D0086700

MODERN HYDRONIC HEATING
for
Residential and Light Commercial Buildings
2nd Edition

To MVCC Library
John Siegenthaler
1/28/04

John Siegenthaler, P.E.
Mohawk Valley Community College
Utica, New York

THOMSON
DELMAR LEARNING™

Australia Canada Mexico Singapore Spain United Kingdom United States

Modern Hydronic Heating for Residential and Light Commercial Buildings, 2nd Edition

John Siegenthaler, P.E.

Vice President,
Technology and Trades SBU:
Alar Elken

Editorial Director:
Sandy Clark

Acquisitions Editor:
James DeVoe

Senior Development Editor:
John Fisher

Marketing Director:
Cyndi Eichelman

Channel Manager:
Fair Huntoon

Marketing Coordinator:
Erin Coffin

Production Director:
Mary Ellen Black

Production Manager:
Larry Main

Production Editor:
Ruth Fisher

Editorial Assistant:
Katherine Bevington

Printed in the United States of America
2 3 4 5 XX 05 04 03

For more information contact
Delmar Learning
Executive Woods
5 Maxwell Drive, PO Box 8007,
Clifton Park, NY 12065-8007
Or find us on the World Wide Web at
www.delmarlearning.com

Library of Congress
Cataloging-in-Publication Data:

Siegenthaler, John.
Modern hydronic heating for residential and light commercial buildings / John Siegenthaler. -- 2nd ed.
p. cm.
Includes bibliographical references and index
ISBN: 0-7668-1637-0
1. Hot-water heating. I. Title

TH7511.S542003
697'.4--dc22

2003055673

NOTICE TO THE READER

Publisher does not warrant or guarantee any of the products described herein or perform any independent analysis in connection with any of the product information contained herein. Publisher does not assume, and expressly disclaims, any obligation to obtain and include information other than that provided to it by the manufacturer.

The reader is expressly warned to consider and adopt all safety precautions that might be indicated by the activities herein and to avoid all potential hazards. By following the instructions contained herein, the reader willingly assumes all risks in connection with such instructions.

The publisher makes no representation or warranties of any kind, including but not limited to, the warranties of fitness for particular purpose or merchantability, nor are any such representations implied with respect to the material set forth herein, and the publisher takes no responsibility with respect to such material. The publisher shall not be liable for any special, consequential, or exemplary damages resulting, in whole or part, from the readers' use of, or reliance upon, this material.

CONTENTS

CHAPTER 10 HYDRONIC RADIANT PANEL HEATING

CHAPTER 11 DISTRIBUTION PIPING SYSTEMS

CHAPTER 12 EXPANSION TANKS

PREFACE

This book, like its predecessor, is about the state of the art in hydronic (water-based) heating systems for residential and light commercial buildings. It was written to provide comprehensive, up-to-date, and unbiased information for heating technology students as well as heating professionals. It is aimed at those willing to work with their heads as well as their hands.

During my 25 years working as an HVAC engineer, it has become apparent that an information gap exists for those interested in upgrading their knowledge of designing and installing high quality hydronic heating systems for smaller buildings. I believe that residential and small commercial buildings deserve better heating systems than they often get. After all, even small heating systems affect the comfort and well being of numerous people over many years.

Much of what has been written on hydronics is aimed at engineers, and intended for use in larger commercial or institutional buildings. It is often impractical to scale such large system designs for use in smaller buildings. Manufacturers' information, while often well done, seldom takes the heating professional through the entire design process. Since most heating systems for smaller buildings are nonengineered, the installing contractor often has sole responsibility for ensuring that the components are properly matched, sized, and installed. In the absence of proper training, this can lead to the same mistakes being repeated over and over. It also leads to design stagnation, where installers are only willing to deal with a certain type of equipment or system design, regardless of the situation with which they are faced. Many overlook the profitable opportunities offered through creative use of modern hydronic heating hardware and design methods.

Since the first edition of this text was released in 1995, growth in the hydronic heating market for residential and light commercial buildings has been phenomenal. The hydronic radiant panel heating market, in particular, has grown by over 1000 percent during the last decade.

With this growth has come a variety of new products and design methods. The second edition has been extensively updated and expanded to include the latest information on both hardware and design procedures.

ORGANIZATION

The first chapters of the book acquaint the reader with the fundamental physical processes involved in hydronic heating. Topics such as basic heat transfer, heating load calculations, and properties of fluids are discussed. A good understanding of these basics enhances the technician's design and troubleshooting skills.

Later chapters use the fundamental principles for overall system design. The reader is referred back to relevant sections of earlier chapters during the design process to reinforce the importance of the basics. Case studies at the end of some of the later chapters show complete system piping and control wiring diagrams.

Chapter 1 provides an overview of hydronic heating, and the basic concepts involved. It emphasizes *comfort* as the ultimate goal of the heating professional. It encourages an attitude of craftsmanship and professionalism.

Chapter 2 discusses heating load calculations. Experience indicates that such calculations can be a stumbling block for students, as well as for those in the trade who want to jump right into design and layout without first determining what the system needs to provide. Without proper load information, *any* type of heating system can fail to deliver the required comfort. A complete method for determining design heating loads is presented, including new worksheets. The second edition includes new information, allowing for quick comparisons of various fuel options.

Chapter 3 surveys a wide spectrum of hydronic heat sources. These include conventional gas and oil-fired boilers, as well as newer devices such as hydronic heat pumps, modulating/condensing boilers, thermal storage systems, and solar energy collectors. The second edition includes new material on the use of wood-fired heat sources. Emphasis is placed on matching the temperature and flow requirements of the heat source with those of the distribution system.

Chapter 4 describes, in simple terms, the physical properties of water including specific heat, density, viscosity, and solubility of air. Two very important equations that relate heat and heat flow to water temperature and flow rate are introduced.

Chapter 5 is a "show and tell" chapter covering the proper use of tubing, fittings, and valves in hydronic systems. The use of copper tubing as well as materials such as PEX and PEX-AL-PEX is discussed. Common valves are illustrated and their proper use is emphasized. Several specialty fittings and valves are also discussed. The second edition provides additional application schematics for the hardware discussed.

Chapter 6 is the key analytical chapter of the text. It introduces a new method for calculating the head loss of a fluid as it flows through a piping system. This new concept, called hydraulic resistance, builds an analogy between fluid flow in piping circuits and the principles of current, voltage, and resistance in electrical circuits. A method is developed that allows the reader to analyze both simple and complex piping circuits. The material presented allows the reader to construct a system resistance curve for both simple and complex piping systems.

Chapter 7 presents both qualitative and quantitative information on circulators for smaller hydronic systems. The pump curve is introduced and used to properly match a circulator to a piping system. The intersection of the pump curve of a candidate circulator and system resistance curve of the piping system reveals the operating flow rate of the circulator/piping system combination. Quantitative and graphical methods are shown for finding this point. The chapter also places strong emphasis on what circulator cavitation is, and how to avoid it. The second edition presents the latest information on specialty circulators with built-in check valves as well as applications for variable speed circulators.

Chapter 8 surveys several types of hydronic heat emitters, including finned-tube baseboard convectors, fan-coils, panel radiators, and radiant baseboards. The advantages and disadvantages of each type are discussed. Performance and sizing information is also given.

Chapter 9 is a major chapter dealing with control components and systems for hydronic heating. This area of hydronic heating has grown immensely since the first edition was published. The second edition has correspondingly expanded to present the latest hardware and design methods. It now delves deeper into the control theory as well as practical applications of the latest hardware. Control systems for multiload/multitemperature systems are discussed. Ladder diagrams are used as a framework on which to design such control systems. The chapter concludes with a look at the future of control systems for hydronic heating, including discussion of adaptive control, Internet accessible controls, and communication among various control devices.

Chapter 10 is devoted entirely to hydronic radiant panel heating. The second edition includes extensively revised installation graphics and design information for radiant floors as well as radiant walls and ceilings. The latest hardware options are discussed. A completely revised step-by-step design method is presented with new data for thin-slab as well as tube and plate systems. This chapter equips hydronic heating professionals to work within this rapidly expanding segment of the hydronic heating market.

Chapter 11 covers several types of distribution piping configurations, including series loop, one-pipe diverter tee, multizone, two-pipe reverse-return, and primary/secondary systems. The second edition presents revised schematics of these piping systems. New material has been included on home run distribution systems using PEX or PEX-AL-PEX tubing, as well as the increasingly popular method of primary/secondary piping. The fundamental analytical methods presented in Chapters 6 and 7 are now called upon to design each type of piping system. The chapter concludes with examples of state-of-the-art distribution systems for multiload/multitemperature systems.

Chapters 12 and 13 deal with the specialized topics of expansion tanks and air removal. Both survey the latest types of hardware, and show how to properly select it.

Chapter 14 covers several miscellaneous topics. Auxiliary loads such as indirect water heating, pool/spa heating, and intermittent garage heating are discussed. New material has been included to assist in sizing indirect domestic water heaters. Extensive material on the design and application of hydronic snow- and ice-melting (SIM) systems has also been added. Other new topics include buffer tanks and minitube distribution systems.

HYDRONICS DESIGN STUDIO

This second edition includes a CD-ROM with the Academic Version of newly released software called the Hydronics Design Studio. This Windows®-based software is built around the analytical methods presented in the text. It allows rapid simulation of flow in user-defined piping systems, including those with multiple parallel branch circuits. It also includes modules for room heat loss estimating, expansion tank sizing, and detailed simulation and sizing of series-loop baseboard systems.

Several examples of how the software can be used to evaluate long numerical calculations are given. References to the software are marked with a disc icon in the margin of the text.

A professional version of the Hydronics Design Studio with increased capabilities is also available. Visit the website www.hydronicpros.com for additional information.

OTHER FEATURES OF THE SECOND EDITION

- Hundreds of *new* and *standardized* schematic drawings for both piping and electrical control systems. These drawings are presented in a consistent format throughout the text rather than simply reprinted from several inconsistent sources. An appendix of the graphic symbols is provided for reference.

- Updated photographs and illustrations of the latest in hydronic heating hardware, as well as work in progress as hydronic systems are installed.
- A full description of all variables, along with the required units, is given whenever an equation is introduced. All inputs are expressed in the customary English units used in the trade.
- Many example calculations organized in the form of a situation statement, solution procedure, and discussion of the results.

Much of the material in this text has been classroom tested at Mohawk Valley Community College in Utica, New York. The text and software has formed the basis of a one semester course dedicated solely to hydronic heating. Student input has been helpful in determining the appropriate presentation level of the material, and in fine-tuning the software.

ACKNOWLEDGMENTS

My deepest gratitude goes to longtime associate Mario Restive of Mohawk Valley Community College who diligently worked to help create the Hydronics Design Studio. Without his skill, this crucial component of the text would not exist.

I would like to thank all members of the production team, especially John Fisher of Delmar, and Vince Potenza of SoundLightMind, for their diligence and patience as the second edition has taken form. It has been a long road, but I'm confident the reader will benefit immensely from the extensive revision efforts that are now complete.

John Siegenthaler, P.E.

FUNDAMENTAL CONCEPTS 1

OBJECTIVES

After studying this chapter, you should be able to:

- Describe the advantages of hydronic heating
- Define heat and describe how it is measured
- Describe three methods by which heat travels
- Explain thermal equilibrium within a hydronic heating system
- Define four basic hydronic subsystems
- Explain the differences between a radiator and a convector
- Explain the differences between an open-loop and a closed-loop hydronic system
- Summarize the basic components of a hydronic heating system and explain how they operate

1.1 WHAT IS A HYDRONIC HEATING SYSTEM?

Hydronic heating systems use water (or water-based solutions) to move thermal energy from where it is produced to where it is needed. The water within the system is neither the source of the heat nor its destination; only its "conveyor belt." **Thermal energy** is absorbed by the water at a heat source, conveyed by the water through the distribution piping, and finally released into a heated space by a **heat emitter.** Ideally, the same water remains in the system year after year.

Water has many characteristics that make it ideal for such an application. It is readily available, nontoxic, nonflammable, and has one of the highest heat storage abilities of any material known to man. All three states of water (solid, liquid, and vapor) are used for various building heating and cooling applications. The modern hydronic systems discussed in this book make use of the *liquid state only.*

The practical temperature range for water in residential and light commercial buildings is from about 50° to 250 °F. At the upper end of this range the water is maintained in a liquid state by system pressurization. The lower end of the range can be extended well below 32 °F by the addition of antifreeze. Such a solution is called **brine.** It would be used in special applications such as hydronic snowmelting.

Hydronic heating began as an outgrowth of steam heating. Early hydronic systems relied on the **buoyancy** of hot water to circulate water between the boiler and the heat emitters. Because of its lower density, hot water would rise upward from a boiler through supply pipes into heat emitters. After releasing heat, the now slightly heavier water flowed downward back to the boiler. Such "gravity hot water" systems required careful pipe sizing and installation since buoyancy-driven flows are rather weak, and their designs were significantly limited in comparison to what can be done with modern methods. The emergence of electrically powered **circulators** made it possible to move water at higher flow rates through much more elaborate piping systems.

Modern hydronics technology enables heat to be delivered precisely when and where it is needed. Hundreds of system configurations are possible, each capable of meeting the exact comfort requirements of its owner. Some may be as simple as a tank-type water heater connected to a loop of flexible plastic tubing for warming a bathroom floor. Others may use two or more boilers operated in stages, releasing their heat through an assortment of heat emitters. The same boiler(s) may also provide the building's **domestic hot water.** They might even heat the swimming pool or melt snow as it falls on the driveway. Well-designed and properly installed hydronic systems provide unsurpassed comfort and fuel efficiency for the life of the building.

1.2 ADVANTAGES OF HYDRONIC HEATING

This section discusses several advantages of hydronic heating. Among these are:

- Comfort
- Energy savings
- Design flexibility
- Clean operation
- Quiet operation
- Noninvasive installation

Comfort

Appropriately, the first advantage listed is comfort. *Providing comfort should be the primary objective of any heating system designer or installer.* Unfortunately, this objective is too often compromised by other factors, the most common of which is cost. Even small residential heating systems affect the health, productivity, and general contentment of several people for many years. It only makes sense to plan and install them accordingly.

The average building owner does not spend much time thinking about the consequences of the heating system he selects. Many view such systems as a necessary but uninteresting part of a building. When construction budgets are tightened, it is often the heating system that is compromised to save money for other, more impressive amenities.

Heating professionals should take the time to discuss comfort as well as price with their clients before decisions on system type are made. Often people who have lived with uncomfortable heating systems simply do not realize what they have been missing. In retrospect, many would welcome the opportunity to have truly comfortable buildings, and would willingly spend more money, if necessary, to achieve it.

Maintaining comfort is not a matter of supplying heat to the body. Instead, it is a matter of controlling how the body loses heat. A normal adult engaged in light activity generates heat at a rate of approximately 400 **British thermal units** per hour. In a typical indoor environment about 48 percent of this heat is released by **thermal radiation** to colder surfaces, 30 percent is released by **convection** to surrounding air, and 22 percent is released by evaporation from the skin. A small amount of heat may also be released by **conduction** from skin or clothing surfaces in contact with cooler objects. When interior conditions allow heat to leave a person's body at the same rate it is generated, that person feels comfortable. If heat is released faster or slower than the rate it is produced, some degree of discomfort is felt.

The interior environment significantly affects the processes by which the body loses heat. For example, most people will not be comfortable in a room containing many cool surfaces such as large windows, even if the room's air temperature is 70 °F. For optimum comfort, the interior environment must provide the proper balance of air temperature, average surface temperature, and relative humidity to accommodate the various processes through which the body releases heat.

Properly designed hydronic systems control both the air temperature and surface temperatures of rooms to maintain optimal comfort. Modern controls can maintain room air temperature to within ± 1 °F of the desired **setpoint temperature.** Heat emitters such as radiant floors or radiant ceilings raise the average surface temperature of rooms. Since the human body is especially responsive to radiant heat loss, these heat emitters significantly enhance comfort. Comfortable humidity levels are also easier to maintain in hydronically heated buildings.

Several factors such as activity level, age, and general health determine what is a comfortable environment for a given individual. When several people are living or working in a common environment, any one of them might feel too hot, too cold, or just right. Heating systems that allow various "zones" of a building to be maintained at different temperatures can adapt to the comfort needs of several individuals. This is called **zoning.** Although both forced-air and hydronic heating systems can be zoned, the latter is usually much simpler and easier to control.

Energy Savings

Ideally, a building's rate of heat loss would not be affected by how that heat is replaced. However, experience has shown that otherwise identical buildings can have significantly different rates of heat loss based on the types of heating systems installed. Buildings with hydronic heating systems have consistently shown lower heating energy use than equivalent structures with forced-air heating systems.

A number of factors contribute to this finding. One is that hydronic systems do not affect room air pressure while operating. Small changes in room air pressure occur when the blower of a forced-air heating system is operating. Increased room air pressure often results from the lack of an adequate return air path from the rooms back to the furnace. This condition drives heated air out through every small crack, hole, or other opening in the exterior surfaces of the room. A study that compared several hundred homes, some with central forced-air systems, others with baseboard convectors, found air leakage rates averaged 26 percent higher and energy usage averaged 40 percent greater in homes with forced-air heating.

Another factor affecting building energy use is air temperature **stratification;** or the tendency of warm air to rise toward the ceiling while cool air settles to the floor. In extreme situations, the difference in air temperature from floor to ceiling can exceed 20 °F. Stratification tends to be worsened by high ceilings, poor air circulation, and heating systems that supply air into rooms at high temperatures. Maintaining comfortable air temperatures in the occupied areas of rooms plagued with a high degree of temperature stratification leads to significantly higher air temperatures near the ceiling as shown in Figure 1–1. The hot air increases heat loss through the ceiling.

Hydronic systems that transfer the majority of their heat by thermal radiation reduce air temperature stratification, and thus reduce heat loss through ceilings. Comfort can often be maintained at lower air temperatures when a space is radiantly heated. This leads to further energy savings. Zoned hydronic systems provide the potential for unoccupied rooms to be kept at lower temperatures, which also lowers heat loss and reduces fuel consumption.

Design Flexibility

Hydronic heating offers almost unlimited possibilities to accommodate the comfort needs, usage, aesthetic tastes, and budget constraints of just about any building. A single

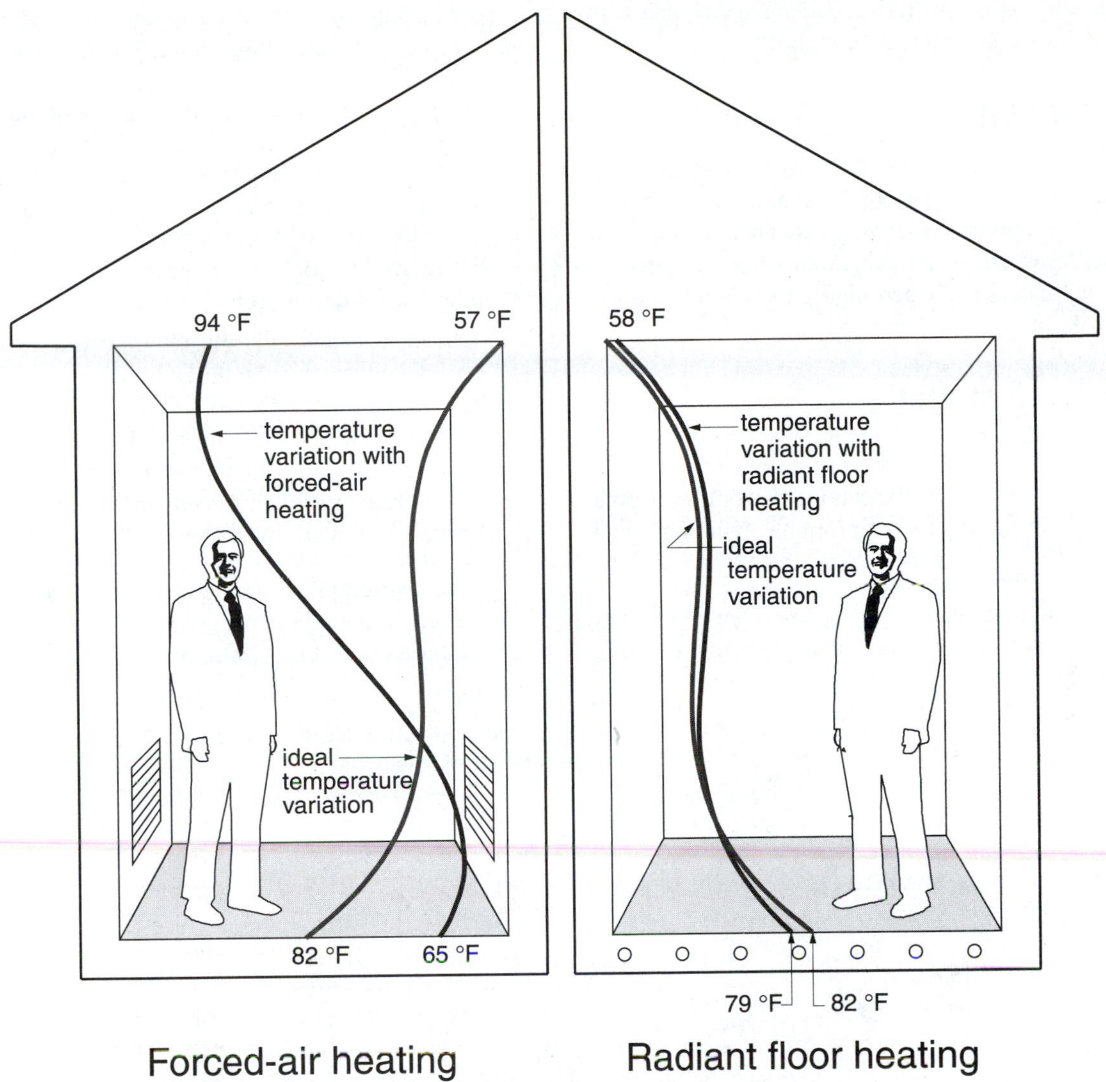

Figure 1–1 Comparison of air temperatures from floor to ceiling for forced air heating (left), and radiant floor heating (right).

system can supply **space heating,** domestic hot water, and specialty **loads** such as pool heating. Such systems reduce installation costs because redundant components such as multiple heat sources, exhaust systems, electrical hookups, and fuel supply components are eliminated.

The heating loads of some buildings are best served through use of different heat emitters. For example, it is common for hydronic radiant heating to be used on the first floor of a house while the second floor rooms are heated using panel radiators or fin-tube baseboard. Commercial and industrial buildings often require heat emitters that are more resistant to physical damage in comparison to those used in residential buildings. Modern hydronics technology makes it easy to combine different heat emitters into the same system.

Clean Operation

A common complaint about forced-air heating is its ability to move dust and other airborne particles such as pollen and smoke throughout a building. In buildings where air-filtering equipment is either low quality or poorly maintained, dust streaks around ceiling and wall diffusers are often evident. Eventually duct systems require internal cleaning to remove dust and dirt that has accumulated over several years of operation.

In contrast, few hydronic systems involve forced-air circulation. Those that do create room air circulation rather than building air circulation. This reduces the dispersal of airborne particles, which is a major benefit in situations where air cleanliness is imperative, such as for people with allergies, health care facilities, or laboratories.

Quiet Operation

A properly designed and installed hydronic system can operate with virtually undetectable sound levels in the occupied areas of a home. Modern systems that use constant circulation with variable water temperature minimize expansion noises that can occur when high temperature water is injected directly into a room temperature heat emitter.

Noninvasive Installation

Consider the difficulty encountered when ducts have to be concealed out of sight within a typical house. The best that can be done in many situations is to encase the ducting in exposed soffits. Such situations often lead to compromises in duct sizing and/or placement.

By comparison, hydronic heating systems are easily integrated into the structure of most small buildings without compromising their structure or the aesthetic character of the space. The underlying reason for this is the high **heat capacity** of water. A given volume of water can absorb over 3400 times more heat as the same volume of air for the same temperature change. The volume of water that must be moved through a building to deliver a certain amount of heat is only about 0.03 percent that of air! This greatly reduces the size of the distribution "conduit."

For example, a 3/4-inch diameter tube carrying water at 6 gallons per minute around a hydronic system operating with a 20 °F temperature drop transports as much heat as a 14-inch by 8-inch duct carrying 130 °F air at 1000 feet per minute. Figure 1–2 depicts these two options side by side. Sawing into the floor joists to accommodate the 14-inch by 8-inch duct would destroy their structural integrity. By contrast, smaller hydronic tubing is easily routed through the framing, especially if it happens to be one of several flexible tube products now available.

If the distribution system will be insulated, which is now a code requirement in many areas, considerably less material is required to insulate the tubing compared to the ducting.

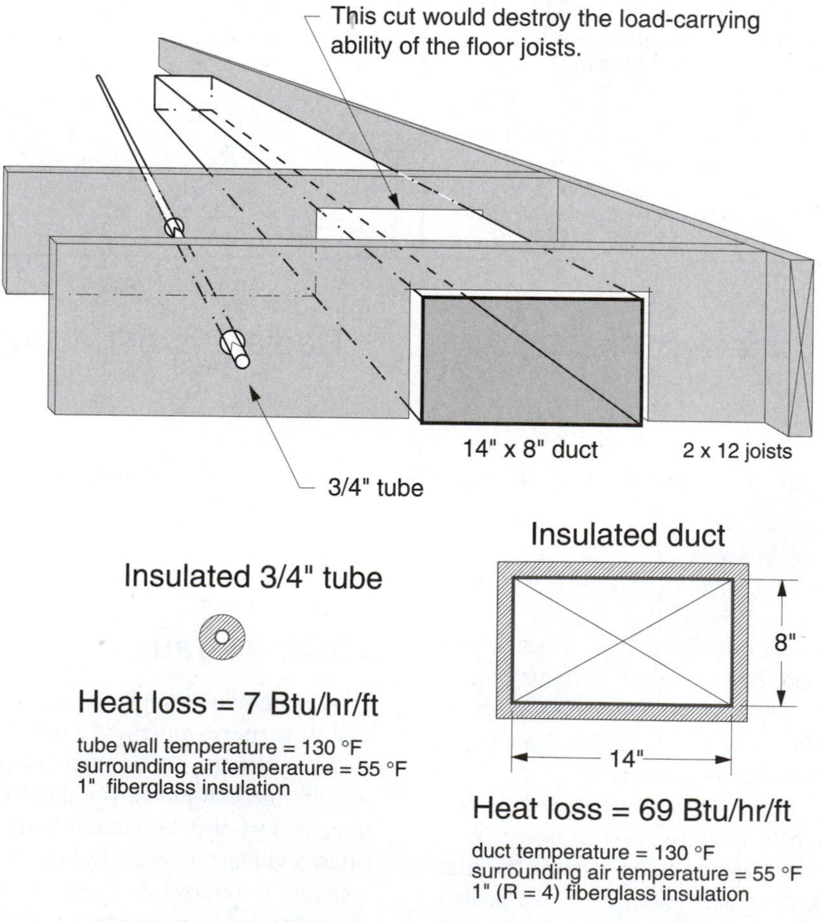

Figure 1–2 A 3/4-inch diameter tube in a hydronic system is capable of transporting heat at the same rate as a 14-inch by 8-inch duct in a forced-air system. Notice that the ducting also will require considerable more insulation material. If the water in the tube and the air in the duct are the same temperature, there is far greater heat loss from the duct because of its surface area.

Even when insulated with the same material, the heat loss of the 14-inch by 8-inch duct is almost *ten times greater* than that of the 3/4-inch tube as shown in Figure 1–2.

Hydronic systems using small flexible tubing are much easier to retrofit into existing buildings than is ducting. The tubing can be routed through closed framing spaces much like electrical cable. A typical retrofit strategy is to run 3/8-inch or 1/2-inch flexible tubing from a central **manifold** to the heat emitter in each room. This approach also allows the option of maintaining different temperatures in each room. The concept is shown in Figure 1–3.

For buildings where utility space is minimal, small wall-hung boilers can often be mounted in a closet. In many cases, these compact boilers supply the building's domestic hot water as well as its heat. The entire system might occupy less than 10 square feet of floor area.

1.3 HEAT AND HEAT TRANSFER

Before attempting to design any type of heating system, it is crucial to understand the entity being manipulated: heat.

What we commonly call heat can also be described as energy in thermal form. Other forms of energy such as electrical, chemical, mechanical, and nuclear can be converted into heat through various processes and devices. All heating systems consist of devices that convert one form of energy into another.

Heat is our perception of atomic vibrations within a material. Our means of expressing the intensity of these vibrations is called temperature. The more intense the vibrations, the greater the temperature of the material, and the greater its heat content. Any material above absolute zero temperature (–460 °F) contains some amount of heat.

There are several units for expressing a quantity of heat. In North America, the most commonly used unit of heat is the British thermal unit (Btu). A Btu is defined as the amount of heat required to raise 1 pound of water by 1 °F.

Heat always moves from an area of higher temperature to an area of lower temperature. In hydronic systems, this occurs at several locations. For systems using a boiler as their heat source, the heat moves from the hot gases in the combustion chamber to the cooler metal walls of the boiler's heat exchanger. The heat continues through the metal walls of the heat exchanger into the cooler water within. After being transported to a heat emitter on a flowing "conveyor belt" of water, the heat passes through the metal walls of the heat emitter into the still cooler air and objects of a room. Finally, heat moves through the exposed surfaces of a room into the outside air. Figure 1–4 depicts this heat transfer. In every instance heat moved from an area of higher temperature to an area of lower temperature. *Without this temperature difference there is no heat transfer.*

The rate of heat transfer from one location to another is governed by several factors. In this book the rate of heat transfer is expressed in British thermal units per hour, abbreviated as Btu/hr, or Btuh. It is very important to distinguish between the *quantity* of heat present in an object (measured in Btu), and the *rate* at which heat moves in or out of the object (measured in Btu/hr). These terms are often misquoted by people, including those in the heating trade.

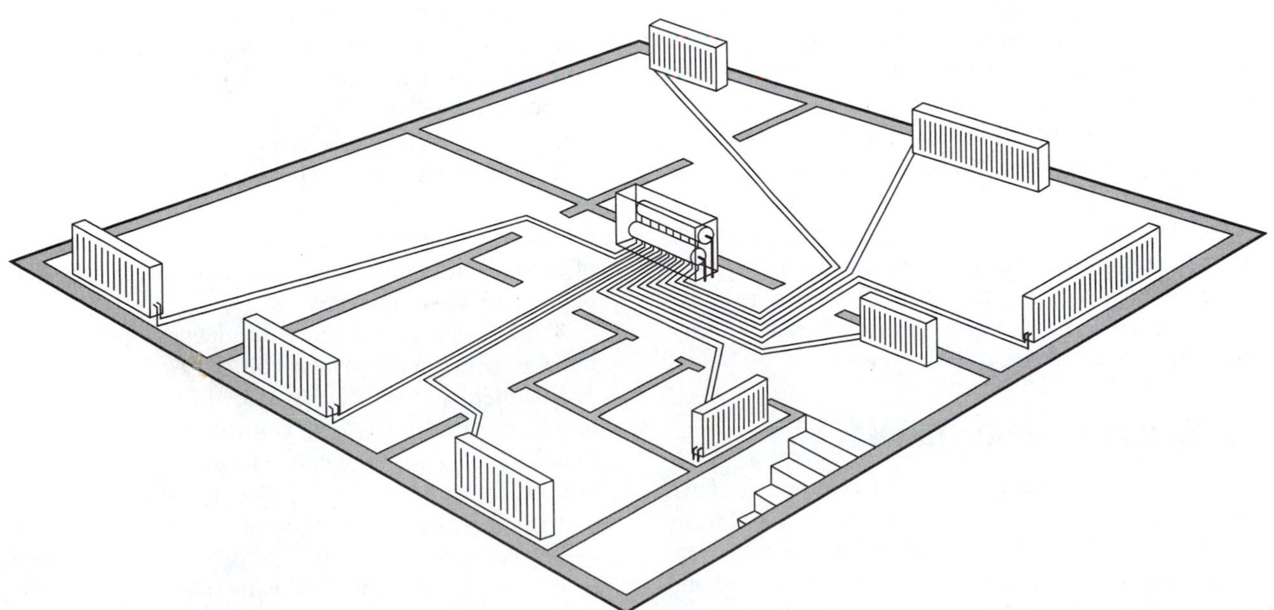

Figure 1–3 Separate supply and return tubing from each of several radiators to a central manifold system. This "homerun" distribution system is especially well-suited for retrofit jobs.

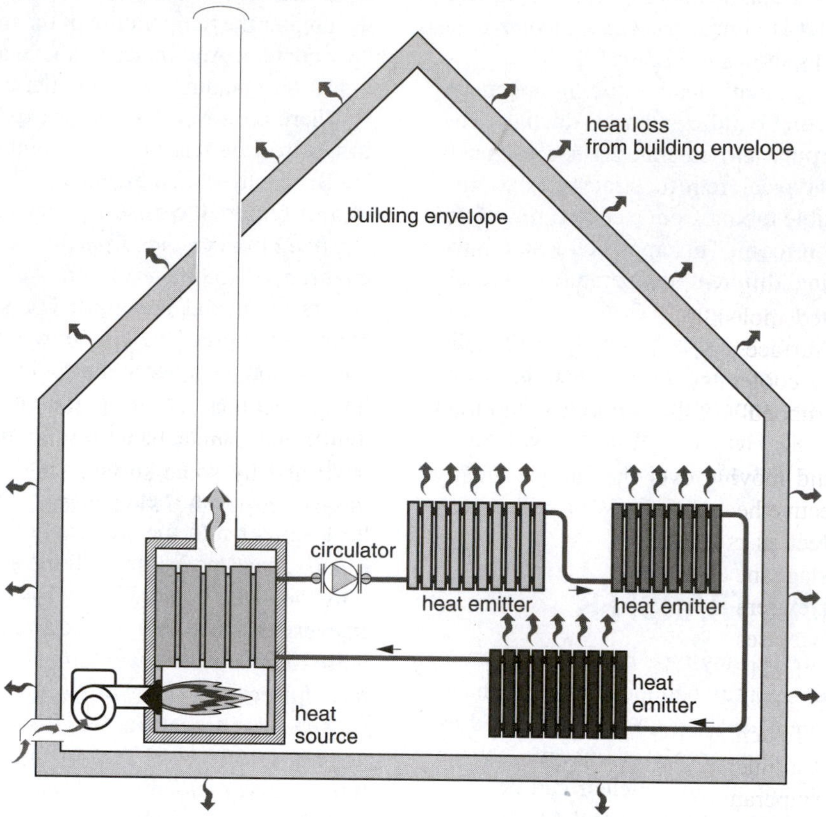

Figure 1–4 Heat movement with a hydronic heating system and the building it serves.

One factor that affects the rate of heat transfer is the temperature difference between where the heat is, and where it is going. *Temperature difference is the driving force that causes heat to move.* The greater the temperature difference, the faster heat flows. In most instances, the rate of heat transfer through a material is directly proportional to the temperature difference across the material. Thus, if we could double the temperature difference across a material, we would also double the rate of heat transfer through it.

Another factor affecting the rate of heat transfer is the type of material through which the heat moves. Some materials, such as copper and aluminum, allow heat to move through very quickly. Other materials, such as polyurethane foam, greatly inhibit the rate of heat transfer.

Three Modes of Heat Transfer

Thus far, two important principles of heat transfer have been discussed. First, heat moves from an area of higher temperature to an area of lower temperature. Secondly, the rate of heat transfer depends on temperature difference and the type of material. To gain a more detailed understanding of how the thermal components of a hydronic system work, we need to classify heat transfer into three modes: conduction, convection, and thermal radiation.

Conduction is the type of heat transfer that occurs through solid materials. Recall that heat has already been described in terms of atomic vibrations. Heat transfer by conduction can be thought of as a dispersal of these vibrations from a source of heat, out across trillions of atoms that are bonded together to form a solid material. The index that denotes how well a material transfers heat is called its **thermal conductivity.** The higher a material's thermal conductivity, the faster heat can pass through it. Heat moving from the inner surface of a pipe to its outer surface is an example of conduction heat transfer. Heat moving through a concrete basement wall is another.

The rate of heat transfer by conduction is directly proportional to both the temperature difference across the material and its thermal conductivity. It is inversely proportional to the thickness of a material. Thus, if one were to double the thickness of a material while maintaining the same temperature difference between its sides, the rate of heat transfer through the material would be cut in half.

In some locations within hydronic systems, designers try to enhance conduction, while in other locations they try to minimize it. For example, the higher the thermal conductivity of a flooring material installed over a heated floor slab, the better it delivers heat to the building. In contrast, the slower the rate of conduction through the insulation of a hot water storage tank, the better it retains heat. Equations for calculating heat flow by conduction are presented in Chapter 2, Heating Load Estimates.

Convection heat transfer occurs when a fluid at some temperature moves along a surface at a different temperature. The term *fluid* can refer to a gas or a liquid.

Consider the example of water at 100 °F flowing along a surface at 150 °F. The cooler water molecules contacting the warmer surface absorb heat from that surface. These molecules are constantly being churned about as the water moves along. The heated molecules are constantly being swept away from the surface into the bulk of the water stream, and replaced by cooler molecules. One can think of the heat as being "scrubbed" off the surface by the flowing water.

The speed of the fluid moving over the surface greatly affects the rate of convective heat transfer. We've all experienced the increased effect as cool air is blown across our skin rather than lying stagnant against it. Although the air temperature may not be extremely cold, the speed it moves over our skin greatly increases the rate of convective heat transfer. Although it may feel like the air is very cold, we are actually sensing the *rate of heat loss* from our skin rather than the air temperature. To achieve the same cooling sensation in calm air requires a much lower air temperature (the so-called "wind-chill" temperature).

When fluid motion is caused by either a circulator (for water) or a blower (for air), the resulting heat transfer is more specifically called **forced convection.** When buoyancy differences within the fluid cause it to move along a surface, the heat transfer is more specifically called **natural convection.** Generally, heat moves much slower by natural convection than by forced convection. The warm air currents rising from the fin-tube element in Figure 1–5 are an example of natural convection. The heat transferred to the air pushed along by the blower is an example of forced convection.

Some hydronic heat emitters, such a finned-tube baseboard, are designed to release the majority of their heat output to the surrounding air by natural convection. Such devices are appropriately called "convectors." Heat emitters that use fans or blowers to force air through a heat exchanger are usually called **fan-coils** or **air handlers.**

Thermal radiation is probably the least understood mode of heat transfer. Just like visible light, thermal radiation is **electromagnetic energy.** It travels outward from its source in straight lines, at the speed of light (186,000 miles per second), and cannot bend around corners, although it can be reflected by some surfaces. Unlike conduction or convection, thermal radiation needs no material to transfer heat from one location to another.

Consider a person sitting a few feet away from a campfire on a cold winter day. If pointed toward the fire their face probably feels warm, even though the air around them is cold. This sensation is the result of thermal radiation emitted by the fire traveling through the air and being absorbed by their exposed skin. The air between the fire and the person is not heated as the thermal radiation passes through it. Likewise, thermal radiation emitted from the warm surface of a heat emitter passes through the air in a room without directly heating that air. When the thermal radiation strikes another surface in the room, most of it is absorbed. At *that instant,* the energy carried by the thermal radiation becomes heat.

The main difference between thermal radiation and visible light is the wavelength of the radiation. Anyone who has watched molten metal cool has noticed how the bright

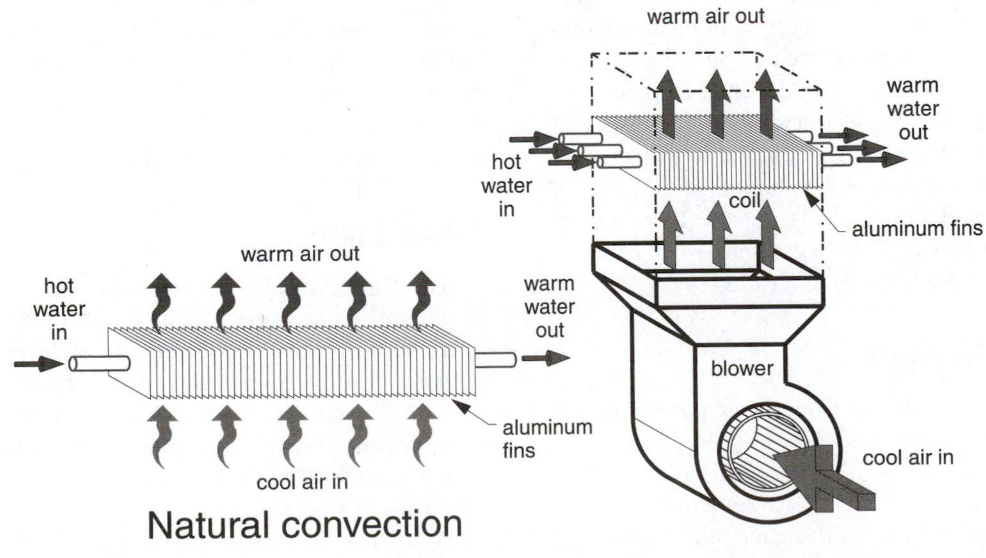

Figure 1–5 Two types of convective heat transfer.

orange color eventually fades to duller shades of red, until finally the metal's surface no longer glows. As the surface of the metal cools below about 970 °F, our eyes can no longer detect visible light from the surface. Our skin however, still senses that the surface is very hot. Though unseen, thermal radiation in the infrared portion of the electromagnetic spectrum is still being strongly emitted by the metal's surface.

Any surface continually emits thermal radiation to any cooler surface within sight of it. The surface of a heat emitter that is warmer than our skin or clothing surfaces transfers heat to us by thermal radiation. Likewise, our skin and clothing give off thermal radiation to any surrounding surfaces at lower temperatures.

The term **mean radiant temperature** describes the average temperature of all surfaces within a room. As the mean radiant temperature of a room increases, the air temperature required to maintain comfort decreases. In this situation heat released from the body by radiation is reduced, so the amount released by convection must increase. The converse is also true. This is why a person in a room with an air temperature of 70 °F may still feel cool if surrounded by cold surfaces, such as large windows, on a winter night.

As thermal radiation strikes an opaque surface, part of it is absorbed as heat and part is reflected away from the surface. The percentage of incoming radiation that is absorbed or reflected is determined by the optical characteristics of the surface and the wavelength of the radiation. Most interior building surfaces absorb the majority of thermal radiation that strikes them. The small percentage that is reflected typically strikes another surface where most of it will be absorbed, and so on. Very little, if any, thermal radiation emitted by warm surfaces in a room escapes from the room. The rate thermal radiation transfers heat between two surfaces depends upon their temperatures, an optical property of each surface called *emissivity,* and the angle between the surfaces.

Hydronic heat emitters deliver a portion of their heat output to the room by convection and the rest by radiation. The percentages delivered by each mode depends on many factors such as surface orientation, shape, type of finish, and surface temperature. When a heat emitter transfers over 50 percent of its heat output by radiation, it is called a **radiant panel.** Heated floors, walls, and ceilings are all examples of radiant panels. They are discussed in detail in Chapter 10, Hydronic Radiant Panel Heating.

Thermal Equilibrium

If a material is not gaining or losing heat and remains in a single physical state (solid, liquid, or gas), its temperature does not change. This is also true if the material happens to be gaining heat from one object while simultaneously releasing heat to another object at the same rate. These principles have many practical applications in hydronic heating. For example, if you observed the operation of a hydronic heating system for an hour and found no change in the temperature of the water leaving the boiler, although it was firing continuously, what could you conclude? Answer: Since there is no change in the water's temperature, it did not undergo any net gain or loss of heat. The rate the boiler injected heat into the water was the same as the rate the heat emitters extracted heat from the water. Under these conditions, the system is in **thermal equilibrium.** *All heating systems will inherently attempt to find a point of thermal equilibrium and remain in operation at that point.* The goal of the system designer is to ensure that thermal equilibrium is established at conditions that properly heat the building, and do not adversely affect the operation or longevity of the system's components.

1.4 FOUR BASIC HYDRONIC SUBSYSTEMS

In a hydronic heating system, water is heated by a **heat source** and conveyed by means of a **distribution system** to heat emitters where it is released to the building. A control system regulates these elements in an attempt to keep the rate of heat delivery very close to the rate of building heat loss.

The overall hydronic system thus consists of four interrelated subsystems. The:

1. Heat Source
2. Distribution System
3. The Heat Emitters
4. The Control System

This section shows how these subsystems are connected to form a simple hydronic heating system. The discussion begins with the concept of a simple piping loop. Several essential components are described and then situated in the loop. Finally, a composite schematic shows all components in their preferred locations relative to each other. After reading this section, you will have a basic understanding of what the major components of a hydronic system are and what they do. The details of proper component selection and sizing are covered in later chapters.

Basic Loop

This simplest hydronic system can be described as a loop, or piping circuit. If the circuit is sealed off from the atmosphere at all locations (as is true for most modern hydronic systems) it is called a **closed-loop system.** If the circuit is open to the atmosphere at *any point,* it is called an **open-loop system.** Figure 1–6 shows the simplest form of a piping circuit. The basic components are represented by schematic symbols.

When there is a demand for heat, water flow in the circuit is established by the circulator. The water carries heat from the heat source to the heat emitter where the heat is released into the heated space. Think of the flowing water as a "conveyor belt" for the heat.

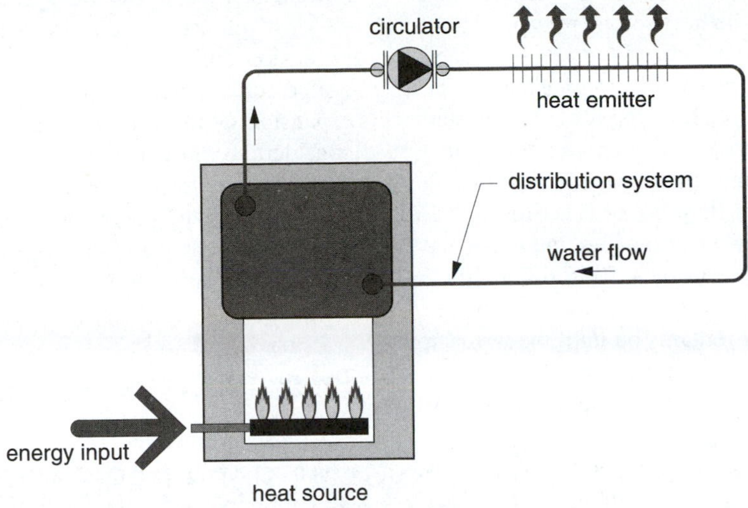

Figure 1–6 The basic hydronic heating circuit.

In an ideal application, the rate of heat production in the heat source exactly matches the rate of heat release by the heat emitters. Ideally, this heat flow also matches the rate of heat loss from the building. Unfortunately, in real systems such ideal conditions seldom exist.

For example, it is fairly common to select a boiler with a heating capacity greater than the building's rate of heat loss on the coldest day of the year. During a mild day, a boiler selected by this method can deliver heat to the building much faster than the building loses heat. Continuous operation of the boiler under such circumstances would quickly overheat the building. Obviously some method of controlling the system's heat delivery is needed.

Temperature Controls

Figure 1–7 adds two simple control devices, the **room thermostat** and the **temperature-limiting control** (often called an **aquastat**), to the basic system. The room thermostat determines when the building requires heat based on its setpoint temperature and the current indoor air temperature. When the indoor air temperature drops slightly below the thermostat's setpoint, the switch contacts inside the thermostat close. This in turn signals other electrical circuits in the system to turn on the circulator and "enable" the heat source to produce heat.

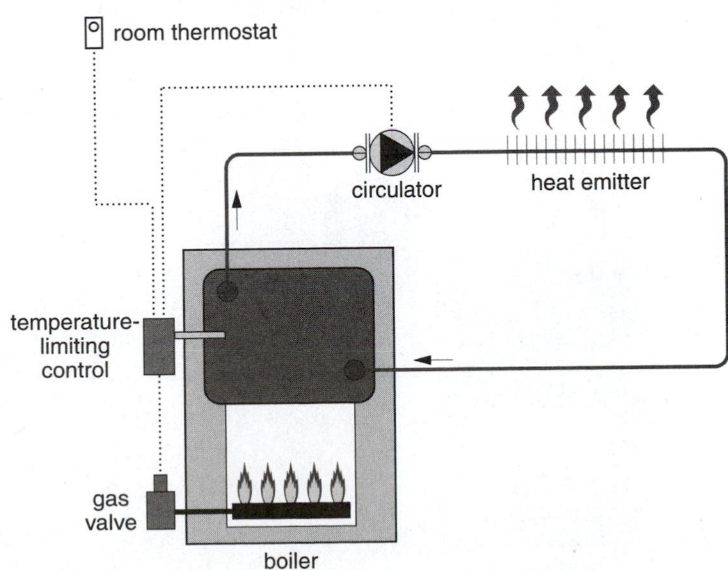

Figure 1–7 Adding a room thermostat and heat source temperature limiting control to the system.

The temperature limiting control (aquastat) ensures that the water temperature within the heat source remains within a predetermined range while the demand for heat is present. It does so by turning the heat-producing components of the heat source on and off as needed. Like the room thermostat, the temperature limiting control is a temperature-operated switch. For example, assume the temperature-limiting control on a typical gas- or oil-fired boiler is set for 160 °F with a 10 °F **differential.** If the water temperature inside the boiler drops to 150 °F (160 °F setpoint – 10 °F differential), a switch contact in the temperature-limiting control closes to operate the burner. The burner remains on until the water in the boiler reaches its upper limit temperature of 160 °F, or the heat demand is no longer present (e.g. the room thermostat is "satisfied").

Together the room thermostat and temperature limit control provide the necessary start/stop signals to the heat source and circulator to reasonably match the heat output to the heating requirement of the building. Such components have been used on literally millions of systems for several decades. These devices, as well as several more sophisticated controls, will be discussed in Chapter 9, Control Strategies, Components, and Systems.

Expansion Tank

As water is heated, it expands. This increase in volume is an extremely powerful but predictable characteristic that must be accommodated in any type of closed-loop hydronic system. Figure 1–8 shows an **expansion tank** added to the basic system. The tank contains a captive volume of air. As the heated water expands, it pushes into the tank and slightly compresses the captive air volume. As a result, the system pressure rises slightly. As the water cools, its volume decreases, allowing the compressed air to reexpand, and the system pressure returns to its original value. This process repeats itself each time the system heats up and cools off.

Most modern hydronic systems use diaphragm-type expansion tanks. Such tanks contain their captive air in a sealed chamber. Older hydronic systems often used expansion tanks without diaphragms. Such tanks had to be considerably larger than modern diaphragm-type tanks. They also had to be mounted higher than the heat source.

The sizing and placement of the system's expansion tank are crucial to proper system operation. Both are discussed in Chapter 12, Expansion Tanks.

Pressure-Relief Valve

Consider the fate of a closed-loop hydronic system in which a defective control fails to turn off the heat source after its upper temperature limit has been reached. As the water gets hotter and hotter, system pressure steadily increases due to the water's expansion. This pressure could eventually exceed the pressure rating of the weakest component in the system. Most residential system components have pressure ratings of at least 60 psi, and may withstand two or more times that pressure before bursting. The consequences of a system component bursting at such high pressures and temperatures could be devastating. For this reason, *all closed-loop hydronic systems must be protected by a* **pressure-relief valve.** This is a universal requirement of all mechanical codes in North America.

Pressure-relief valves are designed and labeled to open at a specific pressure. Residential and light commercial systems typically have pressure-relief valves rated to open at 30 pounds per square inch (psi) pressure. In systems with boilers, the pressure-relief valve is almost always attached directly to the boiler. A pressure-relief valve has been added to the heat source in Figure 1–9. They are discussed in more detail in Chapter 5, Piping, Fittings, and Valves.

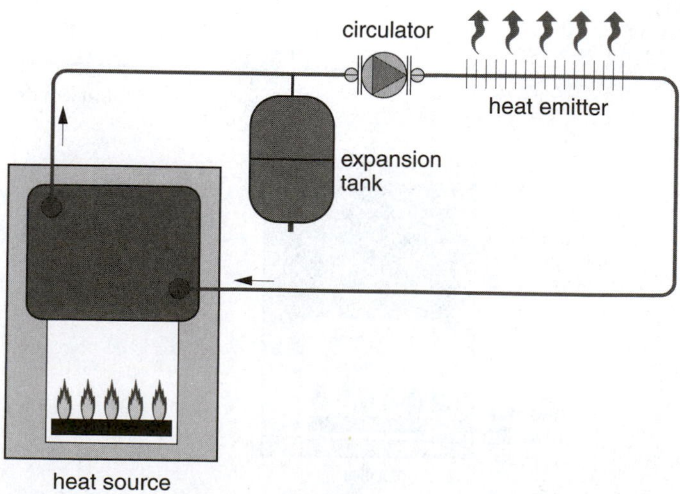

Figure 1–8 Adding an expansion tank to the basic circuit.

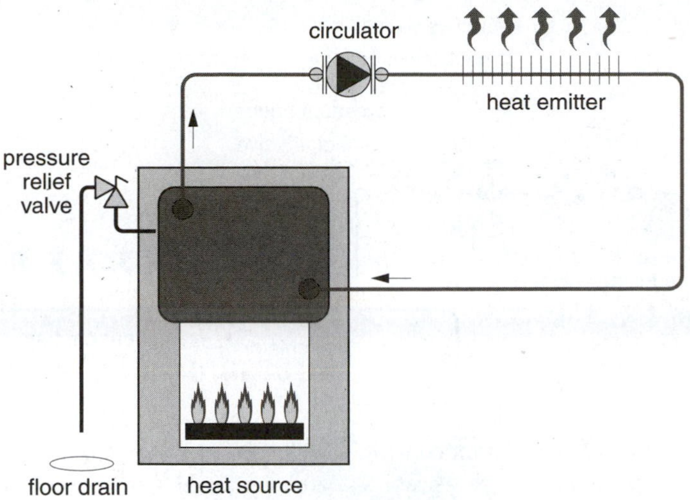

Figure 1–9 Adding a pressure relief valve to the heat source.

Make-up Water System

Most closed-loop hydronic systems experience minor water losses over time due to evaporation from valve packings, pump seals, air vents, and other components. These losses are normal and must be replaced to maintain adequate system pressure. The common method for replacing the water is through a **make-up water system** consisting of a **pressure-reducing valve, backflow preventer,** pressure gauge, and shutoff valves.

Because the pressure in a municipal water main or private water system is often higher than the pressure-relief valve setting in a hydronic system, such water sources cannot be directly connected to the loop. A pressure-reducing valve, also known as a **feed water valve,** is used to reduce and maintain a constant *minimum* pressure in the system. This valve allows water into the system whenever the pressure on the outlet side of the valve drops below the valve's pressure setting. This often occurs as air is vented from the system at start-up or after servicing. Most pressure-reducing valves have an adjustable pressure setting. Determining the proper setting is covered in Chapter 12, Expansion Tanks.

The backflow preventer does just what its name implies. It stops any water that has entered the system from returning and possibly contaminating the potable water supply system. Most municipal codes require such a device on any heating system connected to a public water supply. Even in the absence of code requirements, installing a backflow preventer is a wise decision, especially if any antifreeze or corrosion inhibitors are or ever might be used in the system.

The shutoff valves are installed to allow the system to be isolated from its water source, and to allow components between the shutoff valves to be isolated if they need to be serviced. An optional "fast-fill" valve is sometimes installed in parallel with the pressure-reducing valve so water can be rapidly added to the system. This is particularly beneficial when filling larger systems. The components that constitute the make-up water system are shown in Figure 1–10.

Also shown in the lower portion of Figure 1–10 are valves that allow most of the air in the system to be eliminated as it is filled with water. The valve near the boiler inlet is a shutoff valve. The valve upstream of it is a drain valve (also called a **hose bib**). Together, these valves are called **purging valves.** The shutoff valve is closed just before water is rapidly added to the system through the make-up water subsystem. Pressure then forces water through the distribution system at a high flow rate—high enough that most air is pushed ahead of the water toward the open drain valve. This process is called **purging** the system. It is maintained until no further air exits the drain valve. At that point, most of the air originally in the system will have been expelled through the drain valve.

Flow-check Valve

Another component commonly used in hydronic systems is a **flow-check valve.** This valve can serve one or two purposes depending upon the system it is installed in.

In single-loop systems, the flow-check valve prevents hot water in the boiler from slowly circulating through the distribution system when the circulator is off. Whenever a device containing hot water is part of an unblocked piping path, the potential for such flow exists. If not prevented, such **thermosiphoning** allows heat to "leak" away from the system in an uncontrolled manner, often ending up where it is not desired.

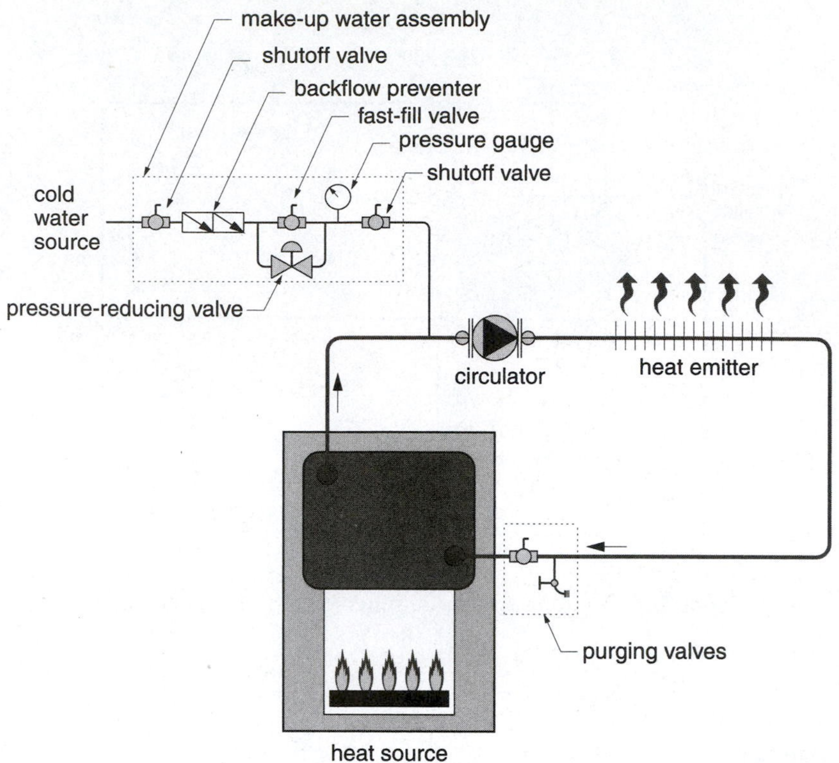

Figure 1–10 Adding a make-up assembly.

In multizone systems using circulators, a flow-check valve is installed in each zone circuit to eliminate thermosiphoning, as well as flow reversal through inactive zone circuits. Figure 1–11 shows the preferred placement of a flow-check valve in a single-loop system. Placement of flow-check valves in multizone systems is covered in Chapter 5, Piping, Fittings, and Valves, and Chapter 11, Distribution Piping Systems.

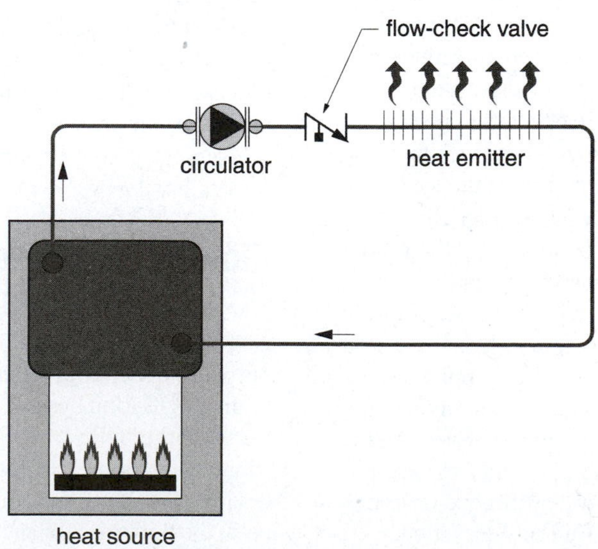

Figure 1–11 Adding a flow check valve to the basic circuit.

Air Separator

An **air separator** is designed to separate air from water and eject it from the system. Modern air separators create regions of reduced pressure as water passes through. The lowered pressure causes dissolved gases in the water to form into bubbles. Once formed, these bubbles are guided upward into a collection chamber where an automatic **air vent** expels them from the system.

The process of separating air from water is enhanced as the water is heated. For best results, the air separator should be located where water temperatures are the highest—in the supply pipe from the heat source. An air separator is shown in Figure 1–12. Air elimination is covered in detail in Chapter 13, Air Removal, Filling, and Purging.

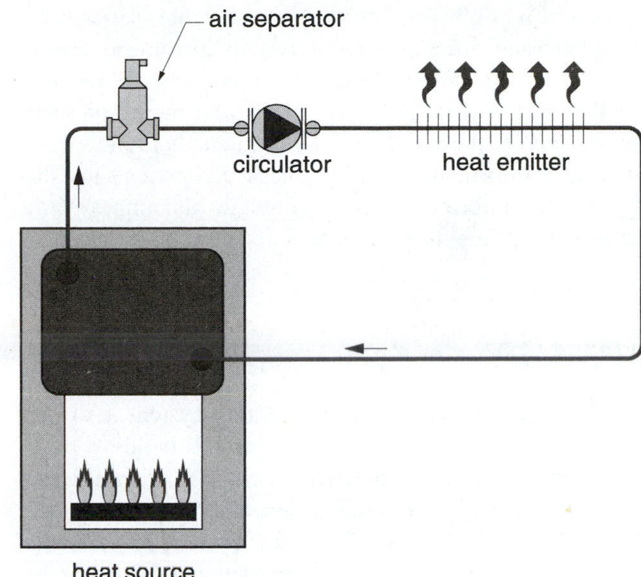

Figure 1–12 Adding an air separator to the basic circuit.

1.5 THE IMPORTANCE OF SYSTEM DESIGN

Figure 1–13 is a composite drawing showing all the components previously discussed in their proper positions relative to each other. By assembling these components, we have built a simple hydronic heating system. It must be emphasized, however, that just because all the components are present does not guarantee that the system will function properly. Combining these components is not a matter of choosing a favorite product for each and simply connecting them as shown. Major subsystems such as the heat emitters and heat source have temperature and flow requirements that must be properly matched if they are to function together as

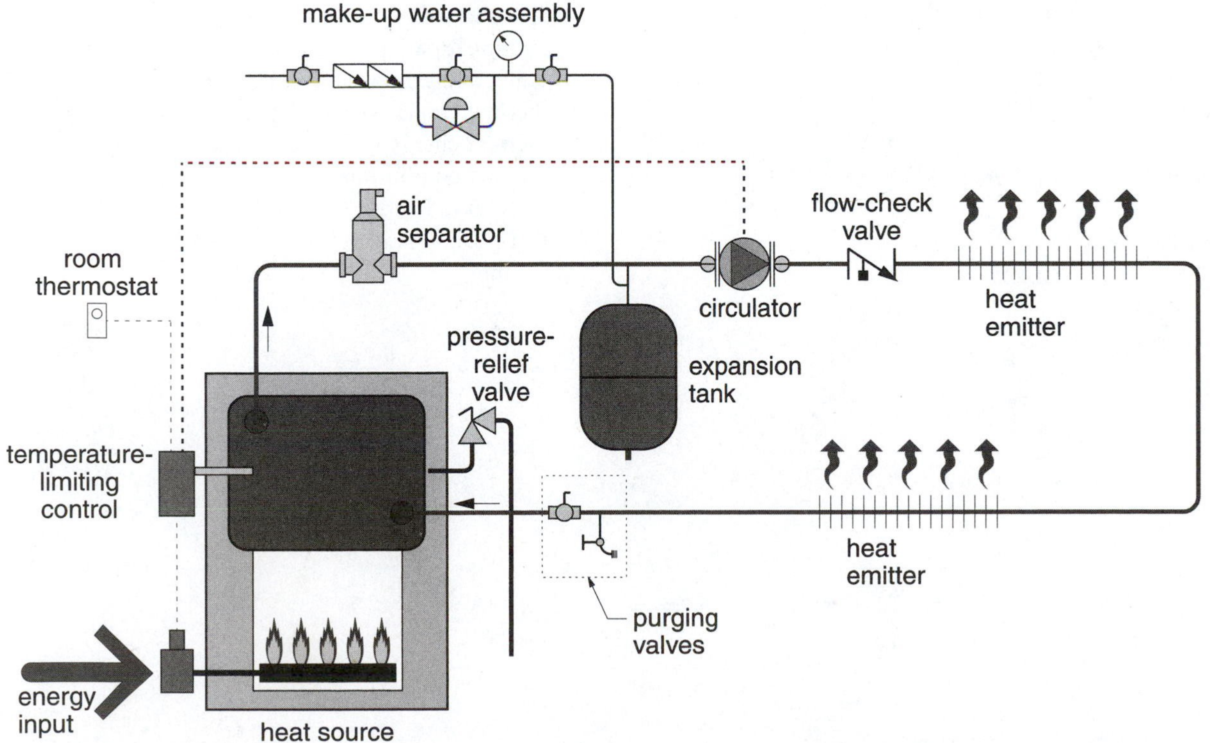

Figure 1–13 Composite drawing showing all the basic hydronic circuit components.

a system. The objective is to achieve a stable, dependable, affordable, and efficient overall system. Failure to respect the operating characteristics of all components will result in installations that underheat, overheat, waste energy, or otherwise disappoint their owners. The chapters that follow present detailed information on all the basic components just discussed. This information is essential in planning systems that deliver optimal performance.

SUMMARY

The primary objective of any heating system is to provide the highest possible comfort for the building occupants. Other important objectives include making efficient use of fuel, providing quiet and dependable operation, and maximizing the service life of all system components. Hydronic heating systems offer tremendous *potential* for attaining all these objectives. The key to turning potential into reality is a solid understanding of the processes and components introduced in this chapter. A lack of understanding of these basics is often the cause of improper system design, or difficulty in diagnosing the cause of poor system performance. Before proceeding into more detailed aspects of design and installation, take time to test your understanding of these all-important fundamentals.

KEY TERMS

air handler
air separator
air vent
aquastat
backflow preventer
brine
British thermal unit
buoyancy
circulators
closed-loop system
conduction
convection
differential
distribution system
domestic hot water
electromagnetic energy
expansion tank
fan-coil
feed water valve
flow-check valve
forced convection
heat capacity
heat emitter
heat source
hose bib
loads
make-up water system
manifold
mean radiant temperature
natural convection
open-loop system
pressure-reducing valve
pressure-relief valve
purging
purging valves
radiant panel
room thermostat
setpoint temperature
space heating
stratification
temperature-limiting control
thermal conductivity
thermal energy
thermal equilibrium
thermal radiation
thermosiphoning
zoning

CHAPTER 1 QUESTIONS AND EXERCISES

1. Why does ducting in forced-air heating systems have to be so much larger than the tubing in a hydronic heating system of equal heating capacity?
2. Why is it better to surround a person with warm surfaces as opposed to just warm air?
3. What type of heat transfer creates the wind-chill effect we experience during winter?
4. A certain block of material conducts heat at a rate of 100 Btu/hr. One side is maintained at 80 °F and the other at 70 °F. Describe what happens to the rate of heat transfer when the:
   a. 100 °F side is raised to 130 °F
   b. thickness of the block is doubled
   c. 100 °F side is raised to 130 °F, and the thickness is cut in half
5. At approximately what temperature does a surface that is cooling from a high temperature stop emitting visible radiation?
6. How does thermal radiation differ from visible light? In what way are they similar?
7. Describe thermal equilibrium within a hydronic system.
8. What is the function of an aquastat on a hydronic heat source?
9. What is the difference between an open-loop system and a closed-loop system?
10. Are there any closed-loop hydronic heating systems on which a pressure relief valve is not required? Why?
11. List two types of hydronic heat emitters other than finned-tube baseboard convectors.
12. What is the function of a flow-check valve in a single-loop system?
13. What is a "brine"? In what type of hydronic heating application would it be used?
14. Why is it necessary to have a backflow preventer in the make-up water line?
15. What are the customary units for heat and heat flow rate in the United States?
16. What is a "zoned" hydronic system?
17. What made water flow through early hydronic systems before circulators were used?
18. What causes air bubbles to form within an air separator?
19. How are the expansion tanks used in modern hydronic systems different from those used in older systems?
20. What are some common ways small amounts of water leak out of a closed-loop hydronic system?

HEATING LOAD ESTIMATES 2

OBJECTIVES

After studying this chapter, you should be able to:

- Describe what a design heating load is and why it is important to hydronic system design
- Explain the difference between room heating loads and building heating loads
- Determine the thermal envelope of a building
- Calculate the effective total R-value of a building surface
- Estimate infiltration heat loss using the air change method
- Determine the heat loss of foundations and slab floors
- Explain what degree days are and how they are used
- Estimate the annual space heating energy usage of a building
- Estimate annual space heating cost for different fuels

2.1 INTRODUCTION

This chapter shows how to estimate the **design heating load** of individual rooms and entire buildings. Determining these loads is the starting point in designing all hydronic systems. *The care given to this step will directly affect system cost, efficiency, and most importantly, customer satisfaction.*

Even for readers already familiar with heating load estimates, this chapter provides new information and a more streamlined approach. It presents the fundamental tools for estimating the heating load using a minimum of mathematics. It also discusses computer-aided load estimating. The Room Heat Load Estimator module which is part of the Hydronics Design Studio software is also described. An example of a complete heating load estimate for a residential building is given. The chapter concludes with a method for estimating the annual space heating energy usage and cost using various fuels.

2.2 DEFINITION OF DESIGN HEATING LOAD

It is important to understand what a design heating load is before attempting to calculate it. *The design heating load of a building is an estimate of the rate at which a building loses heat during the near minimum outdoor temperature.* This definition contains a number of key words that need further explanation.

First, it must be emphasized that a heating load is a calculated *estimate* of the rate of heat loss of a room or building. Because of the hundreds of construction details and thermal imperfections in an object as complex as a building, even a simple house, it is simply not possible to determine its design heating to the nearest Btu/hr. Other factors that add uncertainty to the calculations are:

- Imperfect installation of insulation materials
- Variability in R-value of insulation materials
- Shrinkage of building materials leading to greater air leakage
- Complex heat transfer paths at wall corners and other intersecting surfaces
- Effect of wind direction on building air leakage
- Overall construction quality
- Traffic into and out of the building

Second, the heating load is a *rate* of heat flow from the building to the outside air. It is often misstated, even among heating professionals, as a number of Btus rather than a rate of flow in Btus per hour (Btu/hr). This is like stating the speed of a car in miles rather than miles per hour. Recall from the discussion of thermal equilibrium in Chapter 1 that when the rate of heat flow into a system matches the rate of heat flow out of a system, the temperature of that system remains constant. In the case of a building, *when the rate of heat input to the building equals the rate the building loses heat, the indoor temperature remains constant.*

Finally, the design heating load is estimated assuming the outside temperature is *near* its minimum value. This temperature, called the **97.5% design dry bulb temperature,** is not the absolute minimum temperature for the location. It is the temperature the outside air is at or above during 97.5 percent of the year. Although outside temperatures do occasionally drop below the 97.5 percent design dry bulb temperature, the duration of these low temperature periods is so short that buildings can "coast" through them using heat stored in their thermal mass. Use of the 97.5 percent design temperature for heating system design helps prevent oversizing of the heat source.

Building Heating Load versus Room Heating Load

One method of calculating the heating load of a building yields a single number that represents the heating load of the entire building. This value is called the **building heating load,** and is useful for selecting the heating output of the heat source, as well as for estimating annual heating energy use. However, this single number is not sufficient for designing the heat distribution system. It does not tell the designer how to properly proportion the heat output to each room. Even when the building heating load is properly determined, failure to properly distribute the total heat output can lead to overheating in one room and underheating in another.

One should not assume that room heating loads are proportional to room floor area. A small room with a large window area could have a greater heating load than a large room with smaller windows.

The proper approach is to perform a heat load calculation for each room. Once these individual **room heating loads** are determined, they can simply be added to obtain the building heating load.

Why Bother with Heat Load Calculations?

Would a mason order concrete without first knowing the measurements of the slab to be poured? Would a carpenter begin cutting rafters before knowing the pitch of the roof? Obviously the answer to these questions is no. Yet many so-called "heating professionals" routinely select equipment for a heating system with only a guess as to the building's heating load.

Statements attempting to justify this approach range from: "I haven't got time for doing all those calculations," to "I'd rather be well oversized than get called back on the coldest day of the winter," to "I did a house like this one a couple of years ago, it can't be much different." Instead of listing excuses, let's look at some of the consequences of not properly estimating the building heating load.

- If the estimated heat loss is too low, the building will be uncomfortable during cold weather (a condition few homeowners are willing to tolerate). An expensive call back will eventually result.
- Grossly overestimated heating loads lead to systems that cost more to install. They also cost more to operate due to lower heat source efficiency. The reduced efficiency can waste hundreds, if not thousands of dollars worth of fuel over the life of the system.

Most building owners never know if their heating systems are oversized, and simply accept the fuel usage of these systems as normal. Most simply trust the heating professional they hire to properly size and select equipment in their best interest. Heating professionals who deserve that trust adhere to the following principle:

Before attempting to design a heating system of any sort, always calculate the design heating load of each room in the building using credible methods and data.

The remainder of this chapter shows you how to do just that.

2.3 CONDUCTION HEAT LOSSES

In Chapter 1, Fundamental Concepts, conduction was described as the process by which heat moves through a solid material whenever a temperature difference exists across that material. The rate of conduction depends on the thermal conductivity of the material, as well as the temperature difference across it. The fundamental relationship between these quantities is given in Equation 2.1.

(Equation 2.1)

$$Q = A\left(\frac{k}{\Delta x}\right)(\Delta T)$$

where:

Q = rate of heat transfer through the material (Btu/hr)
k = thermal conductivity of the material (Btu/°F • hr • ft)
Δx = thickness of the material in the direction of heat flow (ft)
ΔT = temperature difference across the material (°F)
A = area heat flows across (ft^2)

Example 2.1: Determine the rate of heat flow by conduction through the panel shown in Figure 2–1.

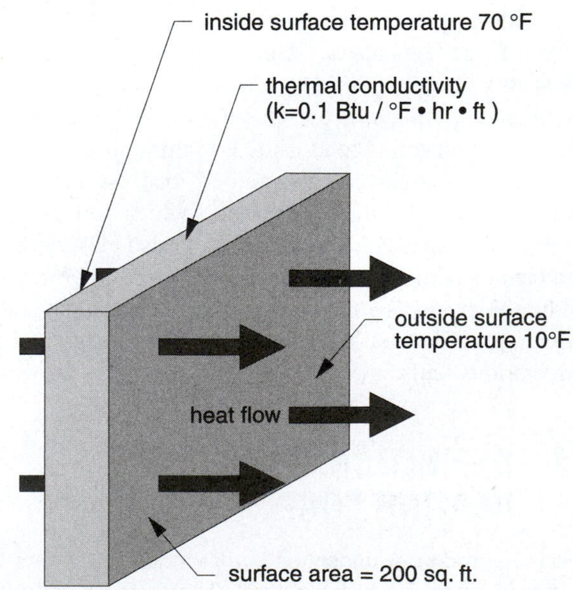

Figure 2–1 Heat flow through a material by conduction.

Solution: Substituting the data into Equation 2.1:

$$Q = A\left(\frac{k}{\Delta x}\right)(\Delta T)$$

$$Q = 200\ \text{ft}^2\left(\frac{0.1\frac{\text{Btu}}{°\text{F} \bullet \text{hr} \bullet \text{ft}}}{0.5\ \text{ft}}\right)(70°\text{F} - 10°\text{F}) = 2400\ \text{Btu/hr}$$

Thermal Resistance of a Material

Since building designers are usually interested in *reducing* the rate of heat flow from a building, it is more convenient to think of a material's resistance to heat flow rather than its thermal conductivity. A material's **thermal resistance** can be defined as its thickness in the direction of heat flow divided by its thermal conductivity.

(Equation 2.2)

$$\text{R-value} = \frac{\text{thickness}}{\text{thermal conductivity}} = \frac{\Delta x}{k}$$

The greater the thermal resistance of a material, the slower heat passes through it when a given temperature differential is maintained across it. Equations 2.1 and 2.2 can be combined to yield an equation that is convenient for estimating heating loads:

(Equation 2.3)

$$Q = \left(\frac{A}{R}\right)(\Delta T)$$

where:

Q = rate of heat transfer through the material (Btu/hr)
ΔT = temperature difference across the material (°F)
R = R-value of the material (°F • ft² • hr/Btu)
A = area across which heat flows (ft²)

Some interesting facts can be demonstrated with this equation. First, *the rate of heat transfer through a given material is directly proportional to the temperature difference (ΔT) maintained across it.* If this temperature difference were doubled, the rate of heat transfer would also double.

Second, the rate of heat transfer through a material is inversely proportional to its R-value. For example, if its R-value were doubled, the rate of heat transfer through the material would be halved (assuming the temperature difference across the material remained constant).

Finally, the larger the object's surface area, the greater the rate of heat transfer through it. For example, a 4-foot by 4-foot window with an area of 16 ft² would transfer heat twice as fast as a 2-foot by 4-foot window with an area of 8 ft².

Example 2.2: Determine the rate of heat transfer through the 6-inch thick wood panel shown in Figure 2–2.

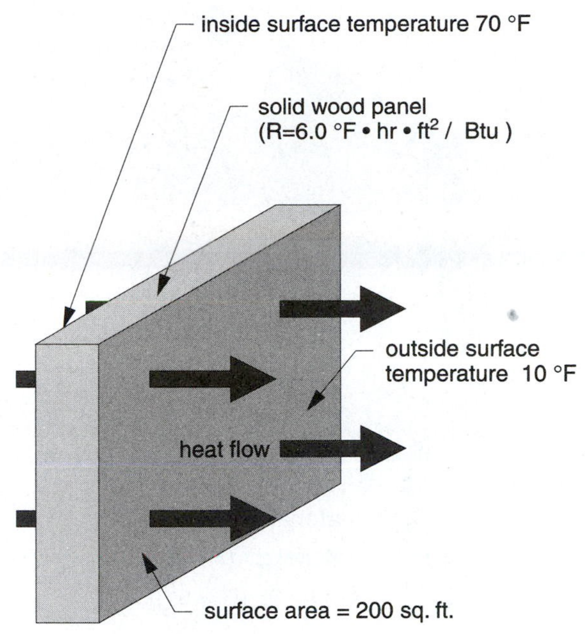

Figure 2–2 Heat flow through a panel by conduction (using R-value).

Solution: Substituting the data into Equation 2.3:

$$Q = \left(\frac{A}{R}\right)(\Delta T) = \left(\frac{200}{6}\right)(70 - 10) = 2000\ \text{Btu/hr}$$

The R-value of a material is also directly proportional to its thickness. If a 1-inch thick panel of polystyrene insulation has an R-value of 5.4, a 2-inch thick panel would have an R-value of 10.8, and a 1/2-inch thick piece would have an R-value of 2.7. This fact is very useful when the R-value of one thickness is known, and the R-value of a different thickness needs to be determined.

The R-value of a material is slightly dependent on the material's temperature. In general, as the temperature of the material is lowered, its R-value increases slightly. For most building materials, the change in R-value is small over the temperature ranges the material typically experiences, and thus may be assumed to remain constant.

Total R-value of an Assembly

Since the walls, floors, and ceilings of buildings are rarely constructed of a single material, it is usually necessary to determine the **total R-value of an assembly** made up of several materials in contact with each other. This is done by adding up the R-values of the individual materials as illustrated in Figure 2–3.

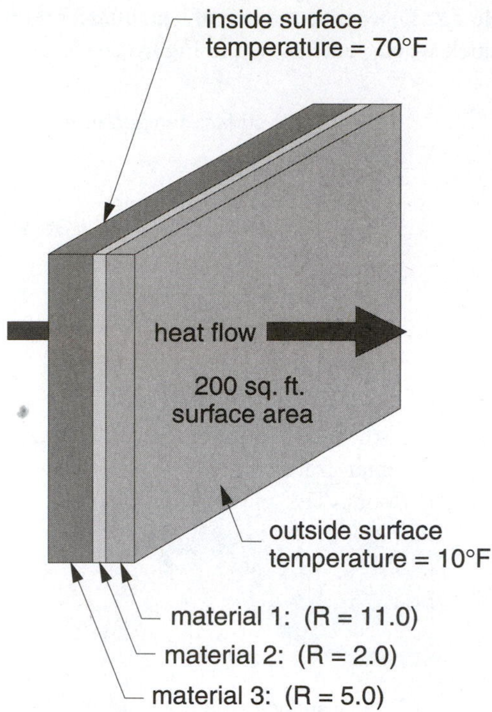

Figure 2–3 Wall assembly consisting of several materials with different R-values.

The total R-value of this assembly is the sum of the individual R-values:

$$R_{\text{total}} = 11.0 + 2.0 + 5.0 = 18.0$$

The heat flow across the panel can now be calculated using the total R-value substituted into Equation 2.3:

$$Q = \left(\frac{A}{R}\right)(\Delta T) = \left(\frac{200}{18}\right)(70 - 10) = 667 \text{ Btu/hr}$$

The R-values of many common building materials can be found in Appendix B. To find the R-values of other materials, consult manufacturer's literature or the latest edition of the *ASHRAE Handbook of Fundamentals.*

Air Film Resistances

In addition to the thermal resistances of the solid materials, there are two additional thermal resistances called the inside and outside **air film resistances.** They represent the insulating effects of thin layers of air that cling to all surfaces. The R-values of these air films are dependent on surface orientation, air movement along the surface, and reflective qualities of the surface. Values for these resistances can also be found in Appendix B. The R-values of air films include the effect of conduction, convection, and radiation heat transfer, but are stated as a conduction-type thermal resistance for simplicity.

The following example illustrates how the total R-value of the wall assembly shown in Figure 2–3 is affected by the inclusion of the air film resistances. Notice that the temperatures used to determine the ΔT in this example are the indoor and outdoor *air* temperatures, rather than the surface temperatures used in the previous examples.

Example 2.3: Determine the rate of heat transfer through the assembly shown in Figure 2–4. Assume still air on the inside of the wall, and 15-mph wind outside. See Appendix B for values of the air film resistances.

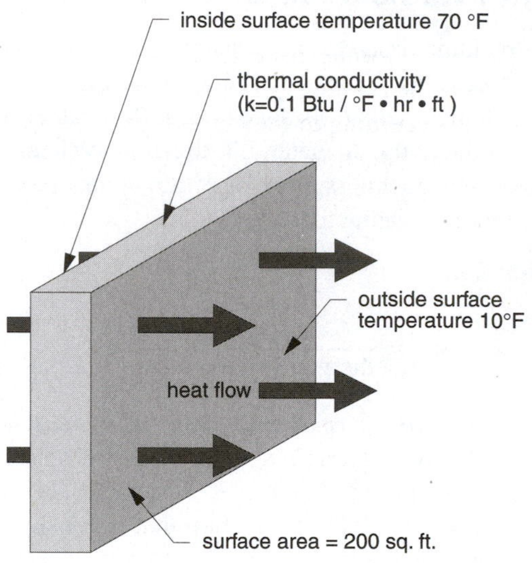

Figure 2–4 Air film resistances add R-value to a wall assembly.

Solution: The total R-value of the assembly is again found by adding the R-values of all materials, including the R-values of the inside and outside air films:

$$R_{\text{total}} = 0.17 + 11.0 + 2.0 + 5.0 + 0.68 = 18.85$$

The rate of heat transfer through the assembly can once again be found using Equation 2.3:

$$Q = \left(\frac{A}{R}\right)(\Delta T) = \left(\frac{200}{18.85}\right)(70 - 10) = 637 \text{ Btu/hr}$$

Effect of Framing Members

The rate of heat flow through assemblies made of several materials stacked together like a sandwich can now be calculated. In such situations, and when edge effects are not considered, the heat flow through the wall is uniform over each square foot of surface area. Unfortunately, few buildings are constructed with walls, ceilings, and other surfaces made of simple stacked layers of materials. This is particularly true of wood-framed buildings with wall studs, ceiling joists, window headers, and other framing members. Wooden framing members that span across the insulation cavity of a wall tend to reduce its **effective total R-value.** This is because the thermal resistance of wood, about 1.0 per inch of thickness, is usually less than the thermal resistance of insulation materials it takes the place of. Thus the rate of conduction through framing materials is usually faster than through the surrounding insulation materials.

It is possible to adjust the R-value of an assembly to include the thermal effects of framing. This requires that the assembly's total R-value be calculated both between the framing, and at the framing. The resulting total R-values are then weighted according to the percentage of solid framing in the assembly. Equation 2.4 can be used to calculate an "equivalent R-value" which can then be applied to the entire panel area.

Equation 2.4:

$$R_{effective} = \frac{R_i \times R_f}{p_f(R_i - R_f) + R_f}$$

where:

$R_{effective}$ = Effective R-value of panel (°F • ft² • hr/Btu)

p_f = percentage of panel occupied by framing (decimal %)

R_f = R-value of panel at framing (°F • ft² • hr/Btu)

R_i = R-value of panel at insulation cavities (°F • ft² • hr/Btu)

Although the percentage of the panel occupied by framing could be calculated for each assembly, such calculations can be tedious, and often result in insignificant changes in the results. Typical wood-framed residential and light commercial construction have from 10 percent and 20 percent of the wall area as solid framing across the insulation cavity. The lower end of this range would be appropriate for 24 inches on-center framing, with insulated headers over windows and doors. The upper end typical of 16 inches on-center framing with solid headers.

To illustrate these calculations, the effective total R-value of the wall assembly shown in Figure 2–5 will be calculated assuming 15 percent of the wall is solid framing.

$$Q = A\left(\frac{k}{\Delta x}\right)(\Delta T)$$

$$Q = A\left(\frac{k}{\Delta x}\right)(\Delta T)$$

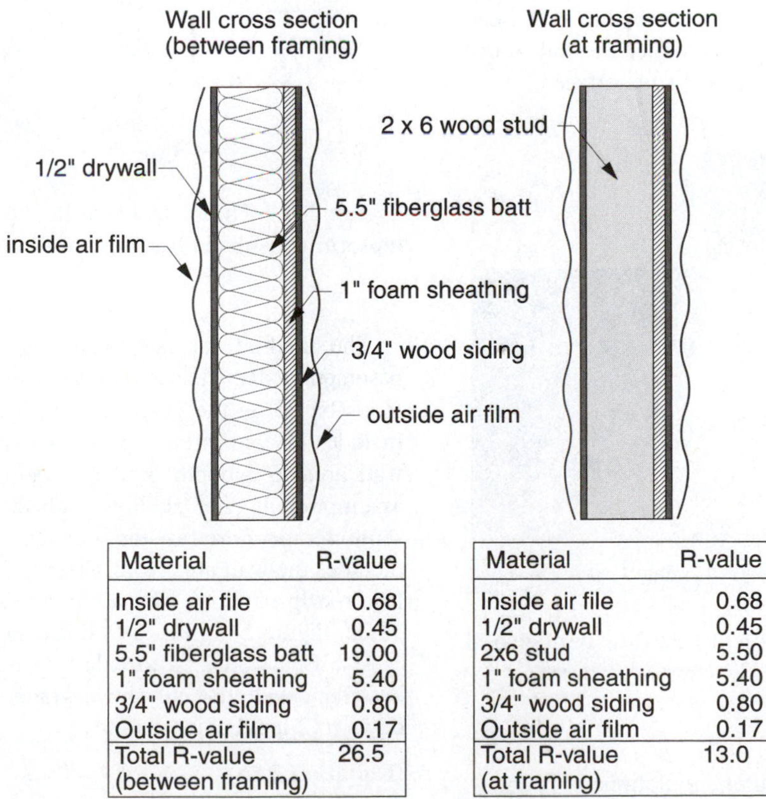

Material	R-value
Inside air file	0.68
1/2" drywall	0.45
5.5" fiberglass batt	19.00
1" foam sheathing	5.40
3/4" wood siding	0.80
Outside air film	0.17
Total R-value (between framing)	26.5

Material	R-value
Inside air file	0.68
1/2" drywall	0.45
2x6 stud	5.50
1" foam sheathing	5.40
3/4" wood siding	0.80
Outside air film	0.17
Total R-value (at framing)	13.0

Figure 2–5 Calculating the effective total R-value of a wall assembly, including the effect of framing members.

To obtain the total heat loss of a room, the effective total R-value of each different **exposed surface** within the room must be determined using a procedure similar to that shown in Figure 2–5. Fortunately, many buildings have the same type of construction for most exposed walls, ceilings, and so on. In such cases, the effective R-value of the assembly need only be calculated once.

It is good practice to make a sketch of the cross section of each exposed assembly (i.e., wall, ceiling, floor), similar to that shown in Figure 2–5. The R-value of each material can then be looked up in Appendix B, and the effective total R-value calculated. The sheets showing the cross sections of the assemblies and the R-values of their materials can be saved in a file, and used as a reference on other buildings with similar construction.

2.4 FOUNDATION HEAT LOSS

Heat flow from a basement or exposed slab edge is determined by complex interactions between the building, the surrounding soil, insulation materials (if present), and the air temperature above grade.

Figure 2–6 shows a computer simulation of heat flow from a basement with an uninsulated heated floor and insulated wall. The contour lines with temperatures labeled in °C are called isotherms. They represent locations having approximately the same temperature. Heat flow is perpendicular to such isotherms in all locations. In this case heat flows downward from the slab, passes under the wall footing, then heads toward the cold exposed soil surface.

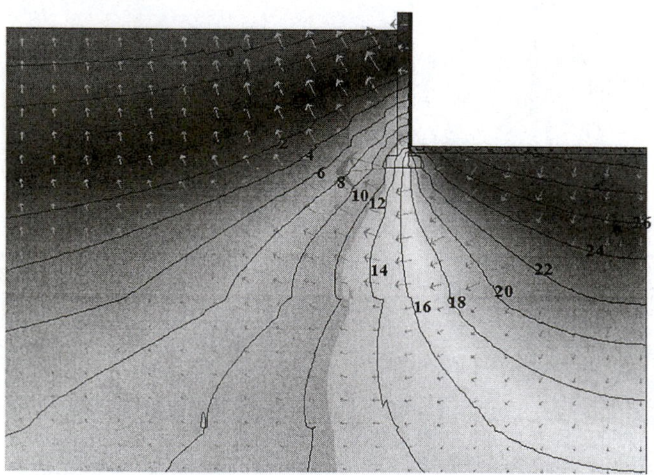

Figure 2–6 Computer simulation of heat flow through a basement wall and adjacent soil. *Courtesy of Beaver Plastics Ltd.*

The results of complex computer simulations of foundation heat loss have been used to develop simpler calculation procedures that can be used for load estimating. One such method is useful for estimating the heat loss of partially buried basement walls, the other for estimating the heat loss from floor slabs.

Heat Loss through Basement Walls

An important factor affecting the heat loss of a basement wall is the height of soil against the outside of the wall. Figure 2–7 illustrates the theoretical heat flow paths through a uniformly insulated basement wall and adjacent soil. Notice that the heat flow paths from the inside basement air to the outside air are longer for the lower portions of the wall. As the path length increases, so does the effective thermal resistance of the soil.

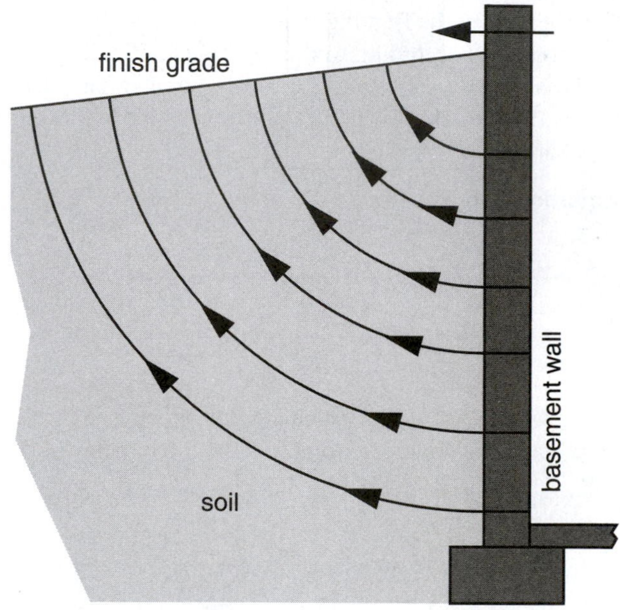

Figure 2–7 Theoretical heat flow paths through a uniformly insulated basement wall.

The method for approximating the heat loss through basement walls is based on dividing the wall into horizontal strips based on the height of finish grade. The upper strip includes all wall area exposed above grade, as well as the wall area to a depth of 2 feet below grade. The implicit assumption is that shallow soils have approximately the same temperature as the outside air. The middle strip includes the wall area from 2 feet to 5 feet below grade. The lower strip includes all wall area deeper than 5 feet below grade. Figure 2–8 illustrates this concept.

Use Equation 2.5a to calculate the effective R-value of basement walls for all above grade areas as well as areas down to 2 feet below grade.

(Equation 2.5a)

$$R_{\text{effective}} = R_{\text{wall}} + R_{\text{insulation}}$$

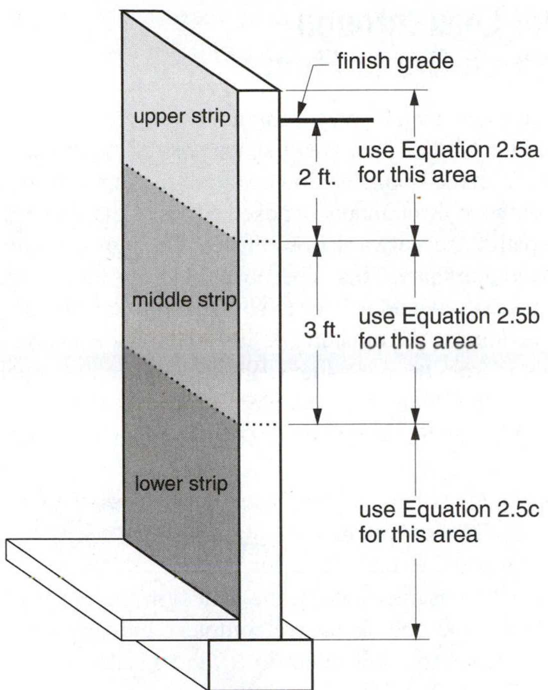

Figure 2–8 Dividing the basement wall into horizontal strips to estimate heat loss.

Use Equation 2.5b to calculate the effective R-value of basement walls for areas from 2 feet to 5 feet below grade.

(Equation 2.5b)

$$R_{\text{effective}} = R_{\text{wall}} + R_{\text{insulation}}$$

Use Equation 2.5c to calculate the effective R-value of basement walls for areas deeper than 5 feet below grade.

(Equation 2.5c)

$$R_{\text{effective}} = 11.3 + 1.13(R_{\text{insulation}})$$

The parameter ($R_{\text{insulation}}$) in Equations 2.5a through 2.5c represents the R-value of any insulation *added* to the foundation wall over the particular strip for which the effective R-value is being determined. The foundation wall itself is assumed to be a masonry or concrete wall between 8 and 12 inches thick.

Once the R-values of each wall strip have been determined, use Equation 2.3 to calculate the heat loss of each area. *The ΔT in Equation 2.3 is the difference between the basement air temperature and the outside air temperature.*

Example 2.4: Estimate the heat loss through a 10-inch thick concrete basement wall 9 feet deep with 6 inches exposed above grade. The wall is 40 feet long, and has 2 inches of extruded polystyrene insulation (R-11) added on the outside over its full depth. The basement air temperature is 70 °F. The outside air temperature is 10 °F. Assume the R-value of the 10-inch concrete wall is 1.0. A drawing of the wall is shown in Figure 2–9.

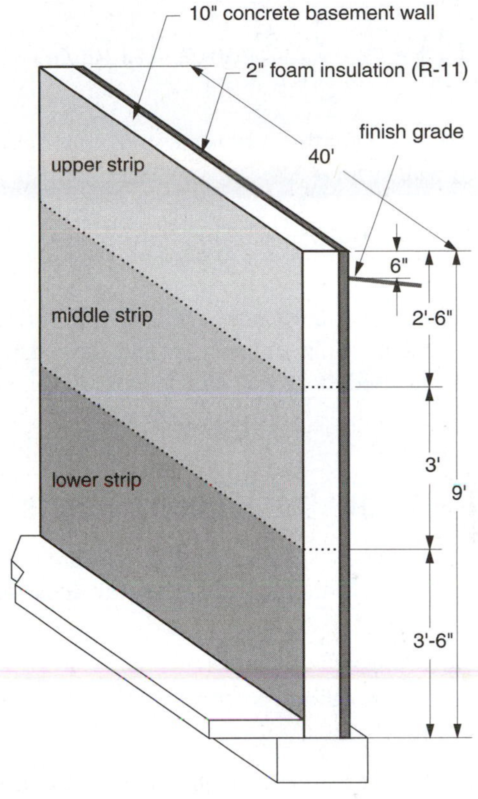

Figure 2–9 Basement wall used in Example 2.4.

Solution: For the upper wall strip, use Equation 2.5a to determine the effective R-value. Then use Equation 2.3 to calculate the heat loss.

$$R = 1.0 + 11 = 12$$
$$A = 40 \text{ ft } (2.5 \text{ ft}) = 100 \text{ ft}^2$$
$$\Delta T = (70\ °\text{F} - 10\ °\text{F}) = 60\ °\text{F}$$

$$Q = \left(\frac{A}{R}\right)(\Delta T) = \left(\frac{100}{12}\right)(60) = 500 \text{ Btu/hr}$$

For the middle wall strip, use Equation 2.5b to determine the effective R-value. Then use Equation 2.3 to calculate the heat loss.

$$R = 7.9 + 1.12\ (R_{\text{added}}) = 7.9 + 1.12\ (11) = 20.2$$
$$A = 40 \text{ ft } (3 \text{ ft}) = 120 \text{ ft}^2$$
$$\Delta T = (70\ °\text{F} - 10\ °\text{F}) = 60\ °\text{F}$$

$$Q = \left(\frac{A}{R}\right)(\Delta T) = \left(\frac{120}{20.2}\right)(60) = 356 \text{ Btu/hr}$$

For the lower wall strip, use Equation 2.5c to calculate the R-value. Then use Equation 2.3 to calculate the heat loss.

$$R = 11.3 + 1.13\,R_{(added)} = 11.3 + 1.13\,(11) = 23.7$$
$$A = 40 \text{ ft } (3.5 \text{ ft}) = 140 \text{ ft}^2$$
$$\Delta T = (70\ °F - 10\ °F) = 60\ °F$$

$$Q = \left(\frac{A}{R}\right)(\Delta T) = \left(\frac{140}{23.7}\right)(60) = 354 \text{ Btu/hr}$$

The total heat loss for this wall is the sum of the heat losses from each wall strip:

$$Q_{total} = 500 + 356 + 354 = 1{,}210 \text{ Btu/hr}$$

Discussion: Note that the same ΔT term (the difference between the basement air temperature and the outside air temperature) was used in the calculations for all wall strips.

Heat Loss through Basement Floors

Equation 2.6 can be used to estimate downward heat loss for basement floor slabs more than 2 feet below finish grade:

(Equation 2.6)

$$Q_{\text{basement floor}} = 0.024(A)(\Delta T)$$

where:

$Q_{\text{basement floor}}$ = rate of heat loss through the basement floor slab (Btu/hr)

A = floor area (ft²)

ΔT = a. (basement air temperature – outside air temperature) (°F) *for unheated slabs*
b. (slab temperature – outside air temperature) (°F) *for heated slabs*

Example 2.5: Determine the rate of heat loss from a 30 ft. by 50 ft. basement floor slab when the outside temperature is 10 °F, and the basement temperature is 65 °F. Assume:

a. the floor has a temperature approximately equal to the inside air temperature.

b. the floor is heated, and has an average temperature of 95 °F.

Solution: Substituting the data into Equation 2.6 for each case:

a: $Q_{\text{basement floor}} = 0.024\ (30 \times 50)\ (65 - 10) = 1980$ Btu/hr

b: $Q_{\text{basement floor}} = 0.024\ (30 \times 50)\ (95 - 10) = 3060$ Btu/hr

Discussion: The effective R-value assumed in Equation 2.6 between the basement floor and the outside air is 1/0.024 = 41.7.

Heat Loss through Slab-on-Grade Floors

Many residential and commercial buildings have concrete floor slabs rather than crawl spaces or full basements. Heat flows from the floor slab downward to the surrounding earth, and outward through any exposed edges of the slab. Of these two paths, the outward flow of heat through the slab edge tends to dominate. This is because the outer edge of the slab is exposed to outside air, or to soil at approximately the same temperature as the outside air. Soil under the slab and several feet in from the perimeter tends to stabilize in temperature once the building is maintained at normal comfort conditions. Hence downward heat losses are relatively small from interior areas of the floor slab. Areas of the floor close to the exposed edges have higher rates of heat loss. This justifies the use of higher R-value insulation near the perimeter of the slab compared to under interior areas.

As with basement walls, the heat flow paths from a slab to the soil and outside air are complex, and vary with foundation geometry, soil conditions, and insulation materials used. Figure 2–10 shows isotherms and arrows indicating the direction of heat flow for a heated slab on grade foundation.

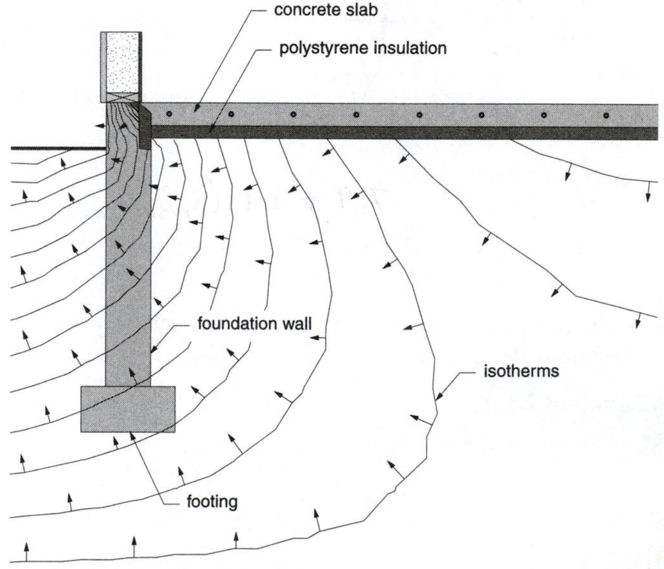

Figure 2–10 Isotherms and arrows indicating the direction of heat flow for a heated slab on grade foundation.

Equation 2.7 is an empirical equation that can be used to estimate the *combined downward and edgewise heat loss* from slab-on-grade floors.

(Equation 2.7)

$$Q_{slab} = \frac{(L)(\Delta T)}{1.21 + 0.214(R_{edge}) + 0.0103(R_{edge})^2}$$

where:

Q_{slab} = rate of heat loss through the slab edge and interior floor area (Btu/hr)

L = *exposed edge length* of the floor slab (ft)

R_{edge} = R-value of edge insulation added to the slab edge (from R-0 to R-11)

ΔT =

a. difference between inside and outside air temperature (°F) *for unheated slabs*

b. difference between the average slab temperature and outside air temperature (°F) *for heated slabs*

Example 2.6: Determine the rate of heat loss from the shaded portion of the unheated floor slab shown in Figure 2–11. The slab has 1.5 inches of extruded polystyrene edge insulation. The inside air temperature is 70 °F, and the outside air temperature is 10 °F.

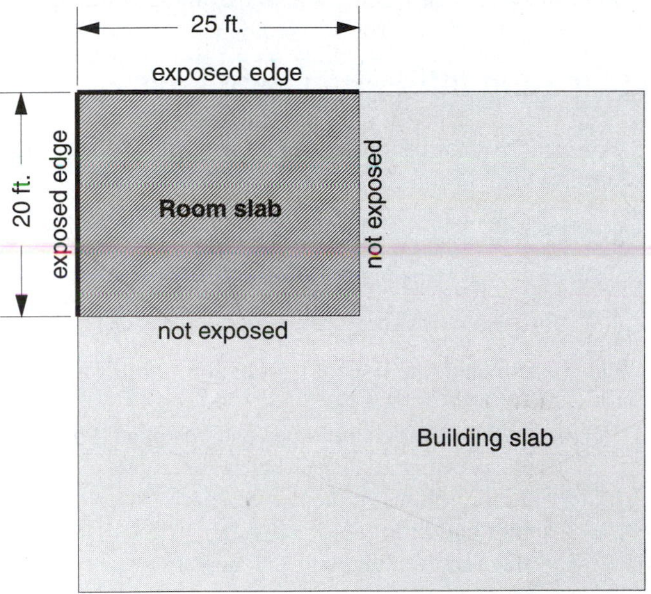

Figure 2–11 Room with exposed slab edge used in Example 2.6.

Solution: Only two sides of the room's slab are *exposed* to the outside. The total linear footage of exposed slab edge is 20 + 25 = 45 feet. The R-value of extruded polystyrene insulation is 5.4 per inch of thickness, therefore the R-value of 1.5 inches is 1.5 × (5.4) = 8.1. Equation 2.7 can now be evaluated:

$$Q_{slab} = \frac{(45)(70-10)}{1.21 + 0.214(8.1) + 0.0103(8.1)^2} = \frac{2700}{3.619} = 746 \text{ Btu/hr}$$

Discussion: This estimate includes heat lost through the exposed edge of the slab, as well as downward heat flow from the shaded interior area.

The Thermal Envelope of a Building:

Heat is lost through all building surfaces that separate heated space from unheated space. Together, these surfaces (walls, windows, ceilings, doors, foundation, etc.) are called the **thermal envelope** of the building. *It is only necessary to calculate heat flow though the surfaces that constitute this thermal envelope.* Walls, ceilings, floors, or other surfaces that separate one heated space from another should *not* be included in these calculations. Figure 2–12 illustrates a part of the thermal envelope for a house.

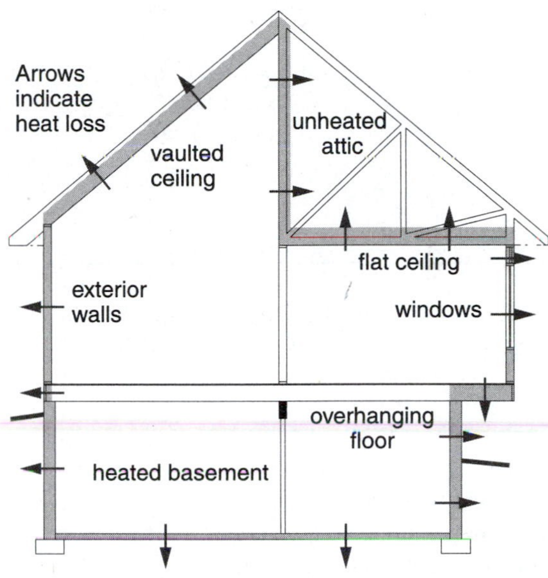

Figure 2–12 A portion of the thermal envelope of a house (shown shaded).

2.5 INFILTRATION HEAT LOSSES

In addition to conduction losses, heat is also carried out of buildings by uncontrolled air leakage. This is called **infiltration heat loss.** The faster air leaks through a building, the faster heat is carried away. The thermal envelope of an average building contains hundreds of imperfections through which air can pass. Some leakage points, such as fireplace flues, exhaust hoods, and visible cracks around windows and doors, are obvious. Others are small and out of sight, but when taken together represent significant amounts of leakage area. These include cracks where walls meet the floor deck, air leakage through electrical outlets, small gaps where pipes pass through floors or ceilings, and many other small imperfections in the thermal envelope.

It would be impossible to assess the location and magnitude of all the air leakage paths in a typical building. Instead, an estimate of air leakage can be made based on the air sealing quality of the building. This approach is known as the **air change method** of estimating infiltration heat loss. Based on a classification of air sealing quality and experience, the designer chooses a *rate* at which the interior air volume of the building (or an individual room) is exchanged with outside air. For example, an air change rate of 1.0 air changes per hour means the entire volume of heated air in a space is replaced with outside air once each hour.

The infiltration rates shown in Figure 2–13 are suggested in *ACCA Manual J.* They are based on a somewhat subjective evaluation of the air sealing quality of the building.

Floor area (sq. ft.)	900 or less	900-1500	1500-2100	over 2100
"Best" quality	0.4	0.4	0.3	0.3
"Average" quality	1.2	1.0	0.8	0.7
"poor" quality	2.2	1.6	1.2	1.0

Add for EACH fireplace:	
"Best" quality	0.1
"Average" quality	0.2
"poor" quality	0.6

Figure 2–13 Air change rates suggested in *ACCA Manual J.*

- *Best* quality air sealing refers to buildings with continuous infiltration barriers such as synthetic housewrap, continuous interior vapor barriers, high-quality windows, well-caulked joints in exterior walls, no combustion air used by appliances, and dampers on all appliance vents.
- *Average* quality air sealing refers to buildings with plastic vapor barriers, average quality windows and doors, major penetrations caulked, noncaulked or taped electrical fixtures, combustion air used by appliances, and dampered appliance vents.
- *Poor* quality air sealing refers to buildings with no air infiltration or vapor barriers, lower quality windows and doors, no caulking at penetrations of thermal envelope, combustion air requirements for heat source, and no dampers on appliances.

These air exchange rates are at best estimates and they vary over a wide range. For example, in a 2000-square foot house, the air leakage rate associated with poor quality air sealing results in four times more infiltration heat loss than the rate associated with best quality air sealing.

Experience is also helpful in estimating infiltration rates. A visit to the building while it is under construction helps give the heating system designer a feel for the air sealing quality being achieved. An inspection visit to an existing building, though not as revealing, can still help the designer assess air sealing quality. Things to look for or consider include:

- Visible cracks around windows and doors
- Poorly installed or maintained weatherstripping on doors and windows
- Slight air motion detected near electrical fixtures
- Smoke that disappears upward into ceiling lighting fixtures
- Deterioration of paint on the downwind side of building
- Poor quality duct system design (especially a lack of proper return air ducting)
- Interior humidity levels (drier buildings generally indicate more air leakage)
- Presence or lack of flue damper on fossil fuel heating system
- Windy versus sheltered building site
- Storm doors and entry vestibules help reduce air infiltration
- The presence of fireplaces (especially those of older masonry construction)
- Casement and awning windows generally have less air leakage than double hung or sliding windows
- Cold air leaking in low in the building generally indicates warm air leaking out higher in the building

Estimating Infiltration Heat Loss

Air leakage rates can be converted into rates of heat loss using Equation 2.8:

(Equation 2.8)

$$Q_i = (0.018)(N)(V)(\Delta T)$$

where:

Q_i = estimated rate of heat loss due to air infiltration (Btu/hr)

N = number of air changes per hour, estimated based on air sealing quality (1/hr)

V = interior volume of the heated space (room or entire building) (ft^3)

0.018 = value representing the heat capacity of air (Btu/ft^3/°F)

ΔT = inside air temperature minus the outside air temperature (°F)

Example 2.7: Determine the rate of heat loss by infiltration for a 30-foot by 50-foot building with 8-foot ceiling height and average quality air sealing. The inside temperature is 70 °F. The outside temperature is 10 °F.

Solution: Referring to Figure 2–13 for a 1,500 square foot house with average air sealing quality, the suggested air leakage rate is 0.8 air changes per hour. Substituting this value along with the other data into Equation 2.7:

$$Q_{infiltration} = (0.018)(0.8)(30 \times 50 \times 8)(70 - 10)$$

$$= 10{,}368 \approx 10{,}400 \text{ Btu/hr}$$

As the conduction heat losses of buildings are reduced, air infiltration becomes a larger percentage of total heat loss. When higher insulation levels are specified for a building, greater efforts at reducing air infiltration should also be undertaken. These include the use of high quality windows and doors, close attention to caulking, sealed infiltration barriers, and so on.

2.6 PUTTING IT ALL TOGETHER

Finding the total design heat loss of a room or building is simply a matter of adding the conduction heat losses to the infiltration heat losses. It is convenient to do this on a room-by-room basis. The resulting numbers can be used to size the heat emitters as the hydronic distribution system is designed. The sum of the room heat losses—the total building design heat loss—determines the required heat output of the heat source.

A worksheet for performing room heat load calculations along with instructions is shown in Figure 2–14. Figure 2–15 is a similar worksheet for use with basement heat loss calculations. To obtain the total heating load of a building with a basement, add the total room loads to the basement load.

2.7 EXAMPLE OF A COMPLETE HEATING LOAD ESTIMATE

This section demonstrates a complete heat loss estimate for an example house having the floor plan shown in Figure 2–16. The room heat load worksheet (Figure 2–14) is used to keep the calculations organized. Other assumed values for these calculations are as follows:

- Unit R-value for all windows: R-3.0
- Unit R-value for exterior door: R-5
- R-value of foundation edge insulation: R-11
- Rate of air infiltration in vestibule and mechanical room: 1 air change/hr
- Rate of air infiltration in all other rooms: 0.5 air change/hr
- Outdoor design temperature: -5 °F
- Desired indoor temperature: 70 °F
- Window height: 4 feet
- Exterior door height: 6 feet 8 inches
- Wall height: 8 feet

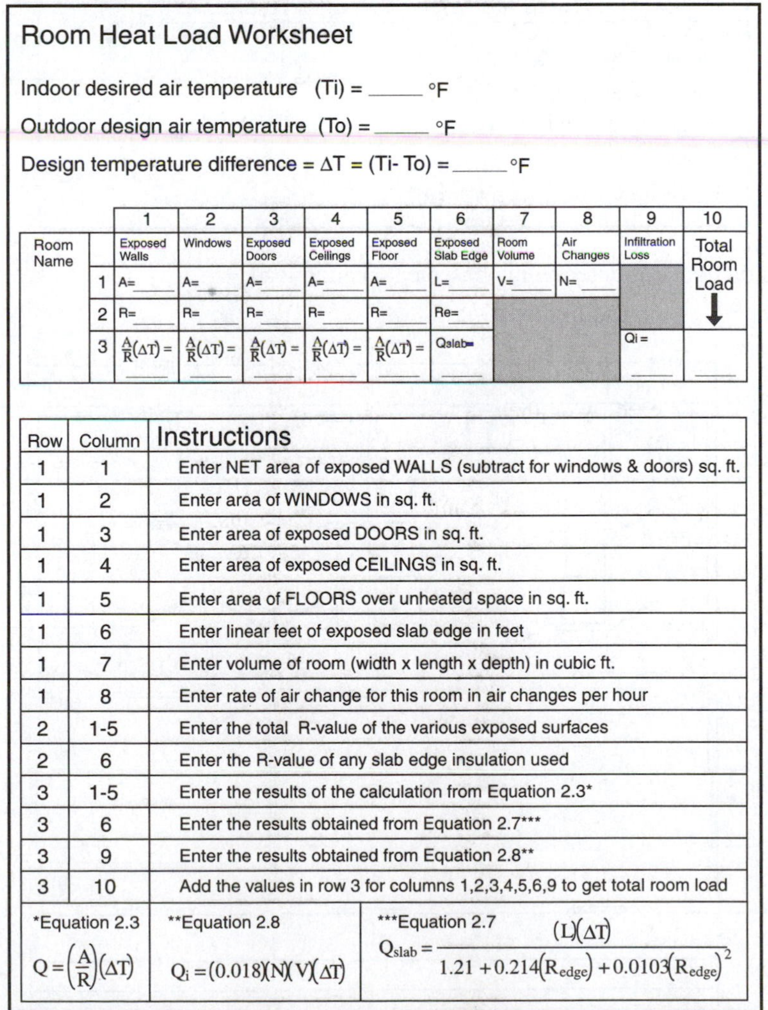

Room Heat Load Worksheet

Indoor desired air temperature (Ti) = _____ °F

Outdoor design air temperature (To) = _____ °F

Design temperature difference = ΔT = (Ti- To) = _____ °F

Room Name		1	2	3	4	5	6	7	8	9	10
		Exposed Walls	Windows	Exposed Doors	Exposed Ceilings	Exposed Floor	Exposed Slab Edge	Room Volume	Air Changes	Infiltration Loss	Total Room Load ↓
	1	A=	A=	A=	A=	A=	L=	V=	N=		
	2	R=	R=	R=	R=	R=	Re=				
	3	$\frac{A}{R}(\Delta T)=$	$\frac{A}{R}(\Delta T)=$	$\frac{A}{R}(\Delta T)=$	$\frac{A}{R}(\Delta T)=$	$\frac{A}{R}(\Delta T)=$	Qslab=			Qi =	

Row	Column	Instructions
1	1	Enter NET area of exposed WALLS (subtract for windows & doors) sq. ft.
1	2	Enter area of WINDOWS in sq. ft.
1	3	Enter area of exposed DOORS in sq. ft.
1	4	Enter area of exposed CEILINGS in sq. ft.
1	5	Enter area of FLOORS over unheated space in sq. ft.
1	6	Enter linear feet of exposed slab edge in feet
1	7	Enter volume of room (width x length x depth) in cubic ft.
1	8	Enter rate of air change for this room in air changes per hour
2	1-5	Enter the total R-value of the various exposed surfaces
2	6	Enter the R-value of any slab edge insulation used
3	1-5	Enter the results of the calculation from Equation 2.3*
3	6	Enter the results obtained from Equation 2.7***
3	9	Enter the results obtained from Equation 2.8**
3	10	Add the values in row 3 for columns 1,2,3,4,5,6,9 to get total room load

*Equation 2.3: $Q=\left(\frac{A}{R}\right)(\Delta T)$

**Equation 2.8: $Q_i=(0.018)(N)(V)(\Delta T)$

***Equation 2.7: $Q_{slab}=\frac{(L)(\Delta T)}{1.21+0.214(R_{edge})+0.0103(R_{edge})^2}$

Figure 2–14 Table and instructions for determining room heating load.

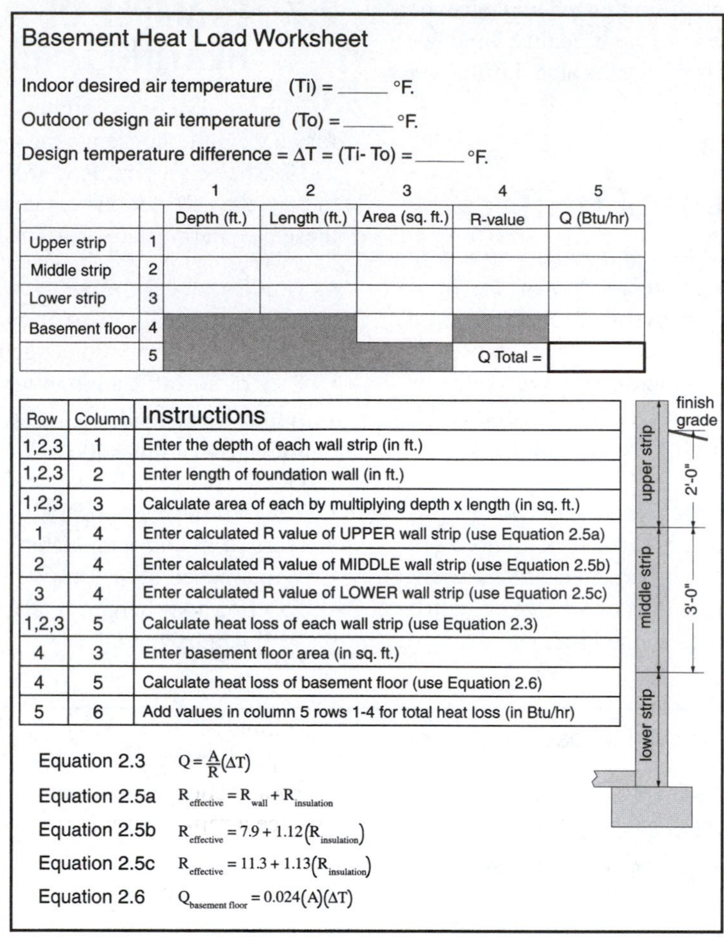

Basement Heat Load Worksheet

Indoor desired air temperature (Ti) = _____ °F.

Outdoor design air temperature (To) = _____ °F.

Design temperature difference = ΔT = (Ti- To) = _____ °F.

		1	2	3	4	5
		Depth (ft.)	Length (ft.)	Area (sq. ft.)	R-value	Q (Btu/hr)
Upper strip	1					
Middle strip	2					
Lower strip	3					
Basement floor	4					
	5				Q Total =	

Row	Column	Instructions
1,2,3	1	Enter the depth of each wall strip (in ft.)
1,2,3	2	Enter length of foundation wall (in ft.)
1,2,3	3	Calculate area of each by multiplying depth x length (in sq. ft.)
1	4	Enter calculated R value of UPPER wall strip (use Equation 2.5a)
2	4	Enter calculated R value of MIDDLE wall strip (use Equation 2.5b)
3	4	Enter calculated R value of LOWER wall strip (use Equation 2.5c)
1,2,3	5	Calculate heat loss of each wall strip (use Equation 2.3)
4	3	Enter basement floor area (in sq. ft.)
4	5	Calculate heat loss of basement floor (use Equation 2.6)
5	6	Add values in column 5 rows 1-4 for total heat loss (in Btu/hr)

Equation 2.3 $Q = \frac{A}{R}(\Delta T)$

Equation 2.5a $R_{effective} = R_{wall} + R_{insulation}$

Equation 2.5b $R_{effective} = 7.9 + 1.12(R_{insulation})$

Equation 2.5c $R_{effective} = 11.3 + 1.13(R_{insulation})$

Equation 2.6 $Q_{basement\ floor} = 0.024(A)(\Delta T)$

Figure 2–15 Table and instructions for determining basement heating load.

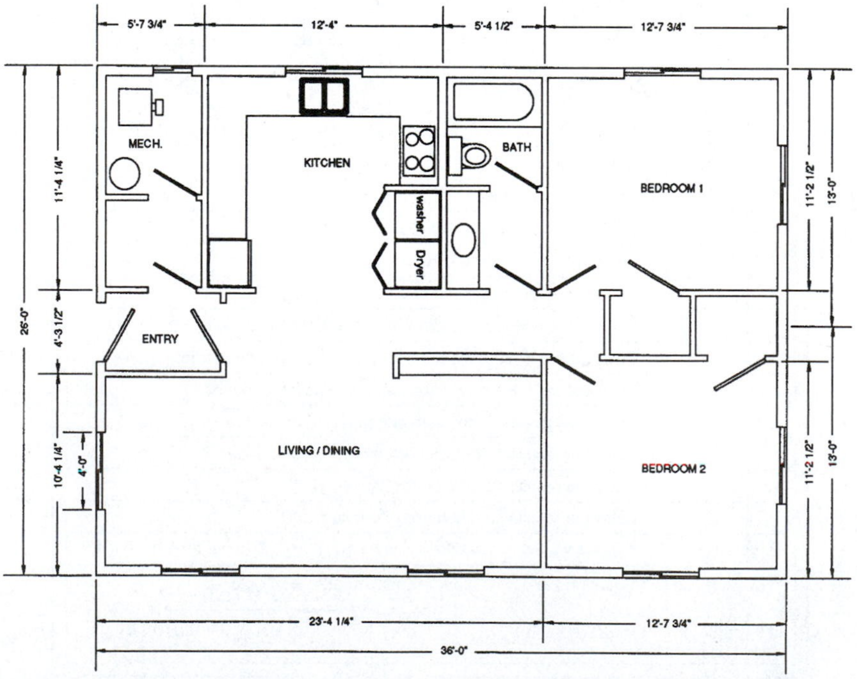

Figure 2–16 Floor plan of example house for which heating load will be estimated.

Determining the Total R-value of the Thermal Envelope Surfaces

Walls: The wall cross section is shown in Figure 2–17. It is made up of 2 × 6 wood framing spaced 24 inches on center with R-21 fiberglass batt insulation in the stud cavities. The inside finish is 1/2-inch drywall. The outside of the wall is sheathed with 1/2-inch plywood, and covered with a Tyvek® infiltration barrier and vinyl siding.

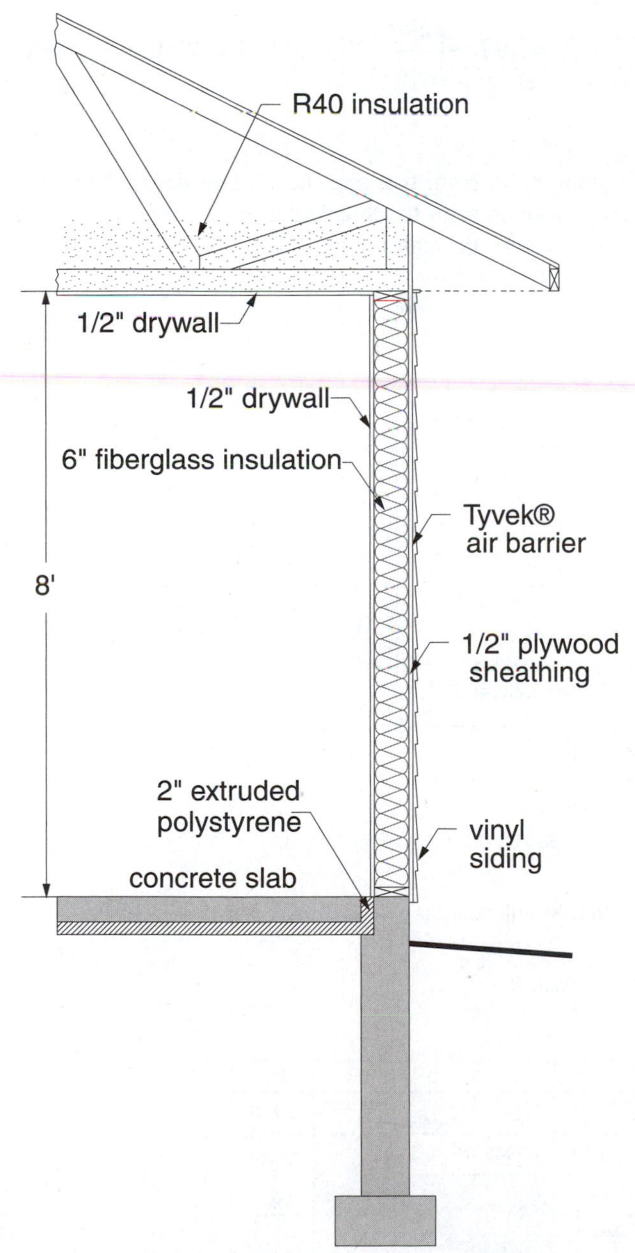

Figure 2–17 Wall cross section for example house.

The R-values of the materials are as follows:

Material	R-value between Framing	R-value at Framing
Inside air film	0.68	0.68
1/2-inch drywall	0.45	0.45
Stud cavity	21.0	5.5
1/2-inch plywood sheathing	0.62	0.62
TYVEK® infiltration barrier	~0	~0
Vinyl siding	0.61	0.61
Outside air film	0.17	0.17
Total	23.53	8.03

Assuming 15 percent of wall is solid framing, the effective total R-value of the wall is:

$$R_{\text{effective}} = \frac{R_i \times R_f}{p_f(R_i - R_f) + R_f} = \frac{23.53 \times 8.03}{0.15(23.53 - 8.03) + 8.03}$$

$$= \frac{188.9}{10.355} = 18.3$$

Ceilings: The ceiling consists of 1/2-inch drywall covered with approximately 12 inches of blown fiberglass insulation. A cross section is shown in Figure 2–18. The roof trusses displace a minor amount of this insulation.

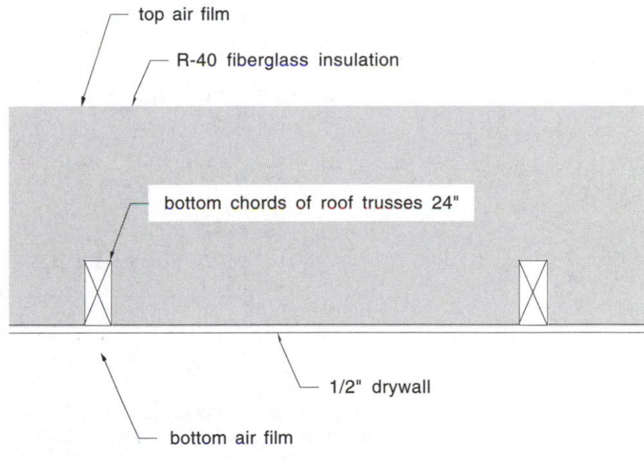

Figure 2–18 Cross section of ceiling in example house.

The R-values of the materials are as follows:

Material	R-value between Framing	R-value at Framing
Bottom air film	0.61	0.61
1/2-inch drywall	0.45	0.45
Insulation	40	28.3
Framing	0	3.5
Top air film	0.61	0.61
Total	41.67	33.47

Assuming 10 percent of ceiling is solid framing, the effective total R-value of the ceiling is:

$$R_{\text{effective}} = \frac{R_i \times R_f}{p_f(R_i - R_f) + R_f} = \frac{41.67 \times 33.47}{0.10(41.67 - 33.47) + 33.47}$$

$$= \frac{1394.7}{34.29} = 40.7$$

The effective R-value is very close to the R-value between framing. The wooden truss chords have minimal effect on the effective total R-value.

Room-by-Room Calculations

Using information from the floor plan and the total R-values of the various surfaces, the design heating load of each room will now be determined. This will be done by breaking out each room from the overall building and determining all necessary information, such as areas of walls, windows, and so on. This information will be entered into the heat load worksheet (Figure 2–14), along with the results of the various equations that have been discussed.

To expedite the process, many of the dimensions are shown on the floor plan of the individual rooms. If such were not the case, the dimensions would have to be estimated using an architectural scale along with the floor plan. As a matter of convenience, the room dimensions are taken to the centerline of the common walls. When working with architectural dimensions, it is best to convert inches and fractions of inches into decimal feet since many dimensions will be added and multiplied together. Example 2.8 illustrates this.

Example 2.8: Convert the dimension 10'–4 1/4" to decimal feet.

Solution:

$$10'\ 4\frac{1}{4}" = 10\text{ ft} + \left(\frac{4.25}{12}\right)\text{ft} = 10\text{ ft.} + 0.35\text{ ft.} = 10.35\text{ ft.}$$

The conversion from feet and inches into decimal feet has been performed prior to the calculation of areas and lengths in Figures 2–19 through Figure 2–23.

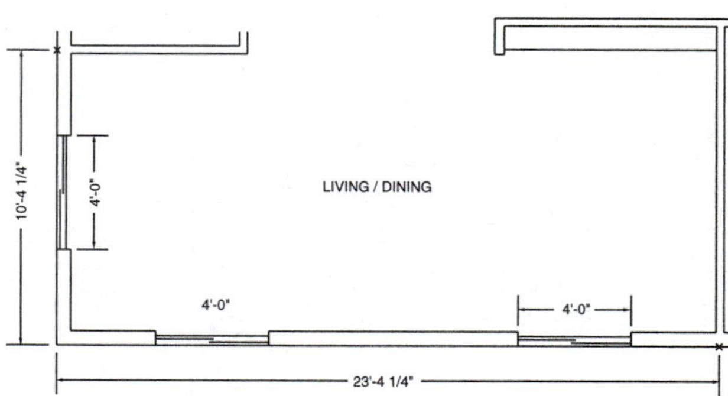

Gross exterior wall area = (10.35 ft. + 23.35 ft.)(8 ft.) = (33.708 ft.)(8 ft.) = 269.7 sq. ft.
Window area = 3 (4 ft • 4 ft) = 48 sq. ft.
Exterior door area = 0 sq. ft.
Net exterior wall area = Gross exterior wall area - Window and door area = 221.7 sq. ft.
Ceiling area = (10.35 ft.)(23.35 ft.) = 241.7 sq. ft.
Exposed slab edge length = (10.35 ft. + 23.35 ft.) = 33.7 linear ft.
Room volume = (10.35 ft.)(23.35 ft.)(8.0 ft.) = 1933.4 cubic ft.

		1	2	3	4	5	6	7	8	9	10
Room Name		Exposed Walls	Windows	Exposed Doors	Exposed Ceilings	Exposed Floor	Exposed Slab Edge	Room Volume	Air Changes	Infiltration Loss	Total Room Load ↓
Living and Dining	1	A= 221.7	A= 48	A= 0	A=241.7	A= 0	L= 33.7	V= 1933	N= 0.5		
	2	R= 18.3	R= 3	R= n/a	R= 40.7	R= n/a	Re= 11				
	3	$\frac{A}{R}(\Delta T)$ = 909	$\frac{A}{R}(\Delta T)$ = 1200	$\frac{A}{R}(\Delta T)$ = 0	$\frac{A}{R}(\Delta T)$ = 445	$\frac{A}{R}(\Delta T)$ = 0	Qslab = 525			Qi= 1305	4384

Figure 2–19 Table with values for living/dining room of example house of Figure 2–16.

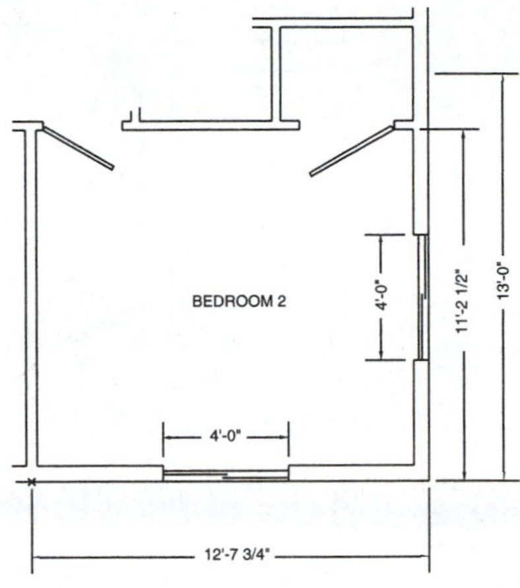

Gross exterior wall area = (12.65 ft. +13.0 ft.)(8 ft.) = (25.65 ft.)(8 ft.) =205.2 sq. ft.
Window area = 2 (4 ft • 4 ft) = 32 sq. ft.
Exterior door area = 0 sq. ft.
Net exterior wall area = Gross exterior wall area – Window and door area =173.2 sq. ft.
Ceiling area = (12.65 ft.)(13.0 ft.) = 164.5 sq. ft.
Exposed slab edge length = (12.65 ft. + 13.0 ft.) = 25.65 linear ft.
Room volume = (12.65 ft.)(13.0 ft.)(8.0 ft.) = 1315.6 cubic ft.

		1	2	3	4	5	6	7	8	9	10
Room Name		Exposed Walls	Windows	Exposed Doors	Exposed Ceilings	Exposed Floor	Exposed Slab Edge	Room Volume	Air Changes	Infiltration Loss	Total Room Load ↓
Bedroom 2	1	A= 173.2	A= 32	A= 0	A= 164.5	A= 0	L= 25.7	V= 1316	N= 0.5		
	2	R= 18.3	R= 3	R= n/a	R= 40.7	R= n/a	Re= 11				
	3	$\frac{A}{R}(\Delta T)$ = 710	$\frac{A}{R}(\Delta T)$ = 800	$\frac{A}{R}(\Delta T)$ = 0	$\frac{A}{R}(\Delta T)$ = 303	$\frac{A}{R}(\Delta T)$ = 0	Qslab = 401			Qi = 888	3102

Figure 2–20 Table with values for Bedroom2 of example house of Figure 2–16.

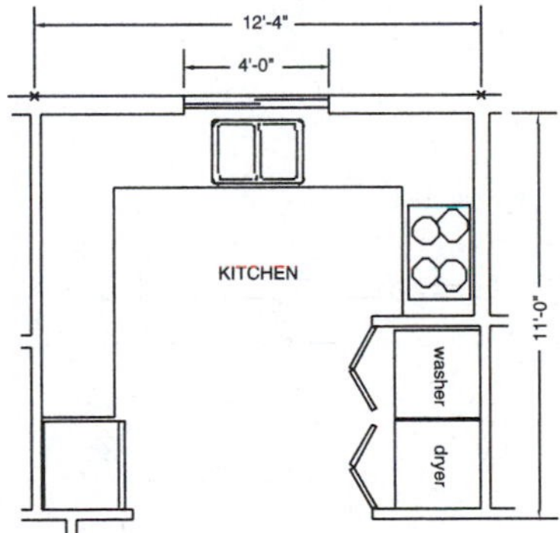

Gross exterior wall area = (12.33 ft.) (8 ft.) = 98.6 sq. ft.
Window area = 1 (4 ft x 4 ft) = 16 sq. ft.
Exterior door area = 0 sq. ft.
Net exterior wall area = Gross exterior wall area - Window and door area =82.6 sq. ft.
Ceiling area = (12.33 ft.) (11.0 ft.) = 135.6 sq. ft.
Exposed slab edge length = 12.33 linear ft.
Room volume = (12.33 ft.) (11.0 ft.) (8.0 ft.) = 1085 cubic ft.

		1	2	3	4	5	6	7	8	9	10
Room Name		Exposed Walls	Windows	Exposed Doors	Exposed Ceilings	Exposed Floor	Exposed Slab Edge	Room Volume	Air Changes	Infiltration Loss	Total Room Load ↓
Kitchen	1	A= 82.6	A= 16	A= 0	A= 135.6	A= 0	L= 12.3	V= 1085	N= 0.5		
	2	R= 18.3	R= 3	R= n/a	R= 40.7	R= n/a	Re= 11				
	3	$\frac{A}{R}(\Delta T)$ = 339	$\frac{A}{R}(\Delta T)$ = 400	$\frac{A}{R}(\Delta T)$ = 0	$\frac{A}{R}(\Delta T)$ = 250	$\frac{A}{R}(\Delta T)$ = 0	Qslab = 401			Qi = 888	2278

Figure 2–21 Table with values for kitchen of example house of Figure 2–16.

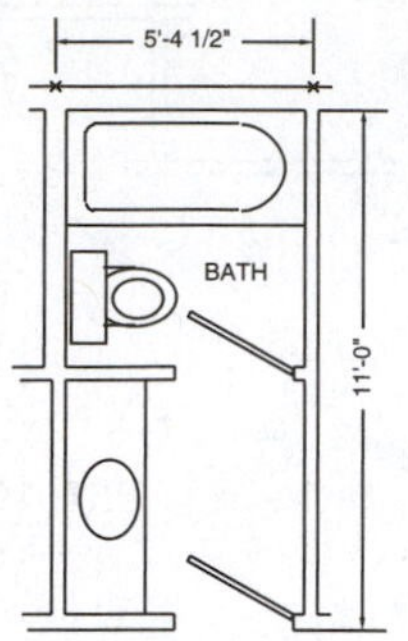

Gross exterior wall area = (5.38 ft.) (8 ft.) = 43.0 sq. ft.
Window area = 0 sq. ft.
Exterior door area = 0 sq. ft.
Net exterior wall area = Gross exterior wall area - Window and door area =43.0 sq. ft.
Ceiling area = (5.38 ft.) (11.0 ft.) = 59.2 sq. ft.
Exposed slab edge length = 5.38 linear ft.
Room volume = (5.38 ft.) (11.0 ft.) (8.0 ft.) = 473 cubic ft.

Room Name		1 Exposed Walls	2 Windows	3 Exposed Doors	4 Exposed Ceilings	5 Exposed Floor	6 Exposed Slab Edge	7 Room Volume	8 Air Changes	9 Infiltration Loss	10 Total Room Load
Bathroom	1	A= 43	A= 0	A= 0	A= 59.2	A= 0	L= 5.38	V= 473	N= 0.5		
	2	R= 18.3	R= 3	R= n/a	R= 40.7	R= n/a	Re= 11				
	3	$\frac{A}{R}(\Delta T)$ = 176	$\frac{A}{R}(\Delta T)$ = 0	$\frac{A}{R}(\Delta T)$ = 0	$\frac{A}{R}(\Delta T)$ = 109	$\frac{A}{R}(\Delta T)$ = 0	Qslab = 84			Qi= 319	688

Figure 2–22 Table with values for bath of example house of Figure 2–16.

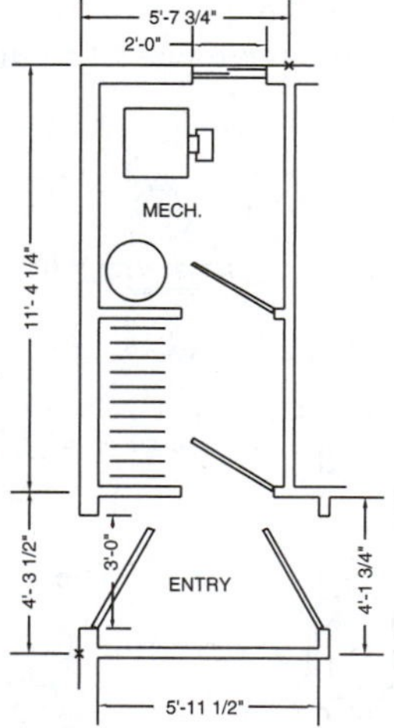

Gross exterior wall area = (4.29 ft. + 11.35 ft. + 5.65 ft.) (8 ft.) = (21.28 ft.) (8 ft.) = 170.2 sq. ft.
Window area = (2 ft. x 4 ft.) = 8 sq. ft.
Exterior door area = (6.66 ft x 3 ft) = 20 sq. ft.
Net exterior wall area = Gross exterior wall area - Window and door area =142.2 sq. ft.
Ceiling area = (4.29 ft. + 11.35 ft.) (5.65 ft.) = (15.64 ft. x 5.65 ft.) = 88.4 sq. ft.
Exposed slab edge length = 21.28 linear ft.
Room volume = (15.64 ft.) (5.65 ft.) (8.0 ft.) = 707 cubic ft.

Room Name		1 Exposed Walls	2 Windows	3 Exposed Doors	4 Exposed Ceilings	5 Exposed Floor	6 Exposed Slab Edge	7 Room Volume	8 Air Changes	9 Infiltration Loss	10 Total Room Load
Mech. room	1	A= 142.2	A= 8	A= 20	A= 88.4	A= 0	L= 21.3	V= 707	N= 1.0		
	2	R= 18.3	R= 3	R= 5	R= 40.7	R= n/a	Re= 11				
	3	$\frac{A}{R}(\Delta T)$ = 583	$\frac{A}{R}(\Delta T)$ = 200	$\frac{A}{R}(\Delta T)$ = 300	$\frac{A}{R}(\Delta T)$ = 163	$\frac{A}{R}(\Delta T)$ = 0	Qslab = 332			Qi= 955	2533

Figure 2–23 Table with values for mechanical room entry of example house of Figure 2–16.

Total Building Heating Load

The total building heating load is obtained by summing the heating loads of each room.

Living/dining room	4,384 Btu/hr
Bedroom 1	3,102 Btu/hr
Bedroom 2	3,102 Btu/hr
Kitchen	2,278 Btu/hr
Bathroom	688 Btu/hr
Mechanical room/entry	2,533 Btu/hr
Total	16,087 Btu/hr

A number of simplifying assumptions were used in this example.

- The rooms were divided at the centerline of the common partitions. This ensures that all the exterior wall area is assigned to individual rooms, and not ignored as being between the rooms.
- Since the two bedrooms are essentially identical in size and construction, it was only necessary to find the heat load of one, then double it when determining the total building load.
- The exterior dimensions of the rooms were used to determine areas, volumes, and so forth. This makes the calculations somewhat conservative because the exterior wall area of a room is slightly greater than the interior wall area.
- A value of 0.5 air changes per hour was selected based on the relatively energy-efficient construction shown on the drawing of the wall cross section. Such a value is consistent with observed rates of air leakage in small, well-sealed houses. A greater air leakage rate was assigned to the entry vestibule since it is a traffic path into or out of the house. This higher rate was also applied to the mechanical room because the combustion-type heat source must have air flow to operate.
- The areas and volumes of the closets between the bedrooms were equally divided up among the two bedrooms. Since a separate heat emitter would not be supplied for each closet, this part of the load is simply included with the bedroom load where the heat emitter would be located.

2.8 COMPUTER-AIDED HEATING LOAD CALCULATIONS

After gathering the data and performing all the calculations needed for a building heating load estimate, it is obvious that a great deal of information goes into finding the results. If the designer then decides to go back and change one or more of the numbers, more time is required to unravel the calculations back to the point of the change, and calculate the new result. Even for people used to handling this amount of information, these calculations are tedious and always subject to errors. The time required to perform accurate heating load estimates is arguably the reason why some "heating professionals" do not do them.

Like many routine design procedures, heating load calculations have made the transition from pencil and paper to desktop computer. The advantages of using a load estimating program are many:

- Rapid (almost instant) calculations to quickly study the effect of changes
- Automated referencing of R-values, and weather data, in some programs
- Significantly less chance of error due to number handling
- Ability to print professional reports for customer presentations
- Ability to store project files for possible use in similar future projects
- Ability to import area and R-value information directly from CAD-based architectural drawings

There are several software packages currently available for estimating the heating loads for residential and light commercial buildings. They vary in cost from under $10 (shareware) to over $400. The higher-cost programs usually offer more features such as material reference files and customized output reports. The lower-cost programs offer no frills output, but can nonetheless yield accurate results.

The Room Heat Load Estimator Module

The Hydronics Design Studio software contains a module called Room Heat Load Estimator that can quickly calculate the design heating load of a building. See Figure 2–24.

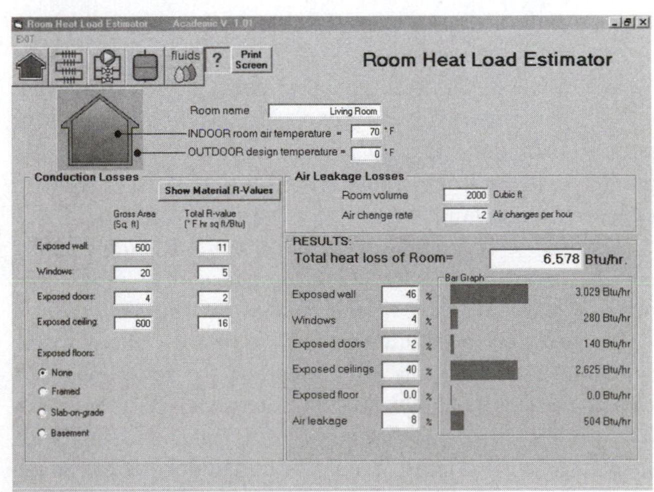

Figure 2–24 Screen shot of the ROOM HEAT LOAD ESTIMATOR module from the Hydronics Design Studio

2.9 ESTIMATING ANNUAL HEATING ENERGY USAGE

Building owners often ask for an estimate of what it will cost to heat their building. Such estimates can be made once the building's design heating load and the seasonal efficiency of the heat source are determined.

Degree Day Method

The seasonal energy used for space heating depends upon the climate where the building is located. The common method of factoring local weather conditions into estimates of heating energy use is the concept of heating **degree days.** The number of heating degree days that accumulate in a 24-hour period is the difference between 65 °F and the average outdoor air temperature during that period (see Equation 2.9). The average temperature is determined by averaging the high and low temperature for the 24-hour period.

(Equation 2.9)

$$DD_{daily} = (65 - T_{ave})$$

where:

DD_{daily} = number of degree days accumulated in a 24-hour period

T_{ave} = average of the high and low outdoor temperatures for the 24-hour period (°F)

The total heating degree days for a month or an entire year is found by totaling the daily heating degree days over the desired time period. Many heating reference books contain tables of monthly and annual heating degree days for many major cities in the United States and Canada. A sample listing is given in Figure 2–25.

CITY	degrees F • days (base 65)
Atlanta, GA	2990
Baltimore, MD	4680
Boston, MA	5630
Burlington, VT	8030
Chicago, IL	6640
Denver, CO	6150
Detroit, MI	6290
Madison, WI	7720
Newark, NJ	4900
New York, NY	5219
Philadelphia, PA	5180
Pittsburgh, PA	5950
Portland, ME	7570
Providence, RI	5950
Seattle, WA	4424
Syracuse, NY	6720
Toronto, ONT	6827

Figure 2–25 Annual heating degree days for some major cities

Heating degree day data has been used to estimate fuel usage for several decades. Such data is often recorded by fuel suppliers, utility companies, and local weather stations. Degree day statistics are also frequently listed in the weather section of newspapers.

When the heating degree day concept was first conceived, fuel was relatively inexpensive, and buildings were poorly insulated. Building heat loss was substantially greater than for comparably sized buildings using modern construction techniques.

Estimating seasonal energy consumption using heating degree days assumes that buildings need heat input from their heating systems whenever the outside temperature drops below 65 °F. While this may be true for many poorly insulated buildings, it is often incorrect for better-insulated buildings. Many energy efficient homes can maintain comfortable interior temperatures, without heat input from their space heating systems, even when outdoor temperatures drop into the 50s. This is a result of greater internal heat gains from appliances, lights, people, and sunlight, combined with much lower rates of heat loss. Internal heat gains eliminate the need for a corresponding amount of heat from the building's heating system. Passive solar buildings are a good example. On sunny days, many require no heat input from their "backup" heating systems, even when outside temperatures are below zero!

Another complicating factor is that some heating degree days occur in months when most heating systems are turned off. This is especially true in northern climates where a few heating degree days are recorded in June, July, and even August. Obviously, if the heating system is turned off, there will be no fuel consumption associated with these degree days. Still, they are included in the seasonal total.

To compensate for the effect of internal heat gains, and warm weather degree days, ASHRAE devised a correction factor known as C_d. This factor is plotted against total heating degree days in Figure 2–26, and will be used to estimate seasonal heating energy use.

The C_d value given by this graph is representative of average conditions over a large number of buildings. The specific C_d value that applies to a given building might be higher or lower than the value shown by the graph. For example, a passive solar building in a sunny climate such as Denver, is likely to have a lower C_d value because it may get a major part of its heating energy from the sun. On the other hand, a building with small windows, in a shaded location, or one

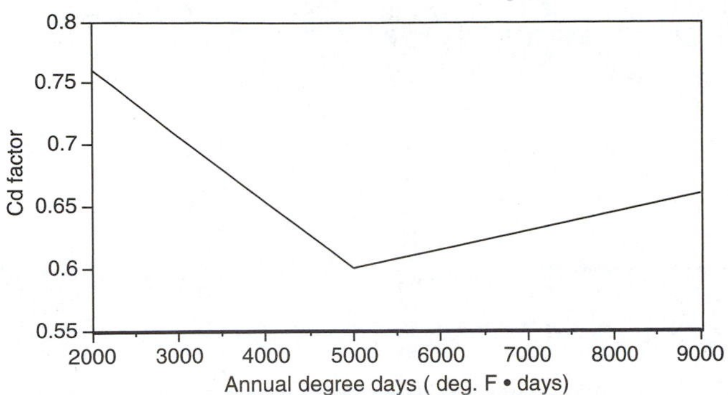

Figure 2–26 Mean values of C_d versus annual heating degree days suggested by ASHRAE. Data source 1989 *ASHRAE Fundamentals Handbook.*

with very little internal heat generation, may have a larger value of C_d. This variation will directly effect the estimate of annual fuel usage.

Equation 2.10 can be used to *estimate* annual heating energy required by the building:

(Equation 2.10)

$$E_{annual} = \frac{(Q_{design})(DD)(24)(C_d)}{1{,}000{,}000(\Delta T_{design})}$$

where:

E_{annual} = estimated annual heating energy *required by the building* **(MMBtu)**

NOTE: 1 MMBtu = 1,000,000 Btu

Q_{design} = design heating load of the building (Btu/hr)

DD = annual total heating degree days at the building location (degree days)

C_d = correction factor from Figure 2–26 (unitless)

ΔT_{design} = design temperature difference at which design heating load was determined (°F)

Example 2.9: A house has a calculated design heating load of 55,000 Btu/hr when its interior temperature is 68 °F, and the outdoor temperature is –10 °F. The house is located in a climate having 7000 annual heating degree days. Estimate the annual heating energy that must be delivered to this building for a typical heating season.

Solution: The *C*d factor for a 7000 heating degree day climate is estimated at 0.63 using Figure 2–26. This and the other given data are then substituted into Equation 2.10:

$$E_{annual} = \frac{(55{,}000)(7000)(24)(0.63)}{1{,}000{,}000[68-(-10)]}$$

$$= 74.6 \text{ MMBtu/heating season}$$

Discussion: The building needs 74.6 MMBtu (or 74,600,000 Btus) delivered to it over a typical heating season. However, this calculation does *not* account for the inefficiencies of the heat source or the distribution system in producing and delivering this heat to the building. These inefficiencies will make the *purchased* energy requirement larger than the *delivered* energy requirement. The lower the efficiency of the heat source, the higher the purchased energy requirement will be relative to the delivered energy requirement. These inefficiencies will be accounted for when estimating annual heating cost in the next section.

Estimating Annual Heating Cost

Knowing how many MMBtus a building requires over a typical heating season is not of much use to most owners. Their main concern is the cost of this energy, as well as how various fuel options compare in cost. To provide this information, the cost of available fuel options must be expressed on a "unit price" basis. The units commonly used are dollars per million Btus *delivered* (abbreviated $/MMBtu). The seasonal efficiency of the heat source and distribution system are factored in when determining the cost of *delivered* energy. Once the unit price of the fuel options has been calculated, a simple multiplication determines the estimated seasonal heating cost in dollars.

The formulas given in Figure 2–27 can be used to convert the purchase costs of several common fuels to a unit price of dollars per million Btus delivered.

Some fuels are priced on "sliding scales." The cost per unit of fuel decreases as the monthly usage increases. This is typical of natural gas pricing from many utilities. In such cases, use a weighted average price per unit of fuel based on usage.

In many cases utilities assess a minimum service charge regardless of the amount of fuel used. One can think of this as "meter rental." Since it has to be paid regardless of how much fuel is purchased, this charge should be deducted from the bill before calculating the purchase price of the fuel. However, such minimum service charges should be considered when weighing the merit of using a utility-supplied fuel versus another fuel that might be available without a minimum monthly service charge. For example, if an owner decides to use natural gas supplied from a utility for space heating, and does not plan to use it for other appliances, the minimum service charges associated with that fuel should be added to the estimated cost of natural gas consumed. The total cost should then be compared to that of other fuel options. Similarly, any known service fees or other fixed costs associated with a given fuel should be included in the total cost associated with that fuel when making comparison.

Example 2.10: Estimate the annual space heating cost of a building requiring 100 MMBtu per year for each of the following fuel options.

- Electric resistance heat at $0.10/Kwhr (kilowatt-hour) and 100% efficiency

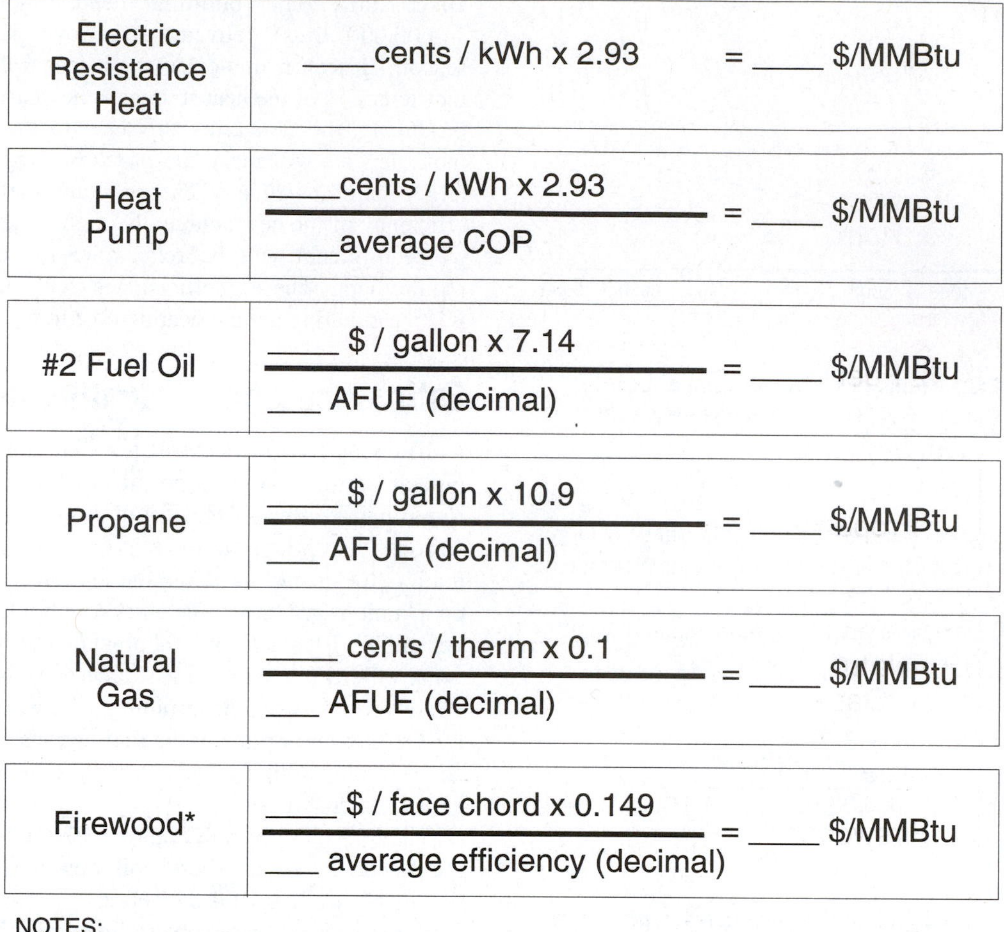

Electric Resistance Heat	____cents / kWh x 2.93	= ____ $/MMBtu
Heat Pump	____cents / kWh x 2.93 / ____average COP	= ____ $/MMBtu
#2 Fuel Oil	____ $ / gallon x 7.14 / ___ AFUE (decimal)	= ____ $/MMBtu
Propane	____ $ / gallon x 10.9 / ___ AFUE (decimal)	= ____ $/MMBtu
Natural Gas	____ cents / therm x 0.1 / ___ AFUE (decimal)	= ____ $/MMBtu
Firewood*	____ $ / face chord x 0.149 / ___ average efficiency (decimal)	= ____ $/MMBtu

NOTES:
1. $/MMBtu = dollars per million Btu of heat delivered to building
2. kWhr = kilowatt • hour = 3413 Btu
3. Average COP = average Coefficient of Performance during heating season (for geothermal heat pump with low temp. distribution system generally 2.5 to 4.0)
4. AFUE = Annual Fuel Utilization Efficiency of appliance (for typical oil or gas boiler use 0.75 to 0.80)

* Assumes a 50/50 mix of maple and beech dried to 20 percent moisture content. Price is for 4 ft. x 8 ft. x 16 inch face chord split and delivered.

Figure 2–27 Equations for determining the unit price of various fuels based on purchase price and efficiency.

- Electrically-operated heat pump with seasonal average COP = 2.5
- #2 fuel oil at $1.00 per gallon and 75% annual heat source/delivery efficiency
- Natural gas at $0.90 per therm and 75% annual heat source/delivery efficiency
- Propane at $1.25 per gallon and 75% annual heat source/delivery efficiency
- Firewood at $45 per face chord and 50 percent annual heat source/delivery efficiency

Solution: Substituting these numbers into the formulas of Figure 2–27 yields the results shown in Figure 2–28.

The total estimated heating cost for the building requiring 100 MMBtu/ season is as follows:

- Electric resistance heat:
  100 MMBtu × 29.3 $/MMBtu = $2930
- System using heat pump:
  100 MMBtu × 11.72 $/MMBtu = $1172
- System using #2 fuel oil:
  100 MMBtu × 9.52 $/MMBtu = $ 952
- System using natural gas:
  100 MMBtu × 12.00 $/MMBtu = $1200
- System using propane:
  100 MMBtu × 18.17 $/MMBtu = $1817
- System using firewood:
  100 MMBtu × 13.41 $/MMBtu = $1341

Electric Resistance Heat	10 cents / kWh x 2.93	= 29.30 $/MMBtu
Heat Pump	10 cents / kWh x 2.93 / 2.5 average COP	= 11.72 $/MMBtu
#2 Fuel Oil	1.00 $ / gallon x 7.14 / 0.75 AFUE (decimal)	= 9.52 $/MMBtu
Propane	1.25 $ / gallon x 10.9 / 0.75 AFUE (decimal)	= 18.17 $/MMBtu
Natural Gas	90 cents / therm x 0.1 / 0.75 AFUE (decimal)	= 12.00 $/MMBtu
Firewood*	45 $ / face chord x 0.149 / 0.50 ave. efficiency (decimal)	= 13.41 $/MMBtu

NOTES:
1. $/MMBtu = dollars per million Btu of heat delivered to building
2. kWhr = kilowatt • hour = 3413 Btu
3. Average COP = average Coefficient of Performance during heating season (for geothermal heat pump with low temp. distribution system generally 2.5 to 4.0)
4. AFUE = Annual Fuel Utilization Efficiency of appliance (for typical oil or gas boiler use 0.75 to 0.80)

* Assumes a 50/50 mix of maple and beech dried to 20 percent moisture content. Price is for 4 ft. x 8 ft. x 16 inch face chord split and delivered.

Figure 2–28 Completed heat load worksheet

Discussion: It's obvious that electric resistance heat, at the cost assumed, is by far the most expensive heating "fuel" option. While the others are significantly lower, their relative standings are based on the assumed pricing and heat source/delivery efficiencies.

SUMMARY

This chapter has presented straightforward methods for estimating the design space heating load of residential and light commercial buildings. This step is an essential part of overall system design. It should never be skipped in favor of immediately laying out piping systems, etc. Many presently installed heating systems that yield marginal comfort testify to the importance of knowing the load before designing the system.

KEY TERMS

air change method
air film resistance
building heating load
degree days
design heating load
effective total R-value
exposed surface
infiltration heat loss
MMBtu
97.5% design dry bulb temperature
room heating load
thermal resistance
thermal envelope
total R-value of an assembly

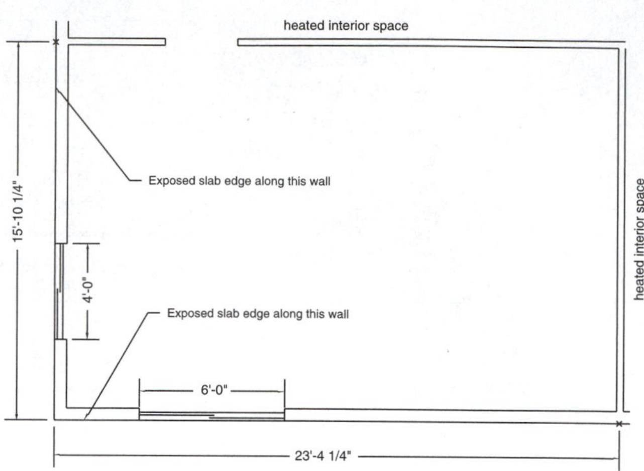

Figure 2–30 Floor slab for exercise 5.

CHAPTER 2 QUESTIONS AND EXERCISES

1. One wall of a room measures 14 ft long and 8 ft high. It contains a window 5 ft wide and 3.5 ft high. The wall has an effective total R-value of 15.5. Find the rate of heat flow through the wall when the inside air temperature is 68 °F and the outside temperature is 5 °F.
2. Using the data in Appendix B, determine the total effective R-value of the wall shown in Figure 2–29. Assume that 15 percent of the wall area is wood framing.
3. The rate of heat flow through a wall is 500 Btu/hr when the inside temperature is 70 °F and the outside temperature is 10 °F. What is the rate of heat flow through the same wall when the outside temperature drops to –10 °F and the inside temperature remains the same?
4. The perimeter of the basement of a house measures 120 ft. The masonry wall is 8 ft tall with 1 ft exposed above grade. The wall is insulated with 2 inches of extruded polystyrene insulation. The basement air temperature is maintained at 65 °F. Determine the rate of heat loss through the basement walls when the outside air temperature is 10 °F. Assume the R-value of the masonry wall is 2.0.
5. Determine the rate of heat loss through the slab-on-grade floor shown in Figure 2–30. The exposed edges are insulated with 1.5 inches of extruded polystyrene. The outside air temperature is 0 °F and the slab is maintained at 90 °F.
6. Using the heat load worksheet in Figure 2–14 determine the design heat loss of the room shown in Figure 2–31. Assume the exterior walls are constructed as shown in Figure 2–29. Assume the outdoor design temperature is –10 °F and the desired indoor temperature is 70 °F.

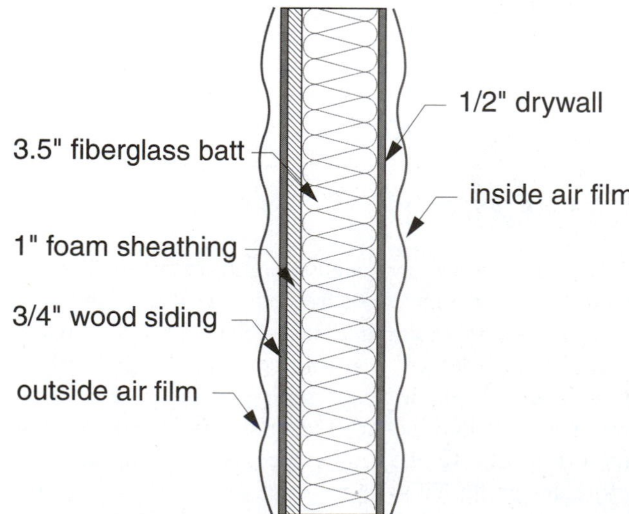

Figure 2–29 Wall cross-section for exercise 2.

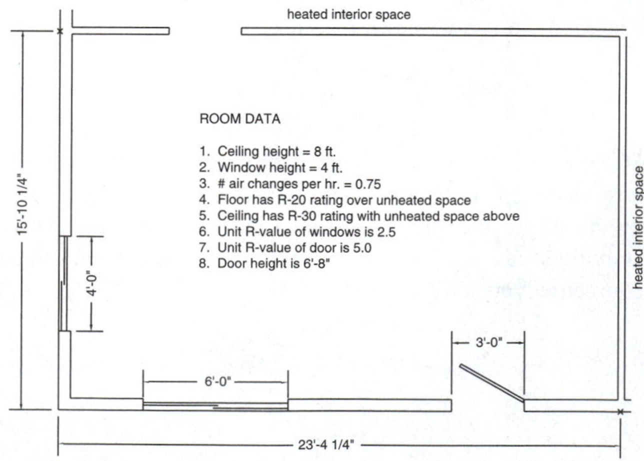

Figure 2–31 Room floor plan for exercise 6.

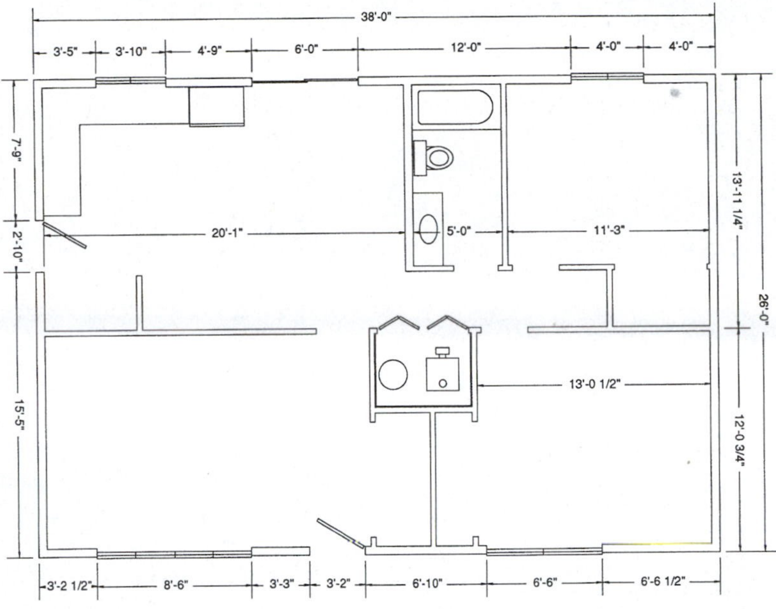

Figure 2–32 House floor plan for exercise 7.

7. Using the heat load worksheet in Figure 2–14, and the construction information used for the example house in section 2.7, determine the design heat loss of the entire house shown in Figure 2–32. Assume all floors are slab on grade. Estimate any dimensions you need by proportioning it to the dimensions shown.
8. Using the Room Heat Load Estimator module in the Hydronics Design Studio, determine the design heat loss of the room described in Question 6. Print the results.
9. Using the Room Heat Load Estimator module in the Hydronics Design Studio, determine the design heat loss of the basement shown in Figure 2–33. Assume an inside temperature of 65 °F and an outside temperature of 0 °F.
10. Using the Room Heat Load Estimator module in the Hydronics Design Studio, recalculate the room load for the example house used in Section 2.7 with the following changes:

    Window R-value = 1.6

    Walls are 2 × 4 studs with 3.5 inches of cellulose insulation

    Ceiling is insulated with 6 inches of blown fiberglass

    Slab edge insulation is 1-inch extruded polystyrene

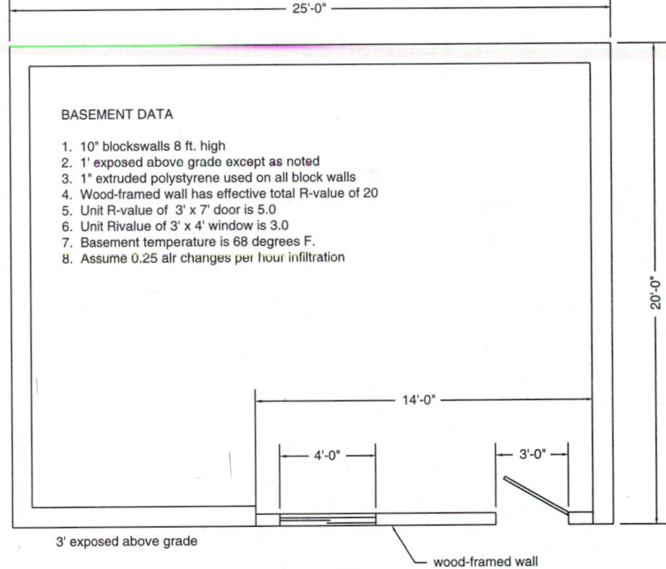

Figure 2–33 Basement plan for exercise 9

HYDRONIC HEAT SOURCES 3

OBJECTIVES

After studying this chapter, you should be able to:

- Evaluate several types of hydronic heat sources for a given application
- Describe several types of boilers and how they are constructed
- Define and work with several boiler performance ratings
- Understand the consequences of excessively oversizing a boiler
- Determine favorable applications for condensing boilers
- Identify the advantages of multiple boiler systems in certain applications
- Discuss the operation of electric thermal storage systems
- Explain the operating principles of a hydronic heat pump
- Predict how operating conditions affect operation of a hydronic heat pump
- Evaluate the suitability of solar energy systems as hydronic heat sources
- Describe the strengths and limitations of wood-fired boilers

3.1 INTRODUCTION

This chapter provides an overview of several types of heat sources suitable for hydronic heating systems. It describes design considerations for each type of heat source, notes the applications where each has certain technical or economic advantages, and illustrates basic schematics for certain types of equipment. Related topics such as exhaust handling, combustion air supply, and methods of piping are also discussed. Several performance ratings for heating output and efficiency are presented with emphasis on using the proper rating for sizing the heat source, as well as predicting its annual energy usage.

This chapter is not meant to be an exhaustive reference on each type of heat source. The details necessary for proper installation of a specific device should be obtained directly from the manufacturer. Sometimes the building code or mechanical equipment codes of a given municipality will influence both equipment selection and installation. For this reason, always check with local codes officials for any specific limitations affecting a proposed hydronic heat source.

Finally, although the heat source is certainly a major piece of equipment, it is still only part of the *system*. Even the best heat source, when misapplied, will yield disappointing performance or fail prematurely. As you read about the different heat sources, pay particular attention to temperature and flow rate limitations that ultimately affect how the heat source interfaces with the remainder of the system.

3.2 CLASSIFICATION OF HYDRONIC HEAT SOURCES

Within this chapter, hydronic heat sources are classified as follows:

- Conventional gas- and oil-fired boilers
- Condensing gas-fired boilers
- Electric resistance boilers
- Electric thermal storage equipment
- Hydronic heat pumps
- Renewable energy heat sources

Of these, conventional gas- and oil-fired boilers are currently used in the vast majority of residential and light commercial hydronic systems in North America. This is largely due to the unit price of natural gas and fuel oil relative to other energy options in the major hydronic markets.

Electrically powered heat sources, such as electric boilers and hydronic heat pumps, have not gained a market share comparable to fossil-fuel boilers to date. However, buildings serviced by electric utilities with low kilowatt-hour rates, or time-of-use rates, can be good candidates for electrically powered heat sources.

Hydronic heat sources that rely on renewable energy currently represent a small portion of the overall market. Nevertheless, when properly applied, they are both technically and economically viable, and will likely gain an increased market share in the future.

3.3 GAS- AND OIL-FIRED BOILER DESIGNS

Most boilers that operate on natural gas or fuel oil can be classified according to their physical construction and heat exchanger material. This section gives a brief description of all major types of gas- and oil-fired boilers.

Cast-iron Sectional Boilers

The cast-iron **sectional boiler** is currently the most common type of boiler used in residential and light commercial buildings in North America. In this type of boiler, water is heated in cast-iron chambers called **sections.** The sections are bolted together to form a boiler block. The hot gases generated in the combustion chamber near the bottom of the boiler rise through cavities between the sections, and transfer heat to the water within. The greater the number of sections, the greater the rate of heat output from the boiler. Figure 3–1 shows a typical sectional **wet-base boiler**.

Most cast-iron sectional boilers for residential and small commercial use are sold as **packaged boilers.** This implies that components such as the circulator, burner, and controls are selected and mounted onto the boiler by the manufacturer. The components selected by the manufacturer assume the boiler will be used in a typical residential system. In some cases, one or more of these components may not be the best choice for a specific application. For example, the small circulator supplied as part of a packaged boiler may not produce the flow rate and pressure differential required by a radiant floor heating system. For this reason, some manufacturers also sell a "bare-bones" version of the boiler without such components. Figure 3–2 shows a typical packaged oil-fired boiler.

Cast-iron sectional boilers are limited to use in closed-loop hydronic systems. Properly designed, these systems rid themselves of dissolved air after a few days of operation. Following this, there is very little further oxidation of the cast iron surfaces. However, *improperly designed or installed systems that allow a continued presence of dissolved air in the system's water will rapidly corrode cast iron.* This can result from undetected (or uncorrected) water leaks, chronic opening of pressure-relief valves, misplaced air vents, and other abnormalities. When properly applied, cast-iron boilers can last for decades. Service lifetimes in excess of 30 years are common.

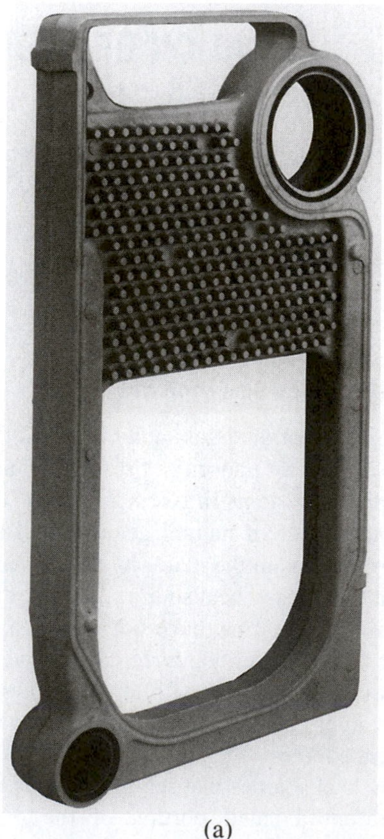

(a)

(b)

Figure 3–1 Example of a (a) wet-base boiler section, and (b) boiler block assembly. *Courtesy of Weil-McLain Corporation.*

Figure 3–2 Example of a packaged cast-iron sectional boiler with factory-installed oil burner, circulator, and controls. *Courtesy of Dunkirk Radiator Corporation.*

The amount of metal in cast-iron sectional boilers makes them relatively heavy. Even a small cast-iron boiler can weigh from 300 to 400 pounds. Such boilers typically contain 10 to 15 gallons of water. The combination of metal and water weight gives cast-iron boilers the ability to absorb a significant amount of heat. Such boilers are said to have a high **thermal mass.**

High thermal mass can be both a desirable and undesirable characteristic. If the boiler is significantly oversized and experiences long off-times between firing cycles, much of the heat stored in the boiler's thermal mass can be carried up the flue by air currents passing through the combustion chamber. This can significantly lower the seasonal efficiency of the boiler. To reduce **off-cycle heat loss**, modern gas-fired high mass boilers are equipped with flue dampers that close whenever the boiler is not firing. Some oil-fired high mass boilers use burners equipped with air shutters that automatically close during the off-cycle for the same purpose. A well-insulated boiler housing, or **jacket** as it is commonly called, also reduces standby heat loss from the boiler.

High thermal mass boilers can also have a favorable effect in systems that are divided into many small, independently controlled zones. The energy stored in the thermal mass helps lengthen the on-cycle as well as the off-cycle. This reduces wear on the burner and improves seasonal efficiency. The desired effect is for a high thermal mass boiler to act as a well-insulated thermal storage device during the off-cycle.

Some manufacturers design the cast-iron sections for vertical passage of flue gases. The sections shown in Figure 3–1 are an example of this geometry. To clean the flue passages in these boilers, the sheet metal top panel, flue connection, and flue gas collector assembly must be removed before a cleaning brush can be maneuvered between the sections.

Another type of cast-iron boiler design uses horizontal flue gas passages between the sections with a hinged door to access both the combustion chamber and the flue gas passages. The exhaust gases usually make two or more passes across the sections before exiting the boiler, allowing more heat to be transferred to the water. Such designs also make it simple to clean the combustion chamber and flue gas passages. An example of a cast-iron sectional boiler with horizontal exhaust gas flow is shown in Figure 3–3.

Figure 3–3 Cast-iron sectional boiler with horizontal flue passages and a hinged combustion chamber door for simplified cleaning. *Courtesy of Buderus Hydronic Systems.*

Steel Fire-tube Boilers

Steel has been used in boiler construction for decades. In a steel **fire-tube boiler,** water surrounds a group of steel tubes through which the hot combustion gases pass. Spiral-shaped baffles known as **turbulators** are inserted into the fire-tubes to increase heat transfer by inducing turbulence and slowing the passage of the exhaust gases. The fire-tubes are welded to steel bulkheads at each end to form the overall heat exchanger assembly.

In some boilers, the tubes are oriented vertically. Figure 3–4 shows an example of a vertical fire-tube boiler. Notice the cut away turbulators within the fire tubes. In this design, the fire-tube assembly sits on top of the combustion chamber.

Other boiler designs are built around horizontal fire tubes. Some boilers route the flue gases through multiple tubes before they reach the flue pipe connection. This allows additional heat to be extracted from the flue gases. In general, multiple-pass horizontal fire-tube boilers will have slightly higher efficiencies compared to single-pass vertical fire-tube boilers. This is due to the greater contact area between the boiler water and fire tubes. Horizontal fire-tube designs also have the advantage of easier tube cleaning through a removable access panel at the front of the boiler. Figure 3–5 shows a cutaway of a three-pass, horizontal fire-tube boiler.

Steel, like cast iron, is also subject to corrosion from system water containing dissolved air. Because of this, steel fire-tube boilers should only be used in closed-loop systems.

Although steel fire-tube boilers generally have less metal weight than cast-iron boilers of similar capacity, they often hold more water. Residential size fire tube boilers often contain 15 to 30 gallons of water. This results in a boiler with a relatively high thermal mass.

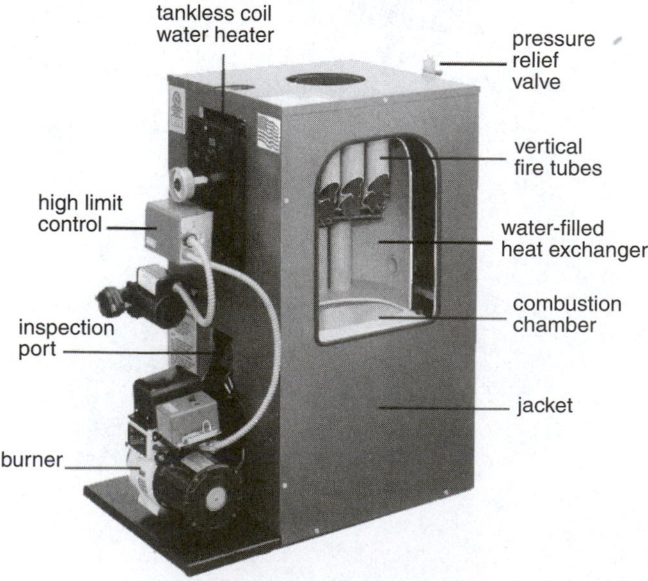

Figure 3–4 Cut-away view of a vertical fire-tube steel boiler. *Courtesy of Columbia Boiler Company.*

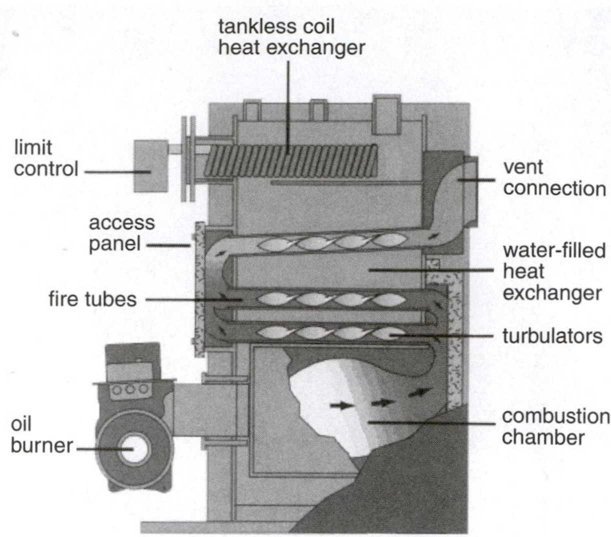

Figure 3–5 Example of a three pass horizontal fire-tube boiler. *Courtesy of Columbia Boiler Company.*

Copper Water-tube Boilers

Some boilers use water-filled copper tubes as their heat exchanger. Design variations include both vertical and horizontal tube arrangements. The copper tubes are usually manufactured with fins that greatly increase the area exposed to the flue gases for improved heat transfer. A finned copper tube heat exchanger is shown in Figure 3–6a. The high thermal conductivity of copper compared to steel or cast iron allows significantly less surface area to yield the same rate of heat transfer. An example of a copper-tube boiler is shown in Figure 3–6b.

The low metal and water content of copper-tube boilers provides minimal thermal mass allowing for very fast warm-up following a cold start. However, low thermal mass provides little storage effect when the load is significantly less than the boiler's heat output. To prevent excessive thermal stress under such conditions, all copper tube boilers must have flow through them while being fired. This requirement must be incorporated into system piping and control design. For example, a copper-tube boiler connected to a distribution system with several zone valves may require "bypass" piping as shown in Figure 3–6c to provide sufficient boiler flow when only one or two zones are operating. In applications where a copper-tube boiler is connected to an extensively zoned/low thermal mass distribution system, an insulated "buffer tank" is often included to prevent short burner cycles. Such applications will be discussed in later chapters.

Many copper-tube boilers use cast-iron tube headers. The internal surfaces of these headers are coated with a heat-fused glass-like material that prevents the water from

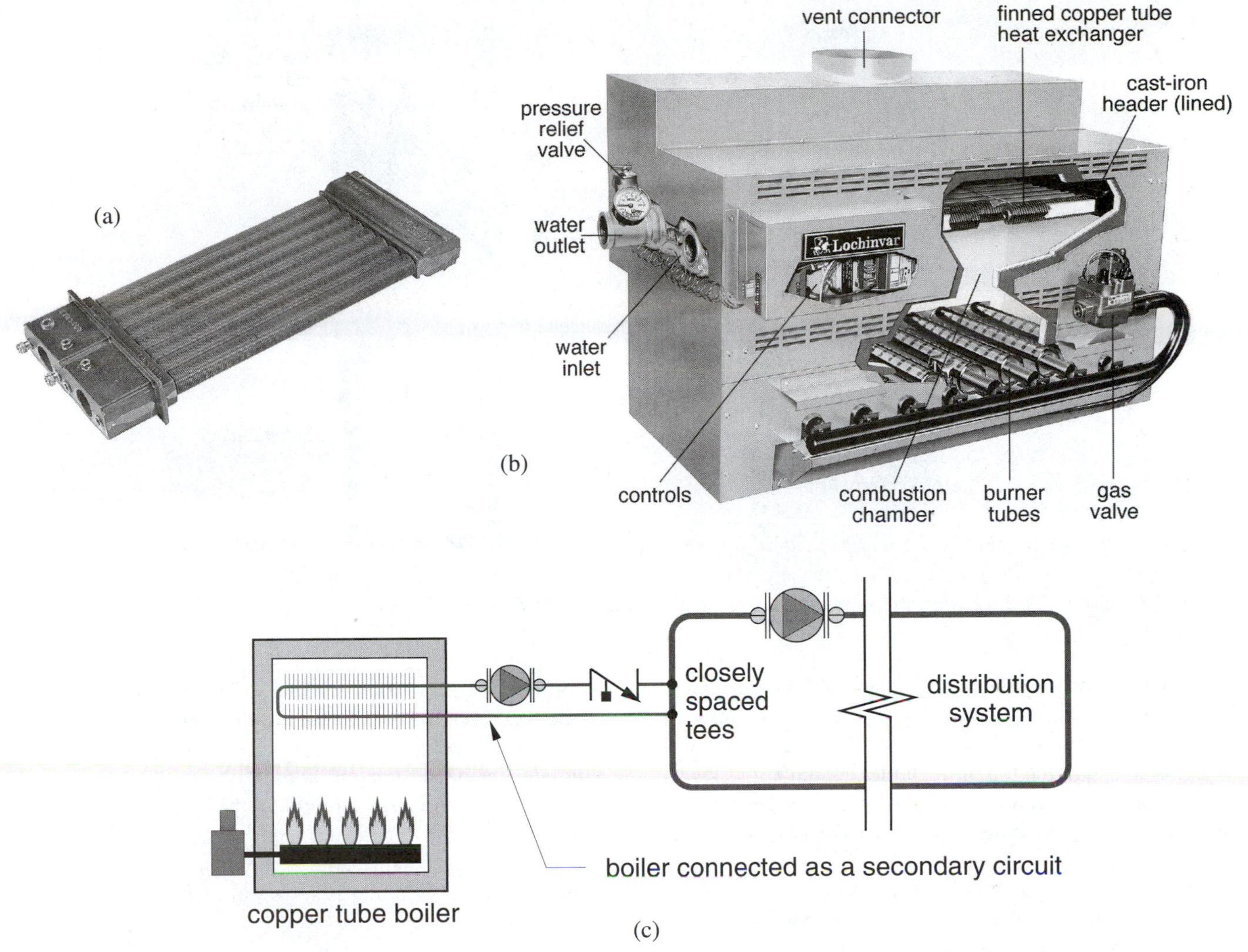

Figure 3–6 (a) Finned copper tube heat exchanger assembly. (b) Cut-away view of a copper water tube boiler. *Courtesy of Lochinvar Corporation.* (c) Piping schematic showing a copper tube boiler connected using Primary/secondary Piping.

contacting the cast iron. This combination of materials allows some copper-tube boilers to operate in open-loop systems in which dissolved oxygen is present in the water. Two examples of such applications would be direct heating of domestic water and swimming pools.

Combined Boiler/Domestic Hot Water Tank Assemblies

A large percentage of residential hydronic systems supply both space heating and domestic hot water. Although several methods have been used to provide domestic hot water from a space heating boiler, it is generally accepted that the most efficient approach uses a storage tank that is indirectly heated by the boiler. Several manufacturers now offer products that combine a boiler and hot water storage tank into a single unit. This approach saves room compared to the use of a separate tank and boiler. It also reduces installation costs because the piping between the boiler and hot water tank is factory installed. Figure 3–7 shows an example of such a product.

Additional Boiler Terminology

Further terminology is used to describe the specifics of how a boiler is designed and constructed.

One approach to boiler design surrounds the combustion chamber with a water-filled heat exchanger. This arrangement is called a wet-base boiler, and it is a common configuration for oil-fired boilers. It has the advantage of exposing more heat exchanger surface to the combustion chamber compared to other designs. The cast-iron sections shown in Figure 3–1 would be used to create a wet-base boiler.

Another common configuration is called a **dry-base boiler.** Such boilers suspend the water-filled heat exchanger above the combustion chamber. Gas-fired cast-iron boilers

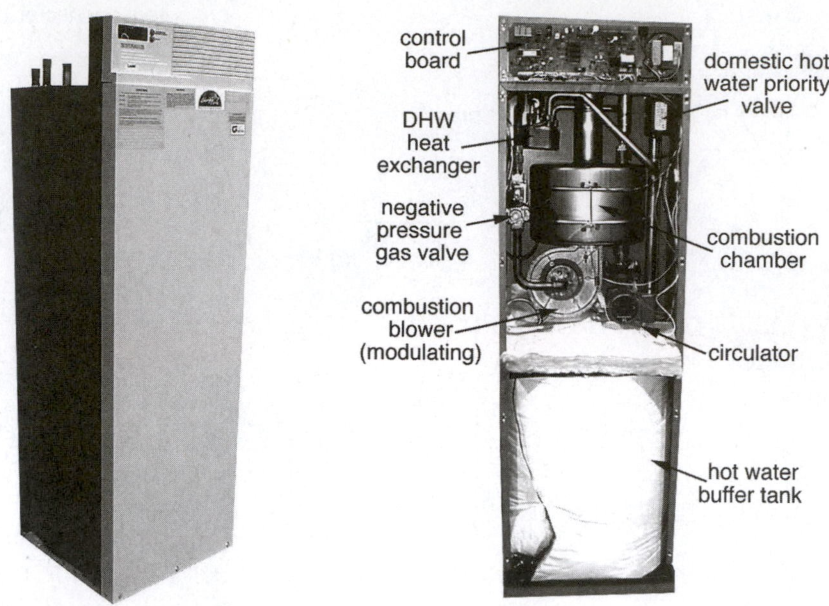

Figure 3–7 Combination boiler and domestic water heater. *Courtesy of Teledyne Laars.*

and oil-fired vertical fire tube boilers are typically designed as dry-base boilers. Hot exhaust gases are drawn upward between boiler sections, or through fire tubes, and enter a flue gas collector at the top of the boiler. Room air is always available at the burner(s). *The use of an automatic flue damper to limit off-cycle draft losses is especially important for this type of boiler.*

A typical cast iron section for a dry-base boiler is shown in Figure 3–8a. Notice the similarity of this section and the upper portion of the wet-base section shown in Figure 3–1. A cutaway of the boiler assembled from these dry-base sections is shown in Figure 3–8b.

A relatively new approach in the gas-fired boiler industry is the wall-hung boiler. It was developed to conserve space in buildings such as apartments or condominiums, and to eliminate the need for a traditional chimney. Wall-hung boilers are usually mounted directly on the interior side of *exterior* walls. Exhaust gases and combustion air travel through a coaxial sleeve that penetrates the wall. An example of a compact wall-hung boiler is shown in Figure 3–9.

Despite their small size, wall-hung boilers are available with heating outputs equal to those of floor-mounted boilers intended for residential and light commercial use. Some can even provide instantaneous domestic water heating as well as space heating.

(a)

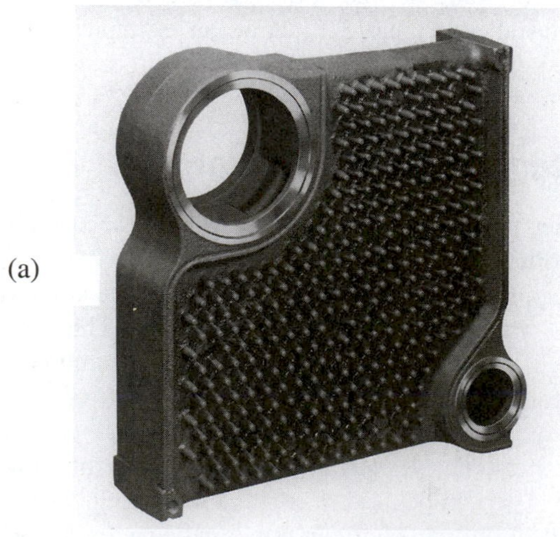

(b)

Figure 3–8 (a) Example of a cast iron section from a dry-base boiler. (b) Construction of a dry-base gas-fired boiler. *Courtesy of Burnham Corporation.*

Figure 3–9a Example of a compact gas-fired wall-hung boiler. *Courtesy of Monitor Products.*

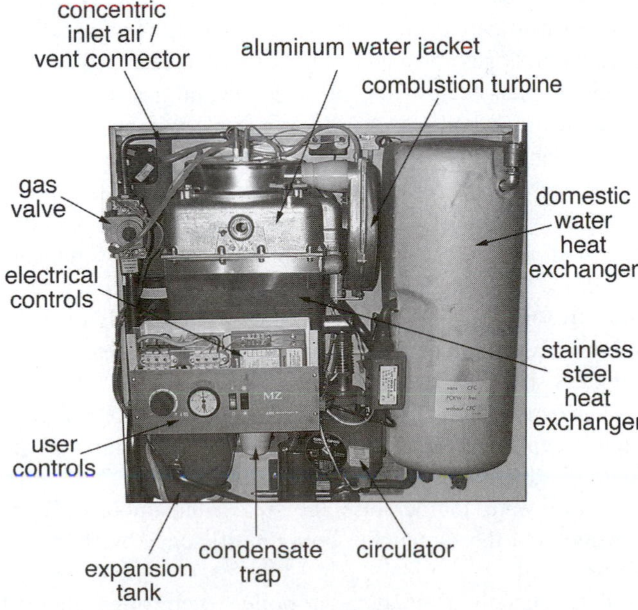

Figure 3–9b Internal construction of boiler shown in Figure 3–9a. *Courtesy of Monitor Products.*

3.4 CONVENTIONAL VERSUS CONDENSING BOILERS

When a hydrocarbon fuel such as natural gas, fuel oil, or propane is burned, water is formed as one of the by-products. In the case of natural gas, the chemical reaction is as follows:

$$CH_4 + 2O_2 \rightarrow CO_2 + 2H_2O$$

Because of the high temperature in the combustion chamber, the water forms as a superheated vapor. As this vapor moves through the boiler's heat exchanger, it is cooled along with the other products of combustion. If sufficient heat is removed from the vapor, it may reach its **dewpoint temperature,** at which point some of the vapor condenses to a liquid. Other compounds in the exhaust stream will also be present in this condensate which make it highly corrosive to any carbon steel or cast-iron surfaces it contacts. This condensate can also cause rapid deterioration of masonry chimneys.

Most boilers currently used in North America were designed with the *intent* that the water vapor produced during combustion would *not* condense within the boiler. Instead, the vapor and the heat it contains, would be carried out of the boiler and through the remainder of the exhaust system before condensing. Boilers designed with this intended mode of operation are called **"conventional" boilers.** *Care must be taken to ensure that conventional boilers do not operate for extended periods during which flue gases are condensing.*

Figure 3–10 shows the relationship between boiler inlet temperature and boiler efficiency. Notice the steep change in the slope of the curve at a temperature of approximately 130 °F. This is a typical dewpoint temperature for a boiler burning natural gas. At higher inlet temperatures, the boiler operates in a **noncondensing mode**. The water vapor passes across the boiler's heat exchanger without condensing. This again, is the *intended* operating mode for a conventional boiler. At temperatures below the dewpoint, a portion of the water vapor condenses on the boiler's heat exchanger. The lower the inlet temperature, the cooler the heat exchanger surface, and the greater the amount of condensate formed. The exact dewpoint temperature depends on the fuel as well as the carbon dioxide content of the exhaust gases. Figure 3–11 shows how the dewpoint varies for gas- and oil-fired boilers.

In the moments following a "cold start," small droplets of condensate form on the heat exchanger of *every boiler.* If the boiler temperature then climbs above dewpoint, this condensate quickly evaporates, and is essentially of no concern. However, if the system prevents the boiler from climbing above the dewpoint temperature, **sustained flue gas condensation** will occur. The latter can quickly and severely damage a boiler not intended to operate in a sustained **condensing mode**.

A good analogy to flue gas condensation in a boiler is seen in the exhaust from a car on a cool day. When the car is first started, the temperature of the exhaust pipe and muffler is

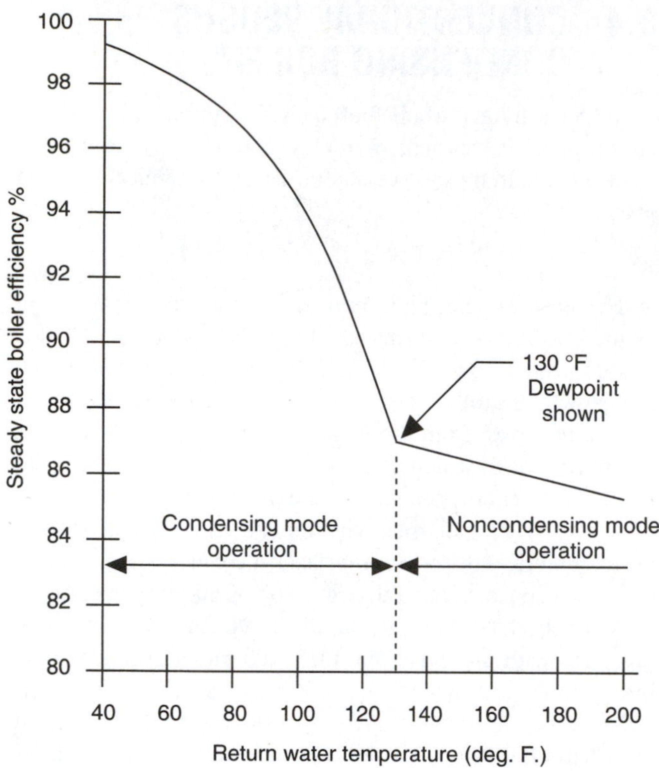

Figure 3–10 Efficiency of a boiler in both condensing and noncondensing modes.

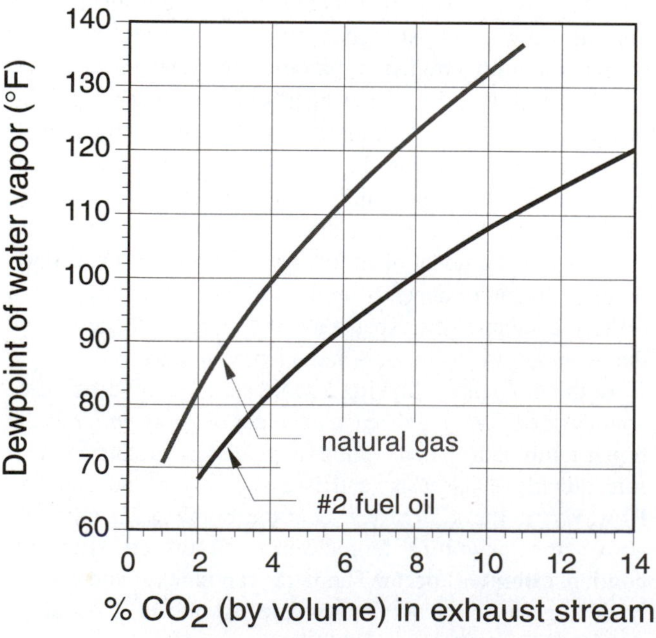

Figure 3–11 Dewpoint of natural gas and #2 fuel oil based on percent CO_2 in exhaust gases.

low enough to cause some of the water vapor produced in the engine to condense by the time it reaches the tailpipe. Water can often be seen dripping from the tailpipe when the car is first started. After the tailpipe warms above the dewpoint temperature, the water vapor no longer condenses, and the dripping stops.

Those who repeatedly drive their car short distances in cold climates often find their tailpipes corrode very quickly. They are succumbing to the same acid-on-steel corrosion reactions experienced in conventional boilers operating with sustained flue gas condensation.

The key to avoiding sustained flue gas condensation is maintaining the boiler inlet temperature above the dewpoint temperature of the exhaust gases. This requires a control that senses the inlet temperature, and, when necessary, prevents the distribution system from extracting heat from the water faster than the boiler can produce this heat. Several methods of accomplishing this are discussed in detail in subsequent chapters.

Condensing Boilers

Following the energy crisis of the early 1970s, higher fuel costs motivated boiler manufacturers to research methods for improving boiler efficiency. The objective was to attain the maximum heat output for each unit of fuel consumed. Analysis of combustion processes reveals that significant amounts of heat can be recaptured from the exhaust stream if the water vapor it contains can be condensed within the boiler.

Under typical operating conditions, each therm (e.g., 100,000 Btus) of natural gas produces approximately 1.15 gallons of water vapor during combustion. If *all* this vapor could be condensed within the boiler, approximately 10,100 Btus of heat would be liberated from the exhaust gases for each therm of gas consumed. This represents about 10 percent of the original energy content of the natural gas, which, in a noncondensing boiler, is wasted as part of the exhaust stream.

Condensing boilers are designed with large heat exchanger surfaces capable of extracting more heat from the exhaust gases compared to the heat exchangers in conventional boilers. With suitable inlet temperature, these heat exchangers can easily cool the exhaust stream below the dewpoint temperature, and thus allow condensation to occur. Figure 3–12 shows an example of a condensing boiler that uses a coiled stainless steel heat exchanger.

The amount of condensate formed within the boiler is highly dependent on the return water temperature of the hydronic distribution system the boiler supplies. The lower the return water temperature, the greater the amount of condensate, and the higher the boiler's efficiency as shown in Figure 3–10.

Early models of condensing boilers were often plagued by corrosion problems. Manufacturers learned from these early experiences, and they have now incorporated new designs and better materials to greatly extend the life of con-

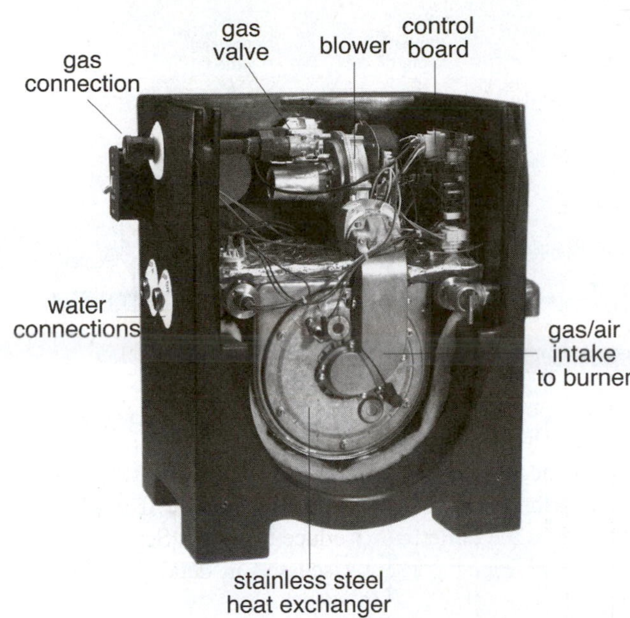

Figure 3–12 Example of a compact condensing boiler with modulating burner. *Courtesy of Heat Transfer Products.*

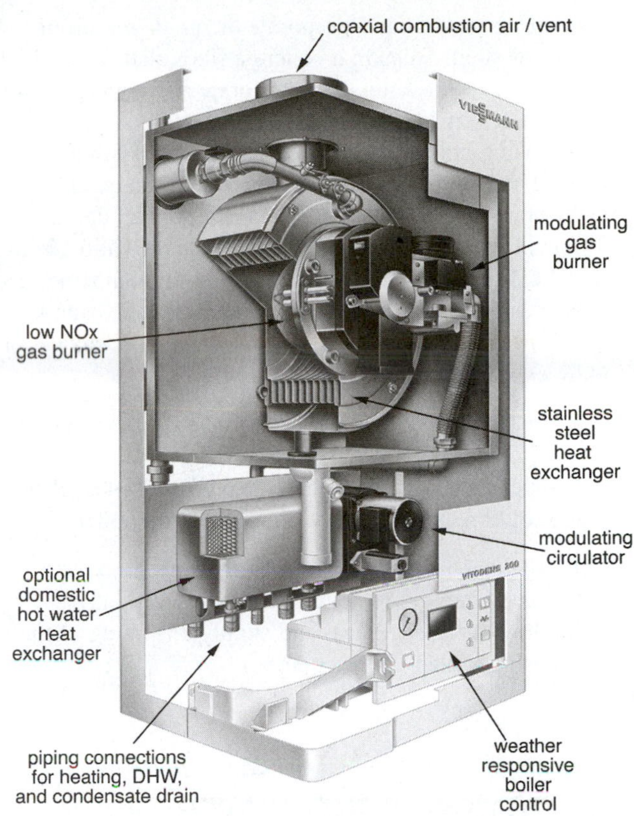

Figure 3–13 Example of a wall hung condensing boiler with modulating burner. *Courtesy of Viessmann Manufacturing.*

densing boilers. Currently available condensing boilers use high grades of stainless steel or proprietary electroplated coatings to effectively eliminate corrosion. Another example of a condensing boiler, this time in a wall-hung configuration, is shown in Figure 3–13.

Oil-fired condensing boilers are not readily available for several reasons. First, the hydrogen content of fuel oil is lower than that of natural gas. This results in less water vapor being formed during combustion and less energy recovery potential. Second, the higher sulfur content of fuel oil leads to higher concentrations of sulfuric acid in the exhaust gas condensate. This increases the risk of corrosion and can cause fouling of the boiler's heat exchanger. Third, soot formation under low temperature operation is more problematic due to oil's higher carbon content.

Conventional versus Condensing: Which Should You Use?

Unfortunately, there is no universal answer to this question. In each situation, the answer depends on the type of hydronic distribution system, installation cost, and energy savings. The following issues should be weighed when considering a condensing boiler.

- Condensing boilers are more expensive than conventional boilers of equal capacity. Their economic viability hinges on recovering this extra cost in fuel savings within a reasonable time. For this to happen, they must be applied in situations where their ability to achieve ultrahigh (95 percent +) efficiency can be realized over most, if not all, operating hours. Low temperature systems such as slab-type radiant floor heating, dedicated domestic water heating, swimming pool heating, and snow melting consistently provide condensing boilers with low inlet water temperatures, allowing them to achieve high efficiency.

- Hydronic systems designed around higher water temperatures (150 ºF or above) may not provide inlet water temperatures low enough for a condensing boiler to actually condense. If used in such a system, a condensing boiler may operate at an efficiency comparable to that of a much less expensive conventional boiler, and thus may *not* be economically justified.

- If a condensing boiler is being considered, the installation location must have provisions for nonattended condensate drainage. In a typical residential system, several gallons of condensate can be formed each day during cold weather, and must be properly disposed of. Most plumbing codes in the United States allow condensate to be routed to the sewer system. A condensate pump may be required depending on the elevation of the boiler relative to the drainage piping. Never assume that condensate will simply evaporate or run down through cracks or small holes in a concrete floor slab. The acidic condensate will eventually deteriorate any concrete with which it comes in contact.

- Condensing boilers cannot be vented into existing flues. The moist/acidic exhaust products can quickly deteriorate clay flues. Masonry chimneys can literally be toppled by extensive, yet undetected, deterioration from moist exhaust gases. Furthermore, the low temperature exhaust stream may not create sufficient draft, even with the assist of power venting. This could create leakage of toxic exhaust gases into the building.
- Only materials listed by the boiler manufacturer should be used for venting. CPVC and AL29-4C stainless steel vent pipe are generally acceptable. Galvanized steel flue pipe should never be used with condensing boilers.
- The existing flue of a dormant chimney (e.g., one not used to vent any other device) can be used as a *chase* through which the exhaust and supply air piping of the condensing boiler can be routed.

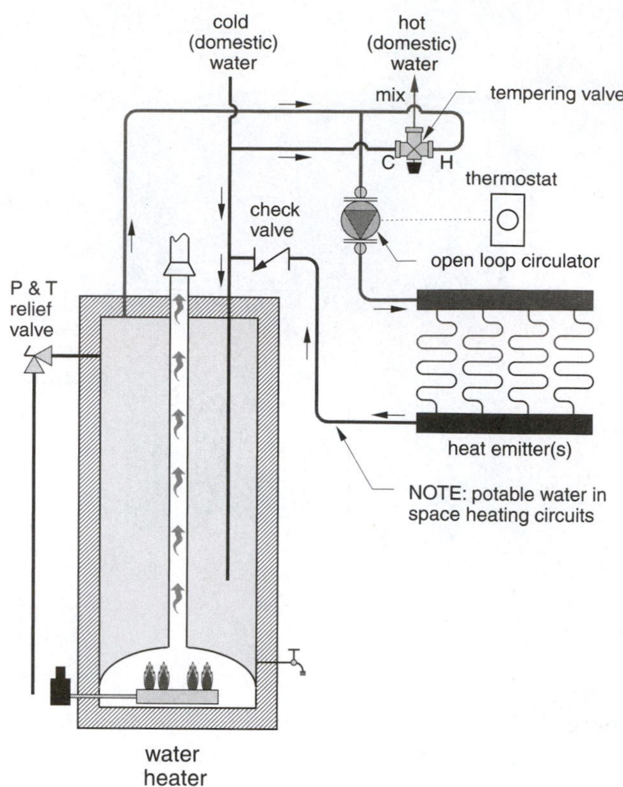

Figure 3–14a Domestic water heater connected as a "duel use" appliance for domestic water heating and space heating. *Note:* Potable water is circulated through the spacing heating distribution system.

3.5 DOMESTIC HOT WATER TANKS AS HYDRONIC HEAT SOURCES

Direct-fired tank-type domestic water heaters have been used as heat sources for hydronic space heating. Their use is generally limited to relatively small spaces with small heating loads such as apartments or residential additions.

Water heaters supplying hydronic space heating loads can be connected as either a "dedicated space heating" device, or as a "dual use" device to supply both space heating and domestic hot water. In the latter category, the space heating distribution system may be isolated from the domestic water, or it may circulate the same potable water that is eventually supplied to fixtures. These options are shown in Figure 3–14a, Figure 3–14b, and Figure 3–14c.

The use of domestic water heaters as hydronic heat sources has created controversy between manufacturers and plumbing code officials. One of the concerns is possible contamination of the potable water from materials used in the space heating portion of the system. Another concern was possible biologic growth in portions of the system where water might lie stagnant for several months during warm weather. Other debated issues include:

- Domestic water heaters are not rated by the ASME (America Society of Mechanical Engineers) as space heating devices. Although they do undergo pressure testing during manufacturing, they are not held to the same standards as boilers.
- When installed so that potable water circulates through space heating circuits, all components in the space heating portion of the system must be compatible with fresh (oxygen containing) water. Ferrous components such as cast-iron circulators, flow-check valves, and unlined expansion tanks would quickly corrode, producing discoloration of domestic water and potential health hazards. Use of bronze or stainless steel circulators and valves to prevent severe oxidation add significant cost to the system. *Lead-based solders such as 50/50 tin lead, commonly used for the nonpotable piping in hydronic heating systems, cannot be used in such systems.*
- Domestic water containing dissolved minerals or other contaminants may precipitate out of solution to foul piping, heat transfer surfaces, circulators, and valves. Such fouling may be difficult or impossible to remove. Aggressive groundwater may lead to corrosion not encountered in closed-loop systems. The use of anticorrosion water treatments is ruled out since they would contaminate water used for domestic purposes.
- Domestic water heating tanks have less heat exchanger area relative to a modern boiler of equivalent capacity. This reduces their operating efficiency, especially if operated at higher temperatures (140 ºF or above). Water heaters also store heated water in direct contact with their flue passage, and are seldom equipped with flue dampers.

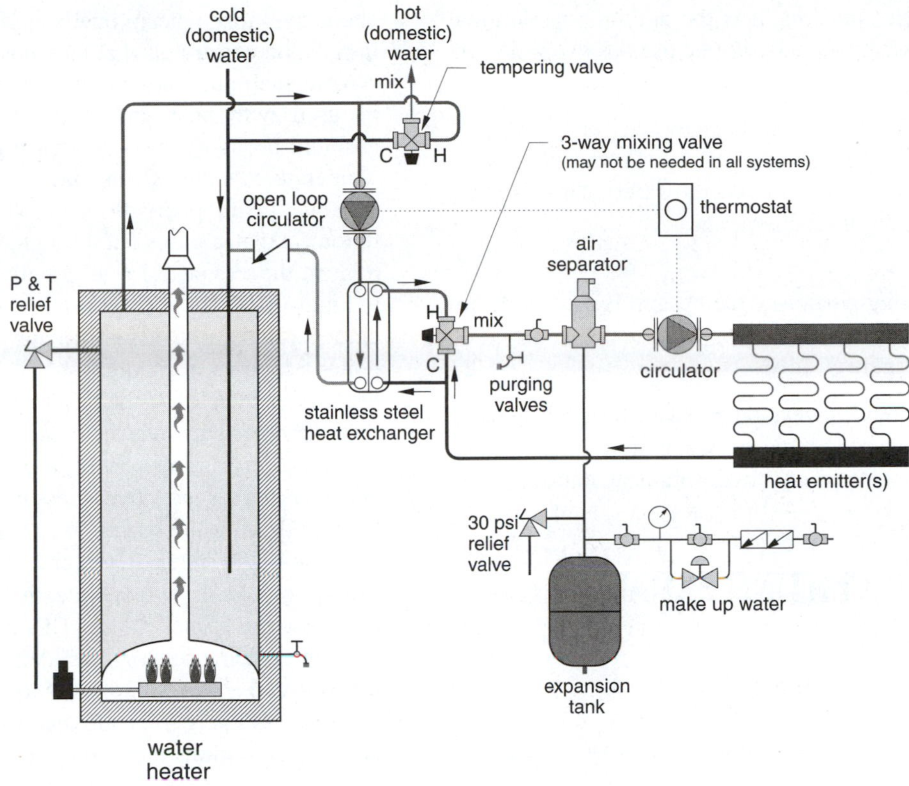

Figure 3–14b Domestic water heater supplies space heating through a heat exchanger.

This increases off-cycle heat loss relative to that of a boiler operating at the same temperature.

- The **heating capacity** of residential water heaters is often substantially lower than that of boilers. Residential gas-fired water heaters are generally capable of producing 30,000 to 50,000 Btu/hr. Oil-fired water heaters have slightly higher capacities depending on the burner nozzle. Although these heat outputs may be adequate for small buildings, they are often inadequate when there are simultaneous demands for both space heating and domestic water heating. For example, during the wake-up period in a typical household, the heating system may have to bring the building temperature up from nighttime setback, while several occupants are also taking showers. In such cases, it is often necessary to temporarily interrupt the space heating load while the tank attempts to supply the domestic water load. When the tank's temperature recovers, the space heating load is turned back on.
- In general, residential water heaters will not last as long as a properly installed boiler. Although their purchase cost is significantly lower, reduced service life can quickly claim the initial savings.

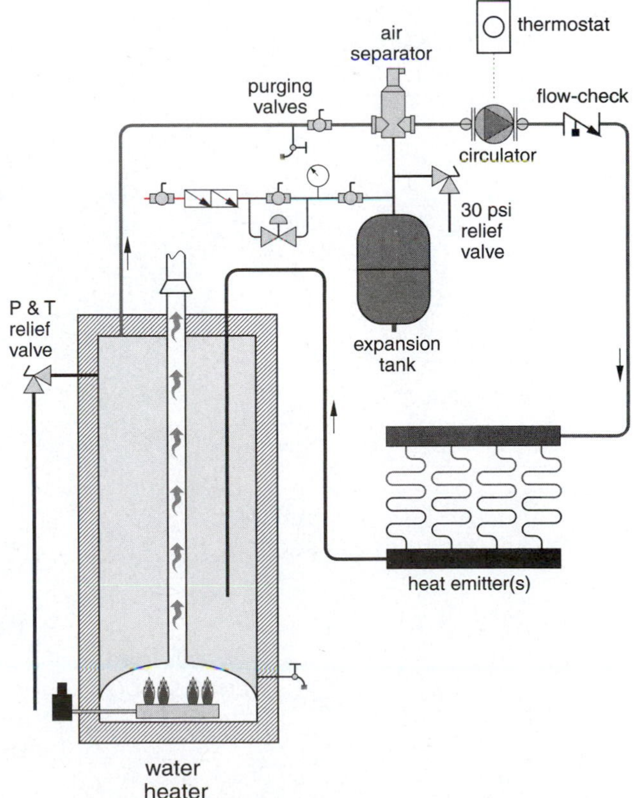

Figure 3–14c Water heater connected as a dedicated heat source.

In light of these concerns, it is the author's opinion that domestic water heaters should only be used as hydronic heat sources when:

A. They are set up as dedicated heat sources, and as such do not supply domestic hot water (as shown in Figure 3–14a). Such setup must include all safety controls that would be present on a boiler.

or

B. The space heating portion of the system is isolated from the domestic water using a suitable heat exchanger (as shown in Figure 3–14b).

Local codes should always be checked for exclusions or special installation requirements when a domestic water heater is being considered as a hydronic heat source.

3.6 POWER VENTING EXHAUST SYSTEMS

The traditional means of removing exhaust gases from a combustion-type heat source is to vent them up a chimney. While this is still the most common approach, there are alternatives that, when properly applied, can help meet unique installation constraints that may preclude chimney venting.

One such alternative is known as **power venting.** It can be used with both gas- and oil-fired boilers, and is an absolute necessity with condensing boilers due to the low temperature of the exhaust gases.

Boilers equipped with power venting use a small blower to force exhaust gases out through an exhaust pipe. On some boilers, the exhaust blower is built into the boiler as shown in Figure 3–15. In other cases, the power venting unit is purchased as a separate assembly, and installed externally to the boiler as shown in Figure 3–16. The vent pipe leading from the blower often exits the building through a sidewall rather than the roof. This reduces costs and simplifies installation.

All power venting systems use a **draft-proving switch** to verify a proper negative pressure in the exhaust system before allowing the burner to fire. In the rare (but possible) event that the exhaust system becomes restricted or the exhaust blower fails to operate, the draft-proving switch will not allow the burner to fire. This makes power venting systems safer than a conventional chimney that cannot automatically stop the burner if draft is lost. Power venting systems are also less likely to be affected by winds or other negative pressure conditions induced within the building.

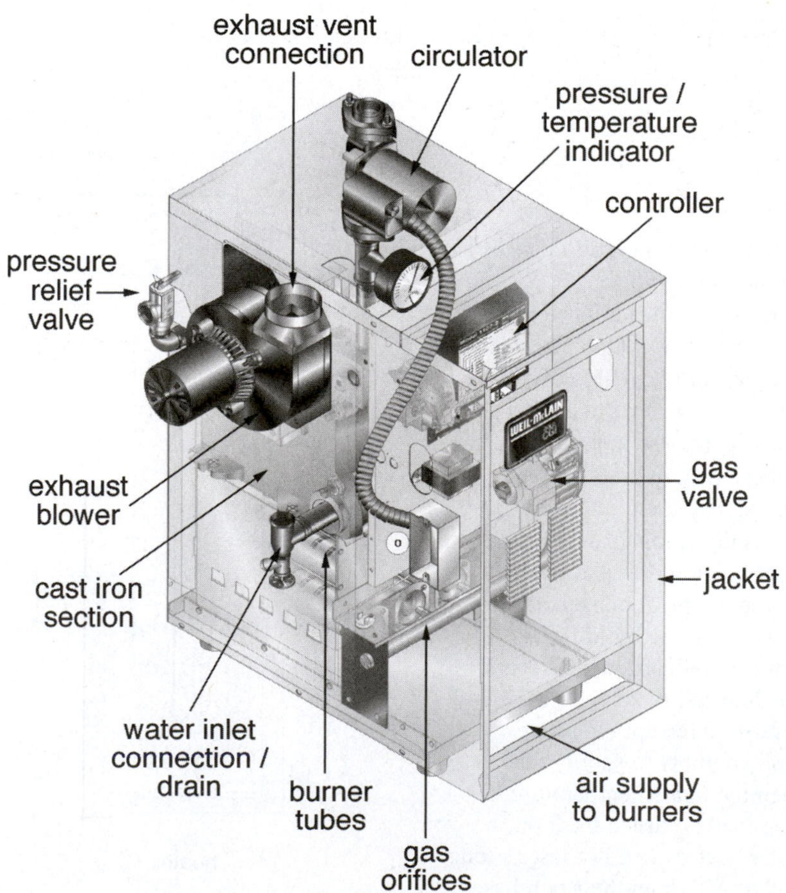

Figure 3–15 Boiler with built-in exhaust blower for side-wall venting. *Courtesy of Weil-McLain.*

(a)

(b)

Figure 3–16 Example of an external power venting system installed on an oil-fired boiler. (a) Interior flue and blower assembly. (b) Outdoor vent terminal. *Courtesy of Tjernlund Corporation.*

When the power venting fan is located in the boiler, the vent pipe from the boiler to the outside is under positive pressure while the boiler is firing. It must be sealed to prevent flue gases from escaping at joints and seams. Only vent pipe approved for positive pressure service should be used. An example of AL29-4C vent pipe designed for positive pressure service is shown in Figure 3–17.

When the power venting fan is at the end of the vent pipe, the latter is under negative pressure and generally does not require sealing.

As always, be sure to use the manufacturer's recommended materials and installation methods.

Direct Vent/Sealed Combustion Systems

A variation of power venting is known as a **direct vent/sealed combustion.** This approach typically uses a coaxial air intake/exhaust assembly through which outside air is drawn for combustion and exhaust gases are routed back outside. A small blower provides the required pressure differential to bring air in and vent exhaust products out.

In a direct vent/sealed combustion system no inside air is used for combustion. Figure 3–18 shows the concept when used in a small wall-hung boiler. Notice the annular passage

Figure 3–17 Positive pressure joining system used with AL29-4C stainless steel venting pipe. *Courtesy of Protech Systems.*

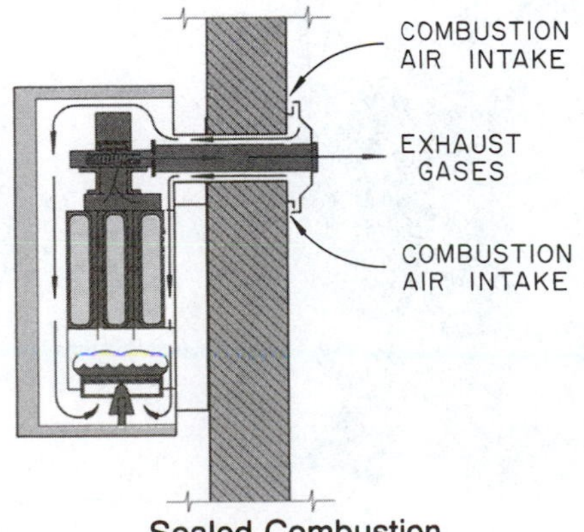

Figure 3–18 A coaxial air intake/exhaust system used with a wall-hung boiler. *Courtesy of Burnham Corporation.*

surrounding the center flue pipe through which outside air is drawn. The incoming air provides a cool metal surface near the wall materials.

Oil-fired boilers can also be configured for sealed combustion. The blower in the oil burner provides the necessary pressure differential to draw in combustion air and vent away exhaust gases. Because of the limited air flow available from the burner, oil-fired sealed combustion boilers generally need to be located close to point where the vent pipe penetrates the wall. An example of an oil-fired sealed combustion boiler is shown in Figure 3–19.

Sealed combustion boilers are especially desirable in buildings that have low air leakage. In such cases, air infiltration alone may not provide sufficient air for proper combustion. A sealed combustion system greatly reduces the possibility that other exhaust fans in the building could cause back drafting of exhaust products from the boiler.

Sealed combustion systems also save energy since outside air rather than heated inside air is used for combustion. In coaxial air intake/exhaust systems such as shown in Figure 3–18, the incoming air is preheated by heat recovered from the exhaust gases prior to reaching the combustion chamber. This improves combustion efficiency.

Many currently available wall-hung boilers use sealed combustion, as do some floor-mounted boilers. An example of the latter is shown in Figure 3–20.

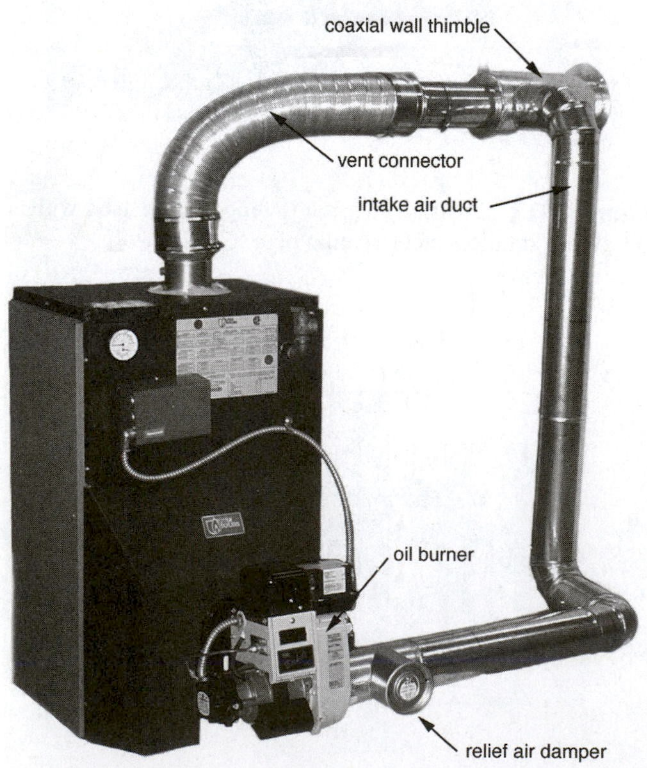

Figure 3–19 Example of a sealed combustion/sidewall vented oil-fired boiler. *Courtesy of Utica Boilers.*

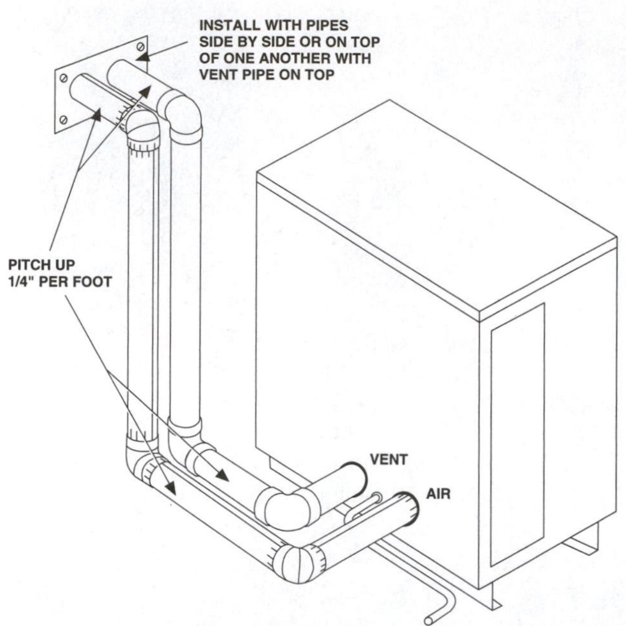

Figure 3–20 Example of a floor-mounted sealed combustion gas-fired boiler. *Courtesy of Weil-McLain.*

3.7 COMBUSTION AIR REQUIREMENTS

Any fuel-burning boiler requires a proper supply of air for safe and efficient combustion, exhaust gas dilution, and boiler room ventilation. Provisions to ensure that such requirements will be met are specified in the latest edition of the **National Fuel Gas Code** (NFPA No. 54/ANSI Z223.1) from the National Fire Protection Association. The provisions of the NFPA code are widely accepted in most state and local building codes.

Under the NFPA code, the requirements for combustion and ventilation air depend on whether the boiler is located in a **confined space** or **unconfined space.** As defined in the code, unconfined space must have a *minimum* of 50 cubic feet of volume per 1000 Btu/hr of gas input rating of *all* gas-fired appliances in that space. Any space not meeting this requirement is considered confined space.

Boilers located in confined space may draw air from adjacent rooms or from outside. When air will be drawn from adjacent rooms, the total volume of the boiler room and the adjacent rooms must meet the above definition for unconfined space. The boiler room must have two openings connecting it with the adjacent space. The *minimum* **free area** of each opening must be 1 square inch per 1,000 Btu/hr of gas input rating of *all* gas-fired equipment within the boiler room, and in no case less than 100 square inches. The top of the upper opening must be within 12 inches of the boiler room ceiling. The bottom of the other opening must be within 12 inches of the floor.

When air is drawn from outside, the size of the required openings vary depending on the type of fuel used and the orientation of supply air ducts. The NFPA code requirements are as follows:

For gas-fired equipment, using *vertical supply air ducting,* or *direct through-the-wall openings:* Provide two openings between the boiler room and the outside, *each* having a minimum *free area* of 1 square inch per 4000 Btu/hr of gas input rating of all gas-fired equipment within the boiler room.

For gas-fired equipment, using *horizontal supply air ducting:* Provide two openings between the boiler room and the outside, *each* having a minimum *free area* of 1 square inch per 2000 Btu/hr of gas input rating of all gas-fired equipment within the boiler room.

For *oil-fired equipment with either horizontal or vertical supply air ducting:* Provide two openings between the boiler room and the outside, each having a minimum *free area* of 1 square inch per 5,000 Btu/hr of all oil-fired equipment in the boiler room.

In all of these cases, one opening, or end of duct, must be within 12 inches of the boiler room ceiling, and the other within 12 inches of the floor. These openings are intended to prevent the boiler room from overheating. They allow natural convection air current to be established to cool the boiler room. They also allow any gases or fumes that might be present in the boiler room to escape.

When ducts are used to bring outside air into the boiler room, their cross-sectional area must equal the free area of the openings to which they connect. The minimum dimension of any rectangular supply air duct allowed by the NFPA code is 3 inches. The NFPA code also specifies that any screening covering these openings cannot be finer than 1/4-inch mesh. The intent is to minimize the chance for the opening to be plugged by grass clippings or other typical debris.

All the above vent area requirements refer to "free area." This is the *unobstructed* area of the opening after accounting for the effects of any screens or louvers covering the opening. Louver manufacturers usually list the free area for each type of louver they offer. The following table lists typical free area to gross area ratios for different types of opening covers. This data should be used for comparison purposes, and not as a substitute for specific data provided by manufacturers.

Type of opening cover:	F/G ratio:
• 1/4-inch screen	0.8
• metal louvers	0.6
• wooden louvers	0.2

Equation 3.1 can be used to determine the required area of the opening or duct.

(Equation 3.1)

$$A_{opening} = \frac{A_{free}}{F/G \text{ ratio}}$$

where:

$A_{opening}$ = required cross-sectional area of each opening or duct (ft²)

A_{free} = free area required by the NFPA code for a given installation (ft²)

F/G ratio = ratio of free area to gross area of the louver or screen as specified by its manufacturer

Example 3.1: The calculated free area required for venting a boiler room is 50 square inches. The exterior of the opening will be covered by metal louvers having a free-to-gross area ratio of 0.6. What is the required gross area of each opening?

Solution: Substituting these numbers in Equation 3.1 and solving yields:

$$A_{opening} = \frac{50}{0.6} = 83.3 \text{ sq. inches}$$

Discussion: The areas required by the code are minimums. Larger areas are also acceptable. The shape of the hole is not specified other than having a minimum dimension of 3 inches in the case of a rectangular opening.

Boilers using sealed combustion also have requirements for ventilation air to cool the space in which they are located. When the sealed combustion boiler is the only combustion appliance in the space, the requirement is to provide two openings, one within 12 inches of the ceiling, the other within 12 inches of the floor. The minimum free area of each opening must be 1 square inch for each 8000 Btu/hr of gas input rating of the sealed combustion boiler. This requirement essentially allows the area of the required openings to be one half that of the openings for nonsealed combustion boilers. The justification is that air flow in such situations is only for ventilation (e.g., cooling) of the boiler room, and not for combustion.

Figure 3–21 shows some possible arrangements for providing combustion/ventilation air. Keep in mind that code requirements can change with time and jurisdiction. Always check with local code officials about any requirements in addition to, or more restrictive than, what is included here.

3.8 BOILER HEATING CAPACITY

The heating capacity of a boiler refers to its rate of useful heat production. In North America, this is usually expressed in Btu/hr or thousands of Btu/hr (MBtu/hr). The word *output* is frequently used as a synonym for heating capacity.

Exactly what portion of the total heat produced by a boiler is useful has led to several different, but standardized, definitions of heating capacity. Each definition contains underlying assumptions about the boiler and the system in which it is used.

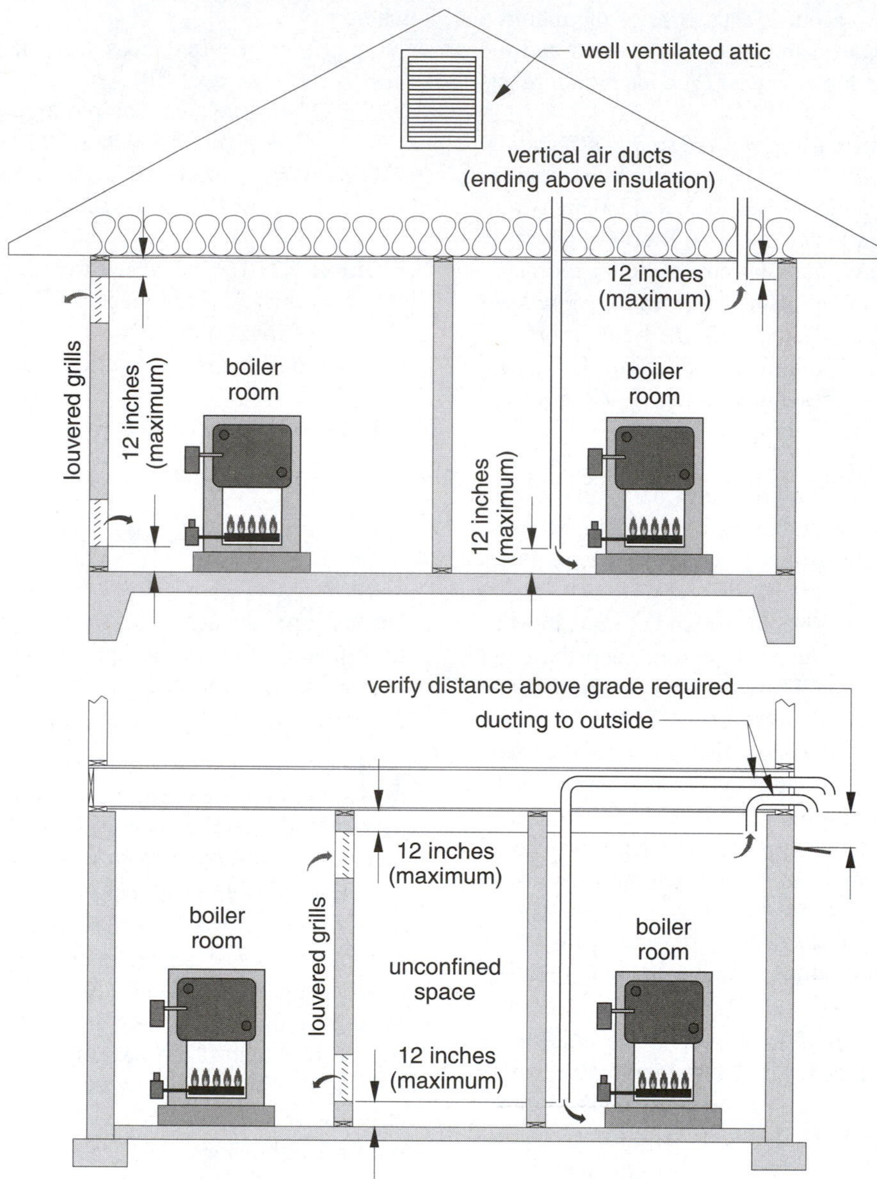

Figure 3–21 Several possible arrangements for combustion/ventilation air supply to a boiler room.

The performance information listed in the boiler manufacturer's literature often refers to one or more of the IBR ratings. IBR stands for the Institute of Boiler and Radiator Manufacturers. This organization was the forerunner of today's Hydronics Institute. The two IBR ratings for boilers capacity are **IBR gross output,** and **IBR net output.**

The IBR Gross Output rating of a boiler is the rate at which heat is transferred to the system's water under steady state operation. Implicit to the definition is the assumption that all heat released from the **boiler jacket** is wasted (e.g., does not offset a portion of the building's heating load).

The IBR Net Output is obtained from the IBR Gross Output rating by multiplying by 0.85. The underlying assumption is that 15 percent of the boiler's gross output is released from piping between the boiler and the intended load, or is needed to warm up the thermal mass of the system following an offcycle. The latter is called the "pickup allowance."

Another index used to describe boiler output is called the **DOE heating capacity.** This rating assumes that all heat losses from the boiler jacket helps to heat the building. The assumption is valid if the boiler is contained within the heated space. DOE heating capacities are only provided for boilers up to 300,000 Btu/hr of fuel input rating.

Because it includes jacket heat losses, the DOE heating capacity of any boiler will always be higher than its IBR Gross Output rating. One or more of these capacity ratings are usually listed on the boiler's data plate and product literature. Figure 3–22 shows a typical table of performance

Boiler Model Number (1)	I=B=R Oil Burner Input (2)		D.O.E. Heating Capacity MBH	I=B=R Net Ratings (3)		Natural Draft Chimney Size	Nozzle Furnished 140 PSIG	A.F.U.E. Rating
	G.P.H.	MBH		Water MBH	Sq. Ft.			
SFH-365	.65	91	79	68.7	458	8×8×15	.60 80°B	86.0
SFH-3100	1.00	140	117	101.7	678	8×8×15	.85 80°B	81.0
SFH-4100	1.00	140	120	104.4	696	8×8×15	.85 80°B	86.0
SFH-3125	1.25	175	144	125.0	834	8×8×15	1.10 60°B	80.0
SFH-4125	1.25	175	149	129.6	864	8×8×15	1.10 80°B	82.5
SFH-5125	1.25	175	151	131.3	875	8×8×15	1.10 80°B	86.0
SFH-4150	1.50	210	175	152.2	1015	8×8×15	1.25 80°B	81.0
SFH-6150	1.50	210	181	157.0	1049	8×8×15	1.25 80°B	86.0
SFH-5175	1.75	245	206	179.1	1194	8×8×15	1.50 80°B	81.5
SFH-5200	2.00	280	231	200.9	1339	8×8×15	1.75 80°B	81.0

Figure 3–22 Typical listing of boiler performance information provided by manufacturer. *Courtesy of Utica Boilers Div. Of ECR International*

information furnished by a boiler manufacturer. This table includes, among other data, the IBR net rating, DOE heating capacity, and oil input rate.

The appropriate capacity rating for use in boiler selection depends on where the boiler and distribution piping are located. If both are located in heated space, the DOE heating capacity is appropriate since losses from the boiler jacket and distribution piping will offset part of the building's heating load. If the boiler is located in unheated space, but the distribution piping is mostly within heated space, the IBR Gross Output rating can be used. If the boiler and distribution piping are both located in unheated space, such as a cold basement, crawl space, or garage, the IBR Net Output rating should be used.

The deduction for piping heat loss assumed in the IBR Net Output Rating is a *fixed* 15 percent of the IBR Gross Output rating. As such, it may not accurately represent piping heat loss in every situation. For example, some state energy codes now mandate that piping carrying hot water through unheated space in new homes must be insulated to specified standards. Such insulation would significantly reduce the piping's heat loss below the 15 percent of boiler output assumed by the IBR Net Output Rating. Using the IBR Net Output Rating to select a boiler in such situations would result in needless oversizing. It would be better to make a separate calculation of piping heat loss, and then assess its effect on required boiler capacity. Such calculations are presented in Chapter 8, Heat Emitters.

3.9 EFFICIENCY OF GAS- AND OIL-FIRED BOILERS

The word *"efficiency"* is often used in describing boiler performance. In its simplest mathematical form, *efficiency is the ratio of a desired output quantity, divided by the necessary input quantity to produce that output.*

(Equation 3.2)

$$\text{Efficiency} = \frac{\text{Desired output quantity}}{\text{Necessary input quantity}}$$

When using Equation 3.2, *both input and output quantities must be measured in the same units.* The resulting efficiency will then be a decimal number between 0 and 1, corresponding to a percentage of 0 to 100. The higher the efficiency of a boiler, the more heat is extracted from a given amount of fuel. All other factors being equal, the higher the efficiency, the lower the boiler's operating cost.

Over the years, several specific definitions of boiler efficiency have been developed. The terms steady **state efficiency, cycle efficiency,** and **annual fuel utilization efficiency** all refer to the ratio of an output quantity divided by an input quantity. However, all are based on different reference conditions. *Using the wrong definition of efficiency can lead to erroneous performance predictions.* Each specific definition of boiler efficiency will be discussed so the appropriate values can be used for their intended purposes.

Steady State Efficiency

A boiler operates at steady state efficiency only when it is being fired continuously under nonvarying conditions. All inputs such as fuel composition, air-to-fuel ratio, air temperature, entering water temperature, and so forth, must remain constant. Under these conditions, the steady state boiler efficiency can be calculated using Equation 3.3:

(Equation 3.3)

$$\text{Steady State Efficiency} = \frac{\text{Heat output rate of boiler (in Btu/hr)}}{\text{Energy input rate (in Btu/hr)}}$$

Example 3.2: A boiler runs continually under steady state conditions for one hour. During this time it consumes 0.55 gallons of #2 fuel oil, and delivers 65,000 Btus of heat.

Determine the boiler's steady state efficiency. *Note:* One gallon of #2 oil = 140,000 Btu **chemical energy content.**

Solution: The fuel input rate of 0.55 gallons per hour is converted into an equivalent energy input rate in Btu/hr by multiplying by 140,000 Btu/gallon. The ratio of output to input is then calculated.

$$\text{Steady State Efficiency} = \frac{\left(65{,}000\frac{\text{Btu}}{\text{hr}}\right)}{\left(0.55\frac{\text{gallon}}{\text{hour}}\right)\left(140{,}000\frac{\text{Btu}}{\text{gallon}}\right)}$$

$$= 0.844 \text{ or } 84.4\%$$

In reality, steady state conditions seldom exist for long in either residential or commercial systems. Theoretically, these conditions would exist only while the heating load of the building was equal to, or exceeded, the heating capacity of the boiler. This usually only occurs for a very small percentage of the heating season, if at all, particularly with boilers that are oversized. Just because a boiler operates at high steady state efficiency under nearly ideal laboratory conditions does not mean this efficiency will be maintained after it is installed. *Erroneous predictions will arise if steady state efficiency values are used in estimating seasonal performance and subsequent operating costs.* This would be like estimating the average gas mileage of a car using only the highway miles per gallon rating.

The steady state efficiency of the boiler can be estimated by dividing its DOE heating capacity by its fuel input rate.

$$\text{Steady State Efficiency} \approx \frac{\text{DOE heating capacity (in Btu/hr)}}{\text{Fuel input rate (in Btu/hr)}}$$

Boiler manufacturers often list the fuel input rate, and DOE heating capacity in their technical literature. For gas-fired boilers, the fuel input rate is listed in Btu/hr or thousands of Btu/hr (MBtu/hr). For oil-fired boilers, the fuel input rate is listed in gallons per hour (gph). For #2 fuel oil, the fuel input rate in Btu /hr is obtained by multiplying the gallons per hour firing rate by 140,000 Btu/gallon.

Combustion Efficiency

The **combustion efficiency** of a boiler is determined by measuring the temperature and carbon dioxide (CO_2) content of the exhaust gases under steady state conditions. These reading are then used to look up the combustion efficiency on a chart based on the fuel being burned. The higher the CO_2 content and the lower the exhaust temperature, the greater the combustion efficiency.

Although attained through different means, combustion efficiency is essentially a field-measured value for steady state efficiency. *Because it is measured under continuous firing conditions, combustion efficiency should not be used to predict seasonal fuel usage.*

Cycle Efficiency

Boilers experience heat losses during each off-cycle that ultimately waste part of the heat produced by the fuel. Another efficiency measurement called cycle efficiency is used to account for these losses. Cycle efficiency is defined as the ratio of the total heat output divided by the chemical energy value of the fuel consumed, measured over an extended period. Mathematically this is expressed as Equation 3.4:

(Equation 3.4)

$$\text{Cycle efficiency} = \frac{\text{Total heat output of boiler over a time period (in Btu)}}{\text{Energy content of fuel consumed over same time period (in Btu)}}$$

Cycle efficiency is always less than steady state efficiency. It is highly dependent on the percent of the time the burner is firing. This percentage is called the **run fraction** of the boiler, and can be calculated using Equation 3.5:

(Equation 3.5)

$$\text{Run Fraction} = \frac{\text{burner on time}}{\text{total elapsed time}}$$

Example 3.3: At a given load condition, the burner of a boiler fires for 5 minutes, then remains off for 20 minutes before restarting. Determine its run fraction under these conditions.

Solution:

$$\text{Run fraction} = \frac{5 \text{ minutes}}{(5 \text{ minutes} + 20 \text{ minutes})} = 0.2 \text{ or } 20\%$$

Equation 3.5 is useful when the actual firing time is known. In other cases, it is necessary to predict what the run fraction would be under certain heating load conditions. This can be done using Equation 3.6.

(Equation 3.6)

$$\text{Run fraction} = \frac{\text{Heating requirement of building (in Btu/hr)}}{\text{Steady state heat output of boiler (in Btu/hr)}}$$

Example 3.4: Find the run fraction of a boiler with an output of 85,000 Btu/hr while supplying a building heating load of 40,000 Btu/hr.

Solution:

$$\text{Run fraction} = \frac{40{,}000 \text{ Btu/hr}}{85{,}000 \text{ Btu/hr}} = 0.47 \text{ or } 47\%$$

Figure 3–23 shows how the cycle efficiency of a boiler decreases as its run fraction decreases. Notice that the decrease in efficiency is relatively small for run fractions from 100 percent down to about 30 percent. However, below 30 percent run fraction cycle efficiency drops rapidly. This implies inefficient use of fuel for the useful heat produced under low run fraction conditions.

Low cycle efficiency can easily occur on days with mild outside temperature. It can also occur on cold days if the building has significant internal heat gain from people, equipment, and sunlight. *Boilers with heating capacities significantly greater than their building's design heating load will operate at needlessly low cycle efficiency even during the coldest weather.* The owner pays the penalty for this oversizing, both in higher initial cost, and, more importantly, greater fuel use over the life of the system.

Example 3.5: A building has a design heating load of 100,000 Btu/hr when the outside temperature is 0 °F, and the inside temperature is maintained at 70 °F. The selected boiler has a heating capacity of 150,000 Btu/hr. Using the graph shown in Figure 3–23, estimate the cycle efficiency when the outside temperature is 40 °F and internal heat gains total 20,000 Btu/hr.

Solution: The heat required of the boiler is the building's current heating load minus the current internal heat gains. This can be calculated as:

$$\text{Heating load} = 100{,}000\left(\frac{70-40}{70-0}\right) - 20{,}000$$

$$= 22{,}860 \text{ Btu/hr}$$

The run fraction required of the boiler can be calculated as:

$$\text{Run fraction} = \frac{22{,}860 \text{ Btu/hr}}{150{,}000 \text{ Btu/hr}} = 0.152 \text{ or } 15.2\%$$

From the graph in Figure 3–23, the corresponding cycle efficiency is about 72 percent. A boiler with a steady state efficiency of 84.5 percent would operate at only 72 percent efficiency under the stated conditions. Keep in mind that spring and fall weather, combined with internal heat gains, can result in several thousand hours of partial load operation during each heating season.

You should consider the following factors in attempting to keep a boiler's cycle efficiency as high as possible.

- *Do not needlessly oversize the boiler.* The author suggests an absolute maximum oversizing factor of 10 percent of the (properly calculated) design heating load. Minimum oversizing allows the boiler to run for longer periods, and thus attain higher efficiency. *An accurate heating load estimate is an absolute necessity before selecting a boiler.*
- In general, boilers with low thermal mass tend to have higher cycle efficiencies due to lesser amounts of heat remaining in the boiler following burner shutdown.
- If a high thermal mass boiler is used, it should be very well insulated by its manufacturer, and it should be equipped with an automatic device that prevents air flow through the boiler during the off cycle. A tight sealing automatic flue damper is especially important on conventional dry base gas-fired boilers.
- A control strategy known as **heat purging** may be incorporated into the system. Flow through the boiler is maintained for several minutes after the burner has shut down to dissipate residual heat into the building, domestic water heater, or another load that can accept the heat. Some boilers are equipped with built-in controls for heat purging. In other cases, a purging system can be assembled from standard control components.
- The use of outdoor reset control will lower boiler water temperature as outside temperature increases, thus reducing its residual heat content and improving efficiency.

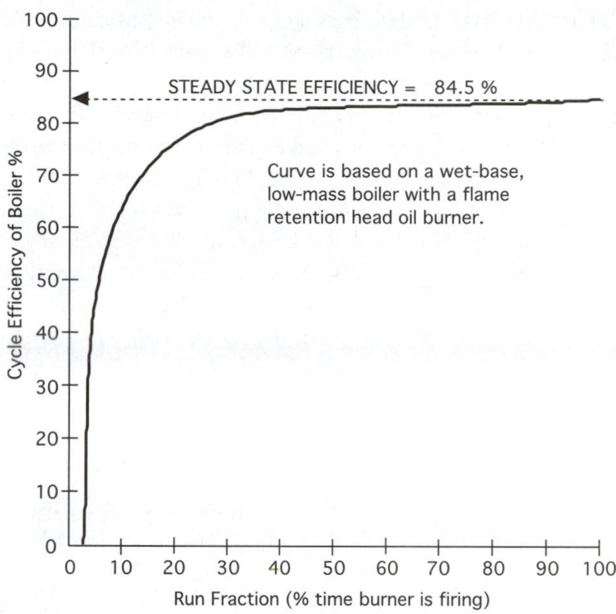

Figure 3–23 Cycle efficiency of a boiler versus run fraction. *Data source: Brookhaven National Laboratory, BNL 60816 & 80-38-HI.*

Annual Fuel Utilization Efficiency (AFUE)

In 1978, the U.S. Department of Energy established a standard for predicting the seasonal energy usage of boilers and furnaces. This standard currently applies to all gas- and oil-fired boilers sold in the United States having fuel input rates up to 300,000 Btu/hr. Its intent is to provide a uniform basis of comparison for boilers used in residential and light commercial systems.

The DOE standard defines a performance indicator called the **Annual Fuel Utilization Efficiency (AFUE)**. The AFUE rating is now the commonly used basis for expressing seasonal boiler efficiency within the United States. Minimum values for AFUE are often cited as part of the heating equipment performance standards in state energy codes.

AFUE values are expressed as a percent similar to steady state efficiency. However, unlike steady state efficiency, or cycle efficiency, AFUE values are intended to account for the many effects associated with part-load operation over an entire heating season. As such, *the AFUE value is intended for use in approximating seasonal fuel usage.* Equation 3.7 can be used for such an estimate:

(Equation 3.7)

$$\text{Estimated seasonal fuel usage} = \frac{\text{Seasonal heating requirement}}{\text{Decimal value of AFUE}}$$

Example 3.6: Assume the calculated seasonal heating requirement of a given home is 60,000,000 Btus (also stated as 60 MMBtus). The home is equipped with a gas-fired boiler having an AFUE rating of 82 percent. The estimated seasonal energy usage of the home is:

$$\text{Estimated seasonal fuel usage} = \frac{60\text{ MMBtu}}{0.82} = 73.2\text{ MMBtu}$$

Discussion: The higher the AFUE, the lower the seasonal fuel usage will be for a given seasonal heating requirement.

The test procedure for establishing a boiler's AFUE uses measured on-cycle and off-cycle performance data as input to a computer program. This program calculates the AFUE rating based on complex engineering principles and assumptions about exhaust gas composition and boiler sizing relative to its load. Those interested in the test procedure for establishing AFUE ratings should consult ASHRAE Standard 103-82.

AFUE values are established assuming the boiler is installed in heated space. As such, all heat losses from the boiler's jacket are considered useful heat output to that space. If the boiler is installed in unheated space, its actual seasonal efficiency will likely be less than its AFUE value.

The AFUE rating standard also assumes the boiler's capacity is approximately 70 percent greater than the design heating load of the building. Boilers sized closer to design load should yield seasonal efficiencies somewhat higher than predicted by their AFUE rating.

Figure 3–24 shows the theoretical effect of boiler oversizing on seasonal efficiency. The numbers are based on computer simulations of a boiler having an efficiency curve as shown in Figure 3–23. Notice the large drop in seasonal efficiency as the boiler size is increased significantly above the design heating load. These results obviously contradict the popular axiom that "bigger is always better."

Boiler capacity / Design heating load	Seasonal boiler efficiency
1.0	68 %
1.5	58 %
2.0	50 %
2.5	45 %
3.0	39 %

Figure 3–24 Decreasing seasonal efficiency as boiler capacity is increased relative to the design heating load. Data is based on a boiler with cycle efficiency described by Figure 3–23, located in a building in Syracuse, New York.

The seasonal efficiency attained by any boiler is always affected by the system and building in which it is installed. Listed AFUE ratings are based on a fixed set of assumptions for these system effects. They will not accurately predict true seasonal efficiency in all cases. Because of this, *AFUE values should only be used for relative boiler comparisons.*

3.10 MULTIPLE BOILER SYSTEMS

When specifying a boiler for a larger home or commercial building, the designer has two options: 1) select a large boiler with sufficient capacity to meet the design heating load, or 2) select two or more smaller boilers that together can meet this design load. The second option, known as a **multiple boiler system,** has a number of advantages including the:

- Higher seasonal efficiency relative to that of a single large boiler.
- Partial heat delivery if one boiler is down for servicing. This is especially important in locations where extreme temperatures can quickly freeze up nonoperational systems.
- Ability to provide elevated output during periods of high domestic hot water demand or snow melting, while still retaining high efficiency under lower space heating loads.
- Potential to eliminate several other dedicated heat sources distributed throughout the building, as well as their associated fuel supplies, electrical hookups, and exhaust systems.
- Smaller/lighter boilers are easier to install than a single larger unit, especially on retrofit jobs. Repair parts are often more readily available for smaller boilers.

In a multiple boiler system, the total output is divided up among two or more boilers. Each boiler constitutes a "stage" of heat input. Only the boilers that are needed to meet the heating load at any given time are fired. This allows the boilers that are being fired to operate at higher run fractions and higher efficiencies.

Consider the situation illustrated in Figure 3–25. Assume two identical buildings each have design heating loads of 300,000 Btu/hr. One building has a single 300,000 Btu/hr boiler. The other has a multiple boiler system consisting of three 100,000 Btu/hr boilers operated as stages. At design load, the single large boiler operates continuously, and thus attains steady state efficiency. Likewise, all three of the smaller boilers operate continuously and at steady state efficiency.

Now consider what happens as the building's load is reduced to 200,000 Btu/hr. The single large boiler operates approximately 40 minutes per hour, or 66 percent run fraction. The efficiency of the boiler drops slightly due to off-cycle heat loss. Meanwhile, in the multiple system, only two of the boilers are operating, *but they operate continuously,* and thus still achieve steady state efficiency.

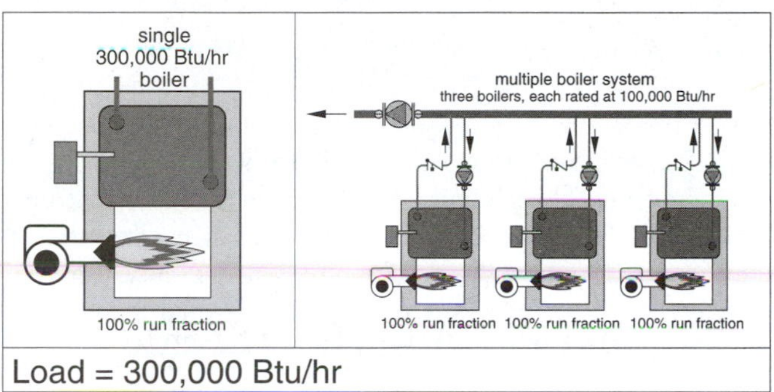

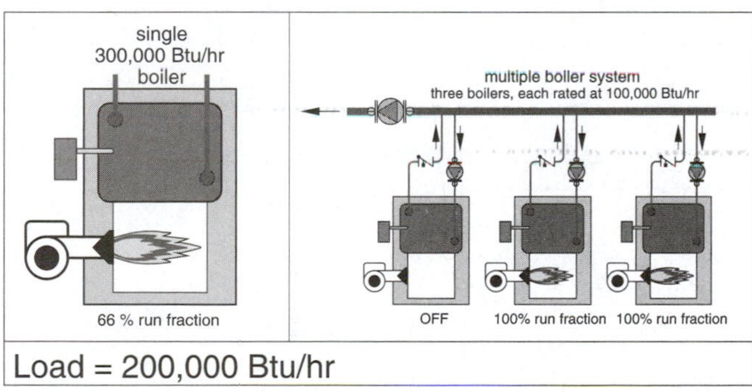

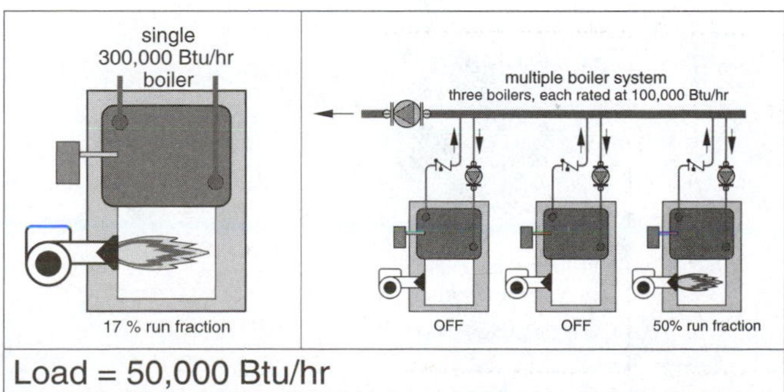

Figure 3–25 Comparison of a single large boiler with a multiple boiler system at three different load conditions.

The difference becomes even more pronounced as the heating load is further reduced. At a load of 50,000 Btu/hr, the single large boiler operates only about 10 minutes each hour, with a run fraction of 17 percent. According to Figure 3–23, its efficiency is significantly lowered under these conditions. In the multiple boiler system, one of the three boilers operates for 30 minutes each hour, or a run fraction of 50 percent. Its operating efficiency is only slightly reduced. The other boilers remain off. Similar scenarios will consistently prove the following: *the smaller the heating load, the higher the efficiency of a multiple boiler system relative to that of a single large boiler.*

The number of boiler stages making up the multiple boiler system also affects seasonal efficiency. Up to a point, the greater the number of boilers, the greater the seasonal efficiency of the multiple boiler system. The more stages available, the better the boiler system can "track" the heating load.

However, studies have shown that efficiency begins to drop off in systems using more than eight to ten boilers. The reason is that a large number of smaller boilers have considerably more jacket surface area than does a single large boiler of equal capacity. The higher heat loss from this extra surface area more than offsets the efficiency gains attained through additional stages.

Modern multiple boiler systems are controlled by microprocessor-based staging controls that perform several functions including:

- Determining the proper water temperature for the system at any given time
- Turning boilers on and off as needed to maintain the proper system supply temperature
- Rotating the boiler firing sequence to accumulate approximately the same running hours on each

Such controls play an important role in maintaining the high efficiency of a multiple boiler system and are discussed in Chapter 9, Control Strategies, Components, and Systems.

The piping of a multiple boiler system is also crucial in achieving high seasonal efficiency. *Boilers should not be piped so that heated water is circulated through unfired boilers.* Doing so allows the unfired boilers to dissipate heat from the flowing water. Some of this heat is carried out the vent by off-cycle air currents. Additional heat is lost by conduction through boiler jackets.

There are a number of ways to prevent hot water from circulating through unfired boilers in a multiple boiler system. One piping method that fulfills this requirement is shown in Figure 3–26.

Each boiler is connected to the main system piping with a pair of closely spaced tees (or "T-drilled" connections). Because of the very small pressure drop between these tees, there is very little motivation for water to flow down through a boiler circuit unless that boiler's circulator is operating. A flow-check valve further assures there will be no flow through the boiler until it is fired and its circulator is operating.

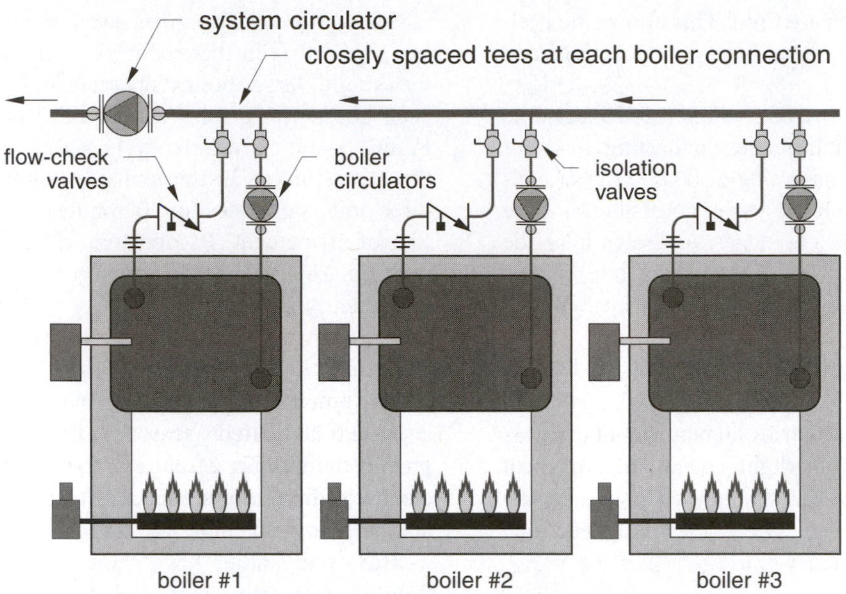

Figure 3–26 Connecting multiple boilers in a series primary/secondary arrangement.

One disadvantage of this approach is that it creates higher entering water temperatures in the downstream boiler(s) when more than one boiler is firing. The higher water temperatures drives up stack temperatures as well as jacket heat loss, decreasing the efficiency of the downstream boilers.

An alternative piping method is shown in Figure 3–27. In this approach, each boiler is piped in parallel with the others. All boilers connect to a common set of headers that in turn connect to the main system piping using closely spaced tees. As with the previous method, each boiler's circulator

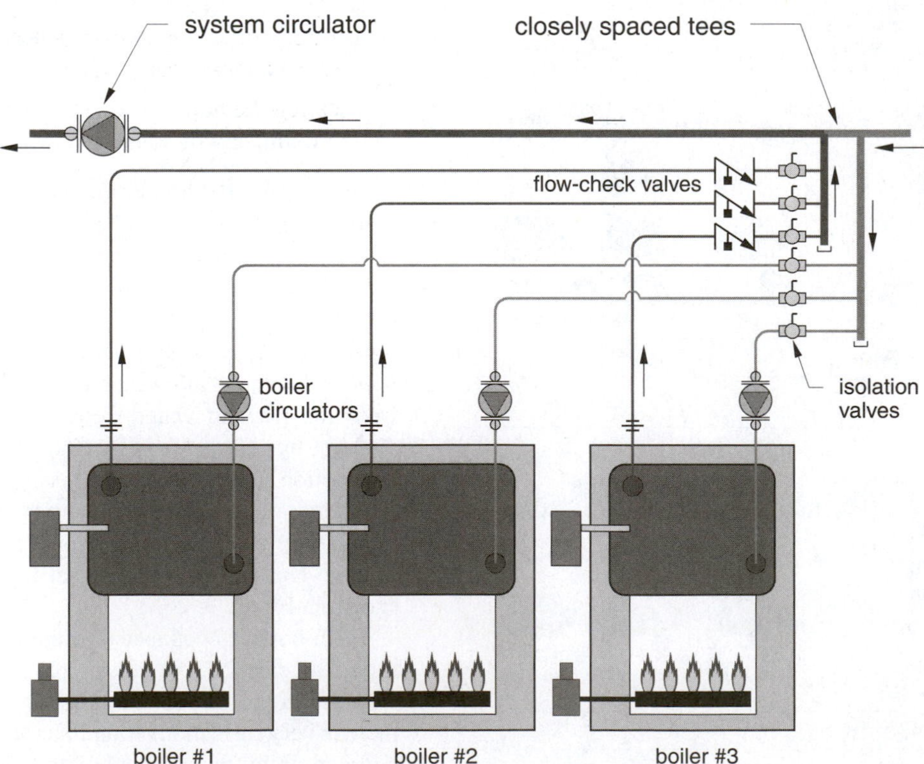

Figure 3–27 Connecting multiple boilers in a parallel primary/secondary arrangement. *Note:* Each boiler operates at the same inlet water temperature.

operates only while the boiler is fired, and thus there is no flow of heated water through unfired boilers. This approach improves system efficiency by providing the lowest water temperature to each boiler. It is used on the system shown in Figure 3–28.

Multiple boiler systems are exceptionally well-suited to homes that have several bathrooms, particularly those having high-flow fixtures. During a high demand for domestic water heating, all stages of a multiple boiler system can go into operation to provide the fastest possible recovery. Other loads in the system may be temporarily turned off during this time. Such "priority" control strategies are discussed in later chapters.

Multiple boiler systems also allow smaller/lighter boilers to be installed in situations where larger boilers simply will not fit. This can be a crucial factor in retrofit applications.

Because of their ability to retain high efficiency under widely varying load conditions, multiple boiler systems are also excellent choices for multiload systems involving snowmelting or pool heating. The author suggests they be considered whenever the total system load exceeds 200,000 Btu/hr.

Figure 3–28 Example of a multiple boiler installation using parallel primary/secondary piping.

High/Low-Fire Boilers

Several manufacturers also offer boilers with two-stage heat output control built into a single enclosure. Such boilers are often called "high/low fired." In some cases, the gas valve on the boiler has two-stage capability (e.g., it is capable of allowing two rates of gas flow to the burners depending on the control signal it receives). In other cases, two separate gas valves and associated burners are installed in the same combustion chamber.

High/low-fire boilers often require less physical space than two separate boilers having the same total capacity. They typically have a single vent connection and are easier to install compared to multiple single-stage boilers.

When used in a multiple boiler system, the stages of high/low-fire boilers must be operated in paired sequences. Modern staging controls can easily be configured for this operating logic.

3.11 ELECTRIC BOILERS

Electric boilers are available as compact wall-hung units. They typically contain two or more immersion-type resistive heating elements mounted in a common enclosure through which water flows. An example of a compact electric boiler is shown in Figure 3–29.

When an electrical current passes through a resistive heating element, the rate of heat output can be calculated using Equation 3.8:

(Equation 3.8)

$$P = iv$$

where:
i = current (amps)
v = voltage (volts)
P = power output (watts)

The following factors allow one to convert the output rating of electrical heating devices (usually expressed in watts or kilowatts) into Btu/hr:

1 watt = 1 w = 3.413 Btu/hr
1 kilowatt = 1,000 watts = 1 kW = 3,413 Btu/hr

The second conversion factor is simply the first multiplied by 1,000. These factors are used so commonly in heating system design that they should be memorized.

With the proper controls, each heating element can be operated independently to provide staged heat output. Only those stages needed to satisfy the load at any given time are operated.

Electric boilers have some advantages as compared to combustion-type boilers. These include the following:

- Since combustion is not involved, no air supply is required for combustion or draft, and no exhaust system is necessary.
- Since there are no flue gases, there is no concern about flue gas condensation.
- On-site fuel storage is not required.
- Periodic maintenance is minimal since there is no soot to remove or fuel filters to replace.
- If used with an outdoor reset control, the water temperature in an electric boiler can be varied by controlling the on-time of the element(s). This would eliminate the need for a mixing device in variable temperature systems.
- Since nearly all buildings have an electric service, the choice of an electric boiler might eliminate the need for a natural gas service and its associated monthly service charge.

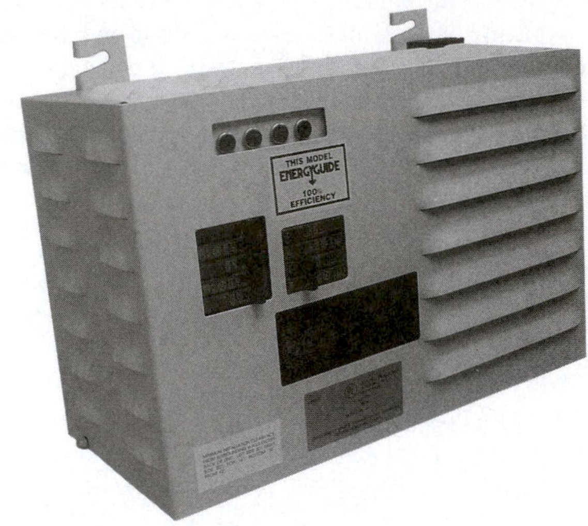

Figure 3–29a Example of a compact wall-hung electric boiler. *Courtesy of Argo Industries.*

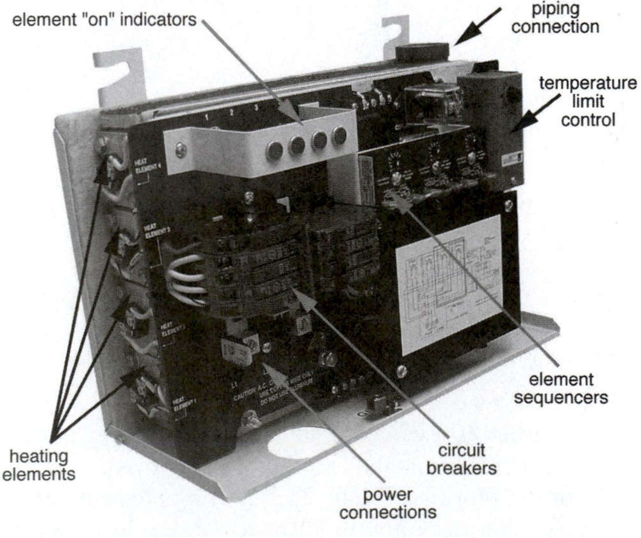

Figure 3–29b Internal construction of boiler shown in Figure 3–29a. *Courtesy of Argo Industries.*

The economic viability of an electric boiler depends on the cost of electricity versus competing fuels. The economics are obviously more favorable in areas where electric utility rates are low. In commercial or industrial buildings where utilities usually impose a "demand charge" as well as an energy charge, the economics of an electric boiler are questionable because their use would significantly increase the building's demand.

Some state energy codes impose more rigid energy conservation standards on electrically heated buildings. Any extra cost associated with meeting these standards should be factored into any economic comparison.

Finally, if an electric boiler is being considered, be sure the building's electric service entrance is adequate to handle the load. This is particularly true in older buildings. A 200-amp/240-VAC service entrance is generally considered minimum for such applications.

3.12 ELECTRIC THERMAL STORAGE (ETS) SYSTEMS

To understand the concept of **electric thermal storage (ETS)**, it is important to understand the situations faced by many electrical utilities. Chief among them is the challenge of meeting the growing demands for electricity without spending hundreds of millions of dollars building new generating facilities. As the national demand for electrical energy increases, all utilities must plan ahead to ensure they can supply the extra energy, especially during critical peak demand periods. A fundamental goal is to reduce peak demand, while simultaneously increasing demand during traditional low-demand periods.

One way to do this is to offer **time-of-use rates.** In their simplest form, time-of-use rates offer customers lower cost electricity during specified "off-peak" periods. Such periods have a set beginning and ending time. In some areas, rates may vary several times during the day, as well as seasonally, during weekends, and even on holidays. The electric meter automatically records usage during each rate period.

Because space heating is usually the dominant electrical load in all-electric buildings, it represents a major opportunity for savings if the necesary energy could be purchased during off-peak periods. This concept requires that the heat produced during these periods be stored for subsequent use during on-peak periods. Since water has excellent heat storage properties, hydronic systems using electric heating elements combined with thermal storage tanks are ideal for such applications.

Classification of ETS Systems

ETS systems are classified according to their ability to meet heating loads using energy purchased and stored during the previous off-peak period. The two common types of ETS system are called **full storage systems** and **partial storage systems.**

In full storage systems, all energy used during the subsequent on-peak period is purchased and stored during the off-peak cycle. The storage tank must hold sufficient heat to supply the building's load through the entire on-peak period, which may be as long as 18 hours. Storage tank volumes of several hundred to more than 1,000 gallons are typical in such applications.

In most full storage systems, the output of the electrical elements, when operating, must be three to four times greater than the design heating load of the building. This is because the electrical elements only have a few off-peak hours in which to transfer enough heat to supply the building for a 24-hour period.

In partial storage systems, only a portion of the on-peak heating energy is purchased and stored during the off-peak period. Once this energy is depleted, the heating elements operate as necessary to maintain comfort. Operating the heating elements during the on-peak period uses electricity at the more expensive rate, and thus this should be minimized. However, because it stores only a portion of heat required for the on-peak heat period, a partial storage system often has fewer heating elements and a smaller storage tank than a full storage system. This approach lowers initial cost in exchange for a somewhat higher operating cost.

Figure 3–30 shows a comparison of hourly electrical demand for a house using nonstorage electric resistance heating, versus a full storage ETS system. Notice the high electrical demand of the ETS system during the off-peak hours when electrical rates are low. This is when the water in a hydronic-based ETS system is being heated. Ideally, sufficient heat for the entire day is purchased during this time. At 7:00 A.M., the heating elements are shut off. During the midday hours, a small amount of electrical energy is needed for components such as pumps, and in some cases, blowers. This demand is very low compared to that of the conventional electric heating system.

The economics of partial storage versus full storage systems are very dependent on the project. An economic analysis that considers the trade-off between initial cost of the equipment, and the locally available on-peak/off-peak rate differential is essential in making an informed decision. The system with the lowest energy cost may not necessarily be the best choice. Instead, one should look for the *lowest total owning and operating cost over the estimated system life.*

ETS System Layout

ETS systems do not use exotic technology. They are composed mostly of standard hydronic components, and designed using fundamental concepts covered in this text. A basic schematic for an ETS system using electric heating elements mounted in a storage tank is shown in Figure 3–31. The temperature of the stored hot water is reduced, when necessary, by a three-way mixing valve before supplying radiant floor heating circuits.

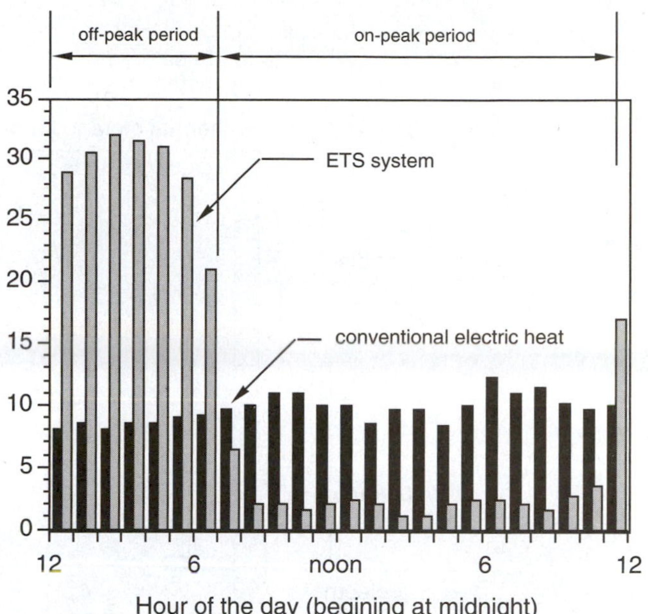

Figure 3–30 Electric demand of full storage ETS system versus conventional electric heating. Data based on a design day in upstate New York.

Storage Tanks for ETS Systems

In both types of ETS systems, the essential component is a properly sized, well-insulated storage tank. Several types of tanks are suitable for such applications including pressure-rated steel tanks and non-pressure-rated polymer tanks. The choice must consider factors such as temperature limit, corrosion, life expectancy, type of insulation, and ease of installation.

Pressure-rated steel tanks are a good choice for closed-loop systems. Although their cost per gallon is usually higher than non-pressure-rated tanks, they allow more flexibility in the remainder of the system design. The cost of such tanks can be reduced by selecting standard size tanks that are manufactured in quantity. For example, tanks designed for bulk storage of liquid propane are mass produced in 500 and 1000 gallons sizes. They are certified for pressures much higher than needed in hydronic heating systems, and come with several threaded openings.

Non-pressure-rated thermoplastic tanks are used in various industrial processes. Some polypropylene tanks have continuous service temperature ratings as high as 220 °F. Such tanks may be used in conjunction with heat exchangers,

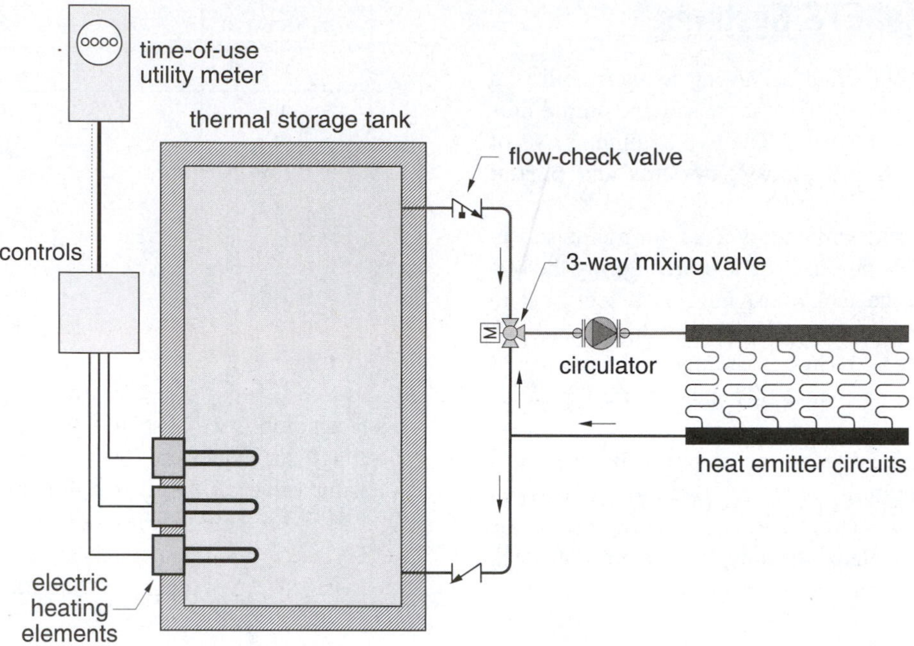

Figure 3–31 Piping schematic of a hydronic ETS system coupled to a radiant floor distribution system.

which allow the remainder of the system to be pressurized. They may also be used in open-loop systems. Rectangular-shaped polymer tanks that will contain hot water must be fully supported at the bottom and along all four sides. Failure to do so will allow excessive bulging of the tank walls. Concentrated loading will also cause the polymer to "creep" over time, permanently deforming the tank.

Although less expensive than pressure-rated tanks, non-pressure-rated thermoplastic tanks introduce many constraints on the remainder of the system design.

All thermal storage tanks must be heavily insulated to limit heat loss. Failure to do so will result in excessive and uncontrolled heat loss and poor overall performance.

3.12 HYDRONIC HEAT PUMPS

A simple definition of a **heat pump** is *a device for converting low temperature heat into higher temperature heat.* The low temperature heat is obtained from some material called the "**source**," and then concentrated and released into another material called the "**sink**."

Categories of Heat Pumps

Heat pumps are classified according to the source from which they extract heat, as well as the sink into which they release it. One that absorbs heat from air is called an air-source heat pump. If it also releases heat into a flowing stream of air, it is called an air-to-air heat pump. Most mechanical contractors who install heating and cooling systems are familiar with air-to-air heat pumps. They have been available since the early 1970s, and are commonly used for heating and cooling in moderate climates.

A heat pump that absorbs heat from a flowing stream of water is called a water-source heat pump. If it then delivers that heat to a flowing stream of air, it is called a water-to-air heat pump. If the heat is delivered to a flowing stream of water, the unit is called a water-to-water heat pump.

In this text, the term "hydronic heat pump" always refers to a heat pump that delivers heat to a flowing stream of water. Although it may gather heat from either air or water, the latter is much more common, especially in what is known as a geothermal heat pump system. Figure 3–32 shows a water-source hydronic heat pump (a.k.a., water-to-water heat pump) that absorbs heat from water flowing through buried tubing, and delivers that heat to a hydronic radiant floor system.

Basic Refrigeration Cycle

The **refrigeration cycle** is the basis of operation of all heat pumps, hydronic or otherwise. During this cycle, a chemical compound called the refrigerant circulates around a closed loop passing through all major components of the heat pump.

To describe how this cycle works, a quantity of refrigerant will be followed through the major components of a water-to-water heat pump. The four stations in the refrigerant cycle are shown in Figure 3–33.

Figure 3–32 A water-to-water heat pump installed as part of a hydronic heating system.

The cycle begins at station 1 as cold liquid refrigerant evaporates within a water-to-refrigerant heat exchanger known as the **evaporator.** Water flowing through the evaporator is supplied from a low temperature source such as a well. Because the liquid refrigerant in the evaporator is colder than this water, it absorbs heat from it, and in the process changes from a liquid to a gas (e.g., evaporates).

At station 2, the gaseous refrigerant flows through the heat pump's compressor where its pressure and temperature are greatly increased. This is where the electrical energy needed to run the compressor is converted to thermal energy, most of which is transferred to the refrigerant.

At station 3 the hot gas travels through another refrigerant-to-water heat exchanger called the **condenser.** Here it releases heat into a flowing steam of water that carries the heat out of the heat pump. As it gives up heat, the refrigerant condenses from a gas to a liquid. Its pressure is still quite high, and its temperature still warm.

To begin the cycle over, the refrigerant must be returned to the same cold liquid state from which this description began. This is accomplished at station 4 where the refrigerant passes through an expansion device (usually called a thermal expansion valve, or TXV) that greatly reduces its pressure. The drop in pressure induces an immediate drop in temperature, and the refrigerant is ready to begin the cycle again.

The refrigerant cycle is not unique to heat pumps. It is used in refrigerators, freezers, room air conditioners, dehumidifiers, water coolers, soda vending machines, and other heat-moving machines. The average person is certainly familiar with these devices, but usually takes for granted how they operate.

Water Sources for Hydronic Heat Pumps

Low temperature **source water** for hydronic heat pumps used for space heating and other applications can come from a variety of sources including wells, lakes, or piping loops

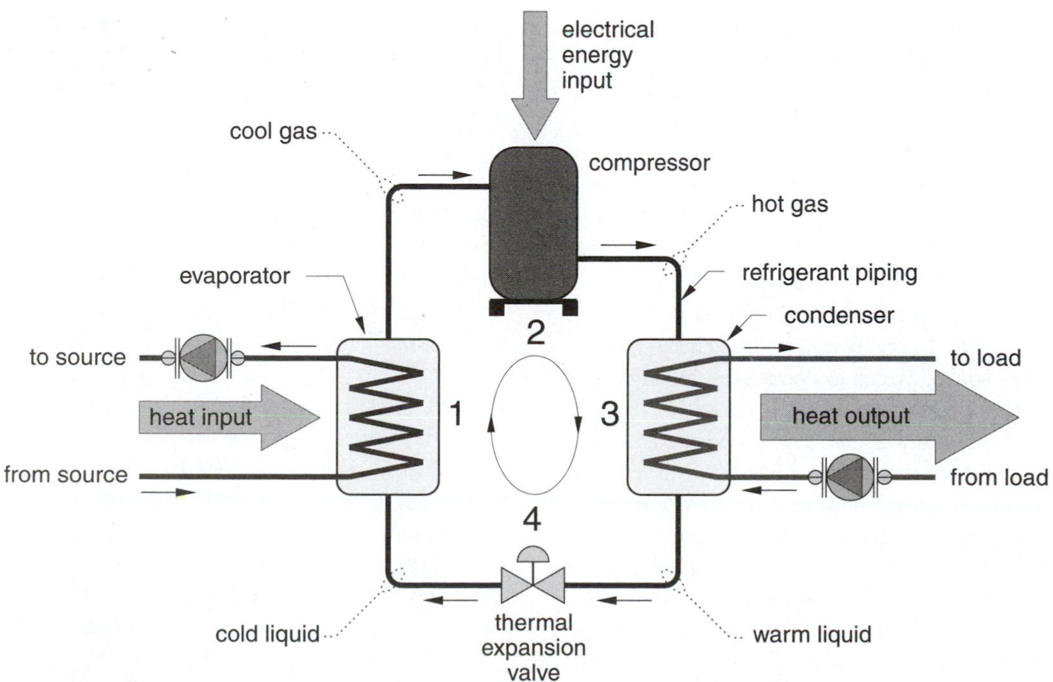

Figure 3–33 The basic refrigerant cycle within a water-to-water heat pump.

buried in the earth. In some applications, heat might even be extracted from a process- or waste-water stream.

Wells, ponds, or lakes are called **open-loop sources** since, at some location, the water they contain is exposed to the atmosphere. Water from such sources can be pumped through the evaporator of a heat pump where heat can be extracted from it. The chilled water can then be returned to its source where it will absorb heat and return to its original temperature.

A piping loop buried in the earth with its ends connects the heat pump is an example of a closed-loop source. The fluid it contains circulates through the loop whenever the heat pump is operating.

Open-Loop Sources

Most hydronic heat pumps require 2 to 3 gallons per minute of water flow through their evaporators, per ton (12,000 Btu/hr) of heating capacity. During cold weather, a typical 4-ton residential system may require several thousand gallons of water to pass through the heat pump's evaporator every day. While this volume of water is readily supplied from a lake or large pond, it can place a tremendous load on a well. If that well cannot keep up with the demand, the heat pump will eventually shut itself off due to insufficient evaporator flow. The building will then be without heat. It is essential that any water source being considered to supply a water-source heat pump can consistently provide the necessary quantity of water. *In the case of a well, always have a certified driller verify the well's sustained recovery rate before committing to its use in supplying a heat pump.*

Since the water must be returned to the environment without contamination, it cannot be chemically treated prior to entering the heat pump. It is imperative that the water quality is suitable to be pumped through the heat pump's evaporator on a long-term basis. Groundwater with high concentrations of calcium, iron, hydrogen sulfide, silts, or other materials may not be suitable for such applications. *Always have the proposed water source analyzed by a competent water testing agency, and the results approved by the heat pump manufacturer, before committing to its use.*

The temperature of the source water supplied to the heat pump is also an important consideration. Most heat pumps have minimum inlet source water temperatures specified by their manufacturer. Typical values range from 36 °F and 40 °F. Minimum temperatures a few degrees above freezing are necessary to prevent ice formation on the cold surfaces of the evaporator. The water temperature in drilled wells usually remains close to the annual average air temperature. Groundwater temperature data is often available from local well drillers.

To summarize, always verify the water quantity, water quality (e.g., levels of impurities), and the likely minimum supply temperature before committing to their use with a water-source heat pump.

Closed-Loop Sources

In situations where open-loop water sources are not available or suitable for use, a closed-loop **earth heat exchanger** can supply low temperature heat to a water-source heat pump. Such a heat exchanger usually consists of several hundred feet of buried plastic pipe through which water, or a mixture of water and nontoxic antifreeze, circulates. As the fluid passes through the heat pump's evaporator, heat is extracted from it. In the process, the temperature of the fluid drops below that of the soil surrounding the earth heat exchanger. The chilled fluid then makes another pass through the earth heat exchanger where it reabsorbs heat from the warmer surrounding soil. This process is illustrated in Figure 3–34.

Once filled and purged of air, no fluid should enter or leave the earth heat exchanger. This closed-loop approach eliminates the need for a well or other source of water at the site. It also eliminates the inevitable scaling or corrosion problems associated with pumping poor quality water through the heat pump.

The type of pipe used for an earth heat exchanger is of critical importance to its longevity. Currently, high-density polyethylene (PE 3408/SDR-11) pipe is the most widely used piping for earth heat exchangers. It is available in a range of sizes and standard coil lengths up to 1000 feet. This pipe can flex to accommodate the stresses imposed by variations in soil temperature and moisture. PVC (polyvinyl chloride) pipe does not have these qualities, and should not be used for buried portions of any earth heat exchanger. All joints in a polyethylene earth heat exchanger must be made using heat fusion. In this process, the two pipe ends are heated to over 500 °F using specialized equipment, then fused together to form a joint stronger than the pipe itself. *There can be no compromise in the materials or methods used to fabricate earth heat exchangers.* When properly installed, they will last for many decades. When improperly installed, they can lead to very expensive corrective actions, or even abandonment of the system.

Figure 3–35 shows a cross section of a typical horizontal earth heat exchanger used by the author on several hydronic heat pump systems. This configuration requires approximately 180 feet of trench (and 720 feet of pipe) per ton of heat pump capacity, based on a minimum fluid supply temperature of 30 °F, when installed in heavy/wet soils in upstate New York. For installations in similar climates but with damp/sandy soils, the earth heat exchanger length would increase to 260 feet of trench (or 1040 feet of pipe) per ton of capacity. The additional piping is required due to the poorer heat transfer properties of sandy soil. These loop lengths were calculated for a specific location (upstate New York), with specific minimum supply temperature to the heat pump (30 °F), and a specific piping geometry (four-pipe square at 5-foot average depth). The length of the earth heat exchanger will vary for other locations, soil types, piping layouts, and operating conditions.

Complete information on designing and installing earth heat exchangers can be found in the references listed at the end of this chapter.

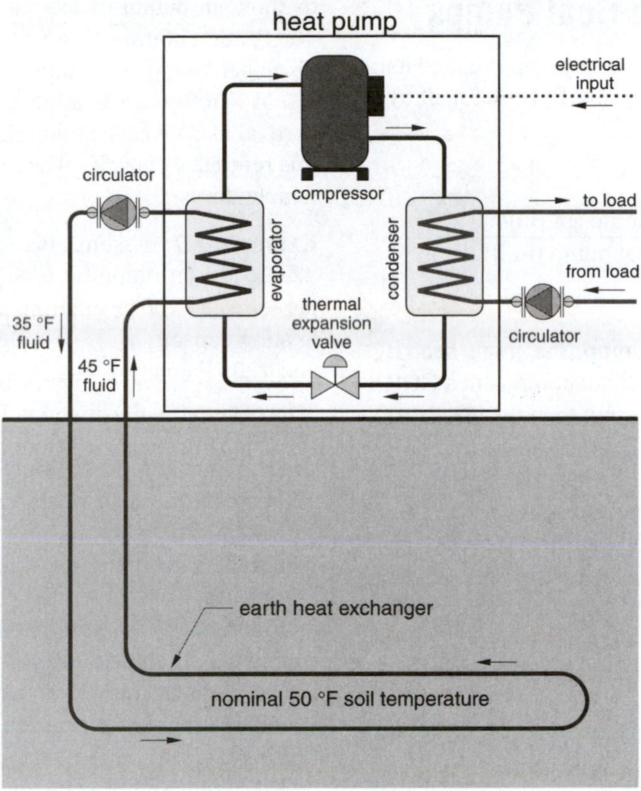

Figure 3–34 Conceptual drawing of how heat is absorbed by the fluid in an earth heat exchanger and extracted by the evaporator of the heat pump.

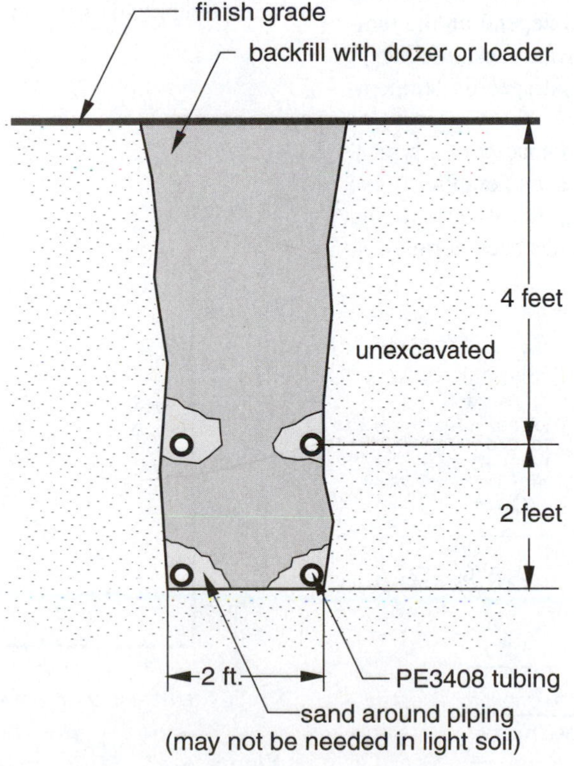

Figure 3–35 Cross section of a "four-pipe square" earth heat exchanger.

Performance of Hydronic Heat Pumps

The efficiency rating of a hydronic heat pump is called its **coefficient of performance (COP).** COP is defined by Equation 3.9.

(Equation 3.9)

$$\text{COP} = \frac{\text{Heat output of heat pump (in Btu/hr)}}{\text{Electrical input to run heat pump (in Btu/hr)}}$$

The higher the COP of a heat pump, the more heat it delivers for a given electrical input. A heat pump with a COP of 3.0 yields three times more heat output than the electrical energy required to operate it. This additional heat is not created by the heat pump, but instead is absorbed from the source water using the refrigeration cycle previously discussed. Water-to-water heat pumps have COPs ranging from 2.0 to upwards of 4.0 depending on the temperatures of the source water as well as the sink water. Because a heat pump can deliver significantly more heat than an electric boiler using the same amount of electricity, its operating cost will be significantly lower.

Figure 3–36 depicts the energy flows in a hydronic heat pump connected to a low temperature water source such as an earth heat exchanger. Notice the relative widths of the arrows representing the two energy flows into the heat pump, and the single arrow representing energy flow out. The wider the arrow, the greater the rate of energy flow.

Both the coefficient of performance and the heating capacity of a water-to-water heat pump depend on the temperature of the source water and sink water. Manufacturers rate the performance of their products at specific temperatures and water flow rates.

Figure 3–37 shows how the heating capacity of a typical water-to-water heat pump varies with the temperatures of the source water and the "load" water (e.g., the water entering the heat pump's condenser). This graph has been normalized to show an output of 1.00 at a source water temperature of 40 °F, and entering load water temperature of 100 °F. For a given heat pump, the output of 1.00 might, for example, represent a reference heating capacity of 30,000 Btu/hr. The vertical axis gives the multiplier that should be used to adjust this reference capacity when the heat pump operates at other temperatures.

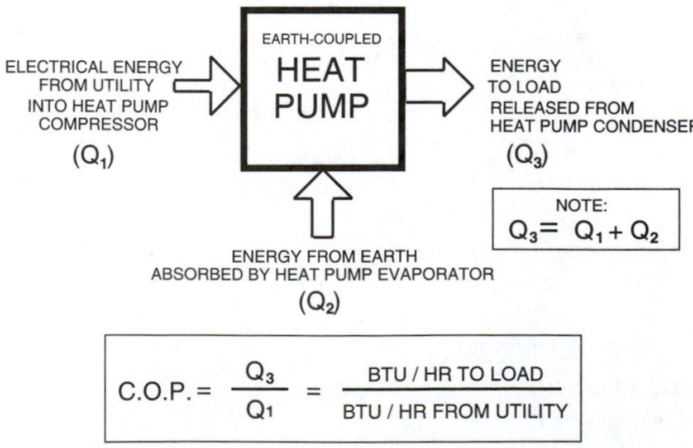

Figure 3–36 Energy flows through a hydronic heat pump.

Example 3.7: Assume that a given heat pump is rated at 32,000 Btu/hr output at a 40 °F entering source water temperature and 100 °F entering load water temperature. Find its heating capacity at 35 °F entering source water temperature and 120 °F entering load water temperature.

Solution: Refer to Figure 3–37. The intersection of the vertical line from 35 °F source temperature with the line for 120 °F entering load water temperature indicates a multiplier of 0.88. The adjusted heating capacity at these conditions would therefore be $(0.88) \times 32{,}000 = 28{,}200$ Btu/hr.

Figure 3–37 shows that heating capacity increases as source water temperature increases, and entering load water temperature decreases. Conversely, the colder the source water, and the warmer the entering load water, the lower the heating capacity of the heat pump.

Like the heating capacity, the COP of a water-to-water heat pump is affected by the source and entering load water temperatures. The higher the source water temperature, and the lower the load water temperature, the higher the COP of the heat pump. The variation in COP due to changes in operating temperatures is shown in Figure 3–38.

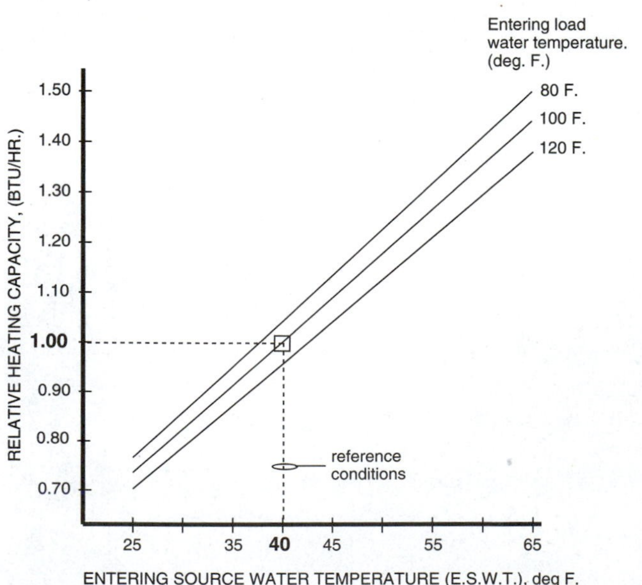

Figure 3–37 Variation in heating capacity of a hydronic heat pump with varying source water and load water temperatures.

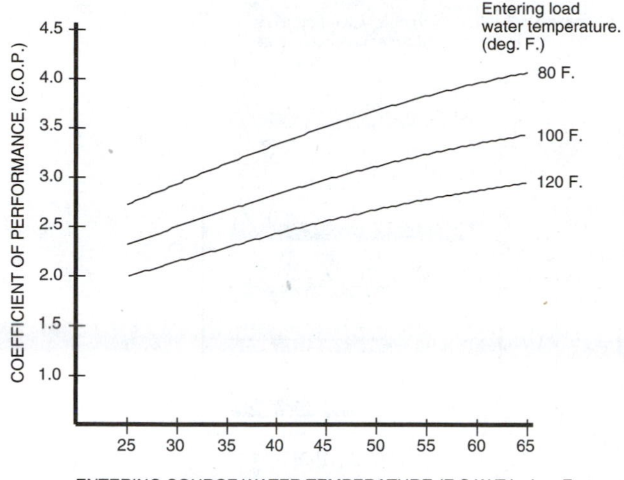

Figure 3–38 Variation in COP of a hydronic heat pump with changes in source water and load water temperature.

3.14 SYSTEM DESIGN CONSIDERATIONS FOR HYDRONIC HEAT PUMPS

Hydronic heat pumps have unique operating requirements that must be addressed during system design. The two primary requirements are sustained water flow, and limitations on supply water temperature to both the evaporator and condenser.

Sustained Water Flow

Whenever a heat pump is operating, refrigerant is removing heat from the water passing through the evaporator, and releasing heat into the water passing through the condenser. If the flow of water through either the evaporator or condenser is stopped or significantly slowed, the heat pump must quickly and automatically shut down to prevent physical damage.

A flow restriction or stoppage in the evaporator can allow ice crystals to form on the cold heat exchanger surfaces separating the water and refrigerant. Since the evaporator contains only a small amount of water, the ice can form rapidly. A "hard freeze" can occur in less than 1 minute of operation without water flow. A number of such hard freezes will eventually rupture the copper tubing in the evaporator and result in a costly repair. Because of this possibility, manufacturers install a temperature-sensing safety switch near the water outlet of the evaporator. This switch monitors the water temperature leaving the evaporator, and stops the compressor if that temperature drops below a preset limit.

Maintaining water flow through the condenser whenever the heat pump is operating is equally important. A flow restriction in this stream can cause the refrigerant head pressure to increase rapidly. Eventually this high pressure will trip a pressure safety switch that shuts down the compressor.

Heat pump manufacturers usually recommend specific minimum flow rates for both the evaporator and condenser. In the absence of such recommendations, the system should provide approximately 3 gpm of water flow per 12,000 Btu/hr of heat transfer in both the evaporator and condenser.

The designer must be careful not to install devices such as zone valves, or three-way mixing valves in series with the heat pump condenser. If such valves are present, they may restrict or totally stop water flow through the condenser causing an automatic shut down of the heat pump's compressor.

One method for combining a hydronic heat pump with traditional zone controls is to install a **buffer tank** between the heat pump and the distribution system as shown in Figure 3–39.

The buffer tank acts as a thermal reservoir between the heat pump and the distribution system. Heat may be added to the tank by the heat pump, and removed by the distribution system at different rates. The heat pump is simply turned on and off as necessary to maintain the buffer tank temperature within a preset temperature range.

When the building requires heat, water from the buffer tank can be circulated through the distribution system regardless of whether the heat pump is on or off. The buffer tank must be well insulated since it contains heated water. Its size depends on the capacity of the heat pump, the temperature differential at which the heat pump will be turned on and off, and the desired minimum running time of the heat pump once it is turned on. Equation 3.10 can be used to determine the minimum volume of the buffer tank for a given set of operating conditions.

(Equation 3.10)

$$V = \frac{t \times (Q_{hp} - Q_{load})}{\Delta T \times 500}$$

where:

V = minimum volume of buffer tank (gallons)
t = desired minimum on-time of heat pump cycle (minutes)
Q_{hp} = output of heat pump while operating (Btu/hr)
Q_{load} = heat flow to load when heat pump is on, if any (Btu/hr)
ΔT = operating temperature differential of tank (°F)

Example 3.8: A hydronic heat pump is rated at 36,000 Btu/hr of output. Determine the necessary buffer tank size to allow the heat pump to run for a minimum of 10 minutes while heating the buffer tank from 95 °F to 120 °F. Assume the load at the time is zero.

Solution: Inserting the data into Equation 3.10 yields:

$$V = \frac{t \times (Q_{hp} - Q_{load})}{\Delta T \times 500} = \frac{10 \times (30{,}000 - 0)}{(120 - 95) \times 500} = 24 \text{ gallons}$$

Discussion: A tank larger than calculated with Equation 3.10 will allow longer on-cycles (minimizing wear on

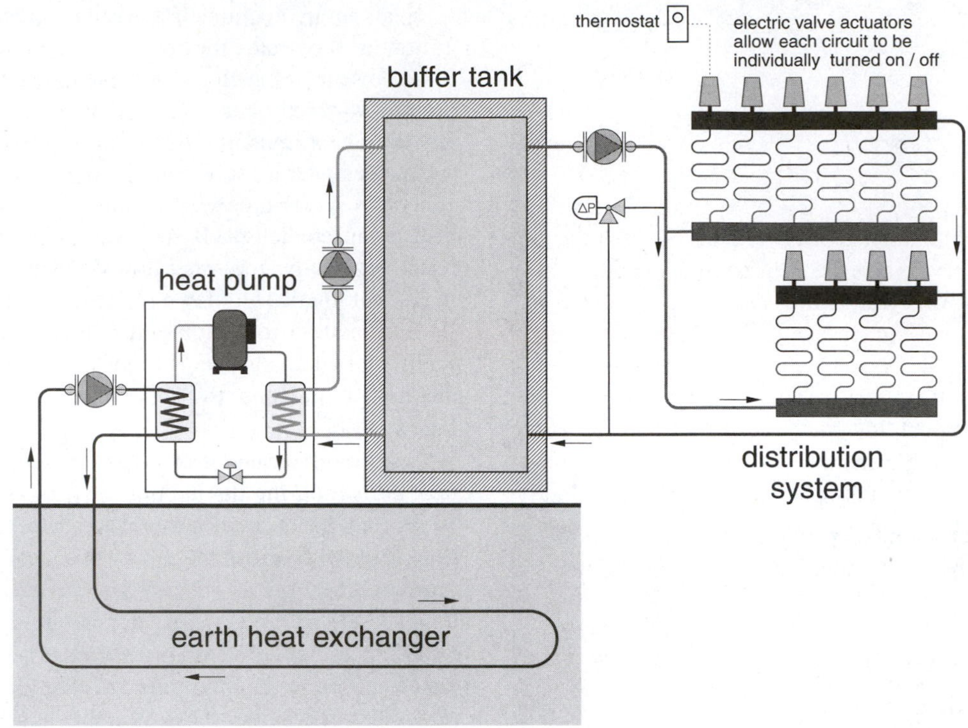

Figure 3–39 Use of a buffer tank to allow for zoned distribution with a hydronic heat pump.

the heat pump components), and/or a narrower range of operating temperature.

Buffer tanks are equally applicable to chilled water cooling applications. As with heating, the buffer tank allows the heat pump to operate independently of the simultaneous load of the distribution system. *If a buffer tank is used to store chilled water, it must be insulated, and have a continuous vapor barrier to prevent moisture migration toward the cold tank surfaces.* Without a vapor barrier, condensation of water vapor on the tank surfaces could present a significant problem.

Water Temperature Limitations

All hydronic heat pumps have a range of acceptable supply water temperatures for both their evaporator and condenser. These temperatures usually depend upon the refrigerant used in the heat pump, as well as the type of expansion device controlling the flow of liquid refrigerant to the evaporator.

The temperature of the water entering the evaporator usually depends on the source. A heat pump supplied from a drilled water well experiences relatively constant water temperature year-round. Well water temperature data is available for specific locations from most water source heat pump manufacturers. Typical values are between 45 °F and 60 °F for the northern half of the United States and Canada. Earth heat exchangers yield fluid temperatures from 25 °F to 70 °F depending on the location, depth, length, and time of year. The lower end of this range can only be used when an antifreeze solution is used in the earth heat exchanger. Such low temperature operation also requires a heat pump equipped with a thermal expansion valve rather than a capillary tube expansion device.

The temperature range of the condenser water also depends on the type of refrigerant used in the heat pump. Heat pumps using R-22 refrigerant have a practical upper temperature limit of 130 °F. R-500 based heat pumps can reach 150 °F to 155 °F, but with some loss of low temperature performance. As previously shown in Figure 3–37 and Figure 3–38, the lower the condenser water temperature, the higher the heating capacity and COP of the heat pump.

The condenser operating temperature is mostly determined by the heat emitters used in the system. Slab-type radiant floor systems are well-suited for use with hydronic heat pumps because they operate with supply water temperatures of 80 °F to 110 °F. By comparison, finned-tube baseboard systems typically require much high water temperatures, and are generally not compatible with hydronic heat pumps.

3.15 RENEWABLE ENERGY HEAT SOURCES

During the late 1970s and early 1980s, America was infatuated with renewable energy sources. Various approaches such as solar heating, wood burning, and wind energy conversion were popular topics both technically and socially. By 1985, popular interest in renewable energy systems began to decline in the face of lower oil prices and dwindling government support. Many lessons, both good and bad, were learned about integrating renewable energy into the American lifestyle and economy.

On the plus side, engineers learned valuable lessons about what worked and what did not work. Products such as solar collectors progressed through various stages of development. Well-designed products evolved, as did good applications.

On the negative side, many concepts, products, and system designs, though well intentioned, did not survive. They succumbed to the inevitable consequences of overcomplexity, use of unsuitable materials, and the reality of having to compete with traditional lower-cost energy sources. The public's zeal for renewable energy was in many cases stunted by market opportunists, unrealistically high pricing, and poorly designed systems that presented major service problems. Still, in light of the rising prices of conventional fuels, renewable energy will likely play an increasingly important role in America's future.

Two types of renewable energy sources, active solar energy systems and wood-fired boilers, are relatively easy to implement as hydronic heat sources.

3.16 ACTIVE SOLAR ENERGY SYSTEMS

Although many different designs for active solar heating systems have been developed, this discussion will be limited to systems using **flat plate solar collectors.** This is the most available type of collector, and has proven to be the most cost-effective in areas of the United States that represent the major hydronic heating markets. Figure 3–40 illustrates the construction of a typical flat plate solar collector.

Individual flat plate solar collectors can be grouped together to form an array. Such an array is often located on a south-facing roof. To maximize wintertime heat output, the array should have a slope angle equal to the local latitude +10 to 15 degrees. A roof-mounted array consisting of six flat plate collectors is shown in Figure 3–41. Ground-mounted collector arrays are also possible, but are usually more expensive due to the need for more extensive rack structures. They are also more subject to accidental damage or vandalism.

Because much of a building's heating load occurs when solar energy is not available, water-filled thermal storage tanks are incorporated into the system to absorb excess solar heat for use during the periods. Heat from the collectors is transferred to the thermal storage tank by circulating tank water through the collectors, or by circulating an antifreeze solution between the collectors and a heat exchanger coupled to the tank.

Closed-Loop/Antifreeze Systems

Figure 3–42 shows a simplified schematic of a closed-loop/antifreeze-type solar energy system.

Since a portion of the collector circuit is exposed to outdoor temperatures, it must be protected against freezing. This is done by filling the collector circuit with a glycol-based antifreeze solution. This solution transports heat from the collector array to the heat exchanger. A second circulator moves water between the storage tank and the heat exchanger. Both circulators operate whenever solar energy is being collected.

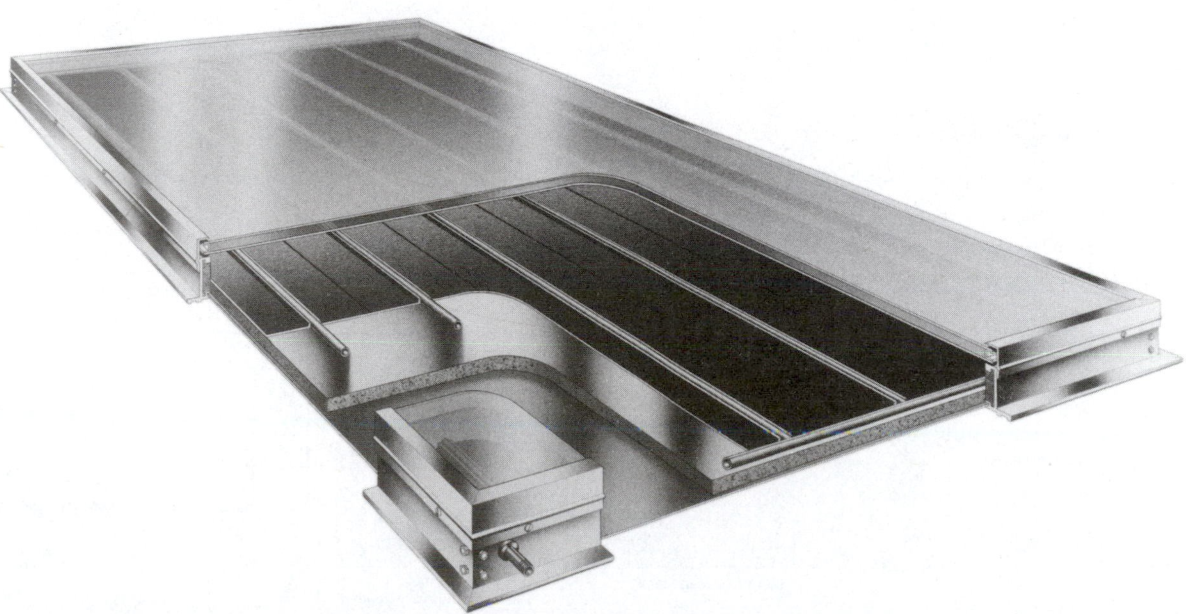

Figure 3–40 Construction of a typical flat plate solar collector. *Courtesy of American Energy Technologies, Inc.*

Figure 3–41 An array of six roof-mounted flat plate solar collectors that supply heat to a hydronic distribution system.

One advantage of an antifreeze-based system is that piping between the collectors and storage tank can be installed in virtually any orientation, inside or outside. This is particularly important in retrofit applications where pipe routing is often more restrictive.

Direct Circulation Systems

There are several advantages to circulating water from the storage tank directly through the collectors. First, the cost of the heat exchanger, antifreeze, and some other components are eliminated.

Also eliminated is the performance penalty associated with using a heat exchanger between the collector array and storage tank. In closed-loop/antifreeze-type systems, the antifreeze temperature is typically 10 to 15 °F warmer than the storage tank when the system is collecting energy. This temperature differential is necessary to move heat across the heat exchanger at a suitable rate. Unfortunately, it also forces the collectors to operate at higher temperatures than would be necessary with a direct circulation system. Higher operating temperature increases collector heat loss and lowers collector efficiency.

Because they don't contain antifreeze, *all direct circulation systems must completely drain the collectors when freezing contains are imminent.* Several methods have been devised to facilitate this draining.

Draindown Systems

One approach, known as a **draindown system,** uses an electronic control to detect near-freezing conditions at the collectors. When necessary, this control operates motorized valves to drain water from the collectors and exposed piping. All collectors and exposed piping must be pitched a minimum of ¼ inch per foot to allow complete drainage. Any failure of the collector temperature sensor, electronic control, or motorized valves can prevent the system from draining. The result can be a costly "hard freeze" that ruptures the collector waterways and/or exposed piping.

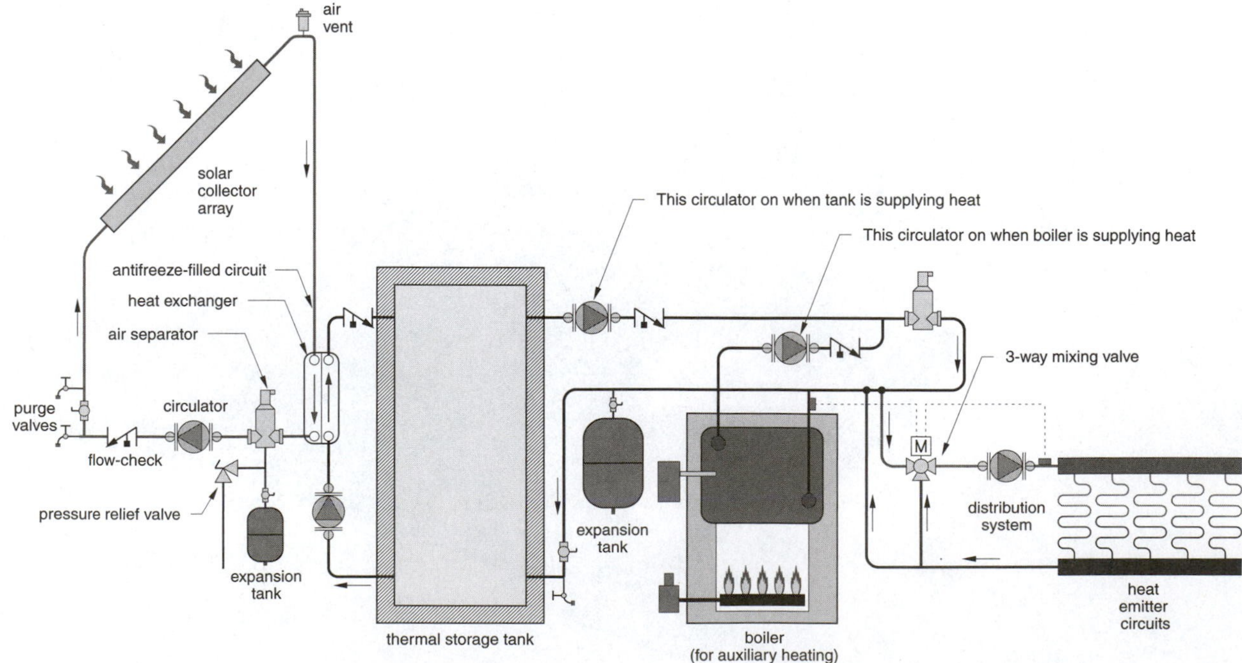

Figure 3–42 Closed-loop/antifreeze-type solar energy system for space heating.

Draindown systems also introduce fresh (oxygen containing), water into the system each time it is refilled following a draindown. Because of this they must be designed as open-loop systems, and cannot contain cast-iron or steel components.

Gravity Drainback Systems

Another approach is known as a **gravity drainback system.** A schematic depicting this approach is shown in Figure 3–43.

In this approach, water drains from the collectors and exposed piping, and runs back to the storage tank whenever the collector circulator stops. No motorized valves or electronic controls are needed. *Again, it is absolutely necessary that the collectors and all exposed piping be pitched a minimum of 1/4 inch per foot toward the storage tank to allow complete drainage.* The collectors and exposed piping are refilled with tank water (rather than fresh water) each time the collector circulator starts. This circulator in a drainback system must be sized to be able to lift the water to the top of the collector array.

Notice the captive volume of air at the top of the storage tank in Figure 3–43. This air quickly replaces the water in the collectors and exposed piping when the collector circulator stops. The air enters the return pipe through an open tee located above the water level in the tank. If properly sized, this air volume can also serve as the expansion tank for both the collector loop and the distribution system.

The return line from the collectors to the storage tank should be sized for a minimum flow velocity of 2 feet/second. This allows the water returning from the collectors to entrain air bubbles and carry them back to the storage tank. As the return pipe fills with water, a siphon is established over the top of the collector loop. Proper sizing of the return pipe also minimizes gurgling noises and increases the water flow rate through the collectors, slightly boosting their efficiency.

The gravity drainback approach has several advantages compared to the other systems. First, since it relies on gravity alone for system drainage, it is not prone to freezing due to failure of a control device. Second, unlike a draindown system, drainback systems recirculate the same water over and over. Since no water leaves the system, fresh water does not keep reentering the system. This allows drainback systems to be designed as closed-loop systems using less expensive cast-iron circulators and other standard closed-loop piping components. Finally, several components associated with the other design approaches are no longer needed, and thus costs are reduced. Of all design approaches, the closed-loop/gravity drainback system is the most preferred by the author.

Design Considerations for Active Solar Systems

An important design objective for any active solar energy system is maintaining collector temperature as low as possible. The lower the collector temperature, the higher its efficiency, and the greater the amount of available solar energy it will collect.

Slab-type radiant floor heating is one of the best hydronic distribution systems for use with active solar collectors. The relatively low water temperatures required by such sys-

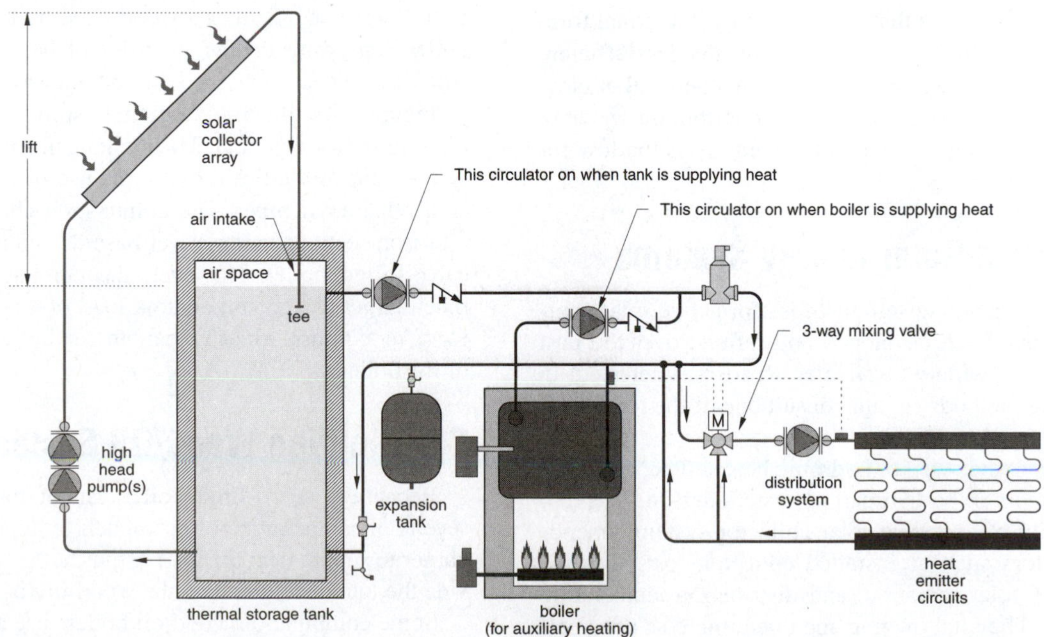

Figure 3–43 Example of a gravity drainback solar energy system for space heating.

tems allow the collectors to operate at high efficiencies, maximizing the amount of solar energy gathered. Conventional hydronic baseboard systems, by comparison, require significantly higher temperature water, and thus are a poor choice for use with solar collectors.

Active solar heating systems usually provide only part of a building's seasonal heating energy. This percentage depends upon location, collector area versus building load, storage tank volume, and many other factors. Practical designs always use some type of "auxiliary" heat source to supply the system when solar energy is not available. Any of the hydronic heat sources discussed in this chapter could potentially be used for this purpose. Since design heating load may occur during inclement weather when no solar energy is available, the auxiliary heat source must be sized to handle the design heating load of the building.

The auxiliary heat source should not maintain the thermal storage tank at temperatures suitable for use by the distribution system during nonsolar periods. Instead, the tank should be allowed to cool down to room temperature during such periods giving up all its transferable heat. Provided the tank is located within heated space, the heat released as the tank cools still offsets a portion of the building's heating load. This energy will eventually be replaced by the collected solar energy, rather than the auxiliary heat source.

The use of auxiliary energy to maintain tank temperature during nonsolar periods can waste solar energy due to delayed collector start-up when solar radiation again becomes available. This delay is caused by the differential temperature control commonly used with active solar heating systems. With this type of control, the collector circulator starts only when collector temperature is 8 to 10 degrees above storage tank temperature.

Another reason for not maintaining the tank temperature using auxiliary energy is that storing energy in thermal form (e.g., as heated water) is thermodynamically less efficient than maintaining this energy in its original chemical or electrical form. Proper design allows the distribution water to bypass the storage tank when its temperature is too low for use by the heat emitters.

Sizing Active Solar Energy Systems

Detailed sizing and design methods for active solar energy systems have been developed and refined over the past decade and are widely available. Readers interested in detailed sizing methods should consult one of the references at the end of the chapter.

As with any other type of hydronic heat source, cost considerations are an important part of system design. The economic viability of an active solar collector system depends on many factors such as installed collector cost, seasonal availability of solar radiation, and distribution temperature requirements. The total owning and operating cost of a solar collector system over its expected life must be compared to those of other potential heat sources.

3.17 WOOD-FIRED BOILERS

Following the energy crisis of the early 1970s, there was a dramatic rise in the number of wood-burning devices being installed for home heating. During this time, wood-burning took on an almost "patriotic" connotation. Many wood-burning boilers intended for use in hydronic heating systems appeared on the market. They ranged from simple firebox heat exchangers, to sophisticated forced-draft boilers capable of producing combustion temperatures in excess of 2000 °F. As was the case with solar energy systems, some wood-burning devices have since disappeared from the market.

The decision to use a wood-fired boiler involves a time commitment much more demanding than for conventional heat sources. It could be said that burning wood for home heating is choosing "a way of life", rather than just choosing a fuel. Many one time wood-burning enthusiasts have lost their zeal for wood burning when these requirements were realized after the system was installed. Some households simply have no one at home during an average workday to tend a wood-fired boiler. Long-term satisfaction with wood-burning under such circumstances is questionable. This observation is not meant to discourage those considering wood-burning boilers, but rather to encourage thorough and realistic comparisons with other options before making a decision.

Types of Wood-fired Boilers

Wood-fired boilers come in a number of designs. Some rely on the natural draft created by a standard chimney, others use a small blower to force air into the combustion chamber. Many boilers that can burn wood can also burn coal if properly set up. These are often called **solid fuel boilers.** Still others can as either a wood-fired boiler, or an oil-fired boiler depending on the intensity of heat output from the wood fire. These are called combination wood/oil boilers.

Figure 3–44 illustrates a modern solid fuel boiler capable of burning wood or coal. This boiler can partially control the heat output of the fire through the use of a thermostatically controlled air damper. The combustion chamber is built of cast-iron sections using a wet base design similar to a typical oil-fired boiler. However, the combustion chamber is much larger to accommodate a load of wood that will burn for several hours. An ash cleanout can be seen at the bottom of the boiler.

Combination Wood/Oil Systems

Because a wood-fired boiler cannot load its own fuel, several manufacturers offer models equipped with an oil burner to supplement the heat output of the wood fire, or provide the total heat output if the wood fire is not maintained.

Some combination wood/oil boilers use a single combustion chamber for both fuels. Such a chamber is of necessity a compromise between the ideal geometry for oil firing, and

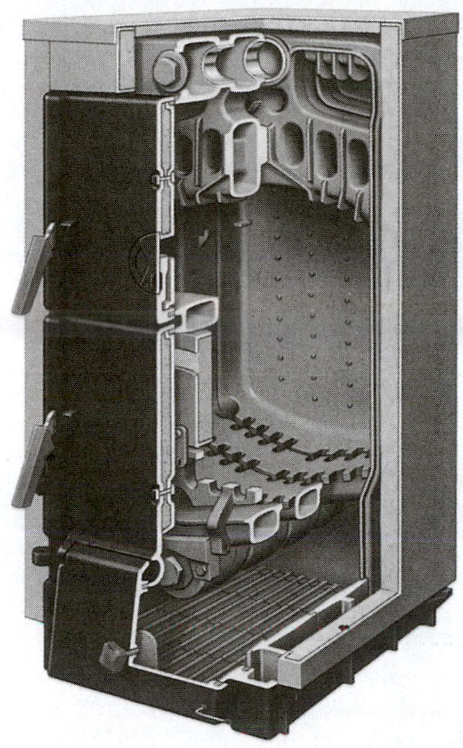

Figure 3–44 Modern solid fuel boiler. *Courtesy of Buderus Hydronic Systems.*

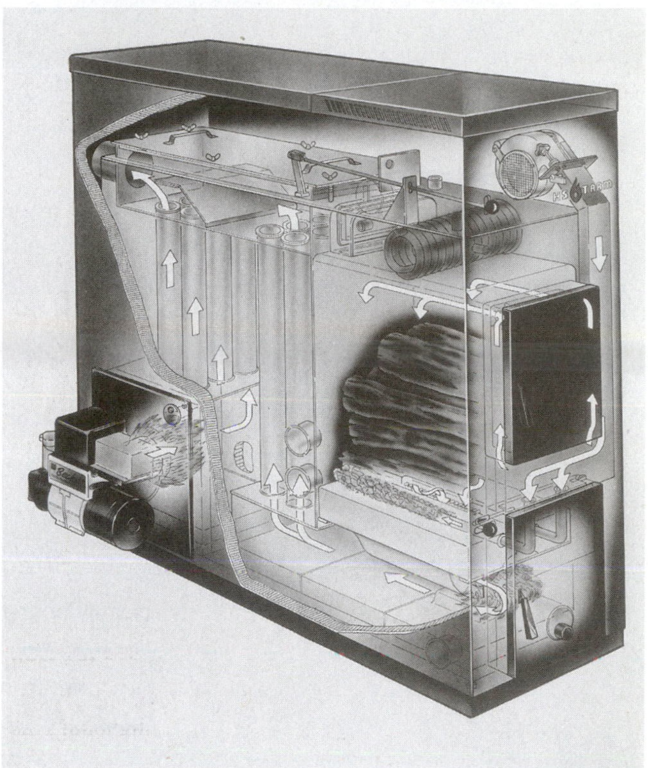

Figure 3–45 Example of a combination wood/oil boiler with separate combustion chambers for each fuel. *Courtesy of HS Tarm.*

the volume needed for wood burning. The efficiency of such a boiler in oil-fired operation generally is not as high as with a modern oil-only boiler. Another concern when an oil burner fires into the same chamber used for burning wood is soot and ash accumulation on the burner's flame head. Such boilers need more frequent cleaning to maintain high burner efficiency. To correct for these, some manufacturers use separate combustion chambers for each fuel. An example of such a combination boiler is shown in Figure 3–45.

Another approach is to install a separate boiler for each fuel. In such a system, each boiler is "optimized" for the fuel it burns. In a properly planned system, the shutdown of one boiler still allows the other to operate. The duel boiler approach is also common for retrofit situations.

There are several ways of piping two boilers into the system. One is to pipe the wood-fired boiler in series with the oil- or gas-fired boiler as shown in Figure 3–46a. The intent is for the wood-fired boiler to handle the heating load until its output decreases due to fuel depletion. At this point (assuming heat is still needed), a temperature control turns on the conventional boiler to supplement and eventually take over for the wood-fired boiler. Although series piping appears simple, it suffers from the fact that heated water always passes through both boilers, even if one of them is not operating. Air currents moving through and around the unfired boiler can absorb heat from this water and carry it up the chimney. Because of this, the author does not recommend the series piping arrangement.

Another possibility is to pipe the wood-fired boiler in parallel with the conventional boiler as shown in Figure 3–46b. Each boiler has its own circulator and check valve. Water only flows through a boiler while it is being fired. This significantly reduces off-cycle heat loss in the unfired boiler. The controls function the same way as with the series arrangement, with the wood-fired boiler having first priority over the conventional boiler provided it maintains a suitable supply temperature. With this arrangement, the circulator on each boiler must be large enough to circulate the water through the entire distribution system. If the controls allow both boilers to operate simultaneously, the system's flow rate would increase significantly. This might lead to excessive flow velocities and associated noise.

A third configuration, shown in Figure 3–46c, makes use of the same piping arrangement preferred for oil- or gas-fired multiple boiler systems. Both boilers are set up as parallel circuits, each with its own circulator and flow-check valve. The parallel circuits are then joined to the main distribution piping using a pair of closely spaced tees. With this approach, flow in the distribution system is unaffected by operation of either boiler circulator. Each boiler also receives the same inlet temperature allowing for maximum efficiency. When a boiler is not firing, its circulator remains off.

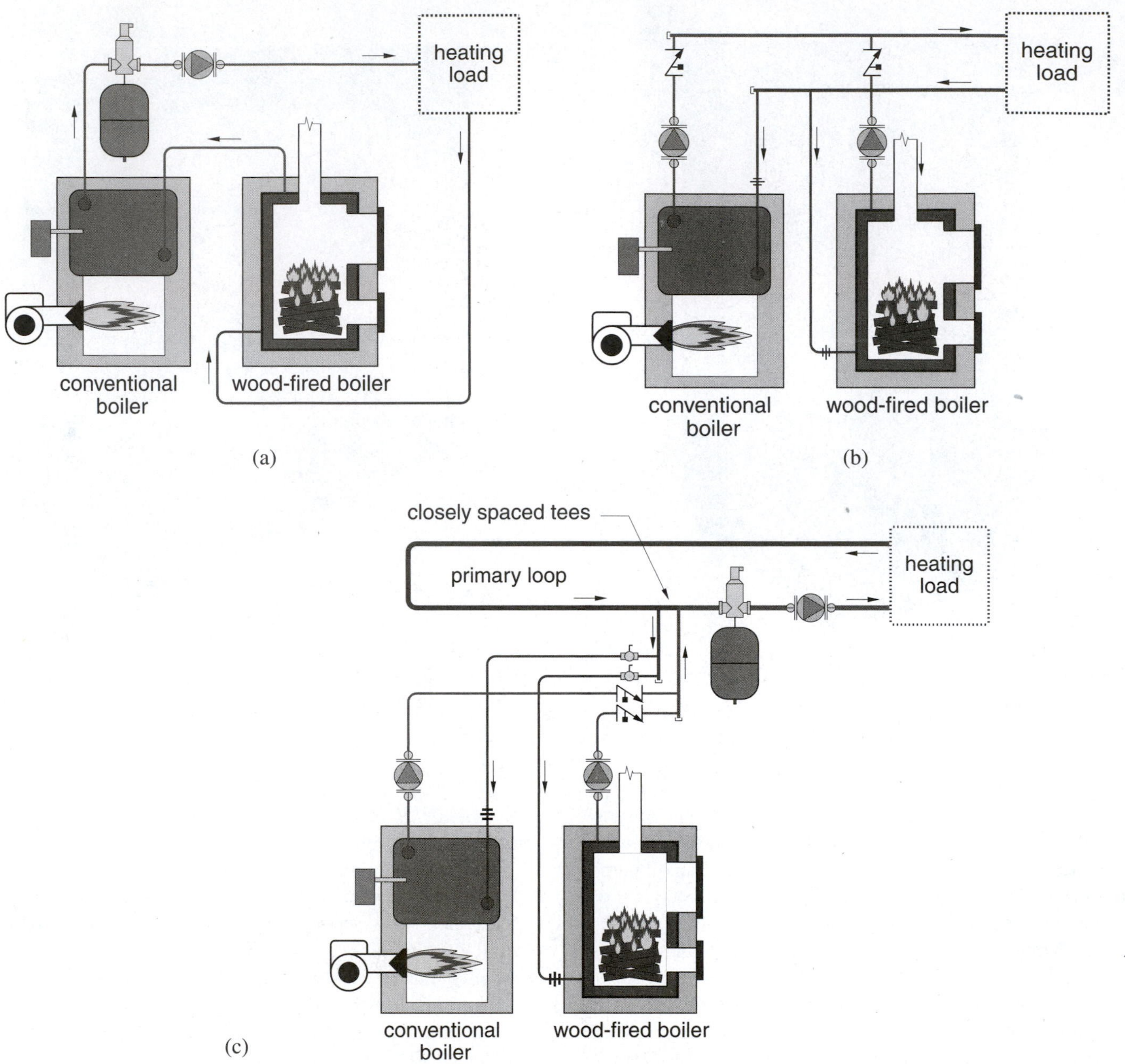

Figure 3–46 (a) Series piping of wood-fired boiler and conventional gas or oil-fired boiler (not recommended). (b) Parallel piping of wood-fired boiler and conventional gas or oil-fired boiler (acceptable). (c) Primary/secondary piping of wood-fired boiler and conventional gas or oil-fired boiler (preferred).

Outdoor Wood-fired Furnaces

Still another option in wood-fired hydronic heat sources is an outdoor wood-fired furnace. These units typically contain a large water-filled heat exchanger that surrounds an equally large firebox. Both are contained in an insulated metal enclosure intended for outside installation. Heated water is conveyed to the building(s) through insulated underground pipes. An example of an outdoor wood-fired furnace is shown in Figure 3–47.

These units are not referred to as "boilers" because their water-filled heat exchangers are vented directly to the atmosphere, and thus cannot be pressurized, as can a boiler. A heat exchanger is often used to separate the nonpressurized boiler circuit from a typical (pressurized) hydronic distribution system as illustrated in Figure 3–48.

Modern units are equipped with automatically controlled blowers and air dampers that provide reasonable control over heat output. The firebox can often hold a large "charge" of wood. Enough that wood might only have to be added

Figure 3–47 Example of an outdoor wood-fired furnace.

once or twice in a 24-hour period, even during cold weather. On larger units, the heat exchanger may hold several hundred gallons of water. This gives the unit considerable thermal mass that can often supply heat to the building for several hours after the fire has dwindled.

Freeze protection is usually accomplished by continuous circulating water between the unit and the building it serves. Antifreeze could be used, but the volume required for a larger unit is considerable, and would add significant cost to the installation. Some installations are planned so the boiler circulator can be powered by an auxiliary electric source such as a generator or battery during an extended power outage.

Outdoor wood-fired furnaces and boilers are obviously more practical in rural locations than in high-density suburbs. They must be located a sufficient distance from buildings so that the smoke they produce does not infiltrate these buildings. Still, they offer the safety benefit of separating the fire from the building. They also eliminate the need to carry wood into a building prior to burning it. In many cases, a single outdoor unit can supply sufficient heat for two or more buildings.

Efficiency of Wood-fired Boilers

The seasonal efficiency of wood-burning boilers is not as well-documented as those of more traditional hydronic heat sources. At present, there are no U.S. government standards or minimum efficiency values for wood-burning boilers. As a guideline, one might expect a quality air-tight wood-fired boiler to achieve a combustion efficiency of 40 percent to 65

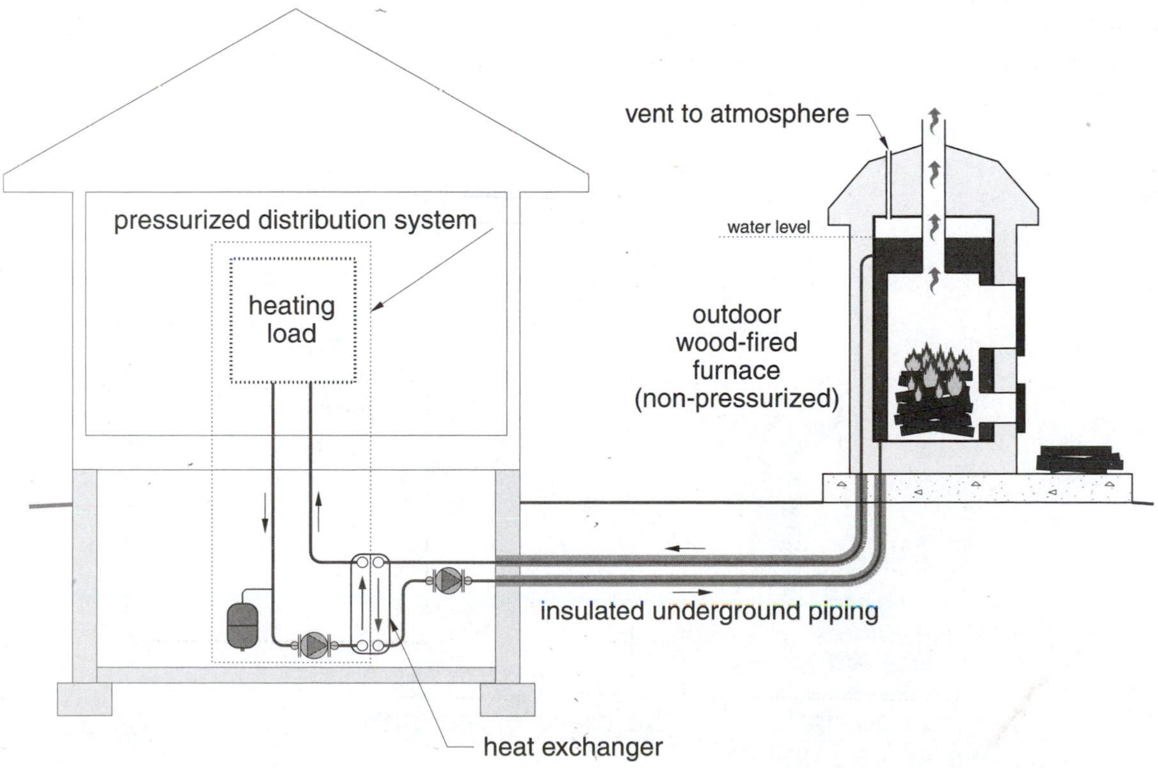

Figure 3–48 Use of a heat exchanger to couple a non-pressurized outdoor wood-fired furnace to a pressurized closely loop hydronic distribution system.

percent under steady operation. Modern forced-draft wood-fired boilers can attain steady state efficiencies in the range of 80 percent. This value however, is highly dependent on the moisture content of the firewood. Use of unseasoned wood, with a high moisture content, could easily cut this figure in half.

The efficiency of a wood-burning boiler is very dependent upon the temperature in the firebox at any given time. To achieve high efficiency, the pyrolytic gases released from the burning wood must themselves ignite and release their chemical energy content. If the temperature of these gases reaches approximately 1100 °F, and sufficient oxygen is available, combustion will occur. If the pyrolytic gases do not ignite due to low flame temperature or insufficient oxygen, they are carried out of the boiler and will likely condense into creosote within the chimney. Forced draft systems, with combustion chamber temperatures approaching 2000 °F, can burn nearly all the pyrolytic gases, and hence essentially eliminate creosote formation. *High combustion temperature is a key to high efficiency and clean operation of wood-fired boilers.*

Another consideration is the ability to regulate the heat output from a wood-fired boiler. Unlike oil, gas, or electrically powered boilers, which can be turned on and off as the heating load requires, wood-fired boilers must essentially burn off the loaded wood before heat output ceases. Many wood-fired boilers have some type of thermostatic air damper for marginal control over the air fed into the firebox. Remember, however, that an oxygen-starved flame smolders away at very low efficiency and increases creosote formation. It is very inefficient for wood-burning boilers to stand by with a smoldering flame while waiting for the next call for heat. If a wood-fired boiler cannot be well-controlled at a low firing rate, its operation may be limited to only the coldest weather.

Forced Combustion Combined with Thermal Storage

One method of minimizing inefficient operation is to completely burn off a charge of firewood at a high firing rate, and simultaneously transfer the heat to a large thermal storage tank. The heat can then be delivered by a hydronic distribution system as needed. This concept is similar to that discussed for electric thermal storage. It allows combustion to occur at a high rate and hence high efficiency. It also reduces the necessary firing time to a few hours each day. The concept is illustrated in Figure 3–49.

The boiler used in such a system must have a heat output significantly greater than the design load of the building. This is necessary because the boiler must produce sufficient heat in a period of a few hours to supply the building load for 24 hours. A forced-draft type boiler is often used to achieve this high firing rate. Such a boiler may only need to be loaded with wood once or twice in a 24-hour period. The size of the thermal storage tank can be calculated using Equation 3.10 discussed in Section 3.14. The cost of this type of system can be substantial because of the high capacity boiler and insulated thermal storage tank. Still, it uses the best available technology and operating strategy to extract maximum heat from the wood, and deliver it to the building in a fully regulated manner.

New technologies will undoubtedly emerge for the improved handling and burning of solid fuels such as wood, and other forms of "biomass." Promising methods such as

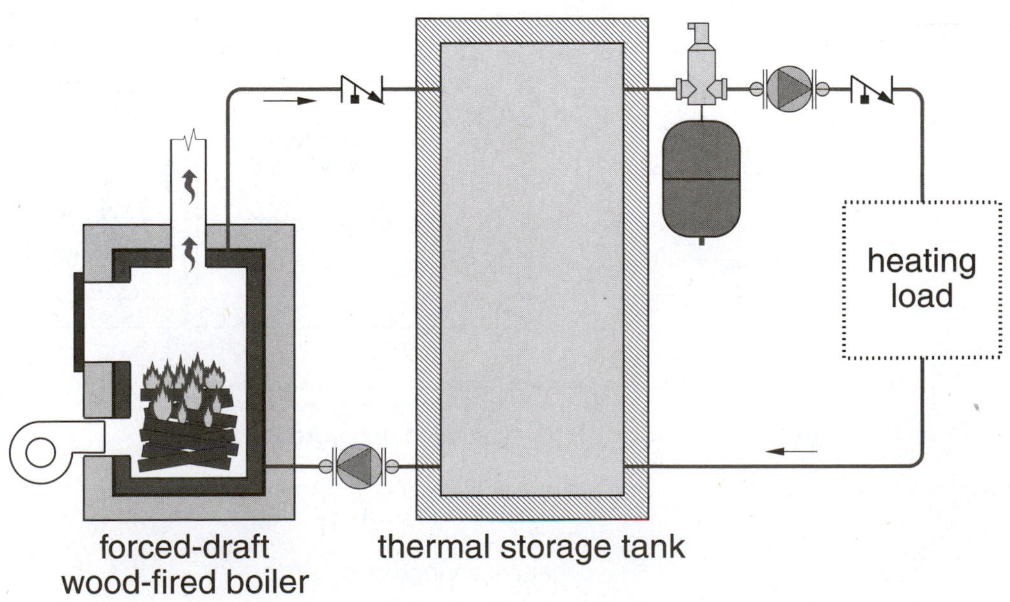

Figure 3–49 Piping schematic of a wood-fired boiler and thermal storage tank.

wood pelletization, in which wood pellets are automatically fed by an auger into a combustion chamber, and wood gasification, which yields a gas suitable for use in a gas-fired boiler, are already in use in large scale industrial systems. Such systems are economically justified at facilities such as paper mills and lumber mills where wood by-products are readily available on site, and can be utilized as fuel. These concepts have also been demonstrated in smaller systems, but presently remain too complex and expensive to compete against traditional heat sources.

SUMMARY

This chapter has examined several types of hydronic heat sources, pointing out some advantages and disadvantages of each. In some cases, the type of heat source used will be specified by the customer. In other cases, it will be dictated by the availability (or perhaps "unavailibility') of certain fuels. Fuel cost will obviously weigh heavily in the decision. In all cases, the designer must consider the effects of the proposed heat source on the distribution system used. Incompatibilities do exist, and are easily conceived by careless "catalog engineering." Always remember that *system* performance, and not just that of the heat source, is what counts.

KEY TERMS

Annual Fuel Utilization Efficiency (AFUE)
boiler jacket
buffer tank
chemical energy content (of fuel)
coefficient of performance (COP)
combustion efficiency
condenser
condensing mode
confined space
conventional boilers
cycle efficiency
dewpoint temperature
direct vent
DOE heating capacity
draft-proving switch
draindown system
dry-base boiler
earth heat exchanger
electric thermal storage (ETS) system
evaporator
fire-tube boiler
flat plate solar collector
free area (of an opening)
full storage system
gravity drainback system
heat pump
heat purging
heating capacity
IBR gross output
IBR net output
jacket
multiple boiler system
National Fuel Gas Code
noncondensing mode
off-cycle heat loss
open-loop source
packaged boiler
partial storage system
power venting
refrigeration cycle
run fraction
sealed combustion
sectional boiler
sections
sink
solid fuel boiler
source water
steady state efficiency
sustainable flue gas condensation
thermal mass (of a boiler)
time-of-use rates
turbulators
unconfined space
wet-base boiler

CHAPTER 3 QUESTIONS AND EXERCISES

1. Describe what is meant by the thermal mass of a boiler. What are some desirable and undesirable traits of low thermal mass boilers versus high thermal mass boilers?
2. What happens when a conventional boiler operates at an inlet water temperature lower than the dewpoint temperature of the exhaust gases?
3. Name some limitations that must be considered if a domestic water heater will be used for space heating.
4. What is the function of a draft-proving switch in a power exhaust system? Describe how it operates.
5. A boiler with a gas input rating of 100,000 Btu/hr will be installed in a basement. Using the NFPA 54 requirements described in Section 3.7, Combustion Air Requirements, determine the required floor area of the basement if it is to be considered unconfined space. Assume the height of the basement is 8 feet.
6. A boiler with a gas input rating of 100,000 Btu/hr will be installed in a basement boiler room. The combustion air will be brought in through horizontal ducts. The ends of the ducts are covered with metal louvers having a free/gross area ratio of 0.6. Determine the size of each duct opening in square inches.
7. A boiler produces 60,000 Btus in 30 minutes. In the process it consumes 0.5 gallons of #2 fuel oil (140,000 Btu per gallon). Determine its average efficiency during this period.
8. In 1 hour, a boiler fires for a total of 12 minutes. Use Figure 3–23 to determine the boiler's cycle efficiency for this period.
9. A house requires 75 MMBtu of heat for an entire heating season. Assume this is provided by an oil-fired boiler with an AFUE of 75 percent. Estimate how many gallons of #2 fuel oil would be required over the season.
10. What is the advantage of *parallel* primary/secondary piping versus *series* primary/secondary piping when connecting boilers in a modular boiler system?
11. An electric boiler outputs 40,000 Btu/hr. It is supplied by a 240-volt circuit. Estimate the electrical current flow to the boiler.
12. Over a 24-hour period, a house has an average heating load of 30,000 Btu/hr. If the house is supplied by a full storage ETS system operating for an off-peak period of 8 hours per night, what is the required total heating capacity of the elements? *Hint:* The entire heat for the 24-hour period must be obtained in the 8-hour off-peak period.
13. A hydronic heat pump has an output of 48,000 Btu/hr. The electrical input rate is 4.5 kW. Determine the COP of the heat pump under these conditions.
14. A hydronic heat pump has an output of 36,000 Btu/hr while its source water enters at 40 °F and its distribution water enters at 100 °F. Using Figure 3–37 estimate its heat output when its source water temperature drops to 35 °F and its load water enters at 110 °F.
15. What is the purpose of a buffer tank when used with a hydronic heat pump supplying several independently controlled zones?
16. Describe some crucial design requirements when matching a hydronic heat pump to a hydronic distribution system. Describe what could go wrong if these requirements are not met.
17. What are some advantages of a drainback solar energy system compared to a draindown system?
18. Why is it necessary to burn wood at high combustion temperatures to achieve high efficiency? What can happen if high temperature combustion does not occur?

FOR FURTHER READING

Ground Source Heat Pump Design

Closed-Loop/Ground Source Heat Pump Systems, Installation Guide. 1988. International Ground Source Heat Pump Association, P.O. Box 1688, Stillwater, OK 74076-1688.

Solar Energy System Sizing

Beckman, et al. *Solar Heating Design by the F-chart Method.* (New York: J. Wiley and Sons, 1977).

Duffie, et al. *Solar Engineering of Thermal Processes.* (New York: Wiley Interscience, 1980).

PROPERTIES OF WATER 4

OBJECTIVES

After studying this chapter, you should be able to:

- Describe several fluid properties relevant to hydronic system design and operation
- Explain why water is an excellent material for transporting heat
- Demonstrate how heat and temperature are related
- Calculate the amount of heat stored in a substance
- Explain the difference between a Btu and a Btu/hr
- Estimate the output of a heat emitter from measurements of temperature and flow rate
- Predict when water will boil based on pressure and temperature
- Explain how a fluid's dynamic viscosity affects its flow rate in a hydronic system
- Describe how air is dissolved into and released from water
- Look up or calculate the values of several fluid properties for use in later chapters

4.1 INTRODUCTION

Water is the essential material in any hydronic heating system. Its properties make possible the high efficiency and comfort associated with such systems. This chapter defines and explains several properties of water that are relevant to hydronic heating. A working understanding of these properties is invaluable to understand the behavior of hydronic systems, and to make design decisions that affect this behavior, as well as for troubleshooting.

4.2 SPECIFIC HEAT AND HEAT CAPACITY

One of the most important properties of water relevant to its use in hydronic systems is **specific heat.** Simply put, *the specific heat of any material is the amount of heat required to raise the temperature of one pound of that material by one degree Fahrenheit (F).* The specific heat of water is approximately 1.0 Btu per pound per degree F. To raise the temperature of 1 pound of water by 1 degree F will require the addition of 1 Btu of heat.

This relationship between temperature, material weight, and energy content also holds true when a substance is cooled. For example, to lower the temperature of 1 pound of water by 1 degree F will require the removal of one Btu of energy.

The specific heat of any material varies slightly with its temperature. For water, this variation is quite small over the temperature range used in most hydronic systems. Figure 4–1 shows this variation. When designing residential or light commercial systems, this slight variation can be safely

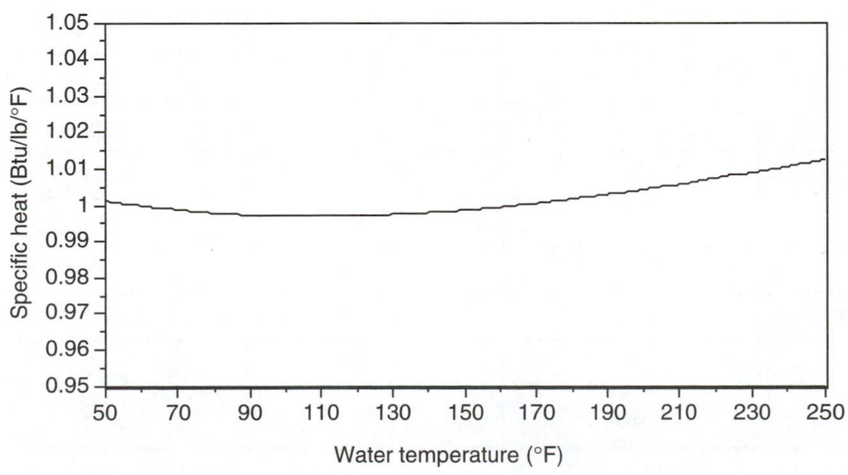

Figure 4–1 The specific heat of water as a function of temperature.

ignored, and the specific heat of water can be considered to remain constant at 1.0 Btu/lb/°F.

The value of a material's specific heat also changes when the material changes phase (i.e., between liquid and vapor, or liquid and solid). For example, the specific heats of ice and water vapor are about 0.48 Btu/lb/°F, and 0.489 Btu/lb/°F respectively. Both values are considerably lower than that of liquid water.

The specific heat of water is high in comparison to other common materials. In fact, water has one of the highest specific heats of any known material. The table in Figure 4–2 compares the specific heats of several common materials. Also listed is the **density** of each material, and the product of specific heat times density, which is called the **heat capacity.**

At first glance, this data may be deceiving. For example, why is the specific heat of air greater than the specific heat of steel? The answer is that although steel has a lower specific heat, it is much denser than air (489 lb/ft^3 for steel versus about 0.074 lb/ft^3 for air). Therefore, the heat capacity of steel, which reflects both its specific heat and density, is much higher than that of air.

By comparing the heat capacity of water and air, it can be shown that *a given volume of water can hold almost 3500 times as much heat as the same volume of air, for the same temperature rise.* This allows a given volume of water to transport much more heat than the same volume of air. Although it is not a scientific term, it could be said that water is about 3500 times more "thermally concentrated" than air.

The Fluid Properties Calculator module in the Hydronics Design Studio software can be used to find the specific heat of water as well as several water-based antifreeze solutions over a wide temperature range.

4.3 SENSIBLE HEAT VERSUS LATENT HEAT

The heat absorbed or released by a material while it remains in a single phase is called **sensible heat.** The word *sensible* means that the presence of the heat can be *sensed* by a temperature change in the material. As a material absorbs heat, its temperature increases. As heat is released, its temperature drops.

This process of sensible heat transfer is not the only way a material can absorb or release heat. If the material changes phase while absorbing heat, its temperature does not change. The heat absorbed in such a process is called **latent heat.** For water to change from a solid (ice) to a liquid requires the absorption of 144 Btu/lb while its temperature remains constant at 32 °F. When water changes from a liquid to a vapor it must absorb approximately 970 Btu/lb while its temperature remains constant. The amount of energy required to vaporize water is very large. This explains why water evaporating from the surface of an object (such as your skin) is a very effective means of cooling it.

4.4 DENSITY

The density of a substance is the number of pounds of the substance needed to fill a volume of one cubic foot. For example, it would take 62.4 pounds of water at 50 °F to fill a 1 cubic foot container, so its density is said to be 62.4 lb/cubic foot.

The density of water is dependent on temperature. Like most substances, as the temperature of water increases, its density decreases. As temperature increases, each molecule of

Material	Specific Heat (Btu / lb. / deg. F.)	Density (lb / cubic ft.)	Heat Capacity (Btu / cubic ft. / deg. F.)
water	1.0	62.4	62.4
concrete	0.21	140	29.4
steel	0.12	489	58.7
wood (fir)	0.65	27	17.6
ice	0.49	57.5	28.2
air	0.24	0.074	0.018
gypsum	0.26	78	20.3
sand	0.1	94.6	9.5
alcohol	0.68	49.3	33.5

Figure 4–2 The specific heats of some common materials.

water requires more space. This expansion effect is extremely powerful and can easily burst metal pipes if not properly accommodated. A graph showing the relationship between temperature and the density of water is shown in Figure 4–3.

This relationship can also be represented by Equation 4.1 for temperatures between 50 °F and 250 °F.

(Equation 4.1)

$$D = 62.56 + 3.413 \bullet 10^{-4}(T) - 6.255 \bullet 10^{-5}(T)^2$$

for: $(50 \leq T \leq 250)$

where:

D = density of water (lb/ft³)

T = water temperature (°F)

The change in density of water or water-based antifreeze solutions directly affects the size of the expansion tank required on all closed-loop hydronic systems. Proper tank sizing requires data on the density of the system fluid both when the system is filled and when it reaches its maximum operating temperature. This subject is covered in detail in Chapter 12, Expansion Tanks.

The Fluid Properties Calculator module in the Hydronics Design Studio software can be used to find the density of water as well as several water-based antifreeze solutions over a wide temperature range.

4.5 SENSIBLE HEAT QUANTITY EQUATION

It is sometimes necessary to determine the *quantity* of heat stored in a given amount of material that undergoes a specific temperature change. Equation 4.2, known as the sensible heat quantity equation can be used for this purpose.

(Equation 4.2)

$$h = wc(\Delta T)$$

where:

h = *quantity of heat* absorbed or released from the material (Btu)

w = weight of the material (lb)

c = specific heat of the material (Btu/lb/°F)

ΔT = temperature change of the material (°F)

Two conversion factors that are needed when using this equation for water are:

- 1 gallon of water weighs approximately 8.33 lbs.
- 1 cubic foot of water weighs approximately 62.4 lbs.

When the first of these values is factored into Equation 4.2, along with the specific heat of water = 1.0 Btu/lb/°F, the resulting equation becomes specific for use with water:

(Equation 4.3)

$$h = 8.33\ v(\Delta T)$$

where:

h = *quantity of heat* absorbed or released from the water (Btus)

v = volume of water involved (gallons)

ΔT = temperature change of the material (°F)

Example 4.1: A storage tank contains 500 gallons of water initially at 70 °F. How much energy must be added or removed from the water to:

a. raise the water temperature to 180 °F?

b. cool the water temperature to 40 °F?

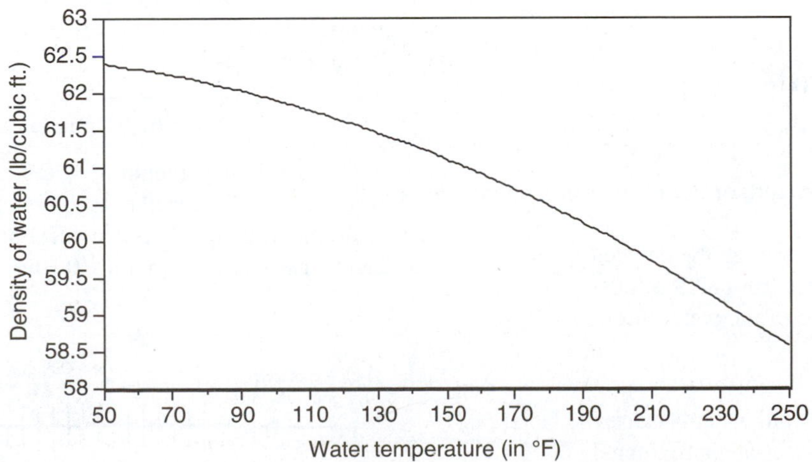

Figure 4–3 The density of water as a function of temperature.

Solution:

a. Using the sensible heat quantity equation for water (Equation 4.2),

h = 8.33(500)(180–70) = 458,000 Btu

b. Using the same sensible heat quantity equation for water (Equation 4.3),

h = 8.33(500)(70–40) = 125,000 Btu

4.6 SENSIBLE HEAT RATE EQUATION

Many times the system designer needs to know the *rate of heat transfer* to or from a stream of water flowing through a device such as a heat source or heat emitter. This can be done by introducing time into the sensible heat quantity equation. The resulting equation is called the sensible heat rate equation:

(Equation 4.4)

$$Q = Wc(\Delta T)$$

where:

Q = *rate of heat transfer* into or out of the fluid stream (Btu/hr)
W = rate of flow of the liquid (lb/hr)
c = specific heat of the liquid (Btu/lb/°F)
ΔT = temperature change of the liquid within the device it exchanges heat with (°F)

For cold water only this equation can be rewritten as:

$$Q = \left(\frac{\text{gallons}}{\text{minute}}\right)\left(\frac{8.33\text{ lb}}{\text{gallon}}\right)\left(\frac{60\text{ minutes}}{\text{hour}}\right)\left(\frac{1\text{ Btu}}{\text{lb} \bullet \text{°F}}\right)(\Delta T\text{ °F})$$

Which simplifies to:

(Equation 4.5)

$$Q = 500 f(\Delta T)$$

where:

Q = *rate of heat transfer* into or out of the water stream (Btu/hr)
f = **flow rate** of water through the device (gpm)
500 = constant rounded off from 8.33 × 60
ΔT = temperature change of the water through the device (°F)

Equation 4.5 is valid for cold water because the factor 500 is based on the density of water at approximately 60 °F (8.33 pound/gallon). The factor 500 is easy to remember when making quick mental calculations for the rate of sensible heat transfer. However, at higher temperatures the density of water decreases slightly. This, in turn, slightly affects its ability to transport heat.

Equation 4.6 is a more general form of Equation 4.5 that allows for variations in both the density and specific heat of the fluid. It can be used for water as well as other fluids.

Equation 4.6

$$Q = (8.01Dc)f(\Delta T)$$

where:

Q = *rate of heat transfer* into or out of the water stream (Btu/hr)
8.01 = a constant based on the units used
D = density of the fluid (lb/cubic ft.)
c = specific heat of the fluid (Btu/lb/°F)
f = flow rate of water through the device (gpm)
ΔT = temperature change of the water through the device (°F)

When using Equation 4.6, the density and specific heat should be based on its average temperature of the process through which the fluid is gaining or losing heat.

Example 4.2: Water flows into a heat emitter at 140 °F, and leaves at 125 °F. The flow rate is 4.5 gpm as shown in Figure 4–4. Calculate the rate of heat transfer from the water to the heat emitter using:

a. Equation 4.5

b. Equation 4.6

Solution: Using Equation 4.5 the rate of heat transfer is:

$$Q = 500(4.5)(140–125) = 33{,}750\text{ Btu/hr}$$

To use Equation 4.6, the density of water at its average temperature of 132 °F must first be estimated using Equation 4.1 or Figure 4–3.

$$D = 61.52\text{ pound/cubic foot}$$

As previously mentioned, the variation in the specific heat of water over the temperature range used in most hydronic systems is very small. For calculation purposes, it can be assumed to remain 1.0 Btu/lb/°F.

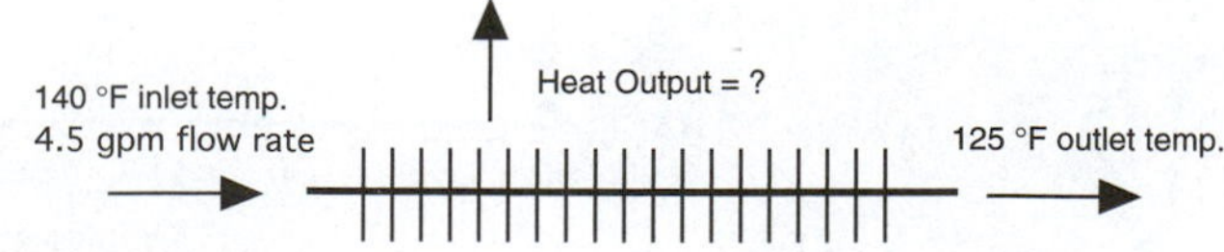

Figure 4–4 Heat emitter for Example 4.2

Substituting these numbers into Equation 4.6 yields:

$$Q = 8.01Dcf(\Delta T) = 8.01(61.52)(1.0)(4.5)(140 - 125)$$
$$= 33{,}262 \text{ Btu/hr}$$

Discussion: Equation 4.6 estimates the rate of heat transfer to be about 1.5 percent less than that estimated by Equation 4.5. This occurs because Equation 4.6 accounts for the slight decrease in the density of water at the elevated temperature, whereas Equation 4.5 does not. The variation between these calculations is arguably small. Equation 4.5 is generally accepted in the hydronics industry for quick estimates of heat transfer to or from a stream of water. However, Equation 4.6 will yield slightly more accurate results when the variation in density and specific heat of the fluid can be implemented into the calculation. Both equations will be used in later sections of this text.

Example 4.3: A heating system is proposed that will deliver 50,000 Btu/hr to a load while circulating only 2.0 gpm of water. What temperature drop will the water have to go through to accomplish this?

Solution: This situation again involves a *rate* of heat transfer from a water stream. Equation 4.5 will be used for the estimate. It will be rearranged to solve for the temperature drop term (ΔT).

$$\Delta T = \frac{Q}{500\,f} = \frac{50{,}000}{500(2.0)} = 50\ °\text{F}$$

Discussion: Although a hydronic distribution system could be designed to operate with a 50 °F temperature drop, this value is much larger than normal. Such a system would require very careful sizing of the heat emitters. This example shows the usefulness of the sensible heat rate equation to quickly evaluate the feasibility of an initial design concept.

4.7 VAPOR PRESSURE AND BOILING POINT

The **vapor pressure** *of a liquid is the minimum pressure that must be applied at the liquid's surface to prevent it from boiling.* If the pressure on the liquid's surface drops below its vapor pressure, the liquid will *instantly* boil. If the pressure is maintained above the vapor pressure, the liquid will remain a liquid, and no boiling will take place.

The vapor pressure of a liquid is strongly dependent on its temperature. The higher the liquid's temperature, the higher its vapor pressure. Stated in other terms: The greater the temperature of a liquid, the more pressure must be exerted on its surface to prevent it from boiling.

Vapor pressure is stated as an **absolute pressure,** and has the units of psia. On the absolute pressure scale, zero pressure represents a complete vacuum. On earth, the weight of the atmosphere exerts a pressure on any liquid surface open to the air. At sea level, this atmospheric pressure is about 14.7 psi. Stated another way, the absolute pressure exerted by the atmosphere at the earth's surface is approximately 14.7 psia. This implies that vapor pressures lower than 14.7 psia would represent partial vacuum conditions from our perspective in the earth's atmosphere.

Did you ever wonder why water boils at 212 °F? What is special about this number? Actually, there is nothing special about it. It just happens that at 212 °F, the vapor pressure of water is 14.7 psia (equal to atmospheric pressure at sea level). Therefore, water boils at 212 °F at elevations near sea level. If a pot of water is carried to an elevation of 5000 ft. above sea level and then heated, its vapor pressure will eventually rise until it equals the lower atmospheric pressure, and boiling will begin at about 202 °F.

Because water has been widely used in many heating and power production systems, its vapor pressure versus temperature characteristics have been well established for many years. This data is available in the form of steam tables. The vapor pressure of water between 50 °F and 250 °F can also be read from the graph in Figure 4–5.

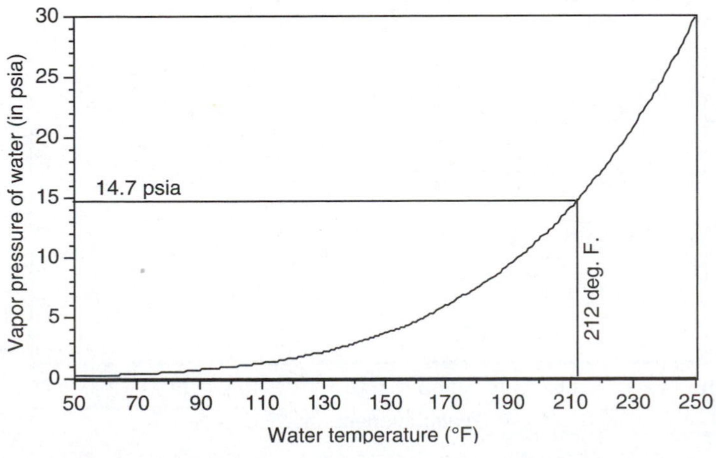

Figure 4–5 Vapor pressure of water as a function of temperature

The relationship shown in Figure 4–5 can also be generated using Equation 4.7 for water temperatures between 50 °F and 250 °F.

(Equation 4.7)

$$P_v = 0.771 - 0.0326(T) + 5.75 \bullet 10^{-4}(T)^2 - 3.9 \bullet 10^{-6}(T)^3 + 1.59 \bullet 10^{-8}(T)^4$$

for: $(50 \le T \le 250)$

where:

P_V = vapor pressure of water (psia)
T = temperature of the water (°F)

Hydronic systems should be designed to ensure that the absolute pressure of the water at all points in the system stays safely above the water's vapor pressure at all times. This prevents problems such as cavitation in pumps and valves.

4.8 VISCOSITY

The **viscosity** of a fluid indicates its resistance to flow. The higher the viscosity of a fluid, the more drag it creates as it flows through pipes, fittings, valves, or any other component in a piping circuit. Higher viscosity fluids also require more pumping power to maintain a given system flow rate compared to fluids with lower viscosities.

Like all other properties discussed in this chapter, the viscosity of water is dependent on temperature. The viscosity of water and water-based antifreeze solutions decreases as their temperature increases. The relationship between the viscosity of water and its temperature is shown in Figure 4–6.

This relationship can also be described by Equation 4.8 for temperatures between 50 °F and 250 °F.

(Equation 4.8)

$$\mu = 0.001834 - 2.73 \bullet 10^{-5}(T) + 1.92 \bullet 10^{-7}(T)^2 - 6.53 \bullet 10^{-10}(T)^3 + 8.58 \bullet 10^{-13}(T)^4$$

for: $(50 \le T \le 250)$

where:

μ = dynamic viscosity of water (lb_{mass}/ft/sec)
T = temperature of the water (°F)

Both Figure 4–6 and Equation 4.8 determine the *dynamic* viscosity of water as a function of temperature. Another type of viscosity, called *kinematic* viscosity is sometimes discussed in other texts on fluid mechanics. However, it is not used in this book.

The physical units of dynamic viscosity can be confusing. They are a mathematical consequence of how dynamic viscosity is defined, and do not lend themselves to a practical interpretation of what the viscosity number actually is. The units used for dynamic viscosity within this book are (lb_{mass}/ft/sec). These units are compatible with all equations used in this book that require viscosity data. The reader is cautioned, however, that several other units for viscosity may be used in other sources of viscosity data (such as manufacturer's literature). *These other units must first be converted to* lb_{mass}/ft/sec) *using the appropriate conversion factors, before being used in any equations in this book.*

Solutions of water and glycol-based antifreeze have significantly higher viscosities than water alone. The greater the concentration of antifreeze, the higher the viscosity of the solution. This increase in viscosity can result in significantly lower flow rates within a hydronic system if it is not accounted for during design.

The Fluid Properties Calculator module in the Hydronics Design Studio software can be used to find the dynamic viscosity of water as well as several water-based antifreeze solutions over a wide temperature range.

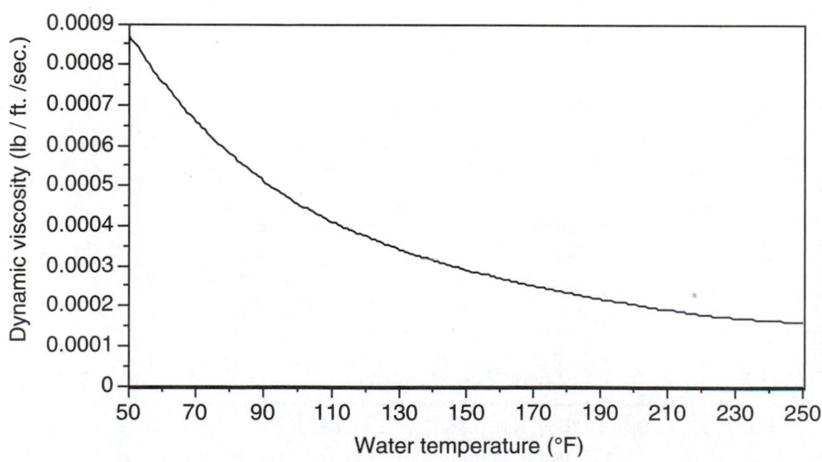

Figure 4–6 Dynamic viscosity of water as a function of temperature

4.9 DISSOLVED AIR IN WATER

Water has the ability to contain air much like a sponge can contain a liquid. Molecules of the various gases that make up air, including oxygen and nitrogen, can exist *in solution* with water molecules. Even when water appears perfectly clear it can contain air in solution. In other words, *just because bubbles are not visible in the water does not mean air is not present.*

When hydronic systems are first filled with water, **dissolved air** is present in the water throughout the system. If allowed to remain in the system, this air can create numerous problems including noise, cavitation, corrosion, and improper flows. A well-designed hydronic system must be able to efficiently and automatically capture dissolved air and expel it. A proper understanding of what affects the air content of water is crucial in designing a method to get rid of it.

The amount of air that can exist in solution with water is strongly dependent on the water's temperature. As water is heated, its ability to hold air in solution rapidly decreases. However, the opposite is also true. As heated water cools, it has a propensity to absorb air. Heating water to release dissolved air is like squeezing a sponge to expel a liquid. Allowing the water to cool is like letting the sponge expand to soak up more liquid. Together these properties can be exploited to help capture and rid air from the system.

The water's pressure also has a marked effect on its ability to hold dissolved air. When the pressure of the water is lowered, its ability to contain dissolved air decreases, and vice versa.

Figure 4–7 shows the maximum amount of air that can be contained (dissolved) in water as a percent of the water's volume. The effects of both temperature and pressure on dissolved air content can be determined from this graph. To understand this graph, first consider a single curve such as the one designated 45 psi. This curve represents a *constant pressure* condition for the water. Notice that as the water temperature increases, the curve descends rapidly. This descent indicates a decrease in the water's ability to hold air

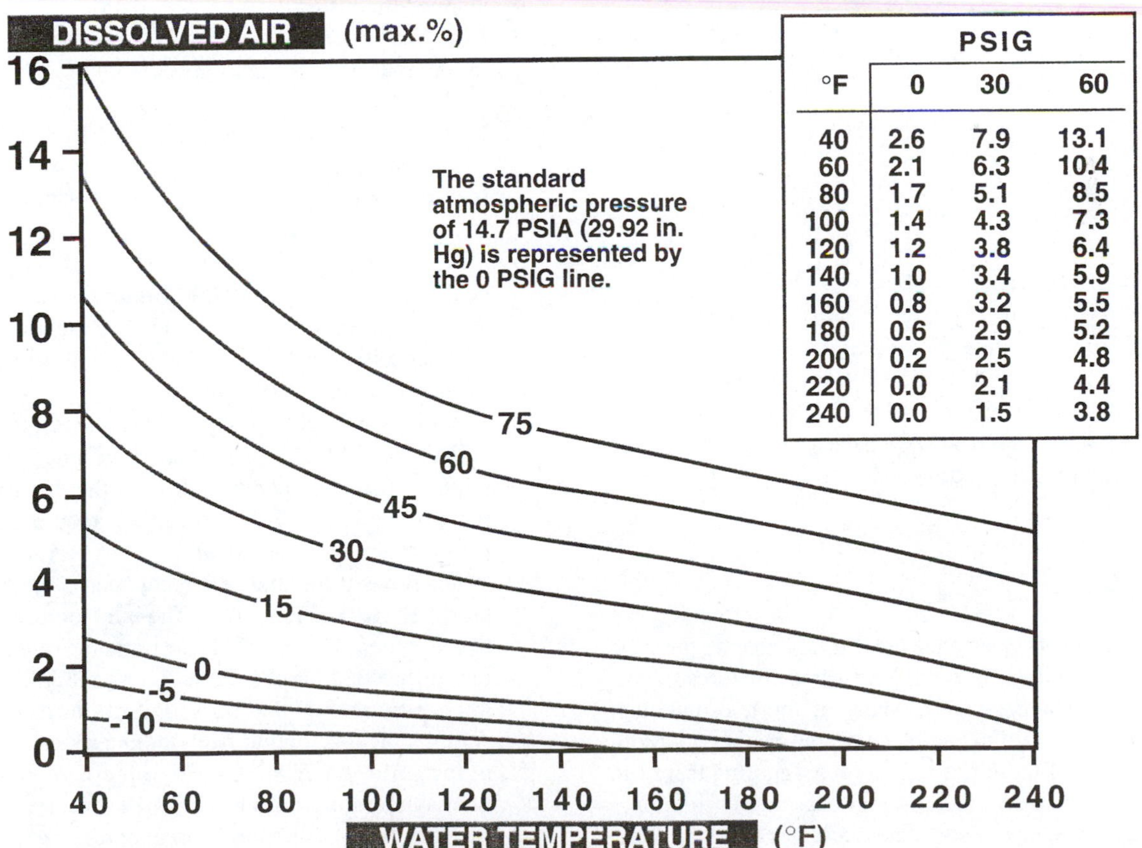

°F	PSIG 0	PSIG 30	PSIG 60
40	2.6	7.9	13.1
60	2.1	6.3	10.4
80	1.7	5.1	8.5
100	1.4	4.3	7.3
120	1.2	3.8	6.4
140	1.0	3.4	5.9
160	0.8	3.2	5.5
180	0.6	2.9	5.2
200	0.2	2.5	4.8
220	0.0	2.1	4.4
240	0.0	1.5	3.8

Figure 4–7 Solubility of air in water at various temperatures and pressures. *Courtesy of Spirotherm Corporation.*

in solution. The exact change in air content can be read off the vertical axis. This same trend holds true for all other constant pressure curves. Now consider what happens if the water temperature remains constant, say at 150 °F, and the pressure is lowered from 90 psi to 15 psi. The lower the pressure, the less air the water can hold in solution.

Keep in mind that Figure 4–7 indicates the *maximum possible* dissolved air content of water at a given temperature and pressure. Properly deaerated hydronic systems will quickly reduce the dissolved air content of their water to a small fraction of one percent. Methods of accomplishing this are discussed in Chapter 13, Air Removal, Filling, and Purging.

4.10 INCOMPRESSIBILITY

When a quantity of water is put under pressure there is very little change in its volume. A pressure increase of 1 psi will cause a volume reduction of only 3.4 *millionths* of the original water volume. This change is so small that it can be ignored in the context of designing hydronic systems. In effect, we can treat water, and most other liquids, as being **incompressible.**

In practical terms this means that liquids cannot be squeezed together. It also implies that if a liquid's flow rate is known at any one point in a closed series piping circuit, it must be the same at all other points, regardless of the pipe size. If the piping circuit contains parallel branch circuits, the flow can divide up through them, but the sum of all branch flow rates must equal the total system flow rate. This simple concept forms the basis for analyzing fluid flow in complex piping systems. It will be applied repeatedly in later chapters of this book.

Incompressibility also implies that liquids can exert tremendous pressure in any type of closed container when they expand due to heating. For example, a closed hydronic piping system completely filled with water, and not equipped with an expansion tank or pressure-relief valve, would quickly burst apart at its weakest component as the water temperature increased.

SUMMARY

The properties of water presented in this chapter describe its heat absorption, boiling point, expansion, flow resistance, and air holding characteristics. Many of the questions that arise during system design can be answered based on a proper understanding of these fluid properties. Ignoring them can lead to serious problems such as low heat output, improper expansion tank size, and inadequate circulator selection.

The sensible heat quantity and rate equations are two of the most often used equations in hydronic system design. You will see them used frequently in later chapters.

KEY TERMS

absolute pressure
density
dissolved air
flow rate
heat capacity
incompressible
latent heat
sensible heat
specific heat
vapor pressure
viscosity

CHAPTER 4 QUESTIONS AND EXERCISES

1. How many gallons of water would be required in a storage tank to absorb 250,000 Btu while undergoing a temperature change from 120 °F to 180 °F?
2. Water enters a baseboard radiator at 180 °F and at a flow rate of 2.0 gpm. Assuming the radiator releases heat into the room at a rate of 20,000 Btu/hr, what is the temperature of the water leaving the radiator?
3. How much pressure must be maintained on water to maintain it as a liquid at 250 °F? State your answer in psia.
4. If you wanted to make water boil at 100 °F, how many psi of vacuum (below atmospheric pressure) would be required? *Hint:* psi vacuum = 14.7 – absolute pressure in psia.
5. Determine the air flow rate (in cubic feet per minute) necessary to transport 40,000 Btu/hr based on a temperature increase from 65 °F to 135 °F. Base the calculation on a specific heat for air of 0.245 Btu/lb/°F and an assumed density of 0.07 lb/ft^3.
6. Assume you need to design a hydronic system that can deliver 80,000 Btu/hr. What flow rate of water is required if the temperature drop of the distribution system is to be 10 °F? What flow rate is required if the temperature drop is to be 20 °F?
7. Water flows through a series-connected baseboard system as shown in Figure 4–8. The inlet temperature to the first baseboard is 165 °F. Determine the outlet temperature of the first baseboard, and both the inlet and outlet temperatures for the second and third baseboards. Assume the heat output from the interconnecting piping is insignificant. *Hint:* Use the approximation that the inlet temperature for a baseboard equals the outlet temperature from the previous baseboard.
8. At 50 °F, 1,000 lbs. of water occupies a volume of 16.026 ft^3. If this same 1,000 lbs. of water is heated to 200 °F, what volume will it occupy?

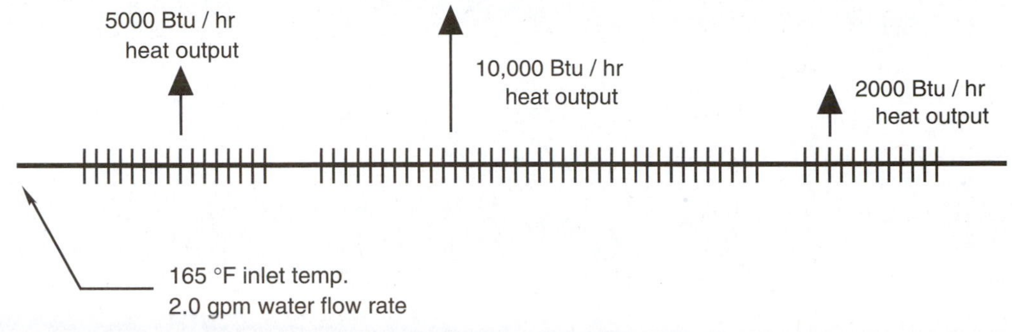

Figure 4–8 Series-connected baseboard convectors for Exercise 7.

9. How much heat is required to raise the temperature of 10 lbs. of ice from 20 °F to 32 °F? How much heat is then required to change the 10 lbs. of ice into 10 lbs. of liquid water at 32 °F? Finally, how much heat is needed to raise the temperature of this water from 32 °F to 150 °F? State all answers in Btus.
10. You are designing a thermal storage system that must supply an average load of 30,000 Btu/hr for 18 consecutive hours. The minimum water temperature usable by the system is 110 °F. Determine the temperature the water must be heated to by the end of the charging period if
    a. the tank volume is 500 gallons
    b. the tank volume is 1,000 gallons
11. A swimming pool contains 18,000 gallons of water. A boiler with an output of 50,000 Btu/hr is operated 8 hours per day to heat the pool. Assuming all the heat is absorbed by the pool water with no losses to the ground or air, how much higher is the temperature of the pool after the 8 hours of heating?
12. An 18,000-gallon swimming pool is to be heated from 70 °F to 85 °F. Assuming an oil-fired boiler with an average efficiency of 80 percent will be used and that all heat is retained by the pool, with no losses to the ground or air. Determine:
    a. the number of Btus required to heat the pool
    b. the number of gallons of #2 oil required by the boiler
    c. at a cost of $1 per gallon of oil, the cost of bringing the pool temperature up
13. A family of four uses 60 gallons of domestic hot water per day. The water enters the water heater at 50 °F and is heated to 140 °F by the time it leaves. Calculate the number of Btus per day used for this process.
14. Recalculate the Btus used for domestic water heating in Question 13 assuming the water is heated to only 120 °F. Estimate the annual energy savings of the reduced water temperature setting assuming the same load exists 365 days per year. Express the answer in MMBtus (1 MMBtu = 1,000,000 Btu).
15. Assume a hydronic system is filled with 50 gallons of water at 50 °F and 30 psi pressure. The water contains the maximum amount of dissolved air it can hold at these conditions. The water is then heated and maintained at 200 °F and 30 psi pressure. Using Figure 4–7, estimate how much air will be expelled from the water.
16. Use the Fluid Properties Calculator module in the Hydronics Design Studio to determine the specific heat of a 50 percent ethylene glycol solution at 100 °F. Also find the specific heat of water at 100 °F. Assume that a flowing stream of each liquid had to deliver 40,000 Btu/hr of heat to a load with a 15 °F drop in temperature. What flow rate would be required for each fluid? Express your answers in pounds per hour.
17. Use the Fluid Properties Calculator module in the Hydronics Design Studio to determine the specific heat and density of a 30 percent ethylene glycol solution at 132.5 °F. Assume this fluid entered a radiator at 140 °F and exited the radiator at 125 °F, at a flow rate of 4.5 gpm. What is the rate of heat transfer from the fluid stream to the radiator? Express your answer in Btu/hr. Compare the results to those of Example 4.2. Explain the difference. *Note:* 1 ft^3 = 7.48 gallons.

PIPINGS, FITTINGS, AND VALVES 5

OBJECTIVES

After studying this chapter, you should be able to:

- Describe several types of piping materials used in hydronic systems
- Compare the temperature and pressure ratings of several piping materials
- Describe several types of piping support devices
- Describe the steps involved in soldering copper tube
- Discuss the issue of oxygen diffusion through polymer pipe
- Identify several common pipe fittings
- Specify the exact type of pipe fitting needed for a given situation
- Explain the use of several specialized pipe fittings
- Calculate the expansion movement of copper pipe
- Select the correct type of valve for a given application
- Identify several specialized valves and explain their use in hydronic systems
- Plan neat and efficient piping installations
- Recognize schematic symbols for piping system components

5.1 INTRODUCTION

This chapter presents basic information on piping, fittings, and valves used in residential and light commercial hydronic systems. It emphasizes the proper selection and installation of these components. The intent is to acquaint the reader with the wide range of available hardware, in preparation for overall system design in later chapters.

Much of the information deals with the benchmark hydronic piping material, copper tubing. Information on common and specialized copper fittings is presented. A short primer on soft soldering is also given.

Newer piping materials such as **cross-linked polyethylene** and **composite tubing** are also discussed. These materials have been successfully used for many years in Europe, and are quickly gaining market share in North America. The versatility they offer in certain situations is unmatched by traditional materials.

This chapter also discusses the intended application of several types of common and specialized valves used in hydronic heating systems.

Tips for professional installation of piping are also given. Small details such as level and plumb piping runs, wiped solder joints, and use of specialty fittings to reduce parts count, distinguish systems that merely work from those that convey professionalism and craftsmanship.

5.2 PIPING MATERIALS

This section compares several piping materials suitable for use in hydronic heating systems. The discussion is limited to those materials that are most practical in today's market. These include copper and **polymer** tubing.

All piping materials have strengths and weaknesses. All have conditions attached to their use that must be followed if the system is to perform as expected and last as long as possible. No single material is ideal for all applications.

Piping material(s) must be selected based on considerations such as temperature and pressure ratings, availability, ease of installation, corrosion resistance, life expectancy, local code acceptance, and cost. The designer should not feel constrained to use a single type of pipe for the entire system. Each material should be applied where it has the ability to overcome obstacles posed by other materials. For example, flexible PEX tubing may be ideal in retrofit situations where use of rigid piping materials would be difficult or impossible. On the other hand, rigid copper tubing has obvious advantages where the piping runs must be straight, or when heavier components will be supported by the tubing. A combination of flexible and rigid piping materials may provide the ideal solution in many systems.

Discussion will be limited to pipe sizes from ⅜ inch to 3 inches nominal diameter. Nearly all the hydronic systems within the scope of this book can be constructed of piping within this range. All analytic methods for piping design to be presented in later chapters, as well as in the Hydronics Design Studio software, also pertain to this range of pipe sizes.

Copper Tubing

Copper water tube was developed in the 1920s to provide an alternative to iron piping in a variety of uses. Its desirable features include:

- Good pressure and temperature rating for typical hydronic applications

- Good resistance to corrosion from water-based system fluids
- Smooth inner walls that offer low flow resistance Lighter than steel or iron piping of equivalent size
- Ability to be joined by the well-known technique of soft soldering

In situations where good heat transfer through the walls of the pipe is desirable, copper is unsurpassed. Examples include heat exchangers or the finned-tube element in baseboard convectors. The high thermal conductivity of copper can also be a disadvantage in situations where heat loss from piping should be minimized.

Copper tubing is available in a wide range of sizes and wall thickness. The type of tubing used in hydronic heating is called copper water tube. In the United States, copper water tube is manufactured according to the ASTM B88 standard. In this category, **pipe size** refers to the **nominal inside diameter** of the tube. The word "nominal" means that the measured inside diameter is *approximately* equal to the stated pipe size. For example, the actual inside diameter of a nominal 3/4" type M copper tube is 0.811 inches.

The outside diameter of copper water tubing is always ⅛ inch larger than the nominal inside diameter. For example, the outside diameter (O.D.) of 3/4-inch type M copper tube is 0.875 inches. This is exactly ⅞ inches, or ⅛ inch larger than the nominal pipe size.

Another category of copper tubing, designated as ACR tubing, is used in refrigeration and air conditioning systems. In this category, pipe size refers to the *outside* diameter of the tubing. ACR tubing is not used in hydronic heating or water distribution systems.

Copper water tube is available in three wall thicknesses designated as types K, L, and M in order of decreasing wall thickness. *The outside diameters of K, L, and M tubing are identical.* This allows all three types of tubing to be compatible with the same **fittings** and valves.

Because the operating pressures of residential and light commercial hydronic heating systems are relatively low, the thinnest-wall copper tubing (type M) is most often used. This wall thickness provides several times the pressure rating of other common hydronic system components, as can be seen in Figure 5–1. In the absence of any building or mechanical codes that require otherwise, type M is the standard for copper tubing used in hydronic heating systems.

Two common hardness grades of copper water tube are available. **Hard drawn tubing** is supplied in straight lengths of 10 and 20 feet. Because of its straightness and strength, hard drawn tubing is the most commonly used type of copper tubing in hydronic systems. So-called **"soft temper" tubing** is annealed during manufacturing to allow it to be formed with simple bending tools. It is useful in situations where awkward angles do not allow proper tubing alignment with standard fittings. Soft temper tubing comes in flat coils having standard lengths of 60 and 100 feet. The minimum wall thickness available in soft temper copper tubing is type L.

Nominal tube size	Inside diameter (inches)	Outside diameter (inches)	Rated working pressure (psi)*
3/8"	0.450	0.500	855/456
1/2"	0.569	0.625	741/395
3/4"	0.811	0.875	611/326
1"	1.055	1.125	506/270
1.25"	1.291	1.375	507/271
1.5"	1.527	1.625	497/265
2"	2.009	2.125	448/239
2.5"	2.495	2.625	411/219
3"	2.981	3.125	380/203

* Pressure ratings are based on service temperatures not exceeding 200 °F. The first number is for drawn (hard) tubing. The number following the / is for annealed (soft temper) tubing.

Figure 5–1 Physical data for selected sizes of type M copper water tube.

Supporting Copper Tubing

Copper tubing must be properly supported to prevent sagging or buckling. On horizontal runs of hard temper tubing, the following maximum support spacings are suggested:

- 1/2-inch and 3/4-inch tube:
  5 feet maximum support spacing
- 1-inch and 1 1/4-inch tube:
  6 feet maximum support spacing
- 1 1/2-inch and 2-inch tube:
  8 feet maximum support spacing

These support spacings do not include any allowance for the extra weight of piping components such as circulators, expansion tanks, or other components that may be supported by the tubing between supports. When such components are present, the piping should be supported immediately adjacent to the component.

On vertical runs, copper tubing should be supported at each floor level, or a maximum of every 10 feet.

A number of different supports are available for small tubing. In residential systems, horizontal piping runs are often supported by plastic coated hangers as shown in Figure 5–2. These simple devices allow the tubing to move slightly as it expands upon heating. The plastic coating reduces squeaking sounds as the tube moves through the hanger. Some pipe hangers have pointed ends that are driven in floor joists. Others have a perforated strap that can be nailed to framing.

Standoff-type piping supports (also called "bell hangers") are used to secure tubing to a solid surface. When tightened, they rigidly clamp the tube in place. This is acceptable for relatively short runs such as found in a typical residential mechanical room, but can result in expansion noises on

Figure 5–2 Examples of plastic-coated metal pipe hanger.

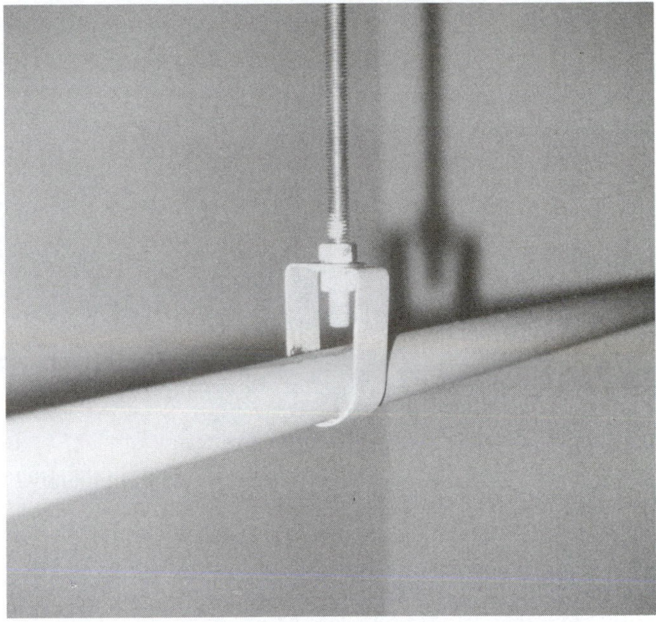

Figure 5–4 Pipe supported by a clevis hanger.

longer runs. Standoff supports can be used to support tubing running at any angle. They also allow sufficient space between the tube and mounting surface to accommodate pipe insulation. See Figure 5–3.

Copper tubing can also be supported at intermediate heights using clevis hangers supported by threaded steel rods attached to the ceiling structure. Clevis hangers are ideal when piping must be supported away from wall or ceiling surfaces. An example of a clevis hanger is shown in Figure 5–4.

In situations where several runs of piping and/or electrical conduit must be run in close proximity, a mounting rail system is a good option. Such systems consist of steel mounting rails to which clamps for various sizes of pipe can be secured. The rails can be mechanically fastened to walls or ceilings, or hung from threaded steel rods. The pipe clamps can easily be moved along the mounting rails before being tightened. Clamps are available with rubber liners that prevent metal-to-metal contact between the clamp and the tubing. Such clamps help dampen vibrations imparted to the piping from components such as circulators. Figure 5–5 shows the versatility of a mounting rail system in accommodating several pipe sizes. Figure 5–6 shows a multiple level mounting rail system supporting overhead piping and electrical conduit in a mechanical room.

Figure 5–3 Example of a stand-off pipe support (also know as a bell hanger).

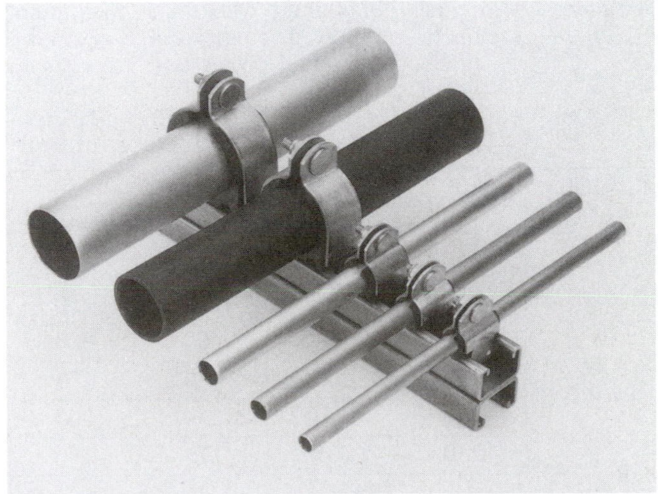

Figure 5–5 Example of a rail-type pipe mounting system. *Courtesy of Hydra-Zorb Corporation.*

Figure 5–6 Piping supported by a suspended rail/clamp system.

Soldering Copper Tubing

The common method of joining copper tubing in hydronic heating systems is soft **soldering** using a 50/50, tin/lead solder. The **working pressure rating** of the resulting joints is dependent on operating temperature and tube size, and can be found in Figure 5–7. These pressure ratings are well above the relief valve settings of most residential and light commercial hydronic systems.

One circumstance in which 50/50 tin/lead solder can not be used is when the tubing carries **domestic water.** *Current plumbing codes do not allow solders containing lead to be used for domestic water service.* One alternative is 95/5 tin/antimony solder. It contains no lead and has higher pressure ratings than 50/50. The melting temperature range of 95/5 solder is 452 °F to 464 °F. This is significantly higher and narrower than the 361 °F to 421 °F range of 50/50 tin/lead solder. Joints made with 95/5 require a longer heating time before the solder will flow. The narrow melting range of 95/5 also means the joint will solidify very quickly when heat is removed. In general, 95/5 solder is more difficult to work with than 50/50 tin/lead, and is often used only for domestic water piping, or when the pressure requirement of the joint cannot be met using 50/50 solder.

Solder type	Service temperature	Maximum allowed pressure (psi) (1/4" to 1" tube)	Maximum allowed pressure (psi) (1.25" to 2" tube)	Maximum allowed pressure (psi) (2.5" to 4" tube)
50/50 tin/lead	100	200	175	150
	150	150	125	100
	200	100	90	75
	250	85	75	50
95/5 tin/ antimony	100	500	400	300
	150	400	350	275
	200	300	250	200
	250	200	175	150

Figure 5–7 Pressure ratings versus operating temperature for soldered joints in copper tubing. *Data provided by the Copper Development Association.*

Soldering Procedure

Proper soldering results in clean, neat, and water-tight joints. The attention given to such joints reflects the professional skills of the installer. *Sloppy joints, even if watertight, indicate poor craftsmanship.* With a little practice, making good soldered joints becomes second nature.

The following procedure describes the steps in making a good soldered joint:

Step 1. Be sure the tube is cut square. Use a wheel cutter or chop saw with a blade suitable for cutting copper tube. Avoid the use of a hacksaw whenever possible. See Figure 5–8.

Step 2. To remove any burrs, ream the end of the tube with the blade of the wheel cutter, rounded file, or other type of deburring tool. See Figure 5–9.

Step 3. Assemble the joint and check it for proper fit and alignment. In most cases, there will be no problem with the fit. However, sometimes a fitting may be damaged or defective. If the fitting wobbles noticeably on the pipe, it should be replaced. Attempting to solder such a joint can result in a leak or poorly aligned joint. See Figure 5–10.

Step 4. Clean the socket of the fitting with a properly sized fitting brush. *Be sure to remove any pieces of the steel brush bristles remaining in the fitting after cleaning it.* These small pieces of steel can create **galvanic corrosion** that eventually could cause a leak in the joint. See Figure 5–11a.

A power cleaning brush driven by a cordless drill as shown in Figure 5–11b is a handy and time-saving alternative to hand brushing when many fittings will be cleaned.

Figure 5–8 Getting a square cut of copper tubing. *Courtesy of the Copper Development Association.*

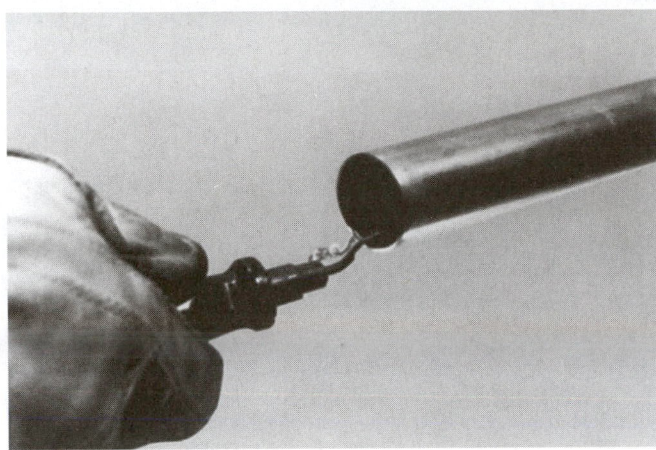

Figure 5–9 Deburring the end of the tube. *Courtesy of the Copper Development Association.*

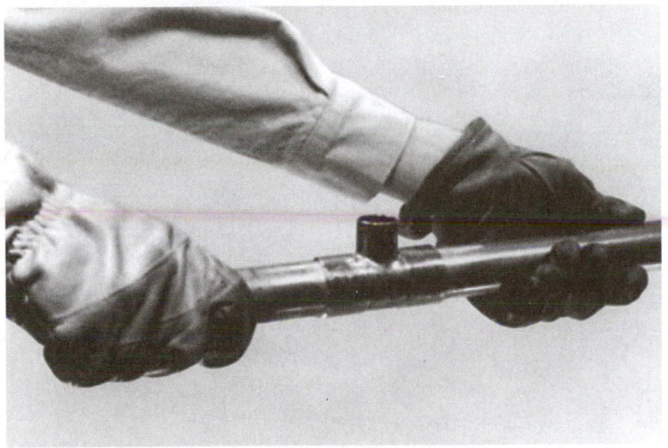

Figure 5–10 Checking the fit of the fitting and tubing. *Courtesy of the Copper Development Association.*

Figure 5–11a Cleaning the socket of the fitting with a steel fitting brush. *Courtesy of the Copper Development Association.*

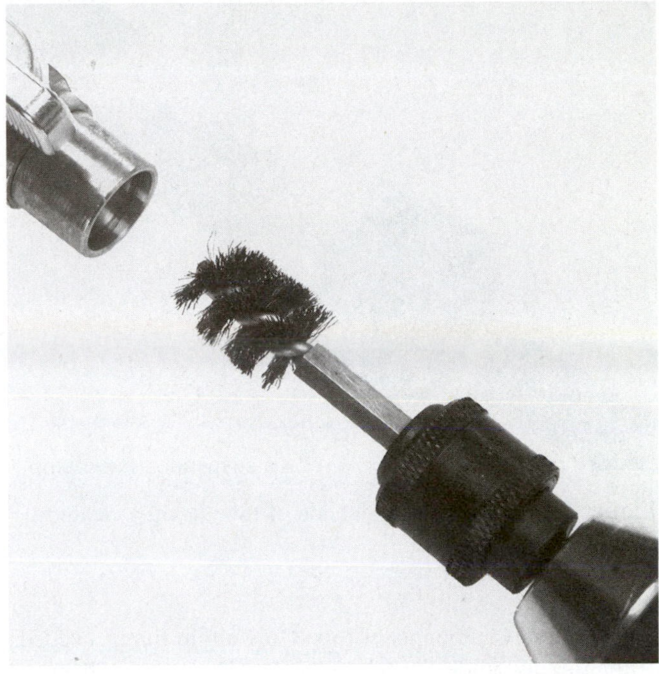

Figure 5–11b Cleaning a fitting socket using a cleaning brush in a cordless drill.

Step 5. Clean the outside of the tube with emery cloth or a power cleaning brush. All oxidation, scale, paint, or dirt should be removed from the tubing surface at least ½ inch farther back than the edge of the fitting socket. The tube should appear bright following cleaning. Be sure to clean all the way around the perimeter of the tube. See Figure 5–12a and Figure 5–12b.

Step 6. Apply paste **flux** to both the fitting socket and the portion of the tube that projects inside the socket. The flux chemically cleans the surface of the copper and helps prevent oxidation of the surfaces when heat is applied. *Always use a flux brush, not your finger, to apply the flux.* Do not

Figure 5–12a Cleaning the outside of the tubing with emery cloth. *Courtesy of the Copper Development Association.*

Figure 5–12b Cleaning outside of tube using a cleaning tool in a cordless drill.

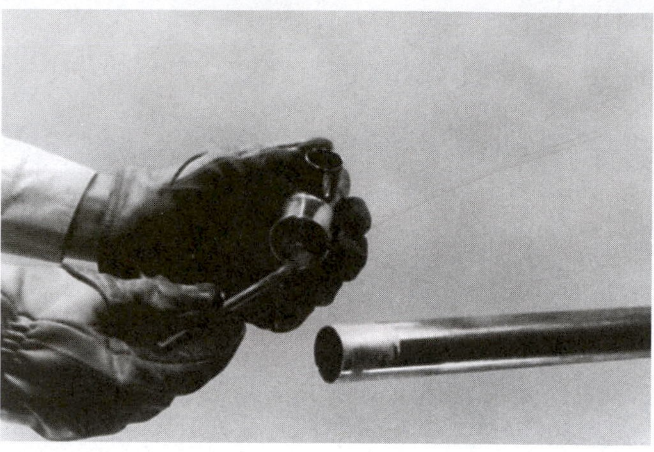

Figure 5–14 Applying paste flux to the inside of the fitting socket. *Courtesy of the Copper Development Association.*

apply excessive amounts of flux. Only a thin film is needed. After fluxing, slide the tube into the fitting and, whenever possible, rotate the fitting once or twice to further spread the flux. Finally, use a rag to remove any excess flux from the surfaces adjacent to the joint before applying heat. See Figure 5–13 and Figure 5–14.

Step 7. Apply heat to the outside of the fitting socket using the torch. Keep the blue tip of the flame just above the surface of the fitting socket as shown in Figure 5–15. Move the flame around the outside of the fitting socket to promote even heating. Heating times differ considerably with the type of torch and gas being used, as well as pipe size, ambient temperature, and type of solder. When the flux begins to sizzle, test the joint by applying the end of the solder wire to the edge of the joint as shown in Figure 5–16. If it sticks but does not melt, continue to apply heat. If the solder immediately melts, the joint is ready to draw in the molten solder by capillary action.

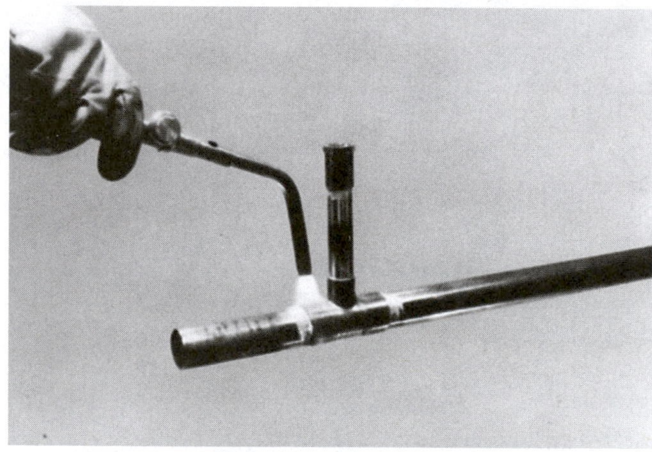

Figure 5–15 Heating the fitting. *Courtesy of the Copper Development Association.*

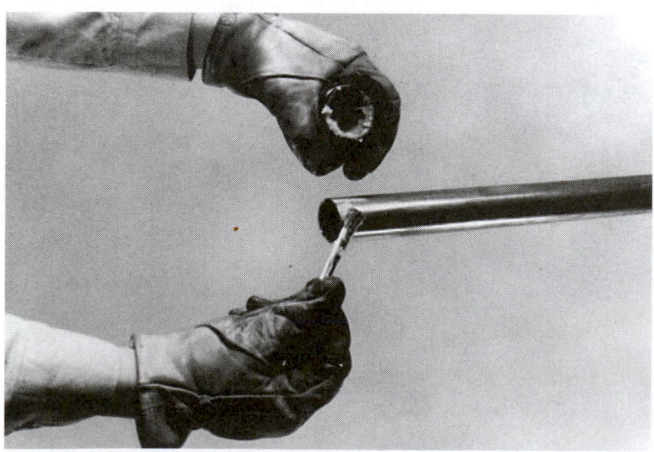

Figure 5–13 Applying paste flux to the outside of the tube. *Courtesy of the Copper Development Association.*

Figure 5–16 Testing the joint for solder melting temperature. *Courtesy of the Copper Development Association.*

Once the solder begins to flow, it can be continually fed into the joint. Only a small amount of solder is needed. An experienced pipe fitter knows by sight how much solder to feed into the joint. An inexperienced person often feeds excessive solder into the joint. The excess solder can solidify into small loose particles inside the pipe. These can be carried around the system by fast moving fluid, and become lodged in valves or circulators. Figure 5–17 lists the approximate length of 1/8-inch solder wire required for joints in tube sizes from 3/8 inch to 3 inches. Loose fitting joints may require more, while tight joints may require less.

A properly heated joint will quickly spread the molten solder around its perimeter. When the solder forms a narrow silver ring around the visible edge of the joint, adequate solder has been applied. If the solder begins to drip off the joint, excess solder is being used. Do not feed any more solder to the joint.

Step 8. After the solder is applied, remove the torch and carefully wipe the perimeter of the joint with a clean cloth rag to remove any excess solder or flux as shown in Figure 5–18. Be careful! All surfaces are still very hot and the solder being wiped is obviously still molten. *Always wear gloves, long-sleeved shirts, and safety goggles when soldering.*

The joint should be allowed to cool naturally before being handled or stressed. Dipping the assembly in water for rapid cooling can cause high stresses in the metal and should be avoided. Cast fittings have been known to crack under such conditions.

After the joint has cooled, one last wiping with a clean cloth moistened with a mild detergent solution will remove any remaining flux. This is an important step because flux residue left on the joint eventually causes unsightly surface oxidation. *Be a professional and clean every joint.*

Copper tube Size (inch)	Approx. length of 1/8" wire solder Required per joint (inches)*
3/8	0.24
1/2	0.40
3/4	0.84
1	1.31
1.25	1.71
1.5	2.27
2	3.65
2.5	4.94
3	6.64

* Based on average joint clearance of 0.005 inch

Figure 5–17 Approximate length of 1/8-inch solder wire required for an average joint.

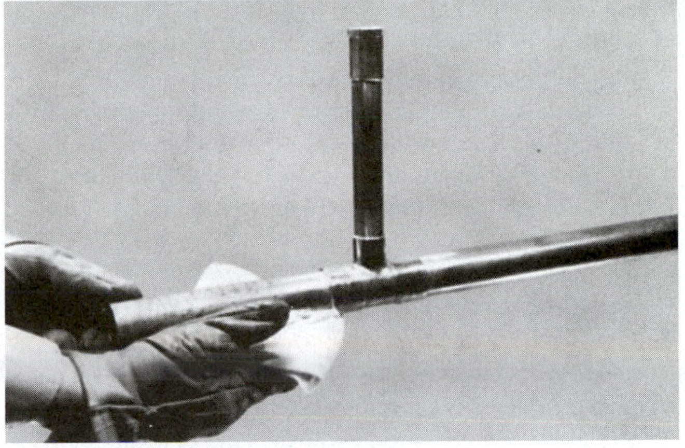

Figure 5–18 Wiping the hot joint. *Courtesy of the Copper Development Association.*

More Tips on Soldering

- Experience shows it is best to make up piping subassemblies on a horizontal working surface whenever possible. The solder flow is more easily controlled, and the work proceeds faster because it is done from a more comfortable position. After cooling, these piping subassemblies can be joined into the overall system.
- When soldering has to take place next to combustible materials, slide a soldering safety pad, as shown in Figure 5–19, between the joint and the combustible materials. This prevents charring wood next to the joint. *Charred wood next to soldered joints does not make a big impression on building owners. Instead, it denotes sloppy workmanship, and disregard for safety.*
- Measure all tubing carefully before cutting. Allow for the take-up of the pipe into the fitting sockets. Figure 5–20 indicates the take-up allowances for copper and brass fittings in pipe sizes from ⅜ inch through 3 inches.

Figure 5–19 Use of a soldering safety pad to prevent charred surfaces. *Courtesy of the Mill Rose Company.*

- Whenever possible, attempt to keep all piping plumb (vertical) or level (horizontal). Even a person with no plumbing background will notice the difference between plumb or level piping and pipes that simply connect point A to point B without regard to appearance. Install temporary piping supports if necessary during piping assembly to hold the pipe in proper alignment for soldering.
- *When soldering valves containing synthetic seals or washers, open the valve to hold the synthetic components away from metal surfaces. Even better, partially disassemble the valve so that only the metal body has to be heated.* This is especially true for zone valves that may not be designed to handle the heat of soldering. *Check the installation instructions before attempting to solder any valve containing nonmetal parts.*
- On fittings having both soldered and threaded connections, make up the soldered joint first. This prevents discoloration or burning of Teflon tape or joint sealing compound during soldering.
- *Always keep one end of the piping assembly open to atmosphere during soldering.* If this is not done the air trapped inside the pipe will increase in pressure due to heating, and blow pinholes through the molten solder before it can solidify. When a piping loop is finally closed in, open a valve or other component near the final joint to allow the heated air to escape.
- Before making up the joint, put a sharp bend in the wire solder at a distance in from the end corresponding to the lengths given in Figure 5–17. This provides a visual indication of how much solder has been fed to the joint.
- Finally, remember that copper is a premium architectural material due to its luster and durability. Take a few extra minutes to clean up any excess flux or shine up oxidized tubing or components. *Your customers will notice it.*

Mechanical Joining of Copper Tubing

Although the most common method of joining copper tubing is soft soldering, there are alternatives. One example is a system in which copper fittings containing elastomer (EPDM) O-rings are mechanically compressed against the tube wall to form pressure-tight joints. An example of such a joint is shown in Figure 5–21.

The electrohydraulic tool used to compress the fittings is capable of exerting a crimping force of 35,000 pounds. This tool, shown in Figure 5–22a, can be fitted with different size jaws to accommodate tube sizes from 1/2 inch to 4 inches. The tool requires about 4 seconds to uniformly compress the fitting (Figure 5–22b). Once the joint is made it cannot be taken apart.

Although the fittings for this type of system are more expensive than standard (solder-type) fittings, the time required for joint preparation is considerably shorter. The tube ends still require reaming to remove any burrs due to

Tube size (inch)	Actual and approx. take-up distance (d) of fitting sockets (inches)		Approximate thread engagement length (x) for NPT threads (inches)
	actual	approximate	
3/8	0.38	3/8	1/4
1/2	0.50	1/2	1/2
3/4	0.75	3/4	1/2
1	0.91	15/16	9/16
1.25	0.97	1	5/8
1.5	1.09	1 1/16	5/8
2	1.34	1 5/16	11/16
2.5	1.47	1 1/2	7/8
3	1.66	1 11/16	1

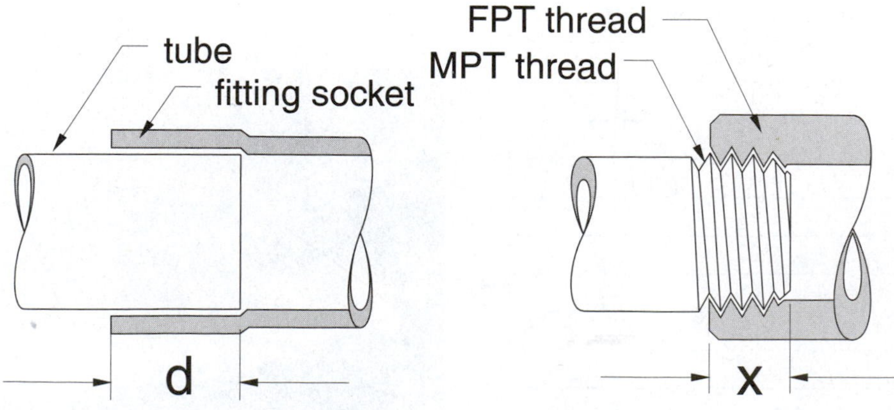

Figure 5–20 Typical take-up allowances for copper and brass fittings.

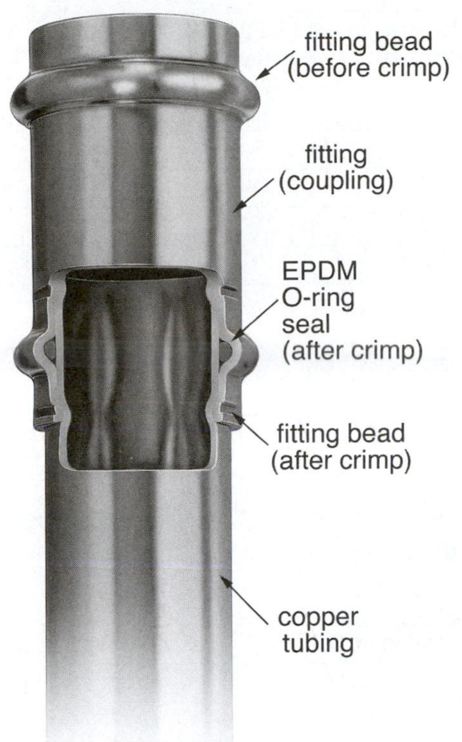

Figure 5–21 Examples of a joint in copper tubing using a mechanically compressed O-ring fitting. *Courtesy of Rigid Tool Co.*

cutting. However, no mechanical cleaning or fluxing is required. This reduces the installation cost of the pressed joint system relative to that of a soldered joint. No torch is required with the pressed joint that arguably is a safety advantage in many environments.

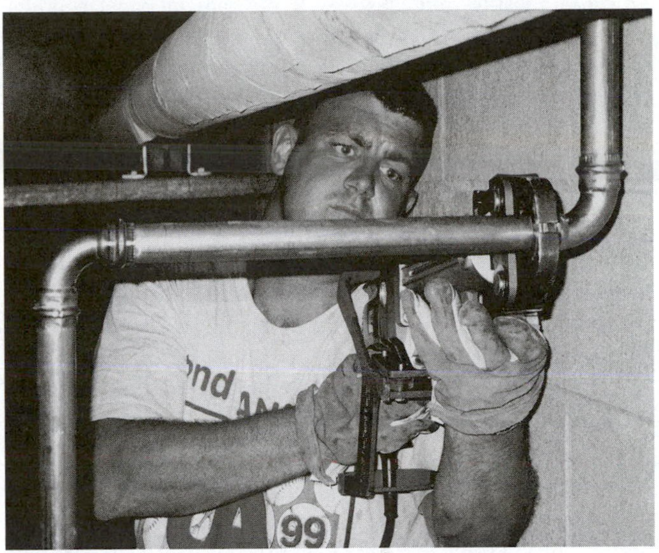

Figure 5–22b Mechanically compressed joint being made. *Courtesy of Rigid Tool Co.*

Figure 5–22a Tooling used for a mechanically compressed joint. The compression tool can be loaded with several different jaws depending on pipe size being joined. *Courtesy of Rigid Tool Co.*

PEX Tubing

Cross-linked polyethylene tubing, also referred to as **PEX tubing,** is a product developed, refined, and extensively used in Europe over the last 35 years. In the mid-1980s the market for PEX tubing in North America began to expand, driven by increasing interest in hydronic radiant panel heating. Worldwide, PEX tubing has proven itself a reliable alternative to metal piping in many hydronic heating applications.

Prior to cross linking, polyethylene molecules are not bonded to each other in a way that prevents their movement. When the polyethylene is sufficiently heated, individual molecules can flow past each other. Eventually the material "melts" and can be reformed in shape while at the elevated temperature. This is a common characteristic of many **thermoplastics** including polyethylene, polybutylene, and polypropylene, all of which can be molded into various shapes or extruded into tubing.

The cross-linking process is the essential step in transforming ordinary high density polyethylene into a **thermoset plastic.** During cross linking, some of the carbon/hydrogen bonds in the polyethylene molecules are broken using various chemical or electoradiative processes. The broken bonds allow adjacent polyethylene molecules to "fuse together" into a three-dimensional network. Once this occurs, the material is classified as a thermoset plastic. It can no longer be melted or reformed. Crosslinking significantly increases the temperature/pressure capabilities of ordinary high-density polyethylene tubing.

PEX tubing is capable of withstanding the repeated stresses experienced through heating and cooling while encased in a rigid material such as concrete. It is also very resistant to chemicals that can cause corrosion or scaling in metal pipe.

PEX tubing is sold in continuous coils ranging from 150 to more than 1000 feet in length (depending on diameter and manufacturer). PEX is available in several diameters suitable for use in hydronic heating applications. The most commonly used nominal tube sizes range from 1/4 inch to 1 inch. Figure 5–23 lists some physical data for PEX tubing that conforms to the widely accepted ASTM F876 standard.

With all plastic tubing there is a tradeoff between operating temperature and allowable operating pressure. The ASTM F876 standard for PEX tubing establishes three simultaneous temperature/pressure ratings based on operation with water. They are listed in Figure 5–24.

The availability of long continuous coils, combined with the ability to bend around moderate curves without kinking, make PEX tubing ideal for use in radiant panel heating systems. It also allows the tubing to run through confined or concealed spaces in buildings where working with rigid copper tubing would be difficult or impossible.

Specialized fittings are available from the tubing manufacturers to transition from PEX tubing to standard metal pipe fittings, both threaded and soldered. Examples of such fittings are shown in Figure 5–25. These fittings allow PEX tubing to be connected to heat emitters, or other system components, as well as used interchangeably with metal pipe.

With reasonable care, PEX tubing is very resistant to kinking during installation. If a kink does occur, it can be easily repaired in the field. The process requires the kinked

Nominal tube size (inch)	Average outside diameter (inch)	Minimum wall thickness (inch)	Minimum burst pressure at 180 °F temperature (psi)
1/4	0.375	0.070	390
3/8	0.500	0.070	275
1/2	0.625	0.070	215
5/8	0.750	0.083	210
3/4	0.875	0.097	210
1	1.125	0.125	210
1.25	1.375	0.153	210
1.5	1.625	0.181	210
2	2.125	0.236	210

Figure 5–23 Physical data for several sizes of PEX tubing conforming to the ASTM F876 standard.

Temperature (°F)	Pressure rating for water (psi)
73.4 °F	160 psi
180 °F	100 psi
200 °F	80 psi

Figure 5–24 Temperature and coincident pressure ratings for PEX tubing conforming to the ASTM F876 standard.

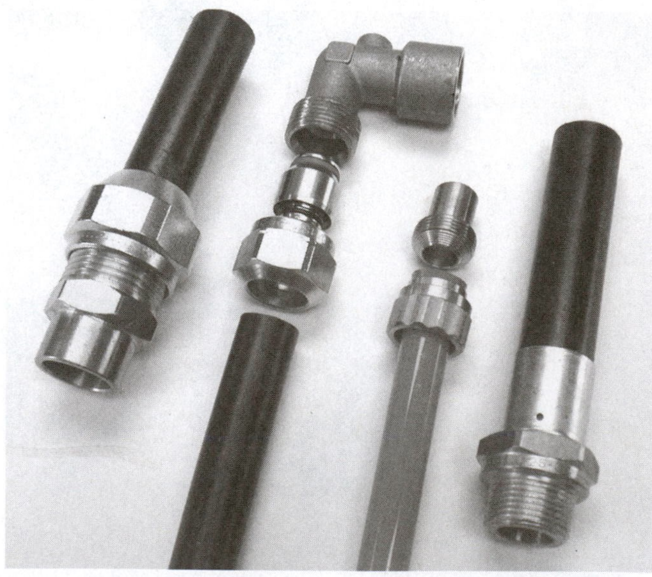

Figure 5–25 Example of fittings to transition from PEX and PEX-AL-PEX to copper tubing.

area to be heated to approximately 275 °F using a hot air gun. At this temperature the crystalline structure of the polyethylene changes to an amorphous state, relieving the stresses created by the kink. The material returns to its normal shape. After cooling, there is virtually no indication that a kink ever existed. Because of this behavior, PEX tubing is said to have a "shape memory."

Composite PEX-AL-PEX Tubing

Another type of tubing well suited for hydronic heating systems is called composite **PEX-AL-PEX tubing.** It consists of three concentric layers bonded together with special adhesives. The inner and outer layers are PEX. The middle layer is longitudinally welded aluminum. A close-up of the cross section of a PEX-AL-PEX tube is shown in Figure 5–26.

The PEX-AL-PEX tubing commonly used in hydronic heating systems conforms to the standard ASTM F1281. Figure 5–27a and Figure 5–27b list some physical properties of PEX-AL-PEX tubing conforming to this standard.

PEX-AL-PEX tubing has temperature/pressure ratings slightly higher than ASTM F876 PEX tubing. This is largely due to the structural characteristics of the aluminum core, which also allows the tubing to hold the shape it is bent to much like soft temper copper tubing. The aluminum core also reduces the expansion movement of PEX-AL-PEX tubing relative to other (all-polymer) tubes.

PEX-AL-PEX tubing cannot be repaired by heating the tube until the PEX reaches its amorphous state. Instead, most kinks can be removed by mechanically reforming the tube with a truing tool.

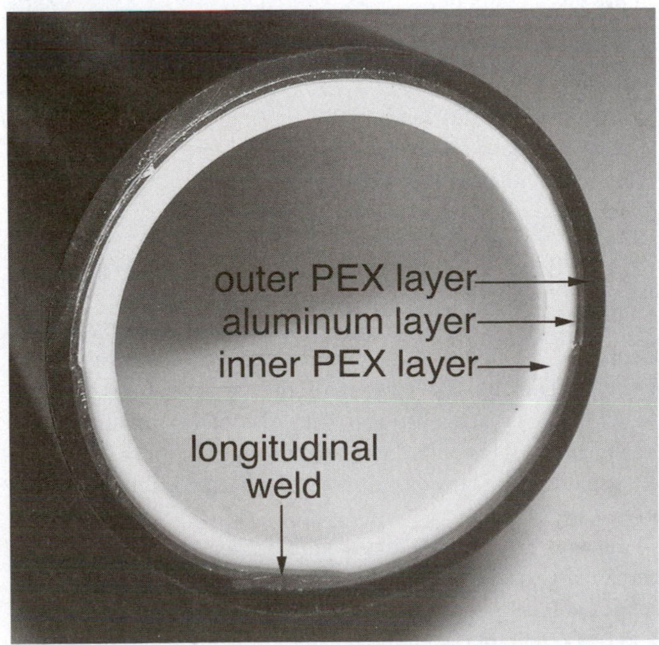

Figure 5–26 Close-up of cross section of PEX-AL-PEX tubing.

Nominal tube size (inch)	Minimum OUTSIDE diameter	Minimum total wall thickness	Minimum aluminum layer thickness (inch)
3/8	12.00 mm/0.472"	1.6 mm/0.063"	0.007
1/2	16.00 mm/0.630"	1.65 mm/0.065"	0.007
5/8	20.00 mm/0.787"	1.90 mm/0.075"	0.010
3/4	25.00 mm/0.984"	2.25 mm/0.089"	0.010
1	32.00 mm/1.26"	2.90 mm/0.114"	0.011

(a)

Temperature (°F)	Pressure rating for water (psi)
73.4 °F	200 psi
140 °F	160 psi
180 °F	125 psi
210 °F	115 psi

(b)

Figure 5–27 (a) Physical data for several sizes of PEX-AL-PEX tubing conforming to the ASTM F1281 standard. (b) Temperature and coincident pressure ratings for PEX-AL-PEX tubing conforming to the ASTM F1281 standard.

Rubber-based Tubing

Another type of tubing available for hydronic heating applications is composed of various layers of rubber-based compounds and reinforcing mesh. Some varieties have temperature/pressure ratings of 200 °F at 100 psi. This type of tubing has been primarily used in hydronic radiant floor heating applications.

A major difference between rubber-based tubing and PEX or PEX-AL-PEX tubing is its flexibility. Most rubber-based tubing can literally be tied into a knot and still not collapse. However, this means the tubing must be properly supported if spanning a distance of more than a few inches. Most rubber-based tubing is joined using a combination of a barbed insert fitting combined with a mechanical compression ring.

Oxygen Diffusion

The molecular structure of nonmetallic tubes such as those made of PEX or rubber compounds is such that oxygen molecules can **diffuse** through their tube walls and be absorbed by the fluid within. The driving force of oxygen diffusion is a lower concentration of dissolved oxygen molecules within the water inside the tube, compared to the concentration outside the tube. The concept is illustrated in Figure 5–28.

The pressure difference between the inside and outside of the tube has very little effect on oxygen diffusion. Diffusion can still occur when the pressure inside the tube is higher than the atmospheric pressure outside the tube. Embedding

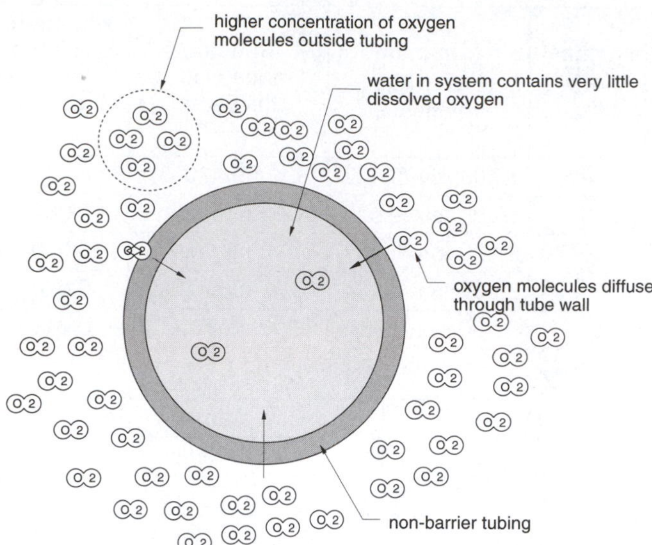

Figure 5-28 Concept of oxygen diffusion through wall of polymer tubing. Note higher concentration of oxygen molecules outside of tube.

the tube in concrete or other solid materials also has very little effect on oxygen diffusion. Oxygen diffusion is also temperature dependent. The higher the operating temperature, the faster oxygen can diffuse through the tube wall.

Any time oxygen enters a hydronic system containing iron or steel components, there is the potential for corrosion. It is therefore important to minimize or eliminate its entry in systems containing iron or steel components. Most manufacturers of polymer or rubber-based tubing intended for hydronic heating applications incorporate an **oxygen diffusion barrier** into their pipe.

The oxygen diffusion barrier commonly used on PEX tubing is a compound called EVOH (ethylene vinyl alcohol). It is applied to the outside of the tube, or laminated between layers of PEX as the tube is manufactured. EVOH acts as a retarder for oxygen diffusion, reducing the rate of diffusion to a level deemed acceptable for long-term service.

Metals such as aluminum are extremely good oxygen barriers. On PEX-AL-PEX tubing, the aluminum core layer acts as the oxygen diffusion barrier. Hence, there is no such thing as "nonbarrier" PEX-AL-PEX tube. Some rubber tubing used in hydronic heating applications include aluminum foil as the oxygen diffusion barrier.

The standard recognized worldwide for oxygen diffusion protection in hydronic heating systems is DIN 4726. This standard allows a *maximum* oxygen diffusion rate of 0.1 milligrams per liter of tubing water content per day. Nearly every manufacturer currently selling polymer tubing for hydronic heating applications offers "barrier" tubing that meets the DIN 4726 oxygen diffusion criteria.

Many tube manufacturers also offer "nonbarrier" tubing. Although intended primarily for use in domestic water plumbing systems, nonbarrier tubing has been successfully used in hydronic systems provided that all steel and iron components are protected from corrosion.

Under the DIN 4726 standard, steel or cast-iron components such as boilers, circulators, and flow-check valves can be used *if isolated from the nonbarrier tubing with a heat exchanger.* Another alternative is the use of oxygen scavenging chemicals in the system. These provisions, while possible, are infrequently used in residential and light commercial systems. Instead, tubing with an oxygen diffusion barrier has become the standard approach when iron or steel components will be used in the system.

There are several means, other than diffusion, by which oxygen can enter a hydronic system. These include fresh water flowing into the system to compensate for minor leaks, misplaced air vents, and improper sizing or placement of expansion tanks. These other forms of oxygen entry can be as much or more of a corrosion threat than oxygen diffusion through nonbarrier tubing. In other words, *the use of tubing with an oxygen diffusion barrier does not guarantee that corrosion will not occur.* However, it is reasonable to assume, with all other factors being equal, that tubing with an oxygen barrier does provide additional protection against corrosion. The slight (if any) additional cost of this protection is minor in comparison to the costs associated with repairing corrosion damage.

5.3 COMMON PIPE FITTINGS

Many of the pipe fittings used in hydronic systems are the same as used in conventional domestic water supply systems. They include:

- Couplings: (standard and reducing)
- Elbows: (90° and 45°)
- Street elbows: (90° and 45°)
- Tees: (standard and reducing)
- Threaded adapters: (male and female)
- Unions

The cross section of most of these common pipe fittings is shown in Figure 5–29.

Fittings for copper tubing are available in either wrought copper or cast brass. Both are suitable for use in hydronic systems. During soldering, wrought copper fittings heat up faster than cast brass fittings. They also have smoother internal surfaces. Of the two, wrought fittings are more typical in the smaller pipe sizes used in residential and light commercial systems.

Fittings are manufactured with solder-type socket ends, threaded ends, or a combination of the two. Standard sockets for solder-type joints create a gap of between 0.002 inch and 0.005 inch between the outside of the tube and the inside of the fitting. This gap allows solder to flow between the tube and fitting by capillary action.

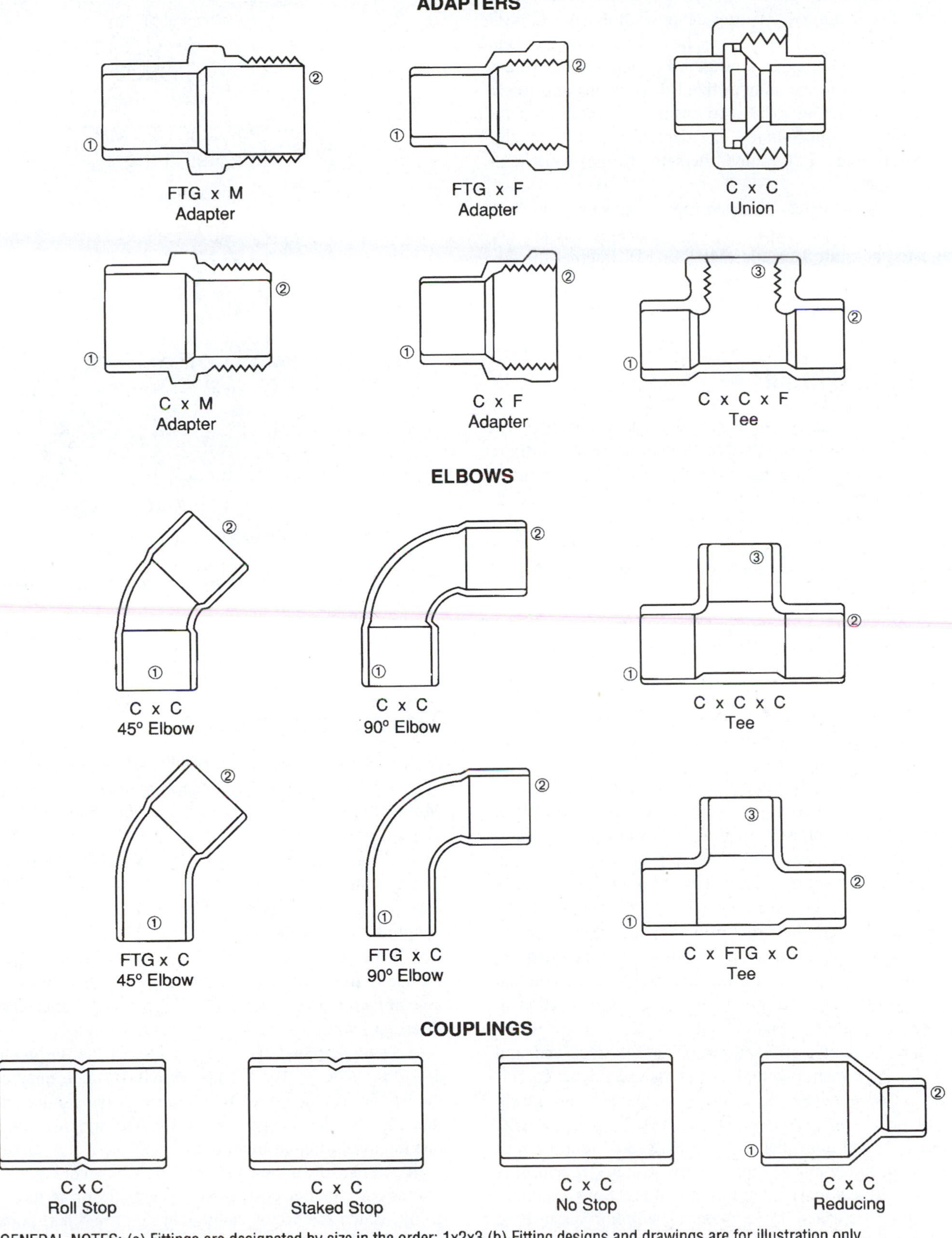

Figure 5–29 Cross section of several common fittings used with copper tubing. *Courtesy of the Copper Development Association.*

The standard designation for solder-type sockets is the letter *C*. For example, a fitting designated as 3/4" C × C indicates both ends of the fitting have solder-type sockets to match 3/4-inch copper tubing. All fitting manufacturers in North America use standardized dimensions and tolerances in constructing fittings to ensure compatibility with tubing. Because the outside diameter of types K, L, and M copper tubing are the same, the same fittings work with all three types.

In the United States, threaded pipe fittings use standardized **national pipe thread** (NPT). External threads are designated as **male threads** (MPT or M). Internal threads are designated as **female threads** (FPT or F).

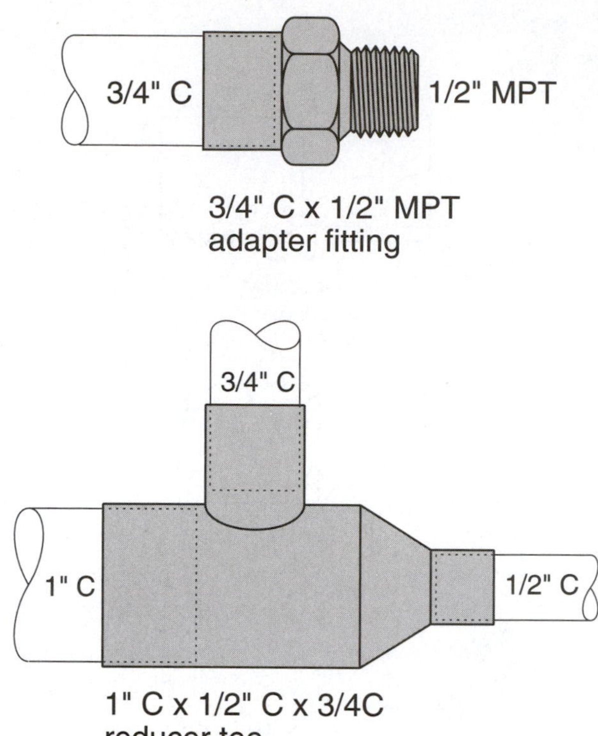

Figure 5–30 Example of proper nomenclature for two reducer fittings.

Specifying Fittings

There are thousands of different fittings available for use with copper tubing. A standard method of describing fittings is needed to ensure that the designer, installer, and supplier all know exactly what type of fitting is being specified. Unfortunately, all people involved do not necessarily adhere to a single standard when describing these fittings. The following minimum information should always be given in describing a fitting:

- Name of fitting (i.e., elbow, tee, union)
- The pipe size(s) of the fitting
- A designation for the type of connections (i.e., solder-type = C, male pipe threads = M, female pipe threads = F)

In the case where all connections on the fitting are the same pipe size, it is only necessary to specify this size once. When a two port fitting such as a coupling or bushing needs to have two different pipe sizes, *always specify the larger size first,* then an "x", followed by the small size. For example, specify a 1" × 3/4" reducer coupling, not a 3/4" × 1" coupling. In the case of a tee, specify the larger of the two end ports, then the size of the other end port, and finally the size of the side port. The designations C, M, and F are included with the nominal size information. An example of the proper specification of two fittings is shown in Figure 5–30.

Although many combinations of fitting sizes and end connections are manufactured, not every imaginable combination can be obtained. As a rule, a single reducing fitting will not vary by more than two pipe sizes. Thus a 1" × 1/2" reducer coupling is available, but a 2" × 3/8" reducer coupling would likely not be, due to very limited demand. It is also very unlikely that all the fitting combinations manufactured will be stocked by a material supplier because of inventory constraints. With experience, installers learn which specialty fittings are frequently used. They often place orders for box quantities of these specialty fittings to prevent job delays.

Tips on Using Fittings

A well-planned piping assembly will use the least number of fittings possible to minimize both labor and materials. Minimal use of fittings also gives the job a more professional appearance.

For example, there are several ways to attach a pressure gauge with a 1/4-inch MPT threaded stem to a 3/4-inch copper pipe. One approach uses a standard tee, a reducer coupling off the side of the tee, and a C × F adapter at the end of the reducer coupling. Two short stubs of copper tubing are required to hold these fittings together. In total, this assembly requires a total of six soldered joints and one threaded joint.

A better approach is to use a special tee designated as 3/4"C × 3/4"C × 1/4"F. This tee has 1/4-inch female pipe threads on its side port. The pressure gauge can be directly threaded into the side port. A total of two soldered joints and one threaded joint are required.

A comparison of these assemblies is shown in Figure 5–31. The slightly higher cost of the special fitting is more than compensated for by the savings in labor and additional fittings. Similar situations occur in mounting devices such as expansion tanks, relief valves, and drain valves. Well-planned use of special fittings can enhance the appearance of the job, and often reduce installation cost.

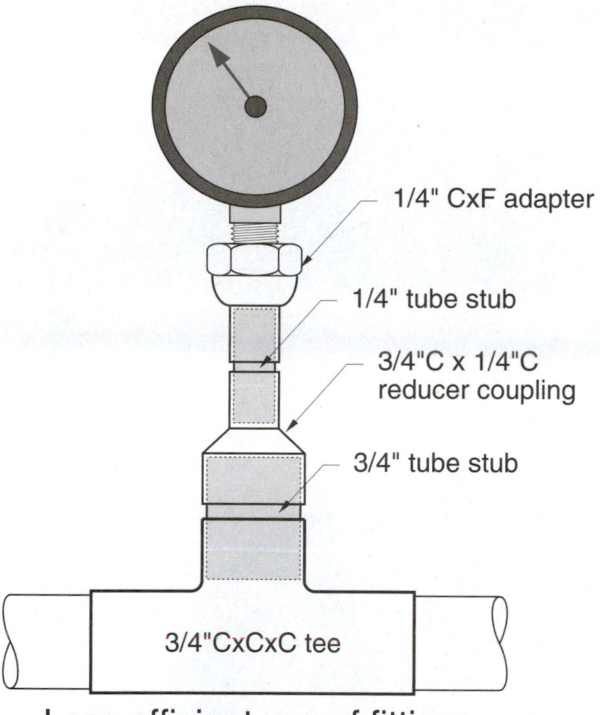

(a)

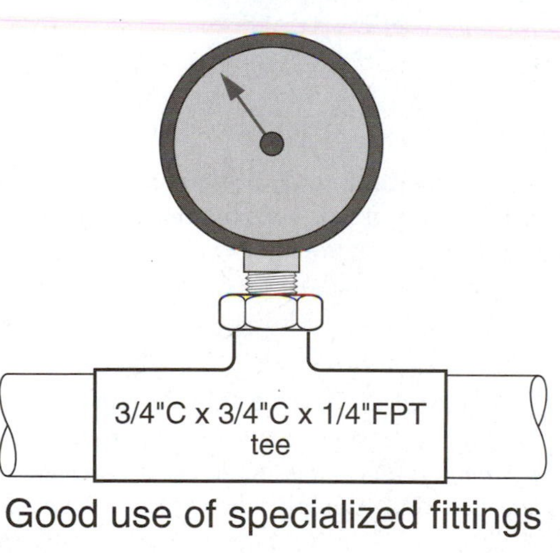

(b)

Figure 5–31 Comparison of fitting options for mounting a pressure gauge to a pipe (a) less efficient use of fittings, (b) good use of a specialty fitting.

5.4 SPECIALIZED FITTINGS

Several specialized fittings have been developed for hydronic heating systems. They include baseboard tees, diverter tees, and dielectric unions. Each is worthy of individual discussion.

Baseboard Tees

A **baseboard tee** (sometimes also called a "vent ell") is commonly used to mount an air vent at high points in a piping circuit. The fitting closely resembles a standard 90-degree elbow, but has an additional 1/8-inch FPT port as shown in Figure 5–32a.

Baseboard tees are available in pipe sizes from 1/2 inch to 1 1/4 inches in both wrought copper and brass, the latter being more common. The fitting's name is derived from its frequent use to mount air vents on the outlet of finned-tube baseboard convectors (Figure 5–32b). These fittings can also be used to mount a threaded temperature sensor, or a 1/8-inch brass drainage plug at a low point in the piping circuit.

(a)

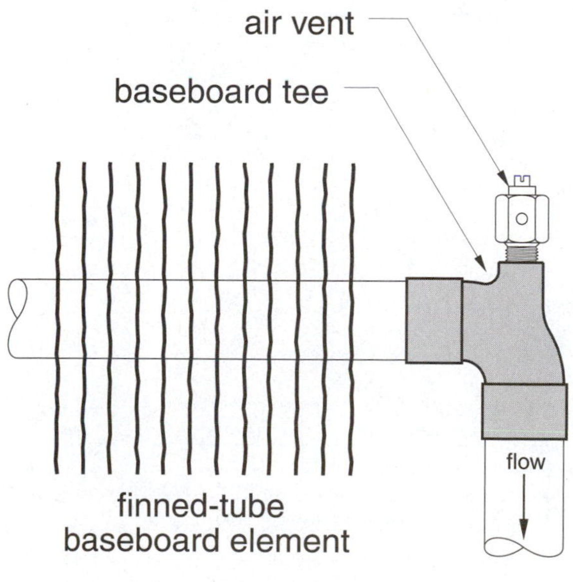

(b)

Figure 5–32 (a) a baseboard tee, (b) typical location for venting air from a baseboard convector.

Diverter Tees

A **diverter tee** is used to create flow through a branch piping path when mounted as shown in Figure 5–33.

When mounted at the beginning of the branch circuit, a diverter tee forces a portion of the entering fluid into the branch. When mounted at the end of the branch circuit, it "pulls" fluid through the branch.

The outside of a diverter tee appears similar to that of a reducing tee, as shown in Figure 5–34a. However, inside there is either a cone-shaped orifice (see Figure 5–34b), or a curved scoop. These internal details create partial obstructions that generate the pressure differentials needed to move flow through the branch piping.

A single diverter tee is often used when the branch piping path has low flow resistance. The diverter tee can be placed at the beginning or end of the branch piping path as shown in Figure 5–35a. If the branch piping path is long, or contains components that create higher flow resistance, two diverter tees can be used to create a higher pressure differential as shown in Figure 5–35b. In either case, *it is crucial that diverter tees be installed in the proper direction within the piping system.* This direction is indicated by an arrow on the side of the fitting, and/or a ring around one end port of the diverter tee. In the latter case, *the fitting should always be installed with the ring facing toward the branch piping path.* The proper orientation of the fitting can also be determined by noting the direction of the cone-shaped orifices in Figure 5–35a and Figure 5–35b.

Within the heating trade, diverter tees are often called **monoflo tees**® after the trademark brand from the Bell and Gossett Company. Another name sometimes used is venturi fitting. This is derived from the pressure-reducing effect of the tee when installed on the return side of the branch piping path.

Diverter tees are extremely useful in designing hydronic systems that allow individual flow control through each heat emitter on a distribution circuit. This is accomplished by regulating the flow rate through the branch piping path using a valve. The design of these so-called **one-pipe systems** is discussed in later chapters.

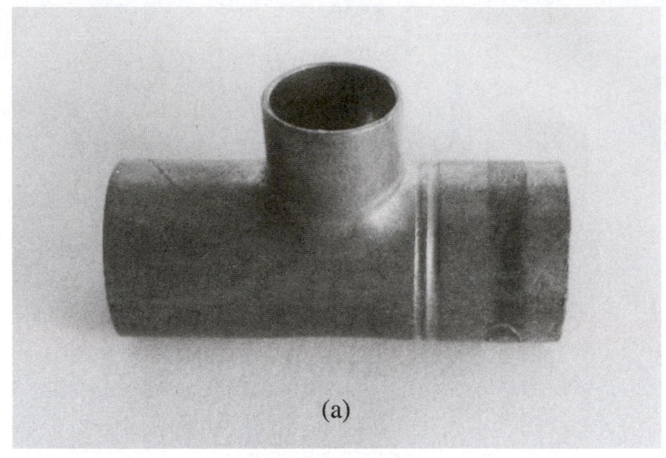

(a)

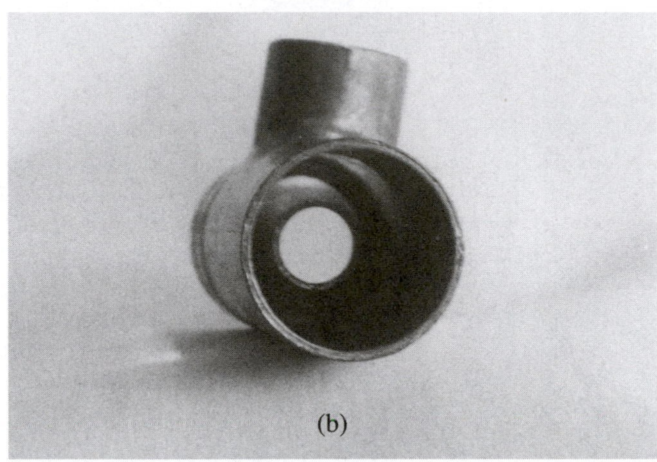

(b)

Figure 5–34 Appearance of a diverter tee (a) side view showing reduced size of side port, (b) end view showing cone-shaped orifice.

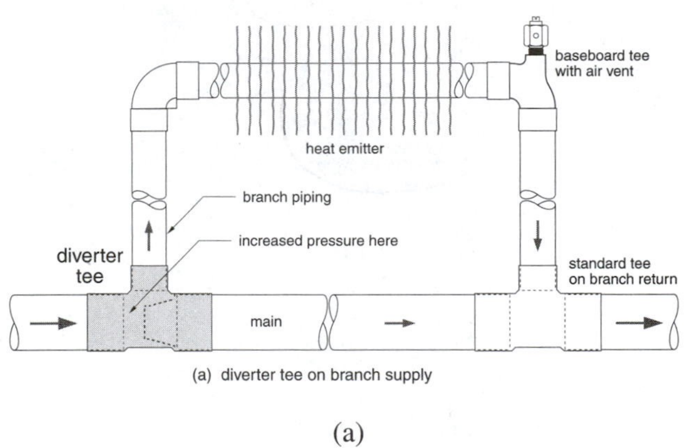

(a)

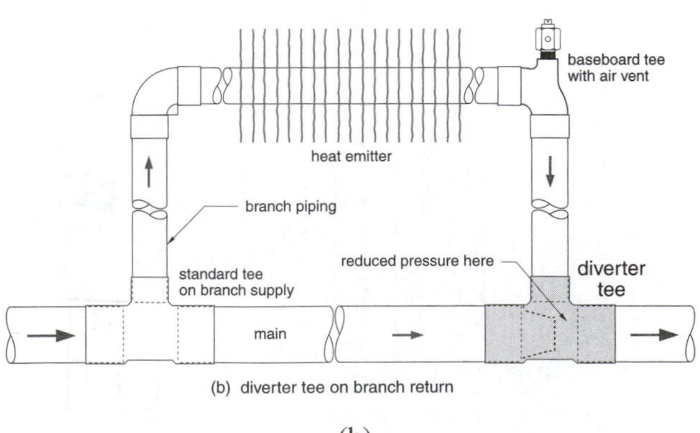

(b)

Figure 5–33 A diverter tee used to create flow in a branch piping circuit (a) diverter tee mounted at the supply of the branch piping path, (b) diverter tee mounted at the return of the branch piping path.

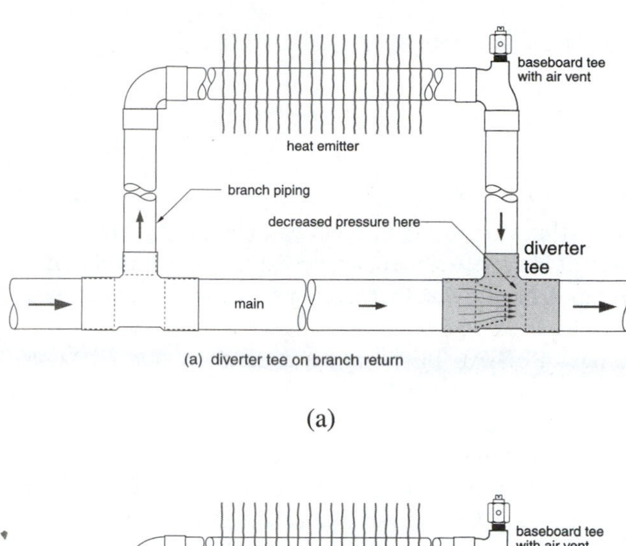

(a)

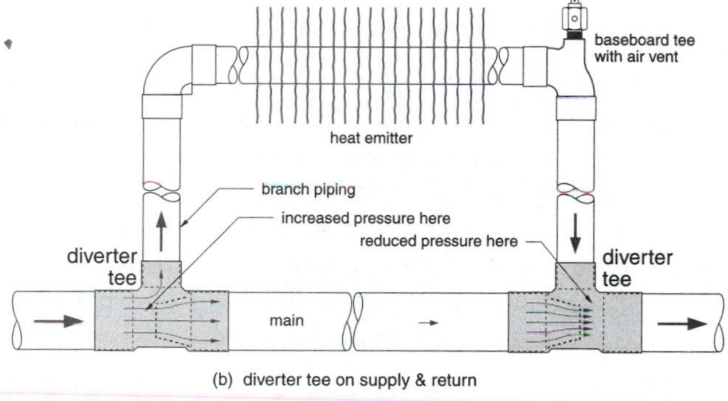

(b)

Figure 5–35 Cross-section of a branch piping path using (a) single diverter tee on return side of branch, (b) two diverter tees. The use of two diverter tees will create a higher flow rate in the branch piping path.

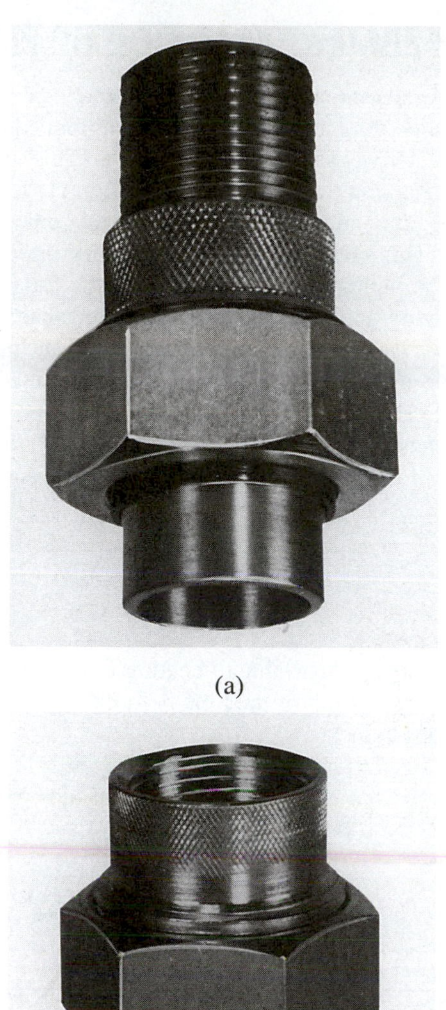

(a)

(b)

Figure 5–36 Examples of dielectric unions (a) with MPT threads on steel side, (b) with FPT threads on steel side. *Courtesy of Watts Regulator Co.*

Dielectric Unions

In many hydronic systems, iron or steel components such as boilers and circulators are connected using copper tubing. If the copper and steel components are in direct contact, and if the system's fluid is slightly conductive, a process called galvanic corrosion will cause the steel and iron components to corrode. To prevent this, it is necessary to place an electrically insulating material called a dielectric between the two dissimilar metals. A **dielectric union** is the most commonly used fitting for this purpose.

At a glance, dielectric unions appear similar to a standard union. Closer inspection, however, shows one side of the union has steel pipe threads, while the other side has a brass socket for soldering to copper tube. The steel and brass parts of the union are prevented from touching each other by a synthetic O-ring and sleeve. Besides preventing galvanic corrosion, dielectric unions also provide for simple disassembly of a piping system, the same as with standard unions. Two available types of dielectric unions are shown in Figure 5–36. A disassembled dielectric union is shown in Figure 5–37.

Figure 5–37 Disassembled dielectric union showing insulating O-ring and sleeve.

5.5 THERMAL EXPANSION OF PIPING

All materials expand when heated and contract when cooled. In many hydronic systems, piping undergoes wide temperature swings as the system operates. This causes significant changes in the length of the piping due to **thermal expansion.** The greater the length of the pipe, and the greater the temperature change, the greater the change in length. If the piping is rigidly mounted, this expansion can cause annoying popping or squeaking sounds each time the pipe heats up or cools down. In extreme cases, the piping can even buckle due to the high thermal stresses that develop when the pipe is prevented from expanding.

The expansion or contraction movement of various types of tubing can be calculated using the following equation:

(Equation 5.1)

$$\Delta L = b(L)(\Delta T)$$

where:

ΔL = change in length due to heating or cooling (inches)
L = original length of the pipe before the temperature change (inches)
ΔT = change in temperature of the pipe (°F)
b = **coefficient of linear expansion** of tube material (see the following)

Tubing material	Coefficient of linear expansion (b) (inch/inch/°F)
Copper	0.0000094
PEX	0.000094
PEX-AL-PEX	0.000013

The smaller the coefficient of linear expansion, the less the tubing expands for a given change in temperature. The linear (lengthwise) expansion of the tube does not depend on the tube's diameter.

Example 5.1: Determine the change in length of a copper tube, 50 feet long, when heated from a room temperature of 65 °F to 200 °F. How does this compare to the change in length of a PEX and PEX-AL-PEX tube having the same length and undergoing the same temperature change?

Solution: The change in length of the copper tube is:

$$\Delta L = \left(0.0000094\frac{\text{inch}}{\text{inch}\cdot{}^\circ\text{F}}\right)\times 50\text{ ft}\times\left(\frac{12\text{ inch}}{1\text{ foot}}\right)\times(200^\circ\text{F}-65^\circ\text{F})$$

$$= 0.76\text{ inch}$$

The change in length for PEX tube is:

$$\Delta L = \left(0.000094\frac{\text{inch}}{\text{inch}\cdot{}^\circ\text{F}}\right)\times 50\text{ ft}\times\left(\frac{12\text{ inch}}{1\text{ foot}}\right)\times(200^\circ\text{F}-65^\circ_{\text{F}})$$

$$= 7.60\text{ inch}$$

The change in length for PEX-AL-PEX tube is:

$$\Delta L = \left(0.000013\frac{\text{inch}}{\text{inch}\cdot{}^\circ\text{F}}\right)\times 50\text{ ft}\times\left(\frac{12\text{ inch}}{1\text{ foot}}\right)\times(200^\circ\text{F}-65^\circ\text{F})$$

$$= 1.05\text{ inch}$$

Discussion: The only difference in these calculations is the value of the coefficient of linear expansion (b) of the different tubes. Notice that PEX tubing expands about ten times as much as the copper tube for the same length and temperature change. PEX-AL-PEX tubing expands about 38 percent more than the copper, but significantly less than PEX. The expansion of the PEX-AL-PEX tubing is largely controlled by the aluminum core layer. Because the values of (b) are very small, be careful to use the correct number of zeros when entering the number into calculations.

Regardless of the tubing used, expansion movement must be accommodated during installation, especially on long straight runs. In many small hydronic systems, the piping is not rigidly supported or contains a sufficient number of elbows and tees to absorb the expansion movement without creating excessive stress in the pipe. Such situations only require piping supports that allow the tubing to expand and contract freely to prevent expansion noises.

When long straight runs of rigid tubing are required, it may be necessary to install an **expansion loop** or an **expansion offset** to absorb the movement. Both are piping configurations capable of absorbing limited amounts of movement without subjecting the pipe to excessive stress. The table and illustration in Figure 5–38 shows some expansion compensating options for copper tubing.

Another alternative is to install an expansion compensator fitting capable of absorbing the movement. These fittings use a flexible "bellows" construction to allow repeated expansion/contraction movement. An example is shown in Figure 5–39.

5.6 COMMON VALVES

There are many types of valves used in hydronic systems. Some are common designs used in all categories of piping including hydronic, steam, plumbing, and pneumatics. Others are designed for a very specific function in a hydronic heating system. The proper use of valves can make the difference between an efficient, quiet, and easily serviced hydronic system and one that wastes energy, creates objectionable noise, or even poses a major safety threat. *It is vital that designers and installers understand the proper selection and placement of several types of valves.*

Most of the common valve types are designed for one of the following duties:

- Component isolation
- Flow regulation

	R/L R=radius of U-bend or coiled loop (inches), L=total legth of offset& return (inches)								
	copper tube size								
expected expansion (inches)	3/8"	1/2"	3/4"	1"	1.25"	1.5"	2"	2.5"	3"
0.5	7/44	8/50	9/59	11/67	12/74	13/80	15/91	16/102	18/111
1.0	10/63	11/70	13/83	15/94	17/104	18/113	21/129	23/144	25/157
1.5	12/77	14/86	16/101	18/115	20/127	22/138	25/158	28/176	30/191
2.0	14/89	16/99	19/117	21/133	23/147	25/160	29/183	32/203	35/222
2.5	16/99	18/111	21/131	24/149	26/165	29/179	33/205	36/227	40/248
3.0	17/109	19/122	23/143	26/163	29/180	31/196	36/224	40/249	43/272
3.5	19/117	21/131	25/155	28/176	31/195	34/212	39/242	43/269	47/293
4.0	20/126	22/140	26/166	30/188	33/208	36/226	41/259	46/288	50/314

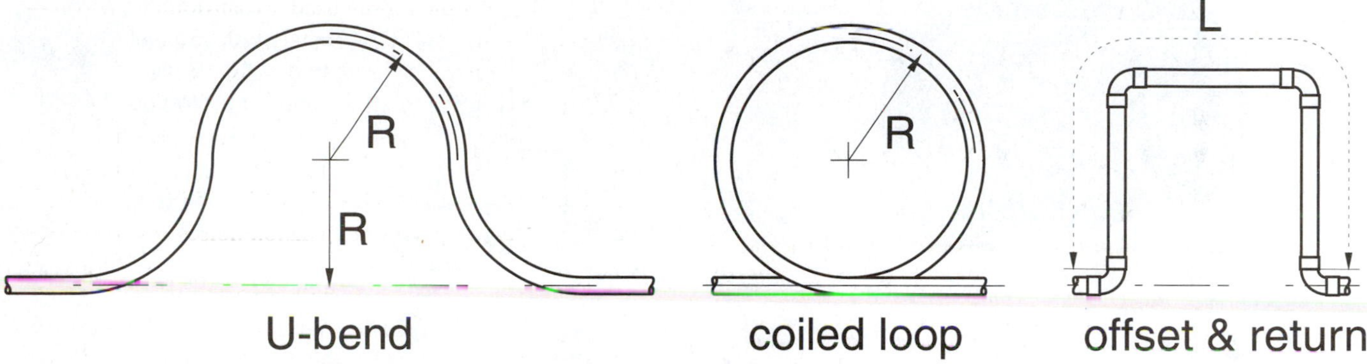

Figure 5–38 Dimensional requirements for expansion loops and expansion offsets for copper tubing.

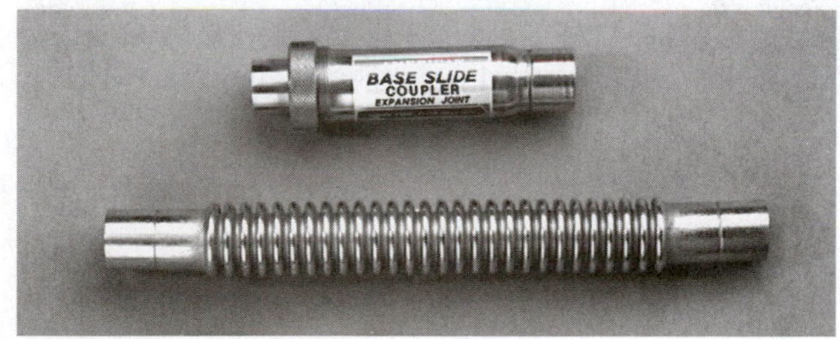

Figure 5–39 Two types of expansion compensators for copper tube. *Courtesy of Amtrol, Inc.*

Component isolation refers to the use of valves to close off the piping connected to a device that may have to be removed or opened for servicing. Examples include circulators, boilers, heat exchangers, and strainers. By placing valves on either side of such components, only minimal amounts of system fluid need to be drained or spilled during servicing. In many cases, other portions of the piping system can remain in operation during such servicing.

Flow regulation requires a valve to set and maintain a given flow rate within a piping system, or portion thereof. Examples include adjusting the flow rates in **parallel piping** branches, or blending hot and cold fluids in a certain proportion. Valves used for flow regulation are specifically designed to remove mechanical energy from the fluid. This causes the fluid's pressure to drop as it passes through the valve. The greater the pressure drop, the slower the flow rate through the valve.

Gate Valves

Gate valves *are designed specifically for component isolation.* As such, they should be fully open or fully closed at all times.

The external appearance and internal design of a typical bronze gate valve are shown in Figure 5–40. When the handwheel is turned clockwise, the wedge moves down out of the "bonnet" chamber until it totally closes off the fluid passage. Near the end of its travel, the wedge seats tightly against the body forming a pressure-tight seal.

In its fully open position, the valve's wedge is completely retracted into the bonnet. In this position, the valve creates very little interference with the fluid stream moving through

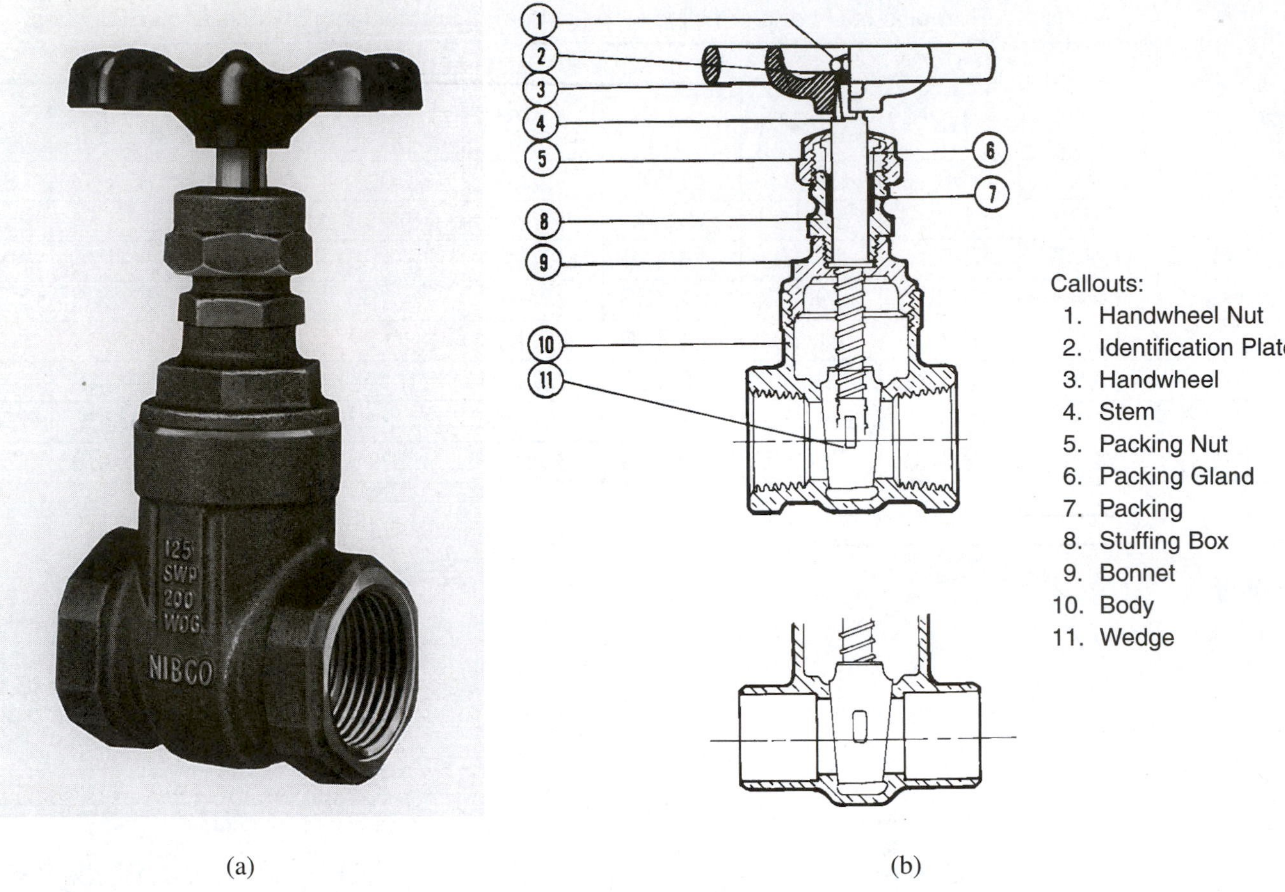

Figure 5–40 A typical bronze gate valve (a) external appearance, (b) internal construction. *Photo and illustration courtesy of NIBCO, Inc., Elkhart, IN.*

it. This results in minimal loss of pumping energy during normal system operation. This is very desirable considering that a typical isolation valve remains open most of its service life.

Gate valves should never be used to regulate flow. If the wedge is set to a partially open position, vibration and chattering are likely to occur. Eventually these conditions will erode the machined metal surfaces, possibly preventing a drip-tight seal when the valve is closed.

Occasionally it becomes necessary to tighten the packing gland of the valve to prevent any minor leakage around the stem. A slight "snugging" of the packing nut is usually sufficient. Never overtighten the packing nut of any valve.

Globe Valves

Globe valves *are specifically designed for flow regulation.* Although their external appearance closely resembles that of a gate valve, there are major internal differences. The fluid's path through a globe valve contains several abrupt changes in direction, as can be seen in Figure 5–41. The fluid enters the lower valve chamber, flows upward through the gap between the orifice and disc, then exits sideways from the upper chamber. The gap between the disc and the orifice determines the pressure drop created by the valve. *Always install globe valves so the fluid flows into the lower body chamber, and upward toward the disc.* Reverse flow through a globe valve can cause unstable flow regulation and noise. All globe valves have an arrow on their body that indicates the proper flow direction through the valve.

Globe valves should never be used for component isolation. The reason is that a fully open globe valve still removes considerably more mechanical energy from the fluid compared to a fully open gate valve. Therefore, during the thousands of operating hours when component isolation is not needed, the globe valve unnecessarily wastes pumping energy. Using a globe valve for component isolation is like driving a car with the brakes partially applied at all times. It may be possible, but it is certainly not efficient.

Angle Valves

An **angle valve** is similar to a globe valve, but with its outlet port rotated 90 degrees relative to its inlet port. This configuration is useful in place of a standard globe valve

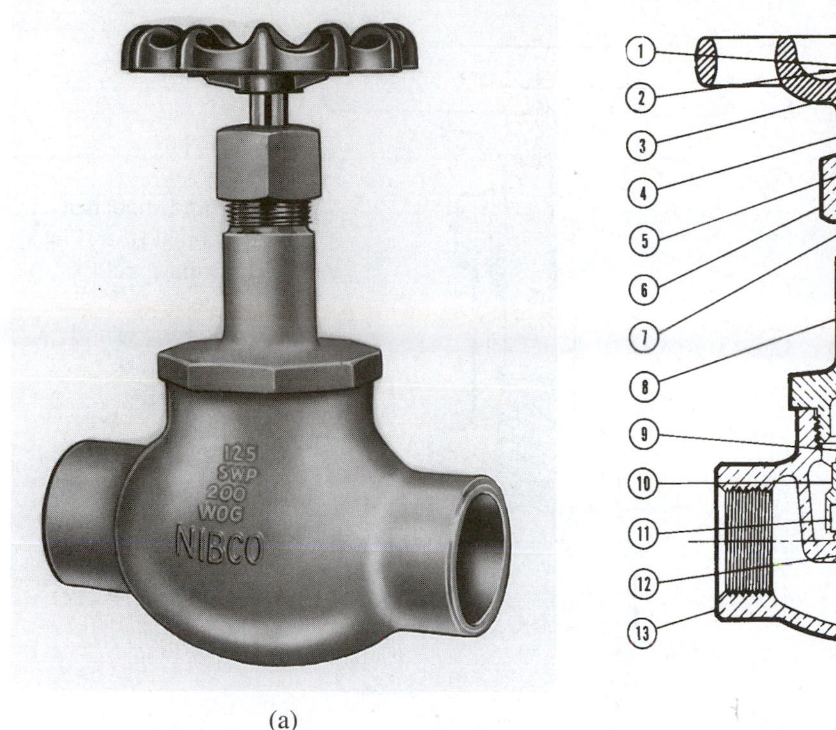

(a)

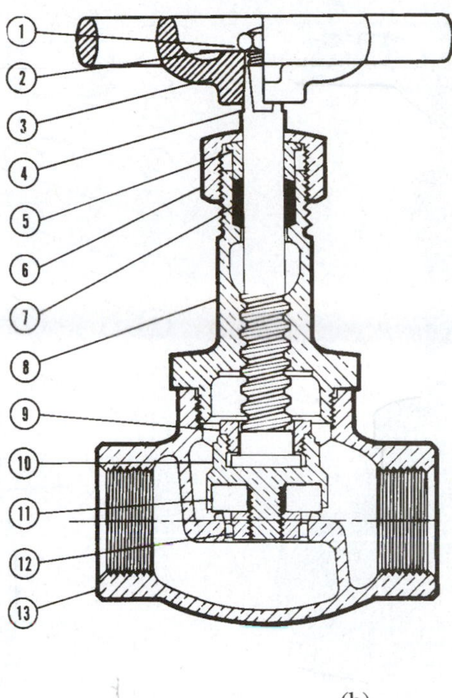

(b)

Figure 5–41 A typical bronze globe valve (a) external appearance, (b) internal construction. *Photo and illustration courtesy of NIBCO, Inc., Elkhart, IN.*

combined with an elbow. Angle valves are frequently used as flow regulation valves on the outlet of heat emitters such as panel radiators or standing cast-iron radiators. Because the path through an angle valve is not as convoluted as that through a globe valve, it creates slightly less pressure drop. However, *angle valves should still not be used solely for component isolation.* As with a globe valve, *be sure to install angle valves so fluid flows toward the bottom of the disc.* An example of an angle valve is shown in Figure 5–42.

A variation on the angle valve, known as a **boiler drain,** is shown in Figure 5–43. This valve allows a standard garden hose to be connected to its outlet port. Boiler drain valves can be placed wherever fluid might need to be drained from the system.

Ball Valves

A **ball valve** uses a machined spherical plug, known as the ball, as its flow control element. The ball has a large hole through its center. As the ball is rotated through 90 degrees of arc, the hole moves from being parallel with the valve ports (its fully open position), to being perpendicular to the ports (its fully closed position). Ball valves are equipped with lever-type handles rather than handwheels. The position of the lever relative to the centerline of the valve indicates the orientation of the hole through the ball. An example of a ball valve is shown in Figure 5–44.

When the size of the hole in the ball is approximately the same as the pipe size, the valve is called a **full port ball valve.** If the size of the hole is smaller than the pipe size, the valve is called a "conventional" or **standard port ball valve.**

Full port ball valves are ideal for component isolation. In their fully open position they create very little flow interference or pressure drop.

It is generally agreed that ball valves can be used for *minor* amounts of flow regulation. However, the valve should never be left in an almost closed position while operating. Doing so can cause high flow velocities through the valve that can eventually erode internal surfaces. Such operating conditions can also cause "crackling" sounds caused by flow cavitation within the valve.

Check Valves

In many hydronic systems it's often necessary to prevent fluid from flowing backwards through a pipe. The **check valve,** in one of its many forms, is the common solution.

Check valves differ in how they prevent reverse flow. The **swing check,** shown in Figure 5–45, contains a disc that is hinged along its upper edge. When fluid moves through the

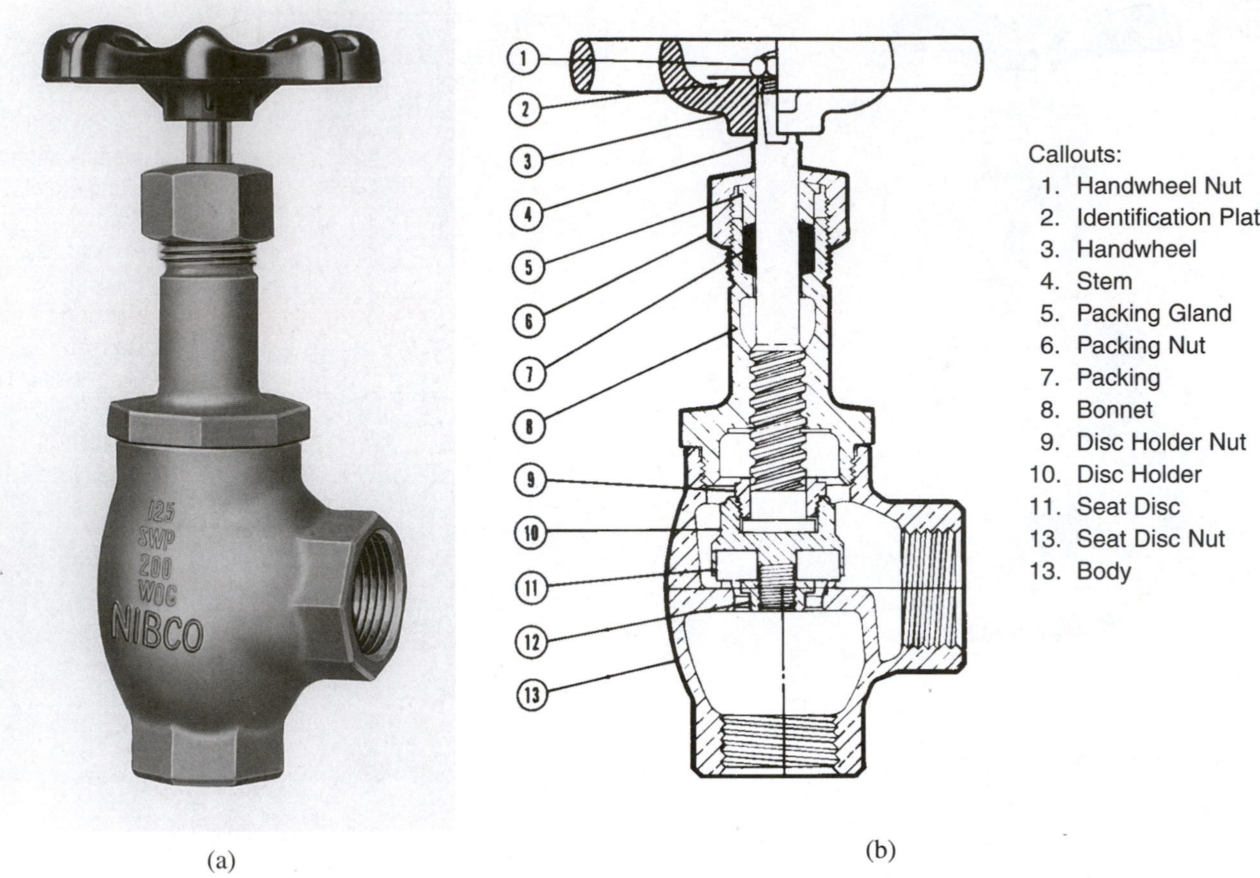

Figure 5–42 A typical angle valve (a) external appearance, (b) internal construction. *Photo and illustration courtesy of NIBCO, Inc., Elkhart, IN.*

Figure 5–43 Boiler drain valve. *Photo courtesy of NIBCO, Inc., Elkhart, IN.*

valves in the allowed direction (as indicated by an arrow on the side of the valve's body), the disc swings up into a chamber out of the direct path of the fluid. The moving fluid stream holds it there. When the fluid stops, or attempts to reverse itself, the disc swings down due to its own weight and seals across the opening of the valve. The greater the back pressure, the tighter the seal.

For proper operation, swing check valves must be installed in horizontal piping with the bonnet of the valve in an upright position. Installation in other orientations can cause erratic operation creating dangerous water hammer effects. It is also important to install swing check valves with *a minimum of twelve pipe diameters of straight pipe upstream of the valve.* This allows turbulence created by upstream components to partially dissipate before the flow enters the valve. Failure to do so can cause the disk inside the valve to rattle. This concept is illustrated in Figure 5–46.

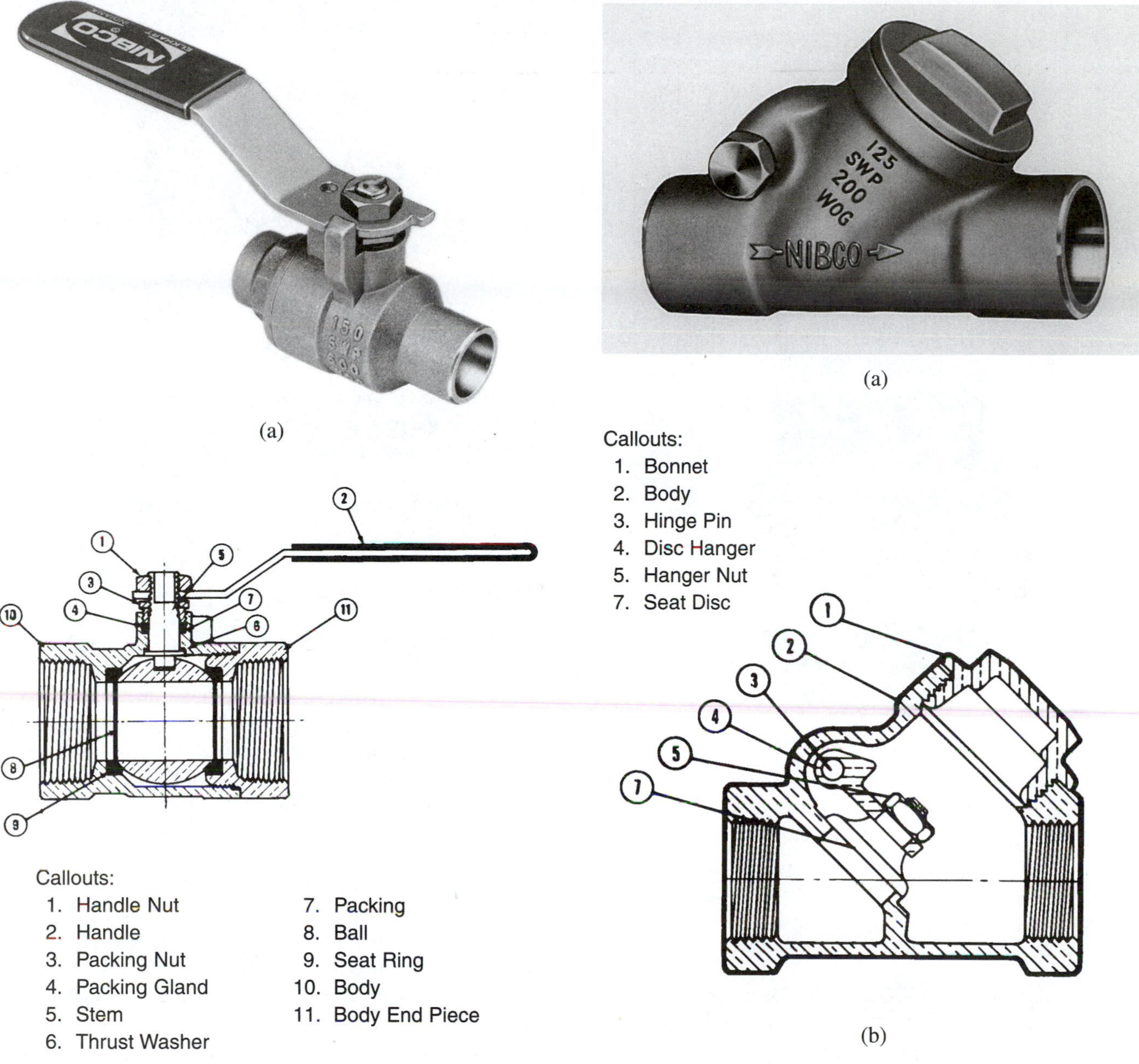

Figure 5–44 (a) External appearance of a full-port ball valve (b) internal construction of a standard port ball valve. *Photo and illustration courtesy of NIBCO, Inc., Elkhart, IN.*

Figure 5–45 A swing-check valve (a) external appearance, (b) internal construction. Notice flow direction arrow on body. *Photo and illustration courtesy of NIBCO, Inc., Elkhart, IN.*

A **spring-loaded check valve** is designed to eliminate the orientation restrictions associated with a swing check. These valves rely on a small internal spring to close the valve's disc whenever fluid is not moving in the intended direction. *The spring action allows the valve to be installed in any orientation.* The force required to compress the spring does, however, create slightly more pressure drop compared to a swing check. As with a swing check, *be sure the valve is installed with the arrow on the side of its body pointing in the desired flow direction.* An example of a spring-loaded check valve is shown in Figure 5–47.

As with swing checks, be sure to install at least twelve pipe diameters of straight pipe upstream of all spring-loaded check valves to dissipate incoming turbulence.

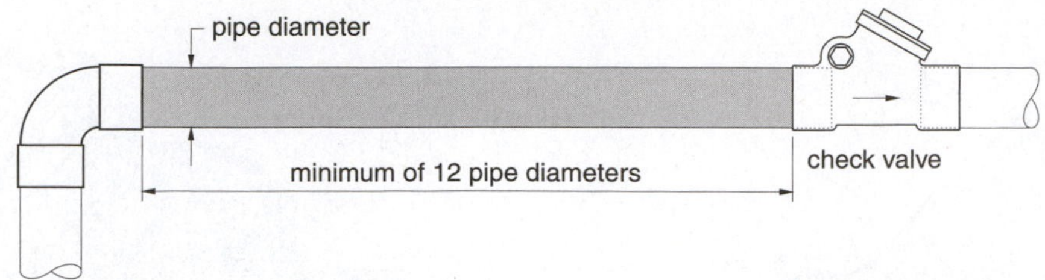

Figure 5–46 Install at least 12 pipe diameters of straight pipe upstream of a check valve to prevent rattling sounds due to turbulence.

(a)

Callouts:

1. Stem
2. Stem Guide
3. Body End
4. Spring
5. Disc
6. Disc Holder
7. Seat Screen
8. Body

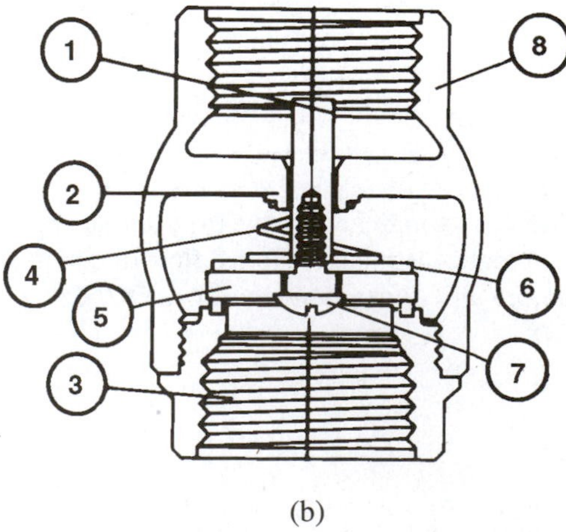

(b)

Figure 5–47 A spring-loaded check valve (a) external appearance, (b) internal construction. Notice flow direction arrow on body. *Photo and illustration courtesy of NIBCO, Inc., Elkhart, IN.*

5.7 SPECIALTY VALVES FOR HYDRONIC APPLICATIONS

Several valves have very specific functions in hydronic systems. Some provide safety protection and are required by plumbing and mechanical codes. Others automatically regulate the temperature and pressure at various points in the system. All these valves will be incorporated into system designs in later chapters.

Boiler Feed Water Valves

A **boiler feed water valve,** or pressure-reducing valve, is used to lower the pressure of water from a domestic water distribution pipe before it enters a hydronic system. This valve is necessary because most buildings have domestic water pressure higher than the relief valve settings of small hydronic systems. The feed water valve allows water to pass through whenever the pressure at its outlet side drops below its pressure setting. In this way, small amounts of water are automatically fed into the system as air is vented out, or as small amounts of fluid are lost through evaporation at valve packings, pump flange gaskets, and other locations. A typical boiler feed water valve is shown in Figure 5–48.

Hydronic systems without automatic feed water valves tend to slowly lose pressure as a result of air venting, and very small water losses through valves packing and pump flange gaskets. The resulting low pressure operation can eventually cause noise and pump cavitation. In multistory buildings, the reduced pressure can also prevent circulation in upper parts of the system. The function of the feed water valve is especially important during the first few days of system operation as dissolved air is released from the fill water and vented out of the system.

The pressure setting of feed water valves can be varied by rotating its center shaft with a screwdriver. Once the desired pressure setting is attained, the shaft position can be secured by tightening the locknut surrounding the shaft. It is often convenient to have a pressure gauge mounted near the outlet of the feed water valve to indicate the current pressure setting. The pressure setting of a feed water valve depends on the height of the system, and will be discussed in Chapter 12, Expansion Tanks.

Most feed water valves also have a lever at the top to allow the valve to be manually opened to quickly fill the

Figure 5–48 Example of a feed water valve. Notice flow direction arrow on body. *Courtesy of Watts Regulator Co.*

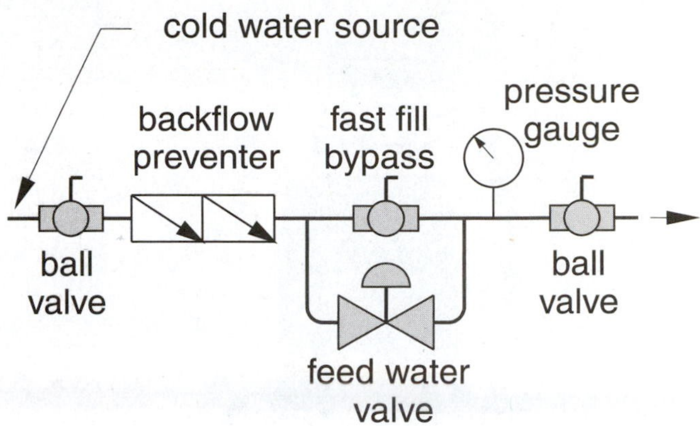

Figure 5–49 Feed water valve installed in a typical make-up water assembly. The bypass ball valve is on the straight pipe to minimize pressure drop and maximize flow rate during fast fill or purging.

system during start-up. This lever effectively disables the pressure-regulating function of the valve when it is lifted. Be sure the lever is returned to its normal position after the system is filled.

The flow through the feed water valve when the fast-fill lever is up is usually sufficient to purge small residential systems (depending on the water pressure available on the inlet side of the valve). On larger systems, it is common practice to install a full port ball valve to provide even faster filling and purging flow. A typical piping arrangement is shown in Figure 5–49. Notice the ball valve is on the straight through piping path, while the feed water valve is installed in the branch piping. This provides the minimum pressure drop through the ball valve for the fastest possible purging.

Many feed water valves are equipped with removable strainers to prevent particulates in the supply water from entering the system. If the strainer becomes clogged, the valve may not supply make-up water to the system. The strainer on the feed water valve should be checked during routine system maintenance. An isolating ball valve on each side of the feed water valve allows it to be quickly isolated so the strainer can be removed with minimal water loss.

When installing a feed water valve, be sure the arrow on the side of the body points *into* the system.

Pressure-Relief Valves

A **pressure-relief valve** *is a code requirement on any type of closed-loop hydronic heating system.* It functions by opening just below a preset pressure rating allowing system fluid to be safely released from the system before dangerously high pressures can develop. The pressure-relief valve is the final means of protection in a situation where all other controls fail to limit heat production. It its absence, some component in the system could potentially explode with devastating results.

Most mechanical codes require pressure-relief valves in any piping assembly that contains a heat source and is capable of being isolated by valves from the rest of the system. Nearly all boilers sold in the United States are shipped with a factory installed ASME-rated pressure-relief valve. The typical pressure rating for relief valves on small boilers is 30 psi. Other hydronic heat sources, such as heat pumps, are usually not equipped with factory-installed pressure-relief valves. A relief valve is also required in any piping circuit supplied with heat through a heat exchanger.

The operation of a pressure-relief valve is simple. When the force exerted on the internal disc equals or exceeds the force generated by the internal spring, the disc lifts off its seat and allows fluid to pass through the valve. The spring force is calibrated for the desired opening pressure of the valve. This rated pressure, along with the maximum heating capacity of the equipment the valve is rated to protect, is stamped onto a permanent tag or plate attached to the valve. A typical relief valve is shown in Figure 5–50.

All pressure-relief valves should have a waste pipe attached to their outlet port. This pipe routes any expelled fluid safely to a drain, or at least down near floor level. The waste pipe must be the same size as the valve's outlet port, with a minimal number of turns and no valves or other means of shutoff. The waste pipe should end 6 inches above the floor or drain to allow for unrestricted flow if necessary. Be sure to check local codes for possible additional requirements on the installation of relief valves.

Figure 5–50 Example of a pressure-relief valve. *Courtesy of Watts Regulator Co.*

Figure 5–51 Example of a small backflow preventer. Note inlet port marking. *Courtesy of Watts Regulator Co.*

Backflow Preventer

Consider a hydronic heating system filled with an antifreeze solution connected to a domestic water supply by way of a feed water valve. If the pressure of the domestic water system suddenly drops due to a rupture of the water main or other reason, the antifreeze solution could be pushed backward by expansion tank pressure into the domestic water system. This could contaminate not only the water in the building, but neighboring buildings as well.

A **backflow preventer** eliminates this possibility. It consists of a pair of check valve assemblies in series, with an intermediate vent port that drains any backflow that migrates between the valve assemblies. *Be sure the valve is installed with the inlet side connected to the domestic water supply.* An example of a small backflow preventer is shown in Figure 5–51.

Most municipal plumbing codes require backflow preventers on all hydronic heating systems that are connected to domestic water piping. Even if local codes do not require this valve, it is good practice to use it, especially if antifreeze or other chemical treatments might be added to the system at a later date. *A check valve is not an acceptable substitute for a backflow preventer.*

Flow-Check Valves

Another variation on the basic check valve is the **flow-check valve.** These valves have a weighted internal plug that drops down over the seat of the valve. The plug's weight is sufficient to prevent hot water below the valve from rising into the distribution system due to its own buoyancy. The weighted plug pops open as soon as the circulator turns on. Flow-checks also prevent reverse flow in multizone hydronic systems.

Flow-checks are available with either two or three ports as shown in Figure 5–52. The three-port body allows the valve to be installed with either a vertical or horizontal inlet pipe. The inlet port that is not used is closed off with a plug.

A small lever or screw at the top of a flow check valve allows it to be manually opened. Doing so allows some hot water flow through the valve in case of a circulator failure. It also allows any air trapped below the valve to rise through the valve as the system is being filled and purged. *The lever does not allow the valve to adjust flow rate.*

Flow checks are available with both cast-iron and bronze bodies. Either type can be used in closed-loop systems. Only bronze-bodied flow checks are suitable for systems where dissolved oxygen is present in the water (such as a system using nonbarrier PEX tubing).

Thermostatic Radiator Valves

Widely used in Europe, **thermostatic radiator valves (TRVs)** can provide precise *room-by-room* temperature control in hydronic heating systems. They are installed in the supply pipe of a heat emitter, or in some cases built into the heat emitter itself. TRVs consist of two parts, the valve body and the thermostatic operator.

The nickel-plated brass valve body is available in pipe sizes from 1/2 inch to 1 inch, in either a straight or angle pattern. In the latter, the outlet port of the valve is rotated 90 degrees relative to the inlet port. Inside the valve is a plug mounted on a spring-loaded shaft. The plug is held in its fully open position by the force of the spring. The fluid pathway through the valve is similar to that of a globe valve, thus making the valve well-suited for accurate flow control.

To close the valve, the shaft must be *pushed* inward against the spring force. No rotation is necessary. Shaft

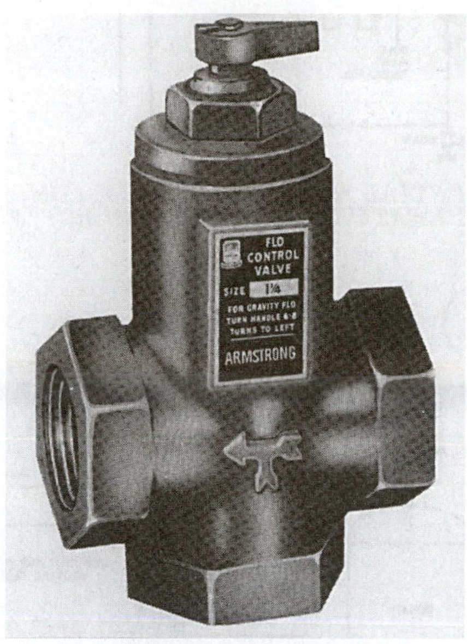

Figure 5–52 Example of flow-check valves (a) two-port design. *Courtesy of Watts Regulator Co.* (b) three-port design. *Courtesy of Armstrong Pumps.* Notice flow direction arrows on valve bodies.

movement can be achieved manually using a knob that threads onto the valve body, or automatically through use of a **thermostatic operator.** Figure 5–53 shows such a thermostatic operator ready to be mounted to a radiator valve.

Several types of thermostatic operators can be matched to radiator valve bodies. The most common configuration attaches the thermostatic operator directly to the valve body. A cutaway of such a combination is shown in Figure 5–54.

The thermostatic operator contains a fluid in a sealed bellows chamber. As the air temperature surrounding the operator increases, the fluid expands inside the bellows which forces the shaft of the valve inward towards its closed position. When the valve is mounted in a pipe supplying a heat emitter, this action decreases water flow, and reduces heat output. As the room air temperature decreases, the fluid contracts, allowing the spring force to slowly reopen the valve plug, and increase heat output from the heat emitter. The combination of the valve and thermostatic operator represents a fully **modulating** temperature control system that continually adjusts flow through the heat emitter to maintain a constant (occupant determined) room temperature.

TRVs are often located close to the heat emitter. The preferred mounting position is with the valve stem in a

Figure 5–53 A thermostatic operator ready to be attached to a radiator valve. *Courtesy of Danfoss, Inc.*

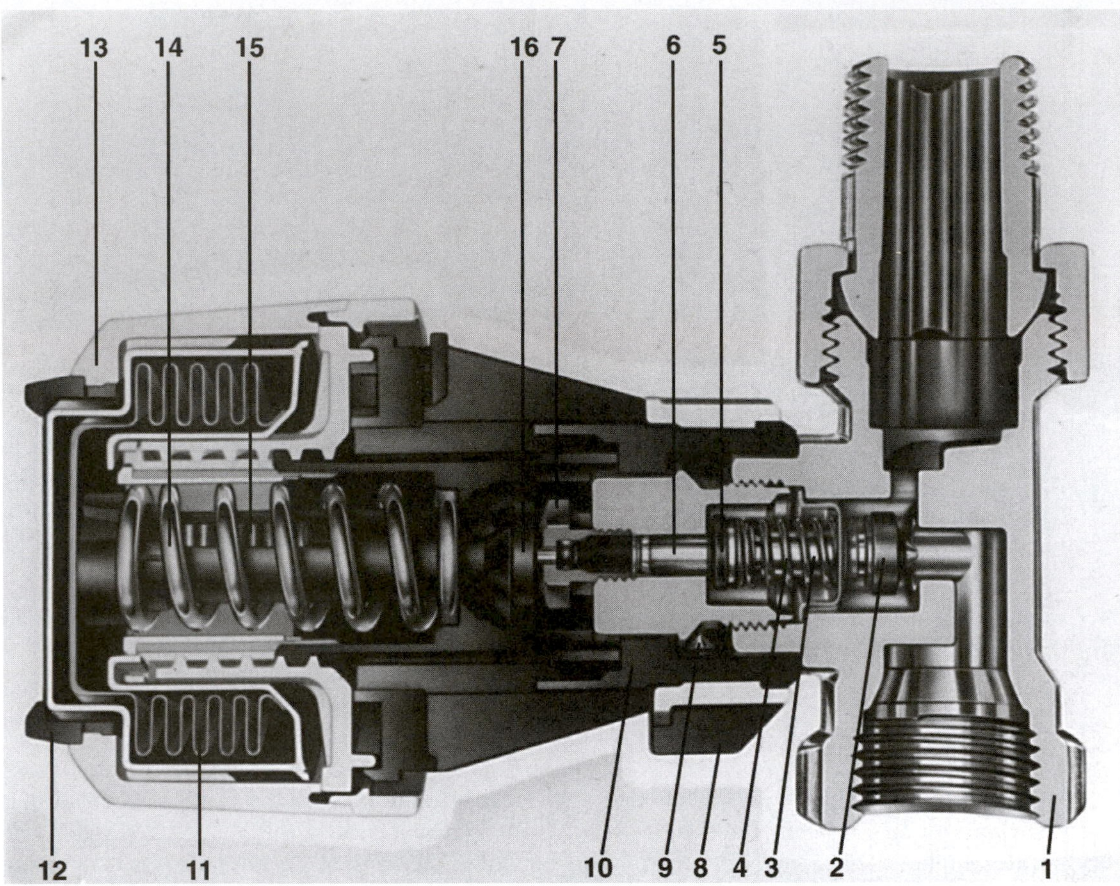

	PART	MATERIAL		PART	MATERIAL
1.	Valve	Nickel plated brass	9.	Allen screw	Steel
2.	Valve disc	EPDM	10.	Socket	POM
3.	Spindle guide	Phosphor-bronze	11.	Bellows	Phosphor-bronze
4.	Spring	Stainless steel	12.	Reference ring	Polyacetal
5.	Back seat washer	EPDM	13.	Handle	ABS
6.	Valve spindle	Brass	14.	Adjustment spring	Steel
7.	Pressure pin	Stainless steel	15.	Safety spring	Steel
8.	Clamping band	Al. alloy	16.	Pressure spindle	Polyamide No. 6
				Capillary tube	Steel

Figure 5–54 Internal construction of a thermostatic operator attached to a straight radiator valve body. *Courtesy of Danfoss, Inc.*

horizontal position with the thermostatic operator facing away from the heat emitter as shown in Figure 5–55. *In no case should the valve be installed so the thermostatic operator is above the heat emitter.* This will cause the valve to close prematurely and greatly limit heat output from the heat emitter. Be sure the TRV is mounted with flow passing through it in the direction indicated by the arrow on the valve body.

For versatility in different applications, most manufacturers of TRVs valves offer operator assemblies with a remote setpoint dial. This allows temperature adjustments to be made at normal thermostat height above the floor. A **capillary tube** runs from the setpoint dial assembly to the operator mounted on the radiator valve. Thermostatic operators can be ordered with capillary tubes up to 12 feet long. Figure 5–56 shows an example of a radiator valve fitted with a remote setpoint adjustment. Care must be taken not to kink or break the capillary tube during installation. *Severing the capillary tube renders the operator useless.*

Thermostatic radiator valves can be used in a variety of piping layouts. One example is combining a TRV with a diverter tee in a "one pipe" distribution system as shown in Figure 5–57.

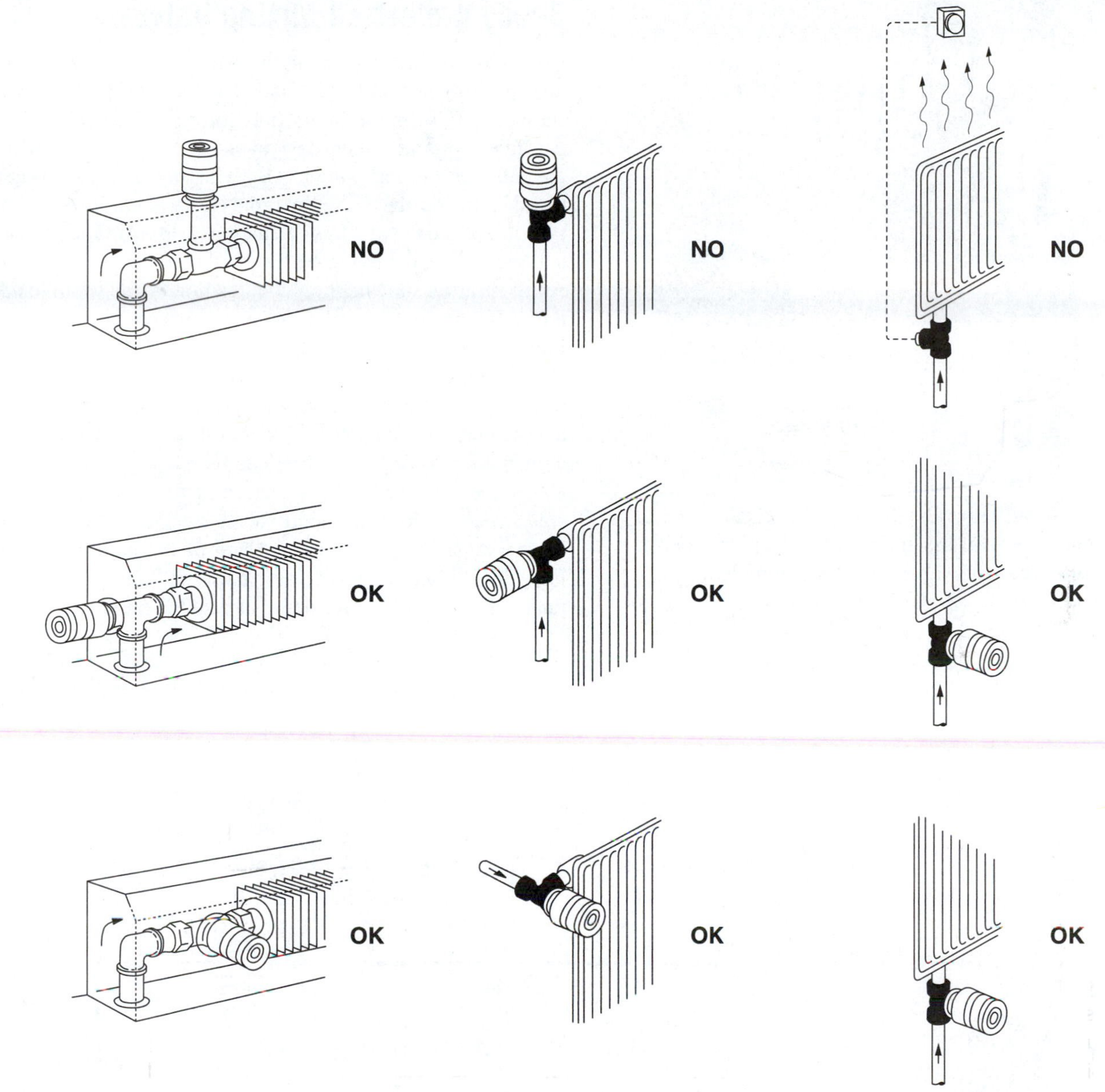

Figure 5–55 Proper and improper mounting locations for a thermostatic operator. Do not mount the operator where it will be directly affected by convective air currents rising from the heat emitter. *Courtesy of ISTEC, Inc., and Danfoss, Inc.*

This arrangement allows each heat emitter to be individually controlled. The TRV on a baseboard in a bedroom might be turned down at night for comfortable sleeping, while the TRV on a bathroom radiator is set to maintain 72 °F. The small circulator would run continuously during the heating season. Heated water would always be available from the boiler (though perhaps at different temperatures depending on the outdoor temperature). This type of distribution system is discussed in detail in Chapter 11, Distribution Piping Systems. With proper design, it offers tremendous versatility and excellent room-by-room temperature control.

Mixing Valves

Some hydronic distribution systems require water temperatures lower than what is available from the heat source. For example, a radiant floor heating system may require water at 110 °F, while the water in a thermal storage tank supplying this load is at 150 °F. In such cases, the lower water temperature is created by blending return water from the distribution system with hot water from the heat source. The flow rate of each fluid stream needed to attain a specific outlet temperature is regulated by a **mixing valve.**

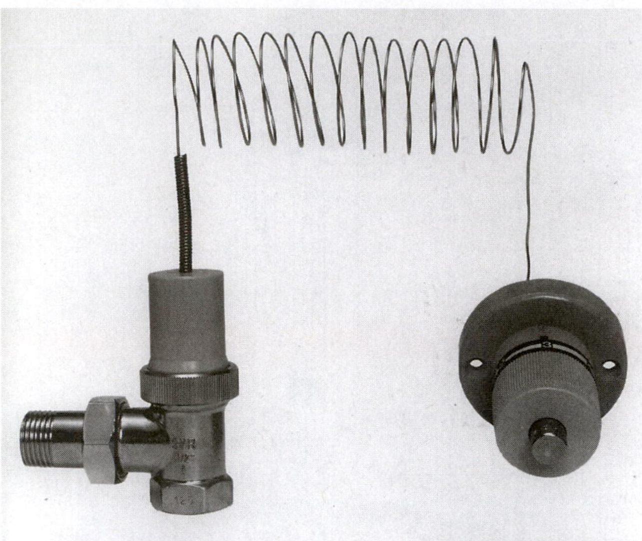

Figure 5–56 A radiator valve fitted with a remote setpoint dial and operator assembly. Notice capillary tubing between setpoint dial and valve operator assembly. *Courtesy of Ammark Corporation.*

3-way Motorized Mixing Valves

All **3-way mixing valves** have three ports. One is the inlet port for heated fluid, another is the inlet port for cool fluid. The third is the outlet port for the mixed fluid. The two inlet ports are often designated as "A" and "B," while the outlet port is designated "AB." The two inlet streams mix together inside the valve in proportions controlled by the position of the valve's flow element (or **spool**). The body of a typical 3-way mixing valve is shown in Figure 5–58. An illustration showing how the position of the spool inside the valve varies the inlet floor proportions is shown in Figure 5–59.

The position of the valve's knob determines the amount of each fluid stream that enters the valve. The flow rate and temperature of each entering stream determines the blended temperature leaving the valve. The blended temperature will always be between the temperature of the entering hot and cold streams. Most three-port valves have an indicator scale on the valve body. The pointed end of the handle or knob indicates the setting of the valve between 0 and 10. The full range of the handle movement is about 90 degrees.

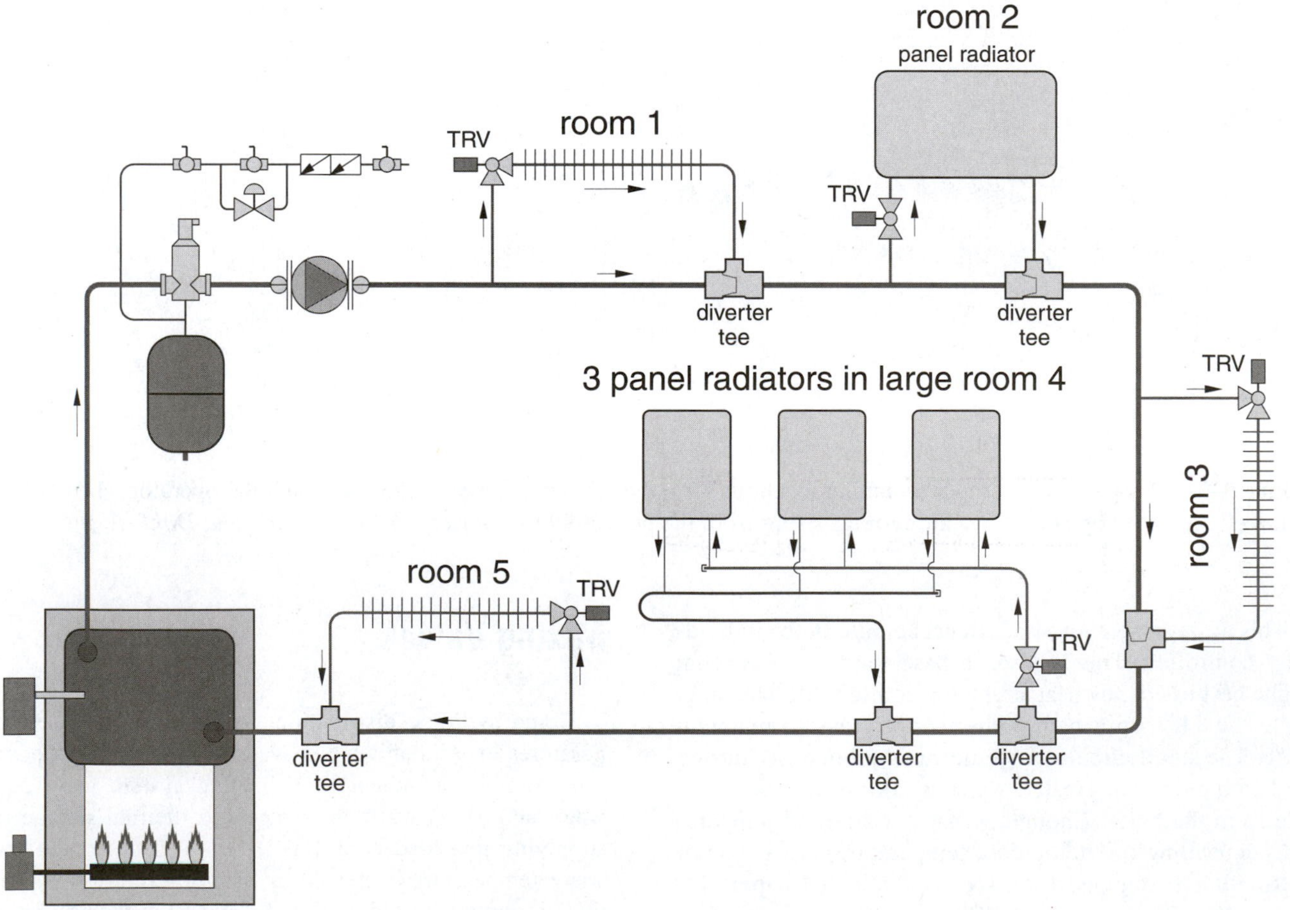

Figure 5–57 Use of thermostatic radiator valves in combination with diverter tees for individual control of branch circuits.

Figure 5–58 A 3-way mixing valve with a manual adjustment knob. Note setting numbers on dial face. *Courtesy of Danfoss, Inc.*

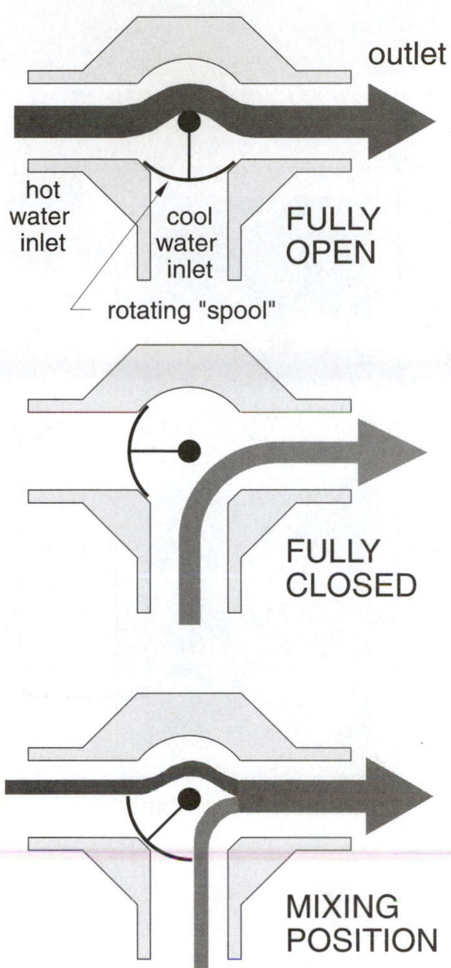

Figure 5–59 Cutaway illustration showing internal construction of a 3-way rotary valve. Spool rotates through approximately 90 degrees of arc.

A basic 3-way mixing valve does not contain an internal thermostatic element. It does not adjust itself to compensate for changes in the temperature of entering fluid streams. For automatic temperature control, the valve must be equipped with a motorized **actuator** operated by an electronic temperature controller. A single 3-way mixing valve can be used when the heat source can accept cool return water from the distribution system without problems such as sustained flue gas condensation (Figure 3–60a). A water heater would be an example of such a heat source.

As discussed in Chapter 3, Hydronic Heat Sources, conventional gas- and oil-fired boilers should not be operated with inlet temperatures that cause sustained flue gas condensation. If a 3-way mixing valve will be used to connect such a boiler to a low temperature distribution system, it must be piped as shown in Figure 3–60b, and operated by a controller that senses boiler inlet temperature.

In this configuration, the controller can limit the rate of hot water flow into the 3-way valve when necessary to prevent the distribution system from extracting heat faster than the boiler can produce it. *Without a controller that senses boiler inlet temperature, a 3-way valve cannot properly protect a conventional boiler from sustained flue gas condensation.*

4-way Mixing Valves

Unlike 3-way mixing valves, **4-way mixing valves** *are specifically designed to couple low temperature distribution systems to conventional boilers (which requires protection from sustained flue gas condensation).* Examples of a 4-way mixing valve for different pipe sizes are shown in Figure 5–61.

In a 4-way mixing valve, a *portion* of the entering hot fluid is blended with a *portion* of entering cool fluid to create the desired supply temperature. The remainder of the hot water is mixed with the remainder of the cool water to boost the inlet temperature to the boiler. This latter function is the main difference between 3-way and a 4-way mixing valves.

Flow through a four-port mixing valve is controlled by a simple rotating vane. As the valve's handle is rotated from one extreme to the other, the vane rotates through an arc of about 90 degrees.

At one end of the travel range, all water returning from the distribution system is routed back to the boiler. At the same time, all hot water entering the valve is routed to the supply side of the distribution system. No mixing takes place when the valve is set to this position. In this position, the valve is allowing the maximum possible heat transfer to the distribution system.

At the other end of the travel range, none of the water returning from the distribution system goes back to the boil-

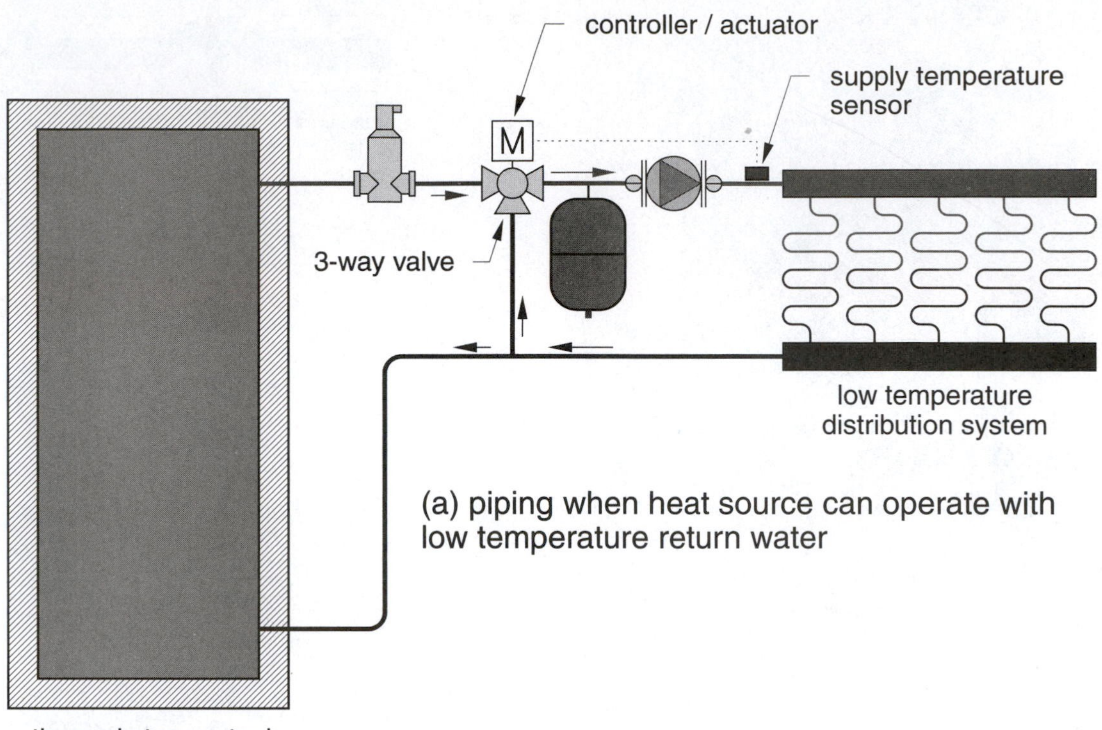

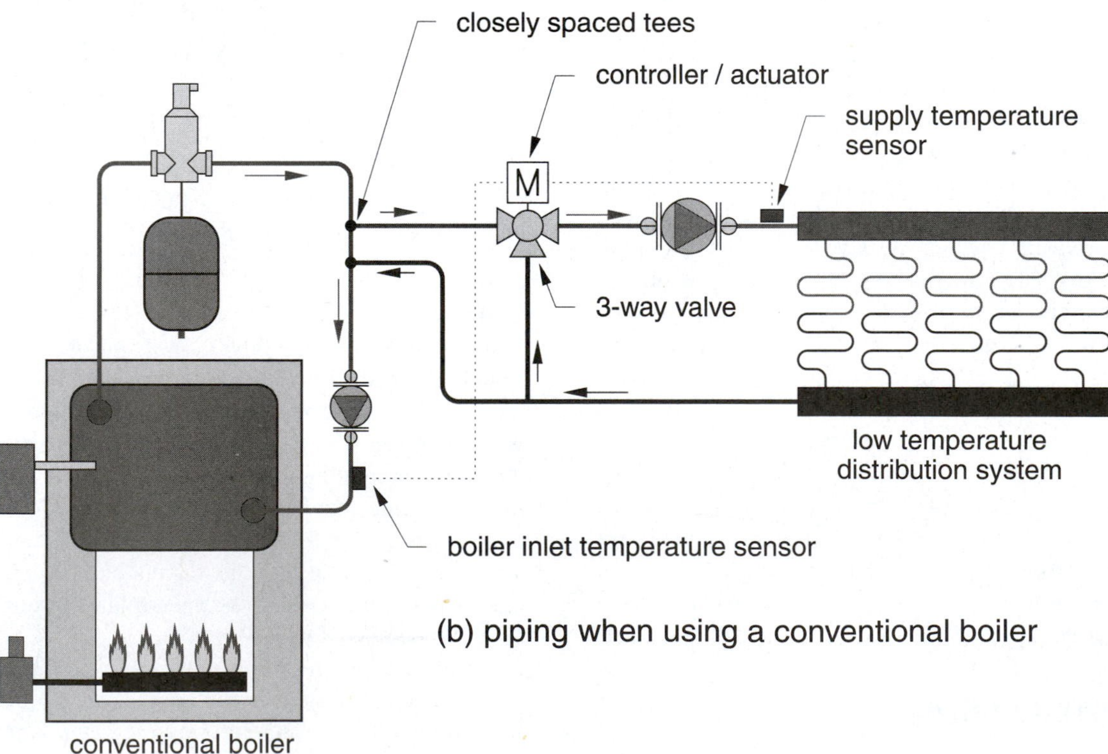

Figure 5–60 Typical piping for a 3-way motorized mixing valve (a) when heat source is not affected by cool return water. (b) When heat source is a conventional boiler that requires protection from sustained flue gas condensation.

Figure 5–61 Examples of a 4-way mixing valve. *Courtesy of tekmar Control Systems.*

er. Instead, all return water is routed back to the supply side of the distribution system. No hot water enters the valve, and hence no mixing takes place. In this position, there is no heat input to the distribution system.

In most cases, the valve operates at an intermediate (mixing) position between these two extremes. These valve setting are illustrated in Figure 5–62.

Like a 3-way mixing valve, 4-way mixing valves do not contain a thermostatic element. For automatic temperature control, they must be equipped with motorized actuators operated by an electronic controller. *A 4-way valve without such an actuator/controller cannot adjust to variations in the temperature of the incoming fluid streams. Thus it cannot provide stable supply temperature control, nor properly protect a conventional boiler from flue gas condensation.* There a very few instances when a 4-way valve should be used without an associated actuator/controller. Figure 5–63 shows a 4-way mixing valve with an actuating motor attached to the valve body.

There are several different ways to pipe 4-way mixing valves. The schematic shown in Figure 5–64a is applicable when the valve is mounted in close proximity to a boiler having low-flow resistance. Flow through the boiler is created by the momentum exchange of water returning to the valve from the distribution system. No circulator is required between the mixing valve and the boiler.

The piping shown in Figure 5–64b is preferred when the heat source supplying the valve has higher flow resistance, or when a boiler requiring a minimum sustained flow rate when firing is used. An additional circulator is used to provide flow in the boiler loop. The valve is coupled to this loop through a pair of closely spaced tees that prevent interference between the circulator in the boiler loop and the circulator in the distribution system. Momentum exchanger of water in the 4-way valve will provide the necessary flow between the valve and the boiler loop.

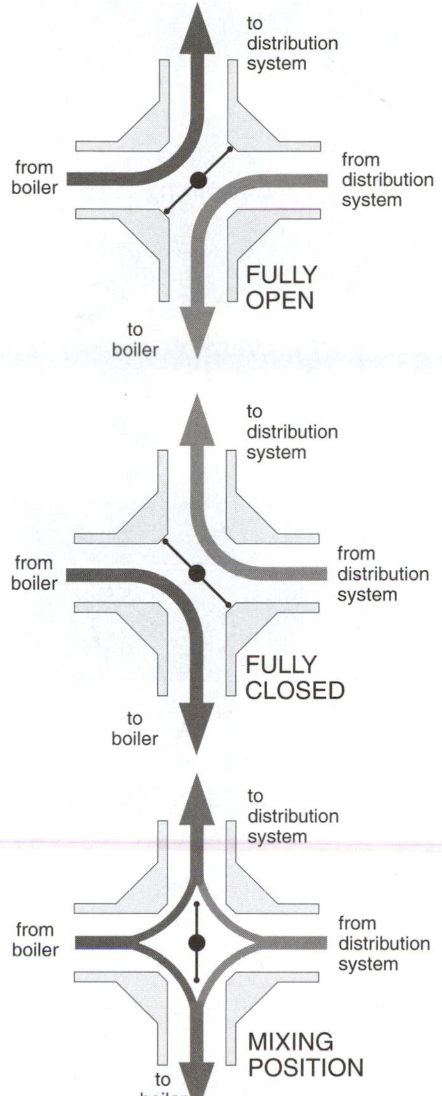

Figure 5–62 Cutaway illustration showing internal construction of a 3-way rotary valve. Internal vane rotates through approximately 90 degrees of arc.

Figure 5–63 A 4-way mixing valve body with actuating motor attached. The actuating motor positions the valve's shaft based on input from an electronic controller.

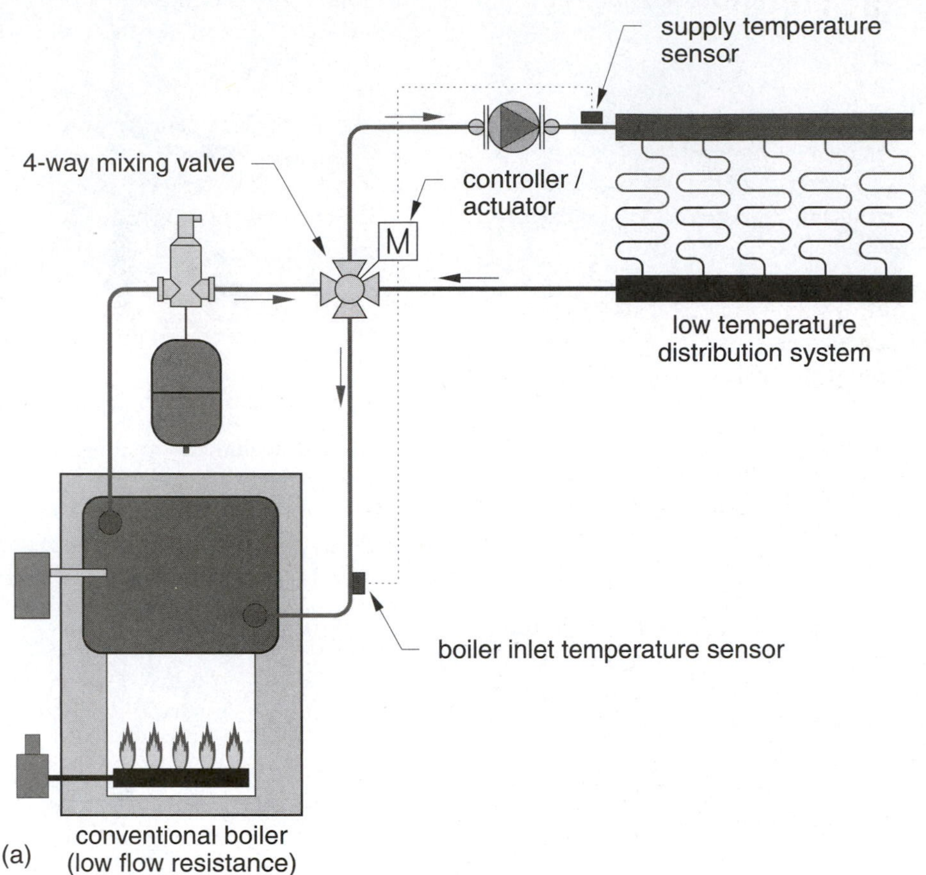

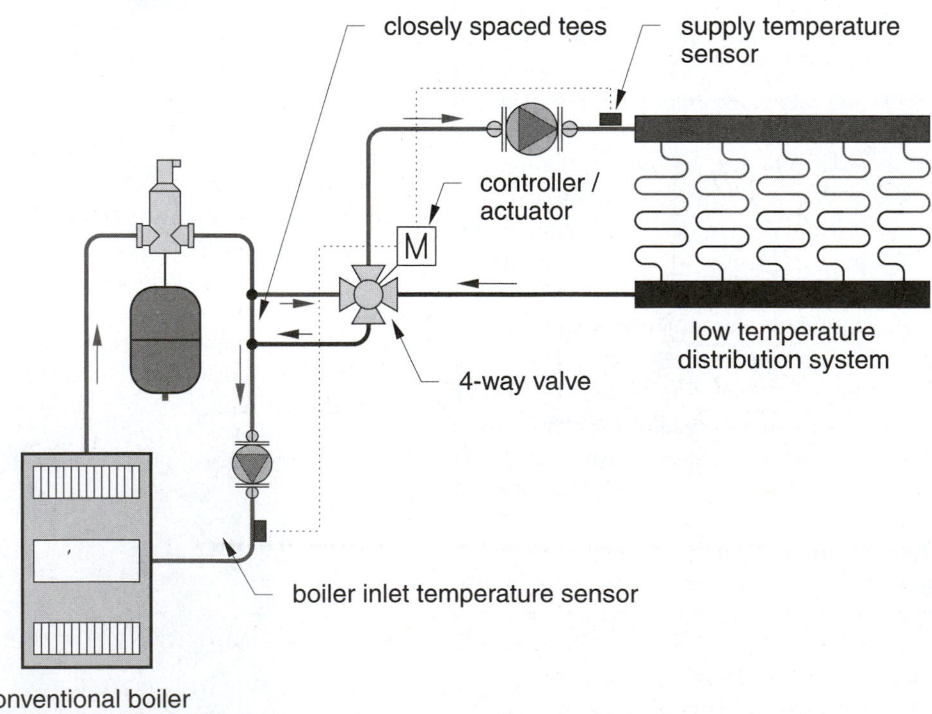

Figure 5–64 Piping options for a motorized 4-way mixing valve. (a) No circulator between 4-way valve and boiler. (b) Valve piped to boiler bypass loop using closely spaced tees.

3-way Thermostatic Mixing Valves

Another type of mixing valve suitable for use in hydronic systems is called a **3-way thermostatic mixing valve.** Unlike the 3-way and 4-way mixing valves just discussed, 3-way thermostatic mixing valves have internal components that automatically adjust the flow of hot and cool water to maintain a set outlet temperature. An example of such a valve is shown in Figure 5–65.

The desired mixed outlet temperature is set on the valve's knob. The two flow streams mix within the valve body and flow past the thermostatic element. This element constantly adjusts the valve's spool to regulate the amount of incoming hot and cool water.

In some systems, the piping of a 3-way thermostatic valve can be identical to that of a 3-way motorized mixing valve. A single 3-way thermostatic valve can be used to provide supply temperature control when the heat source can tolerate low inlet temperatures, and relatively low flow rates (Figure 5–66a). *This is not the case for most conventional boilers.* When a conventional boiler is used as the heat source, a second 3-way thermostatic valve should be used to monitor and boost the boiler inlet temperature when necessary to prevent sustained flue gas condensation. The suggested piping is shown in Figure 5–66b. Notice that a second circulator is now required to provide proper mixing in the thermostatic valve protecting the boiler. Some boiler manufacturers offer models with such a valve already built in.

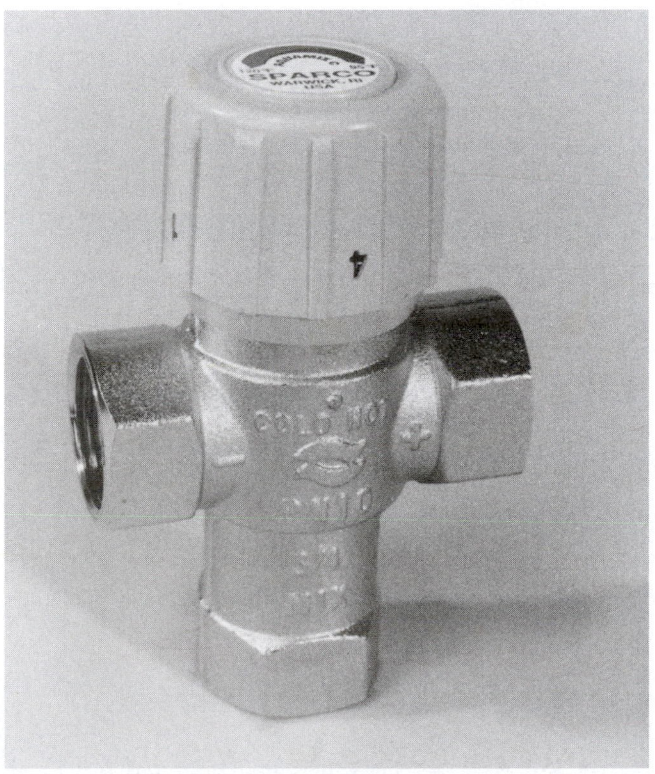

Figure 5–65 Example of a 3-way thermostatic mixing valve. *Courtesy of Sparco, Inc.*

These 3-way thermostatic valves are also available with external temperature-sensing elements. An example is shown in Figure 5–67.

This type of valve functions the same way as the valve with the internal temperature sensing element. The capillary tube connecting the temperature sensing bulb allows the mixed temperature to be sensed a few feet away from the valve body. For the most accurate response, the sensing bulb should be mounted in a well inserted into the mixed flow stream.

2-way Thermostatic Valves

Although technically not classified as a "mixing" valve, 2-way thermostatic valves can still be used to control water temperature in a hydronic distribution system.

An example of a 2-way thermostatically controlled valve is shown in Figure 5–68. Internally the valve is similar to a globe valve, making it a good valve for flow control. The valve body is fitted with a thermostatic operator that regulates the stem position based on temperature. The knob on the actuator sets the desired water temperature at the sensor bulb location.

As the temperature of the sensing bulb increases, the pressure of the working fluid inside the bulb increases. This increased pressure is passed along through a capillary tube to the flexible bellows assembly inside the actuator. The bellows responds by forcing the valve stem toward the closed position. In this way, the actuator modulates the stem position in an attempt to maintain the preset temperature at the sensing bulb.

The **injection mixing** piping typically used with a 2-way thermostatic valve is shown in Figure 5–69. The "flow restrictor" valve is partially closed to create a pressure differential between the injection risers. As the 2-way thermostatic valve begins to open, this pressure differential forces cool water up the return riser and hot water down the supply riser. Mixing begins in the tee where the supply injection riser connects to the distribution system. The sensing bulb should (ideally) be located in a well downstream of the circulator. Here it senses the final mixed temperature, and provides feedback to the thermostatic operator as necessary to maintain a stable mixed temperature.

Zone Valves

Hydronic heating systems often use two or more piping circuits to supply heat to different areas in a building. Such areas are called **zones.** One device often used to allow or prevent flow in a given zone circuit is called a **zone valve.**

Several types of zone valves are available. They all consist of a valve body combined with an electrically powered actuator. The actuator is the device that produces movement of the valve shaft when supplied with an electrical signal. Some actuators use small electric motors and a gear train

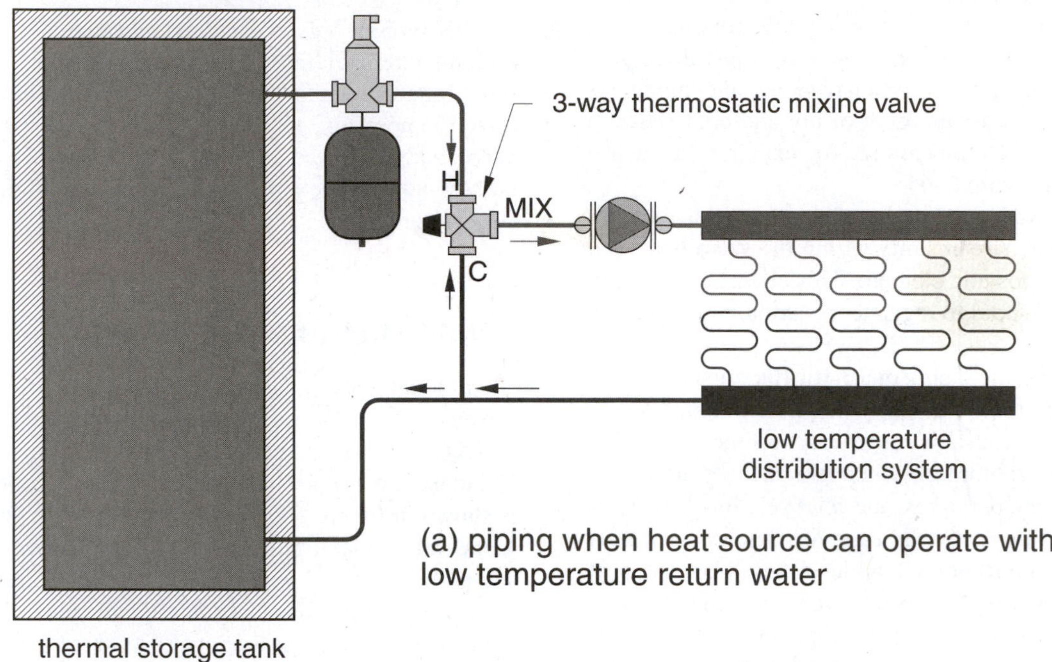

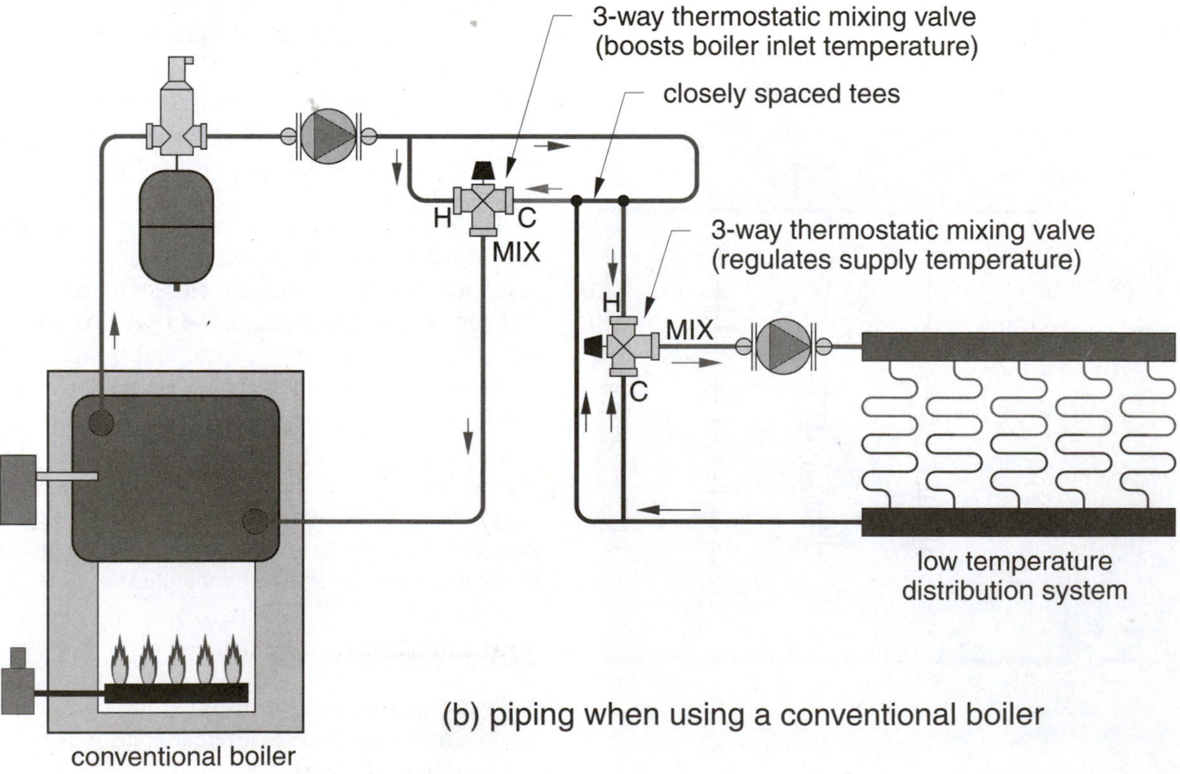

Figure 5–66 Typical piping for a 3-way thermostatic mixing valve. Notice that two mixing valves are required to properly protect a conventional boiler from sustained flue gas condensation when connected to a low temperature/high thermal mass load.

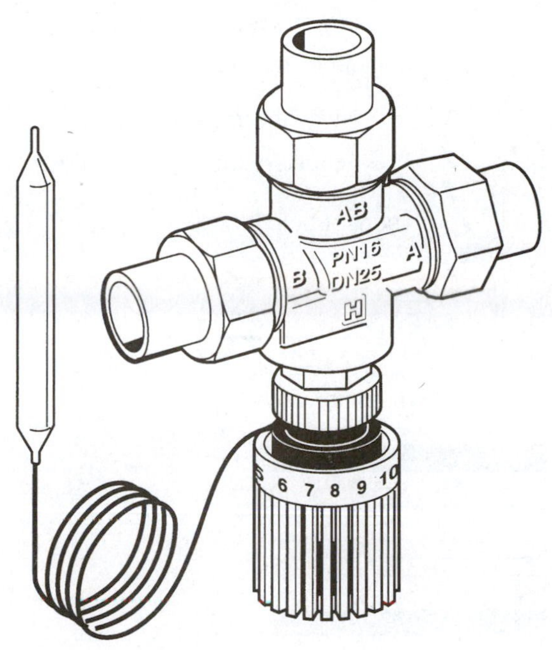

Figure 5–67 Example of a 3-way thermostatic valves using an external sensing bulb and capillary tube. *Courtesy of Honeywell-Braukmann.*

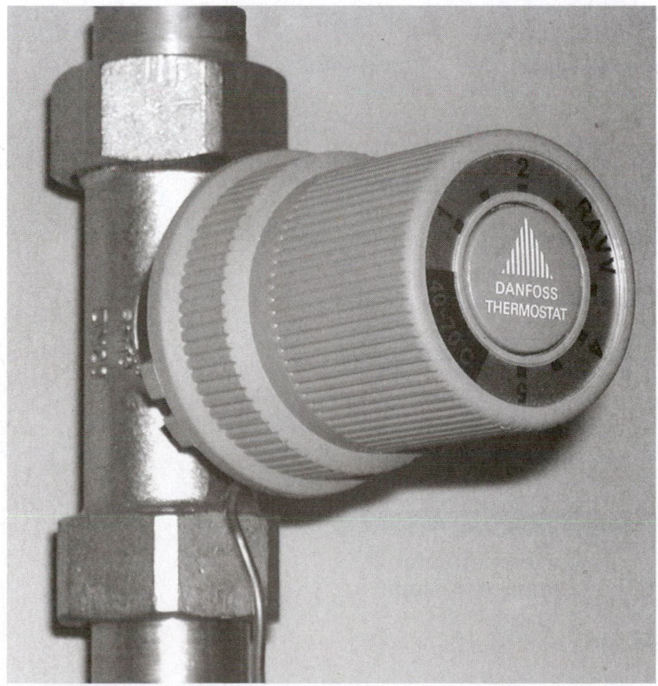

Figure 5–68 Example of a 2-way thermostatic valve. Notice capillary tube leading to sensor bulb. The number set on dial opposite the arrowhead corresponds to a temperature that the valve attempts to maintain.

to produce a rotary motion of the valve shaft. Others use **heat motors** that expand and contract to produce a linear motion that moves the valve's shaft. A zone valve equipped with a gear motor actuator is shown in Figure 5–70a. A zone valve using a heat motor actuator is shown in Figure 5–70b.

Zone valves operate in either a fully open or a fully closed position. *They do not modulate to regulate the flow rate in a piping circuit.* Those with gear motor actuators can move from fully closed to fully open in about 3 seconds. Those with heat motor actuators require 2 and 3 minutes from when the electrical signal is turned on, to when the valve is fully open. Since the response time of a heating system is relatively slow, the slower opening of heat motor type zone valves does not make a noticeable difference in system performance.

Both 2-way and 3-way zone valves are available. The 2-way valve either allows full flow through the valve or totally stops flow. This is the most commonly used zone valve in residential applications.

A schematic for a typical four-zone system using 2-way zone valves is shown in Figure 5–71. Three of the zones are for space heating, the other supplies an indirect water heater. Notice that the zone valves are installed on the *supply* side of each zone circuit. In this location, a closed-zone valve will prevent heat migration due to buoyancy-driven flow of hot water into zone circuits that are supposed to be off. Zone valves mounted on the return side of the zone circuits cannot provide this function.

A 3-way zone valve has an additional port called the bypass. When mounted as shown in Figure 5–72 on page 132, there will always be flow through a 3-way zone valve when the circulator is on. The flow that enters the common port must either pass straight through the valve or out the bypass port. Therefore, 3-way zone valves are useful when flow must be maintained in a certain portion of the piping system. This is often necessary when low mass boilers with constant firing rates are used.

A 3-way zone valve can also be used as a **diverter valve.** In this application, the valve directs the incoming flow into one of two possible outlets. If the valve is deenergized, flow entering the "A" port is directed out the "AB" port. If the valve is energized, flow entering the "B" port is directed out the "AB" port. In some systems, 3-way zone valves functioning as diverter valves are used to direct hot water from the boiler to either the space heating distribution system, or an indirect water heater, but never both at the same time.

Most zone valves for residential and light commercial systems operate on 24 VAC control voltage. This is a common transformer secondary voltage used in many types of the heating and cooling systems. This allows zone valves to be wired with standard thermostat cable. For certain applications, it may be desirable to operate zone valves using standard 120 VAC line voltage. Some manufacturers offer actuators set up for this voltage.

Zone valves are available in two-wire, three-wire, or four-wire configurations. Two wires are always required to operate the valve's actuator. The third, and in some cases fourth,

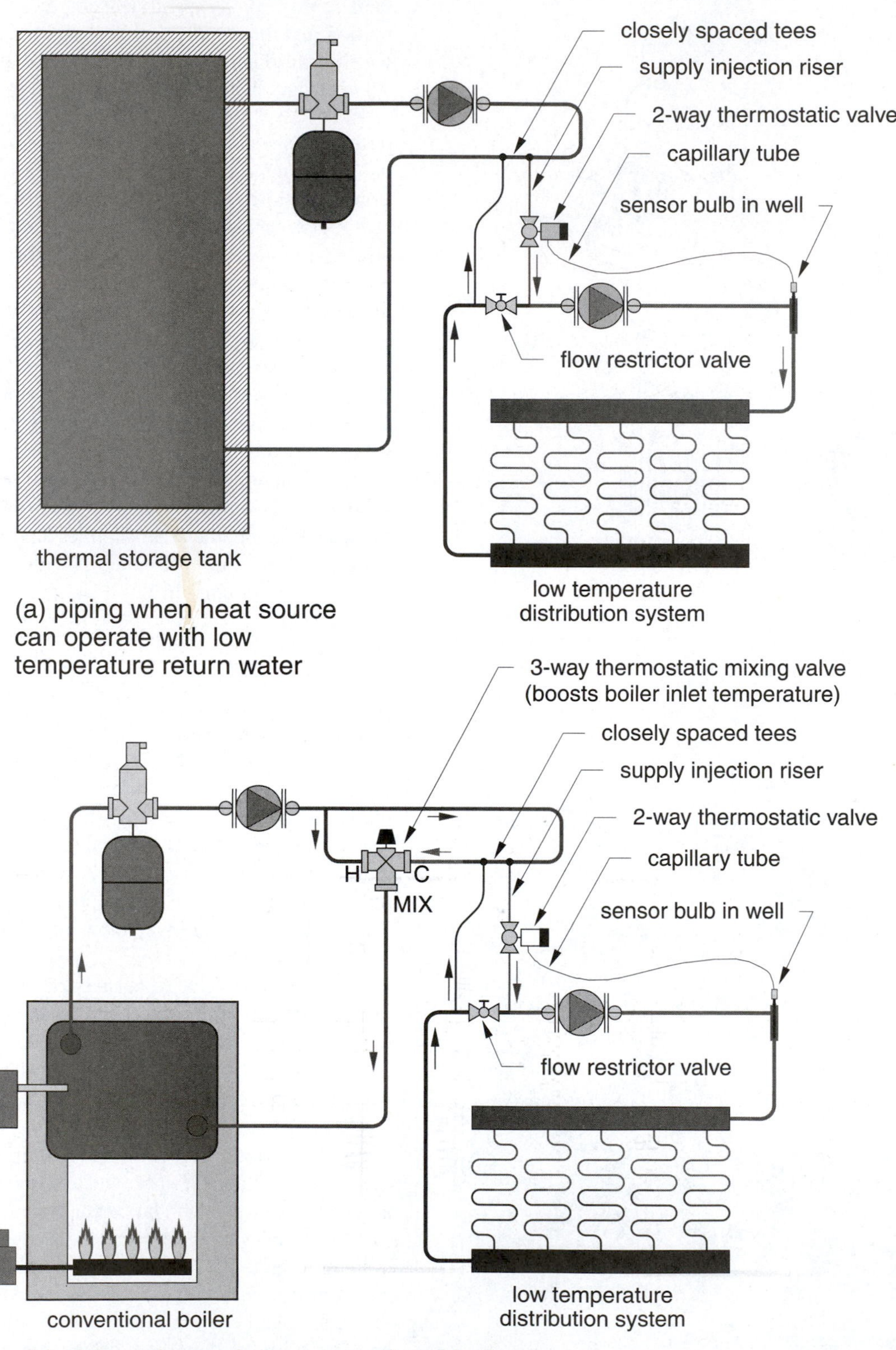

Figure 5–69 Piping a 2-way thermostatic valve for injection mixing control of the supply water temperature. (a) When heat source does not require protection against low inlet water temperature. (b) In combination with a 3-way thermostatic valve to protect a conventional boiler against sustained flue gas condensation.

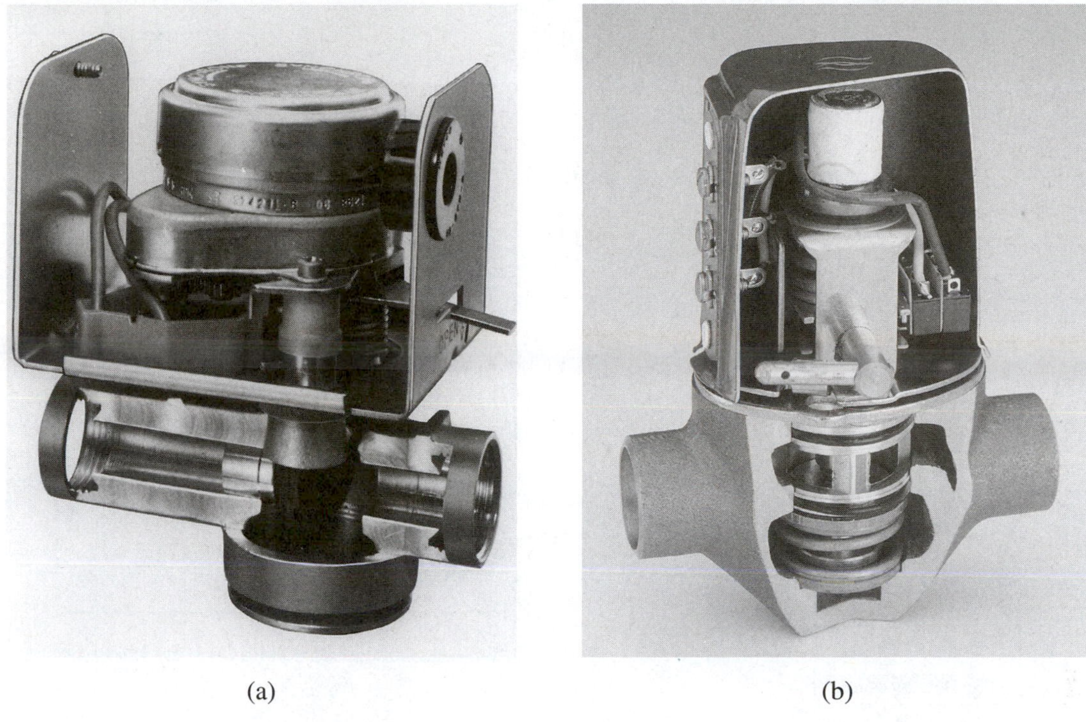

Figure 5–70 Two examples of hydronic zone valves (a) valve equipped with a gear motor actuator. *Courtesy of Erie Manufacturing.* (b) Valve equipped with a heat motor actuator. *Courtesy of Sparco, Inc.*

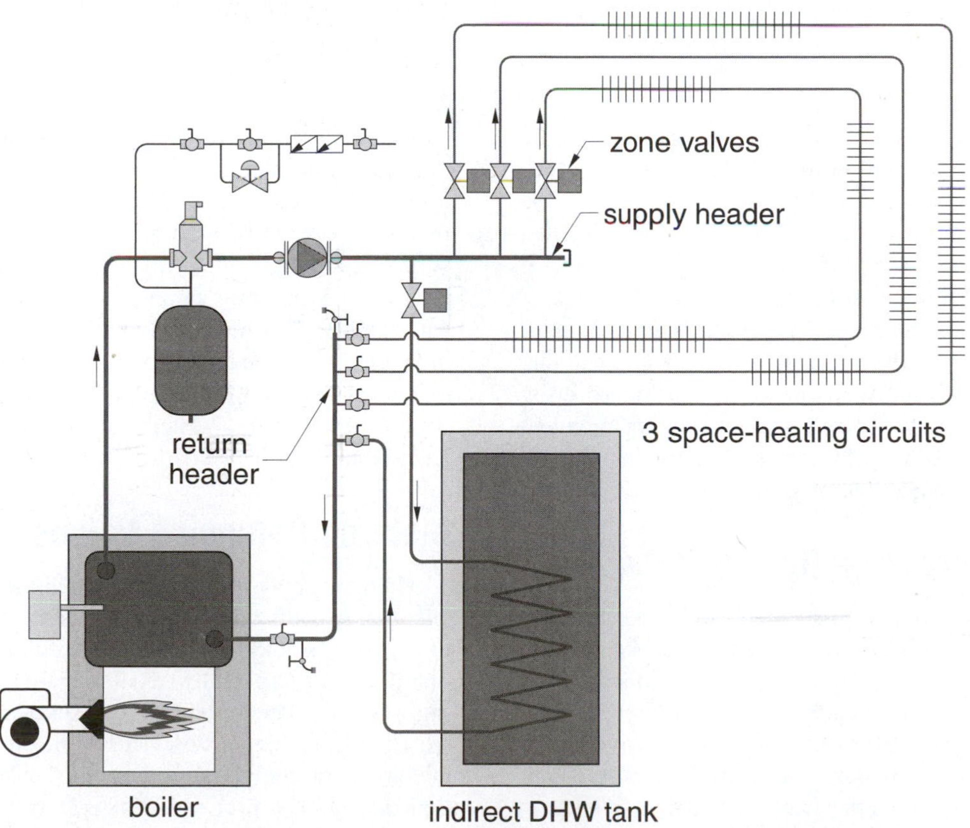

Figure 5–71 Placement of zone valves on the supply side of zone circuits. This placement prevents heat migration into supply side of inactive zone circuits.

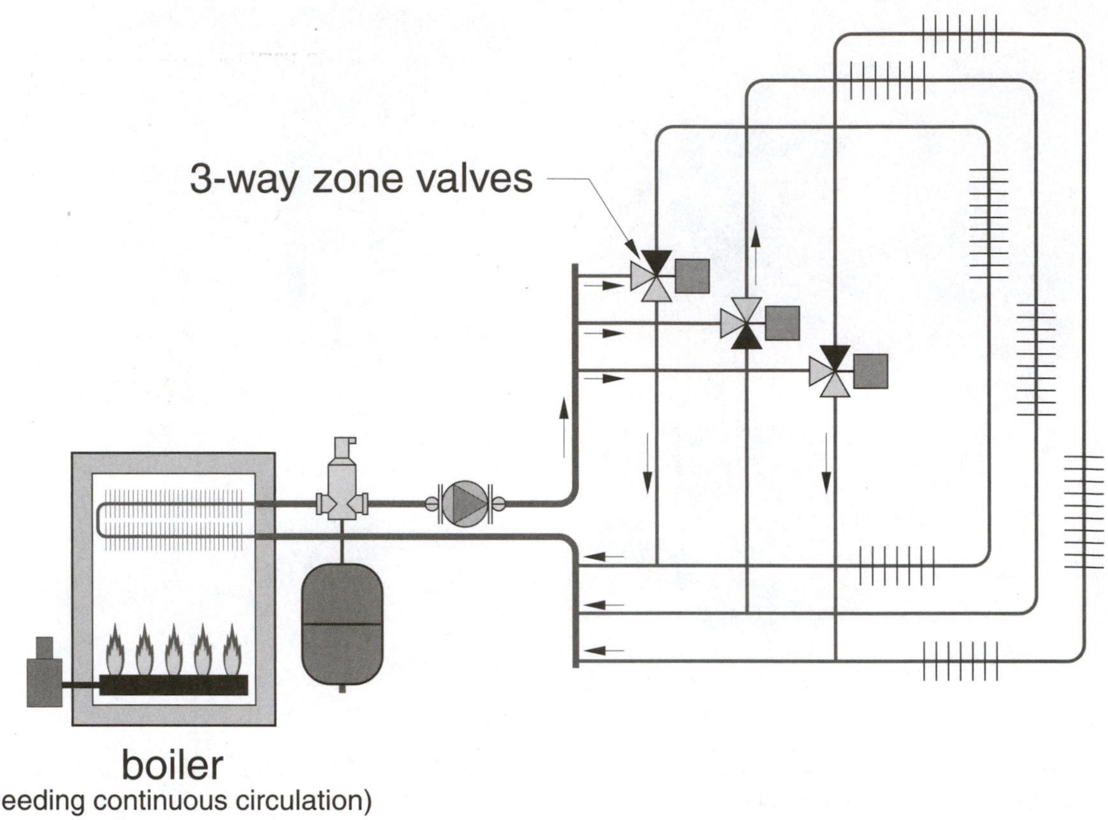

Figure 5–72 Piping schematic showing the use of three-way zone valves. This arrangement maintains flow through boiler regardless of which zones are delivering heat.

wire is connected to an **end switch** in the valve. When the valve reaches its fully open position, the end switch closes. This switch closure is used to signal other portions of the control circuit that a zone is "calling" for heat. More detailed discussion of control wiring for zone valves is presented in Chapter 9, Control Strategies, Components, and Systems.

When soldering a zone valve to copper tubing, the actuator should be removed from the valve body. The disk or ball that controls flow through the valve should be positioned away from the valve seats to minimize heating. Use a hot torch that can quickly heat the sockets of the valve body to minimize the potential for heat damage to other parts of the valve.

Differential Pressure Bypass Valves

Systems that use several 2-way zone valves in combination with a fixed speed circulator can experience undesirable operating conditions when only one or two of the zones are on. The problem that arises is an increase in the differential pressure supplied by the circulator when only one or two zones are operating. The increased pressure can cause excessively high flow rates and flow noise in the active zone circuits.

To prevent this, a **differential pressure bypass valve (DPBV)** should be installed in systems that have more than four zones controlled by 2-way zone valves, or using a circulator larger than 1/25 horsepower. This valve can be set to maintain an almost constant upper limit on the pressure differential across the distribution system. The valve functions by allowing a portion of the flow from the system's circulator to bypass the distribution system, thus limiting the pressure differential the pump can develop. An example of a differential pressure bypass valve is shown in Figure 5–73. The typical placement of a DPBV in a zoned system is shown in Figure 5–74. The details for incorporating DPBVs into systems are discussed in Chapter 11, Distribution Piping Systems.

Metered Balancing Valves

Multizone hydronic systems containing two or more parallel piping paths often require adjustments of the flow rates to properly balance the heating or cooling capacity of each branch. This is especially true in **parallel direct return piping** systems. The piping system shown in Figure 5–75a would produce higher flow rates in the piping paths closer to the circulator compared to those farther out in the distribution system. The proper use of **metered balancing valves** can correct for this situation as depicted in Figure 5–75b.

A metered balancing valve differs from a standard globe valve in three important ways. First, it uses a machined plug

Figure 5–73 Example of a differential pressure bypass valve.

rather than a flat disc as the flow control element. This allows the valve to vary the flow such that heat output from the heat emitter is in closer proportion to stem position. Second, the stem movement per turn of the shaft is less, allowing for adjustments that are more precise. Finally, metered balancing valves have two pressure metering ports that can be used to verify the flow rate through the valve based on the pressure drop across it. A cutaway view of a typical metered balancing valve is shown in Figure 5–76.

The pressure differential across the metered balancing valves can be measured using a digital manometer attached to the valve's pressure taps. This device calculates the flow rate through the valve based on the measured differential pressure across the valve's orifice. It can be easily moved from one valve to the next to balance the entire system. An example of a digital manometer is shown in Figure 5–77.

Metered balancing valves are more intricate and thus more expensive than globe valves. To use them effectively, the digital manometer must also be available.

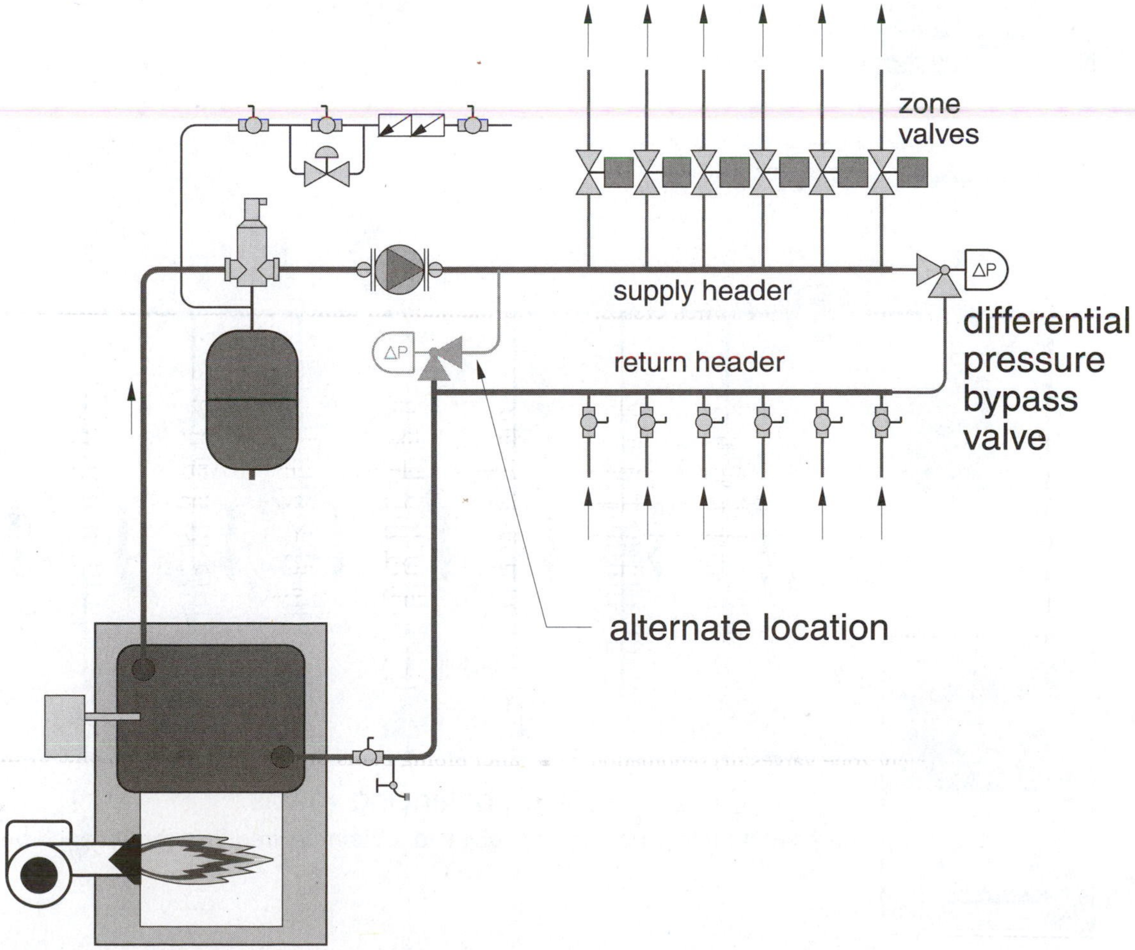

Figure 5–74 Placement options for a differential bypass valve in a system using zone valves.

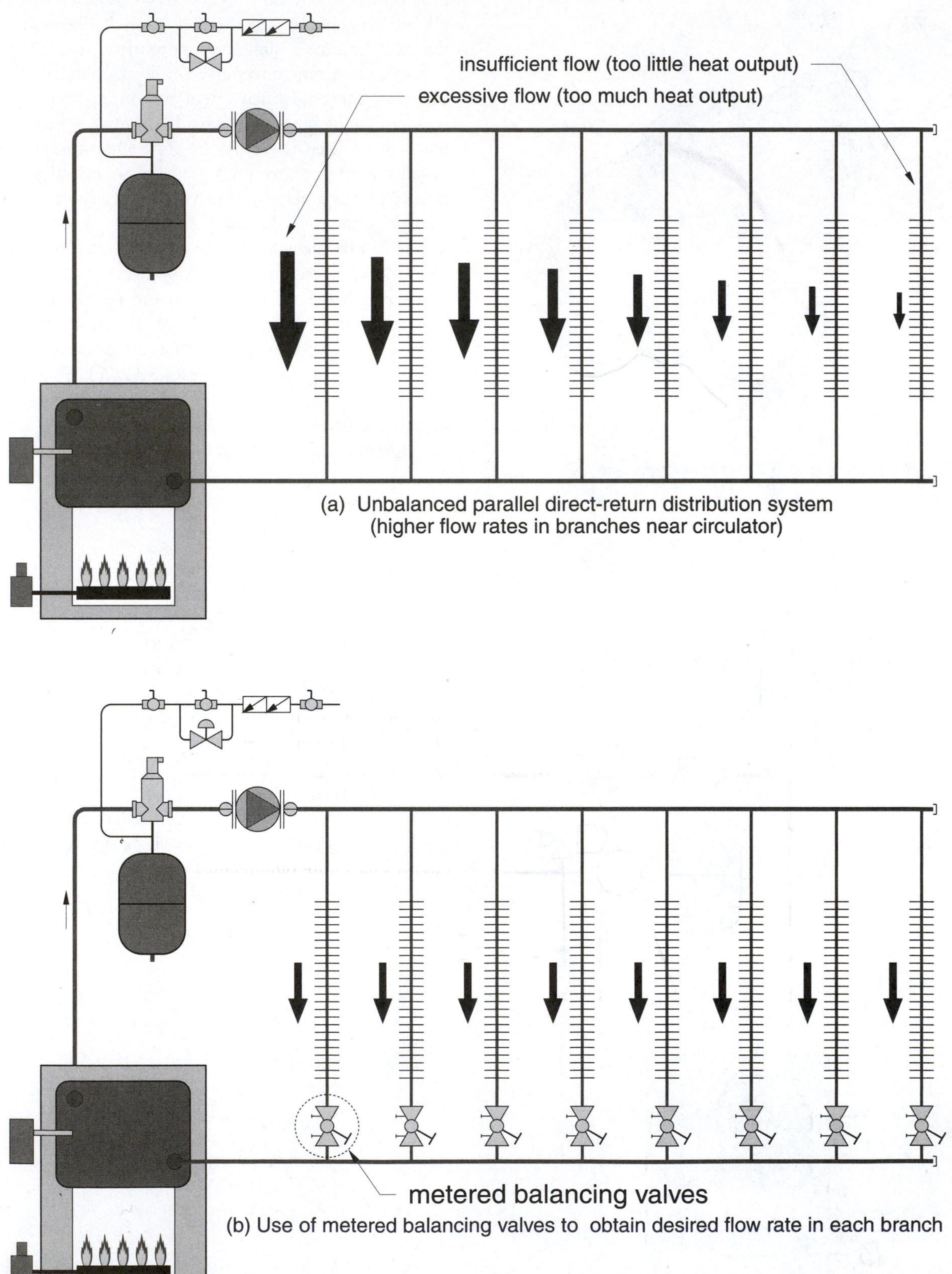

Figure 5–75 Use of metered balancing valves to adjust flow rate in parallel piping paths. (a) Situation without balancing valves, (b) situation after balancing valves are installed and adjusted.

Figure 5–76 Cutaway view of a metered balance valve. *Courtesy of Armstrong pumps.*

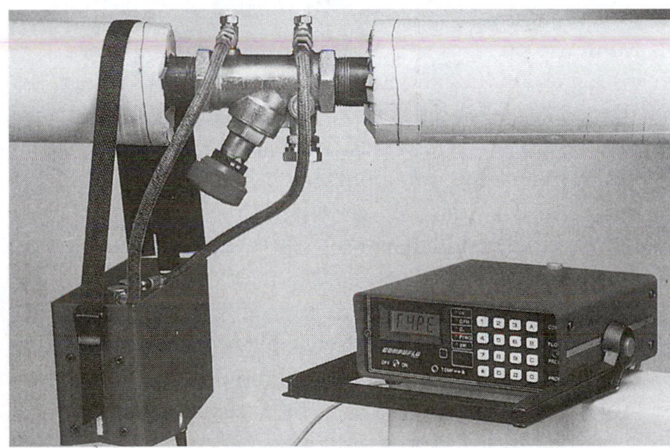

Figure 5–77 Example of a digital manometer system for use with metered balancing valves. *Courtesy of Armstrong Pumps.*

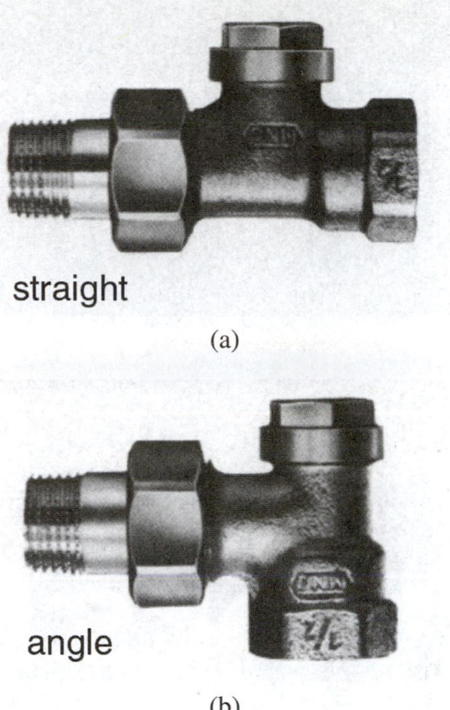

Figure 5–78 Straight and angle versions of a lockshield/balancing valve. *Courtesy of Regin HVAC Products, Inc.*

Lockshield Valves

Another specialty valve created for the hydronics industry provides the ability to isolate, balance, and even drain individual heat emitters. This valve, available in both straight and angle patterns as shown in Figure 5–78, is called a **lockshield valve.**

To adjust the flow setting of the valve, the cap of the valve is removed, and an internal brass plug is closed using an Allen key. A retaining ring that limits the travel of this plug is then adjusted using a screwdriver. When the valve is opened, the plug is rotated open until it contacts the bottom edge of the retaining ring. This design allows the plug to be closed and opened repeatedly without affecting the valve's flow rate setting.

The valve can also be used to drain water from a heat emitter. This is done by screwing a specialized drainage tool onto the valve body as shown in Figure 5–79. The shaft of this tool engages the entire plug/retainer ring assembly and allows it to be positioned so that fluid will drain through the valve. After servicing, the heat emitter can be refilled by pumping fluid back in through this tool. The plug/retainer ring assembly is then screwed back into its original position in the valve body. Lockshield valves are equipped with a union that allows the valve and heat emitter to be quickly separated.

5.9 SCHEMATIC SYMBOLS FOR PIPING COMPONENTS

Piping schematics are the road maps of hydronic system design. They describe the relative placement of dozens (sometimes hundreds) of components that must be assembled to build a smooth operating system. Just as various geographic features are represented by small markings on a map, the components in a piping system can be represented by schematic symbols.

Unfortunately, there is no universal standard for piping symbols to which all those in the hydronics industry sub-

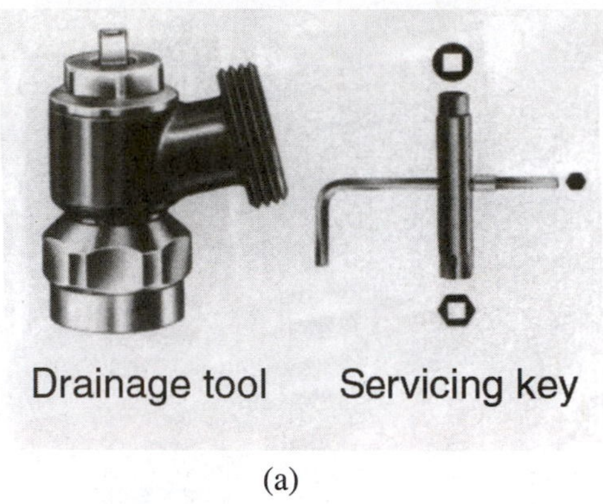

(a)

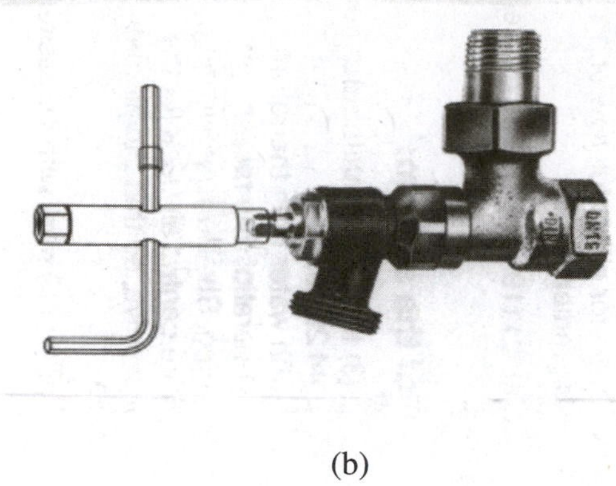
(b)

Figure 5–79 Special tool for draining a heat emitter through a lockshield/balancing valve, (a) drainage tool and key, (b) drainage tool mounted on lockshield/balancing valve. *Courtesy of Regin HVAC Products, Inc.*

scribe. Variations among designers, drafters, etc., are quite common. *Systems designers should establish a consistent set of symbols and identify those symbols using a legend on all drawings in which they are used.* When this is not done, the designer's symbol for a globe valve may be interpreted as a gate valve by the installer and so forth. This can create expensive misunderstandings when the job is bid or installed.

Within this text, all piping schematics will use symbols based on the legend shown in Figure 5–80. Some of the components shown on this legend have not been discussed at this point, but will be in later chapters. This symbol legend is also shown as Appendix A.

The piping schematics shown in this text were all prepared using a simple **computer-aided drafting (CAD)** program. There are many such programs available for both PC and Macintosh computers. The starting point is the creation of a **symbol palette** similar to that shown in Figure 5–80.

Those who will be preparing piping schematics as well as other types of documentation for hydronic heating installations are highly encouraged to use such programs. They allow professional-looking drawings to be quickly prepared for design, installation, and marketing purposes.

5.10 TIPS ON PIPING INSTALLATION

As with any trade, the installation of piping requires an attitude of craftsmanship. Well planned and installed piping systems convey a message of professionalism. The planning of efficient piping layouts is a skill acquired through experience. It is part art and part science. The following guidelines will help in planning and installing good piping layouts.

- As a rule, piping should always be kept vertical (plumb) or horizontal (level). This ensures proper alignment of fitting joints and other components that join into the piping path. The exception is when a pipe must be pitched for drainage. In this case, the pipe should have a minimum slope of 1/4 inch per foot of horizontal run. In either case, use levels and/or plumb bobs to keep the piping runs accurate.
- Plan piping layouts that minimize joints and fittings. If possible, use 20-foot sections of straight pipe on long runs to minimize couplings.
- Minimize changes in the direction of piping. Do not zigzag around other components. This adds to fitting count and makes the system look more complicated than necessary. It also wastes pumping energy.
- Try to align similar components in parallel piping branches. For example, when installing three-zone valves side by side, make sure they are aligned with each other.
- Always consider the need to remove components for servicing. Do not locate serviceable components in small, hard to get to areas, or behind other equipment.
- Always provide support adjacent to components supported by the piping such as circulators, expansion tanks, or other relatively heavy devices.
- Plan piping assemblies that can be put together at a vise or flat working surface to minimize awkward working positions.
- Always allow for the take-up distances of threaded or soldered fittings when measuring and cutting a length of tubing. See the tables and figures in this chapter for the proper take-up allowances.
- In situations where both piping and electrical wiring are in close proximity, *try to keep the piping below the electrical wiring.* This minimizes the chance of water dripping onto electrical components during service, or if a leak should develop.
- Lay out piping runs using the centerline of the piping rather than its edges. This ensures alignment between piping and fixed connection points, especially when different pipe sizes are used.

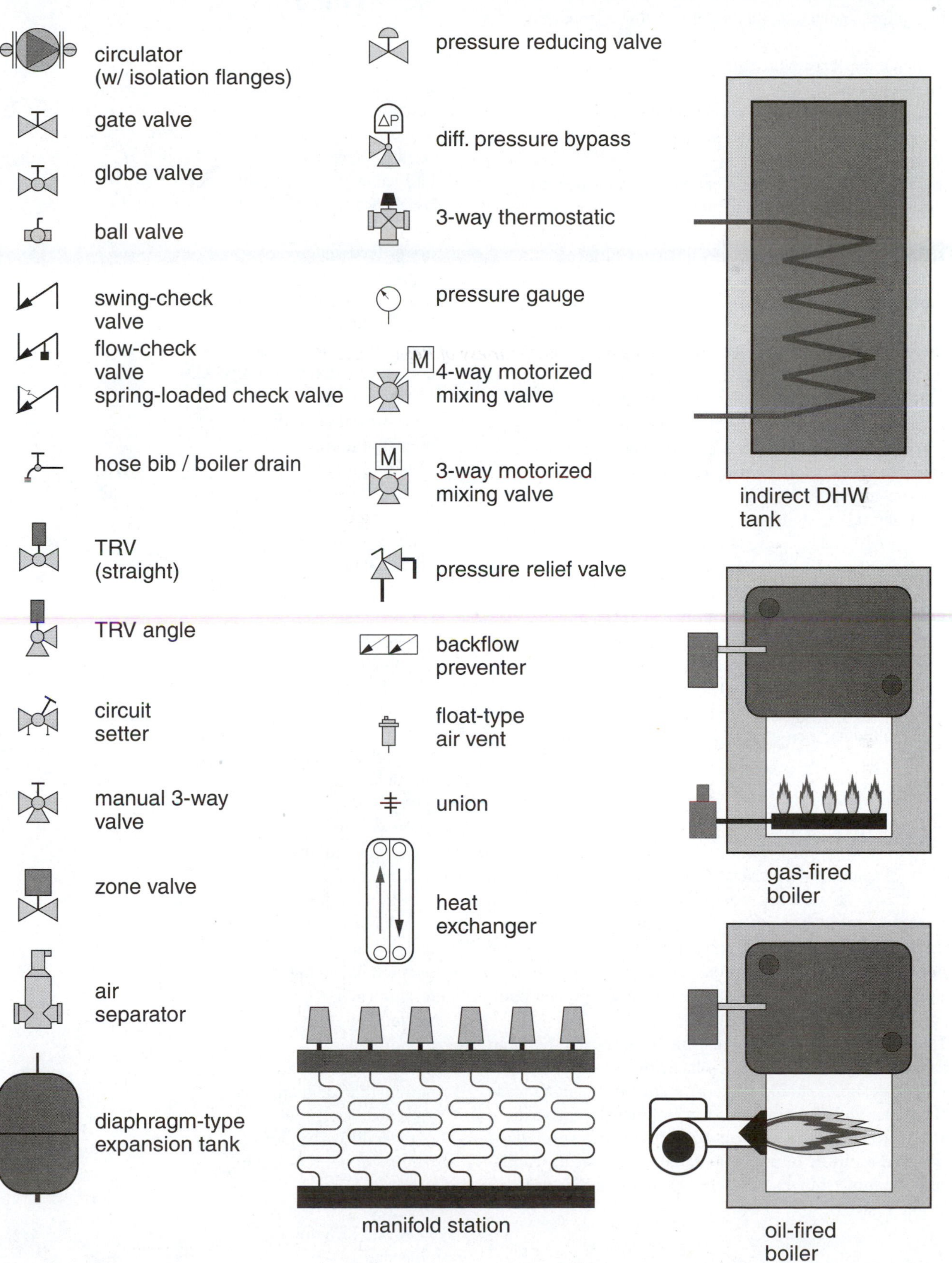

Figure 5–80 Legend of symbols for the piping schematics used in this book

- Plan the piping layout with respect to major pieces of equipment such as the boiler, circulator, mixing valve, and so on. The positions of these devices serve as benchmarks to which interconnecting piping can be referenced.
- Secure all piping to structural surfaces to prevent sagging and to support the weight of components supported by the piping.
- Be sure to allow for piping expansion movement as discussed earlier in this chapter.
- Where insulation will cover the piping, allow sufficient space behind the pipe and any adjacent surface.
- When piping is run through floor joists or other framing members, always drill holes at least 1/4 inch larger than the outside diameter of the piping. This helps prevent the pipe from binding along the framing as it expands and contracts. Use a chalk line during layout to keep the holes precisely aligned. Plan ahead when long piping runs through many floor joists are required.
- The optimum location for a hole through a framing member such as a floor joist is halfway between the bottom edge and top edge. This location minimizes loss of structural strength. *Never notch into the bottom of floor joists, ceiling joists, or trusses to accommodate piping.* Whenever possible, keep the edge of the hole at least two inches in from the edge of framing members to avoid potential punctures from nails or screws.
- Never mount circulators and adjacent piping to surfaces that can amplify vibrations, especially in proximity to living space. If possible, fasten the mounting hardware for circulators and adjacent piping to masonry or concrete walls to reduce the potential for vibration transmission.

5.11 SUMMARY

This chapter has described both basic and specialized piping hardware used for piping hydronic heating systems. The many combinations of piping, fittings, and valves provide almost unlimited flexibility for the creative system designer. Standard piping approaches that might prove cumbersome in certain situations can be replaced by novel combinations of these components.

Unavoidably, not every valve available has been shown or discussed. New variations of the basic designs appear regularly. The designer should make a practice of scanning trade publications and Internet sites for new developments on a regular basis. Keep a file on both basic and specialized valves, and a variety of piping materials. They are the fundamental building blocks of both standard and innovative hydronic systems.

KEY TERMS:

3-way mixing valves
3-way thermostatic mixing valve
4-way mixing valves
actuator
angle valve
backflow preventer
ball valve
baseboard tee
boiler drain
boiler feed water valve
capillary tube
check valve
coefficient of linear expansion
composite tubing
computer-aided drafting (CAD)
copper water tube
cross-linked polyethylene
dielectric union
differential pressure bypass valve (DPBV)
diffuse
diverter tee
diverter valve
domestic water
end switch
expansion loop
expansion offset
female threads
fittings
flow-check valve
flux
full port ball valve
galvanic corrosion
gate valves
globe valves
hard drawn tubing
heat motors
injection mixing
lockshield valve
male threads
metered balancing valves
mixing valve
modulating
monoflo tees®
national pipe thread
nominal inside diameter
one-pipe systems
oxygen diffusion barrier
parallel direct return piping
parallel piping
PEX tubing
PEX-AL-PEX tubing
pipe size
polymer

pressure-relief valve
"soft temper" tubing
soldering
spool
spring-loaded check valve
standard port ball valve
swing check
symbol palette
thermal expansion
thermoplastics
thermoset plastic
thermostatic operator
thermostatic radiator valves (TRV)
working pressure rating
zone valve
zones

CHAPTER 5 QUESTIONS AND EXERCISES

1. Describe the differences between type K, L, and M copper tubing.
2. Describe the difference between copper water tube and ACR tubing.
3. What type of solder should be used for piping conveying domestic water? Why?
4. A straight copper pipe 65 ft. long changes temperature from 65 °F to 180 °F. What is the change in length of the tube? Describe what would happen if the tube was cooled from 65 °F to 40 °F.
5. A hydronic system contains 50 3/4-inch, and 25 1-inch soldered copper joints. Estimate the number of inches of 1/8-in. wire solder required for these joints.
6. Describe a situation where two diverter tee fittings would be preferred to a single diverter tee.
7. Write a standard description for the fittings shown in Figure 5–81.
8. What should be done prior to soldering a valve with a nonmetallic disc or washer?
9. What is the advantage of a four-port mixing valve over a three-port mixing valve when used with a conventional gas- or oil-fired boiler?
10. Describe a situation where the high thermal conductivity of copper tubing is *undesirable.*
11. Describe a situation where a differential pressure bypass valve installed in a multizone hydronic distribution system begins to open.
12. Describe three separate functions of a lockshield/balancing valve.
13. Why should the thermostatic actuator of a radiator valve not be mounted above the heat emitter?
14. Why should a globe valve not be used for component isolation purposes? What type(s) of valves are better suited for this application?
15. What is an advantage of a spring-loaded check valve when compared to a swing check valve? What is a disadvantage?
16. What is the difference between a full port ball valve and conventional port ball valve?
17. Name two applications for a 3-way zone valve.

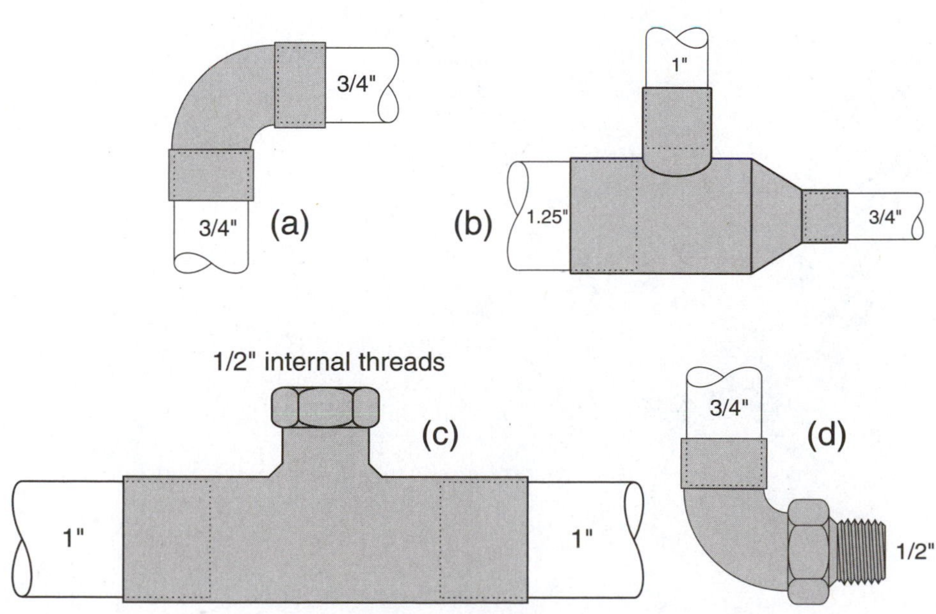

Figure 5–81 Fittings for use in Question 7.

18. Define nominal pipe size.
19. Does the presence of an oxygen diffusion barrier on polymer tubing totally prevent oxygen entry into the system? Justify your answer.
20. What is the function of the end switch in a zone valve?
21. When can a swing check valve offer the same protection as a backflow preventer in the make-up water line of a hydronic system?

FLUID FLOW IN PIPING 6

OBJECTIVES:

After studying this chapter, you should be able to:

- Describe the difference between flow rate and flow velocity
- Explain how the pressure of a liquid varies with its depth
- Define the term *head* and relate it to pressure differential
- Explain what hydraulic resistance is and how it is used
- Use the C_V factor of a valve to calculate pressure drop
- Calculate the hydraulic resistance of piping, fittings, and valves
- Sketch hydraulic resistance diagrams for a given piping layout
- Convert two or more parallel hydraulic resistances into an equivalent resistance
- Reduce complex piping systems into simpler equivalent diagrams
- Plot the system resistance curve of a given piping system
- Explain several considerations that relate to pipe sizing
- Estimate the operating cost of a given piping system

6.1 INTRODUCTION

An understanding of how fluids behave while moving through piping is essential to designing good hydronic systems. This chapter lays the groundwork for describing fluid flow in piping. It introduces new analytical methods for predicting flow rates in all parts of a piping system, even those with complex branches and many components. These methods will become a routine part of overall system design that will be used many times in later chapters.

Unfortunately, experience has shown that many of the people responsible for design and installation of hydronic systems do not understand the principles of fluid flow in piping well enough to scrutinize designs before committing time and materials to their installation. This is especially true of nonstandard projects for which there is no previous example to reference. This often leads to hesitation in trying anything beyond what is already familiar. Many existing hydronic systems could have been better matched to their buildings, installed for less money, or operated at a lower cost if their designer had had some way of evaluating how a slight change in system design would affect performance. This chapter gives designers the analytical tools needed to simulate the behavior of hydronic piping systems *before* committing time and money to their installation.

Many of the concepts in this chapter are drawn from classical fluid mechanics, a subject not unknown for its share of mathematics. However, much effort has gone into distilling this theory into concise and easy to work with design tools.

To further simplify use of this material, the Hydronics Design Studio software contains a module called Hydronic Circuit Simulator that can quickly evaluate the equations and procedures presented in this chapter. This allows the designer to experiment with the many factors involved, and develop a feel for how piping systems behave under a variety of conditions.

6.2 BASIC CONCEPTS OF FLUID MECHANICS

What Is a Fluid?

In a purely technical sense, the word fluid can mean *either* a **liquid**, such as water, or a **gas**, such as air. The science of fluid mechanics includes the study of both liquids and gases. In this book, the word fluid always refers to a liquid unless otherwise stated.

Most liquids, including water, are **incompressible**. This means that any given amount of liquid always occupies the same volume, with the exception of minor changes in volume due to thermal expansion. A simple example would be a long pipe completely filled with water. If another gallon of water were pushed into one end of the pipe, the exact same amount would have to flow out the other end. No additional water can be squeezed into a rigid container (such as a pipe) that is already filled.

By comparison, if the same pipe were filled with a gas such as air, it would be possible to push another gallon of air into the pipe without allowing any air out the other end. The result would be an increase in the pressure of the air within the pipe. Because of this, gases are said to be **compressible fluids**.

The fact that liquids are incompressible makes it easy to describe their motion though closed containers such as piping systems. Two basic terms associated with this motion are **flow rate** and **flow velocity**.

Flow Rate

Flow rate is a measurement of the *volume* of fluid that passes a given location in a pipe in a given period of time. In the North American hydronics industry, the customary units for flow rate are U.S. *gallons per minute* (abbreviated as gpm). Within this book, flow rate is represented in equations by the symbol f. The flow rate of a fluid through a hydronic system has a major effect on that system's thermal performance.

Flow Velocity

The motion of a fluid flowing through a pipe is more complex than that of a solid object moving in a straight path. The speed, or flow velocity of the fluid, is different at different points across the diameter of the pipe.

To visualize the flow velocity of a fluid, think about several small particles that have been positioned across the cross section of a pipe, and released at the same instant to be carried along with the flow. The particles pushed along by the flow would move faster near the center of the pipe than near the edge. The speed of each particle could be represented by an arrow as shown in Figure 6–1a. A curve connecting the tips of all these arrows is called a **velocity profile**. The velocity profile shown in Figure 6–1a is two-dimensional. However, if one imagines this curve rotated around the centerline axis of the pipe, it forms a "nose cone" shape that represents a three-dimensional velocity profile.

The term flow velocity is commonly understood to mean the **average flow velocity** of the fluid as it moves through a pipe. This average velocity would be the velocity that, if present across the entire cross section of the pipe, would result in the same flow rate as that created by the actual velocity profile. The concept of average flow velocity is illustrated in Figure 6–1b. In this book, the term *flow velocity* will always mean the *average flow velocity* in the pipe. The common units for flow velocity in North America are feet per second, abbreviated as either ft/sec or FPS. Within equations, the average flow velocity is represented by the symbol, v.

Equation 6.1 can be used to calculate the average flow velocity associated with a given flow rate in a round pipe.

(Equation 6.1)

$$v = \left(\frac{0.408}{d^2}\right)f$$

where:

v = average flow velocity in the pipe (ft/sec)
f = flow rate through the pipe (gpm)
d = exact inside diameter of pipe (inches)

Example 6.1: What is the average flow velocity of water moving at 6 gpm through a pipe with an inside diameter of 0.811 inches?

Solution: Substituting these numbers into Equation 6.1 yields:

$$v = \left(\frac{0.408}{d^2}\right)f = \left(\frac{0.408}{0.811^2}\right)6 = 3.72 \text{ ft/sec}$$

When selecting a pipe size for a given flow rate, the resulting average flow velocity should be calculated. *A pipe*

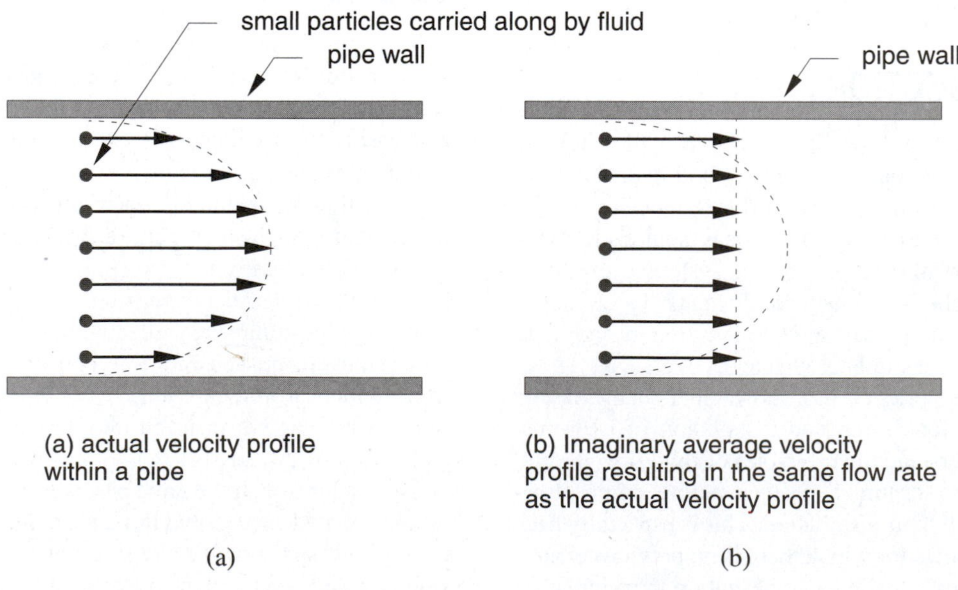

Figure 6–1 (a) Representation of actual velocity profile, (b) equivalent average velocity profile.

size should be selected that keeps the average flow velocity between 2 and 4 feet per second.

The lower end of this velocity range is based on the ability of flowing water to move air bubbles along a vertical pipe. Average flow velocities of 2 feet per second or higher can entrain air bubbles that are in a downward flow. The goal is to route such bubbles to an air separator where they can be collected and discharged from the system.

The upper end of this range (4 feet per second) is based on minimizing noise generated by the flow. Average flow velocities above 4 feet per second can cause flow noise and should be avoided.

Figure 6–2 tabulates the results of applying Equation 6.1 to type M copper tubes in sizes from 3/8 inch to 3 inch. Each equation can be used to calculate the flow velocity associated with a given flow rate. Also given are the flow rates corresponding to average flow velocities of 2 feet per second and 4 feet per second.

Tube size (type M copper)	Velocity versus flow rate equation*	Flow rate at 2 ft/sec (gpm)	Flow rate at 4 ft/sec (gpm)
3/8"	$v = 2.02\,f$	1.0	2.0
1/2"	$v = 1.26\,f$	1.6	3.2
3/4"	$v = 0.62\,f$	3.2	6.5
1"	$v = 0.367\,f$	5.5	10.9
1.25"	$v = 0.245\,f$	8.2	16.3
1.5"	$v = 0.175\,f$	11.4	22.9
2"	$v = 0.101\,f$	19.8	39.6
2.5"	$v = 0.0655\,f$	30.5	61.1
3"	$v = 0.0459\,f$	43.6	87.1

*v = Average flow velocity (feet / second)
f = Flow rate (gpm)

Figure 6–2 Equations relating flow rate (f), and flow velocity (v) for type M copper tube. Flow rates corresponding to flow velocities of 2 feet/second and 4 feet/second are also given.

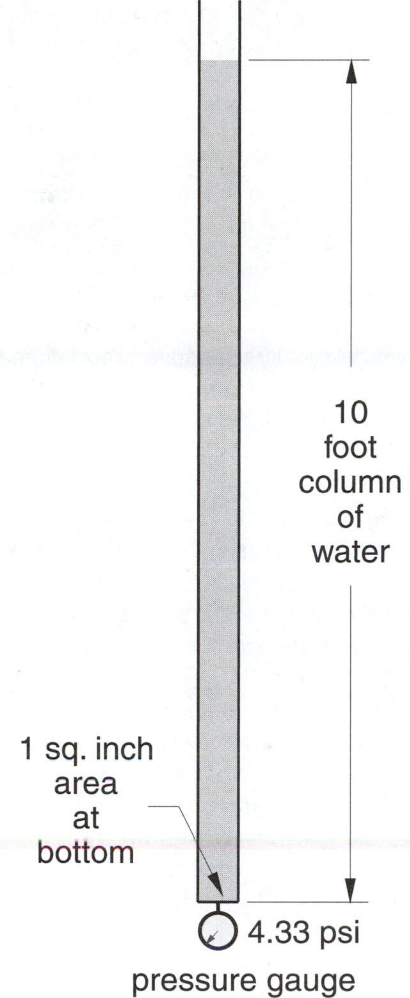

Figure 6–3 Column of water 10 feet high in a pipe with a cross-sectional area of 1 in^2.

Static Pressure of a Liquid

Consider a pipe with a cross-sectional area of 1 square inch, filled with water to a height of 10 feet, as shown in Figure 6–3.

Assuming the water's temperature is 60 °F, its density is about 62.4 lb/ft^3. The weight of the water in the pipe could be calculated by multiplying its volume by its density.

$$\text{Weight} = (\text{volume})\,(\text{density})$$
$$= (\text{cross sectional area})\,(\text{height})\,(\text{density})$$

$$\text{Weight} = (1\text{ in}^2)(10\text{ ft.})\left(62.4\frac{\text{lb}}{\text{ft}^3}\right)\left(\frac{1\text{ ft}^2}{144\text{ in}^2}\right) = 4.33\text{ lb}$$

The water exerts a pressure on the bottom of the pipe due to its weight. This pressure is equal to the weight of the water column divided by the area of the pipe. In this case:

$$\text{Static pressure} = \frac{\text{weight of water}}{\text{area of pipe}} = \frac{4.33\text{ lb}}{1\text{ in}^2}$$
$$= 4.33\frac{\text{lb}}{\text{in}^2} = 4.33\text{ psi}$$

This pressure is called the **static pressure** exerted by the water. The word static means the fluid is not moving. In this case the static pressure at the bottom of the pipe is due solely to the weight of the water. If an accurate pressure gauge were mounted into the bottom of the pipe as shown in Figure 6–3, it would read exactly 4.33 psi.

Next, consider a pipe with a cross-sectional area of 1 *square foot* filled with water to a height of 10 feet. The weight of this column can be calculated as:

$$\text{Weight} = (1\ \text{ft}^2)(10\ \text{ft})\left(62.4\frac{\text{lb}}{\text{ft}^3}\right) = 624\ \text{lb}$$

The static pressure at the bottom of the column would be:

$$\text{Pressure} = \frac{\text{weight of water}}{\text{area of pipe}} = \frac{624\ \text{lb}}{144\ \text{in}^2}$$

$$= 4.33\frac{\text{lb}}{\text{in}^2} = 4.33\ \text{psi}$$

Notice the static pressure at the bottom of the column did not change even with a large change in the diameter of the pipe. This is explained by the fact that *the larger column of water has a proportionally larger area across which to distribute its weight.* Thus the static pressure at a given depth below the surface of a liquid remains exactly the same regardless of the diameter of the pipe.

This same principle remains true for liquids within containers of any size and shape, as shown in Figure 6–4. Scuba divers, for example, feel a static pressure that depends only on their depth below the surface of the water, not on how wide the body of water is.

The static pressure of a liquid increases proportionally from some value at the surface to larger values at greater depths below the surface as shown in Figure 6–5. Again, *the size or shape of the container makes no difference.*

If the top of the system shown in Figure 6–5 were closed off, and a tank containing pressurized air at 5 psi were attached at the top, the static pressure of the liquid at all locations simply increases by 5 psi as shown in Figure 6–6.

From these observations it is possible to develop a relationship between the depth of water within a container and the static pressure it exerts at any point below its surface.

The following approximation is useful for systems containing water:

(Equation 6.2)

$$P_{\text{static}} = \frac{h}{2.31} + P_{\text{surface}}$$

where:

P_{static} = static pressure at a given depth, h, below the surface of the water (psi)

h = depth below the water's surface (ft)

2.31 = unit conversion factor for water at about 60 °F

P_{surface} = any pressure applied at the surface of the water (psi)

If the water is open to the atmosphere at the top of the container, even through a pin-hole size opening, then $P_{\text{surface}} = 0$. If the container is completely closed and filled with water, P_{surface} is the pressure of the water at the top of the container.

The reason Equation 6.2 is an approximation is that the density of water, and thus the static pressure it creates within a container, changes with temperature. To account for

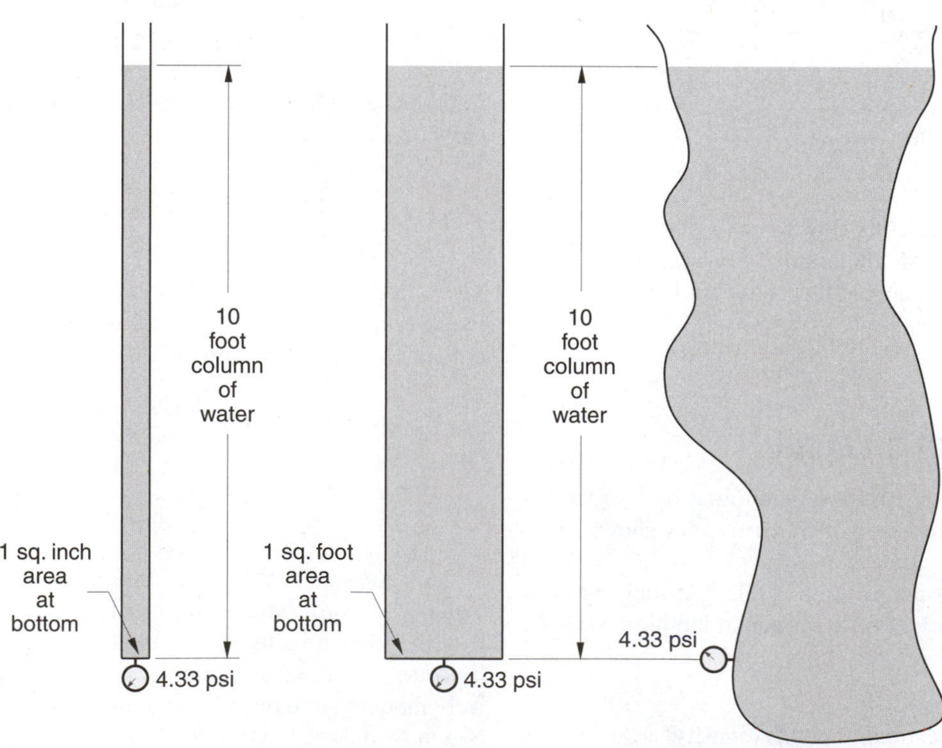

Figure 6–4 The static pressure of a fluid at a given point depends only on depth below the surface, and not on the size or shape of the container.

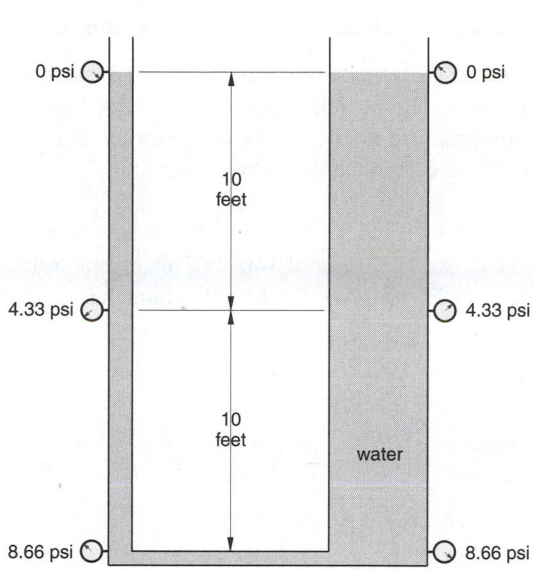

Figure 6–5 Static pressure for various depths below the fluid surface. Notice that the size of the column, or shape of the overall container, has no effect on static pressure.

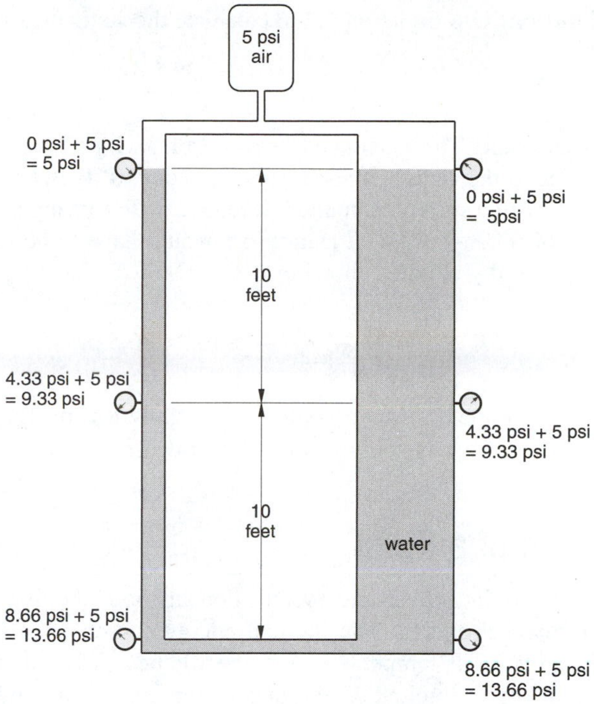

Figure 6–6 The effect of increasing the pressure at top of the piping circuit. All pressures below the surface increase by the same amount.

such changes, and to allow for calculations with liquids other than water, Equation 6.2 can be modified as follows:

(Equation 6.3)

$$P_{static} = \frac{(h)(D)}{144} + P_{surface}$$

where:

P_{static} = static pressure at a given depth, h, below the surface of the liquid (psi)

h = depth below the liquid's surface (ft)

D = density of the liquid (lb/ft^3)

$P_{surface}$ = any pressure applied at the surface of the liquid (psi)

To use Equation 6.3, the density of the liquid must be determined. The density of water over the temperature range of 50 °F to 250 °F can be found in Figure 4–3. The density of several other liquids commonly used in hydronic heating systems can be determined over this same temperature range using the Fluid Properties Calculator module in the Hydronics Design Studio software

Example 6.2: Find the approximate static pressure at the bottom of a hydronic system with an overall height of 65 feet, filled with cool water, and having a pressure of 10 psi at its uppermost point. Refer to Figure 6–7.

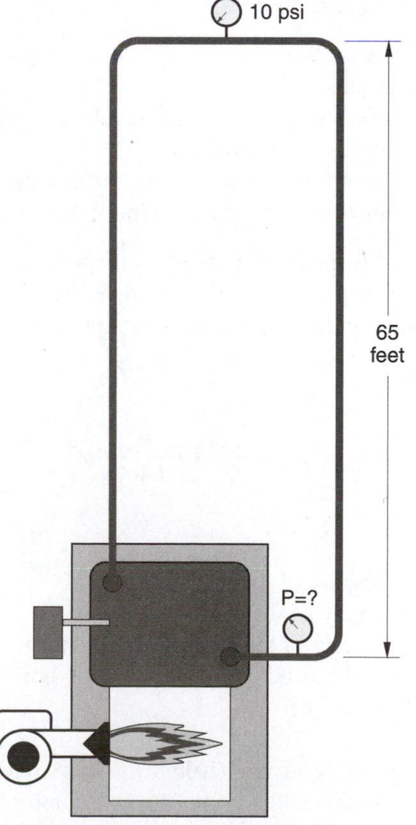

Figure 6–7 Piping system for Example 6.2

Solution: Use Equation 6.2 to calculate the static pressure:

$$P_{static} = \frac{65}{2.31} + 10 = 38.1 \text{ psi}$$

Discussion: The calculated pressure of 38.1 psi would be correct if the water temperature were about 60 °F. If, however, the same system contained water at 250 °F (having a density of 58.754 lb/ft³), Equation 6.3 would have to be used, and would yield the following:

$$P_{static} = \frac{(65)(58.75)}{144} + 10 = 36.5 \text{ psi}$$

Notice the hot water would exert slightly less pressure at the base of the system due to its lower density.

Head of a Fluid

Fluids in a hydronic system contain both thermal and mechanical energy. The thermal energy content of a fluid depends on its temperature and specific heat. We can sense the relative amount of thermal energy in a fluid by its temperature. For example, the hot water leaving a boiler contains more thermal energy than cooler water returning from the distribution system.

The mechanical energy, or head, present in a fluid depends on several factors including its pressure, density, elevation, and velocity at some point in the system. It is helpful to divide head into three categories:

- **Pressure head**, which is the mechanical energy a fluid contains because of its pressure
- **Velocity head**, which the mechanical energy a fluid contains because of its velocity
- **Elevation head**, which is the mechanical energy the fluid contains because of its height in the system

Total head is the sum of the pressure head, velocity head, and elevation head. You could calculate the total head of a fluid at some point in a hydronic system using Equation 6.4:

Equation 6.4

$$\text{Total head} = \frac{P(144)}{D} + \frac{v^2}{64.4} + Z$$

where:

P = the pressure of the fluid (psi)
D = the density of the fluid (lb/ft³)
v = the velocity of the fluid (feet/second)
z = the height of the point above some horizontal reference (feet)

The units for head are (ft•lb/lb). Those having studied physics will recognize the unit of ft•lb as a unit of energy. As such, it can be converted to any other unit of energy such as a Btu. However, engineers long ago chose to cancel the units of pounds (lb.) in the numerator and denominator of this ratio, and express head in the sole remaining unit of feet. To make a distinction between feet as a unit of distance and feet as a unit of fluid energy, the latter can be stated as **feet of head**.

Example 6.3: Calculate the total head of the water at a location in the system which is 3 feet above floor level, where the static pressure is 15 psi, the temperature is 140 °F, and the fluid's velocity is at 4 feet per second.

Solution: Before using Equation 6.4, it is necessary to determine the density of water at 140 °F. This can be found using the graph in Figure 4–3. At 140 °F, water has a density of 61.4 lb/ft³.

Substituting these numbers into Equation 6.4 yields:

$$\text{Total head} = \frac{P(144)}{D} + \frac{V^2}{64.6} + Z = \frac{15(144)}{61.4} + \frac{(4)^2}{64.4} + 3$$

$$= 38.4\frac{\text{ft} \cdot \text{lb}}{\text{lb}} = 38.4 \text{ ft}$$

Liquids have the ability to exchange head energy back and forth between velocity head, pressure head, and elevation head. For example, when a liquid flows through a transition from a smaller to larger pipe size, its velocity decreases, while its pressure increases slightly because some of the velocity head is instantly converted into pressure head.

Relationship between Head and Pressure Change

When head is added to or removed from a liquid in a closed-loop piping system, there will always be an associated pressure change in the fluid. Just as a change in temperature is "evidence" of a gain or loss of thermal energy, a change in pressure is evidence of a gain or loss in head. When head is lost, pressure decreases. When head is added, pressure increases. This concept is illustrated in Figure 6–8.

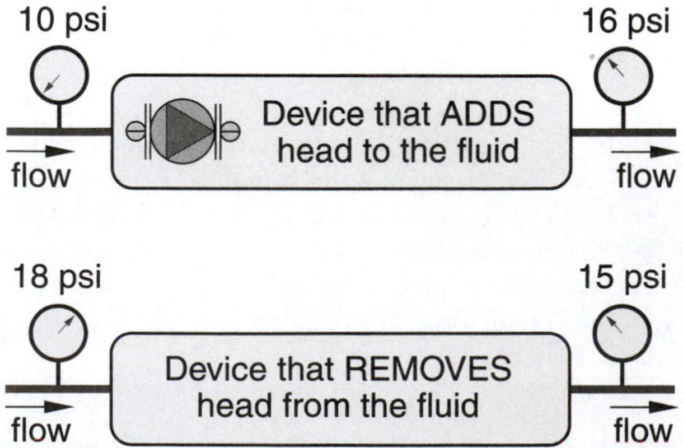

Figure 6–8 Pressure changes associated with head that is added or removed from a flowing fluid.

Using pressure gauges to detect changes in the head of a liquid is like using thermometers to detect changes in the thermal energy content of that liquid. Equation 6.5 can be used to convert the observed change in pressure to the associated gain or loss of head energy. This equation requires the density of the liquid, which depends on its temperature.

(Equation 6.5)

$$H = \frac{144(\Delta P)}{D}$$

where:

H = head added or lost from the liquid (feet of head)

ΔP = pressure change corresponding to the head added or lost (psi)

D = density of the liquid at its corresponding temperature (lb/ft^3)

This equation can also be rearranged to calculate the pressure change associated with a given gain or loss in head:

(Equation 6.6)

$$\Delta P = \frac{HD}{144}$$

Example 6.4: Water at 140 °F flows through a heat emitter. The observed pressure on the inlet side is 20 psi. The observed pressure on the outlet side is 18.5 psi. What is the change in the head energy of the water as it passes through?

Solution: The density of water at 140 °F is 61.4 lb/ft^3. Substituting this and the pressure data into Equation 6.5 yields:

$$H = \frac{144(20 - 18.5)}{61.4} = 3.52 \text{ ft of head}$$

Discussion: Since the pressure decreases in the direction of flow, this is a head loss. If the fluid cools as it passes through the heat emitter, its density slightly increases. However, the variation in density is very small for the 10 °F to 30 °F temperature drop that occurs across a typical heat emitter, and thus it has an insignificant effect on the calculated change in head.

Head Loss Due to Viscous Friction

In an ideal system in which there was no fluid friction, the total head of the fluid would remain constant at all locations. Once set in motion around the circuit by the addition of head energy, such a fluid would continue circulating without further head energy input. There would be no head loss. Such a scenario would constitute perpetual motion, which is impossible.

Whenever a real fluid flows, an energy dissipating effect called **viscous friction** develops both within the fluid stream and along any surfaces the fluid contacts. This friction causes head energy to be converted to (or "dissipated" into) thermal energy.

In the strictest sense, no energy is lost in this process. Instead, a higher quality form of energy (head) is converted to a lower quality form of energy (heat). Before this transformation, the energy in higher quality form (e.g., head) helped move the fluid through the piping system. After being converted to heat by viscous friction, this is no longer possible. If the transformation takes place in a heated space, one could argue that the heat generated by head being dissipated into heat does help heat the space. However, in a well-designed hydronic heating system, the amount of heat produced by head loss is a tiny fraction of the thermal energy carried along by the water.

Viscous friction develops within all fluids, but is greater for fluids with higher viscosities, or when a fluid flows over relatively rough surfaces.

The drag forces associated with viscous friction is what causes fluid molecules closer to a surface to have lower velocities than those farther away from the surface. A familiar example is the slower speed of water near the banks of a stream compared to near the center of the stream. Another is lower wind speeds near the surface of the earth. Within pipes, viscous friction gives rise to a velocity profile similar to that shown in Figure 6–1a.

Because of its incompressibility, a liquid moving at a constant flow rate through a given size pipe cannot slow down as head is dissipated due to viscous friction. Therefore, the head loss must reveal itself as a drop in pressure in the downstream direction of flow as previously discussed. This concept is illustrated in Figure 6–9. Methods for calculating head loss will be detailed in Section 6.4, Hydraulic Resistance of Fittings, Valves, and Other Devices.

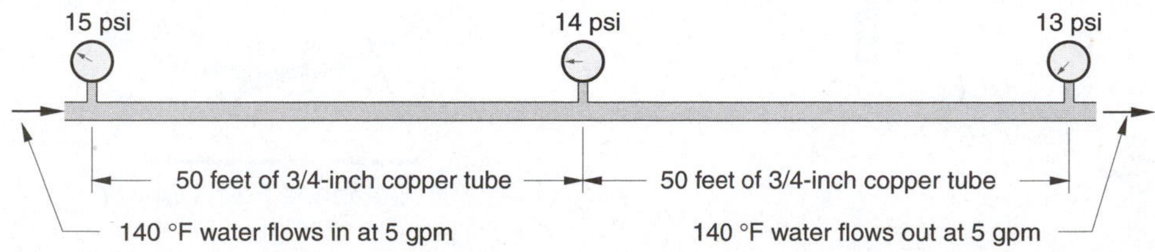

Figure 6–9 Pressure drop in downstream direction associated with head loss created by viscous friction. Note the pipe is horizontal, and thus the static pressure along its length is constant.

Fluid-Filled Piping Loops

Nearly all hydronic heating systems consist of closed piping systems. After all piping work is complete, these circuits are *completely* filled with fluid. During normal operation very little if any fluid enters or leaves the system.

Consider the fluid-filled piping loop shown in Figure 6–10. A static pressure is present at point A due to the weight of the fluid column on the left side of the loop. It might appear that a circulator placed at point A would have to overcome this pressure to lift the water and establish flow around the loop. This, however, is *not* true. The reason is the weight of the fluid on the right side of the loop pushes downward at point B with exactly the same static pressure. Thus, the static pressure on both the inlet and outlet of the circulator are the same. This is true regardless of the pipe size used in the upward flowing portion of the loop versus that in the downward flowing portion. It is also true regardless of the shape, size, or complexity of the loop. Remember, static pressure is not affected by the shape of the container, only the height of the fluid column.

The fluid in the filled loop acts like a Ferris wheel with the same weight in each seat. The weight in the seats that go up exactly balances the weight in the seats that go down. If it were not for friction in the bearings and air resistance, this balanced Ferris wheel would continue to rotate indefinitely once started. Likewise, if it were not for the viscous friction of the fluid, it too would continue to circulate indefinitely within a piping loop.

The following principle can thus be stated:

To maintain a constant rate of circulation in any fluid-filled piping loop, the circulator need only replace the head loss due to viscous friction. This remains true regardless of the shape, height, or pipe size(s) used in the loop.

This principle explains why a small circulator can establish and maintain flow in a filled piping loop, even if the top of the loop is several stories above the circulator, and contains hundreds, or even thousands of gallons of fluid. *Unfortunately, many circulators in hydronic systems are needlessly oversized because this principle is not understood.*

Flow Classifications

Fluid flow in pipes is characterized as being either **laminar** or **turbulent**. Both types of flow can enhance or detract from the performance of hydronic systems depending on where they occur.

To discuss the differences between these types of flow, it is helpful to use the concept of streamlines. A streamline indicates the path taken by an imaginary fluid particle as it moves along a pipe. Streamlines allow one to visualize flow as if the fluid were actually composed of millions of small particles, each with its own unique path.

Laminar flow is characterized by smooth, straight streamlines as the fluid moves through the pipe, as shown in Figure 6–11a. This type of flow is likely to exist when flow velocities are very low. To visualize laminar flow, think of the fluid

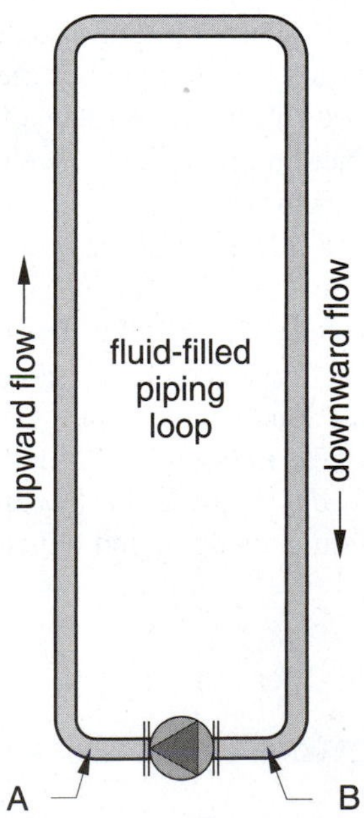

Figure 6–10 A closed loop piping system filled with fluid. The pressure at point A equals the pressure at point B regardless of the height or shape of the loop. To maintain circulation, the pump need only overcome the head loss due to viscous friction.

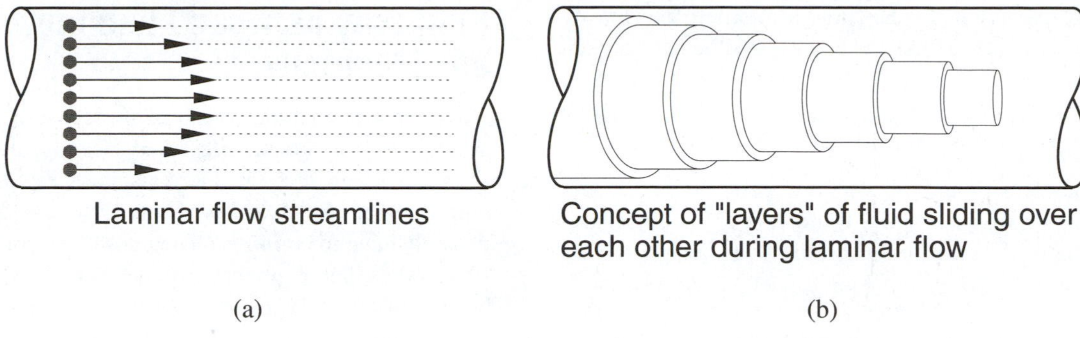

Figure 6–11 The concept of laminar flow. (a) velocity profile within a pipe, (b) concentric layers of fluid sliding over each other.

as if it were separated into many thin layers that resemble concentric cylinders inside a pipe as shown in Figure 6–11b. The closer a given fluid layer is to the inside surface of the pipe, the slower it moves. The faster moving fluid layer tends to slide over the slower moving layers in such a way that mixing between layers is minimal

Laminar flow creates minimal amounts of head loss. This is desirable because the lower head loss means less pumping power will be required to maintain a given flow rate. This is why large pipelines that carry heated or chilled water over hundred or thousands of feet between buildings are often designed to maintain laminar flow.

Another characteristic of laminar flow is that it allows a **boundary layer** to develop near the pipe wall. The boundary layer is a very slow-moving layer of fluid that slides along the inner wall of the tube with very little mixing effect. Such a layer creates a thermal resistance between the pipe wall and the bulk of the fluid stream, reducing the rate of heat transfer between them. Because of this, *laminar flow is undesirable if the objective is to move heat between the pipe wall and fluid stream.*

Turbulent flow is characterized by a vigorous mixing of the fluid as it travels down the pipe. The streamline of a single fluid particle shows it moves in an erratic path that often sweeps in toward the pipe wall, and then back out into the bulk of the fluid stream as illustrated in Figure 6–12. This mixing improves heat transfer because it drastically reduces the thickness and thermal resistance of the fluid's boundary layer. Because of this, heat exchangers, radiators, or any other device that is intended to transfer heat between a fluid and a solid surface should operate with turbulent flow. Fortunately, the conditions present in most small- to medium-size hydronic systems favor the development of turbulent flow.

An undesirable aspect of turbulent flow is that it creates more head loss compared to laminar flow, and therefore requires more pumping power to maintain a given flow rate. However, the increased heat transfer capabilities of turbulent flow are usually far more important than the penalty associated with the increased pumping power.

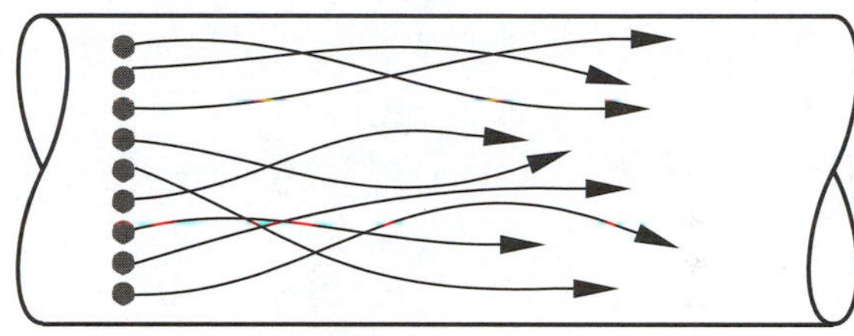

Figure 6–12 Representation of streamlines within turbulent flow.

Reynold's Number

A method exists for predicting if the flow within a pipe will be laminar or turbulent. This method, derived from experimental observations, is based on calculating the **Reynold's number** of the fluid. The calculated value is then compared to a threshold value of 4000. *If the calculated Reynold's number is over 4000, the flow is turbulent.* If it is below 4000, the flow may be either laminar or turbulent. If the Reynold's number is below 2300, the flow will be laminar.

The Reynold's number for flow in a pipe can be calculated using Equation 6.7:

(Equation 6.7)

$$\text{Re}\,\# = \frac{vdD}{\mu}$$

where:

- v = average flow velocity of the fluid (ft/sec)
- d = internal diameter of pipe (ft)
- D = fluid's density (lb/ft^3)
- μ = fluid's dynamic viscosity (lb/ft/sec)

Example 6.5: Determine the Reynold's number of water at 140 °F flowing at 5 gpm through a 3/4-inch type M copper tube. Is this flow laminar or turbulent?

Solution: To calculate the Reynold's number, the density and dynamic viscosity of the water must first be determined. Using the graph in Figure 4–3, the water's density is found to be 61.35 lb/ft^3. Using the graph in Figure 4–6, the water's dynamic viscosity is determined as 0.00032 lb/ft/sec.

(*Note:* The density and dynamic viscosity of water could also have been found using the Fluid Properties Calculator module in the Hydronics Design studio software.)

The inside diameter of a 3/4-inch type M copper tube is found in Figure 5–1 to be 0.811 inches. This must be converted to feet to match the stated units for Equation 6.7:

$$d = (0.811 \text{ in})\left(\frac{1 \text{ ft}}{12 \text{ in.}}\right) = 0.06758 \text{ ft}$$

To get the average flow velocity corresponding to a flow rate of 5 gpm, use the equation for a 3/4-inch copper tube from Figure 6–2:

$$v = (0.62)\,f = (0.62)\,5 = 3.106 \text{ ft/sec}$$

These values can now be substituted into Equation 6.7:

$$\text{Re \#} = \frac{(3.106 \text{ ft/sec})(0.06758 \text{ ft})(61.35 \text{ lb/ft}^3)}{0.00032 \text{ lb/ft/sec}} = 40{,}242$$

Discussion: This value is well above the threshold of 4000, and therefore the flow is turbulent. Note that when the quantities with the stated units are substituted into Equation 6.7, they cancel each other out. Thus, the Reynold's number is a unitless quantity. This must always be true, and serves as a check that the proper units are being used in Equation 6.7.

6.3 ANALYZING FLUID FLOW IN SMOOTH PIPES

This section introduces a method for predicting the head loss in hydronic systems assembled from "smooth" pipe such as copper tubing, PEX, PEX-AL-PEX, or other materials of comparable smoothness. Most residential and light commercial hydronic systems use such tubing. The smooth inside surface of such tubing creates less disturbance of the fluid stream, and therefore reduces head loss.

If major portions of a system will be piped with steel or iron piping, this method will somewhat underestimate head loss, and should not be used without suitable corrections.

This method is based on a combination of established engineering models for fluid flow in pipes, as well as methods borrowed from electrical circuit analysis. It is based on the premise that each component in a piping assembly or circuit has an associated **hydraulic resistance** that defines its ability to dissipate head from the fluid at any given flow rate.

Similarity between Fluid Flow and Electricity

To most people the concepts of electricity and fluid flow seem totally different. Electrical circuits involve concepts such as voltage, current, and resistance. Piping systems deal with flow rate, head, and pressure. What could these systems possibly have in common? The answer lies in the mathematical models that engineers have developed for describing the behavior of both types of systems.

Electrical Device Model

The classic model for describing the operating characteristics of an electrical device is Ohm's law. It is among the first concepts taught in a course on electrical circuits. Mathematically, Ohm's law can be stated as follows:

(Equation 6.8)

$$v = ir$$

where:

v = voltage drop *across* the device (volts)
r = electrical resistance of the device (ohms)
i = current flowing *through* the device (amperes)

Voltage can be thought of as the *driving force* that tries to push electrons through the device. It is measured with a voltmeter across the device as shown in Figure 6–13. Whenever a current flows through a resistance, there is an associated drop in voltage in the direction of current flow. This voltage drop can be calculated using Equation 6.8.

Current is a measurement of the rate of flow of the electrons through the device. Resistance is the ability of the device to oppose the flow of electrons.

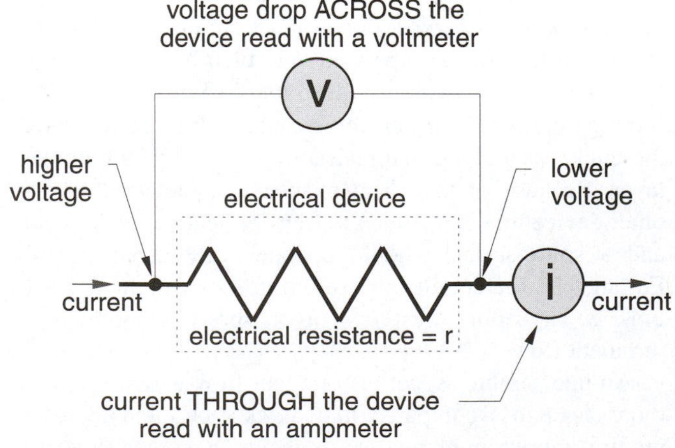

Figure 6–13 The concept of voltage, current, and resistance for an electrical device.

Piping Device Model

For the following discussion, the term *piping device* refers to any component such as a pipe, fitting, or valve through which fluid flows. The relationship between head loss, flow rate, and hydraulic resistance for such a piping device can be described by Equation 6.9:

(Equation 6.9)

$$H_L = r(f)^{1.75}$$

where:

H_L = head loss across a piping device due to viscous friction (ft of head)
r = hydraulic resistance of the piping device
f = flow rate of fluid through the piping device (gpm)
1.75 = exponent (or power) of f

The head loss can be thought of as the mechanical energy used to drive the fluid through the device at a certain flow rate. Recall that head loss reveals itself as a pressure drop across the device. Flow rate, f, is a measurement of the *rate of flow* of the fluid *through* the device. Hydraulic resistance is the ability of the device to oppose this flow. These concepts are illustrated in Figure 6–14.

It is interesting to compare the mathematical equation for electrical devices (Ohm's law) with that for piping devices (Equation 6.9). Notice there is a one-to-one correspondence between the physical quantities in these equations. Head loss *across* a piping device corresponds to voltage drop *across* an electrical device. Electrical current, the indicator of flow rate of electrons *through* an electrical device, corresponds to the flow rate of fluid *through* the piping device. Finally, electrical resistance corresponds to what we now call *hydraulic resistance*. See Figure 6–15.

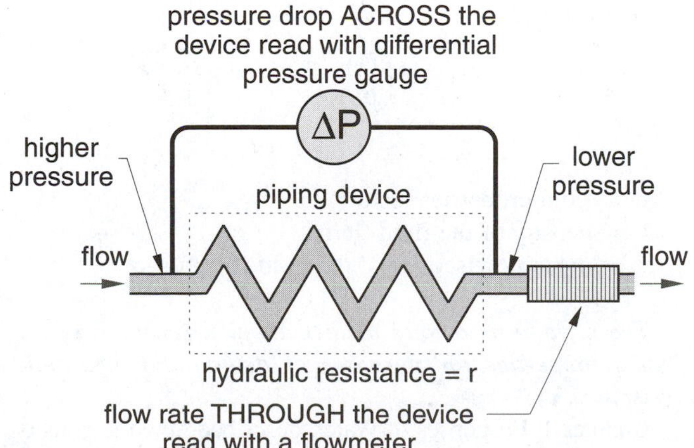

Figure 6–14 Concept of pressure drop (associated with head loss), flow rate, and hydraulic resistance for a piping device.

$$H_L = r_h \times f^{1.75}$$
$$\downarrow \quad \downarrow \quad \downarrow$$
$$V_L = r_e \times \dot{i}$$

H_L = head loss (feet)
V_L = voltage loss (volts)
r_h = hydraulic resistance (ft/gpm^1.75)
r_e = electrical resistance (ohms)
f = flow rate (gpm)
$\dot{i}$ = electrical current (amps)

Figure 6–15 Correspondence of mathematical terms between equations for piping devices and electrical devices.

Still, there is an important difference between these equations. In the electrical device equation, the voltage drop across the device is directly proportional to the current flowing through it. Thus, if the current through the device were doubled, the voltage drop across it would also double. In the case of the piping device model, the head loss across the device is *not* directly proportional to the flow rate through the device. This is because the flow rate is raised to the 1.75 power. If the flow rate through the piping device were doubled, the head loss across it increases by a factor of about 3.4. This difference makes the fluid device equation more mathematically complex, but still manageable. Figure 6–16 shows a comparison between representative graphs of these two equations.

Determining Hydraulic Resistance

Almost any component used in a hydronic piping circuit, be it a length of pipe, fitting, valve, heat emitter, or otherwise, can be represented by a hydraulic resistance. Devices that create relatively little interference with the fluid stream will have low hydraulic resistance. Those that have major restrictions, or twisting paths through which the fluid must flow, will have higher hydraulic resistances. After the hydraulic resistances of all devices in a piping assembly are determined, they can be combined into an **equivalent resistance** similar to how electrical resistances are combined in circuits analysis.

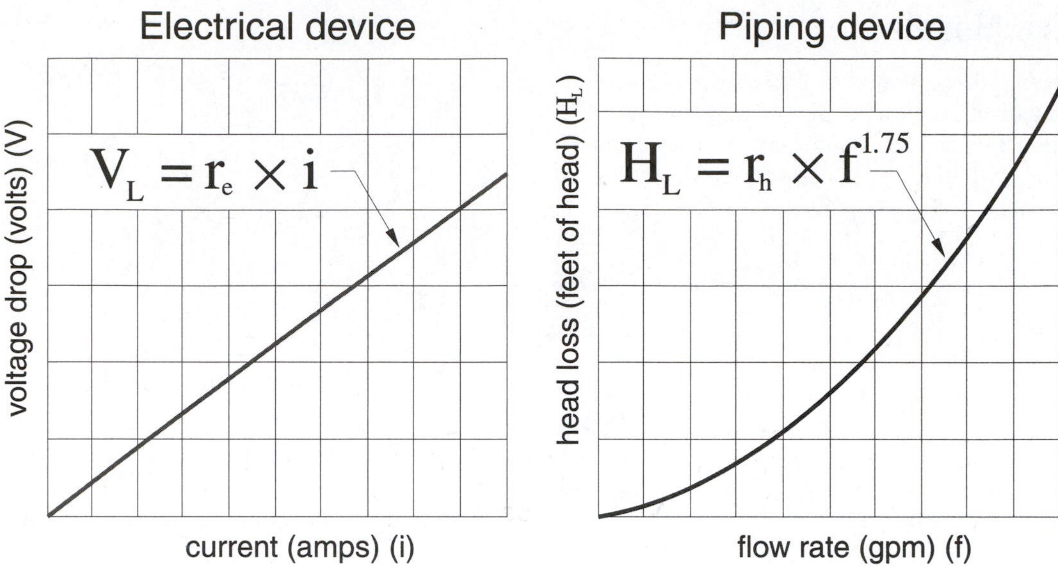

Figure 6–16 Graphical comparison between the equation describing the electrical device (Equation 6.8), and the equation describing the piping device (Equation 6.9).

The hydraulic resistance term, *r*, in Equation 6.9 can be expanded as shown in Equation 6.10:

(Equation 6.10)

$$r = (\alpha c L)$$

therefore:

(Equation 6.11)

$$H_L = r(f)^{1.75} = (\alpha c L)(f)^{1.75}$$

where:

H_L = head loss across a device due to viscous friction (ft of head)

α = **fluid properties factor** based on the fluid's density and viscosity (Equation 6.12)

c = a constant based on the size of the tube hereafter called the **pipe size coefficient** (see table in Figure 6–18)

L = length of pipe, or **equivalent length** of a piping device (ft)

f = flow rate of fluid through the component (gpm)

Equation 6.11 is a specialized relationship derived from the widely used Darcy-Weisbach equation. It incorporates an empirical relationship for the Moody friction factor based on turbulent flow in smooth piping. Because of this empirical factor, *Equation 6.11 is only valid for turbulent flow with Reynold's numbers in the range of 4000 to about 200,000.* Fortunately, this covers most of the operating conditions found in residential and light commercial hydronic systems. At Reynold's numbers above 200,000 the equation gradually begins to underestimate head loss. At a Reynold's number of 300,000, it underestimates head loss by about 6 percent.

The Reynold's number for fluid flow in pipes can be found using Equation 6.7. It is prudent to verify the Reynold's number is between 4,000 and 200,000 when using Equation 6.11 for system design.

Fluid Properties Factor (α)

The fluid properties factor, α (pronounced alpha), which is a factor in Equation 6.10 and Equation 6.11, is defined as follows:

(Equation 6.12)

$$\alpha = \left(\frac{D}{\mu}\right)^{-0.25}$$

where:

α = fluid properties factor

D = density of the fluid (lb/ft^3)

μ = dynamic viscosity of the fluid ($lb_m/ft/sec$)

The α value is a total mathematical description of the fluid's properties for purposes of determining hydraulic resistance.

Chapter 4, Properties of Water, described how the density and dynamic viscosity of fluids vary with temperature. Because it is derived from these properties, the α value is also dependent on temperature. Figure 6–17 is a plot of the α value for water over a temperature range of 50 °F to 250 °F. The Fluid Properties Calculator module in the

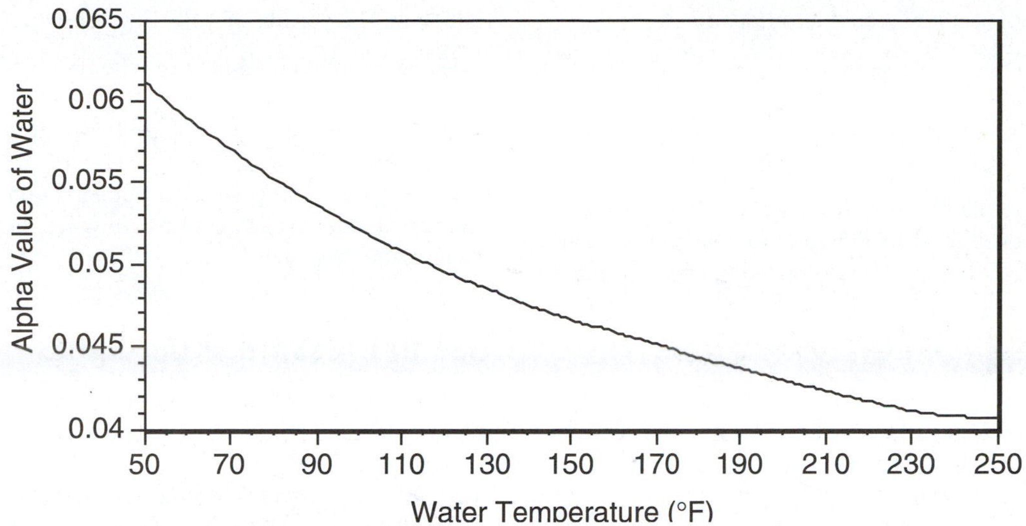

Figure 6–17 Graph of α-value of water for temperatures of 50 °F to 250 °F.

Hydronics Design Studio software can also be used to quickly find density, viscosity, and the fluid properties factor (α) for several fluids commonly used in hydronic heating systems.

The *c* in Equation 6.10 and Equation 6.11 is called the pipe size coefficient. It is a constant for a given tubing size. It incorporates dimensional information such as interior diameter, cross-sectional area, and appropriate unit conversion factors into a single number. A table of pipe size coefficients for several types and sizes of tubing is given in Figure 6–18.

Example 6.6: Using the hydraulic resistance method, determine the pressure drop created when water at 140 °F flows through 100 feet of 3/4-inch type M copper tube at a rate of 5 gpm.

Solution: This situation requires the use of Equation 6.11. First, the value of α for water at 140 °F, as well as the pipe size coefficient (*c* value) for 3/4-inch copper tube, must be determined.

The value of α for water can be found using either Figure 6–17 or Equation 6.12. In this case the latter will be used. The density of water at 140 °F is found from Figure 4–3 to be 61.35 lb/ft³. The dynamic viscosity of water at 140 °F is found from Figure 4–6 to be 0.00032 lb/ft/sec. The α value can now be calculated using Equation 6.12:

$$\alpha = \left(\frac{D}{\mu}\right)^{-0.25} = \left(\frac{61.35}{0.00032}\right)^{-0.25} = (191719)^{-0.25}$$

$$= 0.04779$$

The value of the pipe size coefficient *c* is found from the table of Figure 6–18. For 3/4-inch copper tube, $c = 0.061957$.

Tube	(size & type)	*C* value
3/8"	type M copper	1.0164
1/2"	type M copper	0.33352
3/4"	type M copper	0.061957
1"	type M copper	0.01776
1.25"	type M copper	0.0068082
1.5"	type M copper	0.0030667
2"	type M copper	0.0008331
2.5"	type M copper	0.0002977
3"	type M copper	0.0001278
3/8"	PEX	2.9336
1/2"	PEX	0.786
5/8"	PEX	0.2947
3/4"	PEX	0.14203
1"	PEX	0.04318
1.25"	PEX	0.01668
1.5"	PEX	0.007554
2"	PEX	0.002104
3/8"	PEX-AL-PEX	3.35418
1/2"	PEX-AL-PEX	0.6162
5/8"	PEX-AL-PEX	0.19506
3/4"	PEX-AL-PEX	0.06379
1"	PEX-AL-PEX	0.019718

Figure 6–18 Values of *c* for type M copper, PEX, and PEX-AL-PEX tubing for use in Equation 6.10 and Equation 6.11.

All data can now be substituted into Equation 6.11:

$$H_L = (\alpha c L)(f)^{1.75} = [(0.04779)(0.061957)(100)](5)^{1.75}$$

$$= 4.95 \text{ ft of head}$$

Thus far we have determined the head loss along the pipe, not the pressure drop. The final step is to convert the head loss into a corresponding pressure drop using Equation 6.6:

$$\Delta P = \frac{HD}{144} = \frac{(4.95)(61.35)}{144} = 2.11 \text{ psi}$$

Thus, a pressure gauge at the outlet of the 100 feet of 3/4-inch copper tube should have a reading of 2.11 psi lower than the pressure at the inlet.

Flow Coefficient C_v

Another method for predicting pressure drop due to head loss has been used within the valve industry for many years. It is based on a parameter called the **flow coefficient** (C_v) of the valve.

The C_v is defined as the flow rate of 60 °F water that will create a pressure drop of 1.0 psi through the valve. For example, a valve with a C_v rating of 5.0 would require a flow rate of 5.0 gpm of 60 °F water to create a pressure drop of 1.0 psi across it. Many valve manufacturers list the C_v values of their products in their technical literature. Occasionally C_v values will be listed for devices other than valves. The C_v value is used to describe the setting of balancing valves in the Hydronic Circuit Simulator module in the Hydronics Design Studio software.

The *Cv* value can be used in an equation that estimates pressure drop based on the *square* of the flow rate of fluid through a device. This relationship, given as Equation 6.13, is often used for determining pressure drops through valves.

(Equation 6.13)

$$\Delta P = \left(\frac{D}{62.4}\right)\left(\frac{f}{C_v}\right)^2$$

where:

ΔP = pressure drop across the device (psi)
D = density of the fluid at its operating temperature (lb/ft³)
62.4 = density of water at 60 °F (lb/ft³)
f = flow rate of fluid through the device (gpm)
C_v = known C_v rating of the device (gpm)

Example 6.7: Estimate the pressure drop across a radiator valve having a C_v value of 2.8, when 140 °F water flows through at 4.0 gpm.

Solution: The density of 140 °F water is 61.35 lb/ft³.

Using Equation 6.13 the pressure drop is:

$$\Delta P = \left(\frac{61.35}{62.4}\right)\left(\frac{4.0}{2.8}\right)^2 = 2.0 \text{ psi}$$

For smooth piping such as copper tubing or PEX, the square relationship of Equation 6.13 will slightly overpredict the pressure drop.

Use of Pressure Drop Charts

Many references on piping design contain charts such as the one shown in Figure 6–19. These charts show the relationship between flow rate, flow velocity, and pressure drop due to head loss, for several sizes of copper tubing. They are usually based on water at a temperature of 60 °F, which is typical in domestic (cold) water distribution systems. The reference temperature of 60 °F is, however, not a typical operating condition in hydronic heating systems. The higher temperature water in such systems will have lower density and lower viscosity, resulting in less head loss, and thus smaller pressure drops.

Use of pressure drop charts based on 60 °F water will consistently overestimate the pressure drop in hydronic piping circuits because the water will usually be at higher temperatures. This can result in unnecessary oversizing of the system's circulator(s). Because of this, *these charts should not be used for estimating pressure drop due to head loss in hydronic heating systems*. However, they can still be used for finding the relationship between flow rate and flow velocity.

Example 6.8: Use Equation 6.11 and Equation 6.12 to recalculate the pressure drop for 100 feet of 3/4-inch copper tubing transporting 60 °F water at 5 gpm. Compare the results with those of Example 6.6, and with those obtained from the chart in Figure 6–19.

Solution: Using the graph in Figure 4–3, the density of water at 60 °F is estimated to be 62.3 lb/ft³. Using the graph in Figure 4–6, the dynamic viscosity of water at 60 °F is estimated at 0.00075 lb/ft/sec. The α value can be calculated using Equation 6.12:

$$\alpha = \left(\frac{D}{\mu}\right)^{-0.25} = \left(\frac{62.3}{0.00075}\right)^{-0.25} = (83067)^{-0.25}$$

$$= 0.0589$$

The value of the pipe size coefficient c for 3/4-inch tubing is found in Figure 6–18: $c = 0.061957$.

Substituting these values into Equation 6.11 along with the pipe length and flow rate:

$$H_L = (\alpha c L)(f)^{1.75} = [(0.0589)(0.061957)(100)](5)^{1.75}$$

$$= 6.10 \text{ ft of head}$$

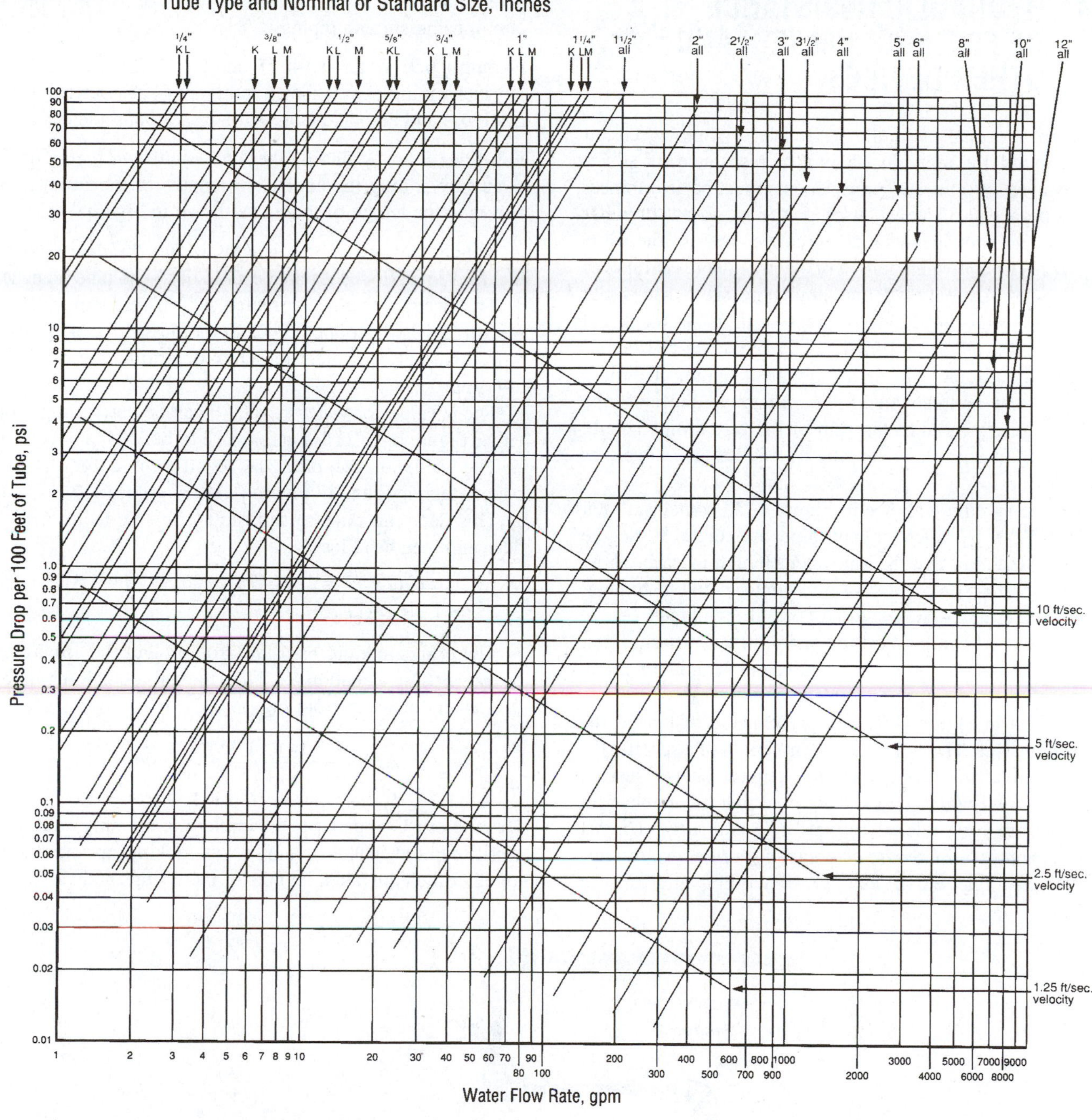

Figure 6–19 Pressure drop chart for copper tubing carrying 60 °F water. *Courtesy of the Copper Development Association.*

The corresponding pressure drop is found using Equation 6.6:

$$\Delta P = \frac{(6.10)(62.3)}{144} = 2.64 \text{ psi}$$

Discussion: The pressure drop for the same piping using 140 °F water in Example 6.6 was 2.11 psi. Comparing these results shows the pressure drop created by 60 °F water to be about 25 percent higher than that created by 140 °F water. This is a significant difference. The chart in Figure 6–19 indicates a pressure drop of about 2.7 psi (for water at 60 °F). This is comparable to the pressure drop found using Equation 6.11 for water at 60 °F.

6.4 HYDRAULIC RESISTANCE OF FITTINGS, VALVES, AND OTHER DEVICES

Before the total hydraulic resistance of a piping circuit can be found, the individual hydraulic resistances of all fittings, valves, or other such components must be determined. One approach is to consider each fitting, valve, or other device as an equivalent length of copper tube of the same pipe size. The equivalent length of a component is defined as the amount of tubing, of the same pipe size, that would produce the same head loss (or pressure drop), as the actual component, at the same flow rate. By replacing all components in the circuit with their equivalent length of piping, the circuit can be treated as if it were a single piece of pipe having a length equal to the sum of the actual pipe length, plus the total equivalent lengths of all fittings, valves, or other devices.

The piping circuit shown in Figure 6–20 contains a total of 58 feet of 3/4-inch copper pipe and eight 90-degree elbows. Assuming that the equivalent length of each elbow is 2 feet of 3/4-inch pipe, the entire circuit could be thought of as if it were 58 + 8(2) = 74 feet of 3/4-inch pipe.

This method can be used for many different fittings and valves. Figure 6–21 gives the equivalent lengths of many commonly used fittings and valves.

The data in Figure 6–21 is representative of typical fittings and valves. The exact equivalent length of a specific fitting or valve will depend on its internal shape, surface roughness, and other factors. Take note that the numbers in the table are for soldered fittings. *The equivalent lengths given in Figure 6–21 should be doubled for threaded fittings.* The values in Figure 6–21 can be used with Equation 6.11 to determine the head loss through any type of hydronic piping circuit using smooth piping.

Example 6.9: Water at 140 °F and 6 gpm flows through the 3/4-inch piping circuit shown in Figure 6–22. Determine the head loss and associated pressure drop around the circuit.

Solution: Start by finding the *total equivalent length* of the circuit by adding the equivalent lengths of the fittings and valves to the length of tubing as shown in Figure 6–23.

Notice that only tubing lengths *within the flow path* are included since flow does not occur in dead-end pipe branches.

The circuit can now be treated as if it were simply 104.8 feet of 3/4-inch copper pipe. Equation 6.11 can be used with this total equivalent length to determine the head loss at 6 gpm.

The α value for water at 140 °F can be determined from either Figure 6–17 or Equation 6.12 to be 0.0478.

The value of the pipe size coefficient c for 3/4-inch copper tube is found in Figure 6–18: $c = 0.061957$

This data can now be substituted into Equation 6.11 to determine the head loss:

$$H_L = (\alpha c L)(f)^{1.75} = [(0.0478)(0.061957)(104.8)](6)^{1.75}$$
$$= 7.14 \text{ ft of head}$$

The corresponding pressure drop around the circuit can be found using Equation 6.6. Note that this equation requires the density of water from Figure 4–3.

$$\Delta P = \frac{(7.14)(61.35)}{144} = 3.04 \text{ psi}$$

Discussion: Compared to Example 6.6, the only additional work required in this example was looking up and adding together the equivalent lengths of the fittings and valves.

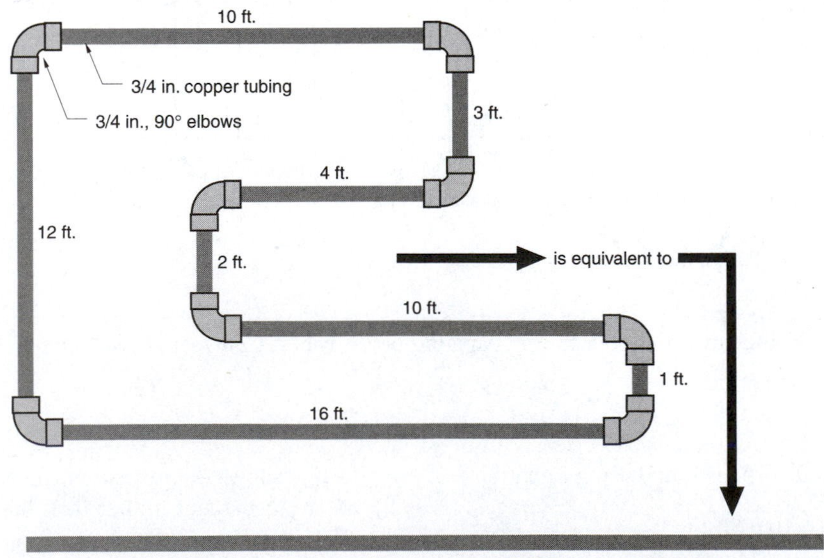

Figure 6–20 Converting a piping circuit containing pipe and fittings to an equivalent length of straight pipe.

Copper Tube Sizes									
Fitting or Valve[1]	**3/8"**	**1/2"**	**3/4"**	**1"**	**1.25"**	**1.5"**	**2"**	**2 1/2"**	**3"**
90-degree elbow	0.5	1.0	2.0	2.5	3.0	4.0	5.5	7.0	9
45-degree elbow	0.35	0.5	0.75	1.0	1.2	1.5	2.0	2.5	3.5
Tee (straight run)	0.2	0.3	0.4	0.45	0.6	0.8	1.0	0.5	1.0
Tee (side port)	2.5	2.0	3.0	4.5	5.5	7.0	9.0	12.0	15
B&G Monoflo® tee[2]	n/a	n/a	70	23.5	25	23	23	n/a	n/a
Reducer coupling	0.2	0.4	0.5	0.6	0.8	1.0	1.3	1.0	1.5
Gate valve	0.35	0.2	0.25	0.3	0.4	0.5	0.7	1.0	1.5
Globe valve	8.5	15.0	20	25	36	46	56	104	130
Angle valve	1.8	3.1	4.7	5.3	7.8	9.4	12.5	23	29
Ball valve[3]	1.8	1.9	2.2	4.3	7.0	6.6	14	0.5	1.0
Swing-check valve	0.95	2.0	3.0	4.5	5.5	6.5	9.0	11	13.0
Flow-check valve[4]	n/a	n/a	83	54	74	57	177	85	98
Butterfly valve	n/a	1.1	2.0	2.7	2.0	2.7	4.5	10	15.5

Footnotes:
1. Data for soldered fittings and valves. For threaded fittings double the listed value
2. Derived from *Cv* values based on no flow through side port of tee
3. Based on a standard port ball valve. Full port valves would have lower equivalent lengths
4. Based on B&G brand "flow control" valves

Figure 6–21 Representative equivalent lengths of common fittings and valves (all values expressed as feet of copper tube of the same nominal size).

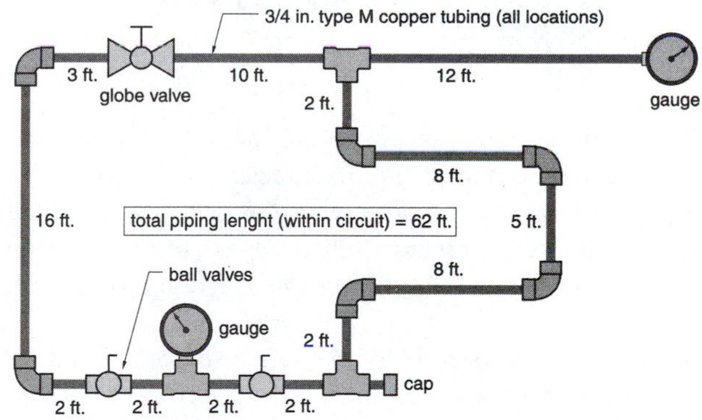

Figure 6–22 Piping assembly for use in Example 6.7.

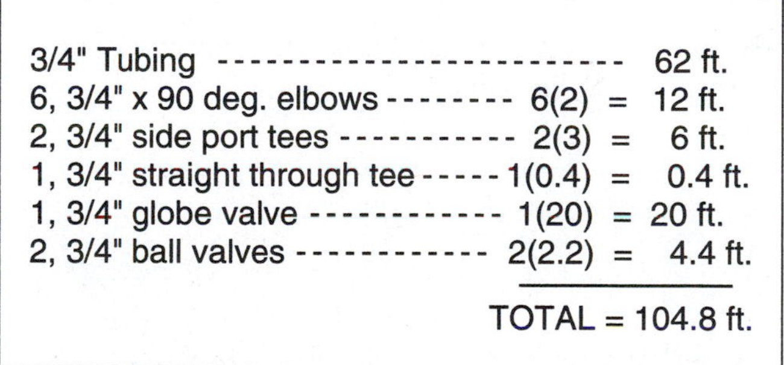

3/4" Tubing ------------------------ 62 ft.
6, 3/4" x 90 deg. elbows -------- 6(2) = 12 ft.
2, 3/4" side port tees ----------- 2(3) = 6 ft.
1, 3/4" straight through tee ----- 1(0.4) = 0.4 ft.
1, 3/4" globe valve ------------ 1(20) = 20 ft.
2, 3/4" ball valves ------------ 2(2.2) = 4.4 ft.
TOTAL = 104.8 ft.

Figure 6–23 Adding the equivalent lengths of the fittings and valves in Figure 6–22.

6.5 THE SYSTEM RESISTANCE CURVE

An examination of Equation 6-11 shows that *for a given piping circuit, operating with a given fluid, at a constant temperature*, the hydraulic resistance term, $\alpha\ c\ L$, remains constant:

(Equation 6.14)

$$H_L = (\alpha c L)(f)^{1.75} = (\text{a constant})(f)^{1.75}$$

Under these conditions, the head loss around the piping circuit depends only on flow rate. Equation 6.14 is a mathematical function that can be graphed by selecting several flow rates, calculating the head loss at each, and plotting the resulting points. Once the points are plotted, a smooth curve could be drawn through them.

Example 6.10: Using the piping circuit and data from Example 6.9, plot several points representing different flow rates and the associated head loss. Draw a smooth curve through these points.

Solution: Start by substituting the appropriate data into Equation 6.11:

$$H_L = (\alpha c L)(f)^{1.75} = [(0.0478)(0.061957)(104.8)](f)^{1.75}$$

Multiply the values of α, *c*, and *L* together to obtain a single number. This value, 0.31, is the hydraulic resistance of the piping circuit shown in Figure 6–22:

(Equation 6.15)

$$H_L = (0.31)(f)^{1.75}$$

Use this equation to generate data for plotting. Simply select a few random flow rates, and substitute each into the equation to find the corresponding head loss. The values used are shown in Figure 6–24.

This data can now be plotted, and a smooth curve drawn through the points as shown in Figure 6–25.

Flow rate (gpm)	Head loss (feet of head)
0	0
3	2.12
6	7.14
9	14.5
12	24.0
15	35.5

Figure 6–24 Values of head loss calculated at several flow rates using Equation 6.15.

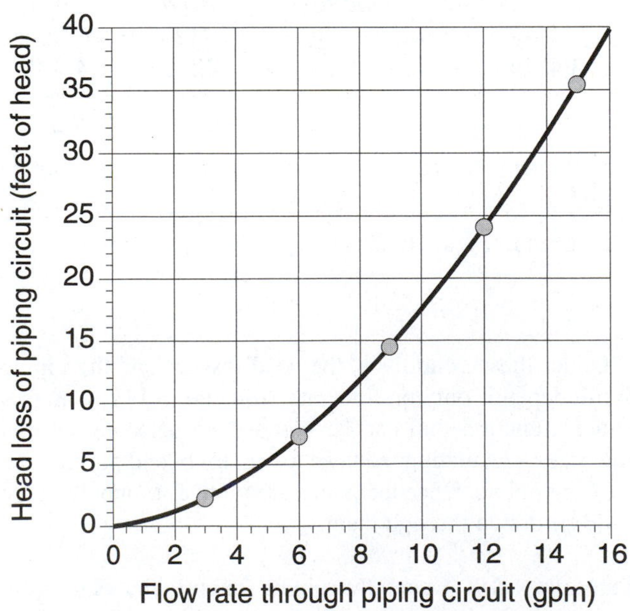

Figure 6–25 Example of a system resistance curve plotted using data from Figure 6–24.

This graph is called a **system resistance curve**, or sometimes just "system curve." It represents the relationship between flow rate and head loss for a given piping circuit using a specific fluid at a given temperature. All piping loops have a unique system curve. It could even be thought of as the analytical "fingerprint" of the loop. Determining the system resistance curve of a piping circuit is an essential step in properly selecting a circulator for that loop.

If additional piping, fittings, or other components were added to the piping circuit, its total equivalent length would increase, as would its hydraulic resistance. This would steepen the system resistance curve.

Figure 6–26 shows several system resistance curves plotted on the same graph. These curves were produced assuming the total equivalent length of the circuit described in Example 6.8 varied from a low of 50 feet to a high of 500 feet. Such changes in length would change the hydraulic resistance, and therefore the "steepness" of the system resistance curve. The hydraulic resistance used in Equation 6.11 to plot each curve is indicated on the graph. Note that the system curve gets steeper as the hydraulic resistance of the circuit increases.

6.6 PIPING COMPONENTS REPRESENTED AS SERIES RESISTORS

The simplest piping assemblies are formed by connecting piping components end to end. In some cases, the collection of tubing, fittings, valves, and other components closes on itself to form a **series circuit**. When this is not the case, the components form a **series piping path**. Examples of each are shown in Figure 6–27.

To visualize the overall hydraulic resistance created by various components of a piping system, each pipe, fitting, valve, or other component can be thought of as a **hydraulic resistor**, and represented with a symbol like that used in electrical schematics. An example of how a series piping path consisting of several components would be represented by hydraulic resistors is shown in Figure 6–28.

As with electrical resistor diagrams, the size of a particular resistor symbol does not indicate the amount of hydraulic resistance it represents. Only numerical values can do this. The orientation of the resistor symbol or the shape of the series piping path also makes no difference. However, *the sequence of hydraulic resistors forming the series piping path or circuit should always match the actual piping component layout they represent.*

A similar representation of piping components using hydraulic resistor symbols can be made for a complete series circuit, as shown in Figure 6–29.

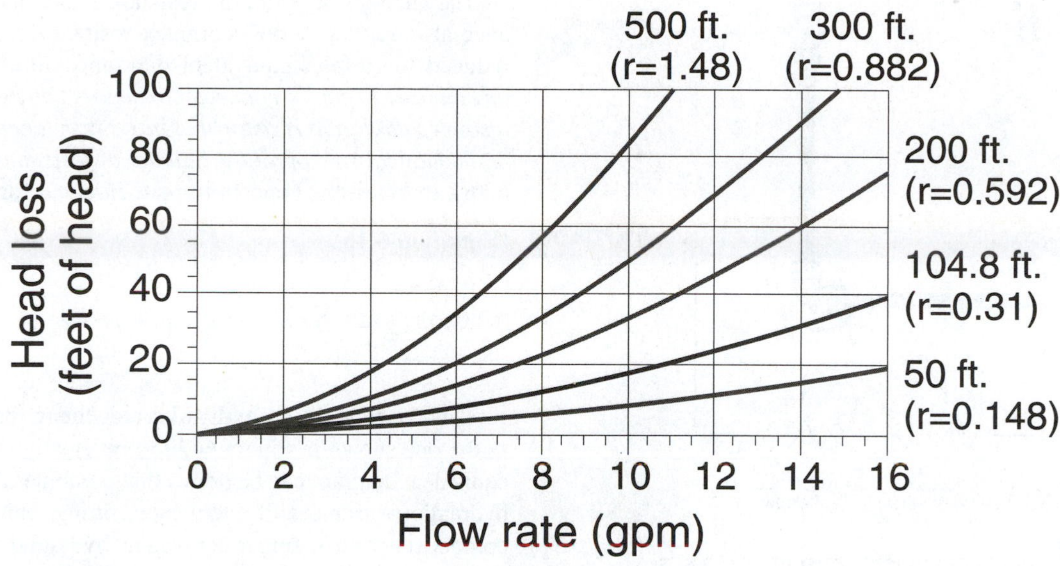

Figure 6–26 System curves for piping circuits having different total equivalent lengths.

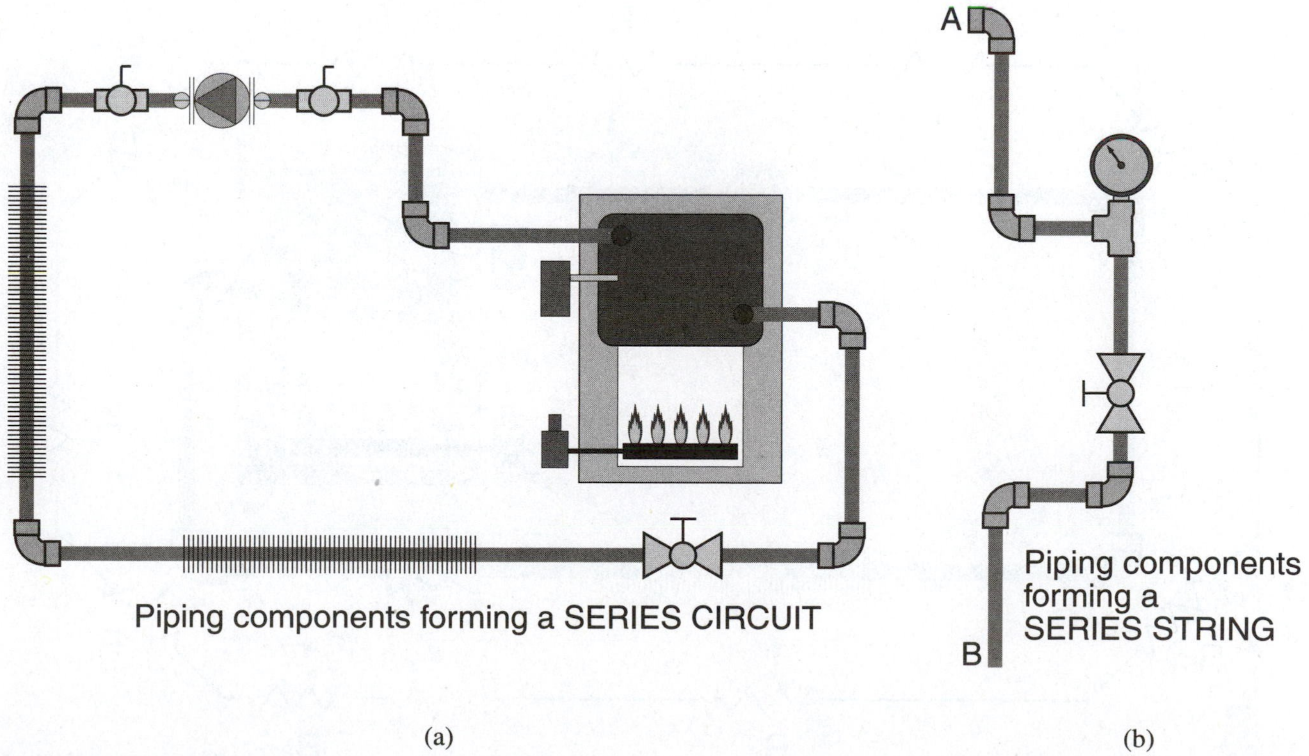

Figure 6–27 Multiple system resistance curves for pipe circuits of various equivalent lengths. Notice that curves get "steeper" as equivalent lengths increase.

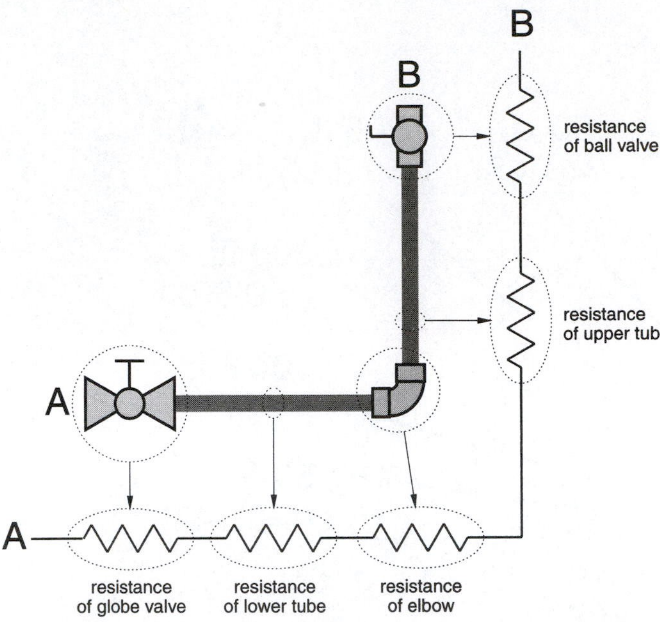

Figure 6–28 Piping components forming (a) a series circuit, (b) a series piping path.

Equivalent Resistance of Series-connected Components

The analogy of hydraulic resistance and electrical resistance also applies to how complex resistor diagrams can be reduced to simpler equivalent diagrams. In either system, *any number of series-connected resistors can be replaced by a single* ***equivalent resistor*** *that has a resistance equal to the mathematical sum of all the individual resistances.* Equation 6.16 expresses this concept in mathematical form.

(Equation 6.16)

$$r_{\text{e series}} = \sum_{i=1}^{n} r_i = (r_1 + r_2 + r_3 + r_4 + \ldots + r_n)$$

This concept allows **hydraulic resistance diagrams** such as the one shown in Figure 6–29 to be greatly simplified. An equivalent diagram can be drawn that combines the individual hydraulic resistances of every pipe, fitting, valve, and other component, into a single equivalent hydraulic resistance as shown in Figure 6–30. The simplified diagram totally represents the hydraulic resistance of *all* the original components, but in a much more compact and easily managed form.

In electrical circuits, the resistance of each resistor is known, and the values are simply added together. In piping

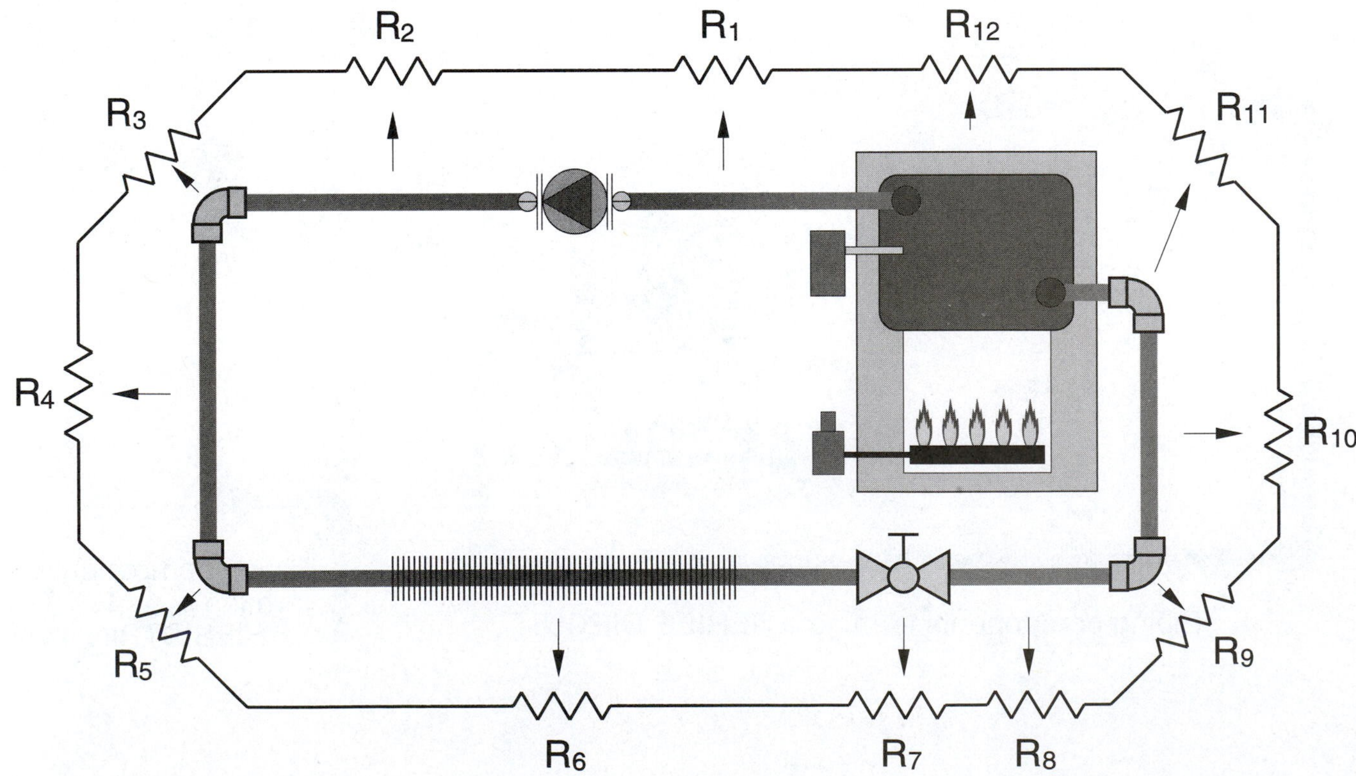

Figure 6–29 Series string of piping components represented as series string of hydraulic resistors.

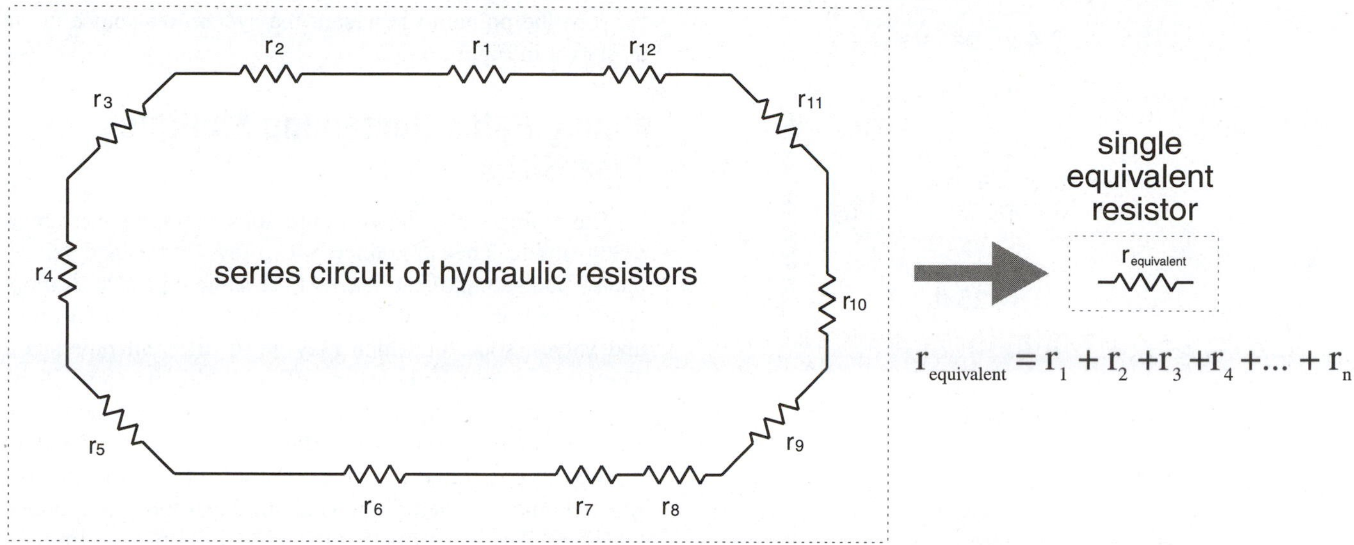

Figure 6–30 Concept of reducing a series circuit of resistors into a single equivalent resistance.

circuits, however, the hydraulic resistance of each component has to be calculated using Equation 6.10, before the values can be added. Although it is certainly possible to do this, it can become quite time-consuming for complex systems.

A more efficient method can be used in the common situation where the piping circuit contains a single size of pipe, fittings, and valves. This method recognizes that the fluid properties factor, α, and the pipe size coefficient, *c*, will be the same for all components, and thus need only be determined once. The only remaining value required for each component is its equivalent length, which can be found in Figure 6–21. Finding the equivalent resistance of the circuit therefore becomes a matter of looking up the equivalent lengths of the fittings and valves, combining these with the total piping length, and multiplying this **total equivalent length** by the common values of α and *c* as shown in Figure 6–31.

Example 6.11: The series piping path shown in Figure 6–32 carries water at 140 °F. Determine the hydraulic resistance of the entire piping path between points A and B (Figure 6–33). Use this hydraulic resistance to sketch a system resistance curve.

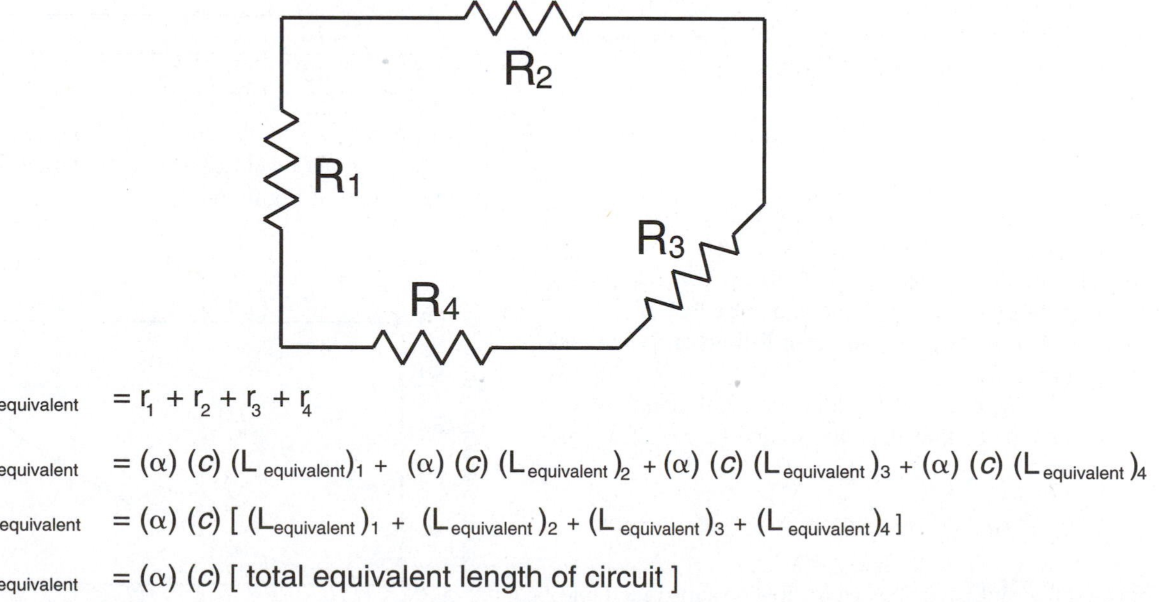

Figure 6–31 Simplified method of adding hydraulic resistances when the piping components are all the same pipe size. Note the value of α and *c* are the same for all components.

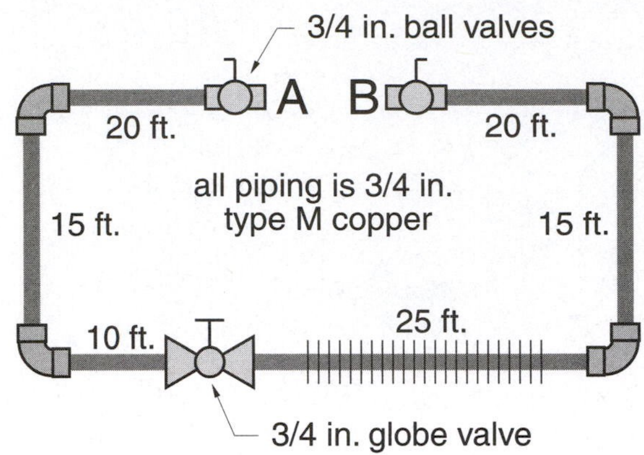

Figure 6–32 The piping system of Example 6.9.

Piping Component				Equivalent Length
105 ft. 3/4 in. copper tubing	=			105 ft.
4, 90 degree x 3/4 in. elbows	=	4(2)	=	8 ft.
2, 3/4 in. ball valves	=	2(2.2)	=	4.4 ft.
1, 3/4 in. globe valve	=	1(20)	=	20 ft.
Total equivalent length	=	L	=	137.4 ft.

Figure 6–33 Adding the equivalent lengths of piping, fittings, and valves in the system shown in Figure 6–32.

Solution: Figure 4–3 and Figure 4–6 can be used to obtain the density and viscosity of water at 140 °F. These values are then substituted into Equation 6.12 to find α:

$$\alpha = \left(\frac{D}{\mu}\right)^{-0.25} = \left(\frac{61.35}{0.00032}\right)^{-0.25} = (191719)^{-0.25} = 0.0478$$

The pipe size coefficient (c) is found in Figure 6–18. For 3/4-inch copper tube $c = 0.061957$.

The only remaining task is to look up the equivalent lengths of all fittings and valves and add these lengths to that of the piping. The equivalent lengths of fittings and valves is found in Figure 6–21.

The values of α, c, and L can now be substituted directly into Equation 6.10 to find the total hydraulic resistance of the piping string:

$$r = (\alpha c L) = (0.04779)(0.061957)(137.4) = 0.407$$

This total hydraulic resistance can now be substituted into Equation 6.11 and used to generate a few points for plotting the system resistance curve as shown in Figure 6–34. The resulting data is plotted and a smooth curve drawn through the points to represent the system resistance curve as shown in Figure 6–35.

Piping Paths Containing Multiple Pipe Sizes

The piping paths shown in previous examples were relatively simple. They contained only a few fittings and valves, and all such components were of the same pipe size. A more typical hydronic piping loop can contain dozens of fittings and valves, some of which may be of different pipe sizes. The total hydraulic resistance of such circuits can still be determined using the basic principles. However, when more than one pipe size is involved, the hydraulic resistance of all piping, fittings, valves, or other components, *of a given pipe size*, should be determined separately, and then added together. This is because each pipe size has a different c value. These calculations can be organized as shown in Equation 6.17.

(Equation 6.17)

$$r_{total} = \alpha\{[c(L_{pipe} + L_{components})]_{size\ 1} + [c(L_{pipe} + L_{components})]_{size\ 2} \ldots + [c(L_{pipe} + L_{component})]_{last\ size}\}$$

$H_L = 0.407(f)^{1.75}$	
Flow rate (f), (gpm)	Head loss H_L, (ft. of head)
0	0
5	6.8
10	22.9
15	46.5

Figure 6–34 Using the equation of the system resistance curve to generate head loss versus flow rate data for the piping system of Figure 6–32.

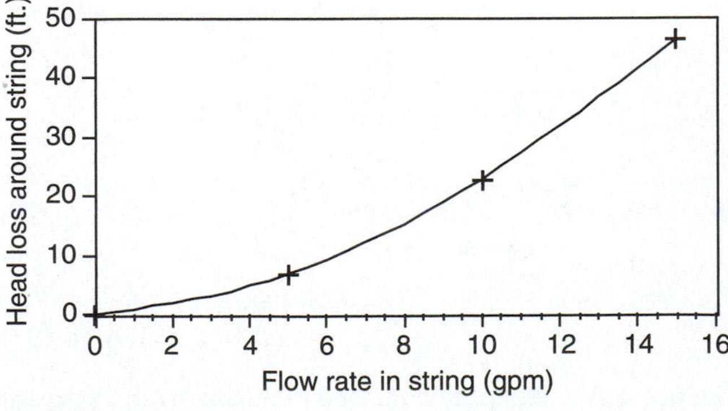

Figure 6–35 The system resistance curve for Example 6.9.

6.7 PARALLEL HYDRAULIC RESISTANCES

Piping systems often contain two or more piping paths connected at common points. Examples include multiple zone distribution circuits that begin and end at the same boiler as shown in Figure 6–36, or radiant floor heating circuits that originate at a common supply manifold, and terminate at a common return manifold, as shown in Figure 6–37. The piping paths that share common points of connection are said to be **piped in parallel**. In such arrangements, the total system flow must divide up among the two or more paths at the common supply point, and then recombine at a common return point.

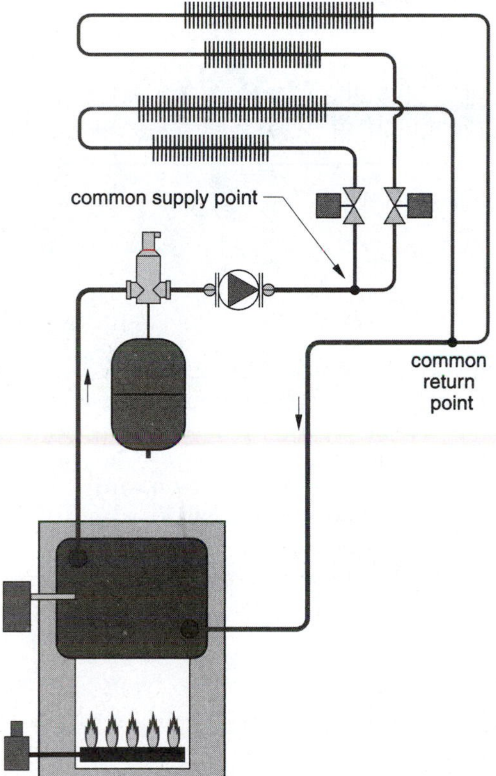

Figure 6–36 Example of parallel piping paths formed by zone circuits.

The analogy of electrical resistance again proves useful for analyzing how the flow divides among two or more parallel piping paths. Figure 6–38 shows a parallel piping assembly along with its associated hydraulic resistor diagram.

In the resistor diagram of Figure 6–38, the combined resistance of the two elbows in the upper piping branch is represented by a single resistor symbol, r_2. This is done as a matter of convenience because it reduces the number of resistor symbols that have to be drawn. Because this resistor now represents both elbows, its numerical value would be doubled. The three pipe segments in the upper branch are also represented by a single resistor symbol, r_3, as are the three in the lower branch, r_8. This simplifying method can be used whenever two or more identical components are configured in series. With experience, it is easy to draw simplified resistor diagrams by combining such resistances together.

The equivalent resistance of the entire piping assembly is a single resistor connected between points A and B. It represents the combined effect of all piping components between these points. To find this equivalent resistance, start by reducing each of the branch piping paths into a single equivalent resistance as shown in Figure 6–39. Since the hydraulic resistors within each branch path are in series with each other, their resistances can be added. This is represented by Equation 6.18a and Equation 6.18b:

(Equation 6.18a)

$$r_{e1} = \alpha c(L_1 + L_2 + L_3 + L_4)$$

(Equation 6.18b)

$$r_{e2} = \alpha c(L_5 + L_6 + L_7 + L_8 + L_9)$$

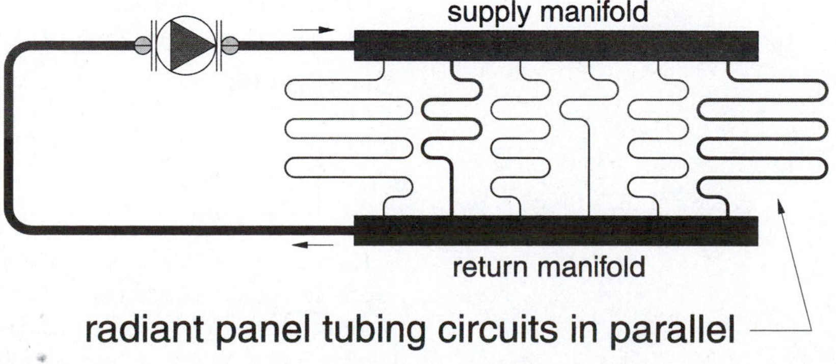

Figure 6–37 Example of parallel piping paths formed by a manifold station.

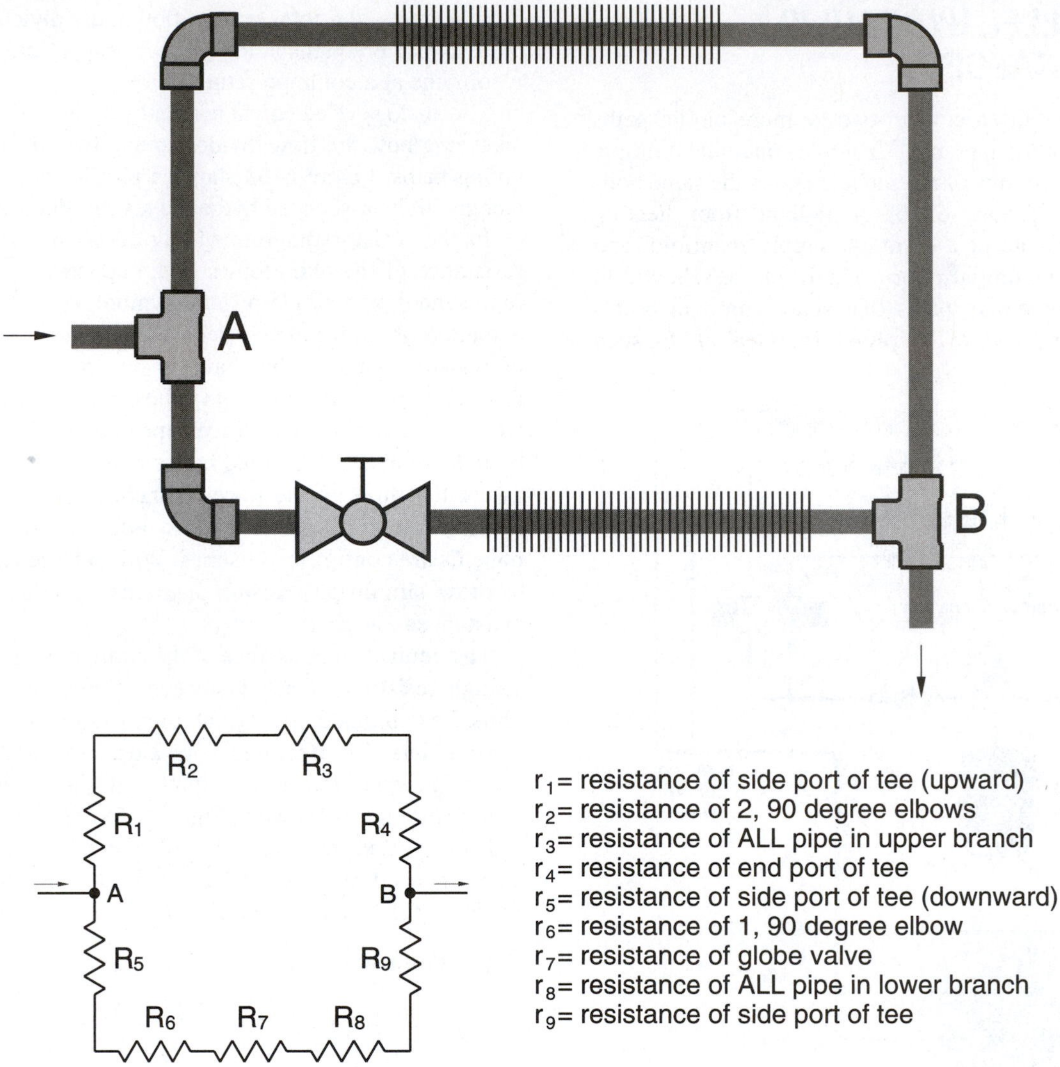

A
B
R2
R3
R1
R4
A
B
R5
R9
R6
R7
R8
r1= resistance of side port of tee (upward)
r2= resistance of 2, 90 degree elbows
r3= resistance of ALL pipe in upper branch
r4= resistance of end port of tee
r5= resistance of side port of tee (downward)
r6= resistance of 1, 90 degree elbow
r7= resistance of globe valve
r8= resistance of ALL pipe in lower branch
r9= resistance of side port of tee

Figure 6–38 Parallel piping assembly with associated hydraulic resistor diagram.

r2
r3
r1
r4
A
B
re1
A
B
A
B
A
B
r5
r9
re2
r6
r7
r8

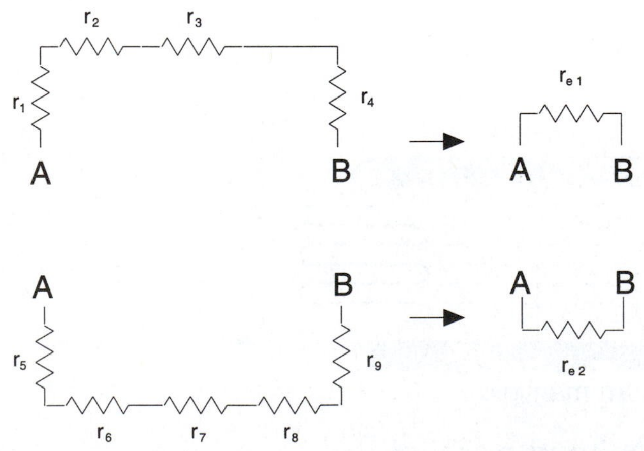

Figure 6–39 Reduce each parallel path to a single equivalent hydraulic resistance.

Once this step is completed, reconnect the two equivalent resistances at points A and B as shown in Figure 6–40.

The arrangement will now be reduced to a single equivalent resistance as shown in Figure 6–41.

The equivalent resistance of two parallel hydraulic resistors is found using Equation 6.19.

(Equation 6.19)

$$R_{\text{equivalent}_{\text{parallel}}} = \left[\left(\frac{1}{r_1}\right)^{0.5714} + \left(\frac{1}{r_2}\right)^{0.5714}\right]^{-1.75}$$

The values of r_1 and r_2 in Equation 6.19 are the hydraulic resistances of the parallel resistors being combined into a single equivalent resistance. This equivalent resistance then represents the entire piping assembly (between points A and B) in Figure 6–38. If this assembly were to be combined with other components to form a more complex piping

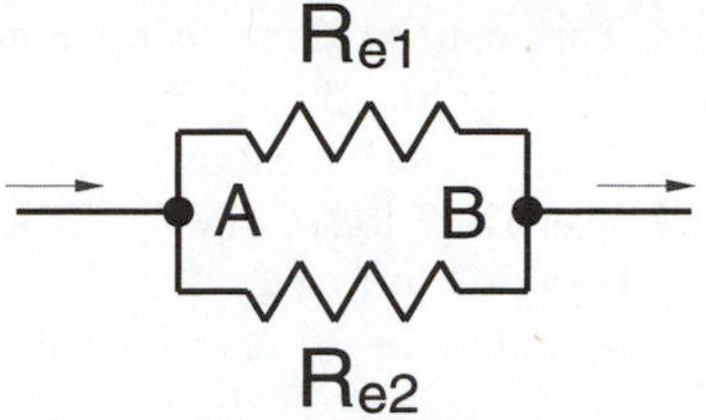

Figure 6–40 A group of two parallel hydraulic resistors.

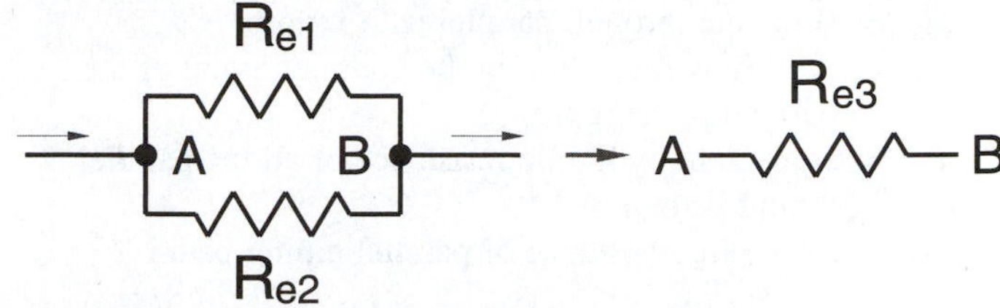

Figure 6–41 Reducing two parallel hydraulic resistances into a single equivalent hydraulic resistance.

system, it could still be represented by this single resistor within the resistor diagram of the overall system.

In some piping systems there are more than two parallel piping strings. An example would be the distribution piping of a radiant floor heating system in which several piping paths originate from a common supply manifold and terminate in a common return manifold. An example is shown in Figure 6–42.

To find the single equivalent hydraulic resistance of this configuration, Equation 6.19 may be extended by one term as follows:

(Equation 6.20)

$$R_{\text{equivalent}_{\text{parallel}}} = \left[\left(\frac{1}{r_1}\right)^{0.5714} + \left(\frac{1}{r_2}\right)^{0.5714} + \left(\frac{1}{r_3}\right)^{0.5714}\right]^{-1.75}$$

This same principle can be extended to *any number* of parallel resistors connected in parallel. Simply add more terms to Equation 6.19 to accommodate all parallel resistances. This can be expressed mathematically for a system with *n* parallel hydraulic resistances as Equation 6.21:

(Equation 6.21)

$$R_{\text{equivalent}_{\text{parallel}}} = \left[\left(\frac{1}{r_1}\right)^{0.5714} + \left(\frac{1}{r_2}\right)^{0.5714} + \left(\frac{1}{r_3}\right)^{0.5714} + \ldots + \left(\frac{1}{r_n}\right)^{0.5714}\right]^{-1.75}$$

Regardless of the number of parallel hydraulic resistances one begins with, the equivalent hydraulic resistor will always be a *single* resistor connected between the common points. This single resistor represents the combined effect of *all* original resistors.

Once the equivalent resistance of a group of parallel hydraulic resistors has been found, the flow rate through each parallel piping path can also be determined using Equation 6.22.

(Equation 6.22)

$$f_i = f_{\text{total}}\left(\frac{R_e}{r_i}\right)^{0.5714}$$

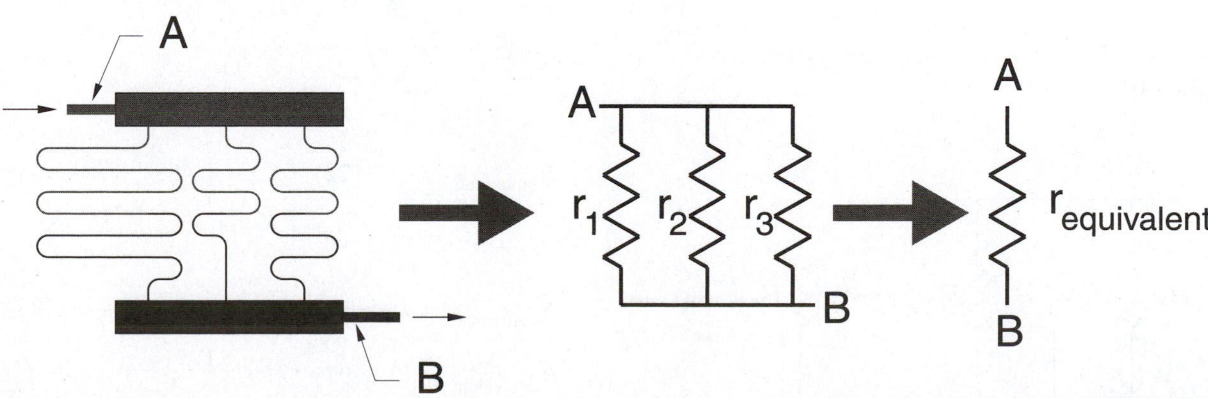

Figure 6–42 Three parallel piping paths represented by hydraulic resistances, which are then combined into a single equivalent hydraulic resistance.

where:

f_i = flow rate through parallel path i (gpm)

f_{total} = *total* flow rate entering the common point of the parallel paths (gpm)

R_e = equivalent hydraulic resistance of all the parallel piping paths

r_i = hydraulic resistance of parallel piping path i

Example 6.12: Using the concept of parallel hydraulic resistors, find the equivalent resistance of the radiant floor distribution system shown in Figure 6–43. All four piping paths consist of 1/2-inch PEX tubing. The average water temperature in the system is 100 °F. To simplify the example, assume there are no valves in the manifolds, and the hydraulic resistances at the points of connection of the piping paths to the manifolds are insignificant. Assuming water enters the supply manifold at 6 gpm, determine the flow rate in each of the branch piping paths.

Solution: The resistor diagram shown in Figure 6–44 represents the assembly as four hydraulic resistors in parallel. After finding the values of these resistors, we can use Equation 6.21 to combine them into a single equivalent resistance.

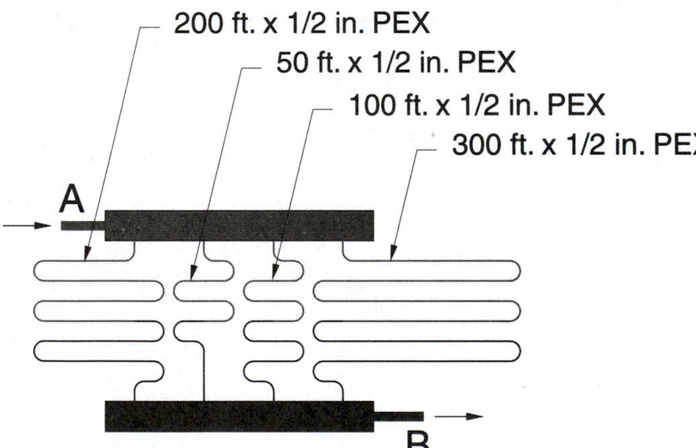

Figure 6–43 Radiant floor distribution system for Example 6.10.

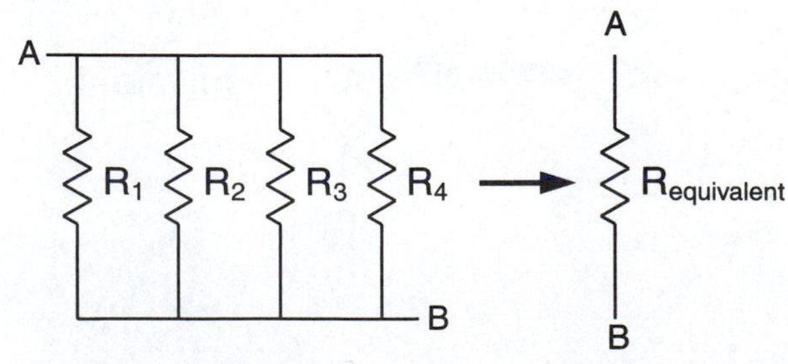

Figure 6–44 Hydraulic resistor diagram for Figure 6–43.

Start by finding the density and dynamic viscosity of water at 100 °F. This can be done using Figure 4–3 and Figure 4–6, or by using the Fluid Properties Calculator module in the Hydronics Design Studio software. The density of 100 °F water is 61.97 lb/ft³. The dynamic viscosity of 100 °F water is 0.0004573 lb/ft/sec.

From these, the value of α can be calculated using Equation 6.12:

$$\alpha = \left(\frac{D}{\mu}\right)^{-0.25} = \left(\frac{61.97}{0.0004573}\right)^{-0.25} = 0.05212$$

The value of c is found in Figure 6–18. For 1/2-inch PEX tube, $c = 0.786$.

The values of α and c will be the same for each piping path. The only remaining quantity needed is the length of each piping path. The lengths along with the values for α and c are now substituted into Equation 6.10:

200-foot piping path:

$$r_1 = \alpha c L_1 = (0.05212)(0.786)(200) = 8.193$$

50-foot piping path:

$$r_2 = \alpha c L_2 = (0.05212)(0.786)(50) = 2.048$$

100-foot piping path:

$$r_3 = \alpha c L_3 = (0.05212)(0.786)(100) = 4.097$$

300-foot piping path:

$$r_4 = \alpha c L_4 = (0.05212)(0.786)(300) = 12.29$$

These resulting values are the hydraulic resistances of each piping path. These resistances can now be substituted into Equation 6.21 and reduced to a single equivalent resistance:

$$R_{equivalent} = \left[\left(\frac{1}{8.193}\right)^{0.5714} + \left(\frac{1}{2.048}\right)^{0.5714} + \left(\frac{1}{4.097}\right)^{0.5714} + \left(\frac{1}{12.29}\right)^{0.5714}\right]^{-1.75}$$

$$R_{equivalent} = [(0.122)^{0.5714} + (0.488)^{0.5714} + (0.244)^{0.5714} + (0.0814)^{0.5714}]^{-1.75}$$

$$R_{equivalent} = [0.3006 + 0.6639 + 0.4467 + 0.2385]^{-1.75}$$

$$R_{equivalent} = [1.6497]^{-1.75}$$

$$R_{equivalent} = 0.416$$

The flow rates in each piping path can now be determined by repeated use of Equation 6.22:

$$f_1 = 6\left(\frac{0.416}{8.193}\right)^{0.5714} = 1.09 \text{ gpm}$$

$$f_2 = 6\left(\frac{0.416}{2.048}\right)^{0.5714} = 2.42 \text{ gpm}$$

$$f_3 = 6\left(\frac{0.416}{4.097}\right)^{0.5714} = 1.62 \text{ gpm}$$

$$f_4 = 6\left(\frac{0.416}{12.29}\right)^{0.5714} = 0.867 \text{ gpm}$$

Discussion: It is worth noting that the equivalent hydraulic resistance of the entire assembly (0.416), is smaller than the smallest individual hydraulic resistance (2.048). This will always be true. In fact, it provides a way of partially checking the results of the calculations.

Another way of checking the results is to add the individual branch flow rates, and verify that the total equals the flow rate entering the common point of the parallel paths:

$$f_{total} = f_1 + f_2 + f_3 + f_4 = 1.09 + 2.42 + 1.62 + 0.867$$
$$= 5.997 \approx 6 \text{ gpm (check)}$$

Note the very slight difference between the calculated total flow rate (5.997 gpm) and the original flow rate (6.0 gpm) is due to rounding off the calculations.

This example also shows that the larger the hydraulic resistance of a piping path, the lower its flow rate. In this example, the longest (300 foot) piping path has the lowest flow rate (0.867 gpm), while the shortest (50 foot) piping path has the highest flow rate (2.42 gpm).

6.8 REDUCING COMPLEX PIPING SYSTEMS

The concept of using series and parallel resistors to represent piping assemblies is a powerful tool when properly applied. An essential concept is that of alternating the use of the series and parallel equivalent resistor procedures to systematically "fold" the complex piping system down to a single equivalent resistance. A system resistance curve can then be produced that will assist in choosing a circulator for the system. The single equivalent resistance can then be "unfolded" back to the individual parallel path resistances to find the flow rates in each path.

The best way to illustrate this concept is to go through an example using a typical hydronic distribution system.

Example 6.13: Use the concepts of series and parallel resistances to reduce the system shown in Figure 6–45 to a single equivalent resistance between points A and B. Find the flow rate in each branch circuit assuming the total system flow rate is 10 gpm.

Solution: Start by sketching a resistor diagram of the system between points A and B. Remember that the hydraulic resistance of two or more identical components can be lumped together and represented by a single resistor symbol. The hydraulic resistor diagram shown in Figure 6–46 uses one resistor symbol to represent the combined resistance of all the tubing segments within a given branch, and another to represent the total resistance of all elbows in the branch. The third resistor symbol in each branch represents the resistance of the zone valve.

The objective is to reduce the resistor diagram of Figure 6–46 down to a single equivalent hydraulic resistance that represents the entire piping system. The first step is to calculate the hydraulic resistances of the tubing, fittings, and zone valves for each parallel piping path. Since all the tubing, fittings, and valves are the same pipe size, this becomes a matter of totaling the equivalent length of all components within a branch, and then multiplying by α and c.

Branch #1: 120 ft tubing + 20(2 ft) elbows
+ 30 ft zone valve equivalent length = 190 ft

Branch #2: 78 ft tubing + 17(2 ft) elbows
+ 30 ft zone valve equivalent length = 142 ft

Branch #3: 165 ft tubing + 26(2 ft) elbows
+ 30 ft zone valve equivalent length = 247 ft

The density of water at 160 °F is determined as 61.02 lb/ft³. The dynamic viscosity of water at 160 °F is determined as 0.00027 lb/ft/sec. These values are then used to find the α value using Equation 6.12:

$$\alpha = \left(\frac{D}{\mu}\right)^{-0.25} = \left(\frac{61.02}{0.00027}\right)^{-0.25} = 0.04586$$

The value of c is found in Figure 6–18. For 3/4-inch smooth tube $c = 0.061957$

The hydraulic resistances of each branch can now be calculated:

Branch #1:

$$R_1 = \alpha c L_1 = (0.04586)(0.061957)(190) = 0.5399$$

Branch #2:

$$R_2 = \alpha c L_2 = (0.04586)(0.061957)(142) = 0.4035$$

Branch #3:

$$R_3 = \alpha c L_3 = (0.04586)(0.061957)(247) = 0.7018$$

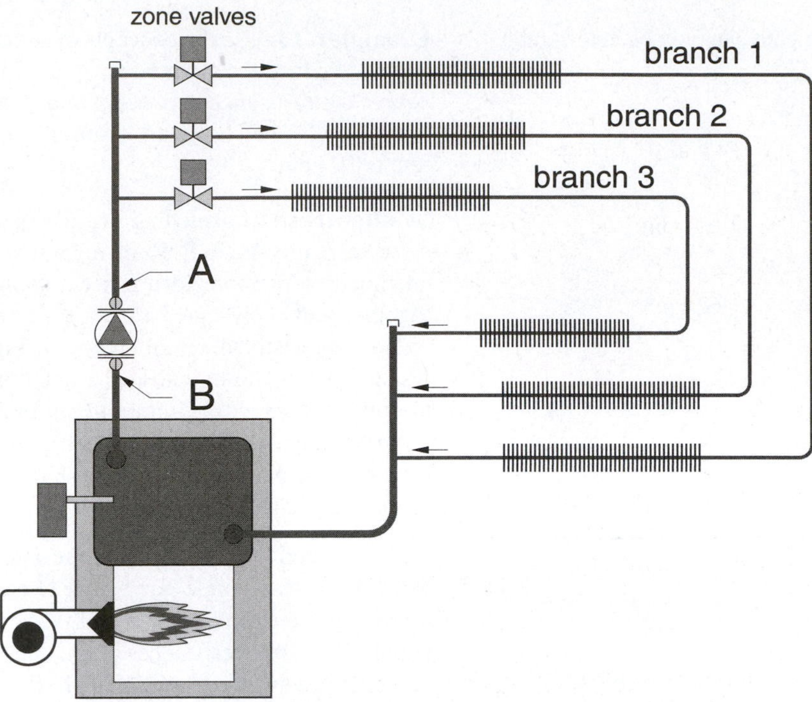

System Description
Three zone system using zone valves and the following fittings:
Branch 1: 120 ft. of 3/4 in. copper tube and 20, 90° elbows
Branch 2: 78 ft. of 3/4 in. copper tube and 17, 90° elbows
Branch 3: 165 ft. of 3/4 in. copper tube and 26, 90° elbows
Each zone valve is assumed to be equivalent to 30 ft. of 3/4 in. copper tube.
The boiler and common header piping is assumed to be equivalent to 20 ft. of 3/4 in. copper tube.
The system operates at an average water temperature of 160 °F.

Figure 6–45 Piping diagram to be analyzed in Example 6.11.

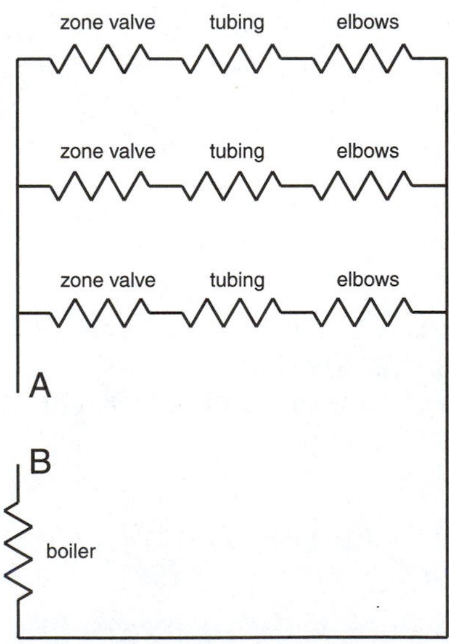

Figure 6–46 Resistor diagram for three-zone system of Example 6.11. Note that only one resistor is shown to represent all the elbows in a branch, and one resistor to represent all tubing segments in a branch.

The resistor diagram can now be reduced as shown in Figure 6–47.

The three parallel resistors can now be reduced to a single equivalent resistance using Equation 6.21.

$$R_{\text{equivalent}} = \left[\left(\frac{1}{0.5399}\right)^{0.5714} + \left(\frac{1}{0.4035}\right)^{0.5714} + \left(\frac{1}{0.7018}\right)^{0.5714}\right]^{-1.75} = 0.07706$$

The resistor diagram can now be further reduced as shown in Figure 6–48.

The boiler is assumed to have an equivalent length equal to 20 feet of 3/4-inch copper tube. Its hydraulic resistance can therefore be calculated as:

$$r_{\text{boiler}} = (\alpha\ c\ L) = (0.04586)(0.061957)(20) = 0.05683$$

The hydraulic resistance of the boiler is in series with the equivalent resistance of the parallel piping paths. Therefore, the boiler's resistance will be added to the equivalent

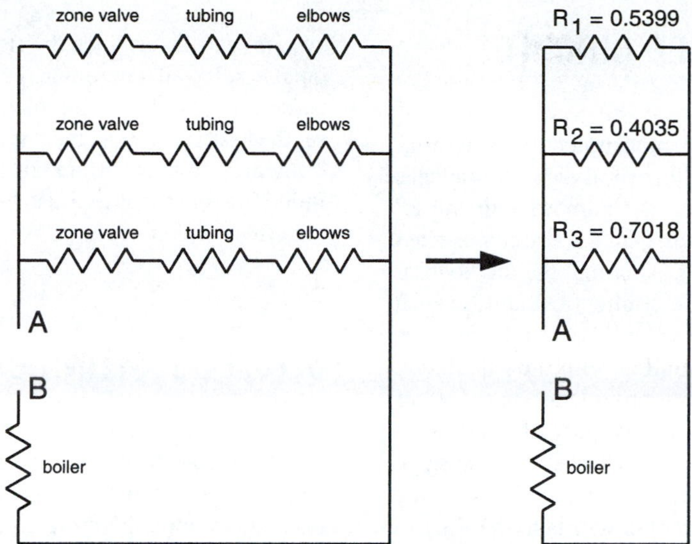

Figure 6–47 First reduction of resistor diagram for Example 6.11.

resistance of the parallel paths to get the overall equivalent resistance of the circuit as depicted in Figure 6–49.

$$r = 0.05683 + 0.07706 = 0.1339$$

Assuming the circulator produces a flow rate of 10 gpm within the system, and all the zone valves are open, we can now find the flow rate in each parallel branch by repeated use of Equation 6.22:

$$f_1 = 10\left(\frac{0.07706}{0.5399}\right)^{0.5714} = 3.29 \text{ gpm}$$

$$f_2 = 10\left(\frac{0.07706}{0.4035}\right)^{0.5714} = 3.88 \text{ gpm}$$

$$f_3 = 10\left(\frac{0.07706}{0.7018}\right)^{0.5714} = 2.83 \text{ gpm}$$

Add the individual branch flows to verify their total equals the system flow rate:

$$f_{total} = f_1 + f_2 + f_3 = 3.29 + 3.88 + 2.83 = 10 \text{ gpm (check)}$$

Discussion: A point worth noting is that the flow rate in any one circuit could be found assuming one or both of the other circuits were closed off at the zone valves. To do this, go back to Equation 6.21 and treat the hydraulic resistance of the closed branch (or branches) as infinite. This has the same effect as simply ignoring the closed zones as parallel resistors. For example, assuming the zone valve on branch #1 was closed, one would find the equivalent resistance of parallel resistors R_2 and R_3 only, and ignore the presence of R_1. The remainder of the procedure would be similar to that shown previously. Knowing the branch flow rates will eventually help in properly sizing the heat emitters in each branch path.

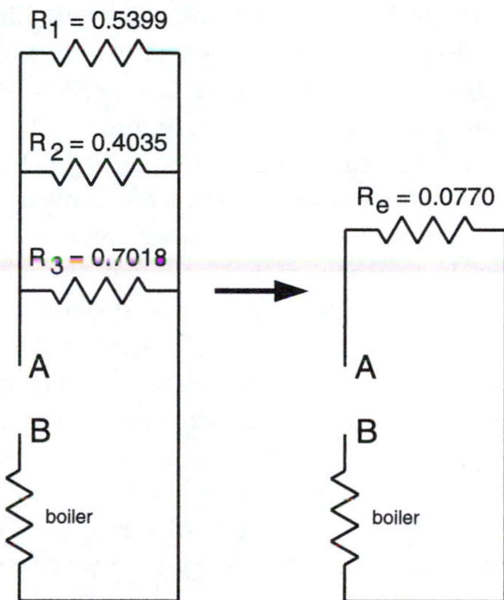

Figure 6–48 Second reduction of resistor diagram for Example 6.11.

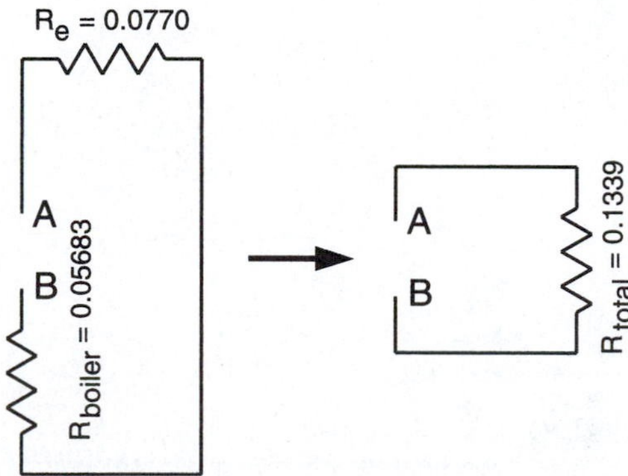

Figure 6–49 Final simplification of resistor diagram into a single hydraulic resistor that represents the entire system between points A and B.

6.9 SOFTWARE-BASED CIRCUIT ANALYSIS

The previous examples have demonstrated that determining the flow rates in systems with several parallel branches requires a considerable amount of "number crunching." Piping systems even more complex than the one represented in Figure 6–45 are often required. Although the methodology demonstrated can be used to determine flow rates in such complex systems, the necessary calculation could requires hours of effort, even for those familiar with the calculation procedures. Such calculations only provide a "snapshot" of the system's performance under a given set of conditions. Any changes, such as use of a different branch pipe size, or the addition of another heat emitter, will require most of the calculations to be redone. Few people will take the time to throughly analyze multiple options when faced with such a task.

The only practical approach in such situations is computer analysis. To that end the **Hydronic Circuit Simulator module** has been developed. A version of this software is included on the Hydronic Design Studio CD-ROM. A screen image from the program is shown in Figure 6–50.

The Hydronic Circuit Simulator uses a graphical user interface to configure a piping system containing up to three parallel branches (Figure 6–51). The user can define the piping components in each branch, as well as the piping between the branches. The user can also select from several system fluids, and dozens of available circulators.

The Hydronic Circuit Simulator uses the same analytical procedures described in this chapter to calculate the flow rates in the parallel branches. To analyze a series circuit, simply set the number of branches to one.

Each time the user changes one of the piping elements, the program recalculates the hydraulic resistance of the system, and determines the flow rates in all branches based on the selected circulator. Methods for the latter are discussed in Chapter 7, Circulating Pumps. The user can also set the hydraulic resistance of balancing valves in each of the branches to determine how the flow rates readjust. Individual branches can also be turned on and off to simulate the effect of zone valves.

6.10 PIPE SIZING CONSIDERATIONS

Several considerations affect the selection of a pipe size for a given application. Among these are pressure loss, flow velocity, potential for erosion damage, operating noise, installation, and operating cost.

Flow Velocity

When selecting a pipe size for a given flow rate, the resulting average flow velocity should be between 2 and 4 feet per second.

The lower end of this velocity range is based on the ability of flowing water to move air bubbles along a vertical pipe. Average flow velocities of 2 feet per second or higher can entrain air bubbles in a downward flow. The objective is to route the bubbles to an air separator where they can be collected and discharged from the system.

The upper end of this range (4 feet per second) is based on minimizing noise generated by the flow. Average flow velocities higher than 4 feet per second cause flow noise and should be avoided.

Figure 6–2 lists the minimum and maximum flow rates corresponding to these flow velocities for type M copper tubing from 3/8 inch to 3 inch nominal diameter.

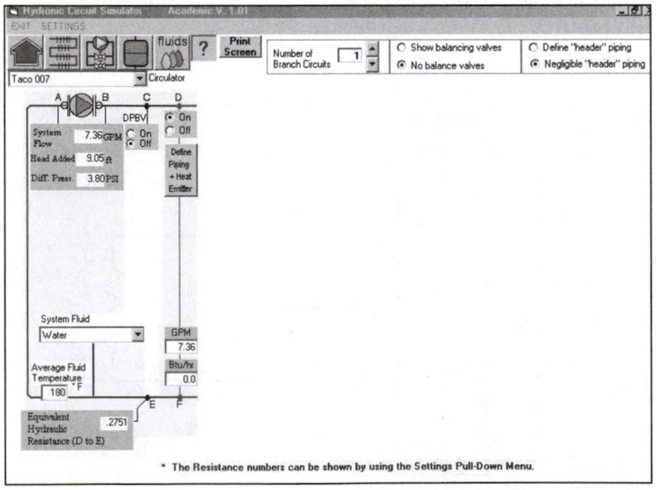

Figure 6–50 User interface screen for the Hydronic Circuit Simulator module in the Hydronics Design Studio software.

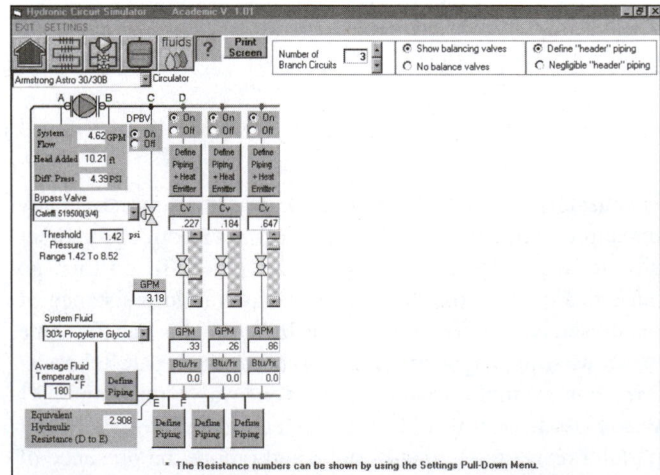

Figure 6–51 User interface screen for the Hydronic Circuit Simulator module in the Hydronics Design Studio software.

Erosion Corrosion

Copper tubing is also subject to another problem when flow velocities extend into the 5 to 8 feet per second range. A condition called **erosion corrosion** can actually scrub metal off the inside wall of a tube or fittings. This effect tends to be localized at tight turns in fittings such as elbows and tees, and has not been known to be a widespread problem except in extreme cases. Still it is prudent to avoid any potential of its occurrence by sizing pipe for lower flow velocities.

Operating Cost

As the flow through a pipe increases, the head loss associated with the flow increases rapidly. Equation 6.9 can be used to show that when the flow rate through a smooth tube is doubled, the head loss increases by a factor of about 3.4. The greater the head loss, the more pumping power is required to maintain a given flow. This in turn can require a larger circulator with a greater electrical power demand, and higher operating cost. Because of this, a piping system should be thought of as having an operating cost as well as an installation cost.

The determination of an optimal pipe size based on minimizing total owning and operating cost can be a complex process. It involves the use of specific installation costs, utility rates, pump efficiency, estimates of hours of operation, rate of inflation of electrical rates, and more. However, since other considerations such as flow velocity often narrow the choice down to two sizes, it is prudent to estimate their operating cost and consider the results when making the final selection.

Equation 6.23 can be used to estimate the theoretical annual operating cost of any piping system transporting a fluid at a certain flow rate and head loss:

(Equation 6.23)

$$E = \frac{(3\times 10^{-6})(D)(H_L)(f)(T)(k)}{\eta_p}$$

where:

E = annual operating cost of the piping system (\$/yr)
D = density of the fluid at its typical operating temperature (lb/ft^3)
H_L = head loss of the piping system at flow rate f (feet of head)
f = flow rate associated with the above head loss (gpm)
T = number of hours per year during which the circulator operates (hr/yr)
k = cost of electrical energy (\$/kwhr)
η_p = efficiency of the motor/pump assembly (decimal percent)

Example 6.14: Determine the operating cost of a piping system that transports 160 °F water at a flow rate of 10 gpm with a head loss of 9 ft, for 4,000 hours per year. Assume that electrical energy costs \$0.10/kwhr, and that the pump/motor combination has an efficiency of 25%.

Solution: The density of water at 160 °F is 61.02 lb/ft^3. Substituting the data into Equation 6.23:

$$E = \frac{(3\times 10^{-6})(61.04)(9)(10)(4000)(0.10)}{0.25} = \$26.37/\text{year}$$

Discussion: Keep in mind that over the life of the system, this annual operating cost will likely add up to several hundreds of dollars.

If Equation 6.23 is combined with Equation 6.11, the head loss term, H_L, can be eliminated, and the following relationship results:

(Equation 6.24)

$$E = \frac{(3\times 10^{-6})(D)(\alpha c L)(f^{2.75})(T)(k)}{\eta_p}$$

where:

E = annual operating cost of the piping system (\$/yr)
D = density of the fluid at its typical operating temperature (lb/ft^3)
α = fluid properties factor for water at a given temperature
c = pipe size factor for the size of copper tube used
L = total equivalent length of a piping path (ft)
f = flow rate associated with the above head loss (gpm)
T = number of hours per year during which the circulator operates (hr/yr)
k = cost of electrical energy (\$/kwhr)
η_p = efficiency of the motor/pump assembly (decimal percent)

This equation can be used to compute the annual operating cost of any piping system using a pump to convey water through copper pipe at a given temperature.

Example 6.15: Assume a hydronic heating system has a total equivalent length of 250 ft of 3/4-in copper tube. It conveys 140 °F water at 6 gpm for 4,000 hours per year. Electrical energy costs \$0.10/kwhr, and the circulator will have an efficiency of 25%. Determine the annual operating cost. Repeat the calculation for the same system using 1-in copper tube.

Solution: From the graph in Figure 4–3, the water's density is 61.35 lb/ft^3. From the graph in Figure 4–6, the water's dynamic viscosity is 0.00032 lb/ft/sec. The α value can be calculated using Equation 6.12:

$$\alpha = \left(\frac{D}{\mu}\right)^{-0.25} = \left(\frac{61.35}{0.00032}\right)^{-0.25} = 0.04779$$

The value of c is found in Table 6–18. For 3/4-inch copper tube $c = 0.061957$.

Substituting all data into Equation 6.24:

$$E = \frac{(3 \times 10^{-6})(61.35)(0.04779)(0.061957)(250)(6^{2.75})(4000)(0.10)}{0.25}$$

$$= \$30.08/\text{year}$$

If the same system were built using 1-inch copper tube, and operated at the same 6 gpm flow rate, only the value of c would change in the above equation. For 1-inch tubing: $c = 0.01776$. The revised operating cost would be:

$$E = \frac{(3 \times 10^{-6})(61.35)(0.04779)(0.01776)(250)(6^{2.75})(4000)(0.10)}{0.25}$$

$$= \$8.62/\text{year}$$

In this particular case, there would be a *theoretical* savings of $21.46 per year if the 1-inch tube were used instead of the 3/4-inch tube. Such a savings could quickly pay for the increased installation cost of the larger tubing. However, in order to obtain this energy savings, a circulator that would produce exactly the same flow rate and head within the larger piping would have to be used. Since only a finite selection of circulators is available, the actual energy savings will depend on the difference in wattage of the smaller circulator used to replace the larger circulator. The savings could be calculated using Equation 6.25.

(Equation 6.25)

$$S = \frac{(w_{PL} - w_{PS})(T)(k)}{1000}$$

where:

S = annual *savings* from being able to use a small circulator ($/yr)

w_{PL} = wattage of the larger circulator (watts)

w_{PS} = wattage of the smaller circulator (watts)

T = number of hours per year during which the circulator operates (hr/yr)

k = cost of electrical energy ($/kwhr)

Example 6.16: By using 1-inch copper tubing rather than 3/4-in copper tubing, a designer finds that a smaller circulator that operates at 85 watts can be used instead of a larger circulator that operates at 200 watts. Estimate the annual savings in using the smaller circulator assuming it will operate for 4,000 hours/year in a location where electricity costs $0.10/kwhr.

Solution: Substituting the data in Equation 6.25:

$$S = \frac{(200 - 85)(4000)(0.10)}{1000} = \$46/\text{year}$$

Discussion: Again it should be emphasized this savings will accumulate year after year. The savings will also be increased each time the cost of electricity increases. Total savings over the life of the system could be several hundreds of dollars.

Selecting a Pipe Size

The following procedure incorporates the previously discussed concepts into a method for choosing an appropriate pipe size:

Step 1. Select a tentative pipe size based on the criteria of keeping the flow rate between 2 and 4 feet per second. Use Figure 6–2 as a guide when working with copper tubing between 3/8" and 3" in size.

Step 2. Estimate the installed cost of the system's piping, or portion thereof, using this tentative pipe size.

Step 3. Estimate the operating cost of the system using Equation 6.24 for the tentative pipe size.

Step 4. Estimate the installed cost of the system's piping, or portion thereof, using the next larger pipe size.

Step 5. Estimate the operating cost of the system using Equation 6.24 for the next larger pipe.

Step 6. If the potential savings in operating cost between Steps 3 and 5 would return the higher installation cost of the large pipe in a reasonable period of time (suggested as 10 years or less), go on to Step 7. If not, use the pipe size from Step 1.

Step 7. Using methods from Chapter 7, Circulating Pumps, select two circulators, one for each pipe size being considered, that will produce the desired flow rate within the system. Assuming that a circulator that is smaller, less expensive, and that uses less power is available for use with the larger pipe size, compare the difference in installation and operating cost of the two circulators. If the savings in installation and operating cost of the small circulator over a period of a few years exceeds the higher installation cost of the larger pipe, use the larger pipe.

Example 6.17: Assume the same pipe systems in Example 6.15. The system's circulator produces 6 gpm, and operates 4000 hours per year using electrical energy purchased at $0.10/kWhr. Assume 3/4-inch copper pipe costs $0.85/ft., and 1-inch copper pipe costs $1.40/ft. Assume the following circulators were selected for each pipe size:

Using 3/4-inch pipe:
Circulator cost = $110 Operating wattage = 150 watts

Using 1-inch pipe:
Circulator cost = $ 65 Operating wattage = 90 watts

Determine which pipe size should be used after considering both technical and economic factors.

Solution

Step 1. Check that the flow velocity in the smaller pipe is equal to or less than the 4 ft/sec limit. From Figure 6–2 one finds that a 3/4-inch copper pipe can carry up to 6.4 gpm and not exceed this limit. Thus 3/4-inch pipe is acceptable at the stated flow rate of 6 gpm, and the potential use of the large pipe size will depend on economic considerations.

Step 2. The installed cost of the 3/4-inch pipe will be (250 ft.)($0.85/ft.) = $212.50.

Step 3. The estimated operating cost using 3/4-inch piping was calculated in Example 6.13 to be $30.08.

Step 4. The installed cost of the 1-inch pipe will be (250 ft.)($1.40/ft.) = $350.00.

Step 5. The estimated operating cost using 1-inch piping was calculated in Example 6.13 to be $8.62.

Step 6. The theoretical savings in operating cost is: $30.08 – $8.62 = $21.46 per year.

The extra installation cost of the 1-inch tubing is: $350 – $212.50 = $137.50.

The payback associated with use of the larger tubing is:

$$\frac{\$137.50}{\$21.46/\text{year}} = 6.4 \text{ years}$$

Since this is a reasonably short payback period, proceed to compare costs and savings associated with the circulators.

Step 7. The savings in purchase cost of the smaller circulator is $110 – $65 = $45.

The *net* increase in system cost using the larger pipe would be the higher cost of the larger pipe minus the savings due to the smaller circulator:

Net increase in system cost = $137.5 – $45 = $92.50

The actual savings in annual operating cost can be calculated using Equation 6.25:

$$E = \frac{(150-90)(4000)(0.10)}{1000} = \$24/\text{year}$$

The time required for operating cost savings to recover the net cost increase would be:

$$\frac{\$92.50}{\$24/\text{year}} = 3.9 \text{ years}$$

Discussion: This example shows a relatively fast return on the extra investment in the large pipe. Furthermore, any increase in electrical rates would reduce this payback period. *Over a system life of perhaps 30 years, the operating cost savings associated with the larger pipe size would return the higher initial investment several times over!* The larger pipe size is well-justified in this case.

SUMMARY

This chapter has laid the foundation for describing fluid flow in piping systems. Many of the fundamentals of fluid mechanics have been condensed and presented in a way especially suited for hydronic heating systems. Terms such as head loss, pressure drop, flow rate, flow velocity, and more have been described. Such terms will be routinely used in later chapters.

The analogy between hydraulic resistance and electrical resistance represents a new approach for analyzing fluid-filled closed-loop piping systems, one that can be a powerful tool when properly applied. This approach will also be used in later chapters dealing with system design.

The reader is strongly encouraged to use the Fluid Properties Calculator module and Hydronic Circuit Simulator module in the Hydronics Design Studio software to expedite many of the calculations demonstrated in this chapter. These modules allow the reader to try many what-if scenarios related to piping system performance. This is an excellent way to gain a solid understanding of the principles in this chapter.

KEY TERMS

average flow velocity
boundary layer
compressible fluids
elevation head
equivalent length
equivalent resistance
equivalent resistor
erosion corrosion
feet of head
flow coefficient
flow rate
flow velocity
fluid properties factor
gas
hydraulic resistance
hydraulic resistance diagrams
hydraulic resistor
Hydronic Circuit Simulator module
incompressible
laminar
liquid
pipe size coefficient
piped in parallel
pressure head
Reynold's number
series circuit
series piping path
static pressure
system resistance curve

total equivalent length
total head
turbulent
velocity head
velocity profile
viscous friction

CHAPTER 6 QUESTIONS AND EXERCISES

1. Find the flow velocities corresponding to the following conditions:
   a. 2.5 gpm flow rate in a 1/2-inch type M copper tube
   b. 10 gpm flow rate in a 3/4-inch type M copper tube
   c. 15 gpm in a tube with an inside diameter of 1.1 inches
2. Use data from Chapter 4, Properties of Water, to determine the value of the fluid properties factor, α, for water at 120 °F.
   Hint: Use Equation 6.12. Compare this to the value obtained using Figure 6–17.
3. What water pressure would be required at street level to push water to the top of a 300-foot building? Express the answer in psi. If the water pressure at the top of the building had to be 50 psi, how would this affect the pressure at the street level?
4. Water at 150 °F moves through a pipe with an internal diameter of 1.3 inches, and at a flow velocity of 3 feet/second. Determine the Reynold's number for this flow. Would the flow be laminar or turbulent under these conditions?
5. Determine the head loss when 200 °F water flows at 10 gpm through 400 feet of 1-inch type M copper tube. Determine the pressure drop (in psi) associated with this head loss.
6. A valve has a C_V rating of 3.5. What will be the pressure drop across the valve when 140 °F water flows through it at 6 gpm?
7. Find the total equivalent length of the piping path shown in Figure 6–52.

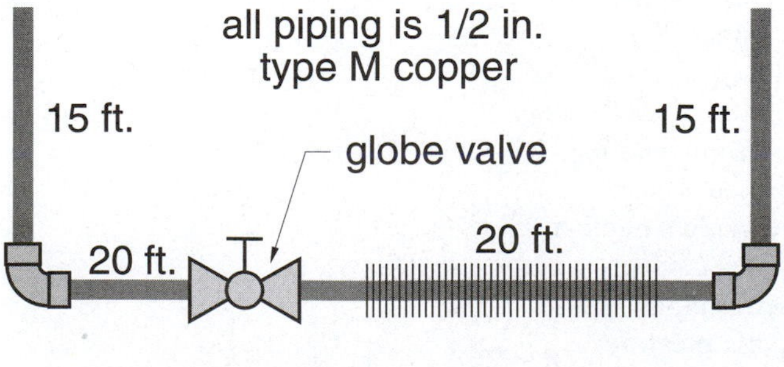

Figure 6–52 Piping assembly for Exercise 7.

8. Find the hydraulic resistance of the piping assembly shown in Exercise 7 assuming water at 120 °F will flow through it.
9. Calculate five data points representing head loss versus flow rate through the piping assembly shown in Exercise 7 for 120 °F water. Plot this data and sketch a system resistance curve.
   Hint: Use Equation 6.11 to generate the necessary head loss versus flow rate data.
10. Use Equation 6.20 to find the equivalent hydraulic resistance of the three parallel hydraulic resistors shown in Figure 6–53.
11. Assuming a flow rate of 8 gpm enters at point A in Figure 6–53, determine the flow rate in each of the three branches. Check your results.
12. Determine the single equivalent resistance between points A and B for the hydraulic resistor diagram shown in Figure 6–54.
13. Draw a hydraulic resistor diagram for the piping assembly shown in Figure 6–55.
    Assume 1-inch copper tubing is used, and that the system operates with 140 °F water:
    a. Using series and parallel resistance concepts, reduce the resistor diagram to a single equivalent resistance.
    b. Assuming flow enters the piping assembly at 15 gpm, find the flow rate in each branch.
    c. Find the flow velocity in each branch.

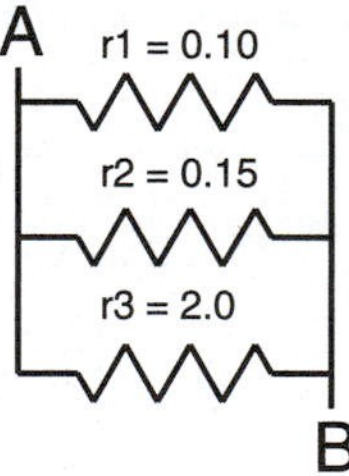

Figure 6–53 Hydraulic resistor diagram for Exercises 10 and 11.

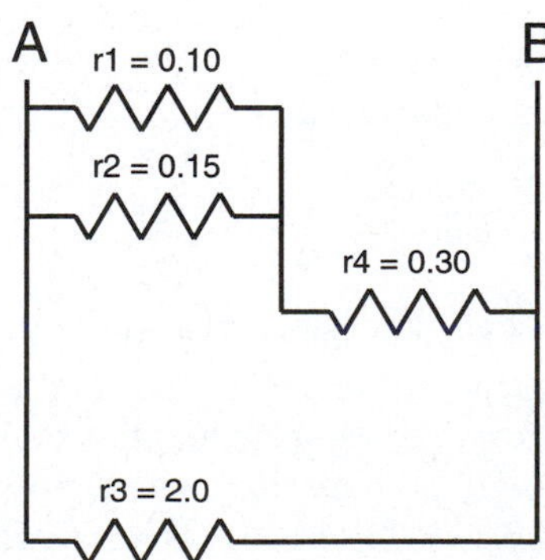

Figure 6–54 Hydraulic resistor diagram for Exercise 12.

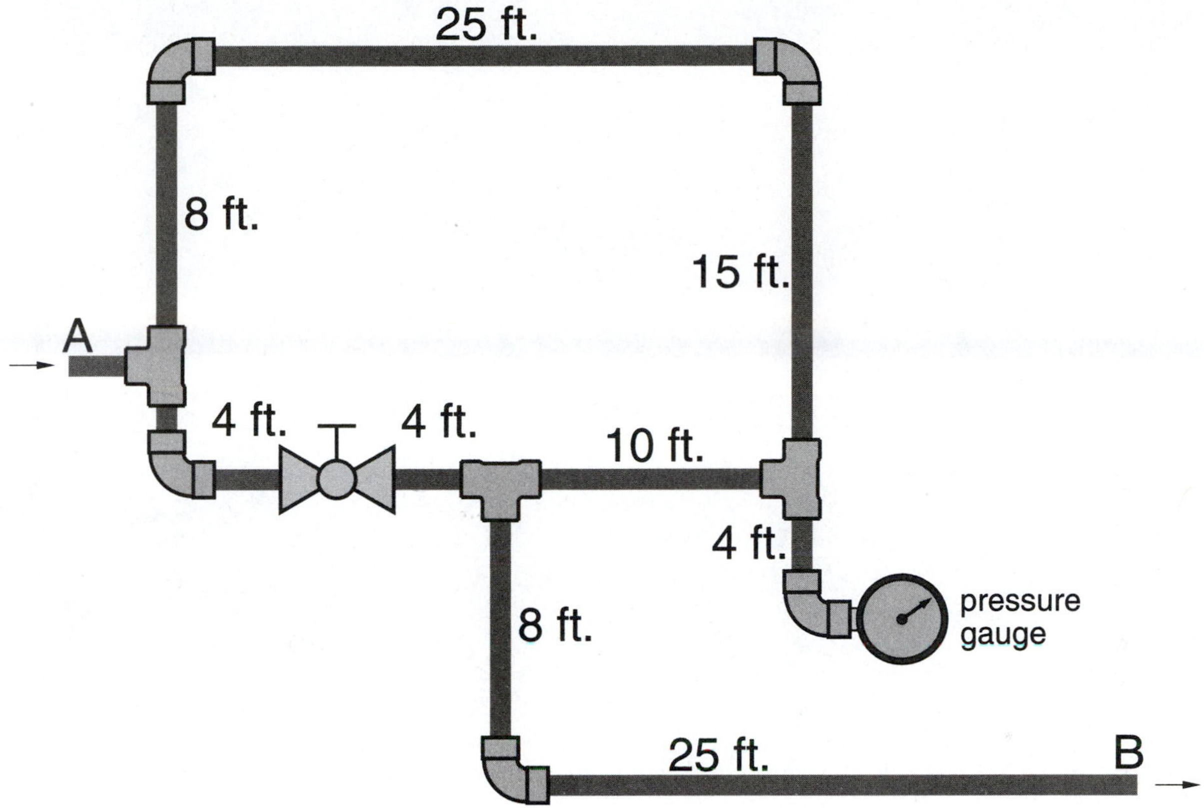

Figure 6–55 Piping assembly for Exercise 13.

14. Use the Fluid Properties Calculator program in the Hydronics Design Studio to find the answer to Exercise 2. Also find the α value for the following fluids and conditions:
    a. 140 °F water
    b. 50 percent propylene glycol solution at 140 °F
    c. 30 percent ethylene glycol solution at 160 °F
15. Use the Hydronic Circuit Simulator module to confirm the answer obtained for Example 6.13.
16. Explain the difference between head loss and pressure drop for a fluid flowing through a piping system.

CIRCULATING PUMPS 7

OBJECTIVES

After studying this chapter, you should be able to:

- Describe different types of circulating pumps used in hydronic systems
- List the main components of a centrifugal pump
- Determine the proper placement of the pump(s) within the system
- Calculate the flow rate a specific circulator will produce in a given piping system
- Estimate the flow rate through a pump from the pressure differential across it
- Work with both graphical and analytical descriptions of pump performance
- Predict the performance of pumps connected in series and parallel
- Explain what cavitation is and how to predict it
- Describe methods of avoiding cavitation
- Select a pump for proper performance and high efficiency
- Describe the use of variable speed pumps in hydronic systems

7.1 INTRODUCTION

The pump is the heart of a hydronic system. Although smaller and less expensive than the heat source, it can prove to be just as vital to proper system performance. The wide variety of circulator pumps available in today's market allows great flexibility in how the overall system is designed and controlled.

This chapter provides an overview of the type of circulating pumps used in small- and medium-size hydronic systems. It goes on to define and illustrate the concept of a pump curve. This curve is combined with the system resistance curve discussed in Chapter 6, Fluid Flow in Piping, to determine the flow rate at which the system will operate. The often-overlooked concepts of series and parallel pumps are discussed from the standpoint of customizing a pumping system to the specific needs of a piping system. The proper placement of the pump in the system is also discussed. Special attention is given to the subject of **cavitation** and its avoidance. Finally, the use of variable speed pumps in hydronic distribution systems is discussed.

7.2 PUMPS FOR HYDRONIC SYSTEMS

Pumps come in a wide variety of designs, sizes, and performance ranges. In closed-loop, fluid-filled hydronic systems, the pump's function is to circulate the system fluid around the piping. No lifting of the fluid is involved in such systems, as discussed in Chapter 6. For this reason, and especially when small fractional horsepower motors are used, pumps are often called **circulator**s. Within this book, the terms circulator and pump are synonymous.

The type of pump commonly used in hydronic systems is known as a **centrifugal pump**. It uses a rotating component called an **impeller** to add mechanical energy (head) to the fluid. Figure 7–1 illustrates the basic construction of a centrifugal pump.

As the impeller rotates, fluid within the center opening (or **eye**) of the impeller is rapidly accelerated through the passageways formed by the impeller vanes between the two impeller disks. The fluid's mechanical energy content is increased as it is accelerated toward the outer edge of the impeller. As it leaves the impeller, the fluid impacts against the inside surface of the chamber surrounding the impeller. This chamber is called the **volute**. Its former speed (or velocity head) is converted to a pressure increase (pressure head). The fluid then flows around the contoured volute and exits through the discharge port.

For this process to be continuous, the rate of fluid entering the pump must be identical to that leaving the pump. Intuition might suggest a pump continuously "sucks" fluid into the eye of its impeller. This is not true. *Water entering a centrifugal pump must be pushed in by system pressure upstream of the inlet port.* This is a very important point, and is often misunderstood. *If the proper conditions are not provided for fluid to be pushed into the inlet port, very undesirable operating characteristics will result.* These are discussed later in the chapter.

Centrifugal pumps can be built with differently shaped volutes while still maintaining the same internal operation. The volute's shape determines how the pump will be connected to the system's piping. Figure 7–2 shows two examples of **inline circulators**. These circulators have their inlet and discharge ports along a common centerline. An inline circulator can be placed into a piping path without need for any lateral offsets between the inlet and discharge ports. *The inline circulator design is by far the*

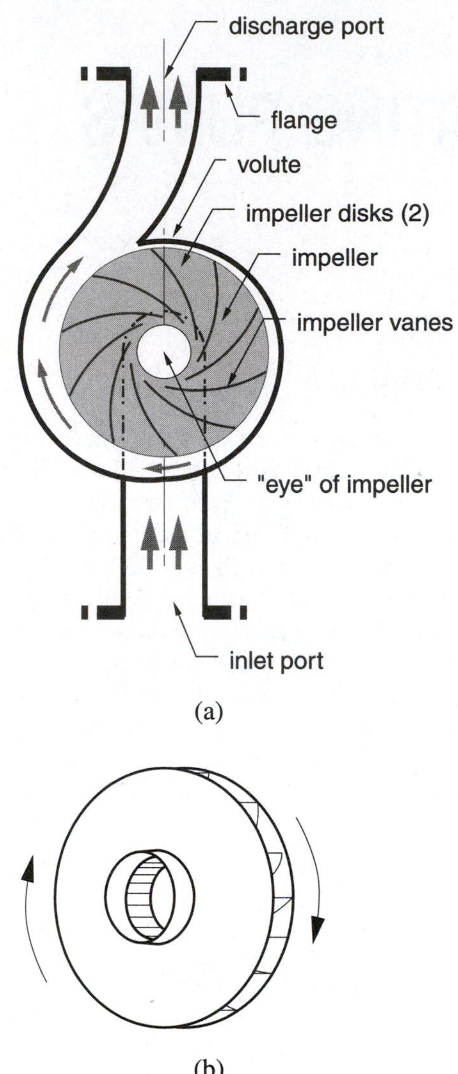

Figure 7–1 (a) Simplified cross section of a centrifugal pump, (b) an impeller.

(a)

(b)

Figure 7–2 Examples of inline circulators. (a) A wet-rotor circulator, (b) a mechanically-coupled three-piece circulator. *Courtesy of Armstrong Pumps.*

most common type used in residential and light commercial hydronic systems.

By modifying the shape of the volute, a design called an **end suction pump** is created. An example of an end suction centrifugal pump is shown in Figure 7–3.

End suction pumps create a 90-degree turn in the system piping. In most cases, an offset between the centerlines of the inlet and discharge piping is also required. End suction pumps are usually floor-mounted, and are more common in larger hydronic systems.

Wet Rotor Circulators

Over the last few decades, a specialized design for small- and medium-sized circulators has been refined by a number of pump manufacturers specifically for use in hydronic systems. This design, known as a **wet-rotor circulator**, combines the rotor, shaft, and impeller into a single assembly that is housed in a chamber filled with system fluid. An example of a wet-rotor circulator is shown in Figure 7–4.

The motor of a wet-rotor circulator is totally cooled and lubricated by the system's fluid. As such, it has no fan or oiling caps. The rotor assembly is supported on ceramic or graphite bushings in the rotor can. These bushings contain no oil, but ride on a thin film of system fluid. The rotor is surrounded by the stator assembly of the motor.

Some advantages of wet rotor circulators are as follows:

- There is no oiling required
- There is no leakage of system fluid due to worn pump seals

Figure 7–3 Example of an end suction centrifugal pump. *Courtesy of Bell and Gossett.*

- Their small size makes them easy to locate and support
- The absence of a cooling fan and external coupling makes for quiet operation
- Several models are available with multiple speed motors
- They are relatively inexpensive due to fewer parts
- They are ideal for applications where limited flow rate and head are required
- They can be **close-coupled** for series pump applications
- Most have permanent split capacitor (PSC) motors that can be operated over a wide speed range by suitable electronic controls

Their disadvantages include:

- The low starting torque of the PSC motors may not be able to free a stuck impeller after a period of prolonged shutdown
- Servicing anything not contained in the external junction box requires opening the wetted part of the pump, resulting in some fluid loss and air entry into the system

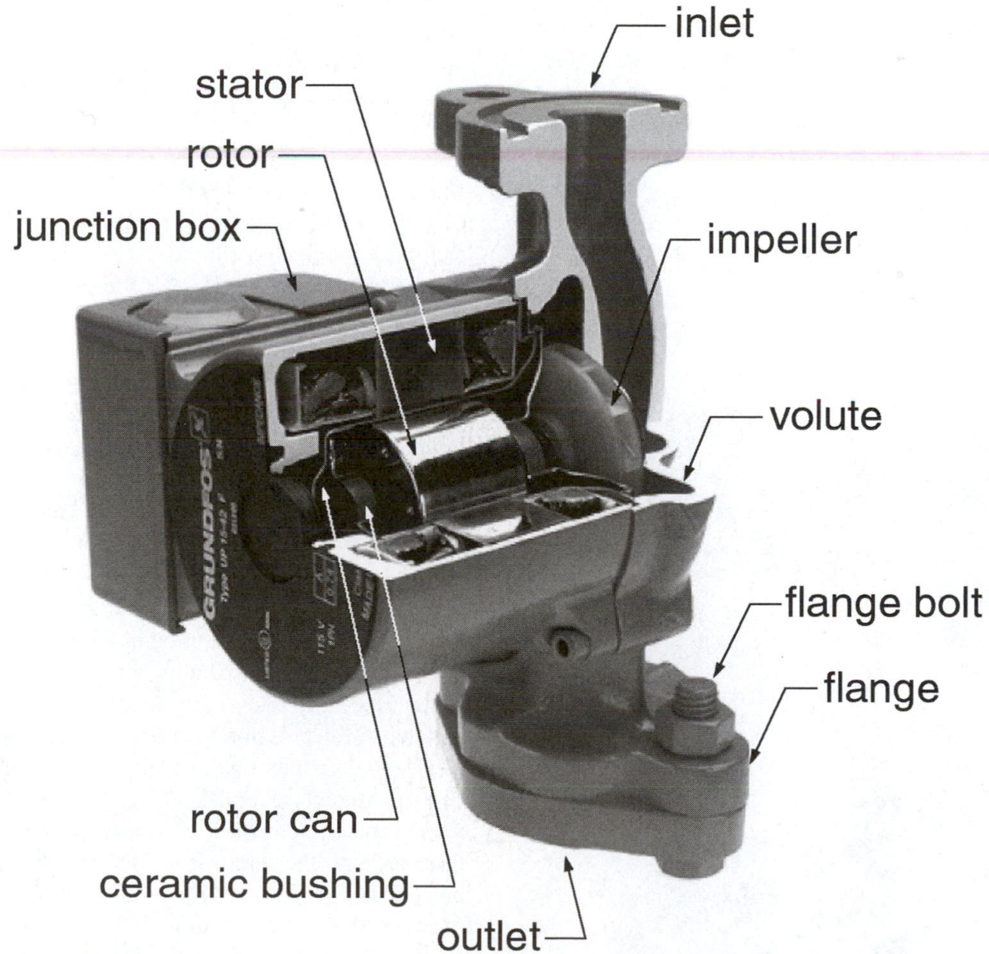

Figure 7–4 Cut-away view of a small wet rotor circulator. *Courtesy of Grundfos Pumps Corp.*

Wet rotor circulators are the most commonly used pumps in modern residential and light commercial hydronic systems. They are available with cast-iron volutes for use in closed hydronic systems, or with bronze or stainless steel volutes for direct contact with domestic water, or other open-loop applications. Most wet rotor circulators have impellers constructed of stainless steel, bronze, or synthetic materials.

Three-piece Circulators

Another common pump design used in small- and medium-sized hydronic systems is called a **three-piece circulator**. It consists of the pump body assembly, coupling assembly, and motor assembly as shown in Figure 7–5.

Unlike wet rotor circulators, the motor of a three-piece circulator is totally separate from the wetted portion of the pump. This allows the motor or coupling to be serviced or replaced without needing to open the piping system. The design of the coupling assembly between the motor and pump shaft varies among manufacturers. A common design employs a spring assembly that absorbs vibration or high torque between the two shafts as the motor starts. The impeller shaft penetrates the volute through **shaft seals** that must limit leakage of system fluid, even under high pressure. Although it is normal for these seals to experience minor fluid losses, the leakage rate of modern pump seals is so small the fluid usually evaporates before it is seen.

The advantages of a three-piece pump design are:

- Potential for longer life if bearings are properly lubricated

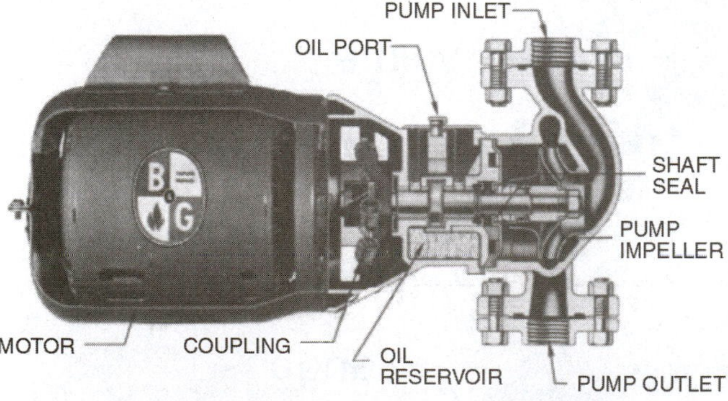

Figure 7–5 Example of a three-piece circulator. *Courtesy of ITT Bell and Gossett.*

- Easy servicing of the motor without need to open wetted portion of pump
- Ability to produce higher starting torque to overcome a stuck impeller condition after a prolonged shutdown
- May use higher efficiency non-PSC motors

The disadvantages include:

- Heavier construction requiring strong supports
- Must be oiled periodically
- More operating noise due to external motor and coupling assembly
- Potential maintenance of mechanical seals and coupling assembly
- Defective or worn shaft seals could allow leakage of system fluid

Like wet rotor circulators, three-piece circulators are available in materials suitable for both open-loop and closed-loop applications.

Pump Mounting Considerations

Most circulators used in smaller hydronic systems are designed to be installed with their shafts in a horizontal position. This removes the thrust load on the bushings due to the weight of the rotor and impeller.

The direction of flow through the pump is usually indicated by an arrow on the side of the volute. *The installer should always check that the pump is installed in the correct flow direction.* As long as the shaft is horizontal, the pump can be mounted with the flow arrow pointing upward, downward, or horizontally. Of these, upward flow is slightly preferred when possible, because it allows the circulator to rapidly clear itself of air bubbles.

In general, the weight of the circulator should not be supported by system piping unless the circulator is relatively light and the piping itself is well-supported adjacent to the circulator. This is especially true for three-piece circulators that create a bending effect on the pipe due to the offset weight of the motor. If the circulator is supported by system piping, the piping should not be rigidly mounted to walls that can transmit vibrations to the building structure. The use of vibration-absorbing mounting brackets is recommended. The piping supporting circulators should also not be mounted to framed partitions adjoining living or sleeping spaces where any noise transmission would be objectionable.

A common practice in many residential installations has been to mount several small circulators that are part of a multi-zone system on a common header assembly directly threaded into the return side of the boiler. Threaded steel or black iron pipe in the 1-inch to 2-inch range is often used to fabricate this header. Although this pipe is strong enough to handle the weight of the circulators, the bending effect imposed on the boiler connection can be very high. For this reason, whenever several circulators are mounted to a common header, the far side of the header should be well-supported to relieve this bending load as shown in Figure 7–6.

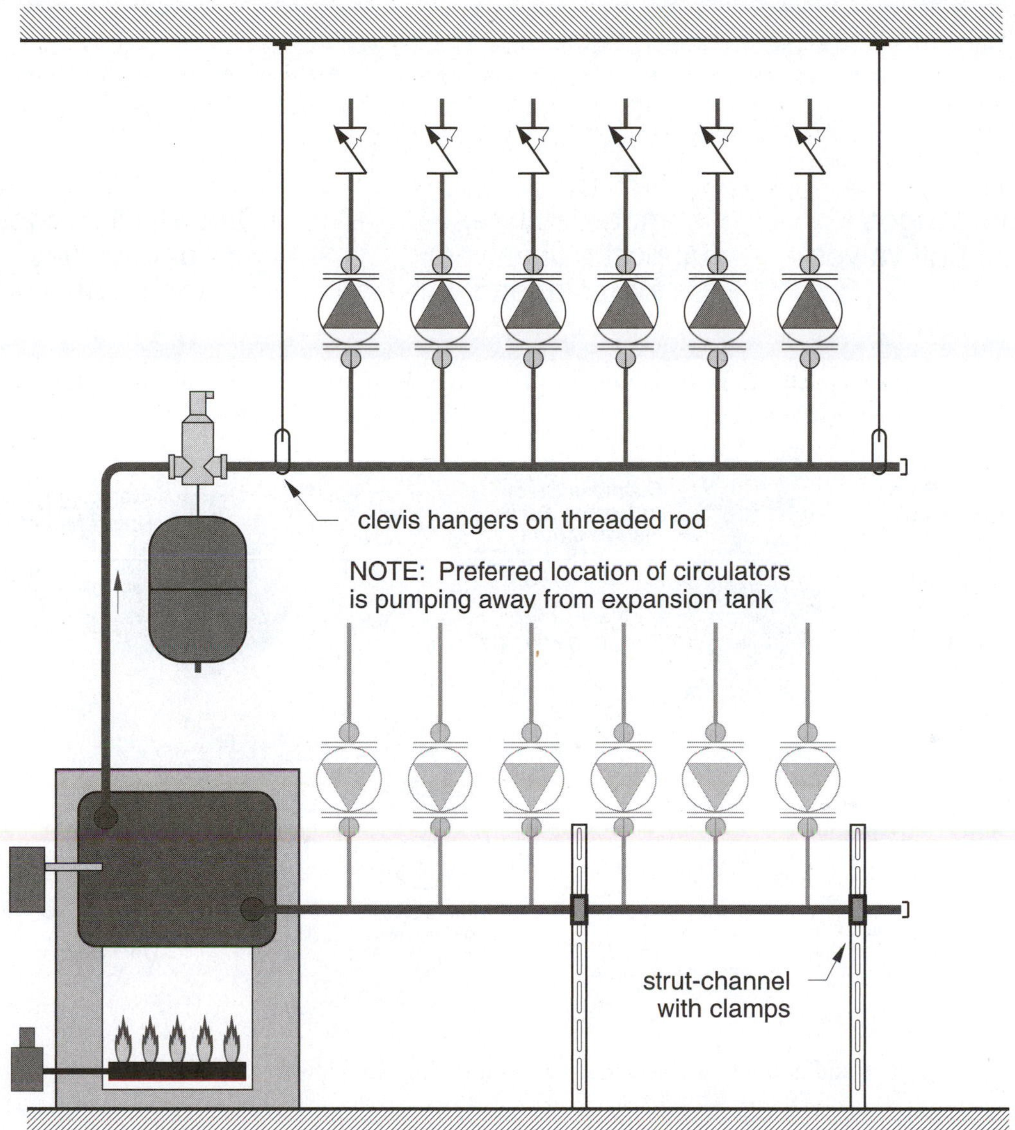

Figure 7–6 Support options for piping connected to circulators.

Connecting the Circulator to the Piping

Unlike fittings and valves, circulators should always be mounted so they can be removed for servicing. The usual method of connecting a circulator to piping is with bolted **flanges**. One side of the flanged joint is an integral part of the pump's volute. The other flange is threaded onto the piping. As the flange bolts are tightened, an O-ring or gasket is compressed between the faces of the flanges to make the seal. For small- and medium-sized circulators, a two-bolt flange is common.

Pump flanges are available in cast iron for closed system applications, as well as bronze or brass for direct contact with domestic water or other open-loop applications. A special type of flange, known as an **isolation flange**, contains a built-in ball valve that can be operated using a screwdriver. With the isolation flanges closed, the circulator may be unbolted and removed from the system. Only the small amount of fluid within the volute will be lost. An alternative method of isolating the pump is to use standard flanges combined with a gate valve or ball valve on each side of the pump. These options are shown in Figure 7–7. *It is highly recommended that every circulator installed in a hydronic system be equipped with one of these means of isolation.*

On some very small circulators, the connection to the piping is made using a half union rather than a flange. The half union threads directly onto the pump's volute. Half unions are available with or without an integral ball valve for isolation. Some small circulators are designed to be directly soldered to copper tubing. Figure 7–8 shows the various flange and half-union options offered by one pump manufacturer.

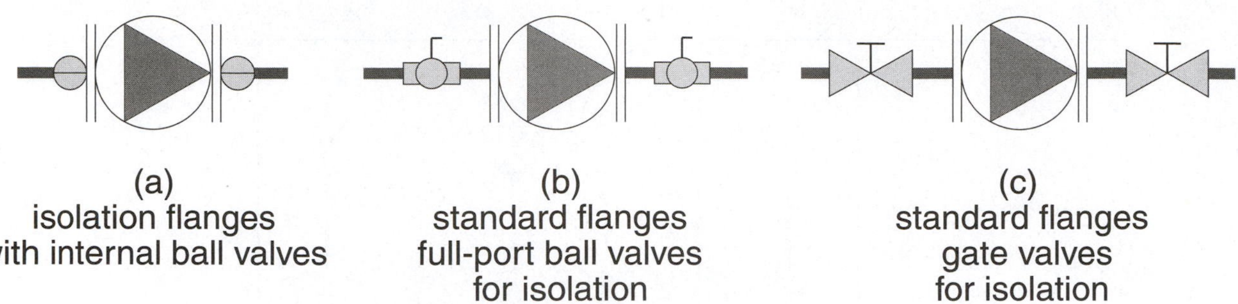

Figure 7–7 Isolating a circulator from system piping using (a) isolation flanges, (b) ball valves, (c) gate valves.

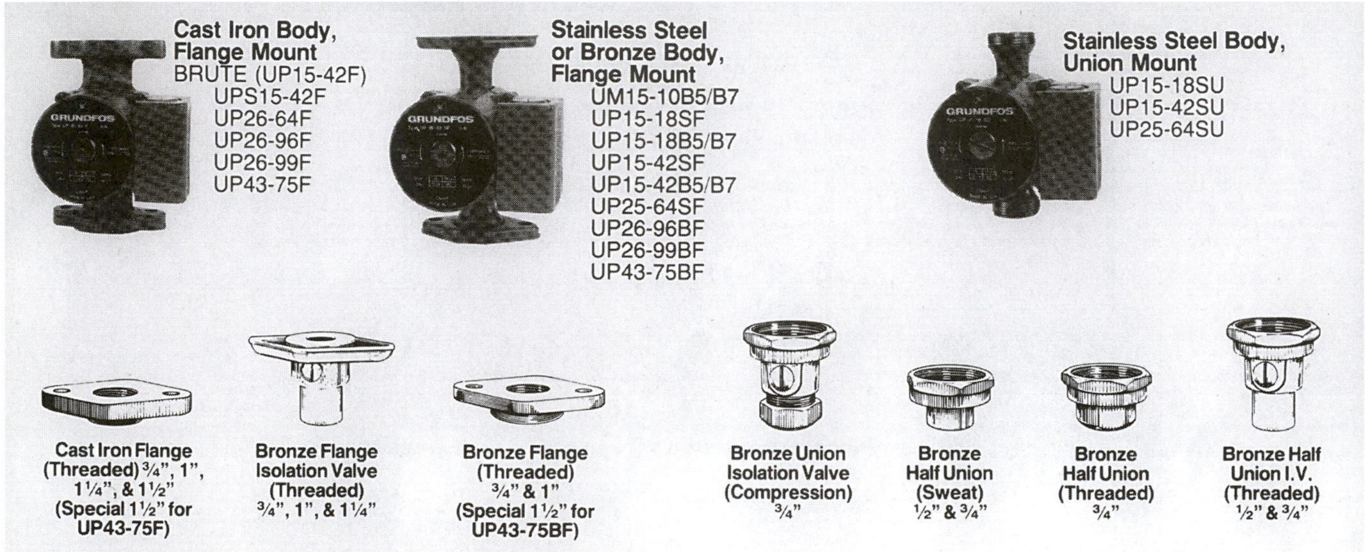

Figure 7–8 Example of flange and half union connections for connecting circulators to piping. *Courtesy of Grundfos Pumps Corp.*

7.3 PLACEMENT OF THE CIRCULATOR WITHIN THE SYSTEM

The location of the circulator(s) relative to the other components in a hydronic system can make the difference between quiet, reliable operation or constant problems. One guiding rule summarizes the situation: *Always install the circulator so that its inlet is close to the connection point of the system's expansion tank.* This principle has been applied for years in commercial hydronic systems, yet is still not universally applied in smaller systems (as it should be).

To understand why this rule should be followed, one needs to consider the interaction between the circulator and expansion tank. In a closed-loop piping system, the amount of fluid, including that in the expansion tank, is fixed. It does not change regardless of whether the circulator is on or off. The upper portion of the expansion tank contains a captive volume of air at some pressure. The only way to change the pressure of this air is to either push more fluid into the tank to compress the air, or to remove fluid from the tank to expand the air. This fluid would have to come from, or go to, some other location within the system. However, since the system's fluid is incompressible, and the amount of fluid in the system is fixed, this cannot happen regardless of whether the circulator is on or off. *The expansion tank thus fixes the pressure of the system's fluid at its point of attachment to the piping.* This is called the **point of no pressure change** within the system.

Consider a horizontal piping circuit filled with fluid and pressurized to some pressure, say 10 psi, as shown in Figure 7–9. When the circulator is off, the pressure is the same (10 psi) throughout the piping circuit. This is indicated by the solid horizontal line shown above the piping.

When the circulator is turned on, it immediately creates a pressure difference between its inlet and discharge ports. *However, the expansion tank still maintains the same (10 psi) fluid pressure at its point of attachment to the system.* The combination of the pressure difference across the circulator, the pressure drop due to head loss in the piping, and the point of no pressure change results in a new pressure distribution as shown by the dashed lines in Figure 7–9.

Notice that the pressure increases in nearly all parts of the circuit when the circulator is turned on. This is desirable because it helps eject air from vents. It also reduces the chance of cavitation. The short segment of piping between the expansion tank and the inlet port of the circulator experiences a slight drop in pressure due to head loss in the piping. The numbers used for pressure in Figure 7–9 are illustrative only. The actual numbers will of course depend on flow rates, fluid properties, and pipe sizes.

Now consider the same system with the expansion tank connected near the *discharge* port of the circulator. The dashed line in Figure 7–10 illustrates the new pressure distribution in the system when the circulator is on.

Again, the point of no pressure change remains at the expansion tank connection. This causes the pressure in most of the system to *decrease* when the circulator is turned on. The pressure at the inlet port has dropped from 10 psi to 2 psi. This situation is not desirable because it reduces the system's ability to expel air. In some cases it can also lead to pump cavitation.

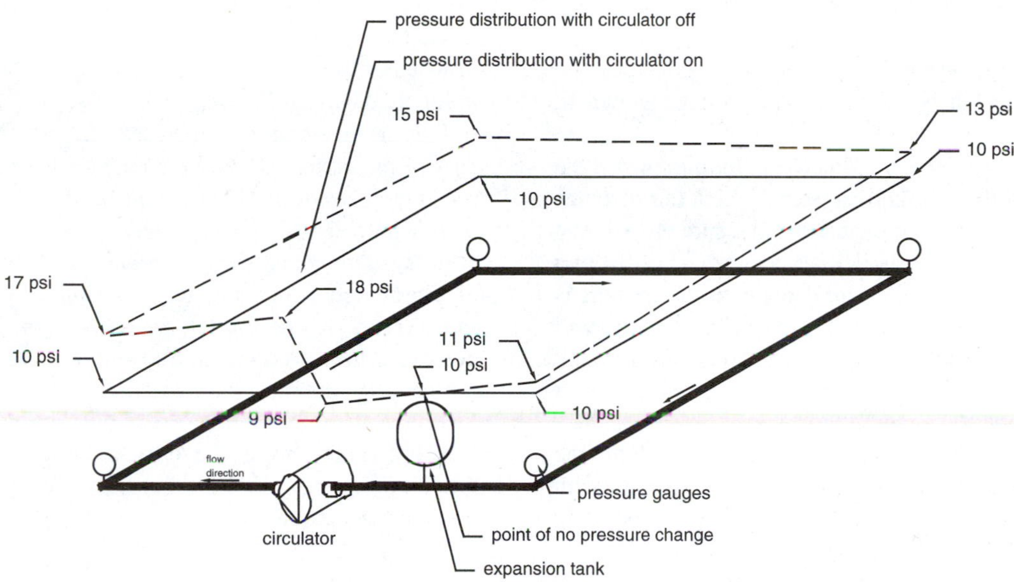

Figure 7–9 Pressure distribution in horizontal piping circuit. Solid line is pressure distribution when circulator is off. Dashed line is pressure distribution when circulator is on. Note expansion tank is near the inlet port of circulator.

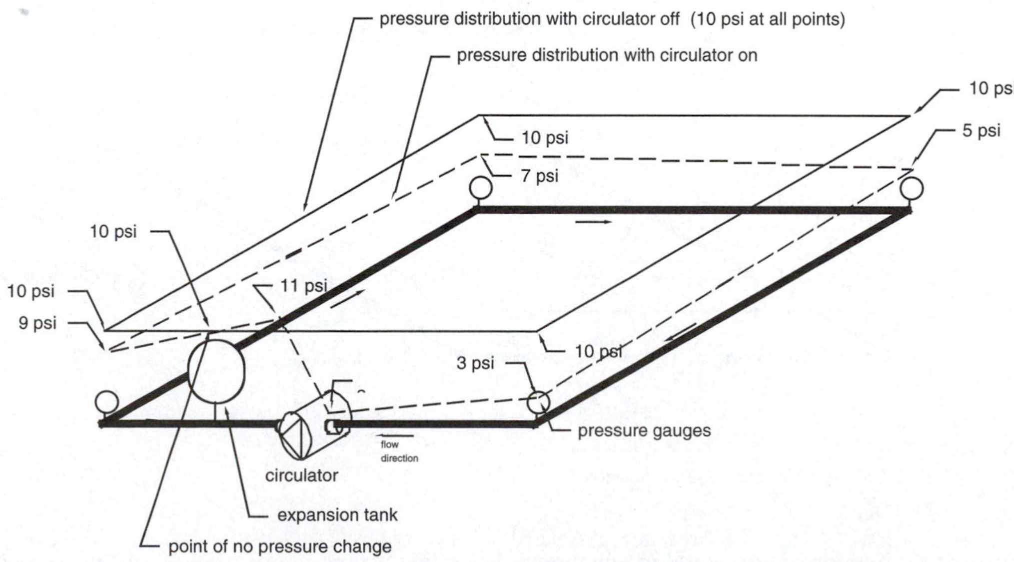

Figure 7–10 Pressure distribution in horizontal piping circuit. Solid line is pressure distribution when circulator is off. Dashed line is pressure distribution with circulator on. Note the expansion tank is near the discharge port of the circulator.

To see how problems can develop, imagine the same system with a static pressurization of only 5 psi instead of the previous 10 psi. When the circulator starts, the same 9 psi differential will be established between its inlet and discharge ports. The pressure profile shown with dashed lines in Figure 7–10 will be shifted downward by 5 psi (10 – 5 = 5 psi). This is shown in Figure 7–11.

Notice the pressure in the piping between the upper right-hand corner of the circuit and the inlet port of the circulator is now *below atmospheric pressure*. If air vents were located in this portion of the circuit, the subatmospheric pressure would actually suck air into the system every time the circulator operates. The circulator is also much more likely to cavitate under these conditions.

Unfortunately, this scenario has occurred in many residential hydronic systems. To see why, consider the piping schematic of a typical hydronic system shown in Figure 7–12.

In this arrangement, the circulator is pumping *toward* the expansion tank. Although there appears to be a fair distance between the circulator and the tank, the pressure drop due to head loss through a typical boiler is very low. Thus, from a pressure drop standpoint, the circulator's discharge port is very close to the expansion tank. This arrangement will cause the system pressure to drop from the expansion tank connection, around the distribution system to the inlet port of the pump, whenever the circulator is on.

One reason circulators were originally located on the return side of the boiler was to allow them to operate with cooler return water. It was believed the lower operating temperature prolonged the life of the packing, seals, and motor. This is not a concern for most currently produced wet rotor circulators that are often rated for *continuous operation* at fluid temperatures up to 230 °F.

Some hydronic systems that are piped as shown in Figure 7–12 have worked fine for years. Others have had problems from the first day they were put into service. Why is it that some systems work and others do not? The answer lies in a number of factors that interact to determine the exact pressure distribution in any given system. These include system height, fluid temperature, pressure drop, and system pressurization. The systems most prone to problems from this type of circulator placement are those with high fluid temperature, low static pressure, low system height, and high pressure drops around the piping circuit. Rather than gamble on whether these factors will work, it is best to arrange the piping as shown in Figure 7–13a.

This arrangement will increase the pressure in nearly all parts of the system when the circulator is on. *Many systems that have chronic problems with air in the piping can be cured with this type of component rearrangement.*

In systems operating at relatively low pressure, the circulator and expansion tank can also be placed at the return side of the boiler as shown in Figure 7–13b. However, this arrangement allows the pressure differential produced by the circulator to add to the static pressure in the boiler. If the static pressure is only a few psi lower than the relief valve setting, the pressure added by the operating circulator may cause the relief valve to open. Obviously, this is unacceptable.

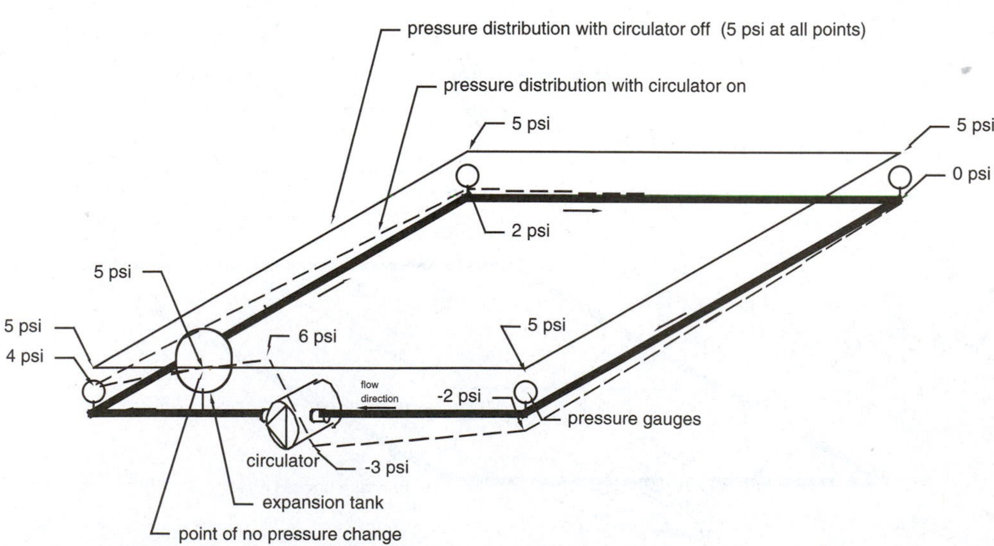

Figure 7–11 Pressure distribution in horizontal piping circuit. Solid line is pressure distribution when circulator is off. Dashed line is pressure distribution with circulator on. Note the expansion tank is near the discharge port of the circulator. Also note that the pressure in some locations is sub-atmospheric when the circulator is on.

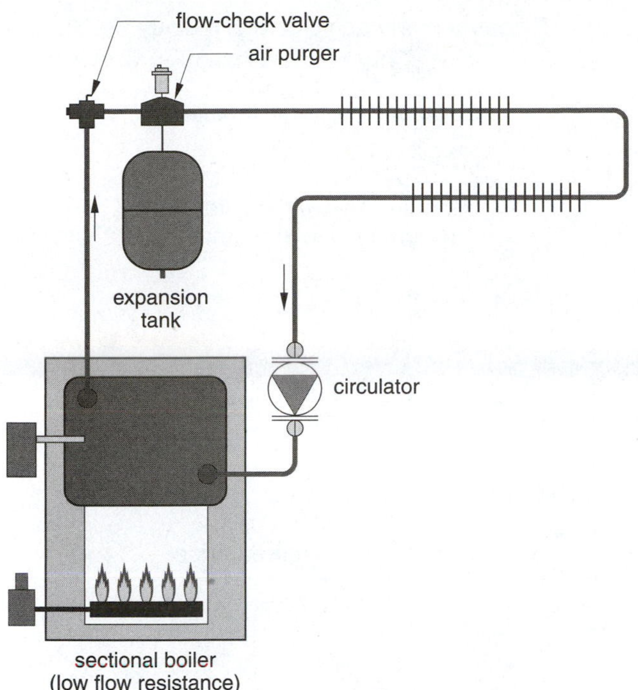

Figure 7–12 Typical piping configuration of a residential hydronic system. Note the placement of the circulator relative to the expansion tank. Expansion tank is near the discharge side of circulator.

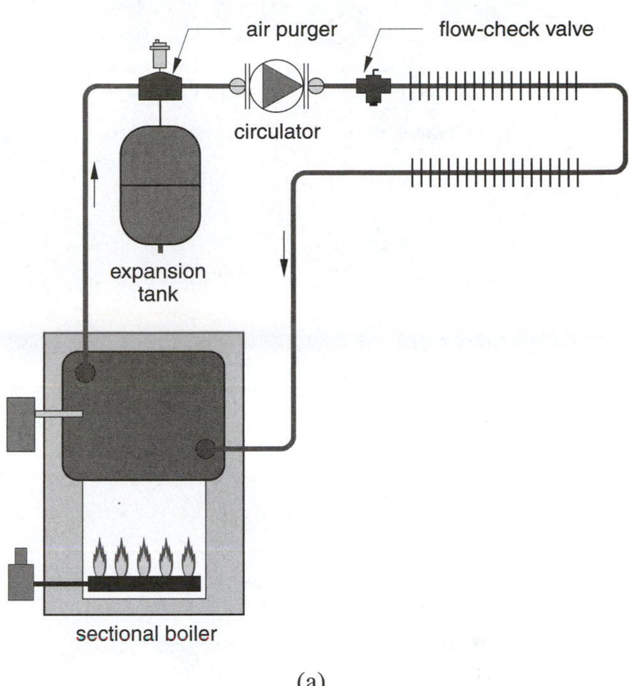

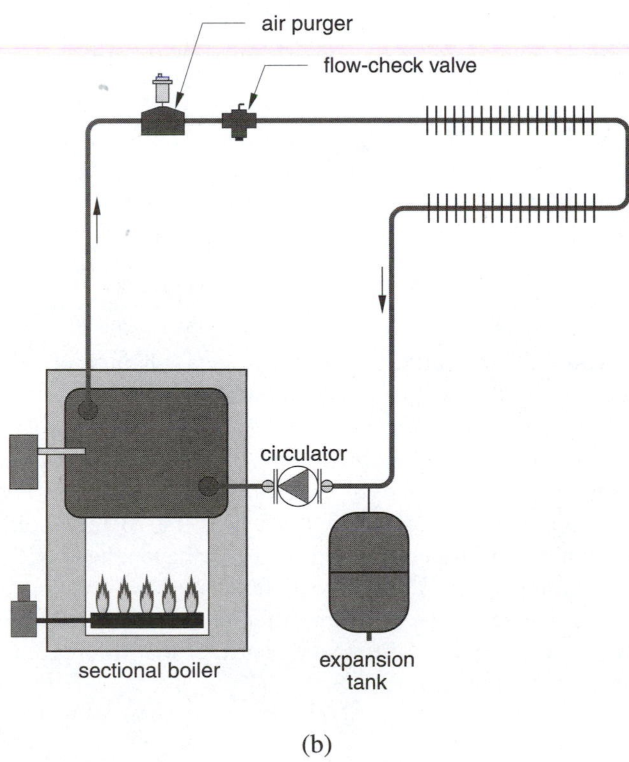

Figure 7–13 (a) One preferred arrangement of circulator and flow check valve relative to expansion tank. Note the expansion tank is located close to the inlet port of the circulator. (b) Another generally acceptable arrangement of circulator and expansion tank.

7.4 PUMP PERFORMANCE

This section presents analytical methods for describing the performance of hydronic circulators. These methods are vital in properly matching the circulator with the flow requirements of a piping system.

Explanation of Pump Head

The **head** produced by a pump is a commonly misunderstood concept. Some references define it as the height to which a pump can lift and maintain a column of water, and others define it as the pressure difference the pump can produce between its inlet and discharge ports. Both definitions are partially correct but somewhat incomplete.

In Chapter 6, Fluid Flow in Piping, head was described as **mechanical energy** present in the fluid. For incompressible fluids such as those used in hydronic systems, the added head (e.g., added mechanical energy) will reveal itself as an increase in fluid pressure between the inlet and discharge ports of the circulator. The extent of this pressure increase depends on the flow rate through the circulator and the density of the fluid being pumped. As the flow rate through a circulator increases, the pressure gain across the circulator decreases as shown in Figure 7–14.

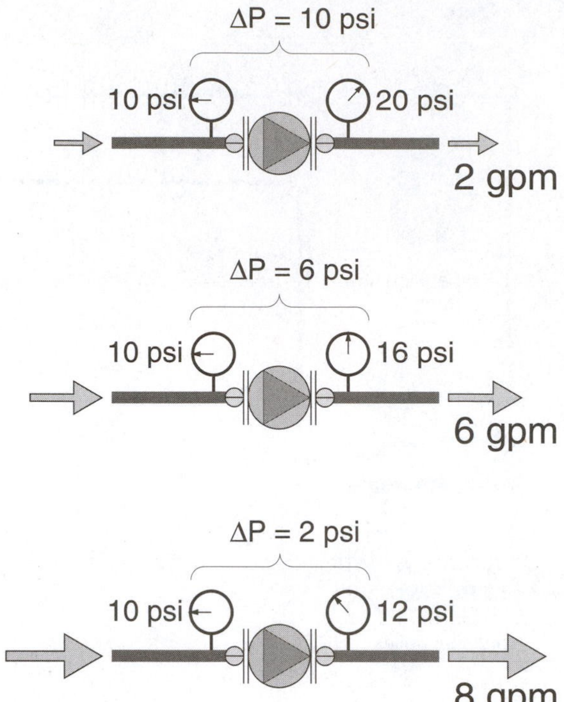

Figure 7–14 As the flow rate through a circulator increases, the pressure difference between the inlet and outlet ports decreases.

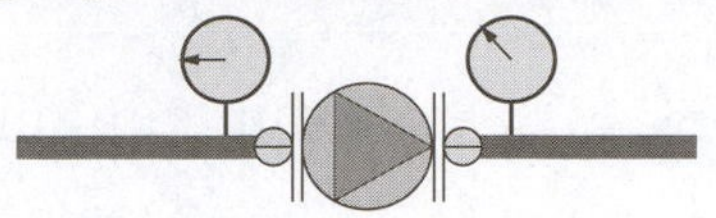

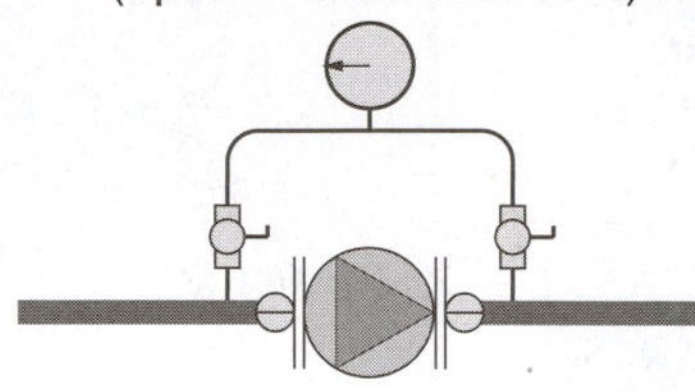

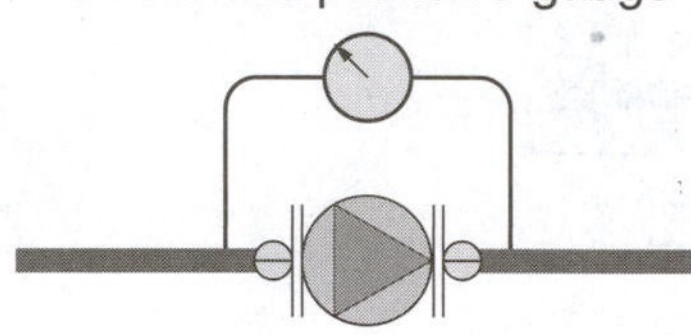

Figure 7–15 Three options for reading the pressure differential across a circulator.

In the United States, the head produced by a circulator is expressed in units of **feet of head.** This unit results from a simplification of the following units:

$$\text{Head} = \frac{\text{ft} \bullet \text{lb}}{\text{lb}} = \text{ft} = \frac{\text{energy}}{\text{lb}}$$

Those having studied physics will recall the unit of ft•lb as a unit of energy. Head may therefore be described as the *mechanical energy added per pound of fluid.*

The head added by a circulator at some given flow rate is essentially independent of the fluid being pumped. For example, a circulator producing a head of 20 feet while pumping 60 °F water at 5 gpm will also have a head of 20 feet while pumping a 50 percent ethylene glycol solution at 5 gpm. However, the pressure differential measured across the circulator will be different because of the difference in the density of the fluids.

Converting between Head and Pressure Difference

The head input to a fluid can be determined from the differential pressure across the circulator. Figure 7–15 shows three options for equipping a circulator with gauges for reading this pressure differential. On some medium-sized circulators, threaded openings are provided in the volute flanges for direct attachment of pressure gauges.

Equation 6.5 can be used to convert the pressure differential produced by a circulator to the head. This calculation requires the determination of the density of the fluid being pumped.

(Equation 6.5)

$$H = \frac{144(\Delta P)}{D}$$

where:

H = head produced by the pump (feet of head)

ΔP = pressure differential between inlet and discharge ports of circulator (psi)

D = density of the fluid being pumped (lb/ft³)

Example 7.1: Based on the pressure gauge readings, determine the head produced by the circulator in the three operating conditions shown in Figure 7–16.

Solution: In all cases, the density of the fluid being pumped needs to be determined. The density of water at 60 °F and 180 °F can be found in Figure 4–3. The density of the 50 percent solution of ethylene glycol was found using the Fluid Properties Calculator module in the Hydronics Design Studio software .

a. (60 °F water): density = 62.31 lb/ft³
b. (180 °F water): density = 60.47 lb/ft³
c. (140 °F ethylene glycol): density = 65.39 lb/ft³

The head produced in each situation can be calculated using Equation 6.5:

a. $$H = \frac{(20.8 - 10)(144)}{62.31} = 25.0 \text{ feet of head}$$

b. $$H = \frac{(15.5 - 5)(144)}{60.47} = 25.0 \text{ feet of head}$$

c. $$H = \frac{(31.35 - 20)(144)}{65.39} = 25.0 \text{ feet of head}$$

Discussion: In all three cases, the pressure rise across the circulator is different, yet the head energy imparted to each fluid is the same.

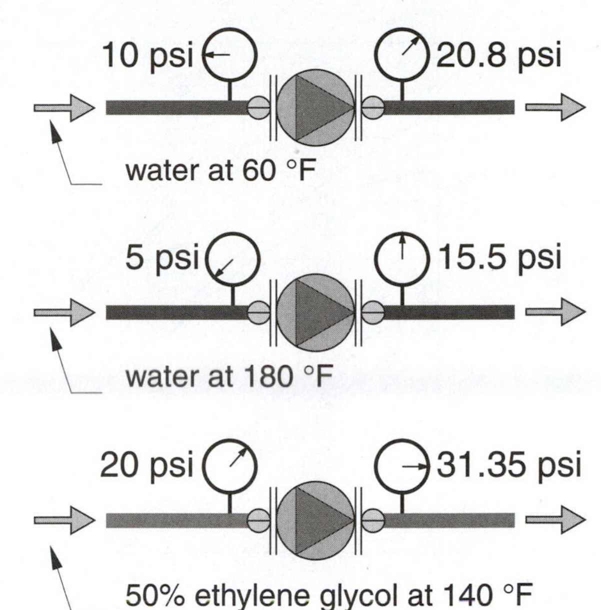

Figure 7–16 Pressure differentials across circulators for Example 7.1

Pump Performance Curves

A graph of a circulator's ability to add head to a fluid at various flow rates is called a **pump curve**. An example of such a graph is shown in Figure 7–17.

Pump curves are developed from test data using water in the temperature range of 60 °F to 80 °F. For fluids with higher viscosities, such as glycol-based antifreeze solutions, there is a very small decrease in head and flow rate capacity of the circulator. However, for the fluids and temperature ranges commonly used in hydronic heating systems this variation is so small that it can be safely ignored. Thus, *for the applications discussed in this text, pump curves may be considered to be independent of the fluid being pumped.* The reader is reminded, however, that the system curve that describes head loss versus flow rate in a piping system is very dependent on fluid properties and temperature as discussed in Chapter 6, Fluid Flow in Piping.

Pump curves are extremely important in matching the performance of a circulator to the flow requirements of a piping system. All circulator manufacturers publish these curves for the circulator models they offer. In many cases, the pump curves for an entire series or family of circulators is plotted on the same set of axes so that performance comparisons can be made. This is illustrated in Figure 7–18.

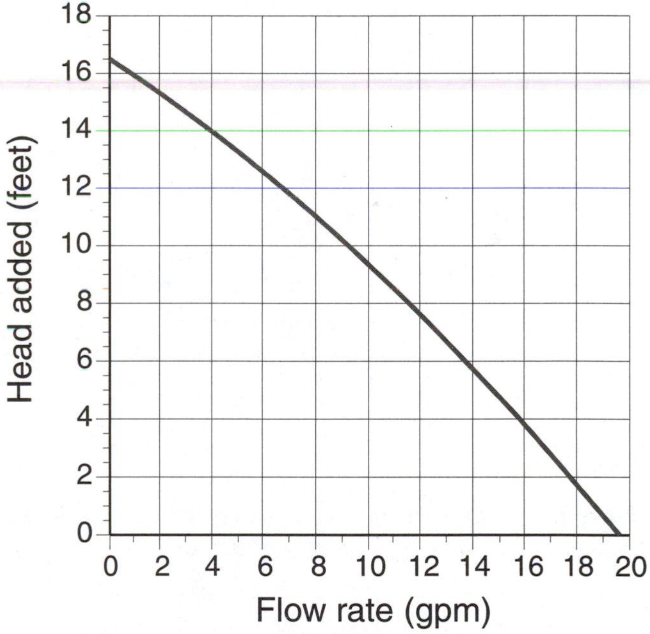

Figure 7–17 Example of a pump curve for a small circulator.

Operating Point of a Pump/Piping System

The system resistance curve of a piping system was discussed in Chapter 6. Such a curve shows the head loss of the piping system as a function of flow rate, for a given fluid at a given temperature. It is a unique graphical description of the hydraulic characteristic of that piping system. An example of a system resistance curve is shown in Figure 7–19.

Notice the pump curve in Figure 7–17 and the system resistance curve in Figure 7–19 have similar quantities on each axis. The system resistance curve shows head *loss* on the vertical axis, whereas the pump curve shows head *added* on its vertical axis. The pump curve can be thought of as the ability of the circulator to *add* mechanical energy to a fluid over a range of flow rates. The system resistance curve can be thought of as the ability of the piping system to *remove*,

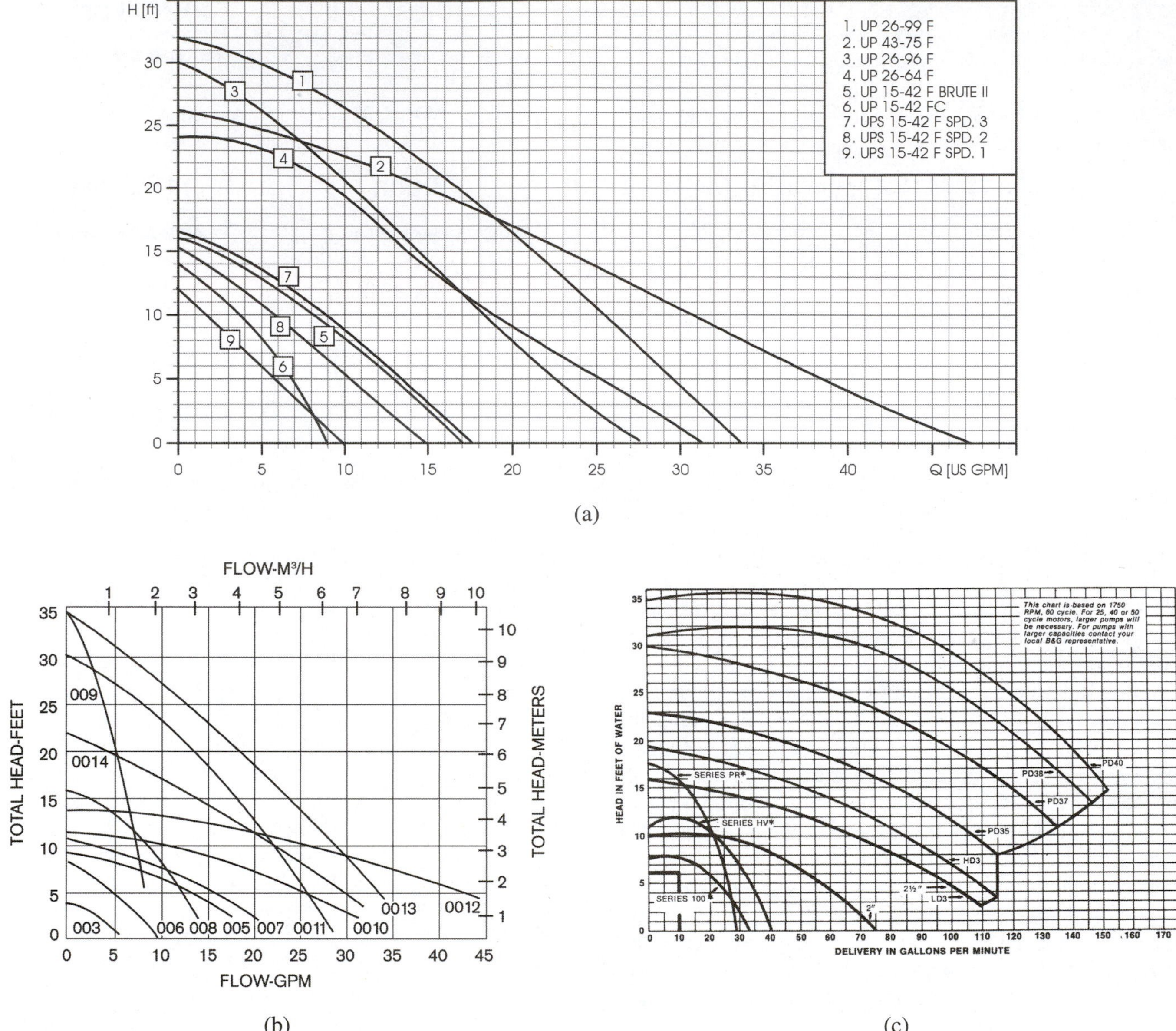

Figure 7–18 Example of several pump curves plotted on a common set of axes. (a) *Courtesy of Grundfos Pumps.* (b) *Courtesy of Taco.* (c) *Courtesy of Bell and Gossett.*

or dissipate, mechanical energy from the fluid over a range of flow rates. In both cases, the gain or loss of mechanical energy is expressed in feet of head.

The first law of thermodynamics states that when the rate of energy input to a system equals the rate of energy removal from the system, the system is in equilibrium, and will continue to operate at those conditions until one of the energy flows is changed. Chapter 1, Fundamental Concepts, discussed the concept of equilibrium for thermal energy flows. It stated that when the rate of heat input to a hydronic system exactly equals the rate of heat dissipation by the system, that system is in thermal equilibrium and temperatures will remain constant.

The concept of equilibrium also holds true for the mechanical energy added or removed from the fluid in a piping system. It can be summarized as follows:

When the rate at which a pump adds head to the fluid in a piping system equals the rate at which the piping system dissipates this head, the flow rate in the system will remain constant.

To find this equilibrium condition, the pump curve and the system resistance curve are plotted on the same set of axes as shown in Figure 7–20. The point at which the pump curve and the system resistance curve intersect is called the **operating point** of the system. This point indicates the condition of mechanical equilibrium of the system's fluid.

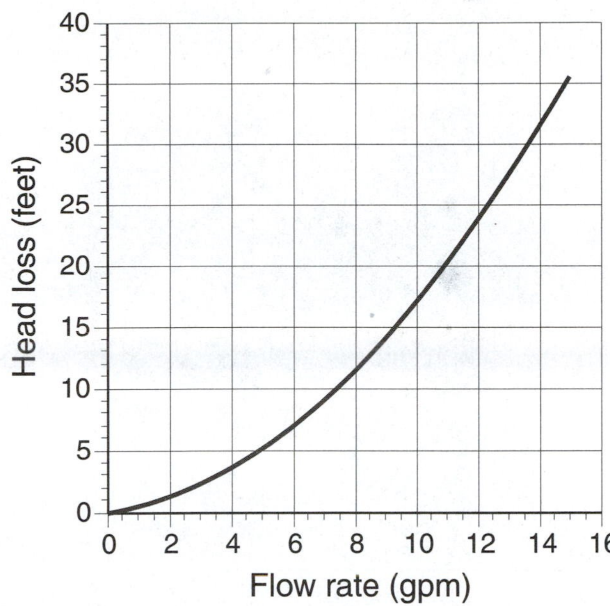

Figure 7–19 Example of a system resistance curve.

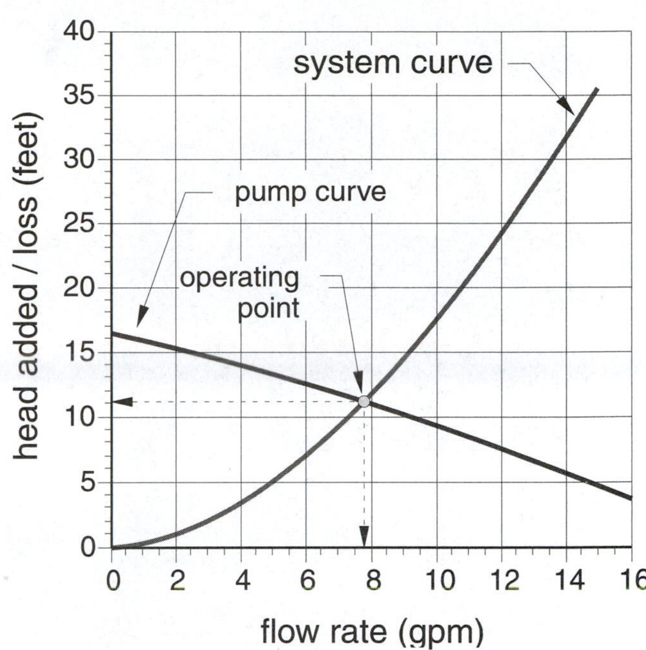

Figure 7–20 A pump curve and system resistance curve plotted on the same axes. The operating point is where the curves cross.

The flow rate at this equilibrium condition can be found by extending a vertical line from the operating point down to the horizontal axis. The head input by the circulator (or head loss by the piping system) can be found by extending a horizontal line from the operating point to the vertical axis. An example is shown in Figure 7–20.

A performance comparison of several "candidate" circulators in a given piping system can be made by plotting their individual pump curves on the same set of axes as the system's resistance curve. The intersection of each circulator's pump curve with the system resistance curve indicates the operating point for that particular circulator. By projecting vertical lines from the operating points down to the horizontal axis, the designer can determine the flow rate each circulator would produce within the system. This concept is illustrated in Figure 7–21.

Notice even though the curves for circulators 1 and 2 are markedly different, they intersect the system curve at almost the same point. Therefore, these two circulators would yield almost identical flow rates of about 7.5 gpm and 7.8 gpm in this piping system. The flow rate produced by circulator 3, about 5 gpm in this case, is considerably lower.

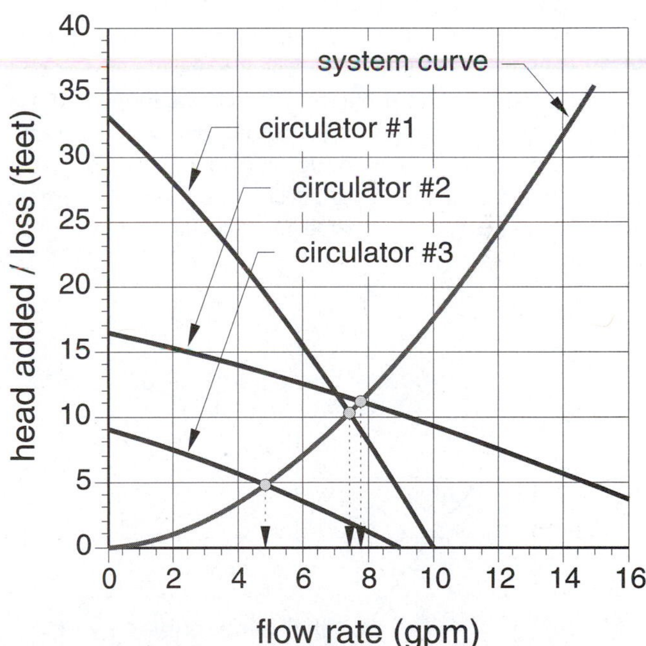

Figure 7–21 Three pump curves plotted along with a system resistance curve. The vertical lines descending from the operating points indicate flow rates produced by each circulator within the piping system described by the system resistance curve.

High Head versus Low Head Circulators

Some circulators are designed to produce relatively high heads at lower flow rates. Others produce lower but relatively stable heads over a wide range of flow rates. These characteristics are fixed by the manufacturer's design of the circulator, in particular the diameter and width of the impeller. Figure 7–22 compares the internal construction of a circulator designed for high heads at low flow rates, versus one designed for lower heads over a wider range of flow rates. Notice the impeller of the high head circulator has a relatively large diameter, but a very small separation between its disks. The low head circulator, on the other hand, has a small diameter impeller with widely separated disks.

Figure 7–22 Comparison between (a) impeller design for a steep pump curve, and (b) impeller design for a flat pump curve. *Courtesy of Taco, Inc.*

The pump curves for each circulator shown in Figure 7–22 are plotted in Figure 7–23. The high head pump is said to have a "steep" pump curve. The other circulator has a "flat" curve. Interestingly, both circulators have the same 1/15th horsepower motor rating, and operate at the same rpm. The design of the impeller has a major effect on how the motor's energy is converted into mechanical energy of the fluid.

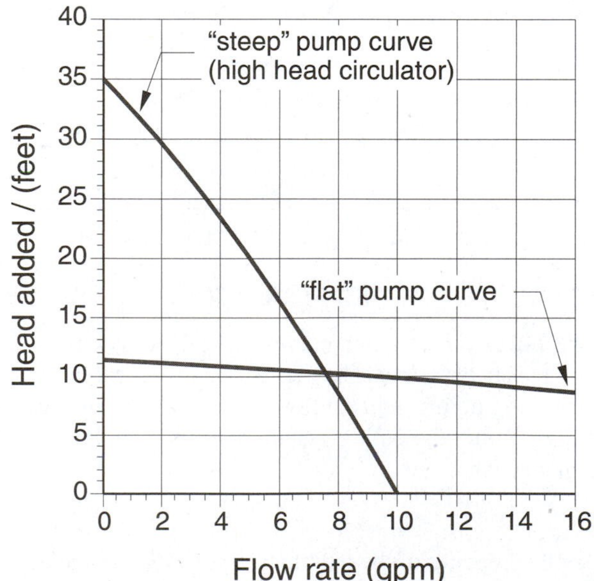

Figure 7–23 Pump curves for circulators shown in Figure 7–22.

Circulators with steep pump curves are intended for systems having high head losses at modest flow rates. Examples would include series piping circuits containing several components with high-flow resistances, earth loop heat exchangers for heat pumps, or radiant floor heating systems using long circuits of small diameter tubing.

Circulators with flat pump curves should be used when it is desirable to maintain a relatively steady pressure differential across the distribution system over a wide range of flow rates. An example of such an application is a multi zone system in which zone valves are used to control flow through individual zone circuits. Depending on the number of zone valves open at a given time, and the shape of the pump curve, the flow rate through the circulator could change considerably. A circulator with a flat pump curve will minimize such changes and thus maintain relatively stable flow rates in the active zones. This is shown in Figure 7–24.

Circulators in Series

Occasionally, a system requires more head than can be supplied by a single circulator. The designer has the option of finding a larger circulator that can supply the necessary head, or possibly using two smaller circulators connected in **series**. The second option is often the more economical choice.

When two identical circulators are connected in series, the resulting pump curve can be found by doubling the head produced by a single circulator at each flow rate.

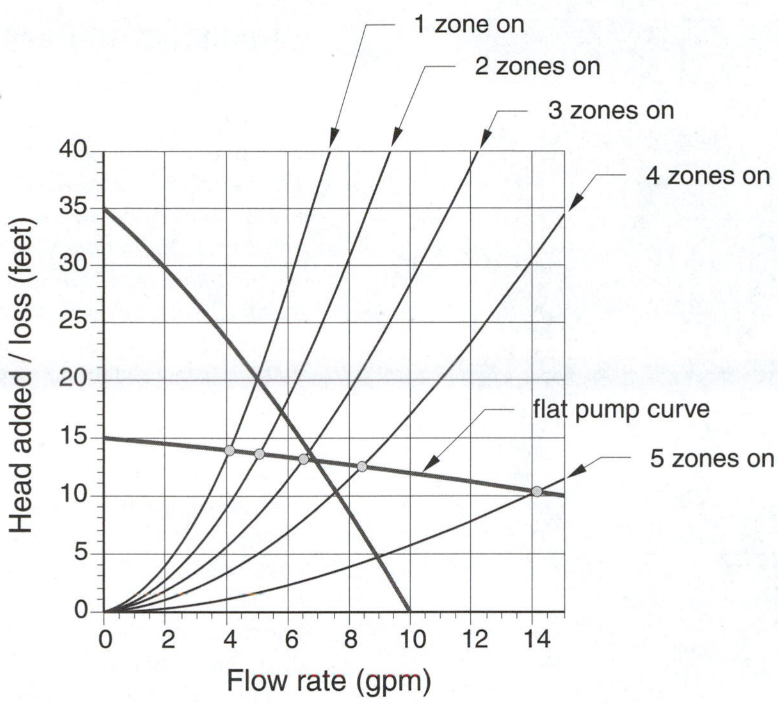

Figure 7–24 Using a circulator with a flat pump curve produces relatively small changes in head and differential pressure as various zone circuits, controlled by zone valves, turn on and off. This is generally desirable.

Figure 7–25 shows the pump curve of a single circulator, as well as two such circulators connected in series. In effect, the two circulators are acting as a multi stage pump with the head added by the first being doubled by the second.

Notice that the flow rate resulting from two circulators in series (about 10.5 gpm) is not double that obtained with the single circulator (about 7.6 gpm). This is typical, and is the result of the curvature of both the pump and system curves. It demonstrates the fallacy of trying to double the flow rate in a system by adding a second circulator in series.

There are two basic ways of piping circulators so they are in series with each other. The first, known as **close coupling**, is accomplished by bolting the discharge flange of the first circulator directly to the inlet flange of the second as shown in Figure 7–26. This is practical for relatively small circulators. It has the advantage of requiring only one set of flanges to connect the resulting assembly to the system piping. *Make sure the flow direction arrow on each circulator is facing the same direction before bolting them together.* The high heads and associated pressure differential resulting from such an assembly makes it critical that the system's expansion tank is located near the inlet port of the first circulator. If this is not done, the likelihood of cavitation is greatly increased.

Another way of installing circulators in series is called a **push/pull** arrangement. In this case, the two circulators are separated by several feet of piping, or a major system component such as the heat exchanger coil of a heat pump. This approach will distribute the pressure differential produced by each circulator more evenly throughout the system.

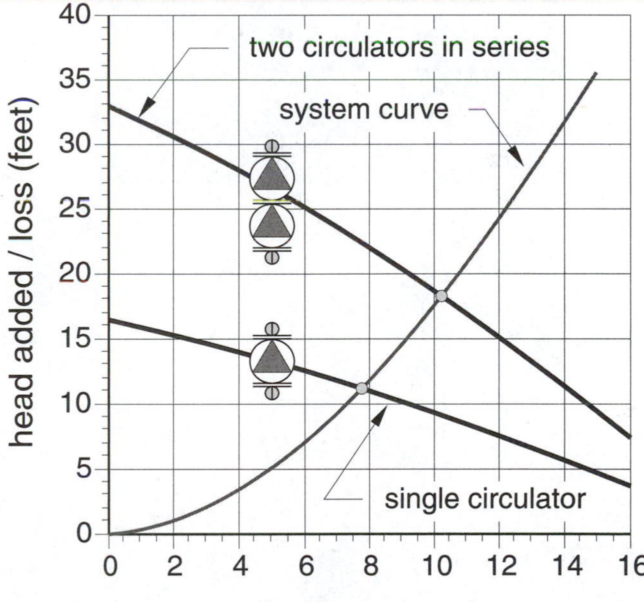

Figure 7–25 Effect of piping two identical circulators in series.

It also can reduce the chances of cavitation relative to the previously described close-coupled arrangement. This configuration is often used for the earth heat exchanger of ground source heat pump systems where a high head is required due to the combined flow resistance of both the heat pump coil and earth heat exchanger piping.

(a)

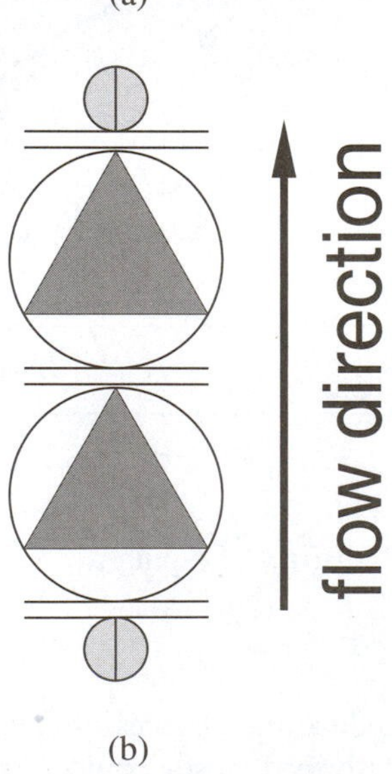

(b)

Figure 7–26 (a) Example of two small circulators in a close-coupled series configuration, (b) schematic representation of this assembly.

Circulators in Parallel

Another way of connecting two or more identical circulators is in parallel. The inlet ports of the circulators are connected to a common header pipe, as are the discharge ports. This arrangement is useful when a relatively high flow rate is required at a modest head. An example would be a parallel distribution system with several zone circuits.

The pump curve for two identical circulators connected in parallel is obtained by doubling the flow rate of a single circulator at each head value. This is illustrated in Figure 7–27. For three identical circulators in parallel, the flow rate would be tripled at each head value, and so on.

System resistance curve #1 in Figure 7–27 is for a piping system with relatively low flow resistance. Point A is the operating point for this piping system using a single circulator. Point B is the operating point using two of the same circulators in parallel. For this particular piping system and circulator combination, using two circulators in parallel increases the flow rate from about 17 gpm to about 26.5 gpm, a significant gain.

Points C and D in Figure 7–27 are the operating points for the same circulator, and parallel circulator combination, in a piping system having a higher head loss characteristic as shown by system resistance curve #2. In this system, adding a second parallel circulator results in a small increase in flow rate from about 7.3 gpm to 8.2 gpm, a very small gain considering twice the pumping power would be used.

These curves demonstrate that the flow rates obtained using multiple circulators in parallel can differ considerably depending on the system resistance curve of the piping system into which they are placed. They also show that *adding a second circulator in parallel will not double the system's flow rate*, even for a system with low head loss characteristics.

Whenever two or more circulators are used in parallel, a check valve should be installed downstream of each circulator as shown in Figure 7–28. This prevents flow reversal through a circulator that may be off because the system needs minimal flow, or because of mechanical failure. Without this check valve, much of the flow produced by the active circulator would simply short-circuit backward through the inactive circulator. To minimize the possibility of chatter, install at least 12 diameters of straight pipe between the discharge ports of the circulators and the inlet of the check valves.

7.5 ANALYTICAL METHODS FOR PUMP PERFORMANCE

The previous sections have shown graphical methods of finding the intersection of a system resistance curve and pump curve. In many situations, it is convenient to be able to find this operating point without having to draw the system and pump curves. This can be done by representing each curve mathematically, and solving for their intersection.

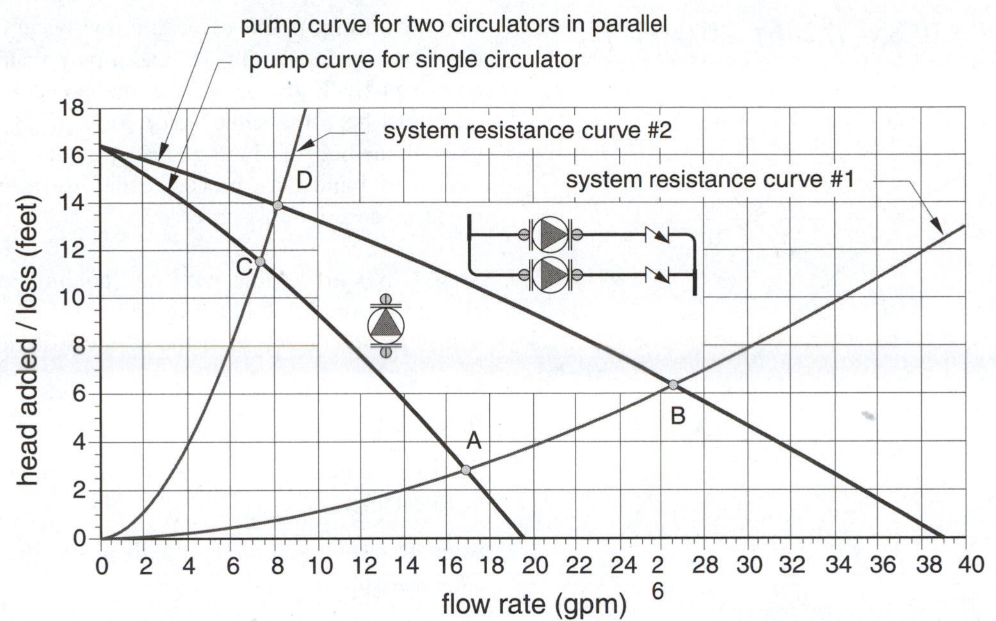

Figure 7–27 Effect of placing two identical circulators in parallel.

The pump curves for most circulators can be accurately represented using Equation 7.1:

(Equation 7.1)

$$H_{circulator} = b_0 + b_1(f) + b_2(f)^2$$

where:

$H_{circulator}$ = head produced by the circulator (feet of head)

f = flow rate through the circulator (gpm)

b_0, b_1, b_2 = constants that are specific for a given pump. These numbers can be positive or negative.

Equation 7.2 is an example of this equation with actual numbers:

(Equation 7.2)

$$H_{circulator} = 10.88 + (-0.206)(f) + (-0.00971)(f)^2$$

When plotted, this equation produces the pump curve shown in Figure 7–29.

The values of the constants b_0, b_1, and b_2 needed for describing a particular pump curve can be determined by mathematical curve fitting. This procedure requires three data points from the pump curve, which represent three flow rates and their associated head gains. The three flow rates constitute three "x" values, and the associated head gains are the associated "y" values. For the best accuracy, these three points should be well spread out on the pump curve.

(a)

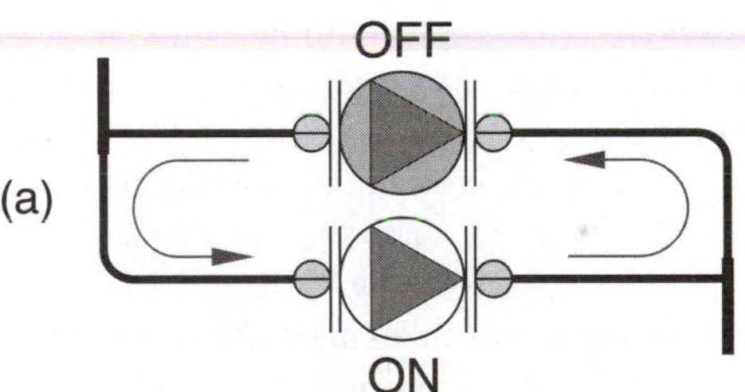

(b)

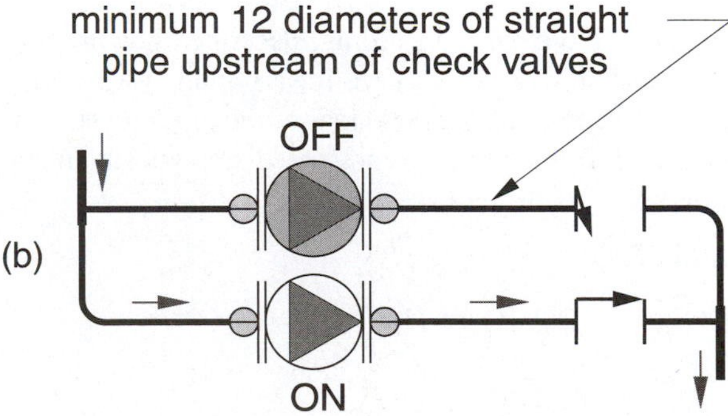

Figure 7–28 (a) Reverse flow occurs through inactive circulator, (b) check valve closes to prevent reverse flow through inactive circulator. Note 12 diameters of straight pipe between outlet of circulator and check valve.

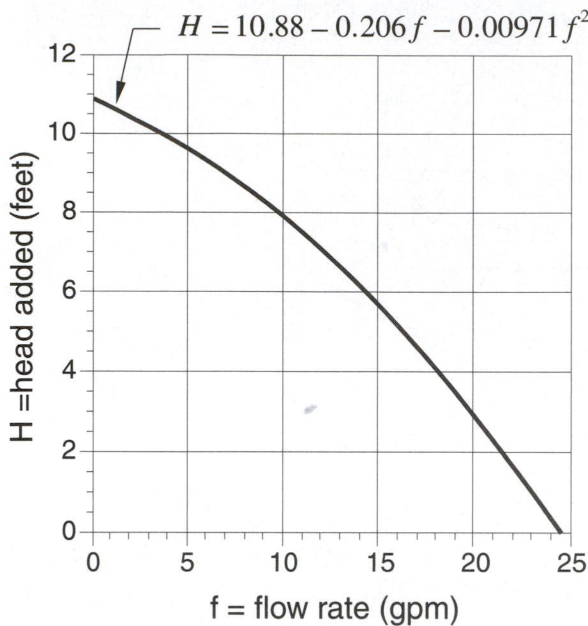

Figure 7–29 Graph of Equation 7.2 representing a pump curve.

Once these three (x,y) points are read from the pump curve, they can be entered into a curve fitting routine such as is included with many computer graphing or spreadsheet programs. Such routines will quickly generate the values of b_0, b_1, and b_2 for a given circulator. Alternatively, the values of b_0, b_1, and b_2 can be manually calculated. Such calculations are quite time-consuming. The procedure for such calculations can be found in mathematical reference books under the topic "linear regression."

Analytical Method for Finding the Operating Point

Chapter 6, Fluid Flow in Piping, described how to construct the system resistance curve for a piping system. This curve, which represents the head loss of the piping system as a function of flow rate, is represented analytically using Equation 6.9.

(Equation 6.9)

$$H_L = r(f)^{1.75}$$

In this equation, the value of r is the total equivalent hydraulic resistance of the piping system. It is the only number needed to describe the exact shape of the system resistance curve. For example, the system curve used in Example 6.10 is represented by Equation 6.15:

(Equation 6.15)

$$H_L = (0.31)(f)^{1.75}$$

The operating point of a piping system such as that described by Equation 6.15, and a pump curve such as that described by Equation 7.2, is the point where these two equations have the same value for $H_{circulator}$ and H_L. The manual method of finding this operating point involves a logical trial and error process called **iteration.** The following steps describe this process:

Step 1. Set up a table with one column for flow rate, f, another for the head added by the circulator, $H_{circulator}$, and a third for the head loss of the piping system, H_L.

Step 2. Make a reasonable first guess for the system's flow rate.

Step 3. Use Equation 7.1 with specific values for b_0, b_1, and b_2 to calculate $H_{circulator}$.

Step 4. Use Equation 6.9 with a specific value of r to calculate H_L.

Step 5. If $H_{circulator}$ is greater than H_L, the estimated flow rate is too low. Increase the value of flow rate f and go back to step 3.

Step 6. If $H_{circulator}$ is less than H_L, the estimated flow rate is too high. Lower the value of flow rate f and go back to step 3.

Step 7. Repeat this procedure until the values of $H_{circulator}$ and H_L are within 0.1 foot or less of each other. The value of flow rate, f, now approximates the actual system flow rate. The head across the circulator is estimated by averaging the last values of $H_{circulator}$ and H_L.

Example 7.2: Find the flow rate corresponding to the pump curve described by Equation 7.2 and the system curve described by Equation 6.15 using a manual iteration procedure.

Solution:

Pump curve: $H_{circulator} = 10.88 + (-0.206)(f) + (-0.00971)(f)^2$

System resistance curve: $H_L = 0.31(f)^{1.75}$

Figure 7–30 shows successive estimates of flow rate with corresponding values of pump head, and system head loss calculated from the two equations. The last estimate of 6.85 gpm showed the pump head and system head loss to be well within 0.1 foot of each other. Hence, this was the last calculation performed.

The head across the pump could now be estimated by averaging the final values of H_{pump} and H_L:

$$\frac{9.01 + 8.99}{2} = 9.00 \text{ ft. of head}$$

Flow rate f (gpm)	Pump head H_{pump} (feet)	System head loss H_L (feet)
4.0	9.9	3.507
6.0	9.29	7.13
8.0	8.61	11.8
7.0	8.96	9.33
6.5	9.13	8.2
6.9	8.99	9.106
6.85	9.01	8.99 ← close enough

Figure 7–30 Iterative solution to find operating point in Example 7.2.

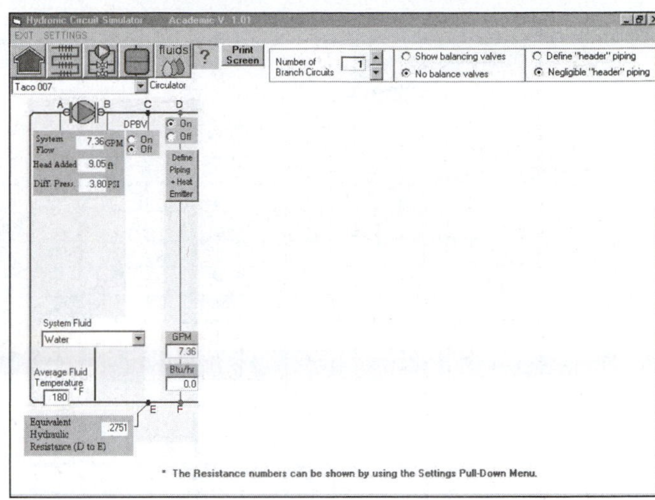

Figure 7–31 User interface screen for Hydronic Circuit Simulator module in the Hydronics Design Studio software.

The Hydronic Circuit Simulator Module

The procedure for finding the operating point of a given piping system/circulator combination is critically important for accurately predicting system performance. Unfortunately, it is often neglected due to the time and level of mathematics involved. This often leads to needlessly oversized or undersized circulators being installed. The cost of correcting such situations can be high.

The **Hydronic Circuit Simulator** module included on the Hydronics Design Studio CD-ROM makes the process of determining the operating point of a given piping system/circulator combination both fast and easy. This module contains a database of the previously discussed b_0, b_1, and b_2 constants for describing the pump curve of over 100 commonly used circulators. It also allows the user to quickly configure the piping system by selecting the piping, fittings, and valves in the system. Each time a change is made to the inputs, the module recalculates the operating point of the currently specified system and displays the resulting flow rate(s). The reader is encouraged to use this module as a routine design tool whenever flow rates in hydronic piping systems need to be determined. A screen shot of the Hydronic Circuit Simulator program is shown in Figure 7–31.

7.6 CIRCULATOR EFFICIENCY

Like any device that converts one form of energy to another, a circulator has an associated conversion efficiency. In Chapter 3, Hydronic Heat Sources, efficiency was defined as the rate of energy output divided by the rate of energy input. In this case, the output is the rate at which the circulator imparts mechanical energy to the fluid. The input is the rate at which energy is supplied to rotate the circulator's shaft.

Equation 7.3 can be used to calculate the **efficiency of a circulator**.

(Equation 7.3)

$$\eta_{circulator} = \frac{f(\Delta P)}{1716(hp)}$$

where:

$\eta_{circulator}$ = efficiency of the circulator
f = flow rate through the circulator (gpm)
ΔP = pressure differential measured across the circulator (psi)
1716 = units conversion constant
hp = horsepower delivered to the circulator shaft by the motor

Although Equation 7.3 is easy to derive from basic physics, it is not necessarily easy to use because measurements of input horsepower to the pump's shaft are difficult to obtain outside of a testing laboratory. Furthermore, the input horsepower to a pump increases with flow rate.

To provide efficiency information to designers, most pump manufacturers plot the efficiency of their medium- and larger-size pumps as a group of contour lines overlaid on the pump curve as shown in Figure 7–32. Unfortunately, such information is often not published for small circulators.

The efficiency of most circulators is highest near the center of the pump curve as shown in Figure 7–33. For the circulator to operate at or near its maximum efficiency, the system curve must pass through this region of maximum efficiency. The author recommends that the operating point (e.g., the intersection between the pump curve and system curve) should fall within the middle third of the pump curve.

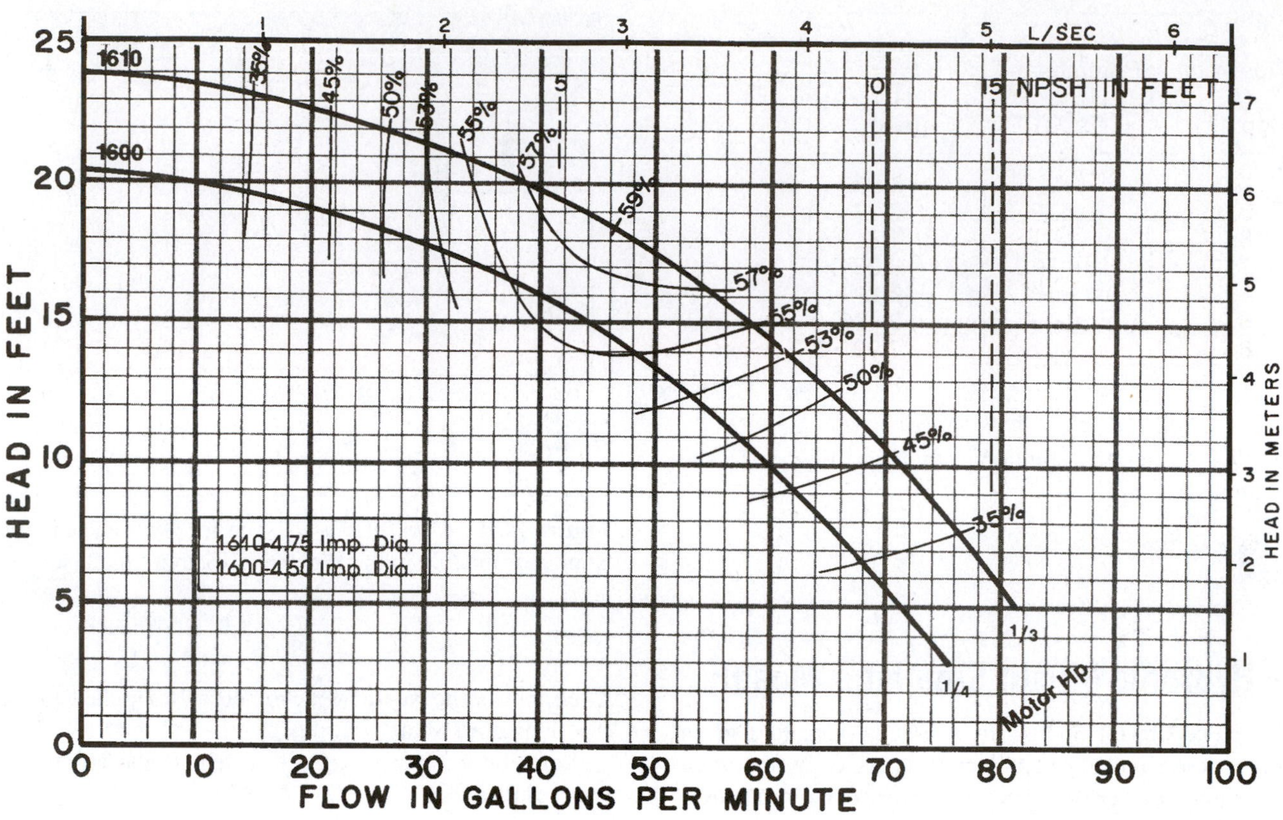

Figure 7–32 Example of efficiency contours plotted with pump curves. Typical of medium and large circulators. *Courtesy of Taco, Inc.*

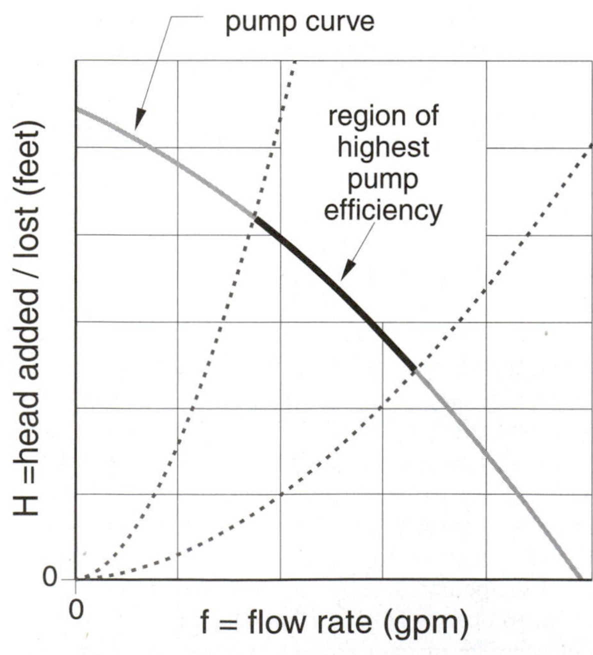

Figure 7–33 Region of highest efficiency for small circulators. For good efficiency, the system curve should cross over the middle 1/3 of the pump curve.

Efficiency of a Pump/Motor Combination

The efficiency obtained using Equation 7.3 is that of the pump only. It does not include any energy losses associated with the electric motor driving the pump. No electric motor is 100 percent efficient in converting electrical energy into mechanical energy. Motor efficiencies vary considerably for different types of construction and different loading conditions. The induction motors used on larger commercial and industrial pumps have efficiencies in the range of 80 percent to 90 percent when loaded between 25 percent and 100 percent of their rated horsepower. The small permanent split capacitor motors used on wet rotor circulators often have efficiencies less than 50 percent. A significant amount of the electrical energy input to the motor is therefore lost before being transferred to the pump shaft as mechanical energy. This energy loss ends up as heat that is dissipated to the circulating fluid or to the surrounding air.

To account for the efficiencies of *both* the pump and the electric motor, Equation 7.3 can be combined with the defining equation of motor efficiency to obtain the following relationship for a pump/motor combination:

(Equation 7.4)

$$\eta_{\text{pump \& motor}} = \frac{0.4344\ f(\Delta P)}{w}$$

where:

$\eta_{\text{pump \& motor}}$ = efficiency of the pump and motor assembly as a whole

f = flow rate through the pump (gpm)

ΔP = pressure differential measured across the pump (psi)

w = input wattage required by the motor (watts)

0.4344 = units conversion factor

The input wattage required by a pump is not a fixed value, but depends on where the pump operates on its curve. Therefore, to make use of Equation 7.4, one needs to know the input wattage *at a particular point on the pump curve*. This data may be available from the pump's manufacturer, or it can be measured using a wattmeter. Figure 7–34 shows the electrical power (wattage) drawn by a small wet rotor circulator over a range of flow rates.

Figure 7–35 shows the pump curve of this same circulator along with its calculated efficiency. Notice that peak efficiency occurs at a flow rate that is below the "knee" of the pump curve. For this particular circulator, it occurs at a flow rate of about 9.3 gpm. Ideally, the intersection of the pump curve and system curve will be at or near this point.

Example 7.3: A small circulator operates at a point where it produces a flow rate of 8.0 gpm of 140 °F water against a head of 9.7 feet. A wattmeter indicates the pump is drawing 85 watts of electrical input. What is the efficiency of the overall motor/pump unit?

Solution: Equation 7.4 can be used, but since it requires the pressure difference across the pump, not the head, Equation 6.6 is first used to convert the head to a pressure difference. The fluid density is also required, and is found from Figure 4–3:

$$\Delta P = \frac{(9.7)(61.35)}{144} = 4.13 \text{ psi}$$

This pressure difference can now be substituted into Equation 7.4 along with the other data:

$$\eta_{\text{pump \& motor}} = \frac{0.4344(8)(4.133)}{85}$$

$$= 0.169 \text{ or } 16.9\% \text{ efficient}$$

Discussion: This value seems surprisingly low, but is in fact typical of most small, wet rotor circulators currently produced. Several factors contribute to this low efficiency. One is the unavoidable decrease in impeller efficiency as its diameter is reduced. Another is the fact that the volute chambers of small circulators do not contain diffuser vanes to guide the fluid off the impeller and out of the volute with minimal turbulence. The economics of producing small circulators simply does not justify marginal efficiency gains at significantly increased production costs.

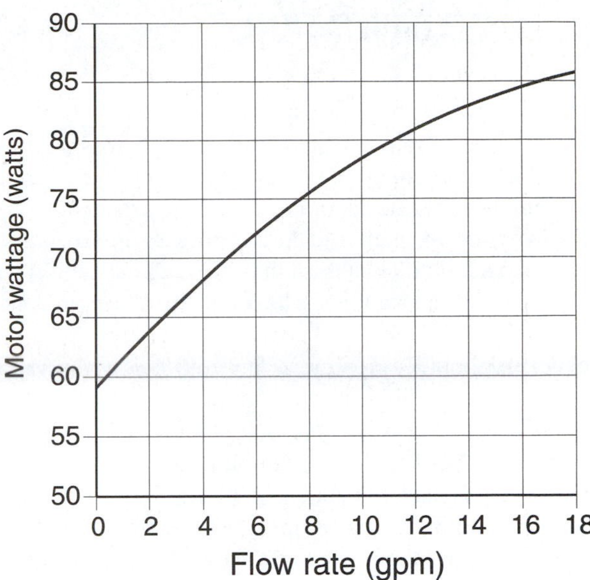

Figure 7–34 Wattage drawn by a small wet rotor circulator.

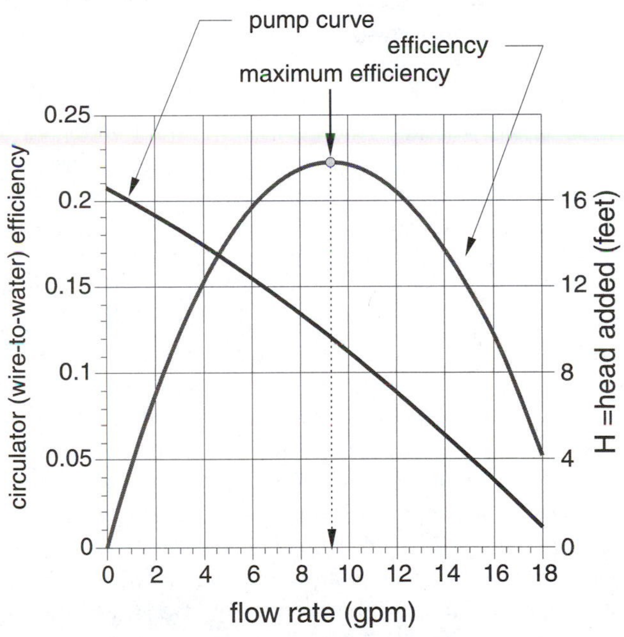

Figure 7–35 Efficiency curve and pump curve for a small circulator. Note region of highest efficiency is within middle 1/3 of pump curve. Efficiency is represented as a decimal percent, i.e., 0.1 = 10%.

7.7 PUMP CAVITATION

Chapter 4, Properties of Water, discussed the fact that liquids will begin boiling at specific combinations of pressure and temperature. The minimum pressure that must be maintained on a liquid to prevent it from boiling at some temperature is called its vapor pressure. The process of boiling involves the formation of vapor pockets in the liquid. These pockets look like bubbles in the liquid, but should not be confused with air bubbles. They will form even in water that has been completely deaerated, instantly appearing whenever the liquid's pressure is lowered below the vapor pressure corresponding to its current temperature. This process is often described as the water "flashing" into vapor.

When vapor pockets form from liquid water, the density inside the vapor pocket is about 1,500 times lower than the surrounding liquid. This is comparable to a single kernel of popcorn expanding to the size of a baseball. If the pressure of the liquid then increases above the vapor pressure, the vapor pockets instantly collapse inward in a process known as **implosion**. Although this sounds relatively harmless, on a microscopic level, vapor pocket implosion is a very violent effect. It is so violent that it can actually rip away material from surrounding surfaces, even hardened metal surfaces.

For a liquid at a given temperature, cavitation will occur *anywhere* the liquid's pressure drops below its vapor pressure. In hydronic systems, the most likely locations for cavitation to occur are in partially closed valves or inside the circulator's impeller.

As liquid flows into the eye of a circulator's impeller, it experiences a rapid drop in pressure. If the pressure at the eye of the impeller drops below the vapor pressure, thousands of vapor pockets instantly form. The vapor pockets are carried outward through the impeller, and as soon as the local fluid pressure increases above the vapor pressure, they implode. This usually occurs near the outer edges of the impeller. The violent implosions create the churning or crackling sounds that are characteristic of a cavitating impeller. A noticeable drop in flow rate and head occurs because the fluid inside the circulator is now partially compressible. If the circulator continues to operate in this mode, the impeller will be severely eroded in as little as a few weeks of operation. Obviously, pump cavitation must be avoided if the system is to have a long, reliable life. Figure 7–36 shows the effects of cavitation damage to a pump impeller.

Figure 7–36 Example of cavitation damage to a pump impeller. *Courtesy of ITT Bell and Gossett.*

Net Positive Suction Head Available

The best way of preventing pump cavitation is to know what operating conditions allow it to occur, and then design the system to avoid these conditions.

Over the years, engineers have standardized a method of predicting the conditions that will cause pump cavitation. This method requires the calculation of a quantity called the **Net Positive Suction Head Available**, or **NPSHA**. The NPSHA is a precise description of the state of the fluid as it enters the pump. It includes the effects of temperature, pressure, velocity, and the fluid's vapor pressure into a single number, which can be calculated using Equation 7.5.

(Equation 7.5)

$$\text{NPSHA} = \frac{v^2}{64.4} + H_e - H_L + (p_s + 14.7 - p_v)\left(\frac{144}{D}\right)$$

where:

NPSHA = net positive suction head available at the pump inlet (ft of head)

v = velocity of the liquid in the pipe entering the pump (ft/sec)

H_e = distance to the liquid's surface above (+) or below (–) the pump inlet (ft)

H_L = head loss due to viscous friction in any piping components between the liquid surface and the inlet of the pump (ft of head)

P_s = *gauge pressure* at the surface of the fluid (psig)

P_v = vapor pressure of the liquid as it enters the pump (psi *absolute*)

If the pressure of the fluid entering the pump is known, it is not necessary to know the pressure at the surface of the liquid, the distance to the liquid's surface, or the head loss due to viscous friction in the piping leading to the pump. These effects will be accounted for in the pressure gauge reading itself. This allows Equation 7.5 to be simplified as follows:

(Equation 7.6)

$$\text{NPSHA} = \frac{v^2}{64.4} + (p_i + 14.7 - p_v)\left(\frac{144}{D}\right)$$

where:

NPSHA = net positive suction head available at the pump inlet (ft of head)

v = velocity of the liquid in the pipe entering the pump (ft/sec)

p_i = gauge pressure of the moving fluid measured at the pump entrance (psig)

p_v = vapor pressure of the liquid as it enters the pump (psi *absolute*)

The NPSHA indicates the difference between the total head of the fluid at the inlet port of the pump and the head at which the fluid will boil. NPSHA is determined by the piping system in which the pump will operate, as well as the properties of the fluid being pumped. It *does not depend on the pump being used.*

Example 7.4: Determine the NPSHA for the piping system shown in Figure 7–37.

a. Assuming the water temperature is 140 °F.

b. Assuming the water temperature is 200 °F.

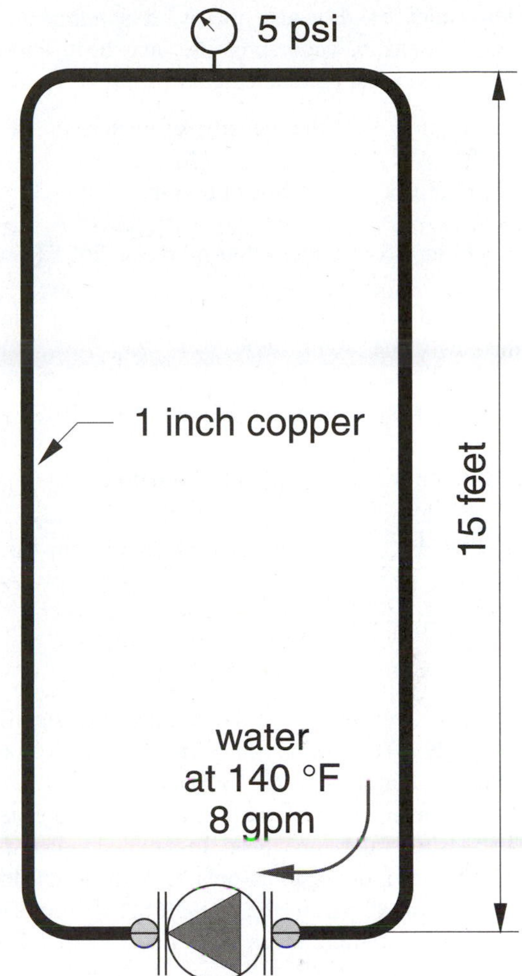

Figure 7–37 Piping system for Example 7.4.

Solution (part a): A number of quantities need to be determined before using Equation 7.5 to calculate the NPSHA:

The velocity of the water in the pipe entering the pump is found using the equation for 1-inch copper tubing from Figure 6–2:

$$v = 0.367\, f = 0.367\,(8) = 2.94 \text{ ft/sec}$$

The distance to the (top) surface of the water as shown in Figure 7–37 is 15 feet. This distance is considered positive because the surface of the water is above the inlet to the pump. Thus, $H_e = 15$ feet.

The head loss from the (top) surface of the water to the inlet of the circulator will be that created by 20 feet of 1-inch type M copper tubing carrying the 140 °F water at 8 gpm. This can be determined using Equation 6.11:

(Equation 6.11)

$$H_L = (\alpha c\, L)(f)^{1.75}$$

The value of α for water at 140 °F is found from Figure 6–17: $\alpha = 0.0475$.

The value of the pipe size factor c for 1-inch type M copper tube is found in Figure 6–18: $c = 0.01776$.

Substituting into Equation 6.11:

$$H_L = [(0.0475)(0.01776)(20)](8)^{1.75} = 0.642 \text{ ft of head}$$

In the case of a closed fluid-filled piping system, the pressure at the surface of the liquid would be the pressure at the top of the piping loop. Therefore, $p_s = 5$ psig.

The vapor pressure of 140 °F water can be read from Figure 4–5: $p_v = 2.9$ psia.

The density of water at 140 °F is found from Figure 4–3: $D = 61.35$ lb/ft³.

Substituting all data into Equation 7.5 yields the following:

$$\text{NPSHA} = \frac{2.94^2}{64.4} + 15 - 0.642 + (5 + 14.7 - 2.9)\left(\frac{144}{61.35}\right)$$

$$= 53.9 \text{ ft. of head}$$

Solution (part b): Because the water's temperature has changed, its density, vapor pressure, and head loss due to viscous friction must be recalculated.

From Figure 4–3, the density of water at 200 °F is 60.15 lb/ft³.

From Figure 4–5, the vapor pressure of 200 °F water is 11.5 psia.

From Figure 6–17, the value of α for 200 °F water is 0.043.

The revised head loss is again found using Equation 6.11 with the new value of α:

$$H_L = [(0.043)(0.01776)(20)](8)^{1.75} = 0.581 \text{ ft of head}$$

Substituting these changes into Equation 7.5:

$$\text{NPSHA} = \frac{2.94^2}{64.4} + 15 - 0.581 + (5 + 14.7 - 11.5)\left(\frac{144}{60.15}\right)$$

$$= 34.2 \text{ ft. of head}$$

Discussion: Increasing the water temperature significantly lowers the NPSHA of an otherwise unchanged system. Other factors that lower the NPSHA of a piping system include decreasing the system's height, lowering its pressure, increasing inlet pipe head loss, or slowing its flow velocity. Of these, the head associated with flow velocity is usually a very small part of the total NPSHA value. This can be seen in this example since the value of $v^2/64.4$ is only 0.134 feet of head.

Net Positive Suction Head Required

Any given pump has a minimum required value of net positive suction head in order to operate without cavitation. This value is called the **Net Positive Suction Head Required**, or **NPSHR**. Its value depends on the design of the pump, as well as the flow rate at which the pump operates. Pump manufacturers determine values of NPSHR by lowering the NPSHA to a pump until cavitation occurs. This is done for several flow rates so a curve may be plotted.

Avoiding pump cavitation is simply a matter of ensuring the NPSHA to a pump always exceeds the NPSHR of the pump under all possible operating conditions.

The higher the NPSHA is relative to the NPSHR, the wider the safety margin against cavitation. A common design practice is to ensure that the NPSHA to the pump is always at least 2 feet of head higher than the NPSHR of a pump.

Example 7.5: The circulator shown in Figure 7–38 has a NPSHR value of 3.5 feet of head. The expected operating point of the pump is at a flow rate of 5 gpm with a 13-foot head gain across the pump. The system operates at a static pressure of 3 psig. Determine if cavitation will occur.

Solution: The pressure differential associated with a head gain of 13 feet can be calculated using Equation 6.6. Note the density of 200 °F water is 60.15 lb/ft³:

$$\Delta P = \frac{HD}{144} = \frac{(13)(60.15)}{144} = 5.43 \text{ psi}$$

Because the expansion tank is (incorrectly) located at the *outlet* of the circulator, this pressure differential will *lower* the pressure of the fluid entering the pump by 5.43 psi.

The pressure at the inlet of the circulator is the static pressure of the system (3 psig), minus the pressure drop due to the head loss around the loop (5.43 psig). The inlet pressure is thus 3 – 5.43 = –2.43 psig. The negative sign indicates that this pressure is below atmospheric pressure. Since the pressure at the inlet of the circulator is known, Equation 7.6 can be used to calculate NPSHA. Again, several pieces of information need to be gathered before substituting into the equation.

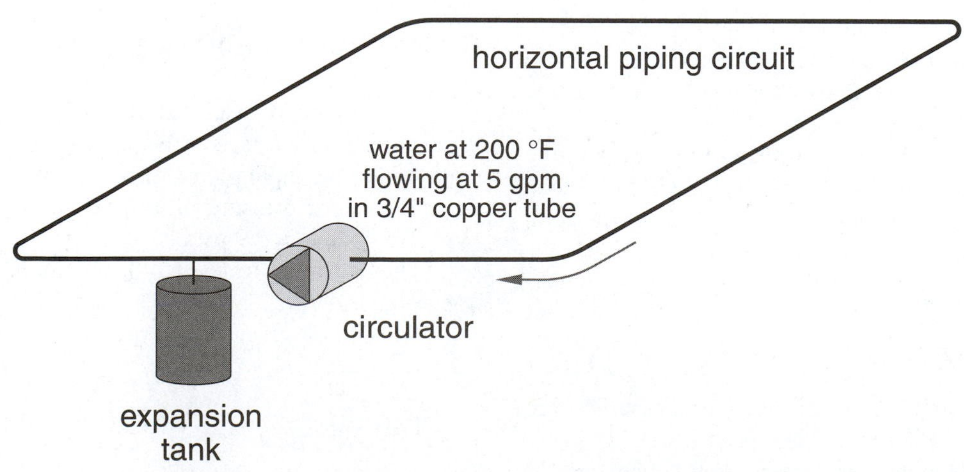

Figure 7–38 Horizontal piping loop for Example 7.5.

The flow velocity corresponding to a flow rate of 5 gpm in a 3/4-inch copper tube is found using the appropriate equation from Figure 6–2:

$$v = 0.6211, f = 0.6211(5) = 3.106 \text{ ft/sec.}$$

The vapor pressure of 200 °F water is 11.5 psia.

Substituting into Equation 7.6:

$$\text{NPSHA} = \frac{3.106^2}{64.4} + (-2.43 + 14.7 - 11.5)\left(\frac{144}{60.15}\right)$$

$$= 1.99 \text{ ft of head}$$

Discussion: *Since the NPSHA of 1.99 feet is less than the NPSHR of 3.5 feet, the pump will cavitate.* In this example, the cavitation was caused by a combination of low static pressure and high water temperature, the latter of which results in high vapor pressure. The incorrectly placed expansion tank also contributed to the problem.

Gaseous Cavitation

Another type of cavitation results from the presence of air in the liquid. The air may be present as small air bubbles entrained by the liquid. Air may also appear inside the vapor pockets formed by vaporous cavitation because the lowered pressure allows some dissolved air to come out of solution. This so-called **gaseous cavitation** is not as violent or damaging as vaporous cavitation. In properly deaerated systems, gaseous cavitation is usually a temporary phenomenon when the system is first put in service.

Guidelines for Avoiding Cavitation

A number of guidelines for avoiding cavitation can be stated based on the previous discussions. These guidelines widen the safety margin against the onset cavitation. Most attempt to make the NPSHA of the system as high as possible. The degree to which each guideline effects the NPSHA of the piping system can be assessed using Equation 7.5 or Equation 7.6.

- Keep the static pressure on the system as high as practical.
- Keep the fluid temperature as low as practical.
- Always place the expansion tank near the *inlet* side of the circulator.
- Keep the circulator low in the system to maximize static pressure at its inlet.
- Do not place any components with high flow resistance (especially flow-regulating valves) near the inlet of the circulator.
- If the system has a static water level, such as a partially filled tank, keep the inlet of the pump as far below this level as possible.
- Be especially careful in the placement of high head pumps or close-coupled series pump combinations because they create greater pressure differentials.
- Provide a straight length of pipe at least 12 pipe diameters long upstream of the pump inlet.
- Install good deareating devices in the system.

The schematics in Figure 7–39 depict some of these guidelines in action.

Correcting Existing Cavitation

Occasionally, an installer or designer may be called upon to correct an existing noisy pump condition. If the pump exhibits the classic signs of cavitation (i.e., churning or crackling sounds, vibration, and poor performance), the problem can probably be corrected by applying one or more of the previous guidelines.

The pump should first be isolated and the impeller inspected for cavitation damage, especially if the problem has been present for some time. Look for signs of abrasion or erosion of metal from the surfaces of the impeller. If evidence of cavitation damage is present, the impeller (or rotor assembly in a wet rotor circulator) should be replaced.

Because every system is different, there is no definite order in which to apply these guidelines in correcting the cavitation problem. In some cases, simple actions such as increasing the system's pressure or lowering its operating temperature will solve the problem. In others, the layout may be so poor that extensive piping modifications may be the only available option. Obviously, corrections that involve only setting changes are preferable to those requiring piping modifications. These should be attempted first.

In cases where cavitation is not severe, increasing the system's static pressure will often correct the problem. This can usually be done by increasing the pressure setting of the boiler feed water valve. Be sure the increased pressure does not cause the pressure-relief valve to open each time the fluid is heated to operating temperature.

If the system experiencing pump cavitation is operating at a relatively high water temperature (above 190 °F), try lowering the boiler aquastat setting by about 20 °F. In this temperature range, the vapor pressure of water decreases rapidly as its temperature is lowered. This may be enough to prevent vapor pockets (the cause of cavitation) from forming. If such an adjustment is made, be sure the system has adequate heat output at the lower operating temperature.

Always check the location of the expansion tank relative to the circulator. If the circulator is pumping toward the expansion tank, consider moving the tank near the inlet of the circulator. This may not only eliminate cavitation, but also improve the ability of the system to rid itself of air.

Also look for major flow restrictions near the circulator inlet. These include throttling valves, plugged strainers, or piping assemblies made up of many closely spaced fittings. If necessary, eliminate these components from the inlet side of the circulator.

Design "inviting" cavitation
(what's wrong)

1. Low system pressure
2. Thottling valve near circulator inlet
3. Expansion tank near circulator outlet
4. High system operating temperature
5. Lack of air separator
6. Turbulent conditions upstream of circulator inlet

Design "discouraging" cavitation
(what's right)

1. Higher system pressure
2. Thottling valve (if needed) near circulator outlet
3. Expansion tank near circulator inlet
4. Lower system operating temperature
5. Use of quality air separator
6. Straight pipe upstream of circulator inlet

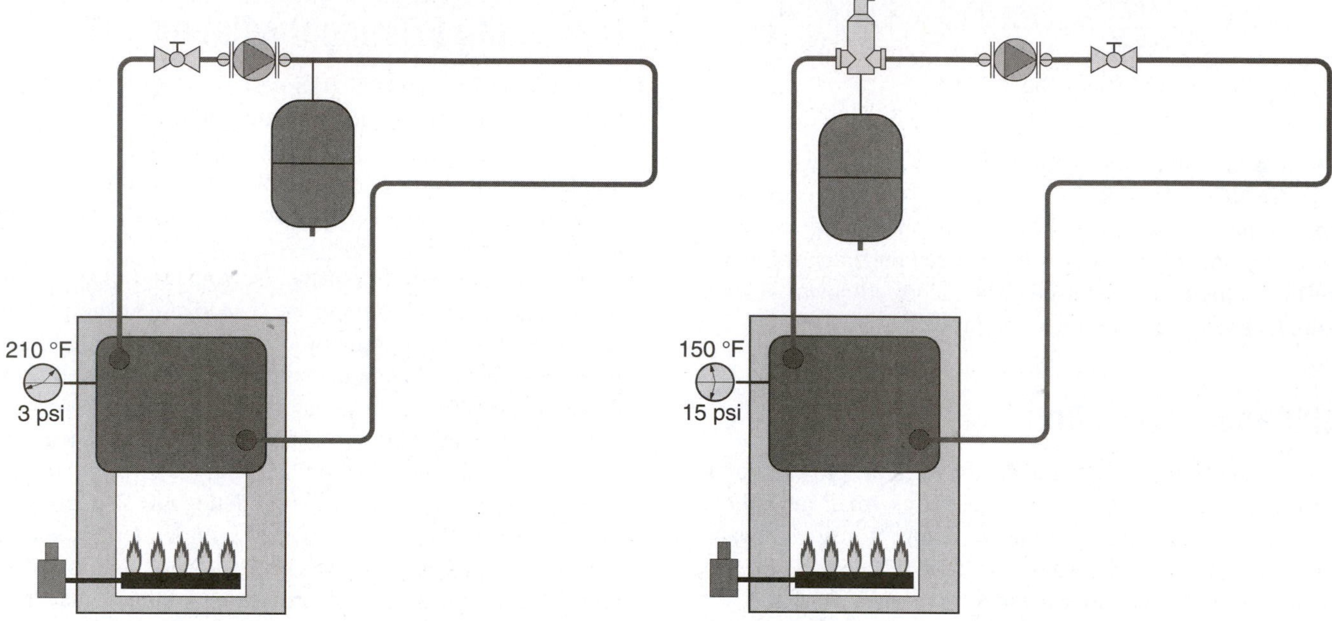

Figure 7–39 Piping schematics depicting good and bad design practices for controlling cavitation.

7.8 VARIABLE SPEED CIRCULATORS

Although they have been used in European hydronic systems for several years, small **variable speed circulators** are relatively new to the North American hydronics industry. Advances in solid state electronics have made it possible to incorporate an electronic speed control subsystem within the junction box of a small, wet rotor circulator. Such subsystems use a combination of AC frequency adjustment and digitally controlled "wave chopping" to regulate the speed of a permanent split capacitor (PSC) circulator motor with relatively good precision and efficiency. An example of a small hydronic circulator with a built-in variable speed drive is shown in Figure 7–40.

The pump curve of a variable speed circulator can vary over a wide range. The curve shifts along the vertical (head) axis as well as the horizontal (flow rate) axis as the motor speed changes. As the speed decreases, the curve shifts inward toward the origin of the graph, and vice versa. This

Figure 7–40 Example of a small circulator with a built-in variable speed drive.

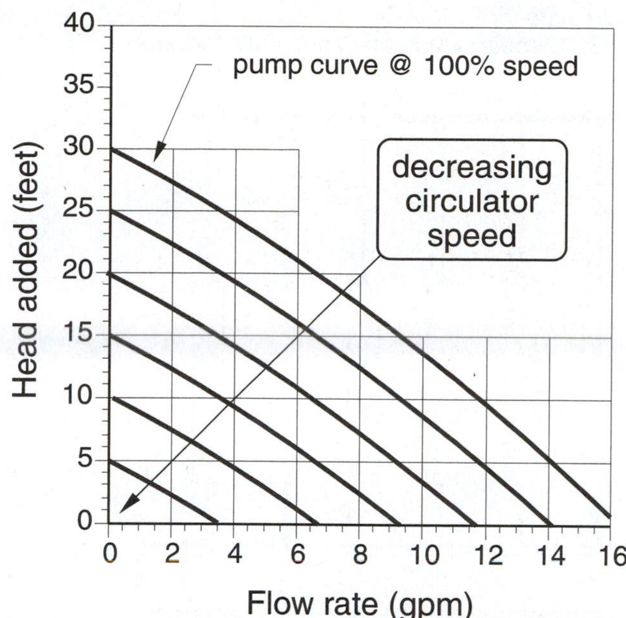

Figure 7–41 Pump curve shifts left and down as the speed of the impeller decreases.

is illustrated in Figure 7–41. Every time the circulator's speed changes, the new pump curve crosses the system resistance curve at a different operating point, and thus establishes a new flow rate in the system.

Most variable speed circulators require a low power **analog control signal** to set their operating speed. One common control signal is a 2- to 10-volt DC voltage (2–10 VDC); another is a 4- to 20-milliamp current (4–20ma). These are standard analog control signals generated by many proportional controllers used in the HVAC industry. Examples include proportional thermostats, temperature controllers, and building automation systems. New controls with these outputs will inevitably become available to facilitate increased use of variable speed circulators.

One excellent application for a variable speed circulator in hydronic systems is as a "**constant pressure circulator**." This refers to the ability of the circulator to maintain a set **differential pressure** over a wide range of flow rates. Systems having several independently controlled zone valves are good candidates for this concept.

If a fixed speed circulator is used in such a system, the operating pump shifts each time a zone circuit opens or closes. This is shown in Figure 7–42. The resulting change in differential pressure causes the flow in the operating zones to increase or decrease. In some systems, the high differential pressure across the circulator when only one or two zone circuits are open causes excessive flow velocity and flow noise. It may even cause the plugs in zone valves that are supposed to be off to lift slightly off their seat and "bleed' hot water into zone circuits where no heat is needed.

The use of a fixed speed circulator with a relatively flat pump curve does reduce the increase in differential pressure as zone circuits close. Another corrective action is the use of a differential pressure bypass valve as discussed in Chapter 5, Pipings, Fittings, and Valves. While both of these measures are currently considered good design practice, they cannot match the benefits of a variable speed circulator capable of maintaining a constant pressure differential on the distribution system regardless of the number of operating zone circuits.

When a zone circuit opens or closes in a system equipped with a variable speed circulator, a **differential pressure transducer** can be used to sense the change in differential pressure across the circulator, and adjusts the analog control signal sent to the circulator. When a zone circuit closes, the speed of the circulator is reduced to prevent any increase in differential pressure due to reduced flow rate. When a zone circuit opens, the speed of the circulator increases. In each case, the objective is to maintain a constant differential pressure on the operating zone circuits. This prevents changes in flow rate that can lead to changes in heat output. Each zone circuit operates totally independent (and unaffected) by the other zone circuits. In such an application, the operating point shifts along a horizontal line representing the intersection of the pump curve (at each speed), and the system resistance curve as shown in Figure 7–43. In effect, one could think of the operating points as shifting along a hypothetical flat pump curve.

Variable speed circulators can also be used as "metering devices" in applications where heat output is regulated by adjusting flow rate. On example is regulating the heat output of an air handler by varying the flow rate of hot water through the heat exchanger "coil," a task that has traditionally used modulating valves.

Another important application of variable speed circulators is injection mixing. Hot water is metered into a circulating distribution system to control the supply temperature and therefore the heat output of the heat emitters. This mixing technique is shown schematically in Figure 7–44 on page 206, and is discussed in detail in Chapter 9, Control Strategies, Components, and Systems.

A significant advantage of varying the speed of the circulator to control flow versus throttling with a valve is that the latter wastes head energy already imparted to the fluid by the fixed speed circulator. This energy originated as electrical input to the circulator motor. Dissipating excess head energy using a valve is akin to driving a car with the gas pedal pushed farther down than necessary and controlling speed by constant application of the brakes. It is possible, but certainly not very energy-efficient.

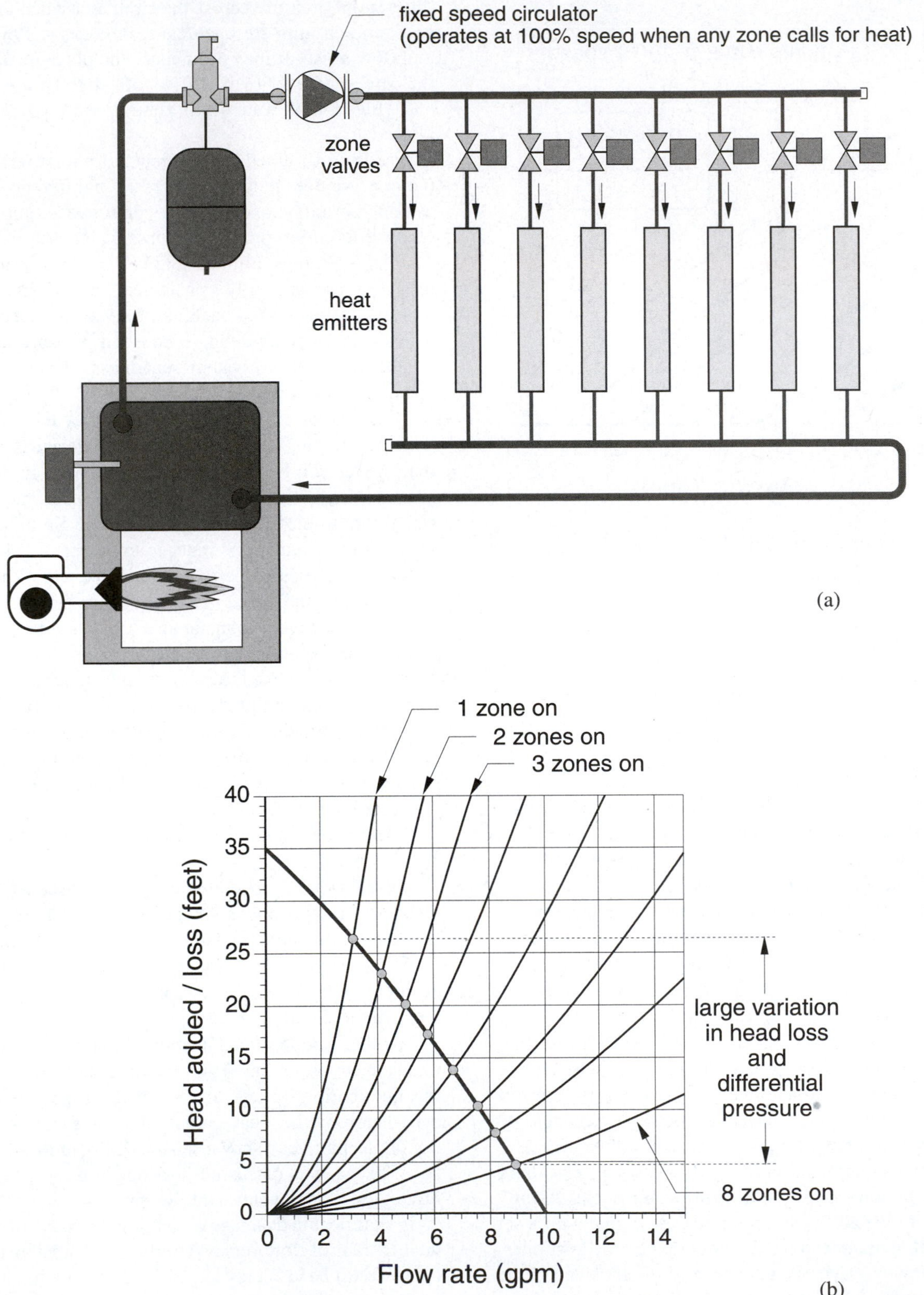

Figure 7–42 (a) A fixed-speed circulator supplying multiple zone circuits controlled by zone valves. (b) Operating point shifts upward along pump curve as the number of operating zones decreases. This increases differential pressure across the circulator.

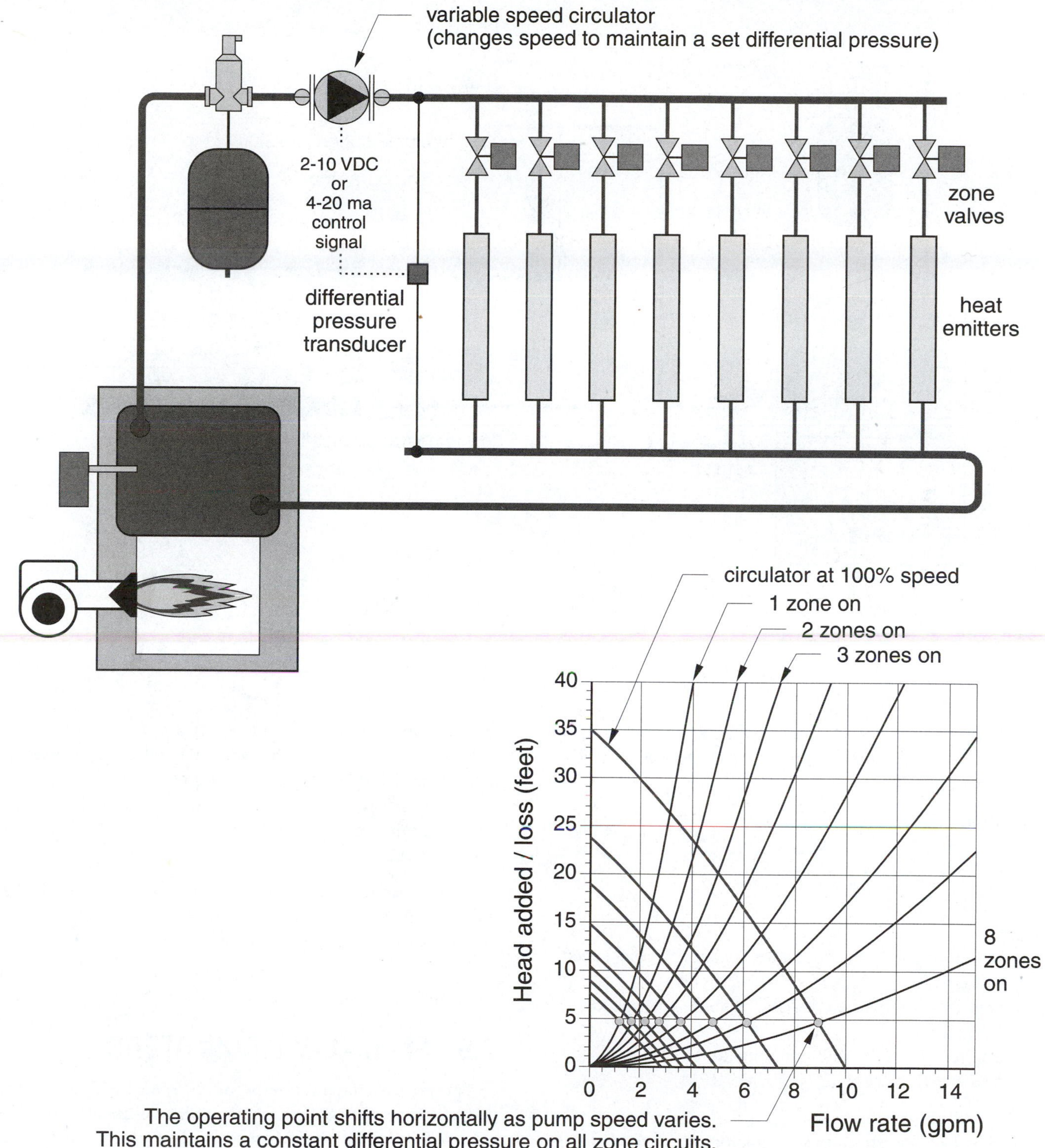

Figure 7–43 A variable speed circulator being controlled to provide constant differential pressure. Note that the operating point shifts horizontally left as zones turn off. Zones that remain on experience no change in differential pressure, and thus their flow rates remain constant.

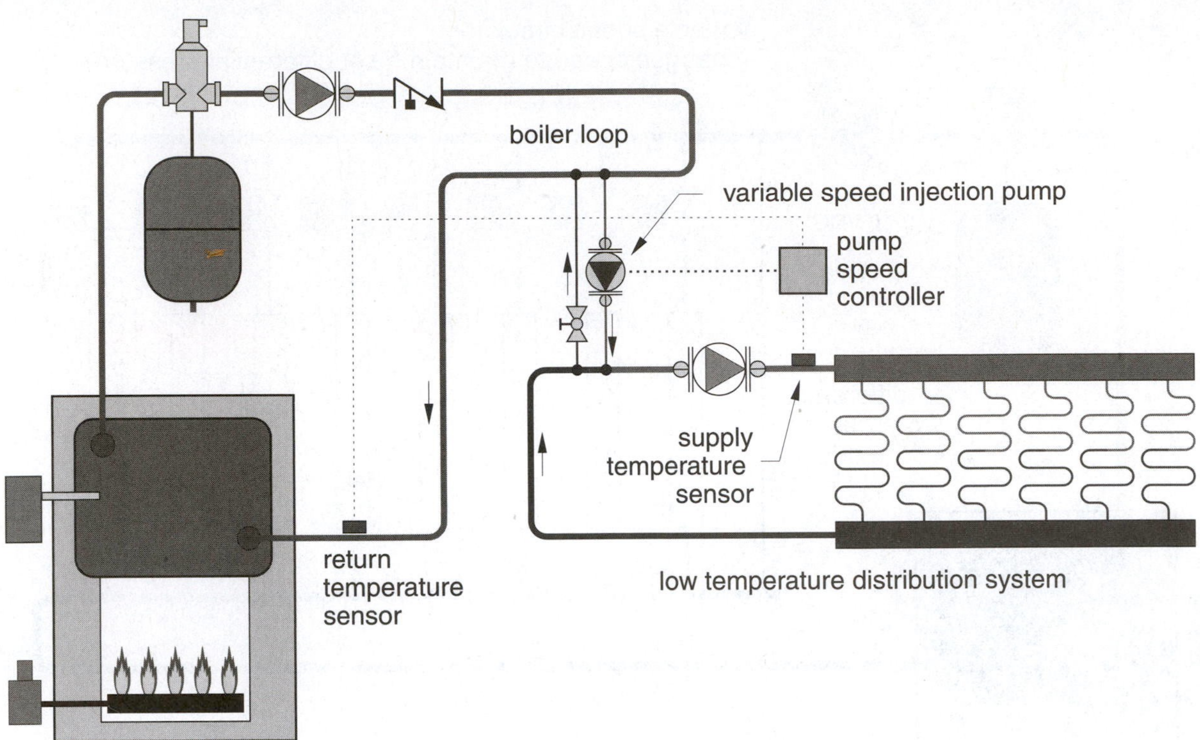

Figure 7–44 Use of a variable speed pump for injection mixing. Injection mixing is discussed in more detail in Chapters 9 and 10.

Theoretically, the power drawn by a pump varies with the cube of the impeller speed. This is expressed mathematically as Equation 7.7:

(Equation 7.7)

$$P_2 = P_1\left(\frac{n_2}{n_1}\right)^3$$

where:

P_1= power drawn by the circulator at some original speed (watts)

P_2= power drawn by the circulator at some new speed (watts)

n_1= original speed of the pump impeller (RPM)

n_2= new speed of the pump impeller (RPM)

Example 7.6: The speed of a circulator is reduced from its full speed of 3600 rpm to 1800 rpm. What is the theoretical reduction in power drawn by the circulator?

Solution: Substituting the original and reduced rpm into Equation 7.7:

$$P_2 = P_1\left(\frac{n_2}{n_1}\right)^3 = P_1\left(\frac{1800}{3600}\right)^3 = P_1(0.125) = \left(\frac{1}{8}\right)P_1$$

Discussion: Regardless of the power drawn by the pump at full speed, only 1/8 of that power is theoretically required at half speed. This demonstrates the significant potential for electrical energy savings when flow rate is controlled by varying pump speed versus the use of throttling valves, especially in larger systems.

As the cost of electrical energy increases, there will be increased incentive for use of variable speed circulators in all types of hydronic heating applications.

7.9 SPECIALTY CIRCULATORS

Several manufacturers now offer "specialty" circulators that combine a basic circulator with additional hardware often needed in hydronic systems. The goal is to reduce installation time by preassembling components, or offering functionally equivalent components, to those that would normally be individually installed at the job site. In most cases this approach also reduces installation cost.

One example is a circulator with built-in circuitry specifically designed to interface with a 24-VAC room thermostat. An example is shown in Figure 7–45. Inside the junction box, this circulator contains a printed circuit board with a transformer, relay, and other electrical components that allow the thermostat to be wired directly to the circulator. When necessary, multiple circulators of this type can be wired together to provide control for a multiple-zone system. This approach eliminates the need for a separate multizone relay center (which is discussed in Chapter 9, Control Strategies, Components, and Systems).

Figure 7–45 Example of a circulator with built-in low voltage circuitry for direct connection to a zone thermostat. *Courtesy of Taco, Inc.*

Circulators with Integral Flow-Checks

A standard circulator that is not operating, can not prevent flow through its volutes due to pressure differentials created elsewhere in the system. Without the installation of a flow-check valve in each zone circuit (as discussed in Chapter 5, Piping, Fittings, and Valves), some of the flow driven by operating zone circulators will reverse through circulators that are off. This allows heated water to pass through zone circuits that are not in need of heat. In general, a flow-check valve needs to be installed in each piping circuit anytime multiple pumps are installed in parallel.

To address this common situation, some pump manufacturers offer circulators with integral flow-check valves. An example is shown in Figure 7–46.

The integral flow-check valve consists of a small spring-loaded valve housed within the inlet or outlet port of the circulator's volute. Like a stand-alone flow-check valve, its purpose is to prevent reverse flow as well as forward heat migration due to differences in fluid density. Use of such a circulator eliminates the need of a field-installed flow-check in each zone circuit. Small circulators with integral flow-check valves are also used to prevent heat migration in some injection mixing systems.

The presence of an integral flow-check reduces the head available from the circulator at any given flow rate. This is reflected in the pump curve supplied by the manufacturer. Be sure to use the proper pump curve when matching the circulator to the piping system.

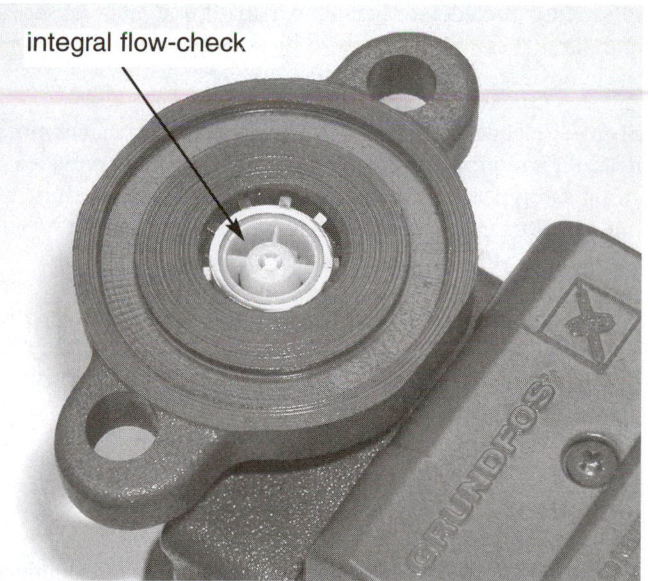

Figure 7–46 Circulator with spring-loaded check valve built into volute. Use of this circulator eliminates the need for separate flow-check valves.

7.10 SELECTING A CIRCULATOR

The ideal circulator for a given application would produce the exact flow rate desired while operating at maximum efficiency, and it would draw the least amount of electrical power. It would also be competitively priced, and built to withstand many years of service. These conditions, although possible, are not likely in all projects because of the following considerations:

- Only a finite number of circulators are available for any given range of flow rate and head requirements. This contrasts with the fact that an unlimited number of piping systems can be designed for which a circulator must be selected.
- Most small circulators are single speed units that operate on a fixed pump curve. These circulators cannot be fine-tuned to the system requirements through impeller changes, motor changes, and so on, as is often possible with larger pumps. The exception is multispeed circulators where each speed represents a different pump curve.
- The choices may be further narrowed by material compatibility. For example, cast-iron circulators should never be used in open-loop hydronic applications because of corrosion.

- Purchasing channels and inventories of spare parts may limit the designer's choices to the product lines of one or two manufacturers.

The selection process must consider these limitations while also attempting to find the best match between the head and flow rate requirements of the system and circulator. A suggested procedure for finding this match is as follows:

Step 1. Determine the desired flow rate in the system based on the thermal and flow requirements of the heat emitters. The relationship between flow rate and heat transport is described by the sensible heat rate equation, and was discussed in Chapter 4, Properties of Water. The flow requirement of various heat emitters will be discussed in Chapter 8, Heat Emitters.

Step 2. Use the system resistance curve or its equation to determine the head loss of the system at the required flow rate. Detailed methods for determining the system resistance curve were given in Chapter 6, Fluid Flow in Piping.

Step 3. Plot a point representing the desired flow rate and associated head loss on the pump curve graph of the circulators being considered. Remember that circulators in series and parallel can be represented by a single pump curve, and thus considered where they offer advantages.

Step 4. Select a circulator that delivers up to 15 percent more head than required at the desired flow rate. This provides a slight safety factor in case the system is installed with more piping, fittings, or valves than what was assumed when constructing the system resistance curve.

Step 5. Finally, the curve of the selected circulator should (ideally) intersect the system resistance curve within the middle third of the pump curve where the circulator's operating efficiency is highest.

The following example will pull together many of the previous information and procedures.

Example 7.7: A piping system is being planned to deliver 35,000 Btu/hr of heat using a design temperature drop of 20 °F. The fluid used will be water at an average temperature of 160 °F. Using the procedures in Chapter 6, Fluid Flow in Piping, the piping loop is estimated to have a total equivalent length of 325 feet of 3/4-inch type M copper tube. Determine the necessary flow rate in the system, and select an appropriate circulator from the family of pump curves shown in Figure 7–18a.

Solution: The "target" flow rate for delivering 80,000 Btu/hr of heat using water and a temperature drop of 20 °F is determined using Equation 4.5:

$$Q = 8.01\ Dcf(\Delta T)$$

where:

Q = *rate of heat transfer* into or out of the water stream (Btu/hr)
8.01 = a constant based on the units used
D = density of the fluid (lb/cubic ft)
c = specific heat of the fluid (Btu/lb/°F)
f = flow rate of water through the device (gpm)
ΔT = temperature change of the water (°F)

From using Figure 4–3, the density of 160 °F water is found to be 60.9 lb/ft^3.

The specific heat of water at 160 °F is 1.00 Btu/lb/°F.

Note: The density and specific heat values could also have been found using the Fluid Properties Calculator module in the Hydronics Design Studio software.

Equation 4.5 is now rearranged to solve for the flow rate:

$$f = \frac{Q}{8.01 \times c \times D \times \Delta T} = \frac{35{,}000}{8.01 \times 1.00 \times 60.9 \times 20}$$

$$= 3.59 \text{ gpm}$$

The head loss of the circuit can now be estimated using Equation 6.11. Doing so requires the value of the fluid properties factor (α), and the pipe size factor (c).

The value of α for water at an average temperature of 160 °F is found in Figure 6–17: $\alpha = 0.046$.

The pipe size factor for 3/4-inch copper tube is found in Figure 6.18: $c = 0.061957$.

These values are combined with the total equivalent length of the circuit and the target flow rate (using Equation 6.11) to estimate the head loss of the system:

$$H_L = (\alpha c L)(f)^{1.75} = 0.046 \times 0.061957 \times 350 \times (3.59)^{1.75}$$

$$= 9.34 \text{ ft}$$

The target operating point of 3.59 gpm and 9.34 feet of head loss is now plotted on the family of pump curves in Figure 7–47. Also plotted are several other random points along the system curve calculated using Equation 6.11 with different flow rates. These points allow the system curve to be sketched on the same graph with the pump curves.

The pump curve for circulator #6 (UP15-42 FC) intersects the system curve almost exactly at the desired operating point, and thus could be used. The curve for circulator 8 (UP15-42 speed 2), although producing slightly more head than necessary, could also be used. In the latter case, the system would operate where the pump curve crosses the system curve at approximately 4.2 gpm.

The Hydronic Circuit Simulator module in the Hydronics Design Studio software could also have been used to quick-

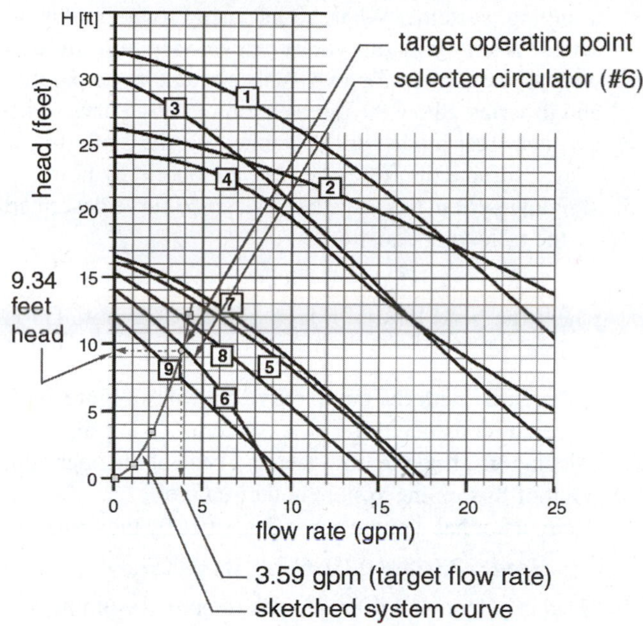

Figure 7–47 Target operating point and system curve sketched on graph of multiple pump curves.

ly estimate the performance of several circulators in the proposed piping system. Because of its internal database of fluid properties and equivalent lengths, it would considerably reduce the time necessary to make these comparisons.

SUMMARY

This chapter has discussed the types of circulators used in hydronic heating systems, as well as methods for describing their performance in a given piping system. The fundamental concept of determining the intersection of the pump curve and system resistance curve will be used in several later chapters as an inherent part of system design. The placement of the circulator so its inlet is near the connection point of the system's expansion tank is also a vital concept to be repeatedly applied in subsequent system design.

By applying the concepts presented in this chapter, the designer can select a circulator that properly supplies the flow rate and head requirements of a piping system, while operating at or near maximum efficiency and without the destructive effects of cavitation.

KEY TERMS

analog control signal
cavitation
centrifugal pump
circulators
close coupling
close-coupled
constant pressure circulator
differential pressure
differential pressure transducer
efficiency of a circulator
end suction pump
eye
feet of head
flanges
gaseous cavitation
head
Hydronic Circuit Simulator module
impeller
implosion
inline circulator
isolation flange
iteration
mechanical energy
net positive suction head available (NPSHA)
net positive suction head required (NPSHR)
operating point
point of no pressure change
pump curve
pumps
push/pull
series
shaft seal(s)
three-piece circulator
variable speed circulator
volute
wet-rotor circulator

CHAPTER 7 QUESTIONS AND EXERCISES

1. Describe the difference between a circulator with a steep pump curve and one with a flat pump curve. Which would be better for use in a multizone system involving zone valves? Why?
2. Describe the difference between the head of a pump and the difference in pressure measured across the pump using pressure gauges. How is this pressure difference converted to a head value?
3. A circulator with a pump curve as shown in Figure 7–48 is installed in a system operating with 140 °F water. Pressure gauges on the inlet and discharge ports have readings as shown. What is the flow rate through the pump?

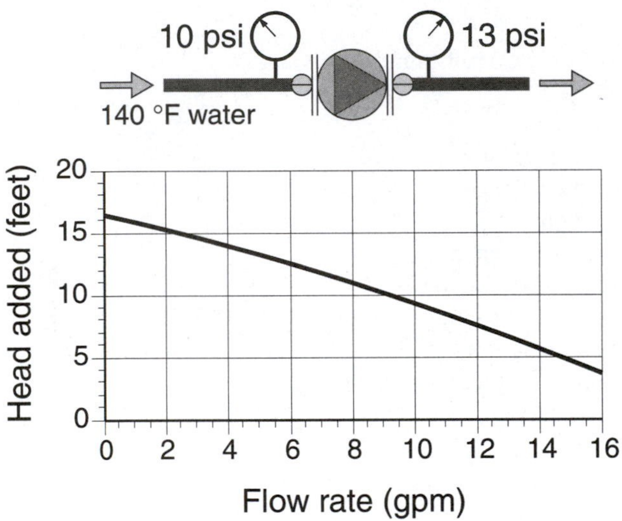

Figure 7–48 Pressure gauge readings and pump curve for Exercise 3

4. Will using two pumps in parallel double the flow rate in a piping system? What about using two pumps in series? Justify your answer by drawing a pump curve as well as the curves for two such circulators in parallel and in series. Show the intersection of these curves with a system resistance curve sketched on the same set of axes. Estimate the flow rates at each operating point.
5. A piping system has a system resistance curve described by the following equation:

$$H_{loss} = 0.85(f)^{1.75}$$

where the head loss is in feet, and the flow rate (f) is in gpm.

A circulator with the pump curve shown in Figure 7–49 is being considered for the system. Using iterative calculations, find the flow rate and head at the operating point of this piping system and circulator.

6. A circulator has a pump curve described by the equation:

$$Head = 7.668 + 0.1086(f) - 0.009857(f)^2$$

Head is in feet, and flow rate f is in gpm. Assuming this pump is used with the same piping system described in Exercise 5, find the flow rate and head at the new operating point using manual iterative calculations.

7. The static pressure in the circulator of a given system is 5 psig when the circulator is off. When the circulator operates, a head of 20 feet is established across it. The fluid being pumped is 140 °F water. Determine the readings of pressure gauges tapped into its inlet and discharge ports when it is running assuming:
   a. The expansion tank is connected at the inlet port of the circulator.
   b. The expansion tank is connected to the discharge port of the circulator.
8. A circulator has a pump curve as shown in Figure 7–49. Sketch the resulting pump curve for the following situations:
   a. Two of these circulators connected in series.
   b. Two of these circulators connected in parallel.

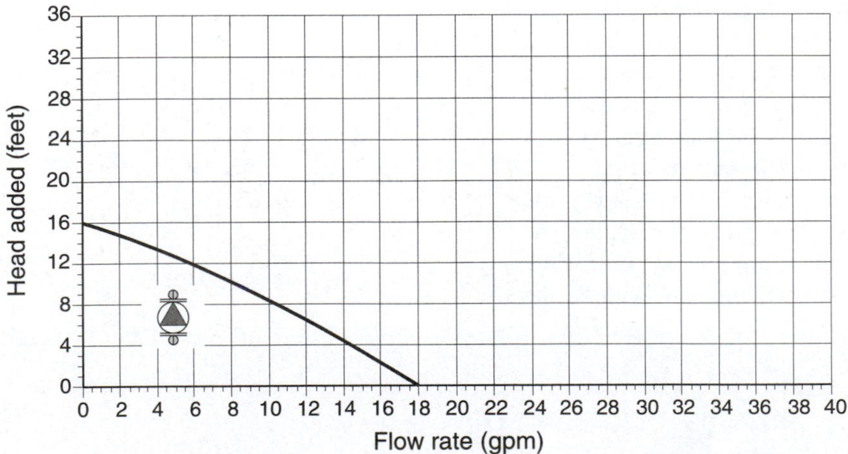

Figure 7–49 Pump curve for Exercise 5

9. A piping system and wet rotor circulator have the head versus flow rate curves shown in Figure 7–50. The measured power consumption of the circulator in this application is 75 watts. What is the overall efficiency of the motor/pump assembly in this application?
10. Determine the NPSHA at the inlet of the pump for the two systems shown in Figure 7–51.
11. Use the Hydonic Circuit Simulator module in the Hydronics Design Studio to estimate the flow rate produced by the following situations in both branches of the piping circuit shown in Figure 7–52.
    a. Using a Taco 007 pump and water at an average system temperature of 140 °F
    b. Using a Bell & Gossett NFR22 circulator and 30 percent propylene glycol at an average temperature of 100 °F
    c. Using a Grundfos UP26-64 and 50 percent ethylene glycol at an average temperature of 160 °F

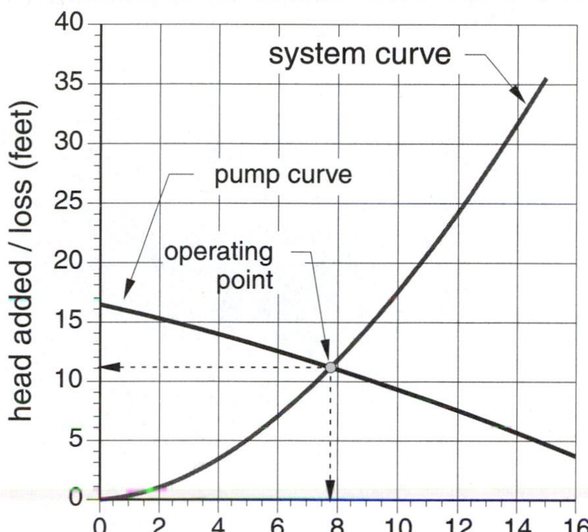

Figure 7–50 Pump curve and system resistance curve for Exercise 9.

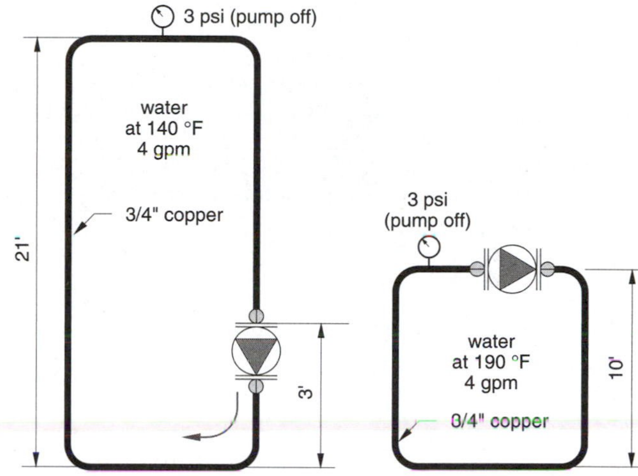

Figure 7–51 Piping diagrams for Exercise 10.

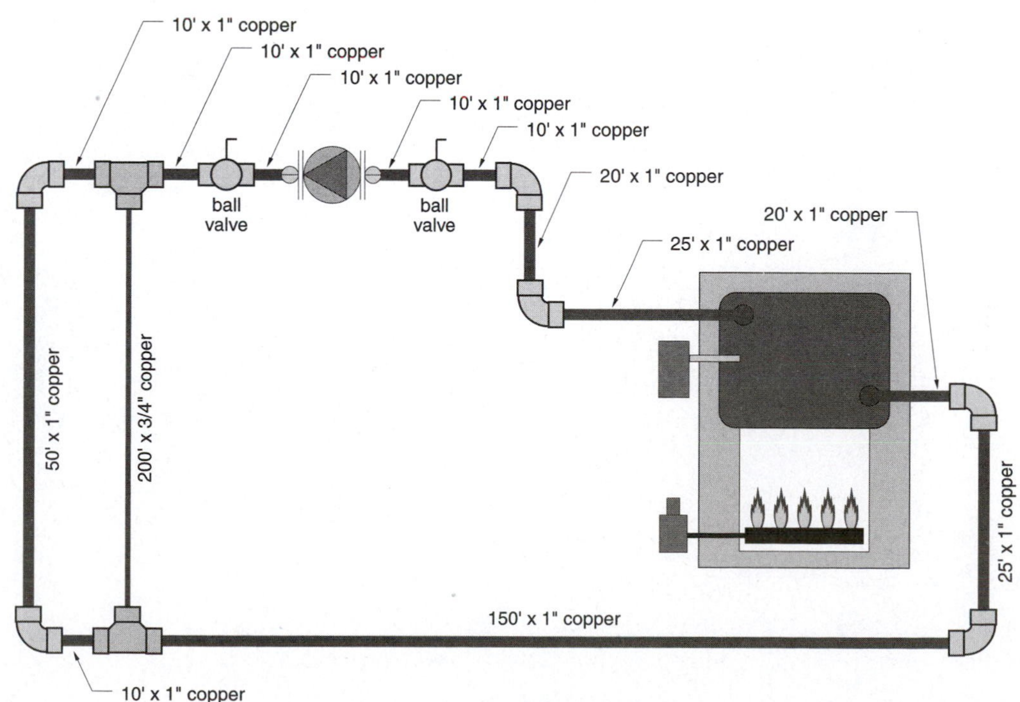

Figure 7–52 Piping system for Exercise 11.

HEAT EMITTERS 8

OBJECTIVES

After studying the chapter, you should be able to:

- Describe several different types of heat emitters
- Identify good applications for various types of heat emitters
- Explain the difference between a radiator and a convector
- Properly size finned-tube baseboard convectors
- Use the Series Baseboard Simulator program in the Hydronics Design Studio software
- Explain the installation methods used for various heat emitters
- Describe different types of panel radiators
- Estimate the performance of heat emitters over a wide range of conditions
- Estimate the rate of heat loss from copper tubing
- Explain the concept of steady state supply temperature

8.1 INTRODUCTION

Hydronic **heat emitters** release heat from the system's fluid into the space to be heated. One distinct advantage of hydronic heating is the wide variety of heat emitters available to suit almost any job requirement. They range from small panel radiators, to underfloor systems that use the entire floor area of a building as the heat transfer surface. The latter type of system is specialized enough to justify a separate Chapter 10, Hydronic Radiant Panel Heating.

The designer must select and size heat emitters for a given project, so that comfort and control are achieved in all areas of the building. Technical, architectural, and economic issues must be considered. The technical issues involve temperature, flow rate, and heat output characteristics of the heat emitter options. Architectural issues include the appearance of the heat emitters in the building, as well as any interference they may create with furniture, door swings, and so on. Economic considerations can greatly expand or restrict heat emitter options. The designer must provide a workable, if not optimal, selection of heat emitters within a given budget. Since most heating systems are designed after the floor plan of the building has been established, the heating system designer is often faced with the challenge of making the system fit the plan. A good knowledge of the available options can lead to creative and efficient solutions.

8.2 CLASSIFICATION OF HEAT EMITTERS

Some heat emitters directly heat the air in a room. They are properly classified as **convectors** because convective heat transfer is the primary means by which heat is released. Other types of heat emitters release the majority of their heat as thermal radiation, and are properly referred to as **radiators**. The balance between convective heat transfer and radiative heat transfer can significantly affect the comfort achieved by various heat emitters. Each type has advantages and disadvantages that need to be understood before making a selection. Often a combination of two or more types provides an ideal match for the requirements or restraints imposed by a given project.

In this chapter the following types of heat emitters will be discussed:

- Finned-tube baseboard convectors
- Fan-coil convectors
- Panel radiators
- Radiant baseboards

These generic categories encompass most of the available types of heat emitters for residential or small commercial hydronic systems. Sometimes a specialized name is used to describe a specific heat emitter. A **kick-space heater**, for example, is a small horizontal **fan-coil** designed to mount under a cabinet or stair tread. A **towel warmer** is a name used for certain styles of panel radiators. These specialties will be discussed within the previously listed groupings.

8.3 FINNED-TUBE BASEBOARD CONVECTORS

The most common hydronic heat emitter used in residential and light commercial systems in North America is the **finned-tube baseboard convector**. It was developed during the 1940s as a flexible, modular, and quickly installed alternative to standing cast-iron radiators. Its widespread use

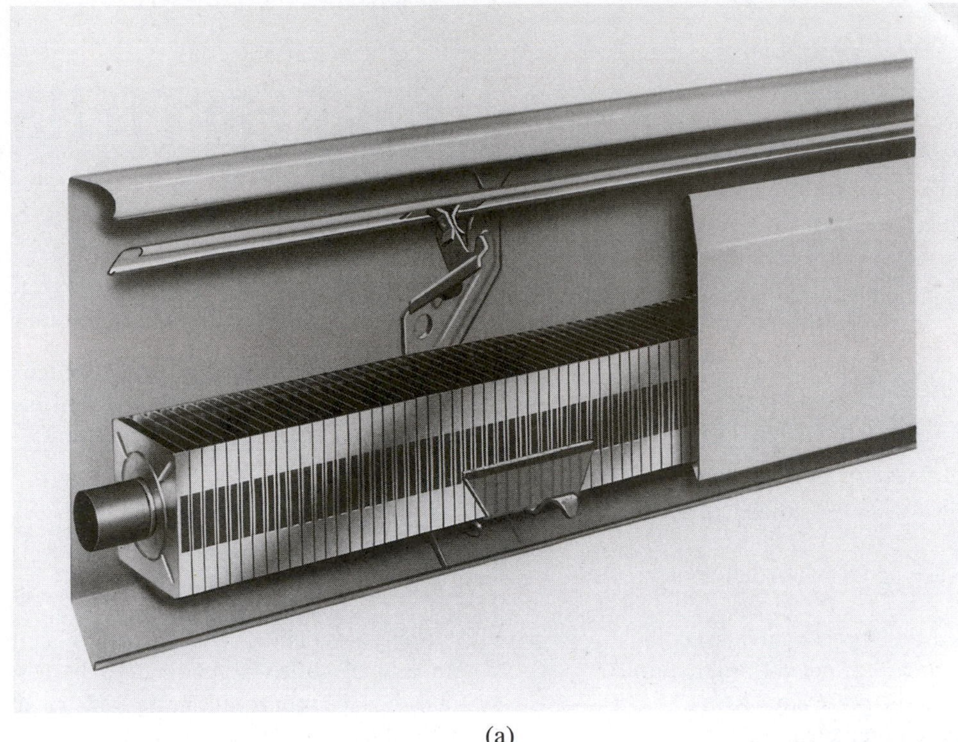

(a)

(b)

Figure 8–1 (a) Cut-away view of a typical residential finned-tube baseboard convector. *Courtesy of Weil-McLain Corporation.* (b) Installed finned-tube baseboard with closed damper.

in smaller hydronic systems is largely due to its price advantage over other heat emitter options.

Figure 8–1 shows a cutaway view of a typical finned-tube baseboard convector. It consists of a sheet metal enclosure containing a copper tube fitted with aluminum fins. The finned tube is often called the **element** of the baseboard, since it is the component from which heat is released.

There are many styles of finned-tube baseboard available. Those intended for residential and light commercial use usually have elements consisting of either 1/2-inch or 3/4-inch copper tubing with rectangular aluminum fins about 2.25 inches to 2.5 inches wide and high. The enclosures are fabricated from steel sheet, and vary from about 6 inches to 8 inches in height. Most manufacturers sell finned-tube baseboard in lengths from 2 feet to 10 feet long in increments of 1 foot. For long runs, combinations of these lengths are joined together.

Larger styles of finned-tube convectors are available, intended for use in larger commercial or institutional buildings. They have elements consisting of either copper or steel pipe in sizes from 3/4 inch up to 1.25 inches. Their fins are larger (3 inches to 5 inches on a side), and are available in both aluminum and steel. These larger convectors can release considerably more heat per foot of length than the smaller residential baseboards. They are also built to withstand the greater physical punishment associated with commercial or institutional usage. Because of their size and cost, they are seldom considered for use in smaller hydronic systems.

In addition to straight lengths of finned-tube baseboard, most manufacturers offer sheet metal enclosure fittings such as end caps, inside and outside corner trim, and joint covers. These allow the baseboard to be fitted to the shape of the room when necessary. They are usually designed to snap into place over the straight length enclosure without the use of fasteners. Figure 8–2 depicts some of the available enclosure trim pieces.

Finned-tube baseboard convectors operate on the principle that heated air rises due to its lower density. This is the same effect that lifts a hot air balloon or makes smoke rise up a chimney. Air in contact with the finned-tube element is heated by natural convection. It rises through the fins, and up through the slot at the top of the enclosure. Cooler air near the floor level is drawn in through a slot at the bottom of the enclosure. As long as the finned-tube element is warmer than the surrounding air, natural convection continues to "pump" air up through the enclosure. Most baseboard enclosures are equipped with a hinged damper over their outlet slot. This can be manually set to limit air flow through the baseboard. When fully closed, heat output is reduced 40 percent to 50 percent.

The upward flow of warm air can negate the effects of downward drafts from exterior walls, and especially windows. Because of this, *finned-tube baseboard is usually placed along exterior walls, and especially under windows.* The circulation of room air due to operation of the finned-tube baseboard is depicted in Figure 8–3.

Heating a room by natural convection of warm air creates both desirable and undesirable effects. On the plus side, the

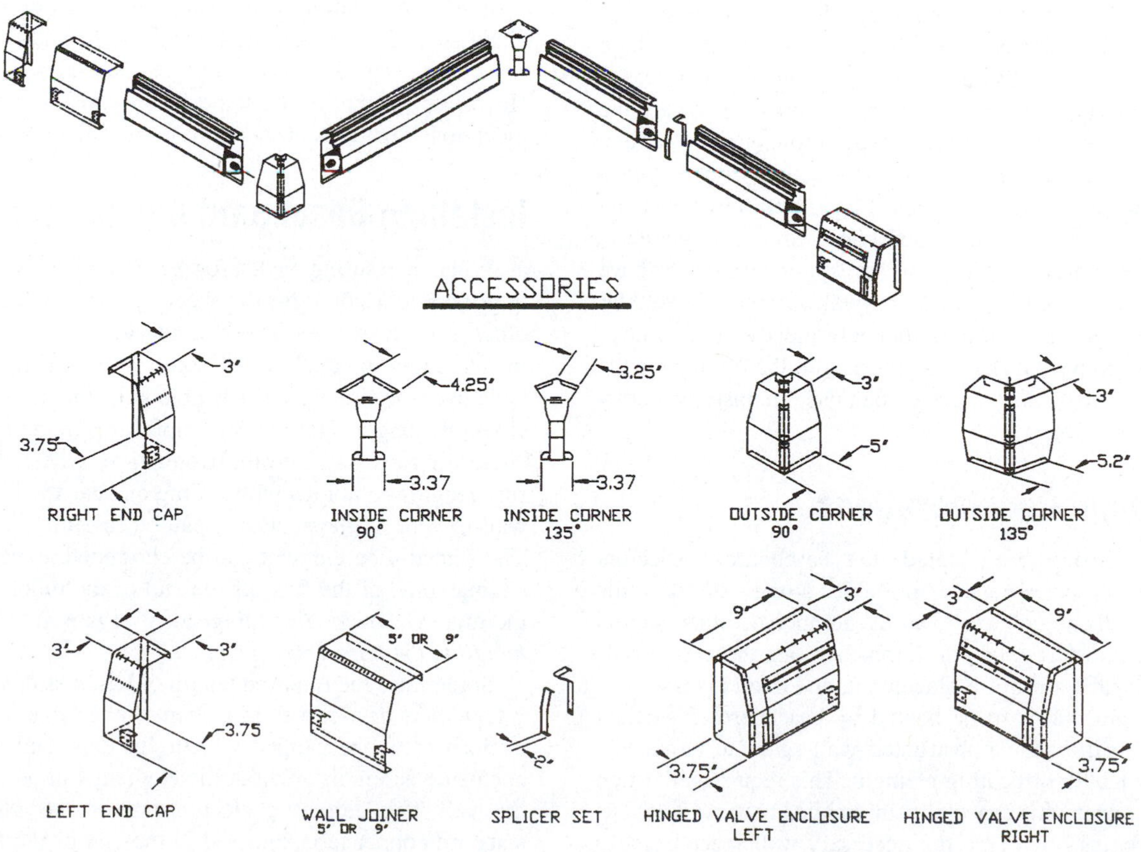

Figure 8–2 Examples of enclosure trim parts for finned-tube baseboard. *Courtesy of Embassy Industries.*

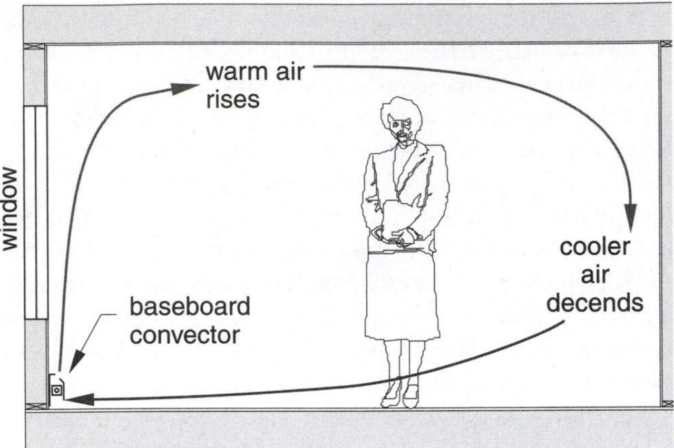

Figure 8–3 Room air circulation established by natural convection currents from finned-tube baseboard.

slow-moving air produces no sound. The rising warm air is also an effective means of limiting condensation of water vapor on cold glass surfaces. It negates the uncomfortable drafts associated with downward movement of cool air along these surfaces. When the baseboard is arranged to blanket an exterior wall with warm air, the wall's surface temperature is increased, improving the comfort in the room.

An undesirable result of heating by natural convection is that warm air tends to stay near the ceiling of the room, while cooler air pools at floor level. This effect, often called **stratification**, is much more noticeable in rooms with high ceilings. It is worsened by convectors operating at high water temperatures. The higher the element temperature, the stronger the rising air currents. The hot air rises toward the ceiling without proper mixing with room air. As a rule, *convectors are not well suited for rooms with high ceilings.*

Anytime air is used to transport heat, dust movement can become a problem. Finned-tube convectors that operate at high temperatures in dusty or smoky environments have been known to produce streaks on wall surfaces above their enclosures. All heat emitters that rely mostly on air movement (e.g., convection) to transfer heat to the room have the potential to create dust streaks when used in dusty or smoky environments.

Placement Considerations

It has already been stated that baseboard convectors should be placed, whenever possible, along exterior walls and especially under windows. A number of other factors also enter into placement decisions. Among these are available wall space, furniture placement, and door swings.

Any room that is to be heated by baseboard convectors must have sufficient unobstructed wall space to mount the necessary length of heating element. This requirement often presents a problem, especially in smaller houses. Kitchens and bathrooms often lack the necessary wall space because of cabinets and fixture placement. However, an estimate of baseboard length must be made before determining if available wall space is adequate. Methods for doing so are presented in the next section.

Heat emitters often compete with furniture for wall space. Most homeowners believe the heating system should be fit around furniture placement. While this may not always be possible, it should be reasonably attempted by the system designer. When interference is likely, it should be brought to the attention of the owner so a compromise can be found. The degree to which heat emitters allow flexibility of interior design can have a big impact on customer satisfaction with the overall system. It is unfortunate that furniture placement is often unnecessarily restricted by a careless placement of heat emitters.

As a minimum, 6 inches of free space should be available in front of any finned-tube baseboard. Without it, heat output will be significantly lowered due to inadequate air circulation. Placing furniture very close to baseboard convectors can also lead to degradation of fabrics, woods, and finishes due to prolonged exposure to relatively hot, low humidity air.

Door swings should also be considered before placing baseboard convectors. Baseboard convectors should not be installed within the arc of the door swing. Their thickness may prevent the door from making its normal swing back to the wall. The closer the baseboard is to the door opening, the worse the problem. Wooden doors that are left open against baseboard can also be discolored, warped, or cracked by prolonged exposure to the warm air.

Every installation can bring its own set of unusual circumstances. It would be impossible to address all situations affecting the placement of baseboard convectors. However, a sampling of some of the most common placement considerations is shown in Figure 8–4.

Installing Baseboard Convectors

When measuring walls for baseboard convectors, it is important to account for the space required by end caps and other trim accessories. These dimensions will vary from one manufacturer to another. In some cases, trim accessories slide over the enclosure for simple adjustment of the overall assembly length. This is very helpful in situations where the enclosure runs wall to wall. In situations where stock enclosure lengths combined with sliding end caps will not allow a wall-to-wall fit, the enclosure can be carefully cut to length. The finned-tube element can be shortened as necessary by sliding some of the fins off the end of the tube, and cutting the tube. *To avoid sliced fingers, always wear gloves when handling the finned-tube element.*

Sometimes the required length of baseboard will end 1 or 2 feet short of a corner of a room. When this happens, the enclosure can be capped with a standard end cap, or the enclosure assembly without the element can be extended to the wall. This dummy enclosure may or may not contain a standard copper tube soldered to the end of the finned-tube element. Its only real purpose is to continue the aesthetics of the baseboard enclosure from wall to wall.

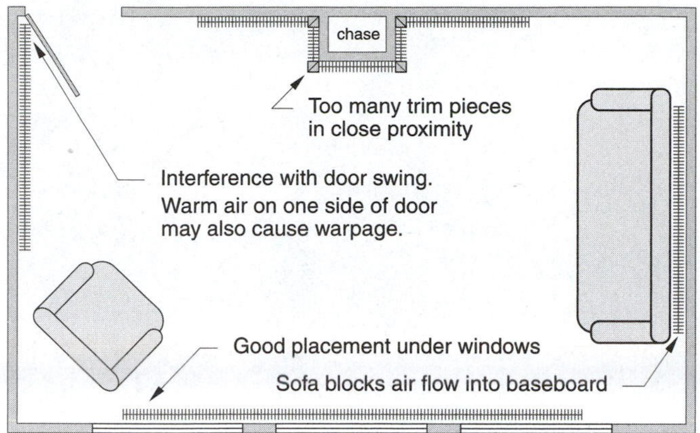

Figure 8–4 Good and bad placement of finned-tube baseboard within a room.

The four basic steps of installing a baseboard convector are:

1. Drill holes for the riser pipes
2. Attach the enclosure to the wall
3. Install the finned-tube element
4. Install the front cover and trim accessories

The holes for the riser pipes should be located so as to avoid drilling into floor joists, girders, wiring, or other plumbing beneath the floor. One approach is to locate a nearby object such as a plumbing stack *that penetrates through the floor*, measure the location of the joists or other objects to this reference object under the floor, then transfer these measurements above the floor to locate the joists. Another method is to drill a small 1/8-inch hole and insert a rigid wire to help locate the hole from under the floor. The distance from the wire to the joists can be measured under the floor and again transferred to the top of the floor. Since most floor joists are located 16 inches or 24 inches on center, the position of adjacent joists can be measured easily. The location of the floor joist, or other obstructions, may require the enclosure, and/or the finned-tube element to be moved one way or the other. Always check this before fastening the enclosure to the wall.

When the centerline of one riser pipe is located, obtain the centerline of the other by measuring the *assembled length* of the heating element, along with any fittings or valves that are attached to it. Locate the other hole, and again *check that it does not fall over a floor joist or other obstruction before boring through.* This method of assembling the finned-tube element, with the required fittings and valves, is simpler than trying to calculate the exact position of the holes while accounting for lengths of the fin-tube element, valves, fittings, and so forth.

The holes for the riser pipes to the finned-tube element should have a diameter about 1/2 inch greater than the nominal pipe size used. This provides a space for the tubing to expand when heated without stressing the pipe. It also allows some tolerance for errors in measurement or alignment when soldering. Because the baseboard element is installed at room temperature, it is close to its minimum length. Because of this, most of the 1/2-inch space created by the larger hole should be located on the side of the tube *away* from the baseboard as shown in Figure 8–5. This allows maximum room for piping expansion. An **escutcheon plate** can be installed on the riser piping to neatly cover the holes through the flooring.

The installer should be careful to mount the enclosure high enough to clear the finish flooring materials. As a rule, the bottom of the enclosure should be at least 1 inch above the subfloor. This allows sufficient space for medium

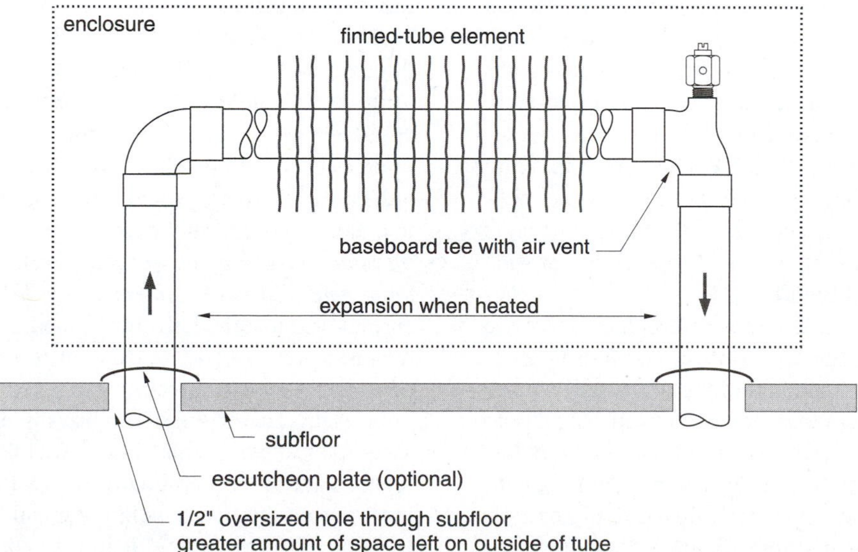

Figure 8–5 Installation of riser piping to a finned-tube baseboard. Note oversized holes through subfloor, and space allowed for element expansion.

thickness carpeting and pad. The enclosure can be nailed or screwed to the wall. On wood-frame walls, the enclosure should be fastened at intervals not exceeding 32 inches (to every other stud in a typical wall). Allow a slight gap with the wall surface at the end of the enclosure to allow end pieces to be snapped in place.

After the enclosure is mounted, the finned-tube assembly can be placed on the cradles of the enclosure and all soldered joints made up. In some cases, the elbows and riser pipes are first soldered to the element, then this assembly is lowered into the enclosure by sliding the riser pipes down through the holes in the floor. In all cases, be sure the element rests properly on the plastic glides that allow the element to expand and contract without creating squeaking sounds.

In some installations, the supply and return piping must exit the same end of a baseboard. This approach is often used when one end of the baseboard cannot be accessed through the floor. This type of installation requires a 180-degree return bend at the end of the finned-tube element. The return pipe then runs back over the top of the finned-tube element. A detail of this return bend fitting, with a tapping for an air vent, is shown in Figure 8–6. The return pipe also requires support within the enclosure. It is usually supported by loops of copper wire wrapped around the brackets inside the enclosure.

To complete the installation, snap the front cover in place, then install the necessary trim pieces. Be sure the damper blade on the enclosure is left in its fully open position.

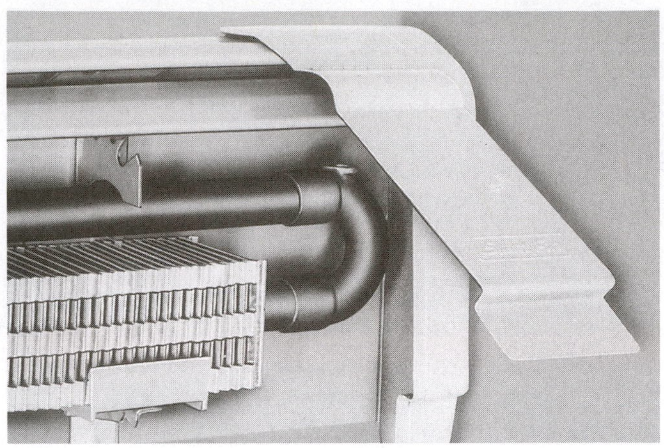

Figure 8–6 Example of a 180° return bend, used when baseboard is supplied and returned form the same end. *Courtesy of Slant/Fin Corp.*

8.4 THERMAL RATINGS AND PERFORMANCE OF FINNED-TUBE BASEBOARD

The heat output of finned-tube baseboard is expressed in Btu/hr *per foot of active element length*. The **active length** of the element is the length of pipe covered with fins. This is usually 3 to 4 inches less than the length of the straight enclosure, not including the end caps.

The heat output of the baseboard is dependent on the fluid temperature in the finned-tube element. Manufacturers publish tables showing the heat output of their baseboard units at several different water temperatures. An example of such a ratings table is shown in Figure 8–7.

These ratings are based on testing conducted by the Hydronics Institute using the IBR Testing and Rating Code for Baseboard Radiation. Output ratings are usually listed for average water temperatures ranging from 220 °F to 150 °F. Any output rating at water temperatures below 150 °F must be estimated using correction factors established by the testing standard, which are listed in Figure 8–8.

Note that these factors multiply the output rating of the baseboard at 150 °F water temperature. They show how the heat output of finned-tube baseboard decreases rapidly at lower water temperatures. For example, the heat output of a baseboard operating with 100 °F average water temperature is only 28 percent of the heat output while operating at 150 °F. For this reason, finned-tube baseboard systems are seldom sized for low temperature operation under design load conditions. However, reduced temperature operation by way of outdoor reset control can be ideal during part-load conditions, and will be discussed Chapter 9, Control Strategies, Components, and Systems.

The ratings in Figure 8–7 are also based on air entering the baseboard at 65 °F and flow rates of either 1 gpm or 4 gpm. The heat output at 4 gpm is obtained by multiplying the heat output at 1 gpm by 1.057. The slightly improved performance is the result of increased convection between the water and the inside surface of the tube.

A footnote below the ratings table indicates the 1 gpm rating should be used unless the flow rate is known to be equal to or greater than 4 gpm. This assumes the flow rate through the baseboard is known. Analytical methods will soon be presented that allow baseboard performance to be adjusted for *any* reasonable flow rate.

Notice the footnote below the thermal ratings table indicating that a 15 percent **heating effect factor** has been included in the thermal ratings. *This means the listed values for heat output are actually 15 percent higher than the measured heat output established by laboratory testing.* The current rating standard for finned-tube baseboard allows manufacturers to add 15 percent to the tested performance provided a note indicating such is listed along with the rating data. The origin of the heating effect factor goes back several decades. At that time it was used to account for the higher output of baseboard convectors placed near floor level rather than at a height typical of standing cast-iron radiators. Unfortunately, it is a factor that has outlived its usefulness. Since the 15 percent higher heat output cannot be documented by laboratory testing, it is inaccurate to select baseboard lengths that assume its presence. *It is the author's*

RATINGS: Fine/Line 15 Series
(Hot water ratings, BTU/HR per linear ft. with 65°F (18.3°C) entering air)

ELEMENT	WATER FLOW	PRESSURE DROP *	140°F†	150°F	160°F	170°F	180°F	190°F	200°F	210°F	215°F	220°F
No. 15-75E Baseboard with 3/4" element	1 GPM	47	290	350	420	480	550	620	680	750	780	820
	4 GPM	525	310	370	440	510	580	660	720	790	820	870
No. 15-50 Baseboard with 1/2" element	1 GPM	260	310	370	430	490	550	610	680	740	770	800
	4 GPM	2880	330	390	450	520	580	640	720	780	810	850

* Millinches per foot.
† Ratings at 140°F determined by multiplying 150°F rating by the I=B=R conversion multiplier of .84.
NOTE: Ratings are for element installed with damper open, with expansion cradles. Ratings are based on active finned length (5" to 6" less than overall length) and include 15% heating effect factor. Use 4 gpm ratings only when flow is known to be equal to or greater than 4 gpm; otherwise, 1 gpm ratings must be used.

Figure 8–7 Example of a thermal rating table for finned-tube baseboard. *Courtesy of Slant/Fin Corp.*

recommendation that baseboard **heat output ratings** *that include a 15 percent heating effect factor be corrected by dividing them by 1.15 before selecting baseboard lengths.* This bases selection on actual tested performance.

Analytical Model for Heat Output per Unit Length

When the rating data shown in Figure 8–7 is combined with the low temperature correction factors given in Figure 8–8, it is possible to plot the heat output of a finned-tube baseboard over a wide range of temperature. This data, and a smooth curve drawn through it, are shown in Figure 8–9.

All types of finned-tube baseboard exhibit a similar curve of heat output versus water temperature. The values on the vertical axis will vary for baseboards with different size elements, but the shape of the curve over the same range of temperature will be very similar.

Average water temperature (deg. F.)	Correction factor to heat output rating at 150 deg. F.
150	1.0
140	0.84
130	0.69
120	0.55
110	0.41
100	0.28

Figure 8–8 Correction factors for heat output ratings of copper tube/aluminum fin baseboard for water temperatures under 150 °F. Based on IBR Testing and Rating Code for Baseboard Radiation.

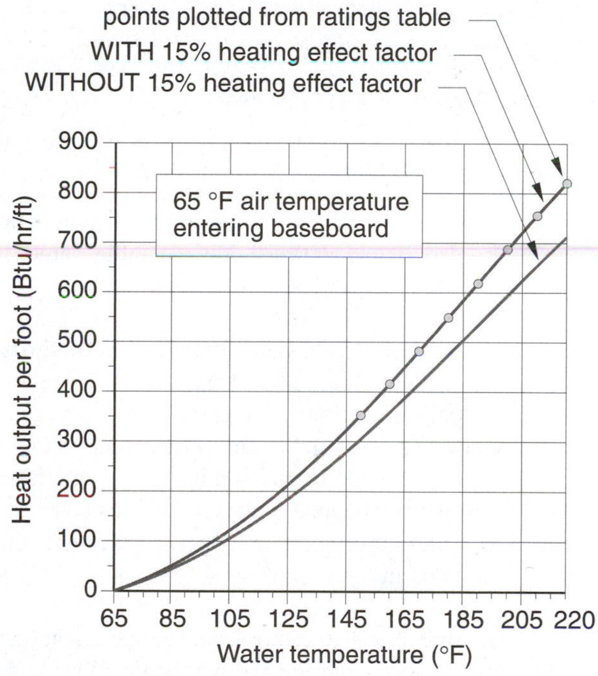

Figure 8–9 Heat output of a finned-tube baseboard for various water temperatures. Based on ratings data from Figure 8–7, and low water temperature corrections from Figure 8–8. The upper curve includes the 15% heating effect factor. The lower curve has the heating effect factor removed.

A generalized way of expressing heat output for a finned-tube baseboard involves two modifications to the data shown in Figure 8–7. First, the heat output data is normalized by dividing each heat output value by the heat output at 200 °F. Second, the heat output is expressed as a function of *temperature difference* between the air temperature entering the baseboard and the water temperature in the baseboard.

This allows heat output to be determined for situations where the baseboard is heating rooms not kept at normal comfort temperature. A factor is also included to account for heat output over a range of flow rates. This factor is based on methods used in the IBR Testing and Rating Code. The result is an empirical equation that gives the heat output of a generic finned-tube baseboard over a wide range of conditions:

(Equation 8.1)

$$q' = \beta(f^{0.04})\ [0.00096865(T_W - T_{air})^{1.4172}]$$

where:

q' = heat output in Btu/hr/ft of finned-tube element length

b = heat output of the baseboard at 200 °F average water temperature, 1 gpm from manufacturer's literature.* (Btu/hr/ft)

f = flow rate of water through the baseboard (gpm)

T_W = water temperature in the baseboard (°F)

T_{air} = air temperature entering the baseboard (°F)

Note: 0.04 and 1.4172 are exponents.

The coefficients and exponents in Equation 8.1 are the result of the empirical derivation.

*If the literature indicates a 15 percent heating effect factor is included in the data, divide the table values by 1.15 to remove it.

Example 8.1: A baseboard has an output of 650 Btu/hr/ft at an average water temperature of 200 °F, entering air temperature of 65 °F, and a flow rate of 1 gpm. Find the output of this baseboard while operating at an average water temperature of 135 °F, flow rate of 3 gpm, and having an entering air temperature of 50 °F. The heat output rating was taken from manufacturer's literature where a 15 percent heating effect factor was included in the ratings data.

Solution: The first step is to remove the heating effect factor by dividing the heat output rating of 650 Btu/hr/ft by 1.15.

$$\beta = \frac{650}{1.15} = 565.2 \text{ Btu/hr/ft}$$

This value can now be substituted along with the other data in Equation 8.1.

$$q' = 565.2(3^{0.04})[0.00096865\ (135 - 50)^{1.4172}]$$

$$= 565.2(1.0449)[0.5254] = 310.3 \text{ Btu/hr/ft}$$

Discussion: Although Equation 8.1 appears complex, it can easily be entered into any spreadsheet program or programmable calculator. It is a powerful analytic tool with the ability to replace dozens of look-up tables that might otherwise be used to show the baseboard's performance at nonstandard operating conditions.

8.5 SIZING FINNED-TUBE BASEBOARD

This section describes two methods for selecting the lengths of finned-tube baseboards connected in a series circuit. The first is a simple but less accurate method suitable for preliminary sizing. The second requires more calculations but also returns more accurate results. After discussing manual sizing procedures, a software module called Series Baseboard Simulator from the Hydronics Design Studio will also be discussed.

Sizing Method #1 (IBR Method)

Preliminary lengths of finned-tube baseboard can be selected by basing the heat output of each baseboard on the *average* temperature of the fluid in the piping circuit. The IBR method as it is often called, customarily takes this to be 10 degrees lower than the outlet temperature of the heat source. The procedure is as follows:

Step 1. Determine the average temperature of the fluid in the circuit:

(Equation 8.2)

$$T_{average} = T_{boiler\ outlet} - 10$$

Step 2. Look up the heat output rating of the baseboard at this temperature from the manufacturer's literature. Interpolate between the listed values if necessary. Again, it is the author's recommendation that the 15 percent heating effect factor be eliminated if it is included in the published ratings. Do so by dividing the listed heat output by 1.15.

(Equation 8.3)

$$\text{Corrected output rating} = \frac{\text{IBR heat output rating (Btu/hr/ft)}}{1.15}$$

Step 3. Divide the design heating load of each room by the heat output per foot of element length number obtained in step 2. Round off the number to the next higher whole foot length of baseboard.

(Equation 8.4)

$$\text{Baseboard length} = \frac{\text{Room design heating load (Btu/hr)}}{\text{Corrected output rating (Btu/hr/ft)}}$$

Example 8.2: A series piping circuit of finned-tube baseboard will be used to heat the rooms having the design heating loads indicated, in the order they are listed. The heat output ratings of the baseboard are given in Figure 8–7. The heating effect factor should be removed from these ratings for purposes of sizing. The boiler outlet temperature is 170 °F. Use sizing method #1 to select a baseboard length for each room.

Design heating load of rooms:

1.	Living room	8,000 Btu/hr
2.	Dining room	5,000 Btu/hr
3.	Bedroom #1	3,500 Btu/hr
4.	Bedroom #2	4,500 Btu/hr
5.	Bathroom	1,500 Btu/hr

Solution:

Step 1. Using Equation 8.2, the average water temperature in the circuit is assumed to be 10 °F less than the boiler outlet temperature:

$$T_{\text{average}} = 170 - 10 = 160\ °\text{F}$$

Step 2. The IBR heat output rating of the baseboard at 160 °F is read from Figure 8–7 as 420 Btu/hr/ft. The heating effect factor, which is noted in the footnotes as being present in these ratings, is removed using Equation 8.3:

$$\text{corrected output rating} = \frac{420}{1.15} = 365.2\ \text{Btu/hr/ft}$$

Step 3. The length of each baseboard is found by dividing the corrected heat output rating value into the design heat load of each room using Equation 8.4, and rounding the result up to the next whole foot length:

1.	Living room	8,000/365.2	=	21.9 or 22 ft
2.	Dining room	5,000/365.2	=	13.7 or 14 ft
3.	Bedroom #1	3,500/365.2	=	9.6 or 10 ft
4.	Bedroom #2	4,500/365.2	=	12.3 or 13 ft
5.	Bathroom	1,500/365.2	=	4.1 or 5 ft

Discussion: Although this method is simple and fast, it can lead to problems if used as the final means of sizing the baseboards. Its weakness is the fact that all baseboard lengths are based on an assumed average water temperature within the circuit. In reality, each successive baseboard in the circuit operates at a lower temperature than the preceding baseboard. *Because this method is based on average temperature, it tends to oversize baseboards near the beginning of the circuit and undersize baseboards near the end. This could lead to overheating some rooms and underheating others.*

Another point worth noting is the average fluid temperature is dependent on the circuit's flow rate. It may be 10 °F less than the boiler outlet temperature, or it may be 5 °F or even 15 °F less. This method does not account for this possibility. *Because of these limitations, sizing method #1 should only be used to make initial estimates of baseboard length.*

Sizing Method #2

The accuracy of baseboard sizing can be improved by accounting for the temperature drop of the fluid as it moves from one baseboard to the next. This allows the length of each baseboard to be based on the fluid temperature at its location in the circuit, rather than an overall average circuit temperature.

This method is based on the use of water as the heat transfer fluid. It is also based on the use of standardized heat output ratings established at an entering air temperature of 65 °F. It requires a reasonable estimate of the flow rate through the baseboard. This can be made using methods from Chapters 6 and 7, or by using the Series Baseboard Simulator module in the Hydronics Design Studio software.

Step 1. Using the rating data from the baseboard manufacturer, make a graph of the heat output of the baseboard versus water temperature, similar to that shown in Figure 8–10. If the ratings data contains the 15 degrees heating effect factor (most do), remove it from each data point using Equation 8.3 *before* plotting the points on the graph. If water temperatures below 150 °F are expected in any of the baseboards, use the data from Figure 8–8 to calculate and plot estimated low temperature performance data. Draw a smooth curve through all the data points. Save this graph for use on future projects using the same baseboard software.

Step 2. Calculate the average water temperature in the first baseboard on the circuit using Equation 8.5:

(Equation 8.5)

$$T_{\text{ave BB}} = T_{\text{in}} - \frac{\text{Room Load}}{1000\,f}$$

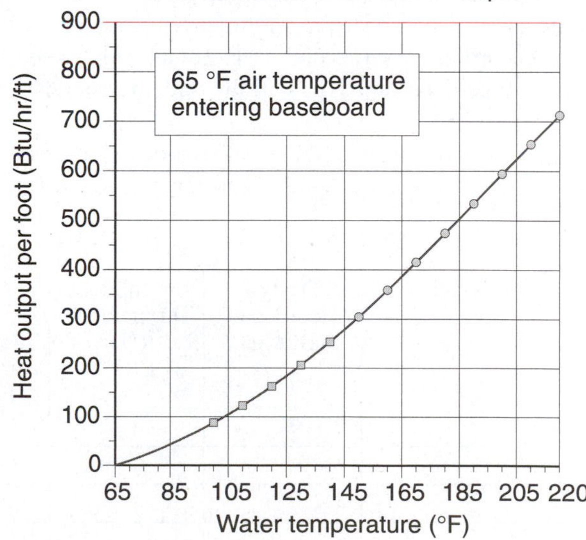

Figure 8–10 Graph of heat output versus water temperature prepared from manufacturer's data and low temperature correction factors from Figure 8–8. The 15% heating effect factor has been removed from the data using Equation 8.3 prior to plotting.

where:

T_{aveBB} = average water temperature in the baseboard (°F)

T_{in} = water temperature entering the baseboard (°F)

Room Load = design heating load of the room (Btu/hr)

f = flow rate of water through the baseboard (gpm)

Step 3. Using the graph prepared in step 1, estimate the heat output of the baseboard at the average water temperature found in step 2. This is illustrated in Figure 8–11. The heat output value read from the graph is designated as q'.

Step 4. Calculate the length of the baseboard by dividing the design heating load of the room by the value of q' found in step 3:

(Equation 8.6)

$$L = \frac{\text{Room Load}}{q'}$$

Round this length off to the next larger whole foot. This length is designated $L_{rounded}$.

Step 5. Calculate the outlet temperature of the baseboard based on its rounded length using Equation 8.7:

(Equation 8.7)

$$T_{out} = T_{in} - \frac{(L_{rounded})(q')}{500\,f}$$

Step 6. Repeat this procedure (steps 2 through 5) using the outlet temperature of the first baseboard as the inlet temperature for the second baseboard. Similarly, use the outlet temperature of each baseboard as the inlet temperature for the next baseboard in the series circuit. Eventually all baseboard lengths will be determined, as will the outlet temperature of the last baseboard.

A good way to organize these calculations is to compile data as shown in Figure 8–12. This table is filled in by working across the top row to obtain the outlet temperature of the first baseboard. This value is then copied to the third column of the second row, and the process is repeated. The last value to be obtained is the outlet temperature of the last baseboard. This table can be extended, if necessary, for circuits with more than ten series-connected baseboards.

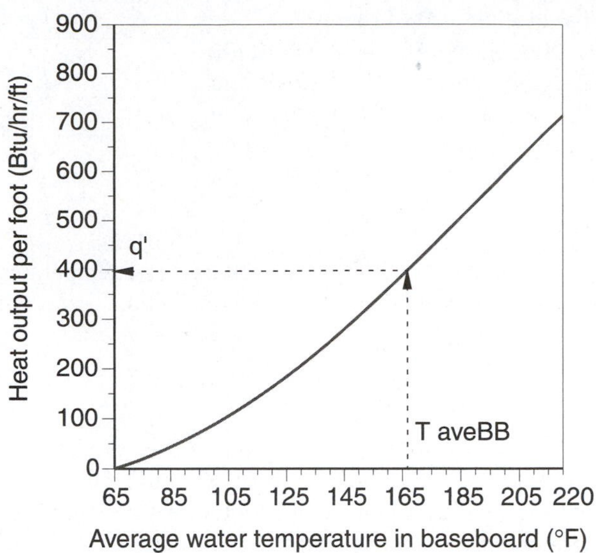

Figure 8–11 Use of heat output graph prepared in step 1.

Example 8.3: Using sizing method #2, repeat the baseboard selection process for the same rooms described in Example 8.2. Assume a water flow rate of 2.25 gpm. Use the graph in Figure 8–13 for heat output ratings. Note that the 15 percent heating effect factor has already been removed from the data shown in Figure 8–13.

Solution: Figure 8–14 shows the results of applying the procedure for sizing method #2. Note the outlet temperature of the first baseboard becomes the inlet temperature to the second baseboard, and so on.

Room Name	Design Heat Load (Btu/hr)	Inlet Temperature (°F)	Average Water Temperature (°F) (Equation 8.5)	Heat Output of Baseboard (from graph)	Calculated Baseboard Length (ft) (Equation 8.6)	Round Baseboard Length (ft)	Outlet Temperature (°F) (Equation 8.7)

Figure 8–12 Table used to organize calculations for baseboard sizing method #2.

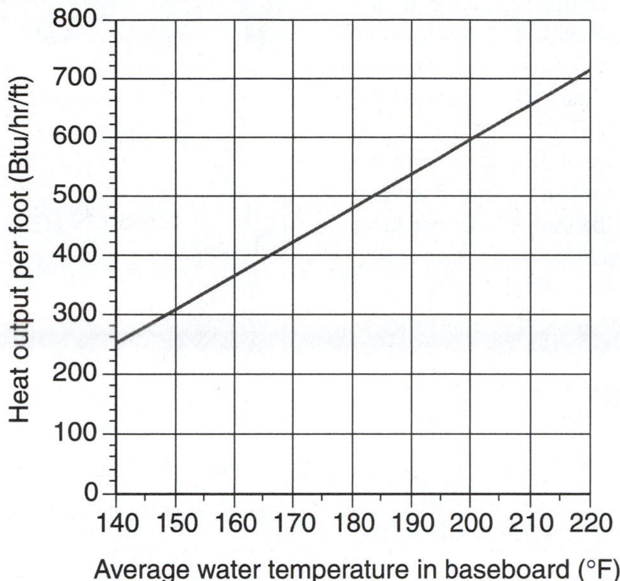

Figure 8–13 Graph of baseboard heat output ratings for Example 8.3. Note: The 15% heating effect factor has been removed from the manufacturer's data prior to plotting this graph.

Room Name	Design Heat Load (Btu/hr)	Inlet Temperature (°F)	Average Water Temperature (°F) (Equation 8.5)	Heat Output of Baseboard (from graph)	Calculated Baseboard Length (ft) (Equation 8.6)	Round Baseboard Length (ft)	Outlet Temperature (°F) (Equation 8.7)
Living	8000	170	166.4	399.5	20.02	20	162.9
Dining	5000	162.9	160.7	366.6	13.6	14	158.3
Bedroom #1	3500	158.3	156.7	343.2	10.2	11	154.9
Bedroom #2	4500	154.9	152.9	321.5	13.99	14	150.9
Bathroom	1500	150.9	150.2	306.6	4.89	5	149.5

Figure 8–14 Table for baseboard sizing method #2 filled in with numbers for Example 8.3.

Discussion: It is interesting to compare the results obtained using sizing method #1 in Example 8.2, with those of sizing method #2 in this example. The best comparison can be made by looking at the *unrounded lengths of baseboard* calculated for each room. The unrounded lengths show the greatest differences between the calculated results, and are given in Figure 8–15.

Note that the length of the first baseboard obtained using method #1 is longer than the length obtained using method #2. Conversely, method #1 calculates a shorter length for the last baseboard on the circuit compared to method #2. These results are due to the use of a single average water temperature in method #1, versus the average temperature *at each baseboard* in method #2. They reveal the tendency of

Room heat load (Btu/hr)	Baseboard lengths using method #1	Baseboard lengths using method #2
8000	21.9 feet	20.02 feet
5000	13.7 feet	13.6 feet
3500	9.6 feet	10.2 feet
4500	12.3 feet	13.9 feet
1500	4.1 feet	4.89 feet

Figure 8–15 Comparison of unrounded baseboard lengths obtained using baseboard sizing method #1 and #2 for the same room heating loads. Method #2 is considered the more accurate.

method #1 to oversize baseboards at the beginning of the circuit and to undersize them near the end.

The greater accuracy of method #2 comes at the expense of more calculations. Furthermore, since the results of the first baseboard affect the input data to the second, and so on, any error in the calculations will invalidate all remaining numbers from that point on. If, for example, the flow rate through the baseboards is changed, the entire table has to be recomputed. Still, the greater accuracy obtained using method #2 justifies its use for final baseboard selection.

An ideal way to automate these calculations is to build the procedure into a spreadsheet program on a PC. This allows long series strings to be calculated error free, and the results of changes in flow rates and inlet temperatures to be obtained almost instantly.

The Series Baseboard Simulator Module

The Hydronics Design Studio contains a module named **Series Baseboard Simulator** that executes a sizing procedure slightly more precise than even sizing method #2. It uses Equation 8.1 as well as additional algorithms and data that allow baseboard performance to be adjusted for different fluids, fluid temperatures, and a wide range of entering air temperatures.

The software version supplied with this book can work with up to three baseboards connected in series. The professional version of the program, discussed at the end of this book, can handle up to twelve baseboards in series. Each baseboard can be selected from a scrollable list. The user can also select the circulator, piping, and fittings from the databases built into the module.

Series Baseboard Simulator determines the flow rate in the specified circuit as well as the rounded length, heat output, inlet temperature, and outlet temperature of each baseboard. The reader is encouraged to experiment with the program to determine the effect of various system design changes on baseboard length. Figure 8–16 shows the user interface for the Series Baseboard Simulator module.

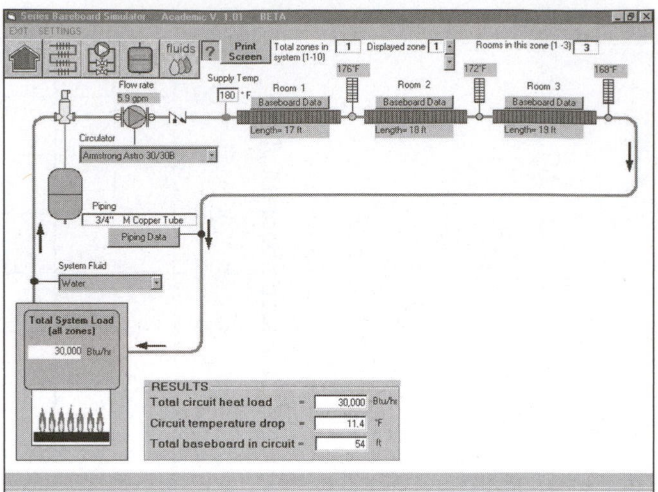

Figure 8–16 Screen shot of the Series Baseboard Simulator module from the Hydronics Design Studio software.

8.6 HYDRONIC FAN-COILS

When rooms lack the wall space needed to mount finned-tube baseboard, one alternative is to use a hydronic fan-coil. These devices come in many different sizes, shapes, and heating capacities. They all have two common components—a finned-tube heat exchanger (often called the "**coil**") through which system fluid flows, and a blower that forces air to flow across the coil.

In some fan-coil units, air is pulled through the coil by the blower. This type of arrangement is called a draw-through design. On other fan-coils, air is blown through the coil. Such units are appropriately called blow-through fan coils. The basic concepts for each are illustrated in Figure 8–17a.

The size and physical orientation of the coil and the blower are largely what distinguish one type of hydronic fan-coil from another. Most hydronic fan-coils used in residential or small commercial systems are designed to blow air directly into the heated space. Some are designed for wall mounting, others are intended to be hung from the ceiling. Some are also designed to be mounted flush with the floor. Another type of fan-coil, often referred to as an **air handler**, is designed for concealed mounting either above or below the occupied space. Air is delivered to the space through a duct system. An example of a small hydronic air handler is shown in Figure 8–17b.

Advantages of Fan-coils

Unlike finned-tube baseboard convectors, fan-coils rely on *forced*-convection to transfer heat from the coil surface to the room air. Force-convection is much more effective than natural convection in transferring heat from a surface to air. This allows the surface area of the coil to be much smaller than that of a finned-tube baseboard element of equivalent capacity. A fan-coil unit can often provide equivalent heating capacity using only a fraction of the wall space required by finned-tube baseboard.

Fan-coils also have very little thermal mass and low water content. They can respond to a call for heat very quickly. This makes them well-suited for spaces that need to be quickly heated from setback temperature conditions, or spaces that are only occasionally heated.

Disadvantages of Fan-coils

The compactness and fast-response characteristics of fan-coils do not come without compromises. First, any type of fan-coil will produce some noise due to the blower opera-

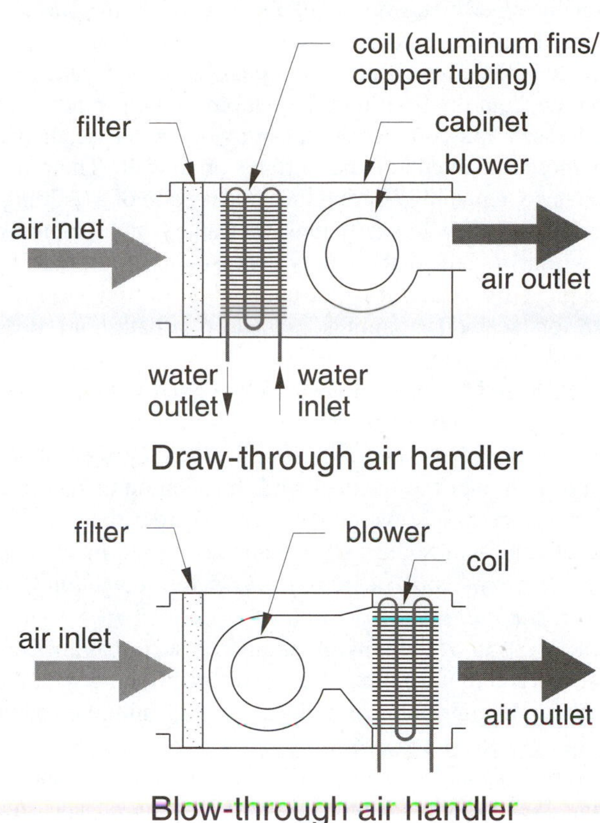

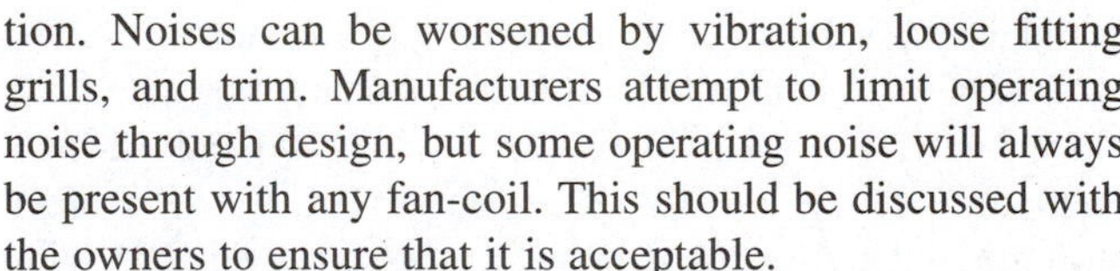

Figure 8–17a Examples of air handlers using "draw through" and "blow through" arrangement of the blower and coil.

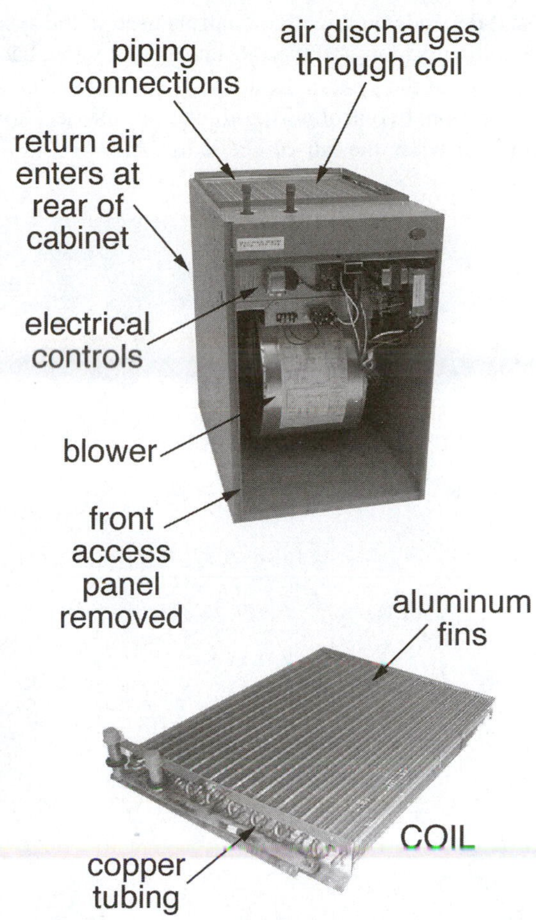

Figure 8–17b Example of a small air handler. *Courtesy of ECR International Corporation.*

tion. Noises can be worsened by vibration, loose fitting grills, and trim. Manufacturers attempt to limit operating noise through design, but some operating noise will always be present with any fan-coil. This should be discussed with the owners to ensure that it is acceptable.

Another factor associated with fan-coils is dust and dust movement. Regardless of how clean a room is kept, some dust is always present in the air and will be carried into the intake of any type of fan-coil. Some fan-coils have intake filters capable of removing most of the dust before it is deposited on the coil surfaces or blown back into the room. When filters are not present, or if the unit is not periodically cleaned, dust accumulation on coil surfaces will reduce the heating capacity, especially if the unit operates at low fluid temperatures.

Fan-coils also require connection to line voltage wiring to operate the blower. Although the amount of power consumed by the blowers in residential size fan-coils is small, the designer must ensure that a source of electricity is available at or near the proposed location of a fan-coil.

Wall-mounted Fan-coils

One type of fan-coil commonly used in residential and light commercial systems is mounted on, or recessed into, a wall or partition as shown in Figure 8–18.

Wall-mounted fan-coils usually draw room air in near the bottom of the unit and discharge it horizontally near the top. Some are designed for installation in short wall spaces such as under windows. Others are designed to fit in tall but narrow wall spaces.

This type of fan-coil is often equipped with controls to adjust the speed of the blower. This allows some degree of control over heat output. On some units, two or three specific speeds can be selected. In others, a solid state speed control allows a wide range of blower speed settings.

Some fan-coils come with an integral thermostat that can turn the blower off once the room reaches its temperature setting. Sometimes a low limit thermostat is also provided to prevent the blower from operating until the proper water temperature is present within the coil. The usefulness of the thermostat depends on the other controls used in the system. For example, most fan-coil thermostats are not wired back to the remainder of the system. In such cases, the installer must provide additional control wiring to operate the heat source and circulator when the fan-coil calls for heat.

(a)

(b)

Figure 8–18 (a) Example of a surface mounted fan-coil convector, (b) Example of a recessed fan-coil convector. *Courtesy of Beacon Morris Division of Mestek, Inc.*

Piping connections to small fan-coils are usually made to 1/2-inch copper tube stubs or FPT connections. In some situations, the piping enters the back of the cabinet, and is thus concealed from view. If this is not possible, piping risers are routed up from the floor into the bottom of the cabinet.

Recessed fan-coils generally provide better aesthetics than those mounted on the surface of a wall. Their use requires good planning during the construction of a building. A properly sized wall cavity must be framed, and the piping and wiring routed to the cavity. The sooner this planning takes place the better. It is not uncommon to find a lack of space for recessed mounting, especially in bathroom areas where plumbing and electrical systems can present interference. Recessed fan-coils are usually designed so that a base-pan assembly can be mounted into the wall cavity before the walls are closed in. The blower/coil assembly then slides into position over the finished wall. Installation of recessed fan-coils in retrofit situations is usually harder than in new construction because the walls are already closed-in. *Be sure to take a careful look at retrofit jobs before committing to recessed fan-coil unit(s).*

The location of the fan-coil should be such that air flow is not blocked by furniture. Discuss the positioning with the owners so that interference problems with furniture or other furnishings can be avoided.

The units should also not blow directly on the occupants of a room. For example, imagine trying to read a newspaper in a reclining chair that is only 2 or 3 feet away from the discharge grill of a wall-mounted fan-coil. The idea is to diffuse the air stream into the room air before it contacts the occupants.

Under-cabinet Fan-coils

Another type of hydronic fan-coil is designed to mount in the 4-inch "kick space" beneath a kitchen or bathroom cabinet. These units, often called kick space heaters, are designed for situations where wall space is not available, particularly kitchens and bathrooms. An example of a kick space heater is shown in Figure 8–19. A typical 12-inch wide unit can yield the heat output of 8 to 10 feet of finned-tube baseboard. Wider units are available with greater heating capacities.

Air flows into the unit through the upper part of the front grill. It is drawn in by a tangential blower mounted at the rear of the unit. The blower reverses the air flow, sending it out through the horizontal coil in the lower part of the unit where it is heated. These types of fan-coils do not have any type of filter, and are somewhat prone to dust accumulation. They should be periodically cleaned with a vacuum cleaner.

Kick space heaters are usually supported on rubber grommets that reduce vibration transfer to the floor or cabinet. Connections are typically 1/2-inch copper tube stubs. An integral junction box is provided for wiring connections. Most under-cabinet heaters are equipped with a low limit aquastat that prevents the blower from operating until the water temperature in the coil reaches a certain temperature.

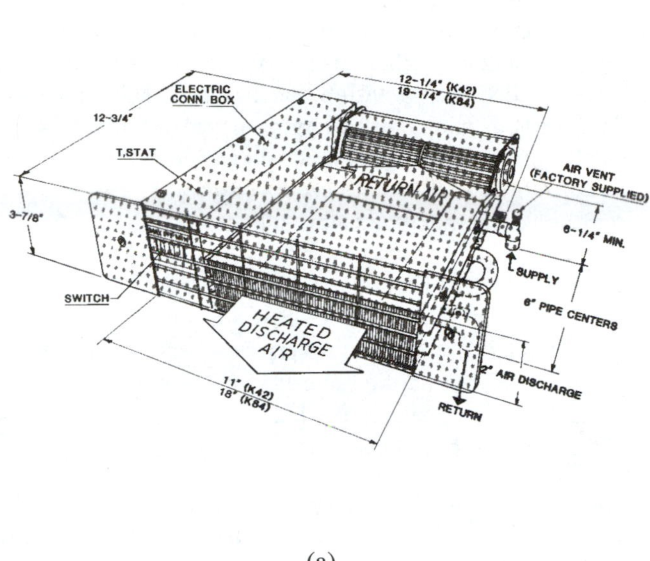

(a)

(b)

Figure 8–19 (a) Example of a small kickspace heater, (b) typical installation under a cabinet.

This setting is determined by the manufacturer and is usually not adjustable. Typical temperatures are 110 °F to 140 °F.

When installing an under-cabinet heater it is convenient to provide an access hatch through the bottom shelf of the cabinet. This can be done by carefully sawing out the bottom shelf over the area where the unit will be mounted and installing wooden ledgers to support the shelf when replaced. Be sure to allow sufficient space to fully access the side-mounted piping and wiring connections. This arrangement permits easy access for cleaning and to the air vent. It also allows the unit to be lifted out if servicing is necessary.

When installing under-cabinet heaters, be careful not to accidentally bend the exposed blades of the blower wheel. The resulting imbalance can cause noticeable vibration and associated noise.

Overhead Fan-coils

Commercial buildings are often designed with overhead piping for mechanical systems. One method of delivering heat from overhead piping without routing it down near floor level is the use an overhead fan-coil, or **unit heater**.

Unit heaters are often used in commercial spaces such as service garages, machine shops, and supermarkets,, where overhead mounting allows them to be out of the way of heavy traffic or rough usage. They are usually supported by threaded steel rods or perforated angle struts attached to structural members such as roof trusses or beams. Figure 8–20 shows a typical unit heater designed for angled horizontal air flow.

Figure 8–20 Example of an overhead unit heater. Note adjustable louvers at front of unit. *Courtesy of Sterling Radiator Division of Mestek, Inc.*

Louvers on the front of the unit allow the heated air stream to be directed toward the floor level at various angles.

Although overhead unit heaters have been used in buildings with relatively high ceilings, such installations often experience stratification in which hot air tends to accumulate near the ceiling while cooler air pools near the floor. Good air circulation is essential to minimize such stratification.

8.7 THERMAL PERFORMANCE OF FAN-COILS

The heat output of a fan-coil is dependent on several factors. Some are fixed by the construction of the fan-coil and cannot be changed. These include the:

- Surface area of the coil
- Size, spacing, and thickness of the fins
- Number of tube passes through the fins
- Air-moving ability of the blower

Other performance factors are dependent on the system into which the fan-coil is installed. These include the:

- Entering water temperature
- Water flow rate through the coil
- Entering air temperature
- Air flow rate through the coil

When selecting a fan-coil, the designer must match the characteristics of the available units with the temperatures and flow rates the system can provide, while also obtaining the required heating capacity. This usually involves a search of manufacturer's performance data in an attempt to find a good match among several simultaneous operating conditions. This data is often provided in the form of tables that list the heat output at several combinations of temperature and flow rate for both the water side and air side of the unit. An example of a thermal performance table for a wall-mounted fan-coil is shown in Figure 8–21.

Figure 8–21 lists the Btu/hr heating capacity of the unit over a wide range of entering water temperatures, two different air flow rates, and two water flow rates. Each heat output value can be thought of as a "snapshot" of heating performance at specific conditions.

Since the heat output of the fan-coil varies continuously over a range of temperatures and flow rates, there is no guarantee it will operate at one of the conditions listed in the table. When it is necessary to estimate the heat output of the unit at conditions other than those listed in the table, a process called **interpolation** can be used. This process is usually not as simple as averaging between the values that are above and below the desired value. Rather it sets up *proportions* to arrive at a more accurate estimate. The best way to illustrate interpolation is through an example.

Example 8.4: The heat output of the fan-coil represented by the data in Figure 8–21 is 3205 Btu/hr at 150 °F entering water temperature, 1 gpm water flow rate, and high fan speed. At 160 °F and the same flow rate the output is listed as 3,580 Btu/hr. Estimate the heat output of the fan-coil at 152 °F water temperature.

Solution: Set up the known and unknown quantities as shown in Figure 8–22, and subtract the quantities connected by the brackets. Form the ratios of these differences, and equate these ratios to each other. Solve the resulting equation for the answer.

Notice the algebraic signs (+ or –) take care of themselves so the answer is positive. Another way of thinking about the problem is that the temperature of 152 is one-fifth of the way between the numbers 150 and 160. This implies the desired value of heat output, x, will be one-fifth of the way between 3205 and 3580. One-fifth of this difference is:

$$(1/5)\,(3{,}580 - 3{,}205) = 75$$

Adding 75 to the lower value of 3205 gives the same answer of 3280 Btu/hr.

In some cases **double interpolation** is required to obtain an estimate of heat output that falls between both the column and row rating points of the table. Again, an example is the easiest way to show how this is done.

Example 8.5: Make an estimate of the heat output of the fan-coil represented by the data in Figure 8–21, at an entering water temperature of 152 °F and a water flow rate of 2.5 gpm.

WH SERIES BTU/HR OUTPUT FOR MYSON FAN CONVECTORS — ENTERING AIR AT 65° F

MODEL	FLOW RATE (GPM)	FAN SPEED	AIR DELIVERY (CFM)	ENTERING WATER TEMPERATURE (°F) 110*	120*	130*	140	150	160	170	180	190	200	210
WH 50	1.0	High	89	1690	2030	2354	2783	3205	3580	3887	4263	4638	5079	5354
		Low	50	1390	1610	2010	2400	2825	3200	3605	4020	4380	4750	5040
	3.0	High	89	1996	2456	2866	3309	3717	4094	4470	4913	5323	5664	6142
		Low	50	1450	1910	2217	2590	2900	3410	3790	4299	4570	4950	5300
WH 90	1.0	High	121	3700	4350	5025	6000	6780	7640	8400	9290	10050	10800	11500
		Low	75	3250	3900	4505	5350	6000	6750	7420	8190	8830	9550	10200
	3.0	High	121	3800	4520	5320	6130	7160	7940	8710	9600	10420	11250	12090
		Low	75	3350	4020	4800	5580	6380	7050	7730	8500	9200	9980	10790

*For Output At These Water Temperatures, Optional 107° Low Temperature Cutoff Required.

Figure 8–21 Thermal performance data for small wall-mounted fan-coil convector. *Courtesy of Myson, Inc.*

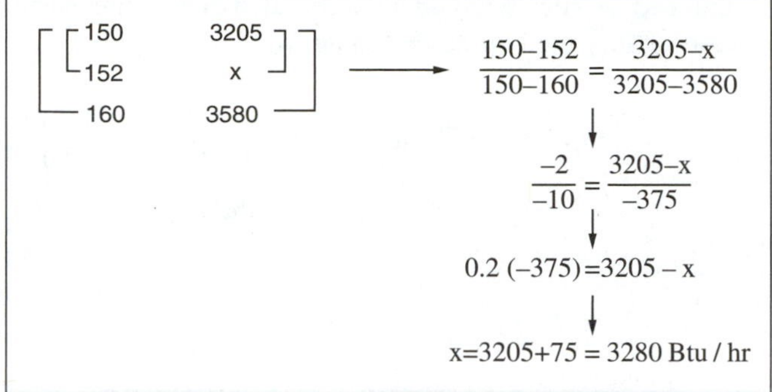

Figure 8–22 Interpolation procedure for Example 8.4.

Solution: Begin by interpolating between the temperatures of 150 °F and 160 °F two times, first for the heat output values listed at a flow rate of 1 gpm, and again for the heat outputs listed at a flow rate of 3 gpm. To finish the calculation, take the results of the first two (temperature) interpolations and interpolate them between the lower flow rate of 1 gpm and the upper flow rate of 3 gpm.

The first temperature interpolation is the same as shown in Example 8.4, with a result of 3280 Btu/hr at 152 °F.

The second interpolation is shown in Figure 8–23.

The final interpolation will be between the flow rates of 1 and 3 gpm, and will use both previously calculated capacities at 152 °F entering water temperature. The final procedure is shown in Figure 8–24.

The estimated capacity of this fan-coil at 152 °F entering water temperature and 2.5 gpm water flow rate is thus 3664 Btu/hr.

Discussion: The final value should always be checked to see if it is reasonable. It should fall between the values listed in the tables both vertically and horizontally.

Another way to perform interpolation is by using a spreadsheet program on a computer. Another useful method, especially if the data will be used repeatedly, is to plot the data on a graph and draw a smooth curve through the points.

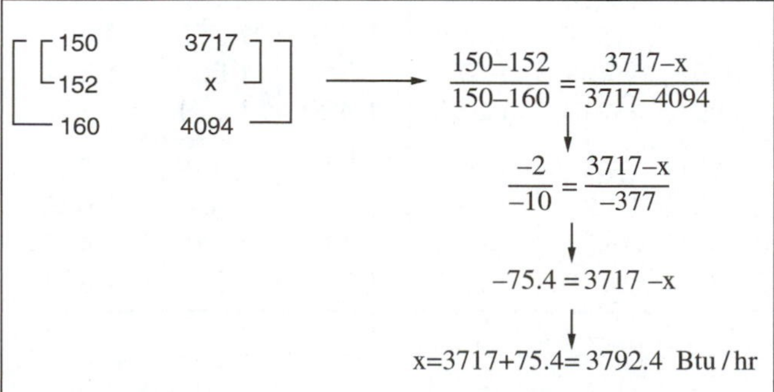

Figure 8–23 Second interpolation procedure for Example 8.5.

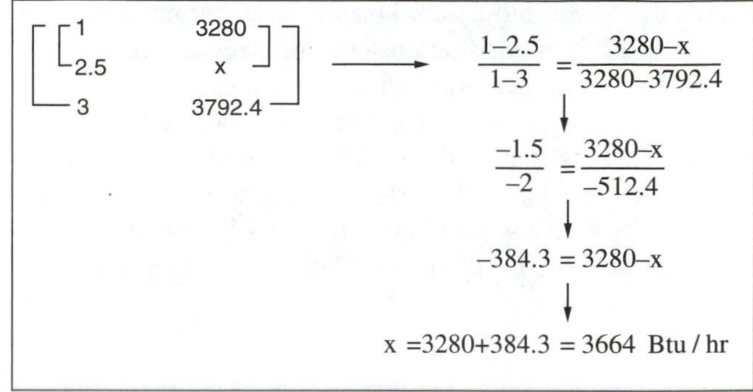

Figure 8–24 Final interpolation procedure for Example 8.5.

The estimated performance between data points is easily read from this curve. Some manufacturers also provide design assistance CD-ROMs capable of providing interpolated output ratings.

Some fan-coil manufacturers list the heat output of a fan-coil unit at a single reference condition consisting of a specified value for the entering temperatures and flow rates of both the water and air. Accompanying this will be one or more tables or graphs of correction factors that can be used to adjust the reference performance up or down for variation in temperature and flow rate of both the entering water and entering air. An example of tables listing these performance correction factors for a unit heater is shown in Figure 8–25.

General Principles of Fan-coil Performance

The heat output of fan-coils can be described in the form of simple principles that can be verified by examining published ratings data. By varying one operating condition such as entering temperature or flow rate, while all others remain fixed, the designer can get a feel for which factors have the greatest impact on heat output. This can be very helpful in evaluating the feasibility of various design options.

Principle #1: The heat output of a fan-coil is approximately proportional to the temperature *difference* between the entering air and entering water.

If, for example, a particular fan-coil can deliver 5000 Btu/hr with a water temperature of 140 °F and room air entering at 65 °F, its output using 180 °F water can be estimated as follows:

$$\text{Estimated output at } 180° = \frac{(180°\text{F} - 65°\text{F})}{(140°\text{F} - 65°\text{F})} \times (5000\text{ Btu/hr}) = 7670\text{ Btu/hr}$$

A proportional relationship between any two quantities will result in a straight line that passes through or very close to the origin when the data is plotted on a graph. Figure 8–26 shows this to be the case when the heat output data from Figure 8–21 is plotted against the difference between the entering water and entering air temperatures.

This concept is useful for estimating heat output when a fan-coil is used in a room kept above or below normal comfort temperatures. For example, if a fan-coil can deliver 5000 Btu/hr with a water temperature of 140 °F and room air entering at 65 °F, its output while heating a garage maintained at 50 °F could be estimated using the same proportionality method described above:

$$\text{Estimated output at 50°F entering air} = \frac{(140°F - 50°F)}{(140°F - 65°F)} \times (5000\ \text{Btu/hr}) = 6000\ \text{Btu/hr}$$

Model No.	Output BTU/ HR*	GPM	Final Air °F	Prssr. Drop FT./H_2O	Motor HP	RPM	Nominal CFM	Outlet FPM	Nom. Amps @ 115VAC
HS-108 A	8,030	.80	91	.80	9 Watt	1550	245	250	.8
	6,800		90			1350	210	215	.8
HS-118A	18,400	1.9	94	2.2	9 Watt	1550	500	500	.8
	15,650		96			1350	420	420	.8
HS-125A	24,800	2.5	102	2.2	1/47	1550	580	590	1.1
	21,230		106			1350	460	450	1.1
HS-136A	35,900	3.6	99	3.0	1/30	1070	850	550	1.1
	32,300		100			900	750	480	1.1
HS-18	13,050	1.3	95	.005	9 Watt	1550	395	395	.8
	11,725		99			1350	350	350	.8
HS-24	17,400	1.8	96	.014	9 Watt	1550	450	450	.8
	15,600		98			1350	380	380	.8
HS-36	26,100	2.7	103	.09	1/47	1550	550	550	1.1
	23,500		103			1350	480	480	1.1
HS-48	34,800	3.5	103	.12	1/30	1070	750	550	1.3*
	31,300		111			900	630	460	1.3*
HS-60	43,600	4.4	105	.17	1/30	1070	900	650	1.3*
	39,200		112			900	700	510	1.3*

(a)

*Hot Water Conversion Factors Based on 200° Entering Water 60° Entering Air 20° Temperature Drop

Entering Air Temperature	Entering Water Temperature—20° Water Temperature Drop										
	100°	120°	140°	160°	180°	200°	220°	240°	260°	280°	300°
30°	0.518	0.666	0.814	0.963	1.12	1.26	1.408	1.555	1.702	1.85	1.997
40°	0.439	0.585	0.731	0.878	1.025	1.172	1.317	1.464	1.609	1.755	1.908
50°	0.361	0.506	0.651	0.796	0.941	1.085	1.231	1.375	1.518	1.663	1.824
60°	0.286	0.429	0.571	0.715	0.857	1.000	1.143	1.286	1.429	1.571	1.717
70°	0.212	0.353	0.494	0.636	0.777	0.918	1.06	1.201	1.342	1.483	1.63
80°	0.140	0.279	0.419	0.558	0.698	0.837	0.977	1.117	1.257	1.397	1.545
90°	0.069	0.207	0.345	0.483	0.621	0.759	0.897	1.035	1.173	1.311	1.462
100°	0	0.137	0.273	0.409	0.546	0.682	0.818	0.955	1.094	1.23	1.371

(b)

Figure 8–25 (a) example of heat output ratings at reference conditions of 200 °F entering water and 60 °F entering air temperature, (b) correction factors for heat outputs at other entering water and air temperatures. *Courtesy of Sterling Radiator Division of Mestek, Inc.*

Although this principle is helpful for quick performance estimates, it should not replace the use of thermal rating data supplied by the manufacturer when available.

Principle #2: Increasing the fluid flow rate through the coil will marginally increase the heating capacity of the fan-coil.

Interestingly, many heating professionals instinctively disagree with this statement. They argue that because the water moves through the coil at a faster speed, it has less time in which to release its heat. However, the time a given particle of water stays inside the fan-coil is irrelevant in a system with continuous circulation. The faster-moving fluid improves convective heat transfer between the fluid and the inner surface of the tubes, resulting in greater heating capacity.

Another way of justifying this principle is to think of the average water temperature in the coil at various flow rates. As the flow rate through the fan-coil is increased, this temperature difference decreases. This means that the average temperature inside the coil increases, and thus the heat output of the coil should increase. Those who claim that a smaller temperature difference between the ingoing and outgoing fluid implies less heat is being released are overlooking the fact that the rate of heat transfer depends on *both* the temperature difference and flow rate. This relationship was described by the sensible heat rate equation of Chapter 4, Properties of Water.

The rate at which heat output increases with flow rate is very "nonlinear." The rate of gain is very fast at low flow rates, but steadily decreases as the flow through the coil increases. This can be seen in Figure 8–27 in which the manufacturer's data for heat output versus water flow rate of a fan-coil is plotted.

Notice the fan-coil has about 50 percent of its maximum heat output capacity at only about 10 percent of its maximum flow rate. The strong curvature of this relationship means that *attempting to control the heat output of a fan-coil by adjusting flow rate can be very tricky*. Very small valve adjustments can create large changes in heat output at low flow rates. However, the same amount of valve adjustment will create almost no change in heat output at higher flow rates. This is also true for other types of hydronic heat emitters, and will be discussed in more detail in Chapter 9, Control Strategies, Components, and Systems.

The nonlinear relationship between flow rate and heat output also implies that attempting to boost heat output from a fan-coil by driving it at an unusually high flow rate will yield very minor gains. The argument against this is further supported when one considers the increased head loss through the coil at higher flow rates.

Principle #3: Increasing the air flow rate across the coil will marginally increase the heat output of the fan-coil.

This principle is also based on forced-convection heat transfer between the exterior coil surfaces and the air stream.

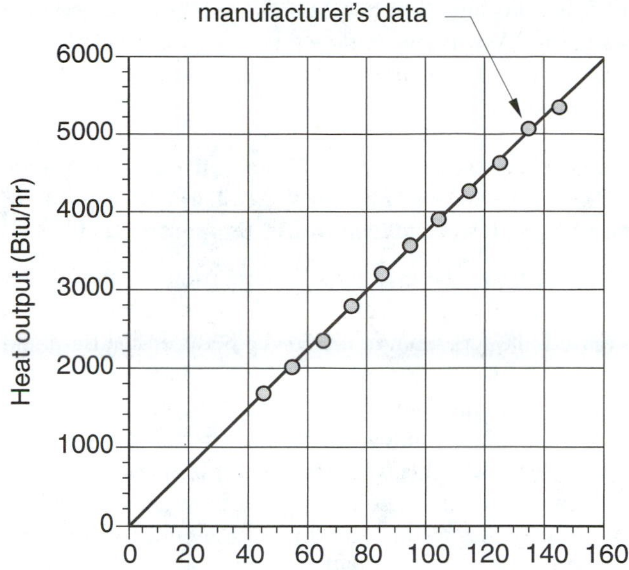

Figure 8–26 Plot of heat output versus temperature difference between entering water and entering air temperature based on data from Figure 8–21. Notice the proportional (straight sloping line) relationship.

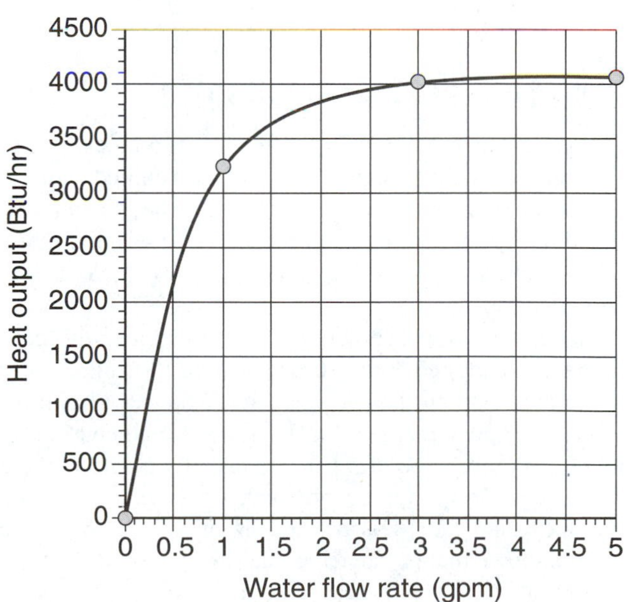

Figure 8–27 Heat output of a small fan-coil at different flow rates and constant entering water and entering air temperatures.

Faster-moving air "scrubs" heat off the coil surface at a greater rate. As with water flow rates, the gain in heat output is very minor above the nominal air flow rate the unit is designed for.

Principle #4: Fan-coils with large coil surfaces and/or multiple tube passes through the coil will be capable of operating at lower entering water temperatures.

Convective heat transfer is also proportional to the contact area between the surface and the fluid. If this contact area is increased, the temperature difference between the fluid and the air streams can be reduced (for a given rate of heat output). This principle can be well-applied in hydronic systems that must operate with low temperature heat sources such as solar collectors or heat pumps. In these applications, the designer may need to use fan-coils with larger coils and/or more tube passes through the fins of the coil, to drive heat from the coil at the required rate. On larger air handlers, manufacturers may offer higher performance coils with up to six rows of tubes as an option. This is usually not true on smaller residential fan-coils. For these products, the only way to increase coil surface area is to install more fan-coils.

Care must also be taken that comfort is not compromised when operating fan-coils at reduced water temperatures. It is critically important to introduce the air stream from such fan-coils into the space so that it mixes with room air before passing by occupants. Failure to do so will usually lead to complaints of "cool air" blowing from the fan-coils, even though the room is maintaining the desired average temperature. *As a rule, the author recommends that fan coil units supplying heat to occupied spaces not be operated with entering water temperatures lower than 130 °F.*

Using Fan-coils for Cooling

Larger commercial buildings often use hydronic distribution systems for supplying both heated and chilled water to fan-coils or air-handler units. This can also be done in residential or small commercial systems if a water-chilling device such as a hydronic heat pump is used as the heating/cooling source.

Only fan-coils/air handlers that are equipped with condensate drip pans and drains should be used for chilled water cooling applications. The drip pan acts as a catch basin for the water droplets that continually form on the coil during cooling operation. The drip pans are usually connected to a drain pipe or floor drain so the condensate can be disposed of at the same rate it is formed. Many of the hydronic fan-coils shown in this chapter are designed only for heating. They do not have condensate pans, and should not be used for chilled water cooling applications. Doing so would result in condensate running out the bottom of the unit causing damage to the structure. An example of a small wall-mounted fan-coil equipped with a drip pan, and thus suitable for chilled water cooling, is shown in Figure 8–28.

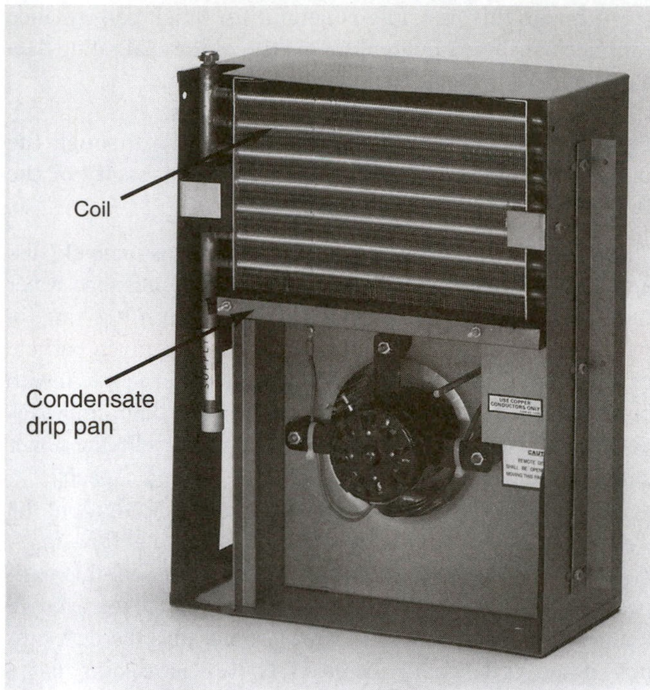

Figure 8–28 Example of a small fan-coil equipped with a drip pan for chilled water cooling applications. Drip pan is located directly under coil. *Courtesy of Burnham Corporation.*

8.8 PANEL RADIATORS

Another type of hydronic heat emitter that is quickly gaining market share in North America is called a **panel radiator**. Having been used in Europe for many years, panel radiators are now available in hundreds of sizes, shapes, colors, and heating capacities to fit different job requirements. They can be used throughout a building or in combination with other types of heat emitters.

Most panel radiators are made of steel. Some are built of preformed steel sheets welded together at their perimeter. Others are constructed of tubular steel components. To prevent corrosion, steel panel radiators should only be used in closed-loop systems. Many panel radiators feature a high-quality fused powder-coat or enamel finish, in several different colors to coordinate with interior design.

Some panel radiators release a significant percentage of their heat as thermal radiation. Such panels typically have a relatively flat front, and only project about 2 inches out from the wall surface to which they are mounted. Their radiant output tends to warm the objects in a room. The lower the water temperature these panels operated with, the higher the percentage of radiant versus convective heat output. This is often desirable because it improves comfort and reduces room air stratification relative to units that release the majority of their heat through convection (e.g., directly heating the room air).

Other panel radiators are designed to release a high percentage of their heat output through convection. Such radiators are equipped with one or more rows of fins that help dissipate heat into the surrounding air. They also tend to have deeper profiles that project 4 to 8 inches from the adjacent wall. This type of panel is well-suited for creating upward air currents to counteract downward drafts from large window areas. They are often operated at relatively high water temperatures that favor convective heat output.

Figure 8–29 shows an example of a **vertical panel radiator**. These panels are designed to be used in tall but narrow wall spaces. They are especially versatile in kitchens, bathrooms, or offices where horizontal wall space is often limited. Lengths in excess of 20 feet are available.

Such radiators are normally mounted with their bottom 6 to 8 inches above the floor. Most hang from mounting brackets fastened to the wall. These brackets, should be fastened directly to wood-framing members or into solid masonry. In new construction with studded walls, wooden backer blocks should be positioned at the proper height and location to accept the fasteners for the brackets. It is important that the heating system designer specifies the position of these framing blocks so they are installed before the wall is closed up.

Piping connections usually consist of 1/2-inch or 3/4-inch risers routed up through the floor, and connecting to threaded opening at the bottom of the radiator. This is one location where steel pipe nipples provide better protection against physical damage, and are preferred over copper tubing. Their use also eliminates any visible soldered joints above the floor. Holes for the risers should be accurately located and neatly drilled about 1/4 inch larger than the outside diameter of the riser pipe. This prevents noise from thermal expansion. The holes can be covered with escutcheon plates for a neat appearance.

The detail shown in Figure 8–30 is useful when flexible tubing such as PEX or PEX-AL-PEX is used to supply a panel radiator. The transition fitting to rigid steel pipe is concealed below the floor. However, the slack left in the flexible tubing allows this fitting to be pulled up through the floor if the panel ever needs to be removed for wall painting or other repair, etc. The steel pipe nipples are often painted to match the radiator. Escutcheon plates cover the oversized holes for a simple/clean finish detail.

To provide for individual heat output control, a thermostatic radiator valve is often mounted in the *supply* pipe of the panel radiator. To facilitate removal of the radiator for wall painting, it is common to mount a shutoff valve in the return pipe. A lockshield valve with integral union, such as discussed in Chapter 5, Piping, Fittings, and Valves. is a good choice for this service. The combination of a radiator valve with integral union in one riser, and a drainable lockshield valve with integral union in the other, allows the panel to be isolated from the system, neatly drained through a hose, and easily disconnected from the piping. It then can be lifted off its brackets for full access to the wall. A representation of these details is shown in Figure 8–31.

Figure 8–29 Example of a vertical panel radiator. *Courtesy of Runtal USA.*

Horizontal panel radiators are also available for wider but shorter wall spaces, such as beneath windows. Again, several widths and lengths are available to match job requirements as shown in Figure 8–32. The mounting and riser piping details are similar to those used with vertical panel radiators.

Freestanding panel radiators are also available. They are usually supported by floor pedestals supplied by the manufacturer as shown in Figure 8–33a. Some are even equipped with a bench seat as shown in Figure 8–33b.

Another variation of panel radiator is known as a towel warmer. These panels are designed for bathrooms or kitchens and provide a space in which to hang one or more towels. In addition to space heating, the simple appeal of a warm, dry towel after a shower or bath has proven very effective in marketing these panels. Hundreds of different styles, sizes, and colors are available. In many cases, towel warmers can be accessorized with garment knobs, mirror segments, and other amenities. When budgets allow, towel warmers are available with chrome, brass, and even 24-karat gold plating. Several types of towel warmers are shown in Figure 8–34.

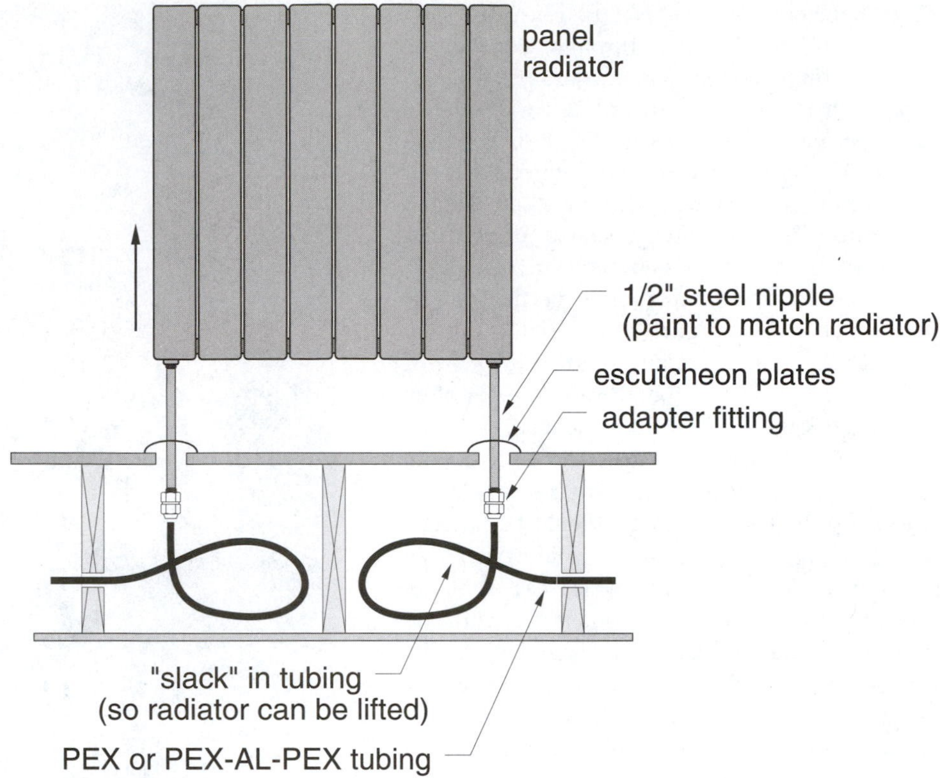

Figure 8–30 Panel radiator installation detail using steel pipe nipples above floor and PEX or PEX-AL-PEX tubing under floor. The tubing slack allows panel to be lifted off its mounting brackets if necessary.

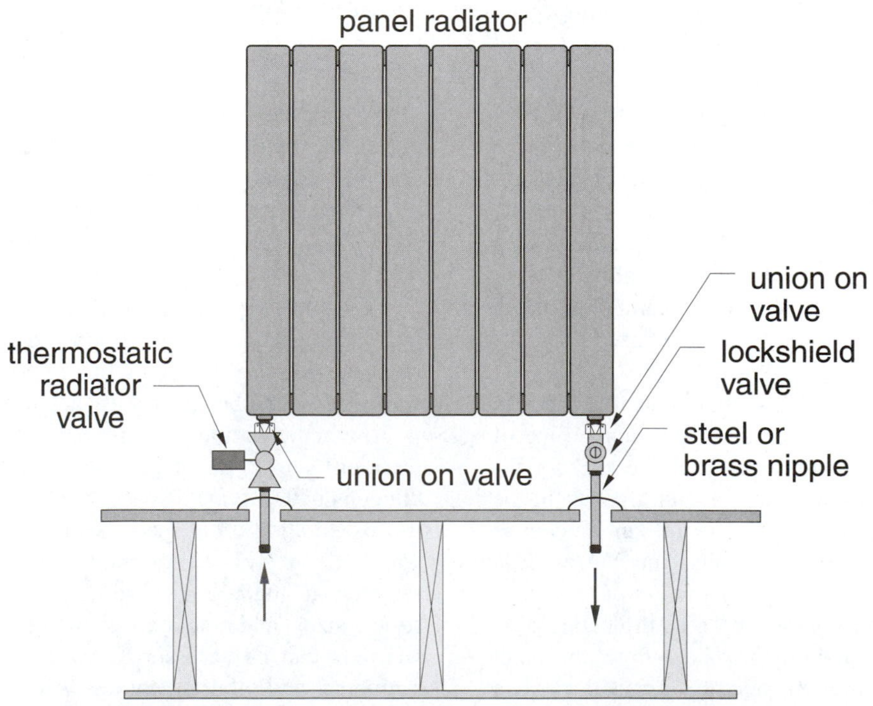

Figure 8–31 Use of a thermostatic radiator valve (TRV) on supply riser, and lockshield valve on return riser of panel radiator. These valves allow heat output control, as well as easy radiator removal if necessary.

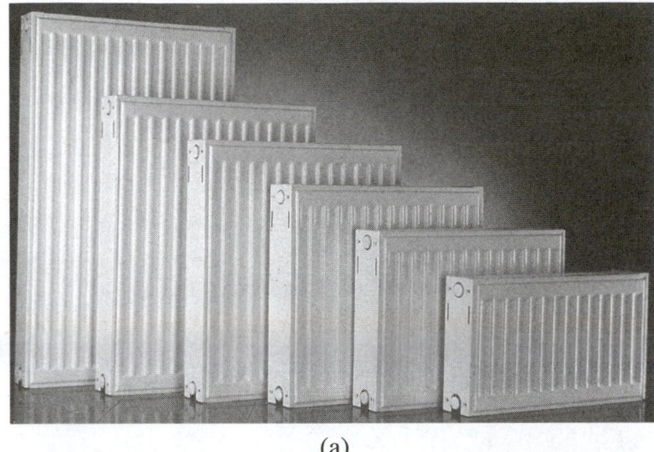

(a)

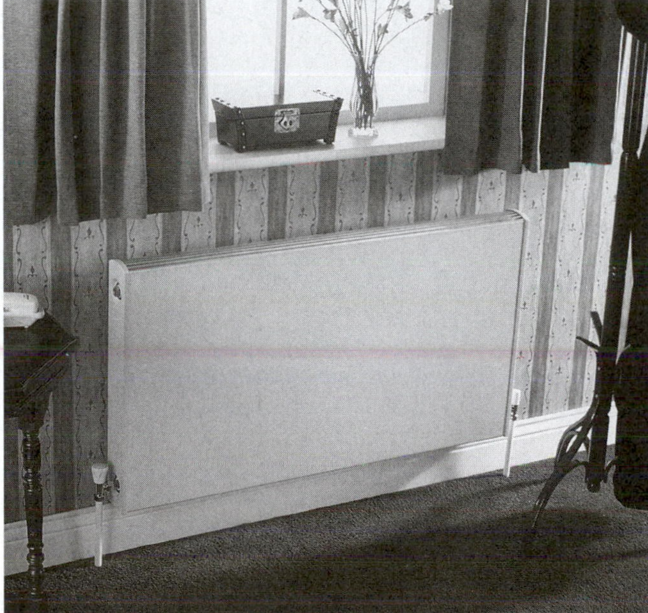

(b)

(c)

Figure 8–32 Example of horizontal panel radiators (a) *Courtesy of Northern Wholesale,* (b) *Courtesy of Myson, Inc.,* (c) *Courtesy of Runtal USA.*

Figure 8–33a Horizontal panel radiator supported by pedestal. *Courtesy of Myson, Inc.*

Figure 8–33b Bench radiator. *Courtesy of Myson, Inc.*

Some panel radiators are available with thermostatic valves already built in as shown in Figure 8–35. This allows the heat output of the panel to be adjusted on a room-by-room basis. If an internal heat gain develops in one room, the thermostatic valve will reduce or stop flow through that panel without affecting the other panels. Be sure to connect the supply and return piping to the correct ports on any panel radiator that contains a thermostatic radiator valve. Reverse flow through the valve often causes flow noise and "banging sounds" as the valve plug closes.

The panel radiator shown in Figure 8–36 contains a chamber that allows water to perform a U-turn when the

(a)

(b)

(c)

Figure 8–34 Examples of towel warmer radiators, (a) *Courtesy of Myson, Inc.*, (b) *Courtesy of Vasco,* (c) *Courtesy of Myson, Inc.*

Figure 8–35 Thermostatic valve built into top right corner of panel radiator.

thermostatic valve is closed. This design allows multiple radiators, each individually controlled by their thermostatic valve, to be connected in a simple series circuit with constant water flow. It gives the same room-by-room temperature control option as provided through other piping options, but with simpler, easier to install piping. This functionality can also be provided by external bypass valves mounted to the panel's inlet and outlet connections. Again, be sure to observe the correct flow direction when connecting piping to any panel radiator using an integral or external bypass valve.

Thermal Performance of Panel Radiators

Thermal ratings for panel radiators are usually given in chart form. For long vertical or horizontal panels the ratings are usually expressed in Btu/hr *per foot of panel length* for various panel widths and entering water temperatures. For compact panel radiators ratings are usually stated in Btu/hr for a given size panel and entering water temperature. In most cases the heat output ratings are based on a

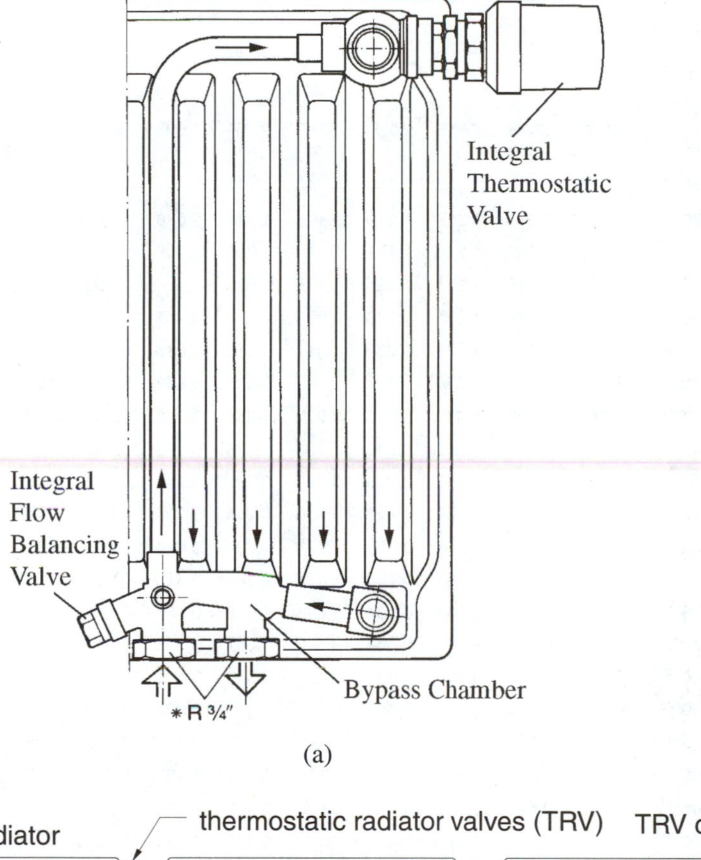

(a)

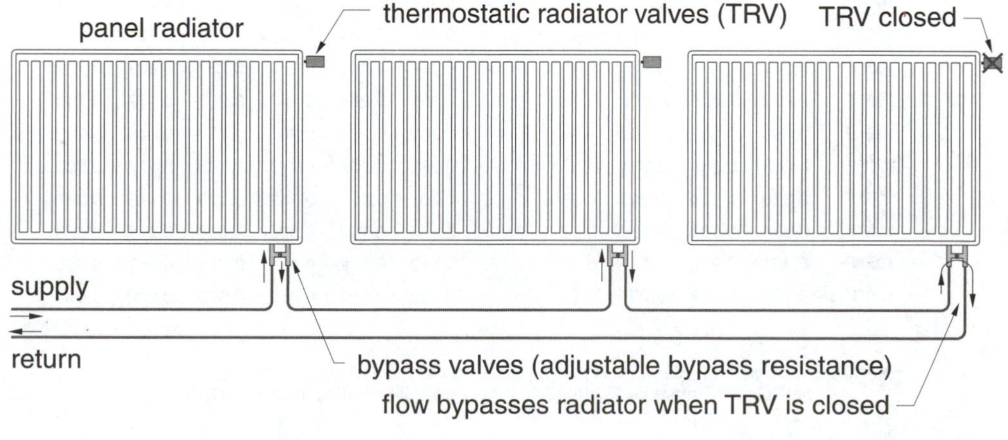

(b)

Figure 8–36 (a) Cross section of panel radiator with integral bypass chamber and thermostatic valve. *Courtesy of Buderus Hydronic Systems, Inc.,* (b) Use of external bypass valves allow individual control of each panel radiator with simple series piping.

BTUH/ft Ratings @ 65°F EAT						
	MODEL	HEIGHT inches	DEPTH inches	215°F	180°F	140°F
PERIMETER STYLE	R-1	2.8	1.6	230	**160**	90
	R-2	5.7	1.6	420	**300**	170
	R-3	8.6	1.6	620	**440**	250
	R-4	11.5	1.6	820	**580**	330
	R-5	14.4	1.6	1020	**720**	410
	R-6	17.3	1.6	1220	**860**	500
WALL PANEL	R-7	20.2	1.6	1430	**1010**	580
	R-8	23.1	1.6	1640	**1160**	660
	R-9	26.0	1.6	1850	**1300**	750
	R-10	29.0	1.6	2060	**1450**	830

(a)

	EAT										
AWT	45°F	50°F	55°F	60°F	65°F	70°F	75°F	80°F	85°F	90°F	95°F
240°F	1.365	1.350	1.304	1.266	1.220	1.171	1.124	1.086	1.039	1	0.953
235°F	1.343	1.305	1.267	1.219	1.171	1.124	1.086	1.038	1	0.952	0.910
230°F	1.305	1.267	1.219	1.171	1.124	1.086	1.038	1	0.952	0.910	0.868
225°F	1.267	1.219	1.171	1.124	1.086	1.038	1	0.952	0.910	0.868	0.826
220°F	1.219	1.171	1.124	1.086	1.038	1	0.952	0.910	0.868	0.826	0.785
215°F	1.171	1.124	1.086	1.038	1	0.952	0.910	0.868	0.826	0.785	0.744
210°F	1.124	1.086	1.038	1	0.952	0.910	0.868	0.826	0.785	0.744	0.704
205°F	1.086	1.038	1	0.952	0.910	0.868	0.826	0.785	0.744	0.704	0.664
200°F	1.038	1	0.952	0.910	0.868	0.826	0.785	0.744	0.704	0.664	0.625
195°F	1	0.952	0.910	0.868	0.826	0.785	0.744	0.704	0.664	0.625	0.587
190°F	0.952	0.910	0.868	0.826	0.785	0.744	0.704	0.664	0.625	0.587	0.549
185°F	0.910	0.868	0.826	0.785	0.744	0.704	0.664	0.625	0.587	0.549	0.511
180°F	0.868	0.826	0.785	0.744	0.704	0.664	0.625	0.587	0.549	0.511	0.474
175°F	0.826	0.785	0.744	0.704	0.664	0.625	0.587	0.549	0.511	0.474	0.438
170°F	0.785	0.744	0.704	0.664	0.625	0.587	0.549	0.511	0.474	0.438	0.403
165°F	0.744	0.704	0.664	0.625	0.587	0.549	0.511	0.474	0.438	0.403	0.369
160°F	0.704	0.664	0.625	0.587	0.549	0.511	0.474	0.438	0.403	0.369	0.334
155°F	0.664	0.625	0.587	0.549	0.511	0.474	0.438	0.403	0.369	0.334	0.301
150°F	0.625	0.587	0.549	0.511	0.474	0.438	0.403	0.369	0.334	0.301	0.269
145°F	0.587	0.549	0.511	0.474	0.438	0.403	0.369	0.334	0.301	0.269	0.237
140°F	0.549	0.511	0.474	0.438	0.403	0.369	0.334	0.301	0.269	0.237	0.207
135°F	0.511	0.474	0.438	0.403	0.369	0.334	0.301	0.269	0.237	0.207	0.177
130°F	0.474	0.438	0.403	0.369	0.334	0.301	0.269	0.237	0.207	0.177	0.149
125°F	0.438	0.403	0.369	0.334	0.301	0.269	0.237	0.207	0.177	0.149	0.122
120°F	0.403	0.369	0.334	0.301	0.269	0.237	0.207	0.177	0.149	0.122	0.096
115°F	0.369	0.334	0.301	0.269	0.237	0.207	0.177	0.149	0.122	0.096	0.071
110°F	0.334	0.301	0.269	0.237	0.207	0.177	0.149	0.122	0.096	0.071	0.050
105°F	0.301	0.269	0.237	0.207	0.177	0.149	0.122	0.096	0.071	0.050	0.030
100°F	0.269	0.237	0.207	0.177	0.149	0.122	0.096	0.071	0.050	0.030	0.011

EXAMPLE: To find the BTUH/ft Rating for an R-6 Panel at 155°F AWT and 65°F EAT,
Multiply the Correction Factor (0.511) by the BTUH/ft Rating at 215°F (1219),
e.g. (0.511) X (1219) = 623 BTUH/ft

(b)

Figure 8–37 (a) Thermal performance ratings given as heat output per foot of panel length for several panel widths (R_1 = 1 tube, R_2 = 2 tubes, etc.). (b) Correction factors for heat output at various room air and entering water temperatures. *Courtesy of Runtal USA.*

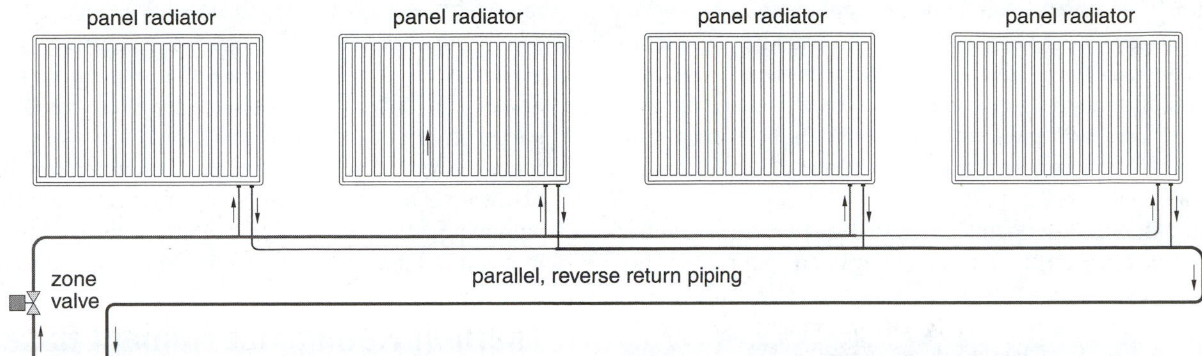

Figure 8–38 Connecting 4 panel radiators in parallel reverse return reduces head loss, and provides the same supply water temperature to each panel.

surrounding air temperature of 65 °F. Correction factors for other entering water temperatures and room air temperatures usually accompany these ratings as shown in Figure 8–37.

When a system requires panel radiators to be piped in series, it is important to account for the drop in fluid temperature as it flows from one panel to the next. A procedure similar to method #2 for sizing baseboard should be used. The size of each panel should be based on the water temperature at its location in the series circuit.

In some situations, it is desirable to keep all the panel radiators the same size. One example of this is where a panel will be placed under each of several identical windows in a room. In these situations, it is preferable to pipe the panel in parallel such that each receives the same supply water temperature. An example of such piping is shown in Figure 8–38.

8.9 RADIANT BASEBOARD

Another relatively new type of hydronic heat emitter to the North American market is referred to as **radiant baseboard**. The active portion of the baseboard consists of an aluminum extrusion that forms a very slim profile about 1 inch wide and 5 inches high. The rear side of the extrusion has two channels that hold either copper or PEX tubing. A cutaway view of a radiant baseboard is shown in Figure 8–39.

Unlike finned-tube baseboards, radiant baseboards have no fins. The majority of the heat output is radiant rather than convective. The radiant output warms the floors and objects near the floor. This enhances thermal comfort, while also reducing temperature stratification from floor to ceiling.

Radiant baseboard is sold in straight lengths of 1 to 10 feet in increments of 6 inches. To join these lengths together, as well as to accommodate corners and other details, manufacturers supply accessories such as compression fittings, mounting brackets, corner assemblies, and filler panels for covering the joints.

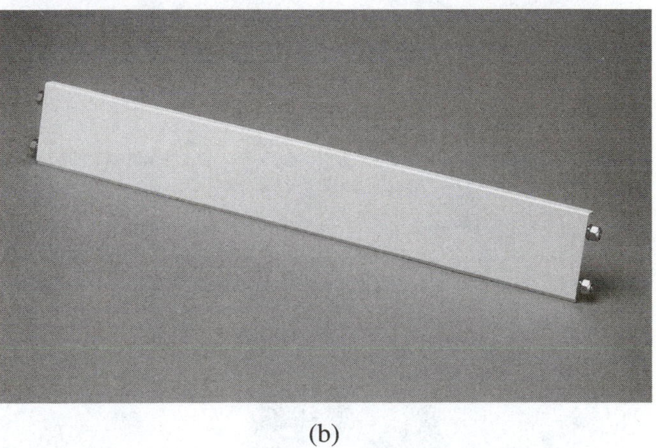

(a)

(b)

Figure 8–39 (a) Cut away of a radiant baseboard consisting of an aluminum extrusion with copper-lined fluid passageways. (b) Complete segment of radiant baseboard with fittings. *Courtesy of Radiant Technology, Inc.*

Radiant baseboard is intended to replace the conventional wooden baseboard at the base of the walls. It is typically installed on all walls in the room. The narrow profile and high-quality enamel finish blend well with most interiors. They are less conspicuous than conventional finned-tube baseboards, and can withstand more physical punishment without bending or denting.

The baseboard segments are held to the wall by clips that are fastened to the wall first. On framed walls, the clips should be fastened into the studs. Once the clips are secured, the baseboard segments simply snap into them.

Connections between baseboard segments are made with compression fittings tightened by wrenches. No soldering is required. The joint is then covered with a filler panel that matches the active baseboard panel. At corners, a short length of PEX tubing with a tight 90-degree turn is used to connect the panels on each wall. The tubing for corners can be purchased preformed, or they can be formed by heating with a hot air gun. Some of these joint details are shown in Figure 8–40.

The double tube design allows both supply and return connections to be made at the same end of the unit. The other end of the baseboard can be at any location in the room.

Radiant baseboards are usually piped using a **home run distribution system** where each baseboard is supplied and returned to a common manifold station as shown in Figure 8–41. Small diameter PEX or PEX-AL-PEX tubing is often used to connect each baseboard to the manifold station. The home run distribution system supplies the same water temperature to each baseboard, limits head loss, and allows room-by-room temperature control provided suitable control hardware is used. Home run distribution systems can be used with many other types of heat emitters and will be discussed in detail in Chapter 11, Distribution Piping Systems.

Thermal Ratings for Radiant Baseboard

Because of its smaller size and surface area, a given length of radiant baseboard has a lower heat output than the same length of finned-tube baseboard. A comparison between the heat output of a typical residential finned-tube baseboard and a 5-inch by 1-inch radiant baseboard is shown in Figure 8–42.

The lower heat output of the radiant baseboard requires a greater length of baseboard. A rule of thumb would be to allow for twice the linear footage of radiant baseboard compared to a typical residential finned-tube baseboard. Be sure each room has sufficient wall space to accommodate the necessary footage of baseboard. If not, some other type of supplemental heat emitter may be required.

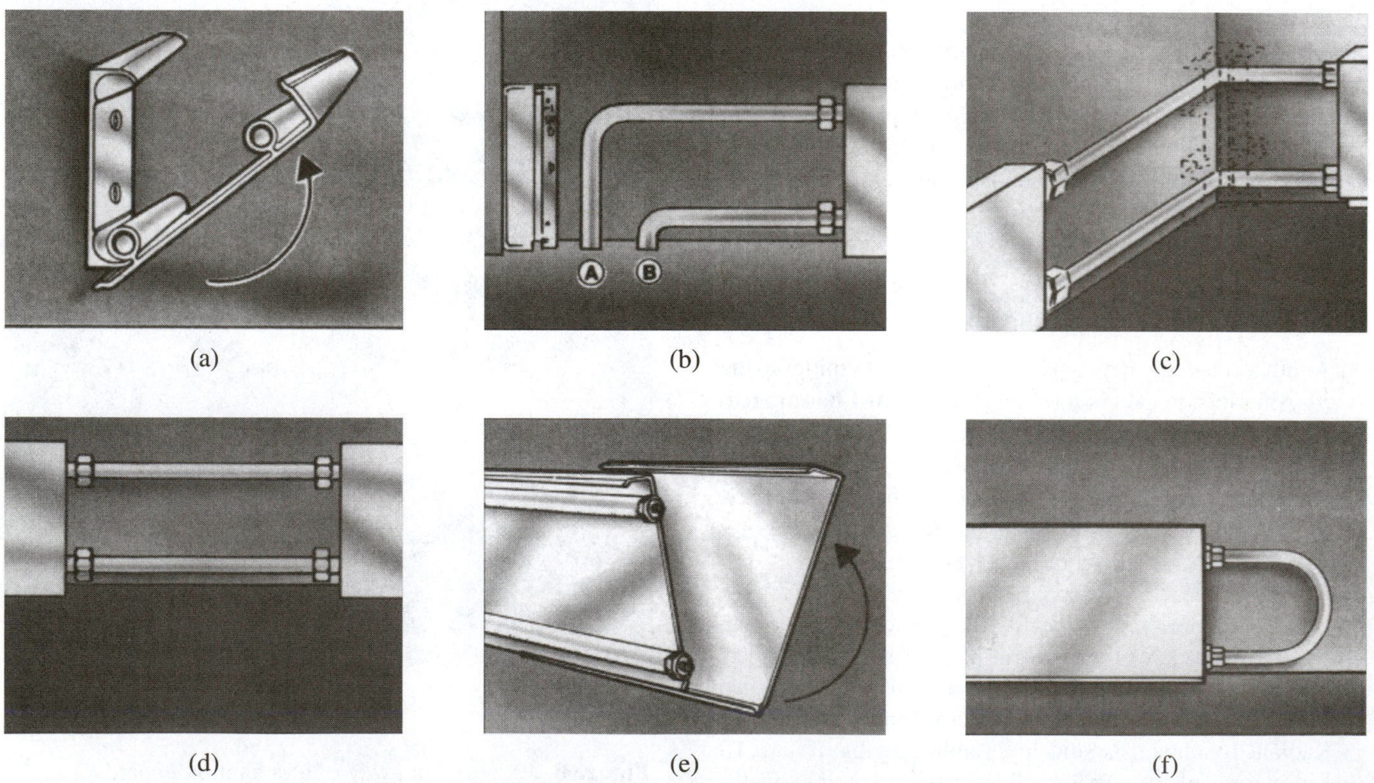

Figure 8–40 Installation details for radiant baseboards: (a) baseboard segment snaps into mounting bracket, (b) PEX tubing penetrates up through floor, (c) preformed PEX tubing at corners, (d) PEX tubing used to connect adjacent baseboards, (e) cover plate snaps over joints, (f) return bend of PEX tubing at end of baseboard. *Courtesy of Radiant Technology, Inc.*

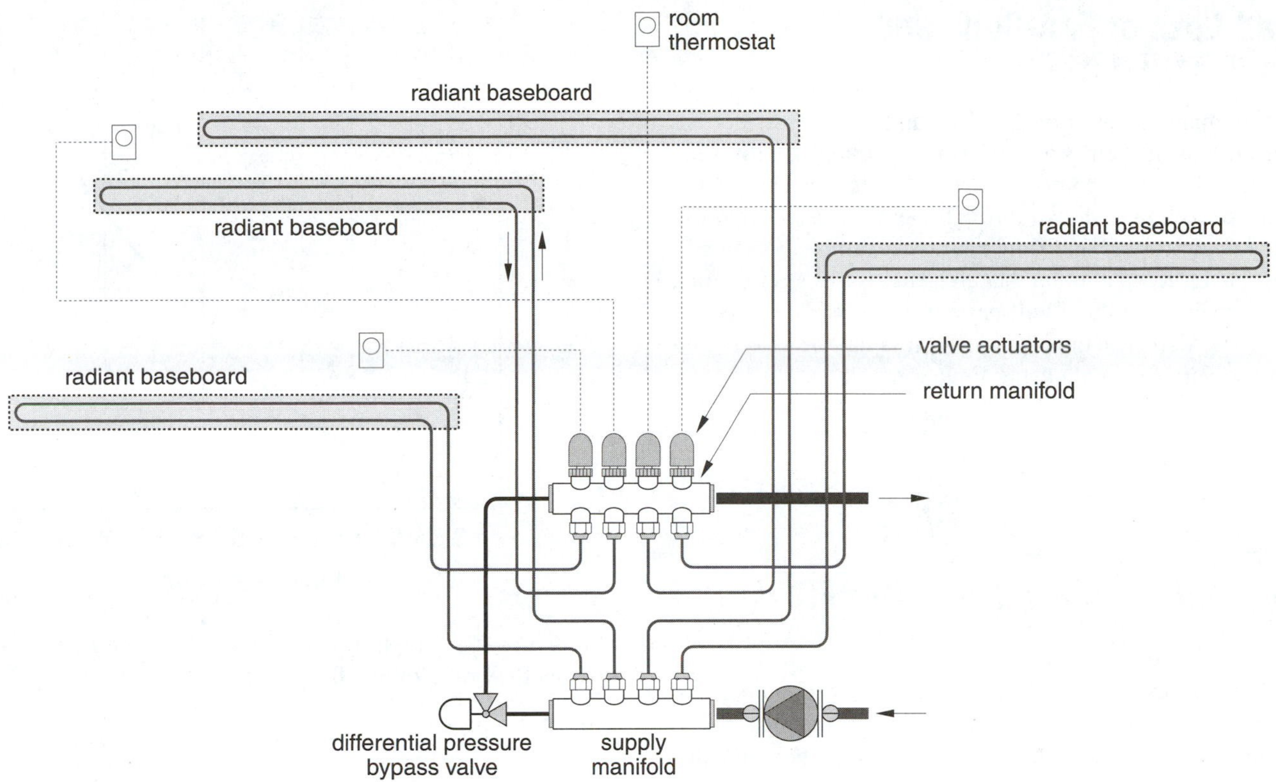

Figure 8–41 Radiant baseboards connected as a homerun distribution system from a common manifold. Flow through each baseboard is individually controlled by room thermostats/valve actuators.

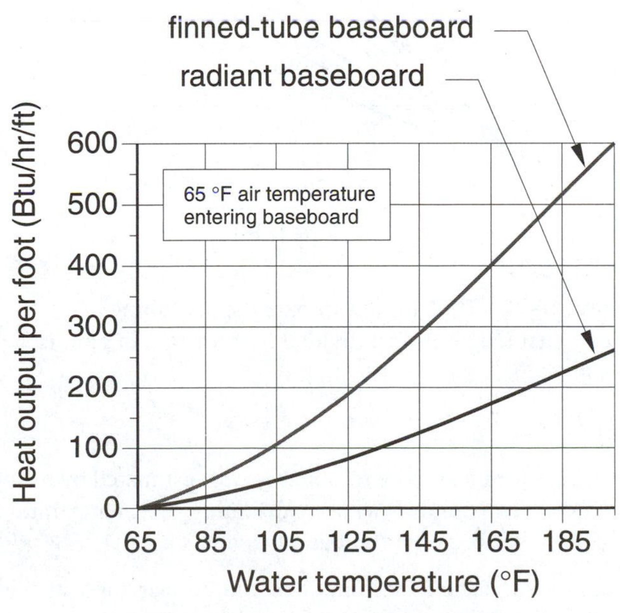

Figure 8–42 Comparative heat output of residential grade finned-tube baseboard and radiant baseboard. Finned-tube output does not include 15% heating effect factor.

8.10 HEAD LOSS OF HEAT EMITTERS

When analyzing piping systems containing heat emitters, it is important to know the head loss they create. The following guidelines describe typical head loss characteristics of each of the heat emitters discussed.

Head Loss of Finned-tube Baseboard

Since the element of a finned-tube baseboard is simply a length of copper tubing with fins attached, the pressure drop will be the same as that of a standard copper tube of the same diameter and length. It can be determined using the methods given in Chapter 6, Fluid Flow in Piping.

The head loss of radiant baseboard can be determined by considering each baseboard to be a tubing circuit approximately twice as long as the baseboard extrusion. This head loss can then be combined with the head loss of the "leader" tubing connecting each radiant baseboard back to the manifold. The head loss of such circuits can also be accurately modeled using the methods of Chapter 6, or through use of the Hydronic Circuit Simulator module in the Hydronics Design Studio software.

Head Loss of Fan-Coils and Panel Radiators

Most manufacturers of fan-coils and panel radiators list the head loss of their products at two or three specific flow rates. If at least two points are listed, the data can be plotted and a smooth curve drawn through them for use in estimating the head loss at other flow rates. An example is shown in Figure 8–43. Keep in mind that all head loss versus flow rate curves for heat emitters will pass through zero head loss at zero flow. Also remember that head loss can be determined from pressure drop data using Equation 6.5:

$$H_{loss} = \frac{144(\Delta P)}{D}$$

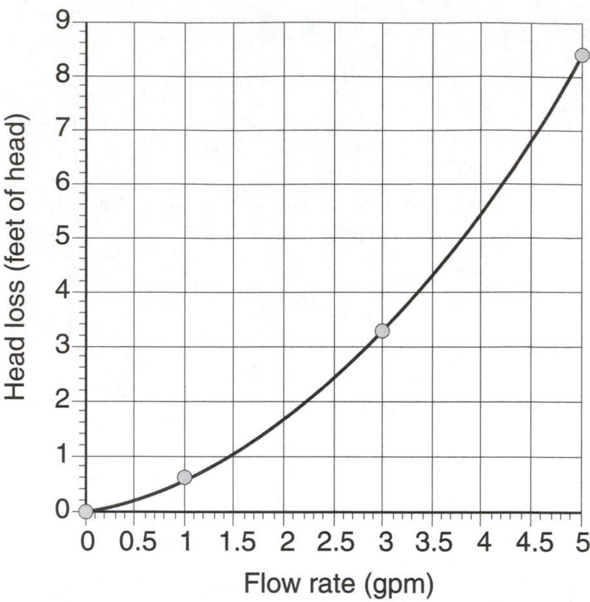

Figure 8–43 Graph of manufacturer's data for head loss versus flow rate for small fan-coil.

8.11 HEAT LOSS FROM COPPER TUBING

Although not specifically intended as a heat emitter, any tube conveying fluid at a temperature greater than that of the surrounding air releases heat. If the tube runs through heated space, this heat offsets a portion of the building's heating load. If the tube runs through unheated space, the heat is lost.

Several issues related to system design can be addressed if the designer has a method for evaluating tubing heat loss. For example, the designer may want to know the heat loss and temperature drop in the tubing between the outlet of the heat source, and the inlet of the first heat emitter. If the system requires long tubing runs through relatively cool spaces, significant heat loss and temperature drop can occur between heat emitters. This will obviously lower the heat output capability of the heat emitters. The designer may also want to know how much heat is released by tubing routed through heated space since this heat partially heats the building.

The heat output of a tube is affected by the material, pipe size, flow rate, fluid, and surrounding air temperature. If the tubing is insulated, its heat loss will be considerably reduced. The complexity of evaluating heat loss under all combinations of these conditions is beyond what can be presented in this book. However, the heat loss of bare copper tube can be evaluated using one of the following two methods:

Method #1: When the tube's length in feet, divided by the flow rate through it in gallons per minute, is less than 20 ft/gpm, the heat loss can be estimated using the graph in Figure 8–44.

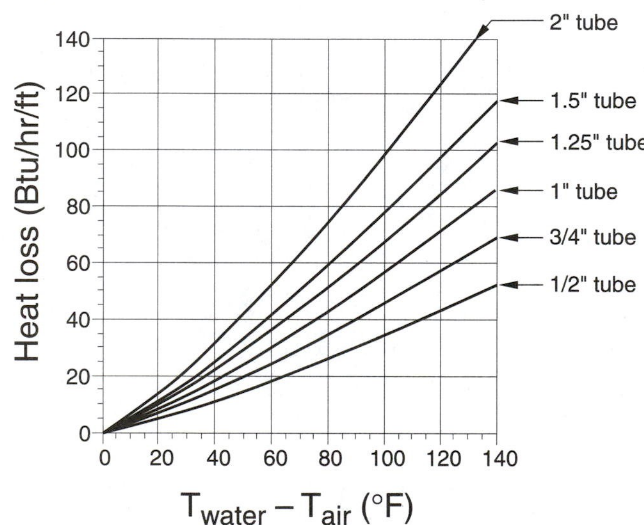

Figure 8–44 Heat loss from bare copper tubing. Used when length in feet divided by flow rate in gpm is less than 20.

This graph is based on extrapolation of copper tube heat loss data published in the *ASHRAE Handbook of Fundamentals*. It plots the heat loss of the tube versus the difference between the fluid inlet temperature and the air temperature surrounding the tube. When the tube is relatively short, or operates at a high flow rate, there is very little temperature drop from the inlet to the outlet. This allows the heat loss to be reasonably well-estimated by using the inlet temperature when calculating the temperature difference between the fluid and the air.

Method #2: When the length of the copper tube in feet divided by the flow rate in gallons per minute exceeds 20 feet/gpm, there will be a greater temperature drop along the tube from inlet to outlet. In such cases, Equation 8.8 can be used to accurately calculate the outlet temperature from the tube.

Equation 8.8

$$T_{out} = T_{air} + \left[(T_{in} - T_{air})^{C1} + \frac{C2 \times L}{(8.021cD) \times f}\right]^{1/C1}$$

where:

T_{out} = temperature of the water leaving the pipe (°F)
T_{air} = temperature of the air surround the pipe (°F)
T_{in} = temperature of the water entering the pipe (°F)
L = length of the pipe (ft)
f = flow rate of water in the pipe (gpm)
Note: $f > 0$
C_1 and C_2 = constants based on tube size and read from Figure 8–45.
c = specific heat of the fluid (at inlet temperature) (Btu/lb/°F)
D = density of the fluid (lb/ft³)

Be sure to copy all the significant figures shown for constants C_1 and C_2 when using them in Equation 8.8. Be sure to include the negative sign associated with each value of C_1.

Once the outlet temperature of the tube is determined, its heat loss can be calculated using Equation 8.9:

(Equation 8.9)

$$Q = (8.021cD)(f)(T_{in} - T_{out})$$

where:

T_{out} = temperature of the water leaving the pipe (°F)
T_{in} = temperature of the water entering the pipe (°F)
f = flow rate of water in the pipe (gpm)
Note: $f > 0$
c = specific heat of the fluid (at inlet temperature) (Btu/lb/°F)
D = density of the fluid (lb/ft³)

Example 8.6: Water enters a 1-inch copper tube at 10 gpm and 160 °F. The tube is 35 feet long and is surrounded by air at 60 °F. Estimate the heat loss of the tube.

Copper tube size	C_1	C_2
1/2"	-0.238285	0.02665
3/4"	-0.237721	0.03695
1"	-0.236284	0.04595
1.25"	-0.235350	0.05475
1.5"	-0.235693	0.06325
2"	-0.235996	0.07985

Figure 8–45 Table of constants (C_1 and C_2) for use in Equation 8.8

Solution: The first step is to check the ratio of the tube length in feet divided by the flow rate in gpm:

$$\frac{\text{Length}}{\text{flowrate}} = \frac{35\text{ feet}}{10\text{ gpm}} = 3.5\text{ ft/gpm}$$

Since this ratio is less than 20, the heat loss can be estimated using the curve for 1-inch bare copper tubing in Figure 8–44. The entering water temperature minus the surrounding air temperature is 160 − 60 = 100 °F. The heat loss per foot of tube is read from the graph as approximately 57 Btu/hr/ft. The total heat loss of the 35 foot tube is thus 35 × 57 = 1995 Btu/hr.

Discussion: At the stated flow rate of 10 gpm, a heat loss of 1995 Btu/hr yields a temperature drop of 0.4 °F as the water flows through the tube. Therefore, the rate of heat loss at the beginning of the tube is approximately the same as at the outlet.

Example 8.7: Water at 200 °F and 1 gpm enters a 105-foot-long run of 3/4-inch bare copper tubing surrounded by 60 °F air. Estimate the heat loss from the tube.

Solution: Dividing the length of the tube by the flow rate in gallons per minute yields a ratio of:

$$105\text{ feet/1 gpm} = 105\text{ feet/gpm}$$

Since this ratio is above 20, Equation 8.8 should be used instead of the graph in Figure 8–45. The density of water at 200 °F is 60.126 lb/ft³. The specific heat is approximately 1.00 Btu/lb/°F. The values of C_1 and C_2 for 3/4-inch copper tube are found in Figure 8–45: C_1 = –0.237721, C_2 = 0.03695. Substituting these into Equation 8.8 yields:

$$T_{out} = 60 + \left[(200 - 60)^{-0.237721} + \frac{0.03695 \times 105}{(8.021 \times 1 \times 60.126) \times 1}\right]^{1/-0.237721}$$

$$T_{out} = 60 + [0.308902 + 0.0080448]^{-4.206612}$$

$$T_{out} = 60 + 125.648$$

$$T_{out} = 185.648 \approx 185.7°F$$

The heat loss is then calculated using Equation 8.9:

$$Q = (8.021 \times 1 \times 60.126)(1)(200 - 185.648)$$

$$= 6922\text{ Btu/hr}$$

Discussion: This is a significant heat loss. Enough to heat a typical room in a house. It points to the need to insulate copper tubing runs that convey hot fluids through cool surroundings.

8.12 SYSTEM OPERATING TEMPERATURE

When turned on, every hydronic heating system tries to establish operating conditions that allow the distribution system to dissipate the full (steady state) heat output of the heat source. If not for the intervention of temperature-limiting controls, every system would eventually stabilize at a supply water temperature where such thermal equilibrium exists. This temperature may or may not provide the proper heat input to the building. Likewise, it may or may not be conducive to safe and efficient operation or long system life.

By adjusting the size, number, or other characteristics of heat emitters in the distribution system, the designer can manipulate the steady state supply water temperature at which the system "wants" to operate. When properly done, this allows both the heat source and distribution system to operate at conditions that are safe, efficient, comfortable, and conducive to long system life. If this tendency to operate at thermal equilibrium is disregarded, the resulting system may attempt to stabilize at a supply temperature that is either unsafe, inefficient, or shortens the life of the heat source.

For example, Figure 8–46 shows a conventional boiler with a steady state output rating of 100,000 Btu/hr connected to a low-temperature radiant floor panel. The temperature-limit control on the boiler is set to 180 °F. The radiant floor panel has been designed so it can dissipate heat at 100,000 Btu/hr when operating with an inlet water temperature of 110 °F. When turned on, this system will eventually stabilize at a supply temperature where the rate of heat output from the heat source matches the rate of heat dissipation by the heat emitter (e.g., the radiant floor panel). Based on the data given, the supply temperature of the boiler will stabilize at 110 °F (e.g., the temperature where the radiant panel can dissipate the full output of the boiler). The fact that the boiler limit control is set to 180 °F is irrelevant. That temperature will only be attained if and when the load on the boiler is significantly reduced or turned off. Barring this, the boiler will not be able to climb above a supply temperature of 110 °F. Such operating conditions will result in sustained flue gas condensation within the boiler. If left unchecked, this will drastically shorten the life of the conventional boiler. The solution is to install a mixing device between the boiler and the low-temperature distribution system. This will be covered in detail in Chapter 9, Control Strategies, Components, and Systems.

Another example of a thermal mismatch would be pairing up residential grade finned-tube baseboard with a condensing boiler as shown in Figure 8–47. Although such a combination is theoretically possible, it will be impractical in most buildings. The length of finned-tube required to allow the condensing boiler to operate at low temperatures where flue gas can condense may be more than what most rooms can accommodate. The alternative of using shorter baseboards and operating the boiler at a higher temperature prevents sustained flue gas condensation and lowers boiler efficiency.

The relationship between supply water temperature and heat output for any hydronic distribution system operating at a fixed flow rate can be represented by a heat output curve, an example of which is shown in Figure 8–48.

To find the **steady state supply temperature** at which the system will attempt to stabilize, first locate the heat

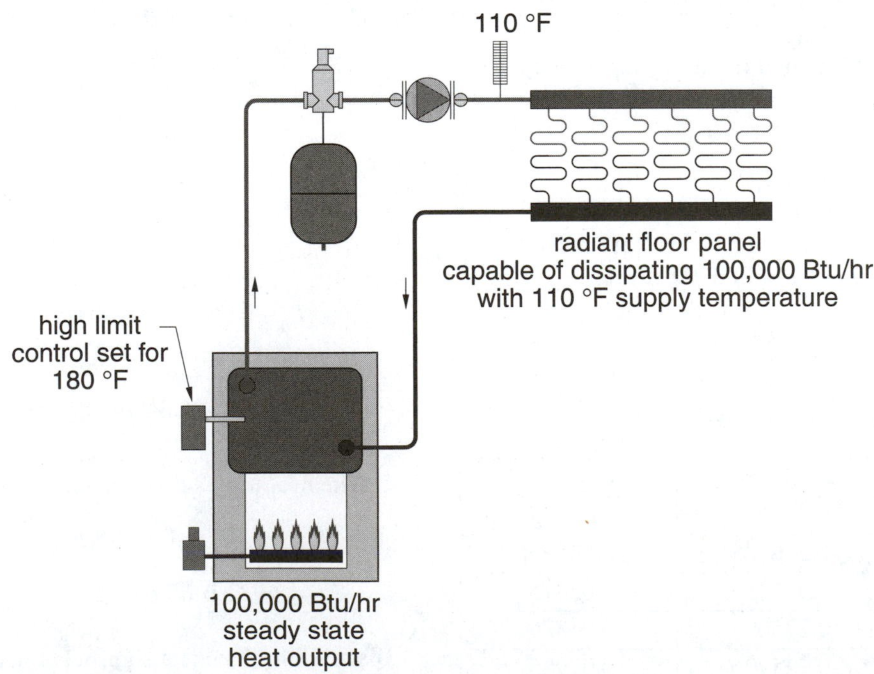

Figure 8–46 Conventional boiler with limit control set to 180 °F connected to low temperature distribution system. Supply water temperature will not climb above 110 °F when this system is at design load.

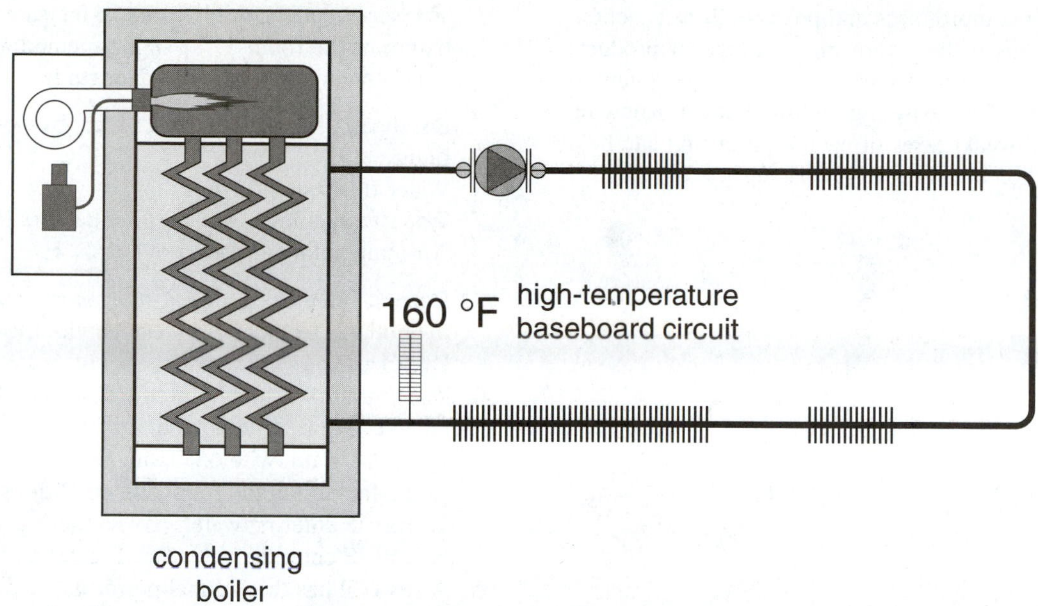

Figure 8–47 Condensing boiler connected to high temperature baseboard circuit. Boiler will not operate at low enough return water temperature to condense flue gases.

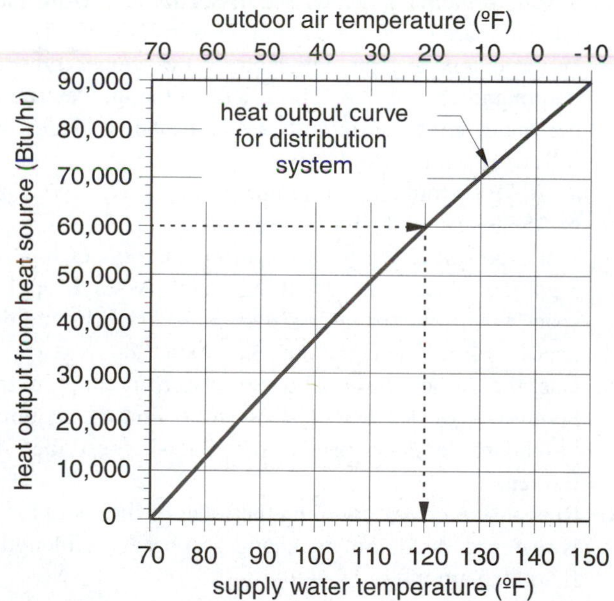

Figure 8–48 Example of a heat output curve for a specific distribution system. If a heat source with an output of 50,000 Btu/hr were used, the steady state supply temperature would be 120 °F.

output rate of the heat source on the vertical axis. Project a horizontal line to the right until it intersects the heat output curve for the distribution system, then downward to the horizontal axis. This supply temperature is where the ability of the distribution system to dissipate heat equals the heat output of the continuously operating heat source.

If the temperature limiting control of the heat source is set *below* this temperature, the heat emitters in the distribution system will not get hot enough to dissipate the full (steady state) output of the heat source. The temperature of the water leaving the heat source will climb as the system operates, eventually reaching the temperature setting of the limit control. At that point, the heat source (burner, compressor, etc.) is turned off. The water temperature leaving the heat source begins to decrease as heat continues to be dissipated by the distribution system. Eventually, the temperature drops to the point where the heat source is turned back on, and the cycle repeats. This is a very common operating mode in many systems during partial load conditions. It can even occur under design load conditions in systems having an oversized heat source.

If the temperature-limiting control on the heat source is set *above* this temperature, the water leaving the heat source will not be able to achieve that temperature setting unless the load is reduced or turned off.

SUMMARY

This chapter has discussed several types of hydronic heat emitters suitable for use in residential and light commercial systems. They have ranged from units having mostly convective heat output such as finned-tube baseboard and fan-coils, to those with a significant percentage of radiant heat output, such as panel radiators and radiant baseboard.

No one type of heat emitter is ideal for all applications. Attempts to use only one type for all kinds of jobs will inevitably lead to compromises and possible dissatisfaction. The designer needs to be aware of the range of products available, and use them as resources to match the system to the specific needs of the building and the budget. Knowing the strengths and weaknesses of each type of unit can help narrow the selection during preliminary design.

KEY TERMS

active length
air handler
coil
convectors
double interpolation
element
escutcheon plate
fan-coil
finned-tube baseboard convector
freestanding panel radiators
heat emitters
heat output ratings
heating effect factor
home run distribution system
interpolation
kick-space heater
panel radiator
radiant baseboard
radiators
Series Baseboard Simulator
steady state supply temperature
stratification
towel warmer
unit heater
vertical panel radiator

CHAPTER 8 QUESTIONS AND EXERCISES

1. A finned-tube baseboard convector has the heat output characteristics given in Figure 8–9 (lower curve). Estimate the heat output of a 1-foot segment of this baseboard operating at an average water temperature of 135 °F, and 3 gpm in a room with 60 °F air near floor level. Use Equation 8.1.
2. The following rooms are to be heated by a series string of finned-tube baseboards:

Room	Design heating load
1. Master bedroom	4,500 Btu/hr
2. Bedroom	3,000 Btu/hr
3. Den	8,000 Btu/hr
4. Living room	11,000 Btu/hr
5. Bathroom	1,500 Btu/hr

The series baseboards will be in the same order as the listing of the rooms. Use sizing method #1, along with the following data, to select a length for each baseboard.

Baseboard thermal ratings given by lower curve in Figure 8.9:
Water flow rate = 1 gpm
Water temperature entering first baseboard = 180 °F
Air temperature at floor level = 65 °F

3. Repeat Exercise 2 using the more accurate sizing method #2. Compare the results with those obtained in Exercise 2.
4. The output of a small fan-coil is rated at 4500 Btu/hr at an entering water temperature of 180 and flow rate of 1 gpm, in a room with 65 °F air. Estimate the heat output of the fan-coil at the following conditions:
   a. 160 °F entering water, 65 °F entering air, 1 gpm
   b. 140 °F entering water, 55 °F entering air, 1 gpm
5. A fan-coil has the thermal performance ratings given in Figure 8–21. Interpolate between the data to estimate the heat output of the unit at 167 °F entering water temperature, 65 °F entering air temperature, and 3 gpm water flow rate.
6. Estimate the output of the fan-coil described in Exercise 5 using principle #1 discussed in Section 8.7. How does this compare to the answer obtained in Exercise 5?
7. Water enters a 1-inch diameter bare copper tube at 9 gpm and 180 °F. The tube is 100 feet long. Determine the outlet temperature and rate of heat loss under the following conditions:
   a. 70 °F surrounding air temperature
   b. 35 °F surrounding air temperature
8. Water at 200 °F enters a 3/4-inch bare copper tube at 3 gpm. The tube is 120 feet long, and passes through a space where the air temperature is 70 °F. Estimate the heat loss from the tube using Equation 8.8.
9. Use the Series Baseboard Simulator module in the Hydronics Design Studio software to find the required baseboard lengths for the room loads described in Exercise 2.
10. Repeat Exercise 9 assuming the order of the room loads is reversed. Are there any changes in baseboard length? If so, explain why.
11. Use the Series Baseboard Simulator module and the room loads stated in Exercise 2 to find the required baseboard lengths under the following assumptions:
    a. The circuit flow rate is 0.5 gpm
    b. The circuit flow rate is 2.0 gpm
    c. The circuit flow rate is 5.0 gpm

    Explain why the baseboard lengths are changing.

CONTROL STRATEGIES, COMPONENTS, AND SYSTEMS 9

OBJECTIVES

After studying this chapter, you should be able to:

- Describe the basic elements of a closed control loop
- Explain the different control outputs and control algorithms used in hydronic heating systems
- Describe the differences between controlling heat output using variable flow versus variable water temperature
- Describe several basic control components such as switches and relays
- Understand the operation of basic electromechanical temperature controls
- Use ladder diagrams to lay out control systems
- Properly connect both three- and four-wire zone valves
- Describe the benefits of outdoor reset control and different ways to implement it
- Explain the operation of several types of mixing valves
- Describe the strengths and limitations of different methods of injection mixing

9.1 INTRODUCTION

The control system is the brain of a hydronic heating system. It determines exactly when and for how long devices such as circulators, burners, and mixing valves will operate. The comfort, efficiency, and longevity of the system are as dependent on the controls as they are on any other component or subsystem.

A well-designed and properly adjusted control system can provide sophisticated operating logic to optimize comfort and efficiency. It will operate unpretentiously behind the scenes, require minimal attention, and help assure a long system life. Conversly, a carelessly designed control system can be a nightmare. The use of high-quality heat sources, circulators, or other components will never compensate for an improperly designed or poorly adjusted control system.

Controls for hydronic heating represent technology that spans several decades. They range from time-proven devices such as the bimetal **room thermostat** to **microprocessor-based controllers** that have the intelligence to optimize system operation for minimum fuel use and maximum comfort. Both high-tech and low-tech controls have their place in modern hydronic systems. With experience, a resourceful system designer will learn to integrate them for maximum reliability and efficiency.

The operation of many heating control systems is based upon the status of various electrical switches at any given time. Some of these switches are manually set, others are automatically operated by temperature, pressure, or other sensed conditions. This chapter discusses many of the basic switch-type controls used in hydronic heating and shows how they interact within an overall system.

Electronic controllers, especially those using microprocessors, have worked their way into nearly all areas of heating and cooling technology over the last decade. Hydronic system designers now have a broad range of such controls with which to work. Many functions that previously required human intervention can now be performed automatically by such controllers. These include automatic adjustments of system water temperature based on weather conditions, automatic shutdown of system pumps during warm weather, even periodic exercising of components such as mixing valves and circulators to prevent setup during nonoperational periods. The accuracy of these controllers also helps improve both comfort and energy efficiency. This chapter will acquaint you with several electronic controllers used in modern hydronic systems.

Many modern hydronic systems use one or more **mixing devices** to control water temperature in various parts of the system. Some of these devices such as 3-way thermostatic valves, and 4-way mixing valves were discussed in Chapter 5, Pipings, Fittings, and Valves. This chapter will go into further detail, and discuss the interaction between the piping components and the electrical/electronic controllers that operate them.

The proper documentation of control systems is essential if they are to be expediently serviced in the future. A universally accepted tool for such documentation is the **ladder diagram**. It will be introduced in fundamental terms, and then expanded to demonstrate its use in more sophisticated systems.

Finally, this chapter discusses how control technology for hydronic heating is likely to progress over the next few years. As good as present-generation controllers are compared to those available only 15 to 20 years ago, there are emerging technologies that will further improve control intelligence and connectivity. Many of the control devices shown and discussed in this chapter did not exist when the first edition of this book was written in 1993. In all likelihood, some of these state-of-the-art devices will be eclipsed by even newer and more advanced devices within a few short years after this edition is released. This is not to imply that present-generation controls are not excellent at what they do. It is simply to point out that control technology is constantly changing. Those heating professionals who wish to avail themselves of all modern hydronic heating has to offer must stay abreast of the most recent developments in control technology.

9.2 CLOSED-LOOP CONTROL SYSTEM FUNDAMENTALS

Hydronic heating systems, like all other HVAC systems, rely on fundamental control techniques to stabilize their operation and optimize their performance. It is imperative that hydronic heating professionals understand the basic control concepts underlying many of the control devices they will specify, install, and service. This section covers those basics, and sets the stage for more detailed discussions of specific control hardware in later sections of the chapter.

Figure 9–1 depicts the elements of a simple **closed control loop** for hydronic heating.

The first device is the **controller**. It may be a simple pressure-operated bellows, or a sophisticated microprocessor-based device. Its function is to read the desired **target value** of the **controlled variable**, as well as to receive **feedback** information from the **sensing element** on the **measured value** of the controlled variable. Any deviation between the target value and the measured value is called **error**. Some controllers used in heating systems receive information from multiple sensors and regulate multiple processes.

Based on these inputs, as well as the **processing algorithms** that are built in or programmed in, the controller generates an **output**. This output may be a simple switch contact closure, a precise DC voltage, or even a variable frequency AC waveform. The output is sent to a **controlled device** that responds by changing the **manipulated variable** (which in a heating system is usually heat flow). The change in the manipulated variable causes a change in the process being controlled (in this case, the heat input to the building). This causes a corresponding change in the controlled variable (in this case, the inside air temperature). The change in the controlled variable is sensed by the sensing element, which provides feedback to the controller.

This entire process is ongoing and simultaneous. The controller receives constant input information and feedback and generates a constant output signal.

Closed-loop control systems can be tuned for very stable operation. The feedback inherent to their design makes them strive to eliminate any error between the target and measured value of the controlled variable.

Control Processing Algorithms

All closed-loop control systems seek to eliminate any error that exists between the target value and sensed value of the controlled variable. The exact procedure by which a controller converts the error present into an output signal is called the processing algorithm. The three common types of processing algorithms used in modern HVAC control systems are **proportional (P)**, **proportional-integral (PI)**, and **proportional-integral-derivative (PID)**.

Proportional (P) Processing

In pure proportional processing, the magnitude of the output signal depends only on the error between the target and measure values of the controlled variable. The greater the deviation (a.k.a. error) between them, the stronger the output signal becomes in an attempt to eliminate this error.

This concept is illustrated in Figure 9–2. The water level in the tank is the controlled variable. Water can run out of the tank at different rates depending on the setting of the drain

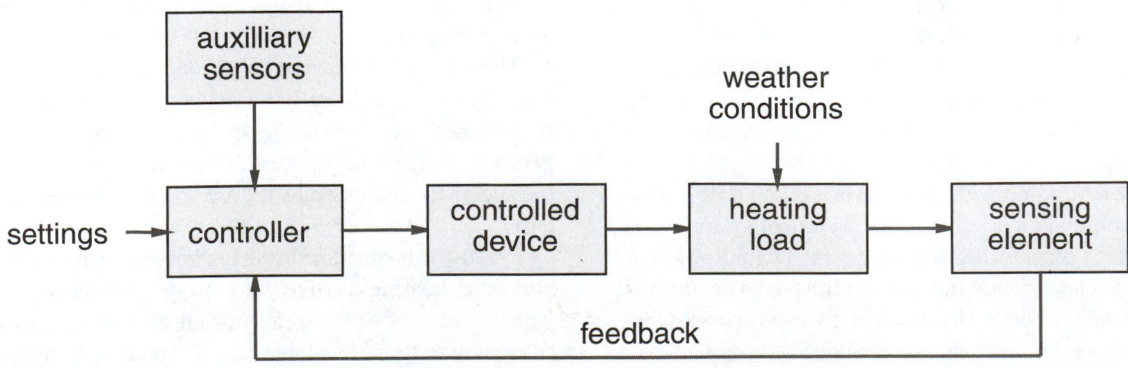

Figure 9–1 Block diagram of a feedback control system for space heating applications.

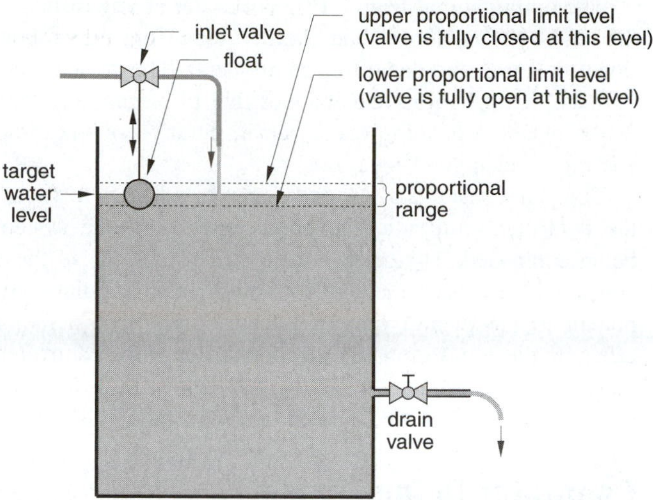

Figure 9–2 Water level in tank is controlled by float operated valve. Rate of water input is proportional to rate of water drainage.

valve. Water is added to the tank by the controlled valve at the top, which is linked to a float that monitors water level. If water leaves the tank faster than the inlet valve admits it, the level drops causing the float to open the control valve to admit water at a higher rate. The farther the float drops, the faster water enters the tank. Similarly, if the water level starts to rise, the float also rises to reduce the water inlet rate. Whenever the water inlet rate equals the water outlet rate, the water level holds constant.

Proportional controllers have what is known as a **proportional band**. This is the range of the controlled variable over which the controlled device varies from fully off to fully on. In the control situation depicted in Figure 9–2, the proportional range is between the upper level, where the inlet valve is completely off, down to the lower level, where the inlet valve is fully open. If conditions are such that the water level drifts above or below these levels, the inlet valve cannot make any further compensation in its attempt to correct this error. Under such conditions, the controlled device is operating outside of its proportional band. In the case of proportional temperature controllers, the proportional band is typically 2 to 4 degrees Fahrenheit.

Any controller that uses purely proportional processing can only maintain the exact target value of the controlled variable under one calibration condition. For example, the water level in the tank will remain at the desired target level only when the outlet flow equals the inlet flow at which the float/valve assembly is calibrated. If the outlet flow is slightly higher than this calibration rate, the float eventually drops to enable a higher inlet flow rate. If the higher flow rate through the tank is within the proportional range of the float/inlet valve, a balance will be reestablished, but at a level somewhat lower than the target level. Likewise, if the outlet flow is slightly less than the calibrated inlet flow, the float rises to reduce the inlet flow, and a new level will be established slightly above the target level. Both situations result in the water level being slightly different from the target value. This effect is called **offset**, and is inherent to all purely proportional controllers. The narrower the proportional band of the controller, the smaller the offset will be. However, proportional bands that are too narrow can lead to erratic control action.

Most mechanical thermostats are pure proportional controllers. They are calibrated by the manufacturer to maintain a target room temperature at some assumed percentage of heat source on-time. When the heating load increases, the room temperature must droop slightly for the thermostat to keep the heat source running a higher percentage of the time.

Although pure proportional processing does have some limitations, it is the dominant mathematical concept underlying nearly all temperature controllers used in HVAC systems.

Proportional-Integral (PI) Processing

Fortunately, the shortcomings of purely proportional processing can be overcome by adding further mathematical sophistication to the processing algorithm. Proportional-integral processing sets the output signal based on *both the amount of error, as well as how long the error has existed.* The latter component of the processing algorithm allows the controller to eliminate the steady state offset inherent with proportional-only control. Because of the mathematics involved, PI processing is only practical in electronic controls. It is now commonly used in many types of electronic temperature controllers including room thermostats.

Proportional-Integral-Derivative (PID) Processing

PID processing adds the ability to monitor the *rate of change of error* to the response provided by a PI controller. This allows the controller to predict what the error is likely to be a short time into the future and adjust the present output accordingly.

PID control is most useful in control systems that must respond to rapid changes in the inputs. Most HVAC systems do not require high-speed response. For this reason, the derivative contribution of the processing algorithm is usually minimal. Still, the presence of some derivative component allows the controller to stabilize to the target value slightly faster than PI control.

Figure 9–3 compares the response of P, PI, and PID control systems to the same initial "error" between measured and target temperatures. Notice how adding the integral function to the processing algorithm eliminates the steady state offset. It is also evident that the system using full PID control stabilizes to the target value slightly faster than does the system using PI control.

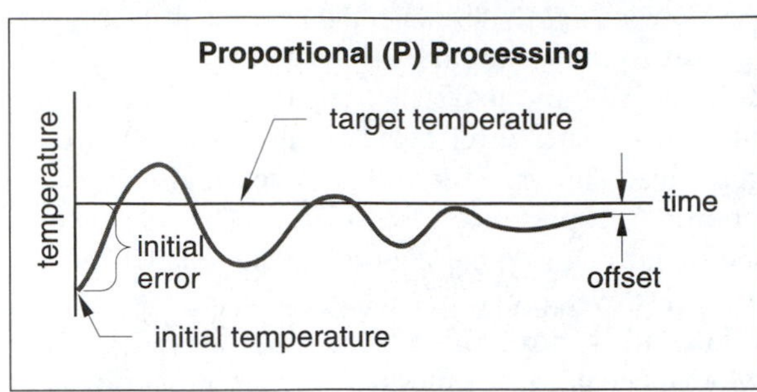

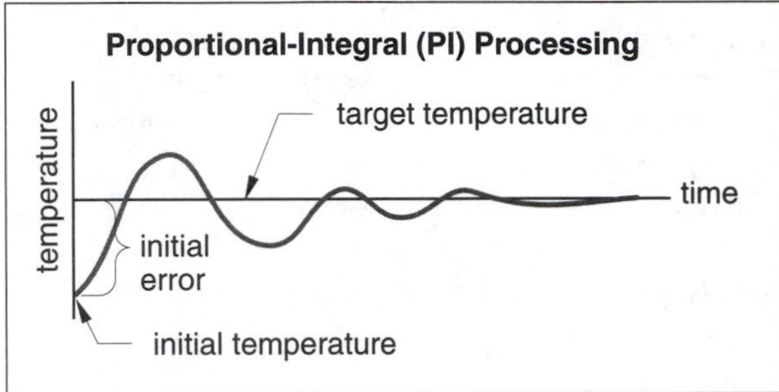

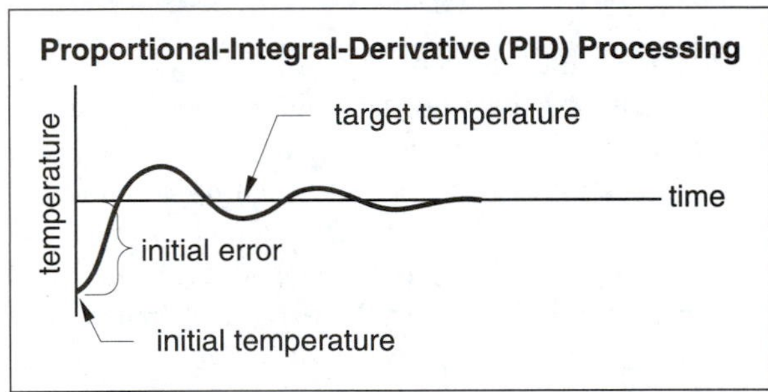

Figure 9–3 Response of a system with an initial temperature error to Proportional (P), Proportional-Integral (PI), and Proportional-Integral-Derivative (PID) control algorithms.

This concept of PID processing can be expressed mathematically as follows:

$$O = c_p(e) + c_i\int(e)dt + c_d\left(\frac{de}{dt}\right) + M$$

where:

O = output signal value at some instant

e = error between the target and measured value of the controlled variable

c_p = gain constant for the proportional contribution to the output

c_i = gain constant for the integral contribution to the output

c_d = gain constant for the derivative contribution to the output

M = a constant representing the output value when the error is zero

The output signal from a PID controller at any instant is the mathematical sum of the contributions from the proportional, integral, and derivative components. The value of any of these contributions can be positive or negative at any given instant depending on the error, duration of the error, and rate of change of the error.

The gain constants c_p, c_i, and c_d can be adjusted to "tune" the controller output to the characteristics of the system being controlled. The greater the value of any one of these constants, the greater the contribution of that particular component to the overall output. Because they have a major effect on the operation and overall stability of the controller, these constants are usually set by the control manufacturer, and can very seldom be adjusted by the installer.

Controller Output Types

A controller can interact several ways with a controlled device. In some cases, the output action is a simple opening or closing of an electrical contact. In other cases, the output action may be a constantly changing voltage or current signal from the controller to the controlled device. This section reviews the controller output types commonly used in hydronic heating control systems.

On/Off Output

In on/off control, the controlled device can only have one of two possible **control states** at any time. If the controlled device is a valve, it can be either fully open or fully closed, but never at some partially open condition. If the controlled device is a heat source, it must be fully on or completely off. On/off output signals are usually generated by the closing and opening of electrical contacts within the controller.

When used to control heating equipment, on/off controllers send heat to the load in pulses rather than as a continuous process. In such applications, all on/off controllers must operate with a **differential**. This is the change in the controlled variable between where the controller output is on to where it is off. In heating systems, the controlled variable is usually the temperature at some location in the system or the building it serves. On some controllers, the differential is below the **setpoint temperature**, as shown in Figure 9–4a. In other controllers, the differential is centered on the setpoint, as shown in Figure 9–4b.

The smaller the differential, the smaller the deviation in temperature from the desired target value. However, if the differential is too small, the controller operates the controlled device in short but frequent cycles. This undesirable effect is called **short-cycling**. In most cases, short-cycling increases the wear on the equipment being controlled. This is especially true of devices such as oil burners, gas valve/ignition components, and compressors used in heat pumps.

If the differential is too wide, there can be large deviations between the target value and measured value of the controlled variable (usually temperature). In some situations this is

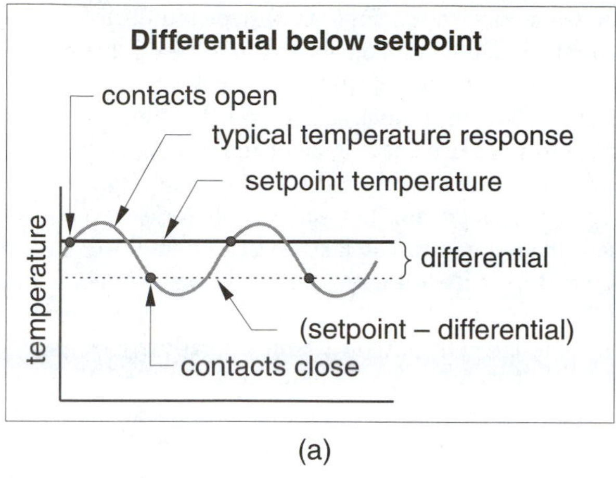

(a)

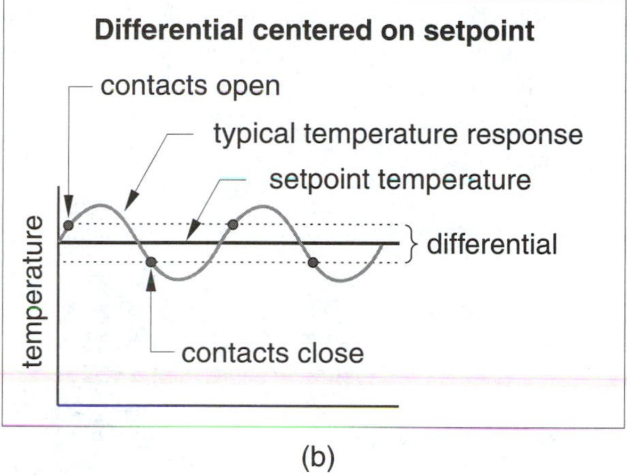

(b)

Figure 9–4 On/off temperature controllers operate with a differential that is either (a) entirely below the setpoint value, or (b) centered on the setpoint value.

acceptable, in others it is not. For example, the large thermal mass of a heated slab-on-grade floor can usually accept variations in water temperature of several degrees Fahrenheit. without creating corresponding swings in room air temperature. However, a room thermostat with a differential of 4 °F or higher, for example, would cause wide swings in room air temperature that would almost certainly lead to complaints.

The term **control differential** refers to the variation in temperature at which the controller changes the on/off status of the output. On some controllers, this is a built-in value and cannot be changed by the designer or installer. On other controllers, the control differential can be adjusted over a wide range.

The term **operating differential** refers to the actual variation in the controlled variable as the system operates. It depends on the combined effect of the controller and the system it is controlling. In heating systems, the operating differential is usually greater than the control differential because of a time lag caused by the thermal mass of the heating system and the building. Figure 9–5 shows the difference between these differentials.

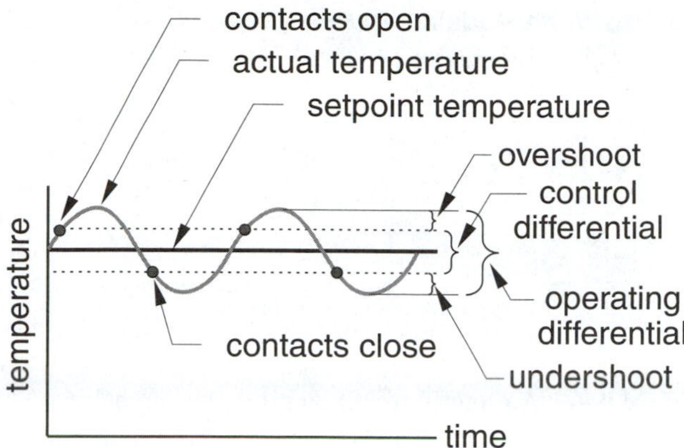

Figure 9–5 The control differential is the desired difference between when controller output turns on and off. Operating differential is the actual temperature range experienced as the process is controlled. The operating differential is wider than the control differential due to overshoot and undershoot.

Pulse Width Modulation Control

When simple on/off control is used in heating systems, energy input begins only when the temperature being controlled drops a certain amount below the desired target value. After the heat source is turned on, the thermal mass of the heating system causes a further drop in temperature until the heat emitters are warm enough to begin raising the room temperature. Heat is added to the room as a result of the room air temperature repeatedly dropping below the desired target value. Over time, the average room temperature becomes slightly less than the desired value. This effect is called **droop**, and is inherent to nonelectronic on/off temperature controls.

Ideally, the heating system would supply enough heat to the room to keep the indoor temperature precisely at the desired value. Doing so requires a control strategy that adjusts the amount of heat supplied to the room based on the error between the desired target temperature and the measured temperature. If the measured temperature is above the target value, the rate of heat input needs to decrease, but not necessarily to zero. Likewise, if the measured temperature is below the target value, the rate of heat input needs to increase, but not necessarily to full design load output. Even when the error is zero, a certain rate of heat input must be maintained to prevent droop.

A control strategy that accomplishes this is called **pulse width modulation (PWM)**. With this approach, the length of the heat input cycle is based on how far the measured temperature is from the target value. The greater the deviation, the longer the heat input is kept on. The concept of pulse width modulation is illustrated in Figure 9–6.

The controller generates the sloping lines internally. Each triangle formed by an up- and down-sloping line represents a preset cycle length of typically 10 to 15 minutes.

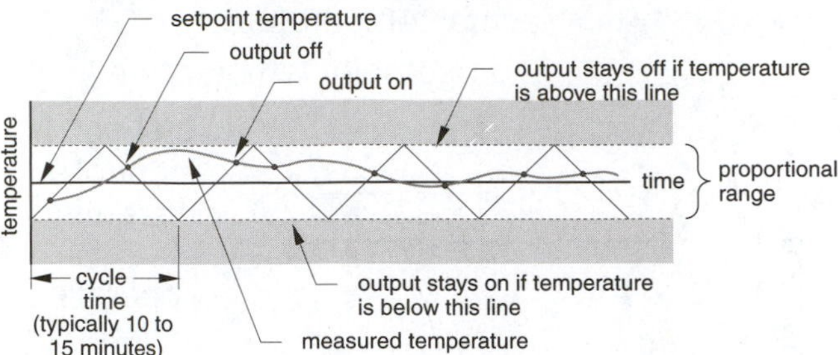

Figure 9–6 An on/off output controlled using pulse width modulation (PWM). Note that the duration of the on versus off time is dependent on the error between the actual temperature and the target (setpoint) temperature.

The controller detects when the sensed value of temperature equals the value represented by the sloping lines. Whenever the measured temperature curve crosses an up-sloping line, the control output (usually an electrical contact) turns on a heat input device such as a heat source, circulator, or zone valve. When the measured temperature crosses a down-sloping line, the output is turned off.

If the measured value drifts into the shaded area above the sloping lines, the output remains off. Similarly, if the output drifts into the shaded area below the sloping lines, the output remains on. Even when the error is zero, the controller keeps the output on 50 percent of the cycle length. This is the essential difference between pulse width modulation and on/off control that allows pulsed heat input to better compensate for continuous building heat loss.

When used to control heat output, pulse width modulation adds energy to the process in proportion to the error present. Most controllers that produce a pulse width modulation output also use integral (I) and in some cases derivative (D) processing to quickly stabilize the system and eliminate droop. When the error is small, the controller gently "nudges" the controlled device in an attempt to eliminate the error without overshooting the desired temperature. When the error is large, the controller operates the controlled device more aggressively to reduce the error as quickly as possible, and eliminate any residual offset. PWM control is particularly effective when controlling air temperature in rooms heated by high mass radiant panel systems.

Floating Control

Another means of providing some heat input to a hydronic distribution system when the error between the target and measured value of the controlled temperature is at or close to zero is called **floating control**. This control output was developed to operate motorized valves and dampers that need to be powered open as well as powered closed. In hydronic heating systems, it is commonly used to drive 3-way and 4-way motorized mixing valves. Figure 9–7 depicts a floating control system in a hydronic mixing application.

One contact in the controller closes to drive the valve's actuating motor in a clockwise (CW) direction. The other contact closes to drive the actuator motor counterclockwise (CCW). *Only one contact can be closed at any time.* The actuating motor driving the valve shaft turns very slowly. Some actuating motors can take up to 3 minutes to rotate the valve shaft over its full travel range. This slow operation is desirable since it allows the sensor ample time to provide feedback to the controller for stable operation.

Notice the **floating zone** in Figure 9–7. When the sensed temperature is within this zone, the valve's actuating motor does not run. The error between the target temperature and actual temperature is small enough to not warrant any correcting action. The ability to hold the valve at a partially open condition is what allows some heat input to the load even when the error is at or close to zero.

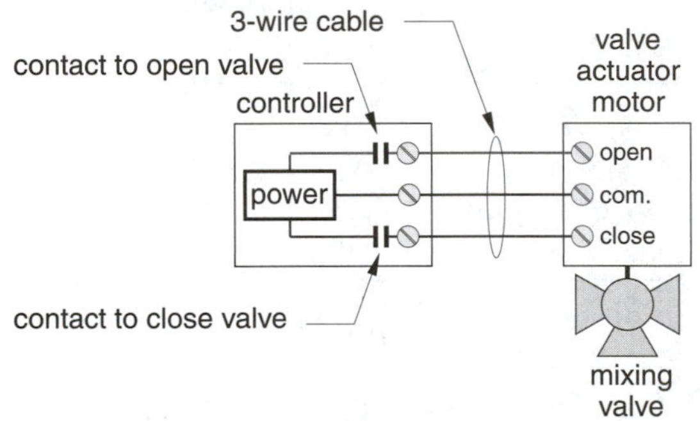

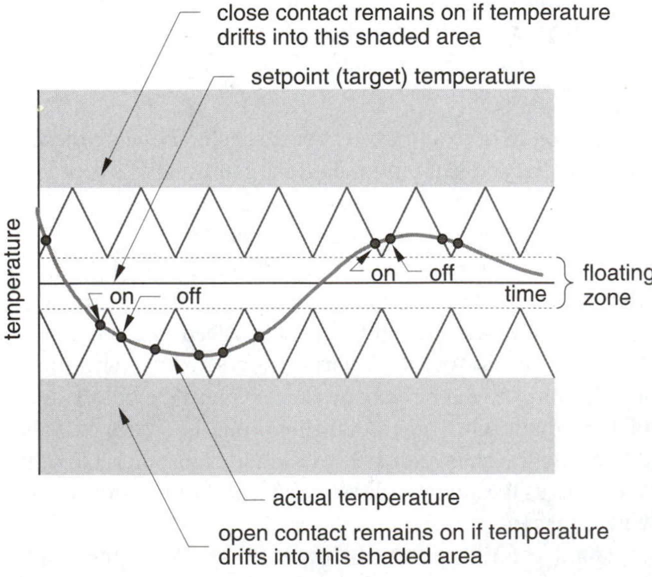

Figure 9–7 Operation of a 3-wire floating control system. Possible control states include opening valve, closing valve, or holding valve in present position. The duration of the on versus off time of either contact is dependent on the error between the actual temperature and the target (setpoint) temperature.

As the sensed temperature drifts above or below the floating zone, the controller responds with a corrective action proportional to the error. The greater the error, the longer the output signal that opens or closes the valve remains on. If the error drifts above or below the PWM zone shown in Figure 9–7, the actuating motor remains on in either the open or closed mode. This usually only occurs while the system is starting up, and a large error is present between the measure and target temperatures.

Floating control is sometimes called **tri-state control** because the output signal can be in any of three operating states at any time (e.g., opening, closing, or hold). It is also sometimes called **three-wire control** because three wires are required between the controller and the actuating motor of the controlled device.

Modulating DC Outputs

On/off control, pulse width modulation control, and floating control all use the opening and closing of electrical contacts to send signals between the controller and controlled device. This type of output is not suitable for all types of devices. For example, none of these outputs can be used to regulate the speed of a motor.

In heating as well as other industrial applications, many controlled devices require a continuous **analog** input signal. Such systems often use a variable voltage between 2 to 10 DC, or a variable current between 4 and 20 milliamps DC as a continuous analog signal between the controller and controlled device. The controller generates a continuous voltage or current within these ranges, and sends that signal to the controlled device. The controlled device responds by operating at a position or speed that is proportional to the input signal. For example, a control signal of 2 volts DC fed to a motor speed control means that the motor should be off. A 10-volt DC signal to the same controller means the motor should be operating at full speed. A signal of 6 volts DC, which is 50 percent of the overall range of 2 to 10 volts, means that the motor should be running at 50 percent of full speed.

The reason these control signals do not begin at zero voltage or current is to prevent electrical interference or "noise" from affecting the controlled device. Wires that run in proximity to other electrical equipment can experience induced voltages and currents due to electrical or magnetic fields. In most situations, raising the starting threshold to 2 volts DC, or 4 milliamps, prevents such interference. However, in some very noisy electrical environments, twisted pair wiring or shielded cable may be necessary between the controller and the controlled device. Consult the control manufacturer's recommendations regarding control wiring in such situations.

Both modulating valves and variable speed circulators are available with 2 to 10 VDC and 4- to 20-milliamp inputs. These inputs are for control only, and do not supply the electrical power to drive the device. For example, the typical wiring of a 2- to 10-VDC modulating valve is shown in Figure 9–8. Notice that 24 volts AC power must also be supplied to the valve to power the motor and the actuator's circuitry. Likewise, a variable speed pump that is controlled by a 2 to 10 VDC or 4- to 20-milliamp signal requires line voltage (120 VAC) to supply operating power.

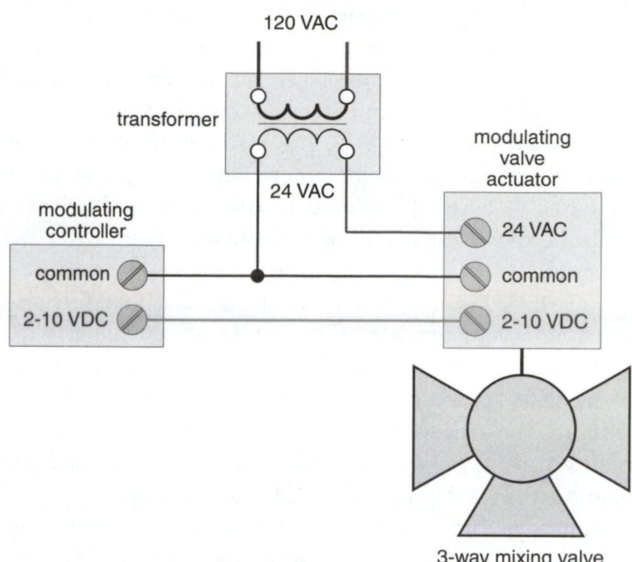

Figure 9–8 Typical wiring for a device controlled by a 2-10 VDC or 4-20 milliamp modulating control signal.

9.3 CONTROLLING THE OUTPUT OF HEAT SOURCES

There are several methods for controlling the heat output of hydronic heat sources. The method used depends on the type of heat source, as well as the total heat output rate. This section discusses the following common methods for controlling heat production:

- On/off control of a single-stage heat source
- Multistage heat production
- Modulating (variable firing rate) heat production

On/Off Control of a Single-Stage Heat Source

Most residential and smaller commercial hydronic systems use a single heat source that has only two operating modes: on or off. Whenever the heat source is on, it produces a fixed rate of heat output (see Chapter 3, Hydronic Heat Sources, for a discussion of the various heat output ratings). When the heat source is off, it produces no heat output. Such heat sources require a simple on/off control signal from their controller. This is typically provided by the closing and opening of an electrical contact.

The total amount of heat added to the system is regulated by turning the heat source on and off. To illustrate this concept, consider the hypothetical building heating load profile shown in Figure 9–9. The load begins at zero, rises at a constant rate to its maximum design value, and then decreases at a steady rate back to half its design value. The time over which this occurs can be assumed to be a few hours.

The shaded rectangles represent the heat inputs from a heat source that provides a fixed rate of heat input while operating. For simplicity, assume that the heating capacity of the heat source equals the design heating load of the building. The heat source is operated by an on/off controller. In this case, assume that controller is a thermostat that monitors the air temperature inside the building.

Notice the height of all the rectangles is the same. This indicates the heat source puts out a fixed of heat flow when operating. The width of the rectangles increases as the load approaches its maximum value. This means the on-time or **duty cycle** of the heat source is increasing. When design load occurs, the rectangle remains uninterrupted. Since the heat source is sized to the design load of the building, it must operate continuously whenever design load conditions exist. As the load decreases, the rectangles become narrower. When the load stabilizes at 50 percent of design load, the width of the rectangles remain constant. Under this condition, the heat source is on 50 percent of the time. When on, its rate of heat output is twice the rate at which the building loses heat.

The shaded area of each rectangle represents the quantity of heat delivery by the heat source during its on cycle. Ideally the area under the load profile, out to some point in time, will exactly equal the area in the shaded rectangles, from the beginning of the graph to that point in time. This implies that the total heat production of the heat source exactly matches the total heat loss of the building over that period of time.

The match between heat output and heating load at low and medium load conditions is certainly not ideal. For example, consider the period of time when the load is very small. Under such conditions, the heat source sends short pulses of heat to the building. The height of these pulses corresponds to the full heating capacity of the heat source, which under partial load conditions may be several times higher than the rate of building heat loss.

In heating systems having low thermal mass, on/off cycling of the heat source is often noticeable and generally undesirable. A forced-air system with a constant speed blower and fixed firing rate is a good example. Under partial load conditions, the furnace produces its full rate of heat output whenever it operates. Since the air in the building has very little thermal mass for absorbing this heat, room temperature increases quickly. When the room thermostat reaches its setpoint temperature, its contacts open and the furnace is turned off. An operating cycle could be as little as 2 or 3 minutes under low load conditions. During this time, the room temperature could increase by several degrees Fahrenheit. The occupants feel this rapid change in air temperature and usually do not like it. When the furnace is shut off, drafts from windows and doors quickly reestablish themselves and comfort rapidly decreases. Eventually the process repeats itself. In effect, the system injects repeated pulses of hot air into the space as its (rather crude) way of attempting to match heat input to heat loss.

Many hydronic systems, especially those having high thermal mass, have an advantage in this respect. The greater thermal mass of the heating distribution system allows it to temporarily absorb some of the surplus heat output generated by the heat source, and spread out its delivery to the building over a period of time. This effect is further enhanced when the distribution circulator operates continuously. The greater the thermal mass of the system, the more stable the room temperatures will remain during partial load conditions.

Radiant floor heating using a 4- to 6-inch thick concrete slab is perhaps the ultimate hydronic system as far as thermal mass is concerned. For example, a 2000 square foot

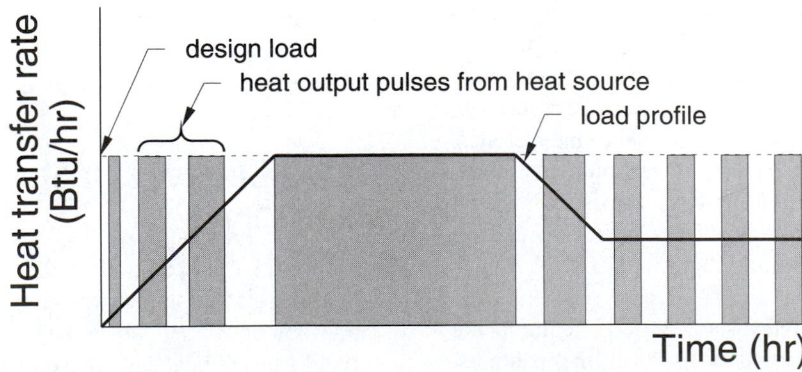

Figure 9–9 Energy input to a hypothetical load profile from a fixed capacity heat source operated by an on/off controller. The output capacity of the heat source is assumed equal to the design load.

floor slab, 4 inches thick, contains over 93,000 pounds of concrete. This thermal mass could absorb about 20,000 Btu while only increasing its temperature 1 °F. The air in the same building (assuming an 8-foot ceiling height) could absorb only about 290 Btu with a temperature rise of 1 °F. The thermal mass of the floor slab is almost 70 times greater than the air. This allows the slab to accept pulses of heat during partial load conditions while delivering a relatively steady heat output to the building. The thermal mass of the system dampens the temperature response of the system. A comparison of temperature swings resulting from pulsed heat input to systems with high and low thermal mass is shown in Figure 9–10.

Multistage Heat Production

Another way of attempting to match the output of the heat source to the building load is with **multistage heat production.** This approach divides the total heating output into a number of increments, or **stages**. As the load increases, these stages of heat input are turned on sequentially. When all stages are on at the same time, the system is at maximum steady state heat output. As the load decreases, stages of heat input are sequentially turned off. The number of stages used in a system can vary from 2 to 8 or more. Larger commercial/industrial systems tend to use more stages than residential systems. The greater the number of stages, the better the match between heat output and load.

Each stage of heat input is usually a separate heat source with a fixed rate of heat output. Each of these heat sources is typically turned on or off by an electrical contact within the staging controller. Some boilers also have two independently controlled burner assemblies that supply heat to the same heat exchanger. They are known as **HI/LO fire boilers,** and are, in effect, a two-stage heat source contained in a single cabinet.

Figure 9–11 illustrates how a two-stage heat source in which each stage represents one-half of design load supplies heat to the same building load profile shown in Figure 9–9.

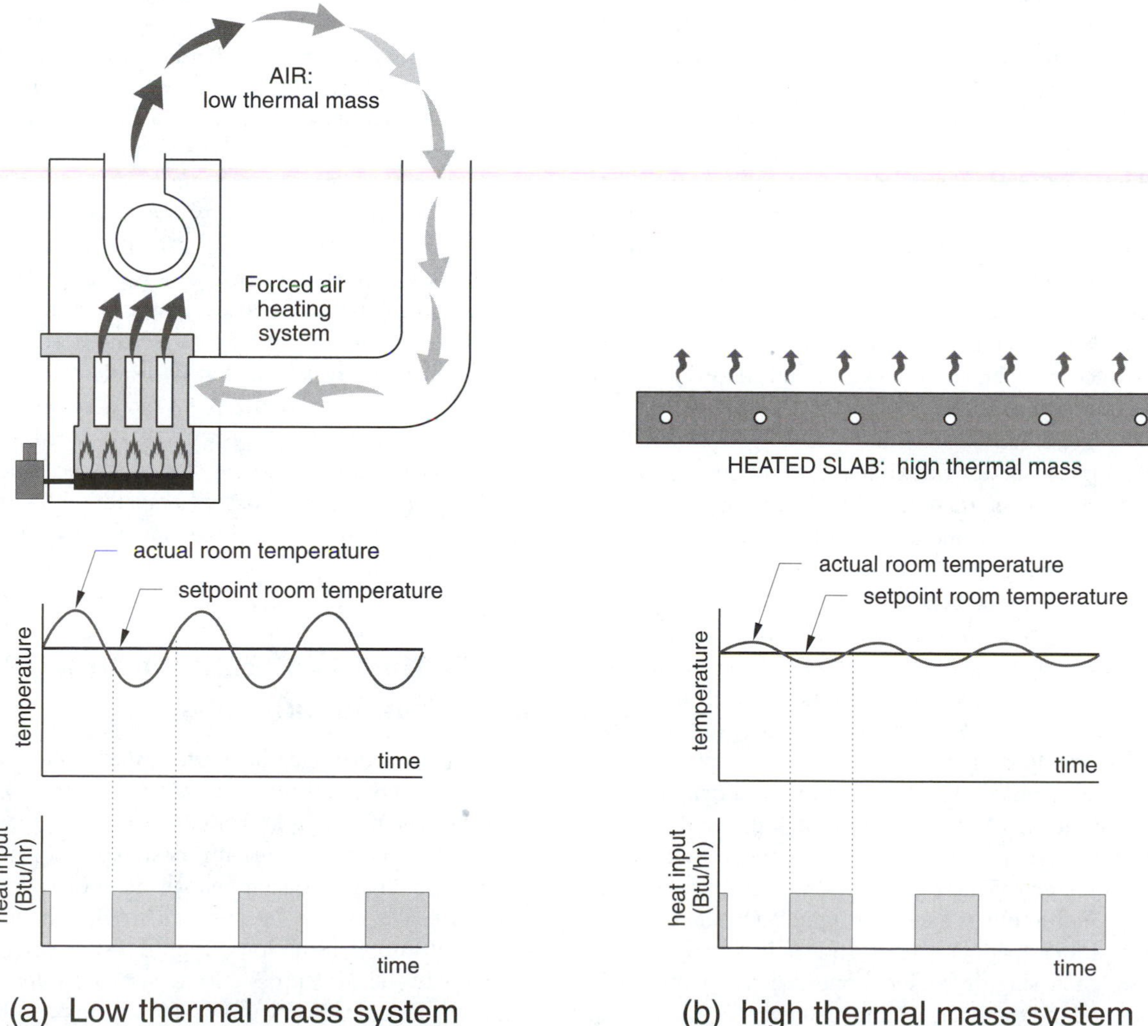

Figure 9–10 Room temperature fluctuations associated with heating systems having low thermal mass and high thermal mass. Note the ability of the high thermal mass system to dampen (e.g., reduce) temperature variations.

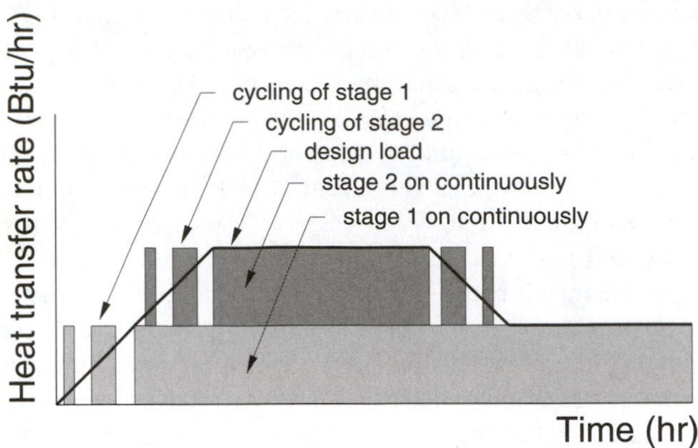

Figure 9–11 Energy input to a hypothetical load profile from a two-stage heat source operated by an on/off controller. The output capacity of each stage is assumed to be 50 percent of design load.

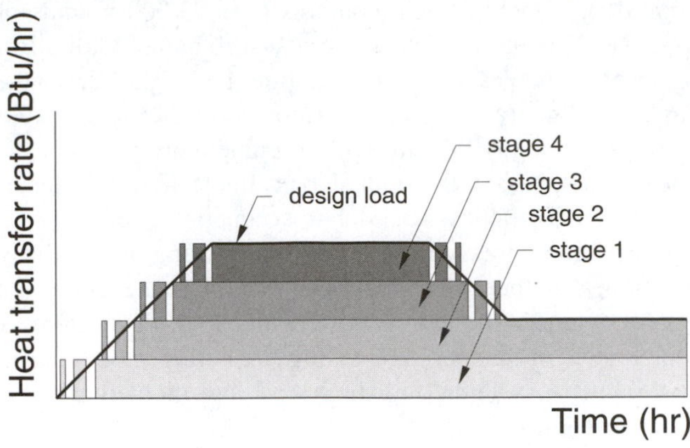

Figure 9–12 Energy input to a hypothetical load profile from a four-stage heat source operated by an on/off controller. The output capacity of each stage is assumed to be 25 percent of design load.

During low load conditions, the first stage of heat output operates intermittently. The width of the rectangles becomes wider as the load increases toward 50 percent of design load. As the load increases above 50 percent of design load, the first stage output remains on continuously, while the second stage operates intermittently to make up the difference between the load and the output of the first stage. Maximum heat output occurs when both stages remain on continuously. This example assumes that the total heat output of both stages equals the design heat load of the building. Any oversizing would necessitate some off-time for the second stage. As the load decreases, the second stage resumes intermittent operation. Since the load profile ends at half of design load, the first stage continues to stay on.

As with single-stage on/off heat input, the total area of the shaded rectangles should equal the area under the load profile up to any point in time. This means that the heat source has supplied the same amount of heat that the building has lost over that period of time.

The notable difference between the two-stage system and the single-stage system is the degree to which the shaded rectangles extend above the load profile, and the amount of "white space" between the rectangle and beneath the load profile. The two-stage system produces less heat in excess of the load. A properly designed multistage heat input system always provides a closer match between heat output and the current heating load. Multiple stage heat sources also help reduce the temperature fluctuations associated with low thermal mass systems. The more stages, the better the match.

An example of a four-stage heat supply system applied to the same building load profile is shown in Figure 9–12. In this case, each stage represents one-quarter of the total heat output, and the total heat output again equals the design heating load of the building.

The practicality of using a multistage heat source system depends on the total heating load, as well as the type of heat source used. For example, small and medium-size houses often have space heating loads low enough that a single gas- or oil-fired boiler is all that is needed. However, two or more stages of heat input may be considered when other large loads such as snowmelting or high capacity domestic water heating are needed. Larger homes and commercial buildings may have space heating loads large enough to consider the use of multistage heat sources such as the multiple boiler systems discussed in Chapter 3, Hydronic Heat Sources. Systems using electric boilers are excellent candidates for multistage heat production. Each electrical element in the boiler can be controlled as a separate stage.

A recent survey found that hydronic heating designers generally consider multiple gas-fired or oil-fired boiler systems when the total system load exceeds 150,000 to 200,000 Btu/hr. In making the decision, the designer must weigh the extra cost associated with a multistage system against the potential fuel savings and more accurate system control.

Modulating (Variable Firing Rate) Heat Production

In an ideal space-heating system, the heat output from the heat source would always equal the rate of heat loss from the building. The heat output from the heat source could take on any value from zero to the full design load of the building. As the load changed due to changes in outdoor conditions or internal heat gain, the heat source would instantly readjust its output to match the load. Figure 9–13 depicts this ideal match between heat production and heating load. The area under the load profile is fully shaded. Unlike with single- or multistage heat production, there are neither "white spaces" under the load profile, nor any instances when heat production exceeds heating load.

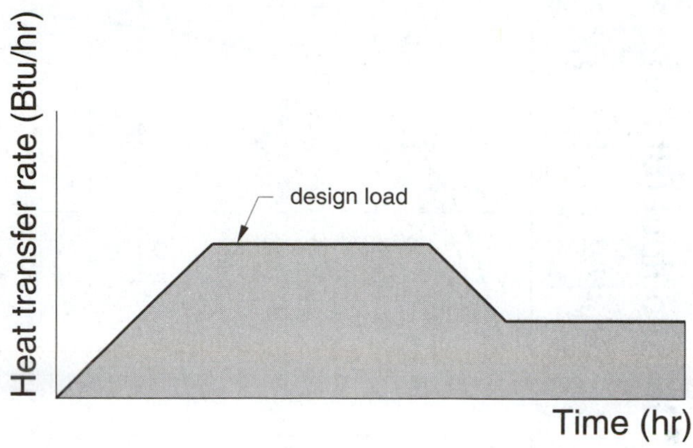

Figure 9–13 "Ideal" match between heat output of a fully modulating heat source and hypothetical load profile.

Although this ideal condition is seldom achieved, several types of modulating boilers are available that approximate it. Until recently, **modulating boilers** were only available in sizes suitable for large commercial or industrial facilities. However, several manufacturers now offer modulating boilers with capacities suitable for smaller residential and light commercial systems. Several more manufacturers are likely to release smaller modulating boilesr in the near future.

Most of the smaller modulating gas-fired boilers that are currently available can reduce their heat output down to 15 to 30 percent of full output. Most use proprietary internal controllers to adjust the rate at which air and gas are supplied to the burner. In some cases the rate of heat output is varied based on the difference in temperature the boiler controller detects between the supply and return side of the distribution system. In other cases, it is based on the ability of the boiler to maintain a set (or calculated) setpoint temperature on the supply side of the distribution system. If the sensed outlet temperature starts to drop, the firing rate is increased.

Because most current generation modulating boilers cannot maintain stable heat outputs lower than about 15 percent of rated capacity, they must cycle from their lowest firing rate to fully off under very low-load conditions. Upon restart, some modulating boilers go to a higher firing rate for a short time then quickly reduce their firing rate based on algorithms used by their internal controllers, as well as feedback from system temperature sensors.

9.4 CONTROLLING THE OUTPUT OF HEAT EMITTERS

Comfort is directly affected by the heat output of the system's heat emitters. The goal of the overall control system is to properly regulate the heat output of the heat emitters. This section describes the fundamental approaches, and lays the groundwork for several types of control techniques and hardware that are discussed in later sections.

There are two fundamental methods of controlling the heat output of heat emitters:

1. Vary the water temperature supplied to the heat emitter while maintaining a constant flow rate through the heat emitter.
2. Vary the flow rate through the heat emitter while maintaining a constant supply water temperature to the heat emitter.

Both approaches have been successfully used in many types of hydronic heating applications over several decades. It is important for system designers to understand the differences, as well as the strengths and weaknesses of each approach.

Variable Water Temperature Control

Chapter 8, Heat Emitters, discussed how the heat output of a heat emitter increases in proportion to the difference between supply water temperature and room air temperature. This can be represented mathematically as follows:

(Equation 9.1a)

$$Q_{output} = k(T_{supply} - T_{room\ air})$$

where:

Q_{output} = rate of heat output from heat emitter (Btu/hr)
k = a constant dependent on the heat emitter(s) used
T_{supply} = fluid temperature supplied to the heat emitter (°F)
$T_{room\ air}$ = air temperature of room where heat emitter is located (°F)

This equation can be rearranged as follows:

(Equation 9.1b)

$$Q_{output} = kT_{supply} - [kT_{room\ air}]$$

For a given heat emitter and room air temperature, the term in the square brackets is a constant. A representative graph of this equation where $k = 100$, and $T_{room\ air} = 70$ °F is shown in Figure 9–14.

Heat output from the heat emitter becomes positive (e.g., heat leaves the heat emitter) only when the supply water temperature exceeds the room air temperature. Since we are not using the heat emitter to cool the room, it is convenient to shift the horizontal axis to the left so it begins at the room air temperature.

For the heat emitter represented in Figure 9–14, an increase of 1 °F in supply water temperature produces an

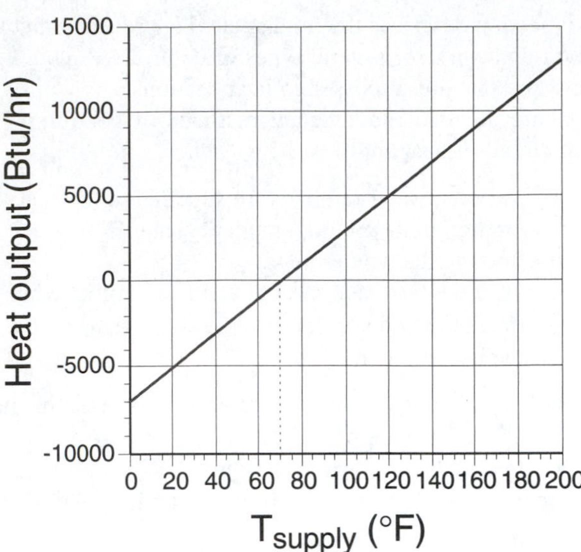

Figure 9–14 Heat output from heat emitter is approximately proportional to difference between supply water temperature and room air temperature. A room temperature of 70 °F is assumed in this graph.

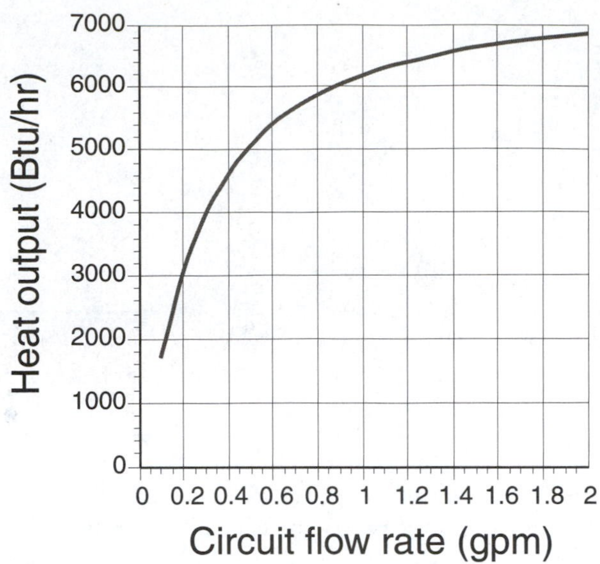

Figure 9–15 Heat output as a function of flow rate for a floor heating circuit supplied with 105 °F water. Circuit is a 300 foot length of 1/2-inch PEX tubing embedded in a 4-inch bare concrete slab with a room temperature of 70 °F.

increase of 100 Btu/hr in heat output, regardless of the starting temperature. Likewise, a decrease of 1 °F in supply water temperature reduces heat output by 100 Btu/hr. The slope of the heat output versus supply water temperature will change depending on the type of heat emitter used, but the linear relationship will remain.

This type of linear relationship between the controlled variable (heat output) and the manipulated variable (supply water temperature) is preferable from a control standpoint. This relationship is relatively easy to mimic when designing both electromechanical and electronic controls. This approach also eliminates some complications that arise when the system flow has to vary to regulate heat output.

Variable Flow Rate Control

The graph in Figure 9–15 shows how the output of a typical floor heating circuit varies as a function of flow rate assuming the supply water temperature is maintained at 105 °F.

Notice how heat output increases rapidly at low flow rates, and much slower at higher flow rates. This characteristic is typical of all hydronic heat emitters. As a rule of thumb, the heat output of a heat emitter reaches approximately 50 percent of its design output when the flow rate reaches approximately 10 percent of design flow. The relationship between the controlled variable, (heat output), and the manipulated variable (flow rate) is very nonlinear. This characteristic makes accurate control more difficult, especially under low-load conditions. To provide accurate control under low-load conditions, the controlled device must be able to maintain stable flow at very low flow rates. Such a device is said to have good **rangeability**.

One way to compensate for rapid increases in heat output at low flow rates is to use a control valve that allows very small changes in flow rate as the valve first begins to open, and larger changes in flow as the valve stem nears its fully open position. Such a valve is said to have an **equal percentage characteristic**.

In valves where the stem moves up and down, the equal percentage characteristic is created by the shape of the valve plug. In some valves, the plug has a specially tapered "logarithmic" shape that allows the gap between the plug and seat to form very slowly as the valve begins to open, then faster as the stem lifts higher. In other valves, the equal percentage characteristic is created by machining a tapered slot into a cylindrical valve plug. This technique is also used with rotary valves. In the case of a ball valve, an insert with a tapered slot is fixed into the orifice of the valve. An example of a ball valve with a special equal percentage insert is shown in Figure 9–16.

The flow rate through an equal percentage valve as a function of its stem position is shown in Figure 9–17. This curve assumes that a constant differential pressure is maintained across the valve as it opens.

When a valve with an equal percentage characteristic is used to control flow to a hydronic heat emitter, the resulting relationship between stem position and heat output is close to proportional, as shown in Figure 9–18.

The very slow increase in flow rate when the valve first begins to open compensates for the rapid rise in heat output from the heat emitter at low flow rates. Likewise, the rapid increase in flow at higher flow rates compensates for the relatively slow rise in heat output at high flow rates. The linear

(a)

(b)

Figure 9–16 (a) A ball valve fitted with flow insert to create an equal percentage flow characteristic. (b) Close-up of equal percentage flow insert with tapered slot orifice. *Courtesy of Belimo (USA), Inc.*

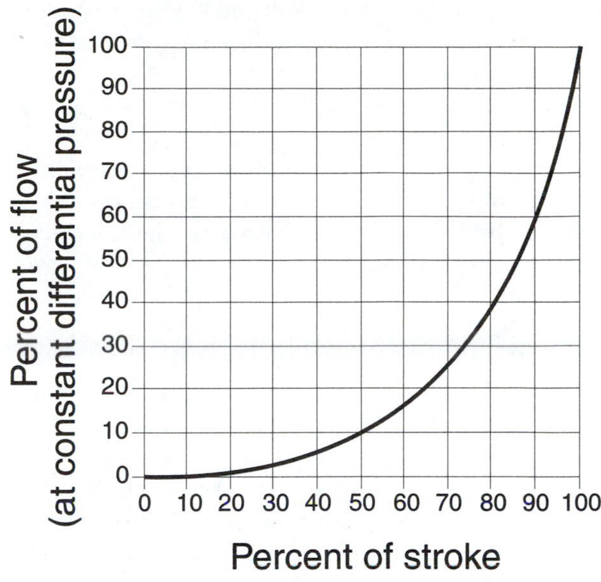

Figure 9–17 Percentage of flow through valve versus percentage of stroke for a valve with an equal percentage characteristic operated at a constant differential pressure.

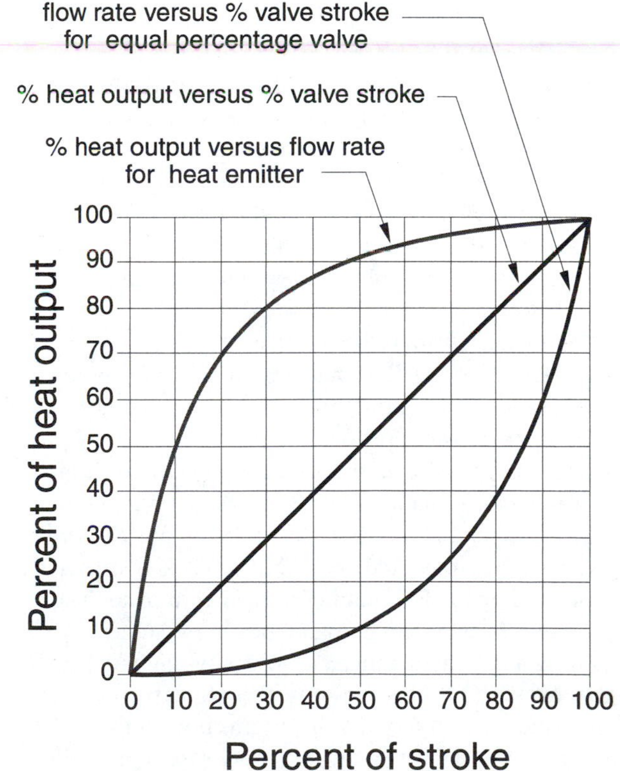

Figure 9–18 Heat output of heat emitter is approximately proportional to percentage of valve stroke when flow through heat emitter is controlled by an equal percentage valve operated at a constant differential pressure. This is a desirable operating characteristic. *Note:* Supply water temperature to heat emitter is assumed constant.

net effect is desirable from a control standpoint. It is especially important under low-load conditions where a **quick opening valve** would make accurate and stable heat output control extremely difficult. In general, an equal percentage valve should be used in any situation where heat output is controlled by varying the flow rate through a heat emitter.

Outdoor Reset Control

An ideal heating distribution system would continually adjust its rate of heat delivery to match the heat loss of a building. This would allow the inside air temperature to remain perfectly stable, regardless of the load on the system.

One means of accomplishing this is to make small but continuous changes to the water temperature supplied to the heat emitters based on the outdoor temperature. As the outdoor temperature drops, and the heating load of the building increases, the water temperature supplied to the heat emitters is increased to create higher heat output. Likewise, as outdoor temperature increases, and the heating load decreases, the water temperature supplied to the heat emitters also decreases as does their heat output. This concept is called **outdoor reset control**.

When properly executed, outdoor reset control is like cruise control for the heating system. It allows just the right amount of heat to be released from the heat emitters to match the current heating load.

There are several benefits associated with outdoor reset control including:

- Stable indoor temperature: When outdoor reset control is properly applied, the water temperature supplied to the heat emitters is just hot enough to satisfy the prevailing load conditions. Excess heat is not being forced into the thermal mass of either the heat emitters or the building materials by "overheated" water. Rooms do not undergo noticeable changes in temperature as is the case when heat input is cycled on and off.
- Near-continuous circulation: Because the water supplied to the heat emitters is just hot enough to meet the prevailing load, the distribution circulator remains on most of the time. This allows heat to be redistributed from areas of low loss to areas of high loss. For example, consider a heated floor area just inside the large overhead doors of a garage. Such areas are susceptible to freezing should the circulator remain off for several hours in cold weather. However, when a properly adjusted outdoor reset controller is used, the circulator remains on most of the time. This allows heat stored in the interior portion of the floor slab to be carried to the locations of higher heat loss.
  In some radiant floor heating systems, continuous circulation also moves heat stored in floor areas covered with higher R-value flooring to areas with lower resistance flooring. This reduces the tendency of the latter to feel cool during what might otherwise be lengthy periods of no flow.
- Reduced expansion noise: The combination of near-continuous circulation and very gradual changes in water temperature minimizes expansion noises from the distribution piping and heat emitters. This is especially beneficial when PEX tubing is used with metal heat transfer plates in radiant panel heating systems.
  During a typical heating season, the piping and heat emitters will likely experience thermal expansion movement similar to that in systems not using outdoor reset control. However, when outdoor reset control is used, the expansion movement takes place over days, even weeks, compared to what might only be seconds in systems that simply turn the flow of hot water on and off. Piping expansion noise is much more noticeable in systems where rapid changes in water temperature occur.
- Reduced thermal shock: The use of outdoor reset control reduces the possibility of thermal shock to both the heat source and the distribution system. Hot boilers are less likely to receive "slugs" of cold water from zones that have not been flowing for several hours. Wood flooring installed over radiant floor panels is less prone to cupping or cracking when undergoing gradual temperature changes.
- Indoor temperature limiting: If water is supplied to the heat emitters at design temperature regardless of the load, occupants can set the thermostat to a high setting, and simply open a window or door to control overheating. Although this sounds like an odd approach to comfort control, it is often done in rental properties where tenants do not pay for their heat. When the water temperature is controlled by outdoor reset control, the water is just hot enough to meet the prevailing load with the windows and doors closed. The use of a "double hung thermostat" to control comfort is no longer viable. This benefit alone has created a strong market for outdoor reset control in rental properties.
- Reduced energy consumption: Outdoor reset control has demonstrated its ability to reduce fuel consumption in both residential and commercial buildings. The exact savings will vary from one project to another. Conservative estimates of 10 to 15 percent are often cited. Figure 9–19 shows the results of field tests in multifamily buildings before and after outdoor reset control was implemented.

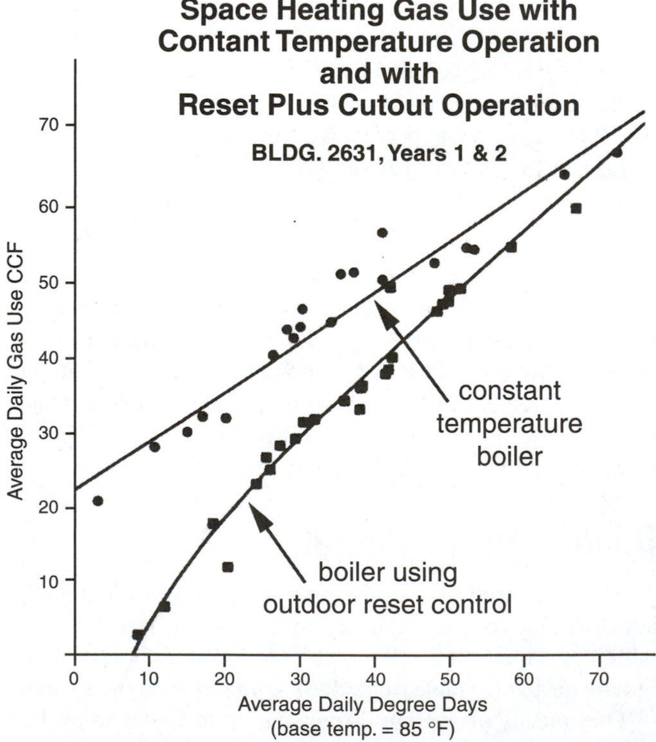

Figure 9–19 Fuel usage for hydronic heating systems with and without boiler reset control. Source: *Home Energy Magazine.*

Basic Theory of Outdoor Reset Control

Outdoor reset control is based on two fundamental heat transfer principles, both of which were discussed in previous chapters.

Principle #1: The rate of heat loss from a building is proportional to the difference between inside and outside air temperatures.

Principle #2: The rate of heat output from a heat emitter is approximately proportional to the difference between the supply water temperature and the inside air temperature.

Example 9.1: A building loses heat at a rate of 80,000 Btu/hr when the inside temperature is 70 °F, and the outdoor design temperature is –5 °F. Express the building's heat loss according to principle #1.

Solution: When one quantity is proportional to another, they can be equated by introducing a constant of proportionality. In this case, that constant can be found by dividing the building's rate of heat loss by the difference between inside and outside air temperature.

(Equation 9.2)

$$Q_{loss} = c_1(T_{inside} - T_{outside})$$

$$c_1 = \frac{80{,}000 \text{ Btu/hr}}{(70°\text{F} - [-5°\text{F}])} = 1067 \text{ Btu/hr/°F}$$

Discussion: In this example, the constant of proportionality (c_1) implies that the building loses 1067 Btu/hr *for each degree Fahrenheit* the outside temperature is below the indoor temperature. The design heat loss of a building divided by the difference between inside and outside air temperature is called the building's **heat loss coefficient**. Each building has a unique heat loss coefficient based on its size and construction.

Example 9.2: A hydronic heating distribution system can release heat at a rate of 80,000 Btu/hr when the supply temperature is 150 °F, and the indoor air temperature is 70 °F. Express the distribution system's heat output according to principle #2.

Solution: This situation is almost identical to that of Example 9.2. Although different quantities are involved, the mathematics are the same.

(Equation 9.3)

$$Q_{output} = c_2(T_{supply} - T_{inside})$$

In this case, the constant of proportionality (c_2) is determined by dividing the heat output of the distribution system by the difference between supply and inside air temperature.

$$c_2 = \frac{80{,}000 \text{ Btu/hr}}{(150°\text{F} - 70°\text{F})} = 1000 \text{ Btu/hr/°F}$$

Discussion: The value of c_2 indicates that this distribution system can release 1000 Btu/hr of heat *for each degree Fahrenheit* the supply water temperature is above the inside air temperature.

When formulas 9.2 and 9.3 are combined, they yield the following relationship:

(Equation 9.4)

$$T_{supply} = T_{indoor} + \left(\frac{c_1}{c_2}\right)(T_{indoor} - T_{outdoor})$$

For the data given in Examples 9.1 and 9.2 this becomes:

$$T_{supply} = 70 + \left(\frac{1067}{1000}\right)(70 - T_{outdoor})$$

Which can be further simplified to:

(Equation 9.5)

$$T_{supply} = 70 + 1.07(70 - T_{outdoor})$$

A graph of Equation 9.5 is shown in Figure 9–20.

The line on this graph is called a **reset line**. It shows what the supply water temperature to the heat distribution system must be to maintain the desired 70 °F indoor temperature for any outside temperature between 70 and –5 °F.

The lower left end of the reset line represents a situation where no heat is needed by the building. For example, if water at 70 °F is supplied to heat emitters in a space that is already at 70 °F air temperature, there will be no heat transfer between the heat emitters and the room air.

The upper right end of the reset line represents design load conditions. This is where the heat output from the distribution system must equal the design heat loss of the building.

Reset Ratio

The slope of the reset line is called the **reset ratio**, and can be calculated as the change in supply water temperature divided by the change in outdoor temperature between any two points on the reset line. The end points of the reset line are typically used to make this calculation.

(Equation 9.6)

$$\text{Reset ratio} = \frac{\Delta T_{supply\ water}}{\Delta T_{outdoor}}$$

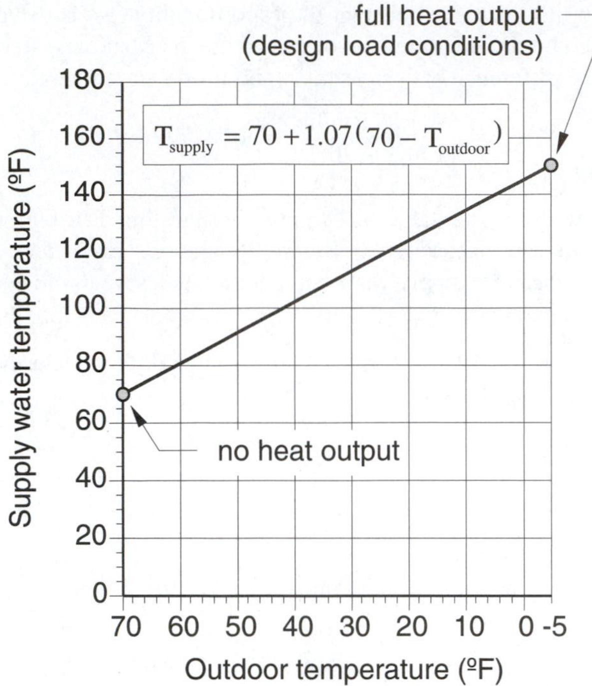

Figure 9–20 Supply water temperture required to allow heat output of a given heat emitter system to match a specified building heating load at various outdoor temperatures. The sloping line is called reset line.

Example 9.3: The building described in Example 9.1 has a design heat loss of 80,000 Btu/hr when the indoor temperature is 70 °F and the outdoor temperature is –5 °F. The heat distribution system described in Example 9.2 can release 80,000 Btu/hr when the supply water temperature is 150 °F and the inside air temperature is 70 °F. Determine the necessary reset ratio for this system.

Solution: The change in water temperature between the no heat input and design load conditions is 150 – 70 = 80 °F. The corresponding change in outdoor temperature is 70 – (–5) = 75 °F. Thus the necessary reset ratio when this heat distribution system is used in this building is:

$$RR = \frac{(150 - 70)}{(70 - (-5))} = \frac{80}{75} = 1.067 \approx 1.07$$

Discussion: Notice that the calculated reset ratio is exactly the same as the mathematical slope of the line shown in Figure 9–20. This allows Equation 9.5 to be written in a slightly more general form as:

(Equation 9.7)

$$T_{supply} = T_{indoor} + RR(T_{indoor} - T_{outdoor})$$

where:

T_{supply} = required supply water temperature to the distribution system (°F)
T_{indoor} = desired indoor air temperature (°F)
$T_{outdoor}$ = current outdoor air temperature (°F)
RR = reset ratio calculated using Equation 9.6

Here are several other facts regarding the reset ratio:

- The greater the value of the reset ratio, the steeper the reset line.
- The reset ratio can be interpreted as the necessary increase in supply water temperature per degree drop in outside temperature.
- The reset ratio is a unitless quantity and does not change when calculated using different temperature units (provided the same temperature units are used in the numerator and denominator of Equation 9.6).
- The value of the reset ratio depends on *both* the heat loss characteristics of the building, and the heat output capability of the heating system in that building.
- Once the reset ratio for a particular building/distribution system has been calculated, it must be entered into the reset controller. This will be discussed in later sections of this chapter.

Figure 9–21 shows some representative reset lines with different reset ratios. In general, distribution systems designed around high water supply temperatures require higher reset ratios. Systems designed around lower water temperatures require low reset ratios.

Compensating for Nonproportional Heat Losses

Theoretically, a heating system in which the supply water temperature is regulated according to outside reset control theory could maintain a set interior temperature while only measuring outside temperature. This can be verified by

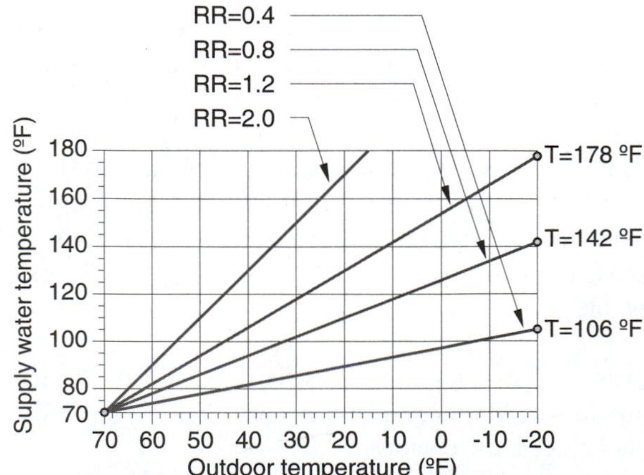

Figure 9–21 Reset lines with different reset ratios (RR). Notice that higher reset ratios result in higher supply water temperatures at a given outdoor temperature.

examining Equation 9.7. Once the desired inside temperature and reset ratio are set, the only remaining variable is outside temperature. This relationship holds true as long as the building's heat loss is strictly proportional to the difference between inside and outside temperature (principle #1 discussed earlier).

Unfortunately, there are several situations that partially invalidate principle #1. The effect of wind speed on building heat loss is one. Heat loss due to air infiltration increases at higher wind speeds. An outdoor reset controller that only measures outdoor air temperature can not sense this condition, and therefore cannot compensate for it.

Another effect that partially invalidates principle #1 is the presence of internal heat gains from sunlight, people, and appliances in the building. For example, solar heat gains through large south-facing windows may at times provide all the heat the building needs, even when it is cold outside. If the outdoor reset control only "looks at" outside temperature, it has no way of knowing that such gains have already provided the necessary heat input to the building. The controller simply goes about providing water to the system at the supply temperature it calculates based on the current outdoor temperature. The result can be significant overheating.

Internal heat gains allow a building to maintain comfortable indoor temperatures when the outdoor temperature equals or exceeds a value known as the **balance point temperature**. *When the outdoor temperature is at the balance temperature, the rate of internal heat gain exactly balances the rate of building heat loss.* The indoor temperature remains at the desired comfort level without any heat input from the heating system.

The balance point temperature of a building can be calculated using Equation 9.8:

(Equation 9.8)

$$T_{\text{balance}} = T_{\text{indoor}} - \left(\frac{Q_{\text{gain}}}{UA_{\text{building}}}\right)$$

where:

T_{balance} = balance point temperature (degree F)
T_{inside} = desired inside air comfort temperature (°F)
Q_{gain} = the rate of internal heat gain (Btu/hr)
UA_{building} = building heat loss coefficient (Btu/hr/°F.)

Example 9.4: Assume a house has a design heat loss of 75,000 Btu/hr when the outside temperature is 0 °F., and the desired inside temperature is 70 °F. The building heat loss coefficient is:

$$\text{building heat loss coefficient} = \frac{75{,}000\ \text{Btu/hr}}{(70°\text{F}-0°\text{F})} = 1071\frac{\text{Btu}}{\text{hr}\cdot°\text{F}}$$

Also assume the total heat output of lights, people, and appliances in the home is 11,000 Btu/hr at some given time. What is the balance point temperature of the house under these conditions?

Solution: Substitute these numbers in Equation 9.8 to yield:

$$T_{\text{balance}} = 70 - \left(\frac{11{,}000}{1071}\right) = 59.7\ °\text{F}$$

Discussion: This building needs no heat input from its heating system until the outside temperature drops below 59.7 °F.

If the rate of internal heat gain drops to zero, the balance point temperature equals the desired inside temperature. The higher the rate of internal heat gain, and the smaller the building heat loss coefficient, the lower the building's balance point temperature becomes.

To compensate for internal heat gains the reset line must be moved downward and to the right on the graph without changing its slope. This displacement of the reset line is called **parallel shifting**. The starting point of the reset line moves along the starting point line as shown in Figure 9–22.

The starting point of the reset line can be adjusted based on the balance point temperature of the building using Equation 9.9

(Equation 9.9)

$$T_{\text{start}} = \frac{T_{\text{inside}} + RR\,(T_{\text{balance}})}{1 + RR}$$

where:

T_{start} = starting point temperature (along starting point line) (°F)
T_{inside} = desired inside air comfort temperature (°F)
RR = reset ratio (determined from Equation 9.6)
T_{balance} = balance point temperature (°F)

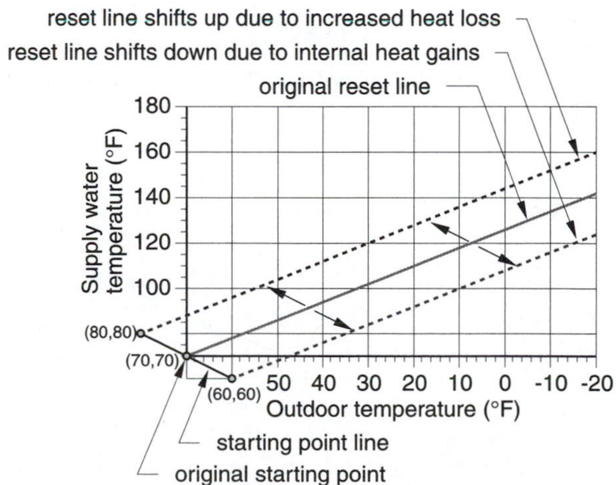

Figure 9–22 Parallel shifting of reset line, down and right to compensate for internal heat gain, up and left to compensate for increase in load due to factors other than drop in outdoor temperture. Note: Starting point of reset line moves along the starting point line to parallel shift the reset line.

Example 9.5: Assume the house used in Example 9.4 has a balance point temperature of 59.7 °F, a desired inside air temperature of 70 °F, and a reset ratio of 0.57. Determine the adjusted starting point temperature of the reset line, accounting for the internal heat gains.

Solution: Substitute the numbers into Equation 9.9:

$$T_{\text{start}} = \frac{70 + 0.57 \times 59.7}{1 + 0.57} = 66.3 \text{ F}$$

Discussion: The reset line with the adjusted starting point temperature of 66.3 °F is shown in Figure 9–23. Notice that the reset ratio has not changed. The reset line has simply been parallel shifted to a new starting point to compensate for the assumed internal heat gains.

The reset line can also be parallel shifted upward and to the left when the heating load becomes higher than expected. For example, suppose a strong wind develops that increases the infiltration heat loss of the building (even though the outside air temperature does not change). If the increase in heating load can be determined, it could simply be treated as a negative internal heat gain in Equation 9.8. Once a revised balance point temperature was calculated, it could be used in Equation 9.9 to calculate a revised starting point for the reset line.

Some outdoor reset controllers can be equipped with indoor temperature sensors. When so configured, they automatically parallel shift the reset line as required to compensate for both internal heat gains and higher than normal heat loss.

If the control does not have an indoor temperature sensor the starting point of the reset line, once calculated, can be manually set using dials or menu-driven inputs on the reset controller. However, because internal heat gains often change, this approach is not as effective as using indoor temperature feedback. Such systems should use an indoor thermostat wired such that it can override the output of the reset controller and prevent further heat input when internal gains push interior temperatures above the desired indoor setpoint.

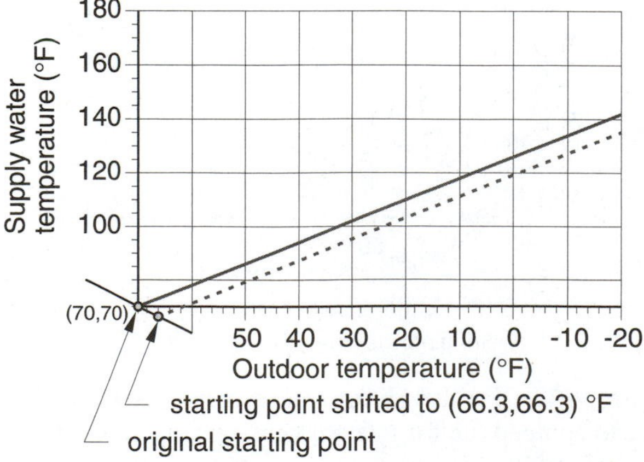

Figure 9–23 A rest line that has been parallel shifted to a starting point of 66.3 °F to compensate for a consistent internal heat gain.

There are two fundamental ways of implementing outdoor reset control in hydronic systems.

1. Boiler Reset Control
2. Mixing Reset Control

Each will be discussed in separate sections:

Boiler Reset Control

Boiler reset control regulates the temperature supplied by the heat source. The colder it gets outside, the hotter the water supplied by the heat source. Although the heat source is often a boiler, the concept is not limited to boilers. It can be used with any heat source capable of delivering water over a range of temperatures.

Boiler reset is somewhat limited when used with a conventional gas- or oil-fired boiler. If the water returning to the boiler is allowed to drop too low, flue gases will condense within the boiler or its venting system as discussed in Chapter 3, Hydronic Heat Sources. To prevent this, most boiler reset controllers have an adjustable **minimum supply temperature**. With this feature, the controller does not allow the boiler to operate below a specified minimum temperature regardless of outdoor temperature. Such an operating strategy is called **partial reset**, and is illustrated in Figure 9–24.

Notice how the lower portion of the sloping line has been "truncated" by the minimum supply temperature setting.

Although partial reset protects a conventional boiler from operating with sustained flue gas condensation, it also prevents the distribution system from operating at theoretically ideal lower water temperature during low-load conditions. Because the supply water temperature to the distribution system is higher than necessary during these times, some means of preventing building overheating is required.

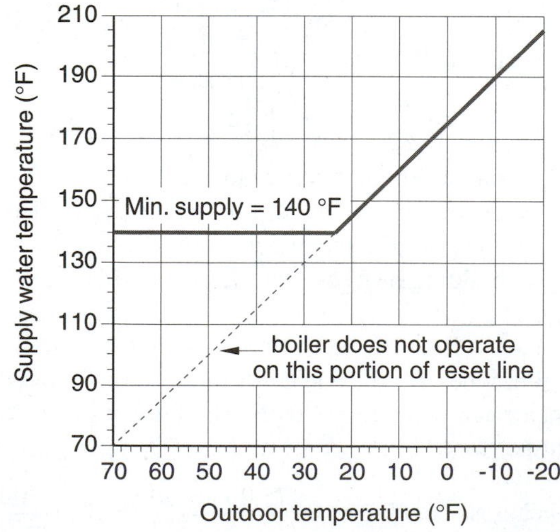

Figure 9–24 Truncation of reset line at a minimum supply temperature of 140 °F. This is known as partial reset control, and is typically done to prevent flue gas condensation in a conventional boiler.

In most systems this is done by turning off a circulator or closing a zone valve whenever the indoor thermostat senses the room is at or above setpoint temperature.

Most boiler reset controllers also have a maximum supply temperature setting. This is used to prevent the heat source from supplying excessively hot water to the distribution system regardless of the outdoor temperature.

Boiler reset controllers have an electrical contact that simply opens or closes to operate the heat source. Like all other on/off temperature control devices, boiler reset controllers use a differential as well as a calculated target temperature to determine when the contacts open and close. On some boiler reset controllers, this differential is centered on the target temperature. The contact operating the boiler opens at one-half the differential above the target supply temperature, and closes at one-half the differential below this temperature. In some reset controllers, the differential is fixed; in others it widens as the outdoor temperature increases. The latter scenario, although allowing wider deviations from the calculated target temperature, also minimizes boiler short-cycling during low load conditions. These concepts are illustrated in Figure 9–25

The implementation of a boiler reset control is shown in Figure 9–26. Notice that the controller has sensors that measure outdoor and supply water temperature.

In most systems, a boiler reset controller does not replace the boiler high limit controller. The latter remains in place as a safety device that can stop the burner from firing should the boiler reset controller ever malfunction. However, during normal operation the boiler reset controller "preempts" operation of the high limit controller, shutting off the burner at lower water temperatures. One can think of the boiler reset controller as the "operating controller" while the standard high limit controller acts as a "safety controller." It is important to set the boiler high limit controller high enough that it will not interfere with normal operation of the boiler reset controller.

Boiler reset control is largely responsible for the energy savings mentioned as one of the benefits of outdoor reset control. These saving are mostly the result of lower heat losses from the boiler.

Mixing Reset Control

Another method of implementing outdoor reset control is by using a mixing device between the heat source and the distribution system. Not surprisingly, this approach is called **mixing reset control**. Several mixing devices are suitable for this purpose and are discussed in detail in Section 9.7. Figure 9–27 shows the general concept of a mixing device that acts as a "bridge" between the boiler loop and the distribution system.

A mixing reset controller outputs one of several types of signals to the mixing device, and continually senses the mixed water temperature supplied to the distribution system.

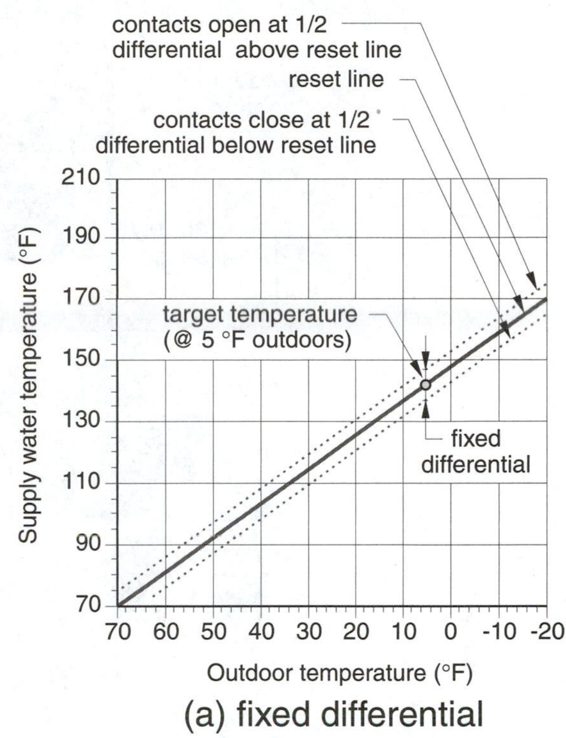

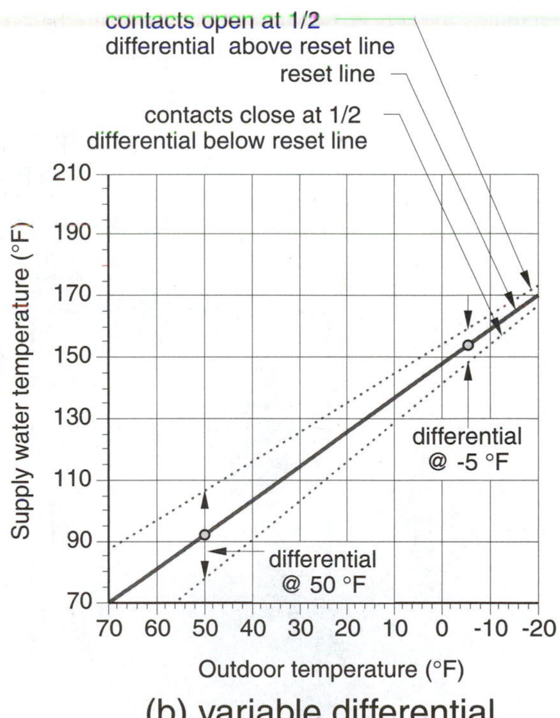

Figure 9–25 Boiler reset controllers are on/off controllers that operate on a differential centered on the reset line (target temperature). (a) fixed differential, (b) variable differential to reduce short cycling of the heat source at low loads.

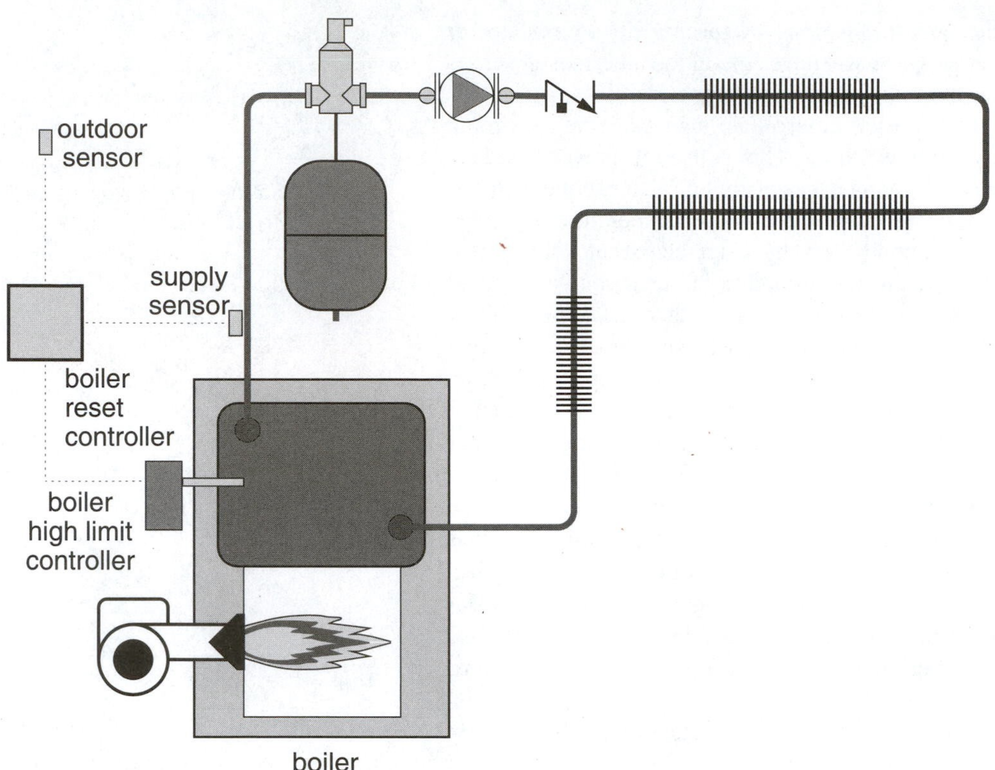

Figure 9–26 Typical installation of a boiler reset controller. The boiler reset controller measures outdoor temperature, then calculates the proper boiler target temperature based on its settings. It operates the burner as necessary to keep the boiler temperature close to the target temperature.

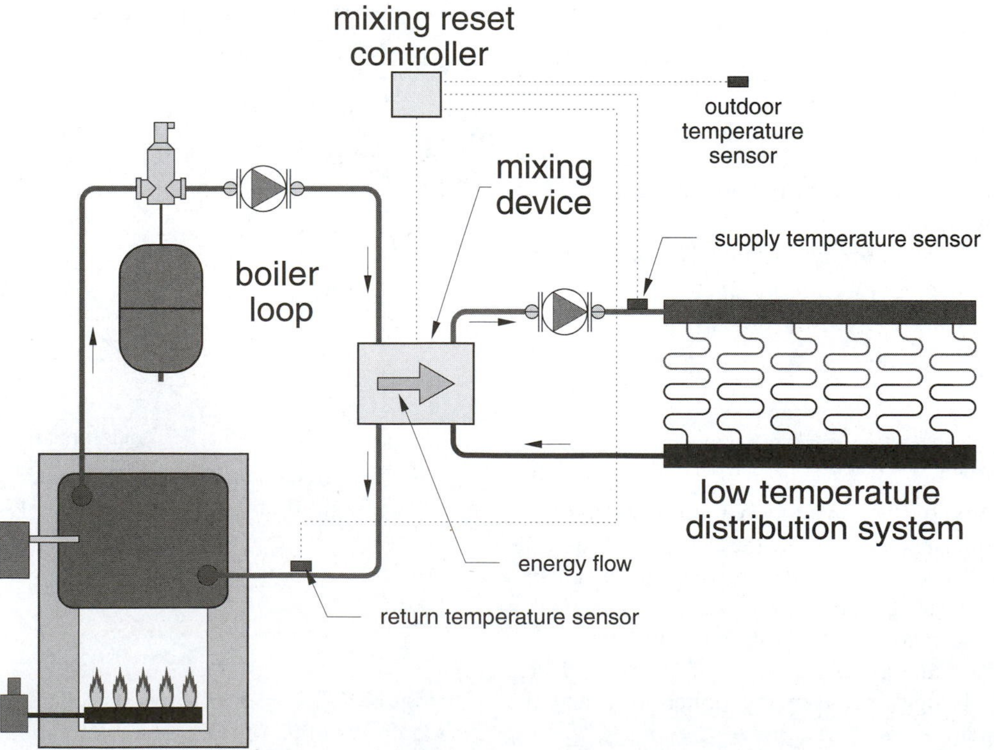

Figure 9–27 The mixing device forms a bridge between the boiler loop and distribution system. All heat reaching the distribution system must pass across this bridge.

Feedback from this sensor allows the controller to zero in on a target supply water temperature that itself is calculated based on outdoor temperature.

Many mixing reset controllers can also protect conventional boilers from low return temperature conditions. To provide this protection, mixing reset controllers must be equipped with an additional sensor to sense the boiler temperature (usually the boiler inlet temperature). When the water temperature entering the boiler approaches the minimum temperature setting, the controller automatically reduces the rate of hot water flow into the mixing device. This prevents the distribution systems from dissipating heat faster than the boiler can produce it.

Mixing reset control allows the water temperature supplied to the distribution system to be reduced all the way down to the room air temperature. This is called **full reset** because the controller can operate the mixing device to produce a water temperature anywhere along the reset line. Full reset is possible when a conventional boiler is used as the heat source *if* the mixing reset controller provides the previously discussed boiler temperature protection.

Combining Boiler Reset and Mixing Reset

It is possible to use both boiler reset and mixing reset in the same system. The boiler supply temperature decreases as the outside temperature increases, as does the water temperature supplied to the distribution system by the mixing device. Such systems increase seasonal boiler efficiency as well as provide the ideal supply water temperature to the distribution system for optimal comfort. Some manufacturers offer controllers capable of handling both functions simultaneously.

Figure 9–28 shows a piping system supplying a circuit of panel radiators, as well as a low temperature radiant floor subsystem. The boiler supply temperature is reset based on the design load requirements of the panel radiators. The mixing device, operated by a mixing reset controller, provides the lower water temperature requirements of the floor heating system. This approach provides partial reset control to the panel radiators (because of the minimum return temperature requirement of the conventional boiler), and full reset of the radiant floor circuits.

Proportional Reset

Many hydronic systems require different water temperatures to be supplied simultaneously to different portions of the distribution system. For example, a residential system might consist of a radiant floor slab on the first floor with panel radiators on the second floor. At design load conditions, the floor circuits may require a supply temperature of 105 °F while the panel radiators require a supply temperature of 150 °F.

Although it is possible to supply these water temperatures all with full outdoor reset, using two independently controlled mixing devices, this approach is not always

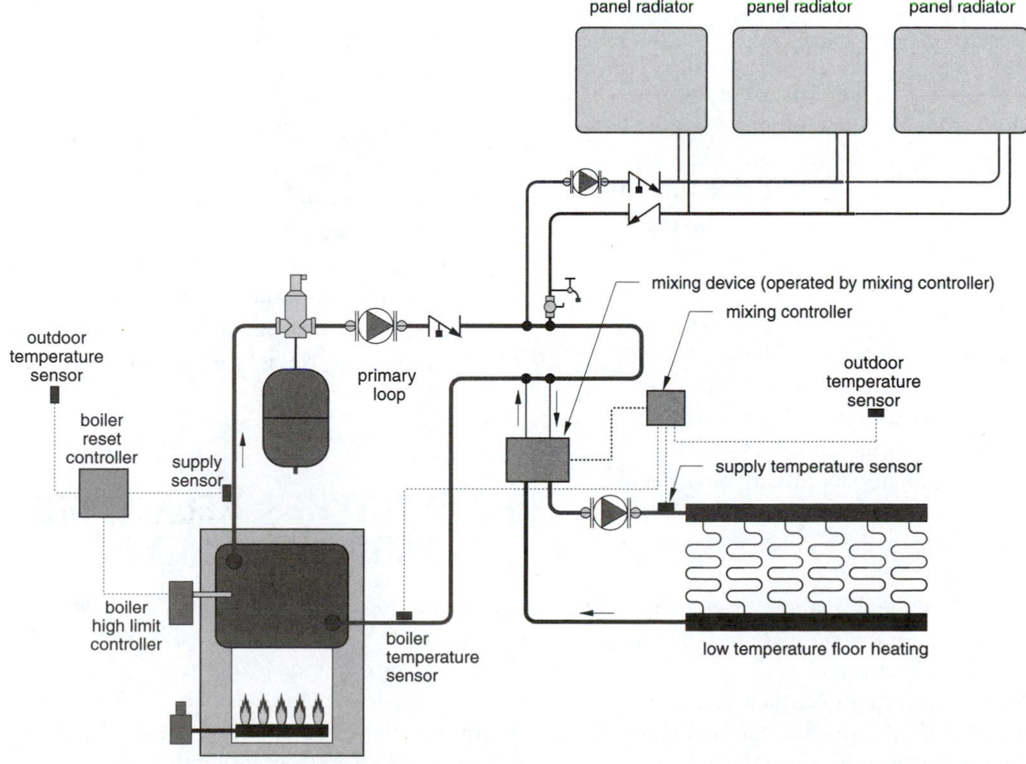

Figure 9–28 Combined use of boiler reset control and mixing reset control.

necessary. When the higher of the two supply temperatures is reset, and the lower temperature is created using a manually set mixing device supplied by the higher water temperature, the lower supply temperature will also be reset. This concept is called **proportional reset**, and is illustrated in Figure 9–29a. The associated piping is shown in Figure 9-29b.

Notice that the lower reset line passes through a supply temperature of 105 °F at the assumed outdoor design temperature of 0 °F. The upper reset line indicates a supply temperature of 150 °F at the same outdoor temperature.

If the higher of the two water temperatures is only partially reset due to minimum boiler temperature requirements, the lower reset line will also be partially reset. Both lines will transition from sloping to flat at the same outdoor temperature. At outside temperatures above this value, flow through the high temperature and low temperature subsystems must be cycled on and off to prevent overheating.

The concept of proportional reset can be extended to produce more than two fully reset or partially reset supply temperatures. In every case, the highest water temperature would be controlled by a reset control, and the lower water temperatures created by simple manually set mixing valves.

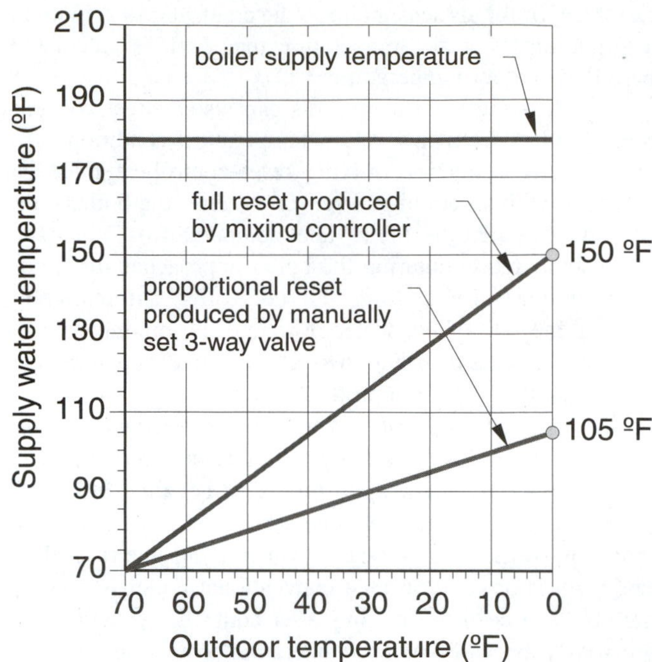

Figure 9–29a Boiler operates as a non-reset (constant temperature) device. Mixing reset control provide steeper reset line. Proportional reset control provides lower reset line.

Limitations of Proportional Reset

Proportional reset is a somewhat underutilized strategy at the present time. When properly applied, it can reduce the cost of the control/mixing hardware necessary to deliver the benefits of reset control in multiple water temperature systems. However, there are limitations that must be respected when applying it. They are as follows:

1. Since any manually set mixing valve only provides fixed flow proportions of the incoming water streams, any variation in the temperature or flow rate of either stream causes a change in the mixed water temperature. Unlike an intelligent mixing device, a manually set mixing valve cannot compensate for such changing inlet conditions. For example, consider a system in which zone valves are installed for individual circuit control. Each time a zone valve opens or closes, the flow and head loss of the distribution system changes. This causes a corresponding change in the flow proportions at the manually set mixing device and a change in the mixed outlet temperature. Under some conditions, there could be wide variations in the mixed temperature.
2. A manually set mixing device cannot protect a conventional boiler from low temperature operation. When such a device provides a low temperature/high thermal mass radiant panel, it should only be used downstream of an intelligent mixing device that senses boiler inlet temperature, and can limit heat transfer to the distribution system when necessary.
3. Proportional reset with a manually set mixing valve is generally not a good idea for spaces that have frequent changes in thermostat settings, especially if that space is heated by a high mass radiant panel. Such an application would experience long recovery time following a setback because the mixed temperature supplied to the circuits would be depressed by the cooler than normal temperature returning from the radiant panel circuits. An intelligent mixing device could compensate for this. A manually set mixing device cannot.

9.5 SWITCHES, RELAYS, AND LADDER DIAGRAMS

This section describes many of the basic electrical switch devices used to build an overall control system. Virtually every heating system, hydronic or otherwise, uses one or more of these switching devices within its control system. A working knowledge of these devices is essential in designing, installing, or troubleshooting hydronic systems. The section also presents ladder diagrams as a fundamental tool for documenting control system wiring.

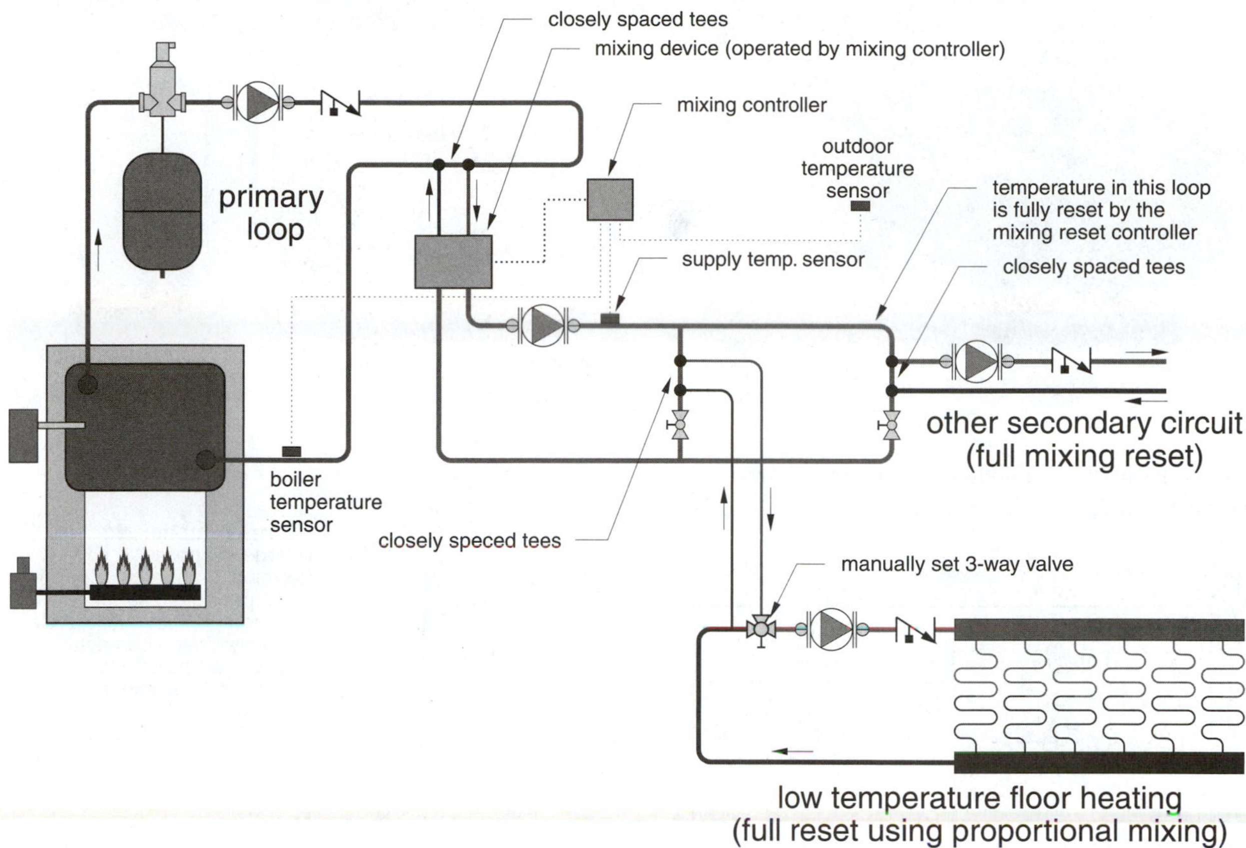

Figure 9–29b Piping used for proportional reset of water temperature to low temperature floor heating circuits using a manually set mixing valve located "downstream" of a mixing reset device.

Switches

The specific **operating modes** of most heating and cooling systems are determined by a specific arrangement of electrical switch **contacts**. Such contacts must be either open or closed at all times. An open contact prevents an electrical signal (e.g., a voltage) from passing a given point in the circuit. A closed contact allows the signal to pass through. These contacts may be part of a manually operated switch or an electrically operated switch called a **relay** or **contactor**. When the settings of the various contacts allow a complete circuit to be formed, a current will flow, and some predetermined control action will take place.

Poles and Throws

Switches and relays are classified according to their **poles** and **throws**. *The number of poles is the number of independent and simultaneous electrical paths through the switch.* Most of the switches used in heating control systems have one, two, or three poles. They are often designated as single pole (SP), double pole (DP), or triple pole (3P).

The number of throws is the number of position settings where a current can pass through the switch. Most switches and relays used in heating systems have either one or two throws, and are called single throw (ST), or double throw (DT) switches.

Figure 9–30 illustrates a typical double pole/double throw (DPDT) toggle switch. Notice the printing on the side of the switch indicates that it is rated for a maximum of 10 amps of current at 250 volts (AC), and 15 amps of current at 125 VAC. *When selecting a switch for a given application, be sure its current and voltage ratings equal or exceed the current and voltage at which it will operate.*

Figure 9–31 shows the schematic symbols used to designate switches having different numbers of poles and throws. On the double and triple pole switches, a dashed line indicates a nonconducting mechanical coupling between the metal blades of the switch. This coupling ensures that all contacts open or close at the same instant.

Figure 9–30 Example of a double pole/double throw (DPDT) toggle switch.

	Single Throw	Double Throw
Single Pole	SPST	SPDT
Double Pole	DPST	DPDT
Triple Pole	3PST	3PDT

Figure 9–31 Schematic symbols used to represent switches having different numbers of poles and throws.

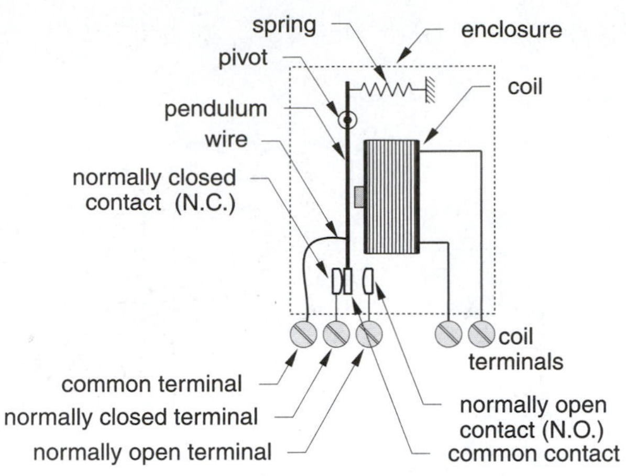

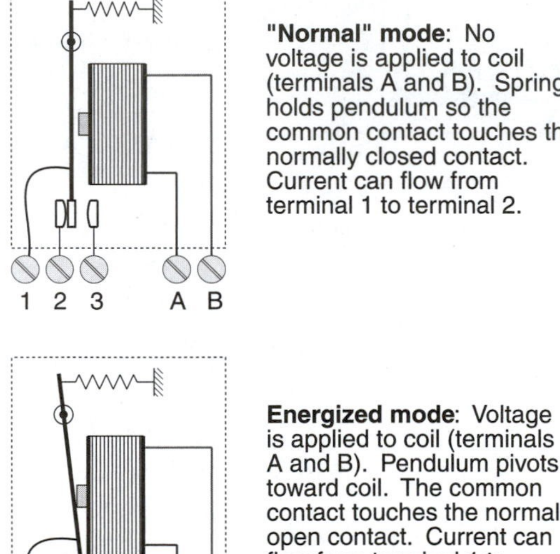

1 2 3 A B

Energized mode: Voltage is applied to coil (terminals A and B). Pendulum pivots toward coil. The common contact touches the normally open contact. Current can flow from terminal 1 to terminal 3.

Figure 9–32 Internal components and operating sequence of a typical relay.

Relays

Relays are simply electrically operated switches. They can be operated from a remote location by a low power electrical signal. They consist of two basic subassemblies: the **coil** and the contacts. Other parts include a spring, pendulum, and terminals. These components and basic operation of a relay are illustrated in Figure 9–32.

The spring holds the contacts in their "normal" position. When the proper voltage is applied to the coil, a magnetic force pulls the common contact attached to the pendulum to the other position, slightly extending the spring in the process. In most relays, this action takes only a few milliseconds. A click can be heard as the contacts move to their other position. As soon as the voltage is removed from the coil, the spring pulls the contacts back to their deenergized (normal) position.

Relay contacts are designated as **normally open (N.O.)**, or **normally closed (N.C.)**. In this context, the word normally means when the coil of the relay is *not energized*. A normally open contact will not allow an electrical signal to pass through while the coil of the relay is off. A normally closed contact will allow the signal to pass when the coil is off. Like switches, relays are also classified according to their poles and throws.

The coil is rated to operate at a specific voltage. In heating systems, the most common coil voltage is 24 VAC.

Relays with coil voltages of 12, 120, and 240 VAC are also available when needed. Relays with coils designed to operate on DC voltages are also available, although less frequently used in heating control systems. Figure 9–33 shows the schematic symbols for the coil and contacts for the relays commonly used in heating control systems.

General Purpose Relays

Some control systems require one or more relays to be installed to perform various switching or isolation functions. The type of relay often used is called a **general purpose relay**. They are built as plug-in modules with clear plastic enclosures. Their external terminals are designed to plug into a **relay socket**. These sockets have either screw terminal or quick-connect tabs for connecting to external wires. Each terminal on the socket is numbered the same as the connecting pin on the relay that connects to that terminal. *This numbering is very important because it allows the sockets to be wired into the control system without the relay being present.* Relay manufacturers provide wiring diagrams that indicate which terminal numbers correspond to the normally closed, normally open, common, and coil terminals. In some cases, the wiring diagram is printed onto the relay enclosure. Figure 9–34 shows a general purpose 3PDT relay along with the socket it mounts into and its wiring diagram. Notice how the terminals are numbered on the bottom of the relay enclosure as well as the relay socket.

Many relay sockets are designed to mount to a flat surface, or to be snapped into an aluminum **DIN rail**. The latter mounting method allows relay sockets to be added or removed quickly and without fasteners. The DIN rail is a standard modular mounting system used in many types of control systems. In addition to relay sockets, many small control devices are designed to be mounted to DIN rails. By mounting a generous length of DIN rail in a control cabinet, relays and other components can be easily added, removed, or moved to a different location. An example of a relay socket mounted to a DIN rail is shown in Figure 9–35.

Time Delay Relays

Often it is necessary to incorporate a time delay between two control events. One example would be allowing a minimum time for pressure to equalize within a refrigeration system before restarting the compressor. Another would be keeping a boiler circulator operating for a few minutes after the boiler has stopped firing to purge out the residual heat. Still another would be allowing the domestic water heating subsystem to be the only load operating for a period of up to 30 minutes once it initiates a call for heat. These and other control functions can be accomplished using **time delay relays**.

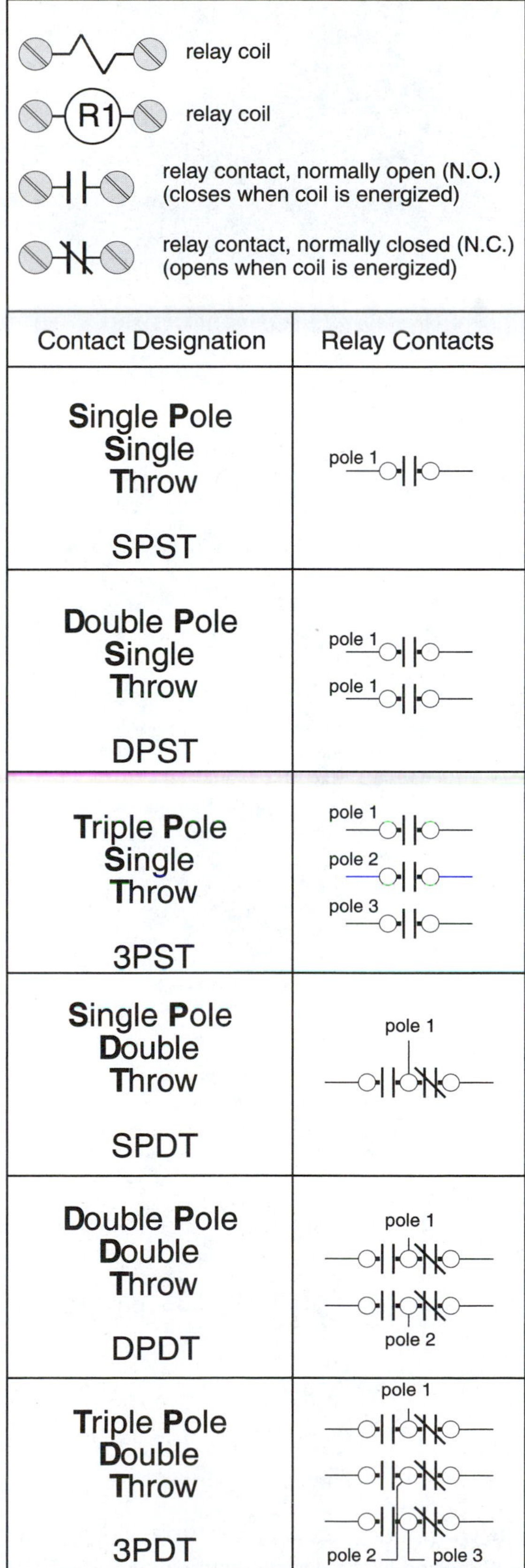

Figure 9–33 Schematic symbols used to represent relays having different numbers of poles and throws.

(a)

(b)

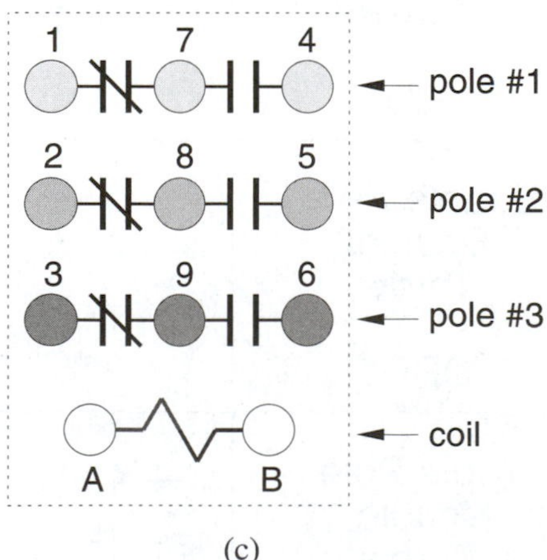

(c)

Figure 9–34 (a) Relay socket for a 3PDT plug-in relay. Note numbers and letters designating terminals. (b) 3PDT plug-in relay in plastic enclosure. (c) Schematic representation of relay associating contacts with terminal numbers. Coil connections identified as A and B.

Time delay relays are available with the following operating modes:

- Delay-on-make
- Delay-on-break
- Interval timing
- Repeat cycle

These operating modes are described using timing charts in Figure 9–36. Of the four operating modes shown, the **delay-on-make** and **delay-on-break** functions are the most commonly used in hydronic heating applications.

The delay-on-make function prevents the contacts from moving until a user-specified time delay period has elapsed from when the input signal is turned on. If input signal is interrupted before the delay period has elapsed, the relay automatically resets the timing circuit to begin from zero the next time the input signal is energized.

One common use of a delay-on-make function is overriding priority control after a specified time has elapsed. For example, assume a designer wants to allow domestic water heating to operate as the only load for a maximum of 30 minutes. All other loads are to be temporarily turned off during this time. However, if the domestic water load is still operating at the end of the 30-minute period, the other loads need to be turned back on. This is desirable to prevent freezing water in remote parts of the space heating distribution system should the domestic water heating load fail in the on position. The wiring necessary for this action is discussed in the next section.

The delay-on-break function is the opposite of the delay-on-make function. The contacts are held in their energized position until a user-specified time has elapsed. The input voltage to the relay is maintained at all times. When an external switch closure is detected across the control switch terminals of the relay, the contacts move to their energized position and remain there. When the external control switch opens, the time delay period begins. After the time delay period has expired, the relay contacts snap back to their "normal" deenergized position. If the control switch closes

Figure 9–35 Relay sockets mounted to DIN rail.

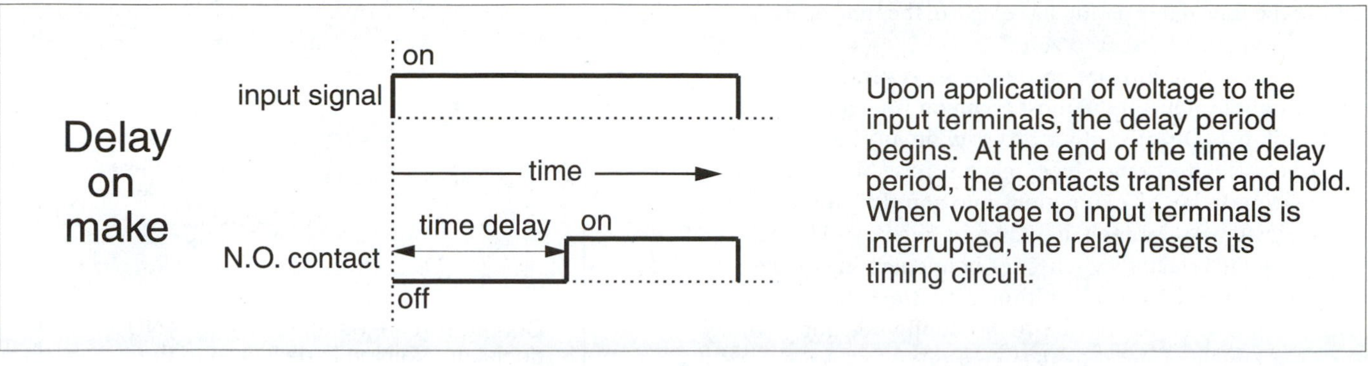

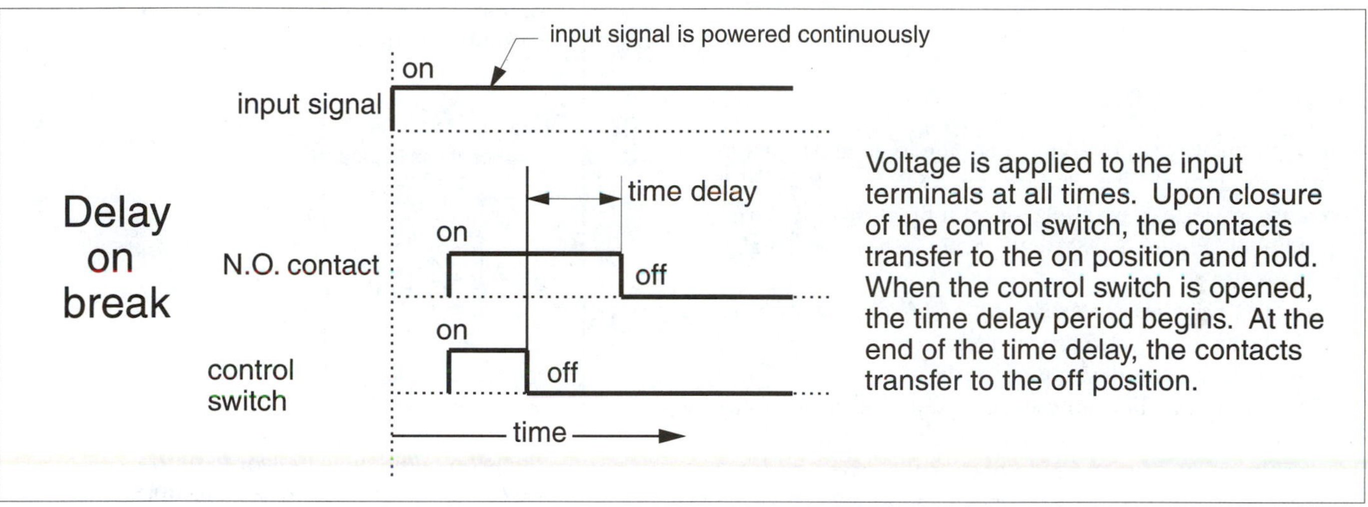

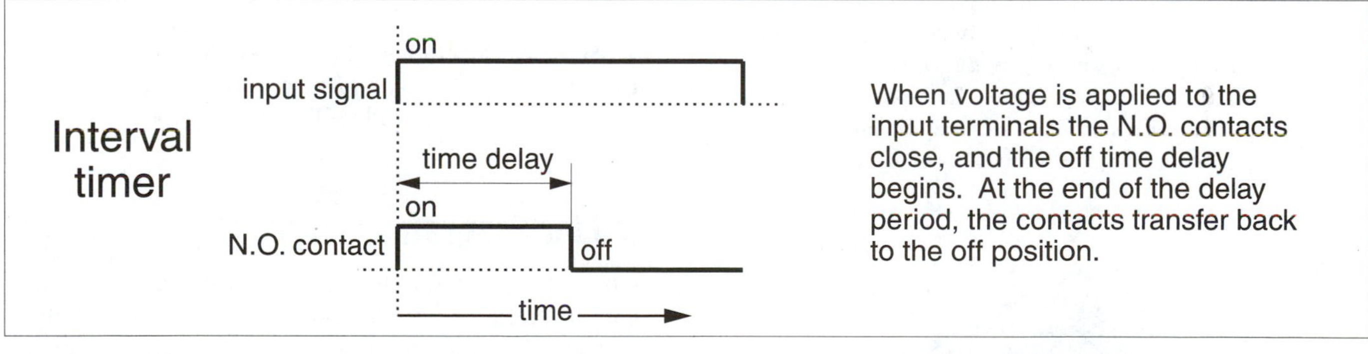

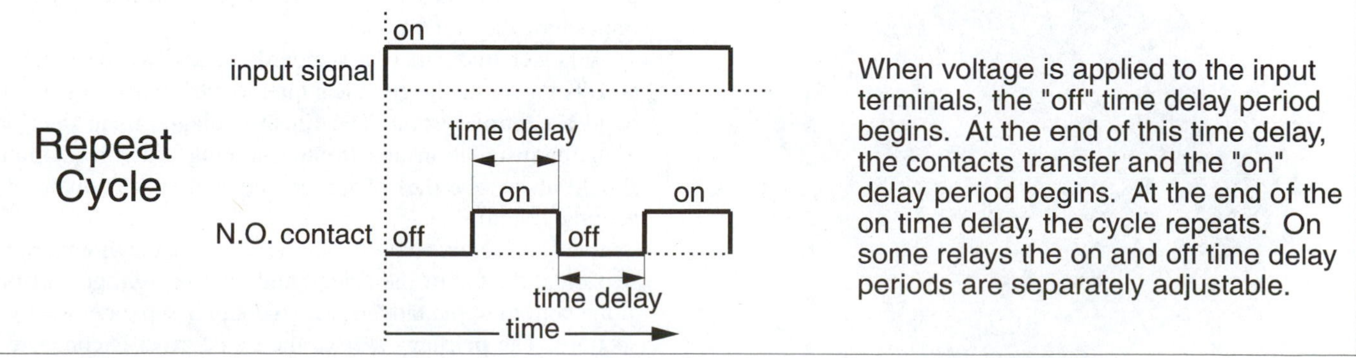

Figure 9–36 Operating modes for time delay relays. Some time delay relays can only provide one of these functions, others can be set to provide any of these functions.

before the time delay period has elapsed, the time delay period is automatically reset to zero.

Time delay relays are available in several pole/throw configurations. One of the most common is the double pole/double throw (DPDT) with a 120-VAC input signal.

Solid state electronics have made it possible to combine several time delay relay functions into a **multifunction time delay device**. An example is shown in Figure 9–37.

Notice the selector switches for both operating mode and time delay range on the multifunction time delay device. These allow a single device to be configured for a wide range of functions and time delay periods.

Hardwired Logic

One of the ways of building operating logic into a control system is through the connecting switch and/or relay contacts in series, parallel, or combinations of series and parallel. Figure 9–38 shows basic series and parallel arrangements.

Contacts in series represent an "AND" decision. For an electrical signal to reach a driven device, all series-connected contacts must be closed simultaneously.

Contacts in parallel represent an "OR" decision. If any one or more of the contacts are closed, the electrical signal crosses the group of switches to operate the driven device.

When switch contacts, relay contacts, and contacts that are part of specialized hydronic controllers are physically wired together in a given manner they create **hardwired logic**. Such logic determines exactly how the control system functions in each of its operating modes. Some operating modes are planned occurrences while others may be fail-safe modes in the event a given component does not respond properly.

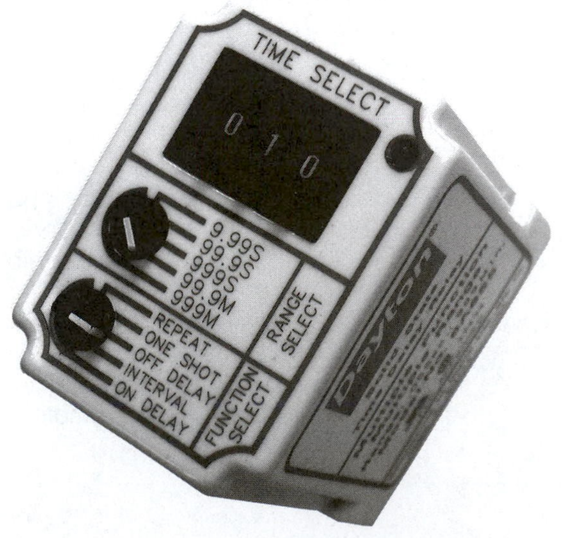

Figure 9–37 Example of a multifunction time delay device.

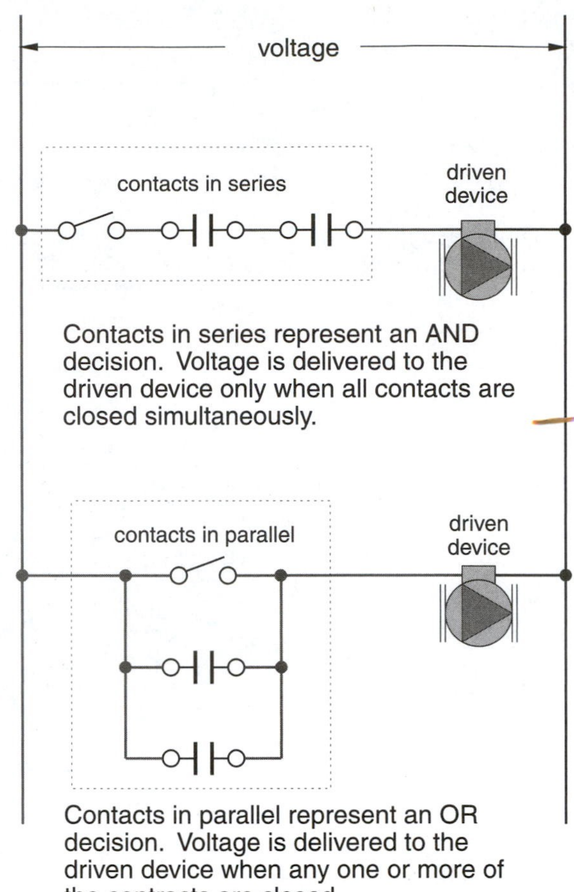

Figure 9–38 Operating logic based on wiring of switches or relay contacts. Contacts in series provide "AND" logic. Contacts in parallel provide "OR" logic.

Ladder Diagrams

It is often necessary to combine several control components to build an overall control system. The way these components are connected to each other as well as to driven devices such as circulators, zone valves, and boilers determines how the system operates.

A ladder diagram is a standard method for developing and documenting the electrical interconnections necessary to build a control system. The finished diagram can then be used for installation and troubleshooting. Such a diagram should always be part of the documentation of a hydronic heating system.

Ladder diagrams have two basic sections, the **line voltage section** at the top of the ladder, and the **low voltage section** at the bottom of the ladder. A **transformer** separates the two sections. The **primary side** of the transformer is connected to line voltage. The **secondary side** of the transformer powers the low voltage section. In North America, most heating and cooling secondary systems operate with a secondary voltage of 24 VAC.

An example of a simple ladder diagram is shown in Figure 9–39. The vertical lines can be thought of as sides of an imaginary ladder. Any horizontal line connected between the two sides is called a **rung** of the ladder. A rung connected across the line voltage section is exposed to 120 VAC. A rung connected across the low voltage section is typically exposed to 24 VAC. The overall ladder diagram is constructed by adding the rungs necessary to create the desired operating modes of the system. The vertical length of the ladder can be extended as necessary to accommodate all the rungs.

Relays are often used to operate line voltage devices such as circulators, oil burners, or blowers based on the action of low voltage components such as thermostats. Ladder diagrams are an ideal way to document how this is done. When a circuit path is completed through a relay coil in the low voltage section of the ladder, one or more contacts of that relay may be used to operate devices in the line voltage section.

Consider a situation in which a line voltage circulator is to be operated by a low voltage switch. Since the circulator needs line voltage to run, it is connected across the line voltage section of the ladder diagram as shown in Figure 9–39. A normally-open relay contact is wired in series with the circulator motor. When this contact is open, the motor is off. To close this contact, the coil of the relay must be energized. This requires a completed circuit path across the low voltage section of the ladder. By wiring the relay coil in series with the switch, the coil is energized when the switch contacts are closed, and deenergized when they are open. The overall operating sequence is as follows: The switch is closed. Low voltage is now applied across the coil of the relay. The energized coil pulls the normally open contacts in the line voltage rung together. Line voltage is applied across the circulator motor to operate it.

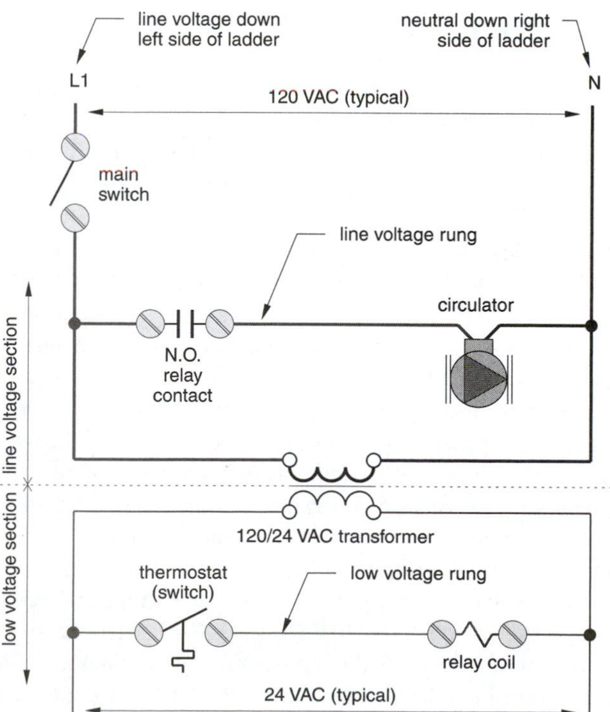

Figure 9–39 Example of a simple ladder diagram to operate a line voltage circulator using a low voltage control circuit.

Although this example is relatively simple, it illustrates the basic use of both the line voltage and low voltage portions of the ladder diagram.

More complex ladder diagrams are developed by placing schematic symbols for additional components into the diagram. Some of these components might be simple switches or relays, others might be special purpose controllers that operate according to how the manufacturer designed them as well as how the adjustable parameters are set. When shown in a ladder diagram, the latter are called **embedded controllers** since they are part of a ladder diagram that documents the overall control system. An example of a ladder diagram with several rungs and an embedded controller is shown in Figure 9–40.

Drawing Ladder Diagrams

Consistency is the key to drawing ladder diagrams that will be useful to many people. The following guidelines are recommended.

- Draw the ladder diagram with line voltage section at the top, and the low voltage section at the bottom.
- Specify the supply voltage and circuit **ampacity** required for the control system represented by the ladder diagram.
- Specify the type and gauge of all wiring to be used for the control system represented by the ladder diagram.
- Always show a main switch capable of completely isolating the ladder diagram from its power source when necessary.
- Use wider lines for line voltage conductors, and thinner lines for low voltage conductors to make the drawing easier to follow.
- When a component symbol appears in the system's piping schematic, and in the ladder diagram, be sure it has identical designations in both drawings (i.e., P1, T2, etc.). This allows for easy cross-referencing.
- Use separate abbreviations for each relay coil and its associated contact(s). Use a designation such as R1-2 to identify pole #2 of relay R1. R2-1 would identify pole #1 of relay R2, and so forth. Without such designations, it is impossible to know which contacts and coils are part of the same relay, especially in complex diagrams.
- It is customary to show switches and relay contacts in their off, or deenergized, positions.
- When several of the same components (i.e., relay contacts, circulators, zone valves, etc.) appear on several

rungs, they should be vertically aligned. This not only makes the drawing look better, it also makes it easier to understand because of consistent symbol placement.

- Although ladder diagrams can be drawn by hand, there are many advantages to creating them using computer drawing software (e.g., CAD). For example, the vertical height of the ladder can be easily extended or shortened as needed to accommodate the necessary rungs. Component symbols and groups of symbols can be easily duplicated and moved to speed the design/drawing process.

Priority Control Strategies

Sometimes the total heating load connected to a hydronic system exceeds the output of the heat source. One way to manage this situation without increasing the size of the heat source is by prioritizing one or more of the loads. **Priority control** is a planned strategy for shedding (e.g., temporarily turning off) one or more loads when a specific condition is detected in the system. This condition might be a specified water temperature at some location, the operation of a specific load, or the occurrence of two or more simultaneous conditions within the system.

A common priority load in many residential hydronic systems is the operation of an indirect domestic water heater. A basic ladder diagram showing how the domestic water heating load is prioritized is shown in Figure 9–41.

When the contacts of the water heater thermostat close, relay coil (R1) is supplied with 24 VAC. The normally open relay contact (R1-1), located in the line voltage section of the ladder diagram, closes to supply line voltage to the tank's circulator. At the same time, a normally closed relay contact (R1-2) opens to break low voltage (24 VAC) to the space-heating thermostats shown in the lower portion of the ladder diagram. Since power to all the space-heating thermostats has now been interrupted, these loads remain off until the priority relay (R1) is deenergized. The full output of the heat source is now directed to the prioritized load. When the priority load is satisfied, relay coil (R1) is deenergized and contact (R1-2) closes to reconnect the space heating thermostats to 24 VAC power.

Although priority control is a useful concept, it can also lead to unexpected problems if not designed for fail-safe operation. For example, imagine what might happen to a system with prioritized domestic water heating if the prioritized load could never be met. This could happen if the tank thermostat failed in the "on" position, or if a temperature sensor was accidentally pulled out of its well in the tank. The control system represented in Figure 9–41 would respond by preventing any of the space-heating circulators from operating. In a cold climate, a couple of days of unattended operation could lead to frozen pipes.

One way to prevent this "priority lockout" is to design the control circuit with priority override. A delay-on-make time delay can be used as shown in Figure 9–42.

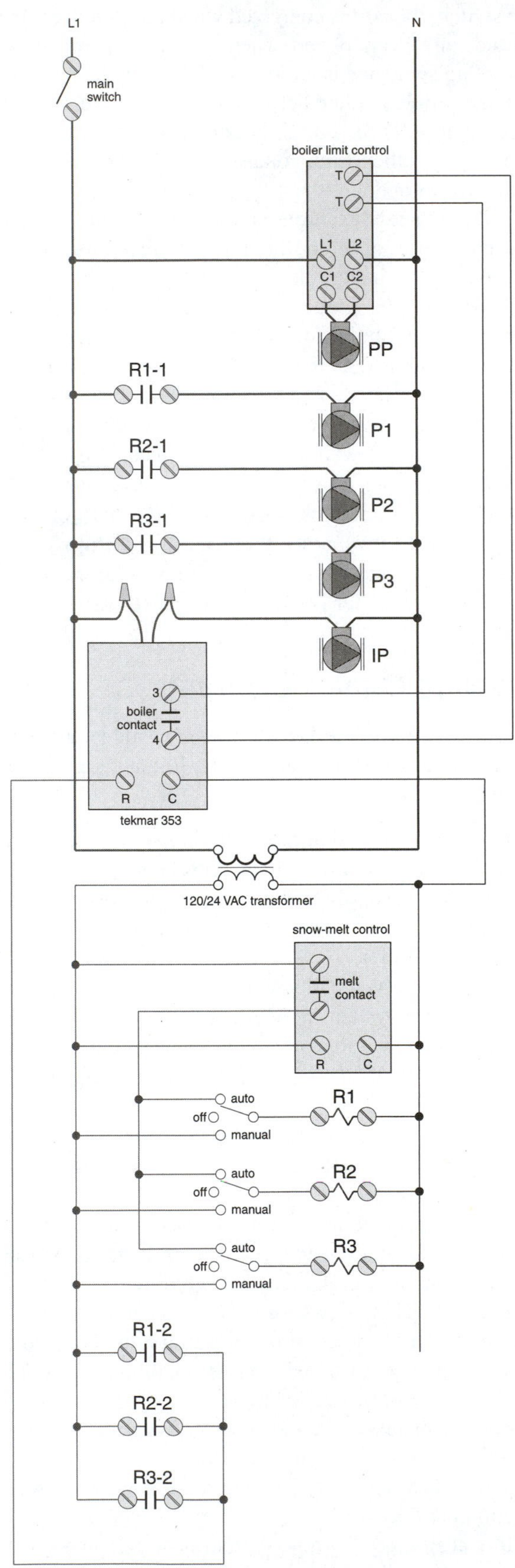

Figure 9–40 Example of a ladder diagram with embedded controllers.

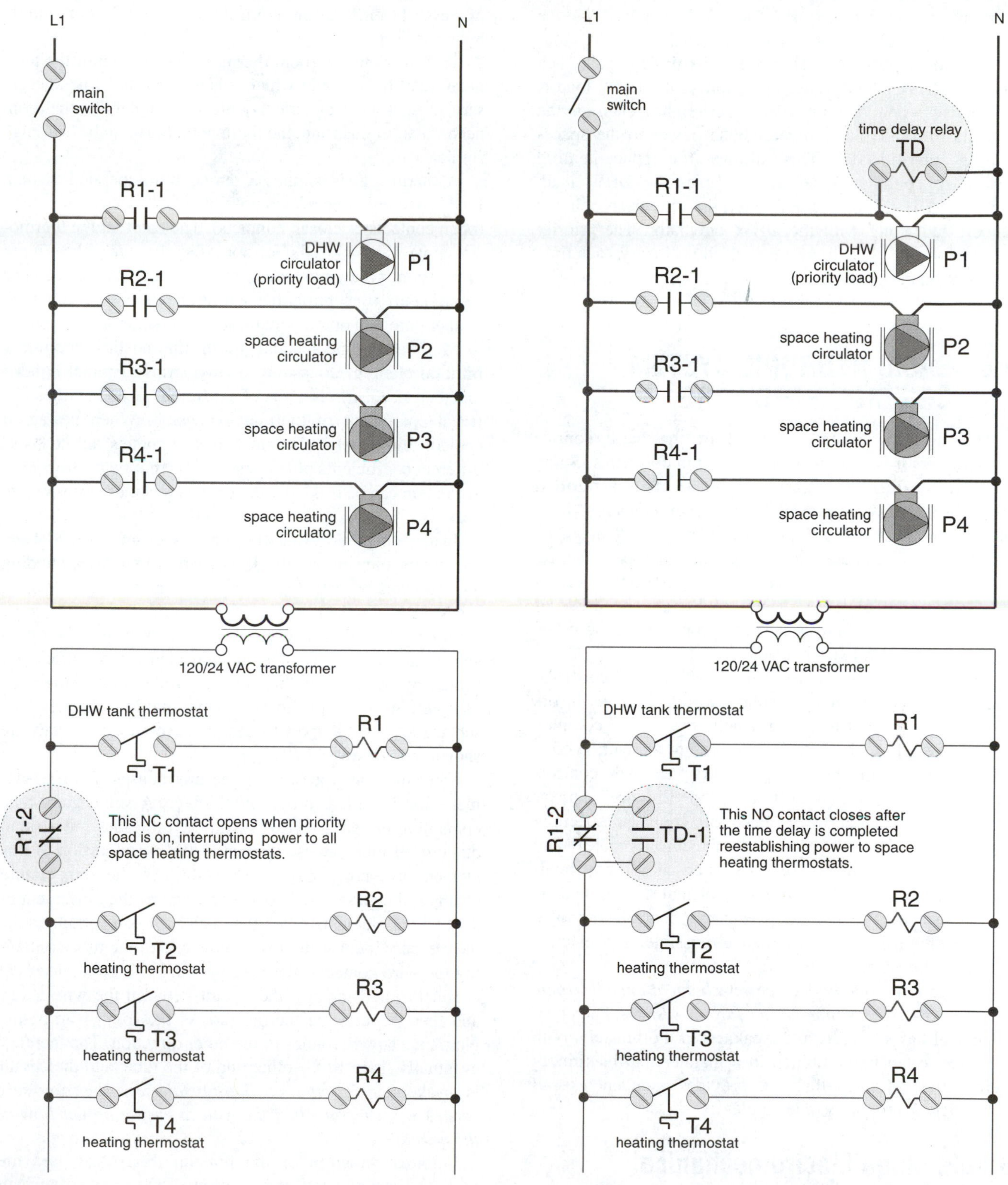

Figure 9–41 Ladder diagram in which the domestic water heating mode is prioritized. All space heating loads are temporarily turned off whenever the domestic water heating load is operating.

Figure 9–42 Ladder diagram showing use of line voltage (delay-on-make) time delay relay to override priority mode after a specified time has elapsed, allowing space heating to operate simultaneously with domestic water heating.

The system goes into the priority mode any time the domestic water heating (DHW) load calls for heat. Notice the time delay relay wired in parallel with the DHW circulator. This delay-on-make relay begins its timing cycle each time the DHW load initiates operation. If the DHW load is still active when the time delay period has elapsed, the contacts in TD-1 close to reestablish power to the space-heating thermostats. This allows the space-heating circuits to operate at the same time as the DHW load. When the DHW load is finally met, the time delay relay is deenergized along with the DHW circulator. The priority override process is set to repeat itself, if necessary, each time the DHW load operates.

9.6 BASIC HYDRONIC SYSTEM CONTROL HARDWARE

This section discusses several of the basic control hardware devices used in modern hydronic systems. Some of these devices are classified as **electromechanical controls**, and have been available for several decades. They use components such as springs, bimetal strips, magnets, and fluid-filled temperature sensing bulbs to open and close electrical contacts based on sensed quantities such as temperature and pressure. There are literally hundreds of electromechanical controls available for specific purposes and applications in hydronic heating.

Other devices are classified as **electronic controls**. They use components such as solid state temperature sensors and microprocessors to perform both simple and complex control tasks. Electronic controls represent the cutting edge of hydronic heating technology. Every year new controls with increased functionality and "intelligence" appear on the market. The ability of electronic controls, especially those using microprocessors, to execute complex and sophisticated control algorithms such as proportional-integral-derivative (PID) processing cannot be matched by electromechanical controls. Electronic controls are also capable of greater accuracy than is possible with electromechanical controls.

Keep in mind that *both electromechanical and electronic controls have a place in modern hydronic systems*. Each type of control has strengths and weaknesses. Fortunately, both types can often be combined in a manner that maximizes their individual strengths, and provides excellent overall performance, reliability, and value.

Single-stage Electromechanical Room Thermostats

No heating control is more familiar by sight and name than a room thermostat. For many homeowners it is the only control they know of, or want to know of, in the entire system. If the house is too cold, they turn it up. If it is too warm, they turn it down. The room thermostat is, however, only one of several temperature control devices, even in simple hydronic systems.

The function of a room thermostat is to monitor the temperature of the room in which it is located, and when necessary, close a set of electrical contacts to signal that the room needs heat. In this state, the thermostat is said to be "calling" for heat.

A thermostat is simply a temperature-operated switch. It can have only one of two possible states at any given time (open contacts or closed contacts). The fact that the thermostat contacts are closed does not affect the *rate* of heat input to the room. Many homeowners do not understand this. The common misconception that setting the thermostat to a high temperature will quickly increase room temperature is proof.

Most electromechanical room thermostats contain a **bimetal element** consisting of two strips of metal bonded together. Each strip is made of a different metal having a different coefficient of thermal expansion. When heated or cooled, this bimetal element bends or rotates due to these different coefficients of expansion. The motion of the bimetal element can be used to pull contacts together or push them apart.

In a heating thermostat, the contacts are arranged so when the bimetal element is sufficiently cooled by the surrounding air, the contacts snap together. This contact closure can be used to start the heat source, open a valve, or initiate some other control action. In a cooling thermostat, just the opposite takes place. When the bimetal element is sufficiently heated by the surrounding air, the contacts snap together, completing a circuit to start the cooling device. The schematic symbols used to designate room thermostats are shown in Figure 9–43.

One common type of room thermostat uses a bimetal element that is coiled into a spiral shape. A small glass tube containing the ends of two flexible wire leads, and a small quantity of mercury, is attached to the outer portion of this element as shown in Figure 9–44. As the temperature changes, the glass tube is slowly rotated by the movement of the bimetal element. When the tube reaches a certain angle, the mercury it contains flows to the end of the tube containing the wire contacts. Because mercury is a good electrical conductor, it completes the circuit between the wire leads, and thus through the thermostat. As the room warms, the glass tube slowly rotates in the other direction. The mercury eventually flows to the other end of the tube, and the circuit is broken. *This type of thermostat must be precisely leveled when mounted on the wall to ensure proper flow of the mercury.*

Another variation of the bimetal thermostat uses the bimetal element itself as a portion of the conducting path through the thermostat. As the room temperature slowly drops, the element brings the contact points closer together. When they are a certain small distance apart, the force created by a magnet snaps the contacts together. This prevents the erratic operation that could occur when minor tempera-

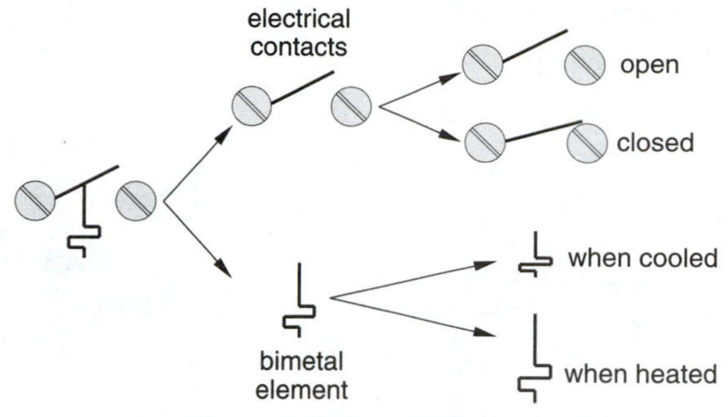

	Standby	Operating	Description
Heating thermostat			Bimetal element contacts when cooled closing contacts. Element expands when heated opening contacts.
Cooling thermostat			Bimetal element expands when heated closing contacts. Element contracts when cooled opening contacts.

Figure 9–43 Schematic symbols used to represent heating and cooling thermostats.

ture fluctuations cause the contacts to bounce between open and closed.

An ideal room thermostat would operate the heating system to maintain the room precisely at a desired setpoint temperature. If the thermostat detected even the slightest drop below the setpoint, it would bring on the heating system. When the temperature of the room increased the slightest amount above the setpoint, it would stop further heat input. Unfortunately, this would result in extremely short and frequent operating cycles of the heating system. Although this is possible with electric resistance heating elements, it would cause extreme wear on boilers and heat pumps. To prevent such short cycling, room thermostats must operate with a differential as discussed in Section 9.2. The differential is the difference in temperature between where the contacts close and where they open.

Figure 9–44 Example of an electromechanical room thermostat. Setting is indicated on upper scale. Actual room temperature is indicated on lower scale.

Some thermostats have an adjustable differential, others are fixed by the manufacturer. The narrower the differential, the greater the number of on/off cycles the heating system will experience during a given period of time. Narrow differentials are beneficial for comfort control *if they can be obtained without creating abnormally short operation of the heat source*. Thermostats are available with differentials as low as 1 °F. Such thermostats are often used with electric resistance heating equipment that is not adversely affected when turned on and off many times an hour. Thermostats that directly operate a combustion type heat source or heat pumps typically use differentials of 2 °F to 4 °F to reduce the number of on/off cycles and thus reduce wear on the equipment.

When a simple on/off thermostat is used, heat input begins only when the room temperature drops a certain amount below the desired setting. The thermal mass of the heating system causes a further drop in temperature while the heat source and heat emitters warm enough to begin raising the room's temperature. This causes an **undershoot** in room temperature as illustrated in Figure 9–45.

If the thermostat contacts remain closed until the room reaches the desired temperature an effect known as **overshoot** occurs. Residual heat stored in the thermal

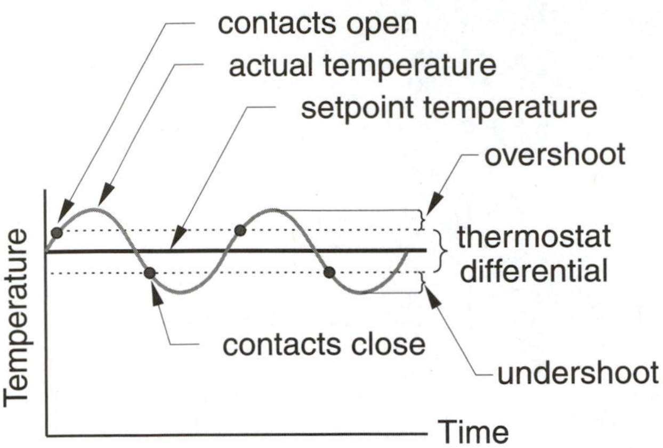

Figure 9–45 Overshoot and undershoot of temperature as thermostat cycles heat on and off.

(a)

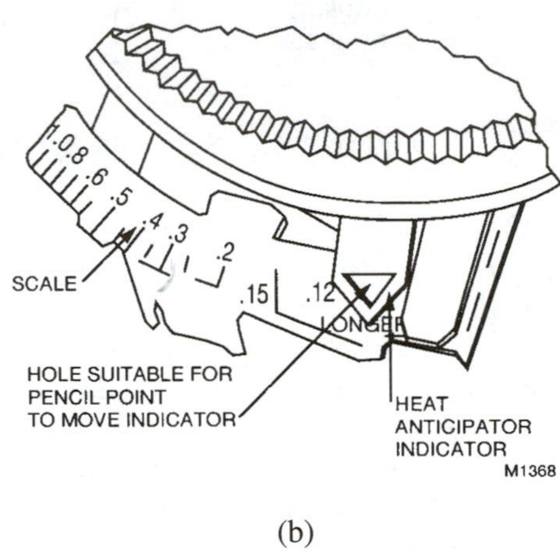

(b)

Figure 9–46 (a) Heat anticipator adjustment on a low voltage room thermostat, (b) close-up of amperage values on anticipator scale. *Courtesy of Honeywell.* The anticipator is set for the amperage draw of the thermostat circuit.

mass of the heat emitters continues to flow into the room causing the air temperature to climb above the desired setting.

To limit overshoot, a **heat anticipator** is built into most electromechanical room thermostats. It consists of a small adjustable resistor that is usually wired in series with the thermostat contacts, and thus has current flowing through it whenever the contacts are closed. The heat given off by the resistor slightly increases the temperature sensed by the bimetel element in the thermostat enclosure. This "fools" the thermostat into opening its contacts before room temperature reaches the setpoint. The residual heat in the heat emitters then causes the air temperature to drift slightly upward with minimal if any overshoot.

The lower the room temperature relative to the setpoint, the longer it takes the anticipator to heat the bimetal element, and the longer the on-cycles become. The anticipator allows the system to increase the amount of heat delivered as the difference between setpoint and actual room temperature increases. However, the longer the on-cycles of the thermostat, the more heat the anticipator releases into the thermostat enclosure. This causes an undesirable effect known as droop in which the room temperature drifts downward as the heating load increases. Standard electromechanical thermostats cannot compensate for droop.

For proper operation, thermostats with series anticipators need to be calibrated for the electrical current that flows through them while operating. This current depends on the circuit the thermostat is wired into, and can be easily measured by connecting an ammeter in series with the thermostat circuit. Once the operating current is read, a small lever or screw inside the thermostat is adjusted to the indicated amperage. Figure 9–46 shows the heat anticipator adjustment on a room thermostat.

In some systems, room thermostats operate devices called **valve actuators** which open and close zone valves. Many of the valve actuators used in radiant panel heating systems use **heat motors**. This type of actuator causes the electrical current through the thermostat circuit to vary depending on the length of time the actuator is on as well as its temperature. The current also varies depending on the number of actuators wired in parallel through a common thermostat. The varying current makes it difficult for a thermostat with a series anticipator to yield consistent performance. One solution is to use a thermostat with a *parallel* anticipator.

Figure 9–47 shows the difference in wiring between thermostats using series and parallel anticipators. In a thermostat with a parallel anticipator, the current passing through the anticipator resistor is independent of the current passing through the thermostat contacts. This allows the anticipator to deliver consistent performance even when the current passing through the thermostat contacts varies. Note that thermostats with parallel anticipators require three wires.

Room thermostats should be mounted on interior walls away from sources of localized heat such as lights, cooking equipment, heat emitters, or direct sunlight. A mounting height of about 5 feet is typical. Choosing a thermostat location should not be considered a trivial matter. *Careless placement can significantly detract from the performance of an otherwise well-installed system.*

Room thermostats can be further classified as low voltage or line voltage devices. The majority of those used in hydronic heating are designed for low voltage operation. They typically switch a circuit operating at 24 VAC. This control voltage is a standard throughout North America. Low voltage thermostats qualify as class 2 devices under the national electrical code, and can be wired with relatively small diameter conductors in the range of 24 to 18 wire gauge.

Line voltage room thermostats are designed for the nominal 120 or 240 VAC line voltage. They are capable of switching line voltage devices such as heating elements, pumps, or fans. All line voltage thermostats have an associated ampacity rating, which is the maximum current the contacts can safely handle. Often two current ratings are given. The greater one is for resistive loads such as heating elements, the lower one for inductive loads such as AC induction motors. When selecting a line voltage thermostat, always check that these ratings meet or exceed the total amperage of all devices that draw their current through the thermostat.

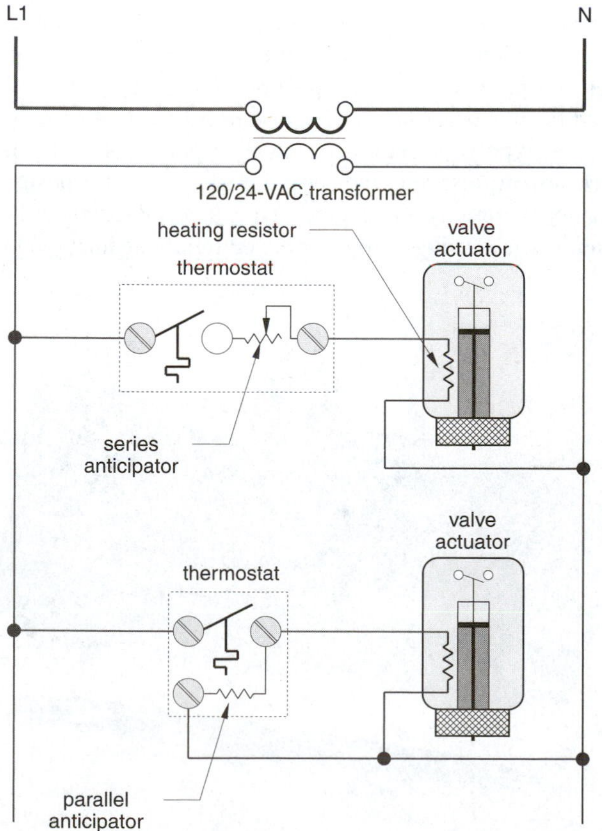

Figure 9–47 Thermostat with a series anticipator resistor and thermostat with a parallel anticipator resistor.

Two-stage Electromechanical Room Thermostats

Occasionally a hydronic system requires two levels of heat input. The first level provides a certain rate of heat input to the space. If the load is such that this input can maintain the setpoint temperature, the other level of heat input is not needed, and therefore not called for. However, if the first level of heat input cannot maintain the room's setpoint temperature, the second level of heat input is called for in an attempt to bring the room temperature back up to the desired setpoint. These levels of heat input are called stages. A thermostat capable of providing two or more levels of heat input is called a multistage thermostat.

In all multistage heating systems, stage one always operates before stage two. For example, in many heat pump systems, stage one of a multistage thermostat operates the compressor that drives the refrigeration cycle discussed in Chapter 3, Hydronic Heat Sources. If the room temperature continues to drop when stage one is operating, stage two of the thermostat operates electric resistance heating elements to provide additional heat input. Because heat from electric resistance elements is significantly more expensive than heat provided by the refrigeration cycle, it will be used only when necessary to prevent the building from becoming uncomfortable.

Another common application for a two-stage thermostat is providing supplemental heat input to a room in which primary heat input is provided by a radiant floor panel. The supplemental heat may come from a baseboard, panel radiator, or other type of heat emitter. However, the supplemental heat emitter will only operate if the radiant floor panel cannot maintain the room's setpoint temperature.

Two-stage room thermostats have two sets of contacts, one for each stage. *The first-stage contacts will always close before the second-stage contacts.* The heating device with operating priority is controlled by the first-stage contacts. The supplemental heating device is controlled by the second-stage contacts.

Two-stage thermostats usually have a fixed temperature differential of 1 to 2 °F between stages. This is called the **interstage differential**. If, for example, this differential was 2 °F and the first-stage contacts closed at a room temperature of 69 °F, the second-stage contacts would close at 67 °F. As the room warms up, the second-stage contacts open first. If the room temperature continued to rise, the first-stage contacts will eventually open.

All multistage electromechanical thermostats create droop as more stages operate. This results from the fact that room temperature must drop below the setpoint before the thermostat can signal for additional heat input. The greater the number of stages the greater the droop.

Electronic Thermostats

One of the first heating controls to enjoy the benefits of electronics was the room thermostat. Although they have been available for over 30 years, the most up-to-date electronic thermostats offer features and benefits that far exceed early models.

Most electronic thermostats, as well as other types of electronic temperature controls, use a **thermistor** as their temperature-sensing element. This small solid state device resembles a glass bead with a diameter of about 0.1 inches, and two wire leads. The type of thermistor commonly used in temperature controllers is called a negative temperature coefficient (NTC) thermistor. The electrical resistance of an NTC thermistor decreases as its temperature increases. This relationship between temperature and resistance is very predictable and repeatable. For some thermistors, a temperature change of 1 °F can result in a change in resistance of over 100 ohms, which is easily detected by the thermostat's circuitry. The small size of a thermistor bead also allows it to respond very quickly to temperature changes.

The most modern electronic thermostats use internal microprocessors and nonvolatile EEPROM memory to store and execute program instructions. Some of these instructions are set by the manufacturer and cannot be changed in the field. Other instructions such as setpoint temperature(s), times to begin and end setback periods, and the type of device the thermostat will control must be entered by the installer, or occupant. Most modern electronic thermostats have internal software that prompts the installer or occupant for information through a liquid crystal display (LCD). An example of a modern microprocessor-based thermostat is shown in Figure 9–48.

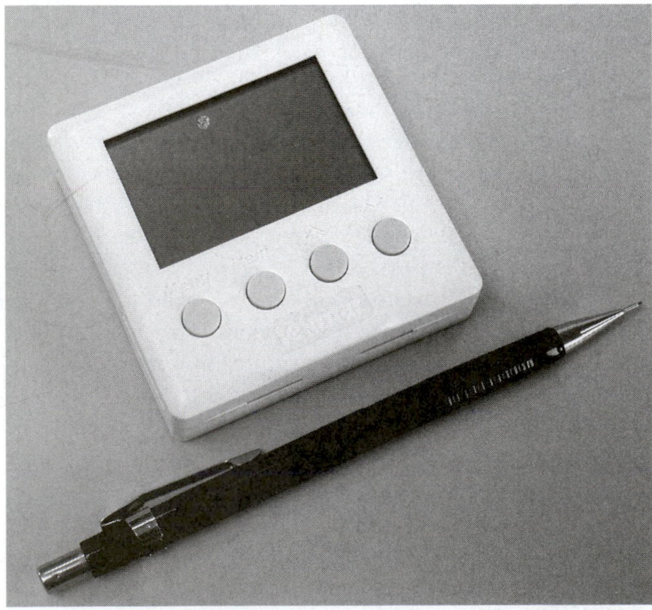

Figure 9–48 Example of a microprocessor-based electronic thermostat.

Once the thermostat has been programmed, it opens and closes its relay contacts in an *attempt* to control the system as desired. The word *attempt* is important because no thermostat can necessarily force the inside air temperature to follow the programmed schedule. For example, imagine a situation in which a thermostat is programmed to begin a 10 °F temperature setback at 11 P.M. and resume normal comfort temperature at 5:30 A.M. the following morning. If the building has very low heat loss, and/or if the heating system has high thermal mass, the inside temperature may not drop more than 2 °F or 3 °F during the setback period, even with no heat input.

Most electronic thermostats draw their very low operating power from the 24 VAC transformer in the controlled device. This allows for both power supply and on/off control through a single pair of wires. Other models operate on batteries that require periodic replacement.

Electronic thermostats generally offer more features and setting options than electromechanical thermostats. This is especially true of microprocessor-based thermostats where menu-driven programming replaces the need for many hardware switches and dials.

Electromechanical Setpoint Controllers

In most hydronic systems, there are several locations where fluid temperatures are measured for safety and control purposes. The opening or closing of electrical contacts is often required when the measured temperature is above or below a predetermined temperature. One type of controller used for this purpose is called a **setpoint controller**.

One type of electromechanical setpoint controller is a **strap-on aquastat** as shown in Figure 9–49a. It is designed to mount directly on a pipe using a metal strap or hose clamp. The sensing bulb is squeezed tightly against the pipe

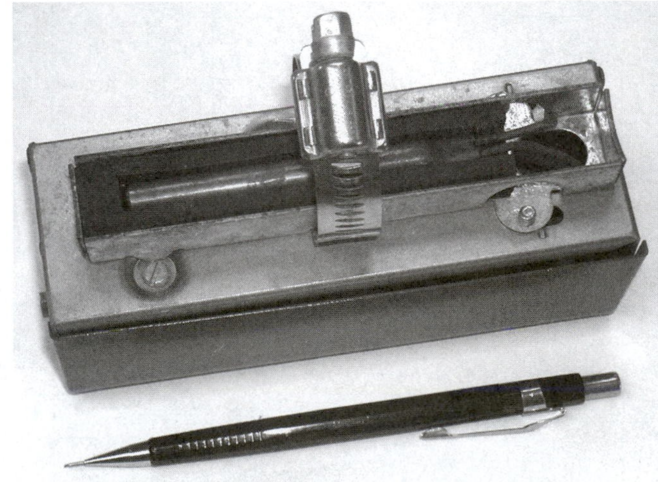

Figure 9–49a Rear side of a strap-on aquastat temperature control. When mounted, the sensor bulb in the center of the cradle is held in contact with surface of pipe.

as the clamp is tightened. This bulb contains a fluid that increases in pressure when heated. The pressure exerted by the fluid's vapor pushes against a bellows assembly inside the control, causing it to expand. This movement opens or closes the electrical contacts in the control. Most strap-on aquastats can switch several amps of current at line voltage. Current and voltage ratings are typically listed on the cover plate. These controls can also be used in low voltage systems.

Another electromechanical setpoint controller is the **remote bulb aquastat** shown in Figure 9–49b. As with the strap-on aquastat, temperature is sensed by a bulb that is connected to a bellow in the aquastat by a capillary tube. However, in this case, the capillary tube can be several feet long allowing the control body and electrical wiring to be located several feet from the location of the sensing bulb. Controls can be ordered with capillary tubes up to 12 feet long. *The length of this capillary tube cannot be changed in the field because the working fluid would be lost.* Any extra length of capillary tubing should be neatly coiled near the control and fastened to something solid to prevent vibration. *Care should be used in routing the capillary tube to prevent kinking.*

In applications where the temperature of a surface is to be measured, the sensing bulb must be tightly secured against that surface using a clamp or tube strap. It should then be covered with insulation to minimize the error from heat loss to the surrounding air.

The most accurate method of sensing a fluid temperature is to immerse the sensing bulb directly in the fluid. Some controls are available with sensing bulbs having MPT threads. These bulbs can be screwed directly into a fitting such as a tee. If the control has to be replaced, some spillage of system fluid is inevitable during removal and replacement of the sensing bulb.

Another option is to mount the sensing bulb into a sensor well. Such wells usually consist of a segment of copper tube slightly larger than the sensing bulb, closed at one end, and open at the other. The open end is equipped with male pipe threads so it can be screwed into a tapping on a tank or a tee. The closed end prevents fluid from entering the well. For best results, the sensing bulb should be coated with the heat-conducting paste often supplied with the well before being inserted into it. This paste improves heat conduction between the bulb and the inside surface of the well, resulting in a faster response to changing temperatures. An example of a sensing well is shown in Figure 9–50.

Most remote bulb aquastats and strap-on aquastats are supplied with a single pole double throw (SPDT) switch. This allows them to make or break a circuit when the desired setpoint temperature is reached. In heating applications, aquastats are usually configured to open a set of electrical contacts to interrupt heat flow to the system when the measured temperature climbs to the setpoint value. The contacts close when the sensing bulb temperature drops through a differential below the setpoint. In most aquastats, this differential can be adjusted over a fairly wide range.

Figure 9–49b A remote bulb aquastat. Note the capillary tube that connects the control's body and sensing bulb. *Courtesy of White Rogers.*

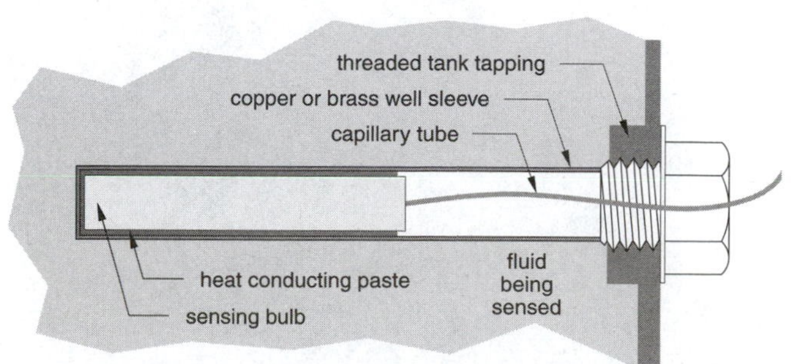

Figure 9–50 A sensor bulb mounted into a sensor well that is immersed in a fluid. Sensor can be removed from well without spillage of fluid. Note the use of heat conducting paste between sensor bulb on inside surface of well.

Electronic (Single-stage) Setpoint Controls

An electronic setpoint control provides on/off contact closure output based on the temperature of its sensor and its programmed setpoint. Like its electromechanical counterpart, output is limited to this simple on/off action. However, several additional features make the control more versatile. These include wider ranges of adjustment for both setpoint and differential, the ability to operate the relay on a rise or drop in temperature, and a digital readout for displaying the sensor temperature. Some electronic setpoint controls can be configured to work in either °F or °C, or to execute a programmable time delay function. Applications for these controls include operating a heat source, pump, diverting valve, or almost any type of electrically driven device at a preset fluid temperature. An example of an electronic, single-stage setpoint control is shown in Figure 9–51.

Electronic setpoint controls also have some *disadvantages* relative to electromechanical controls. Because of the physical size of the relays used, most cannot handle line voltage currents as high as those handled by electromechanical controls. In some cases, the relay contacts are only rated to operate with control voltages up to 24 VAC. If a line voltage device such as a pump or blower is to be controlled, an additional relay will be required between the control and the device.

Electronic setpoint controls tend to be more expensive than electromechanical setpoint controls. This higher cost can be justified when the application requires features such as accuracy, time delay, adjustable differential, and so on. In other applications where accuracy of +/– 3 °F to 5 °F is acceptable, and where rapid temperature changes are not likely, the higher cost may not be justified.

The solid state components used in electronic controls, especially microprocessors, are also more susceptible to damage from voltage spikes. Although manufacturers have improved the ability of these controls to survive moderate voltage spikes, accidental connection of line voltage to low voltage terminals, or a high voltage surge from a nearby lightning strike could still cause damage.

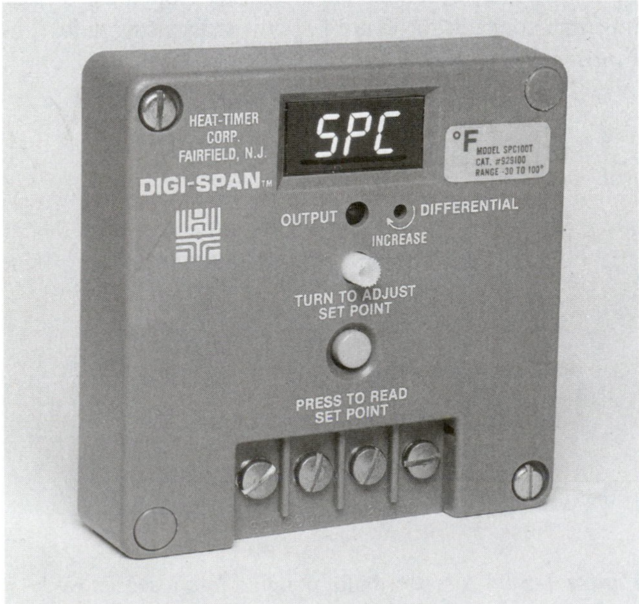

Figure 9–51 Example of an electronic single stage setpoint control. *Courtesy of Heat Timer Corp.*

Electronic (Two-stage) Setpoint Controls

The concept of multistage control has already been discussed. The output of an electronic two-stage setpoint control consists of sequential operation of two switches or relay contacts. The first-stage contacts will always be the first to close and the last to open. The second-stage contacts close when and if necessary to deliver more heating capacity. A two-stage electronic setpoint control allows great flexibility in setting the cut-in temperature of each stage, the differential of each stage, and the interstage differential.

The primary application for two-stage setpoint controls is to activate two stages of heating (or cooling) to meet changing load conditions. The heat sources activated by each stage do not have to be the same type or capacity. The flexible setup allowed by electronic two-stage controls allows the operation of each stage to be matched to the operating characteristics of each controlled device.

Zone Valves

Zone valves are commonly used to allow or prevent flow through two or more independently controlled zone piping circuits. Several types of zone valves are available. The most common types were discussed in Chapter 5, Pipings, Fittings, and Valves. They all consist of a valve body combined with an **actuator**. The actuator is the device that produces movement of the valve shaft when an electrical voltage is applied.

Some actuators use small electric motors combined with gears to produce a rotary motion of the valve shaft. These actuators can fully open or close the valve in about 3 seconds. This is fast enough for rapid heat delivery, yet slow enough to prevent water hammer. Other zone valves use **heat motor actuators** that produce a linear motion when heated by an internal resistor. This type of actuator can take 2 to 3 *minutes* from when the operating voltage is applied to when the valve is fully open.

Although zone valves are available with several different actuator voltages, the vast majority of zone valves used in hydronic heating applications use low voltage (24 VAC) actuators. In many systems, this allows the zone valves to be powered by the same transformer that supplies control voltage to the thermostats and perhaps other low voltage control hardware.

The electrical power required to operate a zone valve is expressed in **VAs (volt-amps)**, and is usually listed on the zone valve or its data sheet. Care must be taken that the VA rating of the transformer supplying several zone valves is greater than the total VAs drawn when all zone valves are operating at the same time. If this is not done, the transformer will be overloaded and eventually burn out. *A rule of thumb is to add the VAs of all zone valves and other devices to be powered, then select a transformer capable of supplying at least 10 VAs more than that total.*

One common approach is to obtain the 24 VAC required by the zone valves from the transformer in the heat source. Before doing this, be sure that the transformer can provide the total VAs required by the low voltage devices in the heat source *in addition to* the VAs required by the zone valves. Manufacturers usually list the maximum **external VA rating** the heat source transformer can safely supply.

In addition to opening the valve body, many zone valves are equipped with internal switches that can be used to signal other components in the system that heat is being called for by a given zone. These switches are called **end switches** because their contacts close when the actuator reaches the end of its travel and the valve is fully open. End switch contacts open as soon as the valve starts to close.

The most common types of zone valve actuators have either three or four terminals or wires. The difference between them is in how the end switch is wired. In a three-wire zone valve, one contact of the switch is wired in parallel with the zone valve motor. In a four-wire zone valve, the end switch is isolated (not connected to any other internal wiring). This difference determines how the zone valve interfaces with the rest of the control system. A simplified representation of both three-wire and four-wire zone valves is shown in Figure 9–52.

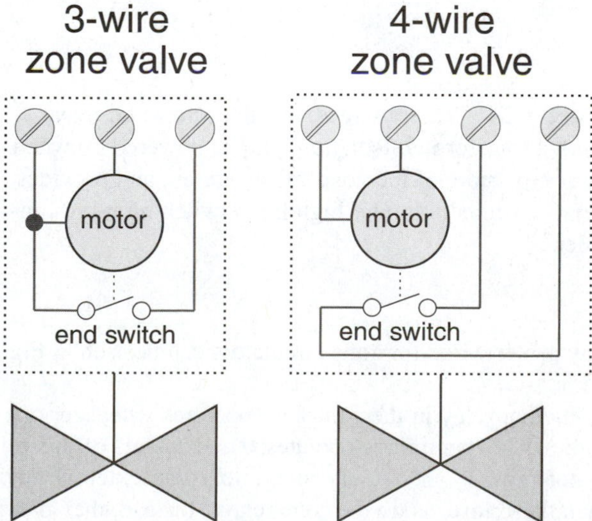

Figure 9–52 Simplified representation of 3-wire and 4-wire zone valves. Note differences in how the end switches are wired.

When a three-wire zone valve is used, the signal passed through the end switch is supplied by the same voltage source that operates the actuator motor. This is provided by a 24-VAC transformer. A schematic showing the wiring of three-wire zone valves and a typical high limit/switching relay control for an oil-fired boiler is shown in Figure 9–53.

The operating sequence of this control system is as follows: When either of the two room thermostats call for heat, the actuator of the associated zone valve is energized, opening the valve. When the actuator reaches the end of its travel (e.g., the valve is fully open), the end switch closes,

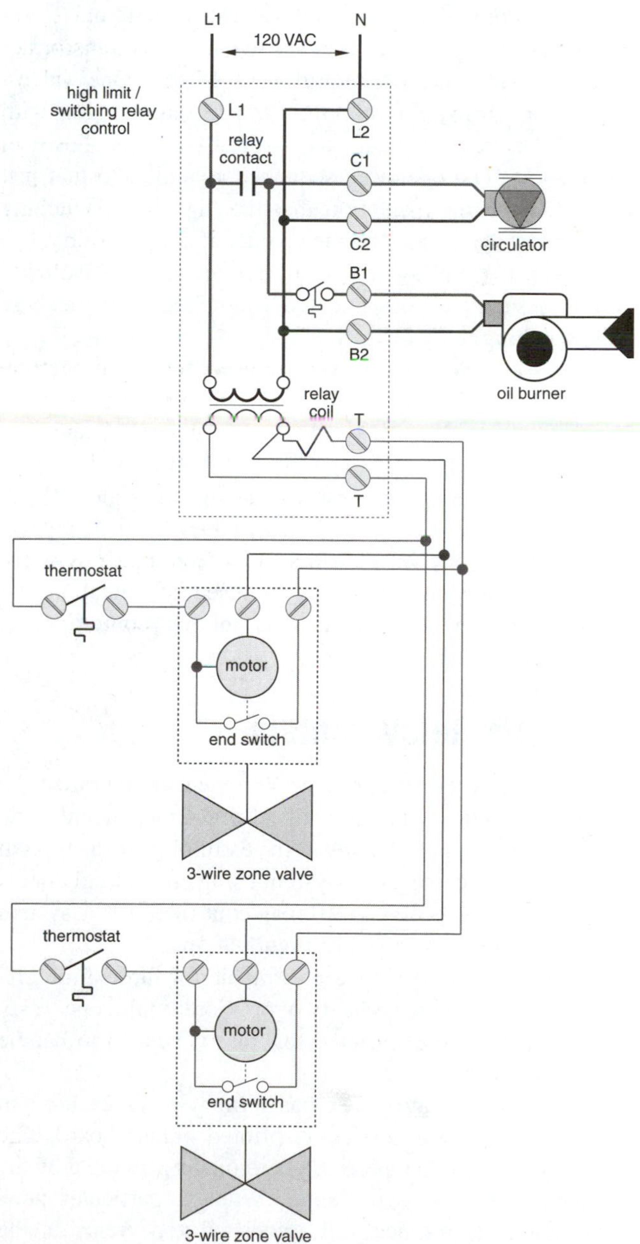

Figure 9–53 Typical wiring of 3-wire zone valves in combination with a boiler high limit/switching relay controller.

completing a circuit across the "T" terminals of the high limit/switching relay control. This control then starts the circulator and burner, and operates as described in Section 9.7. When the thermostat is satisfied, it opens the circuit through the zone valve motor. The burner and circulator stop and the zone valve closes. Burner and circulator operation can be initiated by *any* of the zone valves, and will continue as long as *any* of the thermostats are calling for heat.

If the external VA rating of the transformer in the high limit/switching relay control allows, additional zone valves could be added by simply duplicating the wiring used for one of the zone valves shown. If the external VA rating of the heat source's transformer is not high enough to supply the total VAs required by all zone valves, an external transformer must be used. This relieves the control's transformer from supplying the VA requirement of the zone valves. The typical wiring for this approach in combination with three-wire zone valves and a gas-fired boiler is shown in Figure 9–54. The operating sequence is similar to that just described with the exception that the high limit/switching relay control is activated by the contact of the isolation relay, rather than the end switch of a zone valve. This isolation relay is a separate component not supplied with the high limit/switching relay control.

When four-wire zone valves are used, the isolation relay can be eliminated. In basic systems, the isolated end switches of each zone valve are wired together in parallel, and connected across the "T" terminals of the boiler high limit/switching relay control as shown in Figure 9–55. This allows the circulator and boiler to operate when any one or more of the zone valves is on. In more complex systems, the isolated end switches may be used to signal a heat demand to other devices such as mixing controllers or outdoor reset controllers.

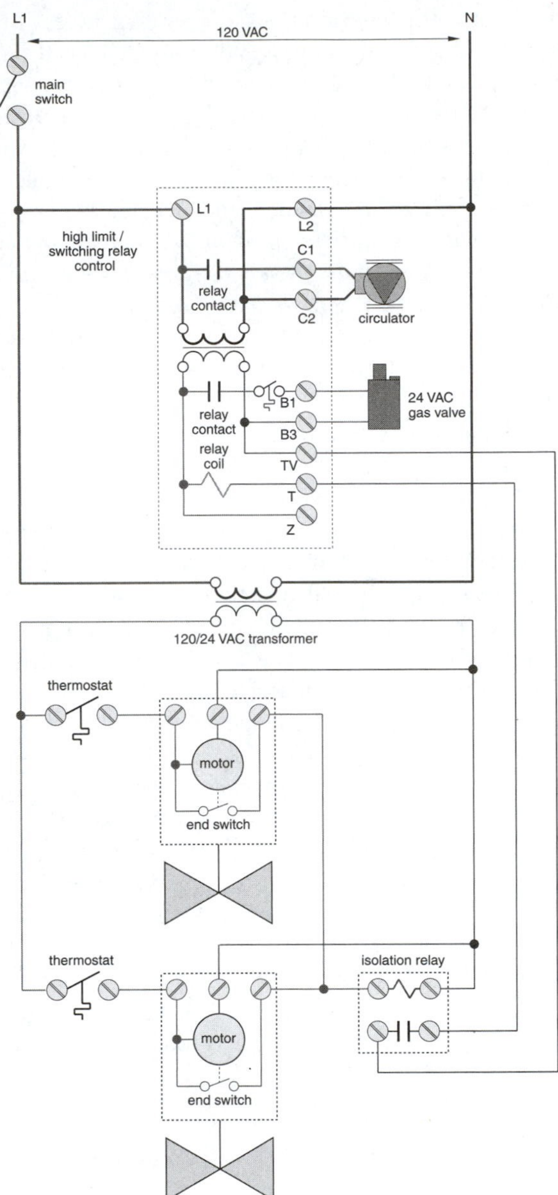

Figure 9–54 Three-wire zone valves powered by an external transformer. Isolation relay is powered up when any one or more of the zone valves are on, and provides operating signal to boiler high limit/switching relay controller.

Multizone Relay Centers

A **multizone relay center** is designed to centralize the wiring and control operation of several independent zone circulators or zone valves. An example of a typical multizone relay center for a system using zone circulators is shown in Figure 9–56a. A similar controller for a system using zone valves is shown in Figure 9–56b.

Multizone relay centers are available for three, four, and six zones. In many cases, two or more multizone relay centers can also be connected together if needed to handle additional zones.

A multizone relay center has terminals for each room thermostat along the top of its printed circuit board. The 24-VAC power for the thermostat circuits is generated by the transformer in the relay center. When a particular zone thermostat calls for heat, an associated zone relay in the relay center is turned on. In systems using zone circulators, this relay provides a line voltage output to drive the associated zone circulator. In systems using zone valves, a 24-VAC output is turned on to operate the associated zone valve. A relay center wired for zone circulators can be seen in Figure 9–56c.

Another relay in the controller operates whenever one or more of the zone relays operates. The contacts of this relay do not provide an output voltage to some external device when they close, as do the zone relays. Instead, they provided an isolated contact closure that can be used to signal other equipment in the system that a heat demand is present. In many systems, this contact is used to turn on the heat source anytime one or more of the zone thermostats calls for

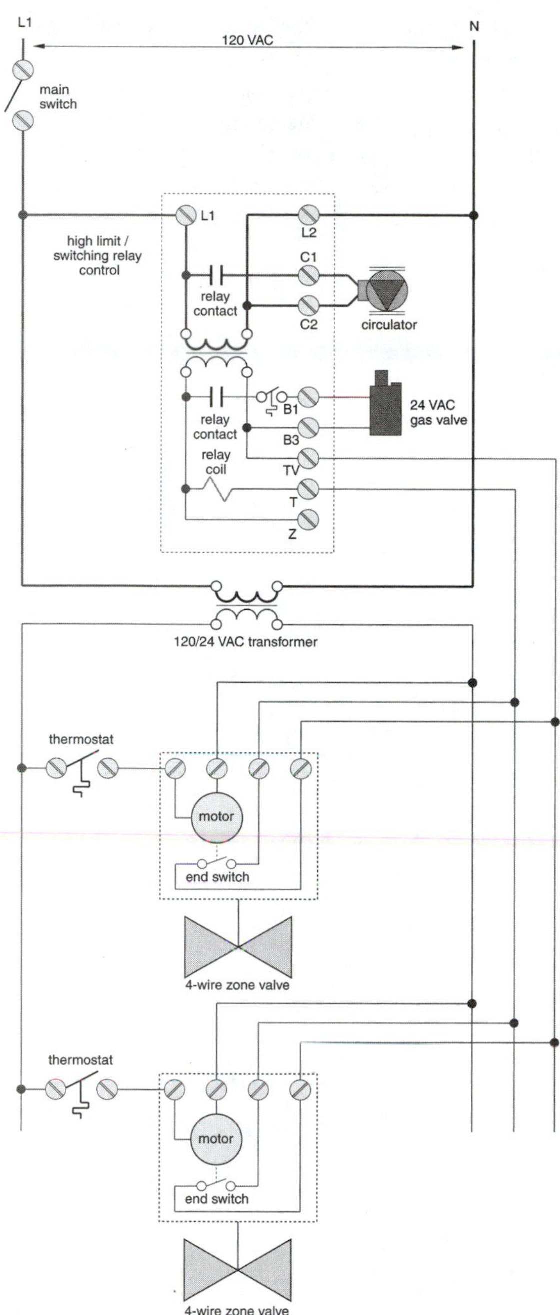

Figure 9–55 Typical wiring of 4-wire zone valves in combination with a boiler high limit/switching relay control.

heat. It can also be wired to turn on a mixing control, outdoor reset control, or some other component(s) in the overall control system.

Most multizone relay centers can be set up to have a **priority zone**. When so configured, the control turns all other zones off whenever the priority zone calls for heat. This allows the full output of the boiler to be available to the load connected to the priority zone.

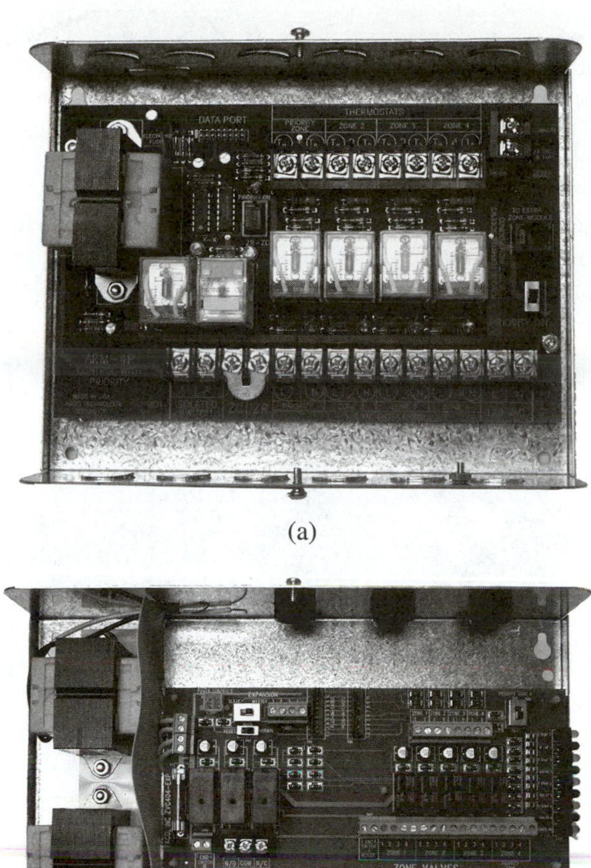

(a)

(b)

(c)

Figure 9–56 (a) Internal view of a multizone relay center designed to operate up to 4 zone circulators. (b) Multizone relay center designed to operate up to 4 zone valves. (c) Relay center wired for zone circulators.

The most common priority load is domestic water heating. Since the full boiler capacity can be directed to domestic water heating, hot water is produced very quickly. This is very desirable in applications that have large demands for domestic hot water, such as homes with large whirlpool tubs. A typical indirectly fired storage water heater, set up with priority control, can produce several times more hot water in a given period of time compared to a residential electric water heating tank.

A schematic representation of a four-zone relay center with domestic water heating as a priority load is shown in Figure 9–57. The same functionality could be assembled using individual components. However, the total labor and material cost would likely be significantly higher than using a preassembled multizone relay center.

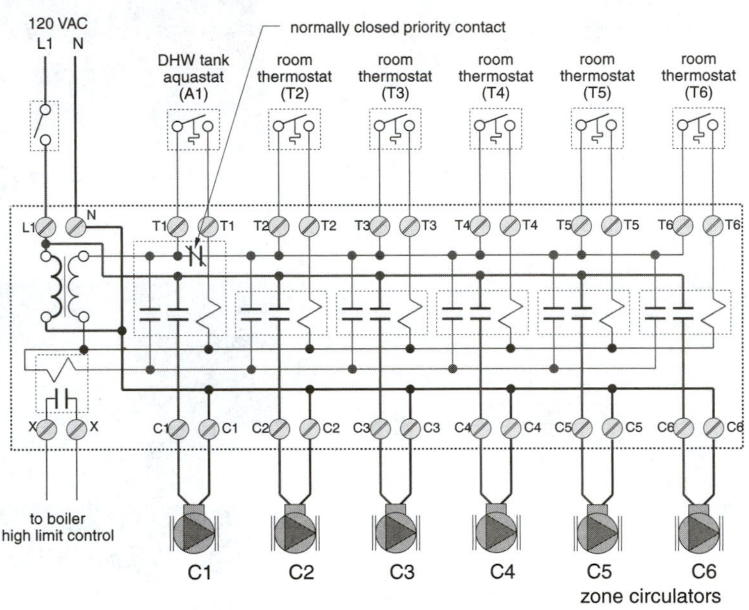

(a) electrical schematic

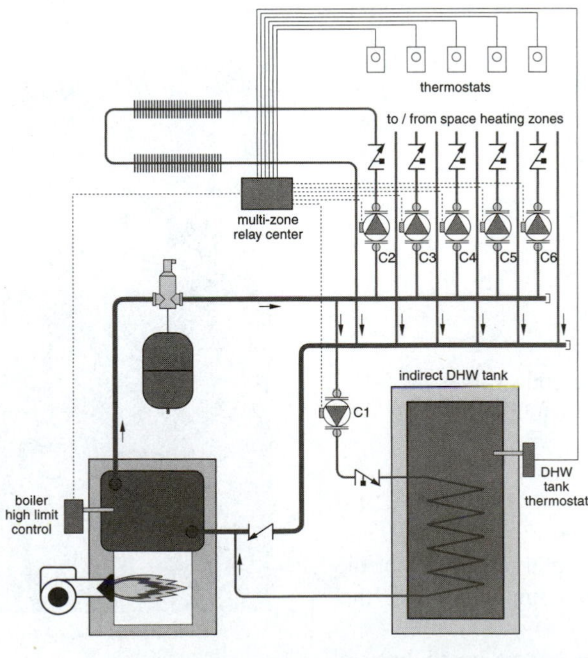

(b) typical piping

Figure 9–57 A 6-zone multizone relay center installed. Notice low voltage thermostat wiring at top, and line voltage circulator wiring at bottom.

9.7 BASIC BOILER CONTROL HARDWARE

This section discusses the basic devices used to control boilers. Some are electromechanical, others are electronic. In most systems, these devices are combined with the system controls discussed in the previous section to form an overall control system.

Combination High Limit/Switching Relay Control

Because many simple residential hydronic systems use a single boiler, single circulator, and single room thermostat, they have virtually identical control requirements. The similarity of these systems allows manufacturers to design a specialized control that can coordinate the operation of all these components. Manufacturers of packaged boilers often equip their boilers with such a control, and usually prewire it to both the burner and circulator. An example of a high limit/switching relay control for an oil-fired boiler is shown in Figure 9–58a. Its circuitry is shown in Figure 9–58b.

This control incorporates several components such as a transformer, relay, and aquastat into a small case that mounts directly onto the boiler. These components are shown inside the dashed lines in Figure 9–58b. The temperature sensing bulb projects from the rear of the case and into a well in the boiler block.

A typical residential system with this control has the following operating sequence:

1. Whenever the system's master switch is closed, line voltage is supplied to the control.
2. When the room thermostat contacts close, a low voltage (24 VAC) circuit energizes the coil of the internal relay.
3. The relay contacts close to supply line voltage to the system circulator. If the boiler is cool, the internal aquastat contacts are also closed, and the burner operates. If the boiler temperature is at or close to the **high limit setting** of the control, the aquastat contacts are open and the burner remains off.
4. If and when the boiler temperature climbs to the high limit setting, the aquastat contacts open to turn off the burner. The circulator, however, continues to run. Such operation is typical when the system load is less than the heating capacity of the boiler.
5. When the boiler temperature decreases to a certain differential below the high limit setting (typically about 8 °F), and the demand for heat continues (as evidenced by the closed thermostat contact across the TT terminals), the burner is turned back on.
6. When the room thermostat contacts open, the circuit through the relay coil is interrupted, and both the burner and circulator are turned off.

Figure 9–58a Example of a boiler high limit/switching relay control. *Courtesy of White Rogers.*

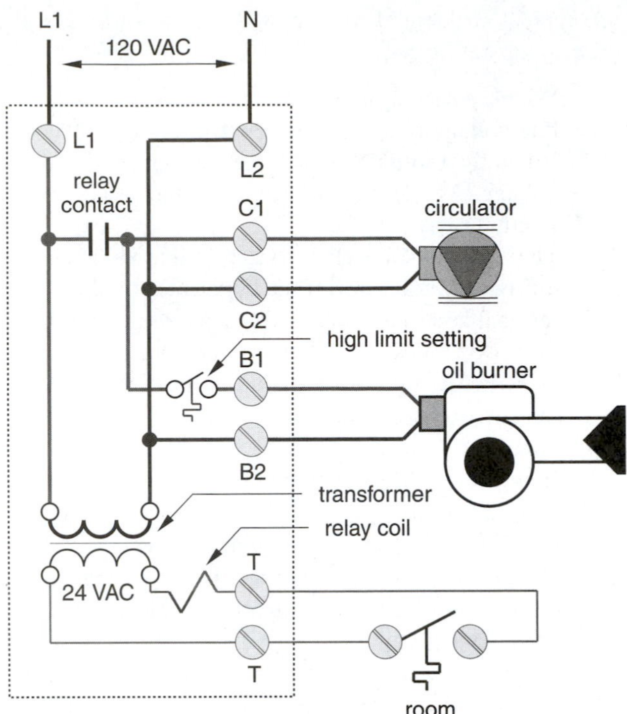

Figure 9–58b Ladder diagram showing internal wiring.

This operation is depicted in Figure 9–59. Notice that boiler temperature eventually drops to room temperature if there is no demand for heat over several hours. A boiler operated in this manner is said to be **demand-fired**. The burner of a demand-fired boiler operates only when there is a demand for heat from the system. The boiler does not maintain a minimum temperature at all times. Most modern boilers are operated in this manner.

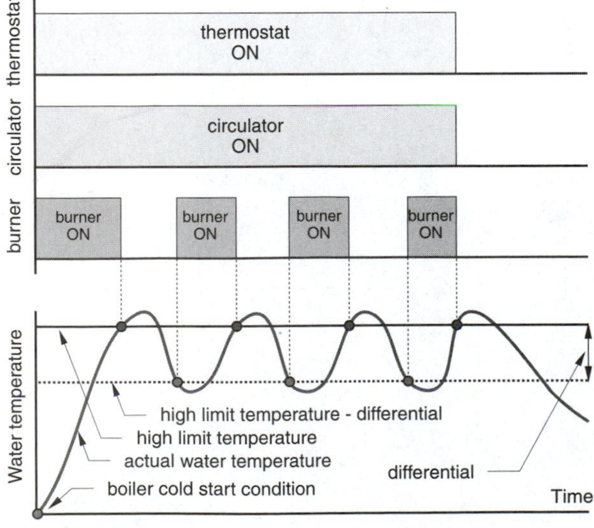

Figure 9–59 Operation of a combination high limit/switching relay control over a period of time.

Triple Action Control

For many years, it was common in residential hydronic systems to produce domestic hot water using a **tankless coil** mounted in the same boiler that provided space heating. Cold water was heated as it passed through this copper coil on its way to a faucet. *Because hot water could be needed at any time, boilers with tankless coils had to maintain a minimum water temperature at all times.*

Due to constant standby heat loss from the boiler, this method of domestic hot water heating is considered inefficient by today's standards. However, many systems that use tankless coils are still operating, and individuals who might service such systems should be familiar with their operation.

Because this type of system was once common, several manufacturers developed controllers specifically for it. This type of control is called a **triple action control**. An example of such a control is shown in Figure 9–60a. Its associated wiring is shown in Figure 9–60b.

The triple action control mounts directly to the boiler. It has a single temperature probe that inserts into a well in the boiler block. The control has two dials for temperature adjustments, one to set the boiler's high limit, the other to set the boiler's low limit. The latter is the minimum temperature the boiler will maintain at all times regardless of load.

The triple action control fires the burner if the room thermostat calls for heat or if the boiler temperature falls below the low limit setting. If the load on the boiler exceeds the boiler's heating capacity, boiler temperature may drop below the low limit setting. When this occurs, the controller temporarily stops the space-heating circulator. The premise is that by shedding the space-heating load, the boiler temperature will increase, and thus be able to maintain sufficient flow of domestic hot water. The space-heating circulator restarts when the boiler temperature recovers its low limit setting plus differential. A sequence showing control operation is given in Figure 9–61.

Manual Reset High Limit Control

Many mechanical codes, particularly those applicable to publicly occupied buildings, require a redundant temperature-limiting control on all boilers. This control must be capable of interrupting boiler heat production should the primary operating control fail in the on position. Most codes also require that the redundant high limit control cannot automatically close its contacts once they have opened. The control designed for this purpose is called a **manual reset high limit (MRHL)**. An example is shown in Figure 9–62

The manual reset limit control should be mounted with its probe sensing the highest water temperature in the system. In some cases, the temperature sensor can be mounted directly into the boiler block. In other cases, the sensor mounts into a tee near the outlet of the boiler as shown in Figure 9–62. The manual reset high limit should be set

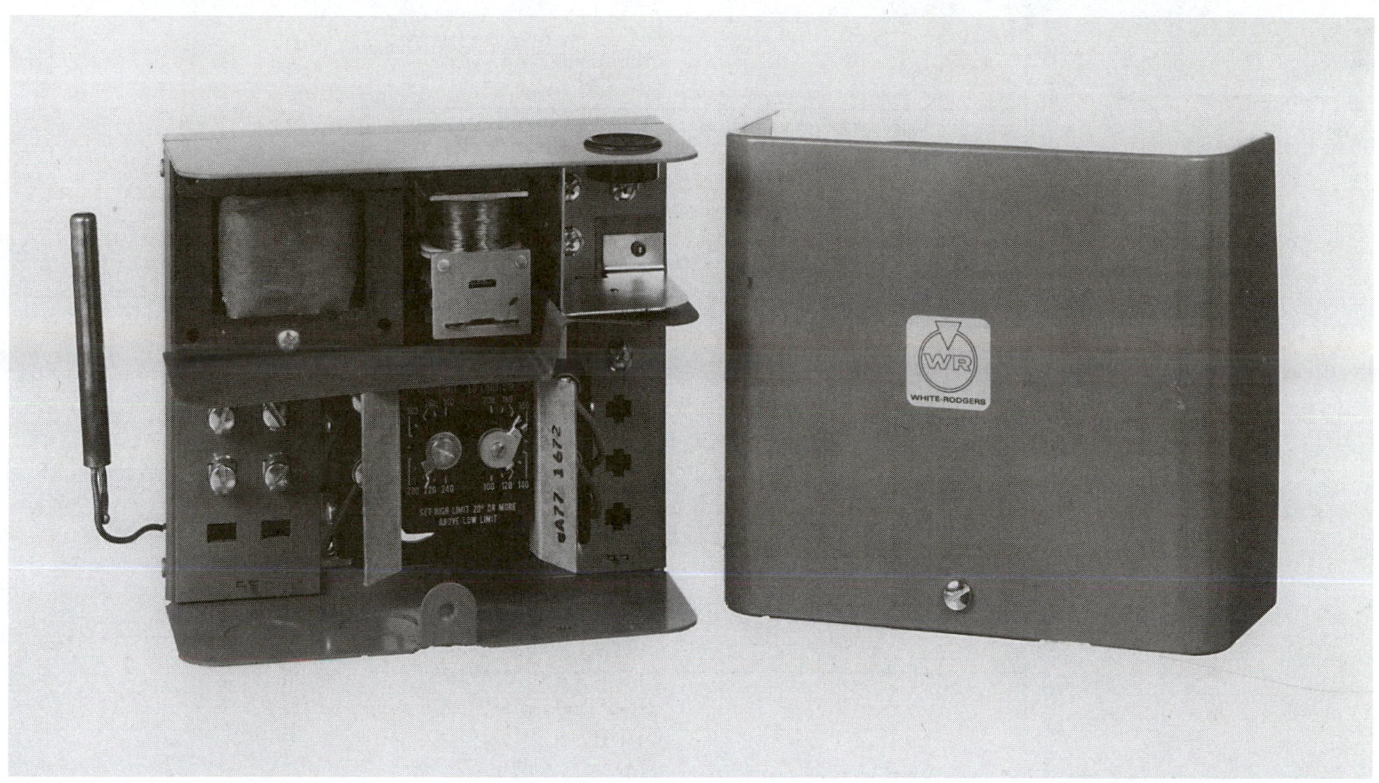

Figure 9–60a Example of a triple action control. Note individual settings for high limit and low limit temperatures. *Courtesy of White Rogers.*

approximately 20 °F above the primary high limit control setting to prevent nuisance tripping due to boiler temperature overshoot. However, most mechanical codes that require this device also stipulate its maximum temperature setting, which should never be exceeded.

Low Water Cutoff Control

Another safety device required on boilers by many codes is a **low water cutoff (LWCO)**. This device uses the conductivity of water to complete a circuit between its probe and the boiler shell. If this circuit is broken due to the water level dropping below the probe, the normally closed contacts inside the LWCO open to interrupt the burner circuit. An example of a low water cutoff control is shown in Figure 9–63.

Low water cutoffs are available for both line voltage and low voltage (24 VAC) circuits. Line voltage controls are commonly used with oil burner circuits, while low voltage controls are commonly used with low voltage gas valve circuits. All low water cutoffs require constant power to operate their internal circuitry. Most also include time delay logic to prevent a nuisance trip due to an air bubble passing over the probe. Always verify any specific requirements for low water cutoff devices with local mechanical codes.

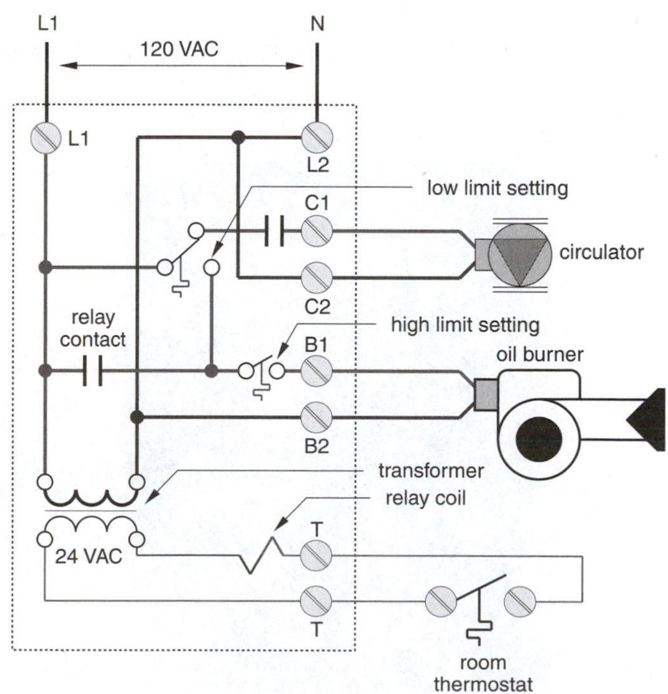

Figure 9–60b Ladder diagram showing internal wiring.

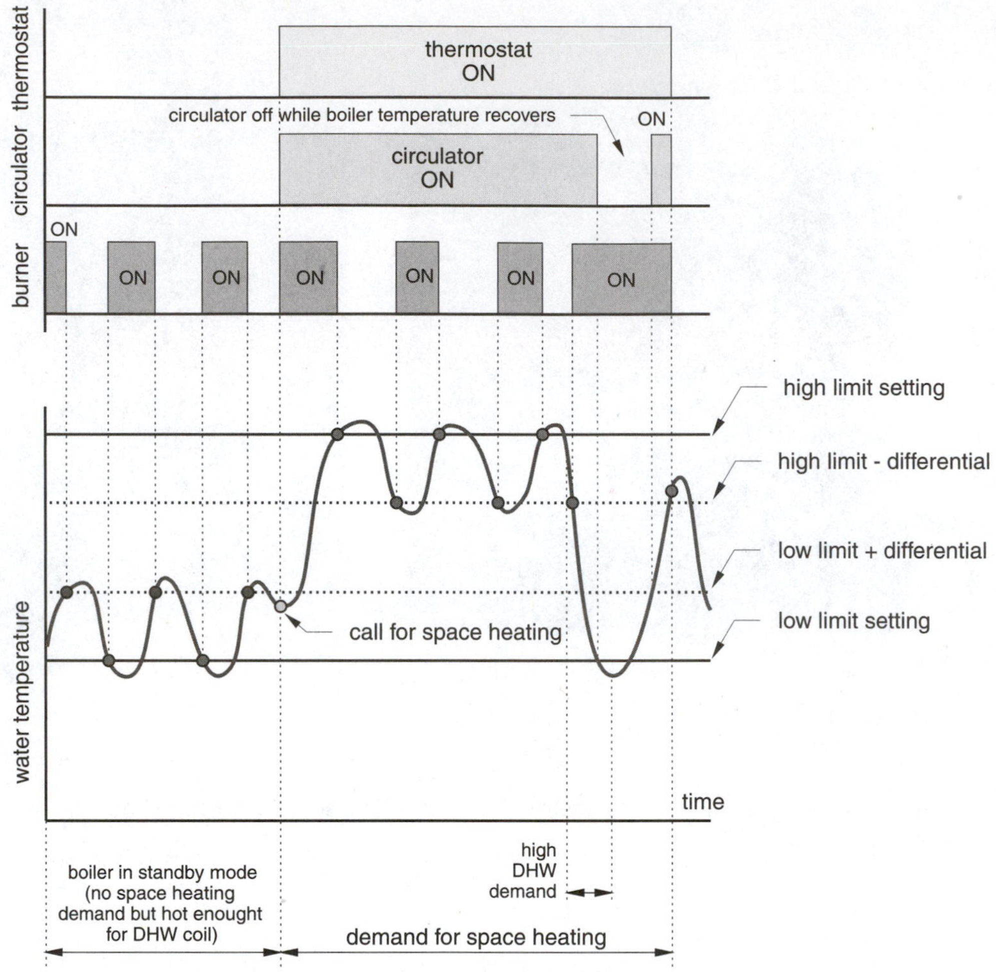

Figure 9–61 Operation of triple action control over a period of time.

Figure 9–62 Example of a manual reset high limit (MRHL) controller. Note the reset button near the bottom of the controller.

Figure 9–63 Example of a low water cut off controller.

Boiler Reset Controllers

The concept of boiler reset control was discussed in Section 9.4. From a hardware standpoint, a modern boiler reset control is an electronic device with an on/off contact that turns the boiler's burner on and off. This on/off action attempts to maintain the outlet temperature of the boiler close to a target temperature that varies according to outdoor temperature. The colder it is outside, the higher the boiler outlet target temperature. Although they are called "boiler" reset controllers, they can also be used with other types of heat sources such as hydronic heat pumps.

Because it uses an on/off output, a boiler reset control must operate with a differential. On some controllers the differential is fixed, and centered on the calculated target water temperature as shown in Figure 9–64. On other controllers, the differential is a function of the outdoor temperature, and becomes wider as the outdoor temperature increases. This approach, which is also illustrated in Figure 9–64, lengthens the boiler on-cycles to minimize short cycling during low load conditions.

A boiler reset control is an operating control, not a safety device. It does not replace the need for a high limit control, or for other safety controls such as a manual reset high limit or low water cutoff that may be required by local mechanical codes.

The normally open output contact of a boiler reset control is often wired in series with other operating and safety controls such as the high limit control(s) and possibly even a thermostat. In this configuration, the boiler reset control simply "preempts" the other limit controls during normal operation. Its contact opens the circuit that keeps the boiler operating at a temperature that is lower than the setting of the limit controls. If, for some reason, the boiler reset control should fail to open its contacts, the other high limit controls will eventually stop the heat source. When used in this way, it is important that the settings of any other limit controls be sufficiently high to avoid interfering with the operation of the boiler reset control.

In some systems the reset control contact is bypassed during domestic water heating so higher temperature water can be supplied to the heat exchanger of an indirect water heater. In this mode, the boiler's high limit control is what limits water temperature. When the domestic water heating cycle is finished, water temperature control is passed back to the boiler reset controller. Some reset controllers have the logic for the domestic water heating mode built in. An example of such a control is shown in Figure 9–65.

When used with a conventional boiler, the boiler reset control must be set for a minimum supply water temperature high enough to prevent flue gas condensation within the boiler during low load conditions. This concept is called partial reset control and is illustrated in Figure 9–24. All modern boiler reset controls have an adjustable minimum boiler temperature that is set using a dial or through a menu-driven programming routine. The controller will not target

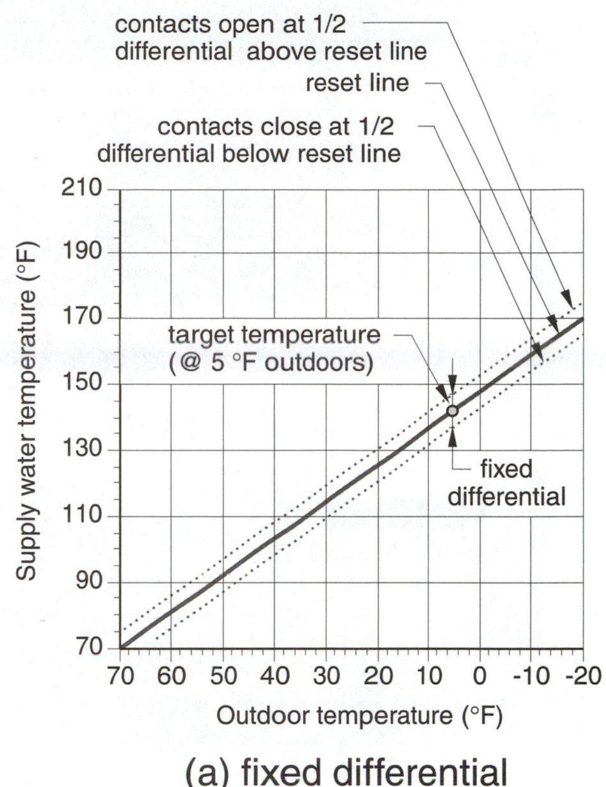

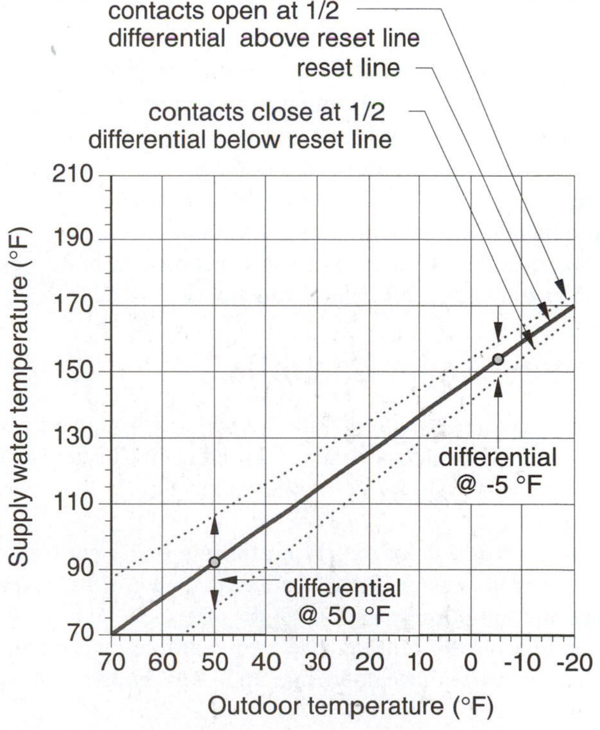

Figure 9–64 Boiler reset controllers operate on a differential centered on the reset line (target temperature). (a) fixed differential, (b) variable differential to reduce short cycling of the heat source at low loads.

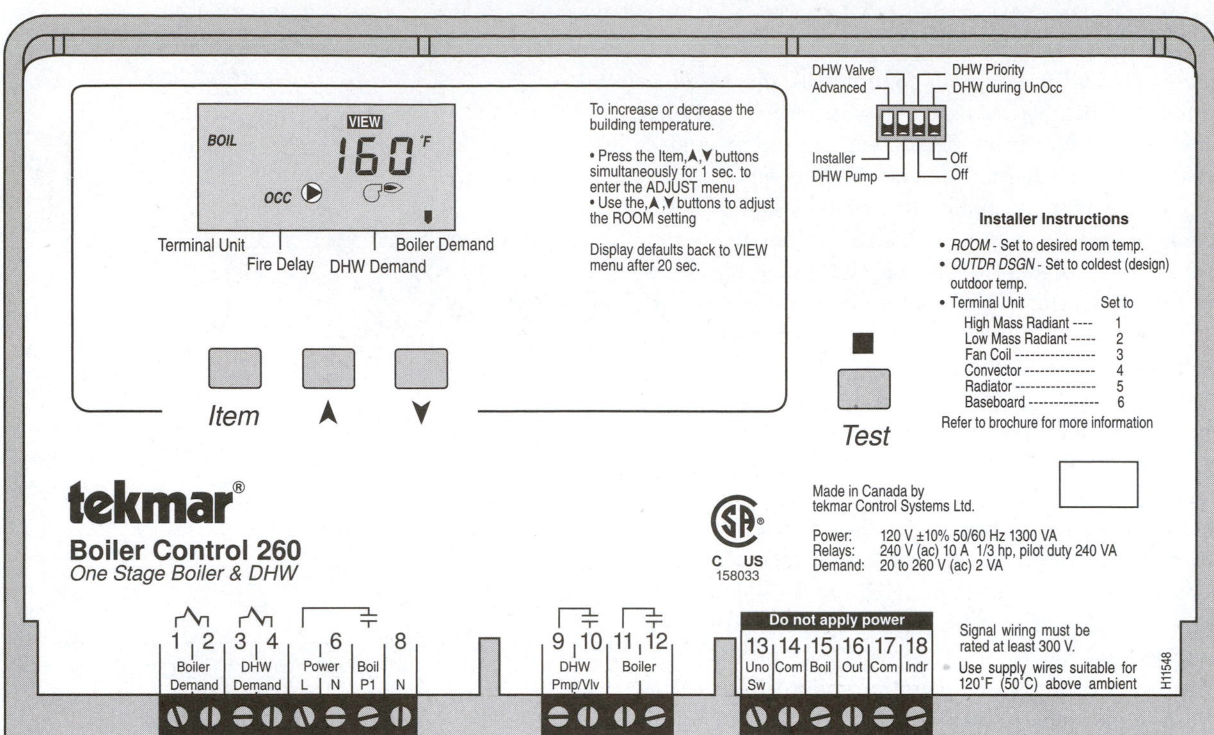

Figure 9–65 Example of a modern boiler reset controller with ability to bypass reset function during calls for domestic water heating. *Courtesy of tekmar Controls.*

supply temperatures below this setting regardless of the outside temperature. The minimum operating temperature varies with the type of boiler and the type of fuel. For most gas-fired and oil-fired boilers, the minimum supply temperature should not be set below 140 °F unless otherwise recommended by the boiler manufacturer.

Some boiler staging controllers incorporate boiler reset logic as one of several control functions.

Multiple Boiler Controllers

The benefits of using a multiple boiler system have been discussed earlier in this chapter, as well as in Chapter 3. The hardware necessary for controlling a multiple boiler system will now be covered.

The current state-of-the-art in **multiple boiler controllers** is a microprocessor-based device that performs several functions to optimize the operation of two to eight boilers depending on the heating load. An example of a multiple boiler controller that can operate up to four boilers is shown in Figure 9–66.

While operating, this controller attempts to keep the water temperature at its supply sensor close to a target temperature. The target temperature depends on the type of "demand" the controller is receiving at the time. This demand is usually a low voltage signal sent to the controller when a circuit is completed through another device such as a thermostat. This voltage is sent to a specific set of terminals on the multiple boiler controller.

One type of demand is known as a **setpoint demand.** When it occurs, the controller targets a fixed supply water temperature that has been previously set or programmed by the installer. *The controller does not adjust this target temperature based on outdoor temperature.* Setpoint demands are commonly invoked when the load is supplied through a heat exchanger. For good heat transfer rates, most heat exchangers require relatively high supply water temperatures in the range of 180 to 200 °F. One example of such a load is an indirect water heater. Another is a heat exchanger used for supplying heat to a snowmelting system or a swimming pool.

The other type of demand is a **space heat demand**, and is invoked when space heating is required. Under this demand, the target temperature is calculated based on outdoor temperature and the reset ratio set on the controller. In this mode, the multiple boiler controller is operating as a boiler reset controller.

When either type of demand is present, the controller monitors the error between the calculated target temperature and the temperature measured at its supply sensor. This error, combined with PID processing, determines the firing sequence of the boilers. After the first boiler is fired, the controller monitors the duration of the error as well as the rate of change of the error to determine if additional boilers need to be fired. If the error is small and being quickly reduced by the first boiler, additional boilers will not be

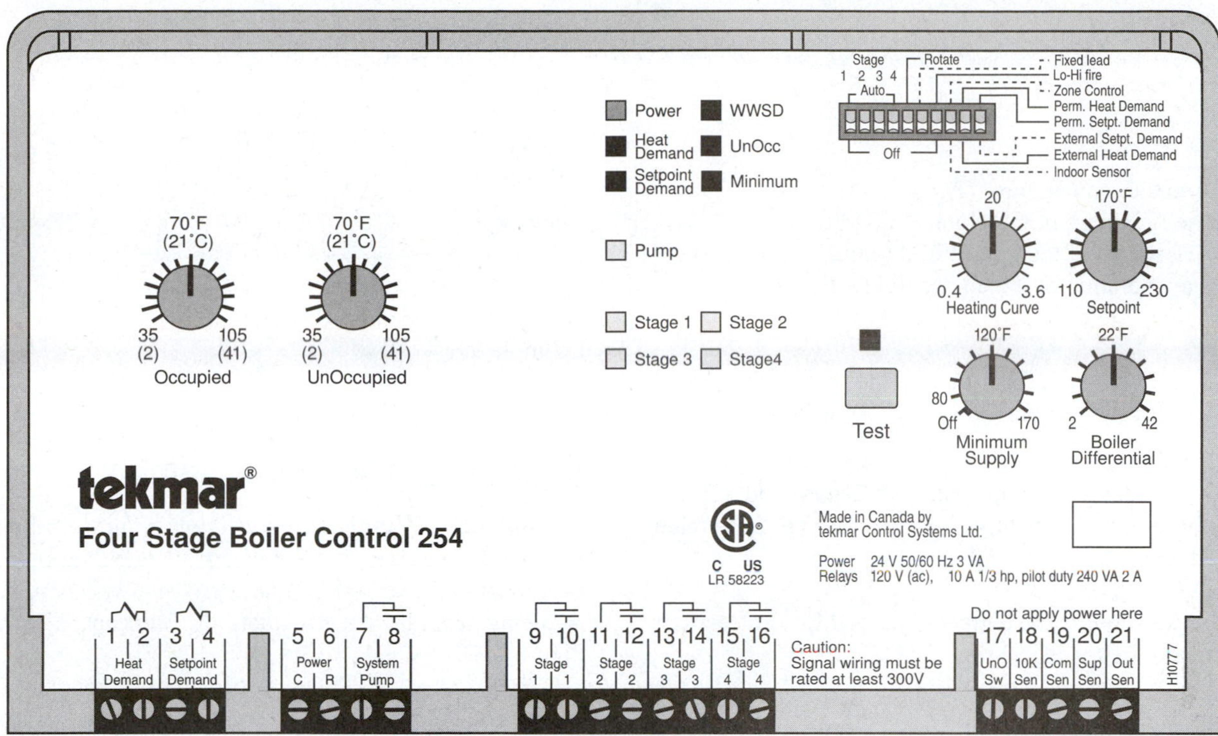

Figure 9–66 Example of a boiler staging controller capable of operating up to 4 boilers. *Courtesy of tekmar Controls.*

fired. If the error is large, or slow to change, one or more additional boilers will be fired.

Most multiple boiler controllers can also be configured to rotate the boiler firing order. The goal is to allocate approximately the same total running time to each boiler so they can be serviced, and ultimately replaced at the same time. **Boiler rotation** is also sometimes called **lead/lag control**.

In some systems, a specific boiler needs to be fired first. This boiler is called the **fixed lead boiler**. In some systems it may be a condensing boiler that operates at very high efficiency when the target supply temperature is low. In other systems, it may be a boiler that needs to fire first to establish the proper draft in the chimney and breaching system. Whatever the case, a switch or program setting on the controller allows a given boiler to serve as the fixed lead. If three or more boilers are used, the other boilers can still have their firing order rotated.

Some multiple boiler controllers can also be configured to operate two or more HI/LO fire boilers. For example, a pair of boilers each having a HI/LO burner represents four stages of heat input. However, the firing sequence is different from the sequence used with four identical boilers. When bringing on additional stages, the controller must fire the HI stage of the first boiler before turning on the LO stage of the second boiler. Most modern multiple boiler controls have this logic built in, and only require the installer to set a switch or program menu item to properly configure the controller.

9.8 MIXING STRATEGIES AND HARDWARE

Earlier in this chapter, the subject of regulating heat output by varying the water temperature supplied to the heat emitters was discussed. The relationship between supply water temperature and heat output is approximately proportional, making this approach preferred from a control standpoint.

This section discusses many of the mixing methods and hardware currently used in hydronic heating. Some approaches use nonelectric thermostat devices, others used sophisticated microprocessor controllers. All have strengths and weakness as well as applications for which they are particularly appropriate.

Purpose(s) of Mixing

The primary purpose of mixing is to control heat output by varying the temperature of the fluid supplied to the heat emitters. This is done by blending higher temperature water from the heat source with cooler water returning from the distribution system. The flow rate and temperature of these two fluid streams determine the blended or "mixed" temperature supplied to the distribution system.

If the flow rate and temperature of two fluid streams that are mixed together are known, the mixed temperature can be calculated using Equation 9.10:

(Equation 9.10)

$$T_{mix} = \frac{(T_{hot} \times f_{hot}) + (T_{cool} \times f_{cool})}{f_{hot} + f_{cool}}$$

where:

T_{mix} = mixed temperature (°F)
T_{hot} = temperature of the hotter fluid (°F)
F_{hot} = flow rate of the hotter fluid (gpm)
T_{cool} = temperature of the cooler fluid (°F)
F_{cool} = flow rate of the cooler fluid (gpm)

The equation applies regardless of the type of fluid used, as long as both streams are the same type of fluid.

Example 9.6: Two streams of water enter a tee as shown in Figure 9–67. The hot stream enters at 2 gpm and 170 °F. The cooler stream enters at 4 gpm and 120 °F. Determine the mixed temperature leaving the tee.

Solution: Substituting the temperatures and flow rates given into Equation 9.10 yields:

$$T_{mix} = \frac{(170 \times 2) + (120 \times 4)}{2 + 4} = \frac{820}{6} = 136.7°\text{ F}$$

Discussion: The mixed temperature always falls between the higher and lower entering temperatures.

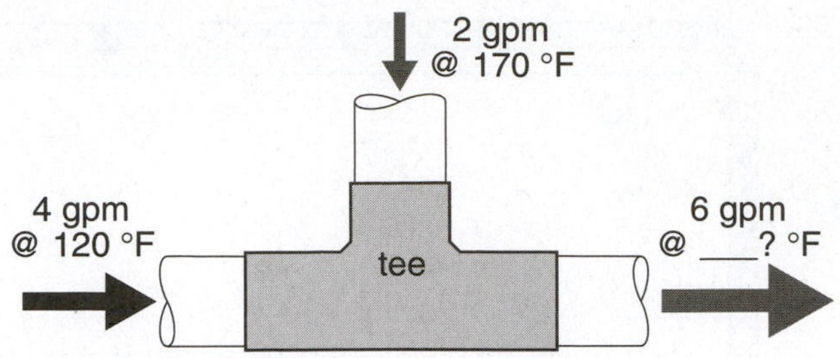

Figure 9–67 Streams of water mixing in a tee for Example 9.6.

Concept of a Mixing Assembly

A **mixing assembly** *is a collection of hardware components and control logic that collectively regulates the rate of heat transfer from the heat source to the distribution system.* A piping schematic that depicts this concept is shown in Figure 9–68.

The gray box in Figure 9–68 can be thought of as a "container" for the various components that make up a specific mixing assembly. It also represents a bridge between where heat is produced, and where it is released by the heat emitters requiring the lower water temperature. All heat that eventually reaches the lower temperature heat

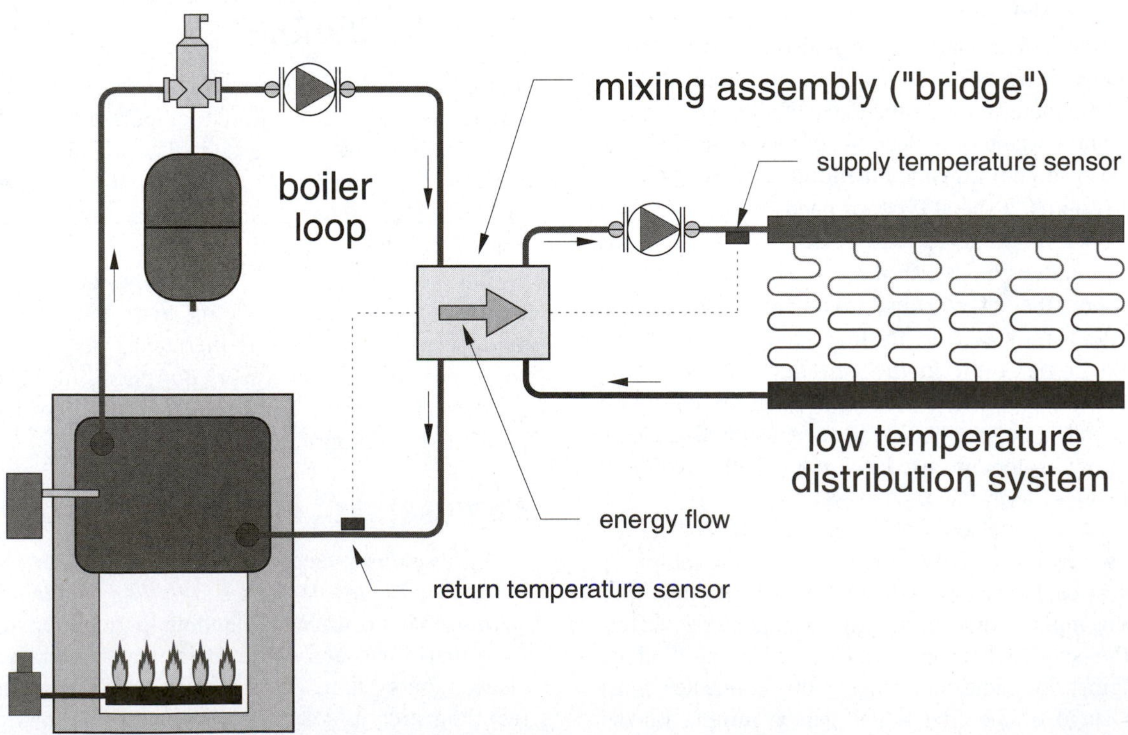

Figure 9–68 Concept of a mixing assembly as a bridge between the higher temperature boiler loop and low temperature distribution system. The mixing assembly acts as a "metering device" to regulate the rate of heat transfer to the lower temperature distribution system.

emitters must pass through the mixing assembly. The piping schematics that follow will show how various components are assembled to make different mixing assemblies.

Boiler Protection

Controlling the supply temperature to the distribution system is not necessarily the only function of the mixing assembly. *In systems using conventional boilers, the mixing device must also protect the boiler from operating at temperatures low enough to cause sustained flue gas condensation.* This function is called **boiler protection**. *Failure to provide it can lead to corrosion and scale formation that drastically shortens the life of the boiler and its venting system.*

Unfortunately, the need to protect conventional boilers from such conditions has too often been viewed as secondary in importance to providing the desired supply temperature. Such disregard for unchecked flue gas condensation has led to serious problems in many systems.

The exact boiler inlet water temperature where condensation problems develop depends on the design of the boiler as well as its operating conditions, as discussed in Chapter 3, Hydronic Heat Sources. It is generally recognized that maintaining return temperatures of 130 °F for gas-fired boilers and 150 °F for oil-fired boilers eliminates the damaging effects of flue-gas condensation. However, these temperatures may not apply to all boilers. Manufacturers should be consulted for the recommended minimum operating temperatures of their boilers.

To properly protect a conventional boiler, the mixing assembly must determine if the distribution system is extracting heat faster than the heat source is generating heat. If heat dissipation exceeds heat production, the first law of thermodynamics demands that fluid temperature in the system must drop until thermal equilibrium is reestablished. If this requires the boiler to operate well below the dewpoint temperature of its flue gases—so be it! Thermodynamics "does not care" if the boiler is condensing; it only cares about balancing heat dissipation with heat production.

All mixing assemblies intended to protect a conventional boiler from sustained flue gas condensation must monitor the boiler temperature. Most monitor boiler inlet temperature. Some monitor boiler outlet temperature and use it along with other information to estimate the boiler inlet temperature. Without this temperature monitoring, the mixing assembly is "blind" to possible low temperature operation that causes flue gas condensation.

Boiler protection is accomplished by reducing the rate of heat flow from the boiler into the distribution system whenever the boiler temperature approaches or falls below a specified minimum operating temperature. This partially "unloads" the boiler from the distribution system allowing its temperature to quickly rise above condensing conditions.

A good analogy to a mixing assembly that provides boiler protection is the clutch in a car. Those who have driven a car with a clutch know that if the clutch is let out too quickly the engine "lugs down." From the standpoint of energy transfer, this is a situation where the drive train is pulling mechanical energy away from the engine faster than the engine can produce it. This condition is quickly sensed by an experienced driver who pushes the clutch pedal in slightly to partially unload the engine, and quickly reestablish a balance between the rates of energy production and energy dissipation. A properly controlled clutch allows the full energy output of the engine to be transferred to the drive train, but does not allow the drive train to pull energy away from the engine faster than it is being produced.

Similarly, a properly controlled mixing device acts as an intelligent "thermal clutch" between the heat source and the distribution system. If heat flows through the mixing device and into the distribution system faster than the boiler can produce it, the boiler temperature immediately starts to drop. When the mixing assembly senses this (by way of the boiler temperature sensor), it immediately responds by reducing the rate of hot water flow into the mixing device, which reduces the rate of heat transfer to the distribution system. A properly controlled mixing assembly allows the full output of the boiler to reach the distribution system, but does not allow the distribution system to pull heat away from the boiler faster than it is being produced.

Mixing Using Manually Set Valves

One simple mixing assembly can be made using manually set valves. Three options are shown in Figure 9–69. The first uses a pair of 2-way valves. The second is a single 3-way mixing valve. The third is a single 4-way mixing valve.

In all three cases, the valves are manually set to a given stem position, and remain in that position as the system operates. Assuming the flow rate in the system in not altered through other means, these mixing assemblies all create a fixed flow rate for both the hot and cool streams entering the mixing point.

If the temperature and flow rates of the hot and cool streams remained constant, so would the mixed temperature leaving the mixing point. Unfortunately, such conditions are very hard to sustain in real systems. The temperature of the hotter fluid varies as the heat source is cycled on and off. The temperature and flow rate of the cooler fluid returning from the distribution system also varies depending on the operating conditions at any given time. Manually set valves cannot sense these changes in temperature and therefore cannot compensate for them. The result is that *the mixed temperature varies whenever the temperature or flow rate of either entering stream changes*. This causes the supply temperature as well as the return temperature to the heat source to vary, in some cases over a wide range. Although this type of mixing may be inexpensive, it does not provide stable supply temperature. It cannot protect a conventional boiler from flue gas condensation. It also lacks the control intelligence to provide mixing reset control. In general, it is of little use in hydronic space heating applications.

Any type of manually set mixing valve should be thought of as a "dumb valve." It has no intelligence to adjust the mixing process to maintain the desired supply and return temperature conditions. Unfortunately, many systems have been installed where false confidence has been placed in manually set valves (3-way and 4-way valves in particular) to provide the proper supply temperature and boiler protection in low temperature radiant heating systems. *A mixing valve without an actuator and controller is no more useful than a car without a driver or a computer without software.*

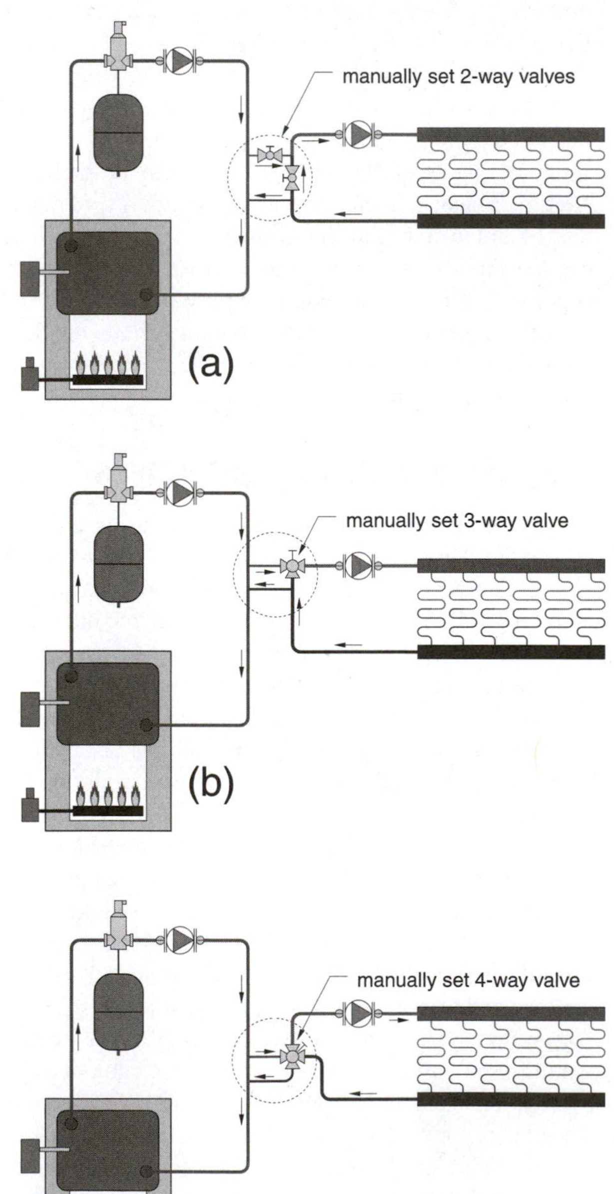

Figure 9–69 Three options for manually-set mixing devices: (a) a pair of 2-way valves, (b) 3-way rotary valve, (c) 4-way rotary valve. None of these options can respond to changes in water temperature or flow rate.

Mixing with 3-way Thermostatic Valves

Chapter 5, Pipings, Fittings, and Valves, discussed 3-way thermostatic mixing valves. These valves have an internal thermostatic element that senses the temperature of the fluid leaving their mixed outlet port. The valve spool automatically adjusts its position to try to maintain the outlet temperature close to the target temperature set on the valve's knob.

A single 3-way thermostatic valve can be used to provide supply temperature control when the heat source can tolerate low inlet temperatures, and relatively low flow rates. This is not the case for most conventional boilers. When a conventional boiler is used as the heat source, a second 3-way thermostatic valve should be used to monitor and boost the boiler inlet temperature when necessary to prevent sustained flue gas condensation. The suggested piping is shown in Figure 9–70. Notice that a second circulator is now required to provide proper mixing in the thermostatic valve protecting the boiler. This valve and circulator also helps maintain flow through the boiler. The latter is desirable when low mass boilers are used.

Mixing with a 3-way Motorized Valve

When a 3-way rotary mixing valve is combined with a motorized actuator and controller, the resulting mixing assembly can perform several functions depending on the capabilities of the controller. Most modern mixing valve controllers provide very accurate supply temperature control. This temperature can be a fixed setpoint temperature, or a temperature that is reset based on outdoor temperature. If the controller is equipped with a boiler temperature sensor, it can also protect a conventional boiler from flue gas condensation when piped as shown in Figure 9–71. Finally, some mixing valve controllers automatically detect when the system has been inactive for a period of a few days and respond by "exercising" certain system components. The distribution circulator is turned on for a few seconds, and the mixing valve is powered fully open, then closed. This reduces the possibility of seizing during long periods of inactivity.

Some controllers used with 3-way motorized valves are built into the actuating motor that attaches to the valve's body as shown in Figure 9–72. Others connect to the actuator motor using a three-wire cable.

Some controllers generate a floating control signal (discussed in Section 9.2). This output operates a 24-VAC motor and gear train assembly in the actuator. The actuator shaft turns very slowly, taking from 2 to 3 minutes to rotate a valve across its full route of travel. Because hydronic systems do not require wide temperature changes over short periods of time, it is not necessary for the valve shaft to rotate quickly. Slower rotation actually helps stabilize the system by allowing adequate time for feedback from the supply sensor located downstream of the mixing valve. The actuator motors are equipped with end switches

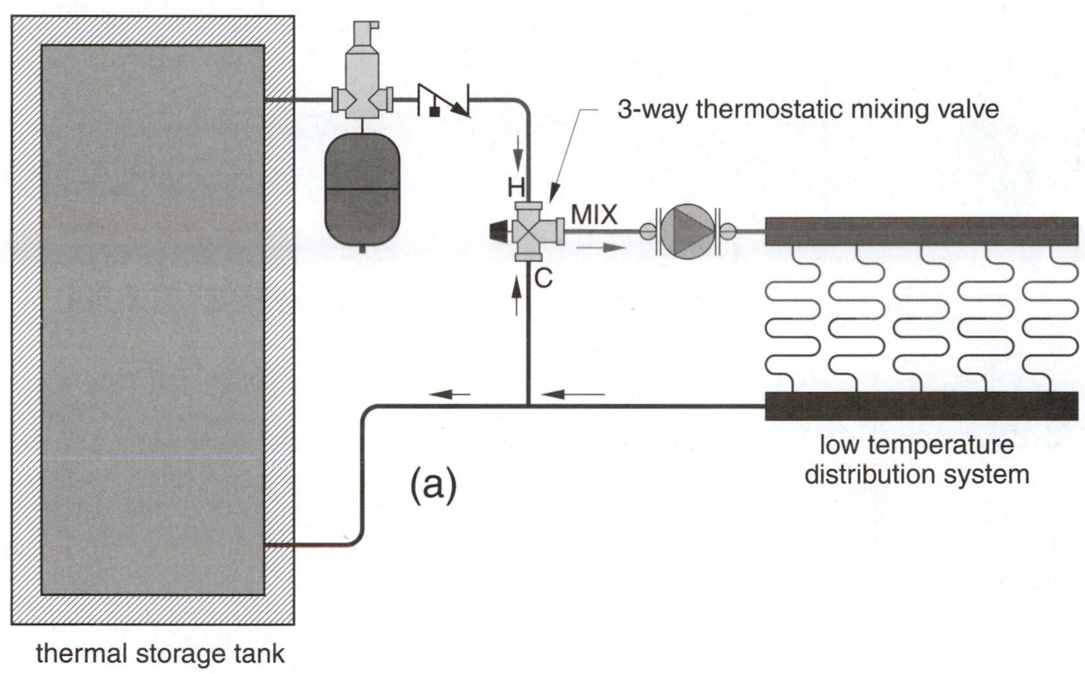

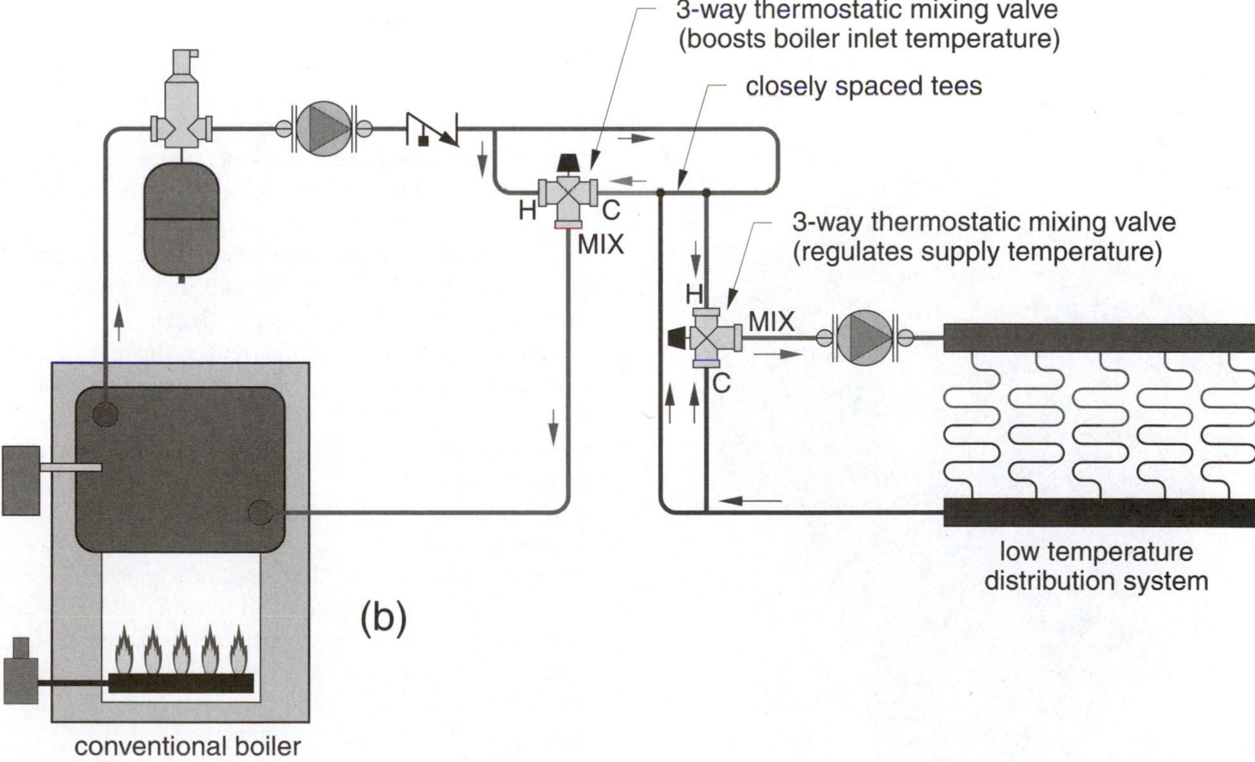

Figure 9–70 Piping for 3-way thermostatic mixing valves: (a) when heat source is not subject to or damaged by flue gas condensation, (b) when heat source is a conventional boiler requiring protection from sustained flue gas condensation.

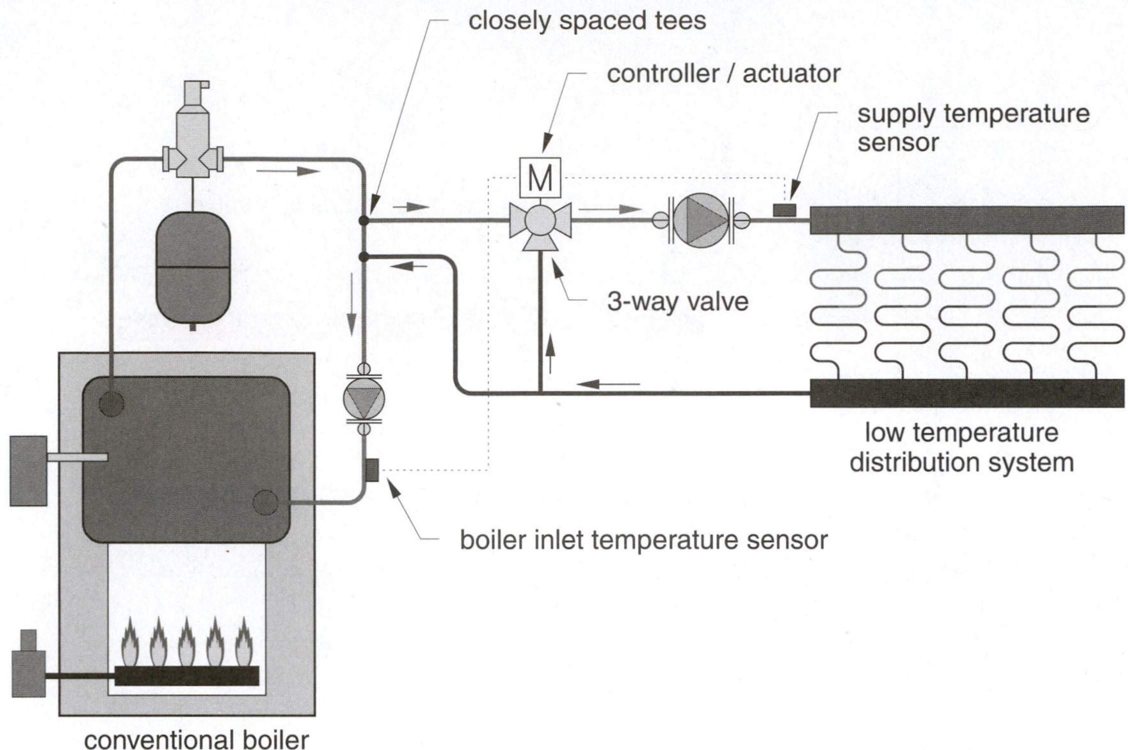

Figure 9–71 Piping for a 3-way motorized mixing valve used in combination with a conventional boiler. Note sensor measuring boiler inlet temperature. This sensor allows the controller to operate valve so that sustained flue gas condensation is prevented.

Figure 9–72 Example of a 3-way valve body with mixing reset controller/actuating motor attached. Sensors attach at left side of controller.

that prevent them from stalling at either end of their travel range. Some also have CAM-operated auxiliary switches that can be used to activate a heat source or to signal other equipment as the valve begins to open.

Other mixing valve controllers generate a continuous 2- to 10-VDC modulating output, and must be matched with actuators designed for this type of signal.

Although less common than either floating or 2- to 10-VDC output, some mixing valve controllers generate a pulse width modulation (PWM) output signal. Such controllers send low voltage electrical pulses to a heat motor type actuator that moves the valve stem. The pulse width is continually adjusted to vary the total energy sent to the heat motor. This in turn determines where the heat motor positions the valve stem. When the pulse width is decreased or stopped, the heat built up in the actuator dissipates, and a spring slowly rotates the valve stem toward the closed position. The housing of the heat motor can get quite warm when the valve is near its fully open position. Full rotation of the valve can take five minutes or more, depending on ambient temperature.

Injection Mixing

The term **injection mixing** applies to any mixing assembly in which hot water is pushed (e.g., injected) into a circulating distribution system as shown in Figure 9–73.

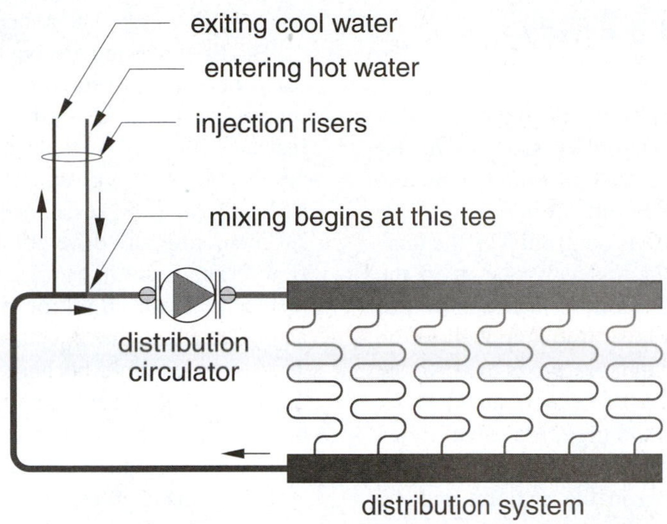

Figure 9–73 Concept of injection mixing. Flow rate of entering hot water always equals flow rate of exiting cool water.

Since the distribution system is completely filled, cooler fluid must exit at the same flow rate hot water is injected. *The greater the rate of hot water injection, the warmer the mixed supply temperature becomes, and the greater the heat output of the distribution system.* This simple concept applies to all forms of injection mixing regardless of the hardware used to implement it.

The injection flow rate needed to establish a given rate of heat transfer into the distribution system can be determined using Equation 9-11.

(Equation 9.11)

$$f_i = \frac{Q}{k \times (T_{hot} - T_{return})}$$

where:

f_i = required injection flow rate (gpm)
Q = rate of heat transfer into distribution system (gpm)
T_{hot} = temperature of fluid being injected (°F)
T_{return} = temperature of fluid returning from distribution system (°F)
k = a number based on the fluid used. The exact value can be calculated using Equation 9.12.

(Equation 9.12)

$$k = 8.012 \times c \times D$$

where:

c = specific heat of the fluid at T_{ave} (Btu/lb/°F)
D = density of the fluid at T_{ave} (lb/ft^3)
T_{ave} = average of the entering and exiting temperatures in the injection risers at design load conditions (°F). For typical applications, T_{ave} can be assumed to be 140 °F, and the following values of k can be used:

- For water: $k = 490$
- For a 30 percent solution of either propylene glycol or ethylene glycol: $k = 479$
- For a 50 percent solution of either propylene glycol or ethylene glycol: $k = 450$

Example 9.7: A radiant floor heating system requires an input of 120,000 Btu/hr at design load conditions. Water at 180 °F is available for injection. The supply temperature of the system at design load conditions is 110 °F, and the temperature drop across the distribution system at design load is 20 °F. What is the required injection flow rate?

Solution: To use Equation 9.11, we need the return temperature of the distribution system. This is the supply temperature minus the temperature drop (110 – 20 = 90 °F). Substituting this temperature and the other data into Equation 9.11 yields:

$$f_i = \frac{Q}{k \times (T_{hot} - T_{return})} = \frac{120{,}000}{490 \times (180 - 90)} = 2.7 \text{ gpm}$$

Discussion: Although the rate of heat delivery is relatively high, the required injection flow rate (2.7 gpm) is quite low. This flow could easily pass through a 1/2-inch tube. This is a direct result of the high temperature difference between the supply and return injection risers ($T_{hot} - T_{return}$). The larger the temperature difference between the injection risers, the smaller the required injection flow rate becomes for a given rate of heat transfer. As you will soon see, this can have a profound impact on the size of the mixing hardware.

Injection Mixing Using a 2-way Thermostatic Valve

One way to control the injection flow rate is with a 2-way thermostatic valve. These valves were discussed in Chapter 5, Piping, Fittings, and Valves. The valve is mounted into the supply injection riser as shown in Figure 9–74.

The hot water injection flow rate is controlled by the thermostatic valve (V1). The knob of this valve is set to the desired supply temperature under design load conditions. The sensing bulb measures the temperature supplied to the distribution system. If the supply temperature is too low, the valve opens farther to allow more hot water into the mix point. If the supply temperature is too high, the valve begins to close allowing less hot water into the mix point.

The greater the deviation between the setpoint and the measured temperature, the more the valve stems moves. Most thermostatic valves have a proportional range of 4 to 6 °F. Half of this range is above the setpoint, and half is below the setpoint. If the supply temperature is more than 2 to 3 °F below setpoint, the valve is fully open. If the supply temperature is more than 2 to 3 °F above setpoint, the valve is fully closed.

In a typical low temperature floor heating system supplied by a conventional boiler the flow rate through the injection valve (V1) is only about 15 to 20 percent of the flow rate in the distribution system. This allows a relatively small valve to regulate a large rate of heat transfer.

The flow restrictor valve (V2) creates the pressure differential necessary to force flow through the injection risers as the injection valve begins to open. The greater the flow resistance created by this valve, the higher the injection flow rate becomes for a given stem position of the injection valve (V1). If the flow resistance is too low, there may not be sufficient pressure differential to move the required injection flow through the injection valve at design load. If the flow resistance is too high, the necessary injection flow will be achieved when the injection valve is only partially open. This is undesirable because it forces the injection valve to operate using a very small percentage of its stem travel under low-load conditions. A valve operating under such conditions is said to have low rangeability as discussed earlier in this chapter.

For stable and accurate control, *the injection valve should be fully open when the maximum injection flow rate occurs under design load conditions*. This requires the valve to be properly sized. It also requires the flow restrictor valve to be properly set.

A common mistake is to select an injection valve based on the size of the injection riser piping. This almost always results in an oversized valve. Instead, the injection valve should be selected based on its flow coefficient (*C*v value). The *C*v value of a valve was discussed in Chapter 6, Fluid Flow in Piping. It is the flow rate (in gpm) of 60 °F water

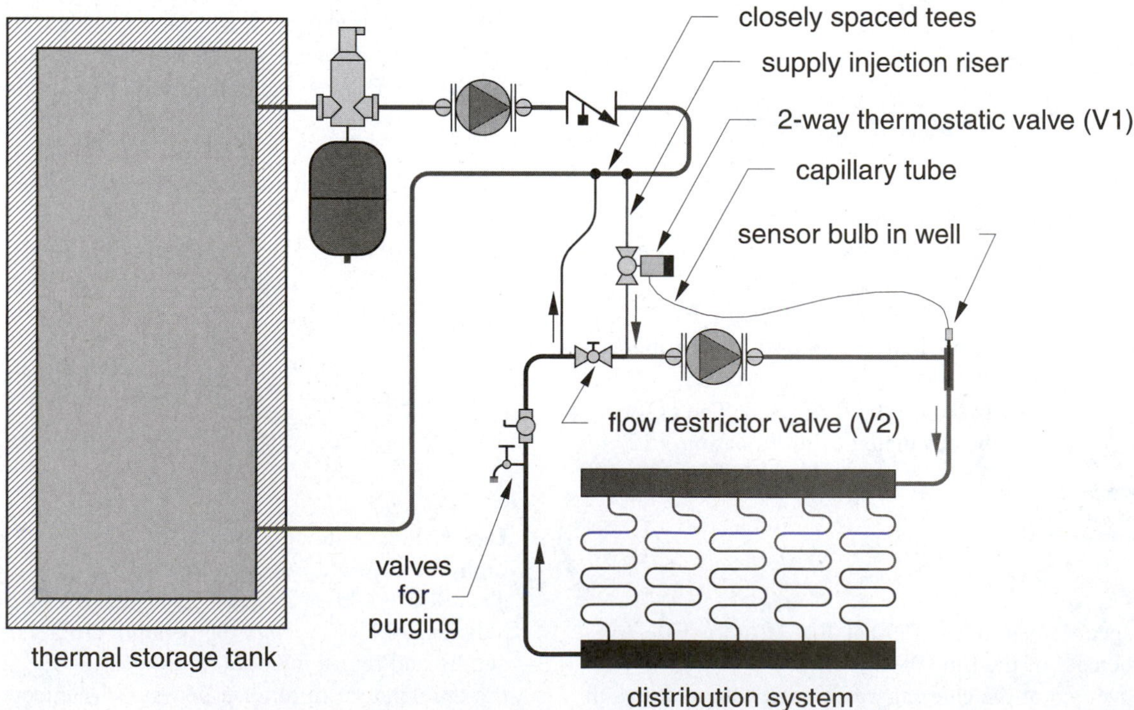

Figure 9–74 Piping for injection mixing using a 2-way thermostatic valve. Note that this piping does not protect a conventional boiler from possible flue gas condensation.

that creates a pressure drop of 1 psi across the valve.

The following procedure can be used to determine the necessary *Cv* of the injection valve as well as the *Cv* of the bypass valve.

Step 1: Use Equation 9.11 to calculate the flow rate through the injection valve at design load conditions.

Step 2: Consult manufacturer's literature to select an injection valve body with a *Cv* value approximately equal to the flow rate calculated in step 1. This will allow a pressure drop of about one psi to develop across the injection valve when it is fully open. Most manufacturers offer 2-way control valves in different pipe sizes, and several *Cv* values for each of these pipe sizes. The pipe size is just a matter of convenience. It is the *Cv* value that determines if the valve will function properly in this application.

Step 3: Calculate the flow rate in the distribution system at design load conditions. This may already be known from previous calculations. If not, it can be estimated using Equation 9.13.

(Equation 9.13)

$$f_s = \frac{Q}{k \times (T_{supply} - T_{return})}$$

where:

f_s = flow rate in the distribution system at design load (gpm)

Q = heat output from distribution system at design load (Btu/hr)

T_{supply} = supply temperature of distribution system at design load (°F)

T_{return} = return temperature from distribution system at design load (°F)

k = a constant based on the fluid.

Use $k = 490$ when the average water temperature in the distribution system is about 140 °F. Use $k = 495$ when the average water temperature in the distribution system is about 100 °F. Use Equation 9.12 to calculate k for other fluids or average temperatures.

Step 4: Calculate the *Cv* setting of the bypass valve using Equation 9.14.

(Equation 9.14)

$$Cv_{bypass} = \frac{0.707 \times f_s \times Cv_i}{f_i}$$

where:

Cv_{bypass} = the necessary *Cv* setting of the bypass valve

f_s = the system flow rate determined in step 3

Cv_i = *Cv* value of the selected injection valve determined in step 2

f_i = the injection flow rate determined in step 1

The *Cv* value of the bypass valve should be thought of as a setting. Several types of valves are available with calibrated *Cv* scales. An example of such a valve is shown in Figure 9–75. Assuming such a valve is used as the bypass valve, all that remains is to set the handle to the *Cv* value calculated in step 4. The injection mixing assembly is now properly sized and adjusted for optimal performance.

Example 9.8: A 2-way thermostatic valve will be used as the injection control device for a system supplying 60,000 Btu/hr to a radiant floor distribution system. Hot water is available from the primary loop at 160 °F. The supply temperature to the floor circuits at design load is 110 °F, with a return temperature of 95 °F. Select an appropriate valve, and determine the Cv setting of the bypass valve.

Solution:

Step 1: Calculate the injection flow rate at design load using Equation 9.11:

$$f_i = \frac{Q}{k \times (T_{hot} - T_{return})} = \frac{60{,}000}{490 \times (160 - 95)} = 1.88 \text{ gpm}$$

Step 2: After surveying manufacturer's literature, an injection valve body with a Cv of 2.0 is selected. The Cv rating is close to the calculated injection flow rate.

Step 3: The flow rate in the distribution system is estimated using Equation 9.13:

$$f_s = \frac{Q}{k \times (T_{supply} - T_{return})} = \frac{60{,}000}{495 \times (110 - 95)} = 8.1 \text{ gpm}$$

Step 4: The *Cv* setting of the bypass valve is calculated using Equation 9.14.

$$Cv_{bypass} = \frac{0.707 \times f_s \times Cv_i}{f_i} = \frac{0.707 \times 8.1 \times 2.0}{1.88}$$

$$= 6.09 \approx 6.1$$

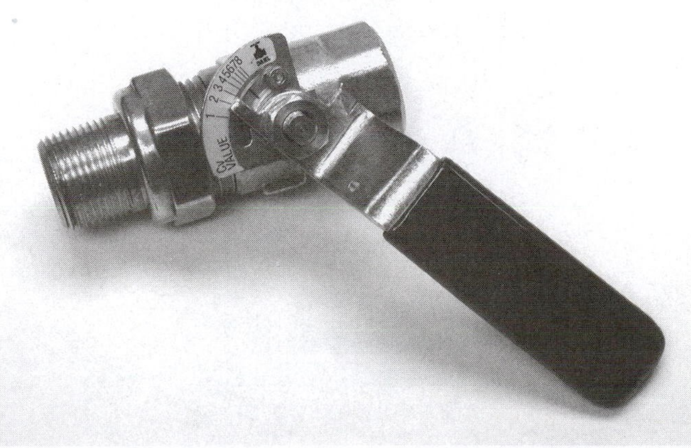

Figure 9–75 Example of a valve that can be set to a specific *Cv* value.

Discussion: Notice that the fluid constant used in Equation 9.13 is 495. This is an appropriate value for the relatively low average water temperature in the distribution system. The bypass valve would be set to a *C*v of 6.1 to properly balance the injection assembly.

Whenever possible, the sensor bulb of the 2-way thermostatic valve should be mounted downstream of the distribution circulator as shown in Figure 9–76. This ensures complete mixing of the hot injection fluid and cool bypass fluid before the mixed flow passes the sensor.

For the most accurate control and fastest response to changing temperatures, the valve's sensor bulb should be in direct contact with the mixed fluid. Some thermostatic valves are supplied with hardware that allows the sensing bulb to be inserted into a tee. A special fitting seals against the capillary tube to prevent leakage. Another possibility is to install a sensor well into a tee, and then slide the sensor bulb into the well. The sensor bulb should fit snugly inside the well. It should also be coated with heat-conducting paste when inserted. If neither of these options can be used, the remaining choice is to strap the sensor bulb to the outside of the pipe downstream of the mixing point. The sensor bulb should be tightly clamped to the pipe with hose clamps, and then completely wrapped in insulation.

When the distribution system cools off, the thermostatic valve will completely open because it is set for a supply temperature significantly higher than room temperature. To prevent any heat migration into the distribution system, the following details are recommended and shown in Figure 9–76:

1. Install the injection risers so there is a drop of at least 18 inches from where the risers connect to the primary loop and where they connect to the distribution system. This creates a **thermal trap** that discourages hot water from migrating downward through the injection risers.
2. Install a flow-check valve in the distribution system downstream of the circulator. The weighted plug in the flow check helps prevent any water movement until the distribution circulator is operating.

Injection mixing using a 2-way thermostatic valve does have some limitations. Since the 2-way thermostatic valve only measures and regulates supply temperature, it does not protect a conventional boiler against flue gas condensation at system start-up or recovery from setback periods. If a conventional boiler is used as the heat source, another temperature-sensing mixing device is required to protect it.

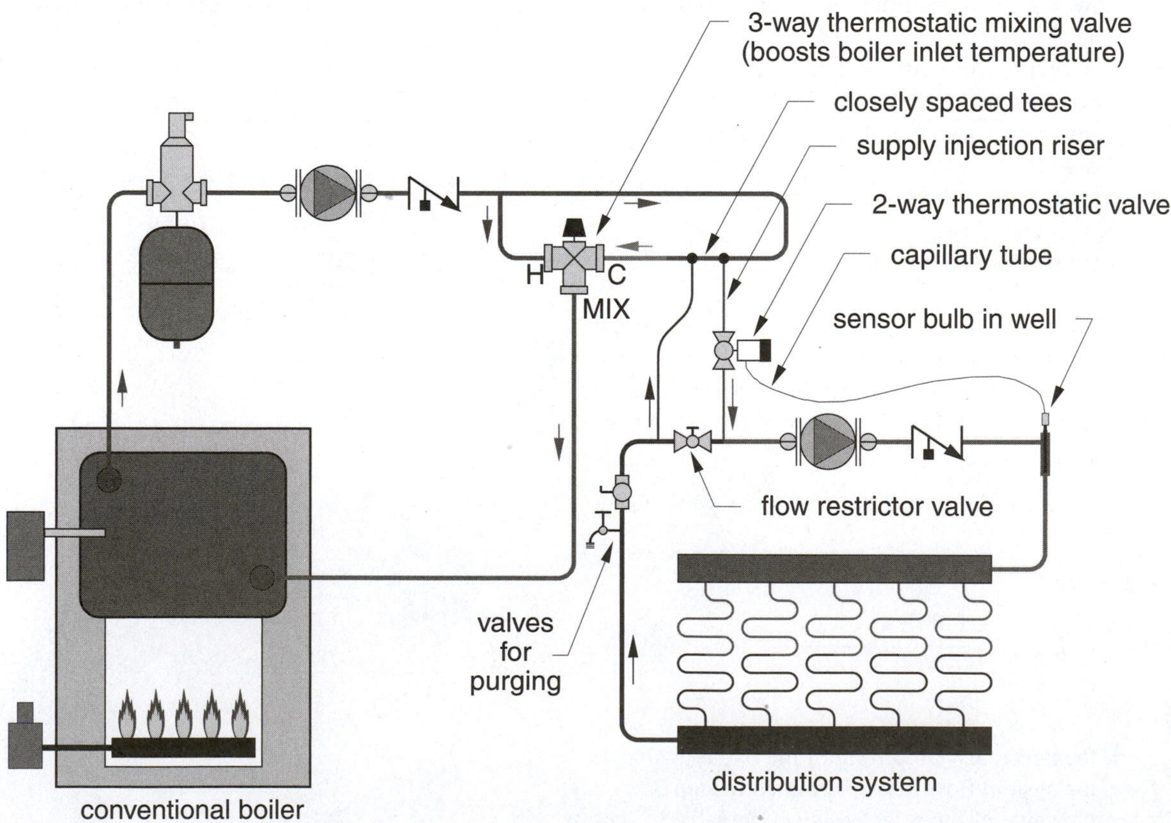

Figure 9–76 Using a 3-way thermostatic valve to provide protection against flue gas condensation when a conventional boiler is used in combination with a 2-way thermostatic injection valve.

One possibility is to use a 3-way thermostat valve set for the minimum allowable boiler inlet temperature, and installed as shown in Figure 9–76. Injection mixing with a 2-way thermostatic valve also does not provide mixing reset control, nor does it allow for constant circulation in the distribution system.

Injection Mixing Using a 2-Way Motorized Valve

The piping used for injection mixing with a 2-way thermostatic valve can also be used for a 2-way motorized valve. These valves are typically operated by electronic controllers that generate a floating output, 2- to 10-VDC output, or a pulse width modulation (PWM) output.

An example of a 2-way motorized valve with a 2- to 10-VDC input that would be suitable for this type of injection mixing is shown in Figure 9–77. A controller capable of operating this valve is shown in Figure 9–78. The piping schematic is shown in Figure 9–79.

Figure 9–77 Example of a 2-way motorized (modulating) valve that can be used for injection mixing applications. The small cylindrical component at the top is the 24-VAC motor which sits on top of the gear train. Wiring terminals can be seen on circuit board. See Figure 9–8 for typical wiring.

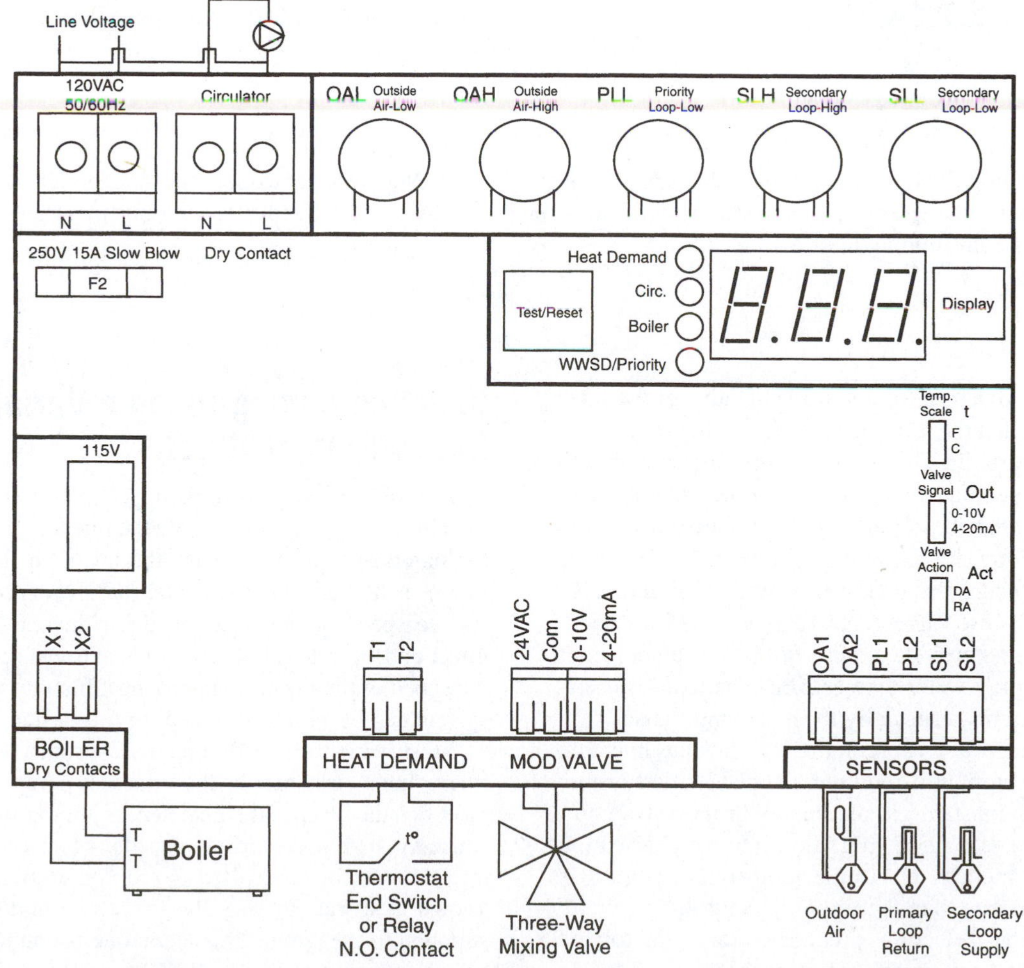

Figure 9–78 Controller designed to operate a 2-way motorized valve for injection mixing applications, as well as a 3-way motorized valve for other mixing methods. *Courtesy of Erie Controls.*

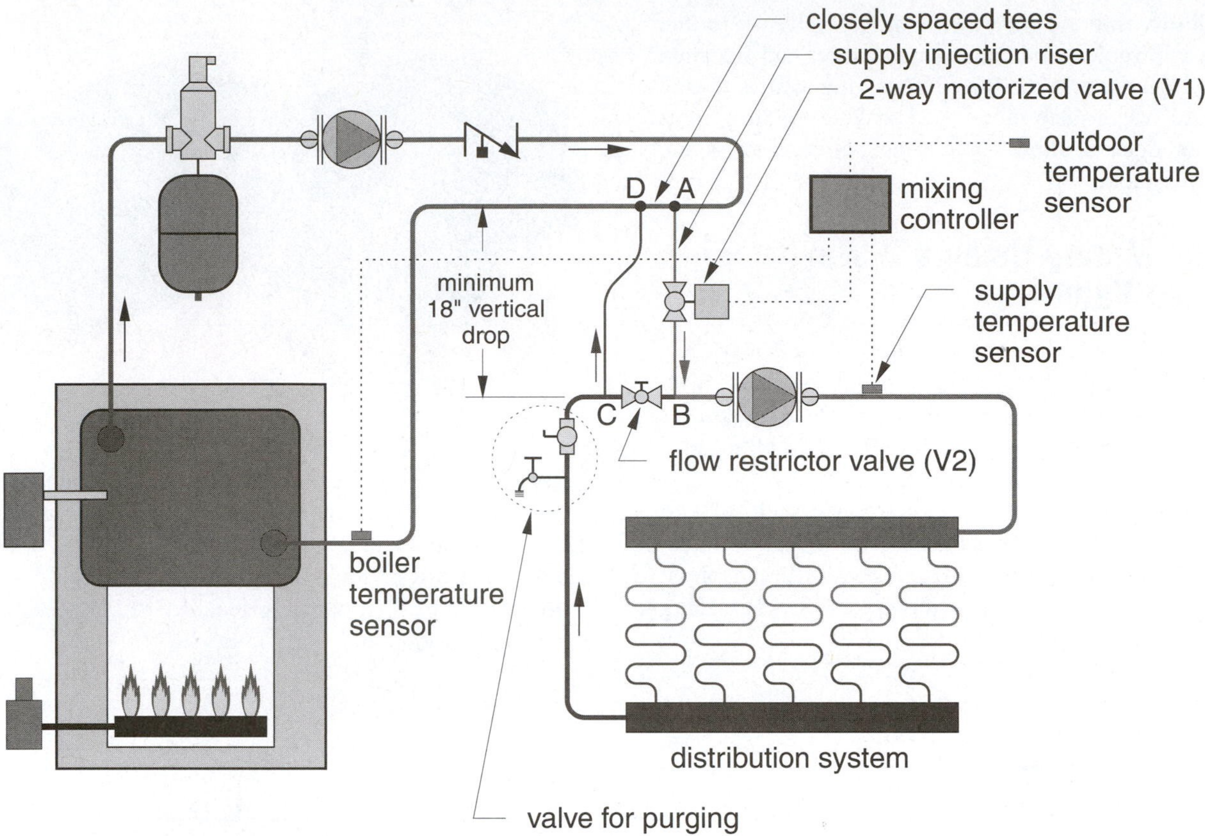

Figure 9–79 Piping for injection mixing with 2-way motorized valve. Note that controller measures boiler inlet temperature, and when necessary, restricts hot water flow through the 2-way motorized valve to prevent sustained flue gas condensation in a conventional boiler.

A significant advantage of a motorized valve over a thermostatic valve is that the motorized valve can be operated by an intelligent controller. This allows the mixing assembly to provide several benefits not available with a thermostatic injection valve. If the controller monitors boiler temperature, it can protect a conventional boiler against low temperature operation by reducing the injection flow through the valve when necessary. Most valve controllers designed for injection mixing can also provide either setpoint or outdoor reset control of the supply water temperature. The latter is desirable in many systems, especially those serving radiant panels. Since the controller can completely shut the injection valve when no heat input is required, the distribution system can operate with constant circulation. With the valve completely shut, there is no heat migration into the distribution system. Finally, modern electronic temperature control is slightly more accurate than thermostatic control.

The sizing and selection procedure for a motorized 2-way valve for injection mixing is identical to that just discussed for a 2-way thermostatic valve. So is the setting of the bypass valve.

Injection Mixing Using a Variable Speed Pump (Direct)

One of the best known forms of injection mixing uses a variable speed pump to regulate the injection flow rate. When the pump runs at low speeds, the injection flow rate is also low, as is the rate of heat transfer into the distribution system. As the speed of the injection pump increases, so does the injection flow rate and the rate of heat transfer. By controlling the speed of the injection pump, heat input to the distribution system can be varied from zero to full design load.

Most variable speed injection systems use a standard line voltage circulator as the injection pump. The controller modifies the electrical signal sent to the pump in a way that causes it to run at various speeds. Most controllers use a microprocessor-controlled solid state switching device known as a triac to vary the frequency and/or wave shape supplied to the pump. This technique is compatible with the permanent split capacitor (PSC) motors commonly used in wet rotor circulators. It does not cause any abnormal wear on the motor.

The variable speed drive circuitry is controlled based on the settings of the controller. Most controllers provide a host of features including the ability to:

- Provide full outdoor reset control of the supply temperature
- Protect a conventional boiler against low temperature operation by reducing injection flow when necessary
- Coordinate the operation of the injection pump, system circulator, and heat source
- Exercise the injection pump and system circulator during inactive periods

Some controllers can also coordinate their operation with other system controls such as those used for zoning or operating multiple boiler systems.

Three examples of variable speed injection controllers are shown in Figure 9–80.

The piping details for variable speed injection mixing with a variable speed pump are shown in Figure 9–81.

When the injection pump is operating, hot water in the primary loop is drawn into the supply injection riser at point A. After passing through the variable speed pump, the hot water is pushed into the tee at point B where it mixes with some of the cooler water from the return side of the distribution system. The remainder of the cooler water leaves the tee at point C, and travels up the return injection riser. *The exiting flow rate of cooler water always equals the entering flow rate of hot water.* When it flows into the tee at point D, it mixes with the bypass flow of hot water. This tee provides a second mixing point that boosts the water temperature entering the boiler.

Note: The injection pump will work equally well when placed in the return injection riser so that it pumps cool water from the distribution system back into the primary loop.

Two piping details are critically important to proper operation of this mixing assembly. Both are shown in Figure 9–81.

Detail #1: The injection risers must drop at least 18 inches between where they connect to the primary loop and where they connect to the distribution system. This forms a thermal trap that discourages hot water from migrating downward into the distribution system when the injection pump is off. Without this trap, the slight pressure drop between the tees from A to D, and from C to B can lead to hot water migration into the distribution system when it is not needed. A standard circulator will not block this flow. Furthermore, a flow-check valve should *not* be installed in either injection riser to prevent this flow because it can cause erratic pump operation ("surging") at low speeds.

Note: The injection risers do not have to drop straight down 18 inches as shown in Figure 9–81.

There can be numerous horizontal offsets in the injection risers if necessary provided the net change in elevation from the primary loop to the distribution loop is at least 18 inches.

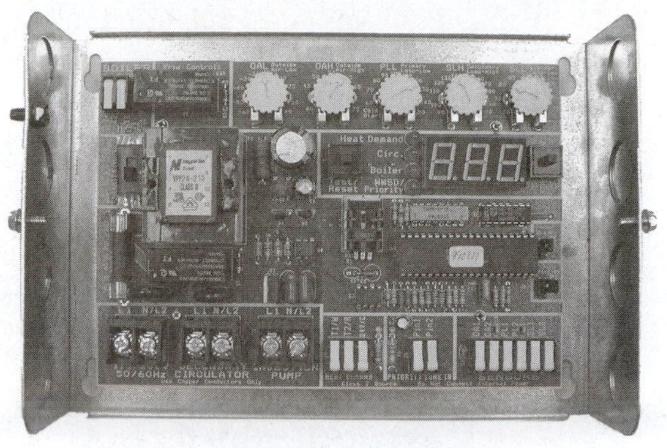

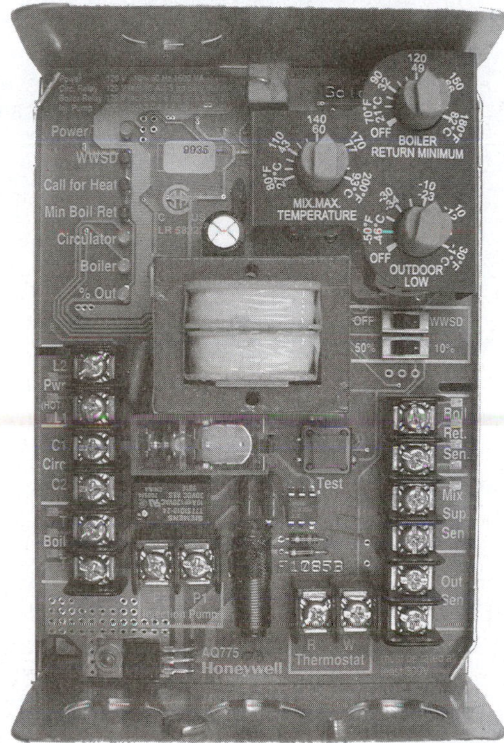

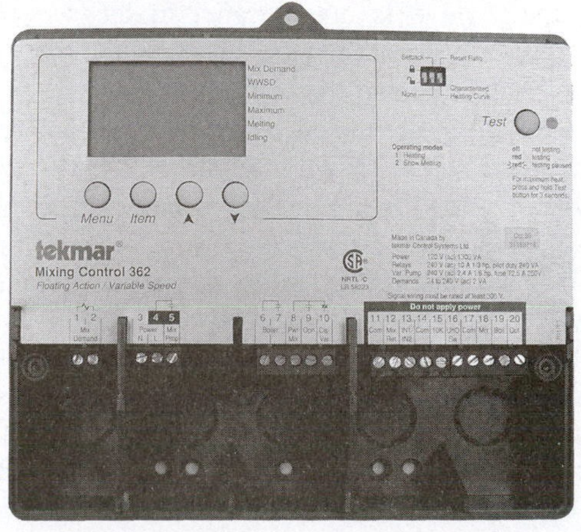

Figure 9–80 Examples of controllers designed to operate a variable speed injection pump.

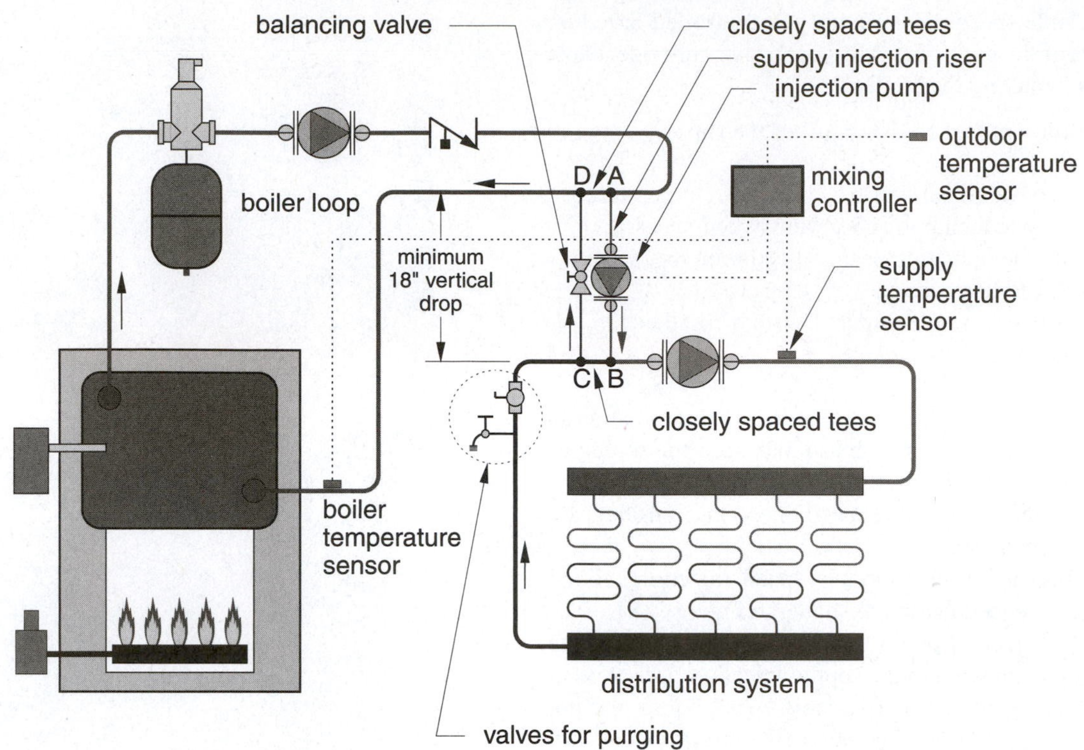

Figure 9–81 Piping for (direct) injection mixing with a variable speed pump.

Detail #2: The tees connecting the injection risers to the primary loop should be as close together as possible. Likewise, the tees connecting the injection risers to the distribution systems should be as close together as possible. The goal is to minimize the pressure drop between the pairs of tees. This minimizes the tendency for uncontrolled hot water migration into the distribution system.

An alternative piping schematic for variable speed injection mixing is shown in Figure 9–82. The injection pump is now located in the return riser rather than the supply riser. The balancing valve is now located in the supply riser. Because of the closely spaced tees coupling the injection risers to the boiler loop and the distribution system, the alternative placement of the injection pump and balancing valve work the same as those shown in Figure 9–81. The injection pump will run cooler using this arrangement. The injection risers also make a turn before connecting to the distribution system. This has no effect on the performance of the mixing assembly. The elevation of the distribution system relative to the injection risers also does not affect operation of the injection mixing assembly, provided the minimum 18-inch deep thermal trap is still provided as shown.

The injection flow rate required at design load is calculated using Equation 9.11. This flow rate does not depend on the injection control device (e.g., 2-way valve versus pump). It is solely determined by the thermodynamics of the situation.

In many residential systems, the injection flow rate is only 1 to 2.5 gallons per minute at design load. The head loss created when this low flow rate passes through 3 to 6 feet of injection riser piping is also very small. The low flow rate and low head loss would suggest that a very small circulator should be selected for the injection pump. Unfortunately, circulators with such limited capacity have virtually no other use in hydronic heating and are simply not available at present.

Instead, the majority of smaller systems using variable speed injection mixing use a standard 1/25 horsepower wet rotor circulator for the injection pump. These circulators are readily available and relatively inexpensive. However, their pumping capacity is far beyond the low flow and head loss requirements of the injection riser piping. This can lead to a situation where the injection pump only operates at a small percentage of its full speed, even under design load conditions. Like an improperly sized control valve, such a pump is said to have low rangeability, and cannot provide the precise control the system is capable of delivering.

This situation can be seen by examining the pump curve of the injection pump and the system curve for the injection riser piping as shown in Figure 9–83a.

If the assumed injection flow rate at design load conditions is 2.0 gpm, the pump speed must be drastically reduced to make the pump curve intersect the system curve at 2.0 gpm. Since this represents design load conditions, the pump will never need to operate at a higher speed. This wastes a large portion of the controller's ability to fine-tune supply temperature by adjusting pump speed.

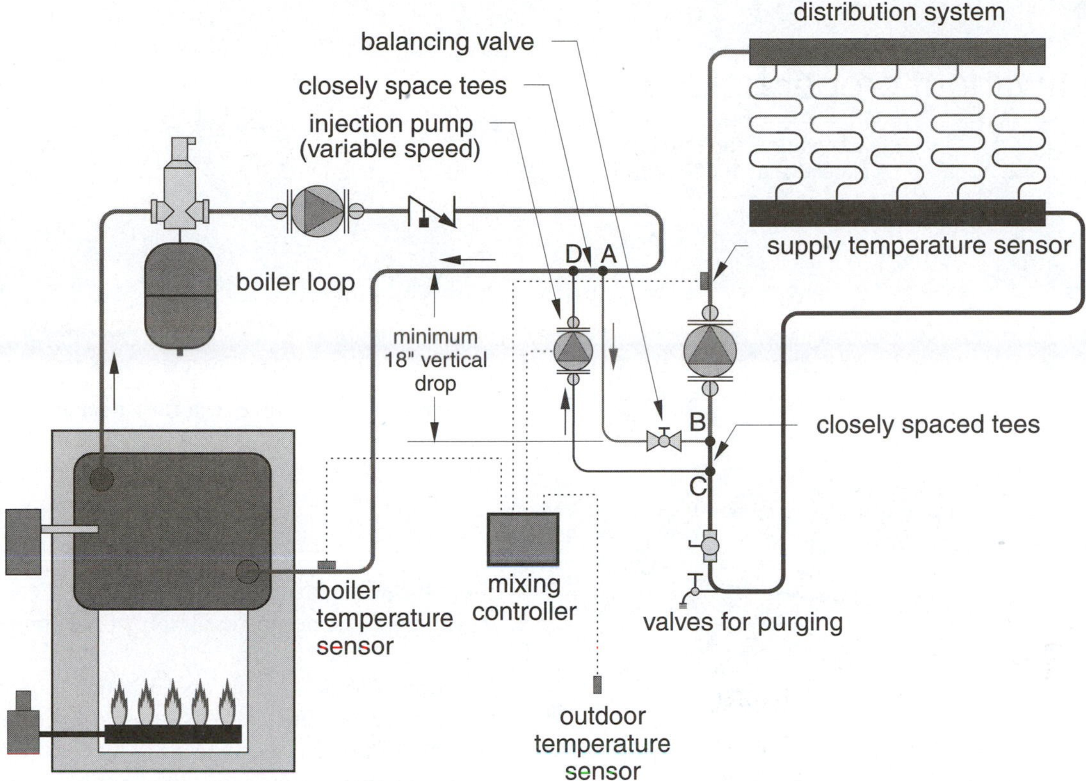

Figure 9–82 Alternate piping for direct injection mixing using a variable speed pump. Notice that injection pump is on the return injection riser and thus operates at a lower temperature.

The solution is to install a balancing valve in one of the injection risers. Its purpose is to add head loss to the injection risers so the injection pump is forced to operate at full speed under design load conditions. As the balancing valve is throttled, the system curve of the injection risers gets progressively steeper. The ideal setting will cause the system curve to intersect the pump curve (when the pump is at full speed) at the desired injection flow rate as shown in Figure 9–83b. This restores the full rangeability of the injection pump, and improves the ability of the injection mixing assembly to fine-tune supply water temperature.

If a flow measuring device is present in one of the injection risers, the procedure for setting the balancing valve is quite simple.

Step 1: Calculate the injection flow rate required at design load conditions using Equation 9.11.

Step 2: With the injection pump running at 100 percent speed, throttle the balancing valve until the indicated flow rate equals the flow rate calculated in step 1. The balancing valve is now properly set.

An example of a combination balancing valve/flow rate indicator is shown in Figure 9–84. This type of device is well-suited for balancing injection mixing systems.

Another approach is to determine the required *C*v of the balancing valve. This assumes that a valve with a calibrated *C*v scale is used.

Step 1: Calculate the injection flow rate required at design load conditions using Equation 9.11

Step 2: Determine the head of the injection pump (operating at full speed) at the injection flow rate calculated in step 1. This head is designated H_p.

Step 3: Calculate the required *C*v setting of the valve using Equation 9.15

(Equation 9.15)

$$Cv_{bv} = \sqrt{\frac{2.39 \times f_i^2}{H_p - r \times f_i^{1.75}}}$$

where:

Cv_{bv} = required Cv setting of the balancing valve

f_i = injection flow rate at design load (gpm)

H_p = head of injection pump when operating at injection flow rate (feet)

r = a constant based on the pipe size of the injection risers*

r = 0.058 for 3/4-inch injection risers

r = 0.25 for 1/2-inch injection riser pipe

* This constant assumes that the total equivalent length of the injection risers (excluding the balancing valve) is approximately 8 feet.

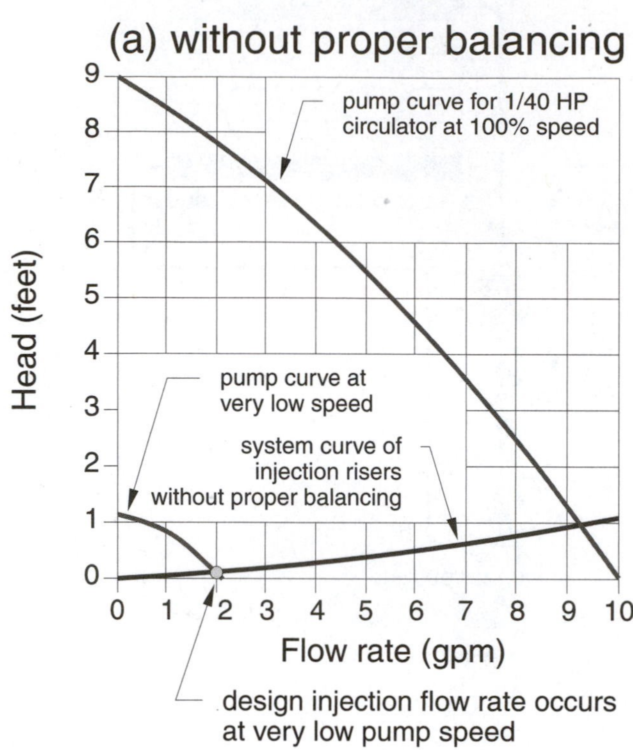

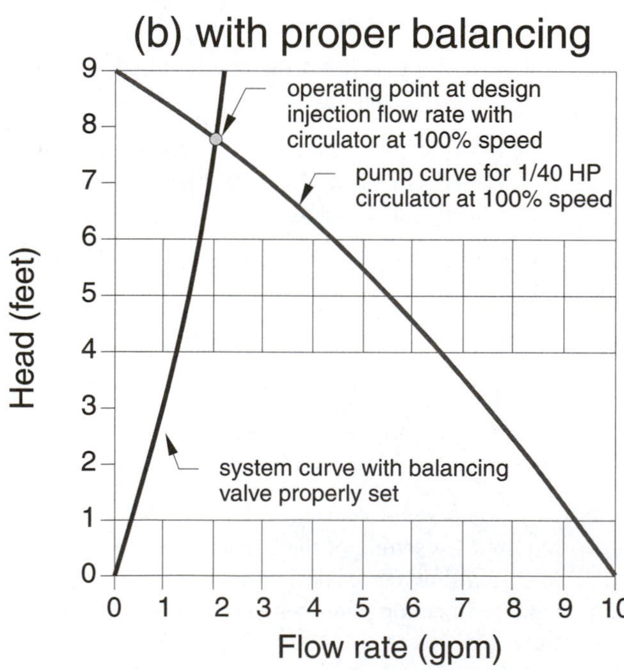

Figure 9–83 (a) Without proper balancing, the injection pump operates at very low speeds, even at design load conditions. This is undesirable. (b) With a properly set balancing valve, the injection pump runs at full speed at design load conditions.

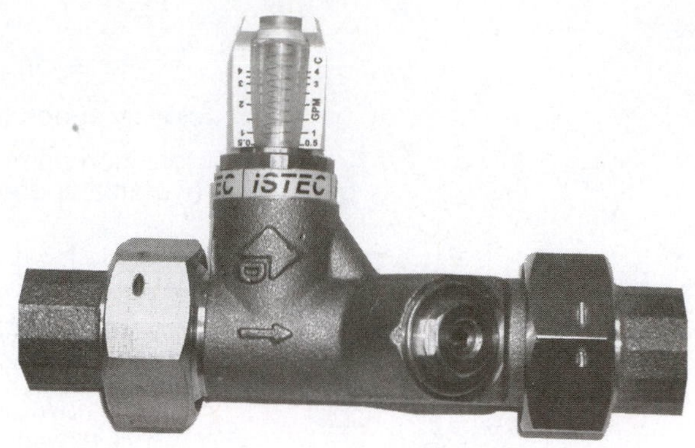

Figure 9–84 Example of a flow balancing valve with built-in flow meter.

Step 4: Set the balancing valve to the *Cv* calculated in step 2. The injection mixing assembly is now properly balanced.

Example 9.9: A variable speed injection mixing system is planned for a radiant heating system with a design load of 100,000 Btu/hr. Hot water is available for injection at 170 °F. At design load conditions, the supply temperature of the radiant panel system is 130 °F with a 20 °F temperature drop. The injection risers will be 1/2-inch copper tubing. The injection pump has the curve shown in Figure 9–85. Determine the required injection flow rate and the proper *Cv* setting of the balancing valve.

Solution: The return temperature of the distribution system will be 130 – 20 = 110 °F.

Step 1: The required injection flow rate at design load is calculated using Equation 9.11.

$$f_i = \frac{Q}{k \times (T_{hot} - T_{return})} = \frac{100{,}000}{490 \times (170 - 110)} = 3.4 \text{ gpm}$$

Step 2: From the pump curve in Figure 9–85, the head loss of the injection pump while operating at 3.4 gpm is approximately 14.4 feet.

Step 3: The *Cv* of the balancing valve is now calculated using Equation 9.15:

$$Cv_{bv} = \sqrt{\frac{2.39 \times f_i^2}{H_p - r \times f_i^{1.75}}} = \sqrt{\frac{2.39 \times (3.4)^2}{14.4 - 0.25 \times (3.4)^{1.75}}} = 1.5$$

Discussion: When the balancing valve is set to the required Cv of 1.5, the system curve of the injection riser piping intersects the pump curve at 3.4 gpm. The injection mixing assembly will provide 3.4 gpm of injection flow when the injection pump is operating at full speed.

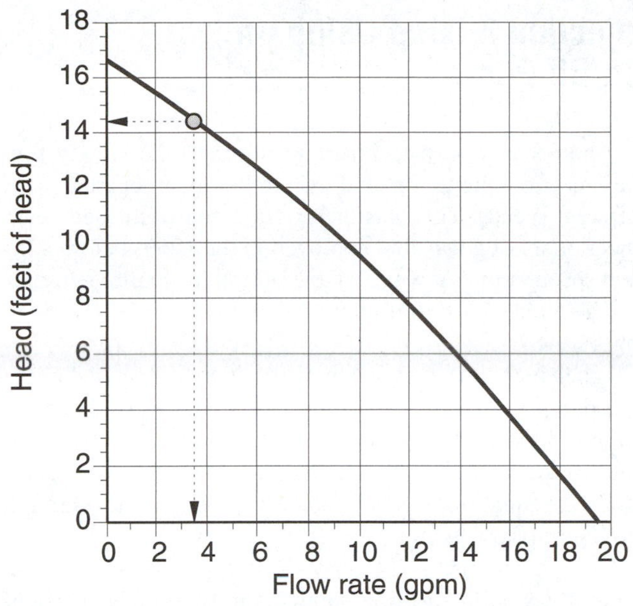

Figure 9–85 Pump curve of injection pump in Example 9.8.

Injection Mixing Using a Variable Speed Pump (Reverse)

The piping shown in Figure 9–81 allows the cool return water from the distribution system to flow up through the return injection riser. This configuration is sometimes called **direct injection mixing**. It provides the largest temperature difference between the supply and return injection risers. This in turn maximizes the heat carried by each gallon per minute of injection flow.

However, there are situations where it is beneficial to reduce the temperature difference between the injection risers. The reduced temperature difference increases the injection flow required to deliver a given rate of heat transfer. This improves the rangeability of the injection pump, and reduces the necessary throttling of the balancing valve.

The piping schematic shown in Figure 9–86 shows the concept of **reverse injection mixing** using a variable speed pump. Notice that the return injection riser now carries water at the mixed supply temperature back to the primary loop. This detail reduces the temperature difference between the injection risers by 10 to 20 °F.

The piping detail where the hot injection flow is delivered to the distribution system is of particular importance with

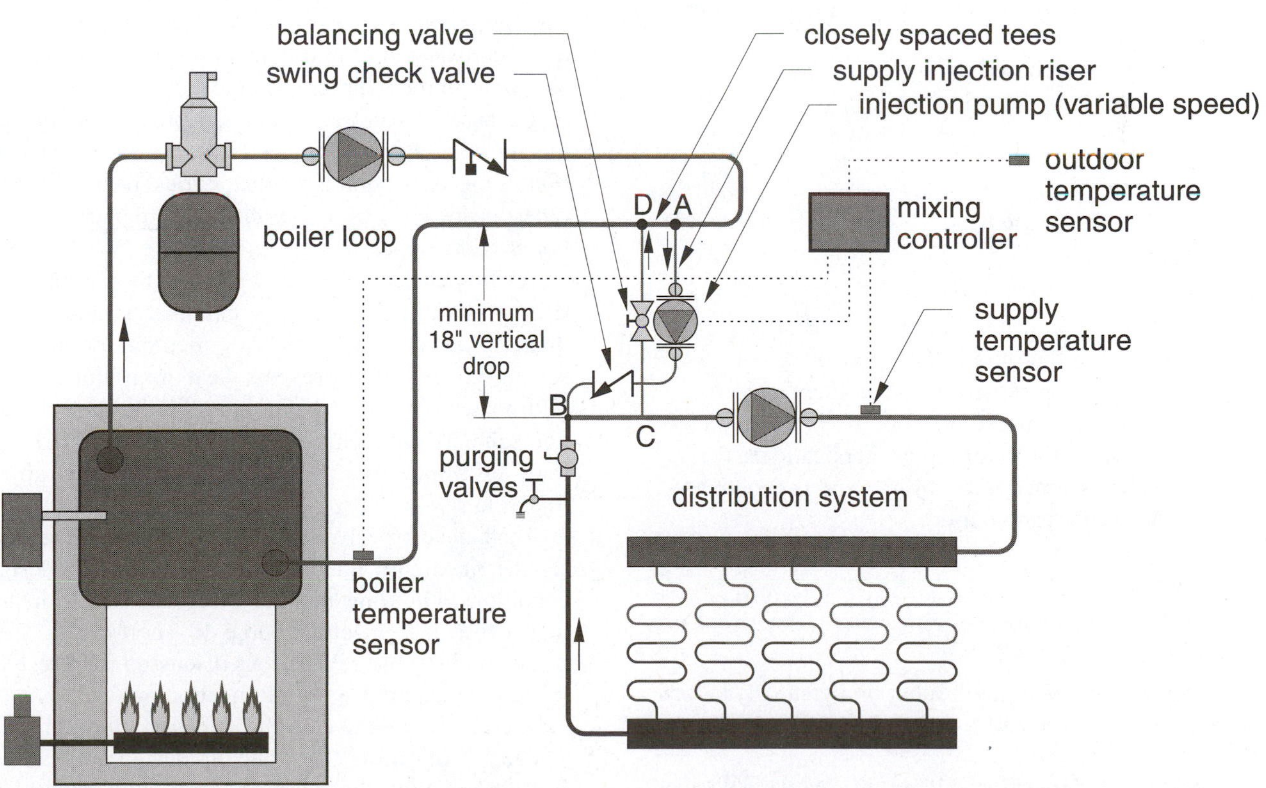

Figure 9–86 Piping for reverse injection mixing using a variable speed pump. Notice that the water in the return injection riser is at the supply temperature of the distribution system rather than the return temperature as in direct injection mixing.

reverse injection mixing. As flow in the distribution system rounds the corner in the tee, it creates a "stagnation pressure" that acts to backseat the swing-check in the supply injection riser. This assists in preventing any heat migration into the distribution system when the injection pump is off. As the injection pump speeds up, it eventually builds sufficient pressure to open the swing-check and force hot water into the distribution system. The net effect is that the pump operates at higher speeds than required for direct injection mixing and thus has better rangeability. This approach also provides excellent protection against heat migration.

Note: Only a swing-check valve should be used in the supply injection riser.

Equation 9.16 can be used to determine the necessary injection flow in a reverse injection system.

(Equation 9.16)

$$f_i = \frac{Q}{k \times (T_{hot} - T_{supply})}$$

where:

f_i = the required injection flow rate (gpm)
Q = the rate of heat transfer into distribution system (gpm)
T_{hot} = temperature of fluid being injected (°F)
T_{supply} = temperature of fluid supplied to distribution system (°F)
k = a constant based on the fluid used.

The exact value can be calculated using Equation 9.12

(Equation 9.12)

$$k = 8.01 \times c \times D$$

where:

c = specific heat of the fluid at T_{ave} (Btu/lb/°F)
D = density of the fluid at T_{ave} (lb/ft³)
T_{ave} = average of the entering and exiting fluid temperatures in the injection risers at design load conditions (°F). For typical applications, T_{ave} can be assumed to be 140 °F, and the following values of k can be used:

- For water: $k = 490$
- For a 30 percent solution of either propylene glycol or ethylene glycol: $k = 479$
- for a 50 percent solution of either propylene glycol or ethylene glycol: $k = 450$

To date, reverse injection mixing using a variable speed pump has been less common than direct injection mixing. Although the piping details are slightly more complicated, it does offer advantages in smaller systems.

Injection Mixing Using an On/Off Pump

Another approach to injection mixing is to simply turn the injection pump on and off rather than operate it at different speeds. This was the original approach used when injection mixing was first deployed in the 1960s before variable speed pumping was a viable option in smaller systems.

The piping for on/off pump injection mixing is shown in Figure 9–87.

When heat is needed by the distribution system, the injection pump is turned on (at full speed). Hot water from the primary loop passes through the supply injection riser and into the distribution system where it mixes with cool return water. An equal flow rate of cooler water flows back through the return riser.

The temperature of the supply water quickly rises several degrees when the injection pump is on. As soon as it is shut off, the supply temperature drops several degrees. Through this cycling, the controller attempts to maintain the *average* supply temperature at or very close to the target value. The target value may be a fixed setpoint temperature, or may change based on outdoor reset logic. Over time, the rate of heat delivery will average out the same as that provided by fully modulating injection mixing.

Because of the relatively quick variations in supply temperature *on/off injection mixing works best with distribution systems that have relatively high thermal mass.* Such systems can absorb the pulses of heat input without creating noticeable variations in room air temperature.

Because it is not a modulating process, *on/off injection mixing does not provide boiler return temperature protection.* If this protection is needed, it must be provided through other means such as a second mixing device that monitors boiler inlet temperature.

A flow-check valve is shown in the supply injection riser of Figure 9–87 to prevent heat migration into the distribution system when the injection pump is off. A second check valve prevents heat migration through the return riser. These valves are possible because the pump is not subject to surging at low pump speeds. If present, they eliminate the need for a thermal trap in the injection risers.

The balancing valve in the return injection riser is used to set the maximum hot water injection rate at design load conditions. Omitting this valve, or not setting it properly, may cause the injection pump to short-cycle. The same setting options and calculations discussed for direct variable speed mixing apply to on/off injection mixing. The goal is to set the balancing valve so the injection pump produces the calculated injection flow rate at design load conditions. With this setting, the injection pump will operate continuously during design load conditions. During partial load conditions, the length of the pump on-cycle decreases in proportion to the load.

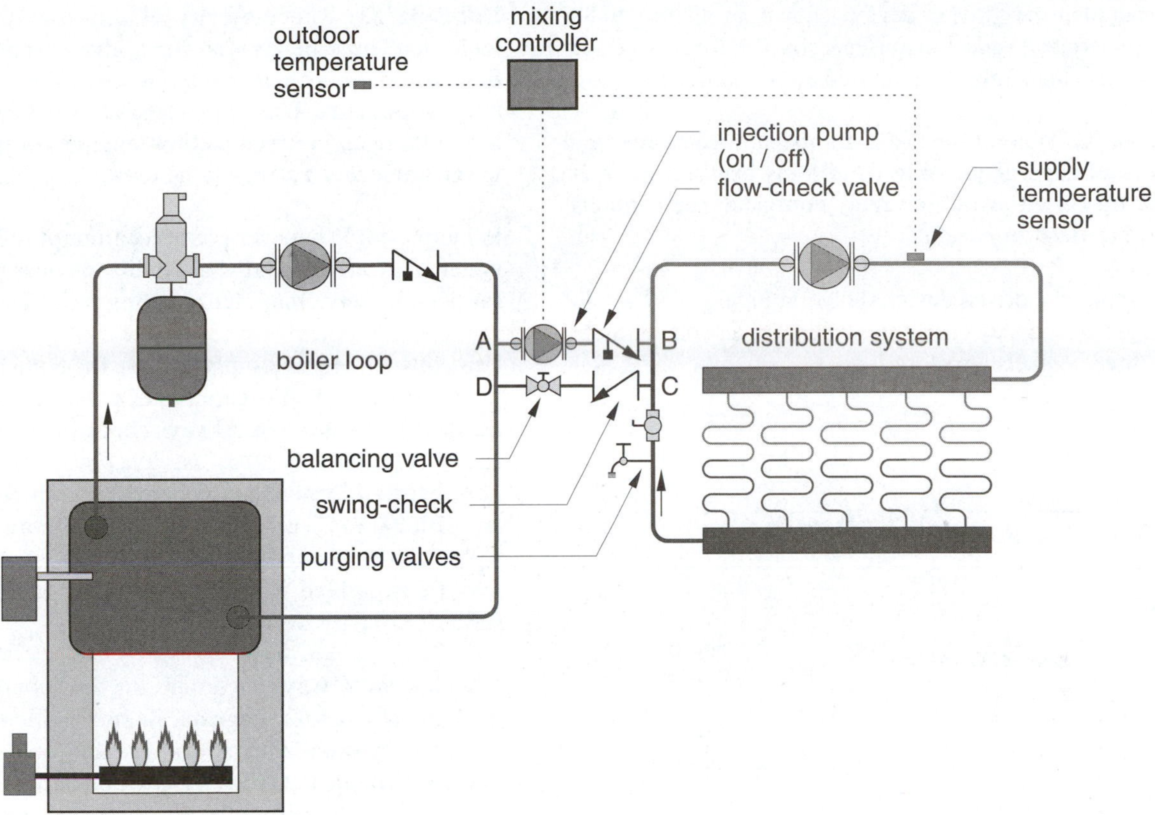

Figure 9–87 Piping schematic for injection mixing using an on/off pump. Note that flow check valve(s) can be used in the injection risers to prevent heat migration. Also note that on/off injection mixing cannot protect a conventional boiler from sustained flue gas condensation.

Some Final Points on Injection Mixing

- One of the strengths of injection mixing is that *only the hot water needed for the mixing process passes through the control device*. Not only does this allow relatively small control hardware to manage significant rates of heat transfer, it also reduces the head losses associated with passing the full flow of the distribution system through the mixing device. Reduced head loss translates into reduced electrical power input to circulators. The life cycle cost savings can be substantial in larger systems.
- The viability of an injection mixing system is directly tied to the temperature difference between the heat source and the distribution system. The greater this temperature difference, the lower the required injection flow, and the smaller the injection hardware. Injection is well-suited to systems where a conventional boiler serves as the heat source for a low temperature distribution system. It is also a very cost-effective alternative to a 3-way or 4-way motorized mixing valve in larger systems.
- All the piping schematics shown for injection mixing show purging valves in the distribution system. These are essential because the closely spaced tees that connect the injection risers to the primary loop do not permit good purging flow in the distribution system.
- Be sure to include the thermal trap detail wherever it has been shown on the injection piping schematics. If omitted, heat migration into the distribution system could become a problem.

9.9 CONTROL SYSTEM DESIGN PRINCIPLES

Entire books have been written describing methods for creating the control logic required for various heating system applications. This text cannot go into such detail. However, by consistently applying a few basic principles, the necessary control logic can usually be created. The principles are as follows:

Principle #1: When a device is to operate only when two or more conditions are met (as evidenced by closed switch or relay contacts), these contacts should be wired in *series*.

For example, the driven device shown in Figure 9–88 will be energized, and remain energized, only if all three series-connected contacts are closed at the same time.

Principle #2: When a device is to operate when any one or more of several possible conditions are met (as evidenced by closed switch or relay contacts), the contacts should be wired in *parallel*.

For example, the driven device shown in Figure 9–89 would be energized, and stay energized, whenever any one or more of the three contacts are closed.

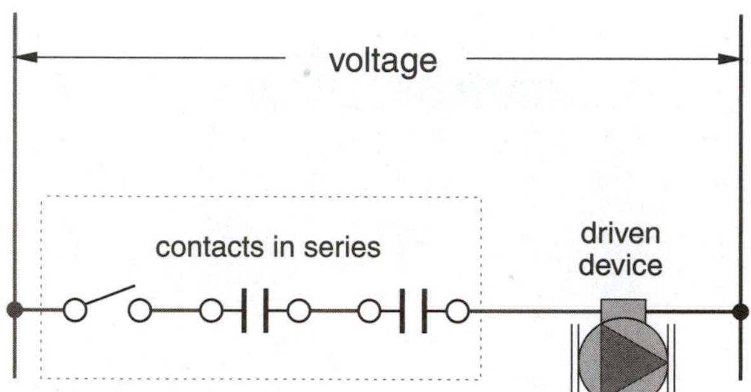

Figure 9–88 Contacts wired in series. All contacts must be closed simultaneously to operate driven device.

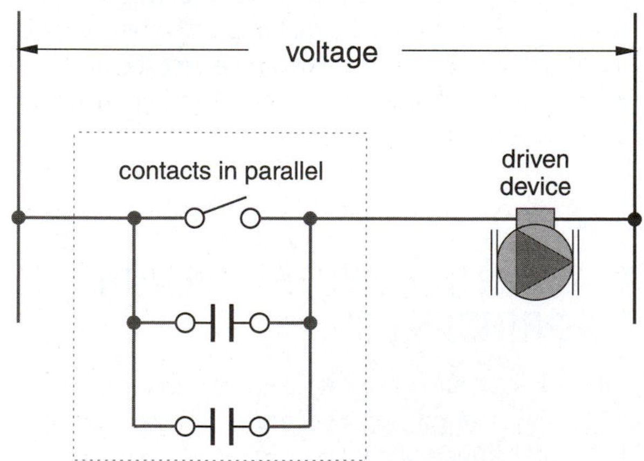

Figure 9–89 Contacts wired in parallel. The driven device is energized when any one or more of the contacts are closed.

Principle #3: The majority of any hardwired control logic that must be created for a given system should be done using low voltage components rather than line voltage components. This approach is less costly because the low voltage and currents allow smaller components with lower amperage ratings to be used.

Principle #4: Whenever possible, attempt to build control systems around readily available devices that can be serviced by any competent heating technician.

Principle #5: Never simply assume the transformer in the heat source control is capable of powering several other controls in the system. Always check for the external VA rating. If none is stated, use a separate transformer to power the additional control devices. Control transformers with 24-VAC secondary voltage are available with VA ratings from about 20 VA through 100 VA. If the system may be expanded in the future, allow at least 20 VA of reserve capacity in the transformer.

Principle #6: Always document the final control system in the form of a ladder diagram. Be sure to include a symbol legend with the diagram, and identify any nonstandard devices. Provide a written description of operation for the system that describes the sequence of each operating mode (i.e., space heating, domestic water heating, pool heating, etc.). This will be demonstrated in the following section. *Lack of good documentation for the control system is a sure recipe for costly problems on the job.*

9.10 EXAMPLE OF A MODERN CONTROL SYSTEM

This section presents an example of a modern control system that manages the operation of a multiload multi-temperature hydronic heating system. Such systems are commonly used in modern residential and light commercial buildings. The system uses relatively basic control components such as a multizone relay center, boiler high limit controller, and an injection mixing controller.

System Description

The example system provides two types of hydronic space heating as well as domestic hot water. Three independent zones of baseboard are controlled using on/off zone circulators. A single zone of radiant floor heating is supplied with a reduced water temperature by a variable speed injection mixing system. Domestic water heating is configured as a priority load. When it is operating, all space heating zones are turned off.

A piping schematic of the example system is shown in Figure 9–90. Notice that all the circulators have designations

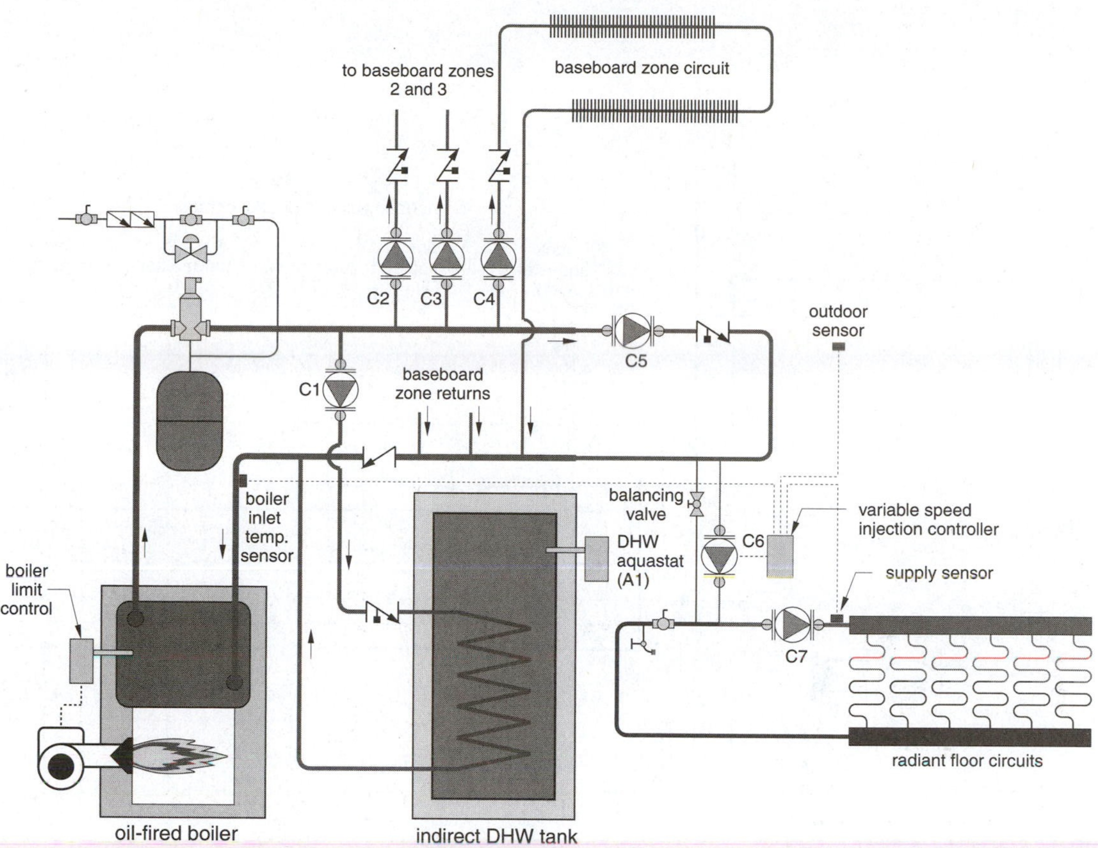

Figure 9–90 Piping schematic for an example system supplying three fin-tube baseboard zones, lower temperature radiant floor circuits through variable speed injection mixing, and an indirect water heater operated as a priority load.

such as C1, C2, etc. This allows them to be cross-referenced between the piping and wiring schematics, as well as referenced in the description of operation.

The electrical schematic for the example system is shown in Figure 9–91. Much of the required control logic is handled by the multizone relay control. All space heating thermostats as well as the DHW tank aquastat are wired to the relay center. All circulators are supplied with line voltage from the outputs of the relay center. The relay center has been configured for priority control of the domestic water heating load. Although not shown in the electrical schematic, this is usually done by removing a jumper or setting a switch inside the relay center. Whenever this load is active, all other line voltage outputs are turned off. A normally open relay contact inside the relay center connects to its (X X) terminals. This contact closes when any of the zone thermostats or the DHW tank aquastat calls for heat. The contact completes the thermostat circuit in the boiler's high limit control allowing the boiler to fire. The wiring shown within the multiload relay center is representative only, and will vary depending on the specific product used. It is shown so the complete circuit can be traced out when following the sequence of operation.

Description of Operation

The following text describes each operating mode of the system using references to the major components identified in the piping and wiring schematics. The description is sequential beginning with the call for heat. As you read the description of operation, refer to the piping and wiring diagrams to find the devices and circuit paths used in each operating mode.

1. Domestic water heating mode. Upon a call for heat from tank aquastat (A1), the zone 1 (priority) relay in the multizone relay center is turned on. This relay provides a line voltage output to the circulator (C1). A normally open relay contact between (X X) terminals in the multiload relay center closes to complete a circuit across the (T T) terminals in the boiler limit control, enabling the boiler to fire. The boiler temperature is regulated by its high limit setting and differential. Hot water from the boiler is pumped through the heat exchanger in the DHW tank where it gives up heat then returns to the boiler. The normally closed priority contact in the

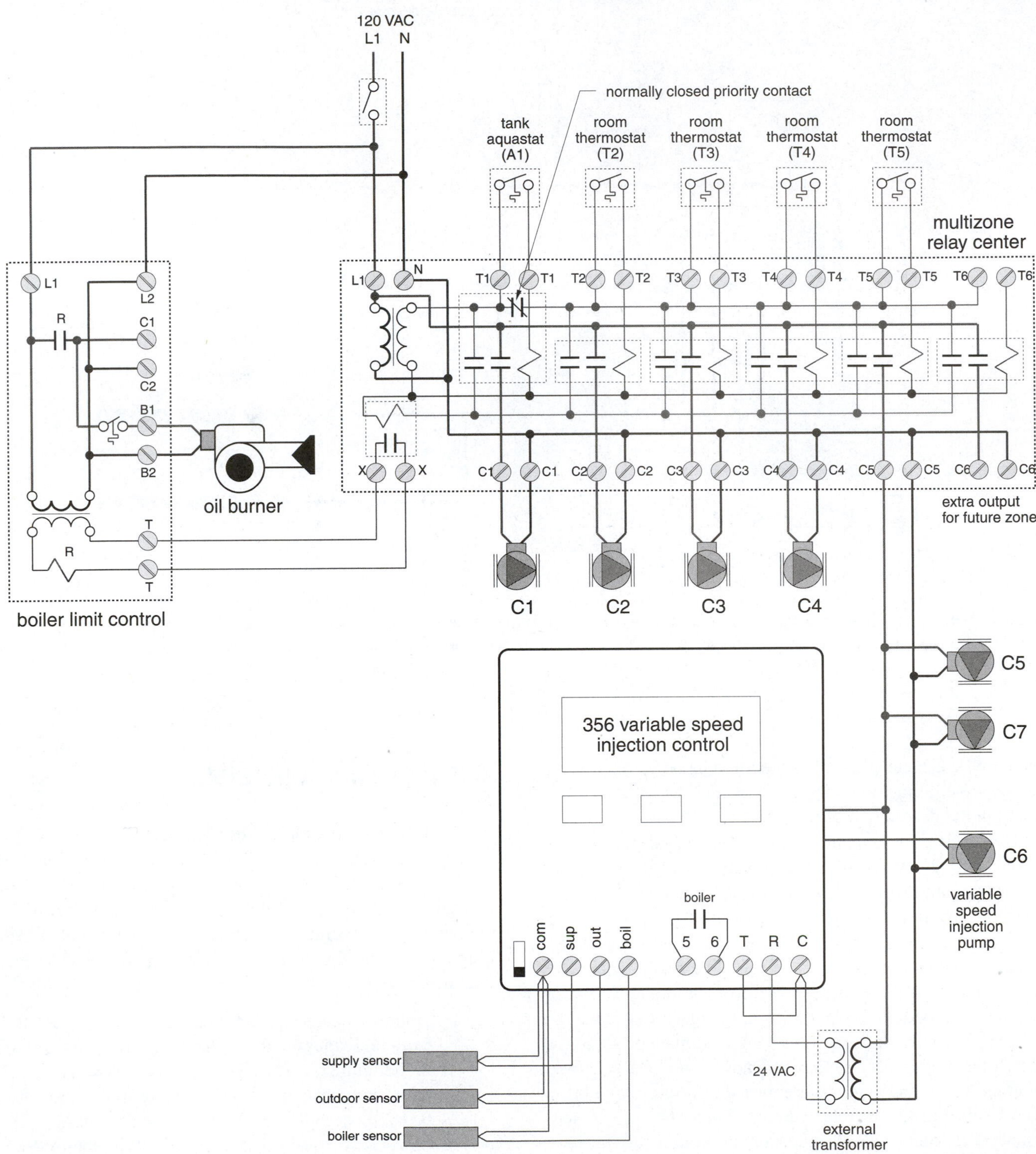

Figure 9–91 A control system for the piping system shown in Figure 9–90. Note the use of a multizone relay center and variable speed injection controller. Also note the cross-referencing of electrical device between the piping and control wiring diagrams.

multiload relay center temporarily turns off all space-heating loads during the DHW heating cycle. The multiload relay center is equipped with built-in priority override that can reestablish space heating if the DHW load is not satisfied within 30 minutes.

2. Baseboard heating mode. When any of the thermostats in the baseboard zones (T2, T3, T4) call for heat, the associated relays in the multiload relay center are turned on. Line voltage is sent to the circulators (C2, C3, and C4) which serve the zones calling for heat. A normally open relay contact between (X X) terminals in the relay center closes to complete a circuit across the (T T) terminals in the boiler limit control enabling the boiler to fire. The boiler temperature is regulated by its high limit setting and differential. This mode of space heating is temporarily interrupted when domestic water heating is operating.

3. Radiant floor heating mode. Upon a call for heating from thermostat (T5), the associated zone relay in the multiload relay center is turned on. It supplies line voltage to circulator (C5), and the distribution circulator (C7). It also supplies line voltage to a 30-VA 120/24-VAC control transformer. This transformer supplies 24 VAC to the 356 variable speed injection mixing controller. When powered up, the controller boots its internal software and begins monitoring the three sensors. If the boiler temperature sensor indicates a low temperature, the injection pump will operate at a very low speed. As the boiler temperature increases, the injection mixing controller ramps up the speed of the injection pump (C6) as necessary to achieve the current target temperature at the supply sensor. The injection mixing controller calculates this target temperature using outdoor reset control. A normally open relay contact between (X X) terminals in the multiload relay center closes to complete a circuit across the (T T) terminals in the boiler limit control enabling the boiler to fire. The boiler temperature is regulated by its high limit setting and differential. This mode of space heating is temporarily interrupted when domestic water heating is operating.

This system is a relatively simple example, and uses basic control devices and operating logic. The same system could be modified in numerous ways to add or remove functionality. For example, a boiler reset controller could be added. This controller might be capable of recognizing when the domestic water heating load is active, and respond by allowing the boiler to climb to its high limit setting during this mode of operation. If the reset controller did not have this functionality, an external relay could be used to bypass the reset control upon a call for domestic water heating.

Another possibility would be to configure the radiant floor distribution circulator (C7) for constant circulation provided the outdoor temperature is below the desired room temperature. Again, this could be accomplished in different ways depending on the specific injection mixing controller selected.

Still another possibility would be to add another circuit of baseboard heating that would serve as supplemental heat for the radiant floor zone. The radiant floor subsystem would be operated by the first stage of a two-stage thermostat. The second stage of that thermostat would operate the supplemental heating circuit.

The functionality of the multiload relay center could also be provided by hardwiring several relays together as shown earlier in this chapter. However, this would likely increase installation cost relative to the approach shown. The capabilities of the multiload relay center are well-used in this system.

It should also be noted that some of the control functions built into the controls were not used. For example, the circulator output on the boiler limit control was not connected. There is no circulator that needs to operate when *any* load calls for heat. Likewise, the sixth zone of the multiload relay center was not needed. It was present because the number of zones multicenter relays are supplied with are typically 3, 4, and 6. If another zone of heat was added in the future, this spare zone connection would prove handy.

9.11 FUTURE OF HYDRONIC HEATING CONTROLS

Hydronic heating technology is constantly improving, especially in the area of controls. New products appear every year that add functionality, reduce cost, boost efficiency, or otherwise improve the benefits offered by modern hydronic heating. Extremely rapid progress in microprocessor-based control technology has had a profound impact on the hydronic heating industry over the past decade. Further advancements in this area are virtually guaranteed. This section discusses control concepts that are very likely to be implemented in residential and light commercial hydronic systems by the year 2005. It is even possible that some are in use by the time you read this.

Communicating Controllers

Many of the current generation controllers discussed in this chapter were designed as stand-alone devices that perform a specific but limited task. Examples include boiler reset controllers, variable speed injection mixing controllers, or multiple boiler controllers. When several such devices are combined in one system, there is often a redundancy of hardware. For example, a boiler reset controller, injection mixing controller, and multiple boiler controller all need to reference (measure) outside air temperature as they operate. With current technology, this requires an outdoor sensor for each controller, even though each measures the same temperature.

Although this is certainly possible, and routinely done, it obviously is more expensive than a system in which a single sensor could be read by several different devices. Similar arguments can be made for other control circuitry such as boiler temperatures sensors, digital displays, power supplies, and even multiple microprocessors.

Several stand-alone controllers operating simultaneously in a system may also interfere with each other if not properly adjusted. For example, if the boiler high limit control is set too low, it may turn off the boiler at a temperature lower than what the reset control has targeted. Likewise, if the boiler temperature is regulated by the reset control during domestic water heating, the recovery rate of the tank may become very low during warm weather.

Future systems will reduce component redundancy and interference by allowing communication between controls. The likely scenario is a simple **two-wire digital bus** that carries signals between controllers. Each device has its own digital address, and thus can be individually polled by other devices. All the controls that need specific information such as boiler temperature, outdoor temperature, or differential pressure would get that information from a single sensing device monitoring that temperature or pressure. This approach also allows the heating system to communicate with other building systems such as those used for cooling, ventilation, lighting, and security. Communication protocols such as LONworks and BACNET are already used for these purposes in **direct digital control (DDC)** systems in larger buildings. Look for this functionality to migrate downward into cost-effective residential applications in the very near future. A concept drawing is shown in Figure 9–92.

Internet accessibility will be an essential element of communicating control systems. Most of the current controllers used in residential and light commercial hydronic heating require a person to be present at the controller whenever an adjustment is required. This will change when the control system can be accessed via the Internet. With proper authorization, a technician could access the control system's Web page and examine a log of how the system has been operating over the last few hours, days, or even weeks. They could also make any necessary adjustments from wherever Internet access is available. Some controllers will have the ability to automatically send a request for service when certain fault conditions are detected. They may even use wireless modems for communication and thus not require dedicated telephone lines.

Adaptive Control

Even sophisticated control algorithms such as PID processing work mostly with **real time inputs** in present-generation controllers. A mixing reset control, for example, bases the calculated target supply water temperature on the outdoor temperature it senses at the time. It may add heat to a building at 7 A.M. while totally oblivious to the likelihood that solar heat gains will overheat that building within the

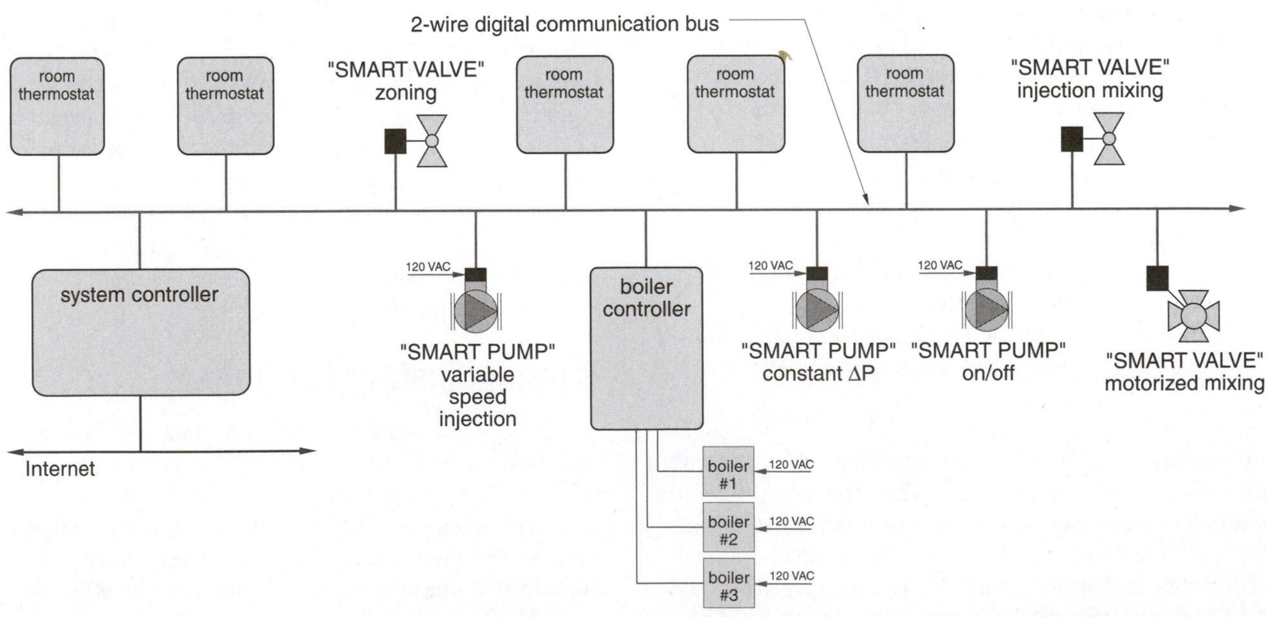

Figure 9–92 Concept of control devices that share information using a 2-wire communications bus. This reduces redundant components such as multiple outdoor temperature sensors. It also allows the control system to better adapt to changing conditions.

next 2 hours. This scenario is likely to repeat itself many times because the controller is not capable of learning how to anticipate and compensate for such situations.

Future controls will use more advanced algorithms, as well as Internet access to learn when such conditions are likely, and automatically modify their actions to compensate. This is type of **fuzzy logic** is currently deployed on a limited basis in some microprocessor-based controllers. Its use and capabilities are sure to increase in future hydronic heating control systems.

User-Definable I/O

Most of the controllers presented in this chapter have input/output (I/O) terminals that must interface with specific devices such as a line voltage pump, 10K thermistor, or a floating actuator on a mixing valve. In some cases, the number of I/O terminals limits the functionality the controller can provide or the type of system in which it can be used. Some applications do not use all the capabilities offered by a given controller. These unused features add cost to the system without improving its performance. Still other systems may need capabilities that a preprogrammed controller cannot provide.

The microprocessors in many current-generation controllers have vast capabilities if the appropriate programming could be created and executed, and the necessary inputs and outputs interfaced to the controller. Future controllers will take advantage of this by offering user-defined inputs and output. For example, in one system, a given pair of terminals on a controller may be configured (through software) to connect to a 10K thermistor temperature sensor. On the next system, that sensor may not be needed. The same terminals may instead be configured to drive a 2- to 10-VDC modulating control valve, or perhaps to the coil of a relay. This ability allows a single mass-produced controller to be programmed to perform exactly as needed in thousands of different applications. The "multiple control box" approach of the present will eventually be replaced with a software-driven universal controller. Functionality will increase, cost will drop, and performance will continue to improve.

Consolidation of Separate Components

As various control strategies become widely accepted, manufacturers are motivated to consolidate the various components involved into a preassembled product. The goal is to reduce installation time as well as to reduce the chance of mismatched components.

One example of such a product is the injection pump block in Figure 9–93. It consolidates the injection pump, distribution pump, and injection mixing controls needed for a typical residential system into a single product that can be quickly installed.

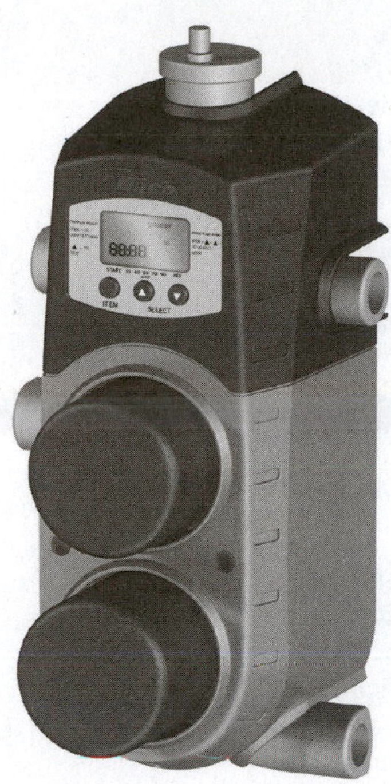

Figure 9–93 Injection pump block which incorporates the injection pump, system circulator, and controls into a preassembled device. *Courtesy of Taco, Inc.*

KEY TERMS

actuator
ampacity
analog signal
balance point temperature
bimetal element
boiler protection
boiler reset control
boiler rotation
closed control loop
coil
contactor
contacts
control differential
control state
controlled device
controlled variable
controller
delay-on-break
delay-on-make
demand-fired
differential
DIN rail
direct digital control (DDC)
direct injection mixing

droop
duty cycle
electromechanical controls
electronic controls
embedded controller
end switch
equal percentage characteristic
error
external VA rating
feedback
fixed lead boiler
floating control
floating zone
full reset
fuzzy logic
general purpose relay
hardwired logic
heat anticipator
heat loss coefficient
heat motor
heat motor actuator
HI/LO fire boilers
high limit setting
injection mixing
Internet accessibility
interstage differential
ladder diagram
lead/lag control
line voltage section
low voltage section
low water cutoff (LWCO)
manipulated variable
manual reset high limit (MRHL)
measured value
microprocessor-based controller
minimum supply temperature
mixing assembly
mixing device
mixing reset control
modulating boilers
multifunction time delay device
multiple boiler controller
multistage heat production
multizone relay center
normally closed (N.C.)
normally open (N.O.)
offset
operating differential
operating mode
outdoor reset control
output
overshoot
parallel shifting
partial reset
poles
primary side
priority control
priority zone
processing algorithm
proportional band
proportional reset
proportional-integral control (PI)
proportional-integral-derivative control (PID)
pulse width modulation (PWM)
quick opening valve
rangeability
real time inputs
relay
relay socket
remote bulb aquastat
reset line
reset ratio
reverse injection mixing
room thermostat
rung
secondary side
sensing element
setpoint controller
setpoint demand
setpoint temperature
short-cycling
space heat demand
stages
strap-on aquastat
tankless coil
target value
thermal trap
thermistor
three-wire control
throws
time delay relay
transformer
triple action control
tri-state control
two-wire digital bus
undershoot
valve actuator
volt-amp (VA)

CHAPTER 9 QUESTIONS AND EXERCISES

1. What is the advantage of proportion-integral (PI) control over proportional only (P) control?
2. A heating system is designed to provide 180 °F water to the distribution system when the outside temperature is –15 °F. The building requires no heat input when the outside temperature is 65 °F. What would be the proper reset ratio for this system?
3. Explain the difference between *floating control* and *2- to 10-VDC modulating control.* Which can be used to operate mixing valves?
4. A building has a design heat loss of 55,000 Btu/hr when the outside temperature is –5 °F and the indoor temperature is 70 °F. Determine the balance point temperature of the building when internal heat gains are 11,000 Btu/hr.
5. Explain why a valve used to regulate heat output by varying the flow rate through a heat emitter should have an equal percentage characteristic.
6. Why do electromechanical thermostats experience droop? How do electronic thermostats compensate for droop?
7. Which is more important when selecting a 2-way injection valve—pipe size or *C*v value? Explain your answer.
8. What must be true of any mixing assembly capable of protecting a conventional boiler from sustained flue gas condensation?
9. What is meant by the term *priority override*? What type of device is commonly used to enable priority override in a control system?
10. What is the difference between a *setpoint demand* and a *space heating demand* to a multiple boiler controller? What is a common application for a setpoint demand?
11. The term *error* refers to the difference between the _____ temperature, and the _____ temperature in a temperature control system.
12. Describe the difference between the reset line for a hydronic system using finned-tube baseboard versus one using a heated floor slab.
13. What is the difference between *partial reset control* and *full reset control*? Which is used for boiler reset of a conventional boiler?
14. Explain the difference between the number of poles and number of throws on a switch. Sketch a schematic symbol for each of the following switches:
    a. a DPST switch
    b. a SPDT switch
15. Explain the effect the thermal mass of a hydronic system has on the operation of the control system and the temperature swings inside the building. Describe a situation where a system with a high thermal mass is desirable. Describe another situation where it is undesirable.
16. Explain the operating principle of electromechanical setpoint controls using a fluid-filled bulb sensor.
17. Complete the ladder diagram shown in Figure 9–94 such that the following control action is achieved. When switch S1 is set in the off position, a red 24-VAC indicator light is on. When the switch is set to the on position, a green 24-VAC indicator light is on, *and* a line voltage circulator is running. Label all components you sketch in the diagram.
18. A control system having the ladder diagram shown in Figure 9–95 is proposed to operate a three-zone hydronic system using a separate circulator for each zone. Identify any electrical errors in this diagram. Describe what would happen if the identified error were present when the system was turned on, or what is unsafe about the error. Sketch out how you would modify the schematic for proper and safe operation of the three-zone system.
19. What is the function of a thermal trap in an injection mixing system?
20. What is an advantage of using a motorized 2-way valve and controller rather than a 2-way thermostatic valve for an injection mixing application.
21. What is the difference between direct and reverse injection mixing? Which is better for smaller systems and why?

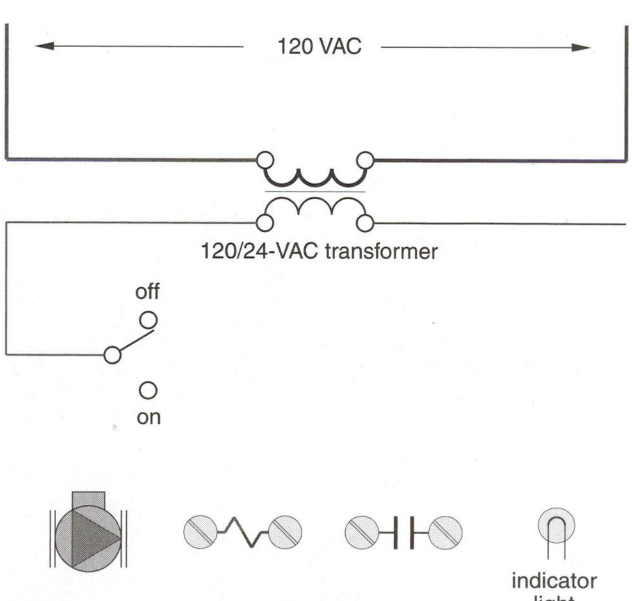

Figure 9–94 Ladder diagram for Exercise 17.

22. An injection mixing system is being designed to transfer 125,000 Btu/hr into a distribution system using water at 185 °F, both under design load conditions. The distribution system has a supply temperature of 105 °F and a temperature drop of 17 °F under design load conditions. Determine the necessary injection flow rate assuming the:
    a. System uses direct injection mixing with a variable speed pump
    b. System uses direct injection mixing with a 2-way modulating valve
    c. System uses reverse injection mixing with a variable speed pump

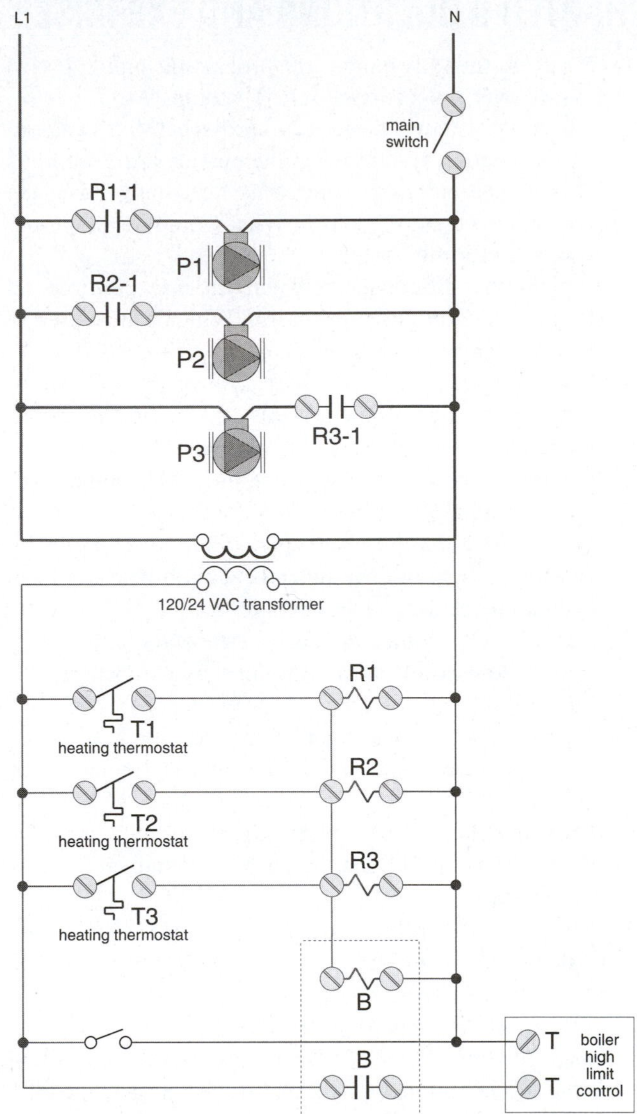

Figure 9–95 Ladder diagram for Exercise 18.

HYDRONIC RADIANT PANEL HEATING 10

OBJECTIVES

After studying this chapter, you should be able to:

- Explain what radiant heating is
- Explain how radiant heating affects thermal comfort
- Describe several types of hydronic radiant panels
- Compare radiant panel heating to other methods of heating
- Discuss the benefits of radiant panel heating
- Identify good opportunities for radiant floor heating
- Identify good opportunities for radiant wall and ceiling heating
- Understand how tubing circuits are placed in radiant panels
- Estimate the thermal performance of several types of radiant panels
- Identify situations where supplemental heating is needed
- Describe the operation and application of several common mixing devices
- Avoid common errors in the design and installation of hydronic radiant panel systems
- Describes methods of producing multiple water temperatures in the same system

10.1 INTRODUCTION

Hydronic radiant panel heating is currently one of the fastest growing segments of the North American HVAC market. During the 1990s, demand for these systems increased over tenfold, and the demand continues to increase each year. The superior comfort and fuel efficiency delivered by radiant panel heating is being discovered and sought after by more and more consumers for use in all types of buildings. Many mechanical contractors have been initiated into the broader field of hydronic heating specifically by the need to install hydronic radiant panel systems.

This chapter describes the general theory of radiant heating as well as the specific details of using hydronics to deliver heat to various types of radiant panels. It will also show how piping and control concepts from previous chapters can be integrated in radiant panel systems.

10.2 WHAT IS RADIANT HEATING?

Radiant heating *is the process of transferring thermal energy from one object to another by* **thermal radiation.** Chapter 1, Fundamental Concepts, briefly described thermal radiation as one of the three modes of heat transfer, along with conduction and convection. Whenever two surfaces are "within sight of each other," and are at different temperatures, thermal radiation travels from the warmer surface to the cooler surface. The rate of energy flow between these surfaces depends on the difference in their temperatures, the distance between the surfaces, the angle between the surfaces, and an optical property of the surfaces known as **emissivity.**

Thermal radiation includes both the visible and infrared portion of the **electromagnetic spectrum.** The wavelengths in these portions of the spectrum are such that any radiation that strikes an opaque surface does not penetrate beyond its surface (unlike X-rays, and gamma rays that can penetrate beyond the surfaces of opaque objects).

All thermal radiation emitted by surfaces at temperatures lower than approximately 970 ºF will be in the infrared portion of the electromagnetic spectrum, and as such cannot be seen by the human eye. All radiant panels discussed in this chapter operate well below this temperature, and thus only emit **infrared thermal radiation.**

Other than the fact that it cannot be seen, infrared thermal radiation behaves similar to visible light. It travels away from the emitting surface at the speed of light (about 186,000 miles per second), and can be partially reflected by some surfaces. Like visible light, infrared thermal radiation travels equally well in any direction. This allows radiant energy to be delivered from heated walls and ceilings as well as from heated floors.

Like other forms of electromagnetic radiation, thermal radiation does not require a media such as a solid, liquid, or gas to move energy from one location to another. This is evidenced by the fact that sunlight passes through approximately 93 million miles of empty space before entering the earth's atmosphere.

Within a room, infrared thermal radiation emitted by a heated surface passes through the air with virtual no absorption of energy. Instead, the radiation is absorbed by the objects in the room. Most nonpolished/nonmetallic surfaces absorb the majority of any thermal radiation that strikes them. The small remaining portion is reflected, often to another surface where further absorption takes place. Thus, very little thermal radiation is reflected out of a room heated by a radiant panel.

It is therefore accurate to state that thermal radiation (more commonly referred to as "radiant heat") warms the objects in a room rather than directly heating the air. This difference is largely what separates radiant heating from convective heating. If the absorbed thermal radiation raises the surface temperature of an object above the room air temperature, some heat will be convected to the room air. Some heat may also be conducted deeper into the object if its interior temperature is lower than its surface temperature.

Infrared thermal radiation is constantly emitted from our skin and clothing surfaces to any cooler surfaces around us. A large portion of the heat generated through metabolism is released from our bodies by infrared thermal radiation. This process greatly affects our thermal comfort, and will be discussed in more detail in Section 10.6.

Finally, it is important to understand that thermal radiation is entirely different from nuclear radiation. The latter is a type of particle radiation emitted by radioactive materials such as uranium and plutonium, and certainly not suited for direct heating of buildings. Unfortunately, many people associate the word *radiation* with nuclear radiation, and thus feel that any kind of radiation is inherently dangerous. This is simply not the case. The thermal radiation emitted by low temperature radiant panels is in no way unhealthy or harmful.

10.3 WHAT IS A HYDRONIC RADIANT PANEL?

A hydronic radiant panel is any object warmed by passing heated water through tubing embedded in or attached to it, and which releases at least 50 percent of that heat to its surroundings as thermal radiation. The heated water is simply the material used to deliver heat to a hydronic radiant panel. If heating cable was embedded in or attached to the same object, one could refer to it as an *electric radiant panel.*

Once heat is transferred to the radiant panel, by any means, its shape, orientation, surface temperature, surface properties, and surroundings determine its thermal output.

The best-known type of hydronic radiant panel is a heated floor. Currently, floors account for over 90 percent of all hydronic radiant panel installations. However, radiant panel heating is by no means limited to floors. There are several established methods of incorporating hydronic tubing into walls and ceilings. Hydronic heating professionals should become familiar with the strengths and limitations of all such panels, and use them where they hold advantages.

Most hydronic radiant panels operate at water temperatures no higher than 200 °F. Many operate with average water temperatures in the range of 80 °F to 160 °F, with surface temperatures in the range of 75 °F to 120 °F.

10.4 THE HISTORY OF RADIANT PANEL HEATING

Heating a room by warming one or more of its surfaces goes as far back as the Roman Empire. In ancient Roman baths, floors and walls were often heated by directing the flow of exhaust gases from wood fires under raised stone floors, and up through hollow wall cavities. Although it is unlikely such systems achieved the even temperature distribution or accurate control possible with modern technology, the intricacy of their construction testifies that they delivered comfort unattainable by simpler means.

The arrival of forced circulation hydronic heating in the United States during the 1930s provided the needed link between producing heat in a boiler and efficiently transporting this heat to a radiant panel. During this time, steel pipe was threaded together to form distribution circuits that were often cast into concrete floor slabs. Some of those systems are still in use today.

The rapid growth in the copper tube market in the 1940s further expanded the possibilities for hydronic radiant panel heating. Architect Frank Lloyd Wright used hydronic radiant panel heating in his well-known Usonian homes. Thousands of systems using copper tube embedded in concrete floors as well as plaster walls and ceilings were constructed during the post-World War II housing boom. Many are still in use today. An example of an early radiant ceiling panel built with copper tubing is shown in Figure 10–1.

During the 1960s and 1970s, the use of hydronic radiant panel heating in America steadily fell victim to market forces that began substituting easier and less expensive heating options in place of proven comfort. As labor costs increased, and central cooling delivered by forced-air systems became readily available, hydronic radiant panel heating systems declined in popularity. During this same time, the publicized failures of some hydronics systems due to corrosion or metal fatigue also discouraged further use.

Fortunately, during this same time new thermoplastic materials were under development in Europe that would eventually bring about a resurgence of interest in hydronic radiant panel heating. Cross-linked polyethylene (PEX) tubing became available on the European market during the early 1970s. This tubing first appeared on the U.S. market during the early 1980s. PEX and other polymer-based tubing materials revolutionized the installation of hydronic radiant panel heating, providing fast installation and a long, reliable service life. The availability of PEX tubing was the spark that rekindled interest in hydronic radiant panel heating. That spark has now grown into a major flame. Over the last 25 years, *billions* of feet of polymer tubing have been installed in hydronic radiant panel heating systems worldwide. The growth of hydronic radiant panel heating in the North American market increased over 1000 percent during the 1990s as shown in Figure 10–2.

Today, hydronic radiant panel heating is being used in all types of buildings including houses, schools, offices, aircraft

Figure 10–1 An early hydronic radiant ceiling panel installation using copper tubing.

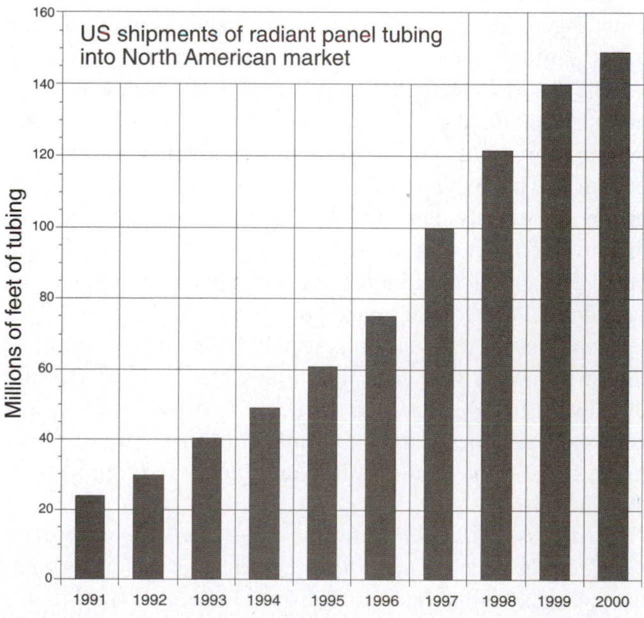

Figure 10–2 Recent sales statistics for tubing used in hydronic radiant panel heating in North America. *Courtesy of the Radiant Panel Association.*

hangers, fire stations, vehicle garages, and more. When combined with modern control technology, hydronic radiant panel heating represents what many, including the author, consider the ultimate form of comfort heating.

10.5 BENEFITS OF RADIANT PANEL HEATING

A properly designed and installed hydronic radiant panel heating system offers many benefits. Some are inherent to radiant heating. Others are inherent to the use of hydronics to deliver heat to the panels. Still others are simply the elimination of undesirable side effects associated with other methods of heating.

These benefits include:

- A system that delivers unsurpassed comfort. Most people who have had an opportunity to compare the comfort offered by a properly installed radiant panel heating system will testify that it delivers comfort superior to other methods of heating. The underlying reasons are discussed in the next section.
- A system that is inconspicuous. Few people enjoy looking at portions of a heating system that, out of necessity, are located within an occupied space. Forced air registers, and even hydronic baseboards convectors can sometimes interfere with furniture placement. In contrast, hydronic radiant panels are totally out of sight within floors, walls, or ceilings.
- A system that is extremely durable. Because radiant panels are literally built into the structure, they are well-protected from physical damage. This is especially true for heated concrete floor slabs, which can withstand punishing traffic in commercial or industrial environments without damage to the embedded tubing.
- A system with virtually no operating noise. A properly designed and installed radiant panel system operates in virtual silence. It does not interfere with the serenity of the space it heats. Modern heat source and circulators operate with minimal noise, and are usually located in a mechanical room that can be acoustically isolated from living spaces. There are no noise-generating fans or blowers located in the occupied spaces.
- A system that minimizes drafts and dust movement. A common complaint about forced-air heating systems is that they create drafts and redistribute dust, pollen, odors, or other airborne pollutants within a building. Radiant panel heating creates very gentle (imperceptible) air motion within individual rooms rather than the building as a whole.
- A system that is compatible with low temperature heat sources. Many types of radiant panels can operate at relatively low water temperatures, even at design load conditions. This allows low temperature heat sources

such as hydronic heat pumps and condensing boilers to supply heat while operating at high efficiency.

- A system that quickly dries floors. Radiant floor heating can quickly dry floors in areas such as bathrooms, entry foyers, and hallways, and therefore reduce the chances of slips and falls. Rapidly dried entry floors also reduce carpet soiling in others areas of the building.
- A system with thermal storage. Some hydronic radiant panels, such as a heated concrete floor slab, have high thermal mass. Such panels can store large quantities of heat, allowing them to deliver a "surge" of heat in situations where large doors are open and cold outside air pours in. Normal comfort is restored very quickly after the doors are closed. This characteristic is particularly desirable in garages or aircraft hangers. The ability of a floor slab to store heat also protects the building against freezing for perhaps two or three days if the heat source or other hardware is nonoperational.
- A system capable of fast response. Not all hydronic radiant panels have high thermal mass. Some are specifically designed for low thermal mass, allowing them to quickly "turn on" and "turn off." Radiant walls and ceilings are usually of low thermal mass construction, and well-suited for situations where comfort needs to be quickly established after a prolonged temperature setback period. They can also respond quickly when significant internal heat gains occur from sunlight, occupants, or other sources.
- A system that is easily zoned. Hydronic radiant panel heating, like other types of hydronic heating, can be easily configured for room-by-room zoning. Sleeping areas can be maintained cool while bathrooms are maintained warm. Heat input can be quickly interrupted to areas experiencing solar heat gains. Rooms that are not frequently used can be maintained at reduced temperatures to reduce energy consumption.
- A system that reduces energy usage. There are many documented examples of fuel savings associated with hydronic radiant panel heating. The underlying reasons include the ability to provide comfort at low interior temperatures, lower operating water temperatures, reduced air temperature stratification, reduced air leakage, and the ability to provide room-by-room zoning. A conservative estimate of fuel savings associated with the use of radiant floor heating relative to convective heating is 15 percent. Some buildings have demonstrated savings in excess of 50 percent.

10.6 PHYSIOLOGY OF RADIANT PANEL HEATING

The most sought after benefit of radiant panel heating is unsurpassed thermal comfort. Such comfort results from the interaction of several factors, not the least of which is heat exchange between the human body and a radiant panel. This section focuses on this interaction.

Our bodies constantly produce heat through a process called *metabolism.* For an adult at rest the rate, of heat production is normally in the range of 300 to 400 Btu/hr. To remain comfortable, this heat must be dissipated at the same rate it is generated. If heat is released faster than it is generated, a person feels uncomfortably cool. Likewise, if heat cannot be released as fast as it is generated, the person feels uncomfortably warm. Under some conditions the rate of heat release from one portion of the body, the feet for example, can become disproportionately high. This will also adversely affect comfort. *Maintaining comfort is thus a matter of providing environmental conditions that allow the body to release heat at the same rate it is produced.* This balance also needs to be maintained without conscious effort on the part of the occupant.

Heat is released from the body through four processes:

- Evaporation of moisture
- Convection of heat to surrounding air
- Conduction of heat to objects in contact with the body
- Radiant heat transfer to cooler surfaces surround the body

Evaporation of water from the body accounts for approximately 25 percent of the heat released from the body during light activity under typical indoor conditions. As the indoor humidity increases, this process becomes less effective as it does during periods of hard physical work or heavy exercise. At other times, evaporation of moisture from the skin can be extremely effective and noticeable. For example, standing in a breeze with wet skin or clothing can produce very noticeable, sometimes even excessive, cooling.

Our bodies also lose heat by convection to cooler surrounding air. About 30 percent of the body's heat production is lost through convection under typical interior conditions.

The speed at which air moves across our skin and clothing surfaces greatly affects convective heat loss. The faster the air moves past the body, the more effective the process becomes. This explains why a room fan that simply increases the air velocity past the body without cooling the air can improve comfort on a hot summer day. The increase in convective heat loss with increasing air speed also explains the "windchill" effect experienced on cool days.

Heat loss by conduction is usually a small percentage of the body's total heat loss. Examples include heat flowing downward from feet placed on a cool floor, or heat loss to a cold car seat on a winter morning. In most interior situations, conduction heat loss is usually less than 5 percent of total heat loss.

Heat loss by thermal radiation is the predominant method by which the body releases heat. Under typical indoor conditions, thermal radiation from skin and clothing surfaces to surrounding surfaces accounts for about 50 percent of total heat loss. The rate of radiant heat transfer from the skin and clothing is very dependent on the surface temperatures in the room. Cool surfaces such as windows can serve as very effective

heat sinks for thermal radiation. They can produce discomfort even when room occupants are surrounded by air temperatures that would otherwise be considered comfortable.

Because a large portion of the body's heat loss is by thermal radiation, comfort is greatly influenced by the temperature of the surfaces in a room. Increasing the temperature of one or more of the larger surfaces can significantly reduce the rate of heat release by thermal radiation from the body. During the heating season, when the objective is to limit the rate of heat loss from the body, moderately warmed room surfaces can greatly improve comfort.

Room Air Temperature Profiles

The priority of the body is to maintain the temperature of its critical organs in the central torso. When placed in an environment that allows heat to be released from its surface faster than it is being produced internally, the body responds by reducing blood flow to its extremities (i.e., hands and feet).

The feet in particular are most affected because they are furthest removed from the central body. To make matters worse, the feet are usually in contact with cooler surfaces that tend to draw heat away by conduction. They also have a greater surface area per unit of mass than do other parts of the body. To put it in mechanical terms, *the feet are like large radiators fastened to cold walls at the far end of a hydronic distribution system.* The heat they receive from the bloodstream is easily dissipated to cool surroundings.

The head, on the other hand, usually has a good supply of heat-carrying blood. It is also insulated, to varying degrees, by hair. This combination allows the head to be comfortable at air temperatures several degrees cooler than those required near the feet. Most people feel comfortable and more alert when air temperatures at head level are in the low to mid-60 °F range.

Figure 10–3 shows the variation of air temperature from floor to ceiling needed for *ideal* comfort at a relaxed activity level. Such a graph is called a **room air temperature profile.** Note that the temperature near the floor needs to be slightly higher than average. This reduces the rate of heat loss from the lower extremities.

The degree to which a heating system provides thermal comfort is in part determined by the room air temperature profile it creates within a space. Profiles similar to that shown in Figure 10–3 are desirable. Unfortunately, many common methods of adding heat to a room result in significant deviations from this ideal.

Figure 10–4 compares typical room air temperature profiles created by three common methods of heat delivery: radiant floor heating, baseboard convectors, and forced air.

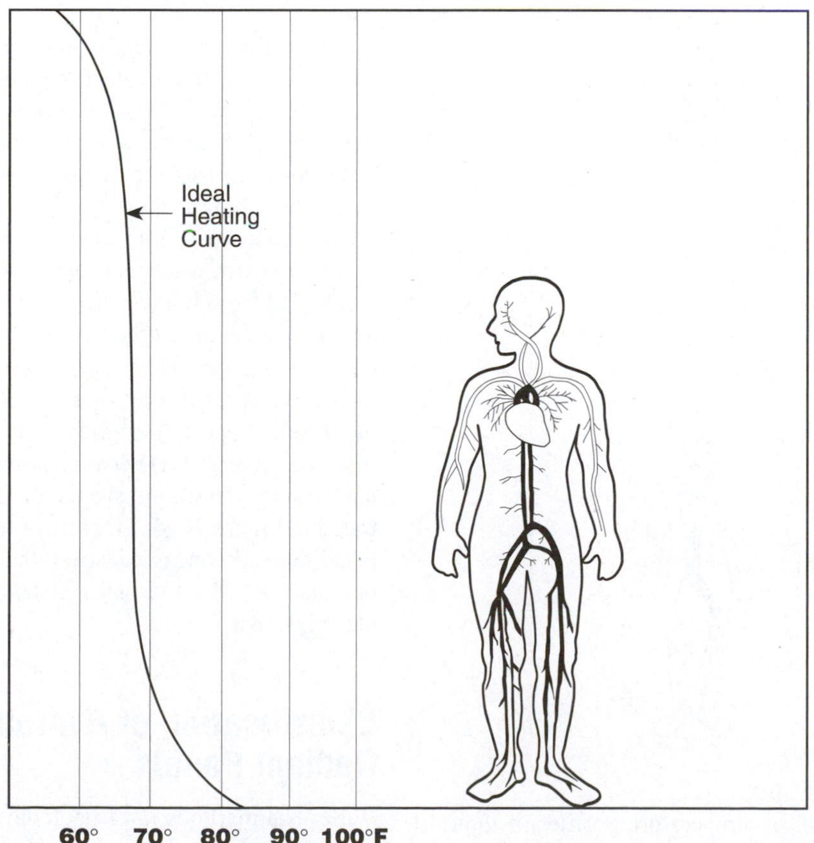

Figure 10–3 Ideal air temperature distribution in a heated room. *Courtesy of Uponor Wirsbo.*

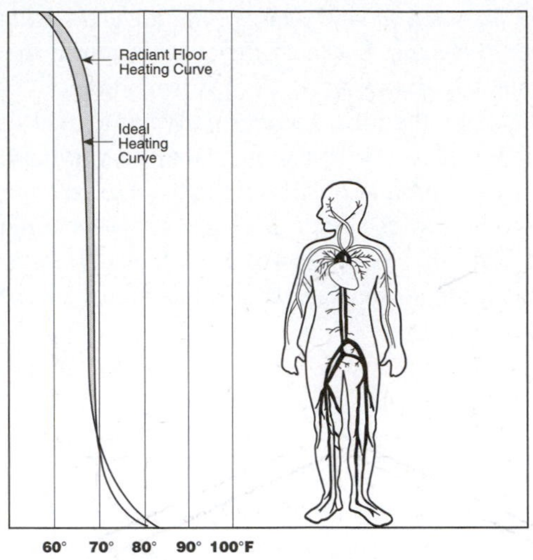

(a)

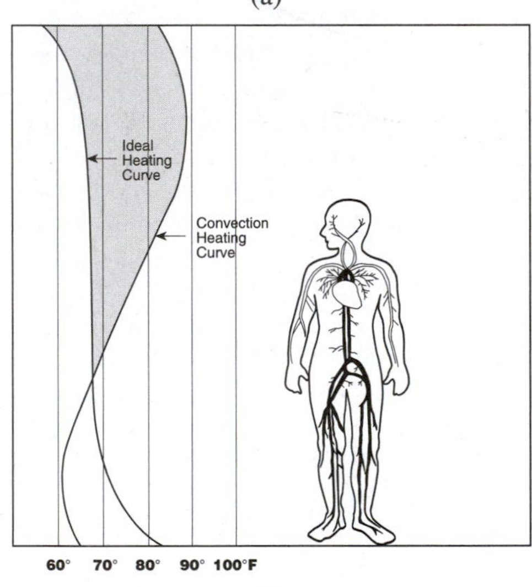

(b)

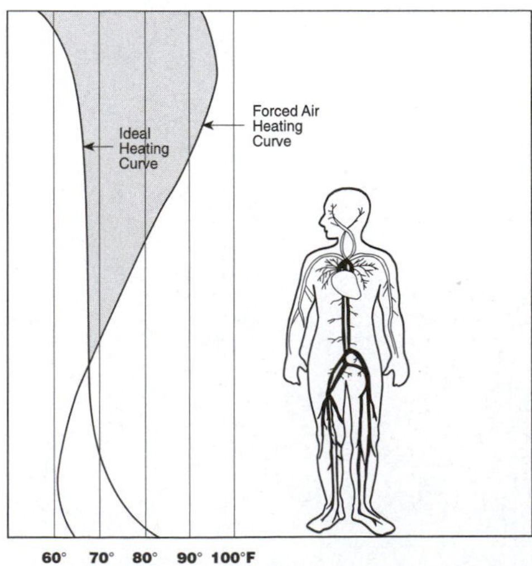

(c)

Figure 10–4 Comparison of temperature profiles of three types of heating delivery systems. (a) Radiant floor heating profile, (b) baseboard convector profile, (c) forced-air profile. *Courtesy of Uponor Wirsbo.*

The profiles for baseboard convectors and forced-air systems reveal lower than ideal temperatures near the floor, as well as higher than ideal temperatures near the ceiling. This is the result of introducing heated air into the space at temperatures significantly higher than the desired room temperature. The "overheated air" quickly rises toward the ceiling, while the higher density cool air collects at floor level, exactly where it is least desired. This undesirable effect is called **air temperature stratification.**

In comparison, radiant floor heating creates a temperature profile very close to the ideal scenario. Since room air is not "overheated," undesirable stratification does not occur.

The room air temperature profiles shown in Figure 10–4 can be influenced by air motion, ceiling height, and insulation levels. Rooms with high rates of air flow and well-designed ducting systems will have less temperature stratification due to vigorous mixing of room air. Rooms with tall ceilings, high register temperatures, and poor air circulation will experience greater temperature stratification.

10.7 METHODS OF HYDRONIC RADIANT PANEL HEATING

All hydronic radiant panel systems discussed in this chapter are created by placing tubing into the floor, walls, or ceiling of a room. In some cases, the tubing is embedded into a poured material such as concrete. In other cases, it is fastened to the surface of a building material. In still other cases, the tubing is suspended in a hollow air cavity such as the space between floor joists.

When warm water is circulated through the tubing, heat is released from it and diffused outward through the materials making up the radiant panel. Most of the heat is directed into the room by the proper selection and placement of insulation. A small portion is released from the rear side of the panel. Properly designed systems allow the heat to diffuse (i.e., spread out) between adjacent tubes so that surface temperature variations are small. These concepts are illustrated in Figure 10–5. The rate of heat flow from the radiant panel depends on its construction, its operating water temperature, and the thermal resistance of any finish material placed over it.

Classification of Hydronic Radiant Panels

Several methods have been developed for incorporating tubing into floors, walls, and ceilings. Within this text, hydronic radiant panel installation methods are classified as follows:

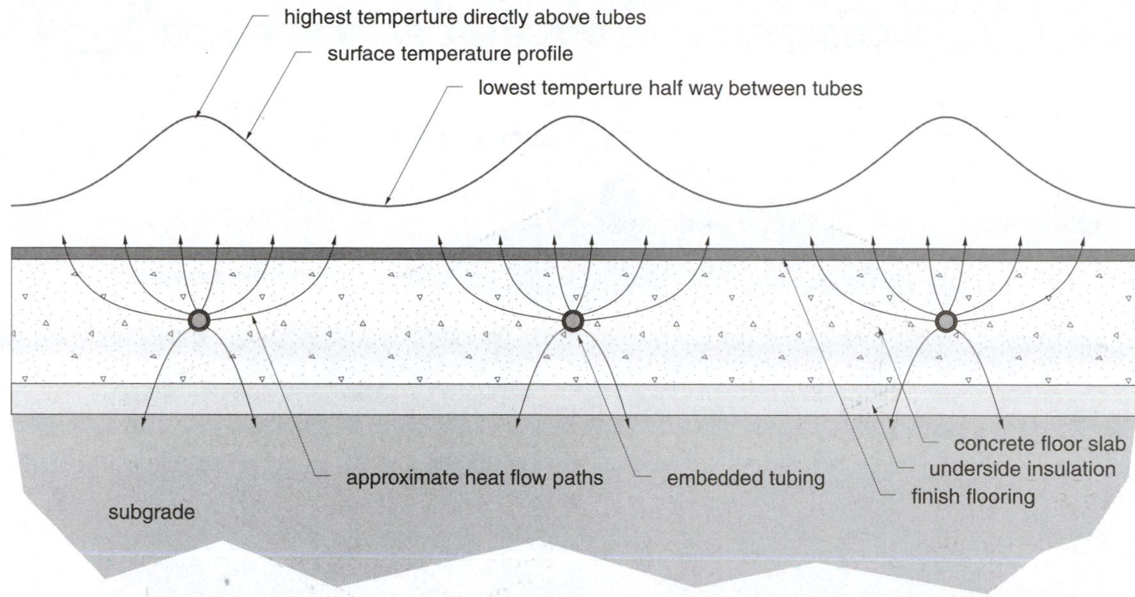

Figure 10–5 Surface temperature profile above a heated concrete slab. Also shown are approximate heat flow paths.

1. Radiant floor panels
   a. Slab-on-grade
   b. Gypsum thin slab
   c. Concrete thin slab
   d. Above floor tube and plate
   e. Below floor tube and plate
   f. Suspended tube
   g. Plateless staple-up
   h. Prefabricated subfloor/underlayment panels
2. Radiant wall panels (tube and plate)
3 Radiant ceiling panels
   a. Tube and plate
   b. Prefabricated panels for suspended ceilings

Each installation method offers technical or economic advantages in certain situations. Each method also has limitations that must be respected by designers and installers. Hydronic heating professionals should learn how to deploy each type of radiant panel in circumstances where they offer advantages. Each installation method is detailed in upcoming sections of this chapter.

10.8 SLAB-ON-GRADE RADIANT FLOORS

The most economical hydronic floor heating systems are those where a concrete slab is already planned as the floor structure. In such situations, the extra cost of making the floor into a radiant panel is limited to the tubing, underslab insulation, and the labor to place these materials. The cross section of a typical concrete slab-on-grade radiant floor is shown in Figure 10–6. A cutaway view is also shown in Figure 10–7.

Slab-on-grade radiant panels usually operate at low water temperatures, especially slabs that have no finish flooring. The thickness of the concrete provides a good "thermal wick" to diffuse heat outward and upward through the slab. A computer simulation of such heat diffusion is shown in Figure 10–8.

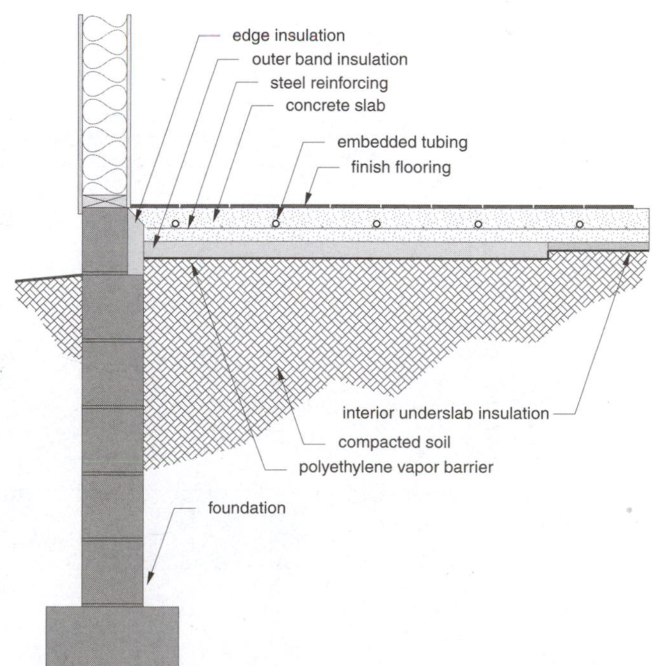

Figure 10–6 Cross-section of a typical slab-on-grade radiant floor.

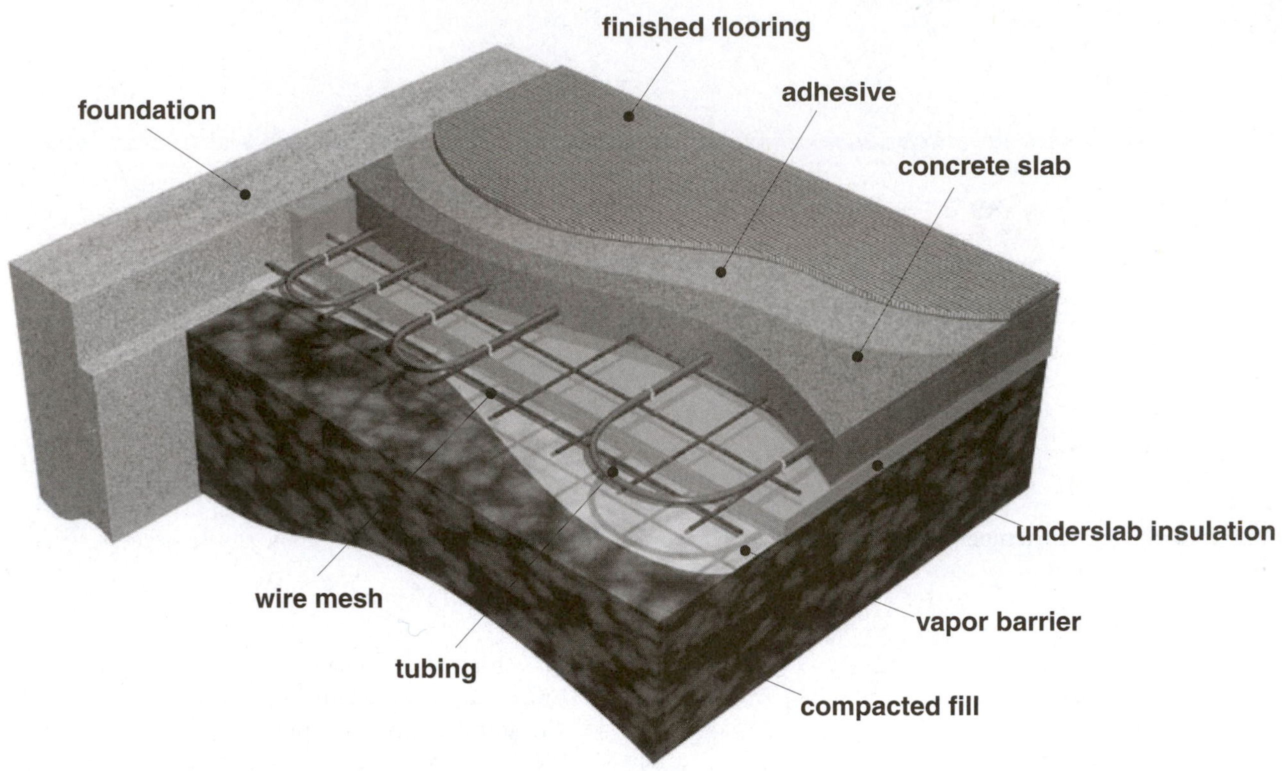

Figure 10–7 Cutaway view of a typical slab-on-grade radiant floor panel. *Courtesy of IPEX, Inc.*

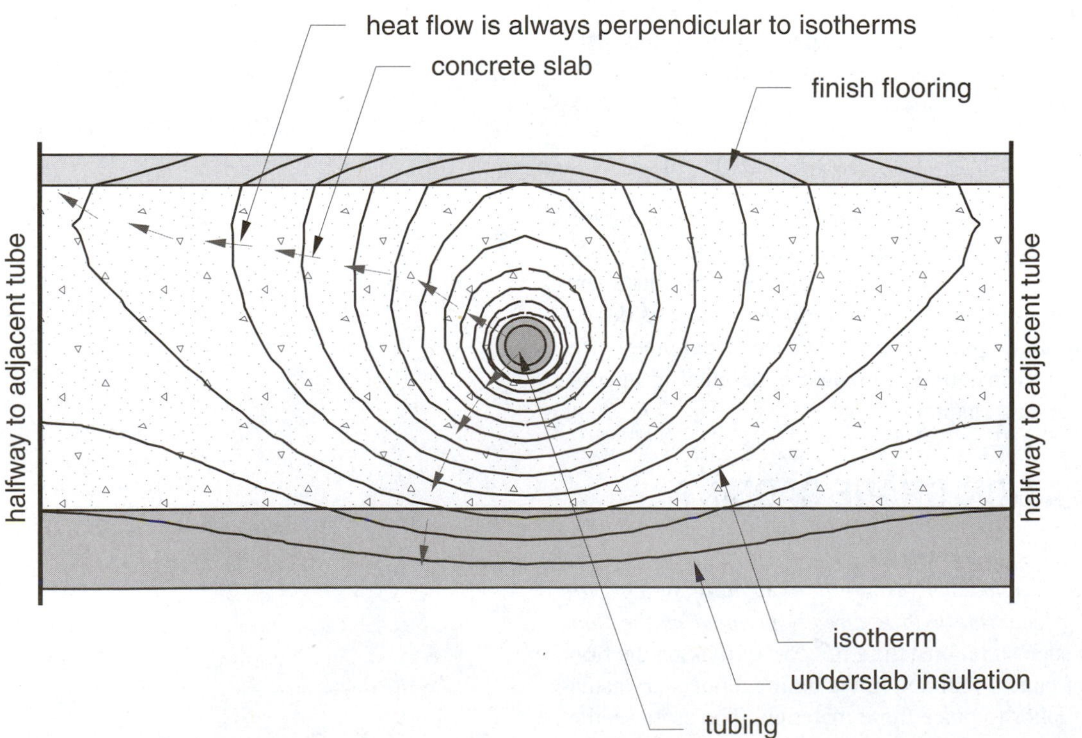

Figure 10–8 Temperature isotherms and heat flow from warm tubing embedded in a concrete slab-on-grade floor.

The curved lines in the cross section are called **isotherms.** Each isotherm is created by connecting points that are at the same temperature. Heat flow at all locations is perpendicular to the isotherms, always in a direction of decreasing temperature. In a sense, isotherms can be thought of as waves of heat moving away from the tubing. Notice how heat moves laterally away from the tubing, then upward toward the surface of the floor. Some heat also moves downward and is lost to the soil beneath the insulation.

Effect of Tube Size

The size of the tubing used in a heated slab has a small effect on the slab's heat output. Larger tubes have more surface area through which heat can pass to the concrete and thus *slightly* increase heat output for a given water temperature. Figure 10–9 shows the predicted thermal performance variations for several different tube diameters.

Tube Depth

The depth of the tube within the slab does affect thermal performance. This effect is illustrated in Figure 10–10.

Ideally the tube should be centered approximately one-third to one-half of the slab depth below its surface. As a practical matter, this is hard to accomplish unless the tubing is carefully tied to steel reinforcing that remains at the proper depth during the pour. In cases where welded wire reinforcing is used, tubing tends to end up near the bottom of the slab because the reinforcing is not lifted properly as the concrete is placed. Sawn **control joints** in the slab may also dictate deeper placement to avoid damaging the tube.

Experience has shown that slab-on-grade radiant floors will still function reasonably well when the tubing is placed near the bottom of a 4- to 6-inch thick slab. Higher water temperatures will be required to maintain a given rate of heat output. The higher operating temperatures tend to lower the efficiency of heat sources such as ground source heat pumps and condensing boilers. They also increase the rate of downward heat loss to the soil. The time required to warm the surface of the slab after a setback period also increases when tubing is placed deeper in the slab. The proper structural functioning of welded wire reinforcing also requires a placement depth between one-third and one-half the slab's thickness.

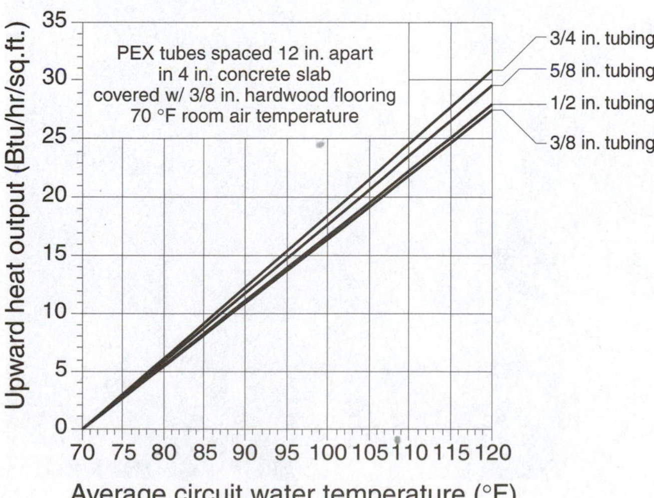

Figure 10–9 Effect of tubing diameter on heat output from a slab-on-grade floor.

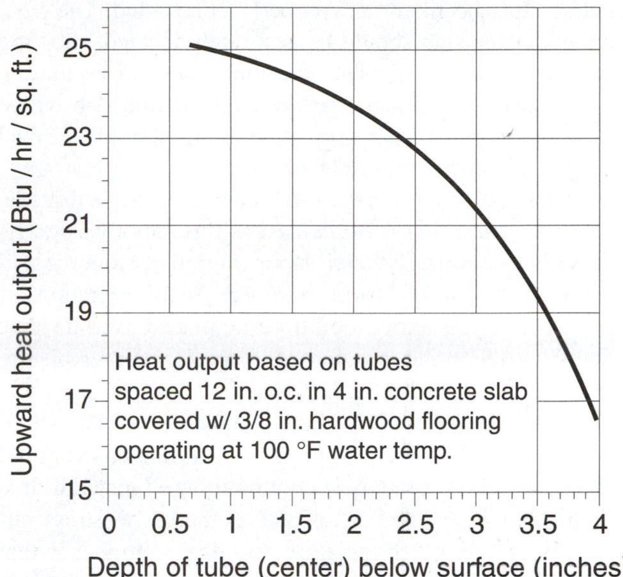

Figure 10–10 Effect of tubing depth on heat output from a slab-on-grade floor.

Slab-on-Grade Installation Procedure

The installation of heating tubing in a slab-on-grade floor is an integral part of the construction sequence of a building. It must be preplanned and closely coordinated with the other trades involved in the construction.

Part of this preplanning is to develop a tubing layout drawing, which is described later in this chapter. *Without an accurate plan showing the location and length of all piping circuits, as well as the location of the manifold stations, even an experienced installer can spend hours attempting installation by trial and error.* The results are still likely to be less than ideal. The installation procedure to be described assumes that a layout plan has been prepared, and thus the installer's main function is to place the tubing according to this plan. The procedure also describes installation of a vapor barrier and underslab insulation. On some projects, these materials may be installed by other contractors. However, the heating contractor should always verify they have been properly installed before installing tubing.

All underslab plumbing or electrical conduit should be in place prior to tubing installation. All underslab utility trenches must be filled and properly compacted. The final grade under the slab should be accurately leveled, allowing for the thickness of the slab and the underslab insulation. This is especially critical near the edges of the slab where thicker insulation is used. Any loose rocks should be raked off to provide a smooth stable surface.

Begin by placing a **vapor barrier** over the subgrade. Polyethylene sheeting is often used as this vapor barrier. Its purpose is to minimize water vapor migration from the soil into the slab during warm weather. Such migration, if unchecked, can discolor wood flooring placed over the slab. In areas with high radon potential, a special radon/moisture barrier may also need to be installed.

Underslab insulation is then installed over the vapor barrier. Although several options are available, **extruded polystyrene foam** is the most commonly used material. It is available in several thicknesses and compressive stress ratings. It does not absorb moisture or outgas compounds that may be harmful to the environment. Designers should verify that local codes allow placement of foam insulation below concrete slabs, especially in areas where termites are present.

The insulation should be neatly placed with all tongue and groove joints interlocked. If the site is windy, place sheets of welded wire fabric and/or wooden planks over the foam sheets as they are placed to avoid wind uplift.

The recommended R-value of underslab insulation varies with geographic location. In the absence of specific codes that require higher values, a *minimum* of R-5 insulation should be installed under all heated slab areas, *other than areas with high structural loading such as footings or bearing pads.* Heat loss is more pronounced near the perimeter of the slab. The R-value of the insulation under the outer 4 feet of the slab is often increased to R-7.5 to R-10. The edge of the slab must also be thermally isolated from the foundation using a minimum of R-7.5 insulation. Figure 10–11 shows 2-inch extruded polystyrene insulation with a tapered upper edge installed on the inside of the foundation wall. The tapered edge allows concrete to be placed close to the foundation wall.

Most concrete slab-on-grade floors are reinforced with welded wire fabric (WWF) or steel rebar. When welded wire fabric is used, it should be placed over the insulation, with adjacent sheets lapped a minimum of 6 inches at all edges and wire tied together.

The next step is to use the tubing layout drawing to locate the **manifold stations** within the building. These are the locations where tubing penetrates up through the slab surface and eventually connects to supply and return manifolds.

Manifold stations are often housed within wall cavities that will eventually be built around them. It is crucial to carefully measure and place the manifold station at the precise location where such walls will eventually be built. A temporary manifold station support is constructed by driving two 3-foot stakes through the insulation and into the soil. A small plywood panel can then be fastened to these stakes, and the manifold mounted to it as shown in Figure 10–12. After the slab is poured and the manifold station is supported within the wall cavity, the stakes can be sawn off flush with the slab surface. Be sure the manifold station faces a direction where it can be accessed when the surrounding wall is completed.

Figure 10–11 Extruded polystyrene edge insulation installed along an exposed foundation wall.

Figure 10–12 Temporary manifold support prior to placing tubing. This support must be accurately placed. Note guide holes for tubing in lower blocking, and sufficient space for tubing bends below the blocking.

After the manifold stations are placed, use the tubing layout drawing to mark of portions of each tube circuit on the foam insulation using spray paint. Corners, return bends, and offsets should be marked as shown in Figure 10–13. This greatly speeds installation and reduces the chance of mistakes as the tubing is placed.

The tubing circuits can now be placed one at a time. Look up the length of the circuit being created on the tubing layout drawing. Select a coil of tubing long enough to create the circuit without splicing. Most tubing sold for radiant panel applications comes with sequential length numbers printed every 3 feet along the tube. By subtracting the smaller number at one end of the coil from the larger number at the other, the length can be quickly determined.

To prevent twisting, the tubing must be unrolled from the coil. *Never pull tubing off the side of a coil.* Most installers make use of an **uncoiler,** such as the one shown in Figure 10–14. This device allows the coil to freely spin as tubing is pulled from it. The uncoiler should be placed several feet away from where the tubing is being laid down. Several feet of tubing should be pulled off the coil to allow ample slack.

Connect the free end of the tubing coil to the supply manifold. Pull some tubing off the coil and proceed along the supply side of the circuit, fastening the tubing along the way. Bends in PEX or PEX-AL-PEX tubing in sizes up to 5/8 inch can be made by hand, taking care not to kink the tubing in the process. Bends in larger diameter tubing or especially tight bends may require a tube bender. Be sure to verify the minimum bend radius allowed with the tubing manufacturer.

Tubing is usually fastened to the welded wire fabric or steel reinforcing using wire twist ties, nylon pull ties, or some type of plastic clip. The manufacturer may require that a specific type of fastener be used to retain the warranty on the tubing. Typical spacing for ties or clips is 24 to 30 inches on straight runs and at two or three locations on return bends. A close-up of tubing secured with wire ties is shown in Figure 10–15.

Figure 10–13 Spray paint used to mark approximate location of corner bends prior to placing tubing. *Courtesy of Harvey Youker.*

Figure 10–14 An uncoiler allows tubing to be pulled off the coil without twisting.

As the tubing is placed, watch for any sharp ends on steel reinforcing. Either bend the reinforcing out of the way or slightly offset the tubing to avoid chafing its surface.

Some tubing suppliers require the installation of **bend supports** where the tubing bends from horizontal to vertical under the manifold stations. Some installation instructions also call for the tubing to be sleeved where it penetrates through the slab surface. Verify the recommended details with the tubing manufacturer.

Control Joints

Larger concrete slabs often require control joints. The purpose of control joints is to intentionally weaken the slab at specific locations so shrinkage cracking occurs along the control joint rather than randomly across the slab. Control

Figure 10–15 Securing PEX tubing to welded wire reinforcing using wire ties.

joints are usually made using a special saw the day after the slab is poured. The depth of the cut is often specified to be approximately 20 percent of the slab's thickness.

Figure 10–16a shows how tubing should be detailed where it passes beneath a sawn control joint. The tubing is covered with a thin plastic sleeve that is centered on the eventual location of the control joint (Figure 10–16b). This sleeve prevents the concrete from bonding directly to the tubing at this location. This minimizes stress on the tubing during any subsequent movement of the slab. Sleeving material is available from most tubing manufacturers for this purpose. The thin-wall sleeving can be slit with a utility knife and wrapped around the tubing after the circuit has been fastened in placed. Many installers mark the location of control joints using spray paint prior to fastening the tubing in place.

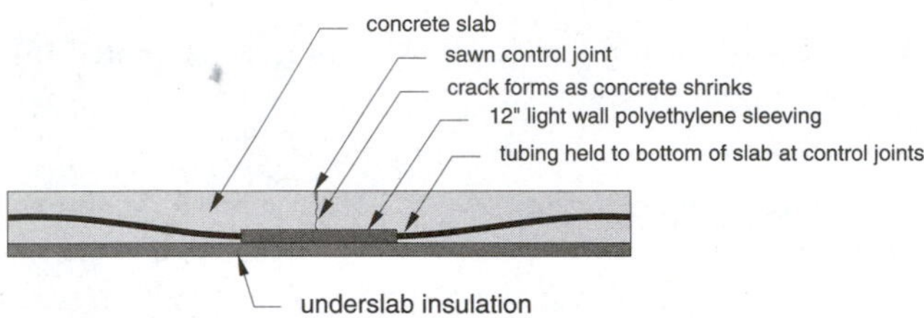

Figure 10–16a Detail where tubing passes under sawn control joints.

Figure 10–16b Thin wall polyethylene sleeving over tubing where it will pass under a sawn control joint. Location of the control joint marked by line painted on underslab insulation.

The tubing should *not* be lifted where it passes beneath control joints. This ensures a wide birth between the tubing and the bottom of the saw cut.

Another important installation detail is to hold tubing at least 8 inches away from any toilet drainage piping rising through the slab. This minimizes heating of the wax ring that is eventually installed on the closet flange.

Pressure Testing

Although leaks due to "defective tubing" are extremely unlikely, it is possible that careless equipment handling on the job site could damage the tubing prior to it being embedded in concrete. To ensure that no leaks are literally cast into concrete, all circuits should be pressure tested prior to the pour.

Pressure testing is best done using compressed air, which is not subject to freezing during cold weather construction. A typical pressure-testing specification requires that all tubing circuits to be pressurized to and maintained at 60 to 75 pounds per square inch (psi) for a period of at least 24 hours.

A convenient approach is to install a pressure gauge on one of the manifolds at each manifold station, along with a Schrader valve for adding air. Plug all remaining manifold ends and vents, then pump air into the assembly with an air compressor. An example of a temporary pressure test setup is shown in Figure 10–17.

Figure 10–17 Temporary installation of pressure gauge and Schrader valves for circuit pressure testing prior to concrete placement.

Once the assembly has reached its test pressure, apply a soapy water solution to all manifold connections to check for any air leaks.

If the pressure drops significantly within a few hours, it is very likely the leak is at one of the manifold connections. Retighten and recheck with soap solution until the system is airtight. If the system will be left unattended for some time, the air pressure should be reduced to about 30 psi for safety. The tubing circuits should remain under pressure while the slab is poured. If a tube were accidentally punctured during the pour, the compressed air would give an immediate indication of the leak. The leak could then be repaired using a coupler specifically designed for that purpose.

Photo Documentation

After all circuits have been placed, several photographs of the tubing installation should be taken. Photos should be taken around each manifold station, as well as in any areas where tubing placement is tight. The use of digital cameras allows such photos to be quickly gathered and stored on disk or CD-ROM. A copy of the photos should be given to the owner for future reference.

Placement of Concrete

There is no standard procedure for placement of concrete slabs. From the standpoint of the tubing, the placement procedure used should minimize heavy traffic over the tubing and ensure that the tubing is placed approximately middepth in the slab (other than where it passes beneath control joints).

Wheelbarrow traffic over tubing under pressure is generally not a problem; however, care should be taken not to pinch the tubing under the nose bar of the wheelbarrow as it is dumped. So-called "power buggies" are too heavy to be driven over the tubing and insulation. Likewise, concrete trucks should never be driven over these materials.

For larger slabs, the use of a concrete pump truck equipped with an aerial boom is the ideal way to efficiently place the concrete with minimal traffic over the tubing. Figure 10–18 shows concrete being placed for a large radiant floor slab using an aerial boom.

As the concrete is placed, lift the welded wire reinforcing and attached tubing to approximately middepth in the slab (other than at control joints). Ideally, one or more workers should be assigned solely to this task. The mesh is pulled up using a lift hook as shown in Figure 10–19. After the coarse (stone) aggregate in the concrete flows under the mesh, it tends to support it quite well.

After the concrete is placed, it is finished in the usual manner. Special care should be taken not to nick the tubing with trowels where it penetrates the slab under the manifold stations. Careless operation of tools such as power trowels can sever tubing circuits and lead to costly repairs.

The installation procedures just described call for good coordination of trades at the job site. It is important for the concrete placement crew to understand details such as depth of the tubing and control joint detailing prior to beginning the pour.

Figure 10–18 Ariel pumping used to efficiently place concrete for a large slab-on-grade heated floor.

Figure 10–19 Mason using a lifting hook to pull welded wire reinforcing and attached tubing to middepth of slab. Note that hook is lifting reinforcing wire rather than directly lifting tubing.

10.9 CONCRETE THIN-SLAB RADIANT FLOORS

There are many buildings where radiant floor heating is desirable, but where slab-on-grade construction is not possible. One of the most common is a wood-framed floor deck in a residential or light commercial building.

One alternative is to install a **concrete thin-slab** radiant panel over the subfloor. A cross-sectional drawing of this approach is shown in Figure 10–20. A cutaway view is shown in Figure 10–21.

Thin-slab systems use the same type of PEX or PEX-AL-PEX tubing as slab-on-grade systems. The tubing is fastened directly to the wood subfloor and then covered with a special concrete mix. Underside insulation is usually installed between the floor framing under the subflooring.

As with slab-on-grade systems, the concrete thin slab provides an effective "thermal wick" allowing heat to diffuse laterally outward away from the tubing. The slab also provides moderate thermal mass to stabilize heat delivery.

Because the slab is thinner than a typical slab-on-grade floor, lateral heat diffusion is slightly less efficient. This is

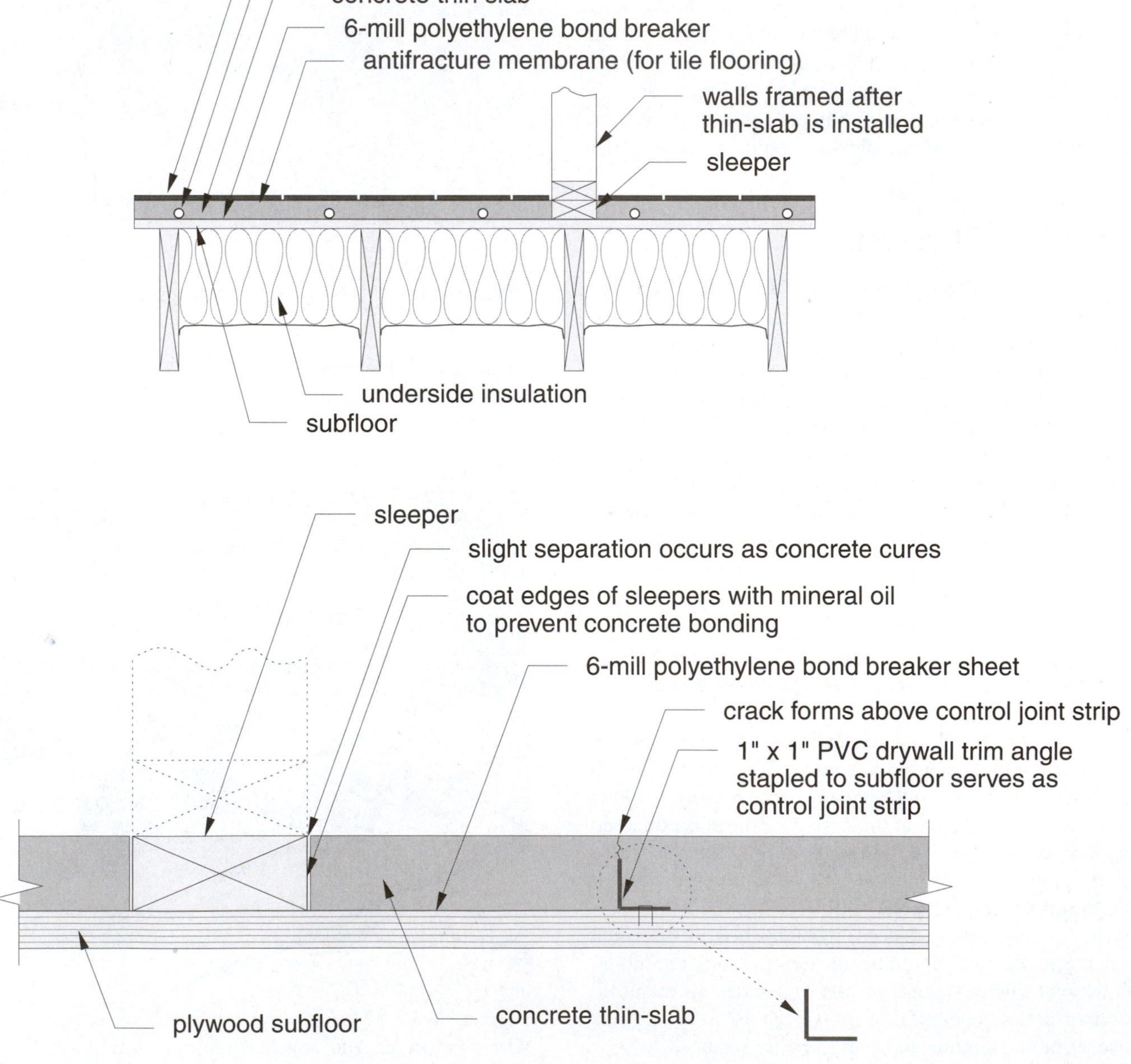

Figure 10–20 Cross-section of a concrete thin-slab radiant floor panel.

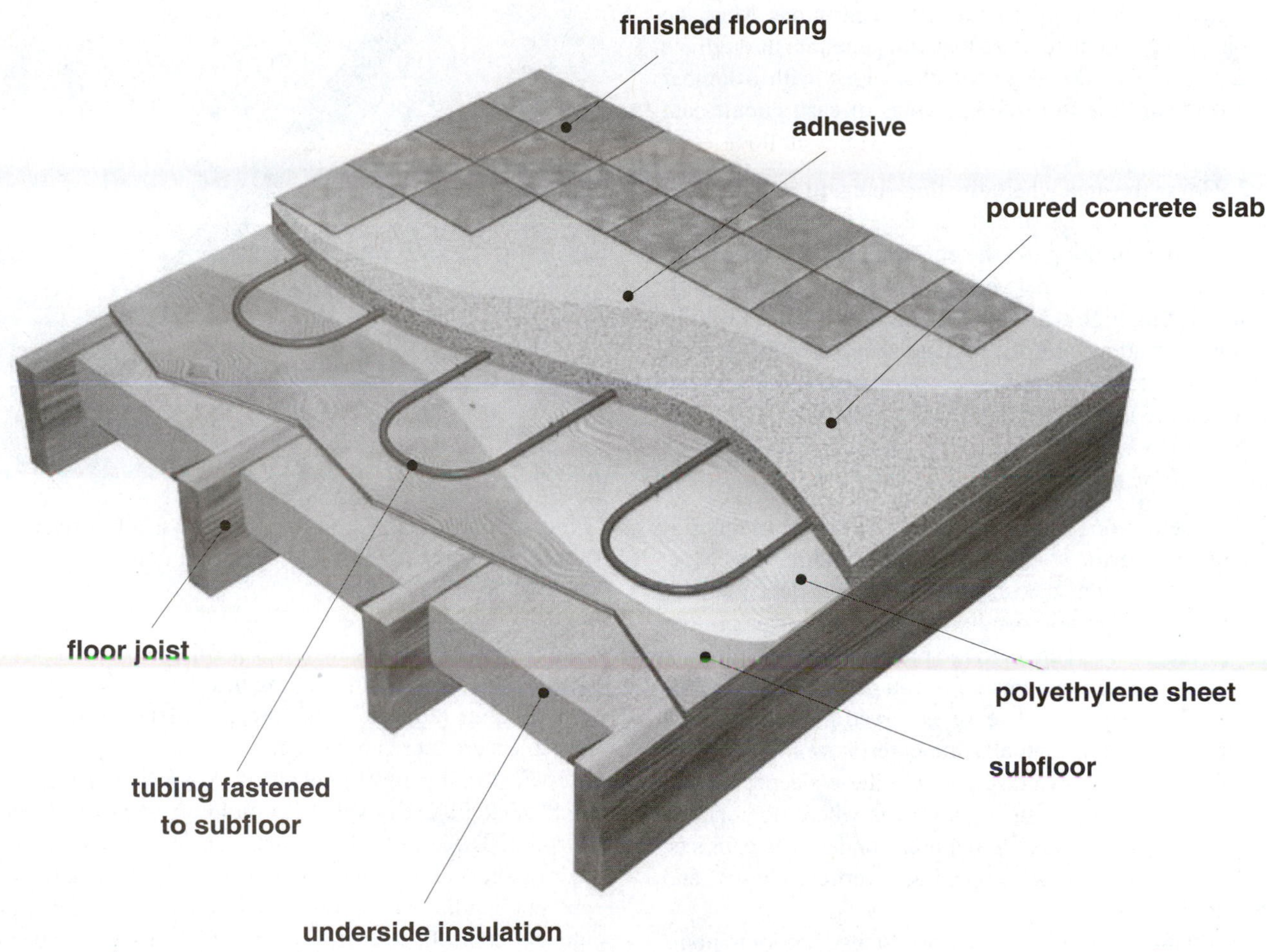

Figure 10–21 Cutaway view of a typical concrete thin-slab radiant floor panel. *Courtesy of IPEX, Inc.*

compensated for by operating the tubing at a slightly higher average water temperature. The thermal performance of concrete thin slabs is discussed in section 10.18.

Thin-slab systems bring a number of issues into the planning process, not the least of which is the added weight of the slab on the floor deck. A typical 1.5-inch-thick concrete thin slab adds about 18 pounds per square foot to the dead loading on a floor. The floor-framing system must be capable of supporting this added load while remaining within code-mandated stress and deflection limits. Although significant, this extra weight can usually be accommodated by adjusting the spacing, depth, or width of floor framing. A competent structural designer or engineer should assess the necessary framing to support the added load. An often-cited maximum deflection for such floors is 1/600th of the floor's clear span under full live loading.

Another consideration is the added height of the thin slab. In most installations, thin slabs add 1.5 inches to 2 inches the height of the floor deck. This affects the height of window and door rough openings, stair riser heights, and rough-in heights for closet flanges. Such adjustments are easily made if the thin slab is planned for when the building is designed. However, these adjustments can be more difficult or even impossible if the decision to use a thin-slab system is made after the building is framed.

Installation Procedure for a Concrete Thin Slab (The Youker System)

The procedure described for creating a concrete thin slab was developed by a radiant heating professional and long-time associate, Harvey Youker. It is called the Youker system, and provides an efficient approach to installing a concrete thin slab in new construction. As with slab-on-grade systems, some portions of the procedure to be described may be delegated to other trades.

Step 1: Begin by sweeping the floor deck clean. Use chalk lines to mark the location of all walls and partitions. Mark the locations of any drainage pipes that will penetrate through the thin slab. Use the tubing layout plan along with a lumber crayon or chalk line to mark key areas of each circuit path such as return bends or offsets directly on the subfloor.

Step 2: Cover the area where the thin slab will be with clear 6-mill polyethylene sheeting. This serves as a **bond breaker** between the subfloor and the concrete. It allows for slight differential movement between the subfloor and the slab without creating high tensile stresses that can crack the concrete. Overlap adjacent polyethylene sheets about one foot, and tape the seams to minimize water absorption into the subfloor before and during the pour. *Never use asphalt saturated roofing felt or other materials that may give off odors when heated for the bond breaker layer.*

Step 3: Install wooden sleepers at all locations of interior partitions or exterior walls. The sleepers should match the thickness of the exterior walls or interior partitions (usually 2 × 4 or 2 × 6). Do not install sleepers across interior door openings. Align the sleepers to the previously snapped chalk lines, which should be visible through the polyethylene sheeting. The sleepers will serve as screed guides when the thin slab is poured. Eventually the exterior walls and partitions will be nailed down directly over these sleepers.

Also install 2 × 4 dams across areas where no concrete will be placed. These include the areas under base cabinets, as well as around stairwell openings, vertical chases, and under shower stalls.

Closet flanges for toilets are easiest to install prior to pouring the thin slab. Be sure the flange is supported with its top flush with the top of the slab. Cover the flange with duct tape to prevent concrete from spilling into it during the pour. If the closet flange is not installed, block off an area large enough to accommodate the piping with a piece of 1.5 inch foam insulation secured to the subfloor. The foam can be easily removed from the slab when the closet flange is installed.

Step 4: Temporarily support the manifold station at the location shown on the tubing layout plan. A small piece of plywood supported on short blocks works well as a temporary support.

Roll out each tubing circuit in a manner similar to that described for slab-on-grade systems with the exception of fastening down the tubing. Use a pneumatic stapler equipped with a depth stop to fasten the tubing to the subfloor. These staplers are usually available from tubing suppliers. They place staples with their crown just touching the top of the tubing, as shown in Figure 10–22. *Never attempt to fasten tubing in place with a stapler not specifically designed for that purpose.* Slide the stapler along the tubing, placing a staple every 24 to 30 inches on straight runs and on each side of return bends. Be sure the tubing lies flat against the subfloor.

Figure 10–22 Close-up of a pneumatically driven staple. Notice the staple does not pinch the tubing.

When all tubing circuits are completed, run a pressure test to ensure there are no leaks.

Step 5: Install control joint strips. A 1 inch × 1 inch × 1/16 inch thick PVC drywall trim molding works well for this purpose. These thin plastic strips can be easy to cut with a pair of snips and stapled to the subfloor. Their purpose is to prevent bonding in the lower two-thirds of the slab, and thus induce controlled cracking that divides the slab into a "mosaic" of smaller pieces. These smaller pieces are better able to adjust to slight movements in the floor deck without developing random cracks.

Place the control joints so the overall slab area is broken into pieces having maximum dimensions of 12 feet. Control joint strips should also be placed across doorways, narrow points in the slab, or any other location where a crack is likely to form. A control joint should also run away from any outside corner formed by adjoining walls.

Step 6: Coat the edges of the sleepers with mineral oil to prevent concrete from bonding to them. This minimizes tensile stress development as the concrete cures, which helps reduce cracking.

A photo of a floor deck prepared for a concrete thin slab using this procedure is shown in Figure 10–23.

The floor is now ready for concrete. The mix proportions suggested for a concrete thin slab are given in Figure 10–24. This mix can be supplied by most batch plants. The small #1A stone aggregate along with the superplasticizer and water-reducing agent yield good flow characteristics and

Figure 10–23 Tubing fastened in place over plywood floor deck that has been previously covered with polyethylene film. Bottom plates of walls fastened in place. *Courtesy of Harvey Youker.*

allow the concrete to fully encase the tubing. This is accomplished without adding access water that can cause shrinkage cracks, loss of strength, and dusty finished surfaces. The Fibermesh® additive provides tensile reinforcement without need for reinforcing steel.

The concrete can be placed using wheelbarrows as shown in Figure 10–25, or in some cases, directly from the chute of the concrete truck. For second floor applications, it can be pumped or transported with a conveyor.

The concrete is screeded flat using the sleepers as depth guides, as shown in Figure 10–26. The slab is then floated and finished in the usual manner. A very smooth finish should be specified in areas that will receive vinyl flooring. Areas that will be covered by ceramic tile or carpet do not require an ultrasmooth finish. After troweling, the slab should be properly cured by maintaining moist conditions at its upper surface.

Mix Design for 1 cubic yard of 3000 psi thin-slab concrete	
Type 1 portland cement	517 pounds
Concrete sand	1639 pounds
#1A (1/4-inch maximum) crushed stone	1485 pounds
Air entrainment agent	4.14 ounce
Hycol (water-reducing agent)	15.5 ounce
Fibermesh	1.5 pounds
Superplasticizer (WRDA-19)	51.7 ounce
Water	About 20 gallons

Figure 10–24 Mix proportions for concrete thin slab

Figure 10–25 "Pea stone" mix concrete poured over tubing. Note plywood scrap beneath nose of wheelbarrow to protect tubing.

Figure 10–26 Concrete is screeded flat using bottom wall plates as screed guides. Note previously poured floor area in background.

Wall framing can begin the following day. Walls can be nailed directly to the previously placed sleepers. Holes for wiring and plumbing can be drilled through the sleepers without need of drilling through concrete. Narrow shrinkage cracks will form between the sleepers and the slab edge, as well as over the control joint strips within 2 or 3 days.

The final step is to install insulation between the floor, framing to minimize downward heat loss. The amount of insulation needed depends on the temperature of the space beneath the floor, and the R-value of any finish floor materials above the thin slab. A conservative rule of thumb is to *provide a minimum of ten times the finish flooring R-value as underside insulation.* This insulation is essential for proper performance, and should never be compromised.

The following minimum R-values are suggested:

- Floors over heated space: R-11
- Floors over partially heated basements: R-19
- Floors over vented crawlspaces or other exposure to ambient conditions: R-30

The insulation can be any suitable product that is relatively easy to install, will stay in place, and will retain its R-value. Fiberglass batts are a good choice. The batts should be neatly tucked up against the underside of the subfloor. If installed in floor framing above a basement, be sure loose fibers from the surface of the batts cannot drift into the space below. If that space is partially heated, kraft paper-faced batts can be used with the facing at the bottom to contain any loose fibers. However, if the space below is not heated, the facing on the bottom of the batt could trap condensed moisture. This situation must be avoided. In such cases, use a vapor permeable layer such as Tyvek® housewrap under the batts. Poly-wrapped fiberglass batts supported by wire stays are another option. Reflective "bubble pack" insulation installed with an adequate air space on each side is yet another possibility. Wiring and plumbing should be installed beneath the floor insulation. This minimizes warming of electrical cables and prevents cold water from being unintentionally heated on it way to the tap.

Lightweight Concrete Slabs

There are several types of **lightweight concrete** currently available. They are primarily used to increase the fire resistance rating and decrease sound transmission through floors. Most formulations use special lightweight aggregates such as expanded shale, vermiculite, and polystyrene beads in place of the usual crushed stone. Depending on the materials and mix proportions used, some lightweight concretes achieve densities as low as 25 pounds per cubic foot. This is approximately one-sixth the density of standard concrete.

Unfortunately, the thermal resistance of most lightweight concrete is substantially higher than that of standard concrete. This is undesirable from the standpoint of diffusing heat away from embedded tubing in a thin-slab radiant panel. The R-value of lightweight concrete of various densities is shown in Figure 10–27.

The greater the R-value of the slab material, the lower the floor surface termperture for a given water temperature in the embedded tubing. This is shown in Figure 10–28.

Higher slab resistance also increases the temperature gradient across the floor surface, from a maximum temperature directly above the tubes, to a minimum temperature halfway between adjacent tubes. This effect is called **striping.** In extreme cases, striping can damage flooring materials such as vinyl tile or carpet. Poor heat diffusion due to low thermal conductivity of the slab material also reduces overall heat output.

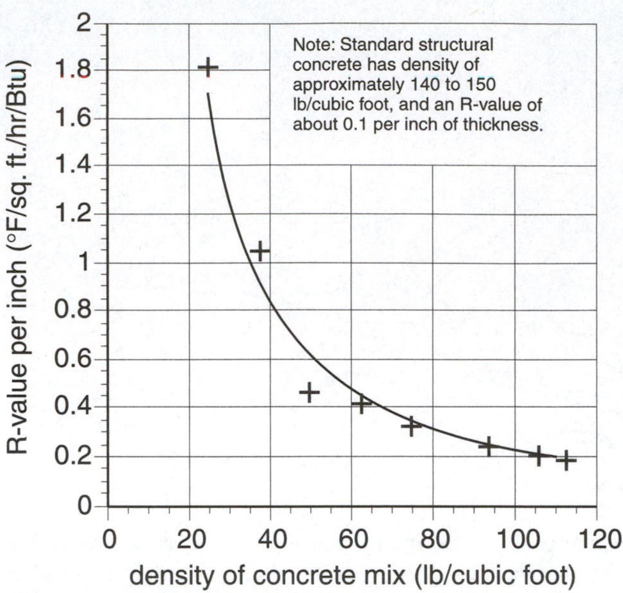

Figure 10–27 Thermal resistance (R-value) versus density of lightweight concrete.

Because of its reduced ability to diffuse heat, lightweight concrete with a density less than 110 pounds per cubic foot is generally not suitable for radiant panels.

10.10 POURED GYPSUM THIN-SLAB RADIANT FLOORS

Poured gypsum underlayments have been used for many years to level uneven floor surfaces, as well as to decrease sound transmission and increase the fire resistance of floor assemblies. They are now widely used for thin-slab radiant floor heating.

Poured gypsum underlayment consists of a mixture of gypsum cement, sand, water, and additives to improve flexibility. These materials are mixed on site and installed by a trained crew. A cross section of a **gypsum thin slab** is shown in Figure 10–29. A cutaway of the same system is shown in Figure 10–30.

The primary advantages of gypsum-based underlayments are the speed and ease of installation relative to concrete. A trained crew of three can install about 2,500 ft² of gypsum-based thin slab in one day. Interior work can usually resume the next day. Gypsum underlayments are also less prone to cracking than is a concrete thin slab. Their ability to decrease air infiltration by sealing small cracks along the base of walls and partitions is also an advantage.

The final compressive strength of gypsum underlayments is typically 2000 to 3000 psi, depending on the mix proportions. This is adequate to support foot and light equipment

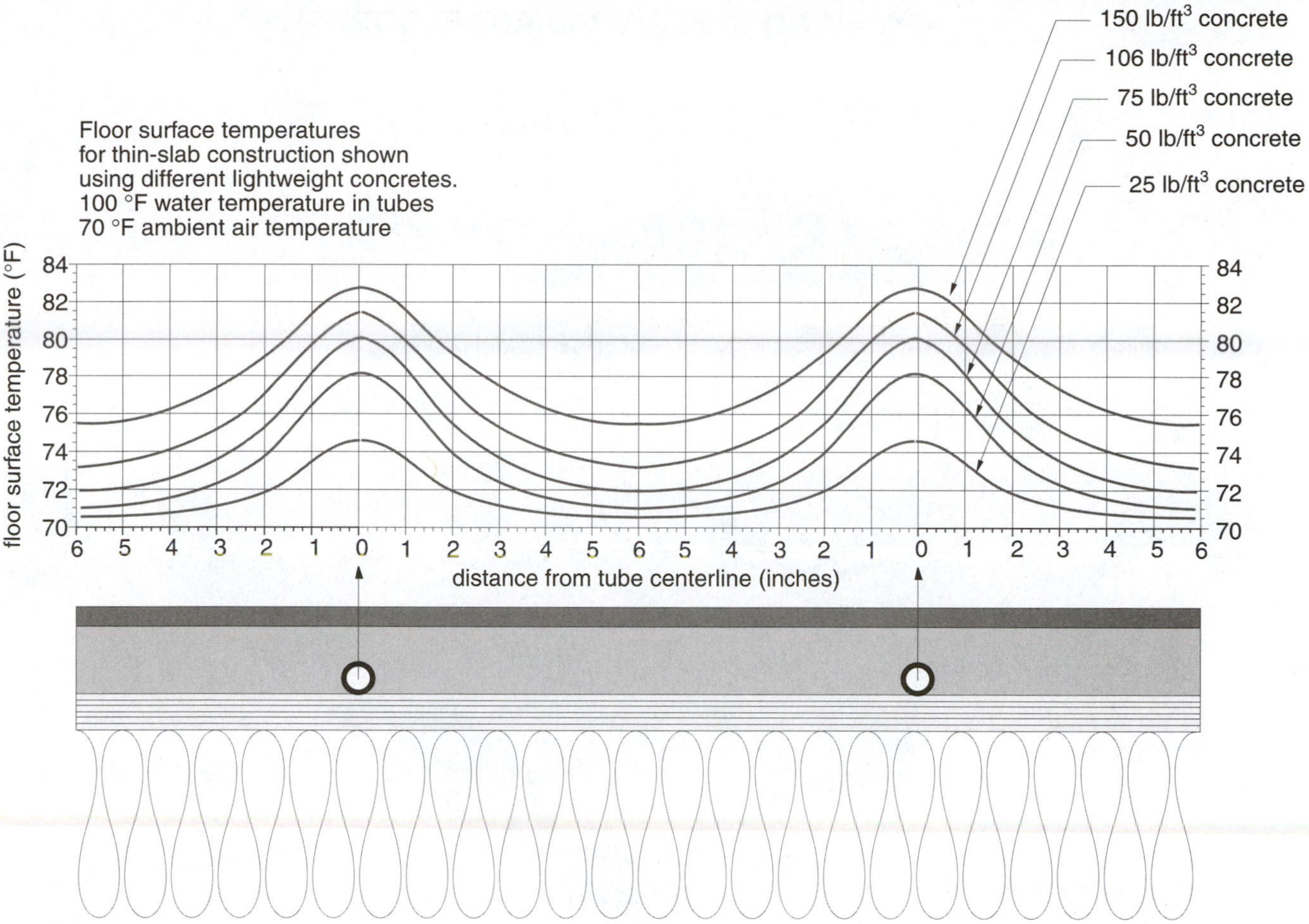

150 lb/ft³ concrete
106 lb/ft³ concrete
75 lb/ft³ concrete
50 lb/ft³ concrete
25 lb/ft³ concrete

Floor surface temperatures
for thin-slab construction shown
using different lightweight concretes.
100 °F water temperature in tubes
70 °F ambient air temperature

floor surface temperature (°F)

distance from tube centerline (inches)

Figure 10–28 Surface temperature profile simulation for thin-slabs made of lightweight concretes of various densities. All temperature profiles assume 100 °F water in tubing. Higher surface temperature profiles result in greater heat output.

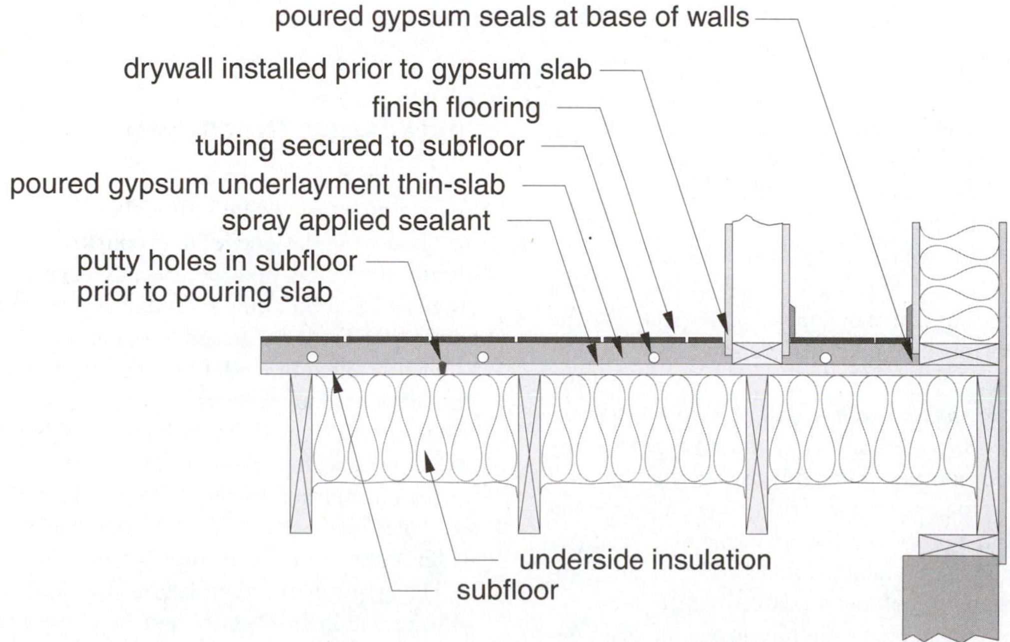

Figure 10–29 Cross-section of a thin-slab radiant floor panel made with poured gypsum underlayment.

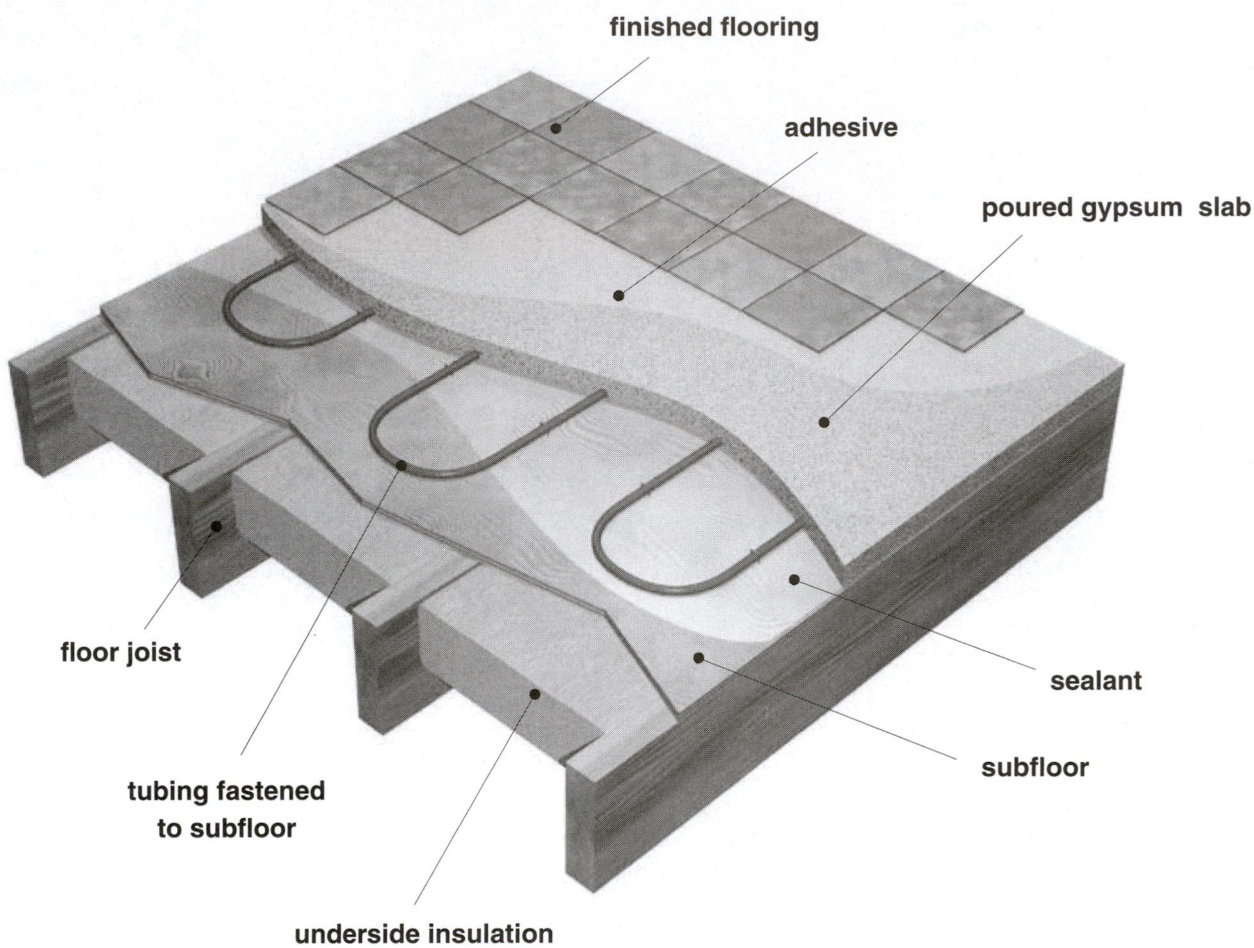

Figure 10–30 Cutaway view of a typical thin-slab radiant floor panel made with poured gypsum underlayment. *Courtesy of IPEX, Inc.*

traffic during construction. Care must be taken, however, not to gouge the completed floor with heavy or sharp objects.

When dry, a 1.5-inch-thick gypsum slab adds about 14.4 lb/ft^2 to the dead loading of the floor. This is slightly less than the 18 lb/ft^2 dead loading of a 1.5-inch concrete thin slab. Still, the added load must be taken into account when the floor framing that will support it is specified. The added height of the thin slab must also be accounted for during building design.

Gypsum underlayments are not designed to function as the permanent wearing surface. They must be covered by some type of finish flooring. After full curing, they can still be damaged by prolonged contact with water. A minor but persistent leak from a plumbing fixture or manifold connection can cause the gypsum slab to soften similar to how gypsum wallboard reacts to prolonged contact with water.

Gypsum underlayments have slightly lower thermal conductivity relative to standard concrete. This results in slightly less heat output under the same operating conditions.

Installation Procedure

In contrast to concrete thin slabs, gypsum thin slabs are installed near the end of the construction sequence, after interior drywall is in place. Because gypsum underlayments are not waterproof, they should never be poured in situations where they will be exposed to precipitation.

The installation begins by stapling the tubing circuits to the subfloor and pressure testing as previously described with concrete thin slabs. Polyethylene bond breaker sheets are not used with gypsum thin slabs. Due to the flow characteristics of gypsum underlayment, it is important to seal off any holes or cracks in the subfloor that exceed 1/8 inch in width. Putty works well for this task.

The gypsum installer begins by spraying the subfloor with a combination sealer and bonding agent as shown in Figure 10–31. This material limits water absorption and improves the bond between the subfloor and slab.

Figure 10–31 Combination sealer/bonding agent sprayed on subflooring prior to pouring thin slab.

The gypsum underlayment is prepared in batches in an outside mixer, then pumped into the building through a hose as shown in Figure 10–32. The material has the consistency of a milkshake, allowing it to self-level as it is poured on the subfloor.

The installer moves back and forth across the room as the material flows from the hose as shown in Figure 10–33. The fluid-like consistency allows it to flow around the tubing, and into any small gaps. This provides a good seal against air infiltration along the lower edge of exterior walls. A wooden float is used to level the material even with the top of the tubing. This layer is called the first "lift," and it is usually firm enough to walk on 2 hours after being poured.

As it cures, the first lift undergoes slight vertical settlement. The top edge of the tubing is just visible at this point. A second lift of the same material is then poured to a minimum depth of 3/4 inch over the top of the tubing as shown in Figure 10–34. A wooden float suspended on two small pins is used to ensure uniform thickness and a flat finished surface.

Figure 10–32 Gypsum-based underlayment is mixed outside and pumped in through a hose.

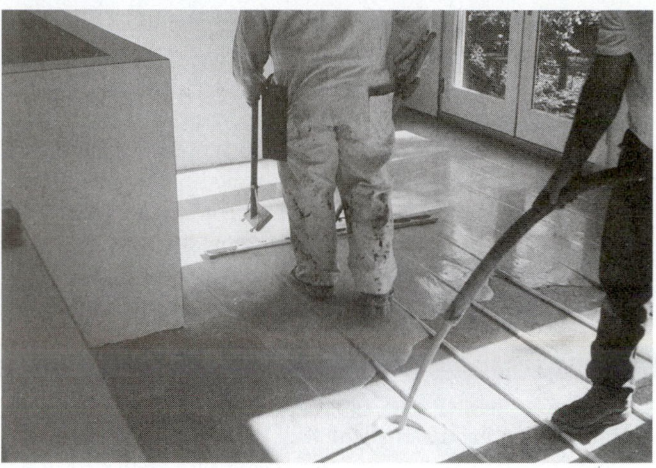

Figure 10–33 Installer places first lift of gypsum-based underlayment. Note its thickness is approximately the same as the tubes's diameter.

Although gypsum slabs can usually be walked on a couple hours after being poured, they can take several days to fully cure. Curing time depends on the temperature and humidity at the job site. If the heating system is operational, a slight amount of heat can be applied to the slab as soon as it sets up to expedite curing. The water temperature in the tubing during curing should never exceed 120 °F.

Gypsum underlayment contains considerably more water than does an equal volume of concrete. Approximately 0.16 gallons of water must eventually evaporate from each

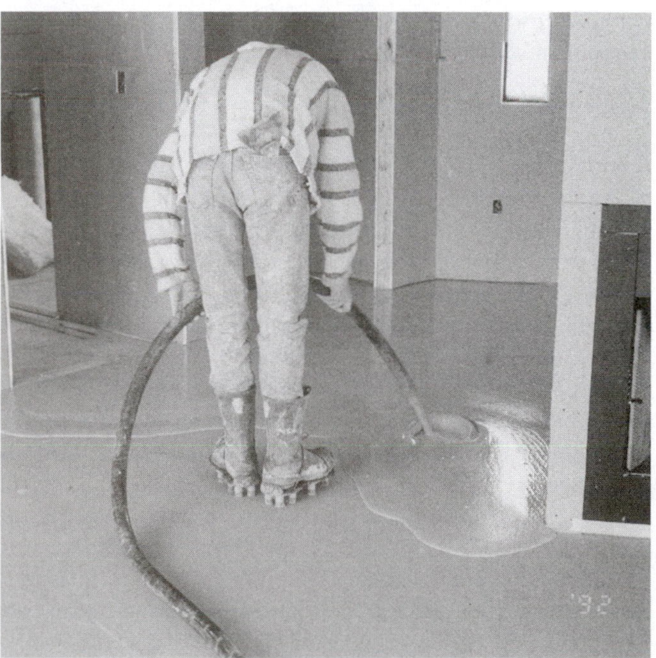

Figure 10–34 Installer places second lift of gypsum-based underlayment over partially hardened first lift.

square foot of floor covered with a 1.5-inch-thick gypsum slab. Therefore, a 2000-square-foot-thin slab pour requires the evaporation of about 320 gallons of water. It is essential to ventilate the space where the slab is poured and to maintain the walls and ceilings warm enough to prevent condensation as this water evaporates, especially if the slab is poured during cold weather. Failure to properly ventilate the space can lead to severe moisture condensation on cool surfaces such as windows, doors, or any uninsulated walls and ceilings.

The moisture content of a gypsum slab should always be checked before installing finish flooring. This can be done by taping down all edges of a 2 foot by 2 foot piece of clear polyethylene sheeting, and verifying that moisture droplets do not accumulate on the underside during the following 48 hours.

Manufacturers often specify that gypsum slabs be sealed with specific products to protect them against moisture from spills or mop water that may penetrate the finish floor. The sealant also binds any fine particles at the surface, which prevents them from working up through carpeting over time. Be sure to follow any sealing recommendations prior to installing the finish flooring.

The final step is to install insulation between the floor framing to minimize downward heat loss. The amount of insulation needed depends on the temperature of the space beneath the floor and the R-value of the finish floor. A conservative rule of thumb is to *provide a minimum of ten times the finish flooring R-value as underside insulation.* This insulation is essential for proper performance, and should never be compromised.

The following minimum R-values are suggested:

- Floors over heated space: R-11
- Floors over partially heated basements: R-19
- Floors over vented crawlspaces or other exposure to ambient conditions: R-30

The insulation can be any suitable product that is relatively easy to install, will stay in place, and will retain its R-value. Fiberglass batts are a good choice. The batts should be neatly tucked up against the underside of the subfloor. If installed in floor framing above a basement, be sure loose fibers from the surface of the batts cannot drift into the space below. If that space is partially heated, kraft paper faced batts can be used with the facing at the bottom to contain any loose fibers. However, if the space below is not heated, the facing on the bottom of the batt could trap condensed moisture. This situation must be avoided. In such cases, use a vapor permeable layer such as Tyvek® housewrap under the batts. Poly-wrapped fiberglass batts supported by wire stays are another option. Reflective "bubble pack" insulation installed with an adequate air space on each side is yet another possibility. Wiring and plumbing should be installed beneath the floor insulation. This minimizes warming of electrical cables and prevents cold water from being unintentionally heated on its way to the tap.

10.11 ABOVE-FLOOR TUBE AND PLATE SYSTEMS

In some buildings, slab or thin-slab-type radiant panels are not possible due to one or more of the following reasons:

- The floor framing cannot support the additional dead loading of a slab.
- The wall framing is already completed and was not adjusted for the added thickness of the slab.
- The finish flooring requires extensive nail penetrations of the subfloor.
- The slab materials are not available or cannot be easily placed.

One alternative is called a **tube and plate system.** Instead of concrete or gypsum, this system uses preformed aluminum **heat transfer plates** to laterally conduct heat away from the tubing.

Systems that locate the heat transfer plate above the subfloor are appropriately called **above-floor tube and plate systems.** If the heat transfer plates are installed beneath the subfloor, the system is called a **below-floor tube and plate system.** The latter type of system is discussed in the next section.

A cross section of an above-floor tube and plate system is shown in Figure 10–35. A cutaway view is shown in Figure 10–36. A topside view is shown in Figure 10–37.

Although there are slight differences from one manufacturer to another, most heat transfer plates are formed from 0.020 inch to 0.025 inch thick aluminum sheets. A typical plate is five or six inches wide and 24 inches long. The groove up the center of the plate wraps around the lower portion of a PEX or PEX-AL-PEX tube. Most plates are formed so the tube snaps tightly into this groove. Good contact between the tube wall and plate is essential to good heat transfer.

The "wings" of the plates conduct heat laterally away from the tube. This part of the plate is typically supported by plywood **sleepers** that have been secured to the subfloor using construction adhesive and mechanical fasteners as shown in Figure 10–35 and Figure 10-36.

The material installed over the plates depends on the type of finish floor planned. If the floor will be covered with vinyl sheet, ceramic tile, carpet, or other materials requiring a smooth substrate, a layer of 3/8-inch plywood is installed over the tube and plates before installing the finish flooring materials. This layer adds some resistance to upper heat flow. It also increases the height of the floor slightly; however, it is essential when a smooth substrate is needed under the finish flooring.

Above-floor tube and plate systems are a good choice where nailed-down wood finish floors are planned. Such flooring can be fastened in place directly over the tube and plates system. This approach minimizes upward thermal resistance between the plane where the heat is laterally

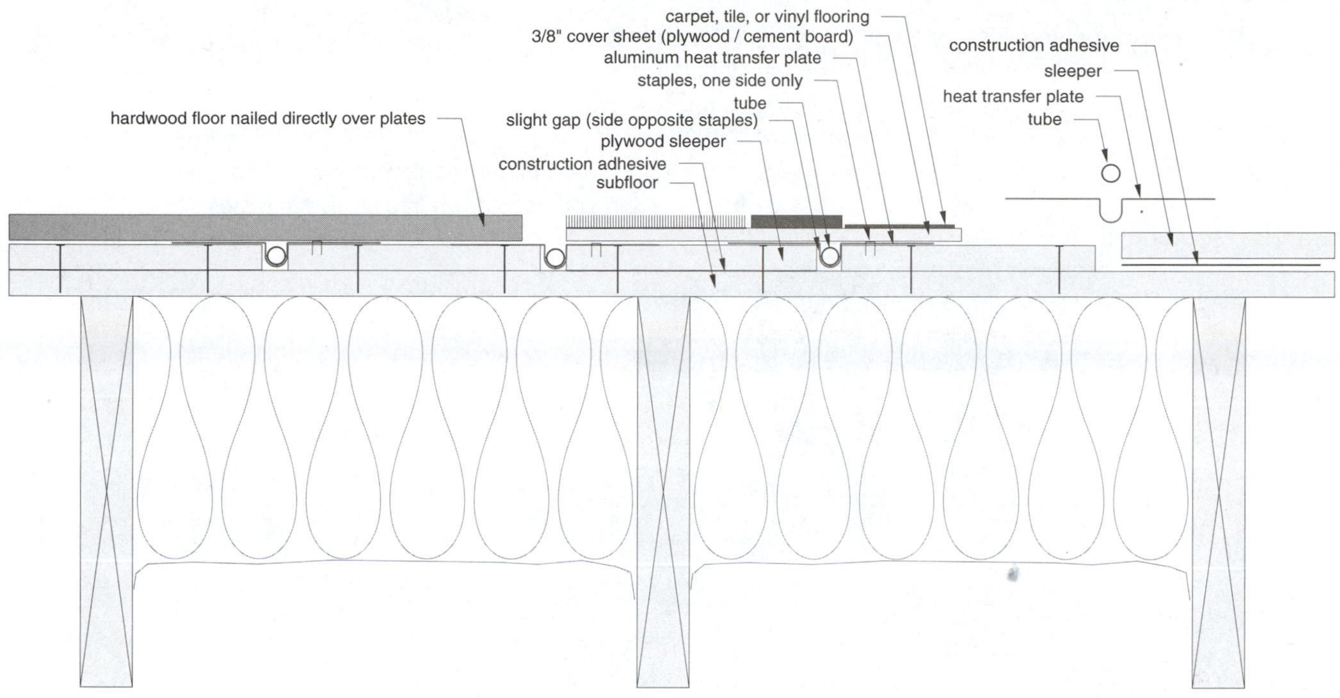

Figure 10–35 Cross section of an above-floor tube and plate radiant panel.

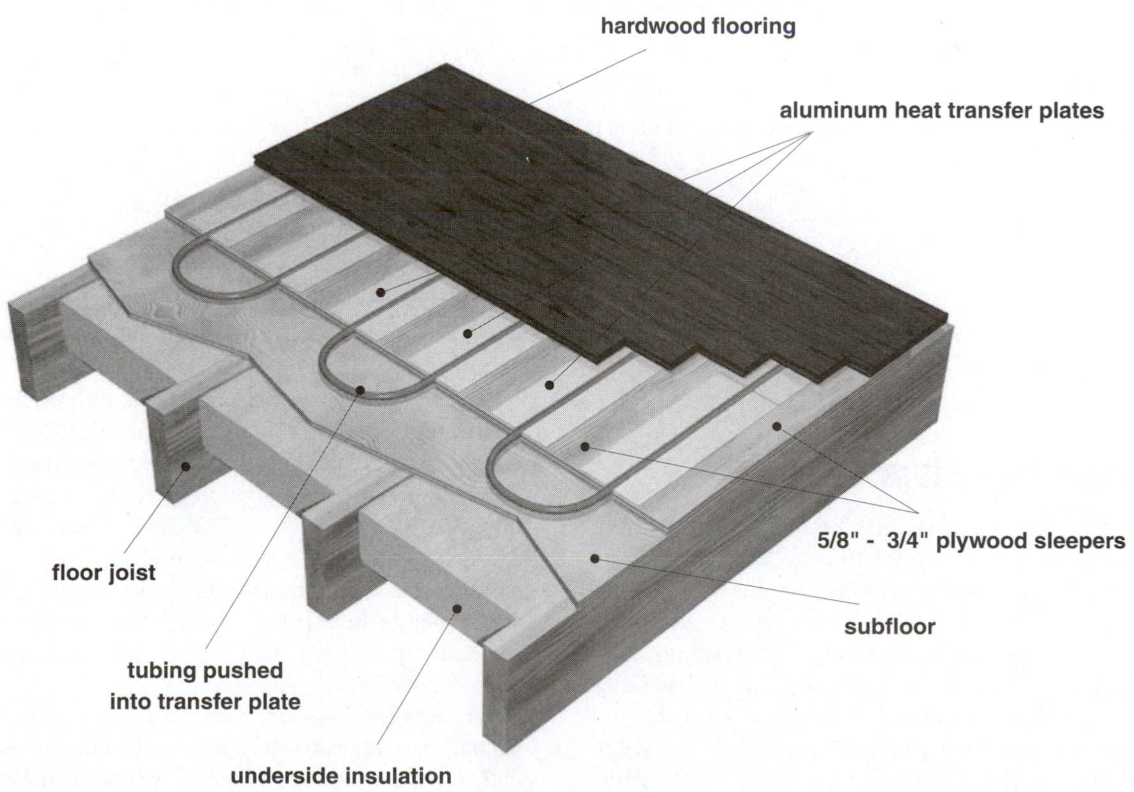

Figure 10–36 Cutaway view of a typical above-floor tube and plate radiant floor panel. *Courtesy of IPEX, Inc.*

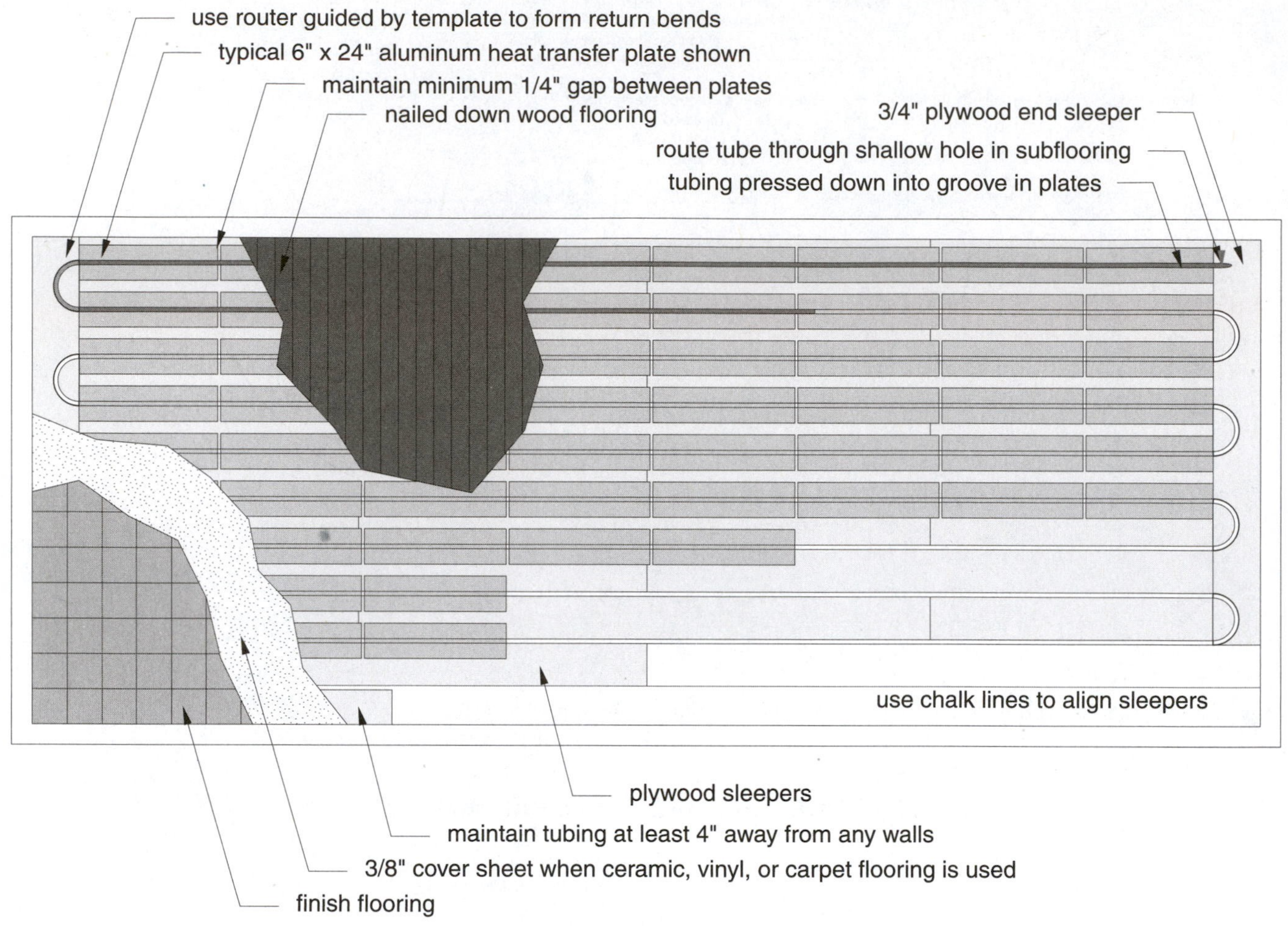

Figure 10–37 Top view of an above floor tube and plate installation with finish flooring options.

dispersed and the top of the finish floor. Lower resistance decreases the necessary water temperature and helps improve overall operating efficiency.

Installation Procedure

As with any radiant panel installation, a circuit layout drawing should be prepared prior to installation. The drawing should show the location of tubing circuits so the sleepers can be properly located. The circuit routing should be kept as simple and straight as possible. Offsets and other complex curves will have to be made using a router, and they do not permit efficient placement of heat transfer plates.

If nailed-down wood flooring will be installed, the sleepers should be placed perpendicular to the direction of the flooring. Determine the preferred direction of the flooring, and plan sleeper installation accordingly. If there is no strong preference on flooring direction, plan the tubing layout to allow the warmest portion of the circuit to run parallel to exterior walls.

Sleepers are typically cut from 3/4-inch plywood. In most applications that use 1/2-inch PEX or PEX-AL-PEX tubing, the sleepers are cut 3/4 inch narrower than the tube spacing. For example, if the tubing will be placed 8 inches on center, the sleepers are sawn to 7.25 inch wide. This creates a 3/4 inch wide by 3/4 inch deep channel between adjacent sleepers that accommodates the groove in the plate. On jobs where a large area will be covered with sleepers, the installer may choose to have the sleepers fabricated ahead of time using a panel saw.

Vacuum the floor deck and snap chalk lines on the floor to mark the edges of the sleepers. Install 3/4-inch plywood end sleepers where return bends will be required. The channels to accommodate tubing at return bends will eventually be cut with a router.

When installing the sleepers, be sure to apply construction adhesive on the bottom side and place them carefully along the chalk lines. Immediately fasten them in place with a pneumatic nailer or screw gun. The construction adhesive helps prevent squeaks in the floor due to subsequent shrinkage.

After the sleepers are installed, the channels for return bends are cut using a two horsepower or larger plunge router fitted with a 3/4-inch cylindrical bit. Construct a template that can be temporarily fastened down to guide the router along a smooth semicircular arc joining the adjacent channels between sleepers. The router can also be used to make any required offsets or other curved paths.

In many installations, the tubing circuit has to eventually pass back across the room. It is easiest to route the tubing down through a shallow hole in the subfloor and then back to the required location through the floor framing. This eliminates a situation where the tubing runs parallel to nailed-down wood flooring and, as such, is more likely to be punctured when the floor is installed.

After all wood fabrication and installation is complete, the floor should again be vacuumed prior to placing the heat transfer plates. Place each plate into the channel between adjacent sleepers, then slide the plate tight to one side of the channel. The slight gap on the opposite side of the plate allows the formed groove to open slightly as the tube is pressed in. Staple the wing on the tight side of the plate to the sleeper with two or three light-gauge staples. Never staple both wings of a plate down since this prevents the groove from expanding. Adjacent plates should be at least 1/4 inch apart to allow for expansion. Some designs are based on wider end-to-end separation of the plates.

When necessary, sheet aluminum heat transfer plates can be cut using a "score and snap" method. Use a utility knife and square to score across the bottom side of the plate and around the convex side of the groove. Place the score line directly over a rigid supporting edge with the groove side up, then snap the plate with a sharp blow. The plate will break along the score line, producing a clean, straight cut without the distortion caused by snips or shears.

The methods used for placing tubing in the plates varies among manufacturers. Some recommend a thin bead of silicone caulk be placed along the bottom of the grooves before pressing in the tubing. The silicone spreads out as the tubing is pressed in place to provide a thin elastic layer that helps minimize expansion sounds as the tubing heats and cools. It is more commonly specified when PEX tubing is used. PEX-AL-PEX tubing has a lower coefficient of expansion and generally does not require the silicone layer. If silicone caulk is used in the plate grooves, it is essential to press the tubing into position as soon as possible before the caulk begins to cure.

Be sure to uncoil sufficient tubing to reach the manifold stations before beginning to press the tubing into the plates. Align the tubing with the grooves and simply walk along the tube to press it in place. If necessary, a wooden block and mallet can also be used to gently tap the tube into the grooves. At return bends, it may be necessary to temporarily hold down the tubing with nail guard plates. These plates should be removed as the cover sheet or wood flooring is installed.

The manifolds can be located under or between the floor framing. They can also be located in vertical stud cavities above the floor. In the latter case, it is convenient to route the tubing down through the bottom plate and then back up through shallow holes in the subfloor.

After all tubing is placed and connected, it should be pressure tested as previously described. Maintain the tubing under pressure as the cover sheet and finish flooring is installed.

Wood flooring should be as dry as possible prior to installation. A moisture level of 4 to 8 percent is suggested. Once the wood is nailed in place, the floor heating circuits should be operated for at least 10 days before sanding and finishing the floor. This helps drive residual moisture from the flooring and sleepers. It also helps the wood to adjust to its environment prior to finish sanding and finishing.

Tube and plate systems, especially those covered by solid sawn wood flooring, should be equipped with an outdoor reset control. As discussed in Chapter 9, Control Strategies, Components, and Sytems, outdoor reset control provides gradual changes in water temperature as outdoor conditions vary. This minimizes expansion noises from the plates. It also minimizes rapid temperature changes in the wood flooring that can induce stress.

Do not install tubing and plates in areas that will be covered by cabinets, islands, or other fixed objects. Such objects block most of the heat output. Food stored in cabinets over heated floors can also spoil faster than normal.

An above-floor tube and plate system adds 1 to 2 pounds per square foot to the floor's dead loading. This is usually insignificant. However, it also adds 3/4 inch to 1 1/8 inch to the floor height. This should be accounted for when planning stair risers, window and door rough openings, and so forth.

The final step is to install insulation between the floor framing to minimize downward heat loss. The amount of insulation needed depends on the temperature of the space beneath the floor, and the R-value of the finish floor. A conservative rule of thumb is to *provide a minimum of ten times the finish flooring R-value as underside insulation.* This insulation is essential for proper performance and should never be compromised.

The following minimum R-values are suggested:

- Floors over heated space: R-11
- Floors over partially heated basements: R-19
- Floors over vented crawlspaces or other exposure to ambient conditions: R-30

The insulation can be any suitable product that is relatively easy to install, will stay in place, and will retain its R-value. Fiberglass batts are a good choice. The batts should be neatly tucked up against the underside of the subfloor. If installed in floor framing above a basement, be sure loose fibers from the surface of the batts cannot drift into the space below. If that space is partially heated, kraft paper-faced batts can be used with the facing at the bottom to contain any loose fibers. However, if the space below is not heated, the facing on the bottom of the batt could trap condensed moisture. This situation must be avoided. In such cases, use a vapor permeable layer such as Tyvek® housewrap under the batts. Poly-wrapped fiberglass batts supported by wire stays are another option. Reflective "bubble pack" insulation installed with an adequate air space on each side is yet another possibility. Wiring and plumbing should be installed beneath the floor insulation. This minimizes warming of electrical cables and prevents cold water from being unintentionally heated on its way to the tap.

10.12 BELOW-FLOOR TUBE AND PLATE SYSTEMS

It is also possible to install aluminum heat transfer plates and tubing below a subfloor. In most cases, the tubing is cradled by the heat transfer plates, which are stapled to the bottom of the subfloor. A cross-sectional view of this system is shown in Figure 10–38. A cutaway view is shown in Figure 10–39.

The feasibility of a below-floor tube and plate system depends on several considerations including:

- Are the spaces between the floor joists readily accessible and free from other mechanical, electrical, or structural components?
- Can the tubing be easily pulled though the floor framing or girders without need for extensive drilling?
- Are there, or will there not be, nails piercing through the underside of the subfloor?
- What finish floor materials are (or will be) installed?
- Are there any materials in an existing floor (such as asphalt-saturated felt) that could outgas if heated?

If the spaces between the floor framing are not relatively free of other obstacles such as wiring, plumbing, or structural bracing, installation of tubing circuits will be slow, difficult, or in some cases nearly impossible. It is crucial to inspect the underside of the floor deck for such obstacles before committing to this type of system.

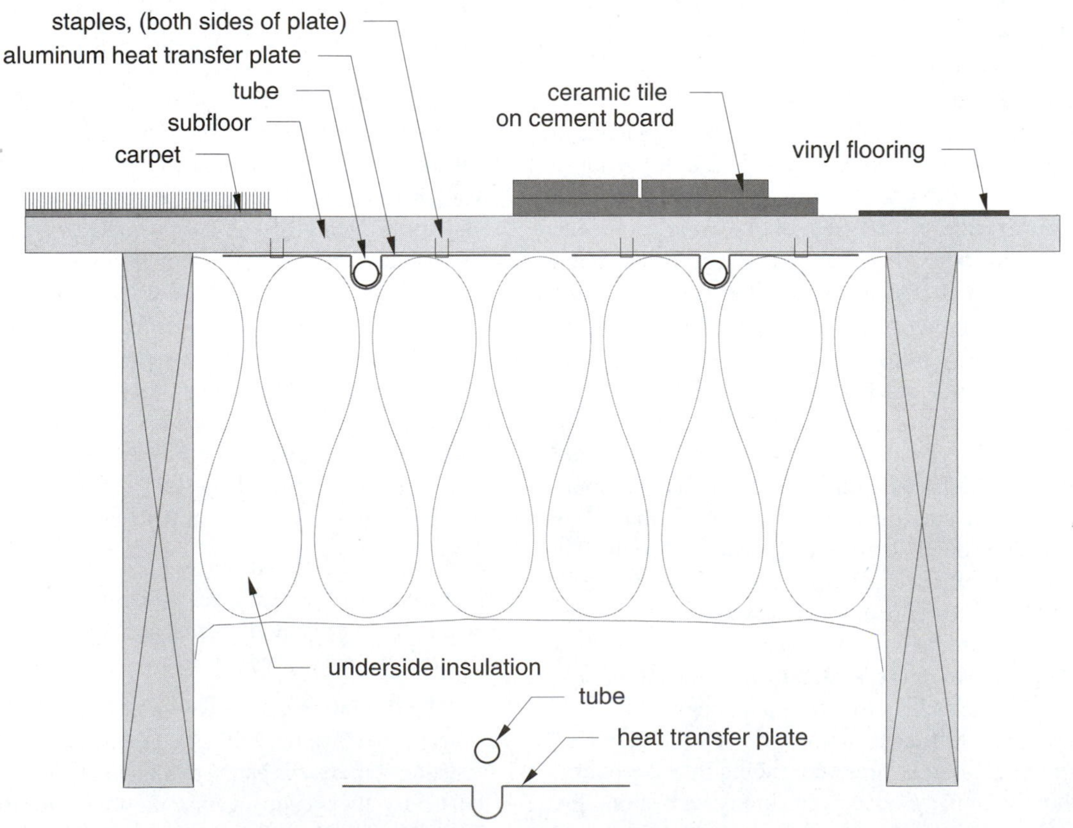

Figure 10–38 Cross-section of a below-floor tube and plate radiant panel.

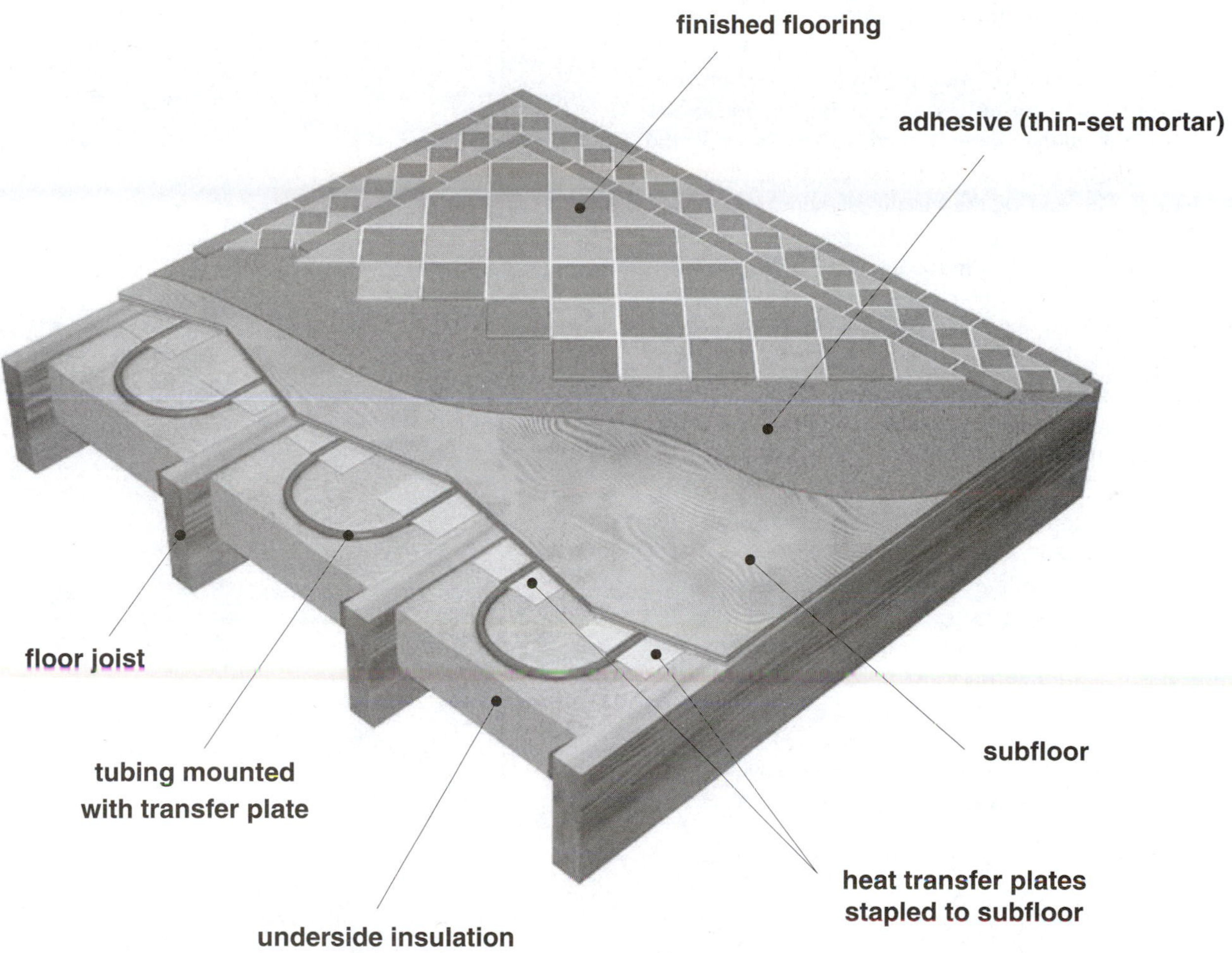

Figure 10–39 Cutaway view of a typical below-floor tube and plate radiant floor panel. *Courtesy of IPEX, Inc.*

The structural design of some buildings can also result in numerous girders crisscrossing the floor deck. Drilling holes for tube routing can be slow. It can also affect the structural integrity of the girders. In general, buildings with complex floor framing are not well-suited to below-floor tube and plate systems.

The presence of nails puncturing through the underside of the subfloor is also a problem. If the nails are already in place, the installer may have to grind them flush with the floor—a very time-consuming process. If the tubing and plates are installed first and nails are then driven through the subfloor, multiple tube punctures are virtually certain.

In retrofit applications, it is essential to determine what materials are currently in place in the floor deck before committing to a below-deck tube and plate system. Although not visible from above or below the floor deck, a layer of asphalt-saturated felt may be installed under hardwood flooring. If heated, it could outgas compounds that create persistent and objectionable odors in the house. Multiple thick layers of wooden flooring are also common in older homes, and they present considerable upward resistance to heat flow.

Installation Procedure

The installation begins by drilling two holes side by side through the floor joists to allow the tubing to be pulled to each of the joist cavities. These holes should be at least

1/2-inch larger than the outside diameter of the tubing (Figure 10–40). They should be kept in a straight line by snapping a chalk line along the joists. The holes should be far enough beneath the subfloor to avoid any nails that might be driven in from above. Never notch the underside of floor framing to accommodate the tubing since this can severely weaken the framing.

The preferred method of pulling tubing into each joist cavity varies among installers and manufacturers. One approach is to thread the tubing through a line of holes in the joists until it reaches the farthest joist cavity as shown in Figure 10–41. The end of the circuit is them routed back through the other set of holes until it reach the manifold location. Tubing is pulled from the uncoiler and then routed up and back along the first joist cavity. The tubing and plates are pushed up tight against the subfloor and stapled in place. Be sure each plate lies tight against the subfloor, and install at least eight staggered staples on each side of a typical two-foot-long plate. A light-gauge pneumatic stapler is well-suited to this task. Allow at least 1/4 inch spacing between ends of adjacent plates. Any excess tube is fed forward toward the next joist cavity. Notice the gentle return bends where the tubing moves from one joist cavity to another.

Be sure the tubing is aligned with the groove in the plates to prevent chaffing. Also, be sure the tubing is free to expand and contract where it passes through holes in the framing. Tubing bends that are tightly bound in place can cause noises as the tubing expands and contracts.

Pressure test the system after all circuits are in place. If a circuit was punctured by a nail, it can be spliced using a coupling supplied by the tube manufacturer.

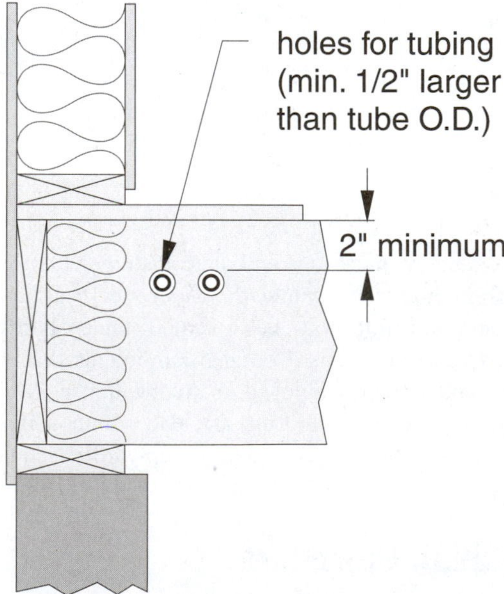

Figure 10–40 Location of holes in floor joists when pulling tubing for a below-floor tube and plate installation.

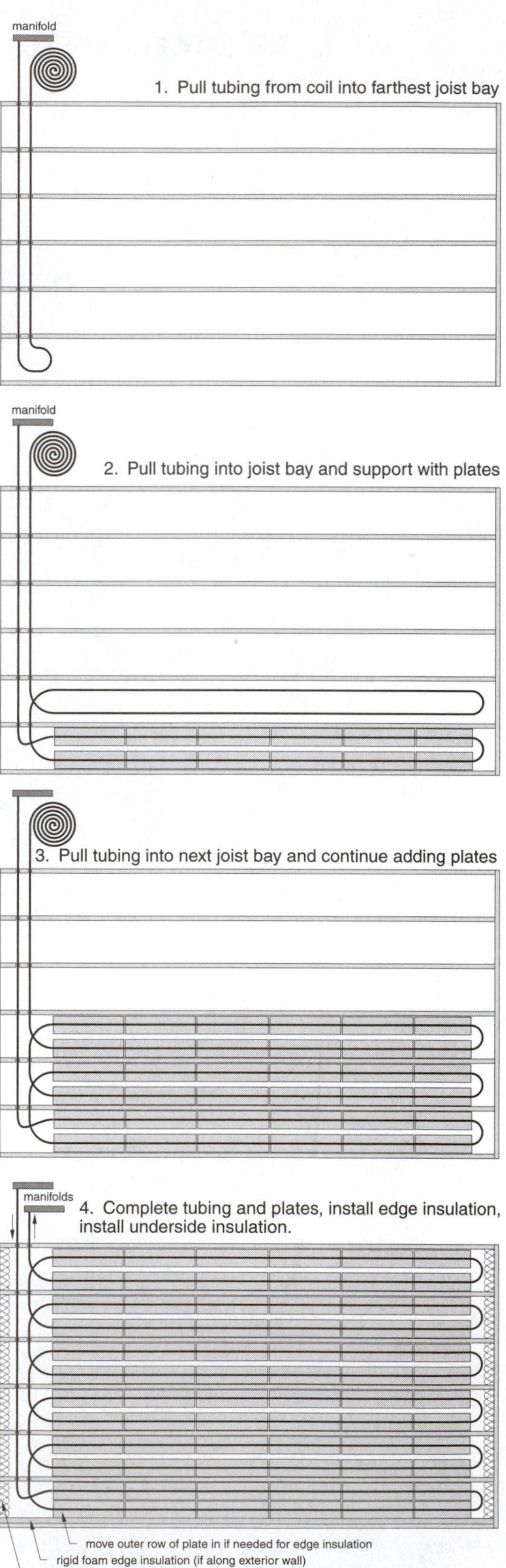

Figure 10–41 Tubing installation sequence for a below-floor tube and plate radiant panel.

The final step is to install insulation between the floor framing to minimize downward heat loss. The amount of insulation needed depends on the temperature of the space beneath the floor, and the R-value of the finish floor. A conservative rule of thumb is to *provide a minimum of ten times the finish flooring R-value as underside insulation.* This insulation is essential for proper performance, and should never be compromised.

The following minimum R-values are suggested:

- Floors over heated space: R-11
- Floors over partially heated basements: R-19
- Floors over vented crawlspaces or other exposure to ambient conditions: R-30

The insulation can be any suitable product that is relatively easy to install, will stay in place, and will retain its R-value. Fiberglass batts are a good choice. The batts should be neatly tucked up against the underside of the subfloor. If installed in floor framing above a basement, be sure loose fibers from the surface of the batts cannot drift into the space below. If that space is partially heated, kraft paper-faced batts can be used with the facing at the bottom to contain any loose fibers. However, if the space below is not heated, the facing on the bottom of the batt could trap condensed moisture. This situation must be avoided. In such cases, use a vapor permeable layer such as Tyvek® housewrap under the batts. Poly-wrapped fiberglass batts supported by wire stays are another option. Reflective "bubble pack" insulation installed with an adequate air space on each side is yet another possibility. Wiring and plumbing should be installed beneath the floor insulation. This minimizes warming of electrical cables and prevents cold water from being unintentionally heated on its way to the tap.

Extruded Underfloor Plates

Below-floor tube and plate systems can also be built using extruded aluminum heat transfer plates. The extruded plates are generally sold in 4- to 8-foot lengths, and are screwed into the underside of the subfloor. The center groove opens to the underside of the floor, allowing the tubing to be installed after the plates are fastened in place. The channel will retain the tubing after it has been tapped in with a mallet. An example of this type of system is shown in Figure 10–42.

10.13 SUSPENDED TUBE SYSTEMS

Another approach to underfloor tube placement is called a suspended tube system. The tubing is suspended in the air space between the floor framing. It may be supported by clips screwed into the subfloor or by light gauge metal hangers nailed between adjacent joists. A representative cross section is shown in Figure 10–43.

Figure 10–42 Aluminum heat transfer plates supporting tubing to the underside of a floor deck. *Courtesy of The Flor Heat Company.*

Suspended tube systems rely on convection and radiation rather than conduction to transport heat from the tube to the underside of the subfloor. The convective air currents depicted in Figure 10–43 help distribute heat across the underside of the subfloor. Reflective insulation under the tubing helps direct radiant output toward the subfloor.

Suspended tube systems operate at relatively high water temperatures in comparison to other types of radiant panels. Supply water temperatures in the range of 160 to 200 °F are typical at design load conditions. These can be supplied by a gas-fired or oil-fired boiler, but are beyond the range of geothermal heat pumps, and impractical for active solar collectors. One advantage of operating at these high water temperatures is that the need for a mixing device between the boiler and distribution system is eliminated. This type of system also has very little thermal mass and thus can warm up or cool down quickly.

Optimal placement of the tubing within the joist cavity has not been fully researched to date. Deeper air cavities would theoretically allow for stronger convection currents that would boost heat output. Heat transfer simulations of two 1/2-inch diameter tubes, suspended as shown in Figure 10–43, suggest an upward heat output of approximately 20 Btu/hr/ square foot with 180 °F water temperature, and 70 °F room air temperature.

Tubing circuits are pulled between the joist cavities in the same manner as for a below-floor tube and plate system. The tubing is supported at intervals of 24 to 30 inches. The supports must allow for movement of the tubing as it expands and contacts. PEX-AL-PEX tubing is especially well-suited for suspended tube systems because it maintains its shape at high temperatures with minimal support.

Tubing should not be run adjacent to drainage traps, cold water pipes, or in direct contact with electrical wiring. Any plumbing traps located in a heated joist bay should be wrapped with insulation to minimize water evaporation.

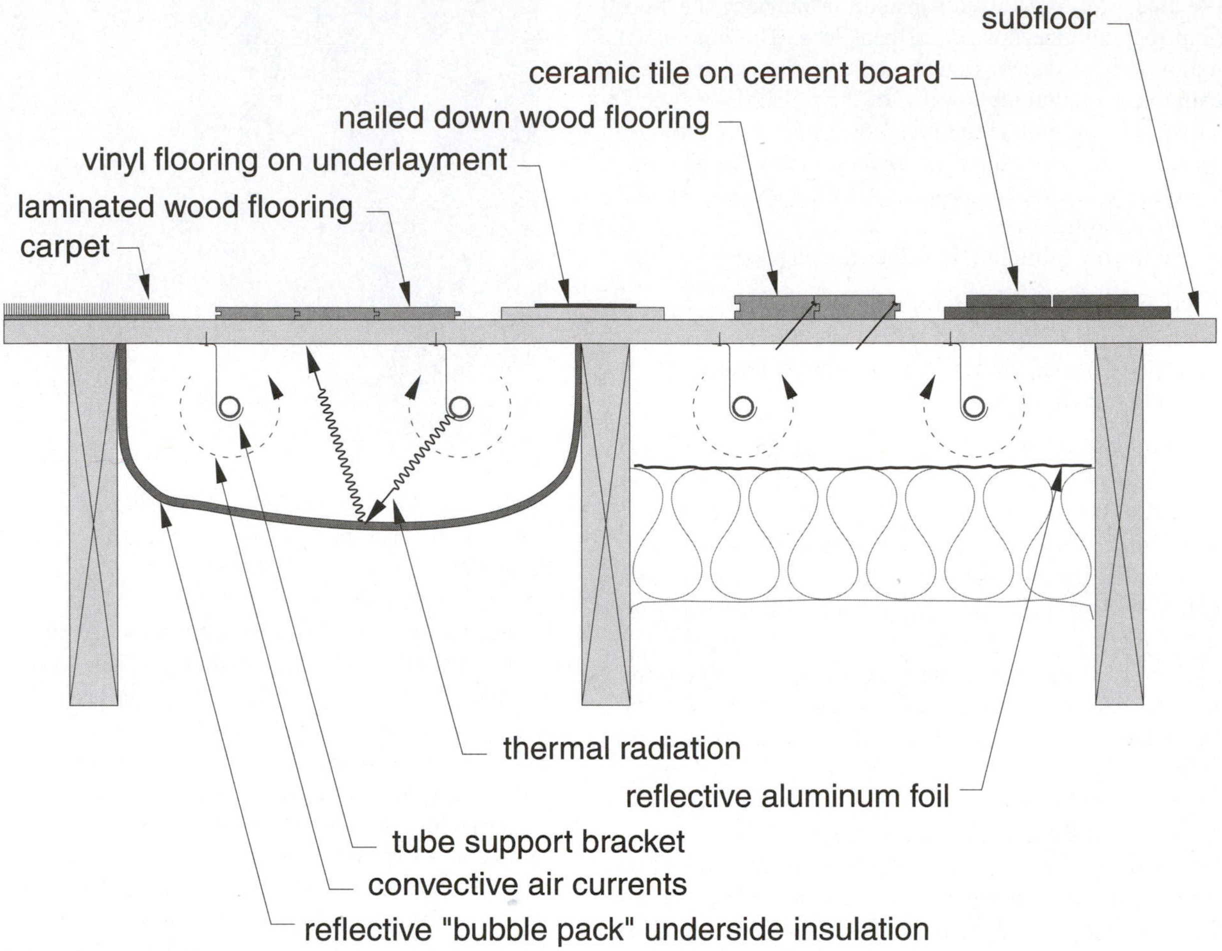

subfloor
ceramic tile on cement board
nailed down wood flooring
vinyl flooring on underlayment
laminated wood flooring
carpet
thermal radiation
reflective aluminum foil
tube support bracket
convective air currents
reflective "bubble pack" underside insulation

Figure 10–43 Cross section of a suspended tube radiant floor panel.

Extended Surface Suspended Tube Systems

A variation on the suspended tube uses slotted aluminum fins clamped to the tubing. These slotted fins extend the surface area over which heat is dissipated from the pipe, thus allowing greater heat output relative to bare tube systems.

An example of such a system is shown in Figure 10–44. The slotted fins are pop-riveted together in tight contact with the tubing. The tubing itself is run perpendicular to the floor joists through sets of holes spaced 18 to 36 inches apart. No clips are needed to support the tubing. Underside insulation is installed allowing a minimum of 2 inches of free air space beneath the tubing and plates. This air space is essential for the development of convective air currents. Published performance data indicates this system can generate heat outputs comparable to those of slab, thin-slab, or other tube and plates systems. Achieving these outputs requires water temperatures in the range of 160 to 200 °F.

Figure 10–44 Installation of slotted aluminum fins on PEX-AL-PEX tubing to form an extended surface suspended tube radiant floor panel. Underside insulation has yet to be placed. *Courtesy of MacDuffco, Inc.*

The elimination of air leakage into and out of the closed joist spaces is crucial to the proper performance of any suspended tube systems. The band joist area at the perimeter of the floor should be tightly sealed against leakage and well insulated. Penetrations of the joist spaces by plumbing pipes, ducts, conduits, or similar hardware should also be well sealed.

Suspended tube systems are well suited for situations where a small area of floor heating is desired in a building that is mostly heated by high temperature heat emitters such as the fin-tube baseboard of panel radiators. A bathroom floor or kitchen floor is a good example. The floor circuit could be supplied with the same water temperature as the higher temperature heat emitters, thus eliminating the need of a mixing device and an additional circulator. This circuit could be controlled as a separate zone using a zone valve or zone circulator.

One possibility is to tie the suspended radiant floor circuit under a bathroom floor in with a towel warmer radiator as shown in Figure 10–45. This provides warm tile floors as well as warm/dry towels. It is a simple, inexpensive approach that allows "spot" application of floor heating in an otherwise nonradiant system.

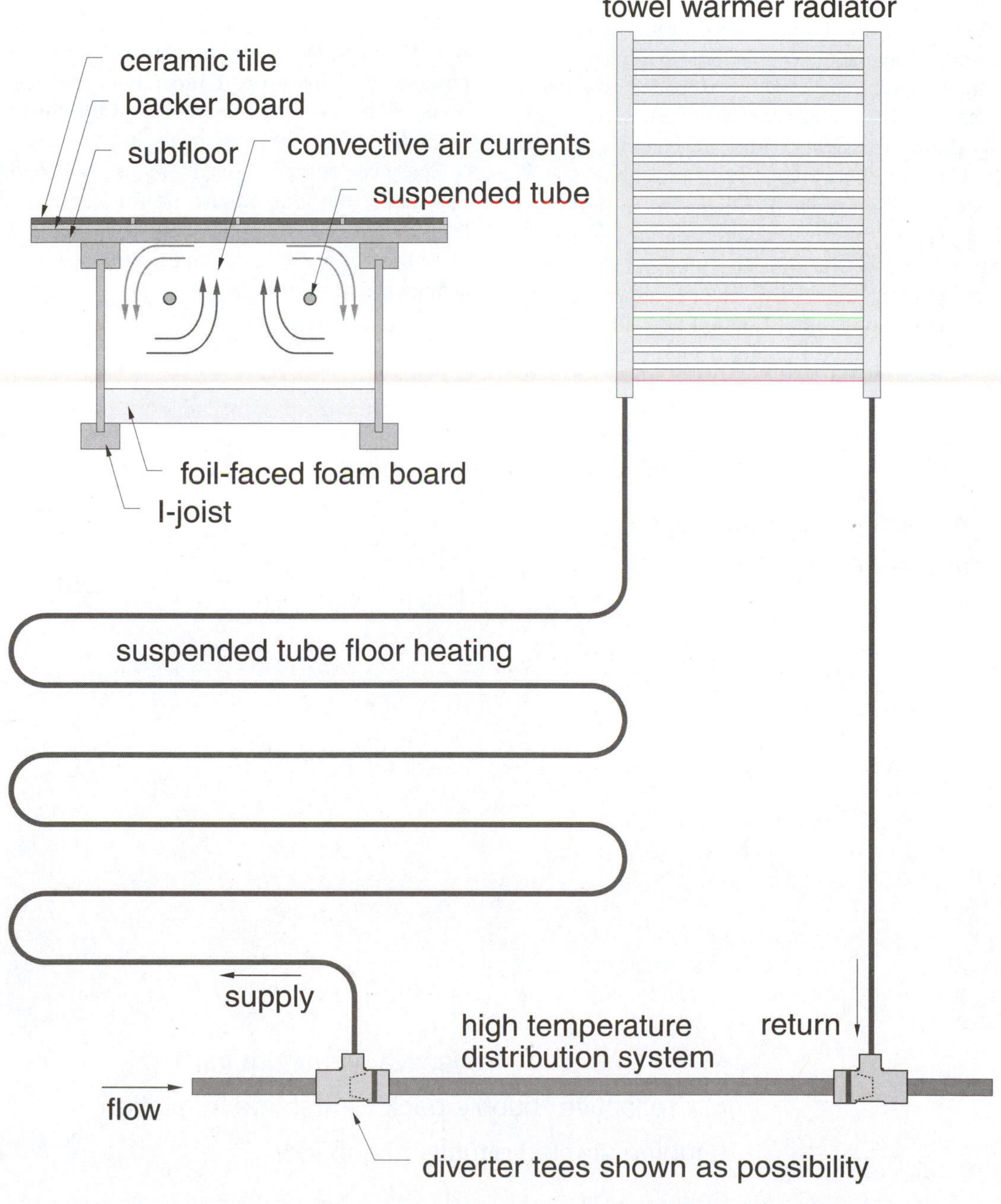

Figure 10–45 Combined use of high temperature suspended tubing circuit (underfloor), with towel warmer radiator.

10.14 PLATELESS STAPLE-UP SYSTEMS

Another approach to hydronic radiant floor heating that has been used in recent years is called a **plateless staple-up system.** The tubing is stapled directly to the bottom of the subfloor without heat transfer plates. A cross-sectional view of this system is shown in Figure 10–46. A cutaway view is shown in Figure 10–47.

In the author's opinion, plateless staple-up systems should be carefully scrutinized, and used only in situations where limited upward heat output (15 Btu/hr/square foot or less) is sufficient.

Because heat transfer plates are not present to disperse heat laterally away from the tubing, the flooring materials directly above the tubing can reach significantly higher temperatures than they would in a plated or suspended tube system. This can lead to high temperature gradients across the floor, especially during transient conditions when hot water begins flowing through tubing stapled to a cool floor deck. These gradients create high thermal stresses in the flooring materials, which can result in cracked flooring, warped flooring, or loss of bond between the subfloor and flooring. High temperature gradients can also discolor certain flooring materials directly above the tubing. Limited lateral heat diffusion also limits the overall heat output from the floor relative to the other systems discussed.

Figure 10–48 shows comparative thermal simulations of two below-floor systems. The simulation models both have 1/2 inch PEX tubing spaced 8 inches apart. The underside insulation is 4 inches of fiberglass pushed up against the subfloor. The flooring is 3/8-inch ceramic tile installed above a 3/4-inch plywood subfloor. One model includes 6-inch wide aluminum heat transfer plates, the other is plateless staple-up.

The surface temperature profile for both systems is plotted for water temperatures of 110 and 140 °F. Notice the relatively flat "plateaus" in the temperature profile for the plated system, versus the steep slopes in the profile for the plateless system. The latter indicates a rapid drop in floor surface temperature as one moves 2 or 3 inches away from the tube. This undesirable effect is known as "striping." It is the direct result of poor heat dispersion away from the tubing. As water temperatures are increased, striping becomes more pronounced. This strong temperature gradient induces stress on the finish flooring materials, and it ultimately may cause premature failure of those materials.

The heat output of each system is proportional to the area under the temperature profile curves. For the conditions simulated, the plated system yields approximately three times more heat output than the plateless system at the same water temperature.

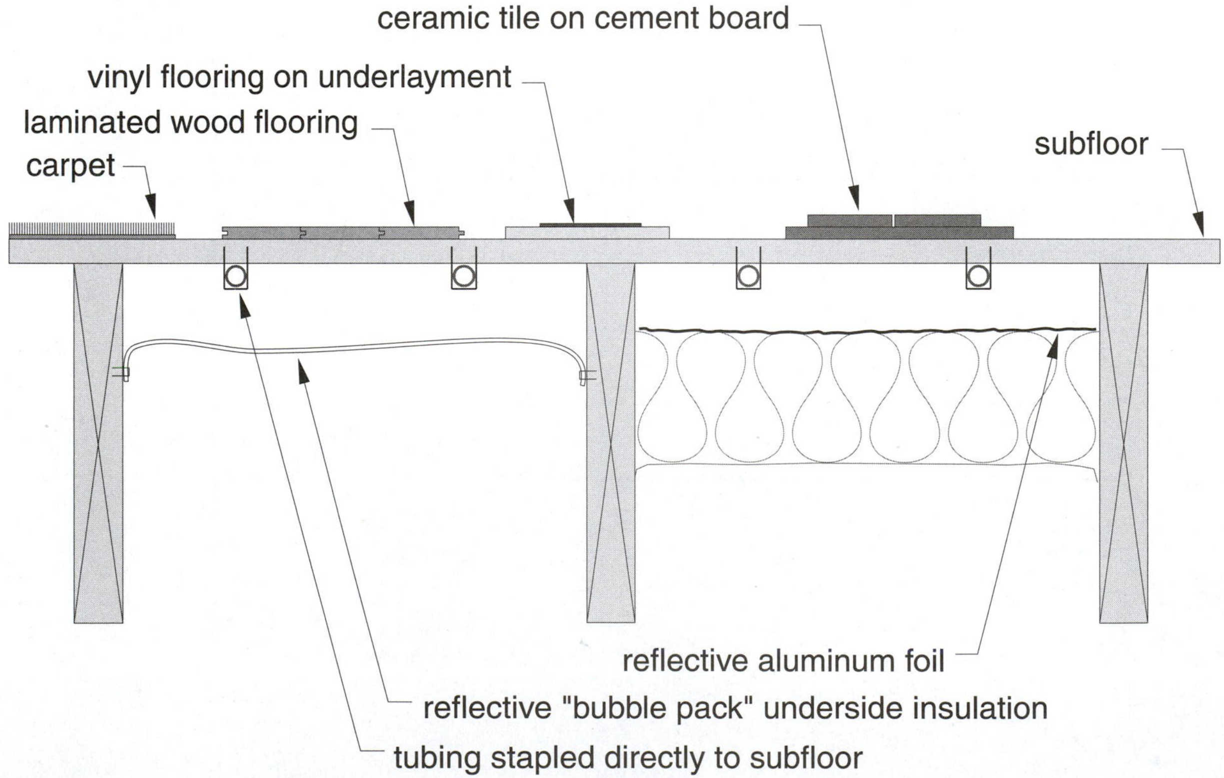

Figure 10–46 Cross section of a plateless staple-up radiant floor panel, with underside insulation options.

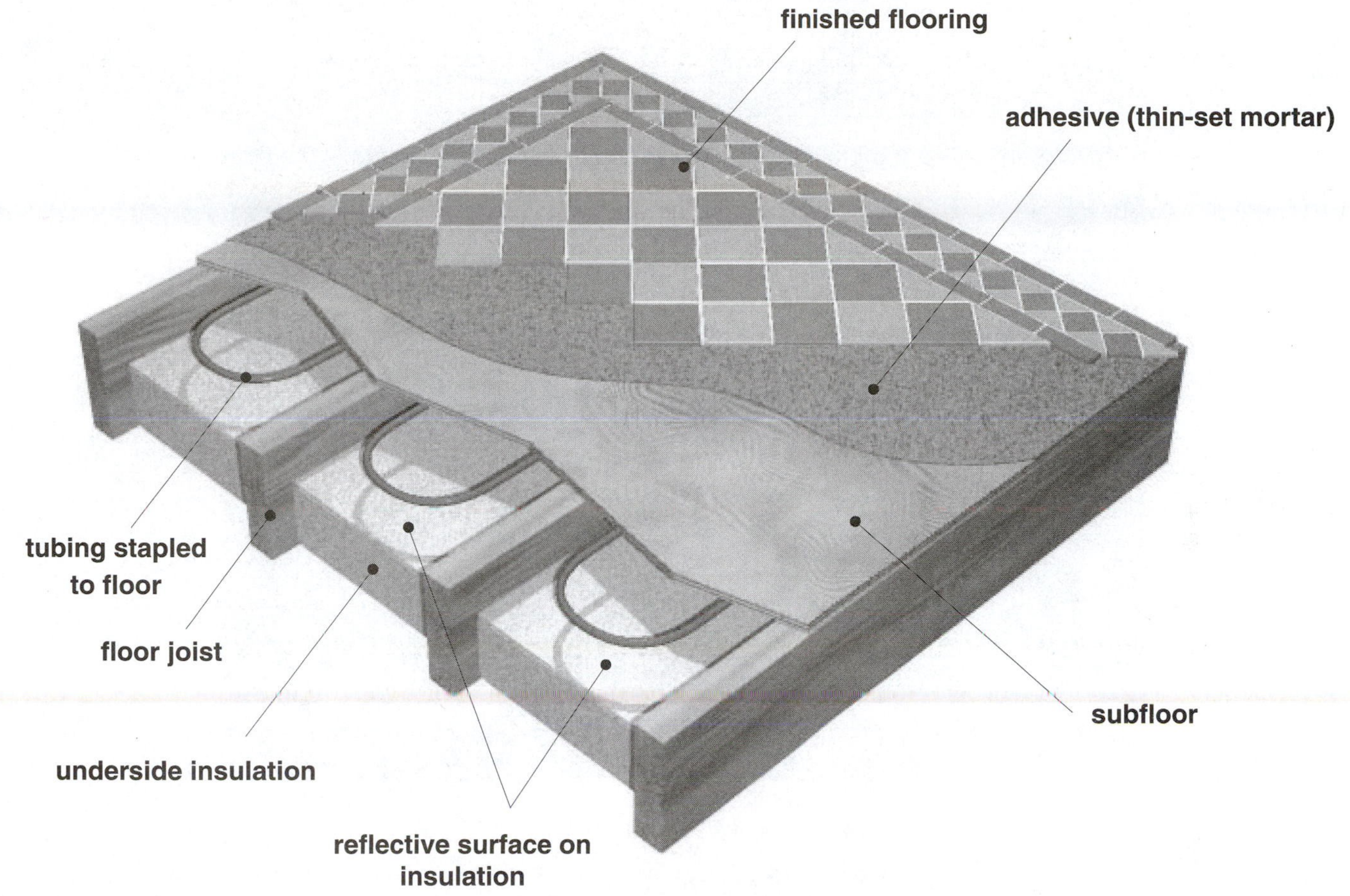

Figure 10–47 Cutaway view of a typical plateless staple-up radiant floor panel. *Courtesy of IPEX, Inc.*

10.15 PREFAB SUBFLOOR/ UNDERLAYMENT PANELS

The use of an aluminum layer to disperse heat away from a tube has also been applied in prefabricated flooring panels. In some cases, these panels serve as the structural subfloor as well as the substrate for the tubing. In other cases, they are fastened to the existing subfloor to form a grooved serpentine path ready to accept tubing.

The product shown in Figure 10–49 is a structural panel that installs directly to floor framing. It is $1\frac{3}{32}$-inch thick, and supplied in 4-foot by 8-foot sheets with various types of circuit routing grooves. A 0.025-inch-thick aluminum layer formed to the groove patterns is bonded to the top of the panel. When the tubing is pressed into the grooves, the result is a low mass radiant floor panel with performance comparable to the previously discussed above-floor tube and plate system. The installation time required for this type of product is significantly less than the previously discussed tube and plate systems.

Another approach bonds plywood strips to a flat layer of aluminum as shown in Figure 10–50. These panels are placed over an existing structural subfloor to form serpentine circuit patterns. Panels are available with both straight grooves and semicircular return bends. Some panels use 1/4-inch PEX tubing, and only add 1/2-inch tubing to the height of the floor. They are particularly well-suited to retrofit applications where minimum gain in floor height is desirable.

After the panels are fastened to the floor, a bead of silicone caulk is laid along the aluminum sheet at the bottom of the groove. The tubing is then uncoiled and driven into the groove with a mallet. The silicone forms an elastic bond between the outer surface of the tube and the aluminum sheet.

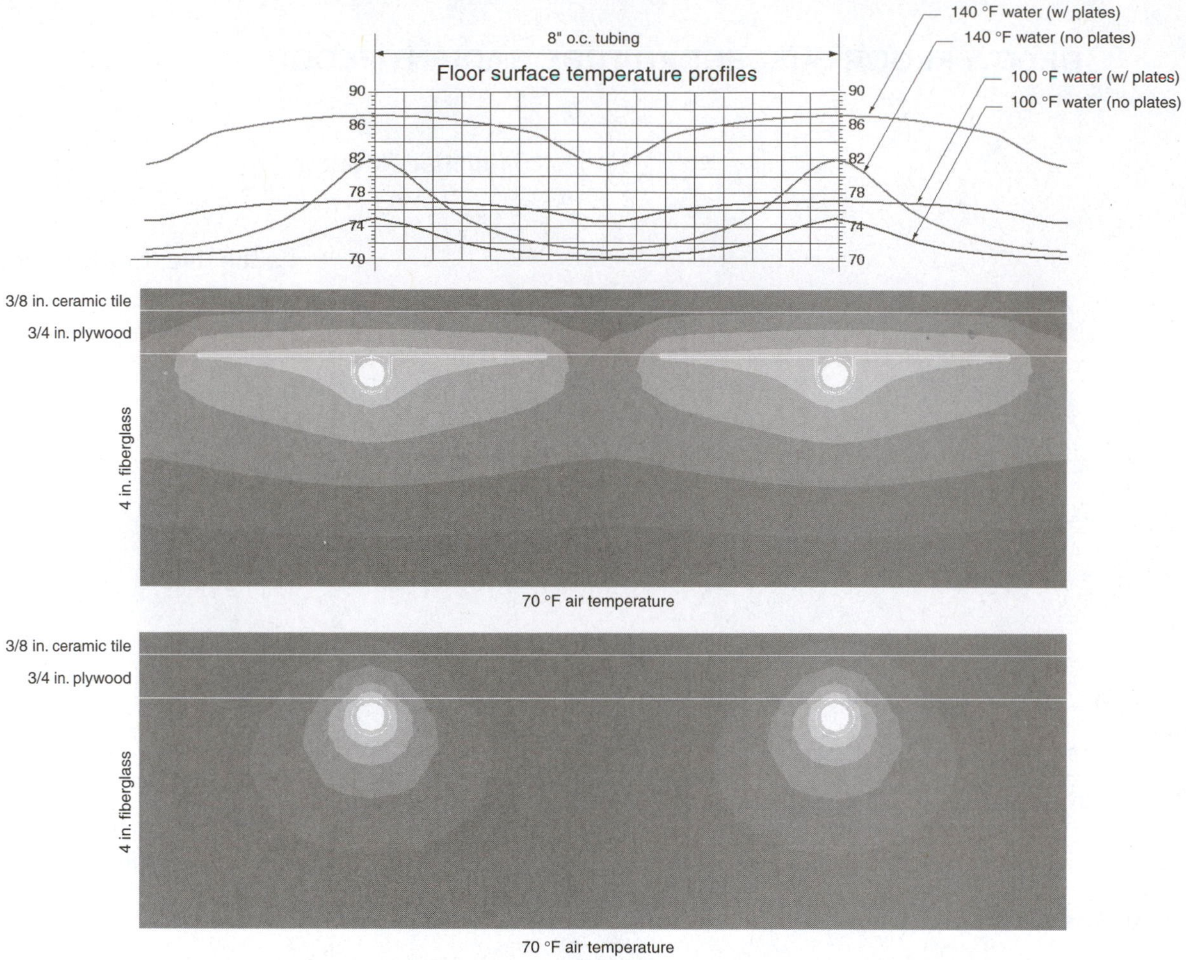

8" o.c. tubing
Floor surface temperature profiles
140 °F water (w/ plates)
140 °F water (no plates)
100 °F water (w/ plates)
100 °F water (no plates)
3/8 in. ceramic tile
3/4 in. plywood
4 in. fiberglass
70 °F air temperature
3/8 in. ceramic tile
3/4 in. plywood
4 in. fiberglass
70 °F air temperature

Figure 10–48 Thermal simulation model of below floor tubing installation with and without 6-inch wide aluminum heat transfer plates on tubing. Surface temperature profiles shown at top.

Figure 10–49 Installation of combined subfloor/radiant panel product. *Coursey of Jeld-Wen, Inc.*

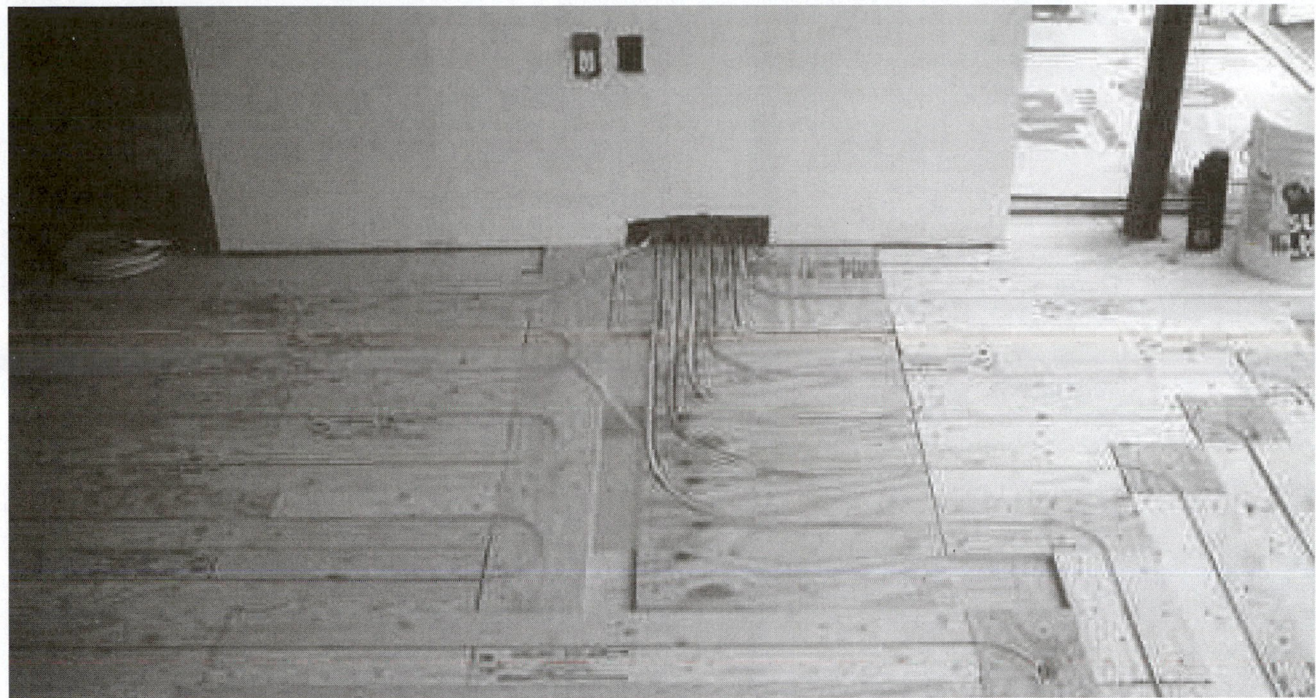

Figure 10–50 Installation of prefabricated radiant underlayment panels. *Courtesy of Foley Mechanical, Inc.*

10.16 RADIANT WALL PANELS

Because radiant heat travels equally well in any direction, it can be delivered from walls and ceilings as well as floors. In some cases, heated walls and ceilings may be possible in circumstances that would rule out floor heating, an example being a room where the owner wants to cover the floor with thick pad and carpet.

There are several advantages to the radiant wall panel system described in this section. They include:

- Fast response: Most radiant walls have very low thermal mass. This is especially true about the wall system described later in this section. Low mass radiant panels can respond quickly to changes in load caused by internal heat gain. They can also quickly reestablish comfort in a room that has been maintained at a setback temperature for several hours.
- Higher heat output: The surface temperature of a radiant wall can be higher than that of a radiant floor without producing discomfort. A typical radiant floor panel is usually not operated above 85 °F average surface temperature to avoid complaints of hot feet. A radiant wall is not subject to these limitations. When operated at an average surface temperature of 110 °F the radiant wall can deliver close to twice the heat output per square foot of a radiant floor.
- Easy to retrofit: Radiant walls are relatively easy to retrofit into existing rooms. The system described later in this section only adds 0.75 to1.25 inches to the thickness of an existing wall. It is also very low in weight and does not present the structural concerns that must be addressed with floor heating systems.
- Not affected by changing floor coverings: Over the life of the building, walls are far less likely than floors to be covered with finish materials or objects having high thermal resistance. However, it is still prudent to consider possible furniture placement relative to a proposed radiant wall. Placing upholstered furniture, an entertainment center, or other large objects against a heated wall could block a substantial portion of its heat output.

The radiant wall system described in this section is shown in cross-section in Figure 10–51. A front view is shown in Figure 10–52.

Notice the foam insulation strips behind the heat transfer plates. These strips function similar to the sleepers in an above-floor tube and plate system. They provide the necessary channel depth to accommodate the plate groove and tubing. However, they also increase the backside R-value of the radiant wall by approximately R-4.5.

A radiant wall panel does not necessarily cover the entire wall. The heated area may only extend 2 to 4 feet above the floor. This allows the radiant output to be directed into the occupied portion of the room where it has the greatest effect on comfort. Because the tubing is installed low on the wall, it is also far less likely to be punctured by nails driven to hang objects on the wall. The drawing in Figure 10–53 shows two options for incorporating the radiant panel into the lower

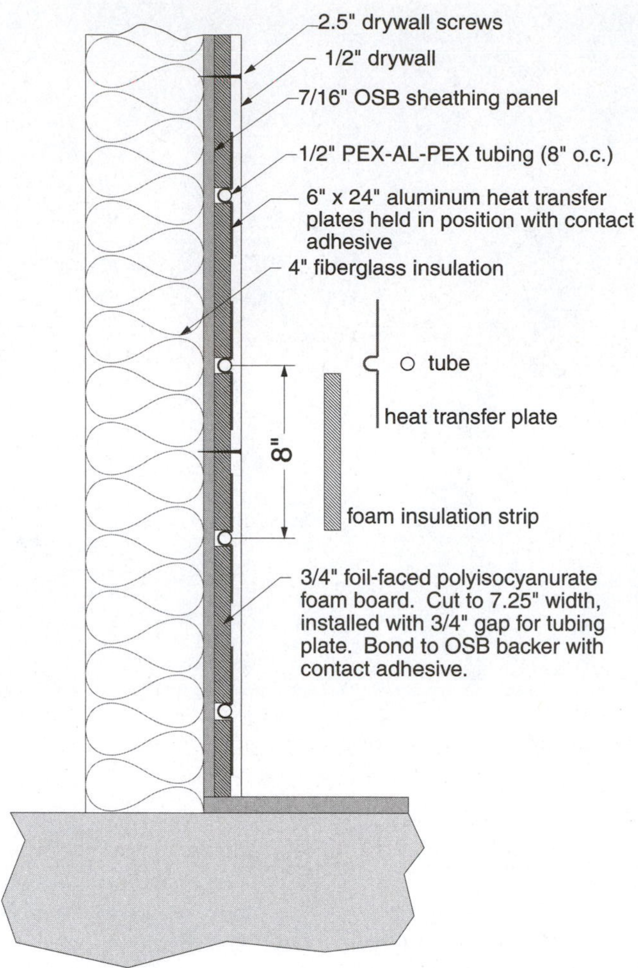

Figure 10–51 Cross section of a low-mass radiant wall panel.

portion of a typical framed wall. In one case, a "chair rail" provides a clean architectural divider between the heated and unheated portions of the wall. The other option uses furring strips to make the finished surfaces of the heated and unheated portions of the wall fall in the same plane.

Installation Procedure

The system to be described requires a flat, smooth surface as a substrate. In new wood-frame construction it can be a layer of 1/2-inch plywood or 7/16-inch oriented strand board (OSB) fastened directly to the studs. In retrofit applications, the existing wall surface can likely serve as the substrate. This substrate should cover the entire area that will become a radiant wall panel.

To "steer" heat output into the room, the wall behind the radiant panel must be well-insulated, especially if it is an exterior wall. In the latter case, the R-value installed behind the heat transfer plates should be approximately 50 percent higher than a normal exterior wall. This keeps the outward heat loss approximately the same as a normal (unheated) exterior wall. The foam insulation strips used as part of this system add about R-5 to the backside R-value of the wall. If the radiant wall is built on an interior wall, the stud cavity should be insulated to R-11.

Once the substrate layer is established, snap horizontal chalk lines on it beginning 3⅝ inches above the floor, and then every 8 inches vertically. These lines mark the top edges of foam insulation strips that will be bonded to the substrate. These strips will support the aluminum heat transfer plates.

The foam insulation strips are made by ripping 3/4-inch-thick foil-faced polyisocyanurate insulation board into 7.25-inch-wide strips with a table saw. The strips are then bonded to the substrate using solvent-based contact adhesive. A light coat of adhesive applied with a paint roller is sufficient. Be sure the strips are accurately aligned with the chalk lines before pressing them to the wall; 3/4-inch-wide wood spacer blocks set along the top edge of each successive row also help keep the foam strips aligned.

To accommodate wall receptacles, 1 1/4-inch-deep plastic junction boxes can be screwed directly to the substrate. The boxes should be centered between adjacent rows of tubing. The front of the box will finish flush with the 1/2-inch drywall. Electrical wiring can be routed out the back of the

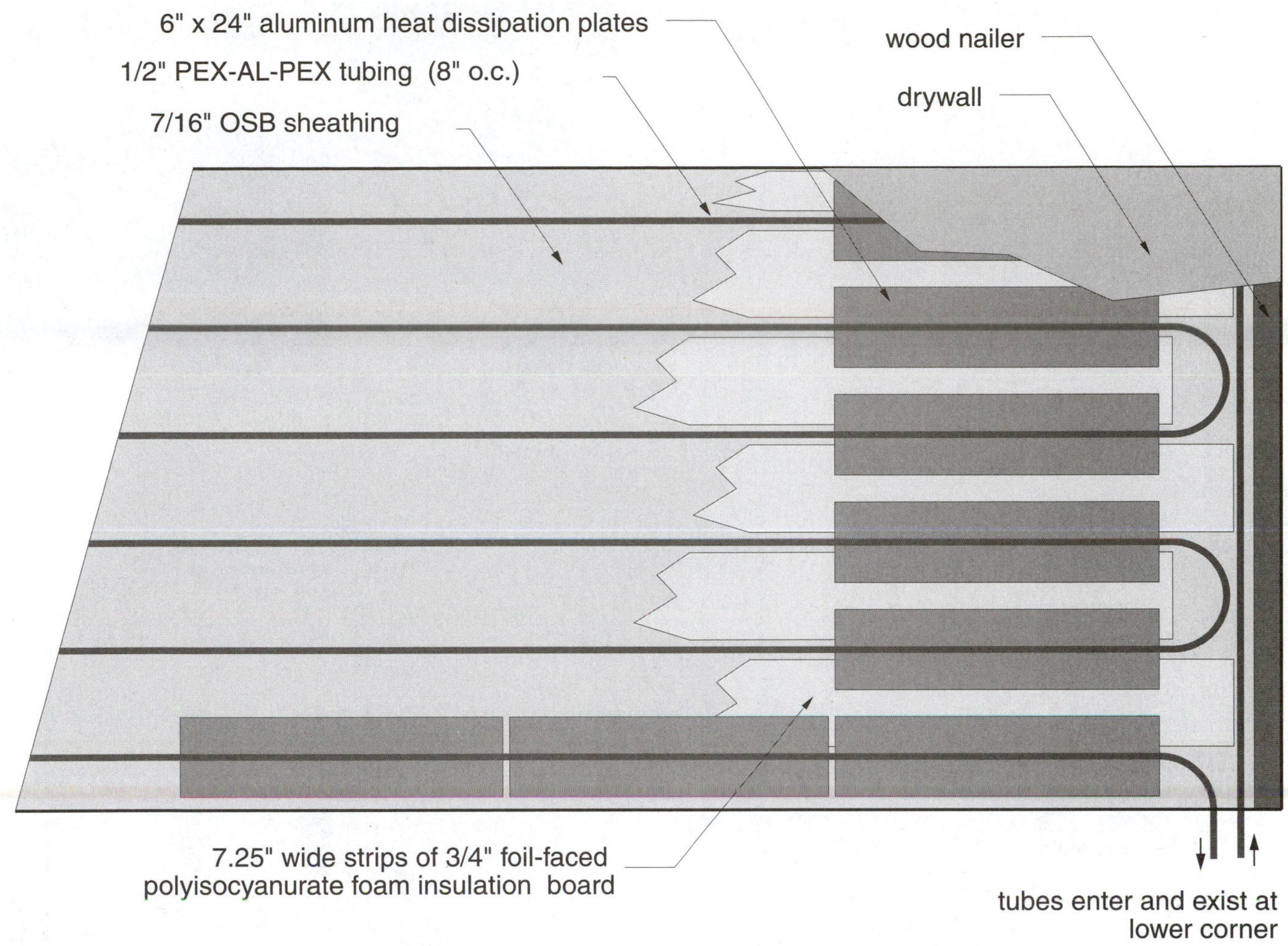

Figure 10–52 Front view of radiant wall panel components.

boxes through the substrate. Do not install heat transfer plates within 4 inches of junction boxes to minimize warming the receptacles and wiring. Mounting the junction boxes horizontally helps keep them a few inches away from the tubing.

At the ends of the wall, the foam strips are cut short to allow space for the return bends. A vertical 3/4-inch by 2-inch wood strip should be installed at each end of the wall to provide a solid nailing surface for the drywall.

The next step is to bond the aluminum heat transfer plates to the foam strips. A light coat of contact adhesive is applied to the back of *one side* of the plate and the matching surface of the foam strips. Minimal bonding strength is sufficient to hold the plates to the foam strips. As the plates are pressed in place, they should be held tight to the strip where the adhesive was applied. This creates a slight gap on the other side of the channel that allows the plate to expand as the tubing is pressed in place.

Besides adding backside R-value, these strips have very low thermal mass. This allows the radiant wall to warm up quickly following a cold start. The aluminum facing on the foam strips also enhances lateral heat diffusion away from the tubing, and provides a good bonding surface for adhesives. The polyisocyanurate foam is thermally stable to temperatures of 250 ºF. This is far higher than any temperature it would experience in this application.

The preferred tubing for this system is 1/2-inch PEX-AL-PEX. It is uncoiled and snapped into the grooves of the plates. A wooden block and rubber mallet can be used to gently tap the tubing into the grooves. Once the tubing is in place, be sure to pressure test the circuit.

The radiant wall is completed by installing 3/8-inch or 1/2-inch drywall directly over the tube and plates as shown in Figure 10–54. 2.5 inch drywall screws can be driven through the drywall and foam strips to catch into the substrate layer. The screws should be driven to a line located halfway between adjacent rows to tubing. *Be very careful not to drive a screw through the tubing near the return bends.*

This radiant wall construction is very scalable. Panels of various sizes can be built in strategic locations. Examples include the walls of a walk-in shower enclosure, the leg space

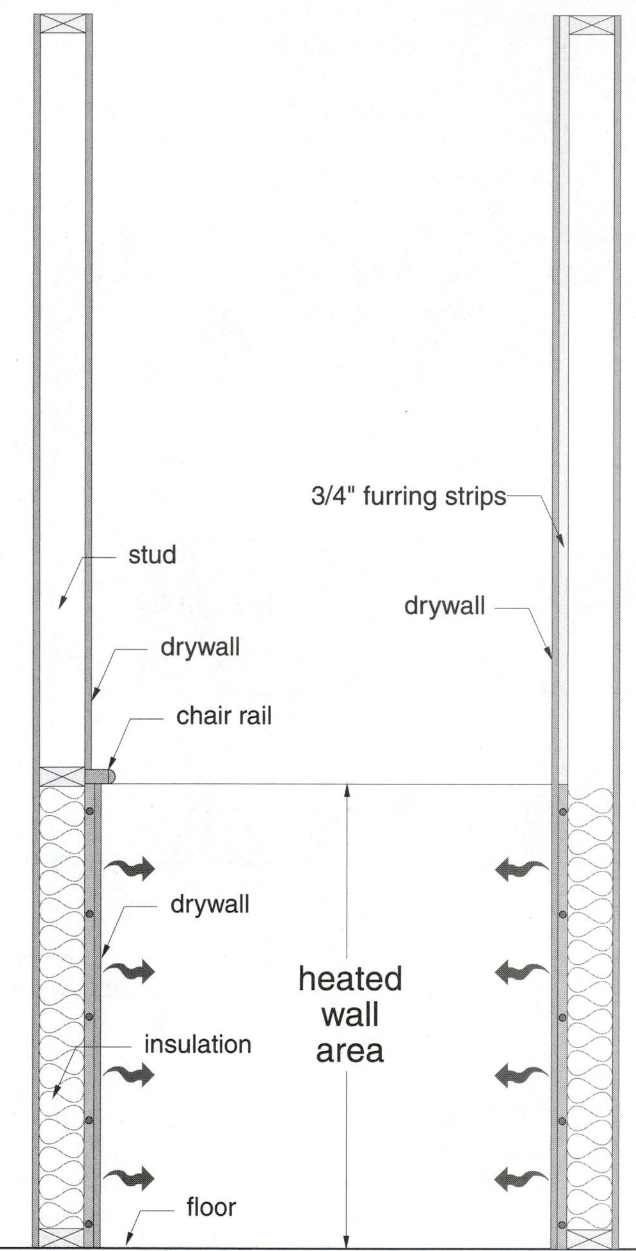

Figure 10–53 Two construction options for a low profile radiant panel.

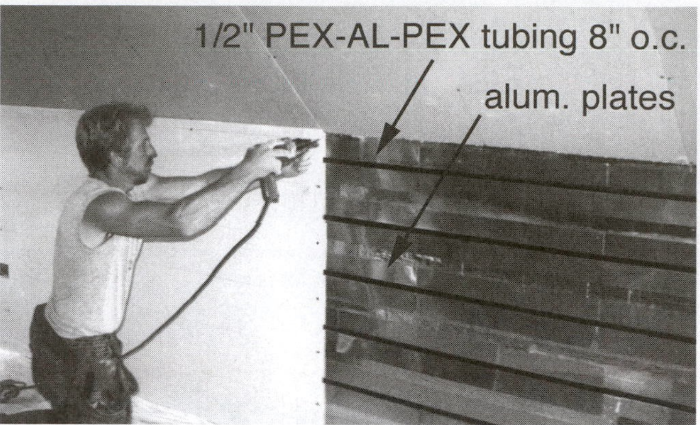

Figure 10–54 Installation of drywall on a low profile radiant wall panel.

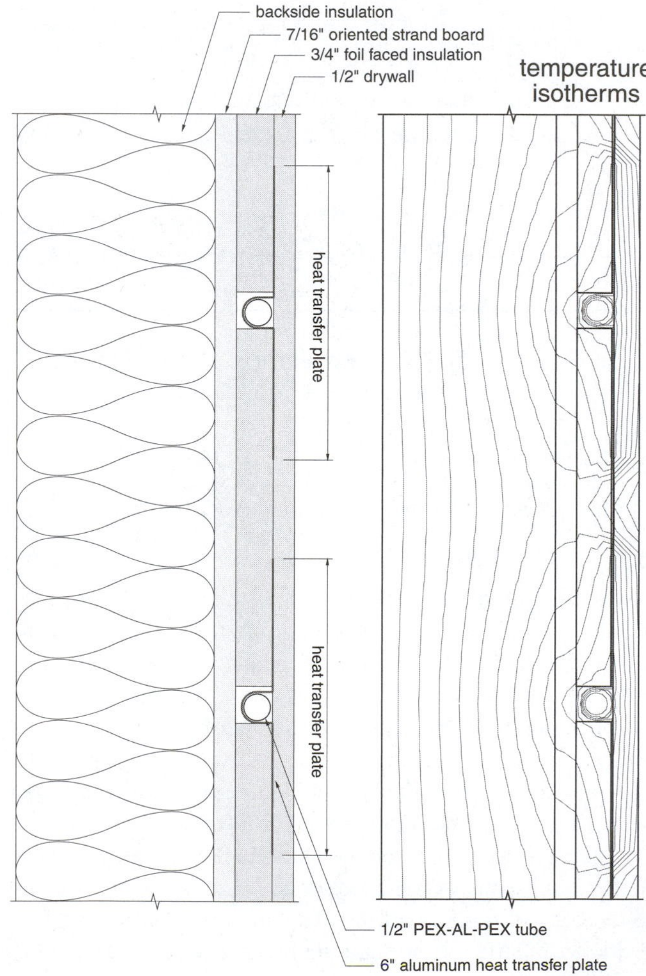

Figure 10–55 Temperature isotherms within a low-mass tube and plate radiant wall panel.

below a breakfast bar, the walls surrounding a dressing area, or the area below the windowsills of an indoor pool enclosure. In high-moisture environments, the wall can be covered with cement board and ceramic tile rather than drywall.

Thermal Performance

The heat transfer through the radiant wall system just described was modeled using computer simulation. Figure 10–55 shows the isotherms in the system under typical steady state operation. The graph in Figure 10–56 gives the theoretical heat output of the wall as a function of the average water temperature minus the room air temperature.

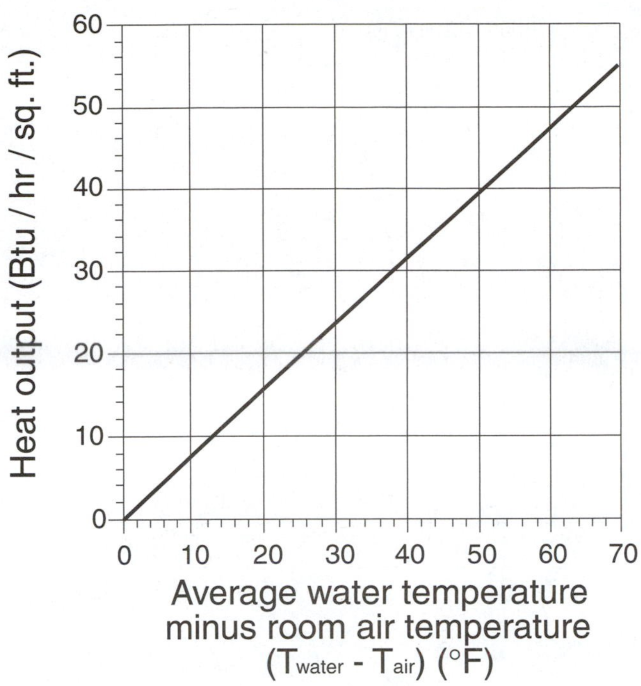

Figure 10–56 Estimated heat output of a low-mass tube and plate radiant wall panel.

10.17 RADIANT CEILING PANELS

Heated ceilings deliver over 90+ percent of their heat output as thermal radiation. In effect, they "shine" thermal radiation down into the room much as a light fixture shines visible light downward. They offer many of the same benefits as radiant walls, including:

- Low thermal mass. Most radiant ceiling panels have low thermal mass allowing them to quickly warm up following a cold start. This characteristic makes them a good choice in rooms where quick recovery from setback conditions is desirable. Equally important is the fact that low thermal mass allow heat input to be quickly suspended when necessary. This can help limit overheating in situations where significant solar heat gain can suddenly occur.
- Higher heat output. Because occupants are not in contact with them, radiant ceilings can be operated at higher surface temperatures than radiant floors. This allows greater heat output per square foot of ceiling. For example, a ceiling operating at a mean surface temperature of 102 ºF will release approximately 55 Btu/hr/sq. ft. into a room maintained at 68 ºF. This is almost 60 percent more heat output than a radiant floor with a mean surface temperature limit of 85 ºF. *Higher heat output per square foot often allows a room to be properly heated without need of covering the entire surface of the ceiling.* This is especially true for the higher temperature ceiling panels discussed later in this section.
- Not affected by changing floor coverings. The ceiling is arguably the least likely surface of a room to ever be covered. Thus, the performance of a heated ceiling is very unlikely to be compromised by changes in floor covering or furniture.
- Warms objects in the room. The radiant energy emitted from a heated ceiling is absorbed by the surfaces in the room below. This includes unobstructed floor area as well as the surfaces of objects in the room. The upward facing surfaces tend to absorb the majority of the radiant energy; the top of beds, tables, and furniture are slightly warmer than the room air temperature. The surface temperature of floors below a radiant ceiling is often 3 to 4 ºF above what it would be in a nonradiantly heated room.
- Easy to Retrofit. Radiant ceilings are usually easier to retrofit into existing rooms than are radiant floors. They add very little weight to the structure, and require minimal loss of headroom.

Hydronic radiant ceiling heating has been used for several decades in the United States. Early versions used copper tubing covered with steel lathe and plaster. An example of an early system was shown in Figure 10–1. Although very labor intensive to construct by current standards, these radiant panels provided good performance. Some are still in use today.

A modern version of a hydronically heated radiant ceiling using aluminum heat transfer plates is shown in Figure 10–57. Other than its orientation, this panel is very similar to the radiant wall panel system described in the previous section. The installation procedure is also more or less the same as for the radiant wall. A photo of this system being installed is shown in Figure 10–58.

Modular Radiant Ceiling Panels

Another type of hydronic radiant ceiling panel consists of a shallow sheet steel or aluminum pan with tubing and insulation attached to its topside. Such panels are designed for installation in "T-bar" ceiling systems or are surface mounted. The metal pan is finished with a baked-on enamel coating on the exposed side. Copper tubing is mechanically attached to the top of this pan. Insulation is then placed over this assembly to limit upward heat flow.

Modular radiant ceiling panels are commonly used to improve comfort near the perimeter area of commercial buildings, especially when large window walls are present. They replace conventional suspended ceiling panels along the exposed perimeter of rooms as shown in Figure 10–59. They are also used in health care facilities to maintain comfort in rooms with high rates of air exchange or in situations where minimal air movement is desirable.

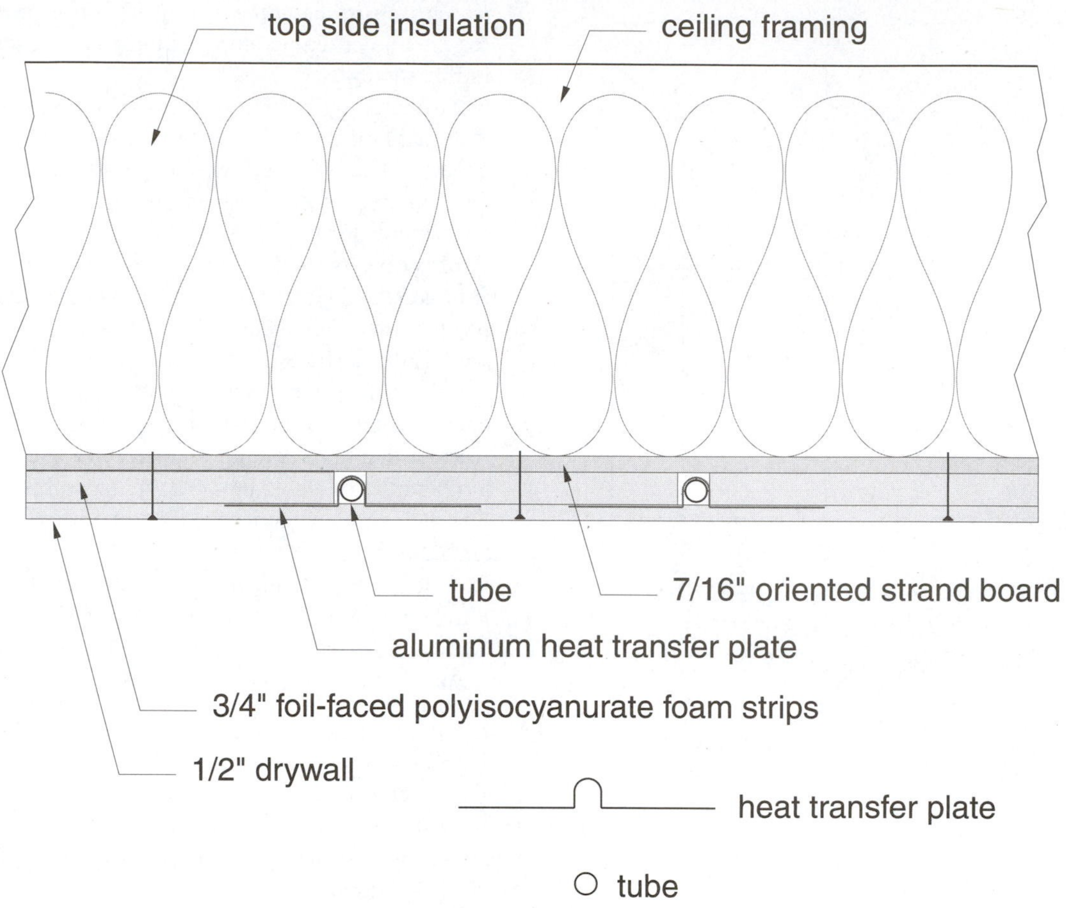

Figure 10–57 Cross section of a low-mass tube and plate radiant ceiling panel.

Figure 10–58 Installation of a low-mass tube and plate radiant ceiling. *Courtesy of Don Jackson.*

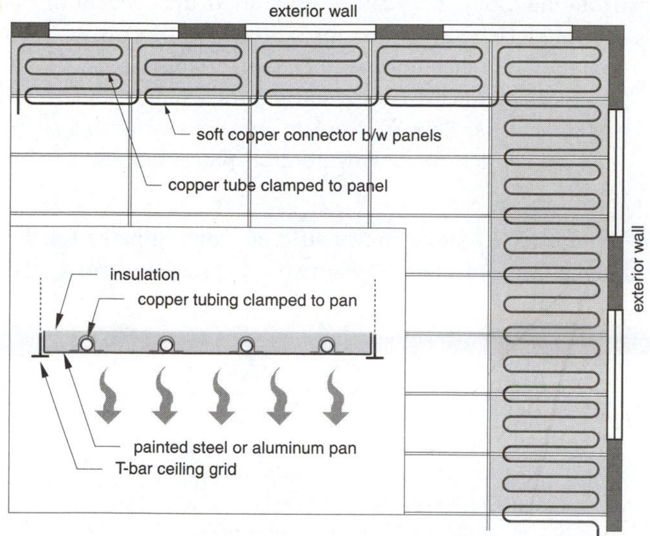

Figure 10–59 Typical placement of prefabricated radiant ceiling panels in a suspended ceiling system.

Thermal Performance of a Heated Ceiling

The heat output of a heating ceiling surface depends on the mean surface temperature of the ceiling and the temperature in the room below. Equation 10.1 can be used to estimate this output:

(Equation 10.1)

$$q = 1.6(T_S - T_R)$$

where:

q = heat output from ceiling (Btu/hr/ft²)
T_s = the mean surface temperature of the ceiling (°F)
T_R = the air temperature in the room (°F)

Example 10.1: What is the required mean surface temperature of a heated ceiling that is required to deliver 45 Btu/hr/ft² to a 68 °F room below?

Solution: Equation 10.1 can be rearranged to solve for the required mean surface temperature:

$$T_S = T_R + \frac{q}{1.6}$$

The given data can then be substituted in and the result calculated:

$$T_S = T_R + \frac{q}{1.6} = 68 + \frac{45}{1.6} = 96.1°\text{F}$$

Discussion: The surface temperature at a given location on the ceiling may be somewhat higher or lower than this calculated *mean* surface temperature. Surface temperature variation depends on the spacing of the tubing as well as the heat dispersion characteristic of the ceiling panel.

For systems involving gypsum ceilings (plaster or drywall), the maximum mean surface temperature should not exceed 120 °F to avoid thermal degradation of the gypsum. Some references also suggest that the maximum mean surface temperature of a heated ceiling not exceed 100 °F when the ceiling height is 8 feet, nor 110 °F when ceiling height is 9 to 12 feet. These limitations are based on thermal comfort rather than material limitations.

The surface temperature of modular (metallic) ceiling panels can be as high as 185 °F. Under such operating conditions, the heat output of the panel can exceed 200 Btu/hr/square feet. Thus, a small perimeter band of panels can often deliver sufficient heat for the entire room.

The surface temperature of a heated ceiling depends on the water temperature in the embedded tubing as well as the other materials used to build the panel. The output of a tube and plate type radiant ceiling panel constructed using 1/2-inch PEX-AL-PEX tubing spaced 8 inches apart as shown in Figure 10–57 is given in Figure 10–60. This graph plots heat output as a function of the difference between the average water temperature and the room air temperature.

The high heat output potential of radiant ceiling panels may allow for situations where it is not necessary to heat the entire ceiling area of a room. A heated perimeter band 4- to 8-feet-wide may provide all the heat output necessary. This reduces the amount of tubing and heat transfer plates needed and therefore lowers cost.

When a radiant ceiling panel is constructed on the bottom of an exposed ceiling, upward heat loss must be minimized. Excessive top side heat lost can raise attic temperatures high enough to melt snow on the roof which often leads to large ice accumulations at the eaves. The upward loss is driven by

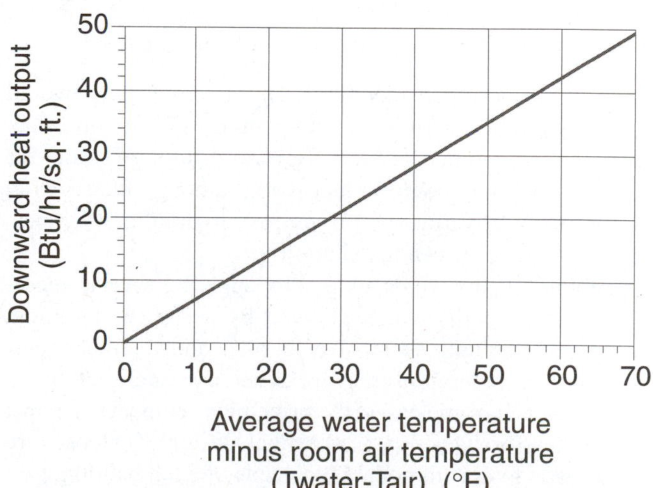

Figure 10–60 Estimated heat output of a tube and plate radiant ceiling panel.

the temperature differential between the heated layer of the ceiling (the horizontal plane of the heat transfer plates), and the air temperature above the insulation. This differential is usually significantly higher than with a nonheated ceiling. *A conservative recommendation is to increase the R-valve above a heated ceiling approximately 50 percent above the R-value planned for a nonheated ceiling.* If the heated ceiling is below heated space, a minimum of R-11 insulation should be installed above it.

10.18 TUBE PLACEMENT CONSIDERATIONS (FLOOR PANELS)

An important part of planning a hydronic radiant floor heating system is deciding how to place individual tubing circuits. The eventual goal is to generate a tubing layout plan that can be easily followed during installation, and serve as permanent documentation of the installation. Experienced installers can attest that such a plan will save hours of frustration and prevent costly errors on the jobsite.

The tubing layout for a given system must account for a number of factors including:

- Dimensions of the spaces
- Design heating loads of the spaces
- Ease of tubing installation
- Any objects that interfere with tubing placement
- Position of the manifold stations
- Position of any control joints
- How the building will be zoned for two or more temperatures
- Types(s) of floor coverings to be used
- Length of the piping circuits
- Horizontal spacing of the tubing
- Position of the exterior walls
- Possibility of fasteners being driven into the floor

With all these considerations, any given designer would probably produce a tubing layout plan that differs somewhat from that of another designer. Several different layouts for the same building could in fact work, although there would likely be some variations in comfort, hydraulic efficiency, and the ease or cost of installation.

Tubing layouts are developed by applying several general principles to the extent allowed by a specific building. Often these principles involve thermal and hydraulic performance considerations that are detailed in later sections of this chapter. In some cases, the principles "compete" against each other for inclusion in the final design. Tradeoffs are often made to accommodate the needs of each building.

- **Principle #1:** *Install the tubing so the warmest fluid is first routed along the exterior wall(s) of the room.* This allows the highest heat output to occur where the losses from the room are greatest. It parallels the concept of placing baseboard convectors at the base of exterior walls. The higher floor surface temperatures help compensate for the cooler surfaces of windows and exterior walls. They also create convective currents that counteract the downward draft present along exterior walls.

Figure 10–61 shows three different routing paths for the tubing in a room having one, two, or three exposed walls.

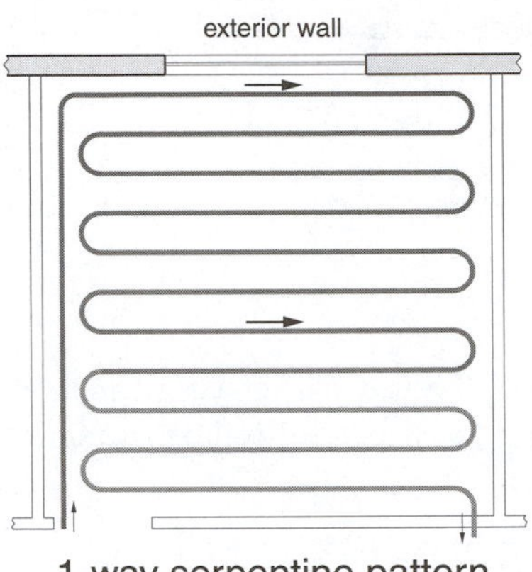

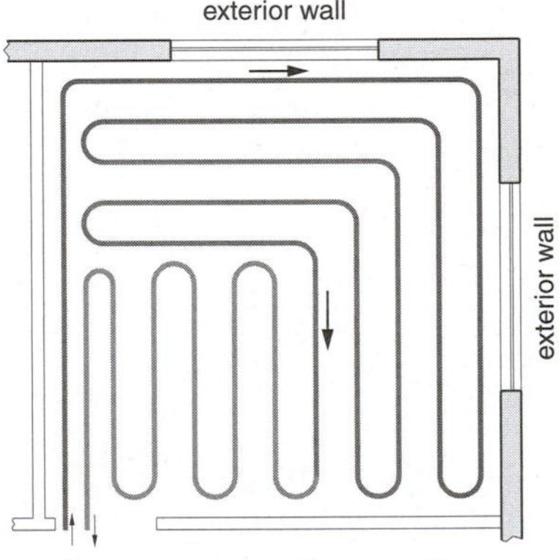

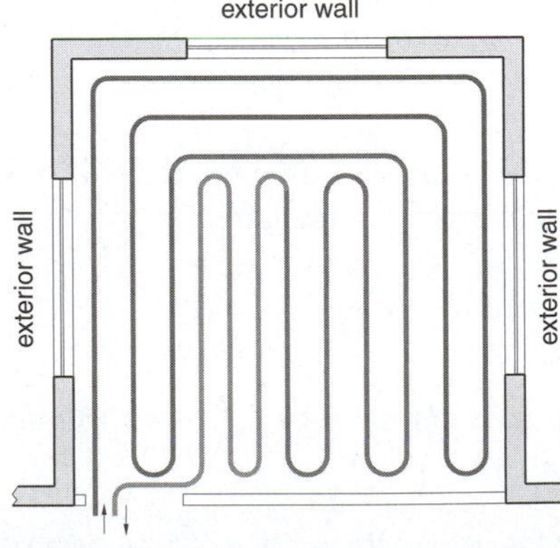

Figure 10–61 Serpentine tubing layout patterns. Note that warmest portion of circuit runs adjacent to exterior walls.

These patterns are called **serpentine patterns** because of how the tubing zigzags across the floor area. Notice in all three cases the warmest water is first routed along the exposed walls. The circuit then works its way back toward the interior area of the room.

These circuit routing patterns work well for rooms in which a single circuit can cover the entire floor area. Larger rooms are often laid out with one or more **perimeter circuits** combined with one or more **interior circuits.** The perimeter circuit(s) usually consist(s) of closely spaced tubing covering a band extending 3 to 4 feet in from exterior walls. The interior circuit typically uses wider tube spacing over the remaining floor area. An example of such a layout is shown in Figure 10–62.

Another circuit routing pattern called a **counterflow spiral** is shown in Figure 10–63. Unlike the three patterns shown in Figure 10–61, this tubing pattern is not designed to bias heat output toward exterior walls. Instead, it is intended to provide the same average heat output over the entire floor area. This pattern is useful for circuits not installed adjacent to exterior surfaces, especially in larger buildings. In most cases, it is used for interior circuits in combination with serpentine perimeter circuits.

- **Principle #2:** *Whenever possible, arrange the tubing to maximize straight runs and minimize return bends and corners.* This speeds installation and reduces the possibility of kinking the tubing. The narrower a room is relative to its length, the more important this becomes. Figure 10–64 illustrates this concept for a room twice as long as it is wide.

- **Principle #3:** *Limit individual circuits to the lengths shown in Figure 10–65:*

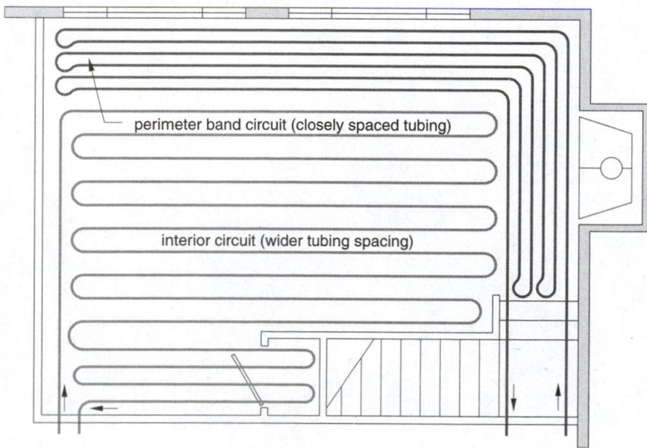

Figure 10–62 Combination of perimeter circuit using closely spaced tubing with an interior circuit using wider tube spacing.

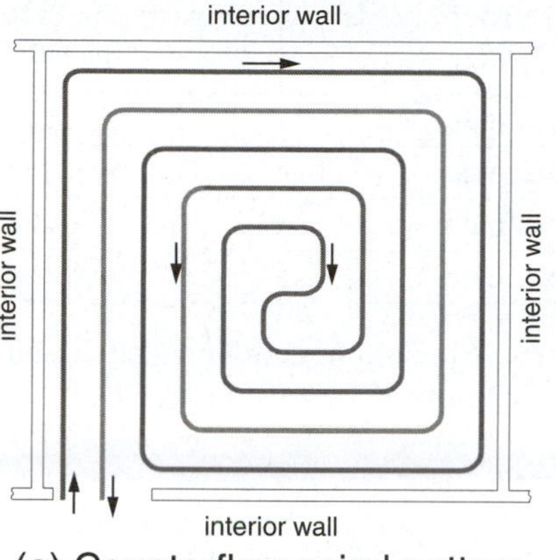

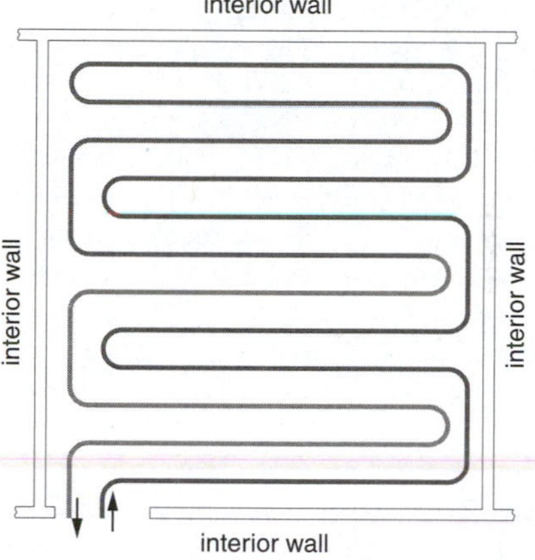

Figure 10–63 Example of a counterflow tubing layout pattern.

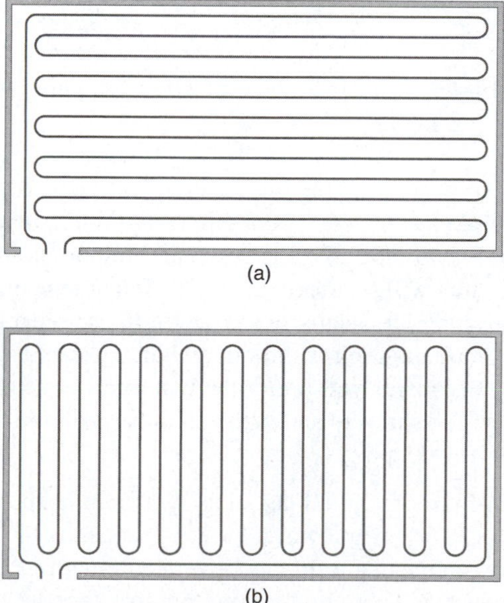

Figure 10–64 Preferred tubing layout (a), maximizes straight runs and minimizes return bends and corners.

Tubing size (nominal ID)	Maximum circuit length
1/4 inch	125 feet
3/8 inch	250 feet
1/2 inch	300 feet
5/8 inch	450 feet
3/4 inch	600 feet

Figure 10–65 Maximum recommended circuit lengths for radiant panel circuits of various tubing sizes.

These lengths limit the head losses of the circuits to reasonable values that can be handled by small circulators.

The circuit length required to cover a certain floor area can be *approximated* by dividing the floor area (ft^2) by the spacing of the tubing (ft). The results of these calculations are shown for several tube spacings in Figure 10–66.

Center-to-center spacing of tubing	Linear feet of tubing required per square foot of floor area
4 inches	3 feet
6 inches	2 feet
8 inches	1.5 feet
9 inches	1.33 feet
12 inches	1.0 feet
15 inches	0.8 feet
18 inches	0.67 feet

Figure 10–66 Linear footage of tubing required for various tube spacings.

- **Principle #4:** *Design circuits so room-by-room flow control is possible.* This is especially important for rooms that will be maintained at different temperatures. For example, it usually makes sense to use separate piping circuits for a bathroom and an adjacent bedroom. Many occupants will prefer the bathroom somewhat warmer for showers and baths, and the bedroom slightly cooler for sleeping.

Room-by-room circuit layout allows room temperature to be controlled in several ways. First, by adjusting the balancing valves on the manifold, the flow rate through each circuit can be altered. This allows some control over heat output and therefore room temperature as discussed in Chapter 9, Control Strategies, Components, and Systems.

Secondly, if frequent changes in individual room temperature are desired, a valve actuator can be attached to each manifold valve. This device opens its associated valve when supplied with a 24-VAC signal through a room thermostat. When this signal is not present, the actuator closes the manifold valve and thus prevents any flow through that circuit.

It is also possible to use nonelectric thermostatic valves to control individual circuits. Such valves are an option to electric thermostats and valve actuators. They are typically located in small boxes mounted into the wall framing. The circuit serving the room passes through the valve body on its way to the floor. An example of such a valve is shown in Figure 10–67.

Designers should remember that individual room thermostats are not needed simply because some rooms need to be maintained at different temperatures. This can usually be handled by circuit flow rate adjustments. Thermostats make sense when *frequent changes* in individual room temperatures are desirable.

This principle is often the dominant concept that determines how a building such as a house is divided into individual circuits. Figure 10–68 shows the concept of room-by-room circuit design for a house.

- **Principle #5:** *Minimize locations where the tubing has to pass beneath control joints.* In many buildings, some tubing will have to cross beneath control joints. When it does, it should be held low in the slab to prevent any possible contact with the saw as discussed in Section 10.8. It should also be protected by a plastic pipe sleeve to relieve stresses that could result if the slab does flex slightly at the control joint.

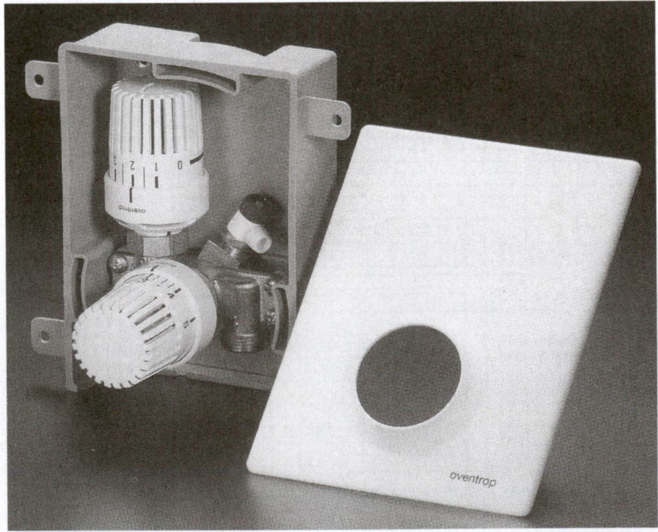

Figure 10–67 Thermostatic valve with recessed wall enclosure used for room by room temperature control of radiant panel circuits. *Courtesy of Oventrop.*

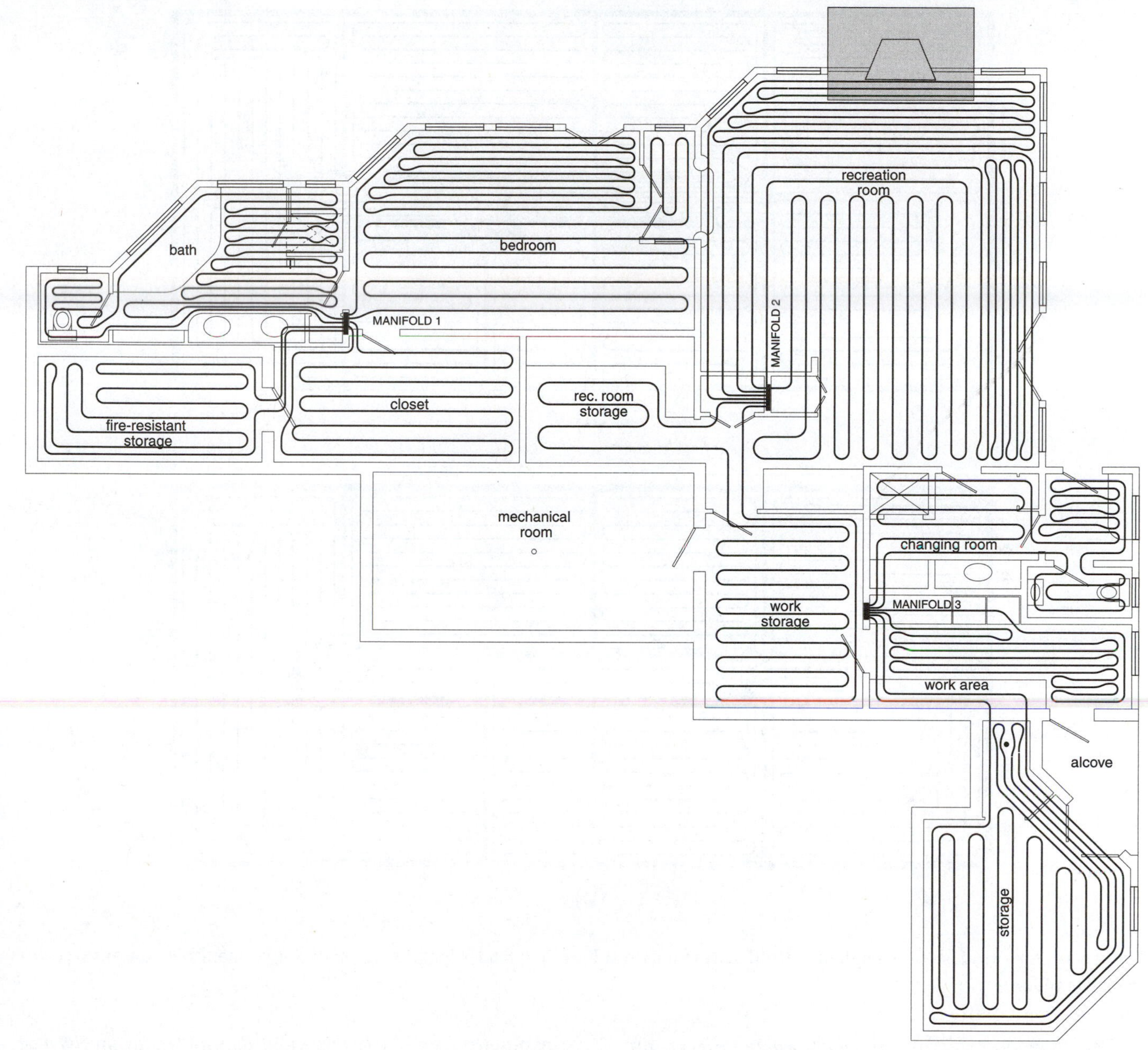

Figure 10–68 Example of room-by-room radiant floor circuit layout.

- **Principle #6:** *For slab-on-grade systems, tubing placement should correspond with the grid formed by welded wire reinforcing whenever possible.* This allows for efficient placement and fastening of tubing. Standard welded wire reinforcing uses a 6-inch by 6-inch grid size. It is also available, although less commonly stocked, in a 4-inch by 4-inch grid size.
- **Principle #7:** *Install manifold stations in central locations to minimize tubing runs between the manifold station and the room(s) to be heated.* Think of the manifold station as the hub of a wheel, with the supply and return pipes of individual circuits as the spokes. By keeping the manifold station centrally located to a group of rooms, long, closely spaced tubing runs called **leaders** can be avoided. This saves tubing and thus reduces cost.

Figure 10–69 shows an example of using two manifold stations rather than one to minimize leader lengths.

- **Principle #8:** *Whenever possible, avoid layouts that require tubes to cross over each other within a slab.* This is an absolute must for thin-slab systems. If unavoidable, crossovers can be made when the tubing is installed in a 4-inch or thicker concrete slab.

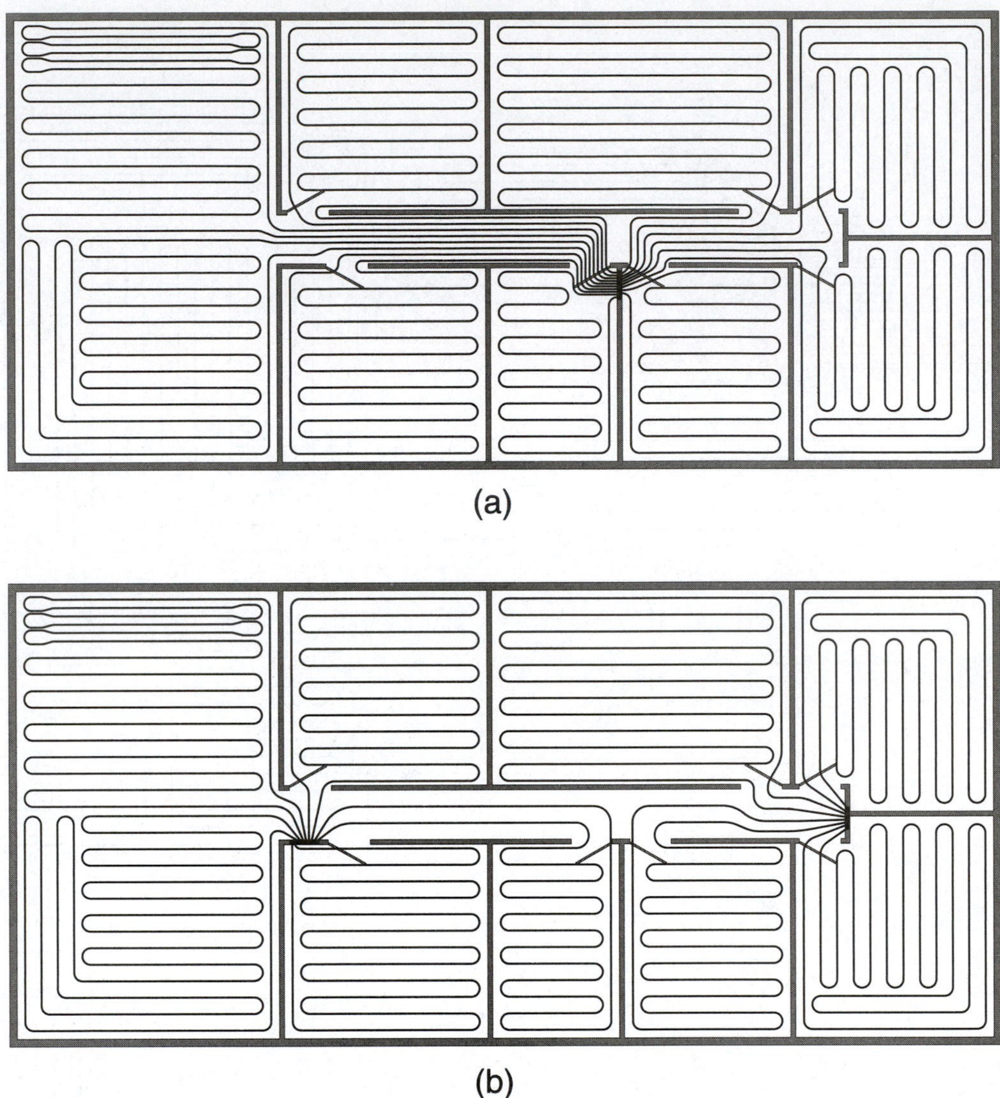

Figure 10–69 (a) Use of a single manifold station may result in long leader lengths. (b) Use of two manifold stations reduces leader lengths.

- **Principle #9:** *Avoid routing tubing where fasteners are likely to penetrate the floor.* This reduces the chance of the tubing being punctured after it is installed. Fastener penetrations are likely beneath interior partitions, under doors with thresholds, or where equipment will be secured to the floor. There is no way to guarantee that a fastener will not penetrate a tube at some point in the life of the building. An accurate layout plan showing the location of the tubing, along with avoiding high probability areas is the best defense.

One solution that has worked on numerous slab-on-grade projects is to coordinate with the framing contractor so that interior partitions over tubed areas are bonded to the slab using construction adhesive rather than mechanical fasteners. This allows simpler layouts and faster installation because the tubing can pass repeatedly beneath interior partitions. *It is vitally important that all trades involved in the project are informed about not penetrating the slab at certain locations.*

Figure 10–70 is a comparison of two different tubing arrangements for a group of rooms with a single exposed wall. In Figure 10-70a, the tubing has been arranged so only one passage beneath the interior partition separating rooms is required. Assuming the location of the crossing is avoided, fasteners could be driven as required to anchor the partition to the slab. However, because the circuit passes from one room to the next in a series, the heat output of the floor in the first room will be the greatest, while that in the last room will be the least. Since the rooms may have nearly identical loads, this could overheat the first room and underheat the last room. This piping arrangement also requires more bends and corners and thus will take more time to install.

Figure 10–70b shows the tubing circuits repeatedly crossing under the partitions that have been secured to the slab with construction adhesive. This arrangement provides a more equitable distribution of heat to each room. It also can be installed faster because of longer straight runs and fewer bends and corners.

Circuit Layout Drawings

An important part of planning a radiant panel system is producing a circuit layout drawing. This drawing shows the routing and flow direction of all tubing circuits as well as the manifold locations. It can save significant time during installation, and resolves a number of issues that otherwise could lead to expensive installation errors. After the project is complete, this circuit layout drawing also provides permanent documentation of the installation. This can help avoid damage in situations where the building undergoes future renovations. It is the author's opinion that a tubing layout drawing should be prepared for every radiant panel installation.

This section describes circuit layout drawings for radiant floor heating systems; however, the same techniques can usually be applied to radiant wall and ceilings. It also assumes the drawings will be produced using **computer-aided drafting (CAD)** software. Several companies now offer relatively inexpensive CAD software for both Windows® and MacOS® computers that is very capable of producing these drawings. Although hand drawn circuit layouts are still possible, they are very time consuming compared to CAD produced drawings, and thus seldom practical.

Circuit layout drawings require an accurate floor plan of the building. Such plans are usually available from the owner or building designer. If the floor plan was prepared using CAD software, it may be possible to obtain a copy of the drawing file from the designer. Such files are now routinely distributed as e-mail attachments. They are usually available in one of several file formats such as HPGL, DXF, DWG, or EPS. These and other file formats can be opened

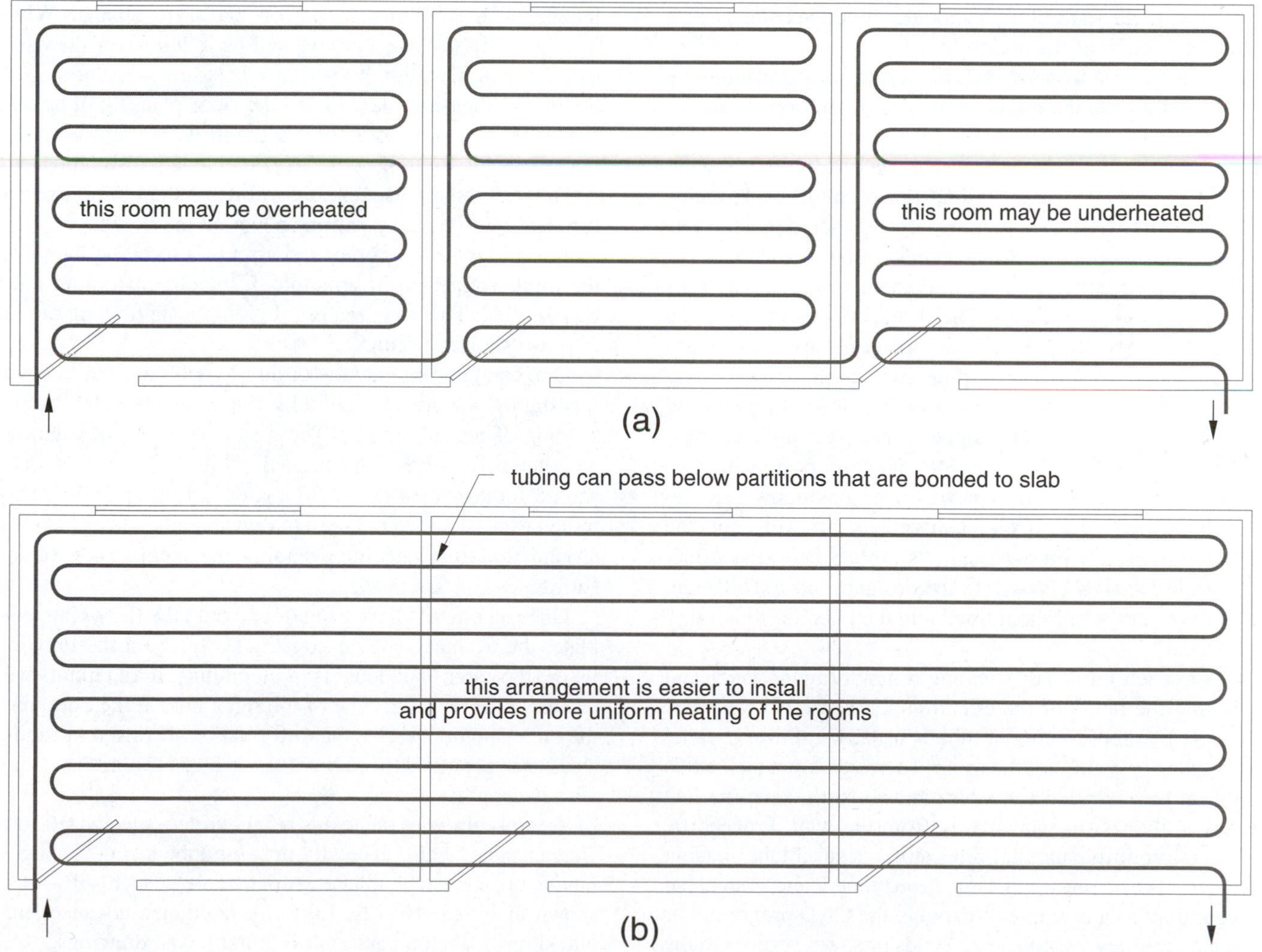

Figure 10–70 Comparison of tubing layouts. (a) Pattern that minimizes routing under partitions to minimize potential puncture from fasteners, (b) preferred pattern allowed by securing partitions to slab with construction adhesive.

by many CAD programs, and used as a "template" upon which the tubing circuits can be drawn.

If the floor plan is not available as a CAD file, radiant heating designers need to recreate a simplified version of the floor plan using their own CAD system. This drawing must be to scale and accurately place all walls, partitions, cabinets, and doorways. It should also show plumbing fixtures such as baths, shower stalls, and toilets. It is usually not necessary to draw in windows or other architectural features that do not directly affect circuit layout. The most common scale for such drawings is 1/4 inch = 1 foot.

Before the circuits can be drawn, it is necessary to determine the required tube spacing for each circuit. This can be done using methods presented later in this chapter, or through use of manufacturer-supplied radiant heating sizing software.

Start the layout drawing by deciding where the manifold station(s) will be located (see Principle #7). The manifold locations serve to anchor the floor circuits. Their location should minimize the distance to the various room circuits, as well as provide easy access when needed. Place a black rectangle on the plan where each manifold station will be located. The scaled length of the rectangle should accurately represent the length of the manifold.

It is preferable to draw each tubing circuit using a unique color. This allows the circuit to be easily seen on the monitor, as well as in color printouts. In many CAD systems, it is also preferable to draw each circuit on a different drawing layer. This makes it easy to isolate each circuit when measuring its length. Assign a unique name to each circuit and its corresponding drawing layer for quick reference.

Drawing circuits involves simultaneous application of the principles discussed earlier. The designer tries to apply the principles "on the fly." As the circuit takes shape, the designer may see a problem developing such as the circuit being too long. In this case, the CAD software allows for quick and easy modifications. If the designer is not sure of the change, most CAD systems allow them to copy and paste the existing drawing elements to a new drawing window prior to making any changes. If the change does not work out, the new drawing window can simply be discarded. This ability makes it possible for designers to try different circuit routings in minimal time, and then decide which layout is best.

Begin each circuit by creating a new drawing layer and assigning the name of the circuit as the name of that layer. The first line representing a tube is usually a drawn parallel to the exterior wall(s) of the room. To avoid carpet tack strips or future holes drilled for cables and so forth, keep the line at least 4 inches (to scale) away from the wall. Proceed by laying out multiple parallel lines at the desired tube spacing distance. These lines can later be trimmed and connected using a fillet tool or semicircle tool in the CAD program. For the best accuracy, avoid square bends or sharp vertices in the tubing pattern. Use a radius drawing tool to make all such changes in direction.

Different CAD systems offer different drawing tools that will affect the manner in which the circuit is drawn. With experience, the designer will learn to optimize the use of these drawing tools to quickly generate accurate drawings. An example of a drawing sequence in which a grid of parallel lines is systematically trimmed and filleted into a final circuit path is shown in Figure 10–71. An exploded view of the completed circuit is shown in Figure 10–72a. This illustrates how the overall circuit can be viewed as simply a collection of lines and arcs. The total length of these lines and arcs is the total circuit length (on the floor plane).

When the circuit has been completed, its length can usually be measured by the CAD software. The exact procedure varies from one CAD system to another. Some require the collection of line segments and arcs that make up the circuit to be joined into a "polyline." Once this is done, the length is instantly calculated by the software. Because each circuit, and only that circuit, is drawn on a separate drawing layer, the circuit's length is simply the sum of the lengths of all line segments and arc on that layer.

Keep in mind that the CAD software only measures the length of the tubing drawn on the floor plane. When installed, slightly more tubing will be required for the risers to the manifold station. Experience has shown that adding 10 feet to the measured length (on the floor plane) will be sufficient for the riser to a typical manifold station located 2 feet above the floor. Part of this added length also represents a safety factor in case some portions of the circuit are not installed in the exact position shown on the drawing.

If the length of a completed circuit is too long based on the limits suggested in principle 3, the circuit routing must be modified. This may require dividing a single long circuit into two shorter circuits.

One can approximate the circuit length required to cover a certain floor area by dividing the floor area (ft^2) by the spacing of the tubing (ft). The results of these calculations are shown for several tube spacings in Figure 10–66. The square footage of a given floor area can be quickly determined using the CAD system to draw a rectangle or polygon around the area, and then reading the area as one of the "attributes" of that shape.

Once all circuits have been drawn onto the floor plan template, the designer should construct a table on the drawing listing the names and lengths of all circuits. If all circuits will use the same size and type of tubing, a note at the bottom of the table should specify this information. If different tubing sizes are used, be sure to note the size of each circuit along with its length.

As this edition of the text is being written, specialized software capable of automatically drawing tube layouts is beginning to appear on the market. An example of such software is shown in Figure 10–72b. Look for continued advances that will simplify the preparation of tubing layout drawings.

Finally, radiant heating professionals should recognize that tubing layout drawings constitute a good marketing tool.

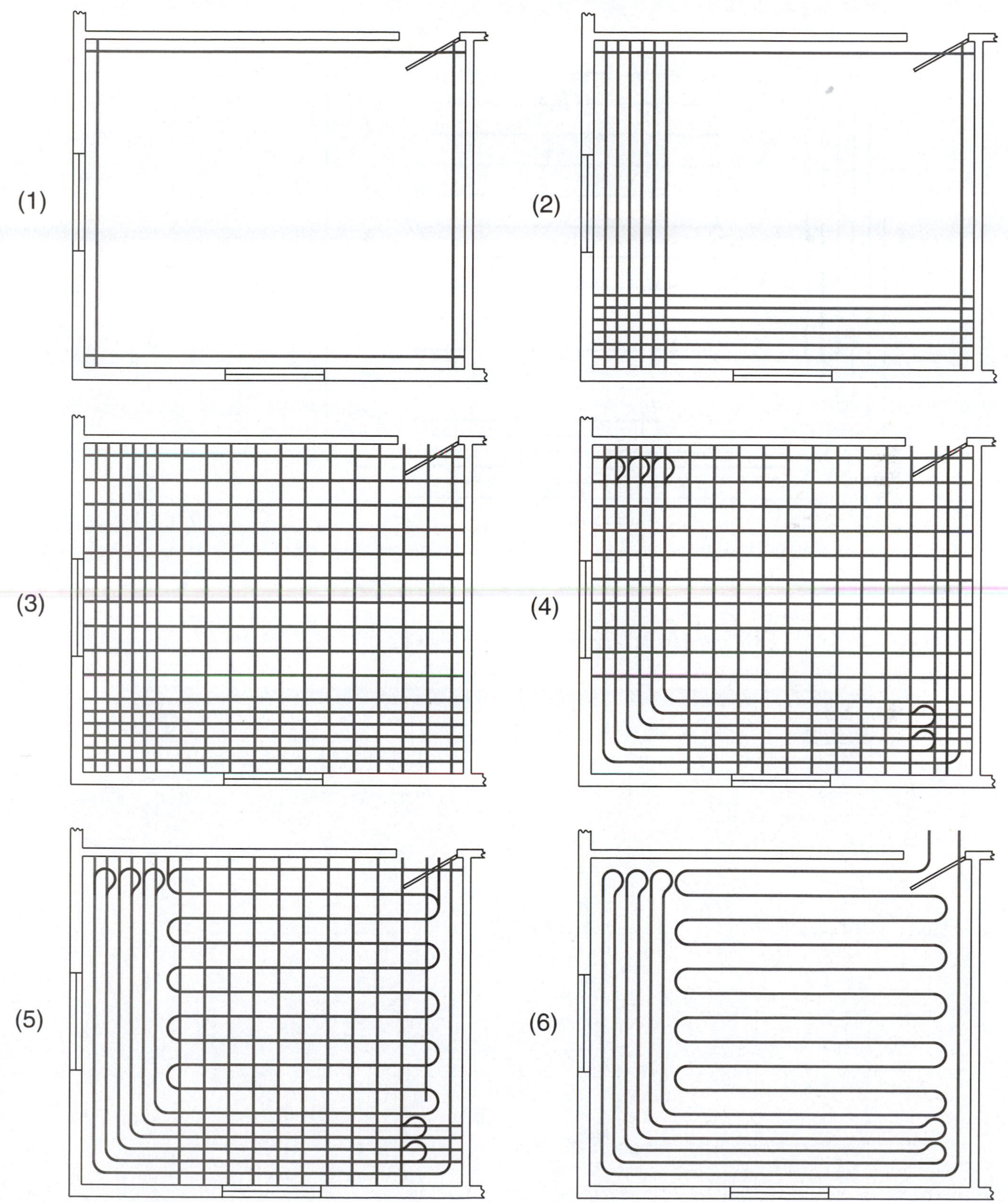

Figure 10–71 One possible approach to drawing a radiant floor circuit using a CAD program.

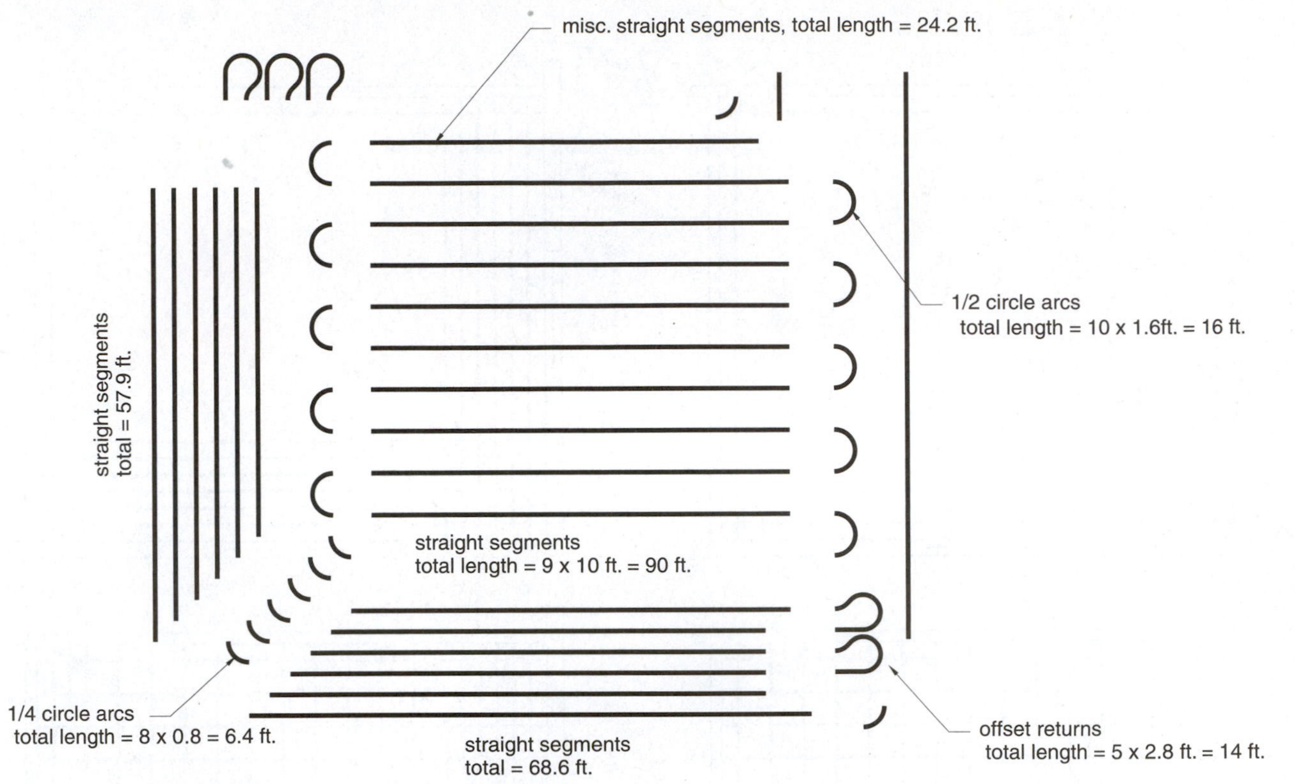

Figure 10–72a How a radiant panel circuit can be thought of for length measuring purposes.

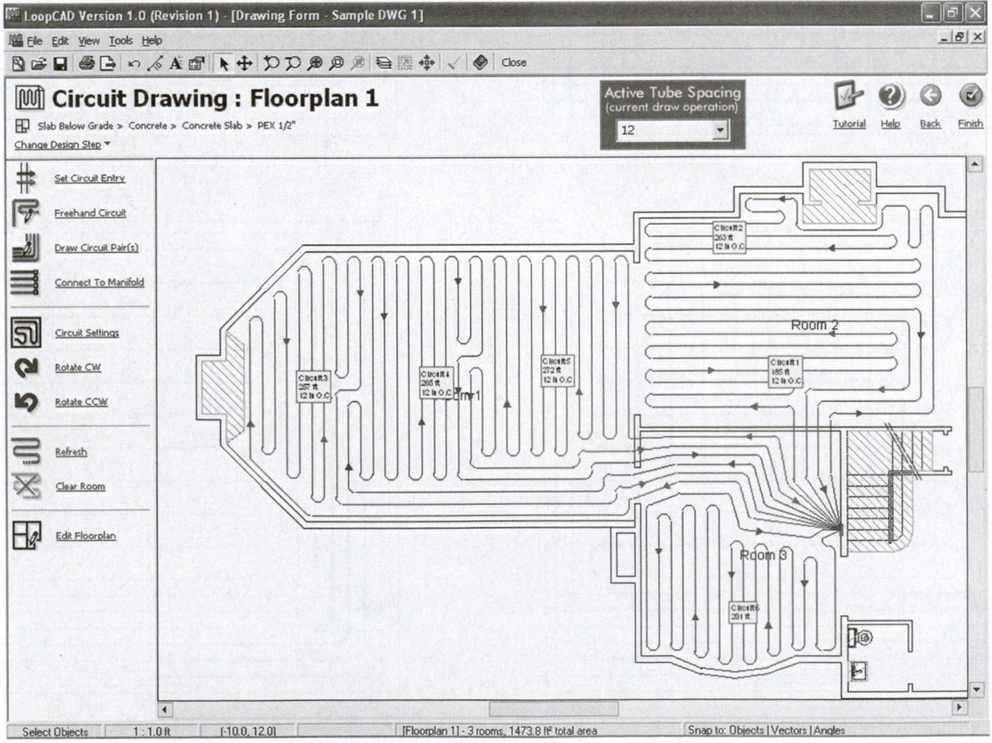

Figure 10–72b Software for automating cirucit layout drawings. *Courtesy of Avenir Software, Inc.*

They demonstrate to the owner that the designer is taking the time to "customize" the layout to the exact needs of the building. They also demonstrate that the system is being accurately documented for future reference. Some radiant heating professionals even frame the circuit layout drawing and mount it in the mechanical room along with the company contact information. Professional workmanship along with a well-documented design usually leads to future referrals.

10.19 RADIANT PANEL CIRCUIT SIZING PROCEDURE

Because hydronic radiant panels are an integral part of a building rather than a manufactured heat emitter assembly, their thermal performance is affected by many factors including:

- Construction of the panel (e.g., slab-on-grade, thin-slab, tube and plate system, hybrid)
- Type of floor covering used (for floor panels)
- Type and amount of the backside insulation
- Size, spacing, and depth of the tubing
- Water temperature and flow rate in the panel circuits
- Surface and air temperatures in the room being heated

The simultaneous interaction of all these factors is difficult (sometimes impossible) to theoretically model. Another complication is that every site-built radiant panel is constructed somewhat differently. For example, if the tubing depth varies in two otherwise identical slab-on-grade systems, so does their performance. The possibilities for building hybrid radiant panels using different materials, dimensions, and installation techniques is virtually unlimited. Thus, a general performance model for every imaginable type of hydronic radiant panel would be very complex and of little use in routine system design.

This section takes a more practical approach. After describing some fundamental concepts, it presents a step-by-step method for sizing the radiant panel circuits for a given building. The selection of available panels includes slab-on-grade, thin-slab, and tube and plate floor systems. The method allows the designer to determine factors such as tube spacing and supply water temperature for each panel circuit based on the heating output it must deliver. It also identifies situations where panel output will not be sufficient to handle the load, and it shows how to determine the amount of supplemental heating needed.

The procedure is presented in a step-by-step manner to demonstrate a logical approach to sizing. Although the required calculations are relatively simple, they can be time consuming for larger projects. As with other design procedures discussed in this text, these procedures and equations could be implemented using software.

Concept of Heat Flux

The useful heat output of a radiant panel is expressed as the rate of heat flow into the room *per unit of area of radiant panel surface.* This is called the **room side heat flux.** In the United States, the common units for room side heat flux are Btu/hr/ft².

The required room side heat flux at design load conditions can be calculated using Equation 10.2:

(Equation 10.2)

$$q' = \frac{Q_R}{A_p}$$

where:

q' = room side heat flux under design load conditions (Btu/hr/ft²)

Q_R = design heat load of room (excluding underside losses of heated floor) (Btu/hr)

A_p = area of radiant panel (ft²)

Example 10.2: The design heat load of a room (excluding the underside losses of the heated floor) is 4400 Btu/hr. The floor in the room is 10 feet wide, 20 feet long, and contains tubing throughout. Determine the room side heat flux requirement of the radiant floor.

Solution: The heated floor area of the room is $10 \times 20 =$ 200 square feet. Substituting numbers into Equation 10.2 and solving yields:

$$q' = \frac{Q_R}{A_p} = \frac{4400\text{ Btu/hr}}{200\text{ ft}^2} = 22\text{ Btu/hr/ft}^2$$

Discussion: If each square foot of floor can deliver 22 Btu/hr to the room, the radiant floor is capable of providing *all* the heat input the room needs during design load conditions. Note that the underside heat loss of the floor was *not* included in the design heat loss (Q_R). This heat loss will be accounted for in another step. Unless otherwise stated, the room side heat flux requirement of a radiant panel is always calculated based on design load conditions. During milder weather, the rate of heat output of the radiant panel decreases (as it would for any heat emitter). This also reduces the room side heat flux.

Available Floor Area

In the case of floor heating, the entire floor area of some rooms may not be available to serve as the radiant panel. Fixed objects such as cabinets or kitchen islands may cover a portion of the floor area. Such items block most upward heat output if placed over heated floors. There is no point in installing tubing under such objects. Premature spoilage has been known to occur in situations where food was stored in cabinets over heated floors.

Heat output can also be blocked by movable objects like padded furniture with floor-length skirting, thick area rugs, or cardboard boxes on the floor. It is crucial to discuss the possibility of such objects with clients prior to designing the system. Ignoring that such objects might be present can lead to serious deficiencies in heat output from otherwise properly designed systems.

The term **available floor area** describes the area of the floor that is heated, and *not* covered by objects capable of blocking upward heat flow. Thus, in the case of floor heating, Equation 10.2 could be restated as follows:

(Equation 10.3)

$$q' = \frac{Q_R}{A_{available}}$$

where:

q' = required upward heat flux at design load (Btu/hr/ft²)

Q_R = design heat loss of room (excluding underside losses of floor) (Btu/hr)

$A_{available}$ = floor area available for heat output (ft²)

Because objects such as furniture and small area rugs are likely to be moved, it still makes sense to install tubing under floor areas where they are likely to be placed.

When calculating available floor area, a conservative approach is to subtract the estimated area covered by thick area rugs, padded furniture with floor-length skirting, or commercial inventory stored directly on the floor from the room's gross floor area. Since this decreases the available floor area, the required upward heat flux increases. In rooms where the value exceeds 35 Btu/hr/ft² the floor panel may not be capable of providing the full heating load of the room. This is due to limitations in floor surface temperature that will now be discussed.

Radiant Panel Surface Temperature Limitations

The room side heat flux delivered by a radiant panel can be limited by physiological or material considerations. In the case of floor heating, *the average surface temperature of the floor should not exceed 85 °F in rooms where prolonged foot contact with the floor is likely.* In areas such as bathrooms, foyers, entries, or the outer perimeter of rooms, the average surface temperature of the floor should not exceed 92 °F. These temperatures are based on avoiding uncomfortably hot feet.

For gypsum-covered wall or ceiling panels, the surface temperature of the panel should not exceed 120 °F. This temperature is based on avoiding long-term thermal degradation of the gypsum board. Modular radiant panels made from steel or aluminum can be operated with surface temperatures up to 185 °F.

Surface temperatures limits produce corresponding limits in heat output. For floor heating, a mean surface temperature of 85 °F will release approximately 35 Btu/hr/ft² into a room maintained at 68 °F. Increasing the mean surface temperature to 92 °F raises heat output to approximately 48 Btu/hr/ft² into a room at 68 °F. A radiant wall with a mean surface temperature of 120 °F releases about 93 Btu/hr/ft² into a room at 68 °F. A radiant ceiling with a mean surface temperature of 120 °F releases about 83 Btu/hr/ft² in a 68 °F room.

Example 10.3: The 14-foot by 16-foot room shown in Figure 10–73 contains a built-in 2-foot by 12-foot cabinet. A sofa with floor-length skirting also covers a 3 by 6-foot area of the floor. The room's design heat loss (excluding downward heat loss of floor) is 4000 Btu/hr. Assuming the room will be entirely heated by the floor, what is the room side heat flux requirement?

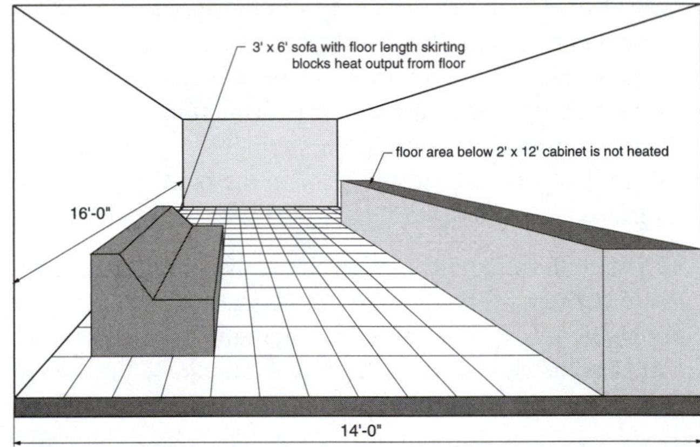

Figure 10–73 Available floor area is floor area that is heated and not blocked by furniture or built in cabinets.

Solution: The available floor area is: 14 × 16 – 2 × 12 – 3 × 6 = 182 sq. ft. Combining this with the design heating load through Equation 10.3 yields:

$$q' = \frac{Q_R}{A_{available}} = \frac{4000 \text{ Btu/hr}}{182 \text{ ft}^2} = 22.0 \text{ Btu/hr/ft}^2$$

Discussion: This required upward heat flux is well under 35 Btu/hr/ft², and therefore the floor will operate well below the maximum surface temperature of 85 °F, even at design load conditions.

Flooring Covering Considerations

Most people would understand that draping a thick wool blanket over a cast-iron radiator would greatly reduce its heat output. However, the same people may not understand that laying a thick carpet over a heated floor will have the same effect. They simply are not accustomed to thinking of the floor as the radiator.

Floor coverings have a major effect on the thermal performance of a radiant floor heating system. This effect must be accounted for when the system is designed. In some cases, the owner's desire to use certain floor coverings may even eliminate radiant floor heating as an option.

From the standpoint of heating, the ideal floor covering is no covering. Bare concrete slab floors are ideal situations for floor heating. Painted, stained, or patterned concrete surfaces all perform essentially the same as a bare slab.

When floor coverings are present, they affect heat output in two ways. First, floor coverings with low R-values allow greater upward heat flow from a given radiant floor operating at a given average water temperature. Viewed another way, *the greater the R-values of the floor covering(s), the higher the average water temperature in the tubing circuits must be to create the desired rate of upward heat flow.*

Secondly, the R-value of floor coverings affects the variation in surface temperature between the tubes. Higher R-value coverings will reduce the variation between the highest floor surface temperature directly above the tube and the lowest surface temperature halfway between the tubes. This effect is shown in Figure 10–74.

It should be noted that although the higher R-value carpeted floor covering creates less variation in surface temperature between the tubes, it also greatly reduces the heat output from the floor. The areas under the surface temperature profile curves in Figure 10–74 are proportional to heat output. Notice that the area under the curve for the ceramic tile floor is considerably greater than the area under the curve for the carpet-covered floor.

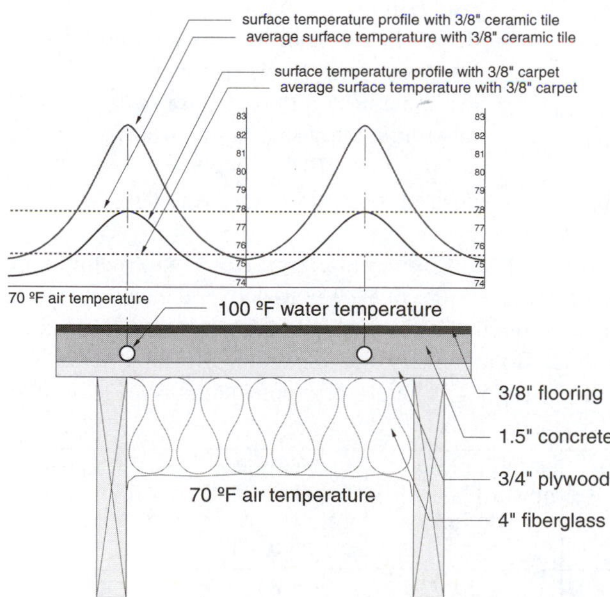

Figure 10–74 Surface temperature profiles, and average floor surface temperature for thin-slab radiant panel covered with 3/8" ceramic tile, or, 3/8" carpet. More heat is released at the same water temperature when the flooring has lower R-value.

A table listing some R-values of floor coverings (and covering systems) is given in Figure 10–75.

Floor covering description	Total R-value of floor covering, or floor covering group (°F hr ft²/Btu)
Bare slab	0
Asphalt or rubber tile	0.05
1/8-inch sheet vinyl	0.20
3/8-inch ceramic tile (thin set)	0.30
1/4-inch carpet (nylon) (glue down)	0.95
1/4-inch carpet over 1/4-inch rubber pad	1.44
3/4-inch marble	0.42
3/8-inch hardwood	0.52
1/2-inch hardwood	0.70
3/4-inch hardwood	1.05

Figure 10–75 Thermal resistance of various floor coverings.

Step-by-Step Circuit Sizing Procedure

The following procedure can be used to determine the tube size, tube spacing, and supply water temperature for several types of radiant floor constructions. The circuits are designed on a room-by-room basis. Examples of the calculations are given in each step.

The calculations in this procedure can be kept organized using a worksheet. That sheet will contain many columns. Examples of parts of such a worksheet will be shown at various stages in the procedure. The full worksheet will be presented at the end of the procedure.

Step 1: Determine the design heating load of each room.
The first step in sizing any heating system is to determine the design heating load of each room in the building. Methods and data for doing so are available in Chapter 2, Heating Load Estimates.

Note: In this procedure, the design heating load of the room should not include the heat loss from the underside of the heated floor. This loss will be calculated separately in a later step. The symbol Q_R will be used to designate the design heat loss of a room excluding downward heat loss from the heated floor.

Step 2: Determine the available floor area of each room.
Available floor area is floor area that is heated and is not blocked from releasing heat into the room. *It does* not

include floor area covered by built-in cabinets, furniture with floor-length skirting, thick area rugs, or boxes stored on the floor. The symbol $A_{available}$ will be used to designate the available floor area of each room.

Step 3: Determine the required upward heat flux for each room.
This is done using Equation 10.3:

$$q' = \frac{Q_R}{A_{available}}$$

where:

q' = required upward heat flux under design load conditions (Btu/hr/ft²)

Q_R = design heat loss of room (excluding underside losses) (Btu/hr)

$A_{available}$ = floor area available for heat output (ft²)

Step 4: Establish a maximum allowable floor surface temperature for each room.

A maximum floor surface temperature of 85 °F is recommended for rooms with prolonged foot contact. In bathrooms, hallways, and entry foyers, this temperature may be increased as high as 92 °F.

Step 5: Determine the average floor surface temperature needed to deliver the required upward heat flux to each room.
This temperature can be calculated using Equation 10.4.

(Equation 10.4)

$$T_S = T_{Room} + 0.5(q')$$

where:

T_s = required average floor surface temperature (°F)

T_{room} = desired room air temperature (°F)

q' = required upward heat flux for the room (Btu/hr/ft²)

Example 10.4: A room heated by a radiant floor is to be maintained at 67 °F at design load conditions. The required upward heat flux has been calculated as 29 Btu/hr/ft². What average floor surface temperature is required to provide this heat input?

Solution: Substituting the numbers into Equation 10.4 and solving yields:

$$T_S = T_{Room} + 0.5(q') = 67 + 0.5(29) = 81.5°F$$

Discussion: This temperature is the *average* floor surface temperature. Areas of the floor closer to the embedded tubing will have slightly higher surface temperature. Likewise, areas several inches away from a tube will have slightly lower surface temperatures. This is shown in the surface temperature profiles of Figure 10–74.

A sample of what the first seven columns of a worksheet to organize the preceding calculations might look like, along with some sample data, is shown in Figure 10–76.

Step 6: Determine if the room requires supplemental heat.

If the required floor surface temperature in step 5 is greater than the maximum allowable surface temperature from step 4, then supplemental heat is required for that room. Identify all such rooms.

The amount of supplemental heat required for a given room can be calculated using Equation 10.5:

(Equation 10.5)

$$Q_{sup} = Q_R - A_{available}\left(\frac{T_{floor\ max} - T_{Room}}{0.5}\right)$$

where:

Q_{sup} = required supplemental heat input at design load (Btu/hr)

Q_R = design heat loss of room (excluding downward loss from floor) (Btu/hr)

$A_{available}$ = available heated floor area of room (ft²)

$T_{floor\ max}$ = maximum allowable floor surface temperature (°F)

T_{Room} = desired room air temperature (°F)

Example 10.5: Room 3 in Figure 10–76 will require a floor surface temperature of 90.9 °F to deliver the design load heat input through the floor. Assuming prolonged foot contact with the floor is likely, the maximum allowable surface temperature is 85 °F. Therefore, supplemental heating is needed.

Room name	Design heat loss (Btu/hr)	Available floor area (ft²)	Room temp. (°F)	Required upward heat flux (Btu/hr/ft²)	Maximum allowable floor surface temp. (°F)	Required floor surface temp. (°F)
Room 1	4000	200	70	20	85	80
Room 2	8000	350	70	22.9	85	81.5
Room 3	5000	120	70	41.7	85	90.9
Room 4	3000	120	70	25	85	82.5

Figure 10–76 Using a table to keep calculations organized.

The amount needed is calculated using Equation 10.5 assuming the desired room temperature is 70 °F.

$$Q_{sup} = Q_R - A_{available}\left(\frac{T_{floor\ max} - T_{Room}}{0.5}\right)$$

$$= 5000 - 120\left(\frac{85 - 70}{0.5}\right) = 1400 \text{ Btu/hr}$$

Discussion: In this room, the radiant floor panel provides 3600 Btu/hr of the 5000 Btu/hr total design load. The remainder is provided by supplemental heat. The latter could be delivered through another hydronic heat emitter supplied by the same heat source. It might also be supplied from a totally separate heating device.

For rooms requiring supplemental heat, the **maximum allowed upward heat flux** from the floor should be recalculated using Equation 10.5a:

(Equation 10.5a)

$$q'_{limit} = \frac{(Q_R - Q_{sup})}{A_{available}}$$

where:

q'_{limit} = maximum allowed upward heat flux (Btu/hr/ft²)

Q_{sup} = required supplemental heat input to room at design load (Btu/hr)

Q_R = design heat loss of room (Btu/hr)

$A_{available}$ = available heated floor area of room (Ft²)

NOTE: The value of q'_{limit} rather than q' should be used in all subsequent calculations for the room (steps 7 and on) that reference q'.

Example 10.6: Room 3 in Figure 10–76 required 1400 Btu/hr of supplemental heat. What is the limiting value of upward heat flux?

Solution: Substituting values into Equation 10.5 and solving yields:

$$q'_{limit} = \frac{(Q_R - Q_{sup})}{A_{available}} = \frac{(5000 - 1400)}{120} = 30 \text{ Btu/hr/ft}^2$$

Discussion: The limiting heat flux of 30 Btu/hr/ft² is based on not exceeding an average floor surface temperature of 85 °F. Again, use q'_{limit} rather than q' in all subsequent calculations for the room that reference q'.

Step 7: Determine the tube spacing based on the required upward heat flux requirement of each room.

Suggested tube spacings are listed in the table in Figure 10–77. These spacings are conservative; they limit variations in floor surface temperature to acceptable ranges for residential occupancy. Spacing for industrial floors, where greater variations of floor surface temperature are acceptable, could be 3 inches wider. In some industrial applications, it is even possible to use tube spacing of 24 inches.

Required upward heat flux (Btu/hr/ft²)	Spacing if 1/2" tubing is used (inches)	Spacing if 5/8" tubing is used (inches)	Spacing if 3/4" tubing is used (inches)
<10	15	18	18
10-20	12	12	15
20-30	9	9	12
30-40+	6	6	9

Figure 10–77 Suggested tube spacings for radiant floor panel circuits.

Step 8: Estimate the total length of tubing for each room.

The length of tubing required in each room can be estimated using Equation 10.6.

(Equation 10.6)

$$L = \frac{12(A_{available})}{S}$$

where:

$A_{available}$ = available heated floor area of room (ft²)

S = tube spacing (inches)

Example 10.7: A room has an available floor area of 260 ft². The tube spacing selected is 9 inches. Estimate the amount of tubing needed in the room.

Solution: Substituting the numbers into Equation 10.6 and solving yields:

$$L = \frac{12(A_{available})}{S} = \frac{12(260)}{9} = 347 \text{ feet}$$

Discussion: Equation 10.6 proportions tubing to floor area assuming straight runs of tubing "wall to wall." It does not precisely account for the length of return bends or risers to wall-mounted manifold stations. The next step will add length to the circuit to account for leaders from the tubing circuit on the floor of the room to the manifold station.

Step 9: Estimate the leader length required for the circuit to reach the manifold station.

The leader length of a circuit is the additional tubing needed to connect the floor circuit to a manifold station as shown in Figure 10–78.

Estimating the leader length requires the designer to know where the manifold station the circuit connects to is located. Based on this location, the designer estimates the

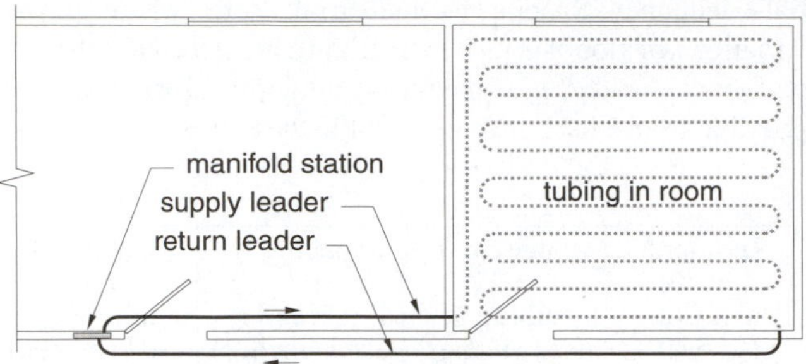

Figure 10–78 Concept of the leader of a radiant floor circuit.

additional tubing needed and *adds it to the length of tubing for the room. Note: Additional circuit leaders may be needed based on the results of step 10.*

Step 10: Determine if multiple tube circuits are needed in the room.

In many rooms, a single tube circuit will suffice. However, in larger rooms with close tube spacing, the length of a single circuit (including its leader length from step 9) may be excessive. The length of a given circuit, including its leaders, should be limited to the values given in Figure 10–65.

If the total circuit length, including leaders, determined in steps 8 and 9 exceeds these maximum circuit length values, divide the total tubing length by the maximum circuit length value and round *up* to the next whole number. This determines the number of tube circuits needed in the room.

Divide the total "in-room" tubing length by the number of circuits, then add an estimated leader length to each circuit. This will be the estimated length of the circuits for that room. In the unlikely but possible event that these circuit lengths still exceed the maximum circuit length allowed for a given tube size, divide the total in-room tubing into one additional circuit and add estimated leader lengths. Be sure to record all circuit lengths, including estimated leader lengths. These lengths will be needed when estimating the head loss of the circuits.

Example 10.8: A room needs 280 feet of 1/2-inch tubing. The estimated leader length (supply + return) is 80 feet. Can this work as a single circuit?

Solution: The total length of a single circuit would be the 280 feet of in-room tubing, plus the 80 feet estimated leader length, or 360 feet. This is longer than the maximum recommended circuit length of 300 feet for 1/2-inch tubing. Therefore the in-room tubing should be divided into two equal circuits of 280/2 = 140 feet. Assume the estimated leader length *for each* is 75 feet. Now the total length of each circuit is 140 + 75 = 215 feet, which is acceptable.

Example 10.9: A room requires 850 feet of 1/2 inch tubing. How many tubing circuits are necessary, and what is the approximate length of each of these circuits? Assume the leader length of each circuit is 60 feet.

Solution: The recommended maximum circuit length for 1/2 inch tubing is 300 feet. Dividing this into 850 total feet yields 850/300 = 2.8. Rounding this up yields three circuits. Dividing 850 by 3 yields circuit lengths of 283 feet. Now assume the leader length for each circuit is 60 feet. The total estimated length of each circuit is now 283 + 60 = 343 feet, which is too long for a 1/2-inch tube. To create equal but shorter circuits, divide the in-room length into 4 rather than 3; that is, 850/4 = 213 feet. Assuming nominal 60-foot leader lengths for each circuit yields a total estimated circuit length of 213 + 60 = 273 feet, which is acceptable.

Discussion: This example demonstrates that long leader lengths can add significant tubing to the system that is "unproductive" since it is not installed into the radiant panel. Long leaders increase both material cost and circuit head loss, both of which are undesirable. A good designer always attempts to minimize leader lengths. In some cases, this requires the use of additional manifold stations. The designer should also attempt to keep the length of circuits installed in the same room reasonably close to each other.

Step 11: Calculate the average water temperature in the tube circuit(s) based on floor covering and tube spacing.

Use Equation 10.7 along with the data given in the table of Figure 10–76 to determine the average water temperature in the tube circuits.

(Equation 10.7)

$$\overline{T}_w = T_{Room} + \frac{(q')}{k}$$

where:

T_w(bar) = average water temperature required in tubing at design load (°F)

T_{Room} = desired room air temperature (°F)

q' = required upward heat flux (Btu/hr/ft²)

k = constant based on tube spacing and R-value of floor covering from the table in Figure 10–79.

Slab-on-grade floors	(k) For 6-inch tube spacing	(k) For 9-inch tube spacing	(k) For 12-inch tube spacing	(k) For 18-inch tube spacing	(k) For 24-inch tube spacing
Rf = 0	1.13	1.0	0.882	0.69	0.563
Rf = 0.5	0.71	0.67	0.62	0.52	0.436
Rf = 1.0	0.53	0.50	0.47	0.41	0.355
Rf = 1.5	0.42	0.40	0.38	0.34	0.30
Rf = 2.0	0.34	0.33	0.32	0.29	0.263

Figure 10–79 (a) k-values for slab-on-grade floor circuits.

Thin-slab floors	(k) For 6-inch tube spacing	(k) For 9-inch tube spacing	(k) For 12-inch tube spacing
Rf = 0	1.2	0.97	0.96
Rf = 0.5	0.74	0.66	0.65
Rf = 1.0	0.54	0.49	0.44
Rf = 1.5	0.43	0.40	0.36
Rf = 2.0	0.35	0.33	0.31

Figure 10–79 (b) k-values for thin-slab floor circuits.

Above-floor tube and plate*	(k) For 8-inch tube spacing	(k) For 12-inch tube spacing
Rf = 0.5	0.58	0.55
Rf = 1.0	0.48	0.44
Rf = 1.5	0.41	0.36
Rf = 2.0	0.33	0.30

* All values assume 6-inch wide aluminum plates installed at indicated tube spacing. The value of Rf must include the R-value of any cover sheet over plates, plus the R-value of any finish flooring materials.

Figure 10–79 (c) k-values for above floor tube and plate floor circuits.

Below-floor tube and plate**	(k) For 8-inch tube spacing	(k) For 12-inch tube spacing
Rf = 0	0.52	0.36
Rf = 0.5	0.42	0.32
Rf = 1.0	0.34	0.24
Rf = 1.5	0.29	0.21
Rf = 2.0	0.26	0.19

** All values assume 6-inch wide aluminum plates installed below 3/4-inch plywood subfloor at indicated tube spacing. Rf is the total R-value of floor coverings installed on top of 3/4-inch plywood.

Figure 10–79 (d) k-values for below floor tube and plate circuits.

The data used to estimate the average circuit water temperature assumes the presence of good underside insulation, specifically that the downward heat loss is limited to no more than 10 percent of the upward heat output. This will typically be the case when the insulation recommendations described in previous sections of this chapter are followed.

Example 10.10: A slab-on-grade floor panel with tubing spaced at 12-inch centers must create a required upward heat flux of 25 Btu/hr/ft². The floor coverings have a total R-value of 0.75. The room is to be maintained at 70 °F. Determine the average water temperature for the circuits:

Solution: The finish floor R-value of 0.75 is not listed in the table for slab-on-grade systems. Therefore one must interpolate between the value of k for R = 0.5 and R = 1.0. In this case, the interpolation is a simple average: (0.62 + 0.47)/2 = 0.545. Substituting into Equation 10.7 yields:

$$T_w = T_{Room} + \frac{(q')}{k} = 70 + \frac{(25)}{.545} = 116°F$$

Discussion: The temperature calculated is the *average* water temperature required in the circuit. This will eventually be converted to a supply temperature based on the selected temperature drop of the circuits. Using Equation 10.7 and the data in Figure 10–79, it is possible to estimate the average circuit water temperature for many heated floor systems having floor covering resistances between 0 and 2.0. Interpolate when necessary for various floor covering resistances and tube spacings.

It is emphasized that heat output from floors having floor covering resistances over 2.0 will be very marginal. The term "floor warming" is perhaps more appropriate than floor heating.

Step 12: Estimate the downward heat loss from the floor panel.

The method used depends on whether the floor is slab-on-grade or built into a suspended floor deck (i.e., thin-slab or tube and plate systems).

For slab-on-grade floors use Equation 10.8:

(Equation 10.8)

$$Q_d = mL[q'(Rf + 0.5) + T_{Room} - T_{outdoor}]$$

where:

Q_d = downward and edge heat loss of slab (Btu/hr)
m = a value based on the edge insulation used (see Figure 10–80)
L = length of exposed slab edge (ft)
q' = required upward heat flux at design load (Btu/hr/ft²)
Rf = R-value of floor covering(s) (°F hr ft²/Btu)
T_{room} = room temperature at design load (°F)
$T_{outside}$ = outdoor design temperature (°F)

R-value of edge insulation (°F hr ft² / Btu)	Value of m
0	0.83
5	0.39
7.5	0.30
10	0.23

Figure 10–80 m-values for different R-values of slab edge insulation.

Example 10.11: Determine the downward and edge loss from a heated slab on grade with 20 feet of exposed length, R-7.5 edge, and underside insulation. The slab releases 30 Btu/hr/ft² into a room maintained at 70 °F at design load. The floor covering has an R-value of 0.75. The outdoor design temperature is 0 °F.

Solution: Looking up the value of m in Figure 10–80 and substituting, along with the other data, into Equation 10.8 yields:

$$Q_d = mL\,[q'(Rf + 0.5) + T_{Room} - T_{outdoor}]$$

$$= 0.30(20)[30(0.75 + 0.5) + 70 - 0]$$

$$= 645 \text{ Btu/hr}$$

Discussion: Equation 10.8 is a special adaptation of Equation 2.7, and it accounts for both edge and downward heat loss from a heated slab.

For radiant panels built into suspended floor decks (i.e., for thin-slab and tube and plate systems) use Equation 10.9 to estimate downward heat loss.

(Equation 10.9)

$$Q_d = \frac{A_f}{R_d}(q'(Rf + 0.5) + T_{Room} - T_{under})$$

where:

Q_d = downward heat loss (Btu/hr)
A_f = heated floor area (ft²)
R_d = total R-value of insulation installed beneath the radiant panel (°F hr ft²/Btu)
q' = required upward heat flux at design load (Btu/hr/ft²)
Rf = R-value of floor covering(s) (°F hr ft²/Btu)
T_{room} = room temperature at design load (°F)
T_{under} = air temperature under the heat floor (°F)

Example 10.12: A 200-square-foot thin-slab system delivers an upward heat flux of 25 Btu/hr/ft² to a room maintained at 70 °F. The floor covering has a resistance of R = 0.5. The underside of the floor deck is insulated to R-19. The temperature below the deck is 60 °F. Estimate the downward heat loss from the floor.

Solution: Substituting the data into Equation 10.9 and calculating yields:

$$Q_d = \frac{A_f}{R_d}[q'(Rf + 0.5) + T_{Room} - T_{under}]$$

$$= \frac{200}{19}[25(0.5 + 0.5) + 70 - 60] = 368 \text{ Btu/hr}$$

Discussion: Equation 10.9 is a special adaptation of the conduction heat loss equation from Chapter 2, Heating Load Estimates.

Step 13: Determine the required flow rate through each circuit.

This step requires the designer to select a temperature drop for the circuits. For residential and light commercial systems, the temperature drop should be between 10 and 20 °F. For larger commercial or industrial systems, it can be from 20 to 30 °F. The higher the temperature drop, the lower the circuit flow rate; however, excessively high temperature drops can create noticeable variations in floor surface temperature from one side of the room to the other.

Once the temperature drop is selected, calculate the flow rate of each circuit using Equation 10.10.

(Equation 10.10)

$$f = \frac{(Q_R + Q_d)}{(N)(b)(\Delta T)}$$

where:

f = circuit flow rate (gpm)
Q_R = design heat loss of room excluding downward loss) from step 1 (Btu/hr)
Q_d = downward heat loss of radiant panel from step 11 (Btu/hr)
N = number of circuits in the room
ΔT = selected temperature drop of circuits (°F)
b = fluid capacitance factor (for water b = 495, for 30 percent glycol b = 479, for 50 percent glycol b = 450)

Example 10.13: A large room has a design heat loss, excluding downward loss, of 20,000 Btu/hr. The downward heat loss from the edge and bottom of the floor slab is estimated at 1800 Btu/hr using methods from step 12. Because of the size of the room, it is divided into four circuits. The designer has selected a circuit temperature drop of 20 °F. The system operates with water in the tubing. Estimate the flow rate in each circuit.

Solution: Substituting the data into Equation 10.10 yields:

$$f = \frac{(Q_R + Q_d)}{(N)(b)(\Delta T)} = \frac{(20{,}000 + 1800)}{(4)(495)(20)} = 0.55 \text{ gpm/circuit}$$

Discussion: Equation 10.10 is a special adaptation of the sensible heat rate equation discussed in Chapter 4, Properties of Water. The calculated flow rate provides for the upward heat output of the floor, as well as the rear side heat loss, without exceeding the specified circuit temperature drop.

Step 14: Estimate the *supply* water temperature for each circuit.

This is found by adding one-half the circuit temperature drop to the average circuit water temperature determined in step 11. It is expressed as Equation 10.11.

(Equation 10.11)

$$T_{supply} = T_w - \frac{\Delta T}{2}$$

For example, if a circuit has an average water temperature of 105 °F and a temperature drop of 20 °F was selected, the supply temperature would be 105 + (20/2) = 115 °F

Step 15: Calculate the average of the supply temperatures of all circuits.

Add up all the supply temperatures of all circuits and divide by the number of circuits.

Check to see if *all* the supply temperatures fall within +/– 5 °F of the average. If this is the case, then all circuits can be generally handled by a system that supplies a single water temperature. The suggested value of the **system supply temperature** is the average supply temperature plus 2.5 °F.

Individual circuits that have supply temperature requirements slightly lower than the system supply temperature may require a slight reduction in flow rate to reduce output. Circuits with supply temperature requirements no more than 2.5 °F above the system supply temperature are generally able to provide sufficient output due to safety factors inherent in the design procedure.

The designer should identify any circuits with supply temperatures more than 5 °F above or below the average circuit supply temperature. Those circuits with higher supply temperature requirements can often be modified by returning to step 11 and using a closer tube spacing to reduce the supply temperature requirement. Likewise, circuits with supply temperatures more than 5 °F below the average may be candidates for wider tube spacing.

In buildings with similar types of radiant panel construction throughout, it is preferable to adjust tube spacing, when necessary, so all circuits fall within the +/– 5 °F range of the average supply temperature. This allows the system to be designed for a single supply temperature.

When the building uses two or more types of radiant panels with widely different supply temperature requirements, it may not be practical to attempt to configure circuits to a single water temperature. Instead, the system can be designed to supply two or more temperatures. Piping and control techniques for this are discussed in the next section.

Figure 10–81 is a worksheet that can be used to organize the calculations from this procedure for a project with up to twenty rooms. A computer spreadsheet could also be developed based on this worksheet along with the equations and data given in this section.

10.20 SYSTEM PIPING AND TEMPERATURE CONTROL OPTIONS

There are several options for connecting hydronic heat sources to radiant panel distribution systems. The method used depends on the water temperature requirement of the

	A	B	C	D	E	F	G	H	I	J	K	L	M	N	O	P	Q	R	S
	room name	design heat loss (Btu/hr)	available floor area (sq. ft.)	room air temp. (°F)	required upward heat flux (Btu/hr/sq. ft.)	max. allowable floor surface temperature (°F)	required floor surface temperature (°F)	supplemental heat ? (Y/N)	amount of supplemental heat needed (Btu/hr)	maximum allowable upward heat flux (Btu/hr/sq. ft.)	tube spacing selected (inches)	length of tubing in room (feet)	leader length for circuit (feet)	total circuit length (feet)	number of circuits in room	average circuit water temp. (°F)	downward heat loss of panel (Btu/hr)	circuit flow rate (gpm)	circuit supply temp. (°F)
1																			
2																			
3																			
4																			
5																			
6																			
7																			
8																			
9																			
10																			
11																			
12																			
13																			
14																			
15																			
16																			
17																			
18																			
19																			
20																			

Figure 10–81 Suggested spreadsheet table for organizing radiant sizing calculations.

radiant panel(s) as well as any other heat emitters used in the system. It also depends on the operating temperature and flow requirements of the heat source. The methods covered in this section include:

- Direct piping (single temperature system)
- Use of mixing devices
- Multiple temperature systems
- Multiload/multitemperature systems

In most cases, the system piping method and control technique used are mutually dependent. One cannot be used without the other. Many of the mixing valves first discussed in Chapter 5, Piping, Fittings, and Valves, will be applied in the piping schematics shown in this section. Several of the control techniques discussed in Chapter 9, Control Strategies, Components, and Systems, will also be used.

Each approach will be illustrated with a basic piping schematics. *These schematics show the general concept but not necessarily every individual component required by the system or local codes.* The reader is encouraged to cross reference previous chapters on piping and controls to fill in the details for these piping concepts.

Manifold Stations

A common subassembly in radiant panel systems is the manifold station. It consists of one **supply manifold** and one **return manifold.** In most systems, all radiant panel circuits begin at a supply manifold and end at a return manifold. Each circuit is therefore in parallel with the other circuits attached to the same manifold.

The number of piping circuits supplied by a manifold station can range from one to twenty or more depending on the total flow rate required by the circuits and the diameter of the manifold. In residential systems, manifolds of 2 to 8 circuits are common. In commercial or industrial buildings with large open spaces, larger manifold stations supplying 10 to 25 circuits may be used.

The supply and return manifolds are usually supported by a common set of brackets. This assembly can be mounted in several locations. In residential and light commercial buildings with slab floors, it is often mounted in the stud cavity of an interior wall and covered by an access panel. In basements, garages, or other utility spaces where exposed hardware is acceptable, manifold stations are often mounted directly to a wall surface. They can also be mounted horizontally on the underside of floor joists and accessed from the basement. Figure 10–82 shows some of these mounting options.

Some manifolds must be ordered with a given number of circuit connections. Others are assembled on the job site using modular components. Figure 10–83 shows an example of a manifold system consisting of brass sections serving either two or three circuits each. These sections can be joined together to form a larger manifold when necessary.

Figure 10–84 is an example of a modular manifold system assembled from individual high temperature thermo-

(a)

(b)

Figure 10–82 (a) Manifold station mounted in a stud cavity (b) large commercial manifold station mounted to the surface of a concrete wall (c) manifold station mounted to underside of floor joists. *Courtesy of Harvey Youker.*

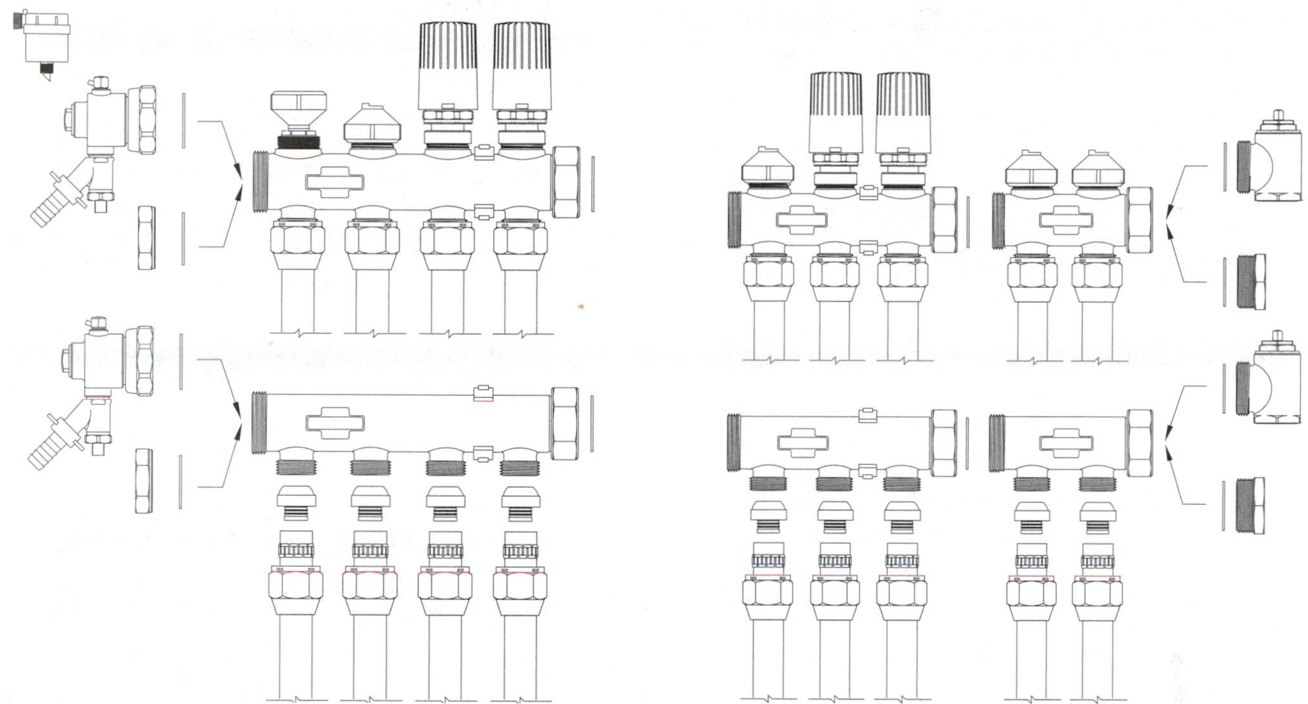

Figure 10–83 Example of a brass manifold system assembled from two, three, and four circuit modules with associated end caps, angle valves, and telestats. *Courtesy of Uponor Wirsbo.*

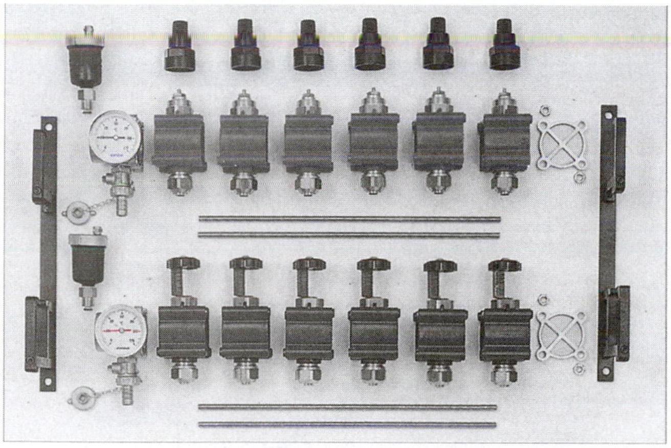

(a)

(b)

(c)

Figure 10–84 Example of a modular thermoplastic manifold system. (a) Individual parts ready for assembly, (b) modules are sealed together using O-rings and held together with threaded rods, (c) the completed manifold station ready to mount. *Courtesy of Euro-tech, Inc.*

plastic components to suit the needs of the system. End caps, drains, and other manifold trim are available from most manufacturers.

Manifolds can be ordered with or without valves. In applications where several circuits of approximately the same length are used in a single zone, and when individual circuit isolation is not necessary, a **valveless manifold** can be used for both the supply and return.

In applications where circuit lengths are significantly different, or when a room-by-room circuit layout is used, one of the manifolds should be equipped with circuit balancing valves. These valves allow the flow rates within individual circuits to be adjusted to increase or decrease heat output.

The most common scenario uses a valved manifold on the return side of the circuits. This allows flow returning from the circuits to pass through the valve seats and then against the valve plugs. This is the proper flow direction for any type of plug-type valve, allowing it to operate quietly and without cavitation.

If each circuit must be capable of being fully isolated, additional valves can be installed on the supply manifold. These valves are strictly for isolation and as such, should be fully open when the system is in normal operation. Figure 10–85 shows schematic representations for some common manifold station configurations with and without valves.

Manifold systems are available with options such as circuit flowmeters, thermometers, vent/drain assemblies, and fittings that allow different types and sizes of tubing to be connected to the manifold.

Many manifold systems also allow electric **valve actuators** to be screwed onto the valves of the return manifold (Figure 10–86). Valve actuators allow a low voltage thermostat circuit to open and close individual circuits. When screwed onto the manifold, they force the spring-loaded valves to their closed position and hold them there. When a given actuator is powered up through a thermostat circuit, it retracts its shaft and allows the manifold valve to open. Valve actuators make it possible to independently (electrically) control flow in each circuit. This is particularly useful when each room needs to be controlled as a separate zone. Two or more valve actuators can also be powered through a common circuit for simultaneous operation.

Sometimes the number of circuits used in a system, or their placement within a building, will require two or more manifold stations. When all circuits can operate at the same supply temperature, multiple manifold stations are piped in *parallel* as shown in Figure 10–87. *Never pipe multiple manifolds in series.* The large pressure and temperature drops that would result from such piping are far beyond what can deliver proper performance.

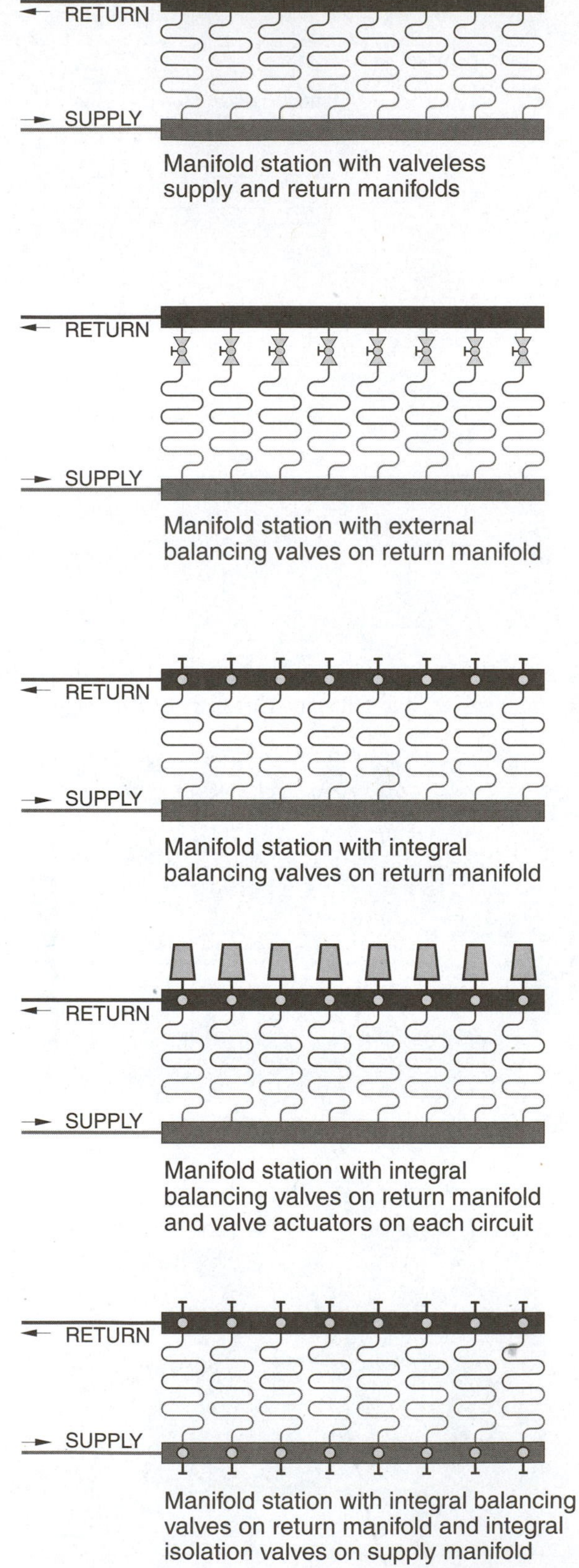

Figure 10–85 Valving options for manifold stations.

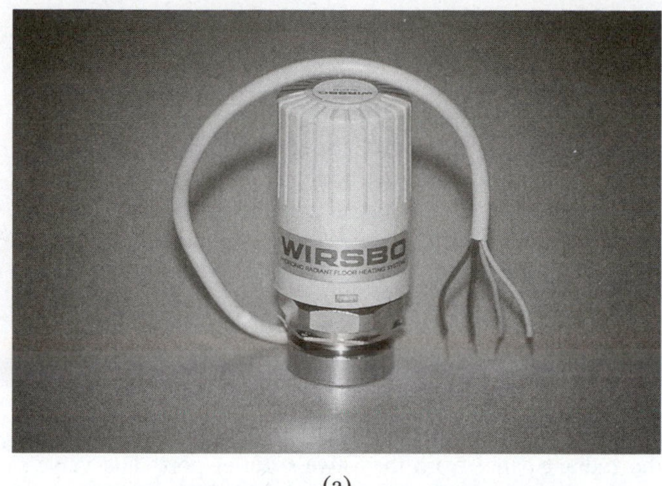

(a)

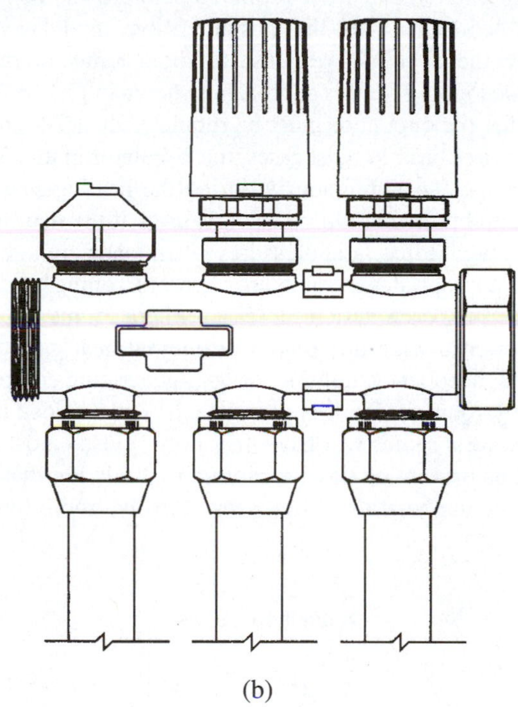

(b)

Figure 10–86 (a) Example of a 24-VAC valve actuator, (b) valve actuators screwed onto a manifold. *Courtesy of Uponor Wirsbo.*

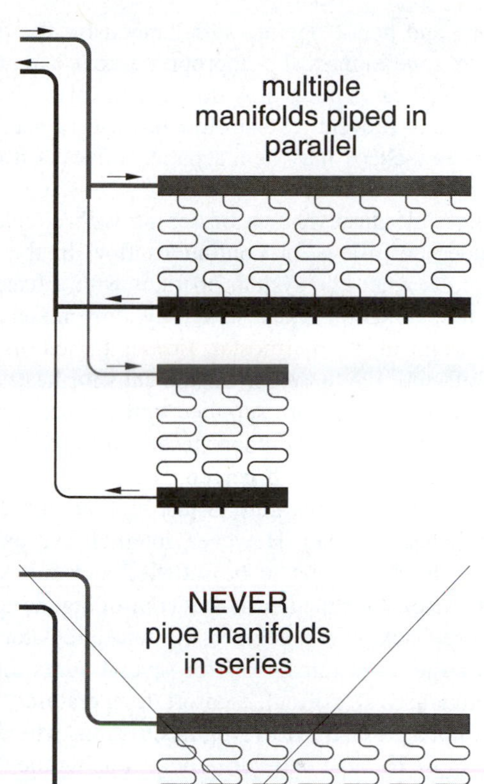

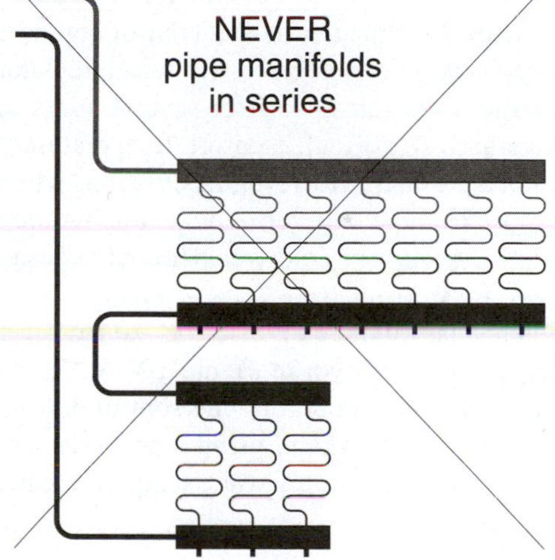

Figure 10–87 Pipe multiple manifold stations in parallel, never in series.

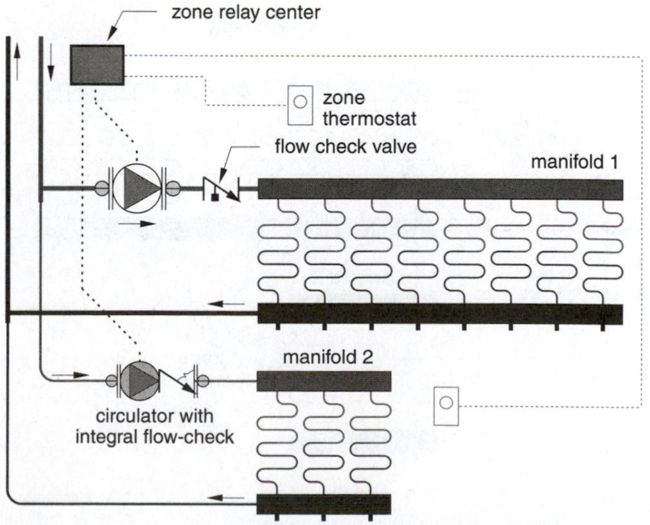

Figure 10–88 Zoning manifold stations using zone circulators.

Zoning Option for Radiant Panel Distribution Systems

Many of the zoning techniques discussed in previous chapters can be used with radiant panel distribution systems. Common approaches include zoning with circulators, zone valves, and electric valve actuators. In some systems, zoning can also be done using nonelectric thermostatic valves.

Figure 10–88 shows how two or more manifold stations can be zoned using circulators. The piping shown assumes that all circuits on a given manifold are located within the same zone and hence operate simultaneously. The thermostat in the zone signals the appropriate zone circulator to operate when heat is needed. A flow-check valve is required in each zone to prevent reverse flow or forward heat migration. The flow-check may be a separate valve, or it may be built into the circulator.

One possible disadvantage of zoning with circulators is that is does not allow for continuous flow in the radiant panel circuits when the system operates with a fixed water temperature. When the space served by a given zone reaches the setpoint of the thermostat, flow is turned off, and it remains off until the thermostat again calls for heat.

If the water temperature supplied to the zones is regulated by a properly adjusted outdoor reset control, the zone circulators tends to run a high percentage of the time. This is because the water temperature is just high enough to meet the current heating load. However, internal heat gain may still cause the thermostat to be satisfied and thus shut off flow. Likewise, when the thermostat controlling the zone circulator is set back several degrees, the zone circulator will be turned off and often remain off for several hours. In buildings maintained at normal comfort temperatures, this is normally not a problem. However, in situations where tubing is routed close to large exterior doors, stopping the flow for several hours during very cold weather could cause water in the tubing to freeze.

Individual manifolds can also be zoned using zone valves. The piping is shown in Figure 10–89. As is the case with circulators, this method of zone control does not allow continuous circulation. Again, using a properly set outdoor reset control to regulate supply water temperature helps keep the zone valves open a high percentage of the time. A differential pressure bypass valve may be needed to prevent high differential pressures when several parallel-piped manifolds are zoned using zone valves.

Flow through individual circuits can also be controlled using valve actuators mounted on one of the manifolds. This approach is very common when room-by-room zoning is desired in residential and light commercial buildings (Figure 10–90).

Individual radiant panel circuits can also be zoned using nonelectric thermostatic valves. These valves are located in small boxes mounted in the stud cavity of an interior partition in each room. The supply side of the circuit is connected to the inlet port of the valve. The circuit continues on to the radiant panel from the valve's outlet port. The valve's thermostatic head projects through a hole in the faceplate of the box where it can sense room temperature and be adjusted by occupants. The thermostatic valve modulates flow through the circuit in response to room temperature. The suggested piping for this approach is shown in Figure 10–91. Note that the circulator must be running whenever any circuit requires heat. In most cases, the circulator in this type of system operates continuously during the heating season. A differential pressure bypass valve is used at the manifold to prevent high differential pressures when loads are light.

The appropriateness of room-by-room zoning should be evaluated for each system. In cases where rooms are likely to experience wide differences in internal heat gain due to sunlight, occupancy, lighting, or heat-generating equipment, room-by-room zoning is appropriate. It is also a good choice when several rooms will have their doors closed most of the time thus preventing passive room air circulation that helps even out temperature differences. Finally, room-by-room

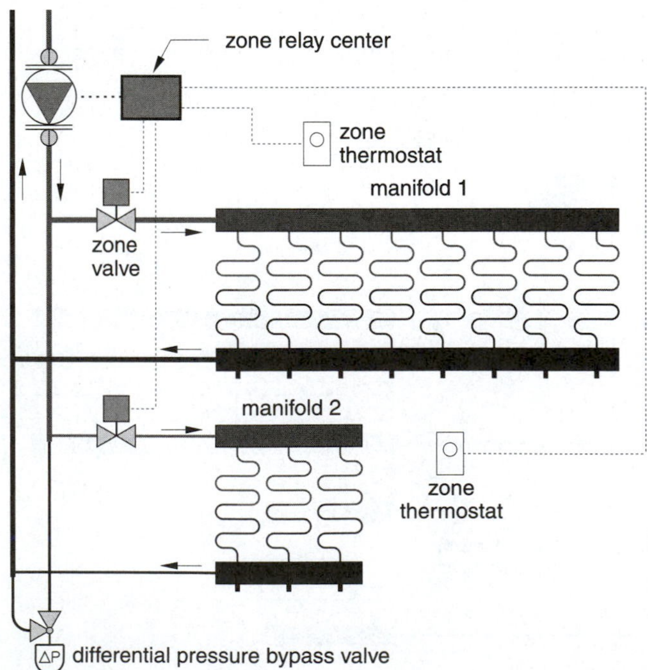

Figure 10–89 Zoning manifold stations using zone valves.

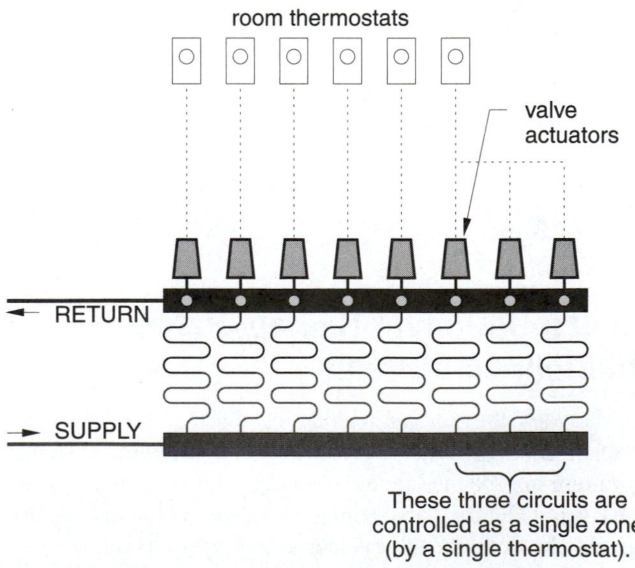

Figure 10–90 Use of electric valve actuators and thermostats for circuit-by-circuit control.

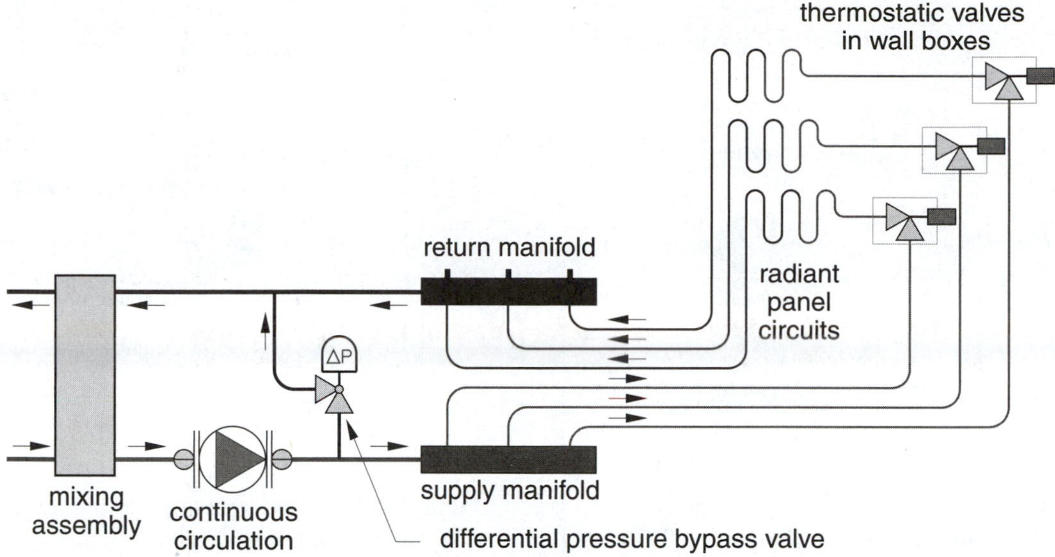

Figure 10–91 Use of non-electric thermostatic valves for circuit-by-circuit control. Note that distribution circulator operates continuously during heating season.

zoning is well-applied when room usage requires frequent changes in room temperature.

Room-by-room zoning is not necessarily justified simply because some rooms need to be maintained at reduced temperatures. This can be accomplished through adjustment of circuit-balancing valves. Reducing the flow through a given radiant panel circuit reduces its heat output and allows the room to stabilize at a slightly lower temperature. One might argue that room-by-room zoning is ultimately the most versatile approach; however, its implementation definitely adds cost to the system, sometimes unnecessarily.

When zoning with zone valves or valve actuators and a constant speed circulator, the designer should evaluate the possible need for a differential pressure bypass valve. These valves were discussed in Chapter 5, Piping, Fittings, and Valves. They open to allow bypass flow when the differential pressure reaches a set value. When properly applied, differential pressure bypass valves prevent high flow velocities that can occur when only one or two zones are operating. They also prevent closed zone valves from being forced open by high differential pressure. When nonelectric thermostatic zone valves are used, the bypass valve prevents **dead heading** the circulator when all zone circuits are closed. *The author's suggestion is to use a differential pressure bypass valve whenever the number of zones exceeds four, and/or the circulator supplying the zones exceeds 1/25 horsepower.* Eventually, the use of variable speed distribution circulators may eliminate the need for differential pressure bypass valves.

It is possible to provide constant circulation and zone control. One approach is shown in Figure 10–92. A 3-way motorized zone valve determines when warm water enters the constantly flowing manifold loop. This valve is a diverting valve, *not* a mixing valve. The water temperature entering the valve must already be at the appropriate temperature. It simply determines if the flow to the supply side of the circuits (port AB of the valve) comes from the return manifold (port B), or the warm water source (port A). Whenever the room thermostat calls for heat, the valve's actuator fully opens port A and fully closes port B.

The operation of zone circulators, zone valves, and valve actuators is typically managed using multizone relay centers. These controls were discussed in Chapter 9, Control Strategies, Components, and Systems. They simplify the wiring of zone thermostats as well as the devices controlling flow in the zone. They also allow the zoned distribution system to "call" for the operation of other devices such as the heat source and mixing controller.

Zone control techniques can also be combined when needed. For example, multiple manifold stations can be controlled using zone circulators, while individual circuits on each of those manifolds can be controlled using valve actuators.

Direct Piping Systems

The simplest method for connecting a heat source to a radiant panel distribution system is direct piping. The concept is shown schematically in Figure 10–93.

This type of piping should only be used with heat sources that can continuously operate at the return water temperatures of the radiant panel, without being damaged by flue gas condensation. Such heat sources include condensing boilers, hydronic heat pumps, electric boilers, some hot water tanks, and solar collectors. *Conventional gas- or oil-fired boilers should not be directly piped to radiant panel heating systems that have return water temperatures below the dewpoint of the boiler's exhaust gases.* To do so will cause sustained condensation of the water vapor on the boiler's heat exchanger

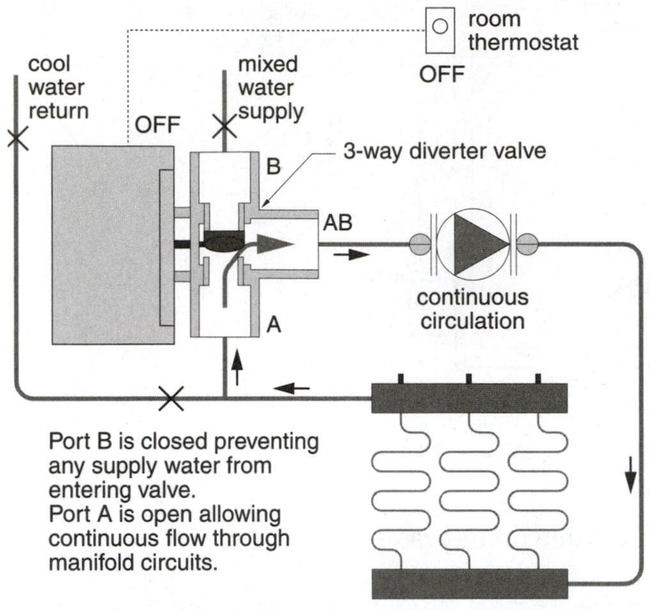

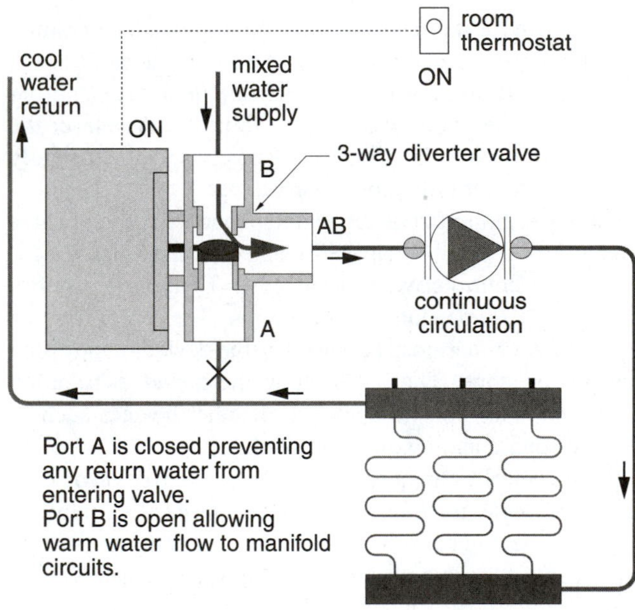

Figure 10–92 Use of 3-way diverter valve to provide zone control of manifold station and constant circulation in radiant panel circuits.

and vent connector piping. The resulting corrosion can cause dangerous leaks in the vent connector piping in a matter of weeks! Several of the piping/control options presented later in this section are specifically piped and controlled to prevent sustained flue gas condensation.

In a direct-piping system, the water temperature supplied to the distribution system is usually controlled by turning the heat source on and off. The water temperature controller may provide a fixed target temperature along with an appropriate differential, or it may operate based on an outdoor

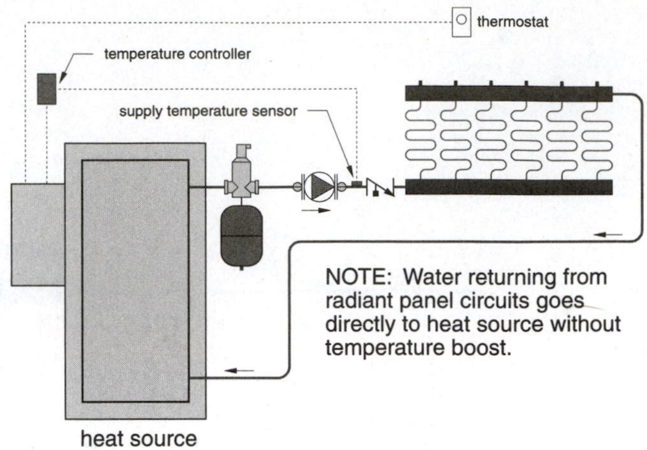

Figure 10–93 Direct piping of heat source to manifold station. No mixing takes place.

reset control. In the latter case, the target supply temperature to the radiant panel circuits varies in response to the outdoor temperature. Outdoor reset control is discussed in detail in Chapter 9, Control Strategies, Components, and Systems.

Care must be taken when designing direct piping systems to ensure there will always be sufficient flow through the heat source. This is especially important when a) the distribution system is zoned, and/or b) a low-mass, fixed-output heat source is used.

Consider, for example, a low-mass copper tube boiler connected to a zoned distribution system as shown in Figure 10–94a. When only one zone is operating, the flow through the boiler may be lower than the minimum flow required by the manufacturer. Operating at such a low flow can create high stresses in the boiler's heat exchanger as well as operating noise. Furthermore, the rate of heat output from the boiler under these conditions may be several times greater than the load. This will lead to **short-cycling** of the burner, which lowers efficiency and ultimately shortens the life of critical components such as the ignition system and gas valve.

The solution for such problems is to install a **buffer tank** between the boiler and the distribution system as shown in Figure 10–94b. This tank provides the additional water volume (thermal mass) to absorb the surplus heat output of the boiler under partial load conditions. It allows the boiler to operate with less frequent but longer cycles, which is desirable. The buffer tank also "uncouples" flow through the heat source from that in the distribution system. Flow-sensitive heat sources such as low-mass boilers and hydronic heat pumps are assured of adequate flow under all loading conditions. The sizing of buffer tanks is covered in Chapter 11, Distribution Piping Systems. *Note that a buffer tank will not ensure that a conventional boiler is protected from flue gas condensation.*

In systems without buffer tanks, the heat source is cycled on and off by a setpoint temperature controller or an outdoor reset controller whenever there is a demand for heat from the

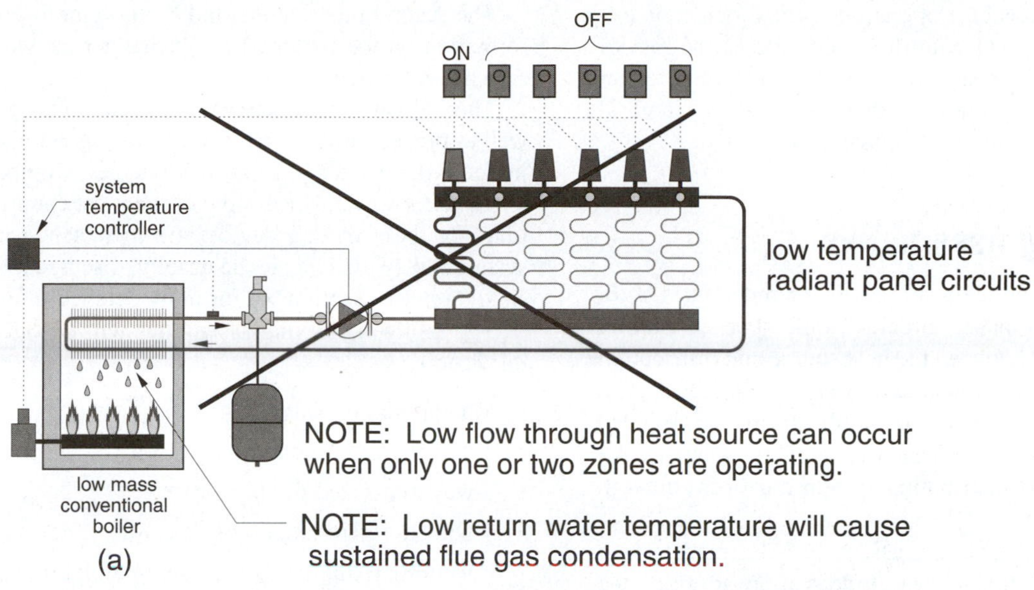

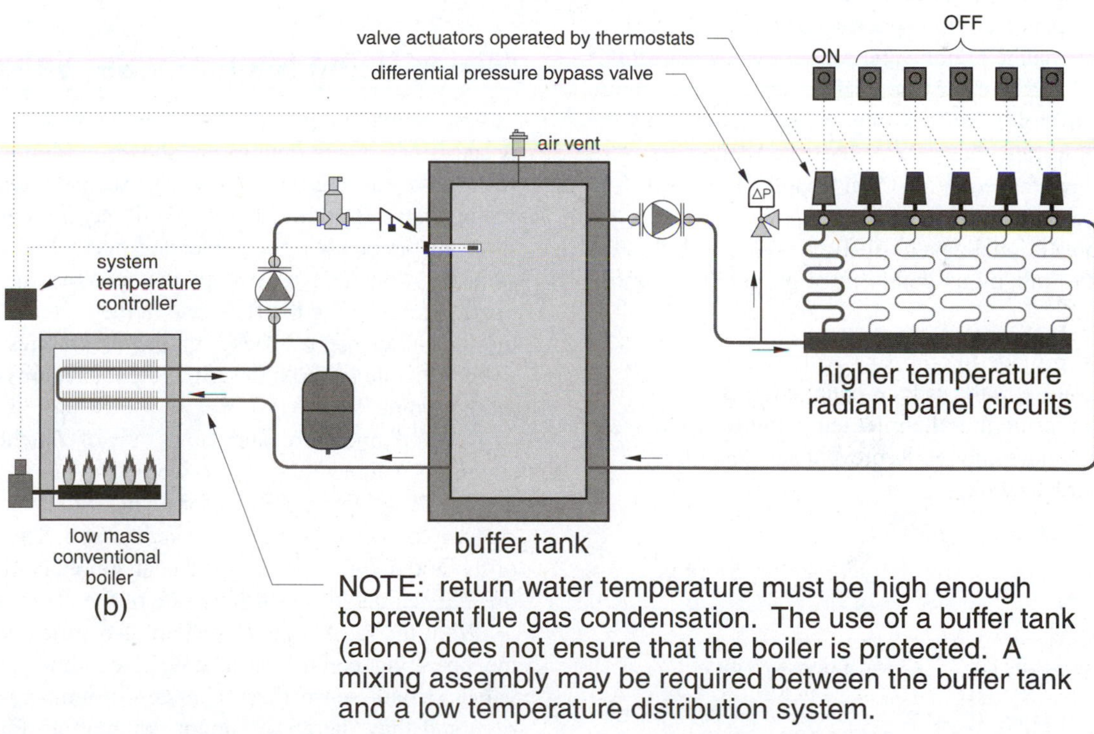

Figure 10–94 (a) Unacceptable connection of manifold station with valve actuators to low mass boiler. (b) Use of buffer tank to ensure proper flow through low mass boiler regardless of which radiant panel circuits are operating.

thermostat. The circulator can be turned on during each demand for heat, or, if outdoor reset control is used, it can operate almost continuously, turning off only if all zones are satisfied. The latter is preferable since it reduces rapid changes in water temperature and evens out heat delivery.

In systems with buffer tanks, the heat source is cycled on and off to maintain the buffer tank temperature within preset limits. These limits may be based on a fixed target temperature and differential, or they may be based on a variable target temperature if an outdoor reset controller is used. The distribution system circulator and valve actuators operate independently of the heat source. Upon a call for heat from any room thermostats, the associated valve actuator opens its valve. When the valve is fully open, the end switch in the actuator closes to signal for operation of the distribution circulator. This piping and control system allows the rate of heat generation at the heat source to be significantly higher than the rate of heat release by the radiant panel circuit(s) without short-cycling the heat source.

Use of Mixing Assemblies

When the heat source must operate at temperatures higher than the radiant panel circuits, some type of **mixing assembly** is needed between the heat source and distribution system. This concept is illustrated in Figure 10–95.

Notice that the mixing assembly forms a "bridge" between the heat source loop and the distribution system. All heat that reaches the distribution system must pass through the mixing assembly.

The purpose of the mixing assembly is to regulate *heat flow* into the distribution system. It does so by adjusting the rate at which hot water passes from the heat source into the distribution system. As hot water flows to the distribution system, an equal volume of cool water from the return side of the radiant panel circuits passes back toward the heat source. This exchange of hot water for cool water constitutes heat transfer. The greater the difference in temperature between the in-going and exiting water streams and the greater the flow rate, the greater the rate of heat transfer. The sensible heat rate equation from Chapter 4, Properties of Water, can be used to calculate this rate.

When a conventional gas- or oil-fired boiler is used as the heat source, the mixing assembly must provide two control actions:

1. It must provide the proper supply water temperature to the radiant panel circuits.
2. It must assure that the inlet temperature to the boiler is high enough to prevent sustained flue gas condensation.

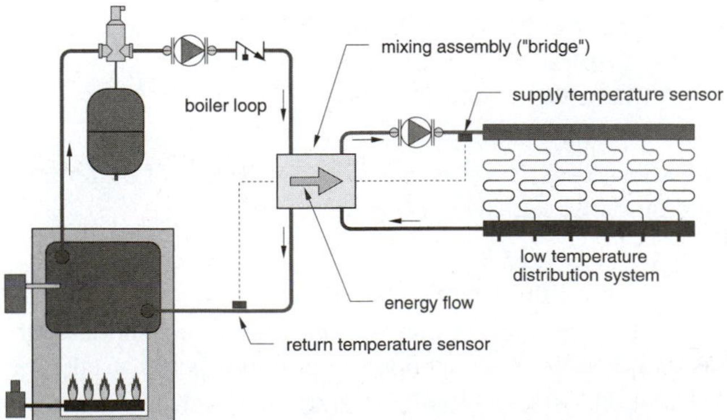

Figure 10–95 The mixing assembly serves as a "bridge" between the boiler loop and the distribution system.

The second function is vitally important to prevent scale formation on the boiler's heat exchanger, as well as corrosion of vent piping.

The individual characteristic of these mixing devices, as well as procedures for sizing and adjusting them, are covered in detail in Chapter 9, Control Strategies, Components, and Systems. However, since some of the most common applications for these mixing devices are in radiant panel heating systems, they will again be briefly discussed. The reader should refer to Section 9.8 for more detailed information.

The following mixing assemblies will be summarized in this section:

- Manual mixing valves
- 3-way thermostatic mixing valves
- 3-way motorized mixing valves
- 4-way motorized mixing valves
- Injection mixing using a 2-way modulating valve
- Injection mixing using a variable speed pump

Mixing Using Manually Set Valves

One simple mixing assembly can be made using manually set valves. Three options are shown in Figure 10–96. The first uses a pair of 2-way valves. The second is a single 3-way mixing valve. The third is a single 4-way mixing valve.

In all three cases, the valves are manually set to a given stem position, and remain in that position as the system operates. Assuming the flow rate in the system is not altered through other means, these mixing assemblies all create a fixed flow rate for both the hot and cool streams entering the mixing point.

If the temperature and flow rate of the hot and cool streams remained constant, so would the mixed temperature leaving the mixing point. Unfortunately, *such conditions are very unlikely in most radiant panel systems.* The temperature of the hotter fluid varies as the heat source is cycled on and off. The temperature and flow rate of the cooler fluid returning from the radiant panel system also varies depending on the operating conditions at any given time. Manually set valves cannot sense these changes in temperature or flow rate, and they therefore cannot compensate for them. The result is that *the mixed temperature varies whenever the temperature or flow rate of either entering stream changes.* This causes the supply temperature as well as the return temperature to the heat source to vary, in some cases over a wide range. Although **manually set mixing valves** may be inexpensive, they do not provide a stable supply temperature. Furthermore, they cannot protect a conventional boiler from flue gas condensation. They also lack the control intelligence to provide mixing reset control. *In general, manually set mixing valves are of little use as the sole means of water temperature control.*

Mixing Using 3-way Thermostatic Valves

A **3-way thermostatic mixing valve** has an internal thermostatic element that senses the temperature of the fluid leaving its mixed outlet port. The valve spool automatically adjusts its position to try to maintain the outlet temperature close to the target temperature set on the valve's knob.

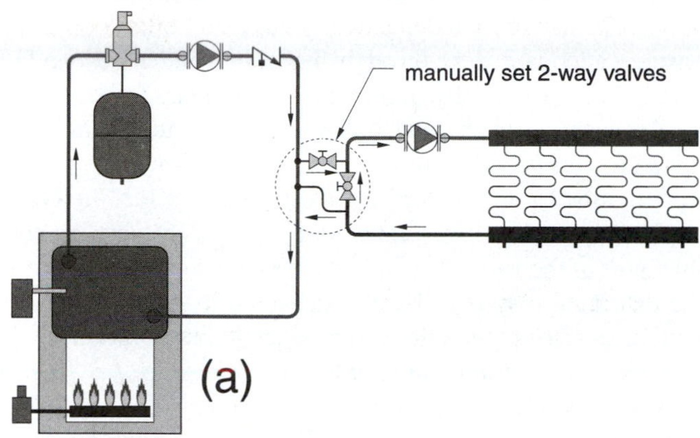

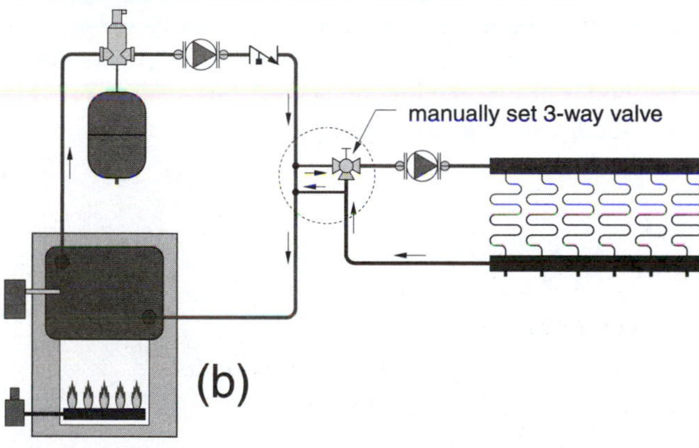

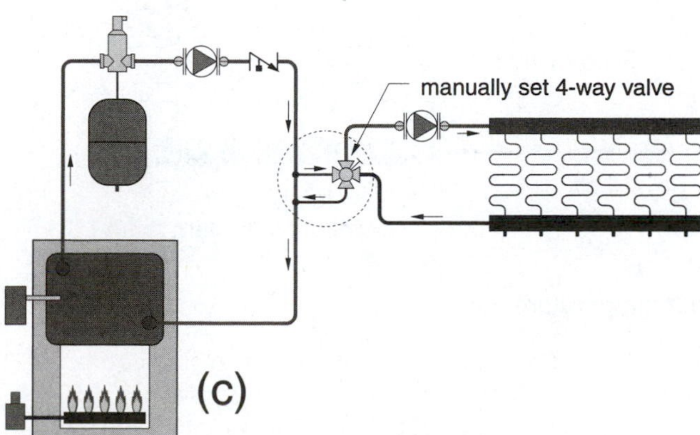

Figure 10–96 Manually-set mixing valve options. (a) Use of two, 2-way valves, (b) use of 3-way valve, (c) use of 4-way valve. Note: None of these options can respond to changing water temperatures or flow rates.

A *single* 3-way thermostatic valve can be used to supply temperature control to radiant panel circuits when the heat source can tolerate low inlet temperatures and relatively low flow rates (Figure 10–97a). *This is* not *the case for most conventional boilers supplying low temperature radiant panel systems.*

When a conventional boiler is used as the heat source, a second 3-way thermostatic valve should be used to monitor and boost the boiler inlet temperature as necessary to prevent sustained flue gas condensation. The suggested piping is shown in Figure 10–97b. Notice that a second circulator is now required to provide proper mixing in the thermostatic valve, protecting the boiler. This valve and circulator also help maintain flow through the boiler. The latter is desirable when low-mass boilers are used.

Three-way thermostatic valves provide a fixed supply temperature to radiant panel circuits. They cannot provide outdoor reset control nor allow for constant circulation in the distribution system without other system modifications.

Mixing Using 3-way Motorized Valves

When a **3-way rotary mixing valve** is combined with a motorized actuator and controller, the resulting mixing assembly can perform several functions depending on the capabilities of the controller.

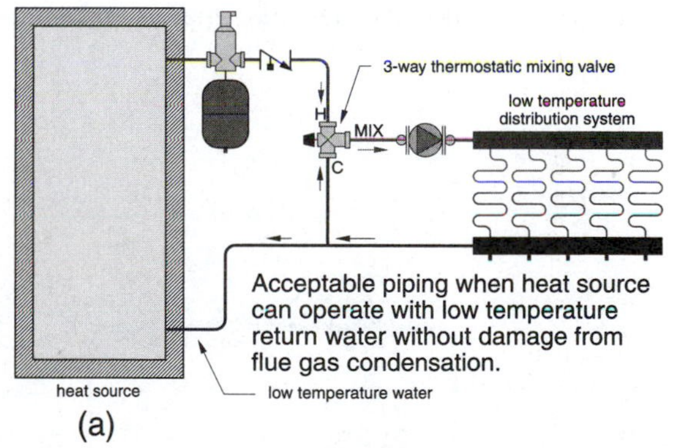

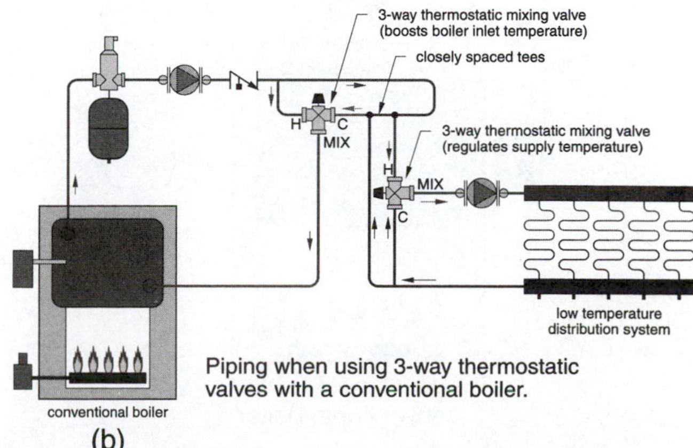

Figure 10–97 (a) Use of a single 3-way thermostatic mixing valve, (b) use of two 3-way thermostatic mixing valves.

Most modern mixing valve controllers provide very accurate supply temperature control. This temperature can be a fixed setpoint temperature, or a temperature that is reset based on outdoor temperature.

When a 3-way motorized mixing valve is used with a conventional gas- or oil-fired boiler and low temperature radiant panels, the controller operating the valve's actuator must be able to monitor the inlet temperature of the boiler. When this temperature is too low, the controller directs the actuator to close the hot water port of the valve as necessary to prevent the boiler from operating with sustained flue gas condensation. In effect, the controller prevents the radiant panel circuits from extracting heat from the water faster than the boiler can generate heat.

The recommended piping for 3-way motorized mixing valves is shown in Figure 10–98.

Mixing Using 4-way Motorized Valves

Four-way motorized mixing valves were specifically developed to regulate water temperature and boost boiler return temperature at the same time. They have a well-established track record in all types of radiant panel systems.

Hot water from the boiler is blended with cooler water returning from the radiant panel circuits at *two* locations within the valve. In the upper chamber, the hot and cool streams blend to produce the desired supply temperature. The remaining portions of the hot and cool streams are blended in the lower valve chamber. The objective of this latter mixing is to boost boiler inlet temperature high enough to prevent sustained flue gas condensation.

To avoid interference between the boiler circulator and the distribution circulator, the 4-way valve should be piped as shown in Figure 10–99. The closely spaced tees uncouple the pressure distribution in the boiler loop from that in the distribution system. The pressure drop established inside the valve will induce the necessary flow of hot water into the valve. The use of an independently pumped boiler loop also ensures adequate flow through the boiler under all conditions. This is especially important for low-mass boilers.

To ensure conventional boilers are protected from sustained flue gas condensation, most modern 4-way valve controllers monitor the boiler inlet temperature. When the controller senses a low inlet temperature, it begins closing the hot inlet port of the valve. This limits the rate of heat transfer into the distribution system. Heat is delivered to the distribution system as fast as the boiler can produce it, but no faster.

A common misunderstanding is that the use of a 4-way mixing valve without an actuator and controller can also properly protect conventional boilers from flue gas condensation. This is *not* true. A manually set 4-way mixing valve has no advantage relative to any other manually set valve or valve set. It cannot react to changing temperatures or flows in the boiler or distribution loops and, thus, cannot provide stable supply or boiler inlet temperature.

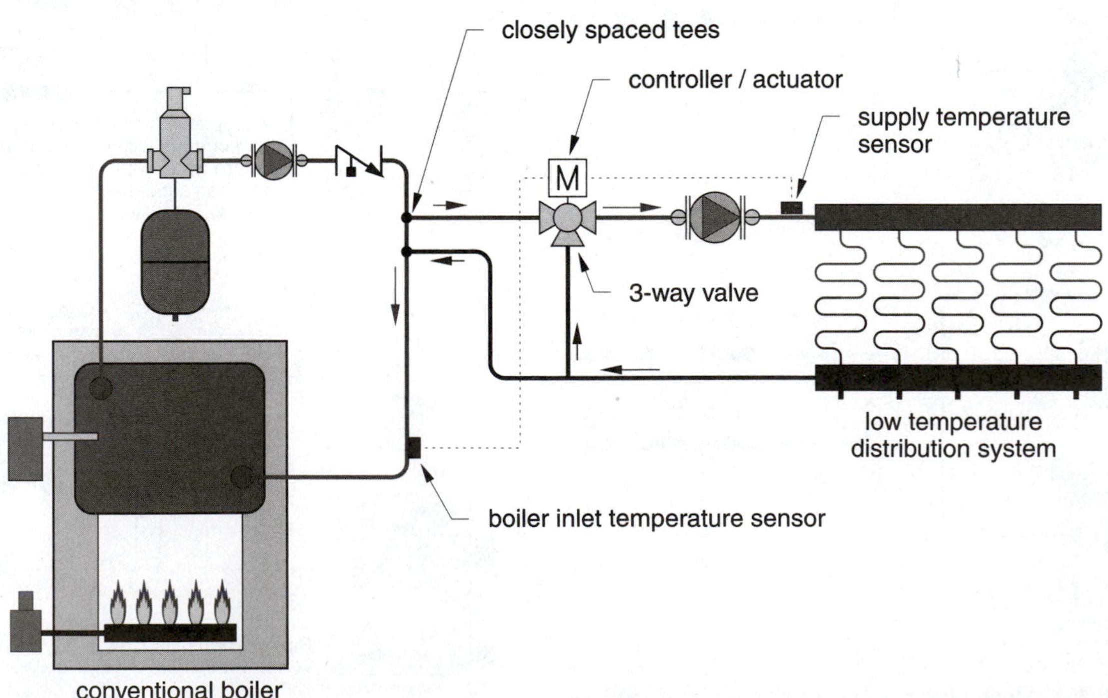

Figure 10–98 Use of 3-way motorized mixing valve with intelligent mixing control that senses supply temperature and boiler inlet temperature.

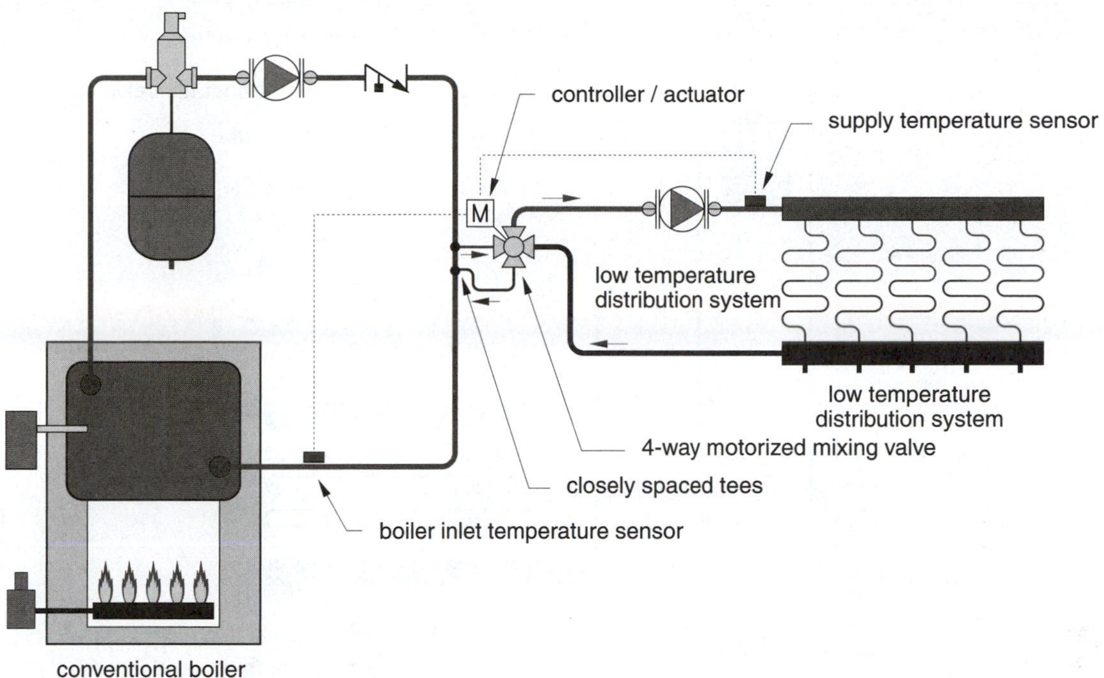

Figure 10–99 Use of 4-way motorized mixing valve with intelligent mixing control that senses supply temperature and boiler inlet temperature.

Injection Mixing

The general concept of injection mixing was discussed in Chapter 9, Control Strategies, Components, and Systems. Injection flow can be regulated by a 2-way valve or an injection pump. The majority of radiant panel systems employing **injection mixing** use either a 2-way *modulating* valve, or a variable speed pump as the **injection control device.** These options are summarized in this section. The reader should refer to Section 8 of Chapter 9 for specific information on sizing and selecting injection-mixing control hardware.

Injection Mixing Using a 2-way Thermostatic Valve

One way to regulate the injection flow rate is with a **2-way thermostatic valve.** The valve is mounted into the supply injection riser as shown in Figure 10–100.

The hot water flow rate is controlled by the thermostatic valve (V1). The knob of this valve is set to the desired supply temperature under design load conditions. The sensing bulb measures the temperature supplied to the distribution system. If the supply temperature is too low, the valve opens farther to allow more hot water into the mix point. If the supply temperature is too high, the valve begins to close, allowing less hot water into the mix point.

In a typical low temperature floor heating system supplied by a conventional boiler, the flow rate through the injection valve (V1) is only about 15 to 20 percent of the flow rate in the distribution system. This allows a relatively small valve to regulate a large rate of heat transfer.

The flow restrictor valve (V2) creates the pressure differential necessary to force flow through the injection risers as the injection valve begins to open. The greater the flow resistance created by this valve, the higher the injection flow rate becomes for a given stem position of the injection valve (V1). For stable and accurate control, *the injection valve should be fully open when the maximum injection flow rate occurs under design load conditions.* This requires the valve to be properly sized. It also requires the flow restrictor valve to be properly set. Refer to Section 9.8 for information on doing so.

Injection mixing using a 2-way thermostatic valve does have limitations. Since the 2-way thermostatic valve only measures and regulates supply temperature, *it does not protect a conventional boiler against flue gas condensation.* If a conventional boiler is used as the heat source, another temperature-sensing mixing device is required to protect it. One possibility is to use a 3-way thermostatic valve set for the minimum allowable boiler inlet temperature and installed as shown in Figure 10–101. Injection mixing with a 2-way thermostatic valve also does not provide mixing reset control nor does it allow for constant circulation in the distribution system.

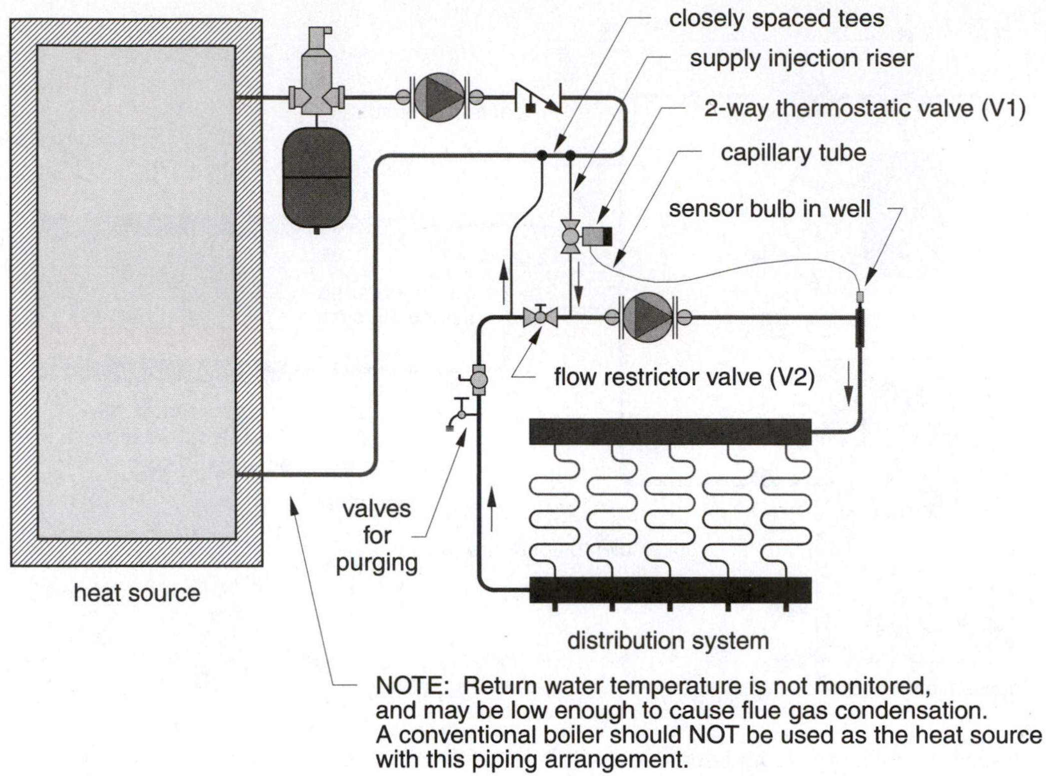

Figure 10–100 Use of 2-way thermostatic valve for injection mixing. Note situation with boiler return temperature.

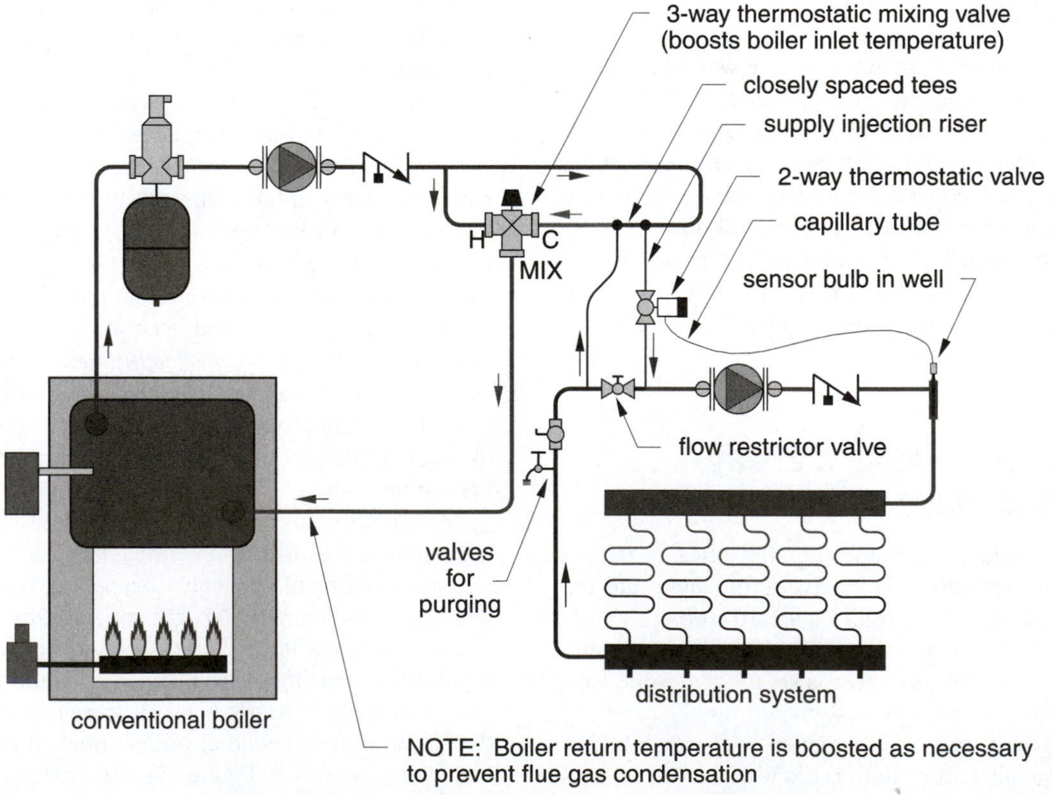

Figure 10–101 Adding a 3-way thermostatic mixing valve to piping shown in Figure 10–100 to boost boiler inlet temperature.

Injection Mixing Using a 2-Way Motorized Valve

The piping used for injection mixing with a 2-way thermostatic valve can also be used for a 2-way motorized valve.

A significant advantage of a motorized valve over a thermostatic valve is that the motorized valve can be operated by an intelligent controller. This allows the mixing assembly to provide several benefits not available with a thermostatic injection valve. If the controller monitors boiler inlet temperature, it can protect a conventional boiler against low temperature operation by reducing the injection flow through the valve when necessary. Most valve controllers designed for injection mixing can also provide either setpoint or outdoor reset control of the supply water temperature. Since the controller can completely shut the injection valve when no heat input is required, the distribution system can also operate with constant circulation.

Injection Mixing Using a Variable Speed Pump

One of the most popular approaches to water temperature control in modern radiant panel systems is injection mixing using a variable speed pump. The piping for this approach is shown in Figure 10–102.

When the injection pump is operating, hot water in the primary loop is drawn into the supply injection riser at point A. After passing through the variable speed pump, the hot water is pushed into the tee at point B where it mixes with some of the cooler water from the return side of the distribution system. The remainder of the cooler water leaves the tee at point C, and travels up the return injection riser. *The exiting flow rate of cooler water always equals the entering flow rate of hot water.* When it flows into the tee at point D, it mixes with the bypass flow of hot water. This tee provides a second mixing point that boosts the water temperature entering the boiler.

Two piping details are critically important to proper operation of this mixing assembly. Both are shown in Figure 10–102:

Detail #1: *The injection risers must drop at least 18 inches between where they connect to the primary loop and where they connect to the distribution system.* This forms a thermal trap that discourages hot water from migrating downward into the distribution system when the injection pump is off. Without this trap, the slight pressure drop between the tees from A to D and from C to B can cause hot water migration into the distribution system when it is not needed. A standard circulator will not block this flow. Furthermore, a flow-check valve should *not* be installed in either injection riser to prevent this flow because it can lead to erratic pump operation ("surging") at low speeds.

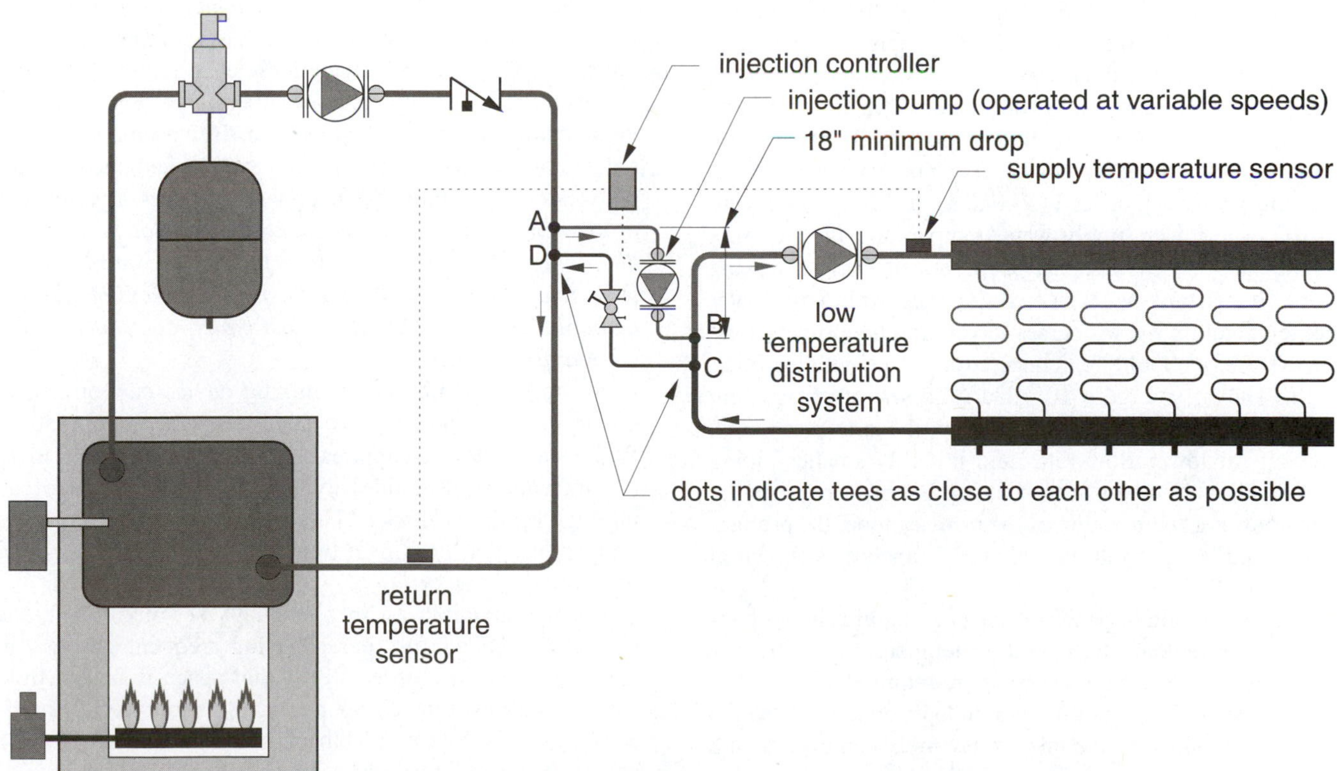

Figure 10–102 Use of variable speed injection mixing with intelligent mixing control that senses supply temperature and boiler inlet temperature.

Note: The injection risers do not have to drop straight down 18 inches. There can be numerous horizontal offsets in the injection risers if necessary provided the net drop in elevation from the primary loop to the distribution loop is at least 18 inches.

Detail #2: *The tees connecting the injection risers to the primary loop should be as close together as possible. Likewise, the tees connecting the injection risers to the distribution systems should be as close together as possible.* The goal is to minimize the pressure drop between the pairs of tees. This minimizes the tendency for uncontrolled hot water migration into the distribution system.

For optimal performance, variable speed injection mixing assemblies must be properly sized and balanced. Information for doing this is given in Section 9.8.

Most variable speed injection mixing controllers can be configured for either setpoint or outdoor reset control of the supply temperature. They also sensor boiler inlet temperature and provide boiler protection when necessary. Some can even coordinate operation of the injection pump, system circulator, and heat source.

Multitemperature Radiant Panel Systems

When two or more types of radiant panel construction are used in the same building, it is often necessary to supply two or more water temperatures simultaneously. This can be done several ways, depending on the heat source used, the operating temperatures of the radiant panels, and the capabilities of the mixing devices.

If one radiant panel subsystem requires water at the same temperature as delivered by the heat source, that subsystem does not require a mixing device. It can be connected directly to the primary loop as a secondary as shown in Figure 10–103. The lower supply water temperature for the other radiant panel subsystem can be produced using a mixing device. If that mixing device can operate with a hot water inlet temperature a few degrees lower than the primary loop temperature, it should be connected in series with the other load as shown in Figure 10–103a. The series arrangement increases the overall temperature drop of the primary loop, allowing for lower flow rates and possibly smaller piping and a smaller circulator. If the mixing assembly requires the same hot water temperature as the primary loop, the primary loop should be split into two parallel branches as shown in Figure 10–103b.

When *both* radiant panel subsystems require supply temperatures lower than the operating temperature of the primary loop, two mixing assemblies are required.

One approach is to use two independently controlled mixing assemblies. If one mixing assembly can operate at a hot water inlet temperature a few degrees lower than the other, series piping, as shown in Figure 10–104a, is preferred. If both mixing assemblies require the same hot water inlet temperature, they should be piped in parallel as shown in Figure 10–104b.

When the series arrangement is used, the designer should calculate the overall temperature drop in the primary loop at design load conditions. If a conventional boiler is used as the heat source, the inlet temperature to that boiler must be maintained high enough to prevent sustained flue gas condensation.

Another approach is to use **submixing.** A single **intelligent mixing device** (one that senses and reacts to water temperature) supplies the highest mixed supply temperature, while other **dumb mixing devices** (that cannot react to changing water temperature) further reduce the water temperature in other parts of the system. An arrangement that uses this concept to provide three mixed temperatures is shown in Figure 10–105a.

Submixing allows multiple water temperatures to be created at a lower cost than when multiple intelligent mixing devices are used. Furthermore, when the "intelligent" mixing device varies the higher mixed water temperature using outdoor reset control, the lower mixed temperatures also track along a reset line. This effect is called **proportional reset.** A representative set of reset lines based on proportional reset is shown in Figure 10–105b on page 400.

There are some limitations to submixing. First, since any manually set mixing valve (i.e., a dumb valve) only provides fixed flow proportions of the incoming water streams, *any variation in the temperature or flow rate of either stream causes a change in the mixed water temperature.* Unlike an intelligent mixing device, a manually set mixing valve cannot compensate for such changing inlet conditions. Changes in the entering flow rate and/or temperatures can cause wide variations in the mixed temperature. In extreme cases, the radiant panel or its coverings could even be damaged by supply water temperatures much higher than necessary. *Because of changing flow rates, submixing should not be used for manifold stations that use valve actuators to control individual circuits. It should also not be used upstream of multiple manifold stations that are individually controlled using circulators or zone valves.*

Secondly, a manually set mixing device cannot protect a conventional boiler from low temperature operation. When such a device supplies a low temperature/high mass radiant panel, it should only be used downstream of an intelligent mixing device. The latter can sense boiler inlet temperature and limit heat transfer to the distribution system when necessary.

Finally, submixing with a manually set mixing valve is not recommended for spaces having frequent changes in thermostat settings, especially if that space is heated by a high mass radiant panel. Such an application would experience slow recovery from setback conditions because of the cooler than normal return water from the panel. An intelligent mixing device could compensate for this. A dumb mixing device cannot.

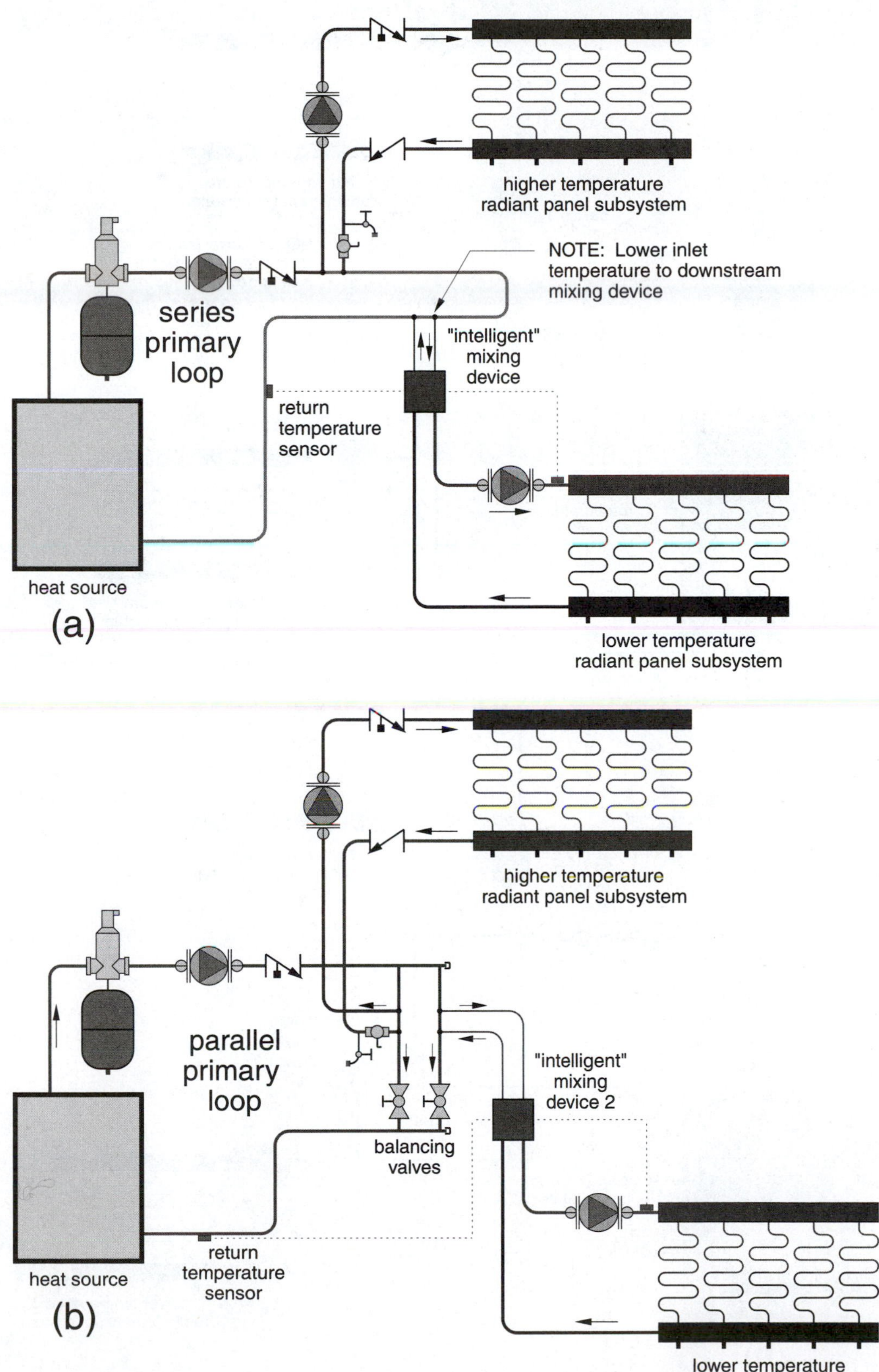

Figure 10–103 Options for providing two different supply water temperatures to two manifold stations when one manifold station *does not* require mixing (a) when downstream mixing device can operate with slightly reduced hot water inlet temperature, (b) when downstream mixing device must operate at the same hot water inlet temperature as upstream manifold.

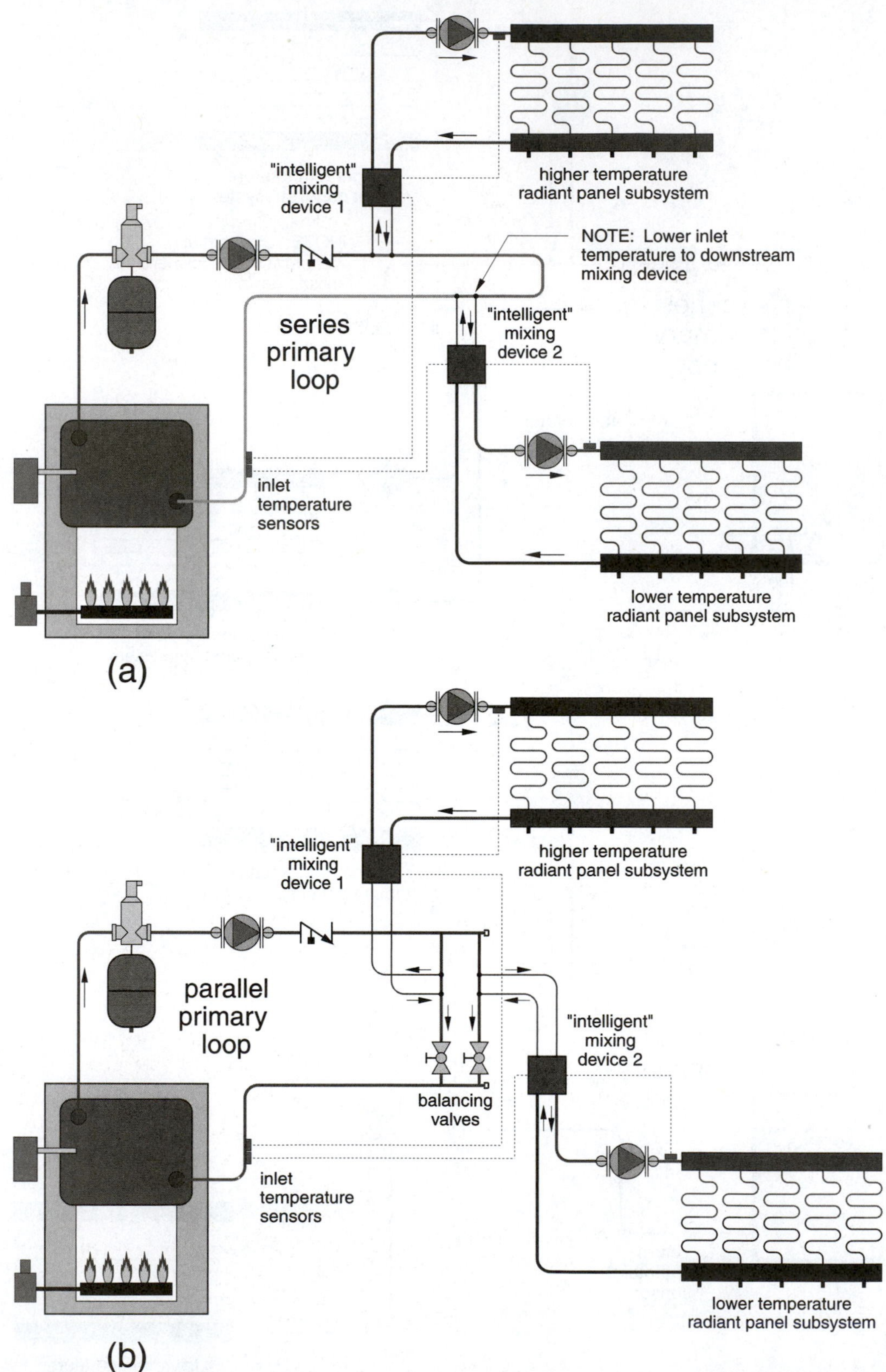

Figure 10–104 Options for providing two different supply water temperatures to two manifold stations when both manifold stations require mixing (a) when downstream mixing device can operate with slightly reduced hot water inlet temperature, (b) when downstream mixing device must operate at the same hot water inlet temperature as upstream mixing device.

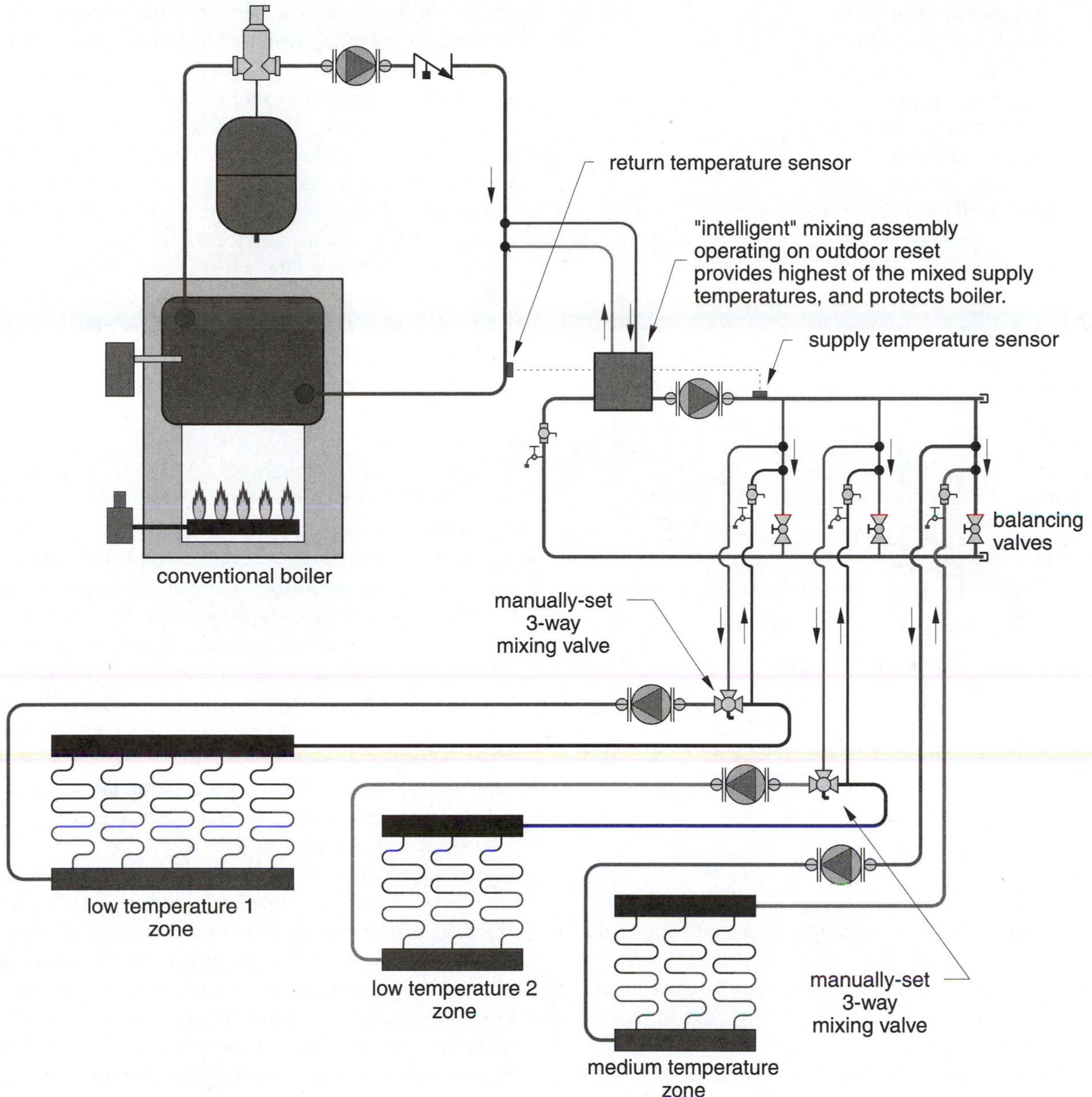

Figure 10–105a Use of submixing to create multiple supply water temperature downstream of an intelligent mixing assembly.

Multiload/multitemperature Systems

Radiant panel heating is often used in a portion of a building along with other types of hydronic heat emitters in the other areas. Ancillary loads such as domestic water heating, snowmelting, and pool heating may also be part of the overall system. The heat source may be a single boiler or a multiple boiler system.

Many of these so-called **multiload/multitemperature systems** can have upwards of a dozen circulators. Some of these circulators may be small while others are relatively large. There can also be an assortment of mixing devices, zone control devices, and specialized hardware, such as heat exchangers. A properly designed system must allow all operating modes to function over a range of conditions, without interference between simultaneously operating circulators, excess temperature drops, flow noise, or sustained flue gas condensation in conventional boilers.

The piping for such systems often uses **primary/secondary tees** between adjoining piping loops. Such an interface consists of a pair of closely spaced tees that isolate the pressure distribution in one circuit from that in the adjoining circuit. Primary/Secondary piping is discussed in detail in Chapter 11, Distribution Piping Systems, but the essential elements are also depicted in many of the drawings in this section.

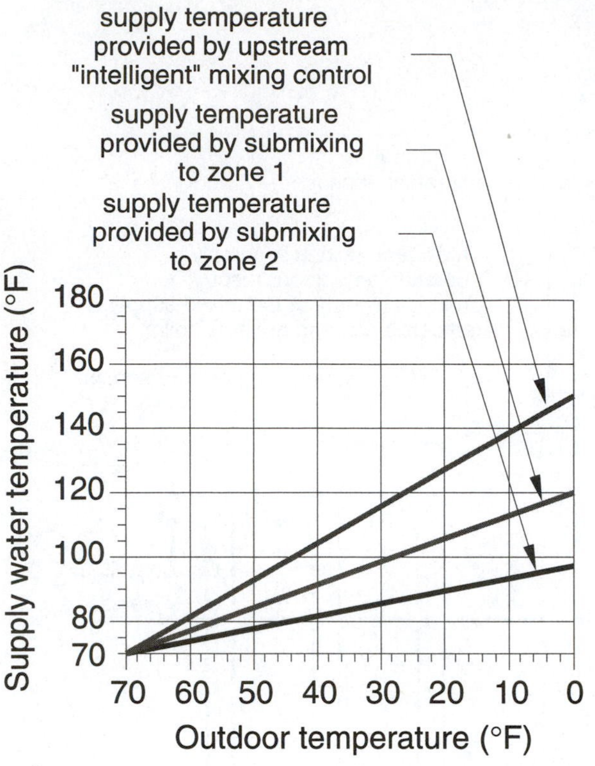

Figure 10–105b Example of reset lines that can be produced by submixing downstream of an intelligent mixing assembly.

Figure 10–106 shows an example of a multiload/multi-temperature system supplying two independent zones of radiant panel heating (both at the same supply temperature), as well as three zones of fin-tube baseboard, and an indirect domestic water heater.

In this piping layout, zone circulators (C1–C5) are assumed to be relatively equal in performance, and as such can be connected to a common header. To minimize interaction among these circulators, the supply and return header piping should be generously sized. The two radiant heating zones are also controlled using zone circulators (C7 and C8). The water temperature supplied to these zones is controlled by a variable speed injection pump (C6). Hot water is supplied to the injection mixing assembly by circulator (C5). Since the circuit powered by (C5) is very short, a flow restricting (globe) valve is installed to prevent (C5) from operating at very high flow rates, which increases amperage and reduces efficiency. The injection risers are connected to the hot water circuit and the radiant distribution piping using pairs of closely spaced tees (e.g.,primary/secondary interfaces). These prevent interference between circulators when the one or both of the radiant heating zones are operating. The injection mixing control monitors both supply and boiler inlet temperature. The latter allows the mixing control to reduce the injection flow rate when necessary to prevent the boiler from operating with sustained flue gas condensation.

Figure 10–107 shows one of several possible control diagrams for this system. A multizone relay center coordinates operation of the zone circulators. It also provides for prioritized operation of the indirect domestic water heater. The injection mixing control, the circulator it drives, as well as circulator (C5) must operate only when one or both of the radiant heating zones call for heat. This is accomplished using two isolation relays wired in parallel with the radiant zone circulators (C7) and (C8). When either of these circulators operates, the associated isolation relay passes line voltage to the injection mixing control, circulator (C5), and the transformer that powers the injection control. Other wiring configurations are possible, depending on the controls used.

Figure 10–108 is a piping schematic for a system supplying radiant floor heating in a small municipal building. The distribution system supplies two manifold stations piped in parallel, and operating at the same supply temperature. Each circuit on both manifolds is equipped with a valve actuator. Some actuators are individually controlled by a room thermostat, others are wired in parallel so they all operate simultaneously from a single thermostat. The building is divided into 10 independently controlled zones. The water temperature to the radiant floor circuits is regulated by a variable speed injection mixing system and is based on an outdoor reset control. Note the use of a differential pressure bypass valve to eliminate high differential pressure across the pump when minimal heating zones are operating.

This system also provides snowmelting for the building's sidewalks. When this operating mode is active, hot water from the primary loop is circulated through the hot side of the heat exchanger. The other side of the heat exchanger and all remaining piping in the snowmelting subsystem is filled with a propylene glycol antifreeze solution. A variable speed injection pump regulates the flow of hot fluid from the heat exchanger into the snowmelting distribution system. A separate air separator, expansion tank, and pressure-relief valve are required on the antifreeze side of the heat exchanger since this portion of the system is pressure isolated from the water-filled portion of the system.

Heat input for both operating modes is supplied by a dual boiler system operated by a staging control. In the space heating mode, the boiler supply temperature is calculated based on an outdoor reset control. In snowmelting mode, the target boiler supply temperature is fixed at 200 °F. This is also the case when both subsystems operate simultaneously.

The control wiring for this system is shown in Figure 10–109 on page 404. Calls for space heating from any of the ten thermostats energize the associated valve actuators through the multizone relay centers. Normally open contacts within the relay centers close to initiate the boiler controller and the distribution circulator. The boiler controller fires one or both boilers depending on the difference between the target boiler temperature and the actual temperature.

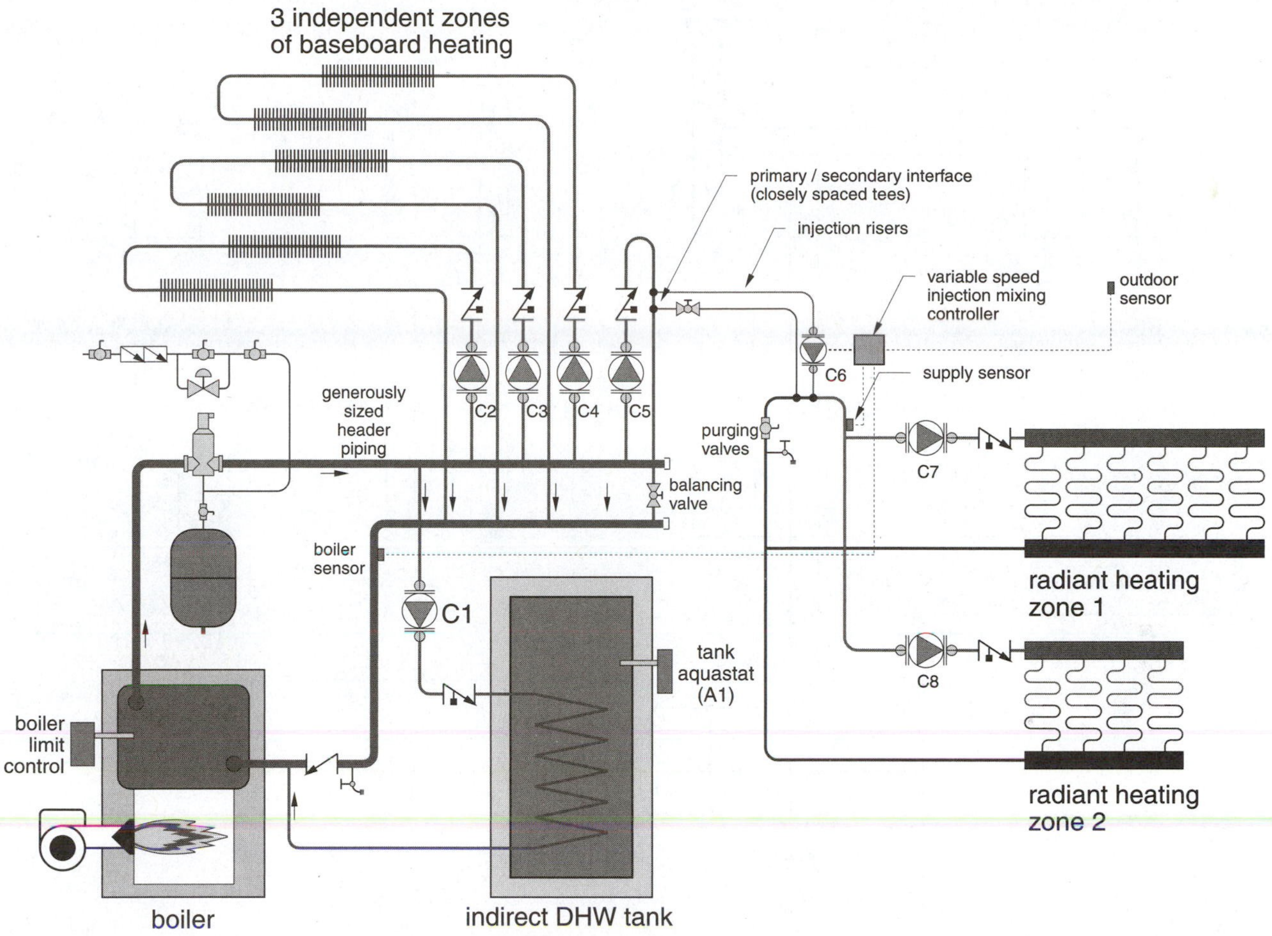

Figure 10–106 Piping schematic of a multiload/multitemperature system combining baseboard heating and radiant panel heating.

Snowmelting can be initiated manually or automatically depending on the setting of the mode switch. In the automatic mode, snowmelting is initiated by a snow detector that senses the presence of precipitation along with simultaneous outdoor temperature less than 38 °F. The contacts of the snow detector close to allow operation of the appropriate circulators and the injection-mixing control.

SUMMARY

Because hydronic radiant panels are an integral part of a building, they require more planning relative to other types of hydronic systems. The time committed to this planning is time well spent. It will return itself many times over.

A final point is that hydronic radiant panel heating systems tends to be "forgiving." Minor differences between calculated loads and estimated heat outputs from radiant panel tend to be insignificant in most projects. *This statement is not meant to endorse careless design or installation,* but rather to calm concerns over obtaining a precise match between calculated heat output and heating load in every room. Remember, the methods used to calculate both the heating load and radiant panel output are subject to inaccuracies. Neither can totally account for all the peculiarities of every building in which these systems are used. These calculations do however serve as a guide to judge the reasonableness of a proposed design and to check for large or unanticipated performance problems.

The technology of hydronic radiant panel heating is constantly evolving. New products and design methods appear each year. The reader is encouraged to contact several manufacturers of flooring heating equipment to stay abreast of the latest materials and design methods.

KEY TERMS

2-way thermostatic valve
3-way rotary mixing valve

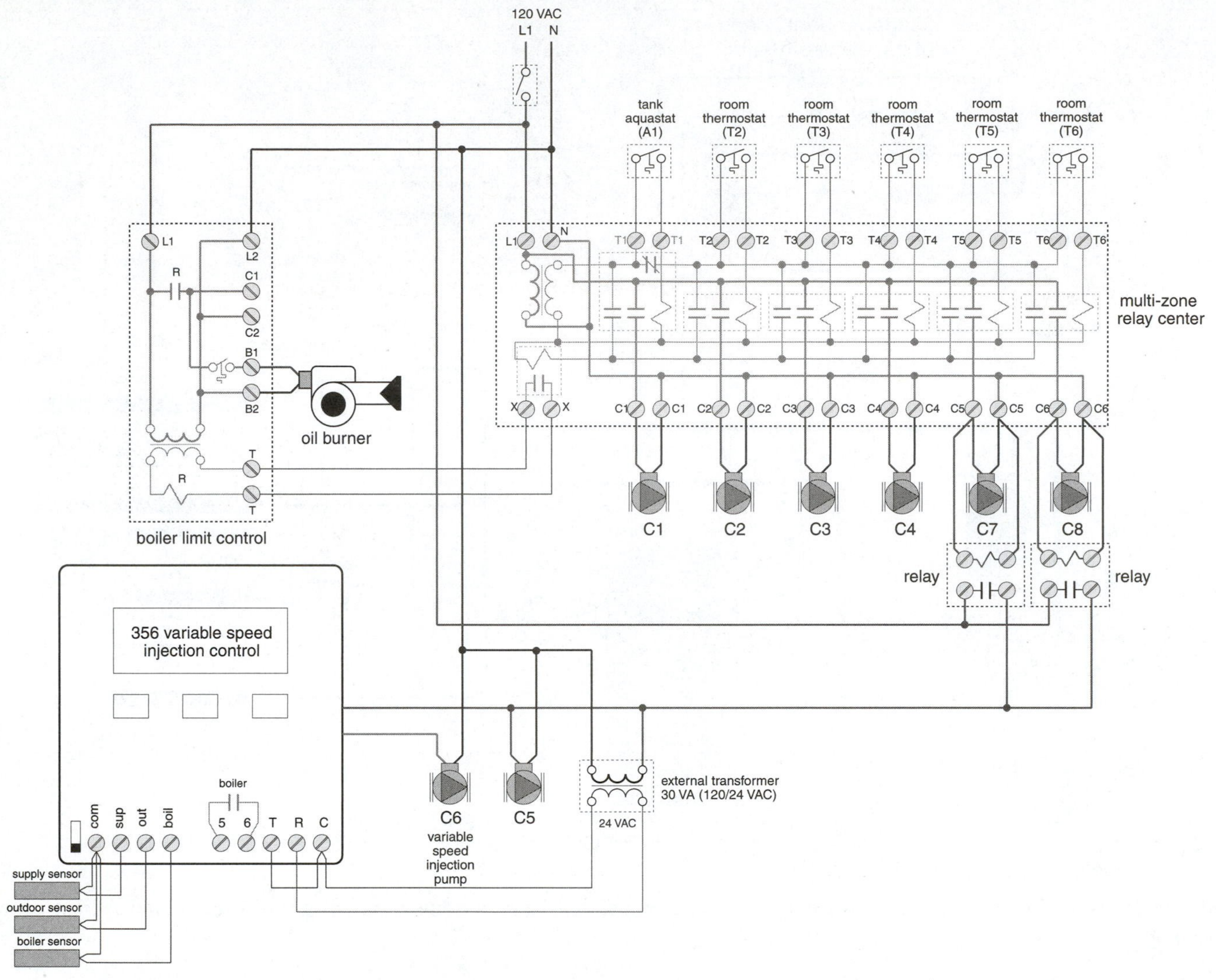

Figure 10–107 Electrical schematic for piping system shown in Figure 10–106.

3-way thermostatic mixing valves
4-way motorized mixing valves
above-floor tube and plate system
air temperature stratification
available floor area
below-floor tube and plate system
bend supports
bond breaker
buffer tank
computer-aided drafting (CAD)
concrete thin-slab
control joints
counterflow spiral
dead heading
dumb mixing devices
electromagnetic spectrum
emissivity
extruded polystyrene foam
gypsum thin slab
heat sinks
heat transfer plates
infrared thermal radiation
injection mixing
intelligent mixing device
interior circuits
isotherms
leaders
lightweight concrete
manifold station
manually set mixing valve
maximum allowed upward heat flux
mixing assembly

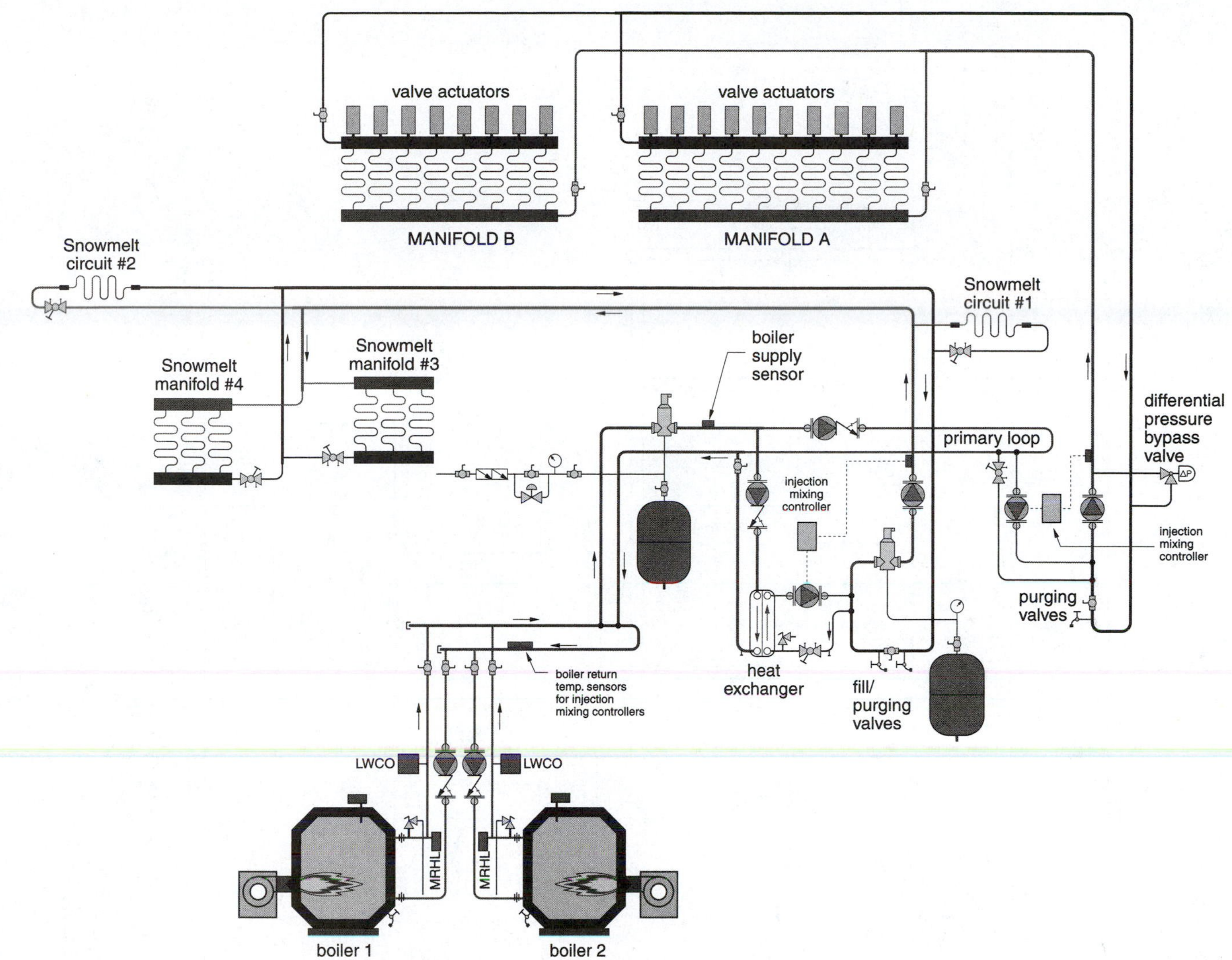

Figure 10–108 Piping schematic of a multiload/multitemperature system combining zoned radiant panel heating with snowmelting.

multiload/multitemperature systems
perimeter circuits
plateless staple-up system
poured gypsum underlayment
primary/secondary tees
proportional reset
radiant heating
return manifold
room air temperature profile
room side heat flux
serpentine patterns
short-cycling (boiler)
sleepers
striping
submixing
supply manifold
suspended tube system
system supply temperature
thermal radiation
tube and plate system
uncoiler
valve actuator
valveless manifold
vapor barrier

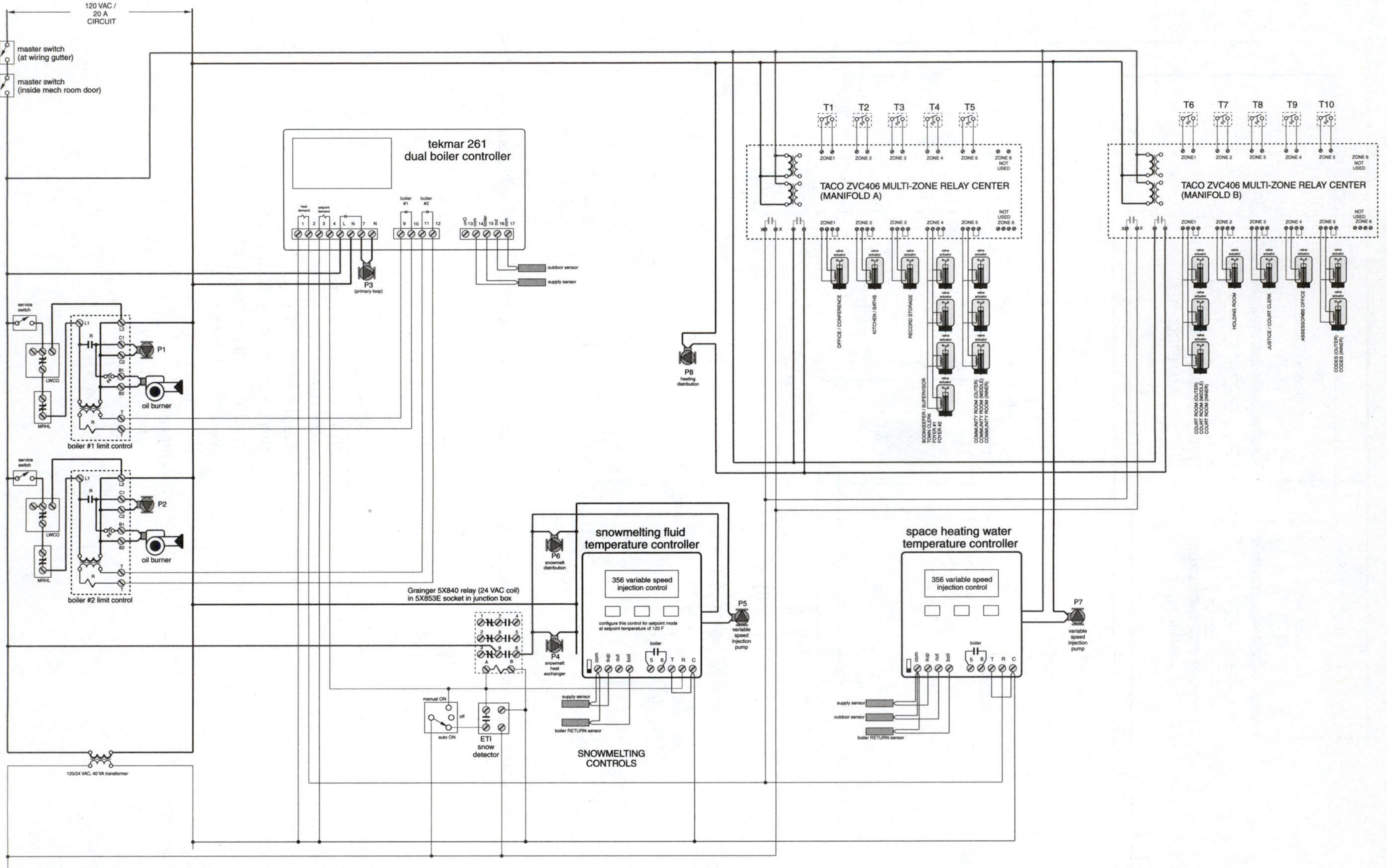

Figure 10–109 Electrical schematic for piping system shown in Figure 10–108.

CHAPTER 10 QUESTIONS AND EXERCISES

1. Why is it desirable to have warmer surfaces and air near the lower extremities of the body?
2. What is an advantage of radiant floor heating compared to finned-tube baseboard convectors in rooms with high ceilings?
3. Describe at least three considerations when locating a manifold station within a building.
4. Why is insulation needed beneath a radiant floor panel installed in a wood-frame floor, even when the space beneath the floor is maintained at normal comfort temperatures?
5. Describe the function of control joints in thin-slab installations.
6. What are some of the architectural considerations (things the building designer needs to know about) associated with using a thin-slab floor heating system?
7. Why are gypsum-based underlayments installed in two lifts rather than a single pour?
8. Why should floor piping circuits route the warmest water near the exterior wall?
9. A kitchen measures 10 feet by 12 feet with an island measuring 3 feet by 5 feet. The kitchen has a design heating load of 4,000 Btu/hr when it is 68 °F inside and 0 °F outside. What is the required average upward heat flux from a floor heating system at design load conditions?
10. Describe what happens to the supply water temperature requirement of a floor heating system when a bare concrete slab floor is covered with 3/8-inch ceramic tile. How does this compare to the supply water temperature requirement if the floor is covered with 1/4-inch carpet?
11. The kitchen described in Exercise 9 has a slab-on-grade floor with one 10-foot wall and one 12-foot wall exposed to the outside. The edges of the slab are insulated with 2 inches of extruded polystyrene insulation. The slab is expected to operate at 105 °F. Estimate the downward and edgewise losses from the slab.

DISTRIBUTION PIPING SYSTEMS 11

OBJECTIVES

After studying this chapter, you should be able to:

- Define eight basic hydronic distribution piping systems
- Assess several considerations when planning a zoned hydronic system
- Discuss the pros and cons of zoning with building owners
- Identify the type of distribution piping used in existing systems
- Explain the iterative processes involved in designing a distribution system
- Assess the suitability of a particular distribution system for a given application
- Use previously discussed analytical tools to design a distribution system
- Explain sizing and selection of differential pressure bypass valves
- Understand the fundamentals of primary/secondary piping
- Identify different types of piping elements in a hybrid distribution system
- Select appropriate hardware for various types of distribution systems.

11.1 INTRODUCTION

The purpose of a hydronic distribution system is to transport heated water to each heat emitter so that it can properly heat its assigned area. The decision of what type of distribution system to use must consider factors such as the:

- Number of heating zones in the building
- Method of controlling zone heat output
- Temperature requirements of the heat emitters
- Head loss characteristics of the heat emitters
- Cost of the piping system
- Size and operating cost of the system's circulator(s)
- Ability to expand the system in the future

This chapter discusses all the standardized piping approaches applicable to residential and light commercial systems. The strengths and weaknesses of each approach are examined. Design procedures are discussed and examples are given.

This chapter also ties together much of the material from earlier chapters into true system design. It is important that the reader be familiar with concepts such as series and parallel piping arrangements, head loss, hydraulic resistance, pumps curves, thermal performance of heat emitters, controls, and other material presented in earlier chapters.

11.2 ZONING CONSIDERATIONS

A heating **zone** *is any area of a building for which indoor air temperature is controlled by a single thermostat (or other type of temperature controller).* A zone may be as small as a single room, or it may be as large as an entire building.

Multizone systems have the *potential* to provide two important benefits not available in single zone systems, namely:

1. The ability to meet individual comfort requirements through independent control of room temperature.
2. The ability to conserve energy by reducing the temperature of unoccupied zones and thus the rate of heat loss from them.

Multizone hydronic heating systems are easy to create, much easier than their forced-air counterparts. Several hardware options exist for implementing zoning in different types of distribution systems. Most have been mentioned in previous chapters, and will be shown in several of the piping schematics in this chapter.

Before selecting a distribution piping system, it is important to evaluate how zoning can best be implemented in the building the piping system will serve. The goal is to maximize the benefits of zoning while at the same time keeping the system affordable, energy efficient, and reliable.

In some cases, hydronic zoning has been "oversold." For example, people sometimes think that the more zones a system has, the better it is designed. This is not necessarily true. Just because it is *possible* to create a separate zone for every room, does not mean it is the best choice. In some cases, the extra piping and control hardware necessary for room-by-room zoning adds complexity and cost without returning tangible benefits. Dividing a system into an excessive number of zones may even create operational problems such as **boiler short-cycling.** It can also add to system service requirements, again without returning any benefit to the owner.

System designers should consider the following when deciding how to best implement zoning.

Effect of Thermal Mass

Occupants sometimes assume that zoned systems are somehow able to lower the temperature of a room (or group of rooms) down to some setback temperature as soon as the thermostat is turned down. Likewise, they also assume that any given room will quickly warm back to normal occupied temperatures as soon as the thermostat is turned up.

Unfortunately, no occupied space can immediately drop in temperature by several degrees in a period of 2 to 3 minutes. Similarly, no (practical) heating system can immediately raise the temperature of an occupied space by several degrees. The time required to cool down or warm up a zone is significantly affected by the thermal mass of the heat emitter(s), and the building itself. The greater the thermal mass of the heating system and building components, the slower the decrease in indoor temperature when heat input to the zone is turned off.

Studies have shown that well-insulated buildings with high thermal mass might only experience a temperature drop of 2 °F to 3 °F over a midwinter nighttime period lasting 8 to 10 hours, *even with their heating system completely off.* If one were to reduce the thermostat setting in such a building from 70 °F to 65 °F at bedtime, the inside air temperature would not decrease to the reduced setpoint before morning. The energy savings potential of such a situation is very limited. If the thermostat were turned down 10 °F or even 20 °F, the results would be the same.

The thermal mass of the heat emitter also affects the recovery period when the setback ends. Low thermal mass heat emitters such as finned-tube convectors can usually restore a room that has been set back 10 °F to normal comfort in perhaps 15 to 30 minutes. High thermal mass systems such as a heated concrete slab-on-grade floor can take several hours to bring a space back to normal comfort temperature following a prolonged setback period. This is illustrated in Figure 11–1 and it shows the simulated heating of both low-mass and high-mass radiant floor panels, starting at an initial temperature of 60 °F, and having a constant heat input of 60,000 Btu/hr.

It is important not to use heat emitters of widely different thermal mass in the same zone (e.g., controlled by the same thermostat) when that zone will operate with frequent changes in setpoint temperature. Because their response times during both cooldown and warm-up are so different, neither type of heat emitter is likely to produce acceptable comfort during the transition periods between stable setpoint temperatures.

It is also important to realize that the energy conservation potential of a high thermal mass-zoned system is very dependent on the *duration* of any reduced thermostat setting. If a zone can be maintained at a reduced temperature for several days, the energy conservation potential can be significant. However, *attempting to create significant temperature changes on a daily basis in zones with high thermal mass heat emitters is usually futile.* The energy savings will be very small, and the potential for temperature overshoot and undershoot will be significant. This characteristic should be explained to occupants, who often do not understand how high mass systems respond to changes in the thermostat settings.

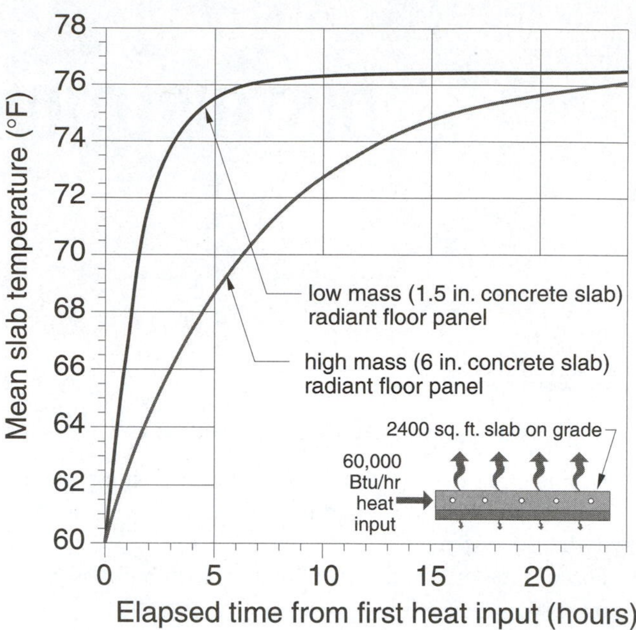

Figure 11–1 Slab temperature versus time for 2400 square foot slabs of different thicknesses while warming from an initial temperature of 60 °F with constant 60,000 Btu/hr input.

Inter-zone Heat Transfer

Another factor affecting zoning performance is **inter-zone heat transfer.** This refers to heat transfer through *interior* partitions as the building attempts to equalize temperature differences from one room to another. If, for example, a particular room is kept at 60 °F, while an adjacent room is kept at 75 °F, heat will flow from the warmer to the cooler room through the wall separating them. This partially defeats attempts to maintain temperature differences by zoning. The greater the thermal resistance of the exterior envelope of the building, the harder it is to maintain significant temperature differences between rooms separated by uninsulated interior partitions.

Zoned areas in which it is desired to maintain temperature differences of 10 °F or more should have insulated partitions separating them from other zones. Doors separating such areas must remain closed when these temperature differences are to be maintained.

Occupant Management of Zoned Systems

Another consideration that helps determine the cost effectiveness of zoning is the willingness of the occupants to regulate the system. It might seem obvious that an owner who is willing to pay more for an extensively zoned system would also be willing to regulate it. However, experience shows this is not always true, especially when nonprogrammable thermostats are used. Today's fast-paced lifestyles mean people often forget to regulate their heating systems, especially when their schedules vary from day to day.

Different rooms within a single zone (e.g., controlled by a single thermostat) and supplied with the same water temperature can still be maintained at different temperatures by adjusting the flow rate through individual heat emitters. This is usually adequate when the goal of the occupants is to maintain several rooms at different (but stable) temperatures. Room-by-room zoning is justified in situations where several rooms are to have both different, and *frequently changing* air temperatures. These distinctions should be carefully discussed with the eventual owners before committing several hundred dollars for extensive zoning controls that may seldom be used.

Effect of Solar Heat Gain

Rooms with large south-facing windows may at times be totally heated by solar heat gains. Rooms in the same building that do not receiving solar gain may still require heat input. Well-planned zoning separates rooms with likely solar heat gain from those without such gain. It also accounts for the fact that the rooms receiving solar heat gain can change over the course of the day as the sun moves across the sky. This type of zoning is crucially important in buildings designed to take significant advantage of solar heat gain.

It is also important to use heat emitters with low thermal mass in areas subject to large and highly variable solar heat gain. The low-mass heat emitters can quickly start and stop heat input as necessary.

Desired Sleeping Comfort

Many people prefer to sleep in cool bedrooms. Good zoning design allows for this possibility. It is common to zone a residential hydronic system so one or more bedrooms are controlled as a single zone. It is also common to create one zone for the master bedroom, with a separate zone for the master bathroom. This allows the bathroom to remain comfortable even when the bedroom is cool.

Effect of Activity Level

Exercise rooms are also good candidates for separate zoning, as are rooms that are frequently unoccupied (guest rooms, hobby rooms, etc.). Rooms where occupants have moderate to high activity levels should be zoned separately from rooms with sedentary occupants.

Zoning of Nonoccupied Areas

It seldom makes sense to maintain normal human thermal comfort requirements in spaces that are infrequently occupied. Basements, if heated, fall into this category, and are normally set up as a separate zone. Garages, if heated, are usually maintained at air temperatures of 45 °F to perhaps 60 °F, and again are usually set up as a separate zone. In cold climates the garage zone is usually designed to operate with antifreeze solutions. This allows for intermittent heating without the risk of freezing should the zone be turned off during very cold weather.

Zoning of Transitory Areas

Entry vestibules with exterior doors and doors into fully heated space are good candidates for separate zoning. In some cases, the goal is to maintain the vestibule at higher temperatures to help buffer interior spaces from cold air gusts when the exterior door is opened. When floor heating is used in such areas, the floor surface temperature is often maintained several degrees above that of other interior areas to help dry tracked-in water and snow.

Effect of Heat Gains from Lighting/Equipment

Rooms that contain heat-producing equipment such as computers, vending machines, or cooking facilities are usually good candidates for separate zoning. It is common for rooms with several electrical appliances to require significantly less heat input than adjacent rooms without such equipment.

Effect of Exterior Wall Exposure

Rooms with minimal exterior exposure will have small heating loads relative to their floor area. When such rooms are controlled as separate zones, overheating can be minimized.

When properly planned, zoning can significantly enhance owner satisfaction while also improving the energy efficiency of the system.

11.3 SYSTEM EQUILIBRIUM

All hydronic systems, regardless of how they are designed, will attempt to stabilize to a unique combination of operating temperatures and flow rate(s).

At these conditions, the system is said to be in **equilibrium**. The system may or may not provide the proper heating of the building even operating at equilibrium. In animated terms, the system "does not care" if proper heating of the building is occurring; it only "cares" about adjusting itself to and maintaining itself at equilibrium conditions.

The task of the designer is to assemble components into a system so equilibrium occurs at values that also allow the building to be properly heated. This needs to be accomplished within an allowable budget, with the highest possible fuel efficiency, and so the resulting system will last as long as possible with minimal maintenance.

Thermal Equilibrium

Thermal equilibrium *occurs when the heat emitters release heat at exactly the same rate the heat source adds heat to the system.* Under such conditions, the fluid temperature at any given point in the system will remain constant.

An analogy to thermal equilibrium between the heat source and heat emitters is that of pouring water into a bucket containing several holes along a vertical line up its side, as shown in Figure 11–2. The holes represent the ability of the heat emitters to release heat into the building. The water being poured into the bucket represents the heat input from the heat source. The height of the water in the bucket, at any time, represents the temperature of the fluid in the distribution system.

When water is slowly poured into the bucket, it can run out the lower hole(s) while the water level rises only a short distance above the bottom of the bucket. As water is poured in faster, the level in the bucket rises until the rate of leakage again balances with the rate of input. Further increases to the water input rate correspond to further rises in the water level. If the input rate is reduced, the water level will drop until equilibrium conditions between input and leakage are again established.

In a similar manner, the fluid temperature in a hydronic system will increase or decrease as necessary to balance the rate of heat input with the rate of heat release. Most control components merely interrupt this process as necessary to prevent wide differences between what the designer thinks the system should be doing and what it naturally wants to do, or to prevent unsafe operating conditions.

True thermal equilibrium is seldom achieved in most hydronic heating systems due to factors such as changing heating loads, on/off operation of the heat source, and the warming and cooling of the system's thermal mass.

Traditional system design is based on assumed steady state operation of components such as the heat source and circulators at "design load conditions." However, during the vast majority of the heating season the system is not at these conditions. A truly well-designed system operates properly at design load conditions, and remains stable and energy efficient during partial load conditions.

Hydraulic Equilibrium

Hydraulic equilibrium *is achieved when the head added by the circulator equals the head dissipated by the flow resistance of the piping system* (Figure 11–3). This balance occurs almost instantly when the circulator is turned on, or whenever flow regulating valves are adjusted. It occurs regardless of how the piping system is designed. Chapter 6, Fluid Flow in Piping, and Chapter 7, Circulating Pumps, showed how to predict where hydraulic equilibrium will occur for any given piping system and circulator combination.

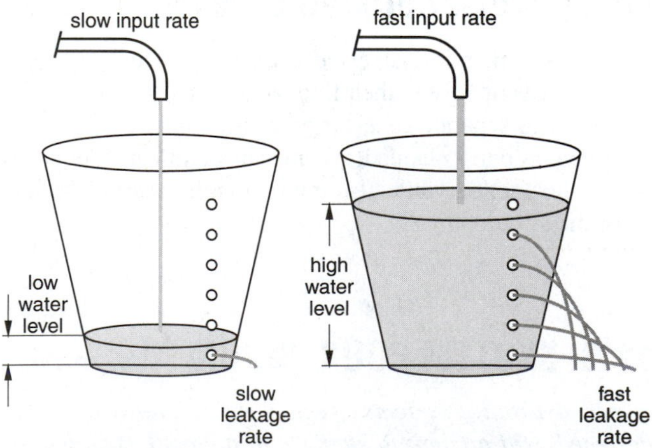

Figure 11–2 Analogy between fluid leaking from a bucket with holes, and thermal equilibrium between a hydronic heat source and the heat emitters of a distribution system.

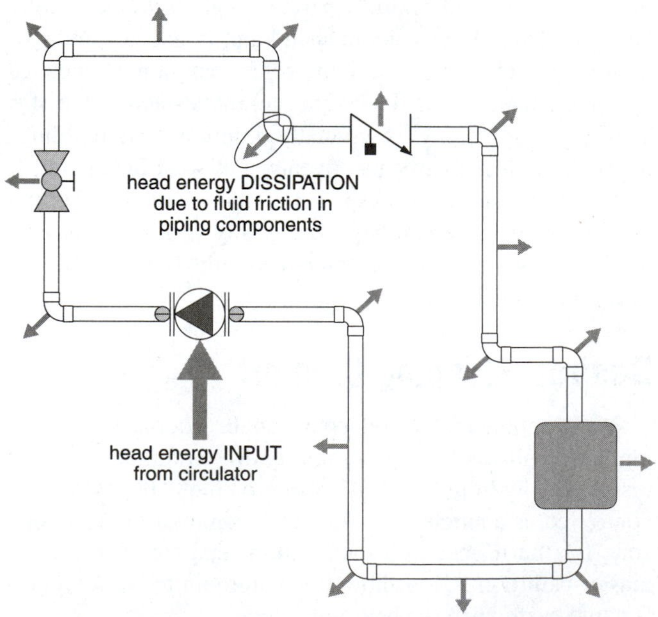

Figure 11–3 Hydraulic equilibrium exists when the rate of head energy input by circulator equals rate of head energy dissipation by piping.

The objective is to design the piping system, and select the circulator, so the operating flow rate is *reasonably close* to flow rate assumed when the heat emitters were selected. This presents a paradox because the heat emitters will often have a significant effect on the flow resistance of the system, and hence its flow rate. To further complicate matters, the heat output of the heat emitters also depends on the flow rate through them.

11.4 THE CONCEPT OF ITERATIVE DESIGN

Situations where the selected equipment depends on the overall performance of the system, yet the system performance depends on the selected equipment, occurs frequently in many areas of engineering. The solution lies in a concept known as **iteration.** Using this concept, the designer repeatedly analyzes both the hydraulic and thermal performance of a system using "candidate" hardware options and operating conditions, until a combination is found that yields acceptable heat distribution.

For hydronic systems the process of iterative design typically goes as follows:

Step 1: Make a tentative selection of heat emitters for the rooms to be heated based on their design heat loads, and on an estimated average water temperature. A common practice is to assume the average water temperature in the system will be about 5 °F to 10 °F lower than the supply temperature from the heat source.

Step 2: The system flow rate for water can be estimated using Equation 11.1:

(Equation 11.1)

$$f_s = \frac{Q}{500(\Delta T)}$$

where:
f_s = initial estimated system flow rate (gpm)
Q = total design heating load served by the system (Btu/hr)
ΔT= intended temperature drop of the system (°F)

Equation 11.1 is a rearranged form of the sensible heat rate equation introduced as Equation 4.4.

When a fluid other than water is used, this equation can be modified as follows:

(Equation 11.2)

$$f_s = \frac{Q}{8.01Dc(\Delta T)}$$

where:
f_s = initial estimated system flow rate (gpm)
Q = total design heating load served by the system (Btu/hr)
ΔT= intended temperature drop of the system (°F)
D = the fluid's density at the average system temperature (lb/ft^3)
c = the fluid's specific heat at the average system temperature (Btu/lb/°F)

If the heat emitters are connected in series, the system flow rate will pass through each heat emitter. If the heat emitters are connected in parallel, the flow rate through each heat emitter can be proportioned based on load using Equation 11.3:

(Equation 11.3)

$$f_i = f_s\left(\frac{Q_i}{Q_t}\right)$$

where:
f_i = estimated flow rate through heat emitter #i (gpm)
f_s = initial estimated system flow rate (gpm)
Q_i= design heat output of heat emitter #i (Btu/hr)
Q_t= total design load of the system (Btu/hr)

Step 3: Select a piping distribution system to connect the heat emitters with the heat source. Several options are shown in later sections of this chapter.

Step 4: Select tube sizes for the series loop or parallel branches based on establishing a flow velocity between 2 and 4 feet per second. Figure 11–4 gives the flow rates that correspond to these flow velocities for various types and sizes of tubing. For other tube types and sizes, use Equation 6.1 to establish the relationship between flow rate and flow velocity, then select a tube size to keep the flow velocity between 2 and 4 feet per second.

Step 5: Using the methods of Chapter 6, Fluid Flow in Piping, the piping system can now be described by a hydraulic resistance diagram. After reducing this diagram to a single equivalent hydraulic resistance, the system curve can be sketched.

Step 6: The pump curves of one or more "candidate" circulators can now be matched against the system curve. Recall from Chapter 7, Circulating Pumps, that the intersection of the system curve and the pump curve yields the operating flow rate for the combination. *It is likely that the system flow rate produced by any given circulator will not exactly match the flow rate estimated in step 2. This is normal and to be expected as part of the design process.*

Step 7: If all heat emitters are in a series circuit with the circulator, this step can be skipped. If some or all of the heat emitters are located in parallel branch circuits, the methods of Chapter 6 can be used to find the flow rate in each branch.

Tubing size/type	Minimum flow rate (based on 2 ft/sec) (gpm)	Maximum flow rate (based on 4 ft/sec) (gpm)
3/8" copper	1.0	2.0
1/2" copper	1.6	3.2
3/4" copper	3.2	6.5
1" copper	5.5	10.9
1.25" copper	8.2	16.3
1.5" copper	11.4	22.9
2" copper	19.8	39.6
2.5" copper	30.5	61.1
3" copper	43.6	87.1
3/8" PEX	0.6	1.3
1/2" PEX	1.2	2.3
5/8" PEX	1.7	3.3
3/4" PEX	2.3	4.6
1" PEX	3.8	7.5
1.25" PEX	5.6	11.2
1.5" PEX	7.8	15.6
2" PEX	13.4	26.8
3/8" PEX-AL-PEX	0.6	1.2
1/2" PEX-AL-PEX	1.2	2.5
5/8" PEX-AL-PEX	2	4.0
3/4" PEX-AL-PEX	3.2	6.4
1" PEX-AL-PEX	5.2	10.4

Figure 11–4 Minimum and maximum flow rates for different types and sizes of tubing, based on flow velocities between 2 and 4 ft./sec.

Note: Steps 5 through 7 can also be handled using the **Hydronic Circuit Simulator** *module in the Hydronics Design Studio software.*

Step 8: The thermal performance of each heat emitter can now be verified at the actual flow rate and entering water temperature present at their location in the system. If the heat output of a given heat emitter is too low, a number of options exist. These include selecting a higher output heat emitter, increasing system operating temperature, or modifying the piping to obtain a higher flow rate. If the output of a heat emitter is more than 10 percent above design load, a lower capacity unit can be used, or, in the case of a parallel system, a balancing valve can be used to reduce its flow rate.

Step 9: Depending on what, if any, changes are made to the system, the designer may need to go back to step 3 and repeat the process of determining the new flow rates in the modified system. *This is iteration.* In most cases only one or two iterations need to be made to find a design that can provide the necessary heat output with the selected heat emitters, piping system, and circulator.

11.5 SINGLE SERIES CIRCUITS

The simplest distribution system is a single piping loop that progresses from the heat source, through each heat emitter, and back to the heat source. System operation is often controlled by a single regulating device such as a room thermostat. This type of distribution system is called a **single series circuit.**

Single series circuits are appropriate for small buildings in which all rooms experience similar changes in heating load. An example of a single series circuit of finned-tube baseboard convectors is shown in Figure 11–5.

Single circuit systems can also be designed around a combination of heat emitters including panel radiators and fan-coils. When different types of heat emitters are used they should have comparable thermal mass.

Because heat input to the entire building is regulated based on the temperature at one location in the building, overheating or underheating of rooms other than where the air temperature is sensed is very possible. *It is crucial that all heat emitters be sized for the load of their respective rooms, and to the water temperature at their location within the piping circuit.* The latter condition is often overlooked when heat emitters are sized based on a single average water temperature. Chapter 8, Heat Emitters, demonstrated the differences in sizing finned-tube baseboard units based on a single average water temperature, versus the temperature of the water at the location of each baseboard. Using the single average water temperature method, the baseboards near the beginning (hottest) part of the circuit tend to be oversized and thus put out too much heat. Those near the end of the circuit tend to be undersized and thus put out too little heat. These tendencies will also be true for other types of heat emitters if sized based on the average water temperature. To prevent this, *the designer must keep track of the fluid temperature as it cools down from one heat emitter to the next.* Begin at the location of the first heat emitter and progress around the circuit treating the outlet temperature of each heat emitter as the inlet temperature to the next.

If a piping segment between heat emitters is longer than 40 feet, or is otherwise expected to have a significant heat loss, the designer should calculate the temperature drop along the pipe and correct the inlet temperature to the next heat emitter accordingly. Methods for estimating pipe heat loss were presented in Chapter 8.

The **Series Baseboard Simulator** module in the Hydronics Design Toolkit software can be used to quickly determine the required lengths of finned-tube baseboard in a series circuit, or portion thereof.

The limiting factors in designing single circuit systems are usually temperature drop and flow resistance. In the United States, series circuits have traditionally been designed around a temperature drop of approximately 20 °F. *However, there is nothing magical about this number.* Systems can be designed to operate properly with both smaller or larger temperature drops.

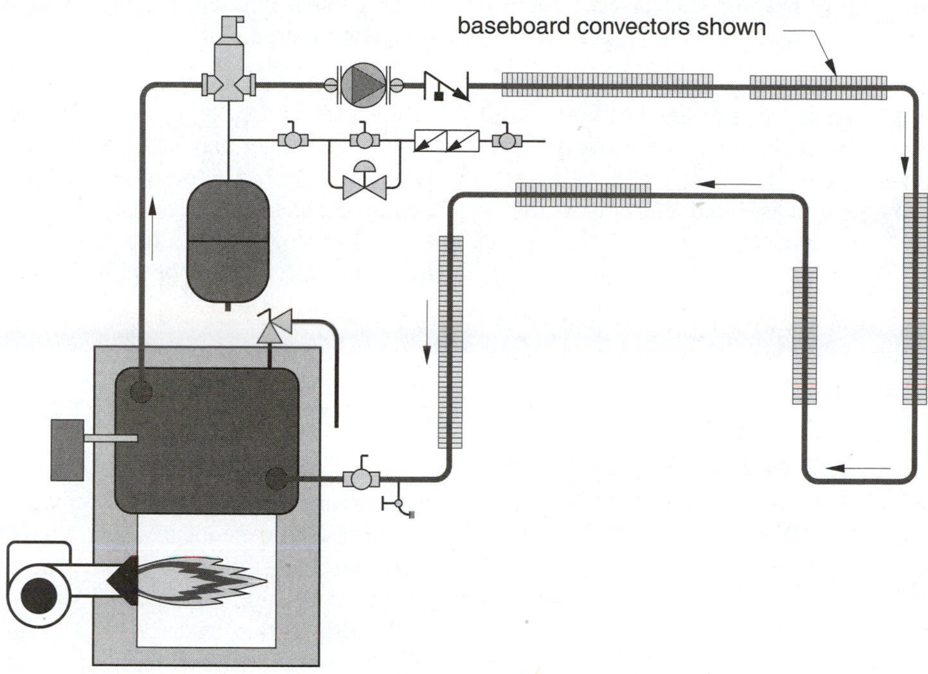

Figure 11–5 Example of a single series circuit distribution system.

In Europe, hydronic distribution systems are routinely operated with temperature drops of 30 °F to 40 °F. The advantage of having a high temperature drop is the flow rate can be relatively low. This in turn allows small diameter piping and a smaller, less power-consuming circulator to be used. The disadvantage of a high temperature drop system is that heat emitters near the end of the series circuit often need to be significantly larger to meet a given load with the lower temperature water. There is also the need of maintaining the return temperature to a conventional boiler high enough to prevent sustained flue gas condensation.

Systems with low temperatures drops usually have relatively high flow rates that may require increased pipe sizes to keep the flow velocity under 4 feet per second. Such systems may also require larger circulators. However, because the heat emitters will operate at higher average temperatures they can usually be smaller compared to those in systems operating at higher temperature drops.

It is impossible to say what the optimum temperature drop is for a single series circuit. The answer depends on the cost of piping, circulators, heat emitters, electricity, and several other factors. The designer may choose to explore different possibilities based on various combinations of supply water temperature and circuit flow rate using the procedures and resources described just ahead.

Series circuits that contain one or more heat emitters with high flow resistance characteristics must be carefully evaluated for potential problems. A heat emitter with high-flow resistance can greatly restrict flow rate through the entire series circuit, limiting its total heat output.

Another potential problem is excessive flow velocity through the small tubes or valves in the restrictive heat emitter. This can create flow noise and eventual erosion corrosion. A high head circulator may be required. High head differentials across a circulator encourage cavitation, especially in high temperature systems. Pumping heads in excess of 30 feet should be carefully scrutinized for cavitation potential. In many cases, a different type of distribution system will prove more appropriate for use with heat emitters having high-flow resistance characteristics.

A definite drawback of a single zone series circuit is that zone control is limited to the control features present on the heat emitters. In the case of finned-tube baseboard, heat output can be reduced up to about 50 percent by closing the dampers on the enclosure. When fan-coils are used, the blower speed can be reduced to slightly limit heat output. Both these methods, however, require manual adjustments to the heat emitters in response to changing load conditions. Many occupants either do not know these adjustments can be made or soon tire of making them. The resulting overheating and underheating is begrudgingly tolerated to the detriment of future referrals for the installer.

Design of Single Series Circuits

The iterative design procedure discussed in Section 11.4 can be adapted to the design of single series circuits. The following is a brief summary of this procedure, specifically tailored to series circuits.

Step 1: Determine the design heating load of each room served by the series circuit.

Step 2: Select a tentative supply temperature and temperature drop for the circuit based on the heat emitters used. For example, for a baseboard circuit supplied by a conventional boiler, the supply temperature is often selected in the range of 170 °F to 190 °F, along with an assumed temperature drop of 20 °F.

Step 3: Use the average water temperature of the system to select tentative sizes of heat emitters for each room. The average water temperature is the supply temperature minus half the assumed temperature drop.

Step 4: Once the tentative sizes of the heat emitters are establish, they can be located within the floor plan of the space being heated. In most cases, the preferred locations will be under windows on exterior walls. However, every building presents different circumstances. It may be necessary to locate some heat emitters in less traditional locations to accommodate furniture or other requirements of the owners.

Step 5: Once the heat emitters are located, sketch a pipe routing path that accommodates both the building construction and heat emitters. Think about where it will be possible (or *not* possible) to route the piping under floors, through partitions, and so on, to access each heat emitter. Once the pipe routing has been determined, estimate the length of the series circuit as well as the number of elbows and other fittings or valves it will likely contain. Do not worry about getting the *exact* length or number of fittings; a reasonable estimate is fine. Be sure to include an allowance for the piping, fittings, and valves in the vicinity of the heat source.

Step 6: Knowing the total heating load served by the circuit, and the estimated temperature drop, use Equation 11.1 or Equation 11.2 to calculate a **target flow rate.**

Step 7: Using the flow rate from step 6, select a tube size for the circuit using the data in Figure 11–4 or Equation 6.1.

Step 8: Use the methods from Chapter 6, Fluid Flow in Piping, to construct a system curve for the series circuit. This curve should include the hydraulic resistance of both the piping and heat emitters. The Hydronic Circuit Simulator module in the Hydronics Design Studio software can also be used for this step.

Step 9: Select a circulator such that the intersection of the system curve and pump curve (e.g., the operating point) falls close to the target flow rate determined in step 6. An operating point flow rate within +/–10 percent of the target flow rate is generally acceptable.

Step 10: Knowing the system flow rate and supply temperature, verify the heat output and sizing of each heat emitter on the circuit. If fin-tube baseboard is used for all heat emitters, the Series Baseboard Simulator module in the Hydronics Design Studio software can be used. If other types of heat emitters are present, their performance and temperature drop will have to be evaluated individually.

It may be necessary to change the size of one or more heat emitters based on the results of this step. If such is the case, the designer may have to iterate the calculations starting with step 3. The designer may also choose to explore options such as reduced supply water temperature, or a larger circuit temperature drop before establishing the final hardware selection and operating conditions.

11.6 SINGLE CIRCUIT/MULTIZONE (ONE-PIPE) SYSTEMS

A variation on the basic series circuit design allows for one or more of the heat emitters connected to a single piping circuit to be independently controlled. This piping arrangement involves the use of **diverter tees** as discussed in Chapter 5, Piping, Fittings, and Valves. These tees are used to divert a portion of the water flowing in the main piping circuit through a branch circuit that includes the heat emitter. An example of the resulting **single circuit/multizone (one-pipe) system** is shown in Figure 11–6.

The diverter tees can be used individually or in pairs. When two diverter tees are used, they set up a greater pressure difference across the branch circuit. This in turn creates a higher flow rate in the branch circuit. Single diverter tee arrangements are often sufficient for low resistance heat emitters such as a few feet of finned-tube baseboard or a panel radiator. Double diverter tees may be needed on heat emitters with higher flow resistances such as fan-coils. They are also suggested to overcome buoyancy forces when the heat emitter is located several feet *below the distribution circuit.* When two diverter tees are used, there should be at least one foot of tube between the tees to allow turbulence created by the upstream tee to partially dissipate before the flow enters the downstream tee.

One of several types of valves can be used to control the flow rate through the branch circuit containing the heat emitter. By regulating the flow, the heat output from the heat emitter can be varied or turned completely off.

The valve can be a simple manually adjusted globe or angle valve adjacent to or even built into the heat emitter. Manual valves used in systems with fixed supply water temperatures will require frequent adjustment if the room is to maintain a constant temperature during changing load conditions. If automatic control of the heat output is desired, a nonelectric thermostatic radiator valve can be used. These valves were discussed in Chapter 5, Piping, Fittings, and Valves.

Systems employing diverter tees are often called **one-pipe systems**. They offer good zoning flexibility at relatively low cost. Any one or more of the heat emitters on the circuit can be selected to have independent temperature control simply by piping it into the circuit using diverter tee(s), and some type of control valve.

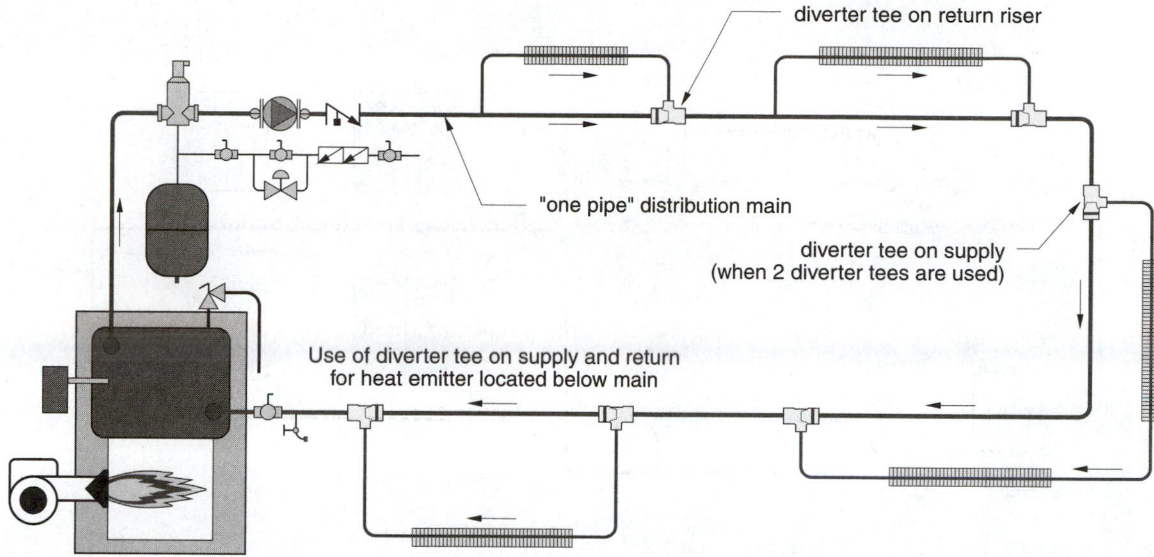

Figure 11–6 Use of one or two diverter tees to induce flow in branch circuits.

Figure 11–7 shows a series circuit system in which one heat emitter is connected using a diverter tee arrangement while the other baseboards are piped in series. This is a convenient way to limit the heat output of one heat emitter without significantly affecting the remainder of the heat emitters. This arrangement could be used, for example, in a guest bedroom that only needs to be heated to normal comfort temperatures a few days each year.

Diverter tees can also be used to supply a piping *subassembly* such as the one shown in Figure 11–8. In this case, the subassembly is a group of four panel radiators, all located in the same heated space. The radiators are connected in a parallel reverse-return arrangement to ensure the same supply temperature and reduce head loss. Parallel reverse return systems will be discussed more in Section 11.9.

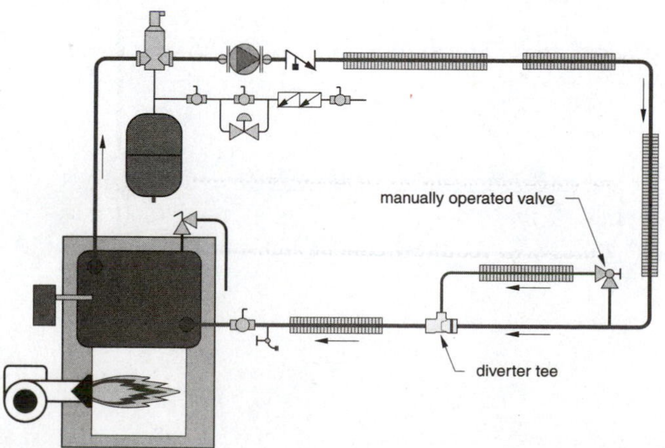

Figure 11–7 Use of diverter tee(s) and a manually adjusted valve to limit heat output of one heat emitter in an otherwise series distribution circuit.

When nonelectric thermostatic radiator valves are used for individual room temperature control, the designer must provide controls to operate the heat source and distribution circulator whenever the building might need heat. One approach is to equip the system with an outdoor reset control that automatically operates the heat source and distribution circulator when the outdoor temperature falls below a preset value. When such a control system is used, the thermostatic valves on the heat emitters serve as temperature-limiting devices for their respective rooms. Since this approach requires constant circulation in the main distribution circuit, all piping should be insulated to minimize heat loss. This concept is shown in Figure 11–9.

Another benefit of one-pipe systems with branch control valves is that they allow some heat emitters to be intentionally oversized without losing the ability to control individual room temperatures. This may be desirable when an infrequently occupied room needs to be quickly warmed following a setback. Intentional oversizing also allows a space to be heated higher than the normal comfort temperature when desired. An example would be when someone wants to quickly warm a bathroom to approximately 75 °F while taking a bath, and then return the room to a normal 68 to 70 °F setting.

Whenever diverter tees and control valves are used to reduce or totally stop water flow to a heat emitter, it is crucial that the heat emitter and the piping leading to it can not freeze during cold weather. Some thermostatic radiator valves have a freeze-proof minimum setting that attempts to maintain the air temperature in the space above freezing, even when the setting knob appears fully closed. If a manual control valve is used, it should be left slightly open during very cold weather.

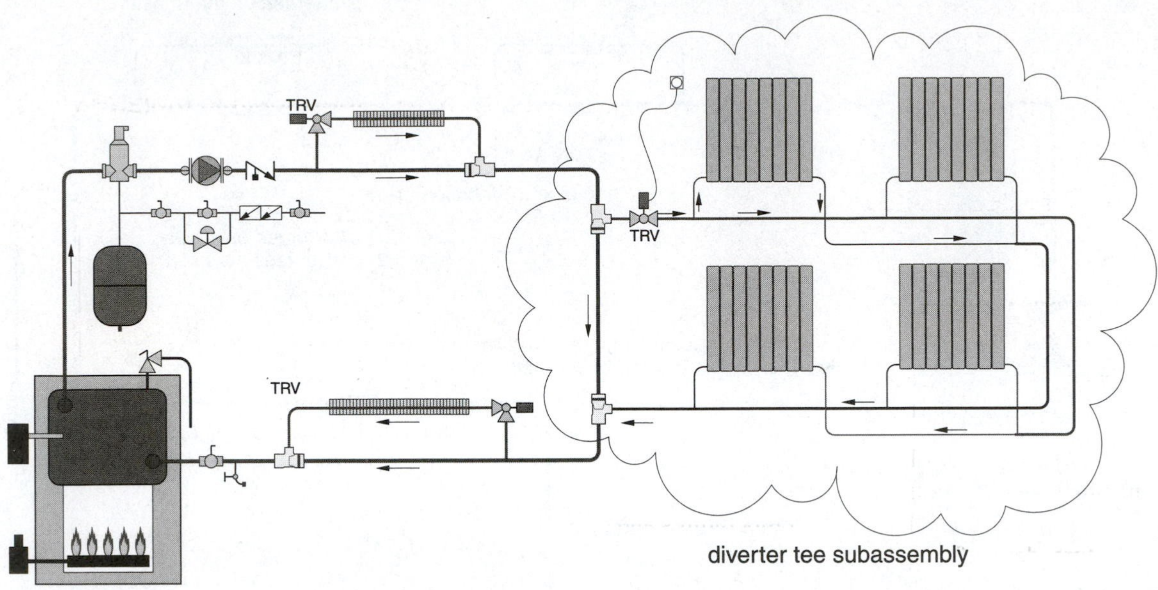

Figure 11–8 Use of diverter tees to route flow through a subassembly of several heat emitters.

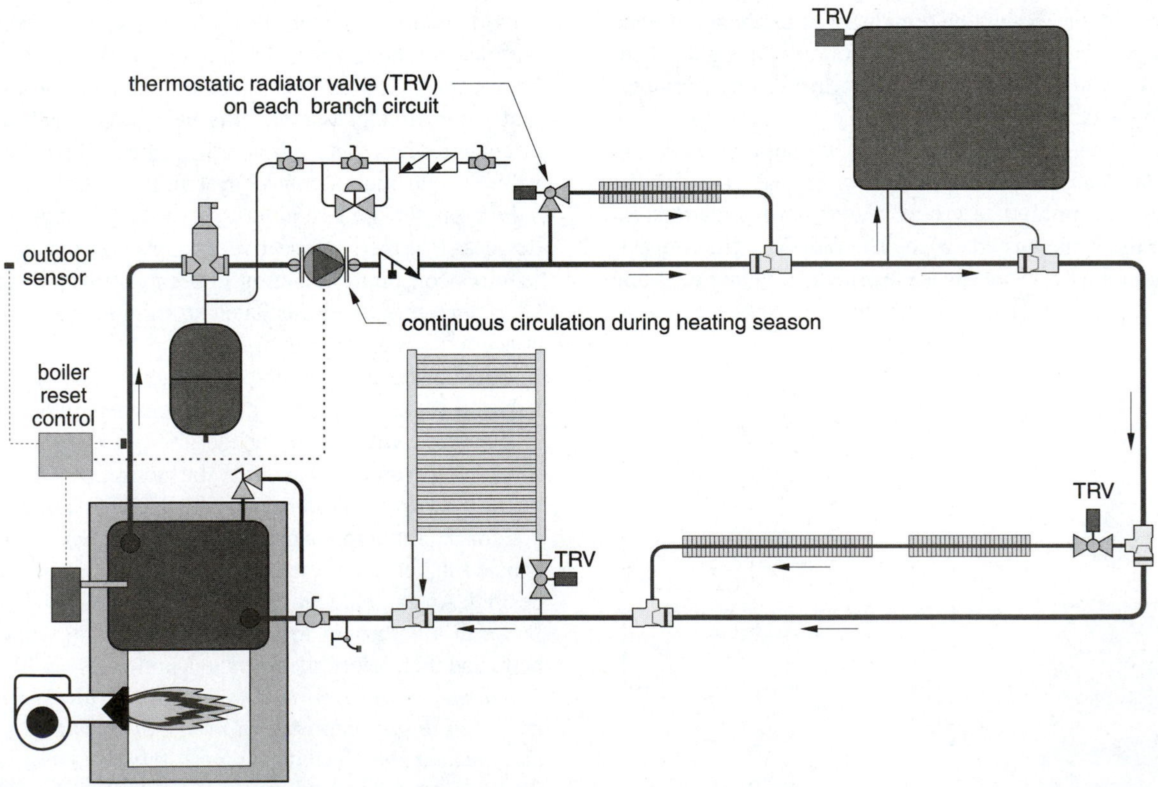

Figure 11–9 Diverter tees and thermostatic radiator valves (TRVs) allow individual heat output control from heat emitters. System also uses outdoor reset control and constant circulation.

If there is a chance the valve may be unintentionally closed, it may be a good idea to remove the handle.

Design Procedure For One-pipe Systems

The iterative design concept discussed in Section 11.4 also applies to the design of one-pipe systems. Additional steps are required due to the more complex piping and parallel circuit branches. The following procedure can be performed manually, or incorporated into software.

Step 1: Determine the design heating load of each room to be served by the one-pipe circuit.

Step 2: Select a tentative supply temperature, and overall circuit temperature drop for the circuit based on the heat source and heat emitters used. The supply temperatures and circuit temperature drops for one-pipe systems are often similar to those used for series circuits.

Step 3: Use the *average* water temperature of the system to select tentative sizes of heat emitters for each room. The average water temperature is the supply temperature minus half the assumed temperature drop.

Step 4: Knowing the total heating load served by the circuit, and the estimated temperature drop, use Equation 11.1 or Equation 11.2 to calculate the target system flow rate.

Step 5: Using the flow rate from step 4, select a tube size for the circuit using the data in Figure 11–4, or Equation 6.1.

Step 6: Once the tentative sizes of the heat emitters are established, they can be located within the floor plan of the space being heated. In most cases, the preferred locations will be under windows on exterior walls. However, it may be necessary to locate some heat emitters in less traditional locations to accommodate furniture or other requirements of the owners.

Step 7: After the heat emitters are located, sketch a pipe routing path that accommodates both the building construction and heat emitters. Think about where it will be possible (and *not* possible) to route the piping under floors, through partitions, and so on, to access each heat emitter.

Also sketch each diverter tee subassembly showing the heat emitter, estimated size and length of the main piping and branch piping, fittings, valves, and the number of diverter tees used to connect it to the main. An example is shown in Figure 11–10. This information is needed to determine the hydraulic resistance of each subassembly.

Step 8: Use the methods described in Chapter 6, Fluid Flow in Piping, to determine the hydraulic resistance of the components in the subassembly. Do this for both the branch circuit and the main piping between the tees. Include the resistance of the diverter tees in the overall resistance of the main pipe. This concept is shown in Figure 11–11.

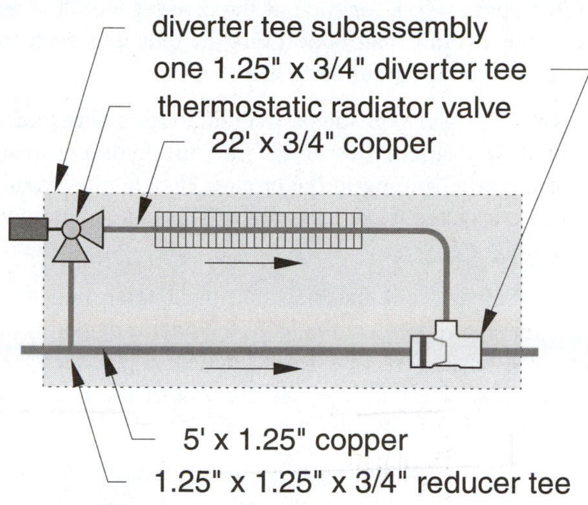

Figure 11–10 Identifying the piping elements in a diverter tee subassembly.

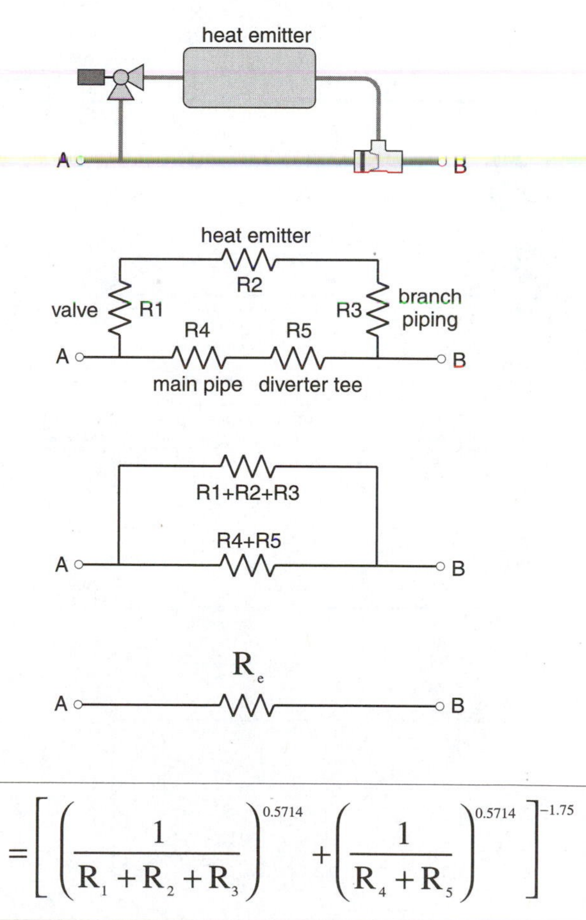

$$R_e = \left[\left(\frac{1}{R_1 + R_2 + R_3} \right)^{0.5714} + \left(\frac{1}{R_4 + R_5} \right)^{0.5714} \right]^{-1.75}$$

Figure 11–11 Reducing a diverter tee subassembly into a single equivalent hydraulic resistance.

Step 9: Repeat step 8 for each of the diverter tee subassemblies in the overall distribution circuit, reducing each to a single equivalent hydraulic resistance.

Step 10: Add the equivalent hydraulic resistances of all diverter tee subassemblies along with the hydraulic resistance of the other piping in the circuit. Use this information to generate a system resistance curve. This concept is shown in Figure 11–12.

Step 11: Select a circulator so that the intersection of the system curve and pump curve (e.g., the operating point) falls close to the target flow rate determined in step 4. An operating point flow rate within +/–10 percent of the target flow rate is generally acceptable.

Step 12: Use the system flow rate, and the parallel hydraulic resistances of each subassembly to calculate the flow rate through each heat emitter.

Step 13: Use the inlet temperature and flow rate to the first heat emitter to calculate its heat output.

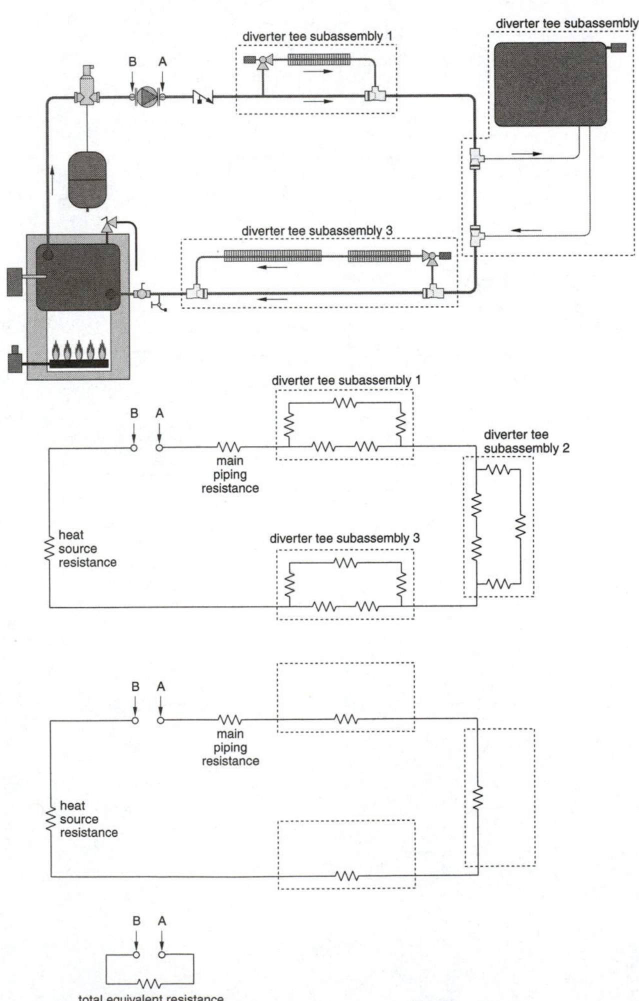

Figure 11–12 Reducing a one-pipe distribution system into a single equivalent hydraulic resistance.

Step 14: Calculate the temperature leaving the diverter tee subassembly using Equation 11.4. Typically this temperature is used as the supply temperature to the next heat emitter. However, if there is a long length of tubing leading to the next heat emitter, it may be necessary to calculate the temperature drop along this tube. Subtract this temperature drop from the outlet temperature of the upstream subassembly to determine the inlet temperature to the next subassembly. Heat loss from base copper tubing can be determined using methods presented in Chapter 8, Heat Emitters.

(Equation 11.4)

$$T_{out} = T_{in} - \frac{Q_i}{500 \times f_s}$$

where:

T_{out} = temperature leaving the subassembly (°F)
T_{in} = temperature of water entering the subassembly (°F)
Q_i = heat removed by the heat emitter (Btu/hr)
f_s = total flow rate entering the subassembly (gpm)
500 = a constant for water*

* For fluids other than water, replace the number 500 with (8.01 x D x c)

where:

D = density of the fluid at average circuit temperature (lb/ft3)
c = specific heat of fluid at average system operating temperature (Btu/lb/°F)
8.01 = a constant

Step 15: Repeat steps 10 through 12 for each diverter tee subassembly, in sequence, until reaching the end of the circuit.

Step 16: After determining the heat output of each heat emitter it may be necessary to make changes to the hardware or operating conditions, and then reevaluate the results by iterating the procedure starting with step 2. The designer may also choose to explore options such as reduced supply water temperature, or a larger circuit temperature drop before establishing the final hardware selection and operating conditions.

11.7 MULTIZONE/MULTICIRCULATOR SYSTEMS

Multiple-zone systems using multiple circulators have been used in both residential and light commercial buildings for many years. They consist of two or more parallel piping loops, each with its own circulator and check valve, connected to a common heat source. Each zone circuit can have several heat emitters. They can be connected in series, or in one of several other piping arrangements described in this chapter. A schematic representation of a four-zone system is shown in Figure 11–13. In this case, three of the zones are used for space heating, while the remaining zone is used for domestic water heating.

Notice the zone supplying the domestic water heater is connected closest to the boiler. While this is not absolutely necessary, it does reduce heat loss from the outer portions of the common piping when only domestic water heating is required, during the summer for example. Also notice that all zone circulators pump away from the system's expansion tank.

Importance of Check Valves

It is crucial that a check valve is used on every zone circuit in this type of system. The check valves prevent water from flowing backward through inactive circuits when other circuits are operating. Without these valves, warm return water flowing backward through inactive zone circuits will cause heat to be released in areas where it is not needed. Reverse flow through an inactive zone circuit is shown in Figure 11–14.

The *type* of check valve used is also important. It should either be a spring-loaded inline check valve as shown in Figure 5-47, or a flow-check valve as shown in Figure 5-52. Either of these valves can prevent both reverse flow *and* undesirable thermosiphoning of hot water through inactive zone circuits. The latter occurs because the higher temperature water in the heat source has a lower density than the cooler water in the distribution piping. Without a valve that imposes a small resistance to forward flow, hot water will migrate upward into inactive zone circuits, causing heating when it is not needed. This effect is especially undesirable during the summer when the boiler operates only for domestic water heating, and heat leaking from the heat emitters adds to the building's cooling load.

Another option is the use of circulators with a built-in spring-loaded check valves such as those shown in Figure 7–46 and Figure 11–15. These circulators eliminate the need for a separate check valve, and are ideal for multizone applications.

There are several advantages to using multiple pumps in zoned systems:

- The head loss and flow rate requirements of a given zone circuit tend to be less than for single circuit systems. This allows the use of smaller piping and circulators. Small wet rotor circulators that draw between 50 and 100 watts are often adequate for zone circuits. Since each circulator operates only when its associated room thermostat is calling for heat, this approach is likely to use less electrical energy than

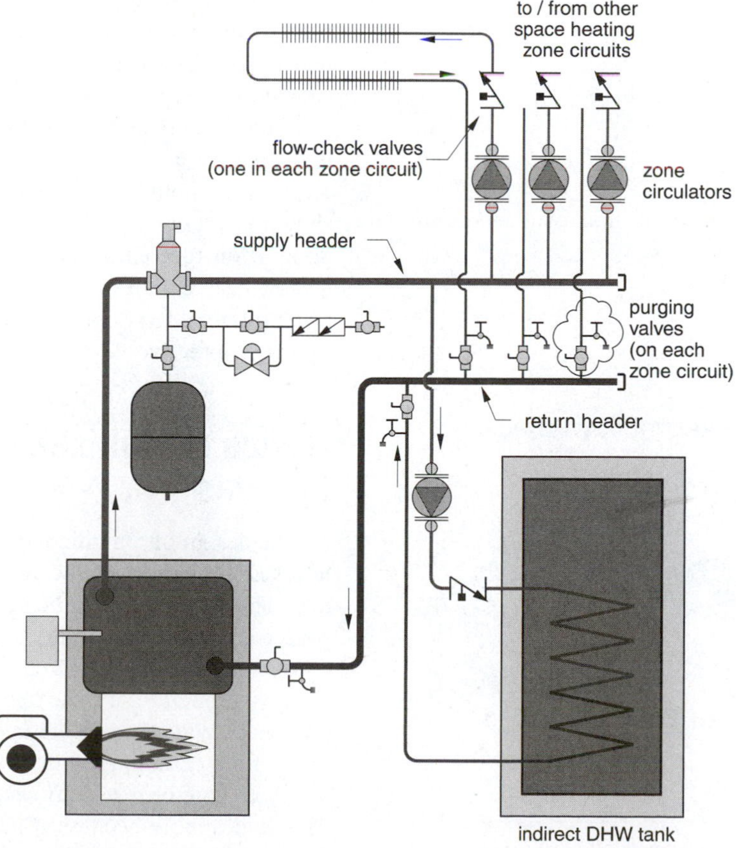

Figure 11–13 Schematic of a four-zone multicirculator system.

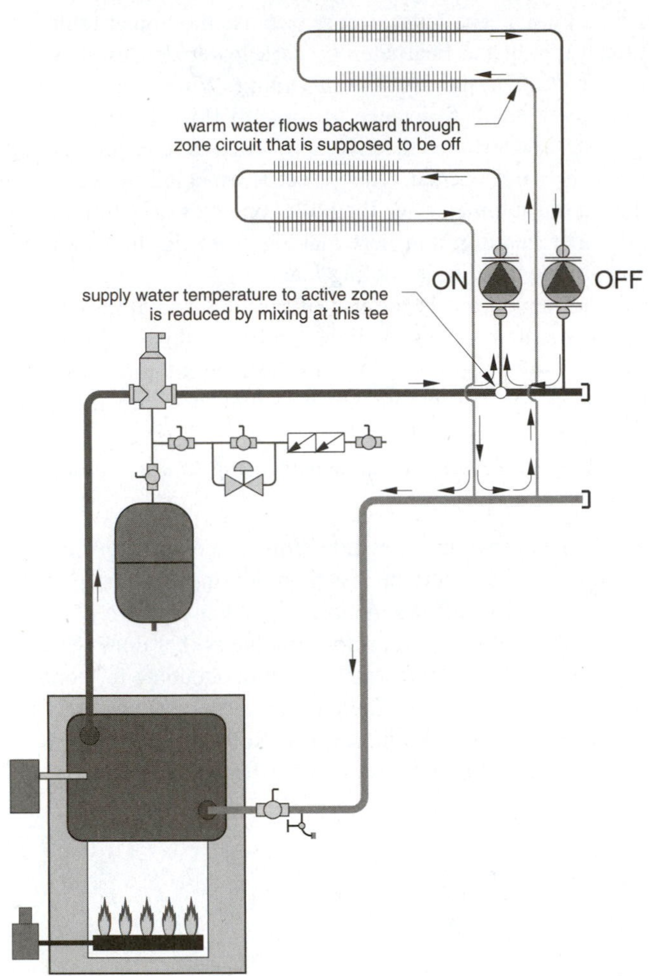

Figure 11–14 Reverse flow through inactive zone circuit when check-valves are not used.

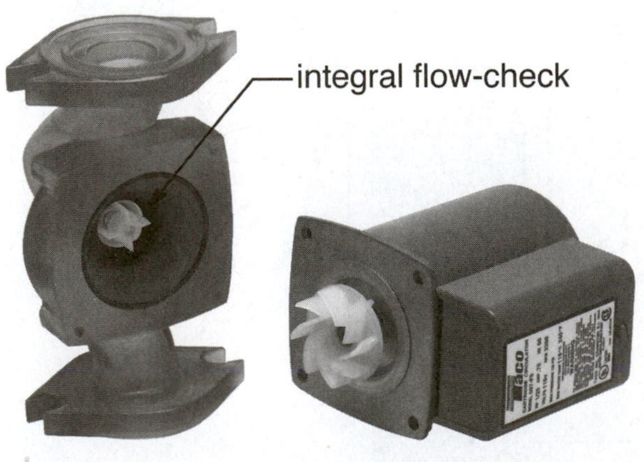

Figure 11–15 Example of a circulator with integral flow-check valve. *Courtesy of Taco, Inc.*

operating a single larger circulator whenever any zone needs heat.

- In a multiple-circulator system, the failure of one circulator only stops heat flow to one of the zone circuits. In a single circulator system, the entire building could be without heat.
- Zone circuits supplied from a common header all operate at the same supply water temperature. This keeps the average water temperature of each circuit higher than if all heat emitters were served by a single series circuit. The higher average water temperature may allow the size of the heat emitters to be reduced.
- If a heat source and common piping with low-flow resistance are used, flow rates in each zone will remain fairly stable regardless of what zone(s) are operating. This allows for better balancing and consistent heat output from each zone circuit. It also eliminates the possibility of flow noise caused by a large circulator pushing water through a single active zone circuit while the other zones are closed off by valves.
- A differential pressure bypass valve, which is often required when zoning using zone valves, is not required when zoning is implemented using circulators.

A Situation to Avoid

Avoid installing parallel-piped circulators of widely differing performance when the **common piping segment** has high-flow resistance. An example of such a situation is shown in Figure 11–16.

If the flow resistance of the common piping segment is high enough, and the maximum head of the smaller circulator is low enough, a pressure distribution similar to that shown in Figure 11–16 could occur. Notice the pressure at point B on the return manifold is greater than at point A. Under these conditions, flow cannot occur from A to B. A primary/secondary piping arrangement, to be discussed later in this chapter, would not allow this situation to develop.

Design of Multicirculator/Multizone Systems

The design of a multizone system using individual circulators is similar to the procedure used for a single circuit system. The difference is in how the flow resistance of the heat source and common piping is treated.

Because the pressure drop across the common piping segment is a function of flow rate, it will change depending on how many zone circuits are operating. When the flow resistance of the common piping is low, the pressure drop across it will change little as various zone circuits turn on and off. This is desirable because it prevents significant changes in flow rate within the active zone circuits regardless of how many zone circuits are operating simultaneously.

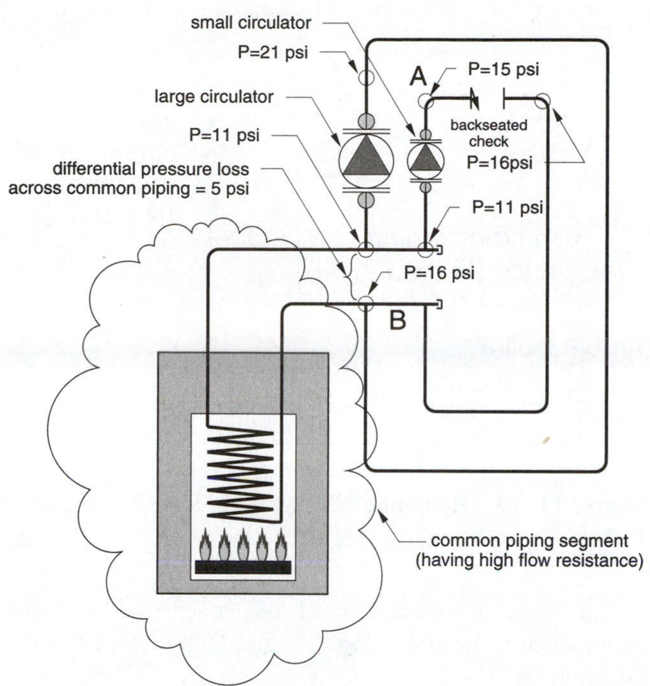

Figure 11–16 System in which flow through small circulator is blocked by the pressure differential produced by a larger circulator.

The conservative design approach is to include the flow resistance of the common piping segment, *assuming all zones are operating,* into the overall resistance of each zone circuit as it is analyzed. This approximates the situation that is likely to occur under design load conditions. This concept is illustrated in Figure 11–17.

Equation 11.5 can be used to modify the hydraulic resistance of the common piping segment to represent full flow conditions:

(Equation 11.5)

$$r_{\text{corrected}} = r_{\text{calculated}}\left(\frac{f_{\text{total}}}{f_{\text{zone}}}\right)$$

where:

$r_{\text{corrected}}$ = hydraulic resistance of the common piping segment at full system flow

$r_{\text{calculated}}$ = hydraulic resistance of the common piping segment as normally calculated.

f_{total} = estimated system flow rate through the heat source when all zones are on (gpm)

f_{zone} = estimated flow rate in the zone circuit being designed (gpm)

Example 11.1: Using manual calculations, estimate the flow rate in the zone circuit shown in Figure 11–18 *when all zones are operating.* The heat source is a boiler with a hydraulic resistance of 0.3. The tentative circulator selection is a Grundfos UP15-42. The zones are intended to have a temperature drop of 15 °F. The design heating load supplied by the boiler is 70,000 Btu/hr. The design load supplied by the zone is 25,000 Btu/hr. The water temperature supplied from the boiler is 160 °F.

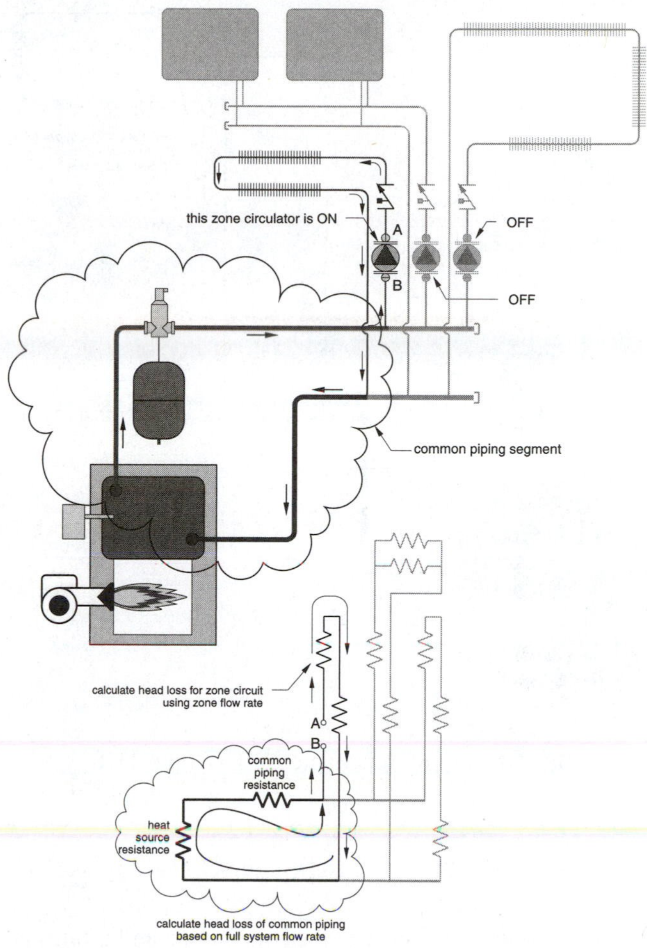

Figure 11–17 Include head loss of the common piping segment at full system flow rate along with the head loss of an individual zone circuit at that zone's flow rate.

Solution: It is necessary to assemble information from Chapter 6, Fluid Flow in Piping, to develop the hydraulic resistance diagram of the zone circuit.

The average water temperature in the zone will be about:

$$T_{\text{ave}} = 160 - \frac{15}{2} = 152.5\ °\text{F}$$

The α value of water at this temperature is estimated from Figure 6–17 as: α = 0.047.

The c-value for 3/4-inch copper tube from Figure 6–18 is: c = 0.061957.

The c-value for the 1.25-inch copper tube from Figure 6–18 is: c = 0.0068082.

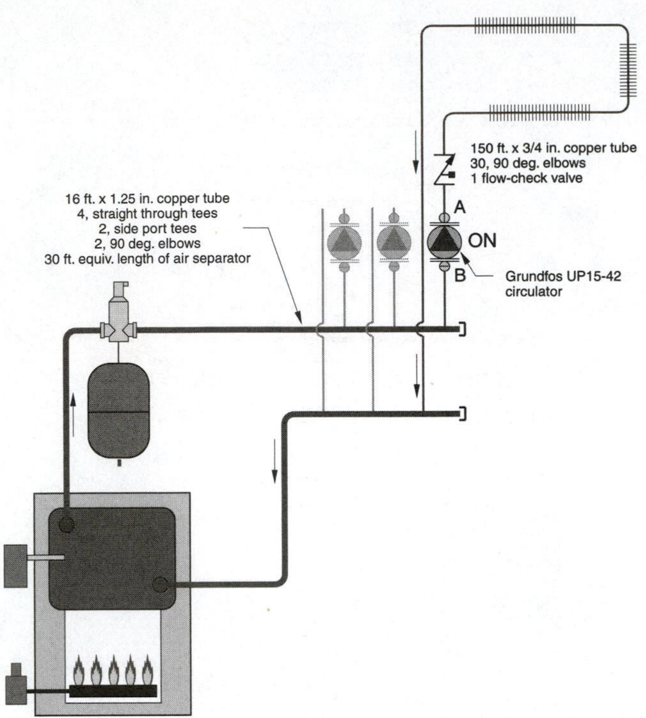

Figure 11–18 Piping schematic for Example 11.1.

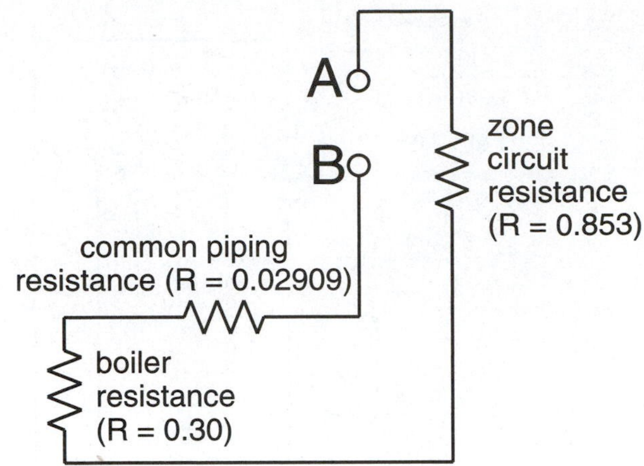

Figure 11–19 Hydraulic resistance diagram for Example 11.1.

The total equivalent length of the zone piping and fittings is: 150 + 30(2) + 83 = 293 ft.

The total equivalent length of the common piping is: 16 + 4(.6) + 2(3) + 2(5.5) + 30 = 65.4 ft.

The hydraulic resistances of the zone circuit and common piping can each be calculated using Equation 6.11: $r = (\alpha\, c\, L)$.

For the zone circuit: $r_1 = \alpha\, c\, L = (0.047)(0.061957)(293) = 0.853$.

For the common piping: $r_2 = \alpha\, c\, L = (0.047)(0.0068082)(65.4) = 0.0209$.

A hydraulic resistor diagram showing these resistances is given in Figure 11–19.

The total system flow rate can be estimated using Equation 11.1.

$$f_{\text{system}} = \frac{70{,}000}{500(15)} = 9.33 \text{ gpm}$$

The target flow rate needed by the individual zone circuit is found using Equation 11.3:

$$f_{\text{zone}} = 9.33\left(\frac{25{,}000}{70{,}000}\right) = 3.33 \text{ gpm}$$

The hydraulic resistance of the boiler and common piping need to be added together and then corrected using Equation 11.5:

$$r_{\text{boiler + common piping}} = r_{\text{boiler}} + r_{\text{common piping}} =$$

$$= 0.3 + 0.0209 = 0.3209$$

$$r_{\text{corrected}} = 0.3209\left(\frac{9.33}{3.33}\right)^2 = 2.52$$

Notice that the *corrected* hydraulic resistance of the common piping segment is considerably greater than the original calculated value. This accounts for the additional head loss when the total flow of all zones passes through the common piping segment.

The total hydraulic resistance of the zone circuit is found by adding the corrected resistance of the boiler plus common piping, to that of the zone piping:

$$r_{\text{zone circuit}} = r_{\text{boiler}} + r_{2\text{corrected}} + r_1 = 2.52 + 0.853 = 3.37$$

The system curve can now be represented as: $H_L = 3.37(f)^{1.75}$.

This system curve is plotted in Figure 11–20 along with the pump curve for the Grundfos UP15-42 circulator to obtain the approximate flow rate in the zone circuit when all zones are operating.

Discussion: Notice the predicted flow rate in the zone (2.36 gpm) is significantly less than the target flow rate for the zone circuit (3.33 gpm) based on obtaining a 15 °F drop. Therefore a larger circulator should be considered. The pump curve for a Grundfos UP26-99 circulator is also shown in Figure 11–20. Notice that it intersects the system curve at about 3.6 gpm. This is slightly higher than the target flow rate and would be acceptable. However, if the designer

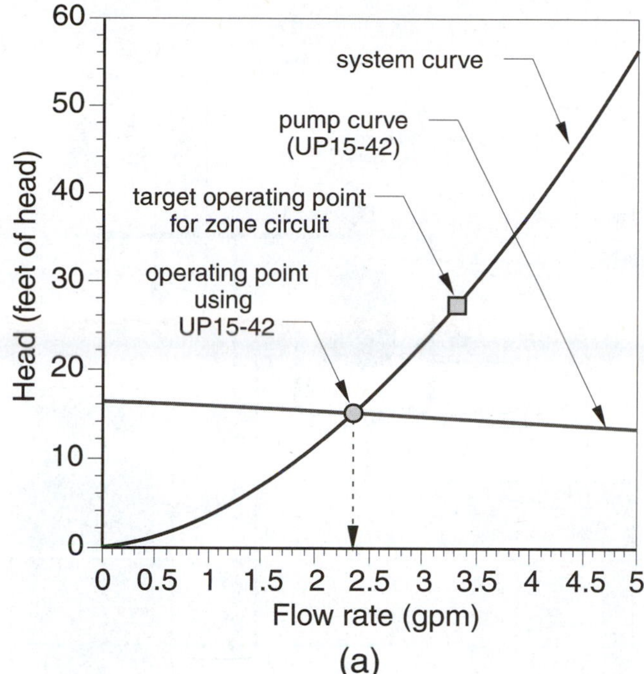

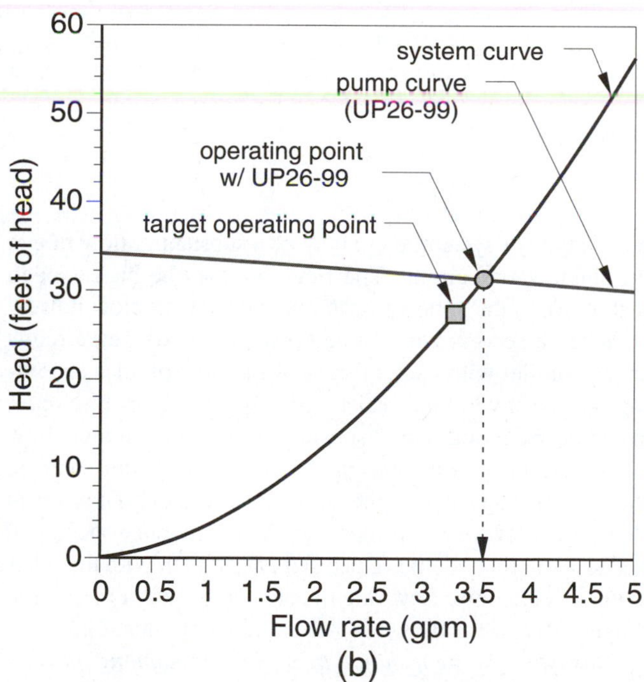

Figure 11–20 System curve and pump curve for Example 11.1.

desired to stay with a smaller circulator they could go on to evaluate the use of larger tubing in the zone to reduce head loss. The equivalent length of the 3/4-inch flow-check (83 feet) is especially high. A design temperature drop of 20 ºF or 25 ºF would also reduce the necessary flow rate in the zone circuit and thus reduce head loss. The Hydronic Circuit Simulator module in the Hydronics Design Studio software could quickly evaluate these options.

Example 11.2: Repeat Example 11.1 assuming a boiler with a hydraulic resistance of 0.002 is used instead of the higher flow resistance boiler.

Solution: The total hydraulic resistance of the boiler and common piping is now:

$r_{boiler+common\ piping} = r_{boiler} + r_{common\ piping} = 0.002 + 0.0209 = 0.0229$

This resistance is corrected for full flow conditions using Equation 11.5:

$$r_{corrected} = 0.0229\left(\frac{9.33}{3.33}\right)^2 = 0.179$$

The total hydraulic resistance of the zone circuit is found by adding the corrected resistance of the boiler plus common piping, to that of the zone piping:

$$r_{zone\ circuit} = (r_{boiler} + r_2)_{corrected} + r_1$$

$$= 0.179 + 0.853 = 1.033$$

The revised system curve and pump curve for the Grundfos UP15-42 pump are shown in Figure 11–21.

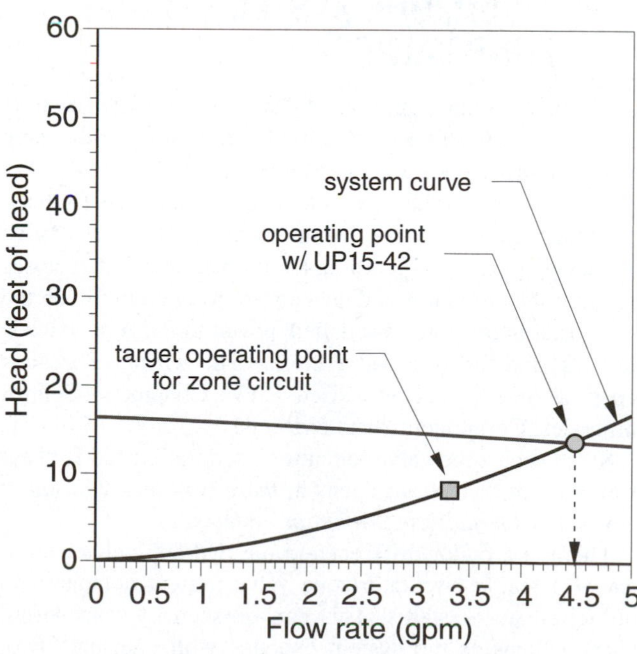

Figure 11–21 System curve and pump curve for Example 11.1 using low head loss boiler.

Discussion: The Grundfos UP15-42 circulator now yields a flow rate of about 4.5 gpm compared to only 2.36 gpm when used with the high-flow resistance boiler of Example 11.1. This significant increase is due solely to the lower flow resistance of the cast-iron sectional boiler. *It demonstrates the need to carefully scrutinize multizone/multicirculator systems using high-flow resistance heat sources.*

The higher flow rate will result in a higher than expected average circuit temperature, which increases heat output. The 4.5 gpm flow rate through the 3/4-inch tubing will not exceed a flow velocity of 4 feet per second, and thus not generate excessive flow noise. In summary, the higher flow rate would improve the zone's thermal performance without creating undesirable operating conditions. The designer could choose to stay with the tentative circulator selection, and accept the added performance. The other option would be to search for a smaller, lower wattage circulator that would yield a flow rate close to the "target" flow rate of 3.33 gpm initially calculated. The latter option, if achievable, would lower the operating cost of the system over its service life.

In summary, multizone/multicirculator systems provide an easy way to deliver hydronic system zoning. To minimize interaction between circulators, the designers should "generously" size the header piping, and plan the common piping segment through which the entire system flow passes for minimal flow resistance. If a heat source with high-flow resistance must be used, it is preferable to pipe it as a secondary circuit using a pair of closely spaced tees as shown in Figure 11–22.

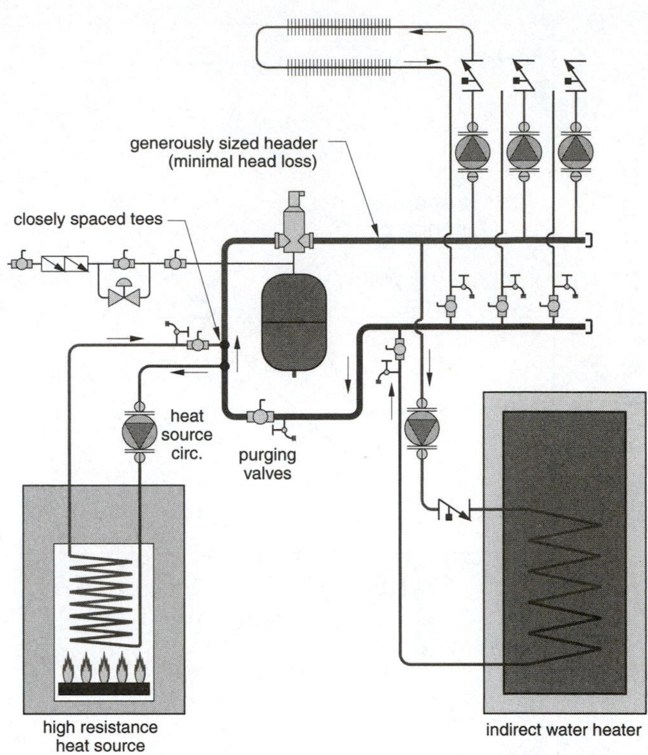

Figure 11–22 Connecting a high flow resistance heat source to generously sized header piping using closely spaced tees.

11.8 MULTIZONE SYSTEMS USING ZONE VALVES

Another common approach to hydronic zoning uses electric zone valves and a single circulator. A piping schematic for this approach is shown in Figure 11–23.

When the room thermostat in a given zone calls for heat, its associated zone valve begins to open. When the valve reaches its fully open position, an internal end switch closes to signal the circulator and heat source to operate. When the thermostat in the zone is satisfied, power to the zone valve is shut off, and the zone valve closes. The wiring of various types of zone valves was discussed in Chapter 9, Control Strategies, Components, and Systems.

Since each zone valve remains closed until a call for heat from a room thermostat opens it, *there is no need for check valves as with multiple-circulator systems.*

There is a major difference in the hydraulic characteristics of a single circulator/zone valve system compared to multicirculator systems. This can be seen by considering what happens as the system operates with one, then two, then three zone valves open at the same time.

With one zone valve open, the system operates as if it were a single circuit. Since a larger pump is likely to be used on this type of system, there may be a substantial flow rate in the single active circuit. The flow rate may be high enough that it causes the zone valve to make noticeable flow noises.

When a second zone valve opens, the two open circuits are in parallel with each other, and the flow divides according to the hydraulic resistance of each branch circuit. Equation 6.20 could be used to predict these branch flow rates. The flow rate through the first zone circuit will be lower than when it was the only active circuit. The *system* flow rate, however, will increase. As additional zones open up, each circuit becomes a parallel circuit path with the other active circuits. The flow rate in each circuit will drop, while the overall system flow rate will continue to increase.

This type of system should be designed assuming all zone circuits are operating simultaneously.

Design Procedure for Multiple Zoned System

Step 1: Determine the design heating load of each zone circuit.

Step 2: Select a target temperature drop for the zone circuits under design load conditions.

Step 3: Determine the target flow rate through each zone circuit using Equation 11.6.

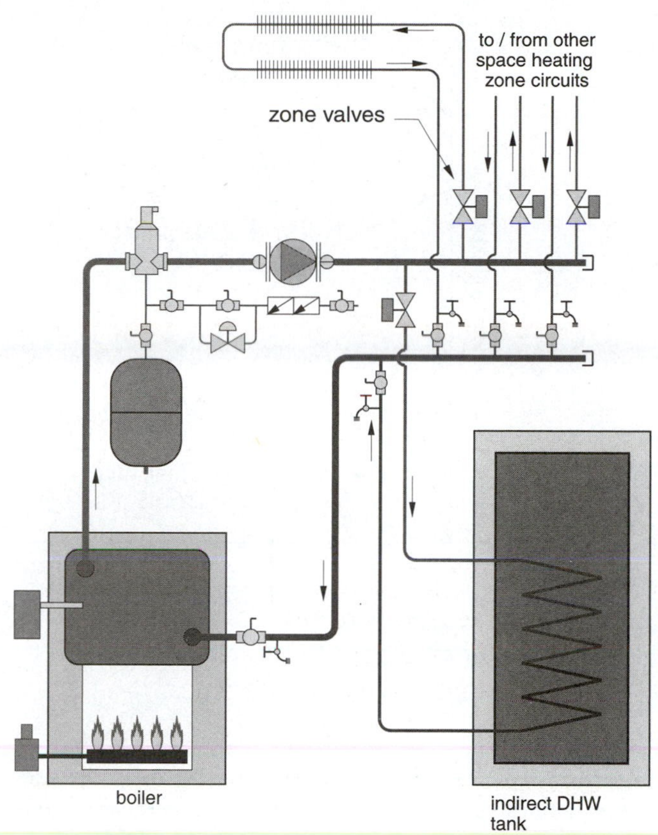

Figure 11–23 Example of a four-zone system using zone valves.

(Equation 11.6)

$$f_{circuit} = \frac{Q_{circuit}}{500(\Delta T)}$$

where:

$f_{circuit}$ = flow rate required in zone circuit (gpm)
$Q_{circuit}$ = design heating load of the circuit (Btu/hr)
ΔT = target temperature drop of the circuit (°F)
500 = a constant for water *

*(for other fluids replace 500 with (8.01 x D x c)

where:

D = fluid density at average circuit temperature (lb/ft3)
c = the specific heat of fluid at average circuit temperature (Btu/lb/°F)

Step 4: Use Equation 11.1 or Equation 11.2 to determine the target flow rate for the *system* based on the total of all zone loads and the temperature drop selected in step 2.

Step 5: Using methods from Chapter 6, Fluid Flow in Piping, determine the hydraulic resistance of each zone circuit.

Step 6: Using methods from Chapter 6, determine the hydraulic resistance of the common piping segment.

Step 7: Use Equation 6.22 to reduce the individual resistances of each zone circuit into a single equivalent resistance. Add this to the hydraulic resistance of the common piping segment to get the overall resistance of the system.

(Equation 6.22 repeated)

$$R_e = \left[\left(\frac{1}{r_1}\right)^{0.5714} + \left(\frac{1}{r_2}\right)^{0.5714} + \left(\frac{1}{r_3}\right)^{0.5714} + \ldots + \left(\frac{1}{r_n}\right)^{0.5714}\right]^{-1.75}$$

where:

R_e = equivalent resistance of n parallel hydraulic resistors
$r_1, r_2, r_3, \ldots, r_n$ = hydraulic resistance of each zone circuit

Step 8: Use Equation 11.7 to plot the system resistance curve for the system.

(Equation 11.7)

$$H_L = r_s(f_s)^{1.75}$$

where:

H_L = head loss of system (feet of head)
r_s = overall hydraulic resistance of system
f_s = total flow rate in system (gpm)

Step 9: Select a candidate circulator and plot its pump curve over the system resistance curve. If the operating point falls within +/– 10 percent of the target system flow rate calculated in step 3, the circulator would typically be acceptable.

Step 10: Use Equation 6.23 to determine the flow rate in each zone circuit based on the system flow rate, and the individual resistances of each circuit. The system flow rate is the flow rate at the operating point found in step 8.

(Equation 6.23 repeated)

$$f_i = f_s\left(\frac{R_e}{r_i}\right)^{0.5714}$$

where:

f_i = flow rate through parallel path i (gpm)
f_s = total system flow rate (gpm)
R_e = equivalent hydraulic resistance of all the parallel piping paths
r_i = hydraulic resistance of parallel piping path i

Step 11: Evaluate the performance of the heat emitters in each zone circuit based on the zone flow rates and the selected supply water temperature. If the heat output of each zone equals or exceeds the design load requirements of the respective zones, there is no need to change the design. However, if the heat output of one or more zones is insufficient to meet design load requirements, changes need to be made. These may include increasing the supply water temperature, reducing the hydraulic resistance of the zone, using a different circulator, or a combination of these changes.

The Hydronic Circuit Simulator module in the Hydronics Design Studio software has the ability to simulate the operation of several parallel zone circuits. A screen shot of the module configured for a three-zone system is shown in Figure 11–24.

The user can designate the number of zone circuits, the piping components in each zone and the common piping segment, the circulator, the fluid, and the supply temperature at which the system operates. The program module instantly finds the hydraulic equilibrium point of the user-specified system and reports all branch flow rates.

Use a Circulator with a Flat Pump Curve

For zone valve type systems, it is important to use a circulator with a relatively "flat" pump curve. The flatter the curve, the more stable the flow rates in each zone circuit as the other circuits cycle on and off. This concept is illustrated in Figure 11–25 for a system having three parallel zone circuits.

When the circulator with the flat curve is used, the differential pressure across the circulator changes very little as zone circuits turn on and off. This allows the flow rate in each active zone circuit to remain relatively stable because the pressure difference between its starting point and ending point stays about the same. If a circulator with a "steep" curve is used, there is a substantial drop in the differential pressure as additional zones turn on. This often leads to significant changes in the flow rate of a given zone depending on what other zones are operating. Another potential problem when using a circulator with a steep pump curve in a zone valve system is excessively high flow velocity through a single open zone.

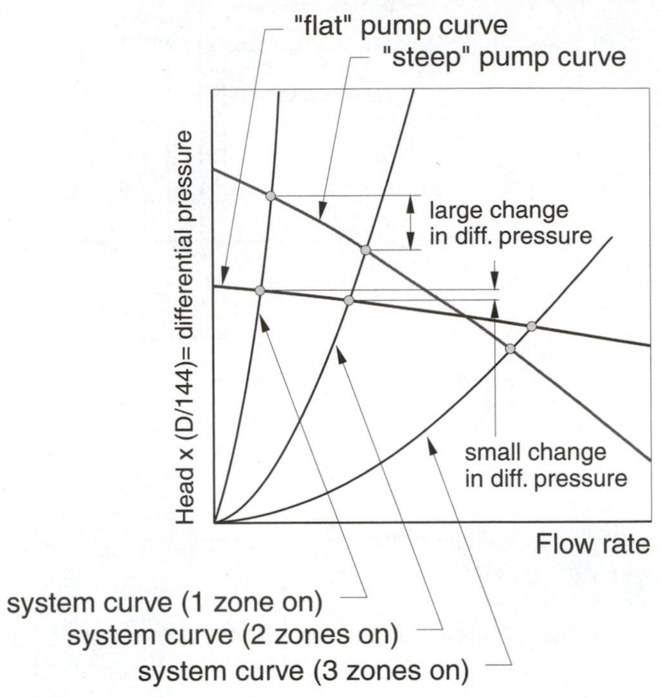

Figure 11–25 The differential pressure between the supply and return header changes based on the number of zone circuits operating. Using a circulator with a "flat" pump curve reduces the change in differential pressure.

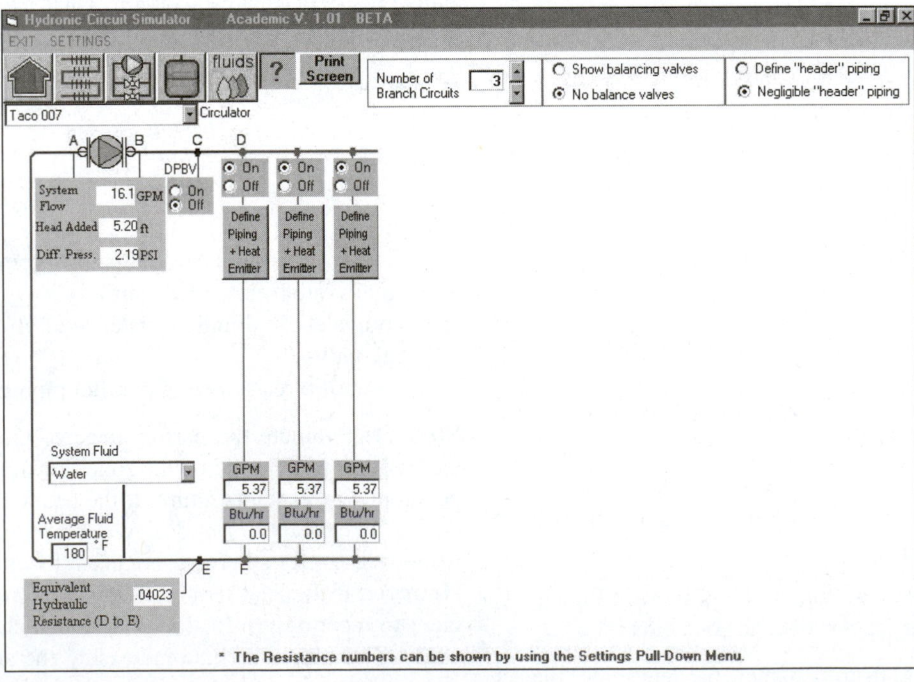

Figure 11–24 Screen shot of the Hydronic Circuit Simulator program module from the Hydronics Design Studio.

Applying Differential Pressure Bypass Valves

Differential pressure bypass valves (DPBV) were discussed in Chapter 5, Piping, Fittings, and Valves. Their purpose is to limit the differential pressure imposed on the distribution system by a constant speed circulator as the system flow rate decreases due to closing zone valves.

There are several undesirable effects associated with not limiting the differential pressure on a system when a small percentage of the total zone circuits are operating. One is an increase in flow rate in the active zone circuits that can lead to **velocity noise.** Beside the obvious acoustical annoyance, high-flow velocities can cause erosion of valve seats as well as other piping components.

Another repercussion of unregulated differential pressure is partial opening of zone valves that are supposed to be off. This leakage allows heated water to migrate into zone circuits that are supposed to be off, which can lead to overheating.

Another potential problem is inadequate flow through low-mass heat sources under low-load conditions. Although this can be corrected in other ways, such as making the heat source a secondary circuit in a primary/secondary system, it is much less likely to need correction in systems using a properly set DPBV. That is because the flow through the bypass valve can be rerouted back through the heat source as shown in Figure 11–26.

Turning the knob on the DPBV adjusts the spring tension holding the plug against the valve seat. The pressure where the plug just starts to lift away from the seat is called the **threshold differential pressure.** Most DPBVs have a calibrated scale for setting the desired threshold differential pressure. When the pressure differential across the valve is less than this threshold, no flow passes through the value. In this situation the valve is "invisible" to the rest of the system.

Above the threshold differential pressure setting the valve maintains a relatively constant differential pressure as flow through it increases. Still, like any piping component, head losses through a DPBV increase with increasing flow.

When a DPBV is installed as shown in Figure 11–26, the pump curve is partially truncated as shown in Figure 11–27. The extent of the truncation depends on the threshold setting of the valve. At differential pressure below the threshold setting the pump's differential pressure curve is unchanged by the presence of the valve.

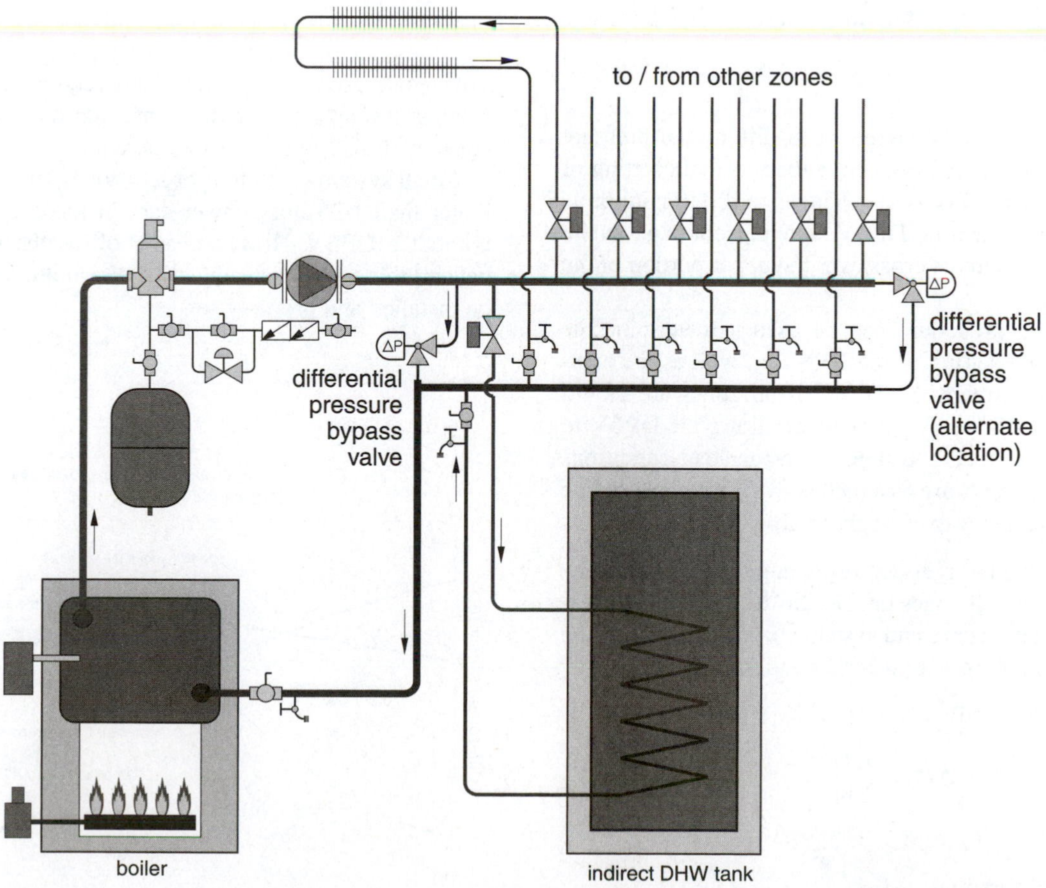

Figure 11–26 Two locations where a differential pressure bypass valve (DPBV) can be installed in a system using zone valves.

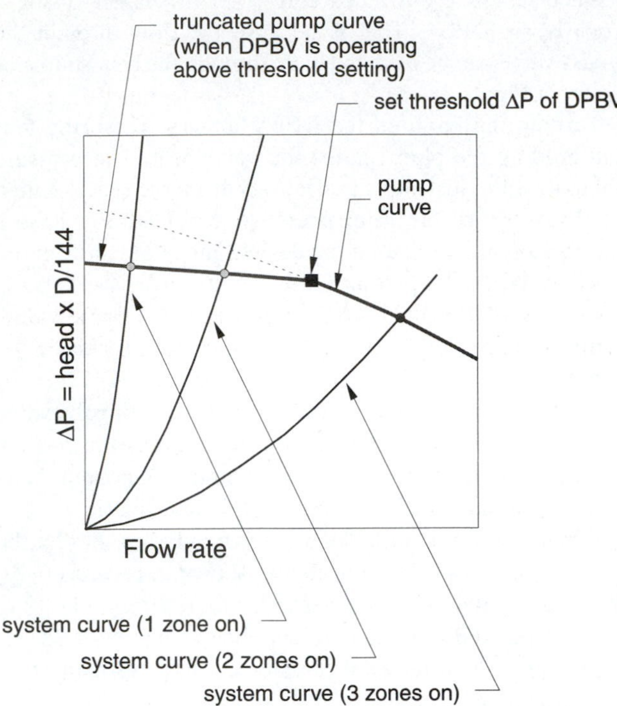

Figure 11–27 The upper portion of the pump curve is "truncated" when the differential pressure bypass valve becomes active.

Notice the very slight increases in differential pressure when the system is operating above the threshold setting of the DPBV. Compare this to the changes in differential pressure that would occur if no DPBV were present (e.g., where the steeper system curves cross over the upper portion of the pump curve).

The following procedure can be used to determine the threshold pressure setting of the DPBV, as well as the maximum flow rate through the DPBV if all zones are closed simultaneously. This setting procedure allows the DPBV to remain "invisible" to the system at design load conditions (when all zone circuits are assumed active), yet come online quickly as the zone's valves begin to close.

Step 1: Determine the pressure differential across the distribution system with all zones on. Do this by finding the intersection of the pump curve and system curve, then convert the head to differential pressure using Equation 6.7 (repeated):

(Equation 6.7 repeated)

$$\Delta P = \frac{HD}{144}$$

where:

ΔP= pressure differential (psi)

H = head at pump curve/system curve intersection (feet)

D = density of fluid being circulated (lb/cubic foot)

Step 2: Set the threshold differential pressure of the valve 0.5 to 1 psi higher than the value determined in step 1. This allows the DPBV to remain fully closed at design load conditions so all circulator flow is directed into the distribution system for maximum heat transfer.

Owing to their European origin, many DPBV valves are calibrated in **bars** or **KPa (Kilo-Pascals)** rather than psi. Use the following conversion factor to convert to psi.

1 psi = 6.895 KPa = 0.068 bar

Step 3: Determine the flow rate through the DPBV when all zones are off. Do this by plotting the differential pressure versus flow rate relationship of the DPBV on the pump's differential pressure curve, and reading the flow rate where the curves intersect (see Figure 11–28).

Step 4: Select a DPBV from manufacturer's literature that is rated to carry the flow determined in step 3.

As a practical matter, most residential and small commercial systems can work with either 3/4-inch or 1-inch size DPBVs. Larger systems may require 1.25-inch size valves.

This sizing/setting procedure is conservative in the sense that the DPBV will begin to open at a relatively high percentage of system load. If the designer's objective is solely to prevent flow noise, the valve's threshold pressure could likely be set higher. In some systems, the DPBV is set while operating by slowly opening the valve from its fully closed position under a minimal loading condition, until excessive flow noise goes away. Such a procedure is acceptable in smaller systems, but does not provide the range of almost constant differential pressure operation.

Small systems with four or less zones, and a circulator no larger than 1/25 horsepower can, in most cases, get along without a DPBV. However, as soon as the system moves beyond this somewhat subjective threshold, a DPBV should be installed and properly set.

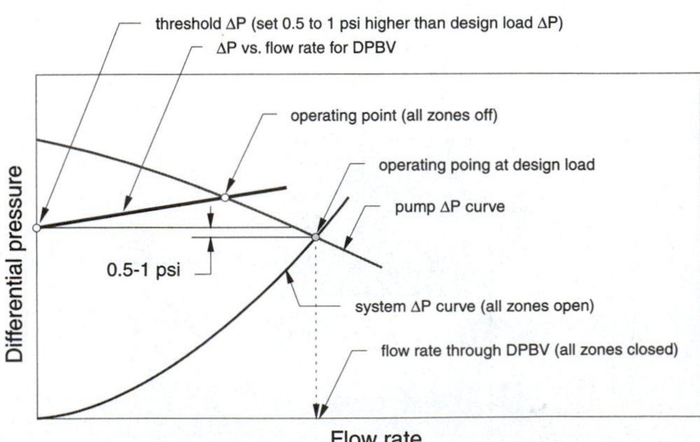

Figure 11–28 Differential pressure versus flow rate relationships when setting a differential pressure bypass valve.

Variable Speed Circulators in Zone Valve Systems

An "ideal circulator" for a zone valve system would yield a perfectly flat pump curve, and thus maintain a constant differential pressure regardless of how much flow is passing through it. This concept is shown in Figure 11–29a. Such a circulator would allow any given zone circuit to turn on or off without affecting flow in the other zone circuits.

Although it is possible to build a circulator with relatively flat pump curves, the curve of a centrifugal pump will always have some drop in head (and hence differential pressure) as flow rate increases.

By varying the speed of the pump, it is possible to shift the pump curve as shown in Figure 11–29b. When the pump speed is decreased, the pump curve shifts down and to the left. When speed is increased the curve shifts up and to the right.

At some specific sets of speeds, the pump curves will cross over the various system curves such that the head (and differential pressure) at the operating points remains constant. This is shown in Figure 11–29c.

This mimics the ideal pump curve scenario shown in Figure 11–29a.

The "intelligence" necessary to keep the pump operating at the right speed in such an application is typically provided by a differential pressure controller. Such a device monitors the pressure difference across the piping mains, compares it to a "target" differential pressure, and outputs a control signal based on the error that exists between the measure and target differential pressure. The concept is shown in Figure 11–29d.

Recently, small circulators have become available with integral variable speed drives as shown in Figure 11–29e. They can be controlled by a differential pressure controller as well as any other controller capable of generating a 2-10VDC/4-20 milliamp signal.

11.9 PARALLEL DIRECT-RETURN SYSTEMS

Parallel distribution systems, also sometimes called **two-pipe systems,** are constructed so that every heat emitter receives water from a common supply pipe and returns it to a common return pipe. A schematic for a **parallel direct-return system** is shown in Figure 11–30.

A parallel direct-return system supplies approximately the same water temperature to each branch circuit. In this respect, it is similar to the multizone/zone valve systems discussed in Section 11.8. However, there are some distinct differences between these two types of distribution systems. One is the length of piping between the beginning and ends of the parallel branch circuits. In the zone valve system, each circuit usually begins from a common supply header and ends at a common return header. The lengths of pipe between the starting points and ending points of zone circuits are very small, usually not more than a few inches. The head loss of such small piping segments is negligible. In a parallel direct-return system, the distance between adjacent parallel branches can be many feet. These longer piping segments can generate significant head loss that must be accounted for when the system is designed. Another difference is that each parallel branch in a parallel direct-return system usually only serves one heat emitter. In a multizone/zone valve system, it is common to have two or more heat emitters on each zone circuit.

The words *direct-return* indicate that the first branch circuit supplied by the hot water main is also the closest to the circulator on the return side of the system (see Figure 11–30). The length of the path from the circulator, through the branch, and back to the heat source has a major effect on the flow rate through the branch. In a situation where all heat emitters have identical hydraulic resistances, the farther out the branch is on the distribution system, the lower its flow rate. The lower the flow rate through a given heat emitter, the lower its heat output. This situation must be accounted for and corrected to avoid serious heat distribution problems.

The balancing valves in each branch are used to correct for the tendency of a direct return system to allow more flow through the branches closer to the circulator, and less flow in the more remote branches. By partially closing the balancing valve on a given branch, the hydraulic resistance of that flow path can be increased to compensate for its distance away from the circulator. If the hydraulic resistance of all branch paths is about the same, the balancing valves closer to the circulator are usually closed more than those farther out in the distribution system.

The amount a given balancing valve must be closed depends on the hydraulic resistance of the heat emitter and the other piping component in the branch, as well as the position of the branch in the system. For example, the valve on a high-flow resistance heat emitter near the circulator may not need to be closed as much as the valve on a low hydraulic resistance branch farther out in the system.

The pipe size of the supply and return mains at any point in the system should be determined by the flow rate present at that point. As the supply main passes by each branch, a portion of its flow is usually routed through that branch. This reduces the flow that continues farther out in the system. This same situation occurs as the return main collects additional flow from each branch it passes on its way back to the heat source. Proper design keeps the flow velocity in all locations under 4 feet/second to avoid flow noise.

Because each branch receives supply water at approximately the same temperature, and since each branch only serves a single heat emitter, all heat emitters can be selected based on the same supply water temperature. This makes parallel direct-return systems well-suited for low temperature applications where temperature drops often must be held to a minimum. It also eliminates the need to correct for temperature drops from one heat emitter to the next as is necessary in series loop and one-pipe distribution systems.

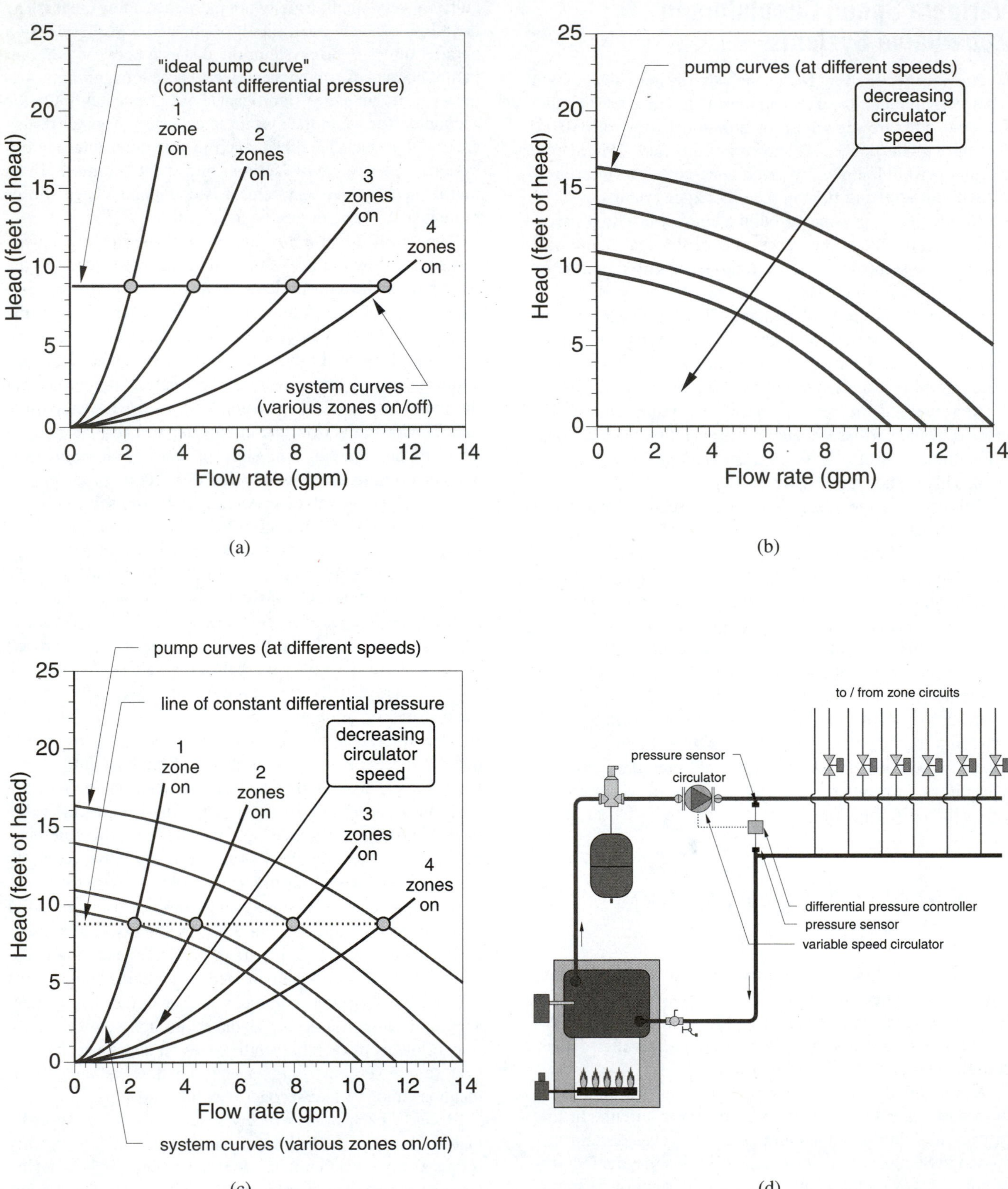

Figure 11–29 By adjusting circulator speed it is possible to maintain a constant differential pressure across the zone circuits regardless of the number of zones operating.

Figure 11–29e Small variable speed circulators are now available. These circulators will likely see widespread usage in zone valve systems in the near future.

Another advantage of parallel systems is that each heat emitter can be separately controlled. The flow through any heat emitter can be modulated or turned on and off by devices such as thermostatic radiator valves, or zone valves.

As the flow through individual heat emitters is reduced or stopped, the *system* flow rate decreases, and the head across the circulator increases. This in turn increases the flow rate in the branches that remain open. If only one or two branches are open, and a constant speed circulator is used, flow noise could develop. To avoid this, a differential pressure bypass valve (DPBV) should be installed and set to prevent the circulator from operating at high pressure differentials. The DPBV can be selected and sized as discussed in Section 11.8.

Variable speed circulators are also well-suited for use in parallel direct-return systems. Such a circulator would typically be configured as a constant differential pressure device. It would speed up or slow down as necessary to maintain a preset pressure differential as the various branch control valves open, close, or modulate. This concept is shown in Figure 11–31.

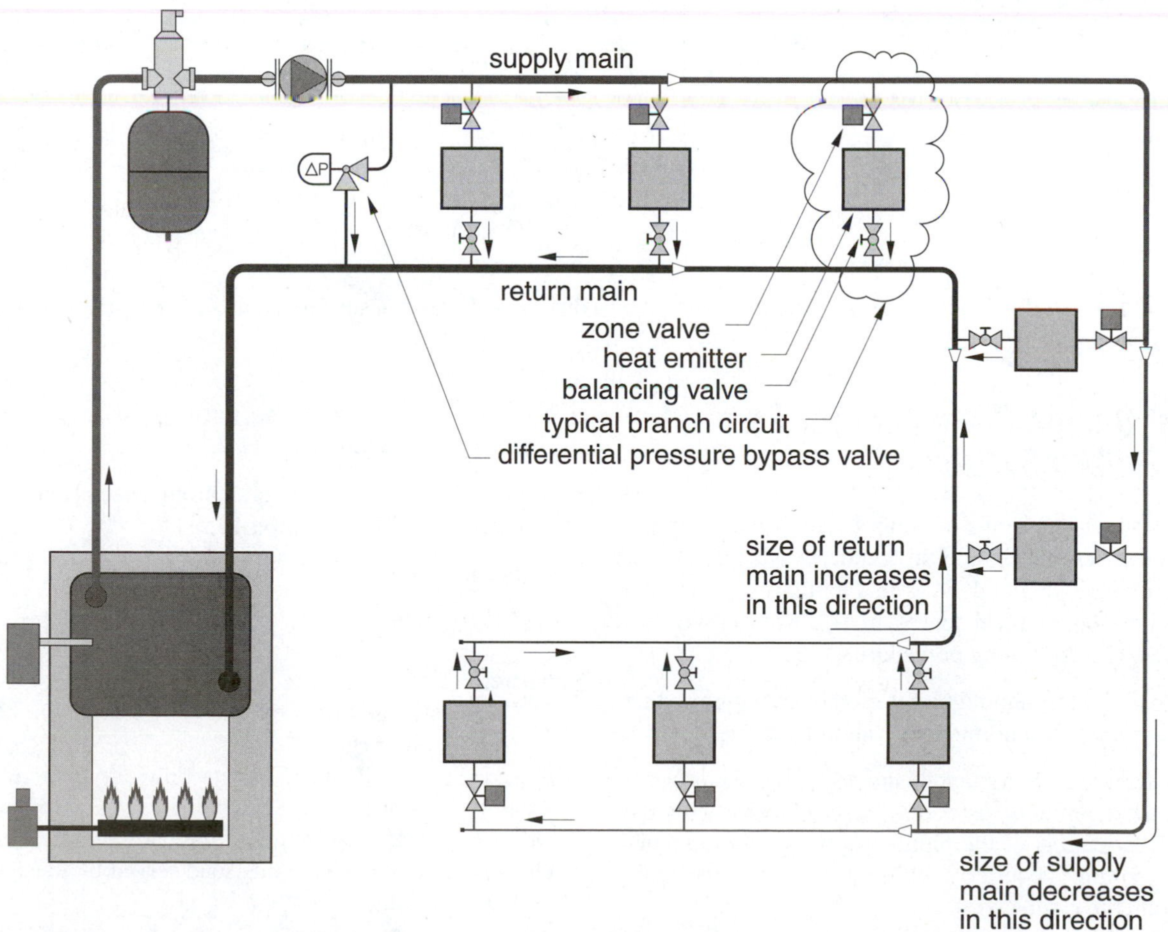

Figure 11–30 Piping schematic of a parallel direct return distribution system. Each branch is individually controlled. Note the use of a differential pressure bypass valve.

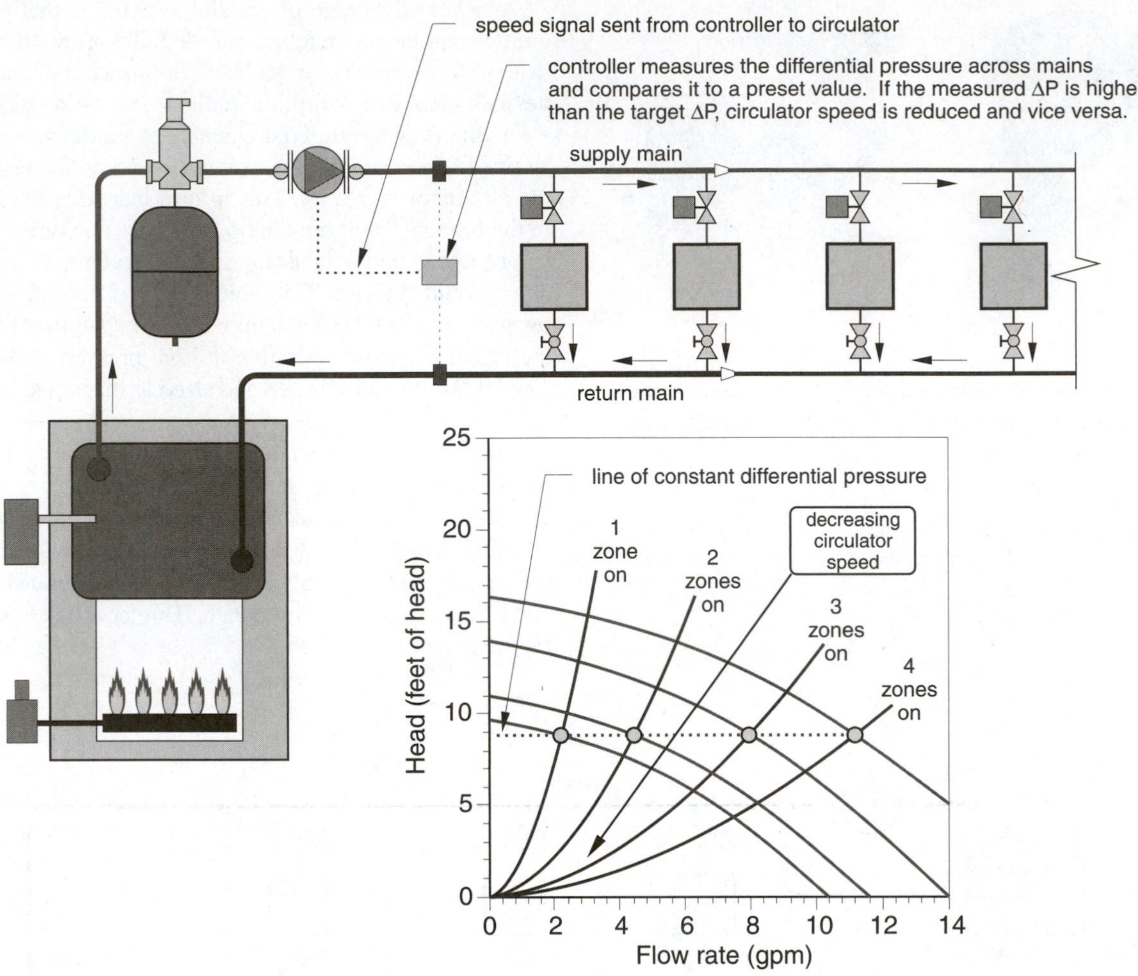

Figure 11–31 Use of a variable speed circulator operated by a differential pressure sensing controller in a parallel direct return distribution system.

Manual Design Procedure for Parallel Direct-Return Systems

Parallel direct-return systems can be analyzed using the concepts of series and parallel hydraulic resistances as discussed in Chapter 6, Fluid Flow in Piping. The following procedure outlines a typical approach. As the reader will see, this can be a relatively long procedure.

Step 1: Heat emitters should be selected based on providing design heat output at a nominal operating flow rate.

Step 2: A tentative piping layout should be sketched for the particular heat emitters used and their placement in the building. The lengths of the piping segments and the number/type of fittings required within the branches and main piping should be estimated.

Step 3: Select a target temperature drop for the system under design load conditions. Suggested temperature drops range from 10 ºF for low water temperature systems, to 25 or 30 ºF when higher water temperatures are used.

Step 4: The *system* flow rate can be estimated using Equation 11.1 or Equation 11.2.

Step 5: The tentative flow rate through each heat emitter can be estimated using Equation 11.8:

(Equation 11.8)

$$f_i = f_{initial}\left(\frac{Q_i}{Q_t}\right)$$

where:

f_i = estimated flow rate through a particular heat emitter (gpm)

$f_{initial}$ = first estimate of system flow rate from step 4 (gpm)

Q_i = design heat output of the particular heat emitter

Q_t = total design heating load served by the system (Btu/hr)

Step 6: Pipe sizes can now be selected based on keeping the flow velocity in all pipes less than or equal to 4 ft/sec. Use the formulas in Chapter 6 to verify flow velocity versus flow rate for various types and sizes of tubing.

Step 7: The pipe sizes, and their estimated lengths, can now be combined with the sketch of the piping layout to construct a hydraulic resistance diagram.

Step 8: The hydraulic resistor diagram can be systematically reduced to a single equivalent resistance using the methods for series and parallel hydraulic resistors from Chapter 6, Fluid Flow in Piping.

Step 9: The single equivalent resistance can be used to construct a system resistance curve. This can be overlaid with the pump curve of a candidate circulator to find the system flow rate.

Step 10: Once the system flow rate is found, Equation 11.1 or Equation 11.2 can be used to verify the temperature drop present under design load conditions. If the temperature difference is significantly *greater* than the initially selected value (e.g., more than 10 percent), a circulator capable of increasing the flow rate is worth considering. Likewise, if the system temperature drop is significantly *lower* than the initial estimate, a smaller circulator should be considered. If the temperature drop is within +/– 10 percent of the initially selected value, the circulator being considered is probably suitable.

Step 11: Knowing the system flow rate, the flow rate through each heat emitter can be found by repeated use of Equation 6.23. When all these flow rates are determined, they should be added up as a check. Their total should equal or be very close to the system flow rate found in step 9. If this does not check out, there is probably a mistake in the calculations.

(Equation 6.23 repeated)

$$f_i = f_s\left(\frac{R_e}{r_i}\right)^{0.5714}$$

where:

f_i = flow rate through parallel path i (gpm)

f_s = total system flow rate (gpm)

R_e= equivalent hydraulic resistance of all the parallel piping paths

r_i = hydraulic resistance of parallel piping path i

Step 12: The flow rates through the individual parallel piping paths and heat emitters should be checked for excessive flow velocity (over 4 ft/sec). These flow rates should also be checked against the tentatively estimated values. If a given flow rate is higher than its initial estimate, it can probably be reduced using a balancing valve. If it is too low, a number of corrective options are available. These include using a large branch piping size or a different heat emitter with a lower flow resistance.

Step 13: After checking the flow rate and expected thermal performance of each heat emitter, and making any changes to the pipe sizes or heat emitters, the hydraulic resistor diagram can be modified to reflect these changes. The modified resistor diagram can then be reanalyzed beginning with step 7. This process can be repeated until the resulting system performs reasonably close to expectations. With practice, this can usually be accomplished within two iterations of the calculations.

The Hydronic Circuit Simulator module in the Hydronics Design Studio software could also be used to simulate the performance of a parallel direct-return circuit. The user can define piping components in both the branches and the main piping segments between the branches. The program module performs the previously discussed design steps anytime one or more of the inputs is changed. Several variations of a proposed system can be analyzed in a few minutes.

11.10 PARALLEL REVERSE-RETURN SYSTEMS

Another method for laying out a parallel system is known as **reverse-return piping.** Figure 11–32 shows a modification of the direct-return system of Figure 11–30, making it into a reverse-return system.

Notice the heat emitter closest the circulator along the supply main is the farthest from the circulator along the return main. Likewise the farthest heat emitter on the supply is the closest on the return. This arrangement helps correct the inherent flow balancing problems of the parallel direct return systems, especially when the heat emitters all have similar hydraulic resistances. Because of this, reverse-return is usually the preferred way of arranging a parallel distribution system.

Reverse-return systems usually require different pipe sizes in various parts of the system. The supply main gets progressively smaller as one moves away from the circulator. The return main gets progressively larger as one moves toward the circulator. The concept is to keep the flow velocity at or below 4 ft/sec to prevent flow noise in all areas of the system.

The ideal arrangement for a **parallel reverse-return system** is to route both the supply and return mains around the perimeter of the area to be heated, as shown in Figure 11–33. If this is not done, there may be a need for considerable extra piping to achieve the reverse return layout as shown in Figure 11–34.

Theoretically, if all branches of a reverse-return system had identical hydraulic resistances, and all pipe size changes in the supply and return mains were symmetrical, each path would operate at the same flow rate. Systems like this are very uncommon. Because of this, a balancing valve should still be installed in each branch of the system for adjusting the flow rate through individual heat emitters in proportion to the loads they supply.

Most of the concepts discussed under parallel direct-return systems are also applicable to reverse-return systems. This design is a good choice for low temperature distribution

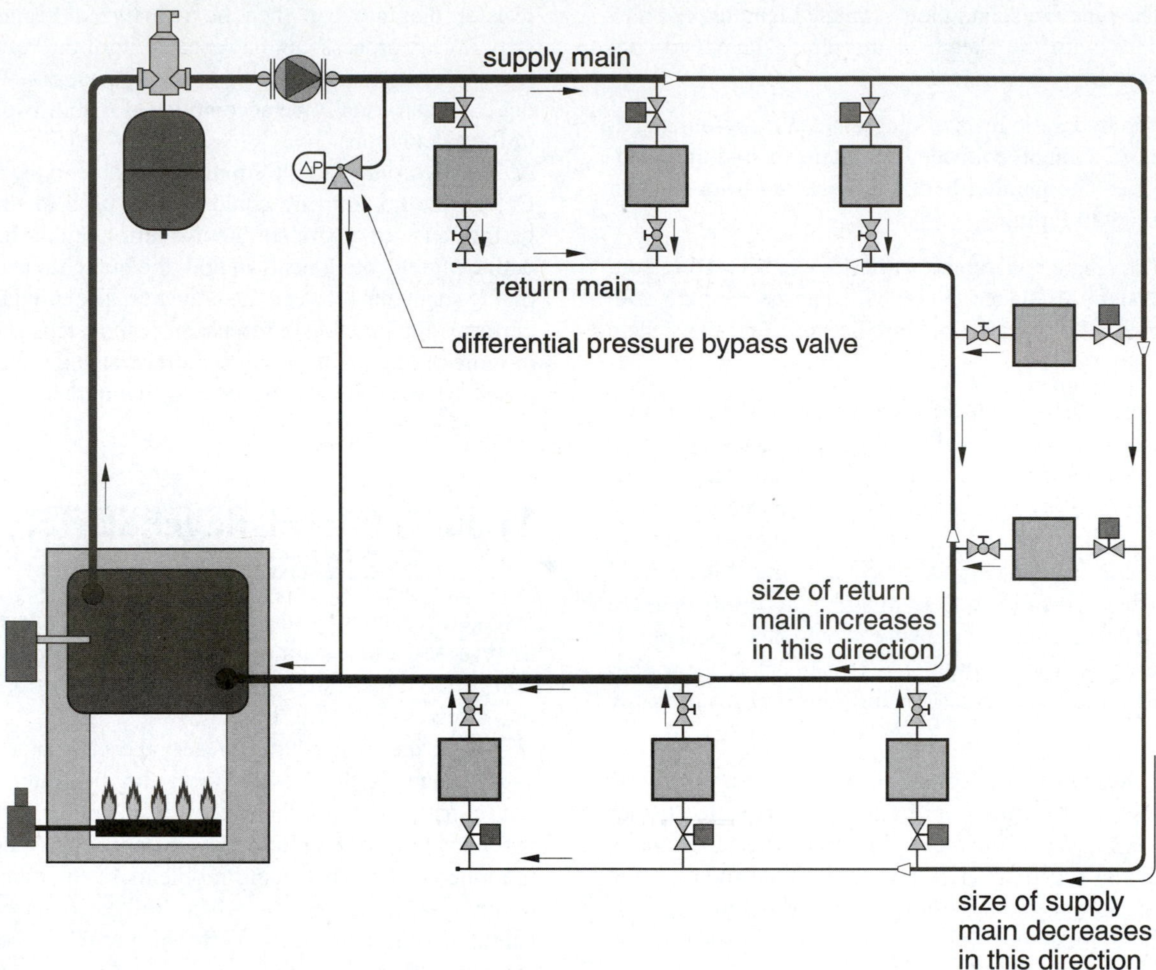

Figure 11–32 Piping schematic of a parallel reverse return distribution system. Each branch is individually controlled. Note the use of a differential pressure bypass valve.

systems. Variable speed circulators are also an option. If a fixed speed circulator is used, a differential pressure bypass valve should be used to reduce the potential for flow noise when only one or two zones are active. This valve would also prevent **dead heading** the circulator if all zone circuits where closed simultaneously.

Interestingly, the standard approach of setting up a hydraulic resistance diagram of the system is not very useful for reverse-return systems. If the number of parallel branches exceeds two, the resulting hydraulic resistor diagram cannot be reduced to a single equivalent resistance using the methods of Chapter 6, Fluid Flow in Piping. The flow rates can still be solved using sophisticated mathematical iteration, but the work is well beyond the scope of this book.

Manual Design Procedure for Parallel Reverse-Return Systems

Since the equivalent resistance methods of Chapter 6 are usually not applicable to reverse-return systems, a somewhat simplified method will be presented. This procedure will result in a conservative (e.g., slightly oversized) circulator being selected. The first four steps of this method are identical to those used for direct-return systems.

Step 1: Heat emitters should be selected based on providing design heat output at a nominal operating flow rate.

Step 2: A tentative piping layout should be sketched for the heat emitters selected and their placement in the building. The lengths of the piping segments and the number and type of fittings required should be estimated.

Step 3: Select a tentative temperature drop for the system under design load conditions. A practical value for parallel systems is 10 to 20 °F. The *system* flow rate can then be estimated using Equation 11.1 (for water), or Equation 11.2 (for other fluids).

(Equation 11.1 repeated)

$$f_{\text{intitial}} = \frac{Q_t}{500(\Delta T)}$$

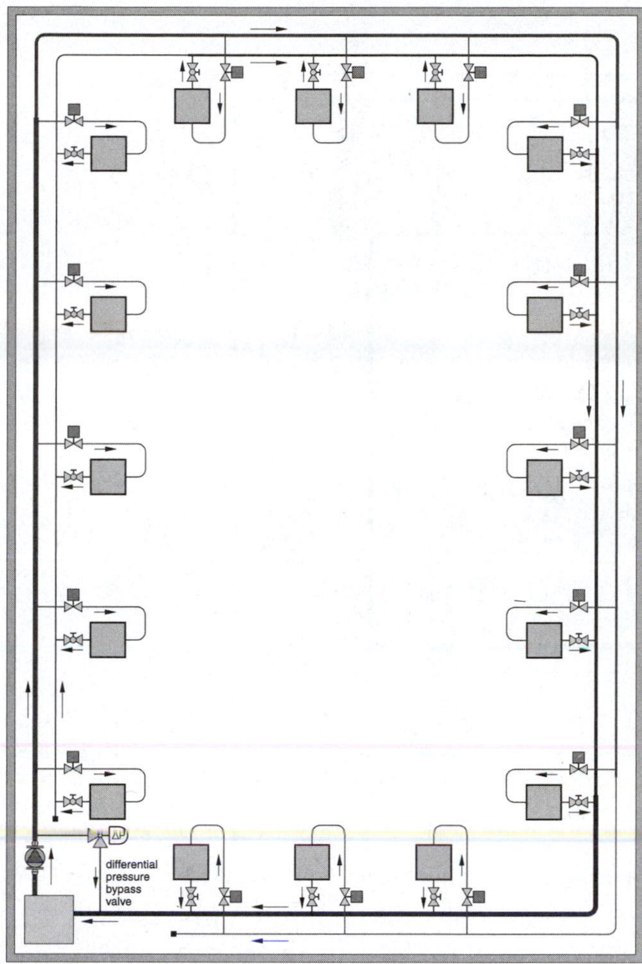

Figure 11–33 Efficient use of parallel reverse-return piping around the perimeter of a building.

where:

$f_{initial}$ = first estimate of system flow rate (gpm)
Q_t = total design heating load served by the system (Btu/hr)
ΔT = selected temperature drop of the system (°F)

Step 4: The tentative flow rate through each heat emitter can be estimated using Equation 11.8:

(Equation 11.8 repeated)

$$f_i = f_{initial}\left(\frac{Q_i}{Q_t}\right)$$

where:

f_i = estimated flow rate through a particular heat emitter (gpm)
$f_{initial}$ = first estimate of *system* flow rate (gpm)
Q_i = design heat output of the particular heat emitter
Q_t = total design heating load served by the system (Btu/hr)

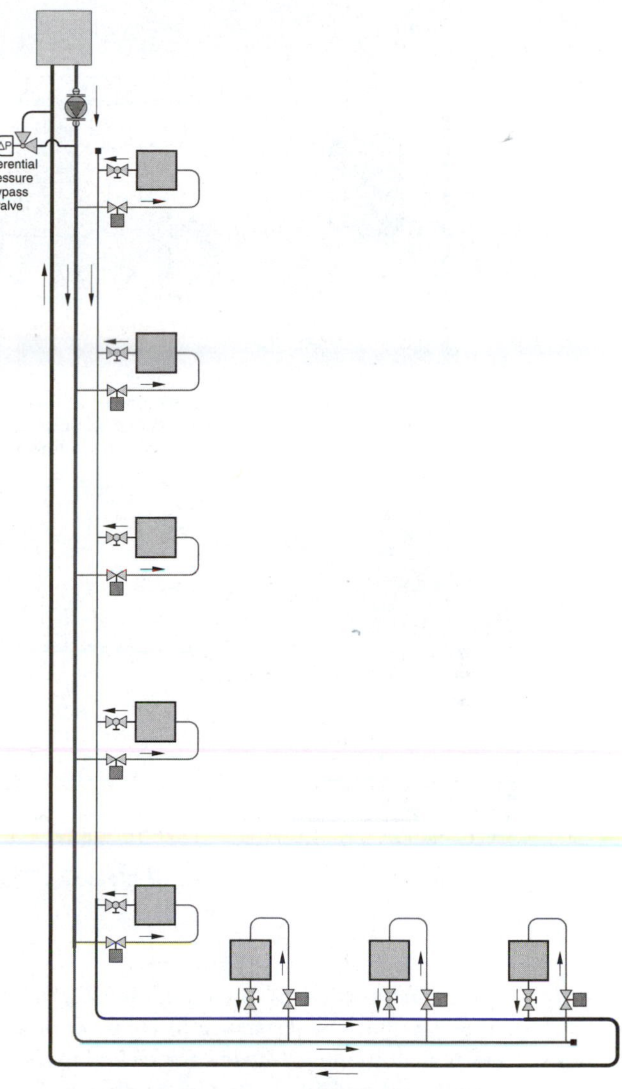

Figure 11–34 Inefficient (but possible) use of parallel direct-return piping in a "dead end" layout.

Step 5: Pipe sizes can now be selected based on keeping the flow velocity in all pipes less than or equal to 4 ft/sec.

Step 6: *Identify the branch with the highest expected flow resistance.* This will often be the branch with the most restrictive heat emitter.

Step 7: Sketch a system diagram showing the expected flow rates (as calculated in step 3) present in all piping segments. An example is illustrated in Figure 11–35.

Step 8: Determine the head loss of the circuit path formed by the branch with the highest resistance and the associated supply and return mains.

Step 9. The head loss obtained in step 9, along with the *full system flow rate,* sets a conservatively high operating point for the circulator. This operating point should fall on or

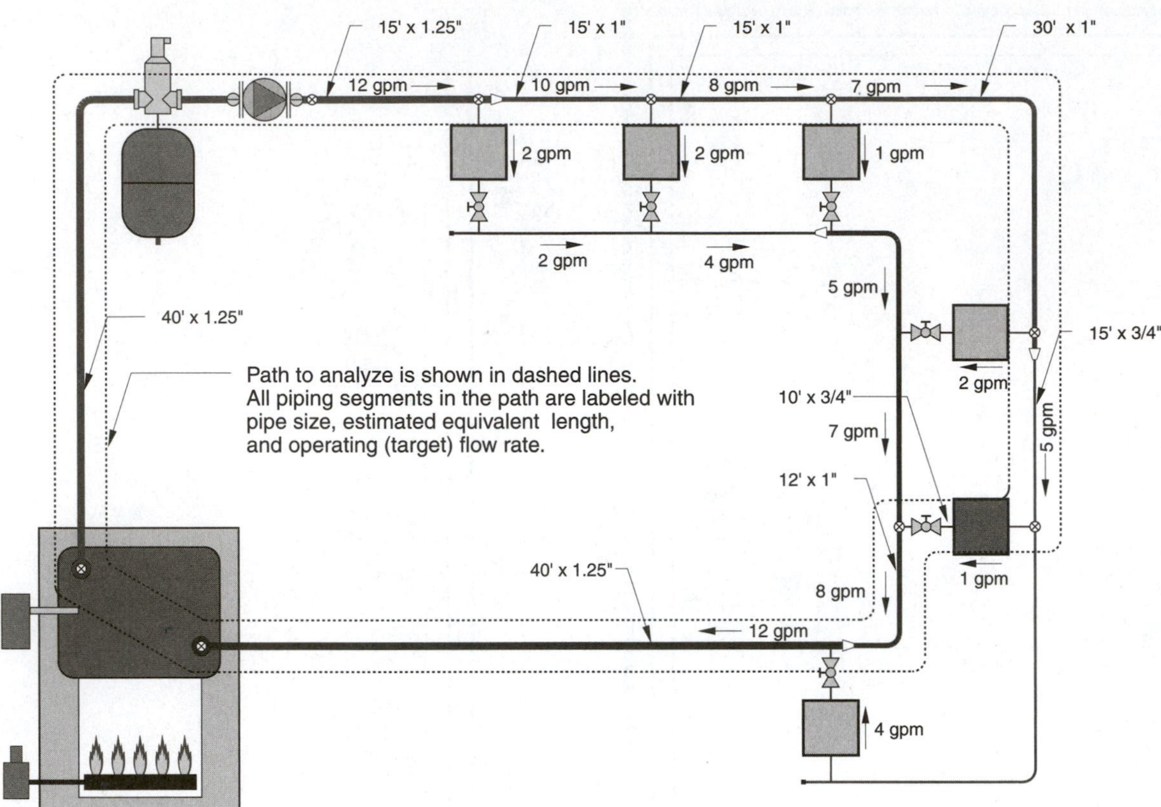

Figure 11–35 Identifying the flow path of greatest resistance when analyzing a parallel reverse-return system.

slightly below the pump curve of circulators suitable for use in the system. Since the flow resistance of the other branches will be lower than the branch used for design purposes, the system will likely operate at flow rates slightly higher than determined by this method.

11.11 HOME RUN DISTRIBUTION SYSTEMS

All of the hydronic distribution systems discussed thus far were developed around the assumption that rigid tubing or pipe would be used to convey heat from the mechanical room to the various heat emitters. They have been successfully used with rigid tubing and pipe for many decades.

Over the last few years, many heating professionals have come to recognize the potential of PEX and PEX-AL-PEX tubing as *universal* piping materials for residential and light commercial hydronic systems. In addition to radiant panel heating applications, the temperature/pressure rating of PEX and PEX-AL-PEX along with their inherent flexibility offers new possibilities for piping traditional higher temperature heat emitters such as cast-iron radiators, fin-tube baseboard, and panel radiators. The most common approach is known as a **home run distribution system,** and is illustrated in Figure 11–36.

In a home run system, a separate supply and return run of small diameter (usually 3/8-inch or 1/2-inch) PEX, or PEX-AL-PEX tubing is routed from a manifold station to each heat emitter. The small flexible tubing can be routed through the framing cavities of the building much like electrical cable. The ability to "fish" tubing through closed framing cavities is a tremendous advantage over rigid tubing, especially in retrofit situations. Home run systems have also proven to be an excellent way of using up the shorter "remnant" lengths of PEX or PEX-AL-PEX tubing often left over from a radiant panel installations.

Besides relatively simple installation, home run systems provide many of the same benefits as other types of parallel distribution systems. They allow the heat output of each room to be individually controlled. They also deliver the same water temperature to each heat emitter, thus simplifying the sizing of the heat emitters.

Home run systems are adaptable to almost any combination of panel radiators, fin-tube convectors, towel warmers, cast-iron radiators, or radiant baseboard. Balancing valves at the manifold station can be used to compensate for the flow resistances of different home run circuits serving different heat emitters.

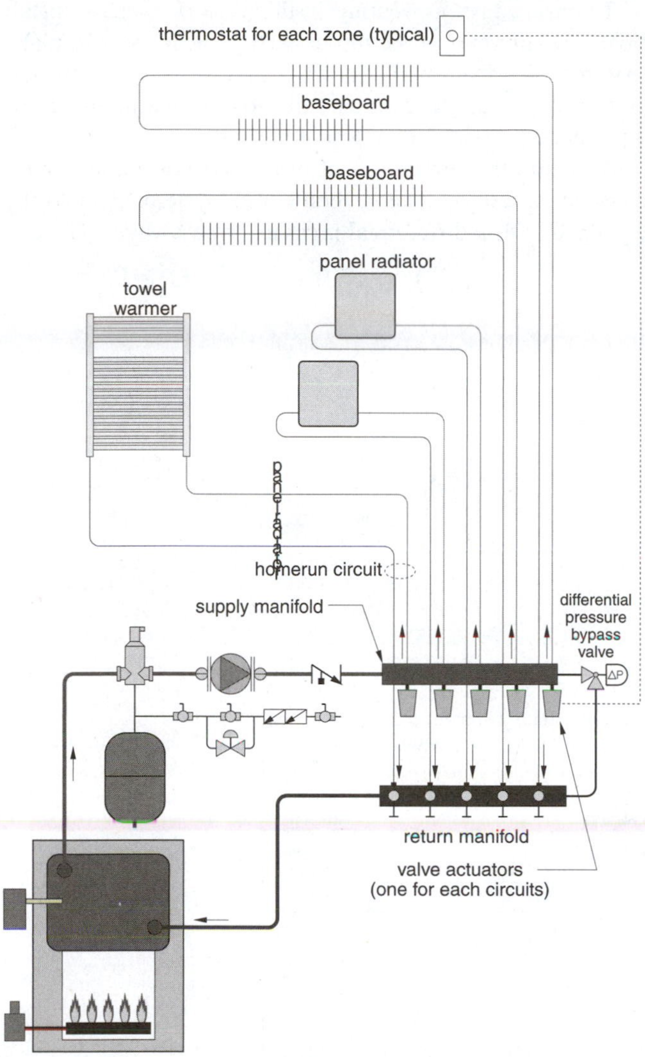

Figure 11–36 Schematic of a home run distribution system using electric valve actuators and thermostats to control flow through individual home run circuits.

Zoning Control Options for Home Run Distribution Systems

Home run systems allow several methods of zoning control. The approach shown in Figure 11–36 uses electric valve actuators to allow or prevent flow in each circuit. These actuators are controlled by 24-VAC thermostats in each room. The end switch in each actuator can be used to call for boiler operation and turn on the circulator. In effect, the manifold and its valve actuators work just like a group of ganged zone valves.

Another approach is to install a thermostatic radiator valve on each heat emitter. These nonelectric valves, which were discussed in Chapter 5, Piping, Fitting, and Valves, modulate the flow through their respective heat emitters to regulate heat output. They are available in various configurations for use with fin-tube baseboard and panel radiators.

Keep in mind that thermostatic radiator valves are nonelectric, and thus can not signal for the boiler to fire or the circulator to run. The distribution circulator must operate whenever heat might be needed by any zone. This is usually accomplished by using a setpoint control that senses outdoor temperature, or a boiler reset control to turn on the distribution circulator and boiler whenever the outdoor temperature is below some starting value (typically about 65 ºF). Figure 11–37 shows the use of the latter in a home run system supplying heat emitters equipped with thermostatic radiator valves.

Stacked Home Run Systems

Imagine a situation where a home run system was to be installed in a tall building such as a three-story house. Significant lengths of tubing would end up as **"leaders"** between the heat emitters on the upper floors and the typical basement mechanical room. One solution is to use a

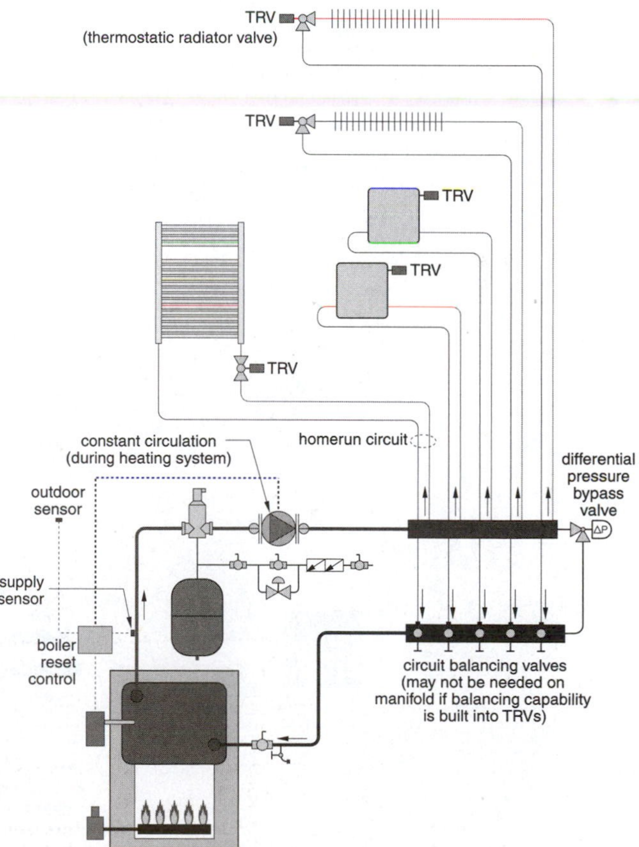

Figure 11–37 Schematic of a home run system using thermostatic radiator valves (TRVs) to regulate flow through each circuit. Boiler is operated by an outdoor reset control. Circulator operates continuously during heating season. Note the use of a differential pressure bypass valve.

"stacked" home run system supplied by a vertical piping risers as shown in Figure 11–38. The designer should look for a central location in the building where the riser piping can be routed from the basement mechanical room to the top floor. Plan the vertical risers so they will be reasonably close to the manifold station on each floor. This approach allows relatively short home run circuits to be routed to each heat emitter from the manifold stations at each floor level.

To ensure fast air venting in this type of system, install float-type air vents at the top of each vertical riser. To make sure they work properly in what may be a relatively tall system, adjust the feed water valve to ensure a minimum of 5 psi static pressure at the top of the system.

As with other systems that use valves for zoning and a constant speed circulator, home run systems should be equipped with a differential pressure bypass valve. The cir-

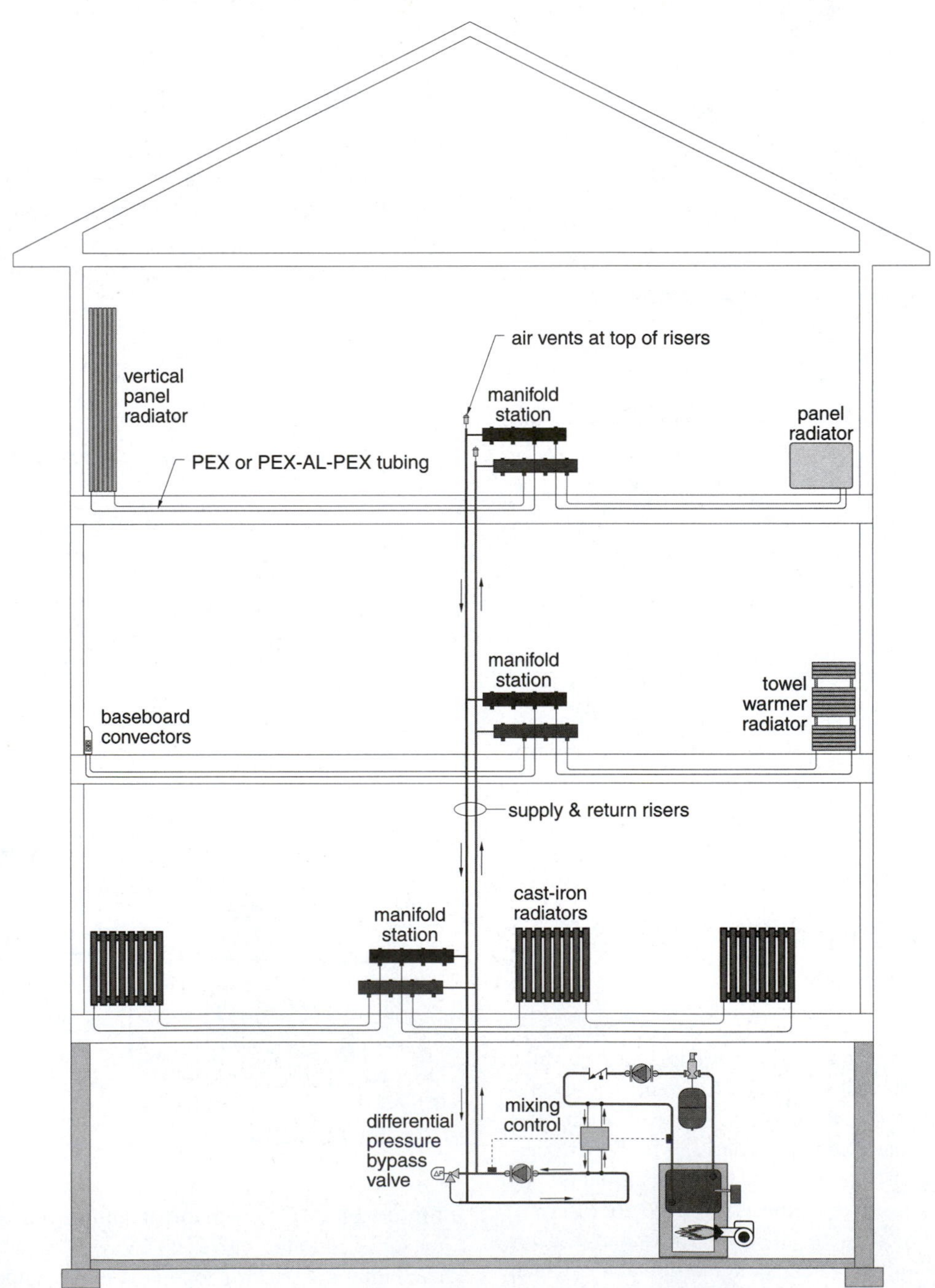

Figure 11–38 Example of a "stacked" home run distribution system in a multi-story building.

culator should also have a relatively "flat" pump curve. Home run systems are also well-suited to variable speed distribution circulators that maintain a constant differential pressure across the manifold station.

Later in the chapter we will show how a **home run subsystem** can also be incorporated within a more extensive distribution system.

Design Procedure for Home Run Systems

From the standpoint of piping "topology," home run distribution systems are very similar to the zone valve systems discussed in Section 11.8. The differences are essentially in the lengths and types of tubing used. The design procedure described in Section 11.8 can thus be used to design a home run system or subsystem. The Hydronic Circuit Simulator module in the Hydronics Design Studio software can also be used to model the flow performance of a home run system.

11.12 PRIMARY/SECONDARY SYSTEMS

The concept of **primary/secondary (P/S) piping** has been around since the 1950s. Until the mid-1980s, it was used mostly for larger commercial heating and chilled water cooling systems. However, the increasingly sophisticated expectations placed on residential and light commercial systems in recent years have prompted designers to look for piping techniques more flexible and expandable than traditional approaches.

To meet this demand, the fundamental concepts of primary/secondary piping have been redeployed and successfully integrated along with modern controls in residential and light commercial systems. The result is an extremely versatile piping technique that now serves as the "backbone" of most modern multiload/multitemperature systems.

A simple example of a simple P/S system is shown in Figure 11–39. The lower circuit is called the **primary loop.** Its function is to convey heated water to one or more locations along its path, and "hand off" some of that heated water to so-called **secondary circuits.** It is common to have two or more secondary circuits supplied from a single primary circuit. The secondary circuit can serve a variety of loads including space heating, domestic water heating, or even snowmelting. These loads usually operate independently of each other based on when their individual controls call for heating. *The circulator in the primary loop must operate whenever any one or more of the secondary circuits is operating.*

The goal underlying the piping configuration of a P/S system is to "uncouple" the pressure differential established by the primary circulator from the pressure differentials established by any of the secondary circulators. *This allows the various secondary circuits to independently turn on and off without affecting the flow rate in any of the other operating secondary circuits, or the flow rate in the primary loop.* In animated terms, each circulator in the system "thinks" it is installed in a totally isolated circuit. It has "no idea" that other piping circuits or circulators are present in the same system. This operating characteristic is highly desirable because it allows for stable operating conditions in all circuits. It also eliminates the possibility of interference between various circulators. Such harmony between simultaneously operating circulators cannot be fully achieved with any of the other systems discussed in this chapter.

To better understand how a P/S system uncouples the pressure dynamics of each circuit from the others, consider the piping system shown in Figure 11–40a. When water flows through the lower piping loop, a pressure drop develops between the tees at points A and B. Since the pressure is higher at point A, this pressure differential tries to induce flow in the upper piping circuit from A to B.

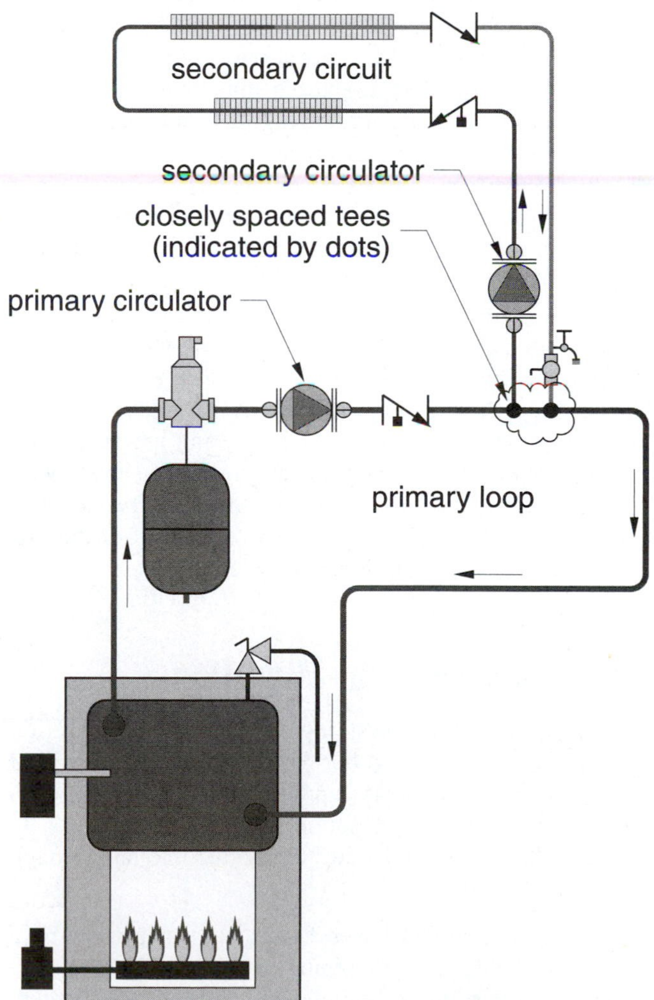

Figure 11–39 Example of a simple primary/secondary distribution system.

The magnitude of the pressure drop from A to B depends on the length and size of the pipe between the tees, as well as the flow rate between them. If the tees are separated by even 2 or 3 feet, the pressure drop between them is relatively small. Small enough that the induced flow in the upper circuit will probably not be able to carry the full design heating load of that circuit. *However, experience has shown that even slow induced flows can carry sufficient heat to cause overheating under mild conditions, or when the spaces served by the circuit do not require heat input.* For this reason, such uncontrolled induced flow must be prevented.

One way to minimize the potential for induced flow is to make the pressure drop between the tees as small as possible. This is done by installing the tees *as close as possible to each other* as shown in Figure 11–40b. Because the pressure drop between such tees is almost zero, there is very little tendency for flow in the primary circuit to induce flow in the secondary circuit. When full design flow in the secondary circuit is desired, a circulator in that circuit is turned on to provide the needed pressure differential.

The use of **closely spaced tees** to minimize the potential for induced flow in secondary circuits is fundamental to all P/S systems. A guideline is to not allow the distance between the centers of these tees to exceed four times the nominal diameter of the primary loop piping. It is also necessary to minimize turbulence both upstream and downstream of the closely spaced tees. This is accomplished by installing at least eight primary pipe diameters of straight pipe upstream up the first tee, and at least four primary pipe diameters of straight pipe downstream of the second tee. These details are shown in Figure 11–41.

The minimal pressure drop associated with a pair of closely spaced tees can also be accomplished using a special primary/secondary fittings as shown in Figure 11–42a. Notice that the side ports of such a fitting are very close together. They have been created using a **T-Drill®** tool. Some installers also use T-Drills® to install two closely spaced copper tubes directly into the tube wall of the primary circuit as shown in Figure 11–42b.

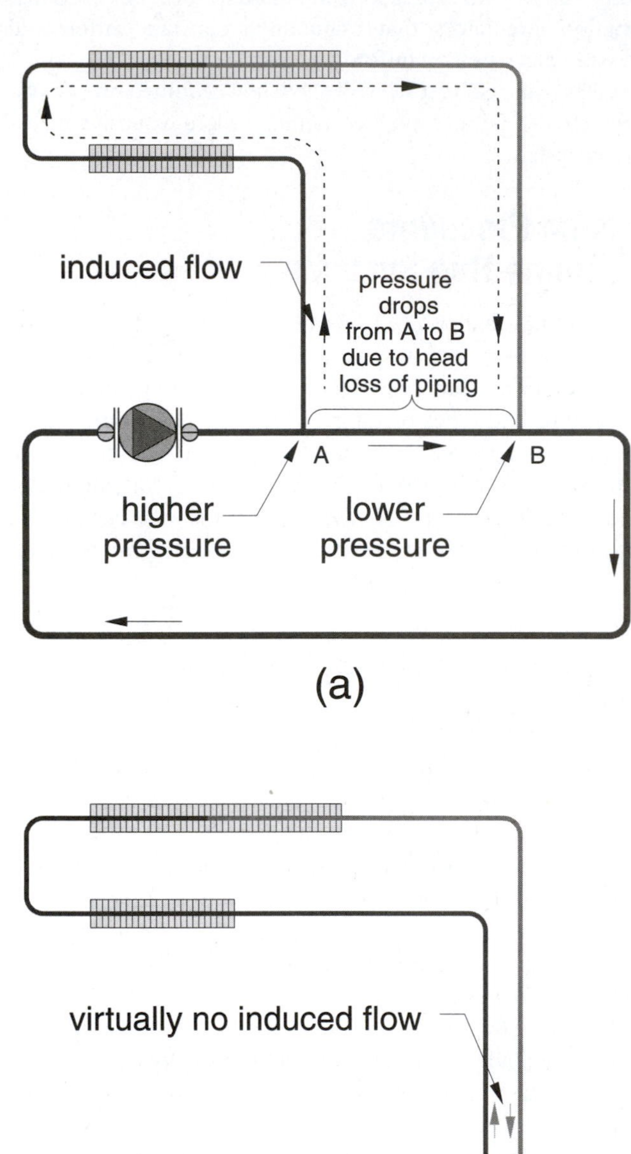

Figure 11–40 (a) Flow is induced in the upper circuit due to pressure drop in piping from point A to point B. (b) By placing tees close together, there is very little tendency for induced flow to occur (due to pressure drop).

Preventing Thermal Migration

Primary/secondary systems also require protection against the tendency for hot water to migrate into and through inactive secondary circuits (e.g., circuits in which the secondary circulator is not operating).

Two factors cause hot water to migrate into and through secondary circuits:

1. The tendency of lower density hot water to rise into the supply side of a piping circuit installed above a hot water source, while higher density cooler water descends through the return side of the same circuit.
2. The fact that the pressure drop between closely spaced tees, although very small, is still not zero.

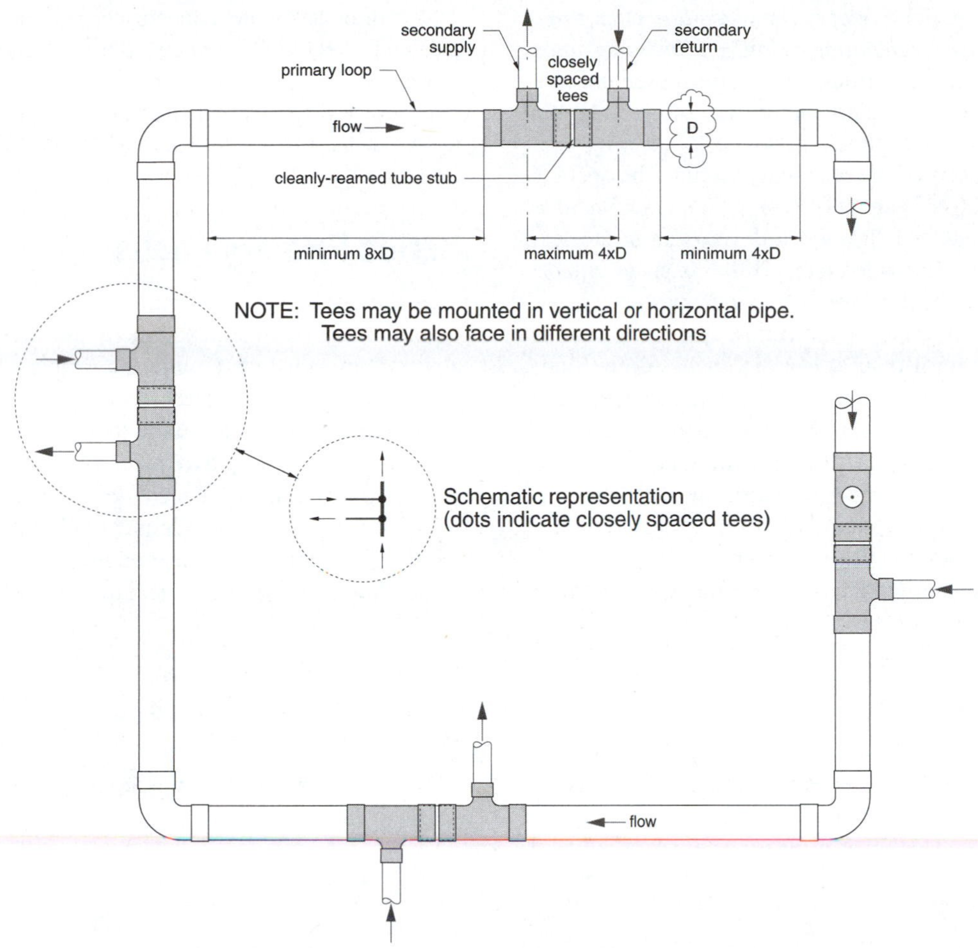

Figure 11–41 Installation details for closely spaced tees in a primary/secondary system. Note the required lengths of straight pipe both upstream and downstream of tees.

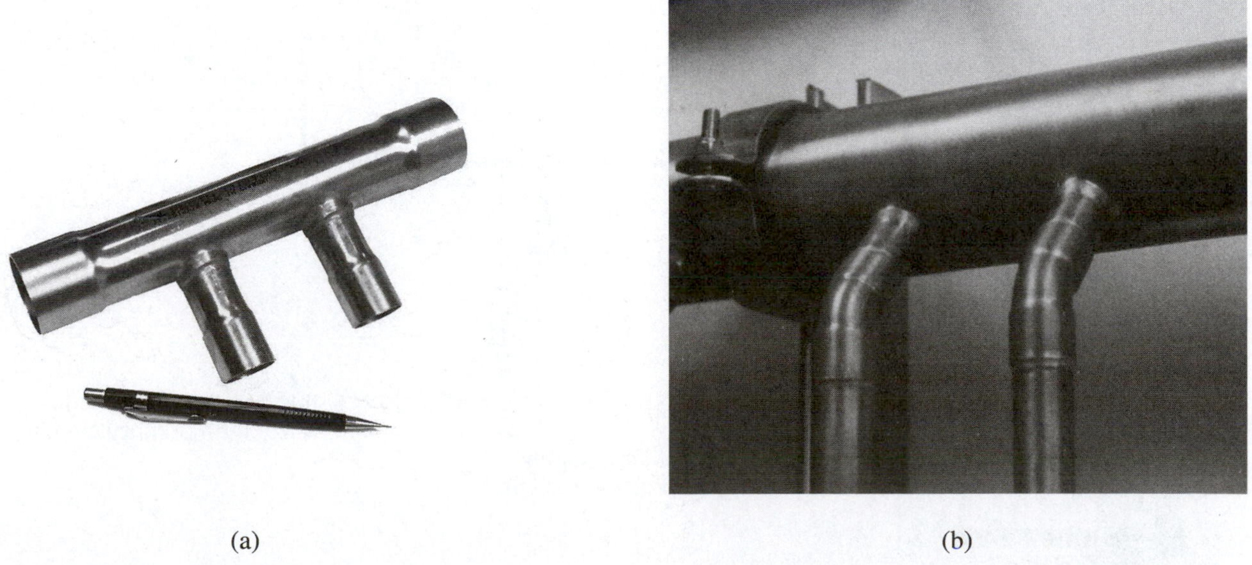

Figure 11–42 (a) Example of a primary/secondary fitting. *Courtesy of Alberta Custom Tee.* (b) Use of a T-Drill® to create a primary/secondary interface.

The supply and return side of every secondary circuit in a P/S system must include detailing to prevent heat migration when its circulator is off. Various options for preventing heat migration into inactive secondary circuits are shown in Figure 11–43. One approach is to install flow-check valves on *both* supply and return risers of the secondary circuit. The opening pressure of these valves is about 0.3 psi, which is sufficient to prevent buoyancy-induced flow from developing in the secondary circuit. A spring-loaded check valve is an acceptable alternative to a flow-check in both these locations.

Two other options exist for the *return riser only.* One is an **underslung thermal trap** at least 18 inches deep. Since hot water "wants" to rise rather than descend, such a piping arrangement "discourages" hot water from migrating into the return side of the secondary circuit. Another option is to install a swing-check on the return side of the secondary circuit. It is important to note that neither the underslung thermal trap or the swing-check valve can prevent hot water from rising into the supply side of the secondary circuit.

It is often desirable to locate a primary loop high within the mechanical room. This allows the secondary circuits to drop down away from the primary loop, even if they eventually must turn and go up in the building. A thermal trap is automatically created in the return side of the secondary circuit whenever it drops at least 18 inches below the primary loop.

Series Primary Loops

By far the most common form of primary/secondary piping is an arrangement where the secondary loads are arranged in sequence around the primary loop as shown in Figure 11–44. The primary loop is now more specifically described as a **series primary loop.**

Series primary loops are ideal when the operating temperature requirements of the secondary circuits vary over a wide range. For example, one secondary circuit might supply relatively high temperature heat emitters such as fin-tube baseboard. Another may serve medium temperature heat emitters

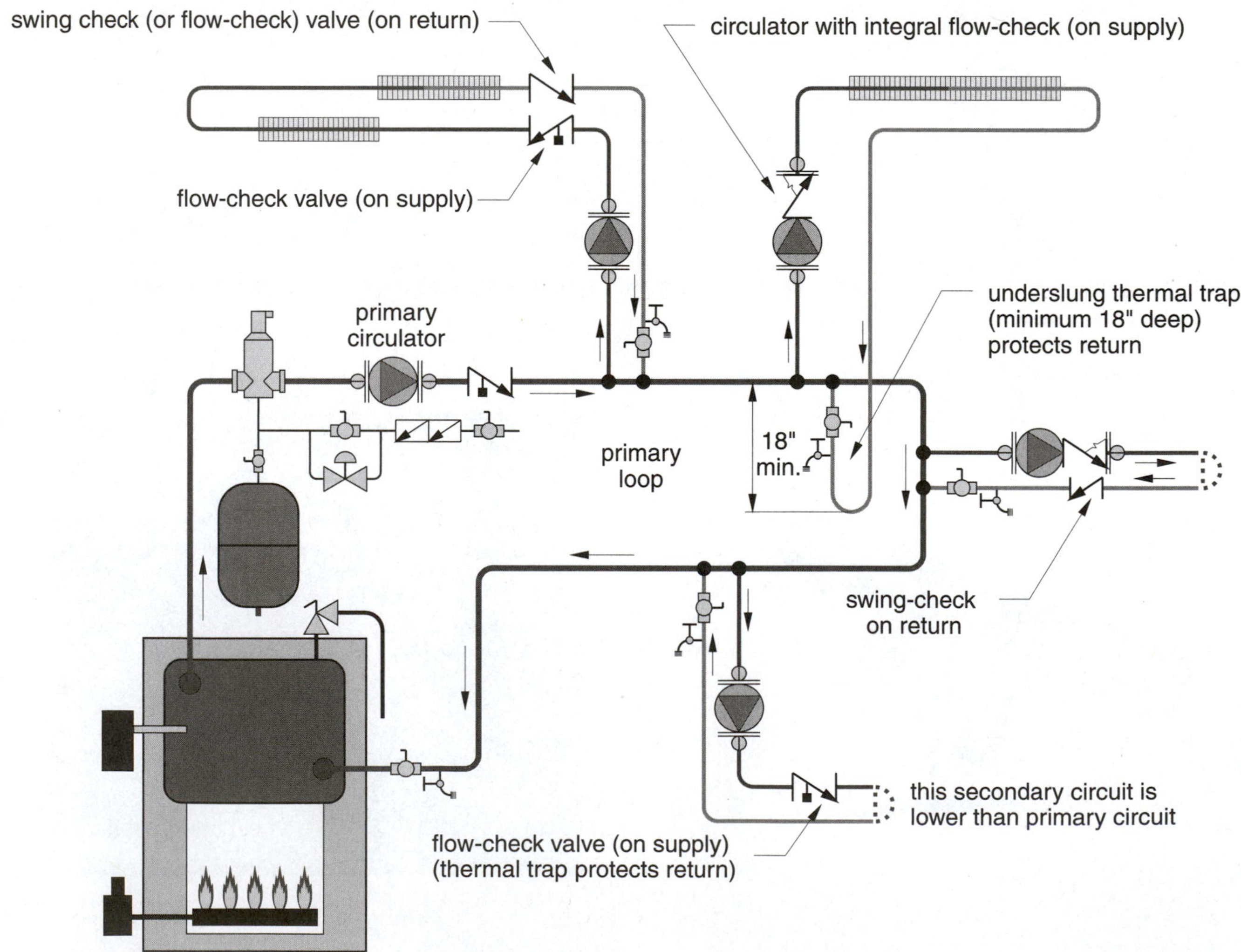

Figure 11–43 Methods of preventing thermal migration into inactive secondary circuits.

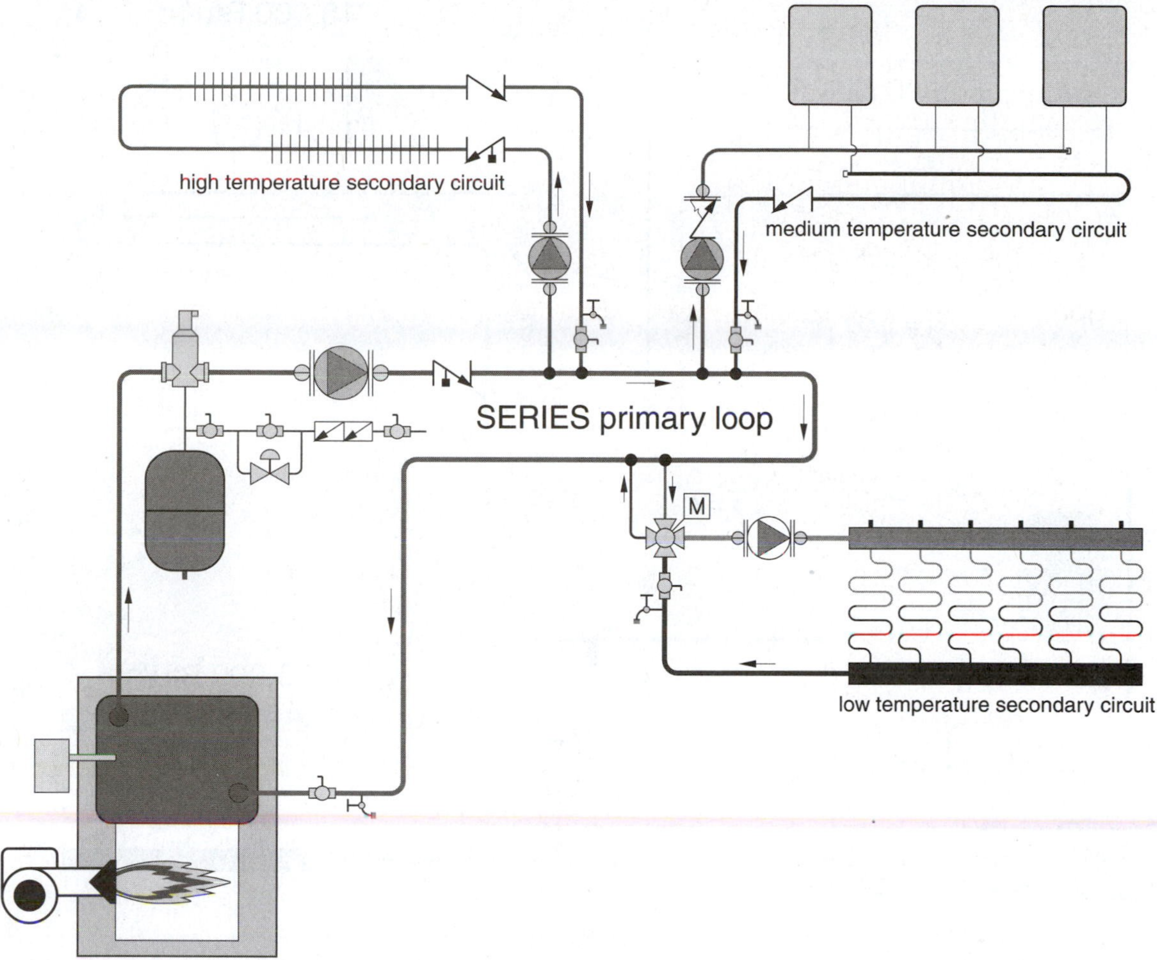

Figure 11–44 Example of a series primary loop. Secondary circuits needing higher supply temperatures placed near beginning of primary loop. Secondary circuits capable of operating at lower temperatures placed near end of primary loop.

such as panel radiators. Still another may serve a low temperature floor heating subsystem through a mixing device.

Higher temperature loads should always be placed near the beginning of the series primary loop, while loads capable of operating at lower water temperatures should be placed near the end of the loop. Such an arrangement works in concert with the drop in water temperature around the primary loop.

The drop in water temperature as the primary loop flow passes an operating secondary circuit can be calculated using Equation 11.9.

(Equation 11.9)

$$\Delta T_p = \frac{Q_s}{500 f_p}$$

where:

ΔT_p = temperature drop in the primary loop as it passes an operating secondary circuit (°F)

Q_s = rate of heat output to the secondary circuit (Btu/hr)

f_p = flow rate in the primary circuit (gpm)

500 = a constant for water

When a fluid other than water is used, replace the factor 500 in Equation 11.9 with (D x c x 8.01)

where:

D = the fluid's density at the average system temperature (lb/ft^3)

c = the fluid's specific heat at the average system temperature (Btu/lb/°F)

To determine the inlet temperature to a given secondary circuit, subtract the temperature drop(s) of any upstream operating secondary circuit(s) from the supply temperature provided by the heat source. This procedure is demonstrated in Example 11.3.

Example 11.3: Determine the water temperatures in the primary loop between the secondary circuits shown in Figure 11–45.

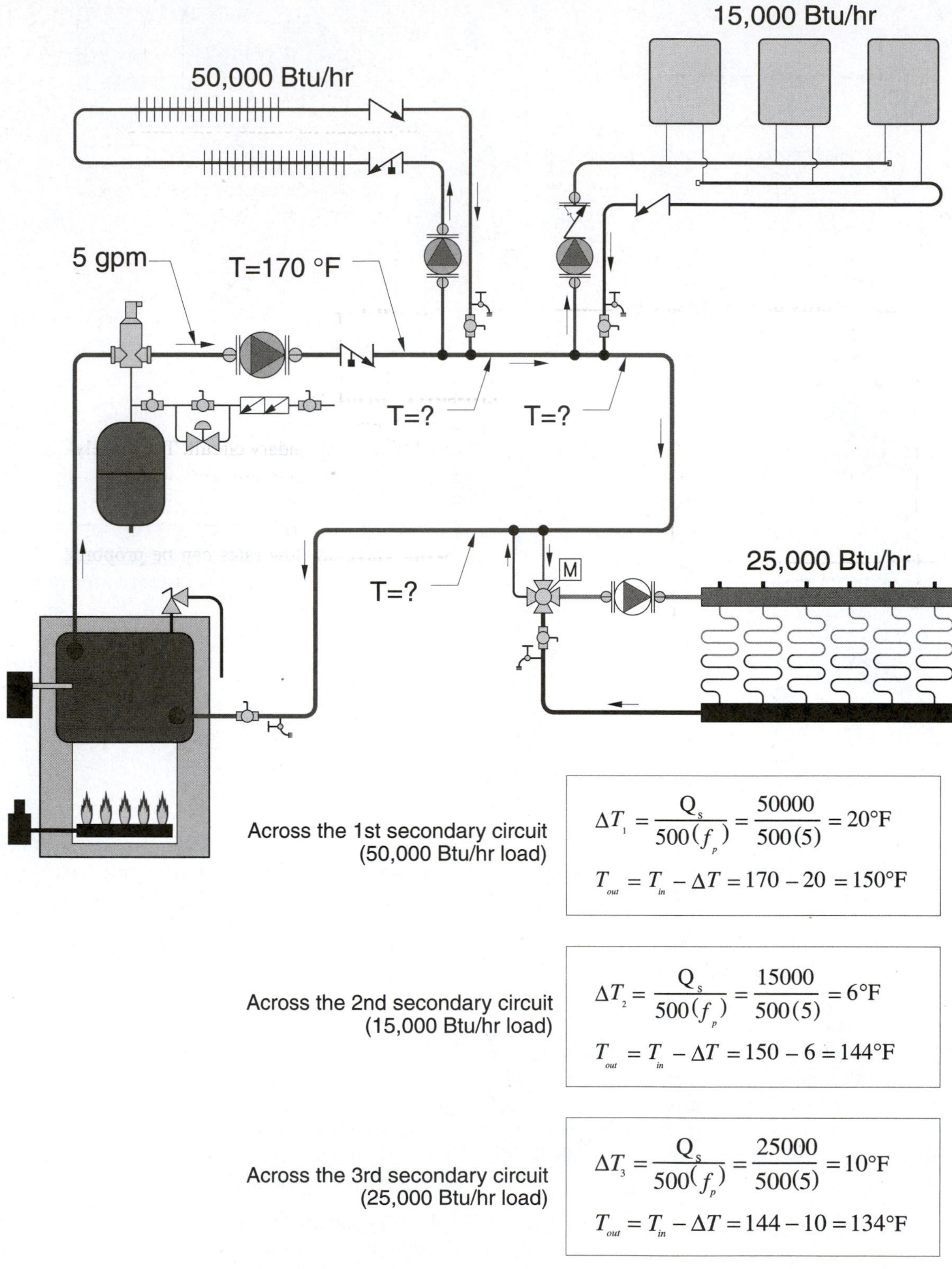

Figure 11–45 Determining the temperature drop across each operating secondary circuit.

Solution: Use Equation 11.9 to determine the temperature drop across each load, then subtract the drop from the inlet temperature to get the outlet temperature.

$$\Delta T_1 = \frac{Q_s}{500(f_p)} = \frac{50000}{500(5)} = 20°F$$

$$\Delta T_2 = \frac{Q_s}{500(f_p)} = \frac{15000}{500(5)} = 6°F$$

$$T_{out} = T_{in} - \Delta T = 150 - 6 = 144°F$$

$$\Delta T_3 = \frac{Q_s}{500(f_p)} = \frac{25000}{500(5)} = 10°F$$

$$T_{out} = T_{in} - \Delta T = 144 - 10 = 134°F$$

Discussion: When a conventional boiler is used as the heat source, the designer should verify that the water temperature returning from the primary loop under design load conditions is high enough to prevent sustained flue gas condensation. This is especially important when the primary loop operates with a large temperature drop.

In some respects, a series primary/secondary system is similar to a one-pipe diverter tee system. Both use a single circuit to transport heated water to several loads. Both can be configured to control each branch circuit as a separate zone. The difference is that a secondary circulator, rather than a diverter tee, is used to create the differential pressure that induces flow in a branch circuit. One advantage of using a secondary circulator is that it can produce far higher flow rates in larger secondary circuits in comparison to those attainable using diverter tees. A secondary circulator also uses electrical energy only when flow is needed in its circuit. A diverter tee, on the other hand, dissipates pumping energy (e.g., head) through its orifice even when its branch circuit is closed off.

Parallel Primary Loops

When several secondary circuits have approximately the same water temperature requirement, they are best supplied from a **parallel primary loop,** as shown in Figure 11–46.

In a parallel primary loop, each secondary circuit is supplied from a pair of closely spaced tees that connect to a **crossover bridge**. The parallel arrangement of the crossover bridges assures the same water temperature is delivered to each secondary circuit. The closely spaced tees ensure there will be no interference between simultaneously operating circulators.

Each crossover bridge should be equipped with a **balancing valve** so flow rates can be proportioned to the load served by each bridge. Equation 11.3 can be used to estimate the flow needed in each crossover bridge.

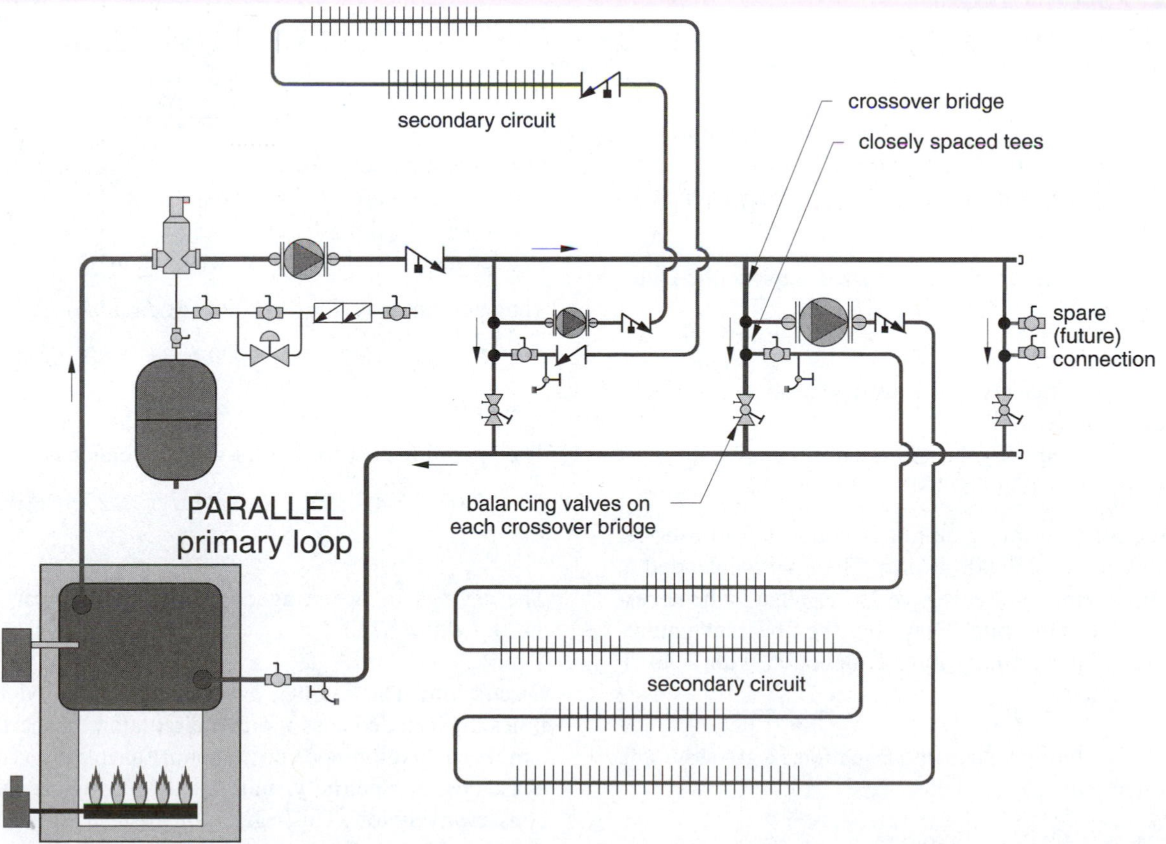

Figure 11–46 Example of a parallel primary loop. Note that each secondary circuit operates at the same supply water temperature.

Parallel primary loops were originally developed for chilled water cooling systems in which all air handlers need to operate within a relatively narrow range of water temperature. They work equally well for heating applications, especially in systems with low supply water temperatures.

Sizing the Primary Circulator

It has been stated that every circulator in a primary/secondary system operates as if it were installed in an isolated circuit. The primary circulator does not assist in moving flow through any of the secondary circuits, or vice versa. The function of the primary loop is simply to convey the output of the heat source to the secondary circuit "pickup" points, while operating at or close to a selected temperature drop.

Contrary to myths that exist in the industry, the primary circulator does *not* necessarily have to be the largest circulator in the system. It may even be the smallest circulator in some systems! The primary circulator also does *not* need to operate a flow rate equal to or greater than the total flow rate of all the secondary circuits that can operate simultaneously.

The flow rate necessary to deliver the full output of the heat source using a specific temperature drop can be found using Equation 11.10:

(Equation 11.10)

$$f_p = \frac{Q_{hs}}{500(\Delta T)}$$

where:

f_p = required flow rate in the primary circuit (gpm)
Q_{hs} = heat output rate of heat source (Btu/hr)
ΔT = intended temperature drop of the primary circuit (°F)
500 = a constant for water

When a fluid other than water is used, replace the factor 500 in Equation 11.10 with (D x c x 8.01).

where:

D = the fluid's density at the average system temperature (lb/ft^3)
c = the fluid's specific heat at the average system temperature (Btu/lb/°F)

Example 11.4: A primary circuit is connected to a boiler having an output of 250,000 Btu/hr. The loop is intended to operate with a temperature drop of 15 °F. What is the necessary primary loop flow rate? How does the flow rate requirement change if the primary loop is operated with a 30 °F temperature drop?

Solution: Substituting values into Equation 11.10 (assuming a 15 °F temperature drop) yields:

$$f_p = \frac{Q_{hs}}{500(\Delta T)} = \frac{250{,}000}{500(15)} = 33.3 \text{ gpm}$$

The required flow rate in the primary loop when the temperature drop is increased from 15 to 30 °F is:

$$f_p = \frac{Q_{hs}}{500(\Delta T)} = \frac{250{,}000}{500(30)} = 16.7 \text{ gpm}$$

Discussion: Doubling the temperature drop of the primary loop cuts the required flow rate in half. Not only does this reduce the required pipe size, it may also reduce the size of the circulator. The smaller circulator will likely operate on lower wattage, and thus reduce the operating cost of the system. Considering that the primary loop circulator must operate whenever any secondary circuit operates, such energy savings over the life of the system can be substantial.

Example 11.5: Assume that operating the primary loop at the lower flow rate in Example 11.4 allows an 80-watt primary loop circulator to be used instead of a 200-watt circulator. Also assume the primary loop circulator operates 3000 hours per year in an area where the present cost of electricity is \$0.10 per kilowatt-hour. What is a conservative estimate for the savings in electrical energy cost over a period of 20 years?

Solution: The operating cost of both circulators can be determined using Equation 11.11:

(Equation 11.11)

$$O = (P)(t)(c)$$

where:

O = operating cost (\$)
P = operating power (kilowatts)
T = operating time (hours)
c = cost of electricity (\$/kilowatt • hour)

The operating cost for the 200-watt circulator is:

$$O = (P)(t)(c) = (0.200\text{kw})(3000 \text{ hr/yr} \times 20 \text{ yr})(\$0.10/\text{kwhr}) = \$1200$$

The operating cost for the 80-watt circulator is:

$$O = (P)(t)(c) = (0.080\text{kw})(3000 \text{ hr/yr} \times 20 \text{ yr})(\$0.10/\text{kwhr}) = \$480$$

The savings in operating cost over the 20-year period is 1200 – 480 = \$720

Discussion: The savings over a 20-year period is several times the installed cost of either circulator. Furthermore, this simple calculation does not account for inflation in the purchase cost of electricity, and thus the savings is likely to be conservatively low. This significant cost savings is also associated with small low-wattage circulators. The savings would be proportionally higher in larger systems using higher wattage circulators.

Once the primary loop flow rate and pipe size are determined, the designer should estimate the head loss of the primary loop based on the piping components it contains. A system resistance curve can be constructed using the methods of Chapter 6, Fluid Flow in Piping. The equivalent lengths of the closely spaced (straight through) tees should be included when constructing the system curve for the primary circuit. However, be sure not to include the resistance of any piping components from the secondary circuits.

A circulator can then be selected by overlaying pump curves for candidate circulators on the system curve to find an operating point that is reasonably close to the target flow rate. The Hydronic Circuit Simulator module in the Hydronics Design Studio software could also be used to model the performance of a proposed primary loop design.

Finally, notice that a primary loop circulator can be selected without reference to the secondary circuits. Likewise, each secondary circulator can be selected based solely on the flow and head loss requirements of the secondary circuit it serves.

A similar procedure can be used to select a circulator for a parallel primary loop. The hydraulic resistance of the parallel crossover bridges and main pipe segments can be modeled using methods from Chapter 6, or by using the Hydronic Circuit Simulator module. Again, the flow rate and head loss requirements of the secondary circuits do not affect the selection process for the primary loop circulator.

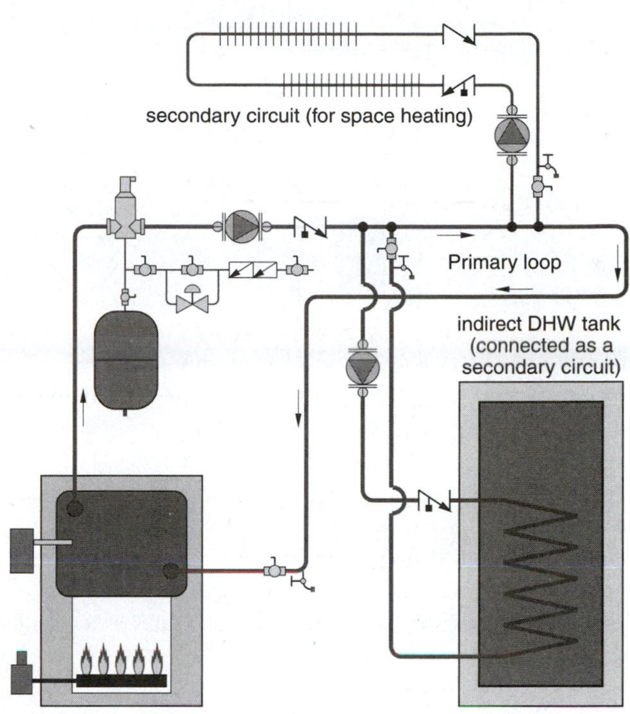

Figure 11–47 Piping an indirect water heater as a secondary circuit.

Indirect Water Heaters in Primary/Secondary Systems

There are two fundamental ways to pipe an indirect water heater into a primary secondary system:

1. As a secondary circuit
2. As a parallel circuit to the primary loop

When piped as a secondary circuit, the indirect water heater should be connected near the beginning of a series primary loop. This allows the hottest water in the system to flow through the tank's heat exchanger for the fastest possible recovery. A representative piping schematic is shown in Figure 11–47.

This piping is appropriate when the indirect tank is located close to the boiler, and thus the primary loop is relatively short, and preferably insulated. The underlying goal is to minimize heat loss from the primary loop piping. If the primary loop were long and/or uninsulated, such heat loss could represent a significant and uncontrolled heat gain to the building. Considering that there would be flow of hot water in the primary loop during warm weather operation of the domestic water heater, such heat loss should be avoided.

It is also crucially important to install a flow-check valve in the DHW secondary circuit to prevent heat migration both into, and out of the indirect tank. Without this valve, hot water from the primary loop may slowly migrate through the tank's heat exchanger while the primary loop supplies hot water to other loads. In some systems, the lack of this flow-check has caused tanks to reach dangerously high water temperatures. Another reason for the flow-check is to prevent heat loss due to reverse flow in the tank's secondary circuit. Without the flow-check installed, the warm water in the heat exchanger will use the secondary circuit piping as a heat sink when both the primary loop circulator and secondary loop circulator are off.

Finally, as with all secondary circuits, it is necessary to run both the primary loop circulator and the secondary circulator whenever the indirect water heater operates. A failure of either circulator would prevent domestic water heating.

The piping for connecting an indirect water heater as a **parallel circuit** is shown in Figure 11–48. With this hybrid piping, the primary loop circulator does not need to operate during the DHW mode. There is also less piping heat loss because hot water does not travel around the entire primary loop while the domestic water is heated.

Because the water heater circuit is in parallel with the primary loop, it is essential to install flow-check valves in both loops to prevent reverse flow through one when the other is active. The flow-checks also prevent forward heat migration from the boiler into inactive circuits. Of the two methods discussed, the author generally prefers to pipe the indirect water heater as a parallel circuit.

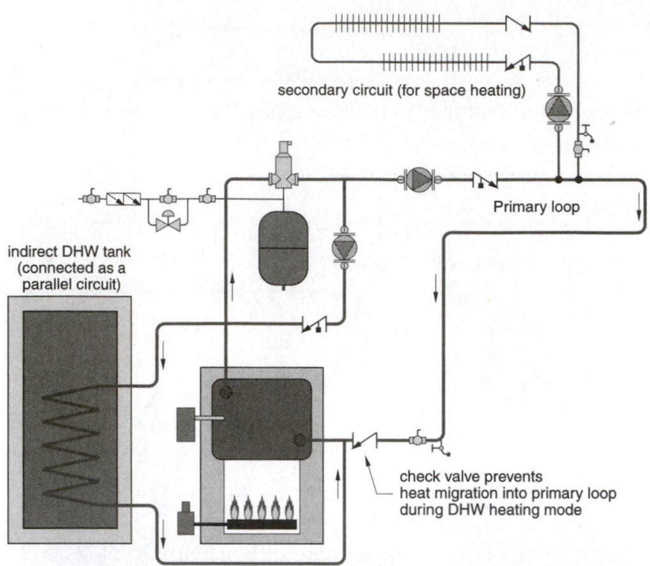

Figure 11–48 Piping an indirect water heater as a parallel circuit.

Purging Provisions for Secondary Circuits

The closely spaced tees connecting a secondary circuit to the primary loop are counterproductive to the desired effect when the system is being purged of air. A strong purging flow in the primary circuit will *not* induce a similar flow in a secondary circuit.

The solution is to install a means of independently purging each secondary circuit. One approach is to install a ball valve and hose bib near the end of each secondary return riser. This detail has been shown on many of the piping schematics in this section. During purging, the ball valve is closed forcing pressurized make-up water in the desired direction through the secondary circuit as air is blown out through the open hose bib. An "all-in-one" purge valve could also be used on the return side of each secondary circuit. On larger systems, the hose bib may be too restrictive for effective purging. In such cases, use a larger ball valve with barbed hose connection as the outlet.

The primary loop should also include valves for purging. Generally, a full port ball valve and hose bib are installed just upstream of the boiler inlet, and used in conjunction with the system's fast-fill valve during purging. The preferred approach is to first purge the primary loop, and then purge each secondary circuit one at a time.

Flow Reversal in Primary/Secondary Systems

It is possible for the flow rate in a secondary circuit to equal or even exceed the flow rate in the primary circuit. When the flow in a secondary circuit equals that in the primary circuit, the flow rate between the closely spaced tees is zero. When the flow rate in a secondary circuit *exceeds* the flow in the primary circuit, the flow reverses between the closely spaced tees. These scenarios are shown in Figure 11–49.

When flow reversal between the tees occurs, the water temperature supplied to the secondary circuit is determined by the proportions of hot water entering the upstream tee, and cooler water returning from the secondary circuit. Equation 11.12 can be used to determine this mix temperature:

(Equation 11.12)

$$T_{mix} = \frac{T_h(f_p) + T_r(f_s - f_p)}{f_s}$$

where:

T_{mix} = temperature supplied to the secondary circuit (°F)
T_h = temperature of fluid in primary loop upstream of first tee (°F)
T_r = temperature of fluid returning from secondary circuit (°F)
f_p = flow rate in primary circuit upstream of first tee (gpm)
f_s = flow rate in secondary circuit (gpm)

Example 11.6: Assume a radiant manifold system is connected as a secondary circuit as shown in Figure 11–50. Using the stated operating conditions, calculate the water temperature supplied to the manifold, and the rate of heat release by the manifold. Verify this rate of heat release by calculating the energy released from the primary loop.

Solution: Substituting the temperature and flow rate data into Equation 11.12 yields the mixed supply temperature to the manifold station.

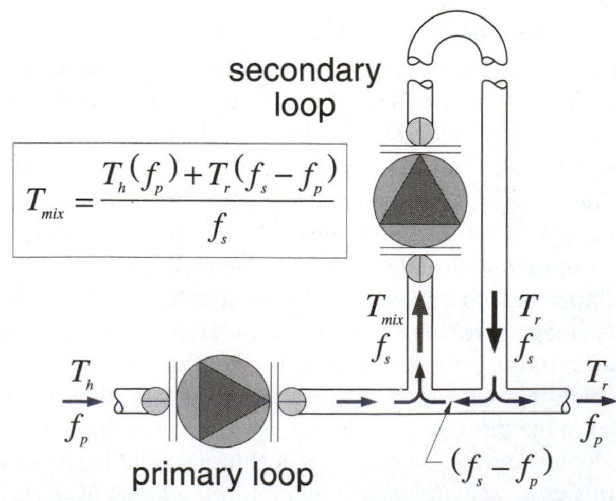

Figure 11–49 Determining the mixed supply temperature of a secondary circuit operating at a flow rate higher than that of the primary circuit.

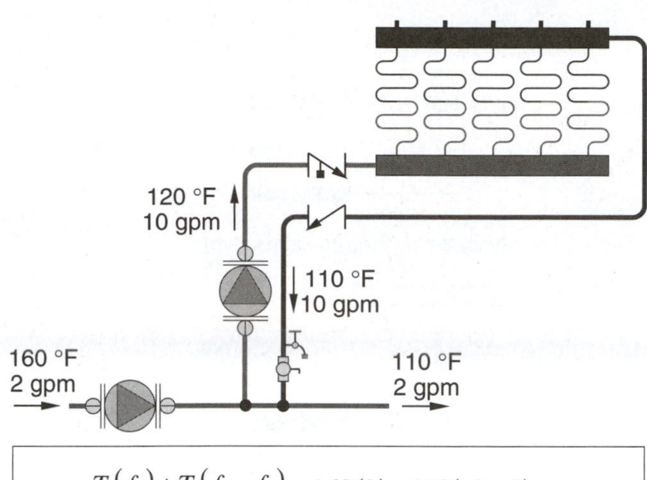

$$T_{mix} = \frac{T_h(f_p) + T_r(f_s - f_p)}{f_s} = \frac{160(2) + 110(10-2)}{10} = 120\,°F$$

Figure 11–50 Piping and operating conditions for Example 11.6.

$$T_{\text{mix}} = \frac{T_{\text{h}}(f_{\text{p}}) + T_{\text{r}}(f_{\text{s}} - f_{\text{p}})}{f_{\text{s}}} = \frac{160(2) + 110(10-2)}{10} = 120°\text{F}$$

The rate of heat release from the manifold station can be determined using the sensible heat rate equation from Chapter 4, Properties of Water.

$$Q = 500 f_{\text{s}}(T_{\text{in}} - T_{\text{out}}) = 500 \times 10 \times (120 - 110) = 50{,}000 \text{ Btu/hr}$$

The rate of heat release from the primary circuit can also be verified using the sensible heat rate equation:

$$Q = 500 f_{\text{p}}(T_{\text{in}} - T_{\text{out}}) = 500 \times 2 \times (160 - 110) = 50{,}000 \text{ Btu/hr}$$

Discussion: Notice that the heat transferred to the manifold station is exactly the same as the heat released from the primary loop. This must be true to satisfy the requirements of the first law of thermodynamics (namely that energy can not be created or destroyed throughout a process). The principle of flow conservation also applies in that the flow rate upstream of the first tees must always equal the flow rate downstream of the second tee. This concept has often been stated as follows: *The flow entering a tee equals the flow leaving the tee.* These simple concepts form the basis for analyzing all flow and temperature exchange processes in hydronic heating.

It is also worth noting that when the secondary circuit flow rate is higher than the primary loop flow rate, the temperature leaving the downstream tees always equals the return temperature of the secondary circuit. In this case, the return temperature of 110 °F is low enough to cause flue gas condensation in a conventional gas-fired boiler, and therefore is not an acceptable system design if such a boiler is used.

To summarize, it is possible for secondary circuit flow rates to exceed primary loop flow rates. Although flow reversal at the primary/secondary interface is not typical in the type of systems described in this text, it does not in any way invalidate the ability of primary/secondary piping to eliminate circulator interference.

Primary Circuit Options

Both series and parallel primary loops have the flexibility of being very short or quite long. For example, the primary loop in a given system may consist of no more than 5 to 10 feet of pipe near the boiler. In such situations, the minimum length of the primary circuit is usually limited by the need to install the necessary piping components and the lengths of straight pipe both upstream and downstream of each set of closely spaced tees.

Another possibility is for a series or parallel primary loop to travel completely around the inside of a building. In such a situation, the secondary circuits are simply tapped into the primary loop where they are located in the building. In this type of system, the primary loop should be well-insulated to minimize extraneous heat gains to the building.

Secondary Circuit Options

Almost any piping circuit that could be connected directly to a boiler or other heat source can also be connected to a primary loop with a pair of closely spaced tees. Some examples of such piping subassemblies are shown in Figure 11–51.

The secondary circuit may be a simple series loop, or a two pipe direct- or reverse-return subcircuit. Another possibility is a home run subsystem as discussed in the previous section.

Several mixing strategies used in conjunction with low temperature distribution systems can also be used with P/S piping. These include any number of mixing devices such as 3-way and 4-way valves, or variable speed injection mixing.

The benefits of a parallel primary/secondary system can also be achieved using a preassembled manifold device such as shown in Figure 11–52.

The chamber on the left side of this specialty manifold serves as the primary/secondary system interface. Two closely spaced holes connect the primary chamber with multiple port supply and return manifolds on the right side of the device. This arrangement assures the same water temperature is supplied to each secondary circuit. The large manifold cross sections, and short distances between connections, create minimum pressure variations along each manifold regardless of which secondary circuits are operating.

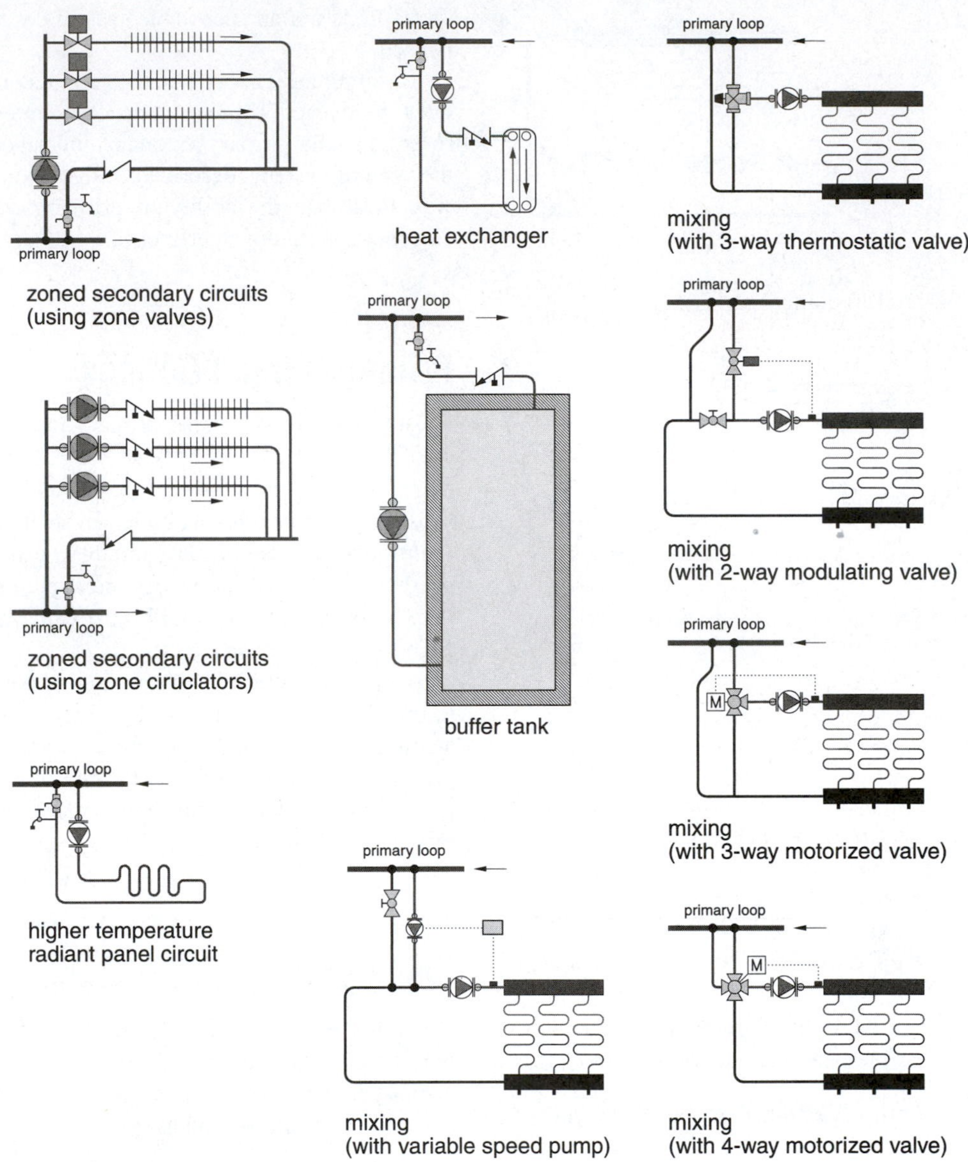

Figure 11–51 Secondary circuit options.

The observant reader may have noticed that all secondary circulators shown in the piping schematics pump *into* their respective secondary circuits. In most primary/secondary systems, the system's expansion tank is located in the primary loop. The primary loop thus becomes the pressure reference point for the secondary circuits. Installing the secondary circulators so they pump into their circuits ensures a pressure increase in those circuits when the secondary circulators operate. This is desirable in that it quiets the circulators, helps prevent cavitation, and helps keep dissolved air in solution until it can be routed back to the deaerator in the primary loop.

11.13 HYBRID DISTRIBUTION SYSTEMS

The diverse loads served by modern residential and light commercial systems often dictate that the distribution system contain **piping elements** from two or more of the basic distribution systems discussed thus far. Figure 11–53 is an example of such a **hybrid distribution system.**

This system uses a combination of primary/secondary piping, parallel piping, and home run piping to supply both high and low temperature space heating loads as well as indirect domestic water heating and snowmelting.

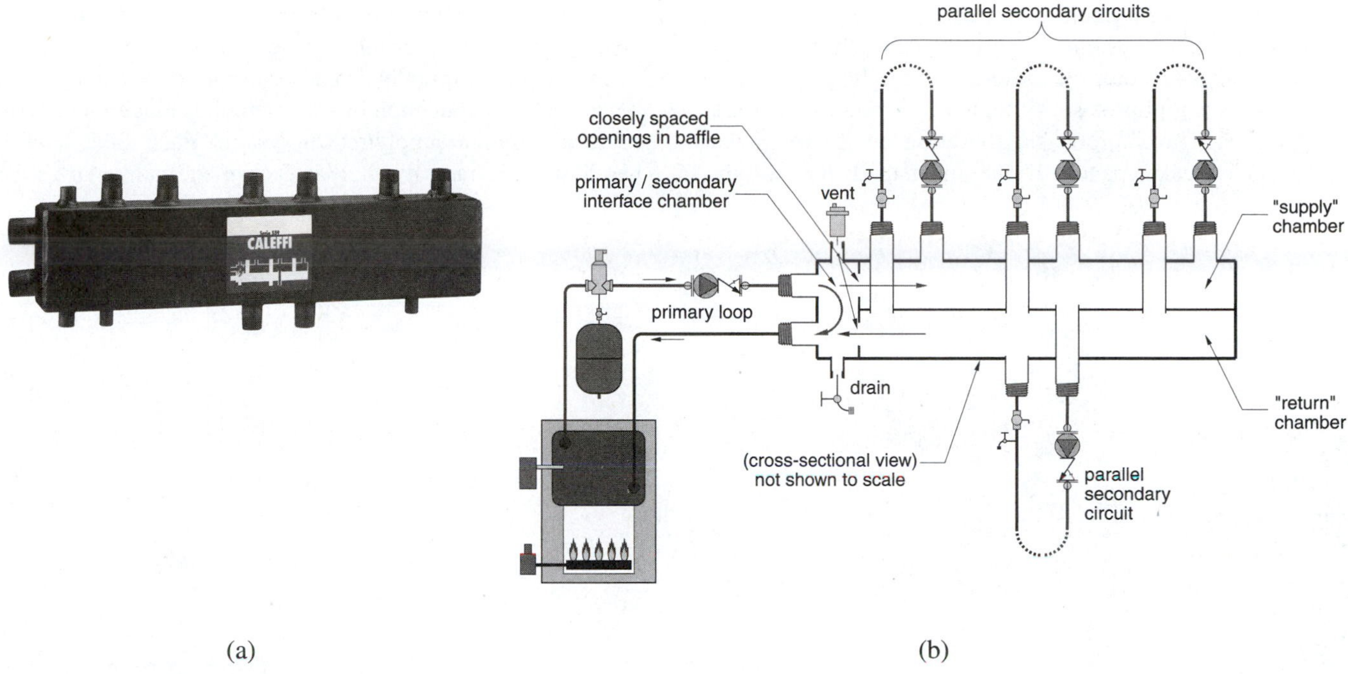

Figure 11–52 (a) Example of a self-contained primary/secondary manifold with parallel secondary circuits. *Courtesy of Caleffi North America.* (b) Cross-sectional drawing showing application of product shown in part (a).

Notice that the two boilers use parallel piping to ensure the same water temperature is delivered to both boilers when both are firing. The circulator associated with each boiler only operates when that boiler is firing. Circulators with integral spring-loaded check valves are used to ensure that there will not be flow reversal through an unfired boiler.

A pair of closely spaced tees connects the boiler header piping to the main distribution piping. This **primary/secondary circuit interface** eliminates the potential for interference between the small boiler circulators and the larger circulators supplying the heating **subsystems.**

The indirect water heater is connected across the supply and return mains as if it were a branch in a parallel direct return system. Downstream of these connections is a similar parallel piping arrangement supplying hot water to a heat exchanger in a snowmelting subsystem.

The fact that the domestic water heating and snowmelting subsystems are piped in parallel allows the possibility of operating both at the same supply temperature. However, in this system, domestic water is operated as a **priority load.** Heat input to all other loads is temporarily suspended while domestic water is being heated. This allows the full output of the multiple boiler system to be directed to the domestic water heating load for the fastest possible recovery. This is a relatively common scenario, especially when the system is installed in a large custom house with large demands for domestic hot water.

Moving to the right, the distribution mains transition into a series primary loop that serves the space heating subsystems. Note that the higher temperature panel radiators are served by a home run subassembly which is the first secondary circuit connected to the primary loop. This ensures the highest possible supply temperature to the panel radiators when they operate.

The manifolds for the lower temperature radiant floor circuits are supplied through a variable speed injection mixing subsystem that is also connected to the primary loop with a pair of closely spaced tees. The injection mixing subsystem can operate with lower water temperatures than the panel radiators, and is therefore connected as the last load on the primary loop.

The radiant floor manifolds each use valve actuators to control the flow through individual circuits. This is an element of multizone/zone valve piping. A differential pressure bypass valve is used to prevent the circulator for the radiant floor circuits from operating at high differential pressures when only one or two circuits are operating.

This system demonstrates the ability to integrate elements of various types of distribution systems into a very functional hybrid distribution system.

If someone were to look at the piping schematic shown in Figure 11–53a prior to reading this section, they might be overwhelmed at the apparent complexity of the system. They might wonder "how does the water know where to go" in such a complex system.

To avoid being overwhelmed, it is best to view such complex systems as a collection of subsystems. A good analogy is the electrical distribution system in a typical house. Although such a system usually contains hundreds of devices such as receptacles, switches, and lights, it is usually viewed as a group of subsystems (e.g., circuits) that are "plugged into" a common supply/return device (e.g., the main circuit breaker panel). If one understands how a light fixture is wired in series with a switch, or how receptacles are wired in parallel, they can extend these concepts while creating multiple circuits. Furthermore, other people with a basic understanding of residential wiring techniques can install, troubleshoot, or modify the system as necessary.

Likewise, complex hydronic distribution systems should be viewed as a collection of subassemblies that are connected to a common supply/return system. Each subassembly has a specific task much like each electrical circuit in a

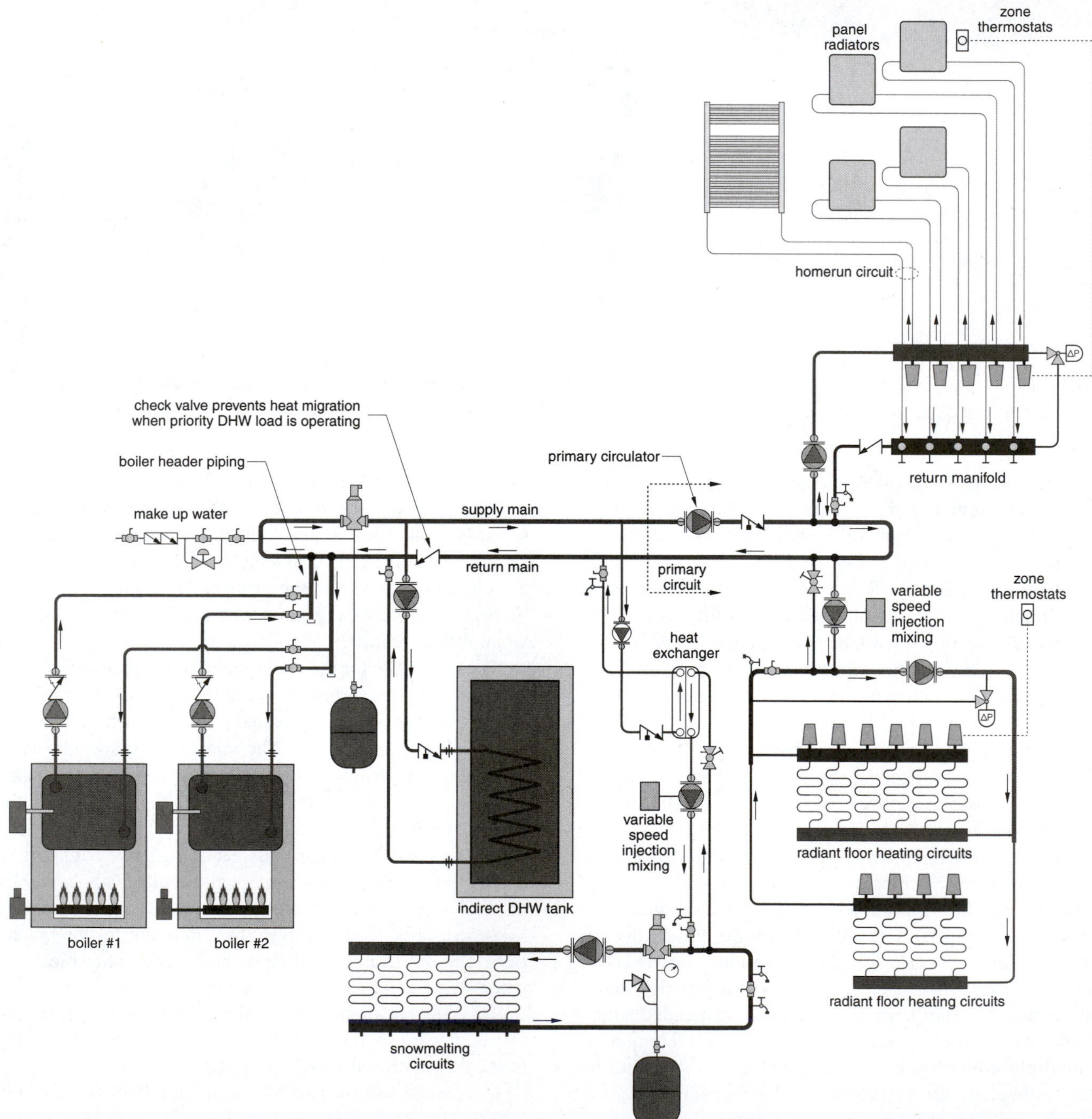

Figure 11–53a Example of a hybrid distribution system serving multiple loads at multiple supply temperatures.

house. This concept is illustrated in Figure 11–53b where the schematic of Figure 11–53a has been disassembled in several subsystems.

To assemble such a system, the designer chooses subsystems based on the loads to be served, then "plugs them into" a common supply/return system (such as a primary loop). This means of conceptualizing complex piping and control systems is an important skill for those who design modern hydronic systems.

The system schematic shown in Figure 11–54 also combines several piping elements and control concepts discussed in earlier chapters.

This system provides:

- Three fully-reset supply water temperatures to various radiant floor heating zones

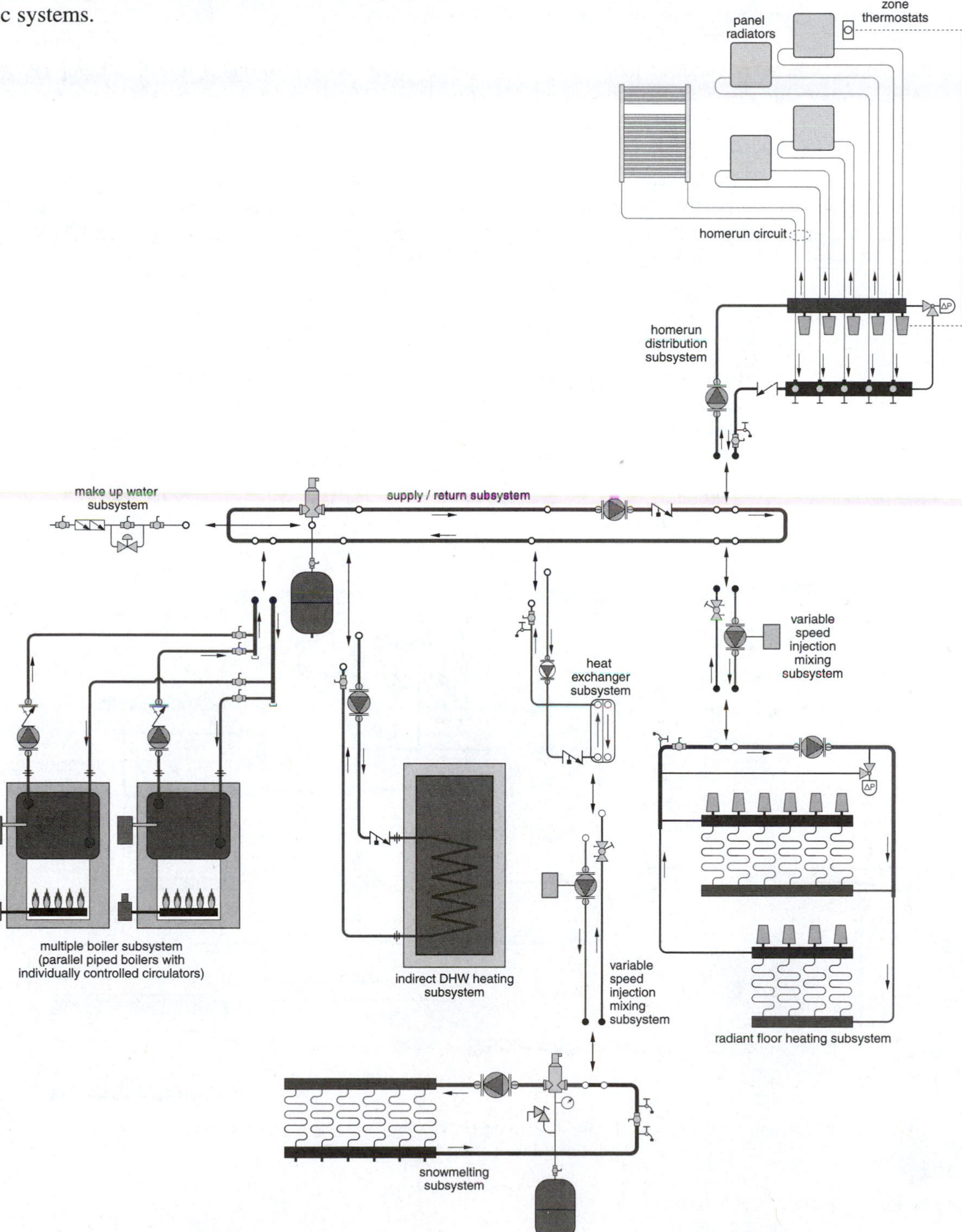

Figure 11–53b Disassembling the distribution system from Figure 11–53a into several subsystems.

- Continuous circulation in the floor heating zones
- Priority domestic water heating
- Full reset control of the condensing boiler

Domestic Water Heating Mode: Upon a call for heat from the thermostat in the indirect water heater, circulator (P1) is turned on. Circulator (P2), if running, is turned off so that the full output of the boiler is directed to the indirect water heater. The target water temperature for the boiler in this mode only is 180 °F.

Space Heating Mode: Whenever outdoor temperature falls below the warm weather shut down temperature of 65 °F the manifold circulators (P7, P8, P9, and P10) are turned on. These circulators provide constant circulation through the floor heating circuits during the heating season.

Upon a call for heat from any of the four zone thermostats, circulator (P2) is turned on. The boiler is enabled to warm to a temperature determined by the setting of the boiler reset controller, and the current outdoor temperature.

If either of the medium temperature floor heating zones calls for heat, the associated zone circulator (P3 or P4) is turned on. Heated water passes by the closely spaced tees in the crossover legs of the active zone circulators, and is "picked up" by the associated secondary circuit. The closely spaced tees prevent the flow rate in the manifold circuits from changing when the zone circulator is turned on.

If either of the low temperature floor heating zones calls for heat, the associated on/off injection pump (P5 or P6) is turned on. These pumps inject heated water into the continuously circulating lower temperature manifold stations. Because the temperature of the injected water is fully reset, the supply temperature to the manifold stations is proportionally reset. The rate of injection determines the reset line along which each lower temperature manifold station operates. This rate is regulated by the setting of the balancing

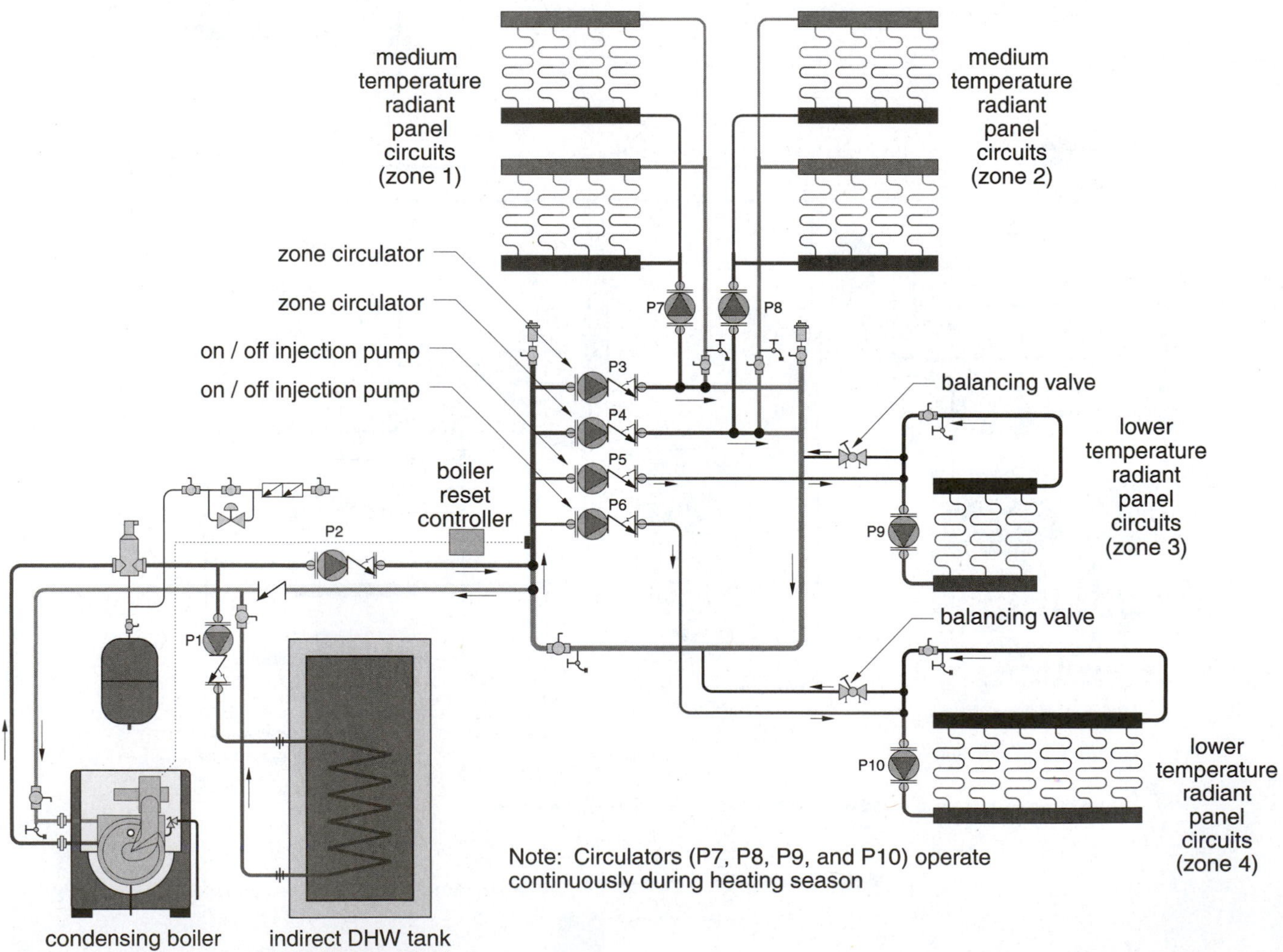

Figure 11–54 Example of a hybrid system containing piping elements from several types of hydronic distribution systems.

Figure 11–55 A portion of the installed piping for the system shown in Figure 11–54. Note the closely spaced tees.

valve associated with each injection pump. Each of the lower temperature manifold stations can therefore operate along a different reset line if necessary.

A portion of the installed piping for this system is shown in Figure 11–55.

Figure 11-56 on page 456 how another example of a hybrid distribution system that contains piping elements from the basic systems discussed earlier in this chapter. Look over this schematic carefully to identify these elements.

SUMMARY

The wide variety of distribution piping systems covered in this chapter allow a hydronic system to be designed for the exact requirements of the building it serves. Study each of them to learn their strengths and weakness. Be ready to deploy them individually or in combination as you design custom heating systems.

KEY TERMS

balancing valve
bars (of pressure)
boiler short-cycling
candidate circulator
closely spaced tees
common piping segment
crossover bridge
dead heading (a circulator)
differential pressure bypass valve (DPBV)
diverter tees
equilibrium
home run distribution system
home run subsystem
hybrid distribution system
hydraulic equilibrium
hydronic circuit simulator
inter-zone heat transfer
iteration
KPa (Kilo-Pascals of pressure)
leaders (tubing)
multizone/multicirculator system
multizone/zone valve system
multizone systems
one-pipe system
parallel circuit
parallel direct-return system
parallel primary loop
parallel reverse-return system
piping elements
piping topology
primary/secondary circuit interface
primary/secondary (P/S) piping
primary loop
priority load
reverse-return piping
secondary circuits
Series Baseboard Simulator
series primary loop
single circuit/multizone (one-pipe) system
single series circuit
stacked home run system
subsystem
target flow rate
T-drill®
thermal equilibrium
threshold differential pressure (of a DPBV)
two-pipe system
underslung thermal trap
velocity noise
zone

CHAPTER 11 QUESTIONS AND EXERCISES

1. A thermometer attached to the supply pipe of a hydronic distribution system shows the temperature of the water rising steadily even after the system has been operating for several minutes. What can you conclude about the heat output of the boiler relative to the heat transferred to the load?

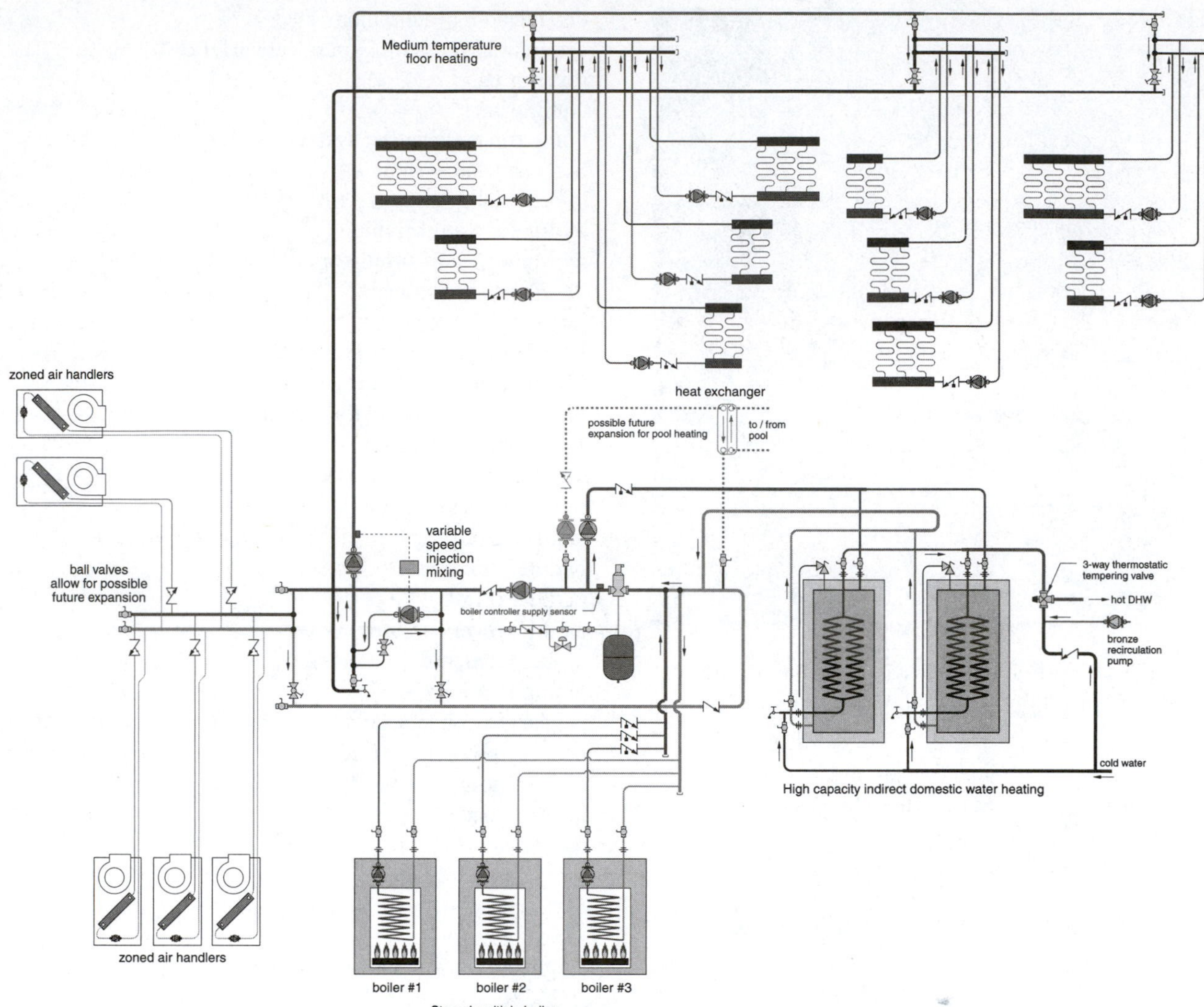

Figure 11–56 Example of a hybrid system using a condensing boiler and piping elements from several types of hydronic distribution systems.

2. When is hydraulic equilibrium established in a hydronic system? Are there certain types of hydronic systems where hydraulic equilibrium will not be established?
3. Will a hydronic heating system, in all cases, properly heat a building when it attains both thermal and hydraulic equilibrium? Explain your answer.
4. Discuss the feasibility of using a setback thermostat with a daily setback schedule in a well-insulated building equipped with a heated 6-inch concrete slab. When would this make sense? When would it not make sense?
5. What is the main advantage of a one-pipe distribution system relative to a single series circuit?
6. A three-zone distribution system is to supply a total load of 90,000 Btu/hr with a design temperature drop of 15 °F. The design heating loads of the three zones are 25,000 Btu/hr, 50,000 Btu/hr, and 15,000 Btu/hr. Determine the system design flow rate and the design flow rate in each of the zone circuits.
7. Describe why check valves are necessary in each zone circuit of a multicirculator zoned system.
8. What is the principal advantage of a reverse-return rather than direct-return parallel distribution system?
9. Why are parallel distribution systems good for use with low temperature heat emitters?
10. What is the purpose of a differential pressure bypass valve in a parallel direct- or reverse-return system. When would it typically operate?
11. What is the advantage of designing a series primary/secondary system for a high temperature drop in the primary circuit? What precaution must be observed if a conventional boiler is used?
12. What circumstances suggest the use of a parallel primary loop rather than a series primary loop?
13. Describe what happens to the flow rate through a particular zone circuit when additional zone circuits (containing zone valves) turn on. What happens to the *system*

flow rate as additional zone circuits open up? Justify your answer.

14. Describe why a circulator with a relatively "flat" pump curve is best suited for multizone systems using zone valves. What can happen if a circulator with a "steep" pump curve is used in such a system?
15. Why is it necessary to "correct" the hydraulic resistance of the common piping segment in a multizone/multicirculator system using Equation 11.5?
16. When the calculated flow rate through a particular parallel branch circuit is significantly lower than the initial design value, what are some things that can be done to increase it?
17. A diverter-tee arrangement operates as shown in Figure 11–57. Calculate the outlet temperature from the diverter tee.

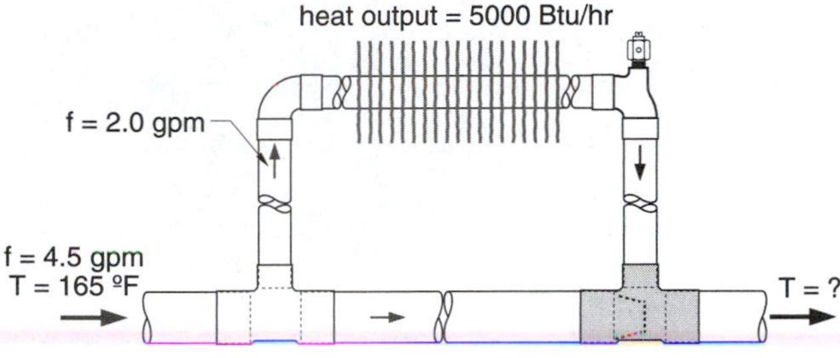

Figure 11–57 Piping and operating conditions for Exercise 17.

18. Examine the piping schematic in Figure 11–56, and identify piping elements from at least three different types of distribution systems.
19. Describe three distinct advantages of a home run distribution system in comparison to a single-series loop system.
20. Describe what happens when the flow rate in a secondary circuit is higher than the flow rate in the primary loop.
21. What is the fundamental reason for placing tees as close as possible at a primary/secondary circuit interface?
22. Why is it necessary to install lengths of straight tubing upstream and downstream of a pair of closely spaced tees in a primary/secondary system?
23. Why is it necessary to use purge valves in each secondary circuit rather than a single purging arrangement in the primary loop?
24. Can an underslung thermal trap protect the supply side of a secondary circuit from thermal migration?
25. Does the primary circulator have to produce a flow rate at least as great as the sum of the secondary circuit flow rates? Why or why not?
26. What are two ways of zoning individual heat emitters in a home run distribution system?
27. Why is it preferable to route the piping for a parallel reverse return system around the perimeter of a building?
28. Why is meant by the term *primary/secondary circuit interface?*
29. What are balancing valves necessary in each crossover bridge of a parallel primary loop?
30. Why can't a swing-check valve prevent heat migration into the supply side of a secondary circuit?

EXPANSION TANKS 12

OBJECTIVES

After studying this chapter, you should be able to:

- Explain the purpose of an expansion tank in a closed-loop hydronic system
- Describe two different types of expansion tanks
- Explain why standard expansion tanks become waterlogged
- Estimate the volume of fluid in a hydronic system
- Determine the proper location of the expansion tank within the system
- Calculate the required expansion tank volume for a given system
- Determine the required air pressurization of a diaphragm-type tank
- Determine the required expansion tank volume in a low temperature system
- Make use of the Expansion Tank Sizer software module
- Avoid incompatibilities between diaphragm materials and system fluids
- Describe how an expansion tank should be mounted

12.1 INTRODUCTION

All liquids used in hydronic heating systems expand when heated. This **thermal expansion** is an unavoidable and extremely powerful fact of nature. Upon heating, each of the trillions of fluid molecules contained in the system become slightly larger. From a macroscopic perspective, one might think there was an increase in the amount of fluid in the system. This is not true. The same molecules simply take up more space when at a higher temperature. The volume of the fluid has increased, but the total mass of the system's fluid has not changed.

For all practical purposes, liquids are **incompressible.** A given number of liquid molecules cannot be compacted or squeezed into a smaller volume without tremendous force. Any container *completely* filled with a liquid and sealed from the atmosphere will experience a rapid increase in pressure as the liquid is heated. If this pressure is allowed to build, the container will burst, in some cases violently.

To prevent this from occurring, *all hydronic heating systems must be equipped with a means of accommodating the volume increase of their fluid as it is heated.* In systems that are open to the atmosphere, such as a nonpressurized thermal storage tank, the volume increase can be accommodated by extra space at the top of the tank. This allows space for the expanding fluid to "park" its extra volume as shown in Figure 12–1.

In a more typical *closed-loop* hydronic system, the extra space is usually provided by a separate chamber called the expansion tank. This tank contains a volume of air that is compressed, somewhat like a spring, as the system's fluid expands against it. This concept is shown in Figure 12–2.

This chapter discusses the types of expansion tanks used for closed hydronic systems. It also gives methods for sizing, pressurizing, and locating the tank in the system. A software module called **Expansion Tank Sizer** in the Hydronics Design Studio software is also introduced as an aid for determining tank size and pressurization.

12.2 STANDARD EXPANSION TANKS

One type of expansion tank used in early hydronic heating systems consisted of a simple open-top drum placed at

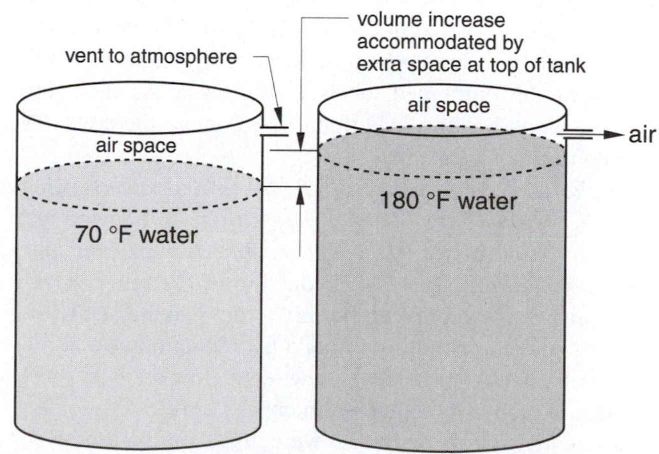

Figure 12–1 Extra space at top of an open tank accommodates expansion of fluid.

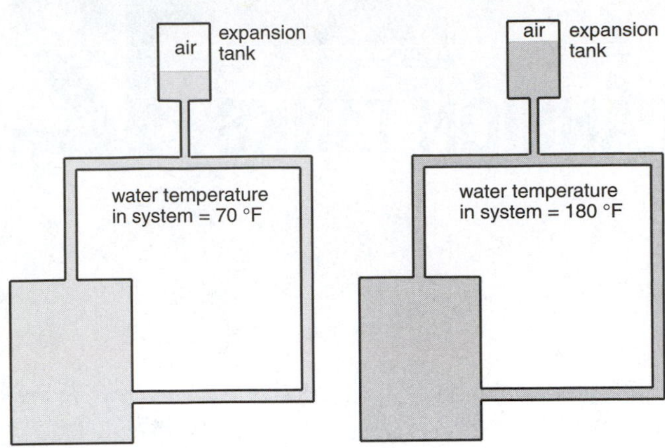

Figure 12–2 Separate expansion tank used to accommodate fluid expansion in a closed-loop hydronic system.

the high point of the system. It served as a chamber into which water would rise when heated. In many cases, the tank was located in the building's attic.

This arrangement presented several problems. First, the open tank allowed fluid to escape from the system by evaporation. The lost water would have to be replaced with fresh water, which was often carried to the attic in buckets. The fresh water contained dissolved oxygen, which caused corrosion of iron-containing components in the system. Second, the height of the tank limited the pressure and thus the upper operating temperature of the system. Finally, since a tank located in a cold attic was often some distance away from the heated parts of the system, it could freeze during cold weather. This could lead to a ruptured tank and a mess in the spaces below. Needless to say, this type of tank is not used in modern hydronic systems.

The next advancement in expansion tank technology was the use of a closed tank located above the boiler as shown in Figure 12–3.

The air in this type of **standard expansion tank** is initially at atmospheric pressure. When the system is filled with fluid, the air is trapped in the tank and partially compressed. The higher the system piping rises above the expansion tank, the more the entrapped air is compressed. As the water expands upon heating, additional fluid enters the tank, further compressing the air.

If the tank is properly sized, the air should reach a pressure about 5 psi below the setting of the system's pressure-relief valve as the water reaches its maximum operating temperature. The 5 psi reduction of the relief valve's rated pressure is to prevent the valve from leaking just below its rated opening condition. This 5 psi reduction also allows the relief valve to be located a few feet below the inlet of the expansion tank as is typical in many systems.

Standard expansion tanks were used on thousands of early hydronic systems. Many are still in service today. These tanks were often hung from the underside of floor joists just above the boiler.

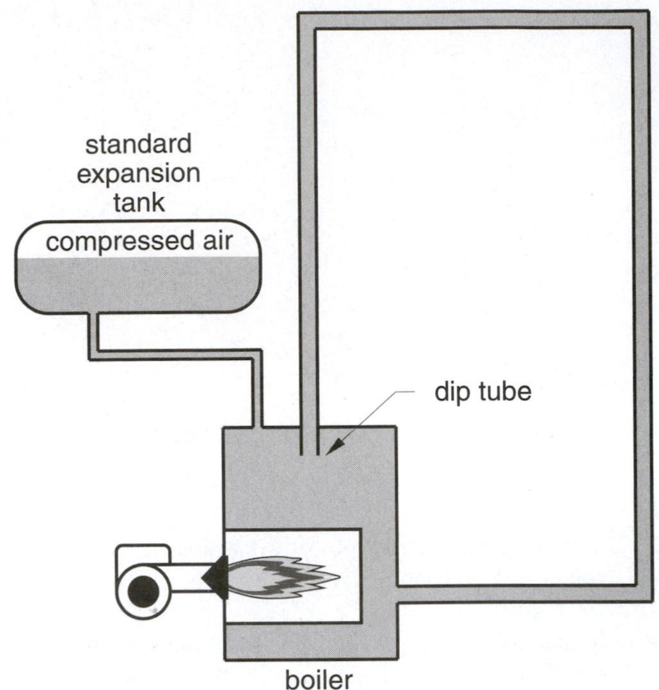

Figure 12–3 Installation of a standard expansion tank.

One inherent problem with standard expansion tanks is that the air and water they contain are in direct contact. As the water in the tank cools off, it has the ability to reabsorb some of the air back into solution. A slight amount of thermosiphoning flow in the piping between the tank and the boiler can carry the cooler water and the dissolved air it contains back to the system. Upon reheating, the dissolved air will again come out of solution, but now in the system piping. Eventually this air will be expelled from the system through air vents. An automatic feed-water valve will then admit small amounts of water to make up for the lost air. The net effect over many heat-up/cooldown cycles is that the air in the expansion tank is replaced by water. Eventually the tank becomes **waterlogged,** which means the tank has become completely filled with water.

When waterlogging occurs, there is no longer a cushion of air for the system's water to expand against. This causes the system's pressure-relief valve to release small amounts of water each time the system heats up. The system's feed water valve replaces this water as soon as the system cools and its pressure decreases. This repeated sequence of events can move many gallons of fresh (oxygen-containing) water through a system during a single heating season. It greatly increases the chance of serious corrosion damage.

Standard expansion tanks typically need to be drained and refilled with air two times a year to prevent the problems associated with waterlogging. Special drain valves that allow air to enter the tank at the same time water is being drained are available for this purpose. These valves also isolate the tank from the rest of the system during this draining operation.

A specialized boiler fitting is often used with standard expansion tanks. This fitting, shown in Figure 12–4, allows air bubbles that accumulate at the top of the boiler to rise into the expansion tank. This boiler fitting contains a short dip tube that extends down below the top of the boiler sections to pick up hot water without air bubbles. This fitting helps return air that may have migrated into the system back to the expansion tank.

The pipe leading from the boiler fitting to a standard expansion tank should be sloped at least 1 inch in 5 feet. This allows the bubbles to rise toward the tank. It should also be at least 10 feet long to minimize heat migration toward the tank. Such heating can cause an additional pressure increase in the tank.

A separate tank fitting is sometimes used to connect the riser pipe from the boiler to the tank. This fitting must be selected based on the diameter of the expansion tank. It helps minimize thermosiphoning flow between the tank water and the water in the system piping.

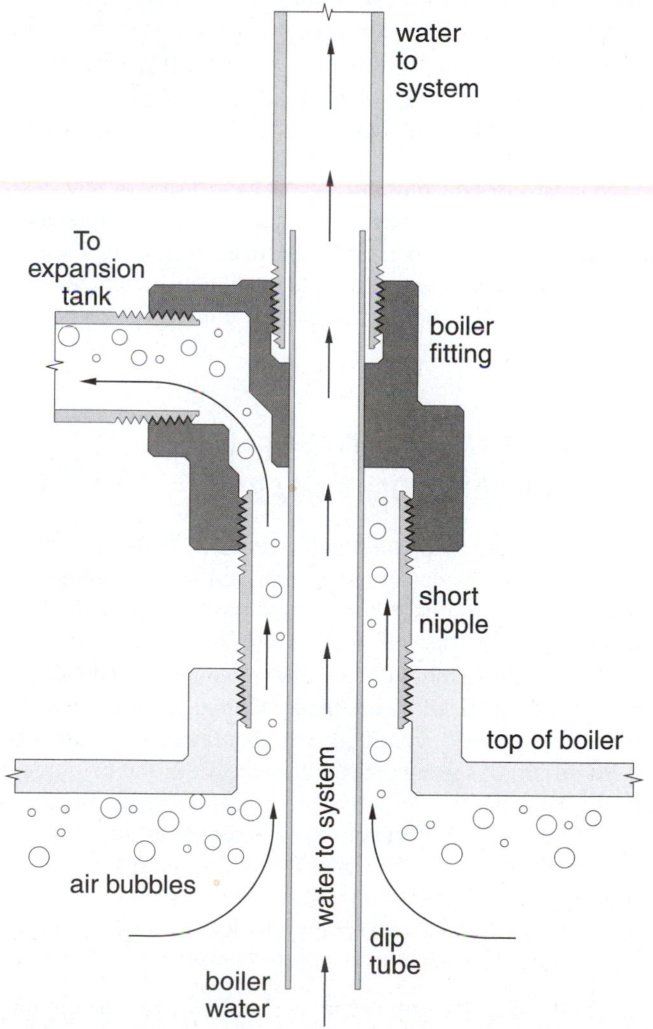

Figure 12–4 Example of a specialized boiler fitting for use with standard expansion tanks.

Sizing Standard Expansion Tanks

The size of a standard expansion tank can be calculated using Equation 12.1:

(Equation 12.1)

$$V_t = \frac{V_s\left(\frac{D_c}{D_h} - 1\right)}{14.7\left(\frac{1}{P_f + 14.7} - \frac{1}{P_{RV} + 9.7}\right)}$$

where:

V_t = minimum required tank volume (gallons)

V_s = fluid volume in the system (gallons) (see Section 12.4)

D_c = density of the fluid at its initial (filling) temperature (lb/ft³)

D_h = density of the fluid at the maximum operating temperature of the system (lb/ft³)

P_f = static pressure of the fluid in the tank when the system is filled (psi)*

P_{RV} = rated pressure of the system's pressure-relief valve (psi)

* The static pressure at the expansion tank connection when the system is filled is based on the height of the system piping above the expansion tank, the type of fluid used, and a customary allowance for 5 psi pressure at the top of the system for proper operation of the air vents. This pressure can be found using Equation 12.2:

(Equation 12.2)

$$P_f = H\left(\frac{D_c}{144}\right) + 5$$

where:

P_f = static pressure of the fluid in the tank when the system is filled (psig)

H = distance from inlet of expansion tank to top of system (ft)

D_c = density of the fluid at its initial (fill) temperature (lb/ft³)

Example 12.1: Calculate the minimum size standard expansion tank for the system shown in Figure 12–5. Water is added to the system at an initial temperature of 60 °F. The maximum operating temperature of the system is 200 °F. The boiler is equipped with a 30 psi rated relief valve. The estimated system volume is 35 gallons.

Solution: Before using Equation 12.1, the pressure in the tank when the system is filled must be found using Equation 12.2. This equation requires the density of the water at 60 °F, which can be read from Figure 4–3, calculated using Equation 4.1, or found using the Fluids Properties Calculator

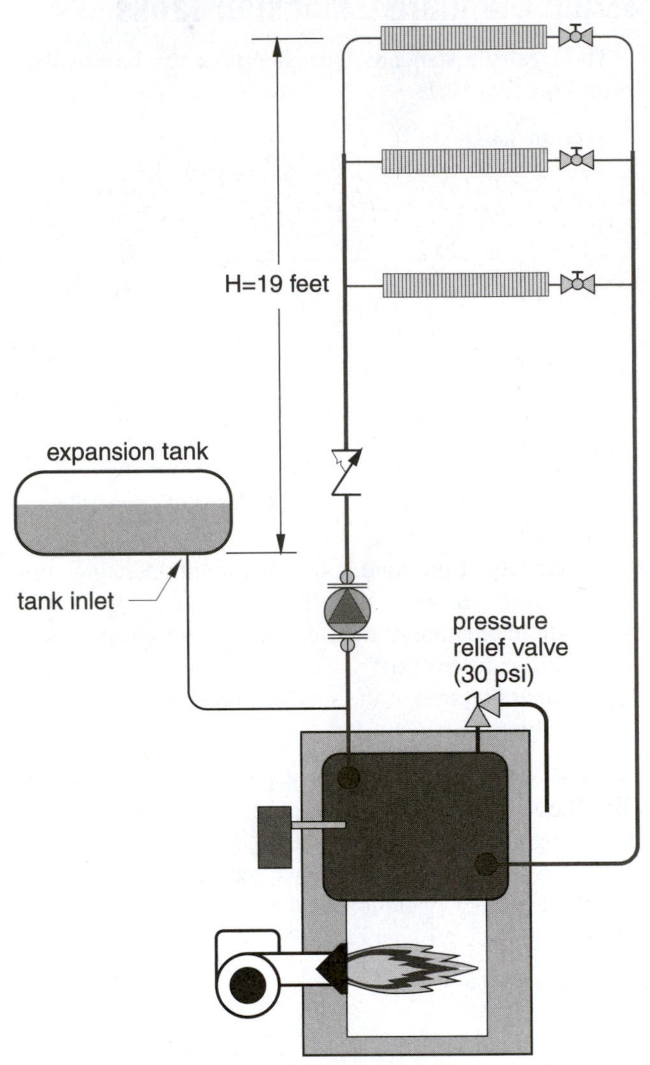

Figure 12–5 System used in Example 12.1.

module in the Hydronics Design Studio software. Since the water's density at the maximum operating temperature of 200 °F is eventually required, it is also found at the same time:

D_c = 62.36 lb/ft³ (water's density at 60 °F)
D_h = 60.13 lb/ft³ (water's density at 200 °F)

The pressure at the inlet of the tank is now calculated using Equation 12.2:

$$P_f = H\left(\frac{D_c}{144}\right) + 5 = 19\left(\frac{62.36}{144}\right) + 5 = 13.23 \text{ psig}$$

This value and the remaining data is now substituted in Equation 12.1:

$$V_t = \frac{35\left(\frac{62.36}{60.13} - 1\right)}{14.7\left(\frac{1}{13.23 + 14.7} - \frac{1}{30 + 9.7}\right)} = \frac{1.298}{0.516} = 8.3 \text{ gallons}$$

Discussion: It is interesting to consider the effect of the height of the expansion tank on its volume requirement. If the tank were located closer to the top of the system, the static pressure caused by fluid in the piping above the tank would decrease. This would result in less initial compression of the air in the tank, and thus more room for fluid. In this particular example, if the tank were moved 10 feet higher, its size could be reduced to 5.1 gallons. A lower maximum fluid temperature would also decrease the tank size required.

It should be noted that Equation 12.1 determines the *minimum* volume of a standard expansion tank. It is certainly possible to use a tank with a larger volume. As the tank's volume increases above its minimum calculated value, the increase in the tank's air pressure as the system's fluid expands is reduced. A larger tank, however, provides no tangible benefit to the performance or longevity of the system. The only reason to use a larger tank would be to use a standard production tank size rather than a custom-made tank.

Standard expansion tanks are seldom used on new residential or light commercial systems. Their maintenance requirements, size, cost, and associated fittings and valves make them much less desirable than the diaphragm-type tanks discussed in the next section.

12.3 DIAPHRAGM-TYPE EXPANSION TANKS

Many of the shortcomings of standard expansion tanks can be avoided by separating the air and water in an expansion tank. During the 1950s, a new type of expansion tank with an internal flexible diaphragm became available. On one side of the diaphragm is a captive volume of air that has been pre-pressurized by the manufacturer. On the other side is a chamber for accommodating the expanded volume of system fluid. As more fluid enters the tank, the diaphragm flexes, allowing the air volume to be compressed. An example of a small **diaphragm-type expansion tank** and a sequence illustrating the movement of its diaphragm is shown in Figure 12–6.

There are several advantages to using diaphragm-type tanks rather than standard expansion tanks:

- Since the diaphragm separates the fluid from the air, the air can not be reabsorbed by the fluid. Diaphragm-type expansion tanks do not have to be periodically drained to prevent waterlogging.

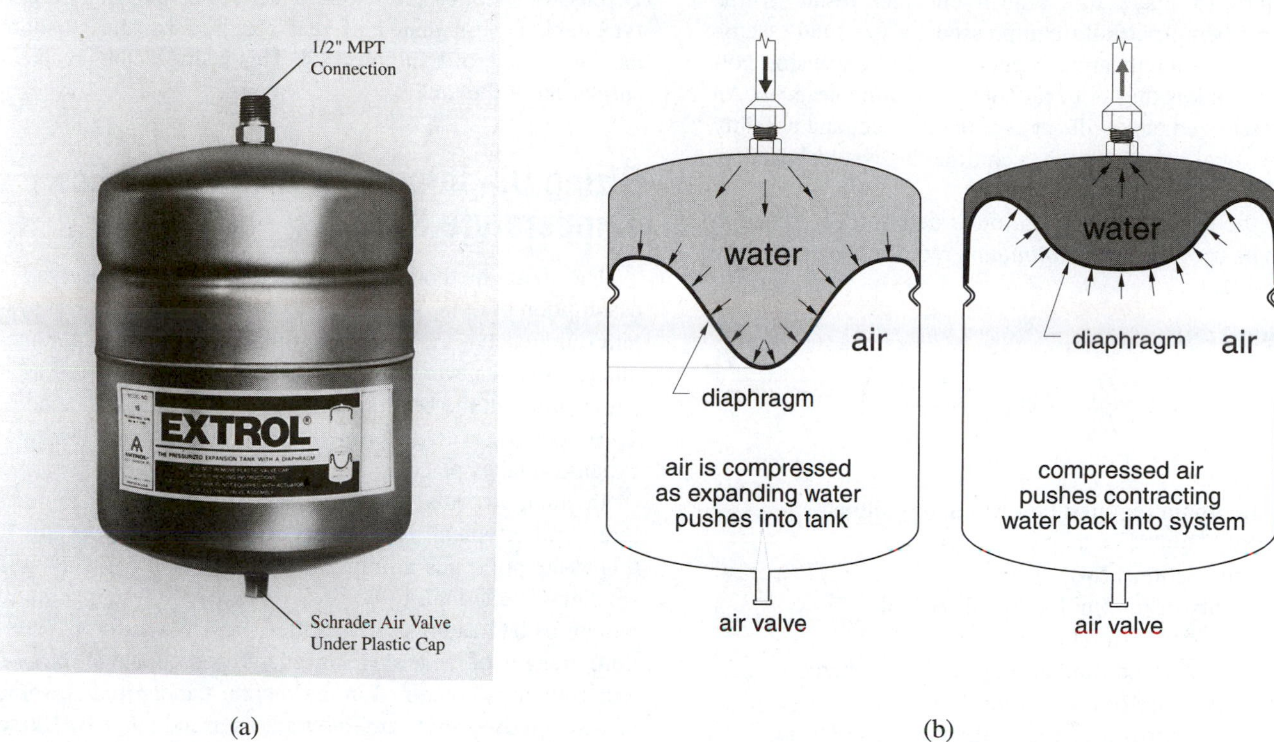

Figure 12–6 (a) Example of a small diaphragm-type expansion tank. *Courtesy of Amtrol, Inc.* (b) Flexing of the diaphragm as additional fluid enters the tank.

- By avoiding waterlogging, the possibility of accelerated corrosion caused by addition of fresh water to make up for relief valve losses is no longer a factor.
- The air pressure in the tank can be adjusted to match the static pressure of the system before the system is filled with water. No significant amount of water enters the tank until the system begins warming up. This results in a significantly smaller and lighter tank.
- Because the air volume is captive, the tank can (theoretically) be mounted in any orientation.
- The special boiler and tank fittings required with standard expansion tanks are no longer needed.

Sizing a Diaphragm-type Expansion Tank

A properly sized diaphragm-type expansion tank should reach a pressure about 5 psi lower than the relief valve setting when the system reaches its maximum operating temperature. The 5 psi safety margin prevents the relief valve from leaking just below its rated opening pressure. It also allows the relief valve to be located slightly below the inlet of the expansion tank.

The first step in sizing a diaphragm-type expansion tank is to determine the proper **air-side pressurization** of the tank, using Equation 12.3.

(Equation 12.3)

$$P_a = H\left(\frac{D_c}{144}\right) + 5$$

where:

P_a = air-side pressurization of the tank (psi)
H = distance from inlet of expansion tank to top of system (ft)
D_c = density of the fluid at its initial (cold) temperature (lb/ft^3)

The proper air-side pressurization is equal to the static fluid pressure at the inlet of the tank, plus an additional 5 psi allowance at the top of the system. *The pressure on the air-side of the diaphragm should be adjusted to the calculated value before fluid is added to the system.* This is done by either adding or removing air through the Schrader valve on the shell of the tank. A small air compressor or bicycle tire pump can be used when air is needed. An *accurate* 0 to 30 psi pressure gauge should be used to check this pressure as it is being adjusted. Because the required pressure may be relatively low, a standard automotive tire gauge is usually not accurate enough.

Proper air-side pressure adjustment ensures the diaphragm will be fully expanded against the shell of the tank when the system is filled with fluid, but before it is heat-

ed. Failure to make this adjustment can result in the diaphragm being partially compressed by the fluid's static pressure *before* any heating occurs. The full expansion volume of the tank is thus not available as the fluid heats up. An underpressurized tank will act as if undersized, and possibly allow the relief valve to open each time the system heats up. This situation must be avoided.

After the air-side pressurization is determined, Equation 12.4 can be used to find the *minimum* required volume of the expansion tank:

(Equation 12.4)

$$V_t = V_s\left(\frac{D_c}{D_h} - 1\right)\left(\frac{P_{RV} + 9.7}{P_{RV} - P_a - 5}\right)$$

where:

V_t = minimum required tank volume (gallons)
V_s = fluid volume in the system (gallons) (see Section 12.4)
D_c = density of the fluid at its initial (cold) temperature (lb/ft³)
D_h = density of the fluid at the maximum operating temperature of the system (lb/ft³)
P_a = air-side pressurization of the tank found using Equation 12.3 (psi)
P_{RV} = rated pressure of the system's pressure-relief valve (psi)

Example 12.2: Determine the minimum size diaphragm-type expansion tank for the system described in Example 12.1. Water is added to the system at an initial temperature of 60 °F. The maximum operating temperature of the system is 200 °F. The boiler is equipped with a 30 psi rated relief valve. The estimated system volume is 35 gallons.

Solution: As in Example 12.1, the density of the water at both 60 °F and 200 °F is required. This information can be read from Figure 4–3, calculated using Equation 4.1, or found using the Fluids program in the Hydronics Design Studio.

D_c = 62.36 lb/ft3
D_h = 60.13 lb/ft3

The proper air-side pressurization of the tank is now calculated using Equation 12.3:

$$P_a = H\left(\frac{D_c}{144}\right) + 5 = 19\left(\frac{62.36}{144}\right) + 5 = 13.23 \text{ psig}$$

This information is now substituted into Equation 12.4:

$$V_t = 35\left(\frac{62.36}{60.13} - 1\right)\left(\frac{30 + 9.7}{30 - 13.23 - 5}\right) = 4.38 \text{ gallons}$$

Discussion: Notice the required volume of the diaphragm-type tank is less than half that required by the standard expansion tank of Example 12.1. This is due to the pre-pressurization of the tank.

Sizing Diaphragm-type Tanks for Low Temperature Systems

The sizing method just discussed has a very conservative assumption built in. It assumes that the *entire volume* of system fluid reaches the maximum fluid temperature simultaneously. This is never the case because the fluid cools as it passes through the heat emitters, or through mixing devices. The cooler the average temperature of the fluid, the less total expansion takes place.

In small systems, operating with single supply temperatures, and a typical temperature drop in the range of 20 °F, this assumption has minimal effect. Seldom will it allow a smaller volume tank to be used. However, low temperature radiant panel heating systems often present a different situation. In many of these systems, a large percentage of the system volume is contained in the radiant panel circuits, some of which only operate at temperatures in the range of 100 to 110 °F. This significantly reduces the total fluid expansion.

Sizing expansion tanks based on the assumption that *all* water in such systems reaches the upper boiler water temperature results in tanks that are needlessly oversized. Although an oversized tank will not cause operating problems, it certainly does add to installation cost.

A more accurate method divides the system water into two portions: That which will operate up to the maximum temperature of the boiler, and that which will operate up to the maximum temperature of the distribution circuits.

Equation 12.5 is a modified version of Equation 12.4 that determines the minimum expansion tank volume in such a low temperature system.

(Equation 12.5)

$$V_t = \left[V_{high}\left(\frac{D_c}{D_{high}} - 1\right) + V_{low}\left(\frac{D_c}{D_{low}} - 1\right)\right] \times \left(\frac{P_{RV} + 9.7}{P_{RV} - P_{air} - 5}\right)$$

where:

V_t = minimum required tank volume (gallons)
V_{high} = volume of fluid contained in the higher temperature portion of the system (gallons)
V_{low} = volume of fluid contained in the lower temperature portion of the system (gallons)
D_c = density of the fluid at its initial (cold) temperature (lb/ft³)
D_{high} = density of the fluid at the maximum operating temperature of the higher temperature portion of the system (lb/ft³)

D_{low} = density of the fluid at the maximum operating temperature of the lower temperature portion of the system (lb/ft³)

P_a = air-side pressurization of the tank found using Equation 12.3 (psig)

P_{RV} = rated pressure of the system's pressure-relief valve (psig)

Example 12.3: Assume a large radiant floor heating system contains 200 gallons of water in the floor circuits, and 15 gallons of water on the boiler side of the mixing device. At design conditions, the water on the boiler side of the mixing devices reaches 180 °F, and that supplied to the radiant floor circuits reaches 110 °F. Assume the air-side pressurization of the tank is 13.5 psi, and the systems relief valve is rated at 30 psi. The system is initially filled with water at 60 °F. Calculate the minimum required volume of a diaphragm-type expansion tank using the:

a. Standard method (Equation 12.4)
b. Modified method (Equation 12.5)

Solution: The densities of the water at the three water temperatures are:

D_c = 62.36 lb/ft³
D_{low} = 61.84 lb/ft³
D_{high} = 60.60 lb/ft³

Using the standard method (Equation 12.4), the minimum expansion tank volume required is:

$$V_t = V_s\left(\frac{D_c}{D_h} - 1\right)\left(\frac{P_{RV} + 9.7}{P_{RV} - P_{air} - 5}\right)$$

$$= 215\left(\frac{62.36}{60.6} - 1\right)\left(\frac{30 + 9.7}{30 - 13.5 - 5}\right) = 21.6 \text{ gallons}$$

Using the modified sizing procedure, the minimum expansion tank volume required is:

$$V_t = \left[15\left(\frac{62.36}{60.6} - 1\right) + 200\left(\frac{62.36}{61.84} - 1\right)\right] \times \left(\frac{30 + 9.7}{30 - 13.5 - 5}\right)$$

= 7.3 gallons

Discussion: Accounting for the fact that only 7 percent of the system water volume reaches the higher temperature, the modified procedure reduces the required expansion tank volume by about 66 percent compared to that calculated by the standard method. Even the modified method is conservative because it assumes that the 200 gallons of water reaches 110 °F at the same time the other 15 gallons reaches 180 °F. In reality, the average system water temperature in the floor circuits is likely to be about 10 °F cooler due to the drop in temperature as heat is released into the floor. The average temperature of the higher temperature water is also less than 180 °F.

Compatibility of the Expansion Tank and System Fluid

It is very important that the diaphragm material used in the tank is chemically compatible with the fluid and/or any dissolved gases in the system. Incompatibilities can result in the diaphragm being slowly dissolved by the fluid. This can create sludge in the system that gums up system components. When the diaphragm eventually ruptures, the tank quickly waterlogs, and the vicious cycle between the relief valve and feedwater valve begins.

Most modern diaphragm-type tanks use a butyl rubber or **EPDM** diaphragm. Although most such tanks are compatible with glycol-based antifreeze solutions, the installer should always verify this compatibility if it is not clearly stated in product specifications.

Another compatibility issue involves the presence of dissolved oxygen in the system water. Most expansion tanks sold for use in hydronic heating systems are intended for installation in closed-loop, oxygen-tight systems. In most such tanks, system water contacts the plain steel tank shell. If the system allows the presence of dissolved oxygen in the water, the shell will corrode and eventually fail. Dissolved oxygen could be present in any type of open-loop system, or in systems using oxygen-permeable tubing. In such cases, an expansion tank rated for use in open-loop systems should be selected. Such tanks often have a polypropylene liner that prevents contact between the system water and the steel tank shell.

Pressure and Temperature Ratings

Expansion tanks have maximum pressure and temperature ratings. This information is stamped on the label of the tank. Typical ratings are 60 psi and 240 °F. These rating are usually adequate for residential and light commercial systems. They must be observed, however, in nontypical applications. For example, if an expansion tank with a 60 psig pressure rating was used on a domestic hot water tank with a 150 psig relief valve, the tank could rupture before the relief valve opened.

Selection, Mounting, and Service

Diaphragm expansion tanks are available in a variety of sizes and shapes. Tanks with volumes from 1 gallon to about 14 gallons are usually adequate for standard residential and light commercial hydronic systems. These tanks are usually equipped with a ½-inch MPT connection from which they are designed to hang vertically. Because of the potential water weight in the tank, the piping supporting the tank should be well-supported.

One commonly used mounting method is to hang a diaphragm-type expansion tank from the bottom of the system's air separator. Many air separators have 1/2" FPT bottom tapping for this purpose. An example of this type of mounting is shown in Figure 12–7a. Notice how the system piping is routed diagonally across the corner space to create clearance for the tank shell.

If necessary for spatial reasons, the tank can also be located several feet away from the point where its connecting piping taps into the system. From a pressure standpoint, the *horizontal* distance between the tank and where the piping from it taps into the system is of no concern. However, lowering the tank inlet several feet below the tap-in point will add static pressure to the tank, and slightly reduce its expansion absorption capability.

When the tank does not hang vertically downward from a well-supported pipe or piping component, the tank itself should be well-supported. Figure 12–7b shows a tank supported by wall brackets.

Although the captive air volume in a diaphragm-type tank would theoretically allow it to function in any orientation, there are other concerns to address whenever the tank is not hung vertically from its inlet connection. One is that air bubbles in the system can be trapped on the water side of the tank shell. The oxygen present in this air will corrode the steel tank shell. Another concern is bending stress on the inlet connection to the tank, especially if the tank is not properly supported. These factors support the author's recommendation that *small diaphragm-type expansion tanks should only be mounted vertically with their inlet connection at the top.*

The pressure in an expansion tank is partially affected by the temperature of the captive air volume. The warmer the shell of the tank, the warmer the captive air inside. Although the pressure increase due to heating is generally compensated for by conservative assumptions in the sizing equations, installation details that minimize tank heating are still desirable. Installing several feet of piping between the tank inlet and the system tap-in point will reduce tank warming. This is especially desirable if the system must operate at water temperatures in excess of 200 °F.

When larger expansion tanks are required, floor mounted models are an option. It is also possible to combine two or more expansion tanks in parallel. The combined volume of the two tanks should at least equal the minimum tank volume calculated with Equation 12.4, Equation 12.5, or the Expansion Tank Sizer software module. Parallel mounted tanks should have their inlet connections at the same height, and have equal air-side pressures. Be sure each tank is properly supported, and that the air valves are accessible in case the air pressure has to be adjusted.

It is also good practice to install an isolation valve near the inlet of any expansion tank. This allows the tank to be easily isolated from the system if it ever needs replacing. A small drainage fitting or valve between the isolation valve and tank is also desirable to relieve any pressure and drain some fluid before disconnecting the tank.

Most diaphragm-type expansion tanks will retain their captive air volumes for decades. However, there is always the possibility that a diaphragm could leak. If this occurs, the tank will eventually lose its air and fill with system fluid.

(a)

(b)

Figure 12–7 (a) Expansion tank hung from its 1/2-inch MPT connection. (b) Expansion tank secured using wall brackets.

Such failure is easy to detect during a routine service check. Simply press in on the stem of the tank's air valve. If a stream of water comes out, the diaphragm has failed and the entire tank needs to be replaced. Gently tapping one's fingers on the side of a tank with an intact diaphragm also yields a hollow sound. If a "thud" sound is heard, it may indicate the tank has filled with water. Check the air valve to be sure.

If a new tank is needed, be sure the air-side pressure is adjusted *before* water is allowed to enter the tank.

12.4 ESTIMATING SYSTEM VOLUME

The sizing equations for both standard and diaphragm-type expansion tanks require an estimate of system volume. In a typical hydronic system, most of the fluid is contained in the piping and heat source. The fluid volume contained in type-M copper, PEX, and PEX-AL-PEX tubing can be calculated by estimating the total length of each tube type/size used in the system, then multiplying these lengths by the appropriate volume factors from Figure 12–8.

For other pipe sizes and materials, Equation 12.6 can be used to estimate the volume factor:

(Equation 12.6)

$$V = 0.04085(d_i)^2$$

where:

V = volume of one foot of the tubing (gallons/foot)

d_i = *exact inside diameter* of the tubing (inches)

The volume of the heat source can vary over a wide range. Small copper tube boilers may only contain one or two gallons of fluid. Cast-iron sectional boilers may hold 10 to 15 gallons. The only reliable source for this data is the manufacturer's specifications.

The volume of any type of storage tank, such as the buffer tank described in Chapter 3, Hydronic Heat Sources, must also be included. Thermal storage systems that employ large *closed* storage tanks will require relatively large expansion tanks.

Most modern heat emitters contain relatively small amounts of water. In the case of either finned-tube or radiant baseboard, the volume can be estimated using the appropriate pipe volume factors in Figure 12–8. Small kick space heaters or wall-mounted fan-coils contain very little fluid, probably not more than 1/4 gallon each. The fluid content of panel radiators varies with their design and size. Standing cast-iron radiators can contain significant quantities of water. Again, manufacturer's specifications are the best source for this data.

NOTE: The volume of the diaphragm-type expansion tank itself is not entered into the system volume. This is because there is no fluid in a properly pressurized diaphragm-type tank until the system is heated and the fluid begins expanding.

Example 12.4: Estimate the volume of the system in Figure 12–9.

Solution: The volume of all piping is estimated using data from Figure 12–8. This is then added to the volume estimates for the heat emitters and boiler.

The total piping volume is:

$$v_{piping} = (55 + 70)(0.06804) + (45 + 40)(0.0454) + (35 + 35)(0.02685) + (25 + 25)(0.01319) = 14.9 \text{ gallons}$$

The total estimated volume of the branch piping and heat emitters is:

$$8(1) = 8 \text{ gallons.}$$

Assume the boiler has a volume of 12 gallons.

The total system volume is the sum of these volumes: 14.9 + 8 + 12 = 34.9 gallons.

Copper (type M)		PEX		PEX-AL-PEX	
	gallons/foot		gallons/foot		gallons/foot
3/8" copper	0.008272	3/8" PEX	0.005294	3/8" PEX-AL-PEX	0.004890
1/2" copper	0.01319	1/2" PEX	0.009609	1/2" PEX-AL-PEX	0.01038
		5/8" PEX	0.01393	5/8" PEX-AL-PEX	0.01658
3/4" copper	0.02685	3/4" PEX	0.01894	3/4" PEX-AL-PEX	0.02654
1" copper	0.0454	1" PEX	0.03128	1" PEX-AL-PEX	0.04351
1.25" copper	0.06804	1.25" PEX	0.04668		
1.5" copper	0.09505	1.5" PEX	0.06516		
2" copper	0.1647	2" PEX	0.1116		
2.5" copper	0.2543				
3" copper	0.3630				

Figure 12–8 Volume factors for tubing.

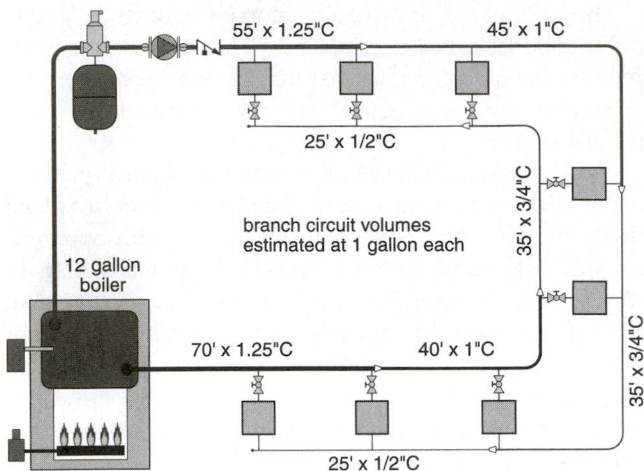

Figure 12–9 Piping diagram of system for Example 12.3.

12.5 THE EXPANSION TANK SIZER SOFTWARE MODULE

The Hydronics Design Studio software contains a module called the Expansion Tank Sizer that can be used to size *diaphragm-type* expansion tanks. The user interface screen is shown in Figure 12–10.

This module allows the user to select up to four different types/sizes of tubing, and enter the associated lengths. Another input is used to enter miscellaneous volumes such as those for the boiler, heat emitters, tanks, and so forth. The user can vary the pressure at the top of the system, the relief valve setting, and the distance from the expansion tank connection to the top of the system. The user can also select from one of several fluids. When a fluid other than water is used, the program corrects for the density differences and estimates the volume of antifreeze required based on its concentration.

The module determines the minimum expansion tank size as well as the proper air-side pressurization. It also reports the estimated system volume, and the volume of antifreeze required (if an antifreeze solution is selected as the system fluid).

12.6 POINT OF NO PRESSURE CHANGE

The placement of the expansion tank relative to the circulator will significantly affect the pressure distribution in the system when it operates. *The expansion tank should always be placed near the inlet port of the circulator.*

To understand why, one needs to consider the interaction between the circulator and expansion tank. In a closed-piping system, the amount of fluid (including that in the expansion tank) is fixed. It does not change regardless of whether the

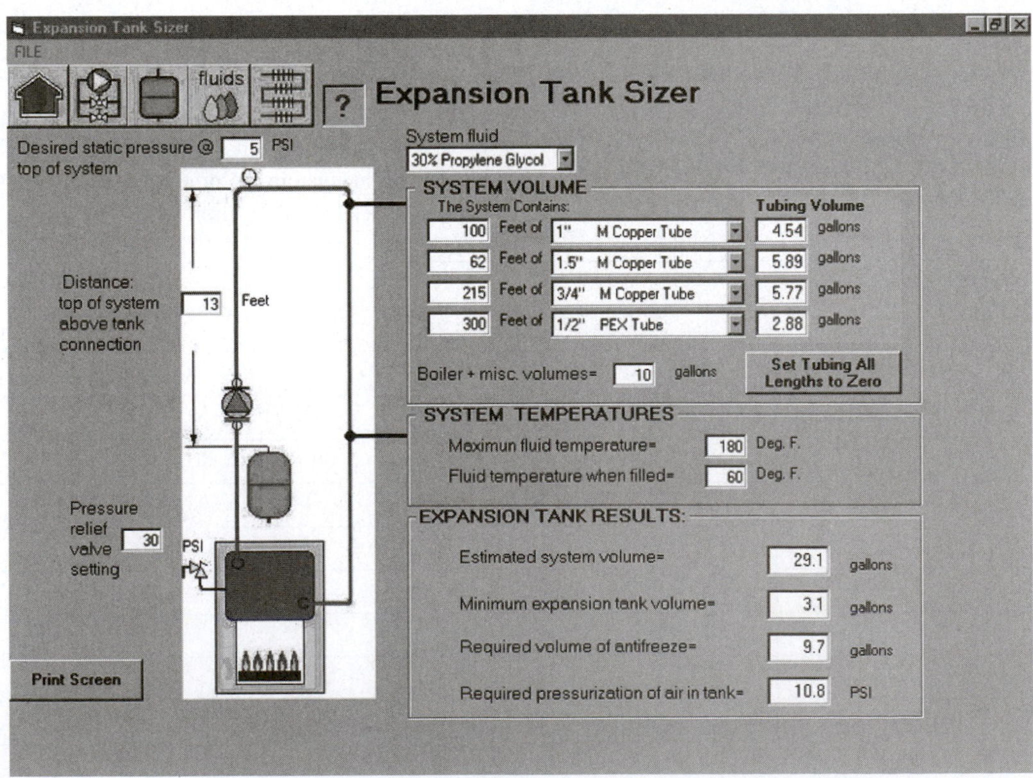

Figure 12–10 Screen shot of the Expansion Tank Sizer module in the Hydronics Design Studio.

circulator is on or off. The expansion tank contains a captive volume of air at some pressure. The only way to change the pressure of this air is to either push more fluid into the tank compressing the air, or to remove fluid from the tank expanding the air. This fluid would have to come from, or go to, some other location in the system. However, since the system's fluid is incompressible, and the amount of fluid in the system is fixed, this cannot happen regardless of whether the circulator is on or off. *The expansion tank thus fixes the pressure of the system's fluid at its point of attachment to the piping.* This is called the **point of no pressure change.**

Now consider a horizontal piping circuit filled with fluid and pressured to approximately 10 psi, as shown in Figure 12–11. When the circulator is off, the pressure is the same (10 psi) throughout the piping circuit. This is indicated by the solid horizontal pressure line shown above the piping. When the circulator is turned on, it immediately creates a pressure differential between its inlet and outlet ports. However, the pressure at the point where the expansion tank is connected to the circuit remains the same. The combination of the pressure differential across the circulator, the flow resistance of the piping, and the point of no pressure change gives rise to a new pressure distribution as shown by the dashed line in Figure 12–11.

Notice how the pressure increases in nearly all parts of the circuit when the circulator is operating. This is desirable because it helps eject air from vents. It also helps keep dissolved air in solution, and minimizes the potential for cavitation at the circulator inlet. The short segment of piping between the expansion tank connection point and the inlet port of the circulator experiences a very slight drop in pressure due to flow resistance in the piping. The numbers used for pressure in Figure 12–11 are illustrative only. The actual numbers will depend on flow rates, fluid properties, and pipe sizes.

Next, consider the same system with the expansion tank (incorrectly) located near the discharge port of the circulator. Figure 12–12 illustrates the pressure distribution that now develops when the circulator is operating.

The point of no pressure change remains at the location where the expansion tank is attached to the system. This causes the pressure in most of the system to *decrease* when the circulator is turned on. The pressure at the inlet port of the circulator has dropped from 10 psi to 2 psi. This situation is not desirable since it reduces the system's ability to expel air. It can also encourage circulator cavitation.

To see how problems can develop, imagine the same system with a static pressurization of only 5 psi. The same 9 psi differential will be established across the circulator when it operates, and the entire pressure profile shown with dashed lines in Figure 12–12 will shift downward by 5 psi as shown in Figure 12–13.

Notice the pressure in the piping between the upper-right corner and the circulator inlet is now *below atmospheric pressure.* If there are air vents, or slightly loose valve packings in this portion of the circuit, air will be sucked *into* the system every time the circulator operates. The absolute pressure at the inlet of the circulator is also lower and thus the potential for vapor cavitation is increased, especially if the system operates at high fluid temperatures.

Unfortunately, many residential hydronic systems have been installed with the circulator pumping toward rather than away from the point where the expansion tank is connected (e.g., the point of no pressure change).

Consider the typical residential system piping arrangement shown in Figure 12–14. In this arrangement, the circulator is pumping toward the expansion tank. The pressure drop created through a typical sectional boiler is very low. Thus, from the standpoint of pressure change, the circulator's outlet is

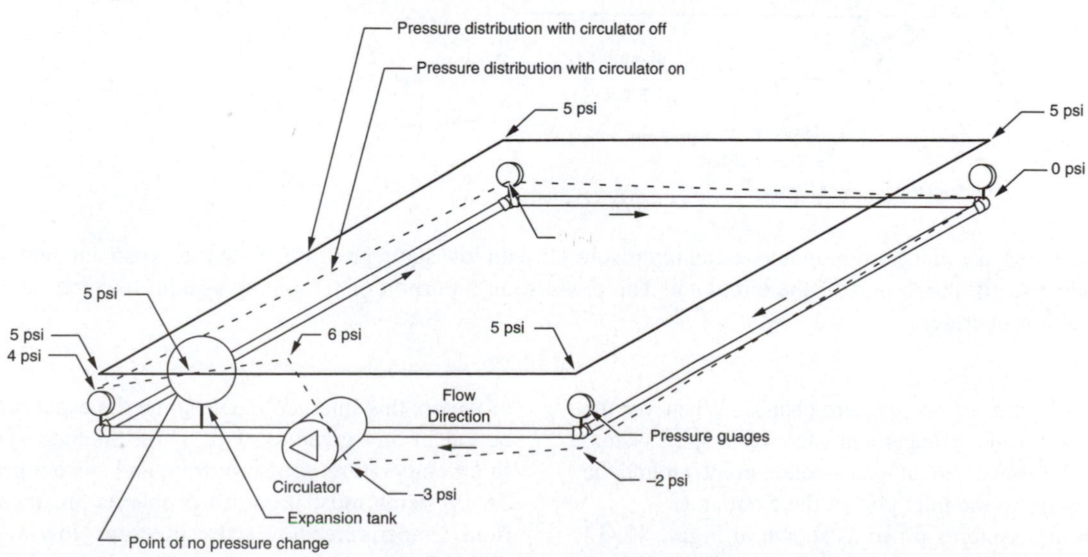

Figure 12–11 Pressure distribution in horizontal piping circuit with circulator on and off. Note the expansion tank is (correctly) located near the inlet port of the circulator.

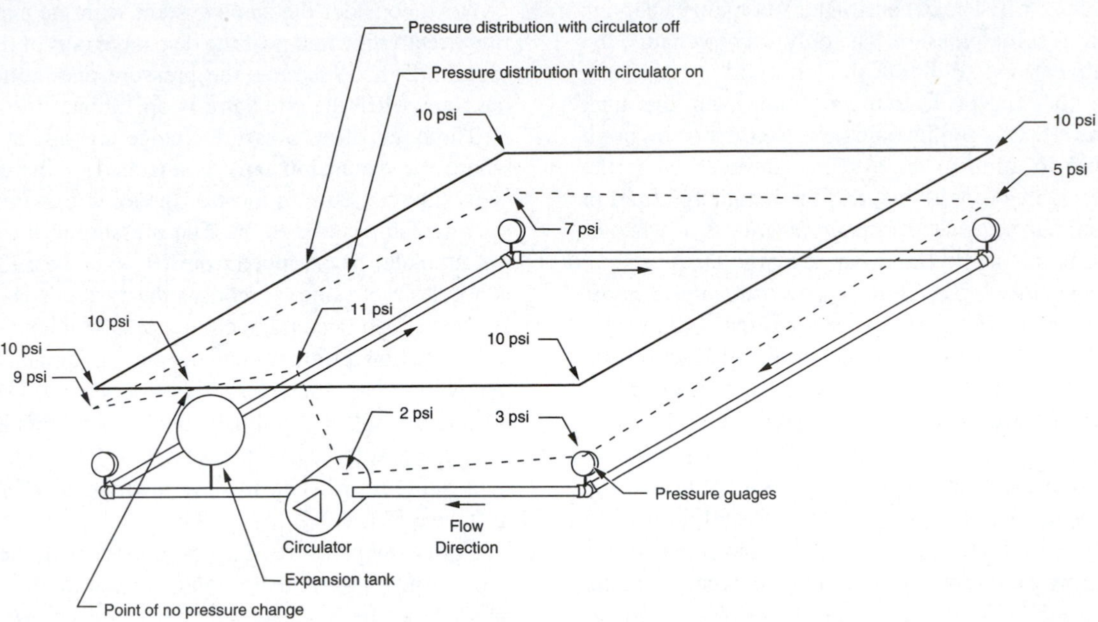

Figure 12–12 Pressure distribution in horizontal piping circuit with circulator on and off. Note the expansion tank is (incorrectly) located near the discharge port of the circulator.

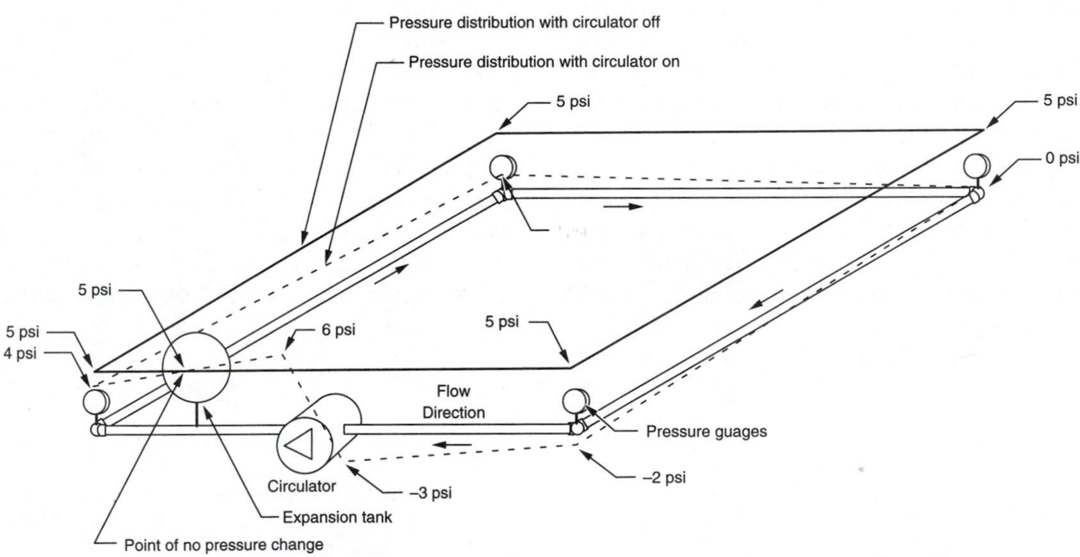

Figure 12–13 Pressure distribution in horizontal piping circuit with low static pressure. Note the expansion tank is (incorrectly) located near the discharge port of the circulator. The pressure in a portion of the piping system becomes subatmospheric when the circulator operates.

very close to the point of no pressure change. Whenever the circulator operates, this arrangement causes a drop in system pressure from the expansion tank connection point, around the distribution system to the inlet port of the circulator.

Some hydronic systems piped as shown in Figure 12–14 have worked fine for years. Others have had problems from the first day they were put into service. Why is it that some systems work and others don't? The answer lies in a number of factors that interact to determine the exact pressure distribution in any given system. These include system height, fluid temperature, pressure drop, and system pressurization. The systems most prone to problems are those with high fluid temperature, low static pressure, low system height, and high pressure drops around the piping circuit.

Rather than gamble on whether problems will occur, it is best to arrange the piping as shown in Figure 12–15. This

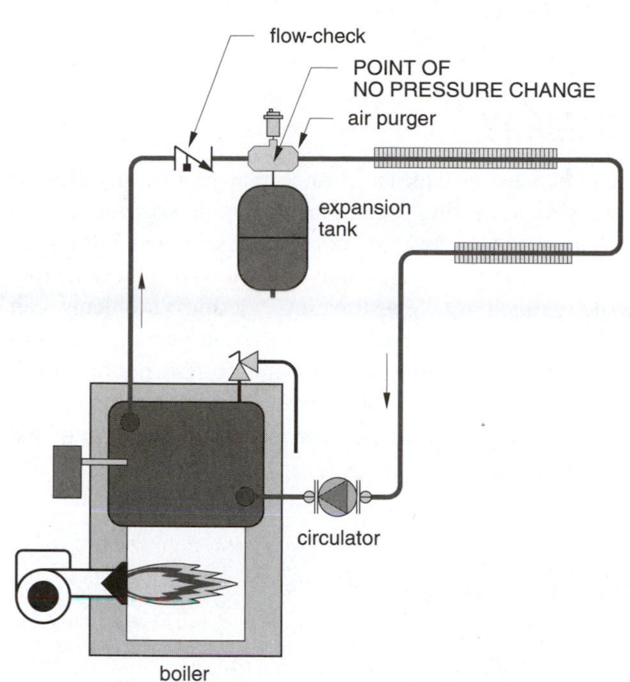

Figure 12–14 Typical piping configuration of a residential boiler. Notice the placement of the expansion tank relative to the circulator.

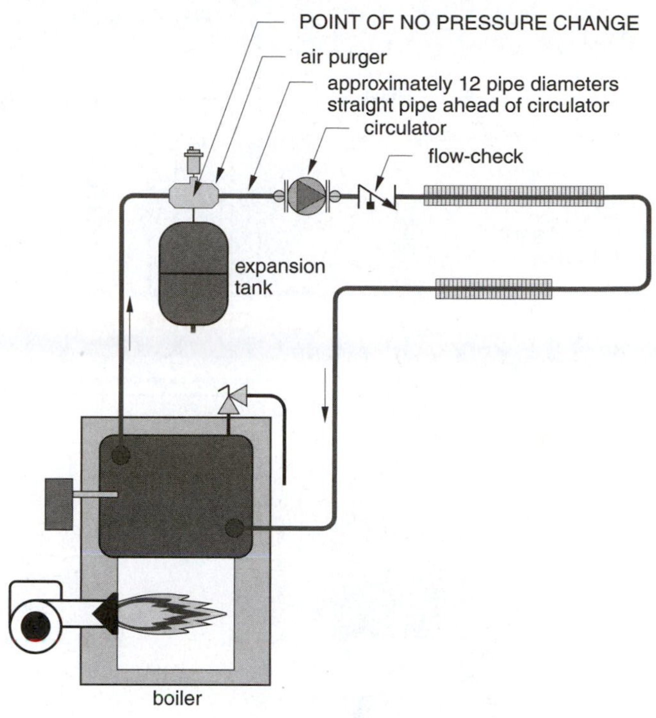

Figure 12–15 Proper arrangement of circulator and flow check valve relative to expansion tank. Notice the inlet port of circulator is now close to the expansion tank connection.

arrangement causes an *increase* in pressure in nearly all parts of the system when the circulator operates.

Many systems with chronic "air problems" can be cured by simply rearranging the piping so the circulator pumps away from the point of no pressure change.

For systems using low-flow resistance heat sources such as sectional boilers, the expansion tank can also be placed at the return side of the heat source as shown in Figure 12–16. This placement allows the tank to operate at slightly cooler temperatures that prolong diaphragm life. *Do not use this arrangement when there will be a significant pressure drop between the point of no pressure change and the circulator inlet.* The greater the pressure drop between these points, the smaller the safety margin against vapor cavitation in the circulator.

When multiple zone circulators are used, the header they connect to should be located close to and just downstream from the point where the expansion tank is connected as shown in Figure 12–17. The header pipe should also be generously sized to minimize the pressure drop between the point of no pressure change and the zone circulators.

In a primary/secondary system, the expansion tank is generally mounted in the primary loop with the primary circulator pumping away from it. *The primary loop serves as the expansion volume for the secondary circuits.* The secondary circulators should pump into their respective circuits as shown in Figure 12–18. This causes the pressure in the secondary circuits to increase when they operate.

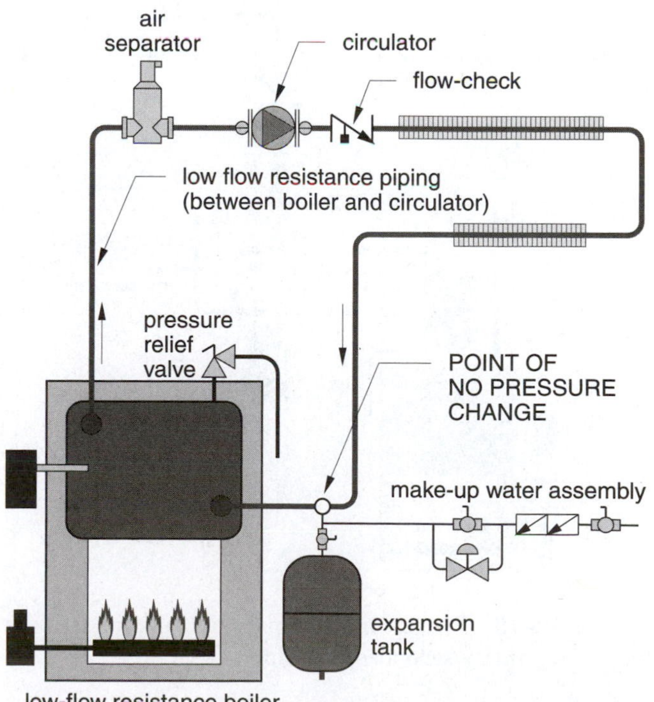

Figure 12–16 Placement of expansion tank near inlet of heat source. Only recommended when boiler and piping upstream of circulator have low flow resistance.

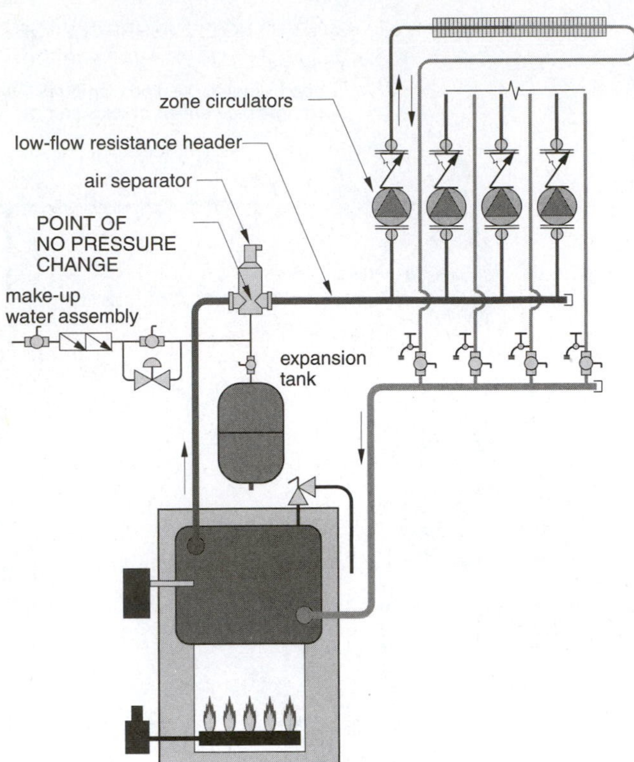

Figure 12–17 Placement of expansion tank in a system using multiple zone circulators.

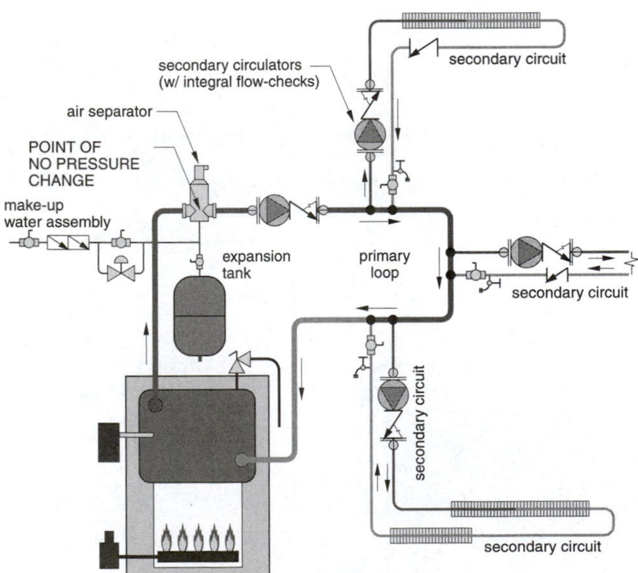

Figure 12–18 Placement of expansion tank in primary loop of a primary/secondary system.

In many systems, the make-up water system is tee'd into the piping connecting the expansion tank to the system. Because this is the point of no pressure change, additional make-up water will not be pulled into the system when the circulator operates. Look for this detail on most of the piping schematics in this text.

SUMMARY

The expansion tank is an important part of any closed-loop hydronic heating system. Its sizing, pressurization, and location within the system are easily determined, but unfortunately often misunderstood or ignored. When these factors are determined by "guestimating," serious problems can develop that lead to unsatisfactory system performance and even premature failure. Those who design and/or install hydronic systems should always review piping diagrams, as well as existing installations, for the proper location of the expansion tank.

KEY TERMS

air-side pressurization
diaphragm-type expansion tank
EPDM
Expansion Tank Sizer
incompressible
point of no pressure change
standard expansion tank
thermal expansion
waterlogged

CHAPTER 12 QUESTIONS AND EXERCISES

1. Explain why air slowly disappears from standard expansion tanks.
2. Can a standard expansion tank be placed lower than the boiler outlet? Justify your answer with a sketch.
3. Calculate the minimum size *standard* expansion tank for a system containing 50 gallons of water where the uppermost system piping is 35 feet above the base of the boiler, and the inlet of the expansion tank is 8 feet above the base of the boiler. The water temperature when the system is filled is 50 °F. The maximum operating temperature of the system is 170 °F.
4. Calculate the minimum size *diaphragm-type* expansion tank for the same system described in Exercise 3. Compare the results with those of Exercise 3.
5. Describe what happens if the diaphragm in an expansion tank ruptures. How could you check the tank to see if it is ruptured?
6. How does heat that migrates from the system piping to the expansion tank affect the pressure of the air in the tank?

7. Why is it necessary to adjust the pressure on the air side of a diaphragm-type expansion tank on each system? What can happen if this is not done, and the pressure in the diaphragm is less than the static fluid pressure at its mounting location?
8. Why should the expansion tank be located near the inlet port of the circulator?
9. How does the use of glycol-based antifreeze affect the required size of an expansion tank? Why is this so?
10. The pressure-relief valve on a hydronic system with a standard expansion tank opens each time the system heats up. How could the expansion tank be causing this situation? What can be done to correct it?
11. An expansion tank is connected 10 feet upstream from the inlet port of a circulator as shown in Figure 12–19. The pipe between the tank connection and the inlet port of the circulator is 3/4-inch copper with 150 °F water flowing through at 8 gpm. Use methods from Chapter 6 to determine the pressure at the inlet port of the circulator assuming the water pressure at the tank connection is 15 psi.

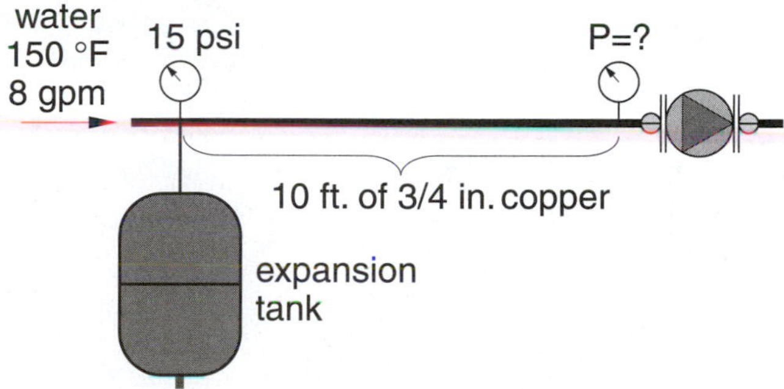

Figure 12–19 Piping arrangement for Exercise 11.

12. In sizing expansion tanks, an allowance for 5 psi fluid pressure at the top of the system is made. Why?
13. A radiant floor heating system contains 125 gallons of water in the floor circuits, and 20 gallons of water on the "hot" side of the mixing device. The boiler and higher temperature piping operate at 190 °F at design load. The floor circuits operate with a supply temperature of 120 °F at design load. The pressure-relief valve is rated at 30 psi, and there is 15 feet of system height above the inlet to the tank with 5 psi pressure at the top of the system. Determine the required expansion tank volume using Equation 12.5. Recalculate the expansion tank volume using Equation 12.4. Why are the results different?
14. Is an expansion tank required in an open-loop hydronic system? How is the expansion of the fluid accommodated in such a system?
15. What is the purpose of the boiler fitting used with standard expansion tanks?
16. Why should expansion tanks be sized so they reach a pressure about 5 psi below the rated pressure of the system's pressure-relief valve, as the system reaches maximum temperature?
17. Look up the exact inside diameters of 3/4-inch and 1-inch Schedule 40 steel pipe in a piping handbook. Calculate the volume of fluid they contain per foot.
18. Use the Expansion Tank Sizer module in the Hydronics Design Studio software to determine the size of a diaphragm-type expansion tank required for the system described in Example 12.3. Resize the tank assuming propylene glycol in concentrations of 30 percent and 50 percent is used as the system fluid. What can you conclude from this?
19. Use the Expansion Tank Sizer module to determine the minimum size of a diaphragm-type expansion tank located 40 feet below the top of a distribution system containing 100 gallons of water. Assume the system is filled with 60 °F water and has a maximum operating temperature of 200 °F.
20. Use the Fluid Properties Calculator module in the Hydronics Design Studio to determine the density of a 30 percent ethylene-glycol solution at 50 °F and 170 °F and repeat Exercise 4 assuming this fluid is used in the system.

AIR REMOVAL, FILLING, AND PURGING 13

OBJECTIVES

After studying this chapter, you should be able to:

- Discuss the different forms in which air can exist in a hydronic system
- Explain the operation of various devices for removing air from a system
- Diagnose some of the problems caused by air in hydronic systems
- Describe where air venting devices should be placed in a system
- Design hydronic systems that will not experience air problems
- Correct chronic air problems in existing systems
- Describe the construction and operation of a purge cart
- Add antifreeze to a system using a purge cart
- Describe the operation of an automatic fluid feeder

13.1 INTRODUCTION

For a hydronic heating system to deliver silent comfort, it must be free of air. If this is not the case, problems ranging from occasional "gurgling" sounds in the pipes, to complete loss of heat output can occur.

Air problems are often bewildering. Just when the problem appears to be fixed, it can recur. Many owners and installers eventually give up, thinking the system's air problem simply cannot be corrected. The installer hopes the same mysterious problem will not occur on the next job. A lack of understanding often prevents the true cause of the problem from being diagnosed, corrected, and most importantly, *avoided* in the future!

At some point, every hydronic system contains a mixture of water and air. This is especially true when the system is first filled and put into service. *If properly designed, a closed-loop hydronic system should rid itself of most air within a few days of initial start-up. The system should then maintain itself virtually air-free throughout its service life.* Systems that experience chronic problems with entrapped air usually contain one or more classic design or installation errors.

This chapter provides the facts of how air gets into the system, how it behaves during system operation, and most importantly, how to get rid of it.

13.2 PROBLEMS CREATED BY ENTRAPPED AIR

Although most people recognize the classic symptom of entrapped air (gurgling or churning sounds in the piping), they are often unaware that entrapped air can lead to other problems including:

- *Accelerated corrosion of iron or steel components due to chemical reactions with the oxygen in the air.* In systems with chronic air problems, corrosion can proceed at several times its normal rate.
- *Loss of heat output in a heat emitter or an entire system due to a large pocket of entrapped air.* This is often called **air binding.** It occurs in a number of ways. One of the most common is when an air pocket displaces water within the volute of a circulator, and thus prevents the circulator from moving fluid through the system. Another possibility is when the circulator cannot lift the water over the top of a piping circuit that is partially filled with air.
- *Reduced circulator head.* Because a mixture of water and air bubbles is a compressible fluid, the circulator is unable to efficiently transfer mechanical energy (e.g., head) to the fluid. This can greatly reduce the system's flow rate and heat output.
- *Improper lubrication of the bushings in wet-rotor circulators.* Because a mixture of water and air bubbles creates a foamlike solution inside the circulator, the fluid film required for proper lubrication of the bushings may not be present. Wet-rotor circulators depend solely on this liquid film for proper lubrication, and can quickly fail if it is not present.
- *Reduced heat transfer within the heat source.* The presence of bubbles along heat transfer surfaces interferes with convective heat transfer, and will reduce the rate of heat output. Air within the heat source heat exchanger can also lead to "hot spots" that add stress and can ultimately cause a failure of the heat exchanger.

13.3 TYPES OF ENTRAPPED AIR

In hydronic systems, air is found in three forms:

- Stationary air pockets
- Entrained air bubbles
- Air dissolved within the fluid

All three forms can exist simultaneously, especially when the system is first put into service. Each exhibits different symptoms in the system.

Stationary Air Pockets

Since air is lighter than water, it tends to migrate toward the high points of the system. These points are not necessarily at the top of the system. Stationary air pockets can form at the top of heat emitters, even those located low in the building. Air pockets also tend to form in horizontal piping runs that turn downward following a horizontal run. A good example is when a pipe is raised to cross over a structural beam, and then dropped to its previous level as shown in Figure 13–1.

When a system is first filled with water, these high points are dead ends for air movement. In some cases, this trapped air can displace several quarts of fluid that eventually must be added to the system. Even after a system is initially purged, stationary air pockets can reform as residual air bubbles merge together and migrate toward high points. This is especially likely in components with low flow velocities where slow-moving fluid is unable to push or drag the air along with it. Examples of such components include large heat emitters, large diameter piping, and storage tanks.

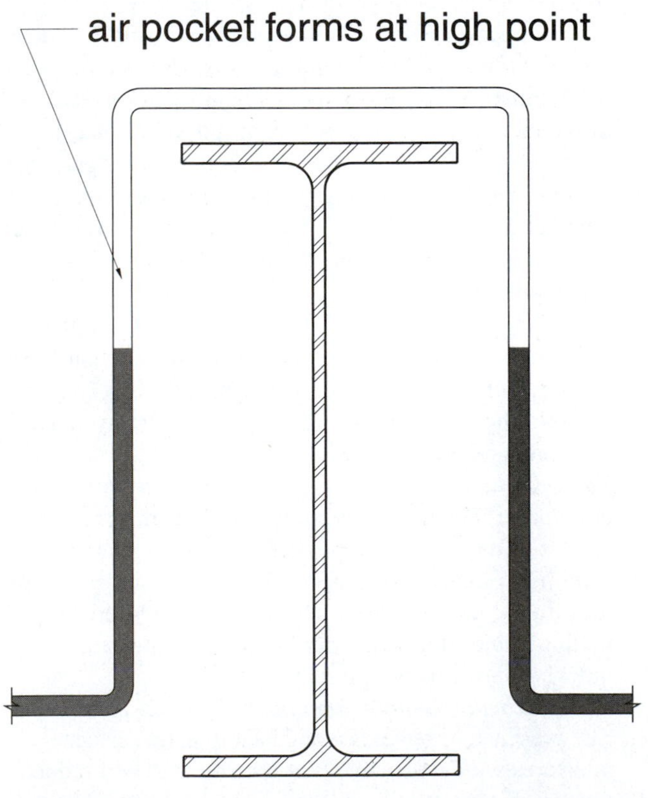

air pocket forms at high point

Figure 13–1 A typical location where a stationary air pocket can form.

Entrained Air Bubbles

When air exists as bubbles, a moving fluid *may* be able to carry them along (entrain them) through the system. **Entrained air** can be both good and bad. It is good from the standpoint of transporting air from remote parts of the system back to a **deareating device;** however, it is bad if the air cannot be separated from the fluid within the deareating device. How well a fluid entrains air is best judged by its ability to move bubbles downward, against their natural tendency to rise. Simply put, *if the fluid moves downward faster than a bubble can rise, it will carry the bubble in its direction of flow.* If air entrainment through a downward flowing pipe is desired, it is crucial that the fluid's flow velocity is greater than the **bubble's rise velocity** as illustrated in Figure 13–2.

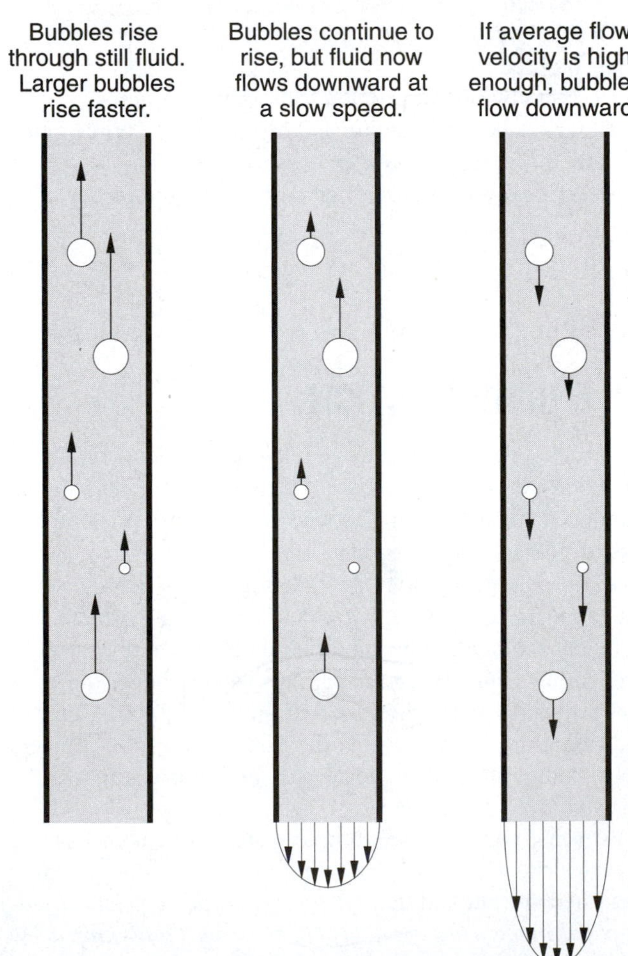

Figure 13–2 The speed of the downward flowing fluid determines if bubbles continue to rise or are pulled along (entrained) in the direction of the flow.

The velocity at which a bubble rises through a fluid depends on the bubble's diameter and density, as well as the density and viscosity of the surrounding fluid. The larger its diameter, the faster the bubble rises. The higher the viscosity of the surrounding fluid, and the lower its density, the slower the bubble rises. Of these factors, the diameter of the bubble has the greatest effect on rise velocity. A bubble half the diameter of another will rise at only one-fourth the velocity through the same fluid. The difference in bubble rise velocity can be observed in a glass filled with a carbonated drink.

One type of bubble often present in hydronic systems is called a **microbubble.** They are so small it is often difficult to see a single microbubble. Dense groups of microbubbles make otherwise clear water appear cloudy. They can often be observed in a glass of water filled from a faucet with an aerator device as shown in Figure 13–3.

Microbubbles can also be formed as dissolved air comes out of solution upon heating, or when a component generates significant turbulence. They have very low rise velocities, and are easily entrained by moving fluids. This characteristic makes it difficult to separate microbubbles from the fluid.

Figure 13–3 Microbubbles slowly rising through a glass of water.

Separation requires the fluid to move through an area of very low velocity and low turbulence, where the microbubbles have sufficient time to rise and merge into larger bubbles.

Unfortunately many hydronic systems have air-separating devices that do not provide sufficiently low flow velocities for efficient microbubble separation. While larger bubbles are captured due to their greater rise velocities, the much smaller microbubbles are swept through before being separated. In some cases, this is due to the design of the air separator. In others it is due to the flow velocity through the air separator being too high. Eventually the microbubbles merge into larger bubbles that can be separated and ejected from the system, but this may take several days of operation.

Air Dissolved within the Fluid

Perhaps the least understood form in which air exists in a hydronic system is as **dissolved air.** Molecules of the gases that make up air including oxygen and nitrogen can exist "in solution" with water molecules. These molecules cannot be seen, even under a microscope. *Although water may appear perfectly clear and free of bubbles, it can still contain a significant quantity of air in solution.*

The amount of air that exists in solution with water is strongly dependent on the water's temperature and pressure. The curves shown in Figure 13–4 show the *maximum* dissolved air content of water as a percentage of volume, over a range of temperature and pressure.

Notice that *as the temperature of water increases, its ability to hold air in solution decreases.* This explains why air bubbles appear on the lower surfaces of a pot of water being heated on a stove. It also explains the formation of microbubbles along the inside surfaces of a boiler's heat

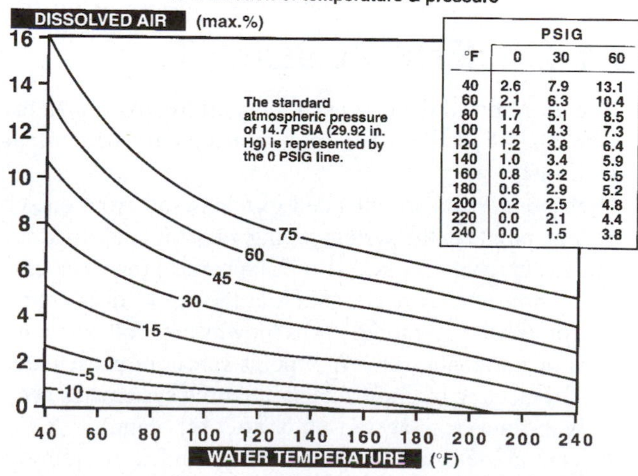

°F	PSIG 0	30	60
40	2.6	7.9	13.1
60	2.1	6.3	10.4
80	1.7	5.1	8.5
100	1.4	4.3	7.3
120	1.2	3.8	6.4
140	1.0	3.4	5.9
160	0.8	3.2	5.5
180	0.6	2.9	5.2
200	0.2	2.5	4.8
220	0.0	2.1	4.4
240	0.0	1.5	3.8

Figure 13–4 Curves showing the maximum solubility of air in water at different temperatures and pressures. *Courtesy Spirotherm, Inc.*

exchanger. As the water nearest these surfaces is heated, some of the air comes out of solution in the form of microbubbles. The opposite is also true. *When heated water is cooled, it will absorb air back into solution.*

The pressure of the water also has a marked effect on its ability to contain dissolved air. *When the pressure of the water is lowered, its ability to contain dissolved air decreases, and vice versa.* This is the reason carbon dioxide bubbles instantly appear when the cap is removed from a bottle of soda. Removing the cap depressurizes the liquid and reduces the fluid's ability to hold carbon dioxide in solution. This results in the instant formation of visible bubbles. Lowered pressure is also one reason air bubbles are more likely to appear in the upper portions of a multistory hydronic system. The lower static pressure in the upper part of the system makes it easier for dissolved air to come out of solution. The greater static pressure near the bottom of the system tends to keep the air in solution.

The ability of water to repeatedly absorb and release air can affect the system in several ways, some good and some bad. For example, the ability of water to absorb air can be very helpful in removing stationary air pockets from otherwise inaccessible portions of the system (such as a high point in the piping not equipped with an air vent), and transporting the air back to a central deareating device. On the negative side, the ability of water to absorb air is also the primary cause of waterlogging in standard expansion tanks as discussed in Chapter 12, Expansion Tanks.

It is always desirable to minimize the dissolved air content of the system's water. This is accomplished by establishing conditions that "encourage" air to come out of solution (e.g., high temperatures and low pressures). When such conditions exist, the resulting microbubbles must be gathered together and ejected from the system. Devices for doing this are discussed in the next section.

13.4 AIR REMOVAL DEVICES

The air removal devices used in small hydronic systems can be classified as either **high point vents** or **central deareators.**

High point vents are intended to release air from one or more high points in the system piping where it tends to accumulate in stationary pockets. Typical locations for such vents are at the top of each heat emitter, at the top of distribution risers, or wherever piping turns downward following an upward or horizontal run. High point vents are particularly useful for ejecting air immediately after the system has been filled with fluid (e.g., at start-up or after servicing).

A central deaerator is a device intended to remove entrained air from a flowing fluid as well as maintain the system at the lowest possible air content. It is usually mounted near the outlet of the heat source, and has the entire system flow passing through it.

There are several types of devices used as high point vents and several others used as central deareators. All types relevant to residential and light commercial systems will be discussed separately.

Manual Air Vents

The simplest type of high point venting device is a **manual air vent.** These components are actually small valves with a metal-to-metal seat. They thread into 1/8-inch or 1/4-inch FPT tappings, and are operated with a screwdriver, square head key, or the edge of a dime. They are sometimes referred to as "coin vents" or "bleeders." When their center screw is rotated, air can move up through the valve seat and exit through a small side opening. An example of a manual air vent is shown in Figure 13–5.

Manual air vents are commonly located at the top of each heat emitter. They can also be mounted into baseboard tees as shown in Figure 13–6, or any fitting with the appropriate threaded tapping. Because of their small size, care should be taken that joint compound or Teflon tape does not cover their small inlet port.

After a manual vent is opened, and any air below it is released, *a small stream of water will continue to flow through the vent until it is closed.* While it is operating, a person should stay near the vent with a can ready to catch any ejected fluid before it stains a carpet or floor. This can be difficult (especially for one person), if all manual vents are left open as the system is filled. It is better to move from one vent to the next, operating them in sequence.

Automatic Air Vents

Another type of small high point venting device is called an **automatic air vent.** An example is shown in Figure 13–7. When the upper cap of this device is open, air can pass through and be expelled. When water reaches the vent, an internal washer or disc swells to seal off the opening. This prevents excessive water spillage if the vent is unattended when

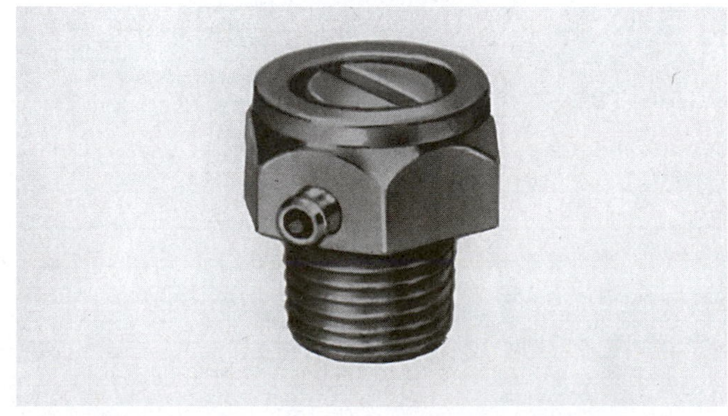

Figure 13–5 Example of a manual air vent. *Courtesy of ITT Corporation.*

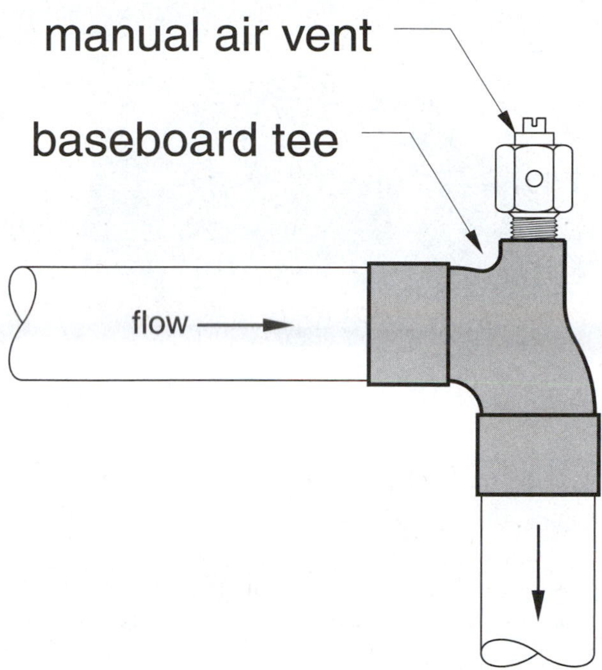

Figure 13–6 Placement of a manual air vent on a baseboard tee.

the system is filled. As more air accumulates beneath the vent, the washer or disc dries out and reopens to expel the air.

The time required for the internal disc or washer to close off the vent is not fast enough to prevent all water loss. Automatic vents may experience a slight water loss from time to time. In some cases installers close off the cap after the system is initially purged to prevent any leakage.

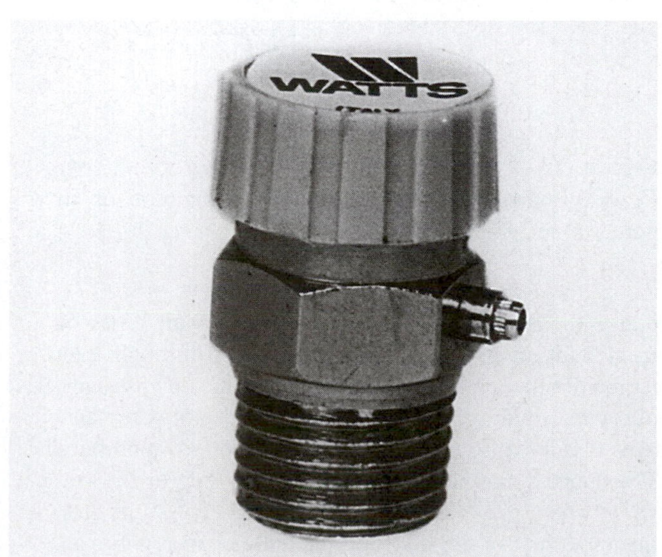

Figure 13–7 Example of an automatic air vent. *Courtesy of Watts Regulator.*

Float-type Air Vents

The need for fully automatic (unattended) venting in locations not easily accessed requires a different type of device. A **float-type air vent** often provides the solution. This device, shown in Figure 13–8, contains an air chamber, a float, and an air valve. When sufficient air accumulates in the chamber, the float drops down and opens the valve at the top of the unit. As air is vented, water enters the chamber and lifts the float to close the air valve.

(a)

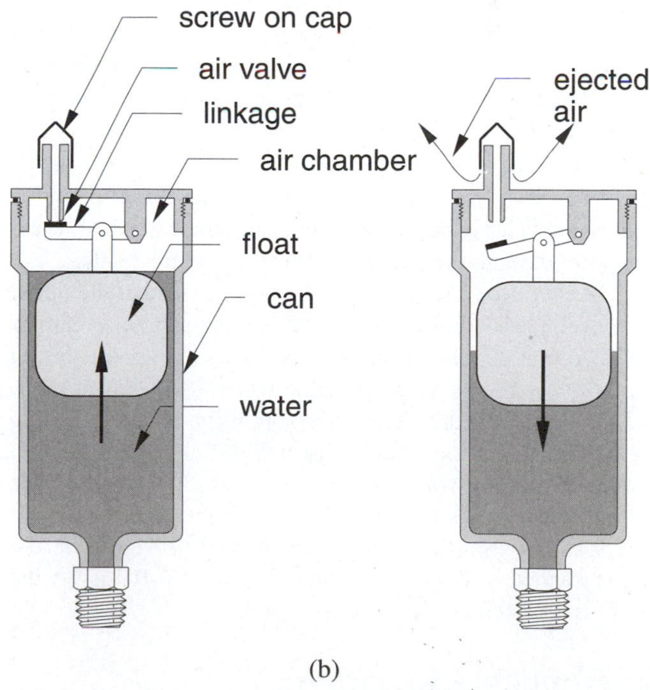

(b)

Figure 13–8 (a) Example of float type air vents. *Courtesy of Maid O' Mist.* (b) Operation of a float type air vent.

Most float-type air vents are equipped with a metal cap that protects the stem of the air valve. *It is important that this cap remains loose when the vent is put into service.* If the cap is tight, air cannot be ejected.

Float-type vents are designed to thread into baseboard tees or reducer tees the same as manual vents. They are available in different sizes and shapes that allow mounting both horizontal and vertical orientations. Low profile designs are available that allow mounting within the enclosures of heat emitters such as finned-tube convectors or fancoils. Some are also available with larger piping connections that allow higher air venting rates when installed on central deareators or at the top of large tanks.

Most automatic and float-type air vents will allow air to enter the system if the fluid pressure at their location falls below atmospheric pressure. This can result from a number of factors, most notably improper placement of the expansion tank relative to the circulator. It is a deceptive problem because it usually only occurs when the circulator is operating. The best way to prevent this from happening is to make sure there is at least 5 psi of positive pressure at the top of the system at all times. This is accomplished by proper setting of the feed water valve as discussed in Chapter 12, Expansion Tanks.

Air Purgers

One type of central deareating device is known as an **air purger** or **air scoop.** This cast-iron component, shown in Figure 13–9a, is equipped with an internal baffle that directs bubbles into a separation chamber where they can rise into a float-type air vent screwed into the top of the chamber. The internal baffle also creates a region of lowered pressure that assists in bringing dissolved air out of solution. An arrow on the side of the air purger indicates the direction it must be installed relative to the flow.

To be effective, air purgers must operate at inlet flow velocities not exceeding 4 feet per second. If the flow velocity is higher, small air bubbles will remain entrained in the flow and be swept through the purger as shown in Figure 13–9b.

The preferred location for an air purger is near the outlet side of the heat source, about 12 pipe diameters upstream of the inlet port of the circulator. This allows the water entering the purger to be as hot as possible while also being at a relatively low pressure. The air's solubility in water is low under these conditions, and thus bubbles are more likely to form. If make-up water is added through the bottom of the purger, air bubbles that may enter with the water are more likely to rise into the upper chamber and be vented before being carried through the system. Figure 13–10 shows the preferred placement of an air purger.

Microbubble Resorbers

Although relatively new to the American hydronics market, **microbubble resorbers** have received widespread praise for their effectiveness in removing all types of air from hydronic systems. As flow passes through them, a region of reduced pressure is created on the downstream side of several small vertical wires. The reduced pressure encourages dissolved air to come out of solution as microbubbles. The bubbles rise along the downstream side of the vertical wires where they are shielded from higher velocity flow. In the upper part of the resorber, the bubbles merge into a larger air volume. A float-operated air valve opens to eject the air when a sufficient amount has accumulated. An example of a microbubble resorber is shown in Figure 13–11.

(a)

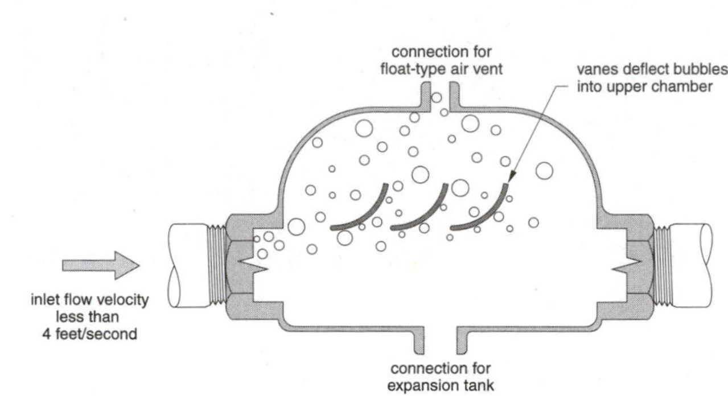

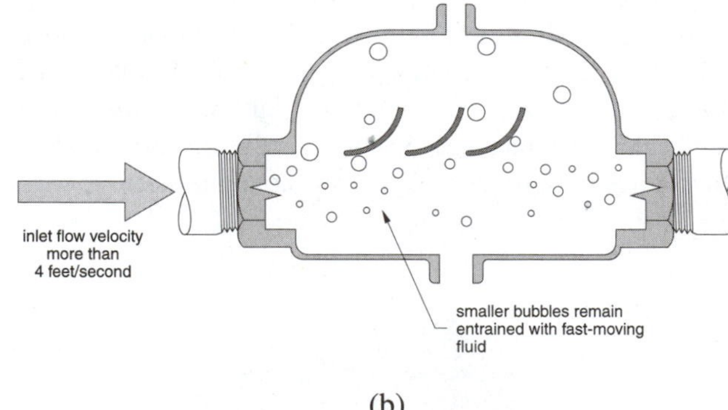

(b)

Figure 13–9 (a) Example of a cast-iron air purger. *Courtesy of Watts Regulator, Co.* (b) Operation of an air purger at different flow velocities.

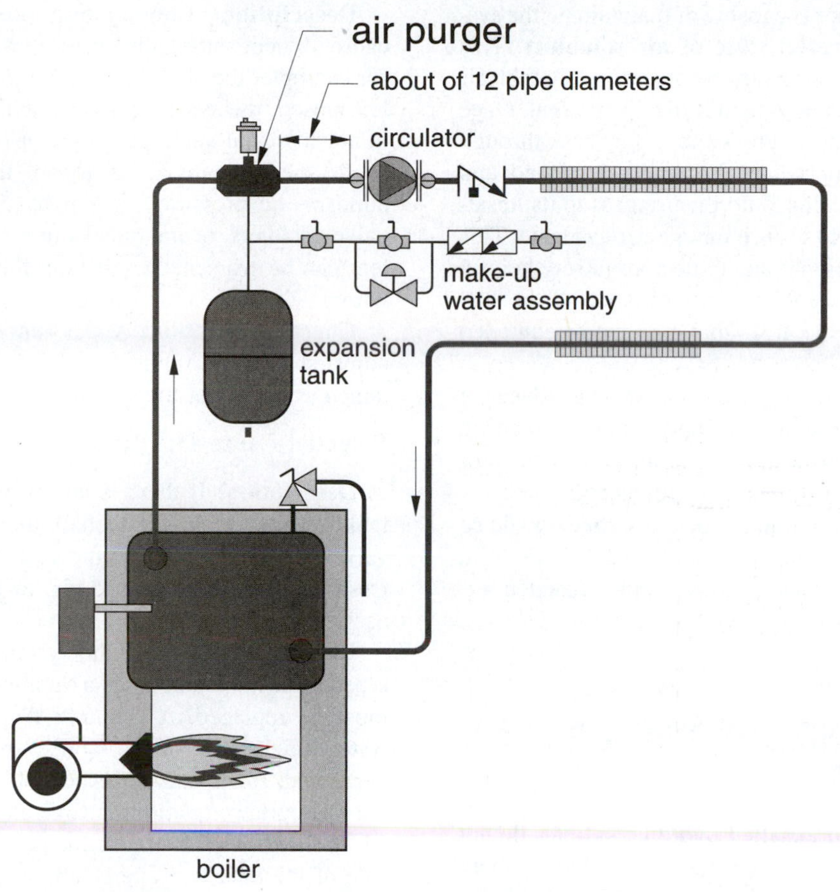

Figure 13–10 Desirable placement of an air purger relative to other components.

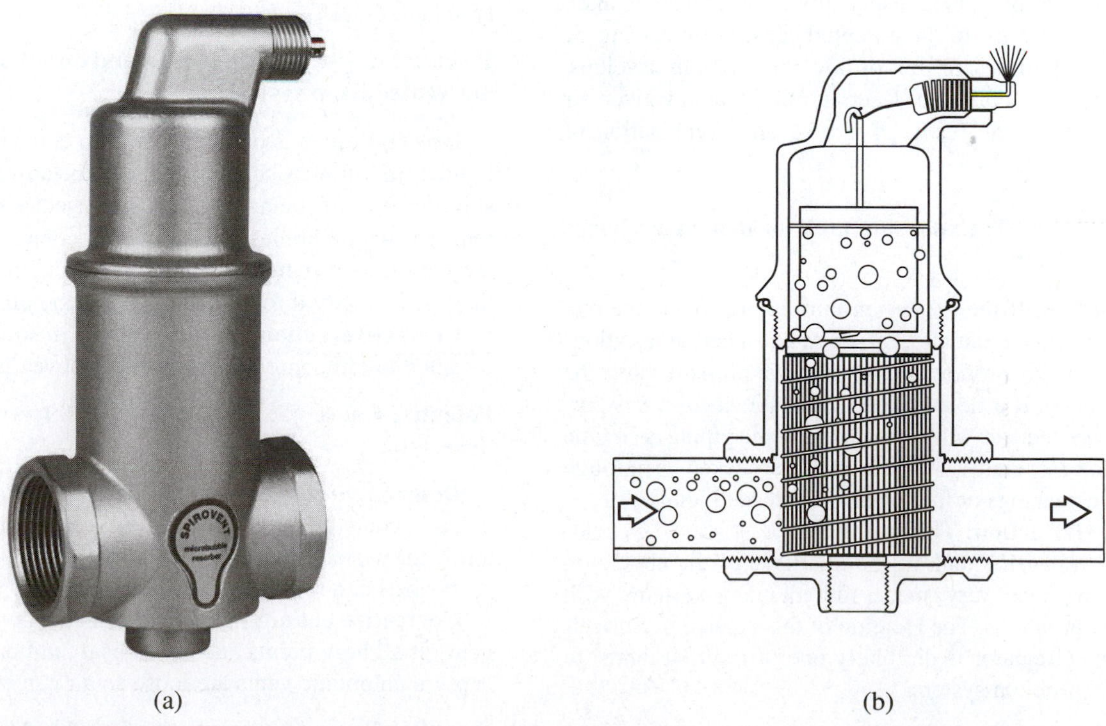

Figure 13–11 Example of a microbubble resorber. (a) External appearance, (b) cut-away view showing internal design. *Courtesy of Spirotherm, Inc.*

A microbubble resorber is capable of maintaining the system fluid in an **unsaturated state of air solubility.** This means the water is always ready to absorb additional air from any areas of the system where it may be present. Once absorbed, the air is transported by system flow back through the microbubble resorber where it can be separated and ejected from the system. The fluid then returns to its unsaturated state and is ready to absorb more air if available. This ability to continually seek out and collect air is very helpful for removing residual air pockets from all areas of the system, especially inaccessible areas that may not be equipped with vents.

Studies have shown microbubble resorbers can reduce the dissolved air content of the system fluid below 0.5 percent. The amount of oxygen present in this small air content is of virtually no significance as far as corrosion is concerned.

The inlet flow velocity to microbubble resorbers should be no higher than 4 feet per second to prevent small bubbles from being wisked through the device without being separated.

13.5 CORRECTING CHRONIC AIR PROBLEMS

Properly designed and installed hydronic systems do not require constant manual air venting. A complaint about recurring air noises is symptomatic of one or more underlying design or installation errors. Many of these errors have little intuitive connection with the problems they create and thus are hard to recognize, especially by a novice installer. What follows is a list of the potential causes of recurring air problems, a short description of how the problem develops, and what to do to correct it. Keep in mind that any given air problem may be the result of almost any combination of these factors.

Potential Cause #1: Expansion tank located on discharge side of circulator.

Description: If the expansion tank is located on the discharge side of the circulator, the pressure in certain locations in the system may drop below atmospheric pressure while the circulator is on. If a float-type air vent is located in this portion of the system, air will be sucked into the piping each time the circulator operates. Air can also be sucked in through loose valve packings or microleaks at threaded joints.

Corrective action: This problem is best solved by relocating the expansion tank to the inlet side of the circulator. This fix has been very successful in many systems with chronic air problems. The location of the expansion tank relative to the circulator is definitely one of the first things to check on a problem system.

Potential Cause #2: Low system pressurization.

Description: Low system pressurization is usually caused by not setting the make-up water system properly. If, for example, the shutoff valve on the make-up water line is left closed, the system pressure will eventually drop due to minor losses at valve packings, or by air removed at vents. When the pressure at the top of the system drops below atmospheric pressure, air can be drawn in through vents, valve packings, or microleaks in threaded joints. This problem can be worsened by having the expansion tank on the wrong side of the circulator as described earlier.

Corrective Action: Make sure the feedwater valve is adjusted so that a minimum static pressure of 5 psig is maintained at the top of the system.

Potential Cause #3: Waterlogged expansion tank.

Description: If there is no air cushion in the expansion tank, or if the tank is too small, the system's pressure-relief valve can open each time the system heats up. This allows fresh water (with dissolved air) to enter the system during each heating cycle.

Corrective action: If the system contains a diaphragm-type expansion tank with a ruptured diaphragm, the tank must be replaced. A ruptured diaphragm is indicated by water flowing out of the tank's air valve when its stem is depressed. If the tank is suspected of being undersized, use the appropriate equations in Chapter 12, Expansion Tanks, to verify the minimum acceptable tank volume. Be sure the air side of the tank is properly pressurized. If the system contains a standard expansion tank, it may require periodic draining to relieve a waterlogged condition. Another approach is to replace the standard tank with a diaphragm-type tank.

Potential Cause #4: Air goes in and out of solution but is not vented from system.

Description: It is possible for air to come out of solution in the form of microbubbles, and then be absorbed back into solution without being captured and ejected from the system. The microbubbles that form as air comes out of solution are not easily captured by common air purgers, especially if the flow velocity through the purger is too high.

Corrective action: A microbubble resorber should be installed to efficiently separate and remove dissolved air.

Potential Cause #5: Lack of high point vents or central deaerator.

Description: High points in the system that do not have either a manual or automatic air vent can persistently collect air. If the system does not have a microbubble resorber, the air pockets can remain in place for weeks.

Corrective action: Install either manual or automatic air vents at all high points and on all heat emitters. Be sure the caps on automatic vents are loose so air can escape.

Potential Cause #6: Unentrained air bubbles.

Description: When persistent gurgling sounds are heard in piping, especially piping with downward flow, it is likely the flow velocity is too low to entrain the air and transport it to a central deaerator. The low flow velocity can be caused by a number of factors. These include but are not limited to significantly oversized piping, insufficient circulator size, or excessively long piping circuits.

Corrective action: All down-flowing piping should be sized to maintain a flow velocity of at least 2 ft/sec to effectively entrain air bubbles. If this is not possible, automatic high point vents must be installed at locations where air can collect.

Potential Cause #7: Air vents or leaks at the top of open-loop systems.

Description: This problem is often associated with hydronic thermal storage systems that have unpressurized tanks. Consider the schematic shown in Figure 13–12. Because this is an open-loop system, the pressure in all piping above the water level in the tank will drop below atmospheric pressure when the circulator stops. Air will try to enter the piping at any point it can (above the water level). Air vents located above the water level are the most likely entry points. However, valve packings, pump flange gaskets, and even microleaks at threaded joints can also admit air.

Corrective action: This type of system should not have air vents in any piping above the water level in the tank. All piping should be designed for a flow rate of at least 2 ft/sec to effectively entrain air bubbles and bring them back to the tank where they can be vented.

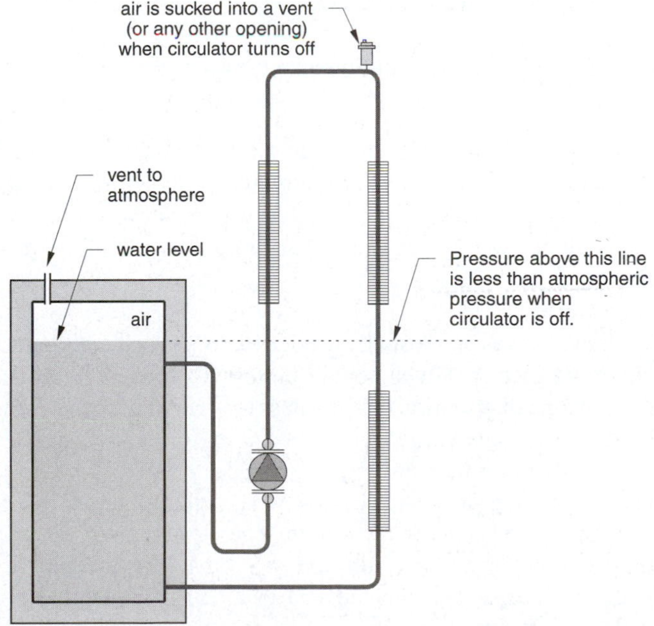

Figure 13–12 Schematic of open loop system showing areas of subatmospheric pressure.

Potential Cause #8: Improper purging during initial start-up.

Description: Inadequate purging of piping can leave large amounts of bulk air in the system. The entrapped air pockets may even be large enough to cause the circulator to not be able to circulate fluid around the system.

Corrective Action: It is essential to remove the bulk air in the system by "forced-water purging" during initial start-up. The next section describes methods for doing so.

13.6 FILLING AND PURGING A SYSTEM

Over the years, many installers have developed their own methods of filling and purging hydronic systems as part of their start-up procedure. This section will show some common approaches, as well as a new approach using a tool called a purge cart.

Gravity Purging

A portion of the air inside an empty hydronic system can be expelled by opening the feed water valve to fill the system, and allowing the air to rise up to and out of the high point vent(s). This approach is known as **gravity purging.** Keep in mind that air pockets will likely form at *all* high points in the system. If all these points are not equipped with a venting device, it may be difficult for the system's circulator to establish flow through the system after filling it, especially if a low head circulator is used. The air pockets may take several minutes to vent through low capacity vents. This method, although simple in concept, is slow in execution, especially if high point vents are not properly located. Figure 13–13 shows some potential problems when gravity purging alone is used.

Forced-water Purging

The amount of air that can be quickly expelled from a hydronic system during filling is substantially increased when the entering fluid has sufficient velocity to entrain air bubbles and carry them to an outlet valve. The greater the water pressure available from a well or water main, the faster **forced-water purging** can push air out of the system.

One method of forced-water purging is to add a boiler drain valve and full port ball valve to the *return* line of the boiler as shown in Figure 13–14. To fill and purge the system, the fast-fill lever at the top of the feed water valve is lifted to admit water at a high flow rate. If the system has a ball valve for bypassing the feed water valve, it should also be opened to allow the fastest possible purging flow. Because the ball valve on the return line to the boiler is closed, the entering water begins to fill the boiler, and then, as the boiler is filled, is forced out through the distribution

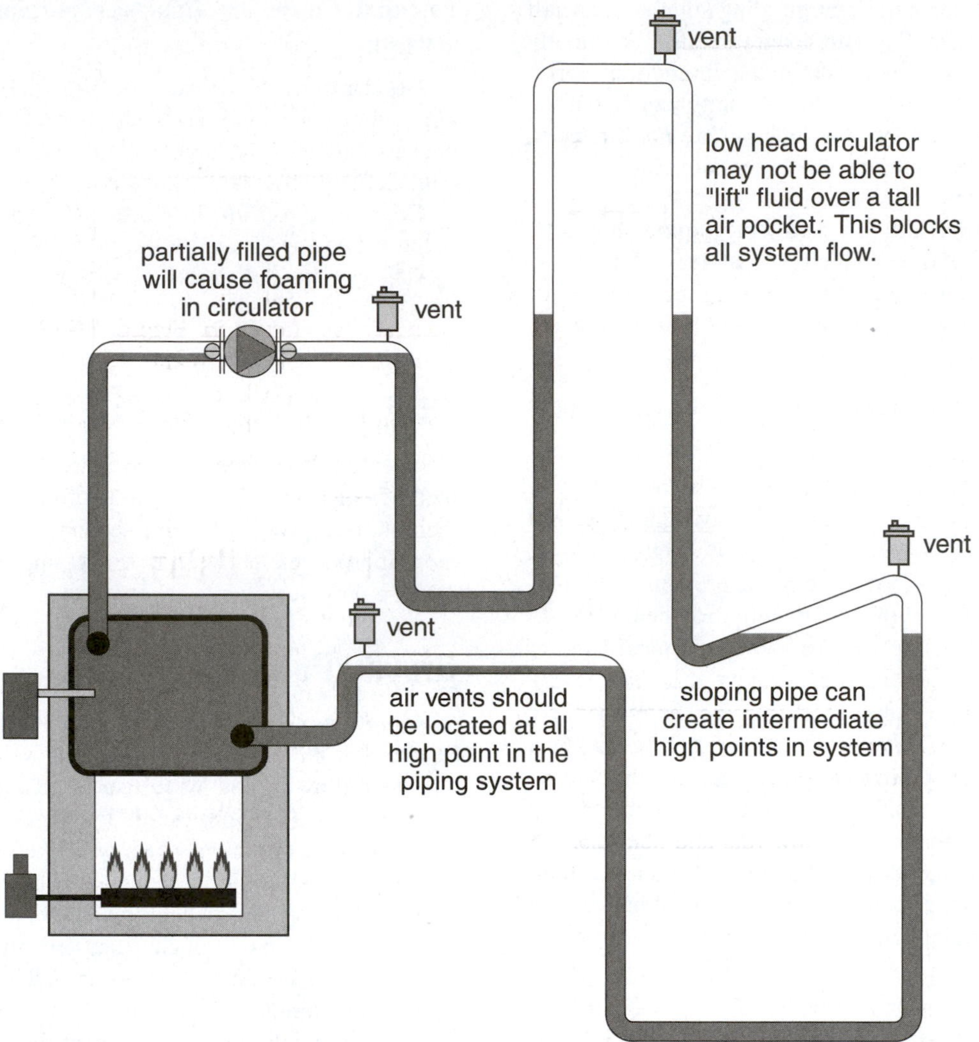

Figure 13–13 Locations where air pockets can form and cause difficulty while gravity purging an empty system.

piping. Most of the air in the boiler will rise into the central deaerator. Some boilers may also have their own air vent.

The purging flow continues through the system and eventually reaches and exits through the outlet drain valve just upstream of the ball valve at the boiler inlet. A hose can be used to route the exiting mixture of air and water to a drain. When the discharge stream is free of visible air bubbles, most of the bulk air will have been purged from the system. At this point, the fast-fill lever is released, and/or the feed water bypass valve is closed, as is the purging outlet valve.

The placement of purging valves near the *inlet* connection of the boiler helps ensure debris such as small solder balls or chips of metal in the piping are flushed out of the system, rather than into the boiler. The reader should note the outlet drain valve *must* be open *before* water is admitted through the feed water valve. If not, the entering water will likely increase the pressure in the system above the pressure-relief valve setting, causing the relief valve to open and defeat the purpose of the high flow rate purging.

A specialty valve that gives the same functionality as the drain valve and ball valve combination is shown in Figure 13–15. This so-called **purge valve** was created to reduce installation time by replacing the drain valve, ball valve, and associated fittings. Several manufacturers now offer such valves.

Purge valves are usually designed to be soldered into either 3/4-inch or 1-inch copper piping. This size is typical for the zone distribution circuits in small- and medium-sized systems. It is generally not a good idea to install a small purge valve in a larger size pipe. Besides creating a bottleneck to purging flow, such a valve would create a larger than necessary head loss as the system operates. In such situations it is better to use a full port ball valve in combination with a high capacity outlet valve. The latter may be equipped with a fitting for a large diameter hose.

To purge a system having parallel branches, it is best to close off all but one branch path at a time. This forces all purging flow through the open branch at a high flow veloci-

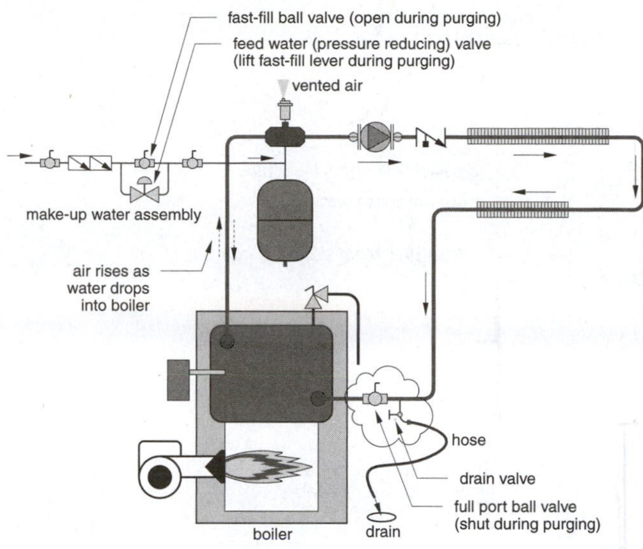

Figure 13–14 Placement of a ball valve and drain valve for system purging.

ty, improving the chances of displacing entrapped air. Proceed from one branch circuit to the next in sequence. After all branches have been purged individually, open them all up at the same time. The reduced equivalent hydraulic resistance will increase system flow rate and help dislodge air pockets in larger piping and components. This technique is especially helpful on radiant floor heating systems having several parallel circuits. The valves on the manifolds can be used to open and close each circuit as needed.

Some installers prefer to install a separate purging valve on the return side of each parallel zone circuit as shown in Figure 13–16. This arrangement allows each circuit to be separately purged during start-up.

It is very important to install purging valves on the return side of any secondary circuit in a primary/secondary system as shown in Figure 13–17. The closely spaced tees that couple the secondary circuit to the primary loop will not induce sufficient purging flow into the secondary circuit, even if there is a strong purging flow in the primary loop.

The author does *not* recommend the installation of a "purging" ball valve between the tees at a primary/secondary interface as shown in Figure 13–17. Although such an arrangement will indeed allow for purging of the secondary circuit, the residual pressure drop of the ball valve between the tees is undesirable. It may induce undesirable flow in the secondary circuit once the system is operational. This must be avoided.

Building and Using a Purge Cart

For about $300 worth of materials, an installer can build a device that makes short work out of filling and purging residential and light commercial hydronic systems. It also serves as a tool for filtering system water, as well as adding antifreeze. This device, referred to as a **purge cart,** consists of a *minimum one horsepower* swimming pool pump mounted on a 30-gallon plastic trash barrel as shown in Figure 13–18.

By installing a union on the pump's intake pipe, the assembled purge cart can be easily disassembled as shown in Figure 13–19.

The piping arrangement that works best with a purge cart is shown in Figure 13–20. The layout is very similar to that shown for use with a purge valve. It does, however, allow the

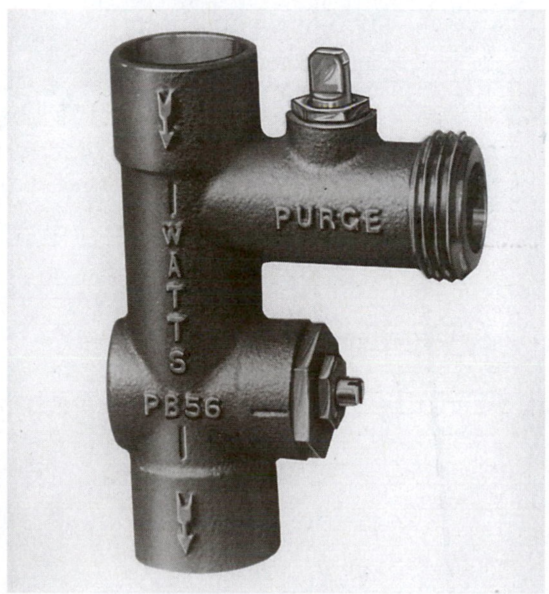

Figure 13–15 Example of a purge valve. *Courtesy of Watts Regulator Co.*

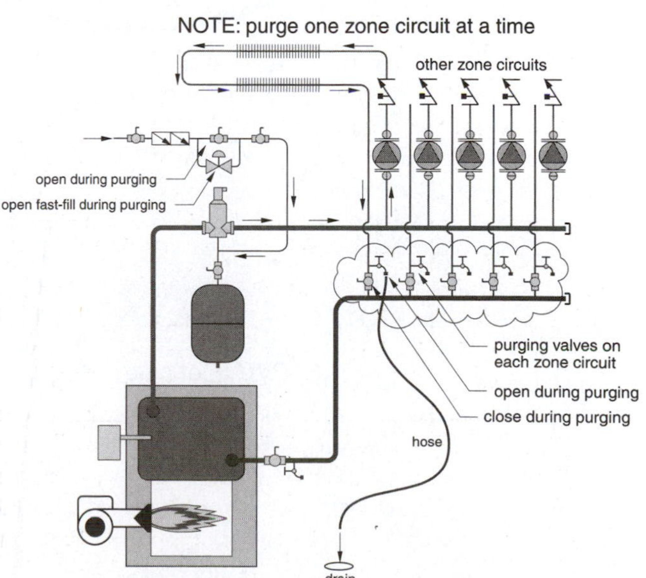

Figure 13–16 Purging valves on each zone circuit.

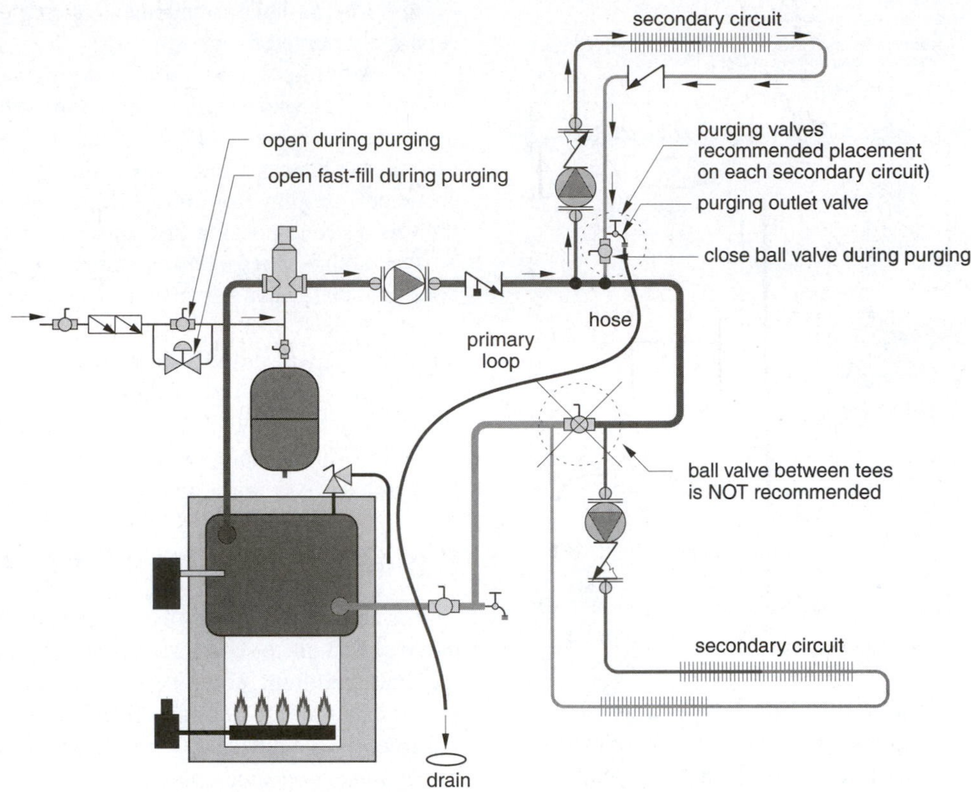

Figure 13–17 Proper use of purging valves on a secondary circuit in a primary secondary system.

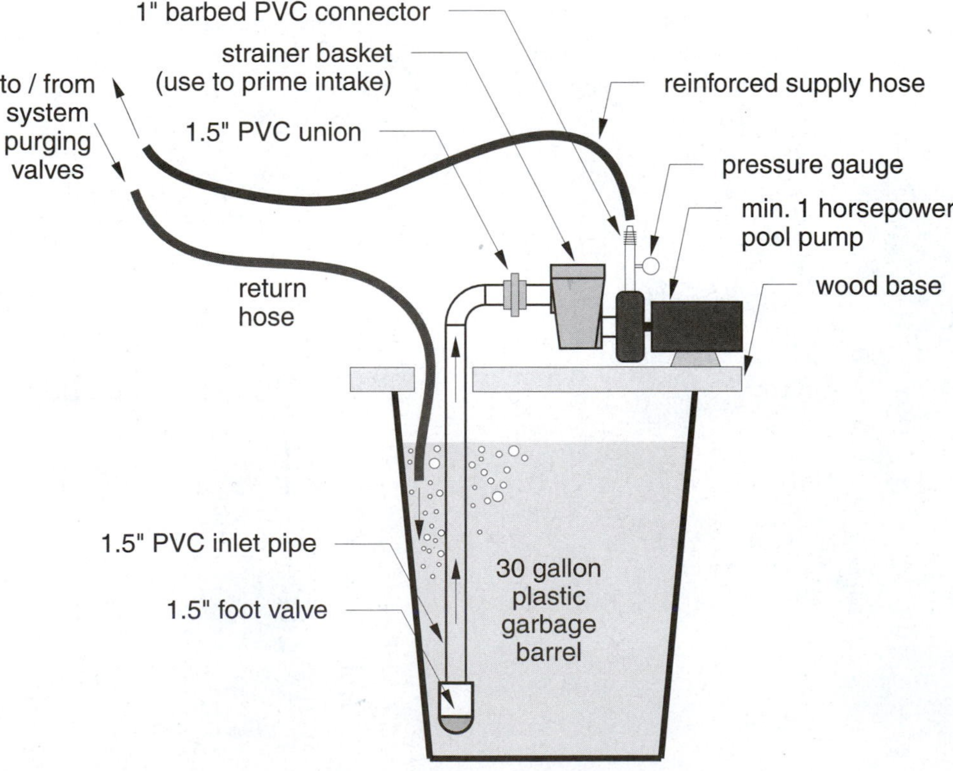

Figure 13–18 The components of a purge cart.

Figure 13–19 Purge cart parts broken down for transportation.

piping system to be purged as two parallel loops—one through the heat source, and the other around the distribution system. The piping connections for the purge cart hoses are usually made using 1-inch ball valves equipped with 1-inch barbed PVC connectors. The reinforced hoses are secured to the barbed connectors using standard hose clamps.

Filling and Purging a System Using a Purge Cart

The following procedure describes how the purge cart is used to fill the system and remove entrapped air bubbles.

Step 1: Before turning on the purge cart pump, make sure all hoses are secured, the suction line of the pool pump is primed, and the inlet and outlet valves on the system are open. The barrel should be almost full of water and at least one parallel branch of the system should be open. Close one of the isolation valves near the circulator so initial purging is through the heat source. Also make sure the return hose is *held* down inside the barrel, or you could get a short but *intense* shower!

Step 2: Turn on the purge cart pump. The fluid level in the barrel will drop rapidly. In 10 to 15 seconds the initially full barrel will be reduced to a minimum pumping level of 4 to 6 inches above the bottom of the barrel. If the system volume is more than about 25 gallons, have a hose or buckets ready to add water to the barrel. Initially the air in the heat source will be pushed ahead of the water and exit through the return hose. After a few seconds of pump operation, a solid water stream will appear in the return hose. *Hold the end of the return hose just below the surface of the fluid in the barrel to prevent air from being entrained down into the fluid.* You may notice some occasional air bubbles exiting the return hose as air pockets are displaced.

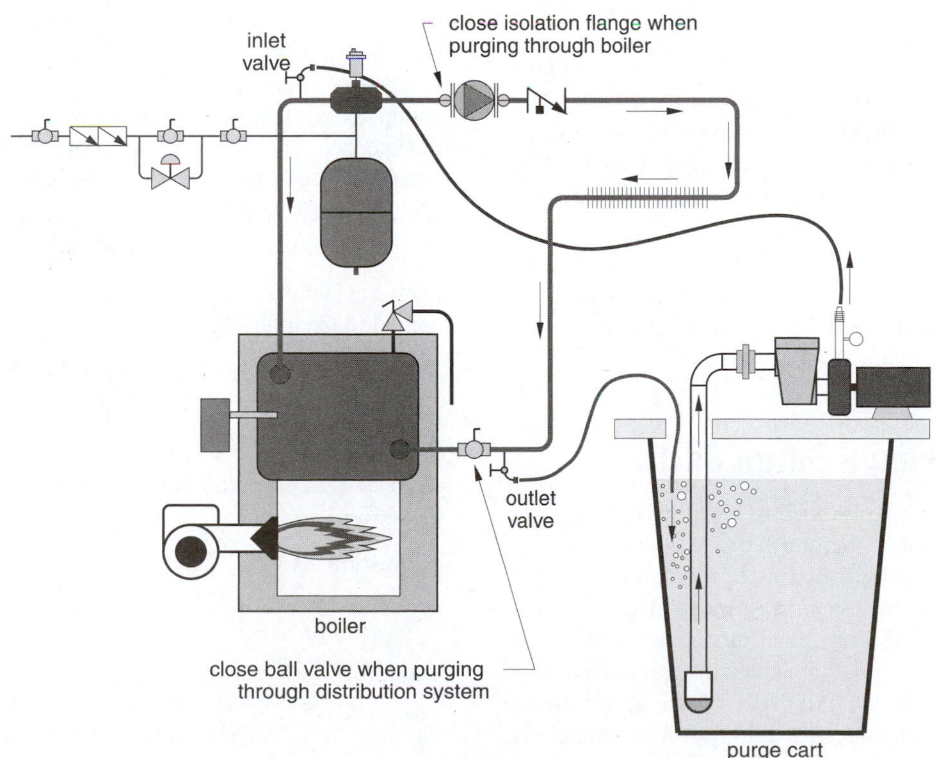

Figure 13–20 Connecting a purge cart to the system.

Step 3: After the heat source has been purged for a minute or so, add water to the barrel until it is almost full. Open the isolation valve on the system's circulator, and close the ball valve near the return connection of the heat source. This allows air and any debris in the distribution system to be purged back to the barrel.

Step 4: Let the purge cart operate for at least 10 minutes while flow is through the distribution system. On multiple zone systems, it is a good idea to open one parallel piping circuit at a time to achieve a high purging flow velocity in each circuit. Be aware that zone valves may make a whining sound due to the high flow rate. If individual zone circulators are used, zones can be isolated by closing the isolation flanges or isolation valves flanking these circulators. After each zone circuit has been purged individually, open all circuits to increase the system flow rate. If the system circulator is located as shown in Figure 13–20, it can also be operated if additional flow is desired. The purge cart, however, usually provides plenty of flow.

Step 5: After the distribution piping has been purged for about 10 minutes, switch the purging flow back through the heat source for a minute or so, then back through the distribution system. Finally, open both parallel paths to allow simultaneous flow through both the heat source and distribution system. Allow the purge cart to operate for another 2 minutes in this mode.

Step 6: The shutdown procedure is simple. Close the *outlet* valve on the system to stop return flow to the barrel. In a few seconds the purge cart pump will have added all the fluid it can to the system as it reaches its maximum (no flow) pressure. This will probably be around 20 psi. Now close the *inlet* valve, and then shut off the purge cart pump. If the system pressure is higher than desired, drain some water from the system. At this point nearly all (nondissolved) air bubbles should be out of the system. The remaining (dissolved) air will be removed by the central deareator over several subsequent heat-up/cooldown cycles. Be sure the feed water valve, or other automatic feed water device is set to add water to the system as the remaining air is collected and vented.

Prefiltering Using a Purge Cart

The purge cart can also be used to prefilter system water to remove some sediments before the water is pumped into the system. This is accomplished by recirculating the fluid through a cartridge filter assembly as depicted in Figure 13–21. A number of different filter cartridges are available with various filtration abilities. It is usually best to start with a coarser filter and work down to smaller particle size limits. Some very fine sediments may still not be completely removed.

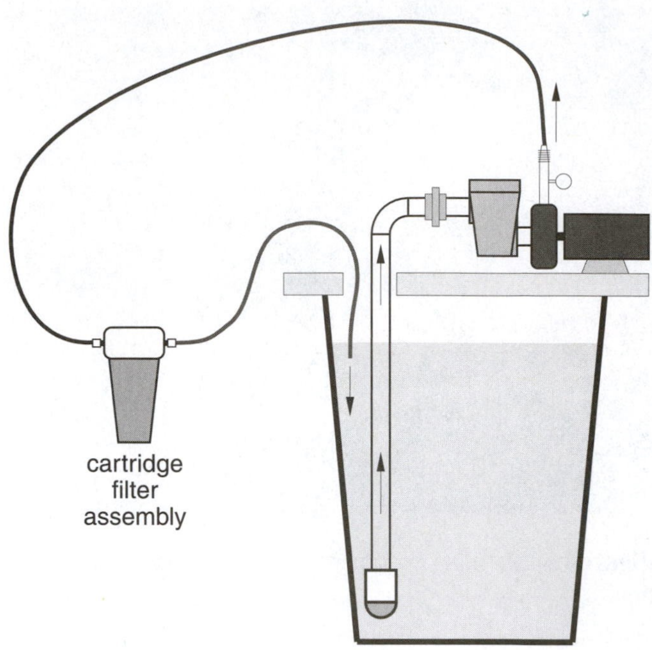

Figure 13–21 Setup for pre-filtering system water using purge cart and cartridge filter.

Adding Antifreeze Using a Purge Cart

The purge cart can also be used to mix and inject antifreeze into a system. Before adding antifreeze, however, it is always advisable to operate the system for a short time with water. This will confirm the presence of any leaks, and the proper functioning of system components. If a leak is found, or a system component must be removed, at least antifreeze will not be lost or spilled. Glycol-based antifreezes tend to leave a slippery film on components and tools that eventually turns sticky. It is much easier to correct any piping problems without this added mess.

Step 1: Calculate the system volume using the methods presented in Chapter 12, Expansion Tanks.

Step 2: If the system uses an automatic feed water system, temporarily turn it off to prevent drained water from being automatically replaced.

Step 3: Calculate the amount of antifreeze required using Equation 13.1:

(Equation 13.1)

$$V_{antifreeze} = (\%)(V_s + V_{b\,min})$$

where:

$V_{antifreeze}$ = volume of antifreeze required (gallons)

$\%$ = desired volume % antifreeze required for a given freeze point (decimal %)*

V_s = calculated volume of the system (gallons)

V_b min = minimum volume of fluid required in the barrel of the purge cart to allow proper pumping (gallons)**

* The Fluid Properties Calculator module in the Hydronics Design Studio software can be used to find the freeze points of several concentrations of both ethylene glycol and propylene glycol.

** The minimum volume of fluid required in the barrel of the purge cart will depend on how the foot valve is located. It should be determined by experimenting with the cart (using water). Reduce the water level in the barrel until the pump begins to draw air into the foot valve. Make sure the return hose is located to minimize turbulence around the foot valve. Using a permanent marker, put a prominent line on the side of the barrel at least 1 inch above this level. This is the minimum operating level of the barrel. Write "Minimum Operating Level" on the outside of the barrel just above this line. Carefully measure the water in the barrel and note its total volume. *Write this minimum operating volume (gallons) on the side of the barrel* as shown in Figure 13–22.

Step 4: *Drain an amount of water from the system equal to the required volume of antifreeze calculated in step 3.* This water should be removed from a low point in the system such as the boiler drain valve.

Step 5: Attach the purge cart hoses to the system as shown in Figure 13–20. Keep both inlet and outlet valves closed.

Step 6: Fill the barrel to its minimum operating level with water, then pour in the required volume of antifreeze. Turn on the purge cart pump. Open the inlet and outlet valves to the system.

Step 7: Follow the same purging procedure described earlier in this section. Allow the mixture to circulate through the system for at least 10 minutes to thoroughly mix the water and antifreeze. Open and close valves as necessary to purge one branch circuit at a time to obtain maximum purging velocity.

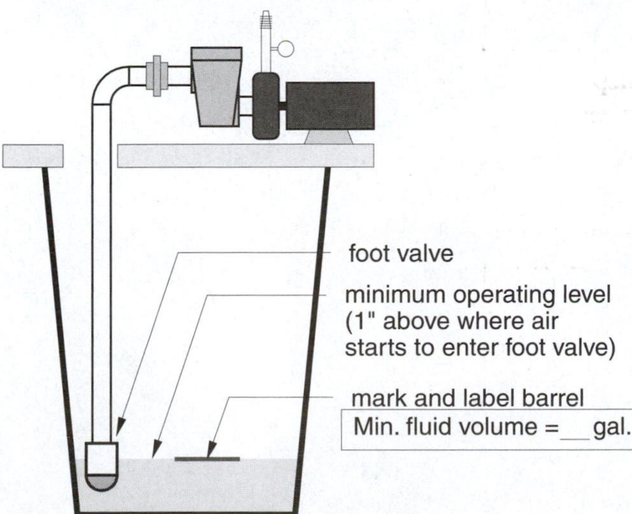

Figure 13–22 Labeling the purge cart barrel for minimum fluid level.

Step 8: When no visible air is being returned from the system to the barrel, close the *outlet* valve on the system. Let the purge cart pressurize the system, then close the inlet valve. If the system pressure is higher than desired, drain some of the fluid back into the barrel through the return hose.

Step 9: The system should now contain the desired concentration of antifreeze and be purged of all but dissolved air. Turn off the purge cart and disconnect the hoses. Reopen the shutoff valve on the make-up water line.

Step 10: The volume of mixed fluid in the barrel should be very close to the minimum operating fluid level from which this procedure began. This mixed fluid should be neatly poured into containers and *labeled.* Be sure to include the exact fluid used, the percentage mixture, and the date the system was filled. This fluid should be saved for the next job *that will use the same fluid and percentage mixture.* It can then be used to make up the minimum operating volume of the barrel. This will reduce the amount of antifreeze used on the next job to just that required by the system. In such a case, Equation 13.1 should now be modified to:

(Equation 13.2)

$$V_{antifreeze} = \frac{(\%)(V_s)}{(1 - \%)}$$

where:

$V_{antifreeze}$ = volume of antifreeze required (gallons)
% = desired volume % antifreeze required for a given freeze point (decimal %)
V_s = calculated volume of the system (gallons)

Other Benefits of a Purge Cart

A purge cart also allows hydronic systems to be filled and operated on construction sites where the building's water supply system may not yet be installed. This is often the case when a structure requires temporary heat during construction. It also provides a means of adding clean water to the system if the water available on the job site is not suitable.

In buildings with a relatively low pressure domestic water system, the purging power of a purge cart will provide faster air removal and less need to manually vent heat emitters or other high points in the piping.

Considering its many uses, it can be a valuable investment for hydronic heating contractors.

13.7 MAKE-UP WATER SYSTEMS

Almost all hydronic systems experience minor water loss over the course of a heating season. These losses often go unnoticed because the water evaporates before visible drops can form. The presence of scale or discoloration near valve stem packings, pump flange gaskets, air vents, or threaded joints usually indicates minor water losses. Eventually such losses allow the system pressure to drop. This may lead to other problems such as cavitation or air entry into the system.

To "make up" for these minor losses it has been customary to install an automatic **make-up water assembly** between the building's cold water plumbing system and the heating system. This assembly usually consists of a pressure-reducing feed water valve, backflow preventer, and at least one isolation valve. Examples of such an assembly are shown on many piping schematics throughout this text, as well as in Figure 13–23. Notice that the feed water (pressure-reducing) valve has been piped to bypass the fast-fill ball valve. This arrangement minimizes the pressure drop through the fast-fill ball valve during purging. Although this is not a crucial detail, it does improve the purging efficiency of the make-up water assembly.

During normal operation, the make-up water assembly automatically feeds small quantities of water to the system to replace the minor water losses, and maintain the pressure set on the feed water valve. Unfortunately a standard make-up water assembly can also be a liability. Since it reacts to a decrease in system pressure, it will continuously feed water if a leak develops in the system. In situations where occupants are not present, or not aware of the leak, considerable water damage to the building can occur. If the leak occurs in a system filled with antifreeze, an automatic make-up water system could eventually add enough water to dilute the solution to the point of freezing. This could lead to further equipment damage.

The hydronics industry continues to debate the merits of leaving an automatic make-up water assembly "active," versus turning it off until the system pressure needs a boost. Arguments can be made both ways. The author's opinion is that an automatic make-up water assembly should be turned off once all portions of the system have been in operation for about 2 weeks. This allows sufficient time for removal of the dissolved air initially in the system. Whenever the system is serviced, the make-up water system should be activated to restore the system to normal pressure. The owner should also be advised of the need to periodically check system pressure and turn on the make-up water system when necessary to maintain pressure.

An alternate means of automatically adding fluid to a hydronic system is shown in Figure 13–24. This **fluid feeder** uses a small diaphragm pump to automatically pump fluid into a system when the pressure drops below the value set on its pressure switch.

The amount of fluid that can be fed into the system is limited by the storage volume of the system feeder. The unit shown in Figure 13–24 holds 6 gallons. It can be placed on the floor or supported by wall brackets. By monitoring the fluid level in the reservoir, it is possible to know how much fluid has been added to the system. If the reservoir is emptied, the fluid feeder provides an electrical contact closure that can activate an alarm, and prevent the feeder pump from further operation. Most fluid feeders can also be used to premix water with glycol or other additives, or to circulate the fluid through a cartridge filter assembly prior to pumping it into the system.

The use of a fluid feeder eliminates the need to pipe a hydronic system to a potable water system. It also eliminates the potential to feed unlimited amounts of water into a sys-

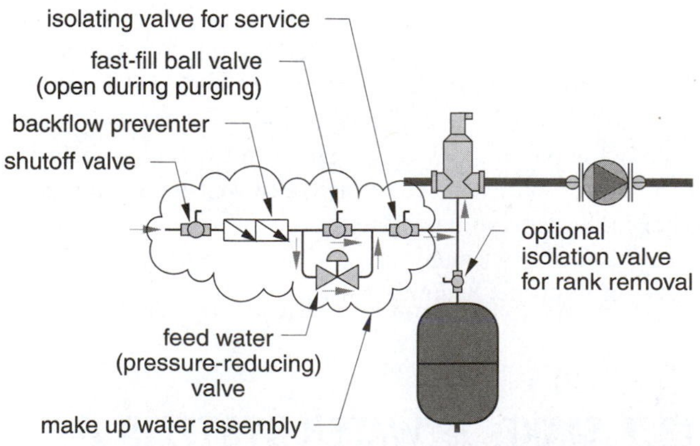

Figure 13–23 The components in a typical make-up water assembly.

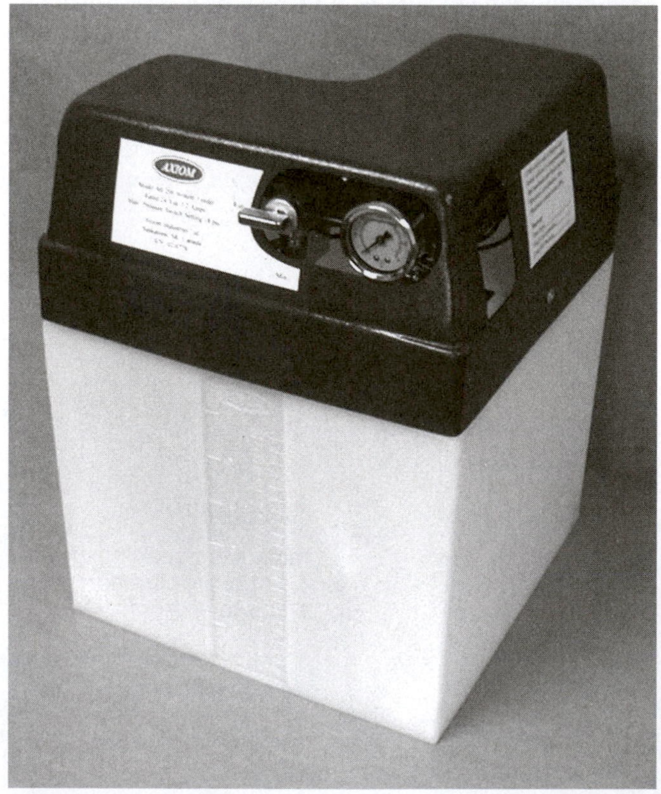

Figure 13–24 Example of an automatic fluid feeder. *Courtesy Axiom Industries Limited.*

tem should a leak develop. Considering these benefits, look for the increasing use of such devices in residential and light commercial hydronic systems.

SUMMARY

This chapter has discussed the various forms in which air can enter and exist in a hydronic heating system. A thorough understanding of how air behaves will allow many of the classic complaints associated with hydronic heating (poor heat output from certain heat emitters and gurgling sounds) to be avoided entirely. This knowledge also gives installers the ability to diagnose and correct previously installed systems exhibiting the classic symptoms of entrapped air. The bottom line is simple: If properly designed and installed, all hydronic systems should automatically rid themselves of almost all internal air after a few days' operation. They should then maintain themselves virtually air-free throughout their service life.

KEY TERMS

air binding
air purger (or air scoop)
automatic air vents
bubble's rise velocity
central deareators
deareating device
dissolved air
entrained air
float-type air vents
fluid feeder
Fluid Properties Calculator
forced-water purging
gravity purging
high point vents
make-up water assembly
manual air vents
microbubble resorber
microbubble
purge cart
purge valve
stationary air pockets
unsaturated state of air solubility

CHAPTER 13 QUESTIONS AND EXERCISES

1. Describe why a waterlogged expansion tank can lead to recurring air problems in a hydronic system.
2. What flow velocity is necessary to entrain air bubbles with the flow in a downward flow direction?
3. Will a bubble rise faster through water or a mixture of water and antifreeze? (*Hint:* Consider the viscosity of each fluid.)
4. Describe why microbubbles form on the inside surfaces of boiler sections as the boiler heats up. How does this affect heat transfer from the boiler wall to the fluid?
5. Describe the difference between an air purger and a microbubble resorber.
6. Describe how a microbubble resorber is able to remove air pockets from remote areas of a system.
7. Why should a 3/4-inch purge valve not be installed in large diameter main piping?
8. A system is filled with water at 60 °F and 15 psi. The water contains the maximum amount of dissolved air it can hold at these conditions. The water is then heated to 160 °F and maintained at 15 psi. How many gallons of air have been given off in the process?
9. What is the best location for a central deaerator in a typical residential or light commercial system? Why?
10. Describe a situation in which air is drawn into system through a vent each time the circulator operates.
11. How would you test an installed diaphragm expansion tank to see if its diaphragm has ruptured?
12. A customer describes air noises in flow passing through the air purger. What might be preventing the air from being separated and vented?
13. Describe a situation where a purge cart would be desirable over the use of forced-water purging through a feed water valve.
14. Does a purge cart remove dissolved air from water? Why?
15. Why is it necessary to hold the end of the return hose to a purge cart below the fluid level in the barrel?
16. A hydronic system consists of a 15-gallon sectional cast-iron boiler, 250 feet of 3/4-inch copper tubing, eight heat emitters holding 0.5 gallons each, and a 2-gallon expansion tank. It is desired to protect the system from freezing to temperatures as low as 0 °F. The minimum usable volume of the purge cart used to fill the system is 4 gallons. How much propylene glycol is needed to produce a 20 percent solution for this system? Use the Fluid Properties Calculator module in the Hydronics Design Studio software to determine the freezing point of this mixture.
17. What is the advantage of piping the feed water (pressure-reducing) valve so that it bypasses the fast-fill ball valve in a make-up water assembly?
18. What are two advantages of using an automatic fluid feeder instead of a make-up water assembly?

AUXILIARY LOADS AND SPECIALIZED APPLICATIONS 14

OBJECTIVES

After studying this chapter, you should be able to:

- Describe how heat exchangers are used in hydronic systems
- Discuss methods of domestic water heating using a hydronic system
- Calculate the energy requirement associated with domestic water heating
- Discuss the advantages of indirect water heaters
- Understand the output ratings used for indirect water heaters
- Modify a tankless water heating system for use with a storage tank
- Describe methods for intermittent garage heating
- Explain how a hydronic system can heat a spa or swimming pool
- Describe installation details for snow- and icemelting (SIM) systems
- Discuss control options available for SIM systems
- Estimate the load associated with a SIM system
- Identify appropriate applications for buffer tanks
- Size a buffer tank for a given application
- Explain the advantages of minitube distribution systems

4.1 INTRODUCTION

Although space heating is the primary load served by most residential and light commercial hydronic systems, they can also provide heat to other **auxiliary loads.** Examples include domestic water heating, intermittent garage heating, pool or spa heating, and snowmelting. Through proper design, a single boiler, or multiple boiler system, can serve as the universal heat source for the multiload system. This approach reduces installation costs relative to using separate heat sources for each load. It also reduces fuel consumption because as the duty cycle of the heat source increases, so does its seasonal efficiency.

The feasibility of supplying both space heating and nonspace heating loads depends on their magnitude, and when they occur. The best results are obtained when nonspace heating loads do not occur at the same time as peak space heating loads. A good example is heating a pool during mild weather when space heating loads are small or nonexistent. If a nonspace heating load is small in comparison to the space heating load, it might be met using the surplus capacity of the heat source, especially during milder fall and spring weather.

Another possibility is to prioritize loads. This approach, described in Chapter 9, Control Strategies, Components, and Systems, temporarily suspends heat input to predetermined low **priority loads** so full boiler capacity can be directed to a high priority load. Once the high priority load is satisfied, heat output is shifted back to the lower priority loads. Such intelligent management of boiler capacity minimizes the total boiler capacity needed.

14.2 HEAT EXCHANGERS

Many auxiliary loads require a transfer of heat from the hot water released by a boiler to some other liquid operating at a lower temperature. Such situations require the use of a **heat exchanger** that prevents the two liquids from mixing, yet allows heat to easily flow from the higher to the lower temperature liquid.

There are several types of heat exchangers used in residential and light commercial hydronic systems. This section discusses the most commonly used types. Later sections in the chapter show how they are integrated into a system.

Heat exchanger selection involves both thermal and hydraulic performance issues. Thermal issues relate to the ability of the heat exchanger to transfer heat from one liquid to another at a specific rate, given the entering temperatures and flow rates. Hydraulic issues relate to the head loss and pressure drop associated with moving the liquids through the heat exchanger at the required flow rates. Failure to address either issue can result in improper selection and poor performance.

Selection must also consider chemical compatibility between the materials in the heat exchanger and the liquids passing through it. For example, some heat exchangers are partially constructed of steel. If the so-called "dead" water within a closed hydronic system is pumped through such a heat exchanger, corrosion due to oxidation is seldom a problem. However, if fresh potable water were pumped through a plain steel heat exchanger, the dissolved oxygen in the water could quickly corrode the steel. Not only would this damage the heat exchanger, it would also contaminate the fresh water. To avoid this, heat exchangers designed to heat (or cool) fresh water must be made of materials such as copper, cupronickel, stainless steel, or have protective coatings that are highly resistant to oxidation. Pool or spa water with high concentrations of chlorine can also be very corrosive to certain materials. Heat exchangers built of 316 stainless steel as well as specific welding/brazing alloys are generally required for such liquids.

One type of heat exchanger often used in modern hydronic systems is called a **flat plate heat exchanger.** An example of such a device is shown in Figure 14–1.

Plate heat exchangers provide a large surface area between the two liquids. This area is distributed among many individual stainless steel plates that fit together like a stack of cards. The overall size of a plate heat exchanger is quite small relative to other heat exchanger designs of similar heat transfer performance. The two liquids exchange heat flow through narrow spaces between the formed plates. If these spaces were numbered, one liquid would flow through spaces 1, 3, 5, and so on. The other would pass through spaces 2, 4, 6, and so on.

The two fluids should flow in opposite directions through the heat exchanger. This **counterflow arrangement** maximizes heat transfer between the two liquids for any given operating conditions (Figure 14–2). When heat exchangers are installed it is vitally important to pipe them for counterflow. Be sure to check manufacturers piping drawings as differences do exist.

Small plate heat exchangers are often further categorized as **brazed plate heat exchangers.** The name stems from the brazing process used to join the perimeter of the plates together. Once brazed, the internal spaces between the plates are not accessible.

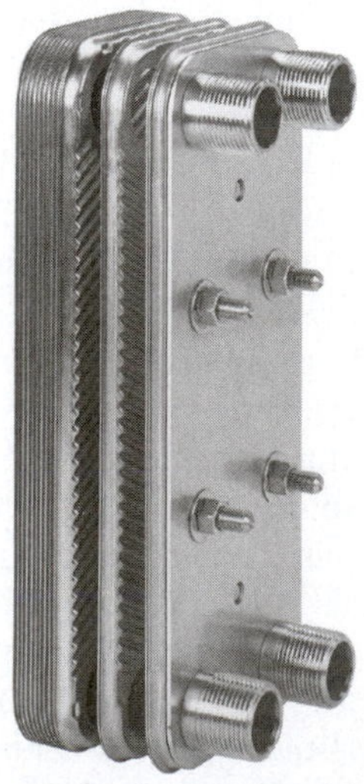

Figure 14–1a Construction of a flat plate heat exchanger. *Courtesy of FlatPlate, Inc.*

Figure 14–1b Installation of a small brazed plate heat exchanger. Note piping is supported adjacent to heat exchanger

Larger plate heat exchangers often hold the plates together using threaded rods and nuts to create mechanical compression. Edge gaskets form the seal between the plates when the nuts are tightened. This type of plate heat exchanger can be disassembled for cleaning if necessary. In some

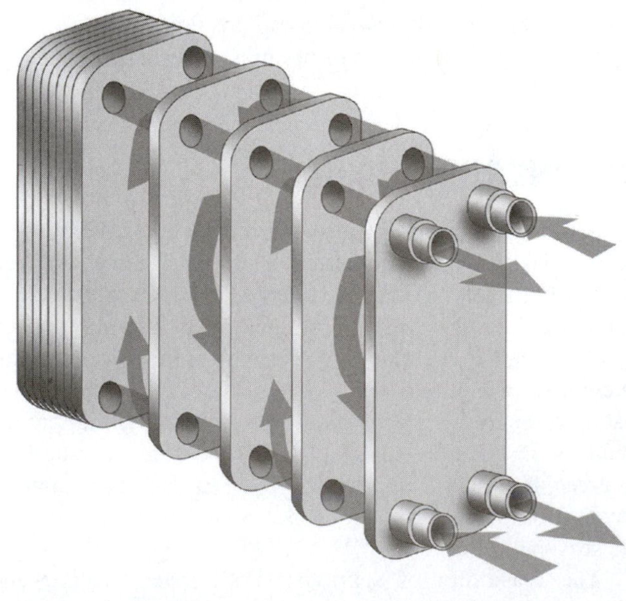

Figure 14–2 Counterflow in a flat plate heat exchanger. Note flow direction in alternating spaces between plates. *Courtesy FlatPlate, Inc.*

instances, additional plates can also be added to increase the heat transfer surface area.

Given the relatively narrow passages inside a plate heat exchanger, it is good practice to install a strainer with a 20-mesh screen upstream of both fluid inlets. These strainers can prevent debris such as balls of solder or metal chips from entering and possibly lodging within the heat exchanger. In older systems, they can prevent precipitants or slug from adhering to the internal surfaces of the heat exchanger. If allowed to occur, such **fouling** can quickly reduce the heat transfer ability of any heat exchanger. Deposits can also cause pitting corrosion. Ball valves should be installed so that the strainers can be isolated and opened for cleaning.

Most brazed plate heat exchangers are sized using charts of thermal performance provided by the manufacturer for typical applications. For more specialized applications, manufacturers generally use their own software to select the appropriate unit based on the operating requirements submitted by the system designer.

Another type of heat exchanger used primarily for hydronic pool heating is shown in Figure 14–3. This design is called a **shell and tube heat exchanger.** The internal construction of such a unit is illustrated in Figure 14–4.

Depending on the design and size of the heat exchanger, boiler water may pass through the tube bundle or through the shell and around the outer surfaces of the tube bundle. For the unit shown in Figure 14–3, boiler water is routed through the tube bundle while pool water is routed through the shell. This type of heat exchanger is made of welded 316 stainless steel to resist the corrosive effects of highly chlorinated pool water.

Figure 14–3 Example of a shell and tube heat exchanger for pool heating applications. *Courtesy of Weil-McLain Corp.*

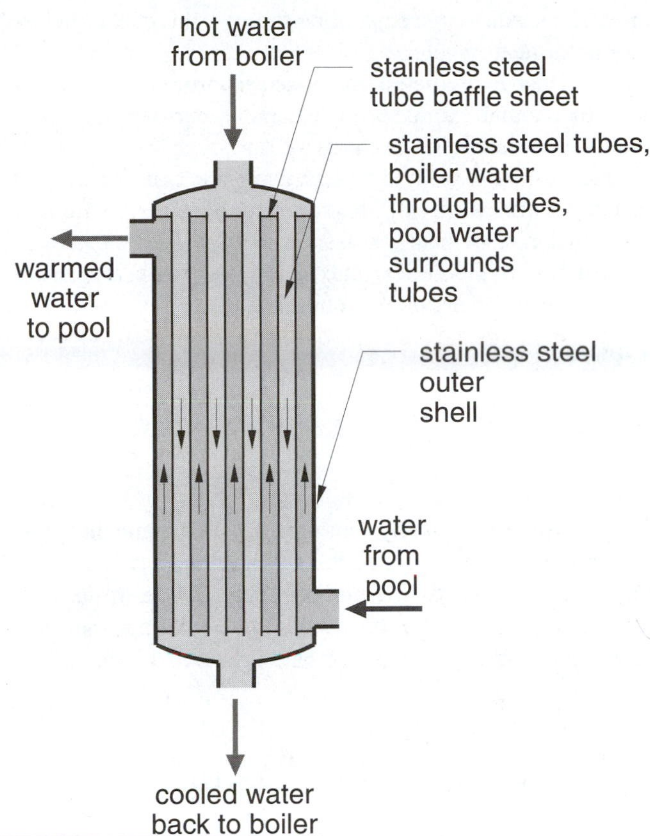

Figure 14–4 Flow paths through tube and shell heat exchanger when used for pool heating.

Another type of heat exchanger commonly found in hydronic systems is called an **internal heat exchanger.** It usually consists of a coil of tubing suspended within a tank. A circulator is used to move water through the coil, and thus heat is transferred to the tube wall by **forced convection.** However, heat released from the outer surface of the heat exchanger is by **natural convection.** Since natural convection is less effective than forced convection in transferring heat, internal heat exchangers require significantly more surface area and/or higher temperature differentials between the entering liquids to deliver thermal performance comparable to "external" heat exchangers. Internal heat exchangers are built into the indirect water heaters discussed in the next section.

Thermal Performance of Heat Exchangers

The rate of heat transfer through a heat exchanger depends on the inlet temperature and flow rate of both fluid streams, as well as the surface area separating them. The greater the temperature difference between the two entering fluids, and the greater their flow rates, the faster a given heat exchanger will move heat from one fluid to another. The larger the surface area separating the two fluids, the greater the rate of heat transfer.

Performance prediction methods vary from one manufacturer to another. Some provide tabular performance data. Others provide graphs or charts.

One common method is based on the concept of heat exchanger **effectiveness.** This term is defined as the ratio of the actual rate of heat transfer across the heat exchanger, divided by the theoretical maximum possible heat transfer for a given set of operating conditions.

(Equation 14.1)

$$\text{Effectiveness} = E = \frac{\text{Actual rate of heat transfer}}{\text{Theoretical maximum possible rate of heat transfer}}$$

Effectiveness is usually determined from a graph that plots it against a parameter called the **capacitance rate ratio.** To determine the capacitance rate ratio, one first determines the capacitance rate (CR) of each fluid entering the heat exchanger using Equation 14.2a and Equation 14.2b.

(Equation 14.2a)

$$CR_{coil} = (f_{coil})(8.01)(c_{coil\ fluid})(D_{coil\ fluid})$$

(Equation 14.2b)

$$CR_{shell} = (f_{shell})(8.01)(c_{shell\ fluid})(D_{shell\ fluid})$$

where:

f_{coil} and f_{shell} = flow rates through the coil and shell respectively (gpm)

$C_{coil\ fluid}$ and $C_{shell\ fluid}$ = specific heats of the coil and shell fluids (Btu/lb/°F)

$D_{coil\ fluid}$ and $D_{shell\ fluid}$ = densities of the coil and shell fluids (lb/ft³)

The capacitance rate ratio is the *smaller* capacitance rate divided by the *larger* capacitance rate:

(Equation 14.3a)

$$\text{If } CR_{coil} \text{ is smaller than } CR_{shell} \text{ then: } CRR = \frac{CR_{coil}}{CR_{shell}}$$

(Equation 14.3b)

$$\text{If } CR_{shell} \text{ is smaller than } CR_{coil} \text{ then: } CRR = \frac{CR_{shell}}{CR_{coil}}$$

Once the capacitance rate ratio is calculated, the effectiveness of the heat exchanger can be read from a graph supplied by the manufacturer. An example of such a graph is shown in Figure 14–5.

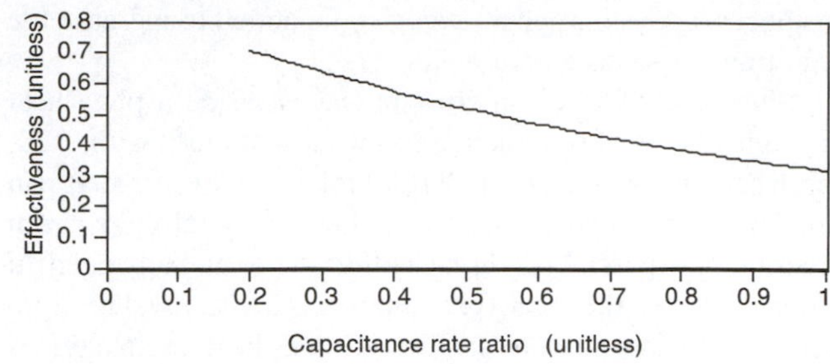

Figure 14–5 Example of a graph of effectiveness versus capacitance rate ratio.

The actual rate of heat transfer across the heat exchanger can now be calculated using Equation 14.4.

(Equation 14.4)

$$Q = E(CR_{min})(T_{hot\ in} - T_{cold\ in})$$

where:

Q = rate of heat transfer across the heat exchanger (Btu/hr)

E = effectiveness from manufacturer's graph (unitless)

CR_{min} = *smaller* of either CR_{shell} or CR_{coil} (Btu/hr/°F)

$T_{hot\ in}$ = *inlet* temperature of the fluid from the heat source (°F)

$T_{cold\ in}$ = *inlet* temperature of the fluid being heated (°F)

Example 14.1: Boiler water enters the shell side of a heat exchanger at 4 gpm and 180 °F. A 50 percent mixture of ethylene glycol and water enters the coil side of the heat exchanger at 3 gpm and 120 °F. Assume that the effectiveness of the heat exchanger is represented by the graph of Figure 14–5. Determine the actual rate of heat transfer across the heat exchanger.

Solution: The Fluid Properties Calculator module in the Hydronics Design Studio software is first used to determine the density and specific heat of both fluids at the indicated entering temperatures:

For 50 percent ethylene glycol:

$C_{coil\ fluid}$ = 0.832 Btu/lb/°F
$D_{coil\ fluid}$ = 65.8 lb/ft³

For water:

$C_{shell\ fluid}$ = 1.002 Btu/lb/°F
$D_{shell\ fluid}$ = 60.5 lb/ft³

The capacitance rate of each fluid stream is now calculated using Equation 14.2a and Equation 14.2b:

CR_{coil} = (3)(8.01)(0.832)(65.8) = 1,316 Btu/hr/°F

CR_{shell} = (4)(8.01)(1.002)(60.5) = 1,942 Btu/hr/°F

Since the capacitance rate of the coil fluid is the smaller value, the capacitance rate ratio is:

$$CRR = \frac{CR_{coil}}{CR_{shell}} = \frac{1316}{1942} = 0.68$$

Entering the graph in Figure 14–5 at 0.68 on the horizontal axis, read up to the curve, then over to the right to get an effectiveness value of approximately 0.44.

The heat transfer across the heat exchanger is now determined using Equation 14.4:

$$Q = E(CR_{min})(T_{hot\ in} - T_{cold\ in})$$

$$= 0.44(1{,}316)(180 - 120) = 37{,}740 \text{ Btu/hr}$$

14.2 DOMESTIC WATER HEATING

Almost all residential and many light commercial buildings require **domestic hot water (DHW).** It is by far the most common auxiliary load to be combined with hydronic space heating. This section presents a simple method for estimating daily and peak hourly DHW heating loads. It also discusses different hardware options for heating domestic water using a hydronic system.

Daily DHW Load Estimates

The *daily* energy requirement for DHW production can be estimated using Equation 14.5:

(Equation 14.5)

$$E_{daily} = (G)(8.33)(T_{hot} - T_{cold})$$

where:

E_{daily} = daily energy required for DHW production (Btu/day)

G = volume of hot water required per day (gallons)

T_{hot} = hot water temperature supplied to the fixtures (°F)

T_{cold} = cold water temperature supplied to the water heater (°F)

The daily usage of domestic hot water is of course heavily dependent on occupancy, living habits, type of water fixtures used, water pressure, and the time of year. The following estimates are suggested as a guideline:

- House: 10 to 20 gallons per day per person
- Office building: 2 gallons per day per person
- Small motel: 35 gallons per day per unit
- Restaurant: 2.4 gallons per average number of meals per day

Example 14.2: Determine the energy used for domestic water heating for a family of four with average usage habits. The cold water temperature averages 55 °F. The supply temperature is set for 125 °F.

Solution: The daily domestic water heating load is estimated using Equation 14.5:

$$E_{daily} = (4 \times 15)(8.33)(125 - 55) = 34{,}990 \text{ Btu/day}$$

The estimated *annual* usage would be the daily usage times *355* days of occupancy per year (assuming the family is away 10 days per year).

$$\left(34{,}990\frac{\text{Btu}}{\text{day}}\right)\left(355\frac{\text{occupied days}}{\text{year}}\right) = 12{,}421{,}000\frac{\text{Btu}}{\text{year}}$$

$$= 12.4\frac{\text{MMBtu}}{\text{year}}$$

Discussion: The cost of providing this energy can be estimated by multiplying the annual energy requirement by the cost of delivered energy in $/MMBtu (see Chapter 2 for determining delivered energy cost on a $/MMBtu basis). For example, at $0.10/kwhr and 100% efficiency, electrical energy has a delivered cost of $29.30/MMBtu. The annual energy cost of providing the domestic hot water in this example would be:

$$\left(12.4\frac{\text{MMBtu}}{\text{year}}\right)\left(29.30\frac{\$}{\text{MMBtu}}\right) = \$363/\text{year}$$

DHW Usage Profiles

The *rate* at which domestic hot water is used varies considerably with the type of occupancy. A typical *residential* domestic hot water **usage profile** is shown in Figure 14–6.

Notice the two distinct periods of high demand: One during the wake-up period, the other during the early evening. The greatest period of usage in this profile is from 7 P.M. to 8 P.M. However, this still represents only 11.6 percent of the total daily DHW demand. For a family using 60 gallons of

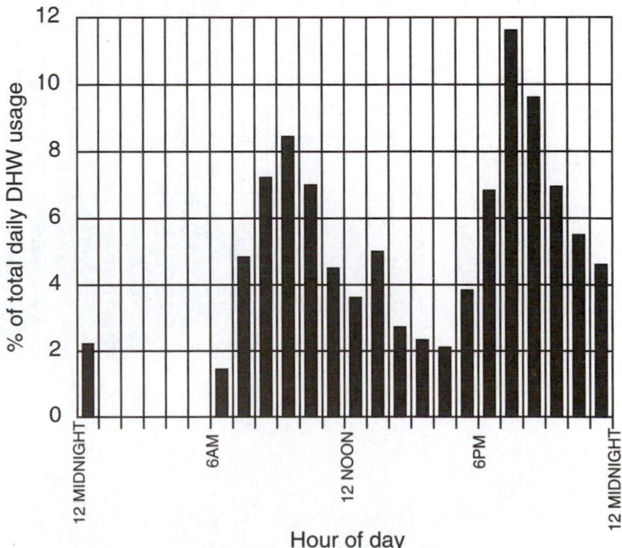

Figure 14–6 Typical usage profile for residential domestic hot water.

domestic hot water per day, this would represent a peak usage of just under 7 gallons per hour. The average *rate* at which energy is required for DHW production during this peak hour (assuming cold water and hot water temperatures of 55 °F and 125 °F, respectively) can be found using Equation 14.5:

$$E_{daily} = [(0.116)(60)](8.33)(125 - 55)$$
$$= 4{,}058 \text{ @ } 4{,}100 \text{ Btu/hr}$$

The first factor in this equation, (0.116), represents the peak demand of 11.6 percent of the total daily water volume. If the heat source could supply an extra 4100 Btu/hr *above the space heating load,* this peak DHW requirement could be met, even on continuous basis if required.

The occurrence of two peak demand periods, one during the wake-up period, the other following the evening meal period, is common in residential buildings. Some households tend to be "high morning users" while others are "high evening users." Figure 14–7 compares these two categories. Notice peak demand for the high evening user is only about 15.5 percent of the total daily demand.

Tankless Water Heaters

The traditional method of providing DHW using a boiler that also supplied space heating is to insert a copper heat exchanger called a **tankless coil** into the upper portion of the boiler. Examples of tankless coils are shown in Figure 14–8. When inserted into a special chamber in the boiler block, these coils are totally surrounded by hot boiler water. A gasketed bulkhead provides a watertight seal at the side of the boiler block. Cold water is drawn through the coil whenever there is a demand for hot water in the building. The entering cold water makes a single pass through the tankless coil on its way to the hot water fixtures.

Since hot water might be required at any time of day, the boiler water surrounding a tankless coil must be kept hot at all times. A triple aquastat control as described in Chapter 9, Control Strategies, Components, and Systems, is used to fire the boiler whenever its water temperature drops below a preset lower limit. During warm weather, when there is no demand for space heating, the boiler water must be kept at a relatively high temperature solely for the purpose of supplying DHW. The standby heat losses associated with this requirement substantially lower fuel efficiency. Values as low as 35 percent have been observed during warm weather operation of tankless coil systems.

Tankless coils can also cause noticeable variations in the temperature of hot water delivered to fixtures, especially if the fixture is operated for several minutes. This is caused by fluctuations in boiler temperature between firing cycles. Since the amount of water contained in the coil is very small, it has very little thermal mass to help absorb energy without a significant temperature change. This problem can be partially compensated for by installing a thermostatic mixing valve on the outlet of the coil.

The inherent inefficiency of maintaining hot water in the boiler at all times, along with the ever-increasing cost of fuel, is making the traditional use of tankless coil water heaters economically obsolete.

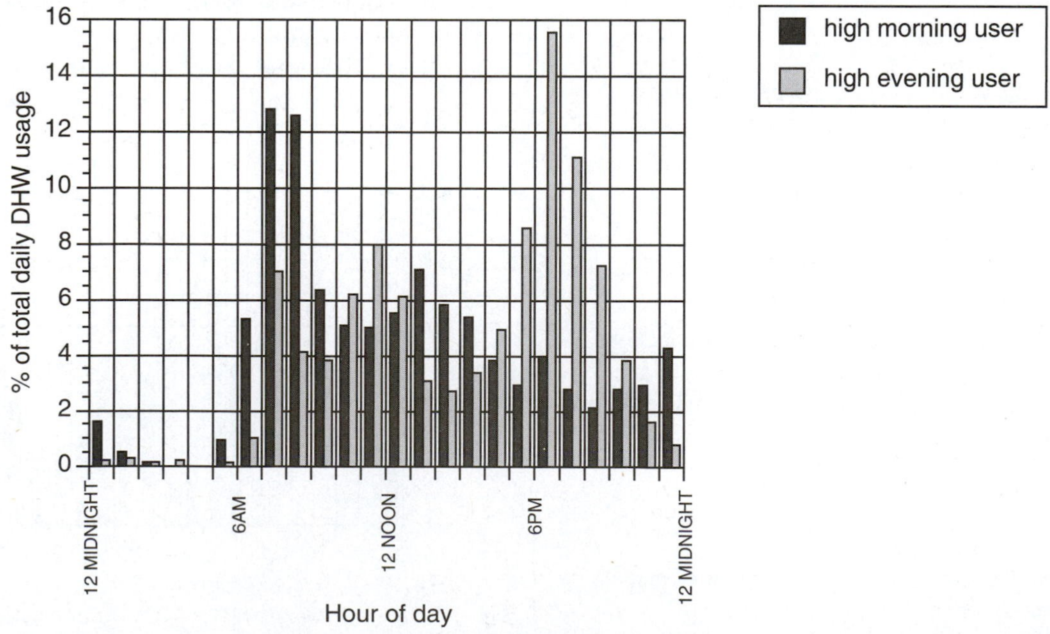

Figure 14–7 Comparison of hourly usage of domestic hot water for a high morning user versus a high evening user. *Source: ASHRAE data.*

(a)

(b)

Figure 14–8 (a) Examples of tankless coil water heaters. *Courtesy of Amtrol, Inc.* (b) A tankless coil being inserted into a boiler. *Courtesy of Weil-McLain Corp.*

Combining a Tankless Coil with a Storage Tank

Some of the disadvantages of traditional tankless coil water heaters can be offset by incorporating a storage tank into the system. A schematic of this concept is shown in Figure 14–9. When used to retrofit a standard tankless coil installation, this modification will significantly improve annual fuel efficiency, as well as reduce temperature fluctuations at fixtures.

For this arrangement to function properly, the triple aquastat control on the boiler must be internally modified to function strictly as a high limit control, or be replaced with a conventional high limit control. The feasibility of converting a triple aquastat control will depend on its make and model. After this modification is made, the boiler should fire only upon a call for heating from the tank's thermostat, or another load in the system. It no longer needs to maintain a constant minimum operating temperature.

Upon a call for domestic water heating from the tank's thermostat, a small circulator moves cooler water from near the bottom of the storage tank through the tankless coil. The heated water from the coil should be returned to a *separate connection* near the top of the storage tank. This ensures it will mix with water in the tank, and thus provide a stable supply temperature to the fixtures. The circulator continues to run until the tank thermostat is satisfied. The larger the volume of the storage tank, and the wider the differential of the tank's setpoint control, the longer the boiler's operating cycle. Longer firing cycles mean higher efficiency. When the storage tank reaches its setpoint temperature, the boiler and tank circulator are shut off. As domestic hot water is drawn from the storage tank, its temperature drops until the tank's setpoint control closes to begin the cycle again.

As an option, the tank circulator can be configured to run for several minutes after the boiler shuts off. This can be done using a delay-on-break time delay relay as discussed in Chapter 9. The temporarily sustained circulation purges residual heat from the boiler, moving it into the storage tank before it is lost up the chimney or through the boiler jacket.

Since the tank circulator moves fresh (oxygenated) water, it must be suitable for use in an open system. Circulators made of stainless steel, bronze, or high temperature resistant plastics are available for such applications. Cast-iron circulators should never be used for this duty.

The tank's thermostat can be a conventional electromechanical control or an electronic control. The sensor bulb or thermistor should be mounted either in direct contact with the inner pressure vessel of an insulated storage tank or, preferably, into a sensor well inserted into the tank water.

If the storage tank used is a conventional electric water heater, the heating elements could be wired and used as a backup DHW heat source in the event the boiler is down for service.

Indirect Water Heaters

Indirect water heaters appeared in North America in the late 1970s as many heating professionals began recognizing the limitations of tankless water heating in the context of modern hydronics technology.

An indirect water heater is an insulated storage tank that does not produce its own heat, as does a typical **direct-fired water heater.** Instead, it uses an internal heat exchanger to transfer heat from a flowing stream of hot boiler water into domestic water. In most (but not all) indirect water heaters,

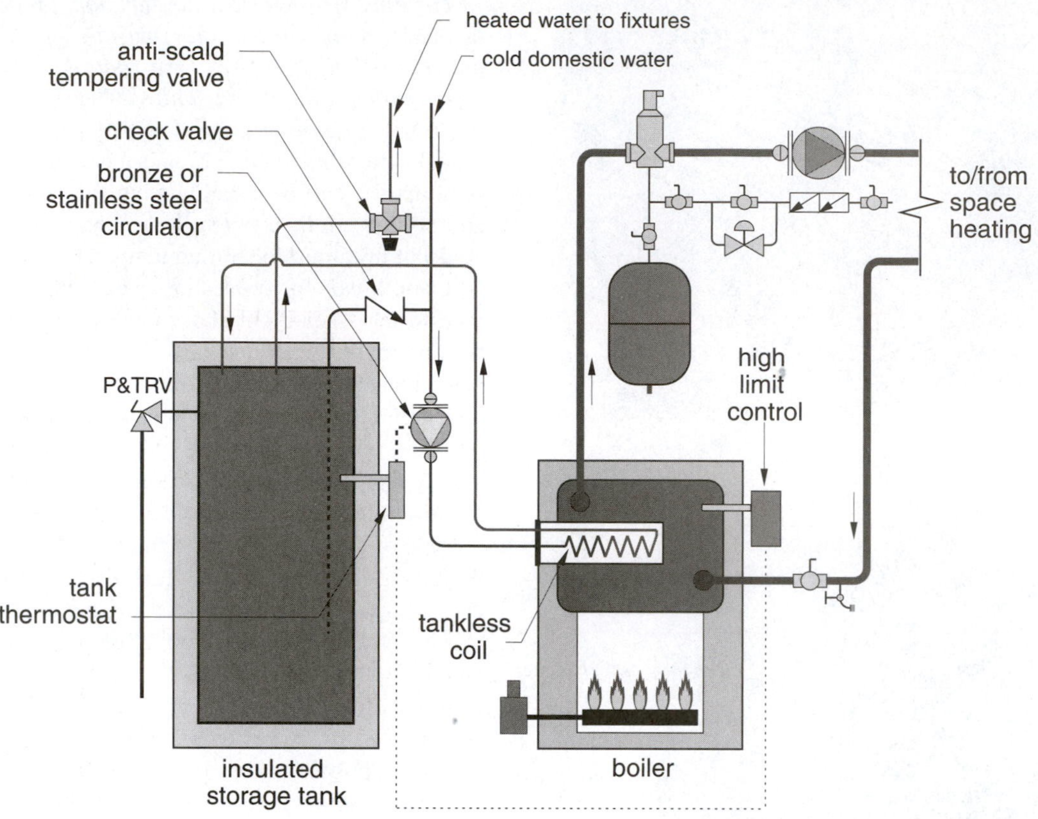

Figure 14-9 Adding a storage tank to a conventional tankless water heating system.

hot boiler water flows through the heat exchanger, while the potable water resides within the tank. Since they are in contact with potable water, the tank and heat exchanger are usually constructed of copper, stainless steel, or lined with a heat-fused porcelain-like glazing.

One example of such a tank is shown in Figure 14–10. The coiled internal heat exchanger is located near the bottom of the tank where it is immersed in the coolest water. This increases the rate of heat transfer from the coil to the potable water. The tank shell and heat exchanger are glazed. A 2-inch or thicker layer of insulation between the tank and external shell keeps heat loss to a minimum.

Another style of indirect water heater is shown in Figure 14–11. In this tank-within-a-tank design, boiler water passes between the steel shell and the inner stainless steel tank. This approach creates a large surface area between the two fluids allowing for rapid heat transfer when suitable rates of heat input are available. It also reduces the head loss associated with circulating boiler water through the tank's heat exchanger.

Another type of indirect water heater is shown in Figure 14–12. This design circulates boiler water through the insulated steel shell, while potable water passes through suspended copper heat exchanger coils. The water in the tank is maintained at a suitable temperature at all times to

Figure 14–10 Example of an indirect water heater with internal coiled heat exchanger. *Courtesy Utica Boiler Div. of ECR International.*

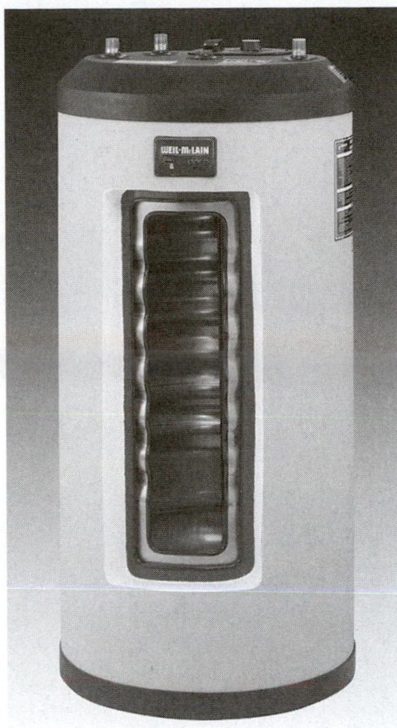

Figure 14–11 Example of an indirect water heater with tank-in-tank design. *Courtesy Weil-McLain Corp.*

provide the thermal storage necessary to meet high demands. The large surface area of the internal coils allows for rapid heat transfer when the tank is connected to a high capacity heat source.

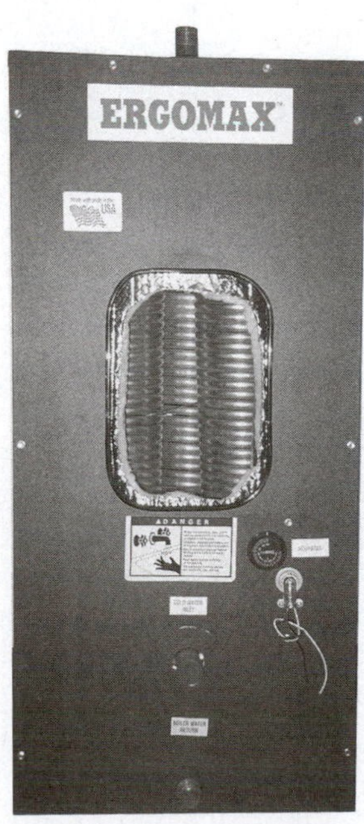

Figure 14–12 Example of an indirect water heater with boiler water in shell and domestic water in suspended coils. *Courtesy Group Thermo, Inc.*

Not all indirect water heaters are vertically oriented cylinders. Some horizontally oriented tanks serve as a platform onto which the boiler is mounted as shown in Figure 14–13. This approach reduces the "footprint" of a typical boiler/indirect water heater installation.

Although there are different ways to pipe an indirect water heater into a system, the goal is always to *get heat from the boiler(s) into the water heater with minimal losses in between.* Whenever possible, indirect water heaters should be located close to boilers to minimize heat loss through interconnecting piping. The piping between the boiler and tank should also be insulated. Remember that hot water will likely be flowing through this piping on the hottest day of the year. The building's cooling system does not need the added load created by heat loss from long lengths of hot uninsulated piping.

One simple and common piping option for an indirect water heater in a multizone system is shown in Figure 14–14. Several other piping arrangements are shown in Figure 11–47, Figure 11–48, Figure 11–53, Figure 11–54, Figure 11–56, as well as other piping schematics throughout this text.

Notice the anti-scald tempering valve shown on the outlet of the tank in Figure 14–14. This valve allows the water in the tank to be heated to a higher temperature than required at the fixtures. When this valve is used in combination with a tank thermostat having a wide differential, the length of the boiler firing cycle can be increased, and the number of firing cycles decreased. This valve is also highly recommended (and pos-

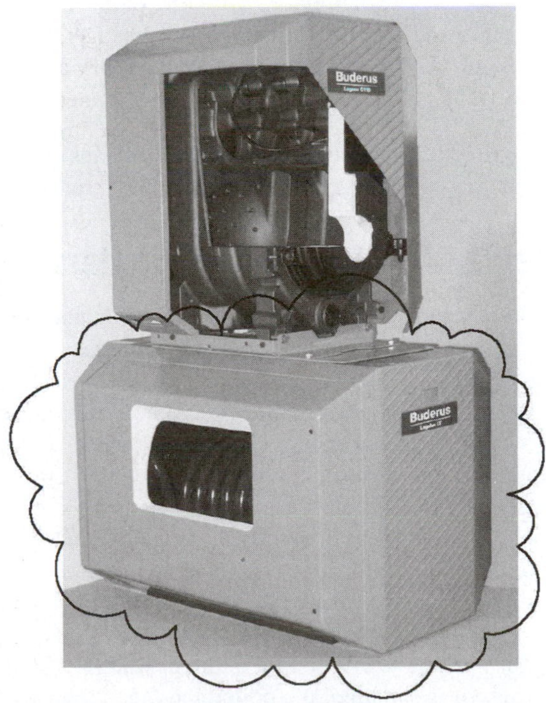

Figure 14–13 Example of an indirect water heater serving as mounting base for boiler. *Courtesy Buderus Hydronic Systems, Inc.*

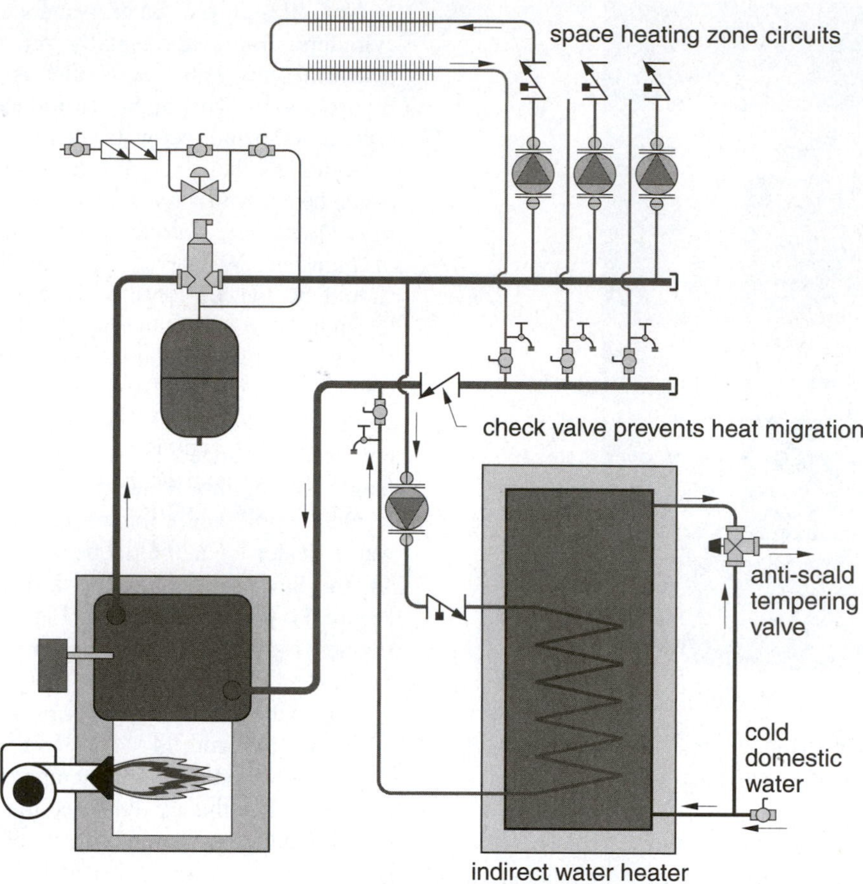

Figure 14–14 Typical piping of an indirect water heater in a multiload hydronic system. Note water heater zone piping is closest to boiler to minimize heat loss.

sibly required by local plumbing code) as a means of preventing the delivery of dangerously hot water to fixtures.

Benefits of Indirect Water Heaters

Indirect water heaters offer several benefits not provided by other means of water heating. These include:

- *The ability to hold 30+ gallons of heated water in reserve to meet high flow rate loads.* Just as a heated concrete slab can deliver high rates of heat output when necessary, an indirect water heater can supply high flow/short duration DHW demands. Because heat is stored in the tank, the rate of heat flow to such demands can be much faster than the heat source can generate heat.
- *The stored water provides the thermal mass necessary to allow the boiler to operate for longer cycles and thus at higher efficiencies.* When the indirect water heater's thermostat is satisfied, the boiler and tank circulator shut off. They can remain off, in some cases for several hours, while the hot water demands of the building are provided directly from the hot water stored in the tank. Standby heat loss from the boiler is substantially reduced since it does not have to remain at a constant elevated temperature, as is the case with a tankless coil.
- *The low pressure drop and relatively large piping connections on most indirect tanks offer far less flow resistance than does a tankless coil.* This allows high flow rates to fixtures without an accompanying drop in water pressure.
- *In sustained flow situations, a properly selected indirect tank set up as a priority load can transfer the full Btu/hr output of the system's boiler(s) into the tank water.* In contrast, most direct-fired water heaters have delivery limitations imposed by the capacity of their heating element or burner. Figure 14–15 compares the sustained DHW flow capability of several common residential water heaters. Relatively low heat input rates are the reason such tanks require large volumes to meet high demand applications. This is especially true for electric water heaters.
- *A modern boiler and indirect water heater can provide DHW at higher efficiencies than conventional direct-*

Heater type	Heat output rate (Btu/hr)	Hot water recovery rate* (gpm/gph)**
Electric (3.8 kW)	12,970	0.35/21
Electric (4.5 kW)	15,360	0.41/25
Electric (6.0 kW)	20,480	0.55/33
Gas-fired (30–50 gallons)	30,000–40,000	0.8–1.1/48–66
* Results are based on cold water temperature of 50 °F, and outlet temperture of 125 °F ** gpm = gallons per minute, gph = gallons per hour		

Figure 14–15 Typical heat input and recovery rates of residential direct-fired water heaters.

fired water heaters. Boiler heat exchangers are larger and designed with better heat transfer geometry than are the heat exchangers in most direct-fired water heaters. Boiler heat exchangers can also be cleaned during their service life to help maintain that efficiency advantage.

- *Indirect tanks are not exposed to the high temperature gradients present in direct-fired tanks.* The high temperature differences between the combustion gases and water create stresses that eventually take their toll on the heat exchanger materials in direct-fired water heaters. The higher temperature heat exchanger surfaces also allow increased precipitation of minerals dissolved in the water. Some of these precipitants bond to the surface of the heat exchanger further reducing its heat transfer efficiency. Direct-fired tanks are also subject to corrosion caused by flue gas condensation, especially when operated at reduced temperatures.

Performance of Indirectly Fired Storage Water Heaters

Most manufacturers of indirect water heaters provide two types of rating information. One of the ratings is called the **first hour rating;** the other is called the **continuous flow rating.**

The first hour rating is the amount of water the tank can deliver within one hour of the initial demand at a specified delivery temperature. The rating assumes the tank is at the specified delivery temperature at the beginning of the demand. The two delivery temperatures that are usually stated are 110 °F and 140 °F.

The first hour rating is influenced by both the storage characteristic of the tank, and its ability to transfer heat to the domestic hot water once the tank's temperature starts to drop. This rating is strongly dependent on the boiler capacity available to the tank, as well as the boiler supply temperature to the heat exchanger. In most cases, the manufacturer provides several first hour ratings along with the associated heat input rate and boiler water supply temperature.

The continuous flow rating is the gallons per minute (or gallons per hour) the tank can deliver under sustained demands with a fixed heat input rate and boiler water temperature. This rating is an indicator of the tank's steady state heat transfer ability rather than its storage capacity.

An example of a manufacturer's rating table showing both ratings is shown in Figure 14–16.

Sizing an Indirect Water Heater

The usage pattern and quantity of DHW needed is difficult to estimate given the variations in lifestyle, schedule, activities, and fixtures/appliances installed in any given building. It is common to estimate the likely demand of DHW during the *peak hour* of a typical day, and base water heater selection on this.

The data given in Figure 14–17 can be used to help estimate a peak hour estimate of DHW usage. This data reflects usage by average occupants drawing water through average fixtures. *It does not necessarily give the usage possible in applications having high-demand luxury fixtures such as multiple head showers, and large whirlpool baths.* Estimates of the DHW demand of such fixtures are best obtained directly from fixture manufacturers.

When selecting an indirect tank, be sure to check manufacturers ratings for the heat transfer capability of the tank's heat exchanger, as well as the necessary inlet water temperature and flow rate needed to achieve that performance. *The author recommends selecting an indirect water heater capable of transferring the full heat output of the system's boiler(s) when operating at a boiler water supply temperature no higher than 200 °F.* When set up as a priority load, such a unit can generate DHW as fast as possible (given the boiler capacity present in the system).

Domestic Water Heating Using an External Heat Exchanger

Another means of generating domestic hot water using a hydronic system is through an external heat exchanger. This approach is especially well-suited to high demand situations where the heat transfer ability of an indirect water heater is not sufficient to handle the load. A piping schematic showing this concept is given in Figure 14–18.

A heat exchanger is installed between the heat source and insulated water storage tank. Upon a demand for heating, boiler water is circulated through one side of the heat exchanger, while domestic water is circulated through the other side. In most cases a cast-iron circulator is used on the boiler side when the system is closed. A bronze or stainless steel circulator must be used between the heat exchanger and storage tank since fresh (oxygenated) water is being conveyed. A strainer should be installed on the domestic water inlet side of the heat exchanger to collect particulates.

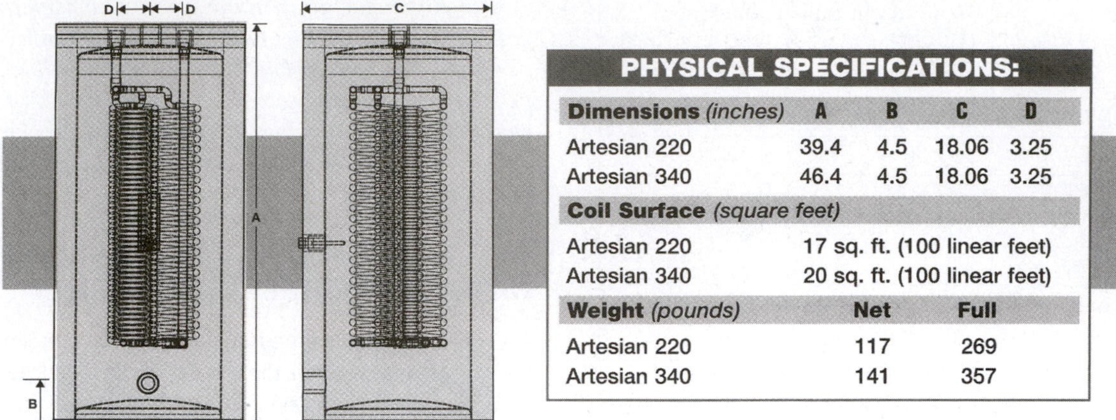

PHYSICAL SPECIFICATIONS:				
Dimensions *(inches)*	**A**	**B**	**C**	**D**
Artesian 220	39.4	4.5	18.06	3.25
Artesian 340	46.4	4.5	18.06	3.25
Coil Surface *(square feet)*				
Artesian 220	17 sq. ft. (100 linear feet)			
Artesian 340	20 sq. ft. (100 linear feet)			
Weight *(pounds)*	**Net**	**Full**		
Artesian 220	117	269		
Artesian 340	141	357		

ARTESIAN

INSTANTANEOUS WATER HEATER — RATINGS & CAPACITIES

BOILER OUTPUT	FLOW RATING AT 140°F			FLOW RATING AT 110°F			MIN. BOILER WATER FLOW
	GAL/HR		GAL/MIN	GAL/HR		GAL/MIN	
MBH	First Hour	Continuous	Continuous	First Hour	Continuous	Continuous	GAL/MIN
ARTESIAN MODEL 220							
50	78	72	1.20	117	108	1.8	2.7 (Taco 007)
70	102	95	1.58	153	142	2.4	2.7 (Taco 007)
90	129	120	2.00	193	180	3.0	2.7 (Taco 007)
110	156	147	2.45	233	220	3.7	3.4 (Taco 007)
140	195	187	3.12	292	280	4.7	5.4 (Taco 007)
160	220	213	3.55	330	320	5.3	6.5 (Taco 007)
ARTESIAN MODEL 340							
160	226	213	3.55	339	320	5.3	4.1 (Taco 007)
190	265	253	4.22	398	380	6.3	5.7 (Taco 007)
220	305	294	4.90	456	440	7.3	6.7 (Taco 0012)
250	343	333	5.55	516	500	8.3	9.0 (Taco 0012)

Test Conditions: 50°F domestic water in, 90° rise and 180°F boiler water in. The max pressure drop of domestic water is 4 psi.

Pressure Ratings: Maximum Allowable Tank Working Pressure – 150 psi. Heat exchangers are tested @ 300 psi.

Specifications and dimensions are subject to change without notice.

Figure 14–16 Example of first hour and continuous domestic hot water ratings of an indirect water heater. *Courtesy of Dunkirk Boiler Division of ECR International*

Usage	High flow fixture (gallons/usage)	Low flow fixture (gallons/usage)
Food preparation	5	3
Hand dishwashing	4	4
Automatic dishwashing	15	15
Clothes washer	32	21
Shower or bath	20	15
Face and hand washing	4	2

Figure 14–17 Estimated usage of domestic hot water for high flow and low flow fixtures in residential applications.

14.4 INTERMITTENT GARAGE HEATING

The option of heating a garage is frequently requested by homeowners in cold and snowy climates. In some cases, the owner's desire is to maintain the garage at a relatively cool temperature sufficient to melt ice from cars, or perhaps to allow the garage to be used as a workshop. In other cases, the intent is to heat the garage only when necessary, and thus minimize fuel usage.

The intended usage of a garage should be considered when selecting the type of heat emitter(s). For example, radiant slab heating is ideal in situations where a relatively constant temperature will be maintained. The warm floors will significantly improve comfort, as well as quickly melt snow and ice from vehicles and dry the floor slab. However, the

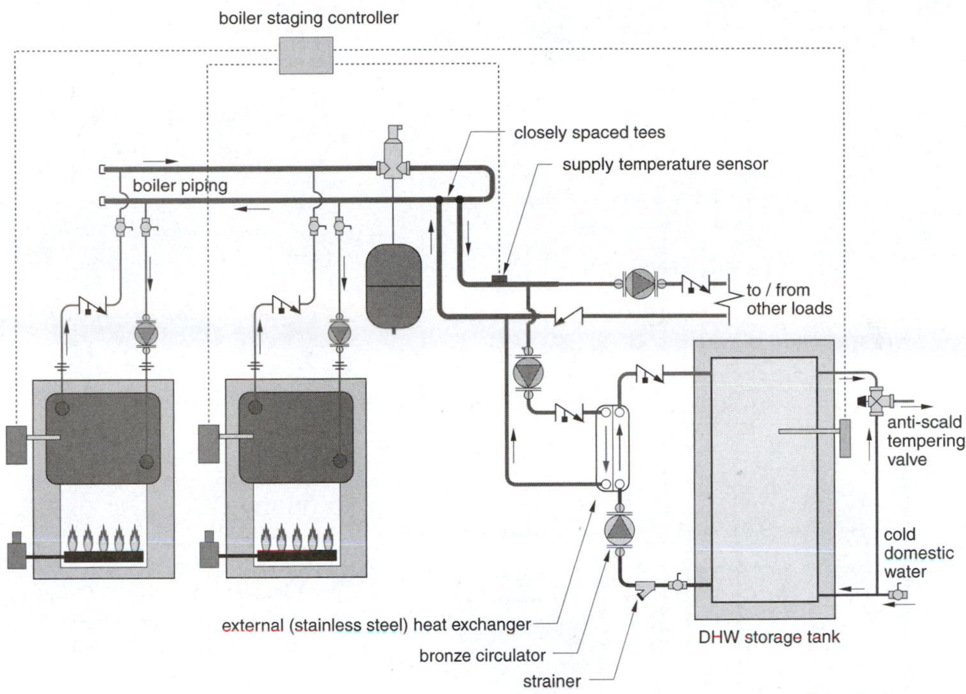

Figure 14–18 Using an external heat exchanger and multiple boiler system for high capacity domestic water heating.

large thermal mass of such systems makes them very slow to react to sudden changes in thermostat setting, and therefore unsuitable for situations where frequent thermostat adjustments will be made. In contrast, an overhead unit heater has very little thermal mass, and can react quickly to changes in thermostat setting. It would be a better choice when the primary goal is to quickly warm the air in a garage.

Arguments can be made both for and against the use of antifreeze in garage heating applications. If the garage is to remain above freezing at all times, one might conclude that antifreeze is not needed. On the other hand, if the heat is to be totally shut off during cold weather, the distribution system will very likely experience subfreezing temperatures and therefore must contain antifreeze.

The author's position is that antifreeze *must* be used in *all* hydronic garage heating applications. Unanticipated events such as power outages or equipment breakdowns can lead to frozen pipes in a matter of hours. A hard freeze can burst piping and create a very expensive repair situation. The extra cost associated with using antifreeze is a small price to pay to be sure the garage heating circuits will not be damaged by freezing temperatures.

If the entire hydronic system will be filled with an antifreeze solution, the garage can be heated by simply creating another zone circuit. This is generally a good approach in smaller systems. However, if the garage zone is relatively small relative to the remainder of the system, the cost of installing and maintaining antifreeze in the entire system may not be justified.

An alternate approach is to separate the garage zone from the rest of the system using a heat exchanger as shown in Figure 14–19.

The heat exchanger is piped into the system as a parallel branch circuit. The piping is arranged so boiler water and the antifreeze solution flow through the heat exchanger in opposite directions (e.g., counterflow).

The garage circuit is now an independent *closed* piping circuit with the heat exchanger as its heat source. As such, it requires most of the usual piping components associated with a closed loop, namely an expansion tank, air separator, make-up water system, and pressure-relief valve. A flow-check valve is not required on the antifreeze filled side of the garage circuit because heat will not be present in the heat exchanger when the garage zone is off. Inlet and outlet purging valves should be included in this circuit if antifreeze will be added using the methods described in Chapter 13, Air Removal, Filling, and Purging.

The author does not recommend the use of a hard-piped automatic make-up water assembly on the antifreeze side of the heat exchanger. The reason is that a small leak in this portion of the system can go unnoticed if an automatic make-up water assembly is used. Eventually the antifreeze solution could be diluted to the point of freezing.

Although not an absolute requirement, an **automatic glycol make-up system** is a good option for the antifreeze filled portion of the system. Such systems are described in Chapter 13, Air Removal, Filling, and Purging.

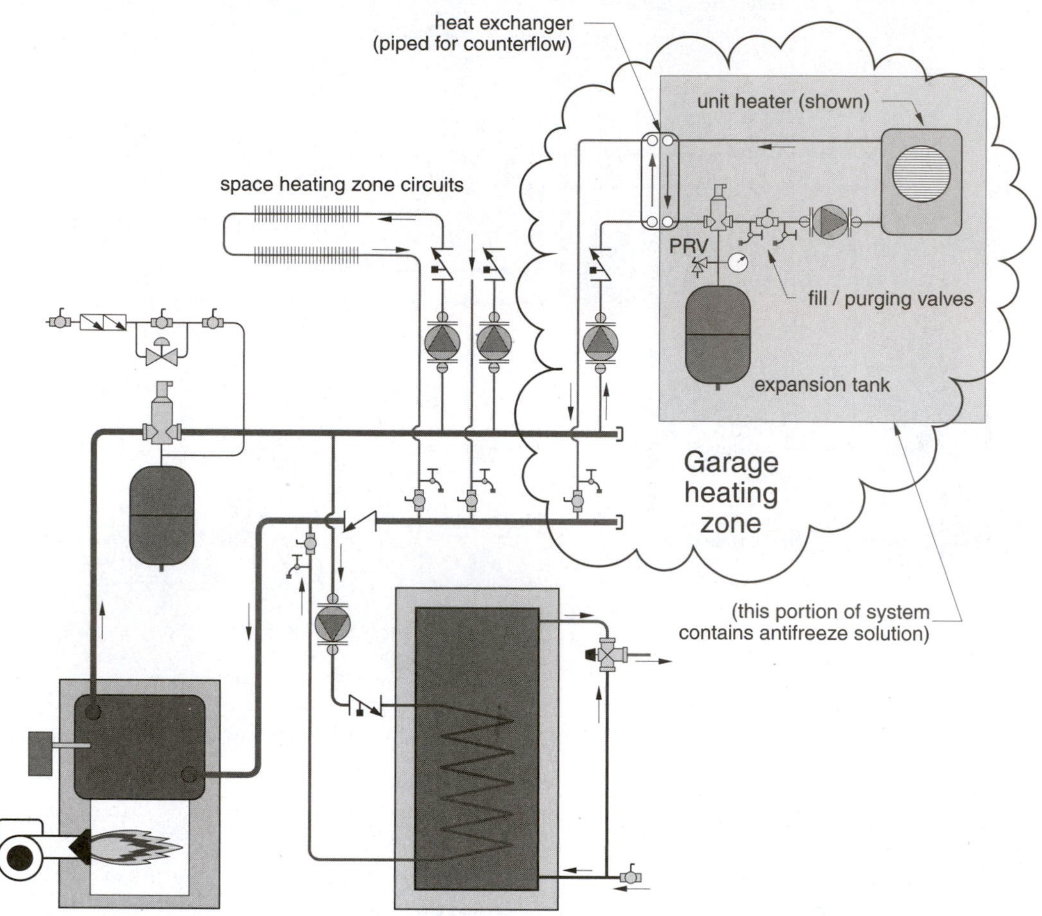

Figure 14–19 Supplying a unit heater in garage zone through a heat exchanger. All piping components outside of continually heated space are filled with an antifreeze solution.

Another option is to intentionally oversize the expansion tank used on the antifreeze side of the system. The extra tank volume provides a reserve of antifreeze solution under pressure, and therefore able to enter the system to make up for minor losses.

Figure 14–20 shows piping that allows a variable speed injection mixing system to regulate the temperature of the antifreeze solution supplied to a heated garage floor. Notice how the injection controller monitors the temperature of the antifreeze solution supplied to the floor circuits, while the variable speed injection pump regulates the flow of hot water supplied to the heat exchanger. The injection controller also monitors the boiler inlet temperature to prevent flue gas condensation in a conventional boiler during start-up of a cold slab. The injection pump is equipped with an integral check valve to prevent migration of hot water from the primary loop through the heat exchanger when garage heating is inactive. This approach has worked well for the author in several systems.

14.5 SPA AND HOT TUB HEATING

The same boiler that supplies space heating can usually be used to supply heat to a spa or hot tub. The most common approach uses a heat exchanger between the space heating system and the spa circuit. This prevents the treated water in the spa from mixing with the boiler water. A conceptual piping schematic is shown in Figure 14–21.

The heat exchanger is piped into the system as a parallel branch circuit. The piping is again arranged so boiler water and spa water flow through the heat exchanger in opposite directions (e.g., counterflow) to maximize the rate of heat transfer.

Only a portion of the spa water leaving the filter needs to be routed through the heat exchanger. This prevents a relatively small heat exchanger from creating an unnecessarily high head loss in the filter circuit which often operates at a high flow rate. Flow through the heat exchanger is regulated by the bypass valve between the tees that lead from the spa

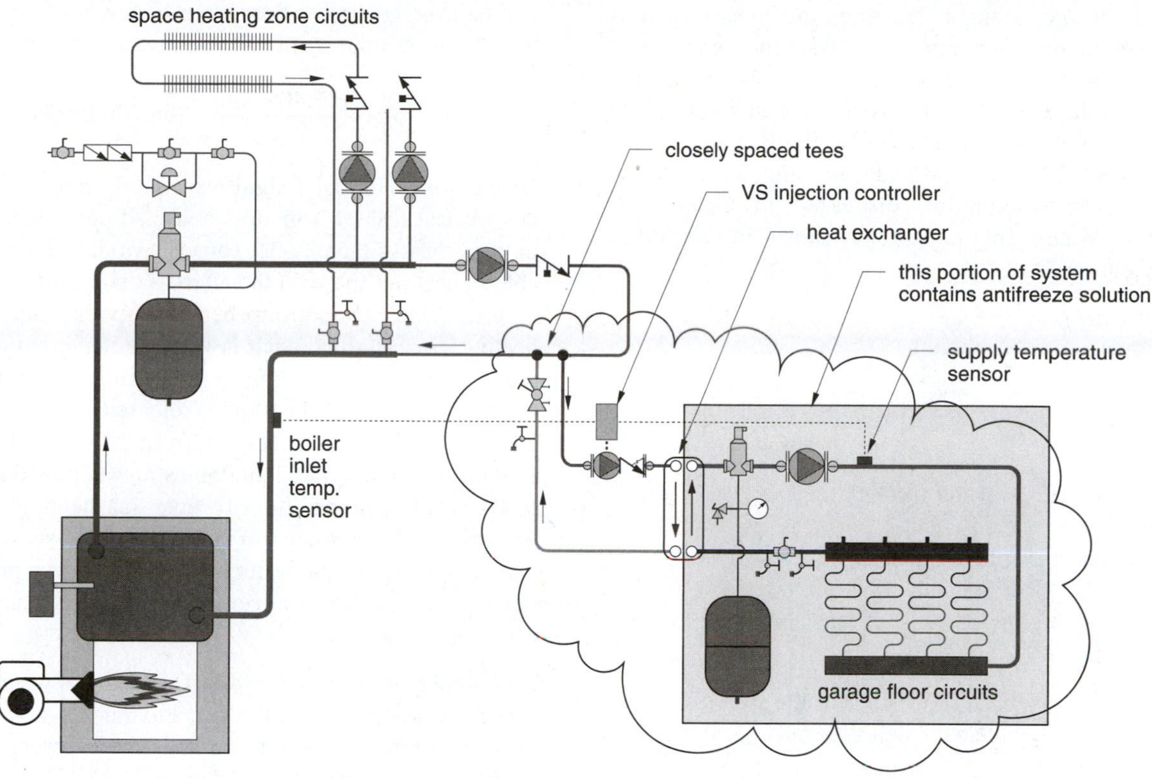

Figure 14–20 Piping schematic showing intermittent garage floor heating supplied through a heat exchanger with heat input regulated by a variable speed injection mixing controller.

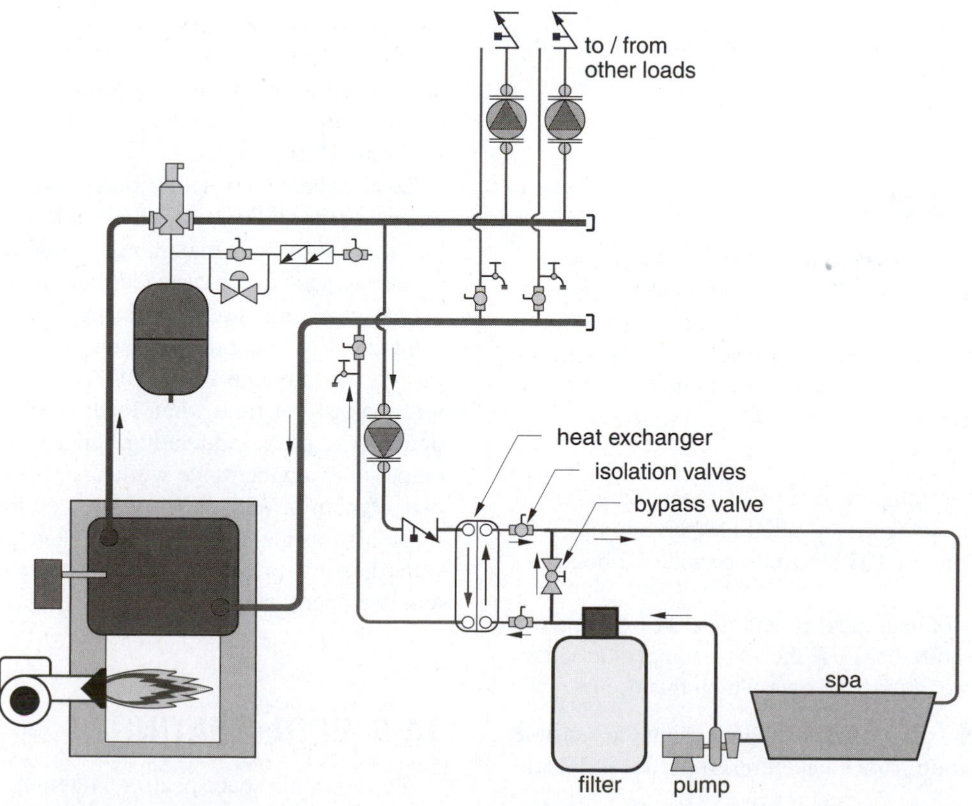

Figure 14–21 Piping schematic showing spa heating zone supplied through stainless steel heat exchanger.

circuit to the heat exchanger. The more this bypass valve is closed, the greater the flow rate of spa water through the heat exchanger. Flow through the heat exchanger should be high enough to produce good heat transfer, yet low enough to minimize the pressure loss caused by the partially closed balancing valve. This flow rate can be estimated using the sensible heat rate equation discussed in Chapter 4, Properties of Water. This equation is shown in rearranged form as Equation 14.6:

(Equation 14.6)

$$f = \frac{Q}{500(\Delta T)}$$

where:

f = flow rate of spa water through the heat exchanger (gpm)

Q = required rate of heat transfer across the heat exchanger (Btu/hr)

ΔT = temperature gain of the spa water through the heat exchanger (°F)

A temperature rise of 5 °F to 10 °F as the spa water passes through the heat exchanger will generally yield efficient heat transfer.

Heat input to the spa can be regulated using a setpoint control to measure the temperature of the water leaving the spa. When this temperature drops below the desired value, the zone circulator and boiler are operated to send hot water through the heat exchanger. A flow switch installed in series with the setpoint control is also recommended to verify flow in the spa circuit before allowing the hydronic heating subsystem to operate.

Heating Loads of Spas

Because spas are usually kept covered and at a reduced temperature when not used, their standby heat loss is relatively small, in some cases only a few hundred Btu/hr. The rate of heat transfer of the spa heating subassembly is mostly determined by the temperature rise required to bring the water to the desired use condition, and what time is allowed for this temperature rise.

Example 14.3: Determine the rate of heat input necessary to bring a 500-gallon spa from an initial temperature of 70 °F to a final temperature of 103 °F over a period of 2 hours.

Solution: The rate of heat input is determined by calculating the total energy required to raise the water temperature, then dividing by the amount of time over which this occurs.

The required energy can be calculated using the sensible heat quantity equation from Chapter 4, Properties of Water.

$$q = (\#\ \text{gallons})(8.33)(T_{final} - T_{initial})$$
$$= (500)(8.33)(103 - 70) = 137{,}400\ \text{Btu}$$

The average rate of heat input is now found by dividing the total heat input by the allowed warm-up time:

$$q = \frac{137{,}400\ \text{Btu}}{2\ \text{hr}} = 68{,}700\ \text{Btu/hr}$$

Discussion: This is a significant load, comparable to the design heat loss of a modest house. If such were the case, most of the available boiler capacity would have to be available to heat the spa with the allotted time. This would entail a "prioritized" relationship between spa heating and space heating. Such an approach is feasible if the building has a high thermal mass heating system. The heat stored in the thermal mass would be slowly released to compensate for the temporary lack of heat input from the boiler. The designer should also be aware that domestic water heating is likely to be an even higher priority than spa heating with most owners. The ladder diagram in Figure 14–22 shows a simple way to make domestic water heating the highest priority, spa heating the next highest priority, and space heating the lowest priority load.

Another possibility is to ask if the owners would accept a longer warm-up time for the spa, and thus lower the required heat input rate. Depending on the time allowed, this could significantly improve the chances of having sufficient boiler capacity to heat the spa and meet space heating requirements without need of priority control.

Given these options, it is very seldom necessary or advisable to size a boiler to the combined total of the spa load and the design space-heating load. Doing so results in a boiler with much higher heat output than necessary during almost all of its operating hours. The resulting drop in seasonal efficiency will be substantial, as discussed in Chapter 3, Hydronic Heat Sources.

If, after being advised of these options, the owner insists on having the ability to rapidly heat the spa, even when other loads are at or near maximum, the designer still has some options. One is to use a staged multiple boiler system rather than a single large boiler. This approach could provide sufficient capacity to handle all loads simultaneously when necessary, yet maintain reasonably high efficiency during the vast majority of time when loads are much lower. Another option is to use a modulating boiler that can adjust its heat output over a wide range while also maintaining reasonably high efficiency. With either of these options, the extra capacity is also available for high demand prioritized domestic water heating, provided the domestic water heating subsystem is properly sized.

14.6 POOL HEATING

The hydronic space heating system in a building can usually be configured so it can also heat a swimming pool in, or adjacent to that building. As with spa heating, a heat

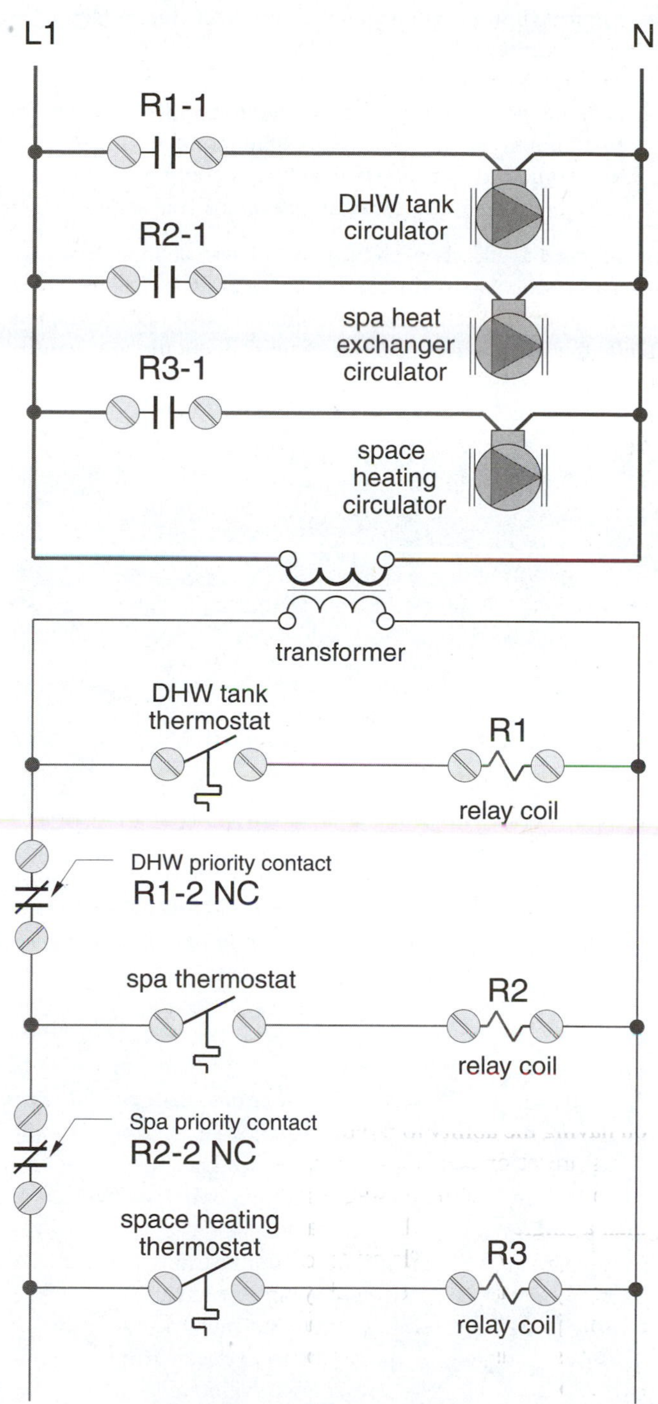

Figure 14–22 Ladder diagram showing three levels of priority control. DHW has highest priority, spa heating second highest priority, space heating lowest priority.

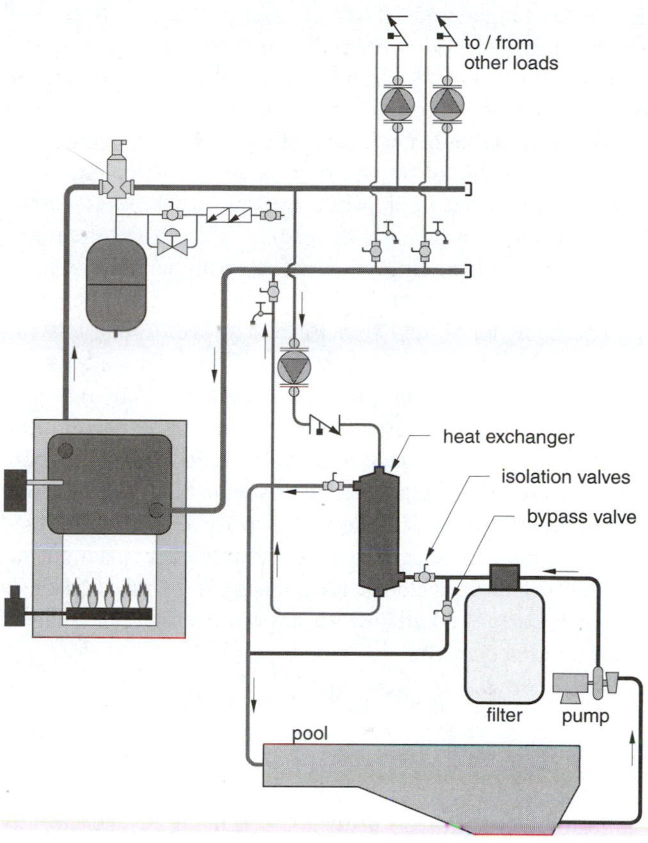

Figure 14–23 Piping schematic showing pool heating supplied through stainless steel heat exchanger. Note that heat exchanger is downstream of pool filter, and piped for counterflow.

exchanger is used to separate the pool water from that in the space heating system. The typical piping for such a system is shown in Figure 14–23.

The feasibility of such a system depends on the magnitude of the pool heating load, and when it occurs relative to the space-heating load. Outdoor pools that are only used from late spring to early fall can usually be heated by the same hydronic heat source that heats the house in cooler weather. This is also true for indoor pools maintained at comfortable temperatures year round. However, outdoor pools that must be kept warm during cold weather will probably have loads high enough to justify extra boiler capacity, or perhaps a staged multiple boiler system.

Many factors affect the heating load of a swimming pool. These include:

- Radiational cooling to the sky (outdoor pools)
- Heat gain from solar radiation (outdoor pools)
- Convective heat loss to the air
- Conduction losses through the sides and bottom of the pool
- Evaporation losses from the water surface

The dominant heat loss from pools is caused by evaporation of water. Each pound of water that evaporates from the surface removes about 960 Btus from the pool. Evaporation losses increase with increasing air velocity across the water surface. They also increase as the relative humidity of the air above the pool decreases. Evaporative heat loss changes almost constantly for outdoor pools.

Estimates range from 30 to well over 100 Btu/hr/ft^2 of pool surface area. *A pool cover will significantly reduce evaporation as well as convective losses, and is highly recommend for any heated pool application.*

The heat exchanger selected for a pool heating application should be capable of transferring the total heat output of the heat source to the pool, while operating within its normal range of temperatures and flow rates. This allows the maximum available heat input rate whenever higher priority loads such as space heating or domestic water heating are satisfied.

The designer should remember that even modest swimming pools have very high thermal mass. This allows the pool to accept heat input, at high rates, almost any time it is available. The high mass also allows heat input to be interrupted for several hours with very little change in pool temperature. These characteristics suggest that pool heating can often be treated as the lowest priority load in a multiload system. Heat can be sent to the pool whenever spacing heating and domestic water heating loads are satisfied. If these higher priority loads then restart, pool heating can be shed, and recommence later.

14.7 HYDRONIC SNOWMELTING

One of the fastest growing segments of the heating market is **hydronic snow- and icemelting systems** (also know as **SIM systems**). Such systems make use of flexible tubing embedded in pavements to melt snow and ice.

SIM systems can be large stand-alone installations, or smaller subassemblies tied into hydronic space heating systems. The potential applications for SIM systems include driveways, walkways, parking areas, exterior steps, patios, wheelchair ramps, loading docks, and any other areas that must be cleared of snow and ice.

Benefits Offered by SIM Systems

Hydronic SIM systems offer several benefits over traditional methods of snow removal. These include:

- The ability to provide fully automatic and unattended snow and ice removal whenever required
- The ability to remove snow without creating banks or piles that often lead to drifting and/or damage to landscaping
- The elimination of sanding and the associated mess created when sand is tracked into buildings, or swept from pavements in spring
- The elimination of salting and its potential damage to landscaping and the surrounding environment
- Less pavement damage due to frost action, salting, or physical damage from plowing, which is especially important for surfaces covered with pavers
- Reduced likelihood of slips, falls, or vehicular accidents due to safer walking and driving surfaces
- Improved property appearance in winter due to lack of snow banks and sand/salt residue
- The ability to use almost any fuel or heat source to provide the energy needed to operate the SIM system. In commercial and industrial installations, it may even be possible to use "waste heat" from cooling or refrigeration processes to supply the SIM system.

Figure 14–24a shows tubing installed in a slope driveway area. Sloped pavements are prime candidates for hydronic SIM systems. Figure 14–24b shows the same pavement while the SIM system is active. The melted patio area shown

Figure 14–24(a) Tubing for snow and icemelting being installed in a sloping driveway. *Courtesy of Clay Thornton.*

Figure 14–24(b) Sloping driveway surface while SIM system is operating. *Courtesy of Clay Thornton.*

in Figure 14–24c demonstrates the ability of hydronic SIM systems to clear complex pavements in heavy snow areas without creating banks or piles.

SIM System Classifications

There are several approaches to designing a hydronic SIM system. They vary in the rate of heat delivery to the surface and the type of controls used to start and stop the melting process. The following SIM system classifications have been developed by the industry.

- **Class 1 SIM systems** are sufficient for most residential walkway and driveway areas. The rate of heat delivery to the surface is typically 80 to 125 Btu/hr/ft^2 depending on the geographic location of the system.

  Class 1 SIM systems typically allow snow to accumulate on the surface during a heavy snowfall, especially if the melting process is not automatically started. This snow layer is beneficial because it acts as an insulator between the heated pavement surface and the outside air, and thus reduces evaporative and convective heat loss. The majority of the heat reaching the surface melts the underside of the snow layer instead of being dissipated in other ways.

- **Class 2 SIM systems** are sufficient for most retail and commercial paved areas. The rate of heat delivery to the surface is generally 125 to 250 Btu/hr/ft^2 depending on geographical location. *Class 2 systems are designed to prevent snow from accumulating on the surface as it falls once the pavement is up to normal operating temperature.*

- **Class 3 SIM systems** have the highest heat delivery rate to the pavement surface, typically 250 to 450 Btu/hr/square foot depending on geographical location. *Class 3 systems are used in areas that must be kept clear of snow and ice at all times, and where the surface should be dry soon after the snow fall ends.* Examples include helicopter pads, toll plazas, sloped pavements in parking areas, and entrances to hospital emergency rooms.

Figure 14–24(c) Courtyard area equipped with SIM system is free of snow and ice. *Courtesy of Clay Thornton.*

Figure 14–25 gives suggested heat delivery rates for all three SIM system classifications in several geographic locations.

Suggested output (Btu/hr/sq ft)			
City	Class 1	Class 2	Class 3
Albuquerque, NM	71	82	167
Amarillo, TX	98	143	241
Boston, MA	107	231	255
Buffalo, NY	80	192	307
Burlington, VT	90	142	244
Caribou, ME	93	138	307
Cheyenne, WY	83	129	425
Chicago, IL	89	165	350
Colorado Springs, CO	63	63	293
Columbus, OH	52	72	253
Detroit, MI	69	140	255
Duluth, MN	114	206	374
Falmouth, MA	93	144	165
Great Falls, MT	112	138	372
Hartford, CT	115	254	260
Lincoln, NE	67	202	246
Memphis, TN	134	144	212
Minneapolis-St. Paul, MN	95	155	254
Mt. Home, ID	50	90	140
New York, NY	121	298	342
Ogden, UT	98	216	217
Oklahoma City, OK	66	81	350
Philadelphia, PA	97	229	263
Pittsburg, PA	89	157	275
Portland, OR	86	97	111
Rapid City, SD	86	102	447
Reno, NV	98	154	155
St. Louis, MO	122	152	198
Salina, KS	85	120	228
Sault Sainte Marie, MI	78	144	213
Seattle-Tacoma, WA	92	128	133
Spokane, WA	87	127	189
Washington, DC	117	121	144
Data source: Section 55.4 *1987 ASHRAE Systems Handbook*			

Figure 14–25 Recommended rates of heat transfer for Class 1, Class 2, and Class 3 SIM systems in various geographical locations.

Melted Pavement Construction Details

This section discusses construction details used to incorporate snow and ice melting in different types of paved surfaces. The designer should always check to see if local codes require variations on the methods presented.

Drainage Provisions

It is very important that all melted pavements be detailed for proper drainage of melt water. The heat delivery rates used with Class 1 and Class 2 systems assume that most of the melted snow is drained from the surface as a liquid rather than being removed by evaporation. If the SIM system is forced to evaporate a high percentage of the melt water, energy usage will be significantly increased.

Improper drainage can also allow melt water to accumulate at low points on the surface, or where melted pavement adjoins nonmelted areas. When melting stops, puddle water can quickly turn to dangerous ice.

Pavements must be sloped toward drains capable of routing the melt water to a drywell, storm sewer, or other discharge means acceptable to local codes. Drainage piping should not run through the heated pavement because the cold water will absorb heat from it. Instead, drainage piping should be routed beneath the underside insulation where it is protected from freezing. When not under the melted pavement, the drainage piping should be either buried below frost level or covered with a layer of extruded polystyrene to control frost penetration.

Trench drains are often installed along the low lines in melted pavements. If the pavement slopes toward a building, the melt water must be intercepted by the trench drain before it can flow into the building. Likewise, be sure melt water running toward a street will be collected by a drain before it contacts unheated pavement. Figure 14–26 shows some examples of pavement drainage concepts.

On some projects, the drainage system will not be installed by the same company that installs the rest of the SIM system. In such cases, be sure the necessary drainage details are clearly documented, communicated, and coordinated with those who are responsible for their installation. The lack of proper drainage can severely compromise an otherwise well-planned SIM system.

Subsurface Conditions

When planning a SIM system, designers should always evaluate the soils under the area to be melted. Failure to address subsoil conditions can lead to unanticipated conditions that could damage both the pavement and embedded tubing.

If the local water table is within 3 feet of the surface, it has the potential to draw heat downward at a high rate as the SIM system operates. High water table situations require drainage. A properly detailed "French drain" constructed around the perimeter of the paved area is a common solution.

If bedrock is present under the melted area, it is important to slope or channel the rock surface so water percolating downward can be drained away. Otherwise, the bedrock may pond water under the melted pavement. It is also important to install at least 1-inch (nominal R-5) extruded polystyrene insulation between the bedrock and upper (heated) layers of the pavement system to reduce downward heat conduction.

Soils containing high amounts of clay or silt are often saturated with water in winter. When these soils freeze, the expanding ice crystals create powerful forces that can easily crack and heave pavements. If such soils are present, the base layer of the pavement system should consist of 6 to 9 inches of 1/2-inch to 1-inch sized crushed stone. The soil surface beneath this layer of crushed stone should be sloped so water will be drained away rather than accumulate within the stones. Be sure a complete drainage path is provided to daylight or a suitable drywell or storm sewer. Also be sure

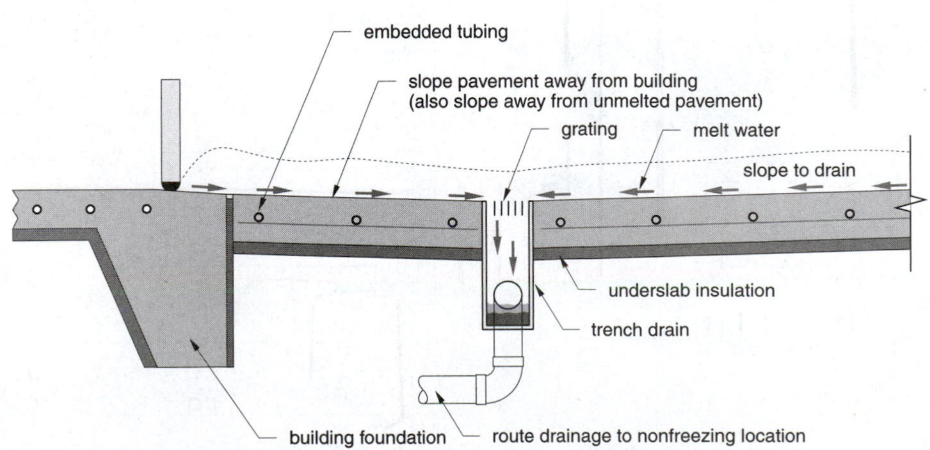

Figure 14–26 Melted pavement slopes toward trench drain.

the water cannot freeze at it passes through that drainage system. This stone layer should be compacted to form a flat and stable surface for the insulation board that will be installed above it.

When pavements will be placed over areas of disturbed or otherwise unstable soil, a geotextile fabric should be incorporated into this base layer. This very strong nondeteriorating fabric helps spread high loads over larger areas to prevent eventual depressions in the pavement. Such depressions could eventually damage embedded tubing.

No snowmelting system can correct for poor pavement design. Be sure to involve knowledgeable professionals in the pavement planning process.

SIM Installation in Concrete Pavement

Figure 14–27 shows the construction details used for a typical SIM system in a concrete driveway or walkway.

If the soil has good drainage characteristics, the base layer of the pavement system generally consists of 6 to 9 inches of compacted gravel. Any moisture percolating down through this base layer will continue downward into the subsoil.

In cold climates, or projects where the pavement will be "idled" at temperatures just above freezing, install a minimum 1-inch (nominal R-5) extruded polystyrene insulation over the compacted gravel base. This insulation greatly reduces downward heat loss. It also shortens the time required for the pavement to reach melting temperature following a cold start. Be sure the insulation board lies flat against the compacted gravel base at all locations so the pavement is fully supported.

The compressive load rating of the insulation should be selected to match loads that may be imposed on the pavement. A 25-psi rated extruded polystyrene insulation board is the minimum rating for pavements subject to light vehicular traffic. If heavier traffic is possible, insulation with a

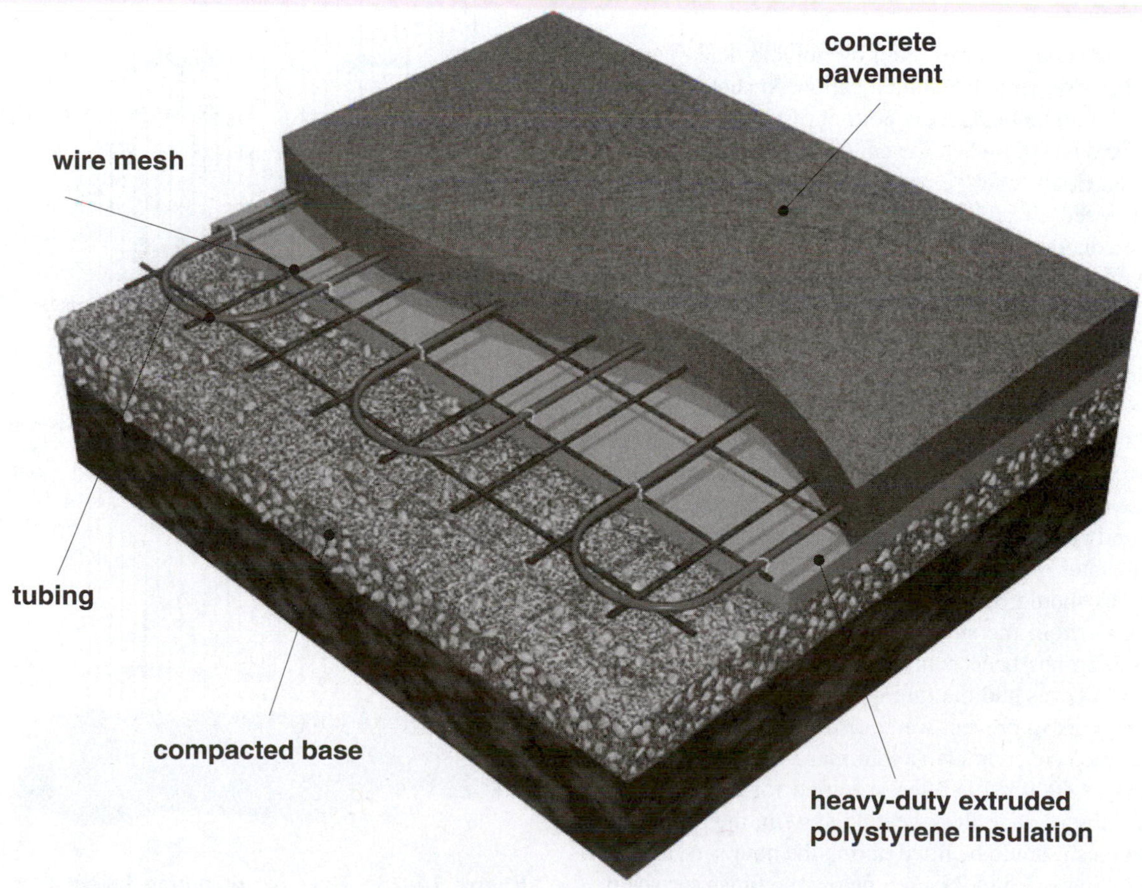

Figure 14–27 Installation of tubing for SIM system in concrete slab pavement. *Courtesy of IPEX Corp.*

compressive load rating of 40 to 60 psi may be required. Manufacturers of extruded polystyrene insulation can provide guidance on selecting the proper compressive load rating for a given pavement application.

Welded wire fabric (WWF), or a grid of steel rebar is now installed over the insulation. Be sure to overlap all sheets of WWF by at least 6 inches and tie them together with wire twist ties.

The tubing can now be secured to the steel reinforcing using wire twist ties spaced 24 to 30 inches apart. *Tube spacing should never exceed 12 inches in Class 1 systems or 9 inches in Class 2 and 3 systems.* Wider spacing can produce uneven melting patterns that may not completely clear the pavement of snow before the melting cycle is complete.

Tube circuits should be planned so as not to exceed the maximum circuit lengths given in Figure 14–35.

The warmest portion of the circuit should be routed in the areas with the highest melting priority. For example, the tire track area of a driveway should have a higher melting priority than the edges of the driveway. Don't install tubing closer than 6 inches to the edge of the pavement.

Figure 14–28 is an example of a tubing layout for a driveway. As with floor heating systems, it is advisable to prepare a tubing layout drawing prior to installing any tubing. The CAD-based methods discussed in Chapter 10, Hydronic Radiant Panel Heating, can be used to quickly develop such a drawing.

The tubing should be protected by polyethylene sleeving wherever it crosses a control joint location. At such locations, the tubing should be held to the bottom of the slab to ensure it will not be damaged when the control joints are sawn.

Once placed, all tubing circuits should be pressure tested with compressed air at 75 psi for at least 24 hours prior to placing the concrete.

In locations where tubing passes through a foundation wall, it should be protected by rigid schedule-40 PVC or steel pipe sleeves. The sleeves provide space around the tubing, allowing it to move without pinching, should the soil or pavement shift vertically. The cut ends of the sleeves should be smoothed to prevent chafing the tubing. The soil beneath the sleeved area should be fully compacted to minimize settling. The sleeves should be cast or tightly sealed into the foundation wall, and pitched slight downward on the exterior side to discourage water intrusion. The soil beneath the outside end of the sleeves should be well-drained so that any water percolating down from the surface cannot accumulate. Finally, once the tubing has been connected, the space between the inside of the sleeves and the tubing should be sealed with an elastomeric caulk to prevent water or insect entry.

Air-entrained concrete with a minimal 28-day compressive stress rating of 4000 psi is often specified for exterior slabs. As with a slab-on-grade floor heating system, the tubing and reinforcing mesh should be lifted during the pour such that the top of the tubing is 1.5 to 2 inches below the finish surface of the pavement. The exception is under control joint locations where the tubing should remain at the bottom of the slab.

Tubing depth is more critical in a snowmelting applications than in radiant floor heating. Leaving the majority of the circuit at the bottom of a typical 6-inch exterior slab will significantly increase the time required to warm the surface to melting temperature. It will also increase the required fluid temperature and create greater downward heat loss.

Asphalt Pavement Installations

Figure 14–29 shows the construction details used for a typical snowmelting system in asphalt paved driveways or walkways. The subgrade, drainage, and insulation layer should be prepared with the same considerations as discussed for a concrete pavement.

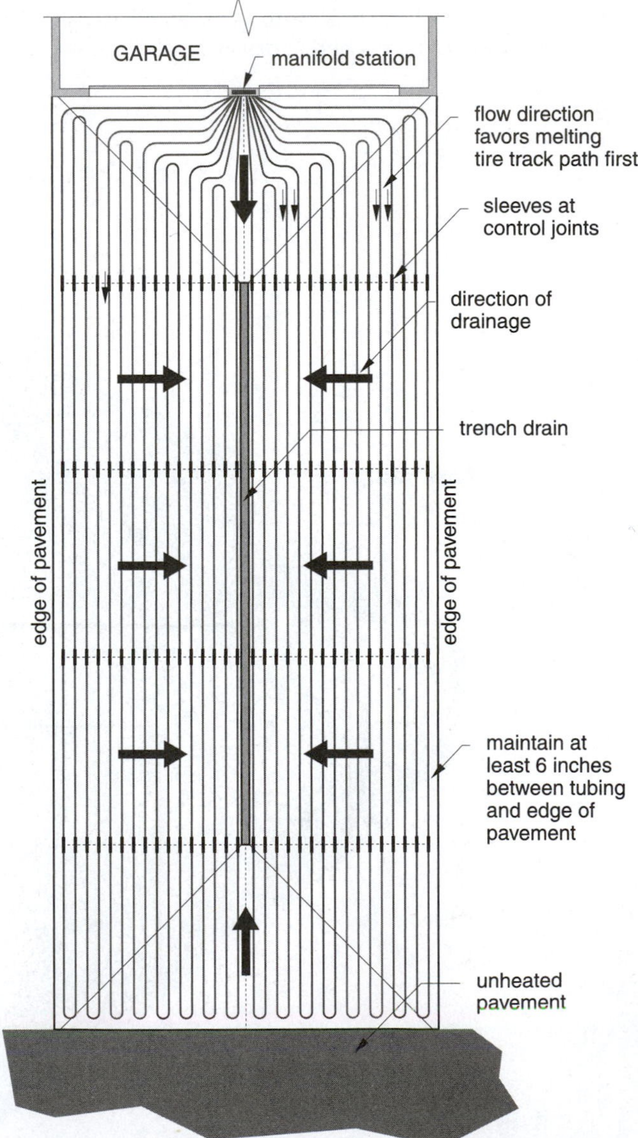

Figure 14–28 Example of tubing layout plan for melted driveway. Note flow direction in circuits and pavement slope.

After this, a layer of sand or stone dust approximately 2 inches thick should be placed on the insulation. This material helps protect the PEX or PEX-AL-PEX tubing from the hot (250 to 350 °F) asphalt when it is placed. A mat of welded wire fabric (WWF) is then laid out over the sand layer. The tubing circuits are again tied to the WWF using wire twist ties, and all circuits pressure tested.

After pressure testing, an additional 2-inch layer of sand or stone dust is placed over the tubing circuits. The circuits should remain under pressure as this material is placed and leveled by hand. Power loaders should not be driven directly over the tubing circuits. After the material is placed and leveled, it should be fully compacted.

Although opinions differ on necessity, some installers prefer to circulate cool water through the tubing circuits as the hot asphalt is placed to further protect them from the elevated temperatures. The final step is to place the asphalt pavement as normal.

SIM Installation for Paver-Covered Surfaces

Surfaces consisting of loosely laid pavers or tiles are easily damaged by conventional methods of snow removal, and therefore well-suited to hydronic snowmelting. Figure 14–30a shows the material assembly used for such installations.

Unlike concrete or asphalt surfaces, water can easily seep down between pavers across the entire melted area. This water cannot be allowed to accumulate under the pavers because subsequent freezing can cause heaving. In areas with low permeability soil, the base layer below the insulation should be detailed for efficient drainage. A 6- to 9-inch deep layer of compacted crushed stone placed over a slightly sloping subgrade grade, and drainage piping that removes the accumulated water to a nonfreezing disposal point is necessary.

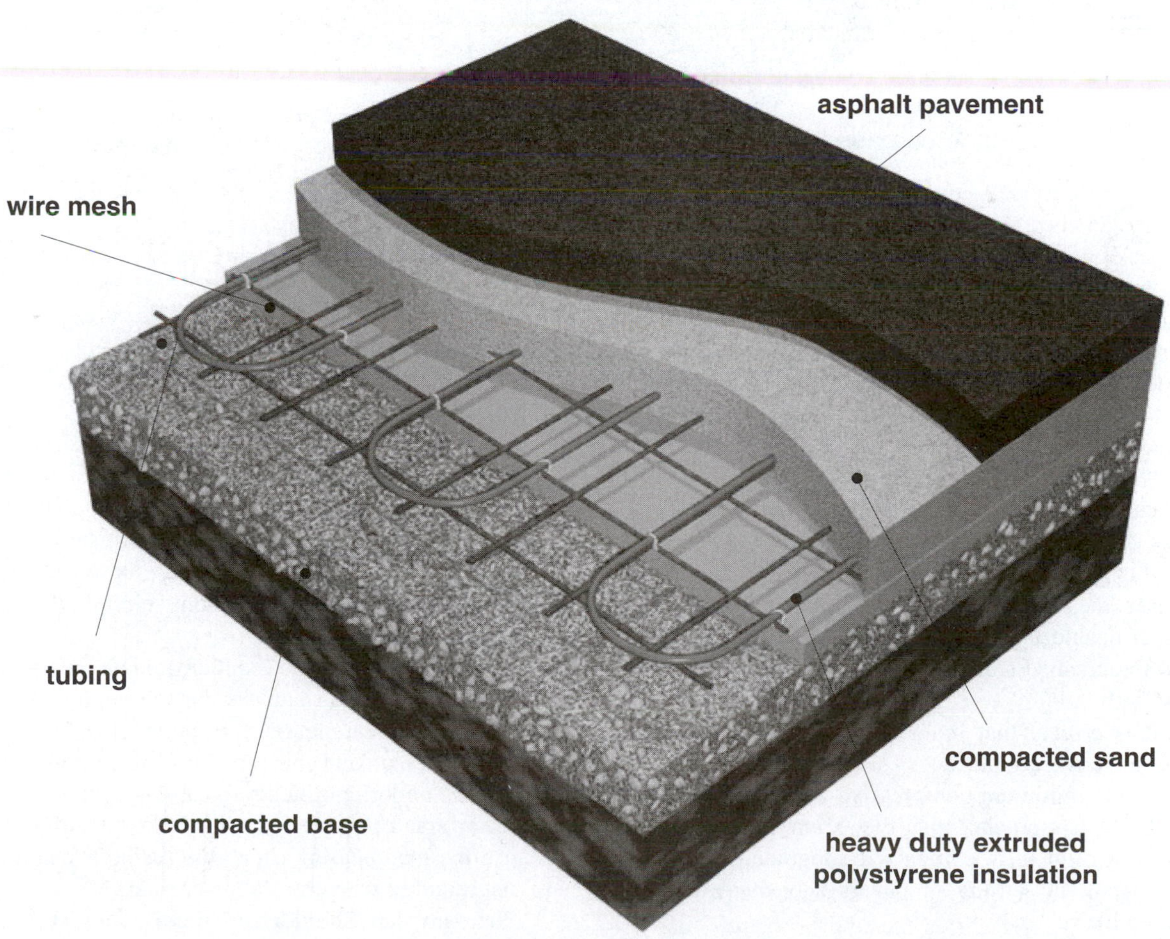

Figure 14–29 Installation of tubing for SIM system in asphalt pavement. *Courtesy of IPEX Corp.*

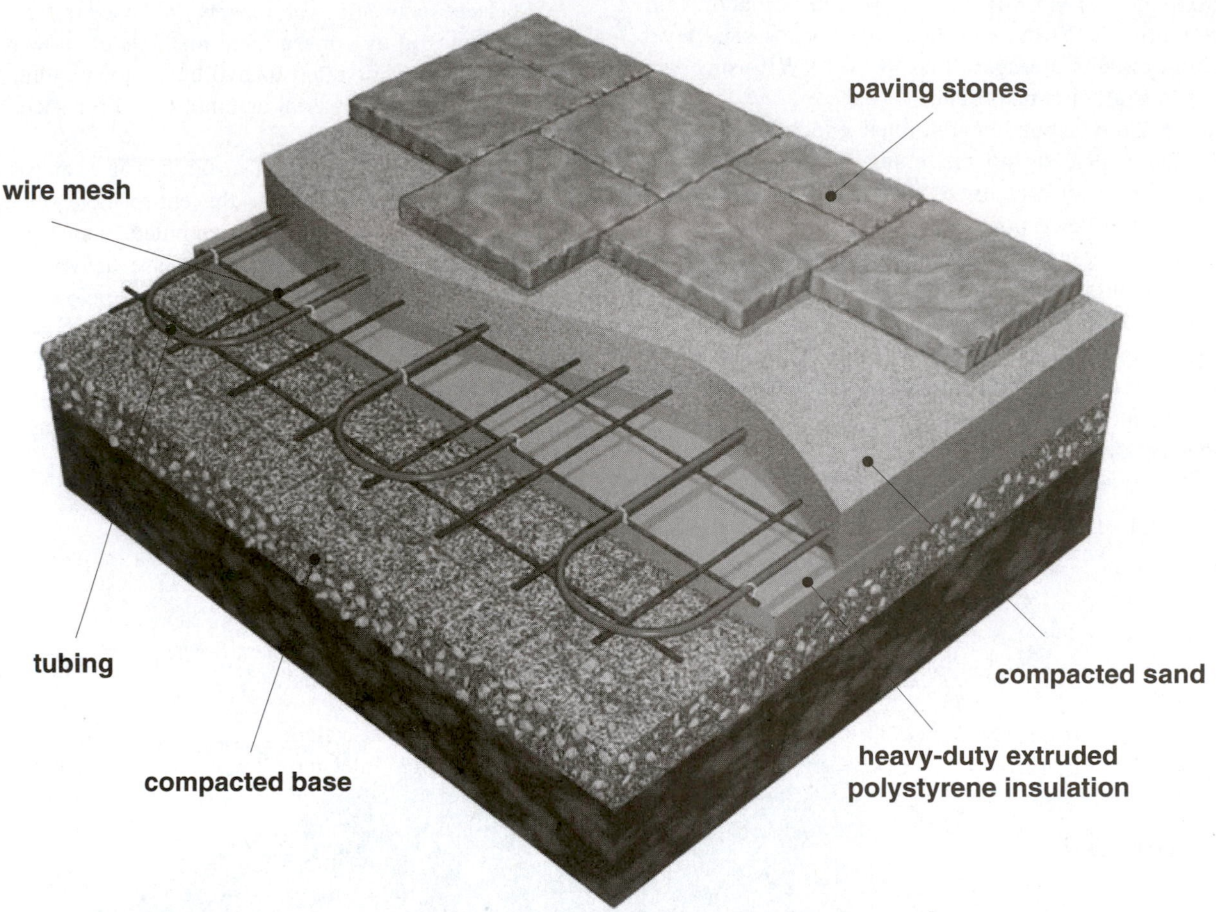

Figure 14–30a Installation of tubing for SIM system in area covered by pavers. *Courtesy of IPEX Corp.*

Extruded polystyrene insulation is impermeable to water. To allow water drainage, nominal 1/2-inch gaps should be left between adjacent sheets of insulation. Alternatively, several sheets of rigid insulation can be stacked and drilled to form a grid of 1-inch diameter holes spaced 12 inches apart. In either case, the gaps or holes should be covered with strips of water-permeable landscape fabric. This allows water to drain into the crushed stone layer without carrying sand or stone dust with it.

Note: It is critical that foam board insulation installed below paver covered surfaces have sufficient bearing strength (60 psi minimum compressive stress rating is suggested). It is also crucial that the foam insulation lies absolutely flat on a fully compacted subgrade. Failure to do so can result in settling of pavers subjected to heavy concentrated loads.

A layer of sand or stone dust about 2 inches thick is now placed over the insulation. A mat of welded wire fabric (WWF) is then laid out over this sand layer. The tubing circuits are tied to the WWF and pressure tested.

Next, an additional 2-inch layer of sand or stone dust is placed over the tubing circuits. The circuits should remain under pressure during the time this material is placed, leveled by hand, and fully compacted. Power loaders should not be driven directly over the tubing circuits. The base layer is now ready for the pavers.

If heavy vehicle traffic is anticipated on the paver covered surface, the use of a sunken concrete sub-slab is recommended. This installation technique is shown in Figure 14–30b. The sunken concrete sub-slab provides additional structural support and helps spread concentrated loads over a greater area of insulation. This load spreading effect will reduce the load bearing capability of the insulation relative to that required when no slab is present.

The sub-slab should be slightly sloped to promote drainage of the water that percolates down between the pavers. This drainage water should be collected in a trench filled with crushed stone and equipped with a drainage pipe as shown in Figure 14–30b. Because the sub-slab will be covered with stone dust, it does not need to be troweled.

System Design and Control Options

There are several ways to design and control SIM systems. They differ in how melting is initiated, as well as how pavement temperature is regulated before, during, and after melting. They also differ in terms of the required hardware.

The approach used must reflect the owner's expectations, the size of the area being melted, and the class of system being designed. The following is a brief discussion of the main issues and options available to address them.

Freeze Prevention Options

Some SIM installations are "dedicated" systems. Snow- and icemelting is the only function of the system. It is common to fill such systems with an antifreeze solution (typically a 30 percent to 50 percent mixture of inhibited propylene glycol and water).

In other systems, snowmelting is of one of several loads served by a common boiler (or multiple boiler system). In these systems, a heat exchanger is used to isolate the antifreeze solution in the snowmelting distribution system from the water in the rest of the system. The plate heat exchangers described in Section 14.2 are commonly used for this service. See Figure 14–31b.

When a SIM system starts during cold weather, the antifreeze solution returning to the heat exchanger can be extremely cold. To avoid the possibility of freezing the heat exchanger, it is good practice to prevent the distribution circulator from operating until heated water from the boiler is flowing through the heat exchanger. This can be done with a setpoint temperature controller such as discussed in Chapter 9, Control Strategies, Components, and Systems. Some snowmelting controllers may have this distribution pump delay feature built in.

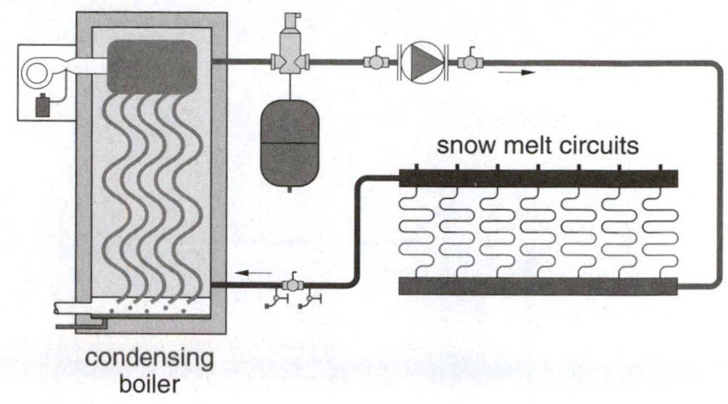

Figure 14–31a Use of a condensing boiler for supplying a dedicated SIM system. No mixing device is required.

Boiler Options

When a conventional boiler supplies heat to a SIM system, that boiler must be protected against sustained flue gas condensation. The system must use a mixing controller that measures boiler inlet temperature, and reduces the rate of heat transfer to the distribution circuits when necessary. The mixing device can be any of those previously discussed for space heating applications as long as boiler inlet temperature protection is provided.

Condensing boilers are also well-suited to the low operating temperatures of hydronic snowmelting systems. When

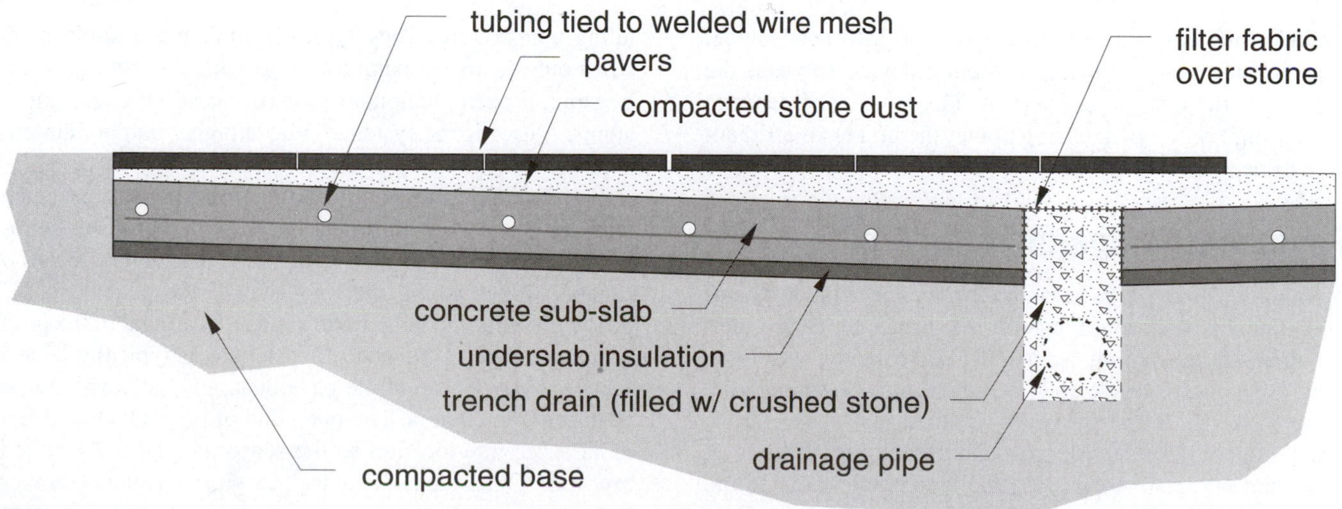

Figure 14–30b Cross-section of SIM system in area covered by pavers, and subject to heavy loads.

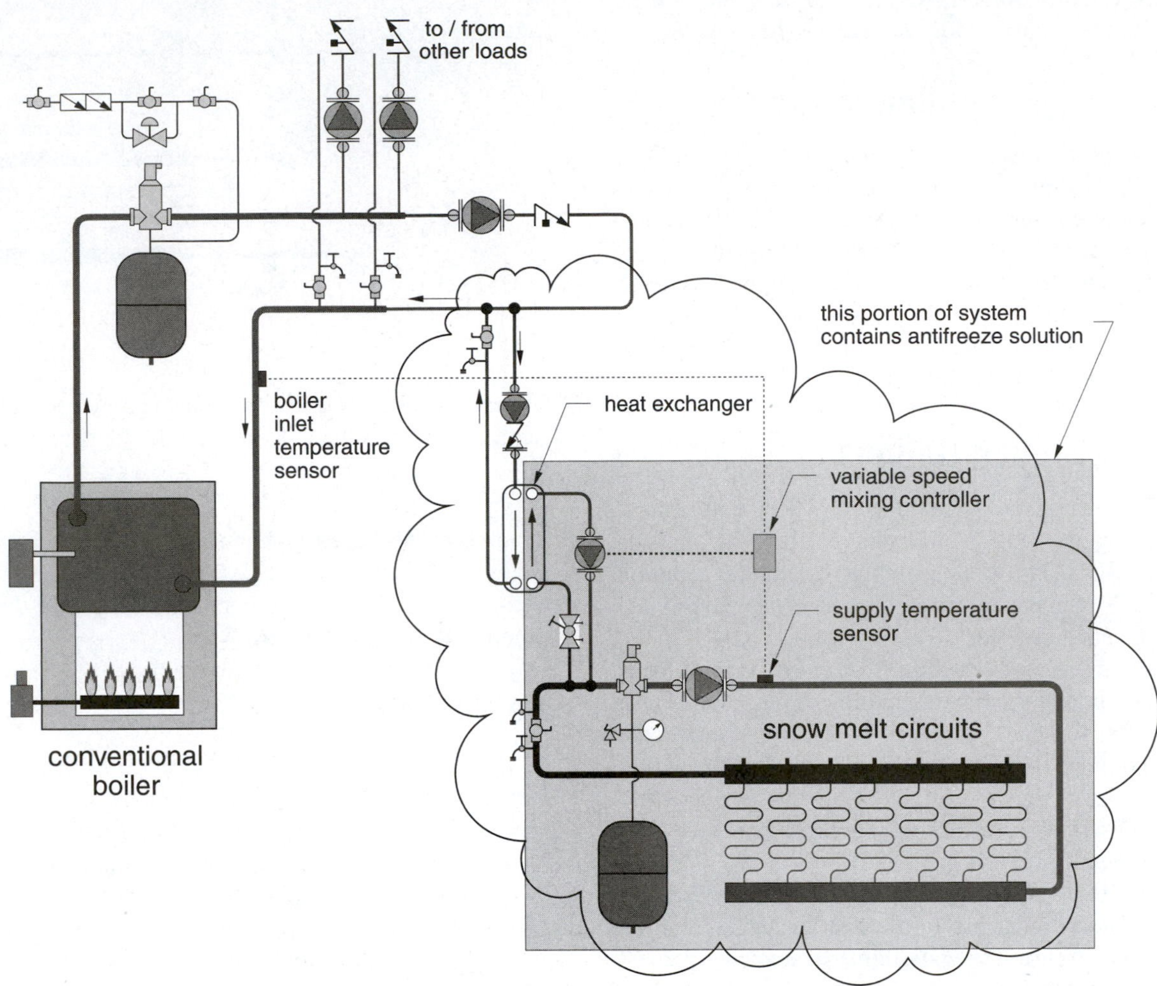

Figure 14–31b A SIM subsystem supplied through a heat exchanger. A mixing device is required to protect the conventional boiler from sustained flue gas condensation.

a condensing boiler is used as a dedicated SIM heat source, it is not necessary to install a mixing device between the boiler and the distribution system. The lower the operating temperature of the condensing boiler, the higher its efficiency will be.

Pavement Temperature Control

When a SIM system starts up from a very cold temperature, it may take several hours for the surface to reach a temperature where melting begins. To decrease this lag time, some SIM controllers **"idle"** the pavement at a temperature close to 32 °F. If the pavement is idled just above 32 °F, it will generally be free of frost and **"black ice."** This is an important safety advantage, especially in public areas. Idling the pavement just below freezing reduces standby heat loss, but still allows rapid warm-up to melting temperature when snow or freezing rain begins falling. Most SIM controllers with idling capability allow the installer to select the desired idling temperature. They typically initiate the idling mode when outside air temperature drops within a few degrees of freezing. Frozen precipitation can occur at these air temperatures. Allowing the system to idle at higher outdoor air temperatures is unnecessary and only wastes fuel.

To idle the pavement, the controller must sense pavement temperature. Some controllers use a small thermistor sensor located within a well in the slab. The well usually consists of capped copper tubing that is cast into the pavement. The position of the slab temperature sensor is crucial to the proper performance of the control system. It is typically located 1 inch below the top of the pavement, and halfway between adjacent tube circuits. The open end of the well should lead to an accessible location so the sensor can be replaced if it ever fails. Other types of controllers sense pavement temperature through the same device that detects precipitation. Always follow any specific installation instructions provided by the manufacturer for the pavement temperature sensor.

On/Off Control Options

SIM systems can be turned on and off by a simple manually operated switch. This is acceptable, provided someone pays attention to the weather and turns the system on and off as needed. If the attending person forgets to turn the system off, it could run indefinitely (or at least until running out of fuel). The cost of unnecessary operation can be high, especially on larger systems. This possibility is the biggest argument against manual start/manual stop control systems.

A somewhat more sophisticated method of control uses manual start combined with automatic shutoff after a set time has elapsed. The operating time is selected by the person turning on the system based on the amount of snow, and previous experience with the system. Overriding conditions such as very cold air temperatures, or sustained air temperatures above freezing may also be used to terminate operation. The goal is to turn the system off as soon as melting is complete and the majority of the surface is dry. The latter is important to prevent the formation of dangerous "black ice."

SIM controllers are available that can automatically detect frozen precipitation on the slab surface and initiate melting operation. They also terminate heat input when melting is complete. Most can also be configured to idle the pavement at a specific temperature when desired.

Some automatic snowmelting systems use a **snow/ice pavement sensor.** Such sensors mount directly into a sleeve that is cast into the pavement. The top of the sensor is flush with the top of the pavement. Such devices detect when frozen precipitation is present and measure the pavement temperature. They must be used with specific controllers that can monitor the signals they produce. An example of a snow/ice pavement sensor is shown in Figure 14–32.

Another type of sensor used in SIM systems is called a **snow switch.** This device measures air temperature and detects the presence of precipitation. It does not however measure pavement temperature. When the outdoor air temperture is below a preset value, typically about 38 °F, and precipitation is falling, the device closes a set of electrical contacts to turn on the SIM system. Snow switches are mounted near, but not on the pavement being melted. They should be located where they will not be tampered with or damaged. An example of a snow switch is shown in Figure 14–33.

Both types of snow detectors have a small electrically heated cell at the top of their housing. The heat generated is just enough to melt any frozen precipitation that falls on a small area of the sensor. The presence of water on the sensor is detected by electrical conductivity.

SIM controllers that monitor pavement temperature tend to reduce fuel usage by allowing the system to maintain the pavement surface just warm enough for sufficient melting. Without pavement temperature sensing, some SIM systems may heat the pavement to a temperature higher than necessary. Although the rate of melting increases, so does pavement heat loss, especially when the pavement is not covered by a layer of snow. Many SIM controllers also have the ability to "lock out" snowmelting operation when outdoor temperatures are above or below preset values.

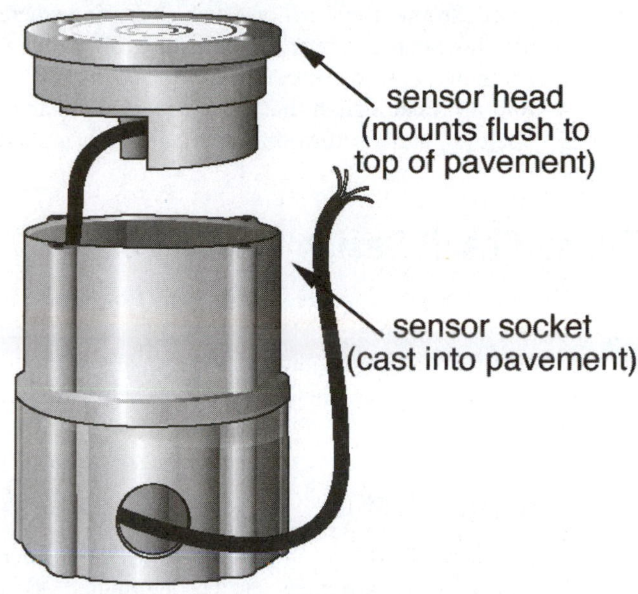

Figure 14–32 Snow/ice pavement sensor that detects precipitation and pavement temperature. *Courtesy tekmar Control Systems.*

With all SIM systems, it is desirable to supply heat to the pavement for a minimum of five hours *after frozen precipitation stops falling.* Some of this time may be needed to melt any accumulation on the surface. The sustained heat input also helps dry the surface to prevent the formation of ice.

Some SIM controllers also provide **ΔT protection** of melted slabs. As the slab starts from a cold temperature, the controller measures the supply and return temperature to the circuits. Heat input is regulated so the temperature drop of the circuits does not exceed a nominal 20 to 30 °F. This reduces thermally induced stresses within the cold slab.

It is the author's recommendation that all snowmelting systems have the option of manually-initiated melting. This

Figure 14–33 Snow switch sensor detects air temperature below 38 °F and presence of precipitation. *Courtesy ETI, Inc.*

provides a work around for situations such as drifting where an automatic snow detector may not be calling for melting, even though snow is present on other portions of the pavement. A wind up timer switch that can be manually started to initiate melting, and yet turn off automatically is a good solution.

Tubing Circuit Design Information

Selecting the proper tube size, spacing, and flow rate is an important part of designing a SIM system. The information that follows gives suggested guidelines and analytical tools that can be used to evaluate the tradeoffs, and help optimize the system.

Flow Requirements

The flow rate required for a SIM circuit to deliver a given amount of heat to the pavement can be determined using Equation 14.7.

(Equation 14.7)

$$f = \frac{Q}{k\,(\Delta T)}$$

where:

f = required flow rate (gpm)

Q = rate of heat output required (Btu/hr)

k = a constant based on the concentration of antifreeze used (see Figure 14–34)

Example 14.4: Determine the circuit flow rate required to deliver 22,000 Btu/hr using a 40 percent solution of propylene glycol in a pavement circuit operating with a 20 °F temperature drop.

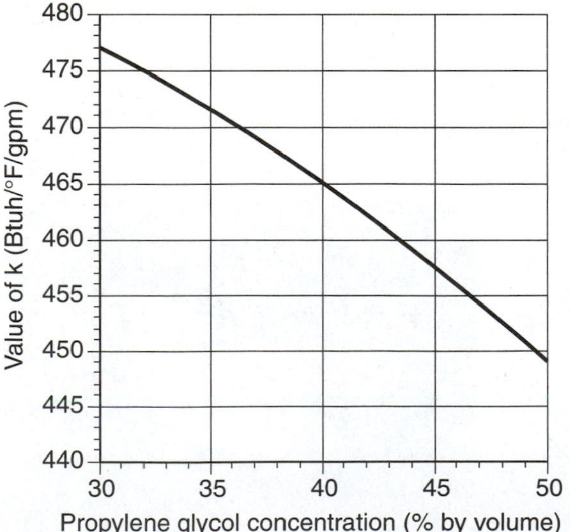

Figure 14–34 Value of constant k (for use in Equation 14.7) for various concentrations of propylene glycol.

	1/2-inch tubing	5/8-inch tubing	3/4-inch tubing	1-inch tubing
Maximum circuit length	210 feet	310 feet	410 feet	530 feet

Figure 14–35 Recommended maximum circuit lengths for use in SIM systems.

Solution: Substituting the values into Equation 14.7:

$$f = \frac{22{,}000}{465 \times 20} = 2.36 \text{ gpm}$$

Discussion: *The rate of heat delivery required for snowmelting is considerably higher than that required for a typical floor heating system.* To deliver more heat without excessive temperature drop, the circuit flow rate must be increased. The use of glycol-based antifreeze solutions, which have lower specific heats than water, further increases the flow rate requirement.

To accommodate higher flow rates, most SIM systems require larger diameter tubing relative to the sizes used in floor heating applications. Residential and small commercial SIM systems typically use 5/8-inch or 3/4-inch tubing. Larger commercial installations often use 1-inch tubing. In areas such as steps, it may not be possible to use 3/4-inch or larger tubing due to bending limitations. The common solution is to use multiple circuits of 1/2-inch or 5/8-inch tubing.

Figure 14–35 lists the maximum suggested circuit lengths for various sizes of PEX and PEX-AL-PEX tubing in SIM applications. The lengths assume that the system operates with a 50 percent propylene glycol solution, and that circuit head loss is equivalent to that of a 300-foot circuit of 1/2-inch PEX or PEX-AL-PEX operating with water. The latter is a common circuit size/length limitation used with floor heating systems.

SIM Circuit Temperature Drop

Most SIM systems are designed for circuit temperature drops of 20 to 25 °F during steady state operating conditions. Temperature drops in excess of 25 °F, when used in standard SIM systems, may result in uneven melting patterns on the pavement surfaces. The areas of the pavement above the warmest portion of the embedded circuits will melt and dry faster than those areas above the cooler portion of the circuits.

Circuit temperature drops in excess of 25 °F can be used provided the flow through the pavement circuits is periodically reversed. This helps even out the overall energy delivery to all portions of the pavement. There are several ways to reverse flow including periodically cycling a 4-way valve from one extreme to the other as shown in Figure 14–36. Operating snowmelting circuits with 25 to 30 °F temperature drops decreases flow requirements and reduces pumping power.

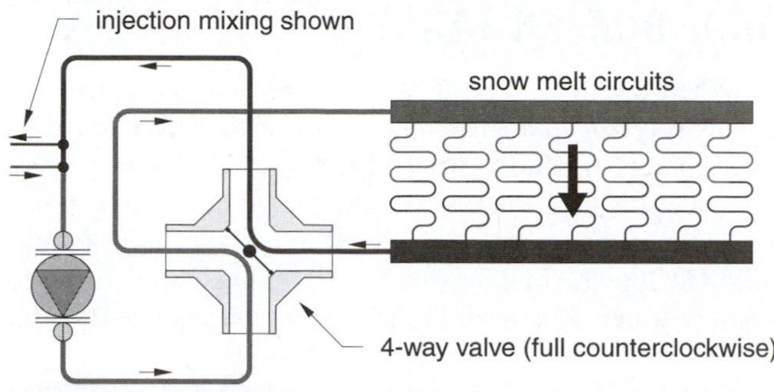

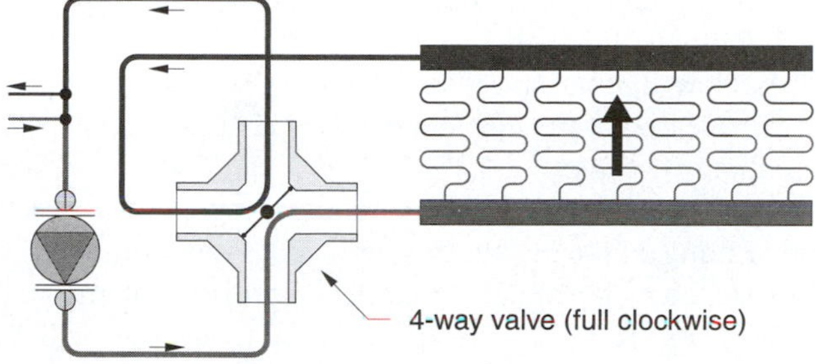

Figure 14–36 Use of 4-way motorized valve for periodic flow reversal through SIM circuits.

Estimating Heat Output

The exact heat output from a heated exterior pavement depends on many factors, many of which can change as the system operates. These include air temperature, wind speed, relative humidity, snow coverage, pavement drainage, tubing placement, average circuit temperature, circuit flow rate, soil temperature, and underside insulation. An exact engineering model that can account for all these factors would be very complex.

Figure 14–37 lists heat outputs and associated average circuit temperatures required to melt snow at a rate of 1/4-inch per hour. Values are given for "snow free" surfaces characteristic of Class 2 and Class 3 systems, as well as snow-covered surfaces characteristic of Class 1 systems. All data assumes tubing installed at 12-inch spacing in a concrete slab with approximately 2 inches of concrete coverage.

The data in Figure 14–37 show that wind speed and air temperatures have major effects on the heat delivery rates and circuit temperatures necessary to maintain snow-free surfaces. Much of the energy delivered to a melted but wet snow-free surface goes into evaporative and convective heat loss rather than just melting snow. Increasing wind speed and decreasing air temperature also increase the required heat delivery rates to snow-covered surfaces, but to a much smaller degree because the snow layer largely blocks convective and evaporative heat loss.

It is the author's opinion that systems requiring melting temperatures in excess of 150 °F are impractical in residential and light commercial systems. To reduce the circuit temperatures required, designers should use closer tube spacing.

Tube Spacing Considerations

Figure 14–38 can be used to estimate the change in heat output of slabs having tubing spacings other than 12 inches. Multiply the slab's estimated heat output using 12-inch spaced tubing by the factor found in Figure 14–38 to estimate the output for closer tube spacings.

All data is for tubing spaced 12-inches apart in concrete slab. Also assumes 80% relative humidity		Air temp. = 0 °F			Air temp. = 10 °F			Air temp. = 20 °F			Air temp. = 30 °F		
		windspeed (mph)			windspeed (mph)			windspeed (mph)			windspeed (mph)		
		5	10	15	5	10	15	5	10	15	5	10	15
to maintain snow-free surface	heat delivery rate (Btu/hr/sq ft)	**308**	**363**	**414**	**275**	**317**	**354**	**241**	**268**	**292**	**204**	**214**	**223**
	Average circuit temperature (°F)	187	215	240	171	192	210	154	167	179	135	140	144
surface covered w/ snow during melting	heat delivery rate (Btu/hr/sq ft)	**208**	**208**	**208**	**202**	**202**	**202**	**195**	**195**	**195**	**189**	**189**	**189**
	Average circuit temperature (°F)	137	137	137	134	134	134	131	131	131	127	127	127

data source: 1999 ASHRAE Applications Handbook

Figure 14–37 Heat delivery rates and average fluid temperatures for SIM circuits using 12-inch spacing. Data given at different air temperature and wind speeds for both snow-free and snow covered melting surfaces. *Data source: ASHRAE 1999 Fundamentals*

Example 14.5: A Class 1 SIM system using tubing spaced at 12 inches can deliver an output of 124 Btu/hr/ft² at a given circuit temperature. Estimate the output at the same average fluid temperature if the tubing is placed at 9-inch spacing.

Solution: The spacing correction factor from Figure 14–38 is 1.15. Thus the estimated output is:

Discussion: Closer tubing spacings obviously increase the

$$Q_{9'' \text{ spacing}} = (124 \text{ Btu/hr/ft}^2)1.15 = 143 \text{ Btu/hr/ft}^2$$

rate of heat delivery to the slab for a given operating temperature. Viewed another way, closer tube spacings allow a given rate of heat delivery to be achieved at lower average circuit temperature. Reducing the average circuit temperature will likely improve the performance of condensing boilers if used as the heat source in a SIM system. It also decreases system heat loss. Closer tube spacing also increases the amount of tubing required, the amount of antifreeze needed, and the size of the expansion tank. In most systems it also increases the number of manifold connections. The tradeoffs in cost versus performance should be evaluated.

The following tube spacings are suggested based on the class of the SIM system. Tube spacings in excess of 12 inches are not recommended because they can create uneven surface melting patterns.

- Class 1 systems: 9 to 12 inches
- Class 2 systems: 6 to 9 inches
- Class 3 systems: 4 to 6 inches

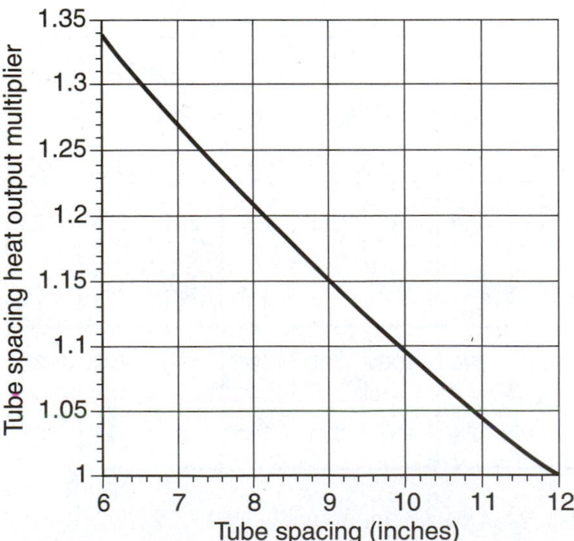

Figure 14–38 SIM circuit output multipliers for tube spacings closer than 12 inches.

14.8 BUFFER TANKS

When low mass on/off heat sources are combined with extensively zoned distribution systems, it is possible for the heat source to **short-cycle** when only one or two zones operate. This happens because the rate of heat production is much greater than the rate of heat release at the heat emitters. The low thermal mass heat source, in combination with the low water content of modern heat emitters and distribution systems, can absorb the excess heat, but only by experiencing a rapid rise in temperature. This causes the heat source to reach its high temperature limit very rapidly. In some systems, heat source on-cycles lasting less than 1 minute have been observed. When the heat source turns off, the low thermal mass components also cool rapidly as heat is released at the heat emitter(s). This leads to undesirably short off-cycles.

Rapid on/off cycling can be tolerated by electric resistance heating elements, but it is very undesirable for components such as gas valves, oil burners, ignition systems, and compressors. The life expectancy of such hardware will be significantly reduced if short-cycling is allowed to persist. Short-cycling also reduces efficiency and raises the concentration of pollutants in the exhaust stream of combustion-type heat sources.

One way of achieving fewer but longer on/off cycles is by adding thermal mass to the distribution system in the form of an insulated **buffer tank.** This thermal mass can "soak up" the excess Btus produced when the heat source operates with minimal loading on the system.

Although there are several possible ways to pipe a buffer tank into a system, not all of them offer the same benefits. For example, piping a buffer tank in series with the heat source and distribution system forces the buffer tank to be heated whenever heat is delivered to any load. In a system containing an indirect water heater, or a heat exchanger for pool heating or snowmelting, this is neither necessary nor desirable. Such loads provide all the thermal mass needed to buffer operation of a low mass heat source while they operate. Furthermore, standby heat loss from a buffer tank that is unnecessarily maintained at elevated temperature only adds to the building's cooling load in summer.

To avoid such situations, the buffer tank should be piped so that it can be bypassed in certain operating modes. One piping arrangement that accomplishes this is shown in Figure 14–39.

Notice how the buffer tank, boiler, and indirect water heater are all connected as secondary circuits to the system's primary loop. The buffer tank can be "online" or "offline" with the primary loop by turning its secondary circulator on and off. This allows the buffer tank to be bypassed during priority domestic water heating, and immediately returned to service when space heating resumes. This piping approach also allows the system to be adapted to future controllers that may include sophisticated algorithms for managing the thermal mass of a buffer tank.

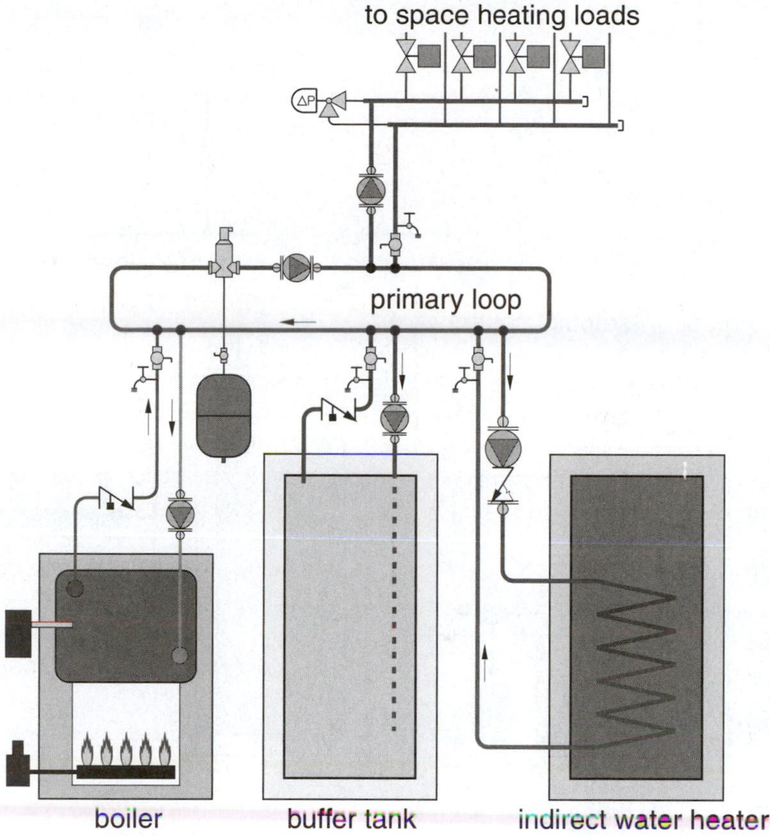

Figure 14–39 Connecting a buffer tank as a secondary circuit to a common primary loop. This piping allows tank to be "online" or "offline" by controlling secondary circulator.

Another piping scenario that allows selective use of the buffer tank is shown in Figure 14–40. A partial control wiring diagram for this system is shown in Figure 14–41. The following is a description of how this system would operate.

Domestic Water Heating Mode: When the indirect water heater requires heat, its aquastat powers up zone relay #1 in the multiload relay center. The 120-VAC output of this relay turns on the 3-way diverter valve directing flow from the boiler to the DHW tank's heat exchanger. With the multizone relay center set for priority operation, all other loads are temporarily turned off. The boiler is fired by passing a 24-VAC signal through the (X, X) terminals on the multizone relay center to the coil of an external (double pole) relay. One contact of this relay is wired in series with the end switch of the diverter valve and the (T, TV) terminals on the boiler limit control. Since both the relay contact and end switch are now closed, there is a complete circuit across the (T, TV) terminals and boiler firing is enabled. The boiler circulator (P1) is turned on by the (C1 C2) terminals on the boiler limit control.

Space Heating Mode: When any of the space heating loads call for heat through the multizone relay center, a 24-VAC signal is passed through the (X, X) terminals to power up the coil of the external relay. One contact pair in the external relay is wired in series with a setpoint control monitoring the temperature in the buffer tank. If the tank temperature falls below the setting of the setpoint control, the contacts in the setpoint control close the circuit between the (T, TV) terminals to fire the boiler. The other set of contacts in the external relay enable the primary loop circulator and buffer tank circulator to operate (except during priority DHW heating mode).

Flow continues through the buffer tank while the boiler is firing, allowing it to absorb heat that is not released to the load. When the buffer tank reaches the high limit setting of the setpoint control, the boiler and its circulator (P1) are turned off. The buffer tank is now in discharge mode providing heat to the primary loop as long as the load persists. When the buffer tank cools to 120 ºF, the boiler and circulator (P1) are turned back on to repeat the cycle.

Sizing a Buffer Tank

The required volume of a buffer tank depends on the rates of heat input and release, as well as the allowed temperature rise of the tank from when the heat source is turned on, to when it is turned off. The greater the tank's volume, and the

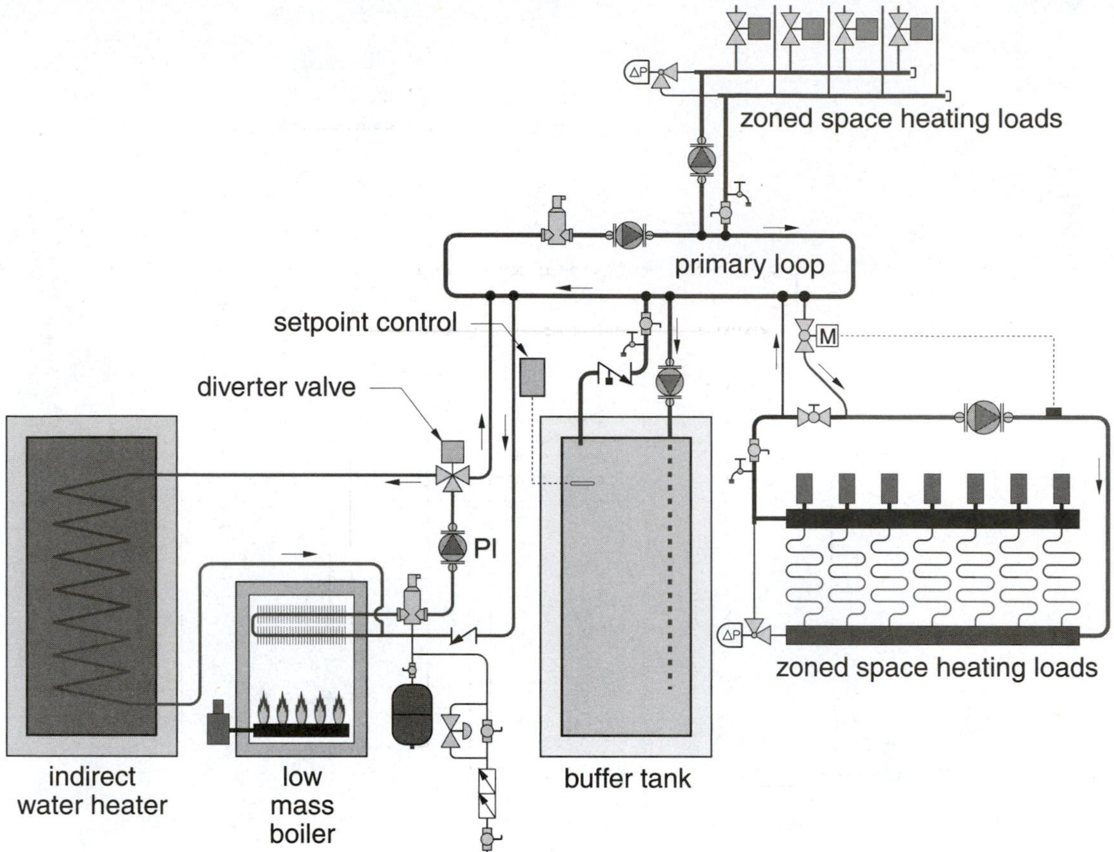

Figure 14–40 Piping that allows buffer tank to be used only when zoned space heating loads are operating.

wider the operating temperature differential, the longer the heat source cycles will be.

Equation 14.8 can be used to calculate the volume necessary given a specified minimum heat source on-time, tank operating differential, and rate of heat transfer:

(Equation 14.8)

$$V = \frac{t(Q_{\text{heat source}} - q_{\text{load}})}{500\,(\Delta T)}$$

where:

V = required volume of the buffer tank (gallons)

t = desired duration of the heat source's "on cycle" (minutes)

$Q_{\text{heat source}}$ = heat output rate of the heat source (Btu/hr)

q_{load} = rate of heat extraction from the tank (Btu/hr)

ΔT = temperature rise of the tank from when the heat source is turned on to when it is turned off (°F)

Example 14.6: It is desired that a boiler operate with a minimum burner on-cycle duration of 10 minutes, while a 75,000 Btu/hr boiler supplies heat to a towel warmer radiator releasing heat at 2000 Btu/hr. The burner turns on when the buffer tank temperature drops to 120 °F, and off when the tank reaches 180 °F. What is the necessary tank buffer tank volume to accomplish this?

Solution: Substituting the numbers into Equation 14.7 yields:

$$V = \frac{t(Q_{\text{heat source}} - q_{\text{load}})}{500(DT)} = \frac{10(75{,}000 - 2{,}000)}{500(180 - 120)}$$

$$= 24.3 \text{ gallons}$$

Discussion: The required volume suggests that a residential water heater tank could probably be used as the buffer tank.

Design Considerations for Buffer Tanks

1. Buffer tanks are seldom needed unless the system contains several small individually controlled loads that have low thermal mass.
2. Heat emitters supplied from a buffer tank will release heat at progressively slower rates as the tank

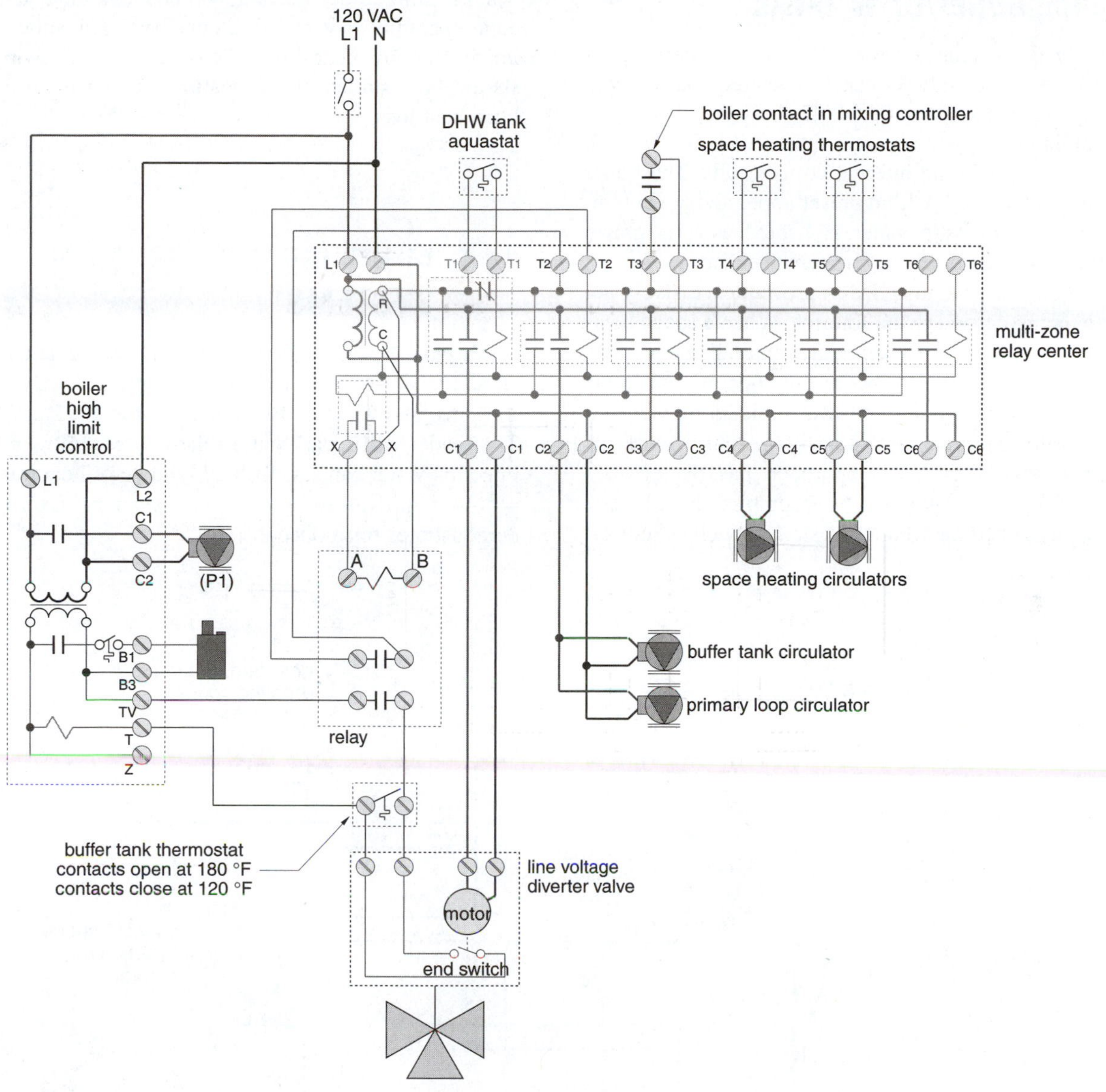

Figure 14–41 Electrical control diagram for system shown in Figure 14–40.

cools during its discharge cycle. To compensate, heat emitters should be sized to provide design load output at the *average* (not maximum) water temperature of the buffer tank.

3. Do not configure the system to bring the buffer tank online during a call for domestic water heating. The thermal mass of the DHW tank itself will buffer the boiler. Not heating the buffer tank during the domestic water heating mode also reduces standby heat loss in warm weather.
4. Do not configure the system to bring the buffer tank online when other high thermal mass loads such as pool/spa heating or snowmelting are active. These loads can usually provide all the buffering effects a low-mass heat source requires.
5. The wider the operating temperature differential of the buffer tank, the fewer (and longer) the heat source cycles will be.
6. Always install a flow-check valve on the outlet of the buffer tank to prevent heat loss due to thermosiphoning during off cycles. A circulator with an integral check valve can also be used for this purpose.
7. Buffer tanks should be well-insulated to minimize standby heat loss. A minimum jacket insulation of R-10 is suggested. Additional insulation may be required on standard DHW tanks to achieve this rating. Whenever possible, the buffer tank should be installed within a heated space.

Combined Buffer/DHW Tanks

Another way to achieve the benefits of a buffering thermal mass is to use an indirect water heater as a buffer tank. The concept is shown in Figure 14–42.

The tank in this system contains boiler water in the shell, and domestic water in the internal copper coils. The boiler fires to maintain the tank within preset upper and lower temperature limits. Domestic water is heated as it is drawn through the copper coils whenever there is a demand. A thermostatic mixing valve is installed to prevent excessively hot water from being carried to the fixtures. When a space-heating load is active, hot water is circulated through the primary loop attached to the tank. The boiler only operates when the buffer tank has discharged to its lower limit setting. Once it starts, the boiler continues firing until the tank reaches its upper temperature limit. This approach has the obvious advantage of using a single tank to accomplish both buffering of a highly zoned distribution system as well as indirect water heating. Since the tank will be maintained at an elevated temperature year round, it must be well-insulated to minimize standby heat loss. Flow-checks and swing-checks should be installed at all locations shown to further minimize heat loss.

14.9 MINITUBE DISTRIBUTION SYSTEMS

Variable speed injection mixing was extensively discussed in Chapters 9, Control Strategies, Components, and Systems, and Chapter 10, Hydronic Radiant Panel Heating. The reader will recall that a relatively low flow rate of hot injection water can establish a high rate of heat transfer into a low temperature distribution system. Example 14.7 demonstrates this concept.

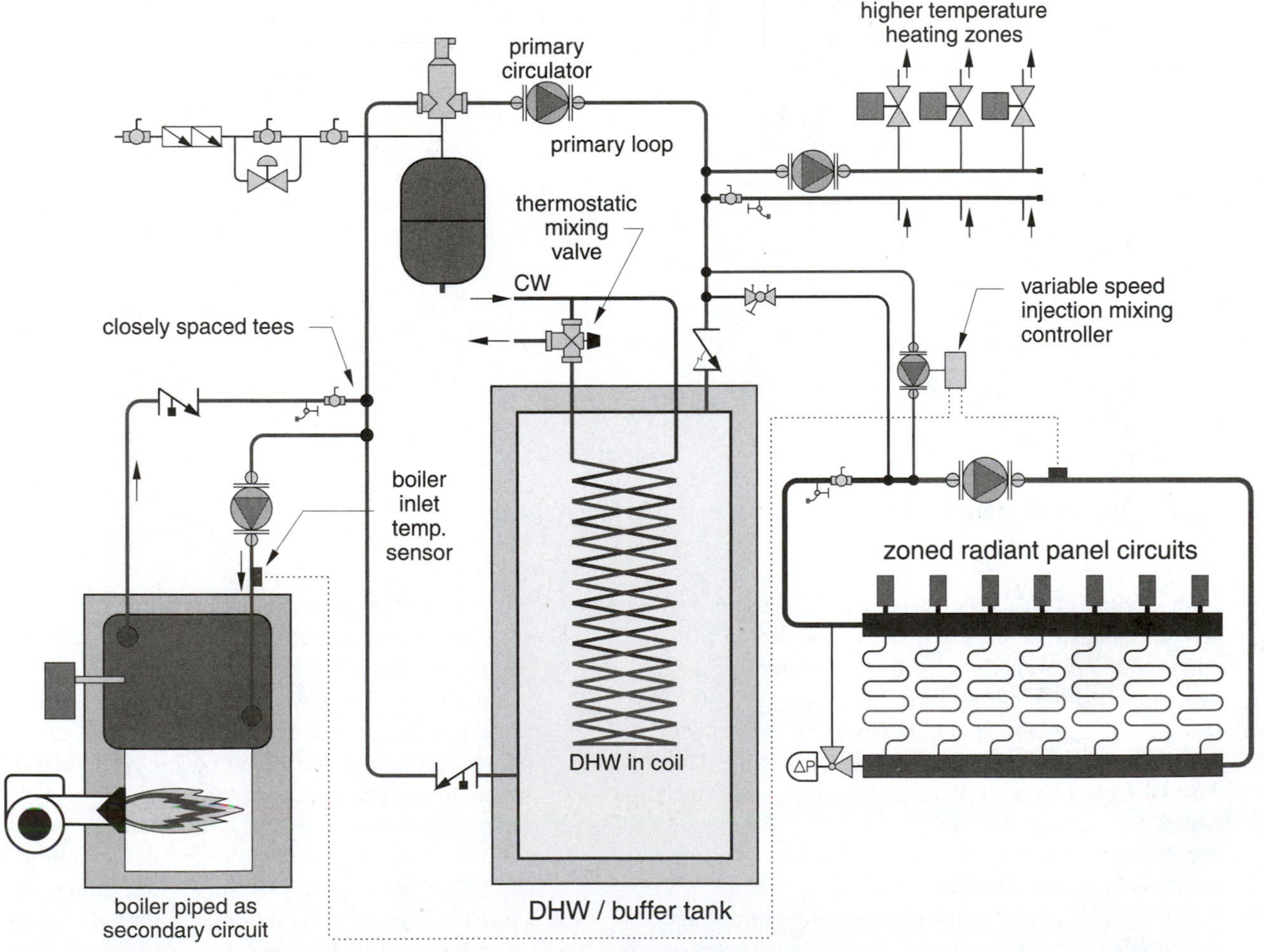

Figure 14–42 Single tank serves as indirect water heater and buffer tank for zoned space heating loads.

Example 14.7: What injection flow rate of 175 °F water is required to transfer heat at 250,000 Btu/hr into a large radiant floor system with a return water temperature of 85 °F?

Solution: Applying the sensible heat rate equation from Chapter 4, Properties of Water, yields:

$$f = \frac{Q}{500(\Delta T)} = \frac{250{,}000}{500(175-85)} = 5.6 \text{ gpm}$$

Discussion: This flow could be handled by a 3/4-inch tube. The large temperature drop between the supply and return injection risers is what enables this relatively low flow rate to deliver such a high rate of heat transfer.

Figure 14–43 shows a piping schematic of a **minitube distribution system** designed to take advantage of the large temperature difference available between hot boiler water and the relatively cool water returning from a heated slab-on-grade floor. Notice the relatively small diameter tubing between the boiler loop and the manifold station. One of these **"minitubes"** carries hot water from the boiler loop to the manifold station, the other carries cool water back to the boiler loop.

Flow through the minitubes is controlled by a variable speed injection pump. The faster the injection pump runs, the greater the proportion of hot water injected at the mix point (A) in the manifold station piping, and the warmer the supply temperature to the distribution circuits becomes. Heat

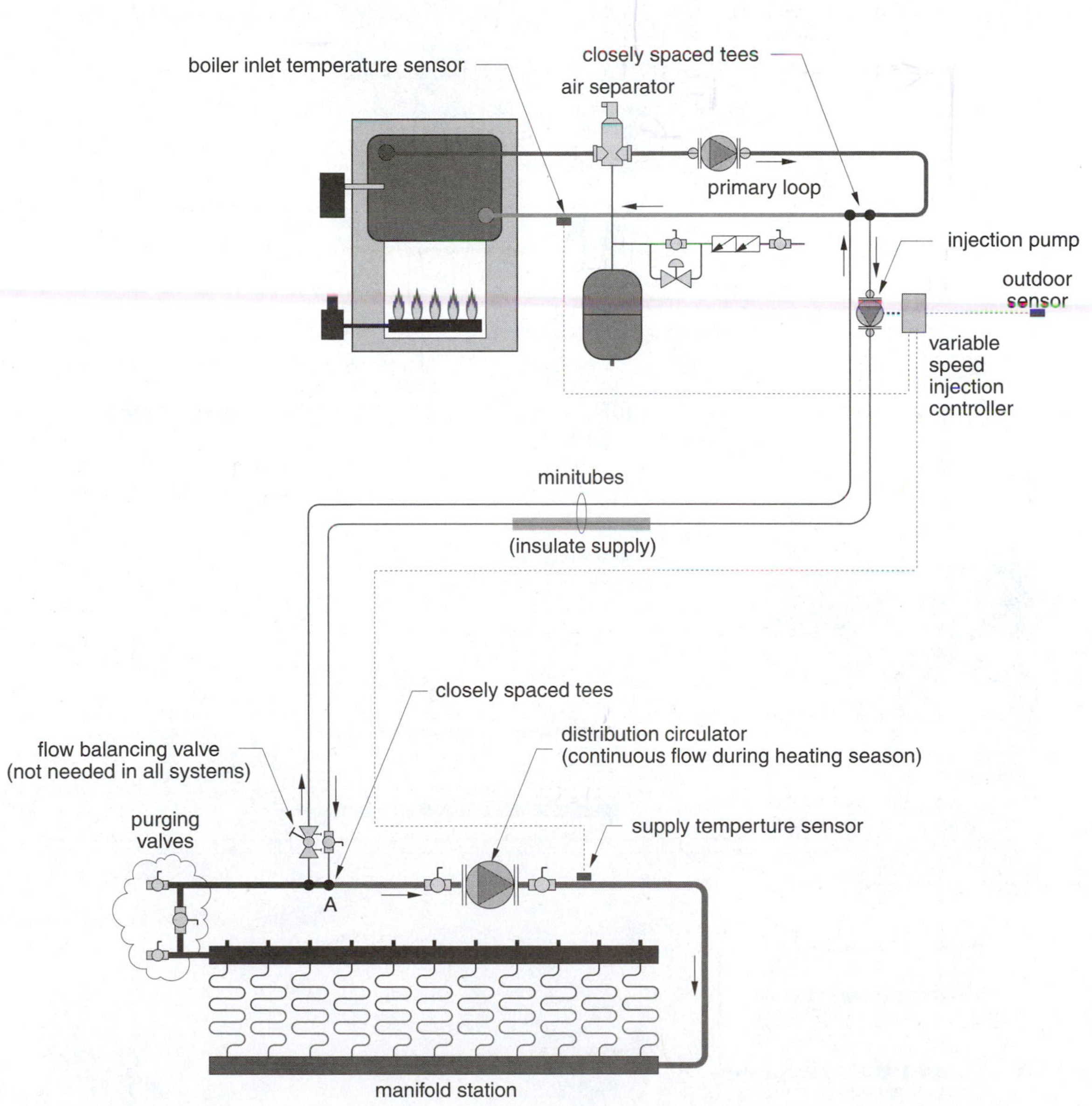

Figure 14–43 Single zone minitube distribution system for radiant floor heating.

input to the manifold station can be regulated from zero (when the injection pump is off) to full design load (when the injection pump is running at full speed).

The supply and return minitubes are coupled to the boiler loop as well as the manifold station piping using pairs of closely spaced tees. These create a primary/secondary interface that prevents the pressure distribution in the minitube subsystem from being influenced by that in either the boiler loop or the manifold station.

Multizone Minitube Systems

When planning a floor heating system for a large building, it is often necessary to locate manifold stations far from the mechanical room. Such situations are well-suited to minitube distribution systems. Using this approach each manifold station has its own variable speed injection mixing system and minitube piping. Each mixing system shares a common primary loop as its source of hot water.

One piping approach for a three-zone system using a multiple boiler array is shown in Figure 14–44. The primary loop is split into three parallel crossovers. The closely spaced tees in each crossover are the beginning and ending points for each minitube circuit. This allows the same water tempera-

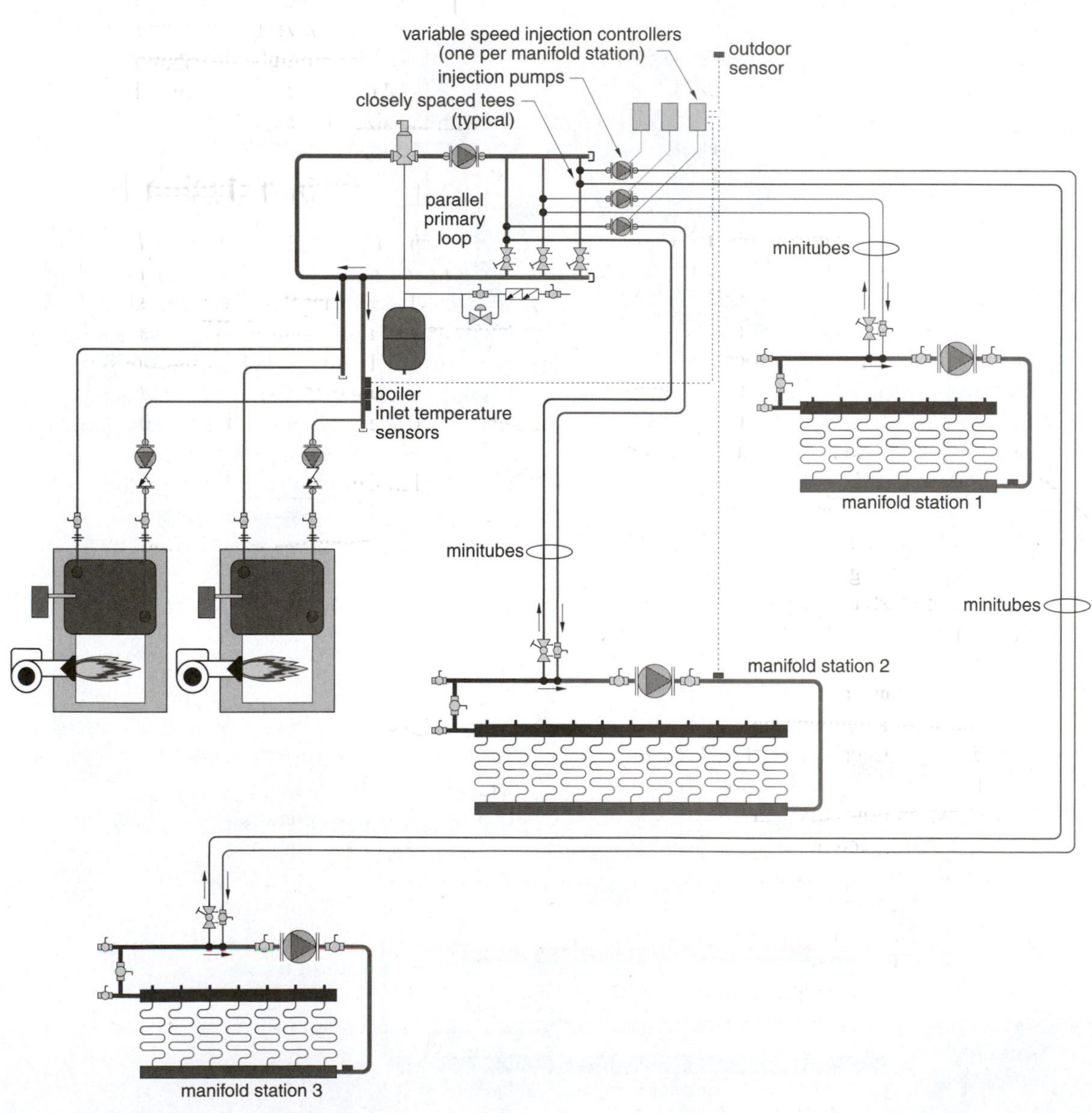

Figure 14–44 Multizone minitube distribution system.

ture to be available to each injection pump. It also eliminates any flow interference between these pumps. The balancing valves in each crossover allow the flow rate in each to be proportioned to the loads served by each manifold station.

Benefits of Minitube Distribution Systems

Minitube distribution systems offer several benefits relative to other methods of supplying remotely located manifold stations:

- *They use much smaller piping between the mechanical room and manifold station.* Besides saving on installation cost, small tubing is also easier to route and support. It reduces system volume, which in turn may reduce expansion tank size, and the required volume of antifreeze.
- *If desired, each manifold station in a minitube system can operate at a different supply water temperature or outdoor reset schedule.* This flexibility is very useful if different areas of the building have different floor coverings or widely varying load characteristics.
- *They allow the distribution circulator at each manifold station to operate continuously during the heating season.* Besides helping to provide stable supply water temperature, constant flow through floor circuits has many benefits in larger buildings. It transports stored heat from internal areas of the slab to localized "heat sinks" such as areas just inside large overhead doors, or the wet floor under a vehicle shedding snow. Should the boiler be inoperable, continuous circulation can help prevent freezing in these high loss areas.
- *Minitube systems improve the rangeability of variable speed injection mixing systems.* When variable speed injection mixing is done using short injection risers, a throttling valve must be installed in the return riser to remove a significant portion of the pump head. This is necessary to force the circulator to operate at full speed under design load conditions, and thus utilize the full range of speed control. In a minitube system, the vast majority of the head loss occurs due to fluid being carried through the minitube piping. If the head loss of the minitube circuit at design flow equals the head added by the injection pump at full speed, it is not necessary to install a balancing valve in the minitube circuit.
- *In some minitube systems it may be possible to use a smaller manifold circulator.* Since the manifold station circulator is not responsible for flow between the mechanical room and manifold station, it may be possible to use a smaller circulator. This reduces both initial cost and life-cycle operating cost.
- *The small minitube piping loses less heat to surrounding air than would larger diameter distribution piping of equivalent heat transport capacity.* Still, prudence dictates that the supply minitube be insulated to minimize heat loss. The return minitube can also be insulated, although energy savings will be minimal given the small surface area and low operating temperature. Obviously any minitubes passing through unheated space should be insulated.
- *The minitube concept is highly scalable.* A small system using 1/2-inch or even 3/8-inch minitube piping may be ideal for larger residential space heating and/or snowmelting applications. It is also well-suited for retrofit applications given the ease of routing small tubing from the mechanical room to distant manifolds. Larger systems using 1.5 to 2-inch minitube piping are capable of transporting several million Btu/hr from a boiler plant to a low temperature load. Such systems are well-suited to large agricultural floor heating installations such as multiple hog or poultry barns supplied from a central boiler plant. Large snowmelting systems using conventional boilers are also good candidates for minitube distribution systems. The cost saving relative to traditional distribution piping increases with the size of the system.

Minitube System Design Details

The circulators at each manifold station in a minitube system should be operated continuously during the heating season. The constant flow past each supply sensor helps stabilize temperature control. With the injection mixing control(s) properly configured for outdoor reset, the water temperature supplied to the distribution circuits is just warm enough to meet the prevailing heating load. Constant flow through the distribution circuit has other benefits as discussed in Chapter 9 and Chapter 10.

If internal heat gains are present in a given zone, and the injection controller has indoor temperature sensing, it may totally stop the injection pump for a time. During this time, it is important to prevent hot water from slowly migrating through the supply minitube and potentially entering the mixing point in the continuous flow manifold system. One solution is to use an injection pump with an integral check valve. Flow-check valves or external spring-loaded check valves are not recommended for this duty because they can create chattering sounds. Another solution is to ensure that the closely spaced tees connecting the minitube to the manifold station piping are at least 18 inches lower than those connecting the minitubes to the primary loop. This drop in elevation creates a thermal trap that discourages heat migration.

When a conventional boiler is used as the heat source in a minitube system, it is important to ensure that it operates at a temperature high enough to prevent sustained flue gas condensation. This is especially true when the load is a large slab with very high thermal mass. Most currently available variable speed injection controllers have the logic necessary to monitor the boiler inlet temperature and slow the injection pump when necessary to prevent the load from extracting heat faster than the boiler is generating it. Be sure to install the boiler temperature sensor as specified by the controller manufacturer to provide this protection.

The author and others have successfully deployed minitube systems in many buildings including municipal highway garages, churches, and large industrial facilities. The largest known installation to date is a 170,000-square foot manufacturing facility having thirteen independently controlled minitube-supplied manifold stations. An example of one of the large manifold stations used in this facility is shown, under construction, in Figure 14–45. Notice the 1-inch copper supply and return minitubes at the top of the manifold, which at design load delivers 500,000 Btu/hr to the floor circuits.

Figure 14–45 Example of large (500,000 Btu/hr) manifold station served by 1-inch minitube supply and return piping.

SUMMARY

The ability to supply several auxiliary loads using the same heat source that provides space heating is a unique feature of hydronic heating systems. It reduces installation costs relative to using separate heat sources for each load. It also increases seasonal boiler efficiency by creating greater run fractions. Furthermore, it reduces the number of fuel supplies, exhaust systems, as well as maintenance requirements.

Wherever possible hydronic system designers should try to maximize the use of the heat source by sharing its heat output as necessary. Prioritized load shedding can almost always provide an acceptable means of managing boiler capacity in the rare event of excess loading.

By making use of the piping and control design information in previous chapters, creative designers can combine space heating with one or more auxiliary loads to create an efficient, cost-effective, and versatile system.

KEY TERMS

ΔT protection
automatic glycol make-up system
auxiliary loads
black ice
brazed plate heat exchanger
buffer tank
capacitance rate ratio
Class 1 SIM system
Class 2 SIM system
Class 3 SIM system
continuous flow rating
counterflow arrangement
direct-fired water heater
domestic hot water (DHW)
effectiveness (of a heat exchanger)
first hour rating
flat plate heat exchanger
forced convection
fouling
heat exchanger
hydronic snow- and icemelting (SIM) systems
idle (a melted pavement)
indirect water heater
internal heat exchanger
minitube
minitube distribution system
natural convection
priority load
shell and tube heat exchanger
short-cycle
snow switch
snow/ice pavement sensor
tankless coil
usage profile

CHAPTER 14 QUESTIONS AND EXERCISES

1. Describe one of the advantages of a flat plate heat exchanger relative to other types of heat exchangers.
2. Describe why all heat exchangers should be piped for counterflow.
3. Estimate the cost of domestic water heating for a typical family of five assuming cold water enters the system at 45 °F and is supplied to the fixtures at 130 °F. Assume the water is heated by an electric water heater in an area where electricity costs $0.12/kilowatt hour.
4. Describe two disadvantages of tankless coil water heaters relative to systems with storage tanks.
5. Why do indirect water heaters have the potential to produce hot water much faster than residential electric water heaters?
6. Describe "priority control" as it applies to domestic water heating in a multiload hydronic system.
7. Why is it a good idea to use antifreeze in garage heating circuits, even if the garage thermostat will be constantly kept above freezing?
8. Water enters the shell of a shell and coil heat exchanger at 175 °F and 10 gpm. The coil has water entering at 135 °F and 8 gpm. The effectiveness of the heat exchanger is given in Figure 14–5. Determine the rate of heat transfer across the heat exchanger. Also determine the outlet temperature on the coil side of the heat exchanger.
9. Describe the operation of a boiler capable of producing 200,000 Btu/hr at 200 °F water temperature when it supplies all its output to an indirect water heater with an internal heat exchanger capable of sinking 150,000 Btu/hr when supplied with 200 °F boiler water. What is the limiting factor on DHW production in such a situation? What could be done to correct this situation?
10. What is an advantage of sensing the temperature of a snowmelting slab versus sensing only air temperature and the presence of precipitation?
11. Name two benefits of "idling" a snowmelting slab.
12. Why is it better to drain melt water from the surface rather than allow the system to evaporate it?
13. Why can't asphalt be placed directly on PEX or PEX-AL-PEX tubing when constructing a SIM system?
14. How do the flow rates used in melted pavement circuits compare to those used in radiant floor heating applications?
15. Why is it recommended that heat be supplied to a melted surface for at least 5 hours after snow stops falling?
16. When is it permissible to use circuit temperature drops in excess of 25 °F for SIM circuits? Describe the hardware details necessary for such a situation.
17. Determine the volume of a buffer tank for a system in which the heat source generates 100,000 Btu/hr and the smallest load releases heat at 2500 Btu/hr. The heat source is to have a minimum on-time of 8 minutes. The buffer tank operating range is from 110 °F to 170 °F.
18. Describe a situation where it is beneficial to avoid routing flow through a buffer tank in a multiload system.
19. Calculate the minitube flow rate needed to transfer 125,000 Btu/hr to a manifold station that supplies 110 °F water to floor heating circuit that in turn operate with a 20 °F temperature drop. Assume water is available from the primary loop at 180 °F. What size minitube piping is needed for this situation? What flow rate and tube size would be needed to supply the manifold station if the 110 °F water was produced in the mechanical room rather than at the manifold station?
20. What are three advantages of using a single boiler or multiple-boiler grouping to supply all the loads in a multiload system versus using a "dedicated" heat source for each load?

SCHEMATIC SYMBOLS A

Schematic Symbols for Piping Components

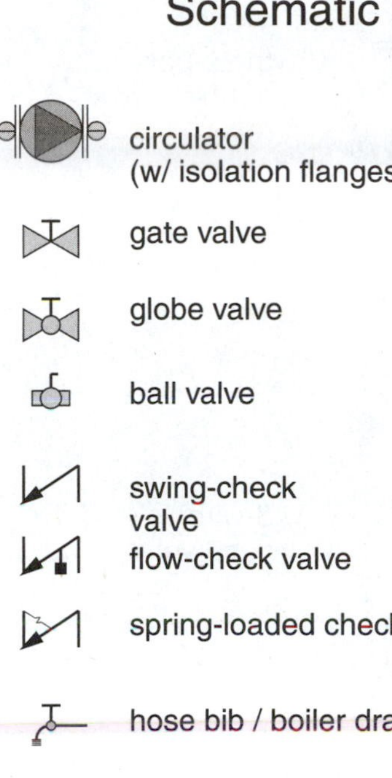

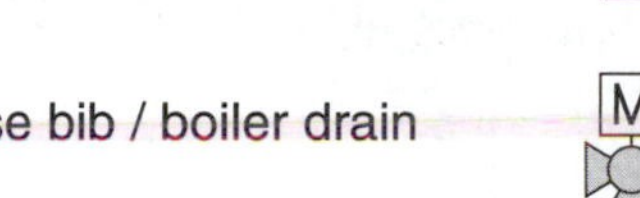

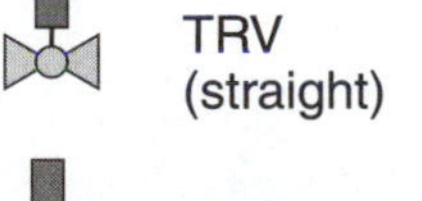

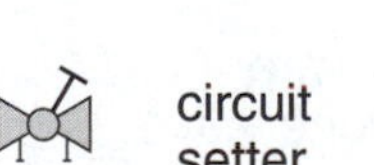

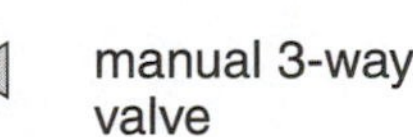

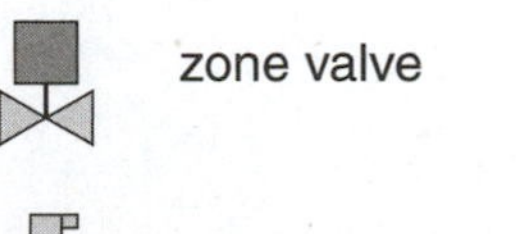

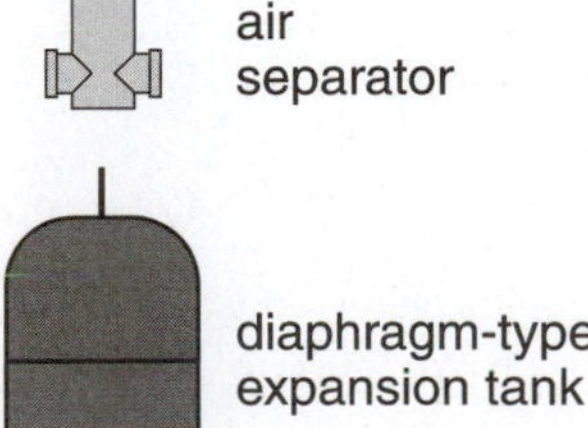

diaphragm-type
expansion tank

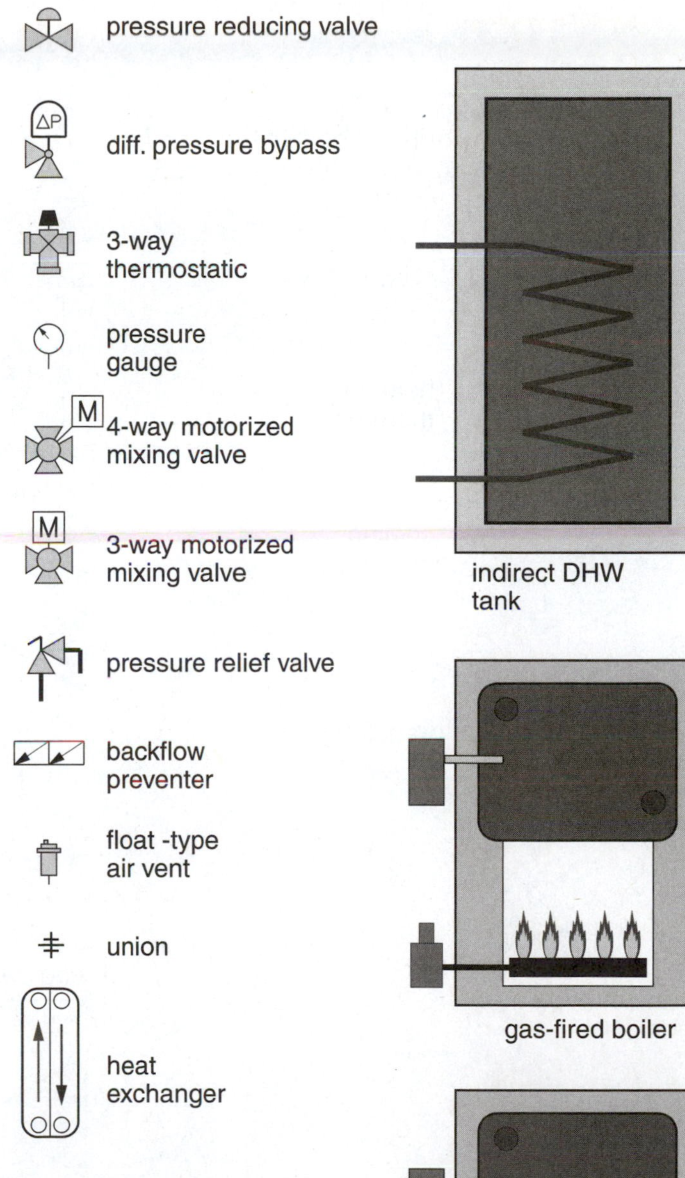

manifold station

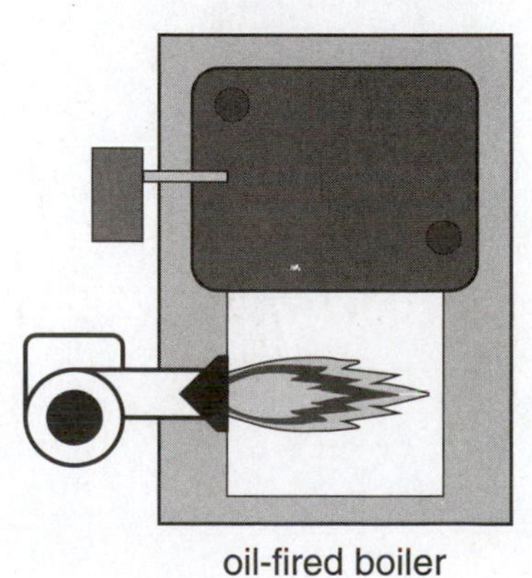

Schematic Symbols for Electrical Components

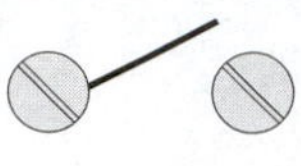
single pole switch

normally-open relay contact

normally-closed relay contact

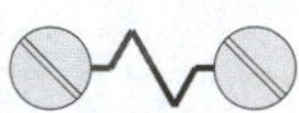
relay coil

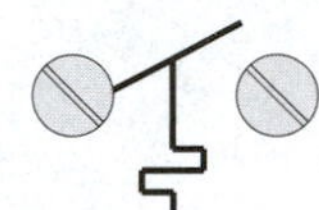
heating thermostat

cooling thermostat

transformer

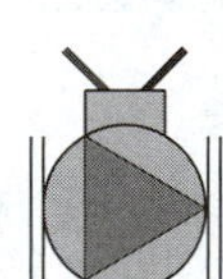
circulator

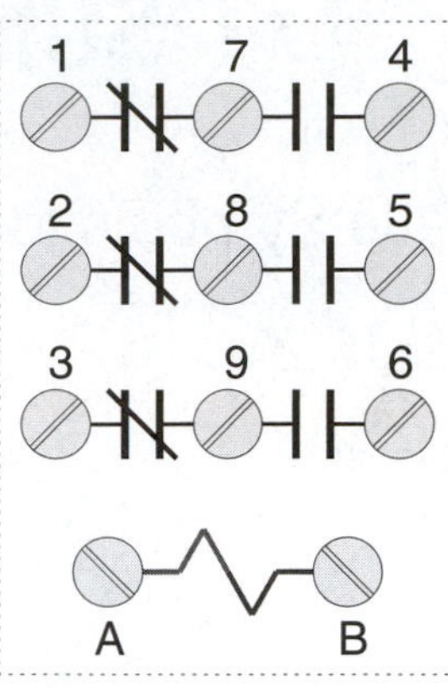

triple-pole double-throw relay

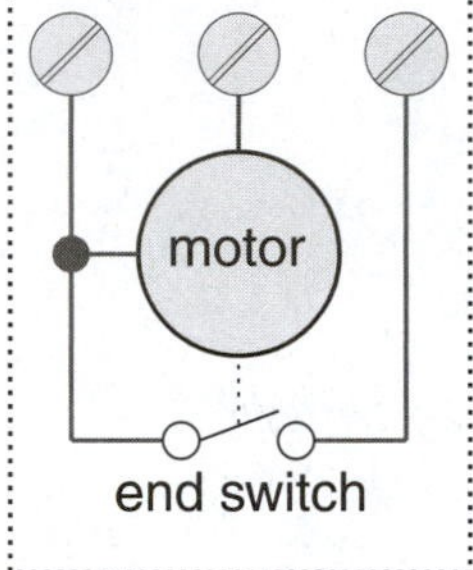

3-wire zone valve

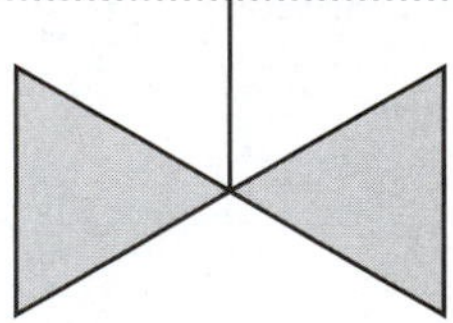

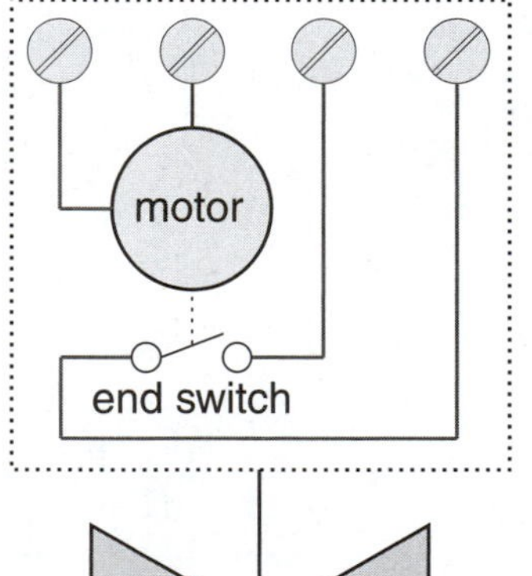

4-wire zone valve

R-VALUES OF COMMON BUILDING MATERIALS B

MATERIAL	R-VALUE*	
INSULATIONS		
Fiberglass batts (standard density)	3.17	per inch
Fiberglass batts (high density)	3.5	per inch
Blown fiberglass	2.45	per inch
Blown cellulose fiber	3.1–3.7	per inch
Foam in place urethane	5.6–6.3	per inch
Expanded polystyrene panels (beadboard)	3.85	per inch
Extruded polystyrene panels	5.4	per inch
Polyicocyanurate panels (aged)	7.2	per inch
Phenolic foram panels (aged)	8.3	per inch
Vermiculite	2.1	per inch
MASONRY AND CONCRETE		
Concrete	0.10	per inch
8-in concrete block	1.11	for stated thickness
w/vermiculite in cores	2.1	for stated thickness
10-in concrete block	1.20	for stated thickness
w/vermiculite in cores	2.9	for stated thickness
12-in concrete block	1.28	for stated thickness
w/vermiculite in cores	3.7	for stated thickness
Common brick	0.2–0.4	per inch
WOOD AND WOOD PANELS		
Softwoods	0.9–1.1	per inch
Hardwoods	0.8–0.94	per inch
Plywood	1.24	per inch
Waferboard or oriented strand board	1.59	per inch
FLOORING		
Carpet (1/4-in nylon level loop)	1.36	for stated thickness
Carpet (1/2-in polyester plush)	1.92	for stated thickness polyurethane
Foam padding (8 lb density)	4.4	per inch
Vinyl tile or sheet flooring (nominal 1/8 in)	0.21	for stated thickness
Ceramic tile	0.6	per inch
MISCELLANEOUS		
Drywall	0.9	per inch
Vinyl clapboard siding	0.61	for all thicknesses
Fiberboard sheathing	2.18	per inch
Building felt (15 lb/100 ft2)	0.06	for stated thickness
Polyolefin housewrap	~0	for all thicknesses
Poly vapor barriers (6-mil)	~0	for stated thickness

* The R-value for a specific thickness of a material may be obtained by multiplying the R-value per inch by the thickness in inches (or fractions of inches). The units on R-value are the standard U.S. units of °F × hr × ft^2/Btu.

The data in this table were taken from a number of sources including the ASHRAE Handbook of Fundamentals and literature from several material suppliers. It represents typical R-values for the various materials. In some cases a range of R-value is stated due to variability of the material. For more extensive data consult the ASHRAE Handbook of Fundamentals or contact the manufacturer of a specific product.

R-VALUES OF AIR FILMS

INSIDE AIR FILMS	R-VALUE*
Horizontal surface w/upward heat flow (ceiling)	0.61
Horizontal surface w/downward heat flow (floor)	0.92
Vertical surface w/horizontal heat flow (wall)	0.68
45-degree sloped surface w/upward heat flow	0.62
OUTSIDE AIR FILMS	
7.5 mph wind on any surface (summer condition)	0.25
15 mph wind on any surface (winter condition)	0.17

* The R-value for a specific thickness of a material may be obtained by multiplying the R-value per inch by the thickness in inches (or fractions of inches). The units on R-value are the standard U.S. units of °F × hr × ft^2/Btu.

USEFUL CONVERSION FACTORS AND DATA

C

ENERGY:
1 Watt hour = 1 whr = 3.413 Btu
1 Kilowatt hour = 1 kwhr = 3,413 Btu
1 Therm = 100,000 Btu
1 MMBtu = 1,000,000 Btu

POWER and HEAT FLOW:
1 kilowatt = 1 kw = 3,413 Btu/hr
1 Ton = 12,000 Btu/hr
1 Horsepower = 0.746 kilowatt = 2546 Btu/hr

LENGTH:
1 Foot = 12 inches = 0.3048 meters

AREA:
1 Square foot = 144 square inches = 0.092903 square meters

VOLUME:
1 Cubic foot = 7.49 gallons = 1728 cubic inches = 0.0028317 cubic meters

PRESSURE:
1 psi = 6894.76 Newtons/square meter = 6894.76 pascal
Absolute pressure (psia) = Gauge pressure (psig) + 14.7

TEMPERATURE:
°C = (°F – 32)/1.8 °F = (°C • 1.8) + 32

CHEMICAL ENERGY CONTENT OF COMMON FUELS:
1 Cubic foot of natural gas = 1020 Btu (chemical energy content)
1 Gallon #2 fuel oil = 140,000 Btu (chemical energy content)
1 Gallon propane = 91,200 Btu (chemical energy content)

GLOSSARY

Above-floor tube and plate system: A radiant floor heating system in which the tubing and heat transfer plates are installed above the subflooring.

Absolute pressure: The pressure of a liquid or gas relative to a pure vacuum. It is often expressed in units of psia (pounds per square inch absolute).

Actuator: An electrically or pneumatically driven device that adjusts the position of a valve stem or damper based on the signal it receives from a controller.

Air binding: The inability of a circulator to dislodge a large air pocket at a high point in a piping system.

Air change method: A traditional method of expressing the rate of air leakage into a building. One air change per hour means the entire volume of air in the building is replaced with outside air each hour.

Air film resistance: The thermal resistance of the air film along a wall, ceiling, or floor surface.

Air handler: A generic name for a device consisting of a blower and a finned-tube coil. It is used to either heat or cool the air passing through it based on the temperature of the fluid circulating through the tubes of the coil.

Air infiltration: The process of outside air entering a building as inside air leaks out of the building.

Air purger (or air scoop, or air separator): A device that separates air bubbles from the moving system fluid and ejects them from the system.

Air temperature stratification: The natural tendency for higher temperature air to rise to the upper portion of a room or interior space while cooler air settles near the floor.

Air vent: A manual or automatic device that releases air bubbles from a piping system.

Air-side pressurization: The pressure on the air side of a diaphragm-type expansion tank before fluid enters the tank. This pressure is adjusted by adding or releasing air through the Schrader valve on the tank.

Ampacity: The maximum current that can be carried by a conductor in a given application as determined by the National Electrical Code.

Analog control signal: An electrical voltage or current that varies over a predetermined range, and typically used to control a modulating device.

Angle valve: A valve with its outlet port rotated 90 degrees from its inlet port.

Annual fuel utilization efficiency (AFUE): An estimate of the seasonal efficiency of a heat source based on a federal government standard.

Aquastat: A device that measures the temperature of a liquid at some point in a system, and opens or closes electrical contacts based on that temperature and its setpoint temperature.

Atmospheric: A term often used to describe gas burners that are designed to operate directly exposed to (unpressurized) room air.

Automatic air vents: Air vents that can automatically eject air as it accumulates within them.

Automatic glycol make-up system: A device that automatically pumps glycol into a hydronic system upon a predetermined drop in system pressure.

Auxiliary loads: Any additional heating load served by a hydronic space heating system.

Available floor area: The floor area in a room that can be used to release heat as part of a floor heating system.

Average flow velocity: The average speed of a fluid particle across a specified cross-section in a piping component.

Backflow preventer: A piping component that contains the functional equivalent of two check valves and a vent port. It prevents any within from a hydronic system from flowing backward and contaminating a fresh water source.

Balance point temperature: The outdoor temperature below which a building requires heat input from its space heating system.

Balancing valve: A valve operated at a partially open setting to achieve a desired flow rate or pressure drop at a given location in a hydronic system.

Ball valve: A valve containing a rotating ball with a hole through it. It can be used for component isolation or limited flow regulation.

Bar: A unit of pressure. 1 bar = 14.504 pounds per square inch (PSI).

Baseboard tee: A special fitting that resembles a 90 degree elbow with an additional threaded tapping. This fitting is often used to mount an air vent at the outlet end of a finned-tube baseboard element.

Below-floor tube and plate system: A radiant floor heating system in which the tubing and metal heat transfer plates are attached to the underside of a subfloor.

Bend supports: Preformed elbows or clips that allow flexible polymer tubing to be bent to a relatively small radius without kinking.

Bimetal element: An assembly composed of two dissimilar metal strips joined at their ends. It creates the movement of electrical contacts in a device such as a room thermostat.

Bin temperature data: A means of reporting the duration of ranges of outside temperature based on the number of hours in a year the temperature is within a given range.

Black ice: A very thin film of ice that forms on pavement surfaces under certain conditions, and is very difficult to see.

Boiler drain: A valve that connects to a hose thread and can be placed at any point in a system requiring drainage.

Boiler feed water valve: Another name for a pressure reducing valve that automatically allows water to flow into a hydronic system if the system pressure drops below a preset value.

Boiler jacket: The outer sheet metal or polymer shell of a boiler.

Boiler protection: A control action intended to prevent a conventional boiler from operating with sustained flue gas condensation.

Boiler reset control: A control action and/or device that automatically changes the upper temperature limit of a boiler based on outdoor temperature.

Boiler rotation: A control technique that varies the firing order of two or more boilers in a multiple boiler system with the goal of equalizing the operating hours of each boiler.

Boiler short-cycling: An undesirable condition during which the burner of a boiler operates for a very short time. It leads to a high number of on/off cycles and premature wear of components.

Bond breaker: A material used to prevent an adhesive bond between two materials.

Boundary layer: A thin layer of fluid that moves slowly along the inside wall of a pipe. The thicker the boundary layer is, the poorer the heat transfer between the bulk of the fluid and the tube wall.

Brazed plate heat exchanger: A compact heat exchanger consisting of parallel stainless steel plates brazed together at their edges.

Brine: A generic term for a solution of water and antifreeze such as propylene glycol, ethylene glycol, or calcium chloride.

British thermal unit: The amount of energy required to raise one pound of water by one degree Fahrenheit, also abbreviated Btu.

Bubble rise velocity: The speed at which a bubble can rise through a liquid. Larger bubbles have greater rise velocities.

Buffer tank: An insulated, water-filled storage tank that forms the link between a heat source and a hydronic distribution system. This tank allows the rate of heat input from the heat source to be different than the rate of heat output to the distribution system.

Building heating load: The rate at which heat must be added to a building to maintain interior comfort.

Buoyancy: The upward force that causes a less dense material to rise above a more dense fluid.

Candidate circulator: A circulator that is being considered as a possible choice for a given flow/head loss situation.

Capacitance rate ratio: The ratio of the smaller fluid capacitance rate through a heat exchanger to the larger capacitance rate.

Capillary tube: A very small diameter tube between a temperature sensing bulb and a controller.

Cavitation: The formation of vapor pockets when the pressure on a liquid drops below its vapor pressure. Cavitation is very undesirable in circulators.

Central deareating device: A device that separates air bubbles from the moving system fluid and ejects them from the system.

Centrifugal pump: An electrically driven device that adds mechanical energy (e.g., head) to a fluid using a rotating impeller. In this text, the word pump refers specifically to a centrifugal pump, and is used synonymously with the word circulator.

Check valve: A valve that limits flow to one direction.

Chemical energy content: The theoretical maximum energy content of a fuel based on its chemical composition.

Circulator: A term that, in this text, is synonymous with the word pump. A device that creates fluid motion in a hydronic system by adding head energy to the fluid.

Class (of SIM system): One of three classes of snowmelting systems (SIMs) based on the rate of their ability to keep the pavement surface free of snow as it is falling.

Close-coupled: Bolting two or more inline circulators together end to end, in the same flow direction, to increase head.

Closed control loop: A control system configuration in which the controlled variable is sensed and the control action modified (if necessary) based on any deviation between the sensed value and target value of the controlled variable. Also know as a feedback control loop.

Closed-loop system: A piping system that is sealed at *all points* from the atmosphere.

Closely spaced tees: Two tees that are installed immediately adjacent to each other to form a primary/secondary connection between two piping circuits.

Coefficient of linear expansion: A number that indicates the change in length of a specified material based on a change in temperature.

Coefficient of performance (COP): The ratio of the rate of heat output of a heat pump to the rate of electrical energy input, *in the same units*. High COPs are desirable.

Coil: A name used to describe a fluid to air heat exchanger mounted within an air handler. It may also refer to the electromagnet device that provides the motive force to move contacts in a relay or contactor.

Combustion efficiency: The efficiency of a heat source in converting the chemical energy content of its fuel into heat, based on measurements of exhaust gas temperature and carbon dioxide content.

Common pipe segment: The piping path in a multiple branch system that is in common with branch circuits and through which the full system flow passes.

Component isolation: The use of valves on all piping connections to a given device that allow that device to be isolated for service if necessary.

Composite tubing: A term often used to describe tubing made of both metal and polymer materials. PEX-AL-PEX is an example of composite tubing.

Compressible fluids: Fluids, usually gases, which can be compressed into smaller volumes when pressure is applied to them.

Compressor: The device in which a refrigerant gas is compressed in volume, and at the same time, increased in temperature.

Computer-aided drafting (CAD): The use of a computer and associated software for preparing technical drawings such as floor plans, and schematics.

Concrete thin-slab: A layer of concrete in the range of one to two inches thick placed over hydronic tubing to create a radiant floor panel.

Condensate: The liquid formed as a gas, such as water vapor, condenses.

Condenser: The device in which a refrigerant gas releases heat, and in doing so changes from a vapor into a liquid.

Condensing mode: Operating a combustion type heat source so that the water vapor produced during combustion condenses on the internal surfaces of the heat exchanger.

Conduction: A natural process whereby heat is transferred by molecular vibrations through a solid or liquid material.

Confined space: Any space not meeting the definition of unconfined space in the National Fuel Gas Code.

Constant pressure circulator: A circulator controlled by a variable speed drive so that it provides a constant differential pressure at all flow rates.

Contactor: Another name for a relay that is rated for relatively high current/line voltage loads.

Contacts: The metal components within a switch or relay that come together or move apart to determine if electrical current will flow through a given pole of the switch or relay.

Continuous flow rating: The rate at which a water heating device can supply heated water at a given temperature on a sustained basis. Often expressed in gallons per hour or gallons per minute.

Control algorithm: The operating instructions or logic that a controller follows.

Control differential: The intended difference in the sensed control variable from where the controller output is turned on to when it is turned of. For example, if a thermostat contact is expected to open at 72 °F and close at 68 °F the control differential is 4 °F.

Control joint: A groove penetrating part way through a concrete slab for the purpose of forcing a crack to occur along a predetermined path.

Control state: The current condition of the output of an on/off controller (e.g., the "ON" state or the "OFF" state).

Controlled device: The hardware device that is being regulated by the controller.

Controlled variable: The physical parameter that is being measured and controlled by the controller. Typical controlled variables in hydronic systems include temperature and pressure.

Controller: The device that monitors the controlled variable and provides control outputs that direct operation of a controlled device.

Convection: A means by which heat is transferred between a solid surface and a fluid such as air or water.

Convector: A generic name for heat emitters that transfer the majority of their heat output by convection.

Copper water tube: Copper tubing specifically manufactured for use in conveying water for domestic use, or within hydronic heating systems.

Counterflow arrangement: A piping arrangement that causes two fluids to flow in opposite directions as they pass through a heat exchanger. This is generally desirable for maximum heat exchange.

Counterflow spiral: A shape for a radiant floor tubing circuit that is intended to evenly distribute heat within a room.

Cross-linked polyethylene: Polyethylene in which long molecular chains have been bonded together to form a stronger and more temperature resistant material. The acronym PEX is often used to represent cross-linked polyethylene.

Crossover bridge: A piping segment that allows water to flow from a supply piping main to a return piping main. It usually contains a pair of closely spaced tees or a heat emitter.

Cycle efficiency: The net efficiency of a heat source accounting for heat losses during off-cycles. Cycle efficiency decreases as run fraction decreases.

Dead heading: An undesirable operating condition in which a circulator is running, but all flow through it is blocked.

Degree day: The difference between the outside average temperature and 65 °F over 24 hours. Daily degree days can be added together to obtain monthly and annual total degree days.

Delay-on-break: An operating mode for a time delay relay in which the contacts maintain their operating state until a set time has elapsed from when the input signal to the relay was turned off.

Delay-on-make: An operating mode for a time delay relay in which the contacts maintain their normal (deenergized) state until a period of time has elapsed from when the input signal to the relay was turned on.

Delta T (ΔT) protection: A control action used in snowmelting systems in which the controller regulates heat input so that the circuit temperature drop does not exceed a set value.

Demand-fired: A term describing a boiler that is only fired when there is a demand for heat (e.g., the boiler does not maintain temperature at all times).

Density: In the context of this book, density refers to the weight of a substance divided by its volume. Some readers will recognize this as being the same as the *specific weight* of a substance. Common English units for density, as defined herein, are lb/ft^3 (pounds per cubic foot).

Design dry bulb temperature (97.5%): The temperature the outside air is at, or above, 97.5 percent of the year. This is the outside temperature at which the design heating load is calculated.

Design heating load: The heating load of a building or room when the outside air temperature is at its design dry bulb value.

Dewpoint temperature: The temperature at which water vapor in a mixture of other gases begins to condense into liquid droplets. Other compounds in a mixture of gases also have an associated dewpoint temperature.

Diaphragm-type expansion tank: An expansion tank that contains a flexible diaphragm that separates the air and water in the tank.

Dielectric union: A union made of steel on one side and brass on the other. A dielectric (electrically insulating) material separates the two metals to prevent galvanic corrosion.

Differential: A *range* of temperature between the point where the electrical contacts of a controller open and when they close. The differential of some controllers is fixed, while on others it can be adjusted.

Differential pressure: The difference in pressure between two points.

Differential pressure bypass valve: A valve designed to allow flow through when the differential pressure across its ports reaches a specific value. It is commonly used in multizone parallel piping systems.

Differential pressure transducer: A device for measuring the differential pressure between two points and generating an analog output signal based on that measurement.

Diffuse: The migration of molecules of one substance through the molecular structure of another material. Oxygen, for example, can slowly diffuse through the molecular structure of certain polymers.

DIN rail: An industry standard mounting rail for relays and other modular controls.

Direct digital control: A control network in which sensors, controllers, and controlled devices all communicate using digital signals sent along a common communication bus.

Direct injection mixing: A piping configuration in which hot water from a heat source is injected into a distribution system, and cool water from the return side of the distribution system flows back to be reheated by the heat source.

Direct vent: A method of exhausting combustion products from a boiler directly through an outside wall.

Direct-fired water heater: A water heater that contains its own heating element or burner.

Dissolved air: Molecules of oxygen, nitrogen, carbon dioxide, and other gases found in air that are present within liquid water.

Distribution system: An assembly of piping, fittings, and valves that conveys heated water from a heat source to heat emitters. It may be a simple piping loop, or an elaborate, multibranch piping system.

Diverter tee: A special fitting resembling a tee with an internal baffle. This fitting is used to force a portion of the flow entering the tee out through a branch circuit connected to its side port.

Diverter valve: A valve (often electrically operated) that directs a single incoming fluid stream in one of two possible directions based on the position of the valve's flow element.

DOE heating capacity: The heat output rating of boilers up to 300,000 Btu/hr based on a federal government rating standard.

Domestic hot water (DHW): Heated water used for purposes of washing, cooking, or other uses. Domestic hot water is considered suitable for human consumption.

Domestic water: Water suitable for human consumption.

Double interpolation: A mathematical method of estimating a value that is known to lie between two other values of each of two independent parameters.

Draft proving switch: A device that measures the negative air pressure in an exhaust system and closes its electrical contacts whenever a safe negative pressure is maintained.

Draindown system: A type of active solar energy system in which freezing is avoided by having all fluid drain out of the collectors and exposed piping *to waste* at the end of each operating cycle.

Droop: The tendency for a controller operating with proportional logic to allow a temperature to stabilize below the desired target value when the rate of heat input to the load increases.

Dry base boiler: A boiler design in which the boiler block rests on top of the combustion chamber.

Dumb mixing devices: A mixing device such as a 3-way valve that does not sense temperature or flow rate, and thus is incapable of reacting to changes in operating conditions.

Duty cycle: The percent of elapsed time in which the device is operating. A 50 percent duty cycle indicates the device is operating 50 percent of the elapsed time.

Earth heat exchanger: Piping buried in the soil for extracting or rejecting heat.

Effective total R-value: The average thermal resistance of a wall, ceiling, floor, etc., after compensating for the presence of framing members.

Effectiveness (of a heat exchanger): The ratio of the actual heat transfer across a heat exchanger to the maximum possible rate of heat transfer. Effectiveness is a unitless number between 0 and 1. Higher values are more desirable.

Efficiency: The ratio of a desired output quantity or rate divided by the necessary input quantity or rate. Both quantities must have the same mathematical units.

Efficiency (of a circulator): The ratio of the rate of mechanical energy imparted to the fluid divided by the rate of energy input to the circulator. When the rate of energy input is the wattage drawn by the circulator motor, this efficiency is often expressed as the "wire-to-wire" efficiency.

Electric thermal storage (ETS) system: A system designed to use off-peak electrical energy to heat water for subsequent use in heating a building. The energy is stored in the form of heated water.

Electromagnetic energy (a.k.a. electromagnetic radiation): Energy that is transferred from one location to another by combined electric and magnetic fields. Examples include visible light, infrared light, microwaves, radio waves, and "radiant heat."

Electromagnetic spectrum: The entire range of frequencies of electromagnetic radiation from low frequency/long wave length radio waves to high frequency/short wave length gamma rays.

Electromechanical controls: Controls that use mechanical force generated by springs, bimetal elements, or pressure diaphragms to open and close electrical contacts.

Electronic control: Controls that use low voltage solid state electronic circuitry including analog electronic circuits and digital electronics that process input signals from sensors and settings to generate output signals.

Element: A name for the finned-tube component in a baseboard convector. The name also applies to the device that creates heat in an electric boiler or water heater.

Elevation head: The head energy within a fluid due to its elevation. The higher the elevation, the greater the elevation head energy content.

Embedded controller: A preassembled controller that performs a specific task, and is part of an overall control system.

Emissivity: A number between zero and one that indicates the ability of a surface to release thermal radiation. Surfaces with high emissivities release thermal radiation at higher rates compared to surfaces with low emissivities.

End suction pump: A pump with an inlet connection perpendicular to the plane of its impeller.

End switch: A switch built into a zone valve that closes its electrical contacts when the valve reaches its fully open position.

Entrained air: Air bubbles that are carried along with the fluid flowing through a piping system.

EPDM: Ethylene Propylene Diene Monomer. A high quality rubber compound resistant to many chemicals.

Equal percentage characteristic: A valve characteristic in which the flow rate through the valve increases slowly as the stem starts to lift, and increases faster the farther the stem lifts. This characteristic gives an equal percentage change over the previous flow for equal increments of stem movement. This characteristic is desirable when heat output is being controlled by varying the flow rate through the heat emitter.

Equilibrium: A "balanced" operating condition in which energy input equals energy output.

Equivalent length: The concept of replacing a particular piping component by a straight length of pipe, of the same diameter, that would yield the same hydraulic resistance as the component being replaced.

Equivalent resistance: The result of mathematically combining two or more hydraulic resistances into a single hydraulic resistance that represents their combined effect.

Erosion corrosion: A process where metal is removed from the inside surfaces of piping or piping components due to excessively high flow velocity.

Error: The difference between a target value and the actual (measured) value of a controlled variable.

Escutcheon plate: A decorative trim piece that slides over a pipe and covers the hole in the surface the pipe penetrates.

Expansion loop: A loop formed within an otherwise straight run of tubing that can absorb the movement in the pipe due to thermal expansion/contraction.

Expansion offset: An offset rectangular path within an otherwise straight run of tubing that can absorb the movement in the pipe due to thermal expansion/contraction.

Expansion tank: A tank specifically designed and sized to accommodate the increased volume of system fluid when that fluid is heated.

Expansion Tank Sizer: A module in the Hydronics Design Studio that assists in sizing diaphragm-type expansion tanks.

Exposed surface: Any building surface exposed to outside air temperature.

External VA rating: The volt-amps that can safely be supplied to an external electrical load from a controller with a built-in transformer.

Extruded polystyrene foam: A type of closed-cell foam insulation board often used for insulation under a concrete slab on grade.

Eye (of an impeller): The center opening where fluid flows into an impeller.

Fan-coil: A heat emitter that contains an internal blower or fan to force air through a finned-tube coil where it absorbs heat.

Feed water valve: A valve that automatically allows cold water to enter a closed hydronic system upon a drop in system pressure.

Feedback: A signal sent from a sensor back to a controller that is regulating the controlled variable measured by the sensor.

Feet of head: The common English units of expressing the head energy of a fluid. The units of feet represent the total mechanical energy content of each pound of fluid, and are derived from simplifying the units of Ft•lb/lb.

Female threads: Pipe threads on the inside diameter of a fitting, valve, or other piping component.

Finish floor resistance: The thermal resistance or R-value of a finish floor material placed on top of the heated subfloor.

Finned tube: A tube with added fins to enhance heat transfer.

Finned-tube baseboard convector: A type of convector that is usually composed of a copper tube with aluminum fins mounted in a steel enclosure located at the base of walls.

Fire-tube boiler: A boiler design where hot combustion products passing through tubes heat water on the outside of the tubes.

First hour rating: The volume of domestic hot water that a given water heater can deliver within the first hour of a draw. It is based on a specified hot water supply temperature and rate of heat input.

Fittings: Components used to attach straight lengths of pipe or tubing. Common fittings include elbows, tees, unions, couplings, reducer couplings, and bushings.

Fixed lead boiler: An operating sequence for a multiple boiler system in which a designated boiler is always the first boiler fired upon a demand for heat.

Flanges: The common means of connecting a circulator to piping. The pump's volute has integral cast flanges on its inlet and discharge ports. These bolt to matching flanges threaded onto the piping. An O-ring provides a pressure tight seal between the flanges.

Flat plate heat exchanger: A modern design for heat exchangers in which the fluids flow through alternating spaces between specially shaped stainless steel plates.

Flat plate solar collectors: Devices that convert solar radiation to heat using a flat, fluid-cooled plate in an insulated and glazed housing.

Floating control: A control output signal used to operate valves and dampers which always has one of three states: opening, closing, or hold the present position. It is also sometimes called "3-wire," or "tri-state" control.

Floating zone: A specified range in which the error between the measured value of a controlled variable and the target value is small enough that no open or close signal is generated by the controller. It is sometimes called the "neutral zone."

Float-type air vents: A type of automatic air vent that uses an internal float to open and close the air venting valve.

Flow-check valve: A valve with a weighted plug that prevents backward flow *and* forward migration of hot water due to buoyancy.

Flow coefficient (C_v): The flow rate, in gallons per minute, of 60 °F water needed to create a pressure drop of one psi across a piping component. C_v values for valves are often listed in manufacturers' specifications.

Flow rate: The volumetric rate of flow of a fluid. For liquids it is often expressed in units of gallons per minute (gpm). For gases, it is often expressed in units of cubic feet per minute (cfm).

Flow regulation: The action of changing the flow rate through a device, often for the purpose of controlling heat output or adjusting a mixed fluid temperature.

Flow velocity: The speed of an imaginary fluid particle at some point in a piping system. Common English units for flow velocity are feet per second.

Fluid feeder: A device that uses a pump to automatically add fluid to a hydronic system whenever the pressure at some location in the system drops below a preset value.

Fluid properties factor (α)**:** A number that combines the fluid properties of density and viscosity of a given fluid into a single index for use in determining hydraulic resistance.

Flux: A paste containing an acid that chemically cleans the surfaces of copper tubing and fittings being joined by soldering.

Forced convection: Heat transfer between a solid surface and a moving fluid where the fluid's motion is created by a fan, blower, or pump.

Forced-water purging: The use of a forced-water stream to entrain air in a piping system and drive it out of a purging valve.

Fouling: A term describing the undesirable accumulation of particulates, minerals, or other debris on the wetted surfaces of a heat exchanger.

Free area: The *unobstructed* area for air to flow through a louvered panel or screen.

Free standing panel radiator: A panel radiator that stands on support brackets secured to the floor rather than being supported by a wall.

Full port ball valve: A ball valve in which the hole through the ball is approximately the same diameter as the pipe size of the valve.

Full reset: The ability of an outdoor reset control to reduce the water temperature supplied to the distribution system all the way from some maximum value at design load conditions down to room temperature at no load conditions.

Full storage systems: A thermal storage system designed to store all the energy needed by the load, for a 24-hour period, during the preceding off-peak period.

Fuzzy logic: Control logic that allows a controller to automatically modify the operating characteristics of the controller based on how the system is responding, and without human intervention.

Galvanic corrosion: Corrosion between dissimilar metals caused by the flow of electrons from a less noble metal (on the galvanic chart) to a more noble metal.

Gaseous cavitation: Cavitation caused by air bubbles entrained in the fluid rather than vapor pockets.

Gate valve: A type of valve designed for component isolation purposes.

Gauge pressure: The pressure of a liquid or gas measured relative to atmospheric pressure at sea level. Common English units for gauge pressure are psig (pounds per square inch gauge), or simply psi.

General purpose relay: A type of relay that can be used for many different control applications.

Globe valve: A type of valve designed for flow regulation purposes.

Gravity drainback system: A type of active solar energy system in which freezing is avoided by having all water drain out of the collectors and exposed piping *into a holding tank* at the end of each operating cycle.

Gravity purging: Filling a hydronic system from the bottom and allowing the air in the piping to rise up to, and out of, high point vents.

Gypsum-based underlayments: A pourable self-leveling mixture containing gypsum-based cement and washed sand. It is specifically intended for use as a thin-slab material.

Gypsum thin slab: A type of construction for radiant floor heating in which a slab of poured gypsum underlayment ranging from 1.25 inch to 2 inch thick is poured over tubing that has been fastened to a subfloor.

Hard drawn tubing: Copper tubing that is sold in straight lengths.

Hardwired logic: Specific wiring connections between components that force a control system to operate in a predetermined manner.

Head: The total mechanical energy content of a fluid at some point in a piping system.

Heat anticipator: An adjustable resistor contained in a room thermostat that preheats the bimetal element to minimize room temperature overshoot.

Heat capacity: A material property indicating the amount of heat required to raise one cubic foot of the material by one degree Fahrenheit. Common English units for heat capacity are Btu/ft^3/°F.

Heat emitter: A generic term for a device that releases heat from a circulating stream of heated water into the space to be heated. Examples include convectors, radiator, and fan-coils.

Heat exchanger: A device that transfers heat from one fluid to another without allowing the two fluids to make physical contact.

Heat flux: The rate of heat flow across a unit of area. The commonEnglish units for heat flux are Btu/hr/ft^2.

Heat loss coefficient: A value that indicates the rate of heat loss (usually for an entire building) per degree of temperature difference between the inside and outside air temperature.

Heat motor actuator: A device that moves the stem of a valve when supplied with a suitable voltage. The necessary movement is produced by the expansion/contraction of a thermal element that is heated by an internal resistor.

Heat output ratings: One of several ratings describing the rate of heat output of a heat source such as a boiler.

Heat pump: A device that uses the refrigeration cycle to move heat from an area of lower temperature to one of higher temperature.

Heat purging: Removing additional heat from a boiler by maintaining circulation through it after the burner has stopped firing.

Heat sink: A material that absorbs heat from a heat emitter or heat exchanger.

Heat source: A device that supplies heat to a space-heating distribution system.

Heat transfer plate: A metal plate that fits over heating tubing and provides lateral heat conduction away from the tubing.

Heating capacity: The rate of heat output of a heat source, usually expressed in Btu/hr.

Heating curve: Another name for a reset line.

Heating effect factor: An allowance of 15 percent extra heat output, beyond the laboratory tested heat output, allowed by the IBR rating standard for finned-tube baseboard convectors.

Hi/Lo fire boiler: A boiler that can produce two rates of heat output based on control input. Hi/Lo boilers typically have two independently controlled burner assemblies.

High limit setting: An upper temperature limit set on a temperature controller.

High point vents: Devices located at the high point(s) of a piping system that can release air that may accumulate at these high points.

Home run distribution system: A distribution system in which each heat emitter has its own supply and return tube routed from a common manifold station.

Horizontal panel radiator: A panel radiator that is wider than it is tall, and has tubes oriented in the horizontal direction.

Hose bib: A common name for a valve that has threads for connecting a common garden hose.

Hybrid distribution system: A distribution system containing more than one type of fundamental distribution piping design.

Hydraulic equilibrium: The condition where the head added by the system's circulator exactly equals the head dissipated by viscous friction of the fluid flowing through the system.

Hydraulic resistance: A means of representing the ability of piping components to remove head energy from a flowing fluid. The greater the hydraulic resistance of a component, the more head loss it creates for a given flow rate.

Hydraulic resistance diagrams: A drawing composed of hydraulic resistors that represents an assembly of piping components.

Hydronic Circuit Simulator: A module in the Hydronics Design Studio software for simulating the flow and heat transfer of a user-defined hydronic piping system.

Hydronic snow- and ice-melting (SIM) system: A system for melting snow and ice by circulating a warm antifreeze solution through an external paving.

IBR gross output: An industry standard rating for boiler heat output that includes the heat transferred to water flowing through the boiler but does not include heat loss from the boiler jacket. This output is expressed in Btu/hr.

IBR net output: The IBR gross output rating minus 15 percent for the assumed heat loss of distribution piping, and delay in heat delivery due to the thermal mass of the boiler. This output is expressed in Btu/hr.

Idling (a melted pavement): To maintain sufficient heat flow to the slab so that its temperature is very close to 32 °F at all times during the winter. Idling a slab allows it to reach melting conditions quickly when frozen precipitation occurs.

Impeller: A rotating disc with curved vanes contained within a centrifugal pump that adds head to a fluid as it is accelerated from its center toward its outer edge.

Implosion: A term describing the rapid and violent collapse of vapor pockets as the pressure around them rises above the vapor pressure of the fluid.

Incompressible: The ability of liquids to transmit large pressures without themselves being compressed into smaller volumes.

Indirect water heater: An insulated storage tank that contains an internal heat exchanger. Hot boiler water is circulated through one side of the heat exchanger, transferring heat to domestic water on the other side.

Infiltration: The unintentional leakage of outside air into heated space.

Infrared thermal radiation: Another name for thermal radiation. It is technically electromagnetic radiation having wavelengths longer than can be seen by the eye.

Injection control device: A device such as a variable speed pump or modulating valve that regulates the rate at which hot water is injected into a circulating distribution system.

Injection mixing: A method of controlling the temperature of a hydronic distribution system by adding (injecting) hot water and simultaneously removing cooler water. The faster the injection rate, the warmer the distribution system becomes and the greater its heat output.

Inline circulator: A circulator constructed such that its inlet and outlet ports lie on a common centerline.

Intelligent mixing device: A mixing device that senses fluid temperature and is able to adjust one or more flow rates based on how the sensed temperature compares to a set target value. Such devices may be electronic or use nonelectric thermostatic actuators.

Interior circuits: A radiant floor tubing circuit that is not run adjacent to exterior (or otherwise exposed) walls.

Internal heat exchanger: A heat exchanger contained within a storage tank and relying on natural convection for heat exchange with the water in the storage tank.

Internet accessibility: The ability of a control system to send or receive information from a remote site using the Internet.

Interpolation: A mathematical method of estimating a numerical value that is know to be between two other stated values. It is based on proportioning.

Inter-zone heat transfer: Heat movement across interior partitions, floors, and ceilings separating building areas that are kept at different temperatures.

Isolation flanges: Special flanges that contain a built-in shutoff valve.

Isotherms: Lines connecting points having the same temperature. Isotherms are frequently used to display the results of finite element analysis of heat transfer through solid materials.

Iteration: A design process that entails making successively refined estimates until a mathematically stable operating condition is determined.

Kick-space heater: A small horizontal fan-coil convector that mounts in the recessed space at the base of a kitchen or bathroom cabinet.

KPa (Kilo-Pascals of pressure): A pressure equal to 1000 newtons of force distributed over one square meter of area.

Ladder diagram: Electrical schematic drawings that are used to design and document a control system.

Laminar flow: A classification of fluid flow where streamlines remain parallel as the fluid moves along the pipe. Laminar flow is more typical at very low flow velocities.

Latent heat: Heat that is added or removed from a substance without any temperature change. This occurs as a material changes phase from solid to liquid, or from liquid to vapor.

Lead/lag control: A control technique that automatically rotates the firing order of boilers in a multiple boiler system in an attempt to keep the operating hours of each boiler approximately equal.

Leaders: The portion of a radiant panel tubing circuit that is not within the radiant panel, but instead connects the tubing within the radiant panel to a manifold station.

Lightweight concrete: Concrete that uses a lightweight aggregate in place of dense stone aggregate. Lightweight aggregates include vermiculite, expanded shale, and polystyrene beads.

Line voltage: A term that usually indicates a nominal 120 VAC relative to ground potential.

Line voltage section: That portion of a ladder diagram containing line voltage devices.

Liquid: A physical state of matter in which a material is pourable, and conforms to the shape of its container. Most liquids are also incompressible.

Load: The heating requirement necessary to maintain a desired temperature in a building, room, or a volume of material such as a tank of water.

Lockshield valve: A valve that mounts on the outlet of a heat emitter and serves as a combination isolation valve, flow balancing valve, and drain valve.

Low voltage: For hydronic systems, most low voltage control circuits operate at 24 VAC.

Low water cutoff (LWCO): A safety control that opens electrical contacts to turn off the heat generating element or burner of a heat source if the water level in a system drops below the level of its probe.

Make-up water assembly: An assembly of components that automatically adds water to a hydronic system to make up for minor losses. It usually consists of a shutoff valve, feed water valve, and backflow preventer.

Male threads: Pipe threads on the outside diameter of pipe or fittings.

Manifold: A piping component that serves as a common beginning or ending point for two or more parallel piping circuits.

Manifold station: A combination of a supply manifold and return manifold that services two or more floor heating circuits.

Manipulated variable: The physical quantity that is being regulated by a control process.

Manometer: A device to measure the pressure difference between two points in a piping system.

Manual air vents: Devices that, when manually opened, allow air to be released (vented) from a piping system.

Manual reset high limit (MRHL): A safety control that opens a normally closed contact when a set temperature is detected at its probe. A button on the device must be pushed to close the contacts once they have opened.

Maximum allowed upward heat flux: The maximum rate of heat transfer from a radiant panel, (often measured in Btu/hr/ft^2), based on limiting the average surface temperature of the panel to a predetermined value.

Mean radiant temperature (MRT): The mathematical average of the temperatures of all wall, ceiling, and floor surfaces in a room. In some definitions of MRT, the temperature of each surface has a weighing factor based on the location in the room at which the MRT is being determined.

Measured value: The value of a controlled physical parameter (such as temperature) as detected by a sensor.

Mechanical energy: The total energy possessed by an object based on its height above the earth, its speed, and in the case of a fluid its pressure. The total mechanical energy content of a fluid is called head.

Metered balancing valves: Specialized valves that allow the flow rate in a branch piping circuit to be precisely adjusted to a desired value.

Microbubble resorber: A type of deaerator that separates and collects microbubbles from the system's fluid. It lowers the dissolved air content of the water, enabling it to absorb any residual air in the system.

Microbubbles: Very small bubbles that give water a cloudy appearance.

Microprocessor: An integrated digital circuit device capable of being electrically programmed to execute specific logical and mathematical operations.

Minimum supply temperature: The lowest temperature allowed as the supply temperature from a boiler being operated by an outdoor reset control. This minimum is typically set to prevent the boiler from operating with sustained flue gas condensation.

Minitube distribution system: A system using a pair of minitubes between a common primary loop and each of the manifold stations.

Minitubes: A pair of small diameter tubes running between a primary loop and a manifold station location for injection mixing control of the water temperature in the manifold circuits.

Mixing assembly: A collection of hardware that constitutes a complete mixing system, and forms a "bridge" between a higher temperature primary loop and a lower temperature distribution system.

Mixing device: Any device in which two or more streams of fluid are intentionally mixed together to create a fluid temperature between the higher and lower entering fluid temperatures.

Mixing reset control: The process of automatically adjusting the supply temperature to a hydronic distribution system based on outdoor reset control logic, and accomplished by mixing hot water from a heat source with cooler return water from the distribution system.

Mixing valve: A valve that blends two fluid streams entering at different temperatures to achieve a desired outlet temperature. The most common types in hydronic heating are 3-way and 4-way mixing valves.

MMbtu: An abbreviation for one million (1,000,000) Btus.

Modulating: A process by which the rate of energy flow is continuously adjusted as necessary to meet the existing load. Common examples include modulating valves, and modulating boilers.

Modulating boilers: A boiler that adjusts its heat output from some maximum value down to a significantly lower level based on the instantaneous demand for heat by the load.

Monoflo tee®: A brand of diverter tees manufactured by the Bell and Gossett Corporation.

Multifunction time delay device: A device that can be configured to provide several common time delay relay functions.

Multiload/multitemperature systems: Hydronic systems capable of simultaneously supplying two or more loads operating at two or more supply water temperatures.

Multiple boiler controller: A controller capable of independently operating two or more boilers as required by the load.

Multistage heat production: A technique for increasing or decreasing the heat output of a heating plant in steps (a.k.a. stages).

Multizone/multicirculator system: A hydronic distribution system that uses a separate circulator for each zone circuit.

Multizone/zone valve system: A hydronic distribution system that uses a separate zone valve for each zone circuit. A single circulator supplies the entire system.

Multizone relay center: A specialized controller consisting of multiple relays and a transformer. This control is used in multizone hydronic systems.

Multizone systems: A hydronic system capable of independently controlling heat input to two or more areas of a building.

National Fuel Gas Code: A model code developed by the National Fire Protection Association that is widely recognized in the U.S. It specifically covers the fuel supply, ventilation, and exhaust requirements of combustion-type heat sources.

National pipe thread (NPT): The standard tapered threads used on piping, fittings, and valves in the U.S.

Natural convection: Heat transfer between a solid surface and a moving fluid when the fluid's motion is created only by natural processes such as the tendency of a warm fluid to rise.

Net positive suction head available (NPSHA): The total head available to push water into a circulator at a given point in a piping system. This indicator is totally determined by the piping system and fluid it contains, and does not depend on the circulator. It is expressed in feet of head. To avoid cavitation, the NPSHA must be equal to or greater than the NPSHR of the selected circulator.

Net positive suction head required (NPSHR): The minimum total head required at the inlet of the circulator to prevent cavitation. This value is specified by pump manufacturers.

Nominal inside diameter: The approximate inside diameter of a pipe or tube.

Noncondensing mode: An operating mode for a combustion-type heat source in which the vapors produced during combustion do not condense (turn to liquid) within the heat source.

Normally closed contacts: Electrical contacts in a relay that remain closed when the relay coil is deenergized.

Normally open contacts: Electrical contacts in a relay that remain open when the relay coil is deenergized.

Off-cycle losses: Heat losses from a heat source that occur when the heat source is not producing heat.

Off-peak periods: Set times when electrical energy is sold at a reduced rate by the utility.

Offset: A residual error between the target value and actual value of a controlled variable associated with proportional only control.

On/off control: Attempting to maintain a setpoint temperature by turning heat delivery on and off.

One-pipe system: A hydronic distribution system that uses diverter tees to connect individual heat emitters to a main distribution piping circuit.

Open-loop source: A source of ground water such as a pond, lake, or well, from which water is supplied to a geothermal heat pump.

Open-loop system: A piping system that conveys fresh water, or is open to the atmosphere at any point.

Operating differential: The actual variation in the controlled variable (often temperature) established by a process using on/off control.

Operating mode: A specific condition or state of the controlled devices in a system that allows it to accomplish a specific objective. Hydronic heating systems can have several operating modes, including off, on, priority domestic water heating, heat purging, etc.

Operating point: The point on a graph where the pump curve intersects the system resistance curve. This point indicates the flow rate at which the system will operate.

Outdoor reset control: A control method that increases the water temperature in a hydronic system as the outdoor temperature drops.

Overshoot: The condition in which a controlled variable (often temperature) exceeds the desired temperature.

Oxygen diffusion barrier: A layer of special compounds or metal within the cross-section that significantly reduce the rate at which oxygen can diffuse through a polymer tube.

Packaged boiler: Boilers that are traditionally supplied with factory-mounted components such as high limit control, circulator, pressure relief valve, and oil or gas burners.

Panel radiator: A metal heat emitter that mounts on a wall and releases a significant percentage of its heat as thermal radiation.

Parallel direct-return piping: A hydronic distribution system consisting of two or more branch piping paths that begin from a common supply pipe and end at a common return pipe. The first branch connected to the supply pipe is also the first to connect to the return pipe.

Parallel piping: When two or more branch piping paths originate from a common starting point and end at a common ending point.

Parallel primary loop: A primary loop in a primary/secondary system having two or more crossovers from a common supply to a common return. A pair of closely spaced tees is installed in each crossover.

Parallel pumps: Two or more pumps that are mounted to a common inlet header and discharge into a common outlet header.

Parallel reverse-return piping: A hydronic distribution system consisting of two or more branch piping paths that begin from a common supply pipe and end at a common return pipe. The first branch connected to the supply pipe is the *last* to connect to the return pipe.

Parallel shifting: Moving a reset line up or down without changing its slope to compensate for internal heat gain, or losses not proportional to the temperature difference between inside and output.

Partial reset: Limiting the lower limit of supply temperature produced by an outdoor reset control. Partial reset is often done to prevent a conventional boiler from operating with sustained flue gas condensation.

Partial storage systems: A thermal storage system that stores only a portion of the energy needed by the load, for a 24-hour period, during the preceding off-peak period.

Perimeter circuits: Radiant floor heating circuits that are routed near the exposed perimeter of a room.

PEX tubing: An acronym for cross-linked polyethylene tubing.

PEX-AL-PEX tubing: A composite tubing having inner and outer layers of PEX, and a center layer of aluminum.

Pickup allowance: A term used to describe extra boiler capacity allowed for quickly heating up a distribution system following an off cycle.

Pipe size: A term that, unless otherwise specified for pipes under 12 inches in diameter, refers to the approximate *inside* diameter of a pipe.

Pipe size factor: A factor designated (c) used to describe all pipe geometry information when determining hydraulic resistance.

Piped in parallel: Two or more piping paths that share a common supply point and return point.

Piping element: Any component through which flow occurs.

Piping topology: The interconnected geometry formed by a specific arrangement of piping, fittings, and valves. Examples of piping topology include 2-pipe reverse return, multizone/multicirculator, and so forth.

Plateless staple-up system: A method of installing radiant floor heating circuits in which tubing is directly stapled to the underside of a subfloor and not equipped with metal heat transfer plates.

Point of no pressure change: The location where an expansion tank is attached to a hydronic system. The pressure at this point remains unchanged regardless of whether the circulator is on or off.

Poles (of a switch): The number of independent electrical circuits that can be simultaneously passed through a switch or relay.

Polymer: A hydrocarbon substance consisting of long molecular chains. Many polymers are commonly called plastics. Common examples include polyethylene, polybutylene, and polypropylene.

Poured gypsum underlayment: A pourable slurry consisting of gypsum cement, sand, and other admixtures that cures into a hard subflooring. Often poured over hydronic tubing that has been stapled to the subfloor.

Power-venting: Using an electrically driven blower rather than a chimney to create the negative pressure required to safely vent combustion products from a heat source to outside air.

Prefabricated subfloor/underlayment panels: A modular, panelized system for installing radiant panels on a floor, wall, or ceiling.

Pressure head: The mechanical energy possessed by a fluid due to its pressure.

Pressure reducing valve: A special valve that reduces the pressure of water as it flows from a water supply system into a hydronic system. This valve is also called a feed-water valve.

Pressure relief valve: A spring-load valve that opens to release fluid from the system whenever the pressure at its location exceeds its rated opening pressure.

Primary/secondary circuit interface: The connection of one hydronic piping loop to another using a pair of closely spaced tees.

Primary/secondary (P/S) piping: A method of interfacing individual (secondary) distribution circuits to a common (primary) circuit using closely spaced tees.

Primary/secondary tees: A pair of closely spaced tees used to form a primary/secondary circuit interface.

Primary loop: A main distribution circuit tied to the system's heat source, and supplying several secondary circuits.

Primary side: The side of a transformer connected to line voltage.

Priority load: A control strategy used in multiload systems in which some loads are temporarily turned off while the priority load is operating.

Processing algorithm: A specific set of programming instructions executed by a controller.

Proportional band: The operating range of a controller in which the output signal or control action is proportional to the error between the target setting and the actual (measured) value of the controlled variable.

Proportional control (P): A control response where the value of the output signal is directly proportional to the error between the target value and the actual (measured) value of the controlled variable.

Proportional reset: A method of producing a reduced supply temperature in a hydronic distribution circuit by supplying a manually set mixing device with a "hot" fluid temperature that itself is reset by an intelligent mixing device.

Proportional-integral control (PI): A control response where the value of the output signal is based on the error between the target value and the actual (measured) value of the controlled variable, as well as how long that error has existed.

Proportional-integral-derivative control (PID): A control response where the value of the output signal is based on the error between the target value and the actual (measured) value of the controlled variable, how long that error has existed, and the time rate of change of error.

Pulse width modulation (PWM): A control algorithm in which the length of the on-cycle varies based on the error between the target value and the actual (measured) value of the controlled variable.

Pump (centrifugal): An electrically driven device that uses a rotating impeller to add mechanical energy (e.g., head) to a fluid. In this text, the word pump refers specifically to a centrifugal pump.

Pump curve: A graph indicating the head energy added to the fluid by the pump versus the flow rate of the fluid through the pump. Such curves are provided by pump manufacturers.

Pump efficiency: The ratio of the rate of head energy transferred to a flowing fluid divided by the rate of mechanical energy input to the shaft of the pump.

Purge cart: An installation tool that uses a high capacity pump to purge air from a system, filter fluid, and help add antifreeze to a system.

Purge valves: A specialized valve or combination of valves that allows easy forced-water purging of a hydronic system or a portion thereof.

Purging: The process of filling a system with water and removing any air within the system.

Quick opening valve: A valve characteristic that allows the flow rate through the valve to increase rapidly as the valve plug first moves away from the seat, and increase slowly as the valve plug nears its fully open position.

Radiant baseboard: An extruded aluminum plate with integral tubes that is fastened to the base of walls, and releases the majority of its heat output as thermal radiation.

Radiant heating: The process by which energy flows by thermal radiation from the surface of an object to the surface of a cooler object.

Radiant panel: A heat emitter that releases 50 percent or more of its heat output as thermal radiation.

Radiator: A commonly used term for hydronic heat emitters that transfer some of their heat output by thermal radiation.

Rangeability: The ability of a control system or device to maintain stable and accurate control under low loading conditions. A control system with high rangeability provides stable control under low loading conditions.

Real time inputs: Information that is instantaneously supplied to a controller from sensors.

Refrigeration cycle: The process of moving heat from a region of lower temperature to one of higher temperature by repeatedly expanding and compressing a compound called the refrigerant.

Relay: An electrically operated switch having a specific number of poles and throws.

Relay socket: The base a relay plugs into, and at which external wires leading to the relay are terminated.

Remote bulb aquastat: An electromechanical temperature controller that uses a capillary tube of several feet in length to connect its sensor bulb to its switch actuator assembly.

Reset line: A straight line having a specific slope on a graph of supply water temperature versus outdoor air temperature.

Reset ratio: The ratio of the change in supply water temperature to the change in outdoor air temperature that a reset control will attempt to maintain. The reset ratio is also the mathematical slope of a reset line.

Return manifold: A metal or thermoplastic chamber to which the return side of one or more heating distribution circuits is attached.

Reverse injection mixing: A piping configuration for injection mixing in which a portion of the fluid at the supply temperature of the distribution system is returned to the primary loop to replace the hot water injected into the distribution system.

Reverse-return piping: A piping configuration in which the branch piping circuit closest to the system circulator on the supply main is farthest from the circulator on the return main. This arrangement helps to naturally balance flow rates through the branch circuits.

Reynold's number: A unitless number that can be used to predict whether flow is laminar or turbulent.

Room air temperature profile: A graph showing how room air temperature varies from floor to ceiling.

Room heating load: The rate at which heat must be added to an individual room to maintain interior comfort.

Room side heat flux: The rate of heat output per unit of area from the room side of a radiant panel. It is usually expressed in Btu/hr/square foot.

Room temperature unit (RTU): A device that measures room and temperature and communicates with a controller in another location.

Room thermostat: An adjustable, temperature-operated switch, mounted in a room, that turns heat delivery on and off to regulate room temperature.

Run fraction: The ratio of the on time of a heat source to the total elapsed time. In general, short run fractions are undesirable.

Rung: A horizontal circuit path run between vertical voltage lines on a ladder diagram.

R-value of a material: A value indicating the thermal resistance of the material. The greater the R-value, the slower heat will conduct through the material, all other conditions being equal.

Sealed combustion: A design for a combustion-type heat source in which all air required for combustion is drawn through sealed ducting from the outside of the building. All exhaust gases are also routed through sealed ducting to the outside of the building.

Secondary circuit: An individual distribution piping circuit with its own circulator that receives heat from the building's primary circuit.

Secondary side: The side of a transformer connected to lower (e.g., control) voltage.

Sectional boiler: A boiler assembled from cast-iron sections.

Sections: The cast-iron assemblies that are joined together to form a sectional boiler block.

Sensible heat: Heat that, when added or removed from a material, is evidenced by a change in the temperature of the material.

Sensing element: A probe used to detect the value of a specific physical quantity such as temperture.

Series Baseboard Simulator: A module in the Hydronics Design Studio capable of simulating the thermal and hydraulic performance of a user-specified series baseboard circuit.

Series (piping circuit): An assembly of pipe and piping components that are connected to form a single closed loop.

Series (piping path): An assembly of pipe and piping components connected end to end to form a single path between two points.

Series primary loop: A primary loop in a primary/secondary piping system to which secondary circuits are connected in sequence, each using a pair of closely spaced tees.

Series pumps: Two or more pumps installed in the same piping circuit, and having the same flow direction.

Serpentine patterns: Patterns for radiant panel circuits in which the tubing path contain numerous 180 degree return bends as it progresses across the panel area.

Setback: Reducing the temperature setting of a setpoint control, such as a room thermostat, to conserve energy.

Setpoint controller: A control that attempts to maintain a preset temperature at some location in a system by controlling the operation of other devices.

Setpoint demand: A demand for boiler operation from a load that needs to be supplied with water at a specific set temperature.

Setpoint temperature: A desired temperature that is to be maintained if possible, and is set on some type of controller such as a thermostat or setpoint controller.

Shaft seal: Metal or synthetic components of a pump that prevent fluid from leaking out where the impeller shaft penetrates the volute.

Shell and tube heat exchanger: A heat exchanger design in which one fluid passes through a group of several parallel tubes while the other fluid passes through the spaces between these tubes.

Short-cycling (boiler): An undesirable operating condition in which the duration of the boiler firing cycles is considered too short for acceptable efficiency or equipment life. Short-cycling often occurs when a boiler is excessively oversized or connected to a highly zoned distribution system.

Single circuit/multizone (one-pipe) system: A distribution piping system that uses diverter tees to connect each heat emitter branch circuit to the main piping circuit.

Single series circuit: A hydronic distribution system in which all heat emitters are connected inline with the distribution piping. The full system flow rate passes through each heat emitter.

Sink (heat): A material into which heat is released.

Slab-on-grade system: A type of radiant floor panel in which tubing is cast into a concrete slab poured on stable soil.

Sleepers: Wooden strips, usually arranged in parallel, and secured to a subfloor.

Smooth pipe: A generic term for tubing with relatively smooth inner walls. Examples include drawn copper tube, polybutylene tubing, and PEX tubing.

Snow switch: A device that automatically turns on a snow-/icemelting system when it detects the possibility of frozen precipitation.

Snow/ice pavement sensor: A device that measure slab temperature as well as detects the possibility of frozen precipitation on the slab surface.

Soft temper tubing: Copper tubing that is sold in coils, and can be bent with relative ease.

Soldering: A method of metallurgically joining copper or copper alloy pipe and components such as fittings and valves using a molten solder alloy fused to the base metal.

Solid fuel boiler: A boiler capable of burning a solid fuel such as wood or coal.

Source (of a heat pump): The material from which the low temperature heat is extracted.

Space-heat demand: The control action whereby a heating system is signaled to deliver heat to a space-heating load.

Space heating: The process of adding heat to a building or room to maintain indoor comfort.

Specific heat: A material property indicating the amount of heat required to raise one pound of the material by one degree Fahrenheit, often expressed in Btu/lb/°F.

Spool (of a valve): An internal mechanism in a mixing valve that slides or rotates to regulate the amount of hot and cool fluid streams entering the valve.

Spring-loaded check valve: A check valve that contains a spring to close the valve regardless of its mounting position.

Stacked home run system: A piping system that uses a home run manifold assembly on each floor level of a multifloor building. These manifolds are connected by a common set of supply and return riser pipes.

Stack effect: The tendency of warm air to move upward in an enclosure, as smoke rises through a chimney.

Stage (control): Using two or more stepped levels of heat input to a load.

Standard expansion tank: An expansion tank that does not have a diaphragm.

Standard port ball valve: A ball valve design where the orifice through the ball is slightly smaller than the diameter of the pipe to which the valve is connected.

Static pressure: The pressure at some point in a piping system measured while the fluid is at rest. Static pressure increases from a minimum at the top of a system to a maximum at the bottom. Static pressure is created by the weight of the fluid itself, plus any extra pressurization applied at the top of the fluid.

Stationary air pockets: Air that accumulates at the high points of piping systems and creates air binding.

Steady state efficiency: The efficiency of converting fuel into heat when a heat source operates continuously with nonvarying input and output conditions.

Strap-on aquastat: A special type of electromechanical setpoint controller designed to strap directly to the outside diameter of a pipe.

Stratification: The natural tendency for a warm fluid to rise above a cooler fluid due to density differences. This term is commonly used to describe warm air rising toward the ceiling of a room while cooler air descends toward the floor.

Striping: The undesirable condition in which the surface of a radiant panel is significantly warmer directly above the tubing.

Submixing: Mixing a warm fluid that has already passed through one mixing device with cooler fluid to achieve an even lower supply temperature.

Subsystem: a collection of components that work together as a system but are contained within a larger overall system.

Supply manifold: A metal or thermoplastic chamber to which the supply side of one or more heating distribution circuits is attached.

Supply water temperature: The temperature of water being supplied to a hydronic distribution system.

Suspended tube system: A type of radiant floor heating system in which tubing is suspended in the air space between floor joists.

Swing check: A check valve with an internal disc that swings on a pivot.

Symbol palette: A collection of symbols representing physical components in schematic drawings (piping and electrical).

System resistance curve: A curve plotted on a graph that indicates the head loss of a piping system versus the flow rate through it.

Tankless coil: A finned copper coil that is inserted into a boiler to provide domestic hot water.

Target flow rate: A calculated flow rate through a piping system based on an expected rate of heat transport, and an assumed circuit temperature drop.

Target value: The desired value of a controlled variable.

T-drill®: A specialized tool for fabricating branch connections in copper tubing.

Temperature drop: The difference between the supply temperature and return temperature in an operating distribution circuit, or part thereof.

Temperature limit control: A controller that prevents the temperature of a heat source from exceeding a set value.

Therm: A quantity of energy equal to 100,000 Btus. The therm is the traditional unit in which natural gas is sold.

Thermal break: The placement of insulation between two materials to prevent rapid heat transfer between them.

Thermal conductivity: A physical property of a material that indicates how fast heat can move through the material by conduction. The higher the thermal conductivity, the faster heat can move through the material.

Thermal energy: Energy in the form of heat. The term thermal energy is used synonymously with the word heat.

Thermal envelope: The shell formed by all building surfaces that separates heated space from unheated space. Exterior walls, windows, and exposed ceilings are all parts of the thermal envelope of a building.

Thermal equilibrium: The condition of a system when the rate of energy loss from the system is exactly the same as the rate of energy input to the system.

Thermal expansion: The natural tendency of materials to expand as their temperature increases.

Thermal mass: The ability of a material or object to store heat due to its mass and specific heat.

Thermal radiation: Heat that is transferred in the infrared portion of the electromagnetic radiation. Thermal radiation cannot be seen by the human eye, but in many other ways behaves like visible light.

Thermal resistance: The tendency of a material to resist heat flow. Insulation materials, for example, tend to have high thermal resistance. Other materials such as copper and glass have low thermal resistance.

Thermal trap: A piping assembly that discourages heat migration due to natural convection when the system circulators are off.

Thermistor: A solid state temperature sensor that changes resistance as its temperature changes.

Thermoplastic: A polymer that can be melted and reformed in shape.

Thermoset plastic: A polymer that has been cross-linked or otherwise chemically altered so that it will not change back to a formable liquid state as will a thermoplastic.

Thermosiphoning: The tendency of heated water to rise vertically within a hydronic piping system, even when the circulator is off.

Thermostatic mixing valve: A mixing valve with an internal temperature control mechanism.

Thermostatic operator: A nonelectric actuator that mounts to a valve and opens and closes the valve as necessary to maintain a set temperature.

Thermostatic radiator valve (TRV): The assembly of a thermostatic operator and a radiator valve.

Three-piece circulator: A circulator in which the wetted parts are separated from the motor by a coupling assembly.

Three-wire control: Another name for floating control in which three wires are required between the controller and actuator.

Threshold differential pressure (of a DPBV): The set pressure at which a differential pressure bypass valve begins to allow flow to pass through itself.

Throws: The number of settings of a switch or relay in which an electrical current can pass through the switch or relay contacts.

Time delay relay: A relay in which the contacts do not open or close at the same time coil voltage is applied or interrupted.

Ton (of capacity): A term representing a rate of energy flow of 12,000 Btu/hr. It is frequently used to describe the capacity of heat pumps and air conditioners.

Total equivalent length: The sum of the equivalent lengths of all piping, fittings, valves, and other components in a piping circuit.

Total head: The total mechanical energy content of a fluid based on its pressure, elevation, and speed at some specified location in a piping system.

Total R-value of an assembly: The overall total thermal resistance of an assembly of materials that are assembled together.

Towel warmer: A specialty radiator designed to have towels draped over it for warming and drying.

Transformer: An electrical component for reducing or increasing voltage in an AC circuit. Its common application in hydronic systems is to reduce line voltage (120 VAC) to control voltage (24 VAC).

Triple action control: A preassembled control that provides high limit, low limit, and circulator control in hydronic systems using tankless water heaters.

Tri-state control: Another name for floating control in which the actuator is either opening the valve, closing the valve, or holding the valve at its current position.

Tube and plate system: A type of radiant panel construction in which tubing is saddled with or snaps into a preformed metal plate intended to laterally disperse heat away from the tubing.

Turbulent (flow): A classification of flow where streamlines repeatedly cross over each other as the fluid flows along the pipe. Turbulent flow is the dominant type of flow in hydronic heating systems.

Two-pipe system: Another name describing a parallel direct-return or parallel reverse-return distribution piping system.

Two-wire digital bus: A wiring method that allows communication between digital devices where each device on the bus has a unique digital address, and only two wires are required.

Uncoiler: A device with a rotating platform that allows coiled tubing to be easily uncoiled without twisting.

Unconfined space: A space in a building that has a minimum of 50 cubic feet of volume per 1000 Btu/hr of gas input rating of all equipment in that space.

Undershoot: When the actual value of a controlled variable drops below the intended lower limit of operation.

Underslung thermal trap: A piping assembly designed to discourage heat migration due to natural convection.

Unit heater: A generic name for a fan-coil that is usually mounted near the ceiling of a room, with downward directed air flow.

Unsaturated state of air solubility: When the amount of dissolved air in the system water is low enough to enable the water to resorb any residual air trapped in other parts of the system.

Usage profile: A graph that shows the quantity of domestic hot water required at any hour of the day.

Valve actuator: A device that physically moves the stem or shaft of a valve using electrical, thermal, or pneumatic input energy.

Valved manifold: A manifold that has a flow regulating valve for each circuit connection.

Valveless manifold: A manifold that does not have flow regulating valves for each circuit connection.

Vapor barrier: A material that has low permeability to vapor flow.

Vapor pressure: The minimum (absolute) pressure that must be maintained on a liquid to prevent it from changing to a vapor. The vapor pressure of most liquids increases as their temperature increases.

Variable speed circulator: A circulator that is operated over a range of speeds using either an internal or external variable speed drive.

Variable speed injection mixing: A method of controlling the temperature in a hydronic heating system by using a variable speed pump to add hot water to a continuously circulating distribution system.

Velocity head: The mechanical energy possessed by a fluid due to its velocity.

Velocity noise: Undesirable sound created when the flow velocity through a pipe or piping component exceeds a certain value.

Velocity profile: A sketch that shows how the velocity of the fluid varies from the centerline of a pipe out to the pipe wall.

Vertical panel radiator: A panel radiator that is taller than it is wide, and has tubes oriented in a vertical direction.

Viscosity: A means of expressing the natural tendency of a fluid to resist flow.

Viscous friction: Internal friction forces that develop within a fluid or between the fluid and a surface due to the viscosity of the fluid. Viscous friction is responsible for loss of head energy.

Volt-amp (VA): A multiplication of volts times amps. A common output rating for control transformers.

Volute: The chamber that surrounds the impeller in a centrifugal pump.

Warm weather shut down (WWSD): A control mode in which the system's circulator is automatically turned off by a controller during warm weather.

Waterlogged: A condition in which the captive air in an expansion tank is eventually replaced by water.

Wet base boiler: A boiler design in which the boiler sections totally surround the combustion chamber.

Wet rotor circulator: A circulator in which the motor's armature and its impeller are an integral unit that is surrounded and cooled by the fluid passing through the circulator.

Working pressure: The allowable upper pressure limit at which a pipe or piping component can safely be used.

Zone: An area of a building for which space heating is controlled by a single thermostat.

Zone valve: Valves used to allow or prevent the flow of hot water from a heat source through individual zone distribution circuits.

Zoning: Dividing a hydronic distribution system into two or more independently controlled distribution circuits.

INDEX

A

B

C

D

E

F

G

H

I

N

O

P

Q

R

S

U

V

W

Y

Z